ELSEVIER'S DICTIONARY OF ZOOLOGY AND GENERAL BIOLOGY

RUSSIAN-ENGLISH and ENGLISH-RUSSIAN

ELSEVIER'S DICTIONARY OF ZOOLOGY AND GENERAL BIOLOGY

RUSSIAN-ENGLISH and ENGLISH-RUSSIAN

compiled by

N.N. SMIRNOV
Moscow, Russia

2004

ELSEVIER

Amsterdam – Boston – Heidelberg – London – New York – Oxford
Paris – San Diego – San Francisco – Singapore – Sydney – Tokyo

ELSEVIER B.V.
Sara Burgerhartstraat 25
P.O. Box 211
1000 AE Amsterdam
The Netherlands

ELSEVIER Inc.
525 B Street, Suite 1900
San Diego, CA 92101-4495
USA

ELSEVIER Ltd
The Boulevard, Langford Lane
Kidlington, Oxford OX5 1GB
UK

ELSEVIER Ltd
84 Theobalds Road
London WC1X 8RR
UK

First edition 2004

Library of Congress Cataloging in Publication Data
A catalog record is available from the Library of Congress.

British Library Cataloguing in Publication Data
A catalogue record is available from the British Library.

ISBN: 0-444-51700-6

♾ The paper used in this publication meets the requirements of ANSI/NISO Z39.48-1992 (Permanence of Paper).
Printed in The Netherlands.

PREFACE

This is the first dictionary which covers all fields of biology, from cytology and molecular biology to behavior and ecology, with special reference to zoology, totaling about 40,000 entries.

Botanical terms (plant names) are not included. For these, the user is referred to "*Elsevier's Dictionary of Botany*" by P. Macura (2002). However, terms relating to plant anatomy and physiology are included, as well as certain generic names of plants, that are necessary for understanding environments and landscapes. Terms relating to aquatic biology (and names of fishes) are for the most part not included either. These can be found in "*Elsevier's Dictionary of Aquatic Biology*" by N.N. and A.N. Smirnov (1999).

Attention is focused on applied fields too, such as pathology, veterinary science, biological aspects of medicine, phytopathology, agriculture, forestry, biotechnology, nature conservation, etc., as well as descriptive terms commonly used in biological literature.

The main part of the dictionary is the Russian-English part, giving in alphabetical order Russian key entries and subentries with their English equivalents. The English-Russian part is an alphabetical listing of English key entries. This part may serve as an index to the Russian-English part: having found an English key entry in the English-Russian part, one can find its combinations in the Russian-English part. Grammatical notes are included as well as indications of scientific fields. Russian words are supplied with the stress indicated after the stressed vowel. Adjectives are in the masculine nominative singular, with the exception of a few which apply only to female objects.

In all stages of the dictionary's preparation the author was aided by A.N. Smirnov, PhD. This and other efforts would not be possible without day-to-day care of Mrs L.A. Smirnova, PhD. The Preface was edited by Mr. A.R. Carpenter (Moscow).

The author hopes that the dictionary will be useful to translators and scientists dealing with various fields of fundamental and applied biology.

Prof. N.N. Smirnov
D.Sc. (Moscow)

ПРЕДИСЛОВИЕ

Предлагаемый словарь – первый, который включает все области биологии, от цитологии и молекулярной биологии до поведения и экологии, причём зоология представлена более подробно. Словарь включает около 40.000 терминов.

Ботанические термины (названия растений) не включены. Их следует искать в "*Elsevier's Dictionary of Botany*", P. Macura (2002). Однако, многие термины по анатомии и физиологии растений включены, также как и некоторые названия растений, необходимые для понимания среды и ландшафтов. Также не включены гидробиологические термины и названия рыб. Они представлены в "*Elsevier's Dictionary of Aquatic Biology*", N.N. and A.N. Smirnov (1999).

Значительное внимание уделено терминам прикладных дисциплин, таких как патология, ветеринария, биологические аспекты медицины, фитопатология, сельское хозяйство, лесное хозяйство, биотехнология, охрана природы и другие, в также описательным терминам, обычно встречающимся в биологической литературе.

Основная часть словаря – русско-английская. В ней представлены в алфавитном порядке гнездовые слова с английскими эквивалентами и их сочетания с определениями, также в алфавитном порядке. Англо-русская часть включает английские термины в алфавитном порядке с их русскими эквивалентами. Она может служить индексом к русско-английской части: найдя в ней английский термин, можно найти его сочетания в русско-английской части. Включены грамматические пометки и указания на дисциплины. В русских терминах указано ударение (после ударной гласной). Прилагательные приведены в единственном числе мужского рода, с исключениями для тех, которые применимы только для объектов женского рода.

На всех этапах подготовки словаря автору оказывал помощь А.Н. Смирнов. Эта, как и другие работы, были бы невозможны без каждодневной заботы Л.А. Смирновой. Английский текст предисловия отредактировал А.Р. Карпентер (Москева).

Составитель надеется, что данный словарь будет полезен широкому кругу переводчиков и специалистов, работающих в различных областях теоретической и прикладной биологии.

Проф. Н.Н. Смирнов
докт. биол. наук (Москва)

ABBREVIATIONS

a	adjective
adv	adverb
agr	agriculture
anat	human and animal anatomy
apic	apiculture
bioch	biochemistry
biom	biometrics
bot	botany
cyt	cytology
ecol	ecology
emb	embryology
ent	entomology
eth	ethology
f	feminine
for	forestry
gen	genetics
geol	geology
helm	helminthology
herp	herpetology
hist	histology
i	interjection
imm	immunology
lab	laboratory techniques
m	masculine
mal	malacology
mbiol	molecular biology
med	medicine
micr	microbiology
myc	mycology
n	neuter
orn	ornitology
pal	palynology
par	parasitology
phys	human and animal physiology
phytop	phytopathology
pl	plural
plsoc	plant sociology
prot	protistology
s	noun
ther	theriology
v	verb
vet	veterinary science
zool	zoology

РУССКО-АНГЛИЙСКИЙ

RUSSIAN-ENGLISH

A

абаксиа'льный *a* abaxial
абапика'льный *a* abapical
аббревиа'ция *f* abbreviation
абдомина'льный *a* abdominal
абду'ктор *m* abductor
абду'кция *f* abduction
аберра'нтный *a* aberrant
аберра'ция *f* aberration
 психи'ческая ~ mental aberration
 хромосо'мная ~ chromosomal aberration
абиогене'з *m* abiogenesis, abiogeny, archigenesis
абиогенети'ческий *a* abiogenetic
абиоге'нный *a* abiogenous
абио'з *m* abiosis
абиоти'ческий *a* abiotic
абиотрофи'я *f* abiotrophy
абиотро'фный *a* abiotrophic
абисса'льный *a* abyssal
аблактиро'вка *f* approach grafting, side grafting, veneer grafting, ablactation
аблакти'ровать *v* graft by approach
абласти'ческий *a* ablastous
абмигра'ция *s* abmigration
абора'льный *a* aboral
абориге'н *m* aborigine, native
абориге'нный *a* aboriginal, native
або'рт *m* abortion
аборти'вный *a* abortive
абортоге'нный *a* abortifacient
абра'зия *s* abrasion
абрико'с *m* apricot (*Armeniaca*)
абсентеи'зм *m eth* absenteism
абсолю'тный *a* absolute
абсорбе'нт *m* absorbent, absorbing agent
абсорби'ровать *v* absorb
абсорби'рующий *a* absorptive
абсорбциоме'тр *m* absorptiometer
абсо'рбция *f* absorption
 вну'тренняя ~, **кише'чная** ~ internal absorption, intestinal absorption
 парентера'льная ~ parenteral absorption
абсце'сс *m* abscess
 ~ **Бецо'льда** Bezold's abscess
 ~ **Бро'ди** Brodie's abscess
авдо'тка *f orn* stone curlew (*Burhinus oedicnemus*)
аве'рсия *f* aversion
 пищева'я ~ food aversion
 приобретё'нная пищева'я ~ learned food aversion
авиауче'т *m* aerotaxation
ави'дность *f imm* avidity
авикуля'рия *f* avicularium
авируле'нтность *f* avirulence
авируле'нтный *a* avirulent
авитамино'з *m* avitaminosis, vitamin deficiency
авифа'уна *f* avifauna
авока'до *n* avocado, alligator pear (*Persea*)
авто- auto-
автога'мия *f* autogamy
автога'мный *a* autogamic, autogamous
автогене'з *m* autogenesis, autogeny
автогенети'ческий *a* autogenetic
автогру'минг *m eth* autogrooming
автоката'лиз *m* autocatalysis
автокине'з *m* autokinesis, autokinesia
автокла'в *m* autoclave, high-pressure steam sterilizer
автоклави'рование *n* autoclaving
автоклави'ровать *v* autoclave
авто'лиз *m* autolysis
автолиза'т *m* autolysate
автолизи'н *m* autolysin
автолити'ческий *a* autolytic
автомати'зм *m* automatism
автома'тия *f* automaticity
автоно'мный *a* autonomous, autonomic
автопарази'т *m* autoparasite
автопаразити'зм *m* autoparasitism
автопартеногене'з *m* autoparthenogenesis
автопои'лка *f* self-filling drinker
авторадиогра'мма *f* autoradiogram, autoradiograph, radio-autogram
авторадиогра'фия *f* radioautography, autoradiography
 ~ **ме'тодом покры'тия** coated autoradiography
 ~ **накло'нных сре'зов** mounted autoradiography
 конта'ктная ~ contact autoradiography
 электро'нно-микроскопи'ческая ~ electron microscopic autoradiography
автори'тмика *f* autorhythmic action
автоспо'ра *f* autospore
автото'мия *f* autotomy
автотропи'зм *m* autotropism
автотро'ф *m* autotroph
автотрофи'я *a* autotrophy

авторо'фность *f* autotrophy
автотро'фный *a* autotrophic
автоферти'льный *a* self-fertile
автофлуоресце'нция *f* autofluorescence
автохо'р *m* autochorous plant, autochore
автохто'нный *a* autochthonous
ага'ва *f* agave *(Agave)*
~ **сиза'левая** sisal agave *(Agave sisal)*
агалакти'я *f* agalactia, agalaxia
ага'мный *a* agamic, agamous
агамео'н *m* agameon
агамо'бий *m* agamobium
агамогене'з *m* agamogenesis
агамогенети'ческий *a* agamogenetic
агамого'ния *f* agamogony
агамоде'м *m* agamodeme
агамосперми'я *f* agamospermy
агамоспиру'ра *f* agamospirura
агамоспо'ра *f* agamospore
ага'р *m* agar, agar-agar
~ **на пивно'м су'сле** beer-wort agar
~ **Сабуро'** Sabouraud agar
~ **с глицери'ном** glycerin agar
~ **с глюко'зой и со'ей** soya glucose agar
~ **с кро'личьей кро'вью** rabbit-blood agar
~ **с ла'ковой кро'вью** laked blood agar, lysed blood agar
~ **с трибутири'ном** tributyrin agar
~ **с фенолфталеи'ном** phenolphthalein agar
~ **Ча'пека** Czapek's agar
ви'смут-сульфи'тный ~ bismuth-sulfite agar
водопепто'нный ~ water peptone agar
глюко'зный ~ glucose agar
глюко'зный ~ Сабуро' Sabouraud's glucose agar
дезоксихо'левый ~ desoxycholate agar
дезоксихо'левый ~ с лакто'зой desoxycholate lactose agar
желати'новый ~ gelatin agar
железосульфи'тный ~ iron agar, iron-sulfite agar
же'лчный ~ bile salt agar
казеи'новый ~ casein agar
карто'фельный ~ potato agar
карто'фельный ~ с декстро'зой potato dextrose agar
косо'й ~ oblique agar, slant agar, slope agar, agar slope
крахмалоаммиа'чный ~ starch-and-ammonia agar
крахма'льный ~ starch agar
кровяно'й ~ blood agar
лактозола'кмусовый ~ lactose-litmus agar
моло'чно-пепто'нный ~ milk peptone agar
мясопепто'нный ~, **МПА** meat infusion agar, beef-extract agar, plain agar
пита'тельный ~ nutrient agar
пита'тельный ~ с цетрими'дом cetrimide agar
полужи'дкий ~ semisolid agar, sloppy agar, soft agar
просто'й ~ plain agar
протеозотрипто'нный ~ proteose tryptone agar
са'харный ~ sugar agar
ско'шенный ~ oblique agar, slant agar, agar slant, slope agar, agar slope
солодо'вый ~ malt agar
су'словый ~ wort agar
твё'рдый ~ stiff agar
трипто'зный ~ tryptose agar
триптоноглюко'зный ~ tryptone-glucose agar
фукси'новый ~ fuchsin agar ~ agar, fuchsin-sulfite agar
яи'чно-желто'чный ~ egg yolk agar
агаро'за *f* agarose
легкопла'вкая ~ low-melting agarose, low-melting point agarose
агарообра'зный *a* agarlike
агарообразу'ющий *a* agar-producing
агастра'льный *a* agastric
агглютина'ция *f* agglutination
~ **жгу'тиков** flagellar agglutination
бактериоге'нная ~ bacteriogenic agglutination
жгу'тиковая ~ flagellar agglutination
иммуˈнная агглютинация immune agglutination, immunoagglutination
мелкозерни'стая ~ "small-flake" agglutination
перекрё'стная ~ cross agglutination
холодо'вая ~ cold agglutination
агглютини'н *m* agglutinin, agglutinating substance
неполноце'нный ~ incomplete agglutinin
агглютини'ровать *v* agglutinate
агглютини'руемый *a* agglutinable
агглютини'рующий *a* agglutinative, agglutinating
агглютиногра'мма *f* agglutinogram

агглютиноскопи'я *f* agglutinoscopy
агглютиноге'н *m* agglutinogen
агглюто'метр *m* agglutometer
агграви'ровать *v* aggravate, malinger
агелиотропи'зм *m* aheliotropism
агене'з *m* agenesis, agenesia
аге'нт *m* agent
~ **желатиниза'ции** gelling agent
активи'рующий ~ activating agent
алкили'рующий ~ alkylating agent
антиви'русный ~ antiviral agent
болезнетво'рный ~ agent of disease
депуринизи'рующий ~ depurinating agent
ингиби'рующий ~ inhibiting agent
инфекцио'нный ~ infectious agent
коагули'рующий ~ coagulating agent
комплексообразу'ющий ~ chelating agent
мембранотро'пный ~ membrane-acting agent
металлохели'рующий ~ metalchelating agent
метили'рующий ~ methylating agent
мутаге'нный ~ mutagenic agent
патоге'нный ~ pathogenic agent
сопряга'ющий ~ coupling agent
трансформи'рующий ~ transforming agent
угнета'ющий ~ inhibiting agent
фазоспецифи'чный ~ phase-specific agent
флокули'рующий ~ flocculating agent
фрагменти'рующий ~ fragmenting agent
хелатообразу'ющий ~ chelating agent
хели'рующий ~ chelating agent
этиологи'ческий ~ etiological agent
аглико'н *m* aglycone
агломера'т *m* agglomerate
агломера'ция *f* agglomeration
городска'я ~ metropolitain agglomeration
аглюко'н *m* aglucone
агматоплои'дия *f* agmatoploidy
агна'ция *f* agnation
агнози'я *f* agnosia
зри'тельная ~ visual agnosia
зри'тельно-простра'нственная ~ visual-spatial agnosia
слове'сно-слухова'я ~ verbal-auditory agnosia
такти'льная ~ tactile agnosia
агонизи'ровать *v* agonize
агони'ст *m* agonist
агонисти'ческий *a* agonistic
аго'ния *f* agony, death struggle, death throes
агранулоци'т *m* agranulocyte
агранулоцито'з *m* agranulocytosis
агра'рный *a* agrarian
агрега'т *m* aggregate
агрега'ция *f* aggregation
~ **автоно'мных органи'змов** irreciprocal aggregation
~ **в моносло'е** monolayer aggregation
необходи'мая ~ essential aggregation
агресси'вность *f* aggressiveness, aggressivity
агресси'вный *a* aggressive
агре'ссия *f* aggression
межсамцо'вая ~ intermale aggression
переадресо'ванная ~ redirected aggression
экологи'ческая ~ ecological aggression
агро- agro-
агробиоло'гия *f* agrobiology
агролесомелиора'ция *f* afforestation amelioration, sylvicultural reclamation, agricultural afforestation, reclamative afforestation, land and forest reclamation
агроло'гия *f* agrology
агроно'м *m* agronomist
агроно'мия *f* agronomics, agronomy
агропочвове'дение *n* agropedology
агростоло'гия *f* agrostology
агроте'хника *f* agricultural technology
агроти'п *m* agrotype
агрофитоцено'з *m* agrophytocenosis
агрохи'мия *f* agrochemistry
агроцено'з *m* agrocenosis
агроценоло'гия *f* agrocenology
агроэкоти'п *m* agroecotype
агу'ти *f ther* agouti *(Dasyprocta)*
адаксиа'льный *a* adaxial
адамантади'н *m* adamantadine
адамантобла'ст *m* adamantoblast
адапта'ция *f* adaptation
~ **гла'за** ocular adaptation
~ **к высоте'** altitude adaptation
~ **к фо'ну** background adaptation
альтруисти'ческая ~ altruistic adaptation
взаи'мная ~ mutual adaptation
вкусова'я ~ gustatory adaptation
диверге'нтная ~ divergent adaptation
дополни'тельная ~ complementary adaptation
зри'тельная ~ visual adaptation
конверге'нтная ~ convergent adaptation
обою'дная ~ mutual adaptation
одновреме'нная ~ simultaneous adaptation
постура'льная ~ postural adaptation

психи'ческая ~ psychic adaptation, psychical adaptation
разносторо'нняя ~ long-range adaptation
светова'я ~ light adaptation
сенсо'рная ~ sensory adaptation
слухова'я ~ acoustic adaptation
темнова'я ~ dark adaptation
ферме'нтная ~ enzymatic adaptation
физиологи'ческая ~ physiologic adaptation, physiological adaptation
функциона'льная ~ functional adaptation
хромати'ческая ~ chromatic adaptation
цветова'я ~ color adaptation
широ'кая ~ long-range adaptation
эпигенети'ческая ~ epigenetic adaptation
адапти'вность *f* adaptability
адапти'вный *a* adaptive
адаптиогене'з *m* adaptiogenesis
ада'птор *m* adaptor
ада'пторный *a* adaptor
адвенти'вный *a* adventive
адвентициа'льный *a* adventitious, adventitial
адгези'вность *f* adhesiveness
адгези'вный *a* adhesive
адгези'н *m* adhesin
адге'зия *f* adhesion, adherence
аллергензави'симая ~ allergen-specific adhesion
гомотипи'ческая ~ homotypic adhesion
иммунная ~ immune adherence, serological adherence
межкле'точная ~ cell-cell adhesion
«физи'ческая» ~ "physical" adherence
аддити'вный *a* additive
адду'ктор *m* adductor
адду'кция а adduction
адеква'тный *a* appropriate
адельфога'мия *f* adelphogamy
адельфопарази'т *m* adelphoharasite
адельфота'ксис *m* adelphotaxis
адени'н *m* adenine
адено- adeno-
адснобла'ст *m* adenoblast
адснови'рус *m* adenovirus
аденогипо'физ *m* adenohypophysis, antehypophysis
аденози'н *m* adenosine
аденозиндифосфа'т *m* adenosine diphosphate
адено'ид *m* adenoid
адено'идный *a* adenoid
аденока'рпный *a* adenocarpous
аденокарцино'ма *f* adenocarcinoma
аденофо'р *m* adenophore
аденоци'т *m* adenocyte
адерми'н *m* adermin
адиактини'ческий *a* adiactinic
адинами'я *f* adynamia
адипоци'т *m* adipocyte
адмира'л *m ent* admiral *(Pyrameis atalanta)*
аднекси'т *m* adnexitis
адолеска'рия *f helm* adolescaria
адора'льный *a* adoral
адренали'н *m* adrenaline
адренерги'ческий *a* adrenergic
адреноблока'тор *m* adrenoreceptor blocking agent
адренокортика'льный *a* adrenocortical
адренокортикотропи'н *m* adrenocorticotropin
АКТГ, адренокортикотропи'н *m* adrenocorticotropin
адреностеро'н *m* adrenosterone
адсорбе'нт *m* adsorbent
адсо'рбция *f* adsorption
адъюва'нт *m* adjuvant
~ **Фре'йнда** Freund's adjuvant
азигомо'рфный *a* azygomorphous
азокарми'н *m* azocarmine
азо'т *m* nitrogen
~ **по Кьельда'лю** Kjeldahl nitrogen
ами'дный ~ amide nitrogen
аммиа'чный ~ ammonia nitrogen
атмосфе'рный ~ air nitrogen, free nitrogen
белко'вый ~ protein nitrogen, albuminous nitrogen
досту'пный ~ assimilable nitrogen, available nitrogen
минерализу'емый ~ mineralizable nitrogen
небелко'вый ~ nonprotein nitrogen
нитра'тный ~ nitrate nitrogen, nitric nitrogen
нитри'тный ~ nitrite nitrogen
оста'точный ~ residual nitrogen
свобо'дный ~ free nitrogen
свя'занный ~ bound nitrogen, fixed nitrogen, arrested nitrogen
сумма'рный ~ total nitrogen
усвоя'емый ~ assimilable nitrogen, available nitrogen
факти'чески усвоя'емый ~ true absorbed nitrogen

азотиза'ция *f* azotization
азоти'рованный *a* azoted
азо'тистый nitrogenous
азо'тный nitrogenous
азотоба'ктер *m* azotobacter
азотобактери'н *m* azotobacterin
азотсодержа'щий nitrogen-bearing
азотфикса'тор *m* nitrogen fixer
азотфикса'ция *f* nitrogen fixation
азотфикси'рующий *a* nitrogen-fixing
азурофили'я *f* azurophilia
азурофи'льный *a* azurophilic, azurophil, azurophile
а'ир *m bot* sweet flag (*Acorus*)
а'ист *m orn* stork (*Ciconia*)
ай-а'й *m ther* aye-aye *(Daubentonia madagascariensis)*
айва' *f bot* quince (*Cydonia*)
ака'нтор *m* acanthor
акантоци'т *s* acanthocyte
акапни'я *f* acapnia
акариа'з, акарио'з *m* acariosis
акариоци'т *m* acaryocyte
акарици'д *m* acaricide
акарици'дный *a* acaricide
акароло'гия *f* acarology
акарофили'я *f* acarophily
акарофи'льный *a* acarophilous
ака'рпный *a* acarpous
аква'риум *m* aquarium
аквариуми'ст *m* aquarist
аквато'рия *f* water area
акине'з *m* akinesis, akinesia
акине'та *f* akinete
акинети'ческий *a* akinetic
акклиматиза'ция *f* acclimatization
акклиматизи'ровать *v a* acclimatize
акклими'роваться *v* acclimate
аккомода'ция *f* accommodation
аккумули'рование *n*, **аккумуля'ция** *f* accumulation
акраниа'льный *a* acranial
акраспедо'тный *a helm* acraspedote
акриди'н *m* acridine
акриди'новый *a* acridinic
акрилами'д *m* acrylamide
акро- acro-
акробла'ст *m* acroblast
акроге'нный *a* acrogenous
акрока'рпный *a* acrocarpous
акролеи'н *m* acrolein
акромега'лия *f* acromegalia, acromegaly
акромиа'льный *a* acromial
акропета'льный *a* acropetal
акроси'ндез *m* acrosyndese, acrosyndesis
акросо'ма *f* acrosome
акроспо'ра *f* acrospore
акроспо'ровый *a* acrosporous
акрофи'т *m* acrophyte
акроцентри'ческий *a* acrocentric
акселера'ция *f* acceleration
аксениза'ция *f* axenization
аксени'ческий *a* axenic
аксо-аксо'нный *a* axo-axonal
аксога'мия *f* axogamy
аксо-дендри'тный *a* axo-dendritic
аксоле'мма *f* axolemma, axilemma
аксоло'тль *m* axolotl (*Ambystoma mexicanum*)
аксо'н *m* axon
аксоне'ма *f* axoneme
аксопла'зма *f* axoplasm
аксопо'дия *f* axopodium
аксо-сомати'ческий *a* axo-somatic
аксости'ль *m* axostyle
актива'тор *m* activator
  ~ **лимфоци'тов** lymphocyte activator
  ~ **плазминоге'на** plasminogen activator
  дистанцио'нный ~ distance activator
  лока'льный ~ local activator
актива'ция *f* activation
  поведе'нческая ~ behavioral arousal
  ко'рковая ~ cortical arousal
активизи'ровать *v* arouse
активи'н *m* activin
акти'вность *f* activity, potency; **проявлять** ~ to display activity
  ~ **«в пустоте'»** *eth* vacuum activity
  ~ **коры' мо'зга** cortical activity
  ~ **супре'ссоров** suppressor activity
  ~ **хе'лперов** helper activity
  бактерици'дная ~ germicidal activity
  биоэлектри'ческая ~ electrobiological activity
  бурстпромо'торная ~ burst-promoting activity
  вы'званная ~ evoked activity
  едини'чная ~ unit activity
  замеща'ющая ~ *eth* substitute activity
  затра'вочная ~ priming activity
  коагули'рующая ~ clotting activity
  ма'тричная ~ template activity
  натрийсберега'ющая ~ sodium-retaining activity

переадресо'ванная ~ *eth* redirection activity
радиопроте'кторная ~ antiradiation activity
рефлекто'рная ~ reflex activity
свё'ртывающая ~ тромбоци'тов platelet coagulant activity
смеще'нная ~ *eth* displacement activity
супрессо'рная ~ suppressor activity
уде'льная ~ specific activity
ферментати'вная ~ enzymatic activity
фузимото'рная ~ fusimotor activity
хе'лперная ~ helper acitivity
акти'вный *a* active
транскрипцио'нно ~ transcriptionally active
акти'н *m* actin
актина'льный *a* actinal
актино- actino-
актинобиоло'гия *f* actinobiology
актиногра'мма *f* actinogram
актино'граф m actinograph
актино'идный *a* **actinoid**
актинока'рпный *a* actinocarpous
актиноме'р *m* actinomere
актиноме'трия f actinometry
актиномико'з *m* actinomycosis
актиномице'т *m* actinomycete
актиномици'н *m* actinomycin
актиномо'рфный *a* actinomorphic, actinomorphous
актинотри'хия *f* actinotrichium
актинотро'ха *f* actinotrocha
актинофа'г *m* actinophage
актогра'мма *f* actogram
акто'граф *m* actograph
актогра'фия *f* actography
актомиози'н *m* actomyosin
актофи'льный *a* actophilous
актуали'зм *m* actualism
акуметри'я *f* acumetry
акупункту'ра *f* acupuncture
акцелери'н *m* accelerin
акселерогра'мма *f* accelerogram
акцепти'ровать *v* accept
акце'птор *m* acceptor
алани'н *m* alanine
алейро'н *m* aleurone
алейро'новый *a* aleuronic
алекси'н *m* alexin
алеуроспо'ра *f* aleurospore
алецита'льный *a* alecithal
ализари'н *m* alizarin, alizarine
ализоге'нный *a* alysogenic
алимента'рный *a* alimentary, nutritional
алисфено'ид *m* alisphenoid
алкало'з *m* alkalosis
алкало'ид *m* alkaloid
алкили'рование *n* alkylation
восстанови'тельное ~ reductive alkylation
алкого'лиз m alcoholysis
алланто'идный *a* allantoid
аллантои'н *m* allantoin
алланто'ис *m* allantois, allantoic sac
аллаутога'мия *f* allautogamia
алле'ли *pl* of allele
мно'жественные ~ multiple alleles
аллели'зм *m* allelism
оппозицио'нный ~ oppositional allelism
аллелобрахиа'льный *a* allelobrachial
аллеогени'я *f* allelogenesis
алломо'рф *m* allelomorph
ло'жный ~ spurious allelomorph
сло'жный аллеломо'рф compound allelomorph
аллеломорфи'зм *m* allelomorphism
ступе'нчатый ~ step allelomorphism
аллеломо'рфный *a* allelomorphic
аллелопати'ческий *a* allelopathic
аллелопа'тия *f* allelopathy
аллелосома'льный, аллелосо'мный *a* allelosomal
аллелота'ксис *m* allelotaxy
аллелоти'п *m* allelotype
аллелохими'ческий *a* allelochemical
алле'ль *m* allele
~ совмести'мости compatibility allele
домина'нтный ~ dominant allele
ине'ртный ~ inactive allele
мно'жественные алле'ли pl multiple alleles
молча'щий ~ blank allele
неакти'вный ~ inactive allele
незапо'лненный ~ blank allele
позицио'нный ~ position allele
потенциа'льный ~ potency allele
пусто'й ~ blank allele
рецесси'вный ~ recessive allele
алле'льный *a* allelic
аллерге'н *m* allergen
пищево'й ~ food-borne allergen
специфи'ческий ~ specific allergen, offending allergen

аллерге'нность *f* allergenicity, allergenic capacity
аллерге'нный *a* allergenic
аллергиза'ция *f* allergization
аллергизи'рующий *a* allergenic
аллерги'н *m* allergin
аллерги'ческий *a* allergic
аллерги'я *f* allergy, allergia
~ **к дома'шней пы'ли** house dust allergy
~ **на пыльцу'** grass pollen allergy, pollen disese, hay fever, pollinosis
~ **на тарака'нов** cockroach allergy
~ **на ужа'ление** sting allergy
~ **неме'дленного ти'па** immediate-type allergy
алимента'рная ~ food allergy
атопи'ческая ~ atopic allergy
бактериа'льная ~ bacterial allergy
бессимпто'мная ~ latent allergy
заме'дленная ~ delayed allergy
ко'жная ~ dermal allergy, skin allergy
конта'ктная ~ contact allergy
лате'нтная ~ latent allergy
лека'рственная ~ drug allergy
мно'жественная ~ multiple allergy
неме'дленная ~ immediate allergy
неспецифи'ческая ~ nonspecific allergy
пеницилли'новая ~ penicillin allergy
пищева'я ~ food allergy
пыльцева'я ~ grass pollen allergy, pollen disese, hay fever, pollinosis
специфи'ческая ~ specific allergy
холодова'я ~ cold allergy
аллерго'з *m* allergosis
аллерго'лог *m* allergologist, allergist
аллерголо'гия *f* allergology, allergiology
аллергопа'тия *f* allergopathy
алле'я *f* avenue
алли'рование *n* allylation
алло- allo-
аллоантиге'н *m* alloantigen, allotypic antigen
сенсибилизи'рующий ~ allostimulating antigen
аллоантиге'нный *a* alloantigenic
аллогами'я *f* allogamy
аллога'мный *a* allogamous
аллогапло'ид *m* allohaploid
аллоге'н *m* allogene
аллогениза'ция *f* allogenization
аллоге'нный *a* allogenic
аллогетеропло'ид *m* alloheteroploid
аллогру'минг *m eth* allogrooming
аллодипло'ид *m* allodiploid
аллодипломоносо'мик *m* allodiplomonosomic
аллоиммуните'т *m* alloimmunity
аллоимму'нный *a* alloimmune
аллойофили'я *f phytop* alloiophylly
аллока'рпия *f* allocarpy
аллока'рпный *a* allocarpous
аллоко'ртекс *m* allocortex
аллоксази'н *m* alloxazin
аллокса'н *m* alloxan
аллометри'ческий *a* allometric, allomorphic
аллометри'я *f* allometry
алломи'ксис *m* allomixis
алломоногетеропло'ид *m* allomonoheteroploid
алломонодипло'ид *m* allomonodiploid
алломорфи'зм *m* allomorphism
алломорфо'з *m* allomorphosis
аллопаралектоти'п *m* alloparalectotype
аллопати'ческий *a* allopathic
аллопа'тия *f* allopathy
аллопатри'ческий *a* allopatric
аллопа'трия *f* allopatrie
аллопла'зма *f* alloplasm
аллоплазмати'ческий *a* alloplasmatic
аллоплоидио'н *m* alloploidion
аллоплоиди'я *f* alloploidy
аллополипло'ид *m* allopolyploid
гено'мный ~ genome allopolyplid
сегме'нтный ~ segmental allopolyploid
аллополиплоиди'я *f* allopolyploidy
аллоритми'я *f* allorhythmia
аллоритми'ческий *a* allorhythmic
аллосемати'ческий *a* allosematic
аллосенсибилиза'ция *f* allosensitization
аллосинде'з, аллосинде'зис *m* allosyndesis
аллосо'ма *f* allosoma
аллосо'мный *a* allosomal
аллостери'ческий *a* allosteric
аллостери'я *f* allostery
аллотетрапло'ид *m* allotetraploid
аллотетраплоиди'я *f* allotetraploidy
аллоти'п *m* allotype
молча'щий ~, **неэкспресси'руемый** ~ latent allotype
аллотипи'рование *n* allotyping
аллотипи'я *f* allotypy
аллотолера'нтность *f* allotolerancy
аллото'п *m* allotope
аллотопи'я *f* allotopia
аллотрансплanta'т *m* allograft

аллотрансплантa'ция *f* allotransplantation, allografting, allogenic transfer
аллотрипло'ид *m* allotriploid
аллотрипло'идный *a* allotriploid
аллотро'фный *a* allotrophic
аллохро'нный *a* allochronic
аллохто'нный *a* allochthonic, allochthonous
аллоцикли'я *f* allocycly
аллоэпито'п *m* alloepitope
аллю'р *m* gait
двухта'ктный ~ two-beat gait
есте'ственный ~natural gait
трехта'ктный ~ three-beat gait
четырёхта'ктный ~ four-beat gait
алый *a* vermilion, scarlet, red
кроцеи'новый ~ crocein scarlet
алыча' *f* cherry plum (*Prunus cerasifera*)
альбатро'с *m* albatross (*Diomedea*)
альбини'зм *m* albinism
альбино'с *m* albino
альбино'сный *a* albinotic
альбуми'н *m* albumin
~ **молока'** milk albumin
неполноце'нный ~ imperfect albumin
сы'вороточный ~ serum albumin
яи'чный ~ egg albumin
альбуми'новый *a* albuminous
альбумино'ид *m* albuminoid
альвеококко'з *m helm* alveococcosis
альвео'ла *f* alveole
альвео'ло-губно'й *a* alveololabial
альвео'ло-зубно'й *a* alveolopalatal
альвео'ло-нё'бный *a* alveolopalatal
альвео'ло-нё'бный *a* alveolopalatal
альвеоля'рный *a* alveolar
альголо'гия *f* algology, phycology
альдола'за *f* aldolase
альдостеро'н *m* aldosterone
альпа'ка *f* alpaca *(Lama pacos)*
альпи'йский *a* alpine
альте'рнанс *m phys* alternans
альтерна'ция *f* alternation
электри'ческая ~ electric alternation, electrical alternation
альтруи'зм *m eth* altruism
эволюцио'нный ~ evolutionary althruism
реципро'кный ~ reciprocal althriusm
альтруисти'ческий *a* altruistic
альфа-разнообра'зие *n* alpha-diveristy
Амазо'ния *f* Amazonian region
амантофи'льный *a* amanthophilous
а'мбер-мута'ция *f* amber-mutation
амбисексуа'льный *a* ambisexual
амблихромати'ческий *a* amblychromatic
амбоце'птор *m* amboceptor
~**бактерио'лиза** bacteriolytic amboceptor
~ **гемо'лиза** hemolytic amboceptor
а'мбра *f* ambergris, ambergrease
амбро'зия ragweed *(Ambrosia elatior)*
амбула'кр *m* ambulacrum
амбулакра'льный *a* ambulacral
амё'ба *f* ameba, amoeba
го'лая ~ naked amoeba *(Gymnamoebia)*
ра'ковинная ~ testate amoeba *(Testacea)*
амебиа'з *m* amebiasis, amoebiasis
~ **пе'чени** hepatic amoebiasis
амёбови'дный, амёбо'идный *a* amoebiform, amoeboid
амебоци'т *m* amoebocyte, amoebocytic cell
амейо'з *m* ameiosis
амелифика'ция *f* amelification
амелобла'ст *m* ameloblast
амелогене'з *m* amelogenesis
аменсали'зм *m* amensalism
амероспо'ра *f* amerospore
аметаболи'ческий *a* ametabolic
аметоптери'н *m* amethopterin
амиди'рование *n* amidation
амидолиз *m* amidolysis
амидоалкили'рование *n* amidoalkylation
амидометили'рование *n* amidomethylation
амикронуклеа'рный *a* amicronucleate
ами'ксис *m* amixis
амикси'я *f* amixia
амило'идный *a* amyloid
амилолити'ческий *a* amylolytic
амилобакте'рия *f* amylobacterium
амилопекти'н *m* amylopectin
амилопла'ст *m* amyloplast
амини'рование *n* amination
восстанови'тельное ~ reductive amination
аминоацили'рование *n* aminoacylation
аминоаци'л-тРНК *m* aminoacyl-tRNA
аминоацилтрансфера'за *f* aminoacyltransferase
аминокислота' *f* amino acid
замени'мая ~ replaceable amino acid, dispensable amino acid
незамени'мая ~ essential amino acid, nonreplaceable amino acid
амино'лиз *m* aminolysis
аминоэтили'рование *n* aminoethylation
амито'з *m* amitosis, amitotic division
амитоти'ческий *a* amitotic

аммоксиди'рование *n* ammoxidation
аммониотели'ческий *a* ammoniotelic
аммонифика'ция *f* ammonification
а'мнион *m* amnion, amnios
амнио'т *m* amniot
амниоти'ческий *a* amniotic
амо'рф *m gen* amorph
амо'рфный *a* amorphous
ампелогра'фия *f* ampelography
амплиту'да *f* amplitude
~ **движе'ний** movement amplitude
~ **цикли'ческих колеба'ний** cyclical amplitude
~ **усло'вий окружа'ющей среды'** environmental range
~ **экологи'ческих усло'вий** ecological range
су'точная ~ daily range
экологи'ческая ~ ecological amplitude, ecological range
амплифика'тор *m* amplificator
амплифика'ция *f* amplification
~ **кло'на** clone amplification
~ **ферме'нта** enzyme amplification
~ **ге'на** gene amplification
а'мпула *f* ampoule, ampule; *anat* ampulla
ампута'ция *f* amputation, ablation, abscission
~ **Бу'нге** Bunge's amputation
~ **Би'ра** Bier's amputation
ампути'ровать *v* amputate, ablate
амфи- amphi-
амфиапоми'кт m amphiapomict
амфиартро'з *m* amphiarthrosis, arthrodial joint, gliding joint, plane joint
амфиа'стер *m cyt* amphiaster
амфиастра'льный *a* amphiastral
амфиатланти'ческий *a* amphiatlantic
амфиби'йный, амфибио'нтный *a* amphibian, amphibious
амфи'бия *f* amphibian (*Amphibia*)
амфибласти'ческий *a* amphiblastic
амфибла'стула *f* amphiblastula
амфиболи'ческий *a* amphibolic
амфибореа'льный *a* amphiboreal
амфиваза'льный *a* amphivasal
амфига'мия *f* amphigamy
амфигапло'ид *m* amphihaploid
амфига'стрий *m* amphigastrium
амфигенети'ческий *a* amphygenetic, amphigonic
амфигони'я *f* amphigony, amphigenesis
амфи'да *f* amphid
амфидельфи'ческий *a* amphidelphic
амфидипло'ид *m* amphidiploid
амфидиплоиди'я *f* amphidiploidy
амфидипло'идный *a* amphidiploid
амфиди'ск *m* amphidisk
амфикарио'н *m* amphikaryon, amphinucleus
амфика'рпный *a* amphicarpous
амфико'н *m* amphicone
амфиконди'льный *a* amphicondylous
амфикриптофи'т *m* amphicryptophyte
амфими'ксис *m* amphimixis
амфими'кт *m* amphimict
амфимикти'ческий *a* amphimictic
амфинуклео'ла *f* amhinucleolus
амфину'клеус *m* amphikaryon
амфио'нт *m* amphiont
амфипати'ческий *a* amphipathic
амфипло'ид *m* amphiploid
амфиплоиди'я *f* amphiploidy
амфипло'идный *a* amphiploid
амфиспе'рмий *m* amphispermium
амфиспо'ра *f* amphispore
амфиспорангиа'тный *a* amphisporangiate
амфисти'лия *f* amphistyly
амфисто'мный *a* amphistomous
амфиталли'чный *a* amphithallic
амфите'на *f* amphitene
амфите'ций *m* amphithecium
амфитоки'я *f* amphytoky, amphoterotoky
амфитрихиа'льный *a* amphitrichous
амфитро'пный *a* amphitropous
амфитро'ха *f* amphitrocha
амфифи'т *m* amphiphyte
амфихромати'зм *m* amphichromatism, amphichromy
амфице'льный *a* amphicoelous
амфици'т *m* amphicyte
амфоге'нный *a* amphogenic
амфоте'рный *a* amphoteric
амфотони'я *f* amphotony
амфофи'льный *a* amphophilic
анабио'з *m* anabiosis
анабио'нт *m* anabiont
анаболи'зм *m* ababolism
анаболи'т *m* anabolite
анаболи'ческий *a* anabolic
анаболи'я *f* anaboly
анавакци'на f anavaccine
анагене'з *m* anagenesis
анагормо'н *m* anahormone
анадро'мный *a* anadromous
анако'нда *f* anaconda *(Eunectes)*

анакроти'ческий *a* anacrotic
анаксиа'льный *a* anaxial
ана'лиз *m* analysis
~ **антия'дерного фа'ктора** ANA assay
~ **бляшкообразова'ния** plaque assay
~ **гла'вных компоне'нт** principal compnent analysis, P.C. analysis
~ **ЕА-розеткообразова'ния** EA-rosette assay
~ **Е-розеткообразова'ния** E-rosette assay
~ **ковариа'нсы** analysis of covariance
~ **компози'ции** compositional analysis
~ **концевы'х групп** end-group analysis
~ **коне'чных результа'тов** end-point analysis
~ **кро'ви** blood count, blood examination
~ **ме'тодом иммобилиза'ции кле'ток** cell attachment assay
~ **непреры'вности** continuity analysis
~ **погре'шностей** error analysis
~ **по Оухтерло'ни** Ouchterlony analysis
~ **розеткообразова'ния** rosette assay
~ **сме'шанного колониеобразова'ния** mixed colony assay
~ **соотве'тствий** correspondence analysis
~ **сопряжё'нности** association analysis
~ **соста'ва семе'й** household analysis
~ **травосто'я** herbage analysis
~ **ферментати'вной гранулосекре'ции** granule enzyme release assay
адгезио'нно-кле'точный ~ cell attachement assay
активацио'нный ~ activation analysis
весово'й ~ gravimetric analysis
вы'борочный ~ sampling analysis
гармони'ческий ~ harmonic analysis, Fourier analysis
гено'мный ~ genome analysis
гравиметри'ческий ~ gravimetric analysis
градие'нтный ~ gradient analysis
гранулометри'ческий ~ granulometric analysis, particle-size analysis
вуме'рный ~ two-dimensional analysis
двухмаркё'рный иммунофлуоресце'нтный ~ double-label immunofluorescence assay
двухсло'йный иммуноферме'нтный ~ two-site enzyme immunoassay
двухстади'йный иммуноферме'нтный ~ two-site immunoenzymometric assay, two-site enzyme immunoassay
дискримина'нтный ~ discriminant analysis
дисперсио'нный ~ analysis of variance, variance analysis
дистанцио'нный ~ *biom* distance analysis
дот-блот ~ dot-blot analysis
ЕА-РОК ~ EA-rosette assay
Е-РОК ~ E-rosette assay
изото'пный ~ isotope analysis
иммуногенети'ческий ~ immunogenotypic analysis
иммунокарбо'новый ~ carbon immunoassay
иммунологи'ческий ~ immunoassay, immunodetection
иммунолюминесце'нтный ~ luminescent immunoassay
иммунорадиометри'ческий ~ immunoradiometric assay
иммунофенотипи'ческий ~ immune marker analysis
иммуноферме'нтный ~ enzymoimmunoassay
иммунофлуоресце'нтный ~ immunofluorescence analysis, fluoroimmunoassay
иммунохими'ческий ~ immune marker analysis
иммунохими'ческий ~ опухолей tumor marker immunoassay
кла'стерный ~ cluster analysis
клини'ческий ~ clinical analysis
ковариацио'нный ~ covariance analysis
кого'ртный ~ cohort analysis
коли'чественный иммунофлуоресце'нтный ~ immunofluorometric assay
комплементацио'нный ~ complementation analysis
компози'ции *ecol* compositional analysis
кондуктометри'ческий ~ conductometric analysis
контро'льный ~ check analysis, control analysis
конформацио'нный ~ conformational analysis
корреляцио'нный ~ correlation analysis
КОЕ-микст ~ mixed colony assay
микроскопи'ческий ~ microscopic examination
микроспектра'льный ~ microspectral analysis

непреры'вный ~ continuous analysis
ка'пельный ~ spot analysis
коли'чественный ~ assay
конкуре'нтно-свя'зывающий ~ competitive binding assay
ла'тексный ~ latex-based assay
многоэта'пный иммуноферме'нтный ~ multilayer enzyme immunoassay
мономаркё'рный иммунофлуоресце'нтный ~ single-label immunofluorescence assay
обме'нно-фа'зовый иммуноферме'нтный ~ phase change enzymoimmunoassay
объё'мный ~ volumetric analysis
полево'й ~ field analysis
приближё'нный ~ approximate analysis
прижи'зненный ~ intravital analysis
пыльцево'й ~ pollen analysis
радиоиммунопреципитацио'нный ~ radioimmune precipitation assay
радиоиммуносорбе'нтный ~ radioimmunosorbent assay
радиоиммунохими'ческий ~ radioimmunochemical analysis
радиометри'ческий ~ radiometric analysis
регрессио'нный ~ regression analysis
рекомбинацио'нный ~ recombination analysis
рентгенострукту'рный ~ X-ray analysis
рестрикцио'нный ~ restriction analysis
седиментацио'нный ~ sedimentation analysis
серологи'ческий ~ serologic assay
сетево'й ~ network analysis
систе'мный ~ system analysis
сонографи'ческий ~ sonographic analysis
сорбцио'нный ~ sorptographic analysis
спектра'льный ~ spectroscopic analysis
спо'рово-пыльцево'й ~ sporo-pollen analysis
структу'рно-группово'й ~ group analysis
твё'рдофазный иммуноферме'нтный ~ enzyme-linked immunosorbent assay
твёрдофа'зный иммунофлуоресце'нтный ~ fluorescent binding immunoassay
тетра'дный ~ tetrad analysis
трансдукцио'нный ~ transduction analysis
фа'кторный ~ factor analysis
флуоресце'нтный ~ fluorescence analysis
хими'ческий ~ chemical analysis
хроматографи'ческий ~ chromatographic analysis
часто'тный ~ frequency analysis

анализа'тор *m* analyzer
~ **ско'рости реа'кции** reaction rate analyzer
~ **усло'вных раздражи'телей** conditioned stimulus analyzer
автомати'ческий ~ autoanalyzer
аминокисло'тный ~ aminoacid analyzer
вестибуля'рный ~ vestibular analyzer
вкусово'й ~ gustatory analyzer, taste analyzer
дви'гательно-кинестети'ческий ~ motor-kinesthetic analyzer
дви'гательный ~ motor analyzer
зри'тельный ~ visual analyzer
ко'жный ~ cutaneous analyzer
межфа'зный ~ interphase analyzer
многокана'льный ~ multichannel analyzer
обоня'тельный ~ olfactory analyzer
проприоцепти'вный ~ proprioceptive analyzer
речедви'гательный ~ speech motor analyzer
слухово'й ~ acoustic analyzer
такти'льный ~ tactile analyzer

аналити'ческий *a* analytic, analytical
ана'лог *m* analog, analogue
~ **основа'ния** *mbiol* base analogue
синтети'ческий ~ гормо'на hormone substitute

аналоги'чный *a* analogous
анало'гия *f* analogy
ана'льно-полово'й *a* anogenital
ана'льный *a* anal
ана'мнез *m* history, anamnesis
ана'мнии *f pl* anamniotes *(Anamnia)*
анамнио'т *m* anamniote
анаморфо'з *m* anamorphosis
анана'с *m* pineapple (*Ananas sativus*)
анаплази'я *f* anaplasia
анаплевроти'ческий *a* anapleurotic
анапофи'з *m* anapophysis
анастомо'з *m* anastomosis
анастомози'ровать *v* anostomoze
анастомози'рующий *a* anastomosing
анастра'льный *a* anastral
анатокси'н *m* anatoxin
ана'том *m* anatomist
анатомический *a* anatomical
анато'мия *f* anatomy

~ **челове'ка** anthropotomy
макроскопи'ческая ~ gross anatomy
микроскопи'ческая ~ microscopical anatomy
описа'тельная ~ descriptive anatomy
патологи'ческая ~ morbid anatomy
прикла'дная ~ applied anatomy
систе'мная ~ systemic anatomy
сравни'тельная ~ comparative anatomy
топографи'ческая ~ topographic anatomy, regional anatomy
ча'стная ~ special anatomy
анатро'пный *a* anatropic
анафа'за *f* anaphase
анафи'за *f* anaphysis
анафилакси'я *f* anaphylaxis
акти'вная ~ active anaphylaxis
возвра'тная ~ recurrent anaphylaxis
приобретё'нная ~ acquired anaphylaxis
анафилакти'ческий *a* anaphylactic
анафилактоге'н *m* anaphylactogen
анафилотокси'н *m* anaphylotoxin
анахоре'з *m* anachoresis
анаэро'б *m* anaerobe, anoxybiont, anaerobion
гнило'стный ~ putrefactive anaerobe
облига'тный ~ strict anaerobe
факультати'вный ~ facultative anaerobe
анаэробио'з *m* anaerobiosis, anoxybiosis
анаэробио'нт *m* anaerobe
анаэро'бия *f* anaerobia
анаэро'бный *a* anaerobic, anoxybiotic
а'нгел *m*: **морско'й ангел** angel shark (Squatina)
ангидробио'з *m* anhydrobiosis
анги'на *f* angina, tonsillitis
~ **Венса'на** Vincent's disease
~ **Гено'ха** Henoch's angina
~ **Дюге'** Duguet's angina
~ **Лю'двига** Ludwig's angina, Ludwig's disease
ангиобла'ст *m* angioblast
ангиогра'фия *f* angiography
ангиока'рпный *a* angiocarpous
ангиоло'гия *f* angiology
ангио'ма *f* angioma
ангиоспори'ческий *a* angiosporous
ангиотензи'н *m* angiotensin
ангиото'мия *f* angiotomy
ангиотони'н *m* angiotonin
анголоцикли'ческий *a* anholocyclic
андро- andro-
андрогамо'н *m* androgamone
андроге'н *m* androgen
андрогене'з *m* androgenesis
андрогенети'ческий *a* androgenetic
андроге'нный *a* androgenic, androgenous
андрогинофо'р *m* androgynophore
андрогони'дий *m* androgonidium
андрого'ний *m* androgonium
андрокли'ний *m* androclinium, clinandrium
андроко'ний *m* androconium
андромерого'ния *f* andromerogony
андропета'льный *a* andropetalous
андросо'ма *f* androsome
андроспо'ра androspore
андроспора'нгий *m* androsporangium
андростадиоло'н *m* androstadiolon
андростеро'н *m* androsterone
андрофи'лл *m* androphyll
андрофи'льный *a* androphilous
андрофо'р *m* androphore
андро'цей *m* androecium
андроци'т *m* androcyte
андроэ'ция *f* androecy
аневри'зма *f* aneurism
аневри'н *m* aneurin, aneurine
анеми'чный *a* anemic
анеми'я *f* anemia
~ **Аддисо'на-Би'рмера** Addison-Biermer's anemia
~ **Кули'** Cooley's anemia
~ **Ледере'ра** Lederer's anemia
~ **Фанко'ни** Fanconi's anemia
анемо- anemo-
анемо'метр *m* wind gauge
анемоспори'ческий *a* anemosporic
анемота'ксис *m* anemotaxis
анемотропи'зм *m* anemotropism
анемофи'льность *f* anemophily
анемофи'льный *a* anemophilous, anemogamous
анемофо'б *m* anemophobe
анемохо'р *m* anemochore
анемохо'рный *a* anemochorous
анемоэнтомофи'льный *a* anemoentomophilous
ане'ргия *f* anergy
анестези'рование *n* anesthetization
анестези'ровать *v* anesthetize
анестези'я *f* anesthesia, anaesthesia, anaesthetization
~ **Куленка'мпффа** Kuhlenkampff's anesthesia
анизо- aniso-

анизогаме'та *f* anisogamete
анизога'мия *f* anisogamy
анизога'мный *a* anisogamous
анизоге'ния *f* anisogeny
анизоге'нный *a* anisogenous
анизогено'мный *a* anisogenomatic
анизода'ктильный *a* anisodactylous
анизодо'нтный *a* anisodont
анизока'рпный *a* anisocarpous
анизомери'я *f* anisomery
анизометри'ческий *a* anisometric
анизопло'ид *m* anisoploid
анизоплоиди'я *f* anisoploidy
анизопло'идный *a* anisoploid
анизополиплоиди'я *f* anisopolyploidy
анизополя'рный *a bot* anisopolar
анизорефле'ксия *f* anisoreflexia
анизоспо'ра *f* anisospore
анизоспори'я *f* anisospory
анизотони'ческий *a* anisotonic
анизотрисоми'я *f* anisotrisomy
анизотропи'я *f* anisotropy
анизотро'пный *a* anisotropic
анизофилли'я *f* anisophylly
анизоцито'з *m* anisocytosis
анимализа'ция *f* animalization
анима'льный *a* animal
анкилостомо'з *m helm* ancylostomiasis
анкофи'льный *a* ancophilous
аннида'ция *f* annidation
аноксибио'з *m* anoxibiosis
анокси'я *f* anoxia
анома'лия *f* anomaly, abnormality
  ~ **Эбште'йна** Ebstein's disease
аноректа'льный *a* anorectal
анэупло'ид m anorthoploid
анортоспира'ль *f* anorthospiral
аносми'я *f* anosmia
ановуля'ция *f* anovulation
аноксеми'я *f* anoxaemia, anoxemia
анома'льный *a* anomalous, abnormal
анортогене'з *m* anorthogenesis
анортопло'ид *m* anorthoploid
анофелеци'д *m* anophelecide
анофелоге'нный *a* anophelogenous
ансери'н *m* anserine
антагони'зм *m* antagonism, antipathy
  конкуре'нтный ~ competitive antagonism
антагони'ст *m* antagonist
антагонисти'ческий *a* antagonistic, antipathic
Антарктогея *f* Antarctogaea
антекли'макс *m* anteclimax
антената'льный *a* antenatal
анте'нна *f* antenna
антенна'льный *a* antennal, antennary
анте'ннула *f* antennule
антенну ля'рный *a* antennular
антеридиа'льный *a* antheridial
антери'дий *m* antheridium, androgametangium
антеридионо'сец *m* antheridiophore
антерозо'ид *m* antherozoid
антеспорофи'лл *m* antesporophyll
анти- anti-
антиаллерге'нный *a* antiallergenic, antiallergic
антиамбоце'птор *m* anti-intermediary body
антиантите'ло *n* antiantibody
антибактериа'льный *a* antibacterial
антибиогра'мма *f* antibiogram
антибио'нт *m* antibiont
антибио'тик *m* antibiotic
антибиоти'ческий *a* antibiotic
антивени'н *a* antivenin
антивирусный *a* antiviral
антивитами'н *m* antivitamin
антивспе'ниватель *m* antifoam
антигельми'нтный *a* antivermicular
антигемолити'ческий *a* antihemolytic
антиге'н *m* antigen
  ~ **АВО** ABO antigen
  антиге'ны групп кро'ви *pl* blood-group antigens
  ~ **Буаве'на** Boivin antigen
  ~ **ви'русной оболо'чки** viral coat antigen
  ~ **гистосовмести'мости** histocompatibility antigen
  ~ **Гудпа'счура** Goodpasture's epitope
  ~ **зо'ны бере'менности** pregnancy-associated antigen, pregnancy zone antigen
  ~ **ка'псулы** capsular antigen, envelope antigen
  ~ **кле'тки-мише'ни** target antigen
  ~ **ко'жи** skin antigen
  ~ **Лью'иса** Lewis antigen
  ~ **окружа'ющей среды'** environmental antigen, natural antigen
  ~ **Па'уля-Бу'ннеля** Paul-Bunnel's antigen
  австрали'йский ~ Australia antigen
  активацио'нный ~ activation antigen
  аллотипи'ческий ~ alloantigen, allotypic antigen
  «аресто'ванный» ~ entrapped antigen

бактериа'льный ~ bacterial antigen
близкоро'дственный ~ shared antigen
блоки'рующий ~ blockading antigen
ви'русный ~ viral antigen
внекле'точный ~ extracellular antigen
волокни'стый ~ fiber antigen
«воскре'сший» ~ recall antigen
высокоиммуноге'нный ~ potent antigen, immunodominant antigen
гетерогенети'ческий ~ heterogenetic antigen
гетерокле'точный ~ heat-labile antigen
группо-специфи'ческий ~ group-specific antigen
далё'кие антиге'ны *pl* distantly related antigens
дерма'льный ~ skin antigen
диагности'ческий ~ diagnostic antigen
дифференциро'вочный ~ differential antigen
жгу'тиковый ~ flagellar antigen
замаскиро'ванный ~ "buried" antigen
иммобилизо'ванный ~ fixed antigen
иммунодомина'нтный ~ immunodominant antigen
инактиви'рованный ~ dead antigen
инкапсули'рованный ~, интегра'льный ~ entrapped antigen
иску'сственный ~ artificial antigen, synthetic antigen
ка'псульный ~ capsular antigen, envelope antigen
карцинофета'льный ~ carcinofetal antigen, oncofetal antigen
кле'точный ~ cellular antigen
клони'рованный ~ cloned antigen
конкури'рующий ~ competing antigen
контро'льный ~ test antigen
корпускуля'рный ~ particulate antigen, particle antigen
кортика'льный ~ cortical-specific antigen
мембра'нный ~ membrane antigen
миело'идный ~ myeloid-lineage antigen
моде'льный ~ model antigen
модифици'рованный ~ modified antigen
моновале'нтный ~ monovalent antigen
монодетермина'нтный ~ monovalent antigen
мо'стиковый ~ bridged antigen
мультиме'рный ~ multimeric antigen
нейробласто'мный ~ neuroblastoma-associated antigen
неочи'щенный ~ crude antigen
нераствори'мый ~ particle antigen, particulate antigen
неро'дственный ~ "nonshared" antigen
опухолеспецифи'ческий ~ tumor-associated antigen, tumor-specific antigen
органоспецифи'ческий ~ organ-specific antigen
оста'точный ~ residue antigen
парциа'льный ~ partial antigen, incomplete antigen
перекрё'стно-реаги'рующие антиге'ны *pl* cross-reacting antigens
пове'рхностный ~ surface antigen
поливале'нтный ~ polyvalent antigen
по'лный ~ complete antigen
полово'й ~ sex-associated antigen
приро'дный ~ natural antigen, environmental antigen
секвестри'рованный ~ sequestered antigen, inaccessible antigen
сенсибилизи'рующий ~ sensitizing antigen
синтети'ческий ~ artificial antigen
сло'жный ~ complex antigen
«слу'щивающийся» ~ "shed" antigen
сомати'ческий ~ "shed" antigen
стадиеспецифи'ческий ~ stage-specific antigen, transitory antigen
сы'вороточный ~ serum-born antigen
термолаби'льный ~ heat-labile antigen
ти'мус-зави'симый ~ thymic-dependent antigen
ти'мус-независи'мый ~ thymic-independent antigen
тканеспецифи'ческий ~ tissue-specific antigen
Т-кле'точный ~ T-cell antigen
Т-кле'точный дифференциро'вочный ~ T-lineage-associated antigen
транзи'торный ~ stage-specific antigen, transitory antigen
трансплантацио'нный ~ transplantation antigen
тромбоцита'рный ~ platelet antigen
трофобласти'ческий ~ trophoblastic antigen
фибро'зный ~, фибрилля'рный ~ fiber antigen
холодово'й ~ cold antigen
цитоплазмати'ческий ~ cytoplasmic antigen

чужеро'дный ~ "nonshared" antigen
экрани'рующий ~ blockading antigen
эмбриона'льный ~ embryonic antigen, embryonic-stage antigen
антигени'я *f* antigeny
антиге'н-метаболи'т *m* metabolic antigen
антиге'н-мише'нь *m* target antigen
антиге'нность *f* antigenicity
антиге'нный *a* antigenic
антиглобули'н m antiglobulin
антигормо'н m antihormone
антидетермина'нта *f*:
внутрицепьева'я ~ *imm* "grove"-type determinant, intrachain determinant
антидо'т *m* antidote
антидро'мный *a* antidromic
а'нтии *f pl* antiae
антиидиоти'п m anti-idiotype
антикатализа'тор *m* anticatalyst
антикоагули'рующий *a* anticoagulative
антикоагуля'нт *m* anticoagulant
антикодо'н *m* anticodon
антикомплемента'рный *a* anticomplementary
антило'па *f* antelope (*Antilopinae*); **ста'до антило'п** herd of antelope
~ **бе'йза** beisa antelope *(Oryx gazella beisa)*
~ **бе'йра** beira antelope *(Dorcatragus megalotis)*
вилоро'гая ~ prong-horned antelope *(Antilocapra americana)*
винторо'гая ~ spiral-horned antelope *(Strepsicerotini)*
ка'рликовая ~ pygmy antelope *(Neotragus)*
лошади'ная ~ **roan antelope, roane antelope** *(Hippotragus equinus)*
четырёхро'гая ~ four-horned antelope *(Tetracerus quadricornis)*
антиме'р *m* antimere
антиметаболи'т *m* antimetabolite, metabolic antagonist
антимикро'бный *a* antimicrobial
антимитоти'ческий *a* antimitotic
антимо'рф *m* antimorph
антимутаге'н *m* antimutagene
антиокисли'тель, антиоксида'нт *m* antioxidant
антипериста'льтика *f* antiperistalsis
антипета'льный *a* antipetal, antipetalous
антипо'рт *m* antiport
антипо'ртер *m* antiporter
антисе'птик *m* antiseptic
антисе'птика *f* antisepsis
антисепти'ческий *a* antiseptic
антисмыслово'й *a mbiol* antisense
антиспазмати'ческий *a* antispasmodic
антисы'воротка *f* antiserum, antibody-containing serum, immune serum, protective serum
~ **к Ia-антиге'нам** anti-Ia antiserum
~ **по'сле реиммуниза'ции** hyperimmune serum
~ **про'тив носи'теля** anticarrier serum
абсорби'рованная ~ absorbed antiserum, exhausted antiserum
аллоге'нная ~ alloserum
антивидова'я ~ antispecies antiserum
высокоави'дная ~, **высокоспецифи'чная** ~ high-avid antiserum
истощё'нная ~ exhausted antiserum
низкоави'дная ~ nonavid antiserum
перекрё'стно-реаги'рующая ~ cross-reactive antiserum
подкла'сс-специфи'ческая ~ subclass-specific antiserum
поливале'нтная ~ multispecific antiserum
поликлона'льная ~ polyclonal antiserum
полиспецифи'ческая ~ multispecific antiserum
термоинактиви'рованная ~ heat-inactivated antiserum
типоспецифи'ческая ~ type-specific antiserum
челове'ческая антиглобули'новая ~ Coombs-type antiserum
чужеро'дная ~ foreign antiserum
антите'ло *n* antibody
~ **ма'тери** maternal antibody
абортоге'нное ~ abortifacient antibody
агглютини'рующее ~ agglutinating antibody
анафилакти'ческое ~ anaphylactic antibody
антиадгезио'нное ~ adhesion-blocking antibody
антигапте'новое ~ antihapten antibody
антиидиотипи'ческое ~ anti-idiotypic antibody, anti-antiself antibody, "internal image" antibody
антиклонотипи'ческое ~ anticlonotypic antibody

антиплацента'рное ~ antiplacental antibody
антире'зусное ~ anti-Rh antibody
антитрансплантацио'нное ~ graft-specific antibody
антиэритроцита'рное ~ antired-cell antibody
«аресто'ванное» ~ entrapped antibody, integral antibody
аутоантиидиотипи'ческое ~ autoanti-idiotypic antibody
биотинили'рованное ~ biotin-conjugated antibody
блоки'рующее ~ blocking antibody, incomplete antibody, univalent antibody
втори'чное ~ secondary antibody, second-layer antibody
гемолити'ческое ~ hemolytic antibody
гетеровале'нтное ~ heteroligating antibody, bifunctional antibody
гетероклити'ческое ~ heteroclytic antibody
гетероспецифи'ческое ~ heteroligating antibody
гиперимму'нное ~ hyperimmune antibody
гомоцитотро'пное ~ homocytotropic antibody
гумора'льное ~ humoral antibody
двухвале'нтное ~ bivalent antibody
домен-специфи'ческое ~ domain-specific antibody
есте'ственное ~ natural antibody, classical antibody
защи'тное ~ protective antibody
идиотипи'ческое ~ idiotypic antibody, idiotype-bearing antibody
иммобилизи'рованное ~ solid-phase antibody, capture antibody
инкапсули'рованное ~ entrapped antibody, integral antibody
кожно-сенсибилизи'рующее ~ skin-sensitizing antibody
комплементсвя'зывающее ~ complement-fixing antibody
лейкоагглютини'рующее ~ leukoagglutinating antibody
локус-специфи'ческое ~ locus-specific antibody
модифици'рованное ~ modified antibody
моновале'нтное ~ monovalent antibody
моноклона'льное ~ monoclonal antibody
мономо'рфное ~ monomorphic antibody
непо'лное ~ incomplete antibody, monovalent antibody
непреципити'рующее ~ nonprecipitating antibody
низкоави'дное ~ low-grade antibody
опсонизи'рующее ~ opsoning antibody
по'зднее ~ late antibody
поликлона'льное ~ polyclonal antibody
полимо'рфное ~ polymorphic antibody
по'лное ~ whole antibody
поствакцина'льное ~ vaccine-induced antibody
преципити'рующее ~ precipitating antibody
приро'дное ~ natural antibody, classical antibody
протекти'вное ~ protective antibody
ра'ннее ~ early antibody
реаги'новое ~ reaginic antibody
реагинოподо'бное ~ reagin-like antibody
рекомбина'нтное ~ recombinant antibody
сайт-специфи'ческое ~ site-directed antibody
серотипи'ческое ~ serotype-specific antibody
сесси'льное ~ sessile antibody, tissue-fixed antibody
следово'е ~ minute antibody
твердофа'зное ~ solid-phase antibody
токсиннейтрализу'ющее ~ antitoxic antibody
хелатизи'рованное ~ chelate-tagged antibody
химе'рное ~ chimeric antibody, recombinant antibody
холодово'е ~ cold-reactive antibody
цитостати'ческое ~ cytostatic antibody
цитофи'льное ~ cytophilic antibody

антителоблоки'рующий *a* antibody-blocking
антителозави'симый *a* antibody-dependent
антитермина'ция *f* antitermination
антитокси'н *m* antitoxin
антитокси'ческий *a* antitoxic
антиферме'нт *m* antienzyme
антифона'льный *a* antiphonal; **антифона'льное пе'ние** antiphonal singing, duetting
антихолинерги'ческий *a* anticholinergic
антиципа'ция *f* anticipation
антогене'з *m* anthogenesis
анто'дий *m* anthodium

антока'рпный *a* anthocarpous
антокли'ний *a* anthoclinium, clinanthium
антоксанти'н *m* anthoxanthin
анто'лиз *m* antholysis
анто'новка *f* antonovka apple tree
антотакси'я f anthotaxy
антофи'льный *a* anthophilous
антофи'т *m* anthophyte
антофо'р *m* anthophore
антоциа'н *m* anthocyan
антоцианиди'н *m* anthocyanidin
антоциани'н *m* anthocyanin
антракно'з *m phytop* anthracnose *(caused by Colletotrichum)*
~ **зла'ков** cereal anthracnose *(caused by Colletotrichum graminicola)*
~ **фасо'ли** common bean anthracnose *(caused by Colletotrichum lindemuthianum)*
се'верный ~ кле'вера northern anthracnose *(caused by Kabatiella caulivora)*
ю'жный ~ кле'вера southern anthracnose *(caused by Colletotrichum trifolii)*
пятни'стый ~ spot anthracnose *(caused by Elsinoe sphaceloma)*
антропометри'ческий *a* anthropometric
антропометри'я *f* anthropometry
антропоморфи'зм *m* anthropomorphism
антропомо'рфный *a* anthropomorphous
антропогене'з *m* anthropogenesis
антропоге'нный *a* anthropogenous, anthropogenic
антропозооно'з *m* anthropozoonosis
антрополо'гия *f* anthropology
антропосфе'ра *f* anthroposphere
антропофи'льный *a* anthropophilous
антропофи'т *m* anthropophyte
антропохо'р *m* anthropochore
антропохо'рный *a* anthropochorous
антропоценоло'гия *f* anthropocenology
антропоцентри'зм *m* anthropocentrism
а'нус *m* anus
анцестра'льный *a* ancestral
анча'р *m* antiaris *(Antiaris)*
анэлектрото'н *m* anelectrotonus
анэ'струс *m* anestrus
анэуга'мия *f* aneugamy
анэупло'ид *m* aneuploid
анэуплоиди'я *f* aneuploidy
анэупло'идный *a* aneuploid
анэуцентри'ческий *a* aneucentric
анча'р *m* upas-tree (*Antiaris*)
анэупло'ид *m* aneuploid
анэуплоиди'я *f* aneuploidy
ао'рта *f* aorta
брюшна'я ~, вентра'льная ~ abdominal aorta, ventral aorta
грудна'я ~ thoracic aorta
дорса'льная ~, спинна'я ~ dorsal aorta
аорта'льный *a* aortic
апалогами'ческий *a* apalogamic
апати'чный *a* torpid
а'пекс *m* apex
апельси'н *m bot* orange (*Citrus sinensis*)
аперту'ра *f bot* aperture
аперту'рный *a* aperturate
апета'льный *a* apetalous
апика'льный *a* apical
апитокси'н *m* apitoxin
аплази'я *f* aplasia
апланогаме'та *f* aplanogamete
апланогамета'нгий *m* aplanogametangium
апланоспо'ра *f* aplanospore
апласти'ческий *a* aplastic
аплацента'рный *a* aplacental
аплоперисто'матный *a* aploperistomatous
аплостемо'нный *a* aplostemonous
апно'э *n* apnoea
апобази'дия *f* apobasidium
апога'мия *f* apogamy
апога'мный *a* apogamic, apogamous
апоге'й *m* acme
апогени'я *f* apogeny
апоги'ния *f* apogyny
аподе'ма *f* apodema, apodeme
апока'рпий *m* apocarp
апока'рпия *f* apocarpy
апока'рпный *a* apocarpous
апокри'нный, апокри'новый *a* apocrine
аполега'мия *f* apolegamy, apolegamic mating
апомейо'з *m* apomeiosis
апоми'ксис *m* apomixis
вегетати'вный ~ apospermy
апоми'кт *m* apomict
апомикти'чный *a* apomictic
апоневро'з *m anat* aponeurosis
висо'чный ~ temporal aponeurosis
апопета'льный *a* apopetalous
апоплазмати'ческий *a* apoplasmatic
апопти'ческий *a* apoptotic
апопто'з *m* apoptosis
апорепре'ссор *m* apo-repressor
апорога'мия *f* aporogamy
апосемати'ческий *a* aposematic

апосперми'я *f* apospermy
апоспори'я *f* apospory
апоспорогони'я *f* aposporogony
апострофи'я *f* apostrophe
апоте'ций *m* apothecium
апотро'пный *a* apotropous
апоферме'нт *m* apoenzyme
апофи'з *m zool* apophysis
апофи'за *m bot* apophysis
апофи'т *m* apophyte
апохромати'ческий *a* apochromatic
апоци'т *m* apocyte
аппара'т *m* apparatus
~ **Ва'рбурга** Warburg's apparatus
~ **Го'льджи** *cyt* Golgi apparatus, Golgi's organ, reticular apparatus of Golgi, canalicular apparatus
~ **движе'ния** locomotor apparatus
~ **для титрова'ния** volumetric apparatus
~ **Сес Гри'на** Seth Green apparatus
~ **Соксле'та** Soxhlet's extraction apparatus
~ **Холде'йна** Haldane's apparatus
ве'беров ~ Weberian apparatus
вестибуля'рный ~ vestibular apparatus
голосово'й ~ vocal apparatus
диффузио'нный ~ diffusion apparatus
жева'тельный ~ masticatory apparatus
замыка'ющий ~ **у'стьиц** *bot* stomatal closing apparatus
запира'тельный ~ **трахе'и** *ent* closing apparatus ~ of trachea
зубно'й ~ dentition
мочеполово'й ~ urogenital apparatus, genitourinary apparatus
опо'рно-дви'гательный ~ locomotor system, locomotor apparatus
пищевари'тельный ~ digestive apparatus
ротово'й ~ *ent* mouth-parts
слухово'й ~ hearing aid
аппарату'ра *f* equipment
регистри'рующая ~ recording equipment
стациона'рная ~ fixed equipment
аппе'ндикс *m* appendix, cecal appendage
аппете'нтный *a* appetitive
аппете'нция *f* appetence
аппети'т *m* appetite
во'лчий ~ cynorexia, bulimia
апплика'тор *m* applicator
аппози'ция *f* apposition
апрессо'рий *m* appressorium
аппрессо'рия *f* appressorium
аппроксима'ция *f* approximation
после'довательная ~ *biom* successive approximation
апте'рий *m* apterium
арахни'диум *m* arachnidium
арахноло'гия *f* arachnology
арбови'рус *m* arbovirus
арбу'з *m bot* water melon (*Citrullus vulgaris*)
арбу'скула *f bot* arbuscle, arbuscule
аргали' *ther* argali *(Ovis ammon)*
аргентофи'льный *a* argentophilic, argentaffin, argyrophil, argyrophilous
аргини'н *m* arginine
аргирофи'льный *a* argentophilic, argentaffin, argyrophil, argyrophilous
арде'лла *f* ardella
ареакти'вность *f* areactivity
иммунологи'ческая ~ immunological areactivity
ареа'л *m* area, range, geographic range
~ **популя'ции** of population
высо'тный ~ altitudinal area
ле'нточный ~ belt area
нагу'льный ~ foraging area
разо'рванный ~, **разъедине'нный** ~ disconnected area, discontinuous area, discontinuous range, breaking-up area
распыле'нный ~ diffuse area
рели'ктовый ~ relic area
совреме'нный ~ present range
соприкаса'ющиеся ареа'лы *pl* adjacent areas, adjacent ranges
сплошно'й ~ continuous area, continuous range
у'зкий ~ restricted area
арео'ла *f* areola, areole, areolation
локуля'рная ~ loculate areola *(of diatoms)*
поро'идная ~ poroid areola *(of diatoms)*
ареоли'рованный *a* areolate
ареоля'рный *a* areolar
арео'метр m areometer
арефле'ксия *f* areflexia
аридиза'ция *f* aridization, exciccation
ари'дность *f* aridity
ари'дный *a* arid
арилалкили'рование n arylalkylation
арили'рование n arylation
ари'ллус *m* aril
ло'жный ~ arillode
аритми'ческий *a* arrhythmic
аритми'я *f* arrhythmia, anisorhythmia
аркообра'зный *a* arciform
аркоце'нтрум *m* arcocentrum

аркти'ческий *a* arctic
арктоальпи'йский *a* arctoalpine, arctic-alpine
Арктоге'я *f* Arctogaea
а'ркулус *m ent* arculus
аро'лий *m ent* arolium
арома'т *m* scent (*of flowers*), aroma, flavour, fragrance
аромати'ческий *a* aromatic
аромо'рфа *f* aromorph
аромoрфо'з *m* aromorphosis
аро'нник *m bot* arum (*Arum*)
арреногенети'ческий *a* arrhenogenic
арренотоки'я *f* arrhenotoky
артерено'л *m* arterenol
артериализа'ция *f* arterialization
артериа'льный *a* arterial
артериовено'зный *a* arteriovenous
артериогра'мма *f* arteriogram
артериогра'ф *m* arteriograph
артериогра'фия *f* arteriography
артерио'ла *f* arteriole
арте'рия *f* artery
~ **стопы'** pedal artery
алланто'идная ~ allantoic artery, umbilical artery
альвеоля'рная ~ alveolar artery
базиля'рная ~ basilar artery
бараба'нная ~ tympanic artery
бе'дренная ~ femoral artery
большеберцо'вая~ tibial artery
брыже'ечная ~ mesenteric artery
вене'чная ~ coronary artery
верхнечелюстна'я ~ maxillary artery
висо'чная ~ temporal artery
возвра'тная ~ recurrent artery
ворси'нчатая ~ choroid artery
восходя'щая ~ ascending artery
выношя'щая ~ efferent artery
гио'идная ~ hyoid artery
глазни'чная ~ ophthalmic artery
гло'точная ~ pharyngeal artery
глубо'кая ~ deep artery
головна'я ~ cephalic artery
горта'нная ~ laryngeal artery
грудна'я ~ thoracic artery
грудоакромиа'льная ~ thoracoacromial artery
губна'я ~ labial artery
диафрагма'льная ~ phrenic artery
дорса'льная ~ dorsal artery
дугообра'зная ~ arcuate artery
жа'берная ~ branchial artery
жева'тельная ~ masseteric artery
желто'чная ~ vitelline artery
желу'дочная ~ gastric artery
желу'дочно-са'льниковая ~ gastroepiploic artery
завитко'вая ~ helicine artery
запира'тельная ~ obturator artery
заты'лочная ~ occipital artery
клубо'чковая ~ glomerular artery
корона'рная ~ coronary artery
ко'жная ~ dermal artery
крема'стерная ~ cremasteric artery
крестцо'вая ~ sacral artery
лё'гочная ~ pulmonary artery
лицева'я ~ facial artery
ло'бная ~ frontal artery
локтева'я ~ ulnar artery
лучева'я ~ radial artery
малоберцо'вая ~ peroneal artery
ма'точная ~ uterine artery
менингеа'льная ~ meningeal artery
мозгова'я ~ cerebral artery
мы'шечная ~ muscular artery
надбло'ковая ~ supratrochlear artery
надглазни'чная ~ supraorbital artery
надлопа'точная ~ suprascapular artery
надпо'чечниковая ~ suprarenal artery
надчре'вная ~ epigastric artery
нё'бная ~ palatine artery
нисходя'щая ~ descending artery
ободо'чная ~ colic artery
плечева'я ~ brachial artery
плю'сневая ~ metatarsal artery
пове'рхностная ~ superficial artery
подвздо'шно-ободо'чная ~ ileocolic artery
подвздо'шно-поясни'чная ~ iliolumbar artery
подглазни'чная ~ infraorbital artery
подключи'чная ~ subclavian artery
подколе'нная ~ popliteal artery
подлопа'точная ~ subscapular artery
подмы'шечная ~ axillary artery
подо'швенная ~ plantar artery
подрё'берная ~ subcostal artery
подъязы'чная ~ sublingual artery
позвоно'чная ~ vertebral artery
полова'я ~ pudendal artery
по'чечная ~ renal artery
поясни'чная ~ lumbal artery
принося'щая ~ afferent artery
пробода'ющая ~ perforating artery

пупо'чная ~ umbilical artery, allantoic artery
пя'стная ~ metacarpal artery
решё'тчатая ~ ethmoid artery
селезё'ночная ~ splenic artery
семеннико'вая ~ testicular artery
сигмови'дная ~ sigmoid artery
скулоглазни'чная ~ zygomatico-orbital artery
слё'зная ~ lacrimal artery
со'нная ~ carotid artery
сонно-бараба'нная ~ caroticotympanic artery
тощекише'чная ~ jejunal artery
углова'я ~ angular artery
ушна'я ~ auricular artery
хвостова'я ~ caudal artery
ше'йная ~ cervical artery
шилососцеви'дная ~ stylomastoid artery
ще'чная ~ buccal artery
щитови'дная ~ thyroid artery
ягоди'чная ~ gluteal artery
язы'чная ~ lingual artery
яи'чковая ~ testicular artery
яи'чниковая ~ ovarian artery

артефа'кт *m* artefact, artifact
артишо'к *m bot* artichoke, globe artichoke, cardoon (*Cynara*)
артро'дия *f* arthrodia, ball-and-socket joint, spheroidal joint
артро'з joint
артрокони'дия *f* arthroconidium
артропоно'зный *a* arthropod-borne
артроспо'ра *f* arthrospore
артростери'гма *f* arthrosterigma
архалла'ксис *m* archallaxis
арха'р *m ther* argali *(Ovis ammon)*
архебакте'рия *f* archaebacterium
архебио'з m archebiosis
архегониа'льный *a* archegonial
архегониено'сец *m* archegoniophore
архего'ний *m* archegonium
архе'йский *a* Archean
архентеро'н *m* archenteron
археофи'т m archeophyte
археоци'т *m* archaeocyte, archeocyte
археспо'ра *f* archespore
археспо'рий *m* archesporium
архети'п *m* archetype
архетоми'я *f* architomy
архика'рп *m* archicarp
архихлами'дный *a* archichlamydeous

архитекту'ра *f* architecture
ландшафтная ~ landscape architecture, landscaping
садо'во-па'рковая ~ gardening architecture

архитоми'я *f* architomy
арче'вник *m* juniper stand
асепа'льный *a* acalycine, acalycinous
асепта'тный *a* aseptate
асе'птика *f* asepsis
асепти'ческий *a* aseptic
асимметри'чный *a* asymmetrical
асимметри'я *f* asymmetry
асимфи'тный *a* asymphytous
асина'псис *m* asynapsis
асинга'мный *a* asyngamic
аси'ндез *m* asynapsis
асинхрони'зм *m* asynchronism
асинхро'нный *a* asynchronous
асистоли'я *f* asystole
аск *m* ascus
аскари'да *f* ascarid *(Ascaris)*
аскаридо'з *m* ascariasis
аскарици'дный *a* ascaricid
аскоге'нный *a* ascogenous
аскогониа'льный *a* ascogonial
аского'ний *m* ascogonium
аскока'рп, **аскока'рпий** *m* ascocarp
дискови'дный ~ ascoma

аско'ма *f* ascoma
аскоспо'ра *f* ascospore
аскостро'ма *f* ascostroma
аскофо'р *m* ascophore
аспараги'н *m* asparagine
аспе'кт *m* aspect
весе'нний ~ vernal aspect
зи'мний ~ hiemal aspect
осе'нний ~ autumnal aspect
сезо'нный ~ seasonal aspect

аспекти'вный *a* aspective, aspecting
аспекти'рование *n* aspection
аспергилле'з *m* aspergillosis
а'спид *m* coral snake (*Elaphidae*)
аспира'ция *f* aspiration
аспири'н *m* aspirin
аспири'ровать *v* aspirate
аспороге'нный *a* asporogenic, asporogenous
ассениза'ция *f* sludge removal
ассимили'ровать *v* assimilate
ассимили'рующий *a* assimilating, assimilatory, assimilative
ассимиля'т *m* assimilate

ассимиляцио'нный *a* assimilative
ассимиля'ция *f* assimilation, anabolism
~ **отхо'дов** waste assimilation
ви'димая ~ apparent assimilation
генети'ческая ~ genetic assimilation
и'стинная ~ net assimilation, nette assimilation, real assimilation
наблюда'емая ~ apparent assimilation
окисли'тельная ~ oxidative assimilation
у'стьичная ~ stomatal assimilation
ассоциати'вный *a bot* associate
ассоциа'ция *f* association
близнецо'вая ~ twin association
«га'лстучная» ~ *bot* "necktie" association
гетероге'нная ~ heterogenic association
гомоге'нная ~ homogenic association
длительнопроизво'дная ~ secondary permanent
заключи'тельная ~ terminal association
за'мкнутая ~ closed association
кли'максовая ~ climax association
коренна'я ~ climax association
кратковре'менная ~ short-timed association, temporary association
негомологи'чная ~ nonhomologous association
одновидова'я ~ homotypical association, homotypical aggregation
откры'тая ~ open association
перви'чная ~ primary association
перехо'дная ~ transitional association
произво'дная ~ derived association, secondary association
расти'тельная ~ plant association
сери'йная ~ subordinate association, associes
танде'мная ~ tandem association
узлова'я ~ climax association, fundamental association
усто'йчивая ~ tolerant association
элемента'рная ~ elementary association
ассоции'рованность *f* association
ассоцио'н *m* associon
ассоциу'ла *f* associule
ассо'ция *f* associes, subordinate association
астаксанти'н *m* astaxanthin
астати'ческий *a* astatic
астели'я *f* astely
астени'ческий *a* asthenic
астери'гмный *a* asterigmate
астогени'я *f* astogeny
а'стра *f bot* starwort (*Aster*)
астробиоло'гия *f* astrobiology
астробла'ст *m* astroblast
астробота'ника *f* astrobotany
астроглиоци'т *m* astrogliocyte, spider cell
плазмати'ческий ~ protoplasmic astrocyte
астросфе'ра *f cyt* astrosphere, astrocenter
астроци'т *m* astrocyte
плазмати'ческий ~ protoplasmic astrocyte
асулька'тный *a pal* asulcate
асфикси'я *f* asphyxia, asphyxiation
асци'т *m* ascite
атави'зм *m* atavism
атависти'ческий *a* atavistic, ancestral
ата'ка *f* attack
ло'жная ~ sham attack
атакова'ть *v* attack
ата'ксия, **атакси'я** *f* ataxia
аталами'ческий *a* athalamous
ателио'з *m* ateliosis
атели'я *f* atelia
ателомити'ческий *a* atelomitic
атипи'чный *a* atypical
атла'нт *m anat* atlas
атлантозаты'лочный *a* atlantooccipital
атмосфе'ра *f* atmosphere
ато'кный *a* atocous
ато'лл *m* atoll
а'том *m* atom
~ **ме'ченый** атом tagged element, tracer element
атони'я *f* atony
атопи'я *f* atopy
атрези'я *f* atresia
атрети'ческий *a* atretic
атриа'льный *a* atrial
атриовентрикуля'рный *a* atrioventricular
атриопо'р *m* atriopore
а'триум *m* atrium
атропи'н *m* atropin, atropine
атро'пный *a* atropous
атрофи'я *f* atrophy
атроцито'з *m* athrocytosis
аттенюа'тор *m* attenuator
аттенюи'рованный *a* attenuated
атракта'нт *m* attractant
АТФ, **аденозинтрифосфат** *m* ATP, adenosine triphosphate
аудиометри'я *f* acumetry
ауксаногра'мма *f* auxanogram
ауксаногра'фия *f* auxanography
ауксано'метр *m* auxanometer
ауксимо'н *m* auximone

аукси'н *m* auxin
ауксоавтотро'ф *m* auxoautotroph
ауксогетеротро'ф *m* auxoheterotroph
ауксоспире'ма *f* auxospireme
ауксоспо'ра *f* auxospore
бокова'я ~ lateral auxospore
интеркаля'рная ~ intercalar auxospore
полуинтеркаля'рная ~ semi-intercalar auxospore
свобо'дная ~ free auxospore
ауксотони'ческий *a* auxotonic
ауксотро'ф *m* auxotroph
ауксотро'фный *a* auxotrophic
ауксохро'м *m* auxochrome
ауксоци'т *m* auxocyte
аулофи'т *m* aulophyte
аурикуля'рия *f* auricularia
аускульта'ция *f* auscultation
аутбри'динг *m* outbreeding
ауги'зм *m* autism
ауткро'ссинг *m* outcrossing
ауто- auto-, self-
аутоагрега'ция *f* self-aggregation
аутоагресси'вность *f* autoaggressiveness
аутоагре'ссия *f* self-aggression
аутоаллергиза'ция *f* autoimmunization
аутоаллерги'я f autoallergy
аутоантагони'зм *m* autoantagonism
аутоантиге'н *m* autoimmune antigen, autostimulating antigen, autoantigen
стимули'рующий ~ autostimulating antigen
аутоантиге'нность *f* autoantigenicity
аутоантиге'нный *a* autoantigenic
аутоантите'ло *n* autoreactive antibody, autoantibody
антиидиотипи'ческое ~ anti-antiself antibody
аутогемагглютини'рующее ~ autohemagglutinin
гемолити'ческое ~ autohemolysin
есте'ственное ~ true autoantibody
аутобази'дия *f* autobasidium
аутобивале'нт *m* autobivalent
аутовакци'на *f* autovaccine, homologous vaccine, corresponding vaccine, personal vaccine
аутога'мия *f* autogamy
аутога'мный *a* autogamic, autogamous
аутогемагглютина'ция *f* autohemagglutination
теплова'я ~ warm autohemagglutination
холодова'я ~ cold autohemagglutination
аутогемагглютини'н *m* autohemagglutinin
аутогемолизи'н *m* autohemolysin
аутогемотерапи'я *f* autotransfusion
аутогене'з *m* autogenesis, autogeny
аутоге'нный *a* autogenic, autogenous
аутогено'мный *a* autogenomatic
аутогетеропло'ид *m* autoheteroploid
аутоде'м *m* autodeme
аутодетермина'нта *f* self-determinant
аутоидиоти'п *m* self-idiotype
аутоиммуниза'ция *f* self-immunization
аутоиммуните'т *m* autoimmunity
аутоимму'нный *a* autoimmune
аутоиммуноге'нность *f* autoimmunogenicity
аутоинва'зия *f* autinvasion
аутоинфе'кция *f* autoinfection, self-infection
аутока'рпия *f* autocarpy
ауто'лиз *m* autolysis
аутологи'ческий, аутологи'чный *a* autologous
аутоми'ксис *m* automixis
аутомутаге'н *m* automutagen
аутоориента'ция *f* autoorientation
аутополипло'ид *m* autopolyploid
аутополиплоиди'я *f* autopolyploidy
аутопси'я *f* autopsy
авторедуплика'ция *f* autoreduplication
аутореплика'ция *f* self-replication
аутосенсибилиза'ция *f* autosensitization
аутоси'напс *m* autosynapsis, autosyndesis
аутоси'ндез *m* autosynapsis, autosyndesis
аутосо'ма *f* autosome
аутосо'мный *a* autosomal
аутотилли'я *f* autotilly
аутотолера'нтность *f* autotolerance
аутото'мия *f* autotomy
аутотрансплanta'т *m* autograft, autotransplant, autoplast
аутотрансплanta'ция *f* autografting, autotransplantation
аутотрансфу'зия *f* autotransfusion
аутофаги'я *f* autophagy
аутофагосома *f* autophagosome
аутофе'н *m* autophene
аутоцито'лиз *m* autocytolysis
аутоцитотокси'чность *f* self-rectricted cytotoxicity
аутоэпито'п *m* autoepitope, self-epitope
аутэкологи'ческий *a* autecological
аутэколо'гия *f* autecology
афаги'я *f* aphagia

афле'бия *f* aphleby
афилакси'я *f* aphylaxis
афилли'я *f* aphylly
афлатокси'н *m* aflatoxin
афоти'ческий *a* aphotic
а'фта *f* aphtha
аффе'кт *m* affect
аффере'нтный *a* afferent
аффилиа'ция *f* affiliation
ахира'льный *a* achiral
ахлоро'з *m* achlorosis
ахроглоби'н *m* achroglobin
ахрома'т m achromat
ахромати'н *m* achromatin
ахромати'новый *a* achromatinic
ахромати'ческий *a* achromatic, achromic
ахромато'лиз *m* achromatolysis
ахроми'я *f* achromia
ахроморетикулоци'т *m* achromoreticulocyte
аце'льный *a* acoelous *(a vetrebra)*
ацентри'ческий *a* acentric
ацетализа'ция *f* acetalation, acetalization
ацетали'рование *n* acetalation, acetalization
ацетили'рование *n* acetylation, acetylization
ацетилхоли'н *m* acetylcholine
ацетоацетили'рование *n* acetoacetylation
ацетоба'ктер *m* acetobacter
ацето'лиз *m* acetolysis
ацетони'рование *n* acetonation
ацетонури'я *f* acetonuria
ацетоксили'рование *n* acetoxylation
ацетоксиметили'рование *n* acetoxymethylation
ацетоэтили'рование *n* acetoethylation
ацефалоци'ста *f* acephalocyst
ацефа'льный *a* acephalous
ацидо'з *m* acidosis
ацидо'лиз *m* acidolysis
ацидотро'фный *a* acidotrophic
ацидофи'т *m* acidophyte
ацидофи'льный *a* acidophilic, acidophile
ацикли'ческий *a* acyclic
аци'кула *f* acicle, aciculum
ациламиноалкили'рование *n* acylaminoalkylation
ациламинометили'рование *n* acylaminomethylation
ацили'рование *n* acylation
ацилоксили'рование *n* acyloxylation
аэротакса'ция *f* aerotaxation
аэре'нхима *f* aerenchyma
аэри'ровать *v* aerate
аэро'б *m* aerobe
  облига'тный ~ obligate aerobe
  факультати'вный ~ facultative aerobe
аэробио'з *m* aerobiosis
аэробиоло'гия *f* aerobiology
аэробио'нт *m* aerobe
аэро'бный *a* aerobic
аэроге'нный *m micr* aerogenic
аэрозо'ль *m* aerosol
аэроморфо'з *m* aeromorphosis
аэропла'нкто'н *m* aeroplankton, anemoplankton
аэроско'п *m* aeroscope
аэрота'ксис *m* aerotaxis
аэротропи'зм *m* aerotropism
аэротро'пный *a* aerotropic
аэротэ'нк *m* aerotank
аэрофи'льный *a* aerophilic, aerophilous
аэрофи'т *m* aerophyte
аэрофотосъе'мка *f* air survey, aerial survey, aerial photographic mapping, aerial photography

Б

бабезио'з *m vet* babesiosis
ба'бка *ent* corduliid dragonfly *(Corduliidae)*
ба'бочка *f* butterfly, moth (*Lepidoptera*)
бе'лая америка'нская ~ fall webworm moth *(Hyphantria cunea)*
дневна'я ~ butterfly
бабуи'н *m* yellow baboon *(Papio cynocephalus)*
баго'р *m* gaff
ба'за *f* base
кормова'я ~ food resources, food supply, food reserve
база'льно-не'рвный *bot* basal-nerved
база'льный *a* basal
база'р *m*: **пти'чий** ~ colony of birds, rookery
бази- basi-
базиальвеоля'рный *a* basialveolar
базиге'н *m* basigene
базидиа'льный *a* basidial
базидиено'сец *m* basidiophore
базидиока'рп *m* basidiocarp
базидио'ла *f* basidiole
базидиомице'т *m* basidium fungus (*Basidiomycetes*)
базидионо'сец *m* basidiophore
базидиоспо'ра *f* basidiospore
базидиофо'р *m* basidiophore
бази'дия *f* basidium
базикони'ческий *a* basiconic
базикраниа'льный *a anat* basicranial
базилатера'льный *a* basilateral
базили'к *m* basil (*Ocimum*)
базиля'рный *a* basilar
базипета'льный *a* basipetal
базиподи'т *m* basipodite
базисфено'ид *anat* basisphenoid
базопени'я *f* basopenia
базофи'л *m* basophil, basophile, basophilic leukocyte
иммунореакти'вный ~ antigenically responsive basophil
базофили'зм *m* basophilism
базофили'я *f* basophilia
базофилобла'ст *m* basophiloblast
базофи'льный *a* basophilic, basophilous, basophil, basiphil
байба'к *m* steppe marmot (*Marmota bobac*)
бакау'т *m bot* lignum-vitae (*Guaiacum officinale*)
баклажа'н *m* eggplant, garden eggplant, aubergine (*Solanum melongera*)
бакла'н *m orn* shag, cormorant, sea raven (*Phalacrocorax*)
большо'й ~ great cormorant, large cormorant, common cormorant *(Phalacrocorax carbo*)
короле'вский ~ king cormorant *(Phalacrocorax albiventer)*
ма'лый ~ pygmy cormorant (*Phalacrocorax pygmaeus*)
че'рный ~ little cormorant (*Phalacrocorax niger*)
бактериа'льный *a* bacterial
бактери'н *m* bacterin
бактериоге'нный *a* bacteriogenous
бактерио'з *m* bacteriosis
бактерио'лиз *m* bacteriolysis
бактериолизи'н *m* bacteriolysin
бактериолити'ческий *a* bacteriolytic
бактериологи'ческий *a* bacteriological
бактериоло'гия *f* bacteriology
бактериори'за *f* bacteriorhiza
бактериоскопи'я *f* bacterioscopy
бактериоста'з *m* bacteriostasis
бактериоста'т *m* bacteriostat
бактериостати'ческий *a* bacteriostatic
бактериофа'г *m* bacteriophage
~ **ля'мбда** lambda bacteriophage
уме'ренный ~ temperate bacteriophage
бактериофаги'я *f* bacteriophagy
бактериофо'бия *f* bacteriophobia
бактекрици'д *m* bactericide
бактерици'дность *f* bactericidal action
бактерици'дный *a* bactericidal, bacillicidal
бакте'рия *f* bacterium
автотро'фная ~ autotrophic bacterium
азотфикси'рующая ~ nitrogen-fixing bacterium
алкалифи'льная ~ alkaliphilic bacterium
анаэро'бная ~ anaerobic bacterium
ароматобразу'ющая ~ aroma-forming bacterium
ацетоге'нная ~ acetogenic bacterium
ацетокласти'ческая ~ acetoclastic bacterium
ацетонбути'ловая ~ acetone-butanol bacterium
ацетонэти'ловая ~ acetone-ethanol bacterium

ацидофи'льная ~ acidophilic bacterium
аэро'бная ~ aerobic bacterium
аэротолера'нтная ~ aerotolerant bacterium
болезнетво'рная ~ malignant bacterium
газообразу'ющая ~ gas-producing bacterium
галофи'льная ~ halophilic bacterium
гетероферментати'вная ~ heterofermentative bacterium
гетеротро'фная ~ heterotrophic bacterium
гни'лостная ~ putrefactive bacterium
гноеро'дная ~ pyogenic bacterium
гомоферментати'вная ~ homofermentative bacterium
грамвариа'бельная ~ gram-variable bacterium
грамотрица'тельная ~ gram-negative bacterium
грамположи'тельная ~ gram-positive bacterium
денитрифицирующая ~ denitrifying bacterium
каталазонегативная ~ catalase-negative
кислотоустойчивая ~ acid-fast bacterium
кише'чная ~ colibacillus (*Escherichia coli*)
клубенько'вая ~ root nodule bacterium, nodule bacterium
колиподо'бная ~ coliform bacterium
колифо'рмная ~ coliform bacterium
коринефо'рмная ~ coryneformic bacterium
масляноки'слая ~ butyric-acid bacterium
люминесци'рующая ~ luminescent bacterium, luminous bacterium, photogenic bacterium
лизоге'нная ~ lysogenic bacterium
маслянокислая ~ butyric-acid bacterium
мезофи'льная ~ mesophilic bacterium
мета'новая ~, **метанокисля'ющая** ~ methane-oxidyzing bacterium
метано'льная ~ methanolic bacterium
метатро'фная ~ metatrophic bacterium
метилотро'фная ~ methylotrophic bacterium
молочноки'слая ~ lactic-acid bacterium, lactobacillus
неферменти'рующая ~ nonfermentative
нитеви'дная~, нитчатая ~ filamentous bacterium, thread bacterium
нитрифици'рующая ~ nitrifying bacterium
осмофильная ~ osmophilic bacterium
палочкови'дная ~ rodlike bacterium, rod-shaped bacterium
патогенная ~ pathogenic bacterium
пигментообразу'ющая ~ chromogenic bacterium
почкующаяся ~ budding bacterium
пропионовоки'слая ~ propionic bacterium
прототро'фная ~ prototrophic bacterium
психрофи'льная ~ cryophilic bacterium
пурпу'рная ~ purple bacterium
рубцовая ~ ruminal bacterium
сапрофи'тная ~ saprophitic bacterium
сахаролити'ческая ~ saccharolytic bacterium
светя'щаяся ~ luminiscent bacterium, luminous bacterium, photogenic bacterium
се'рная ~ sulfur bacterium
скользящая ~ gliding bacterium
слизеобразу'ющая ~ slime-forming bacterium
солеусто'йчивая ~ salt-tolerant bacterium
спорообразу'ющая ~ spore-forming bacterium
стебелько'вая ~ stalked bacterium
сульфатвосстана'вливающая ~, сульфатредуци'рующая ~ sulfate-reducing bacterium
термосто'йкая ~ thermoduric bacterium
термоге'нная ~ thermogenic bacterium
термофи'льная ~ thermophilic bacterium
турно'зная ~ tourne bacterium
уксусноки'слая ~ acetic-acid bacterium
усло'вно-патоге'нная ~ opportunistic pathogenic bacterium
фотосинтезирующая ~ photosynthetic bacterium
хемосинтезирующая ~ chemosynthetic bacterium
хемотрофная ~ chemotrophic bacterium
целлюлозоразрушающая ~ cellulose-fermenting bacterium

бактеро'ид *m* bacteroid
бактеро'идный *a* bacteroid
бактерия'дный *a* microbivorous
бактриа'н *m* Bactrian camel (*Camelus bactrianus*)
бакулови'рус *m* baculovirus
бала'нс *m* balance
азо'тистый ~ nitrogen balance
во'дный ~ water balance
генети'ческий ~ genetic balance, genetical balance

ге'нный ~ genic balance
кислоро'дный ~ oxygen balance
полиге'нный ~ polygenic balance
радиацио'нный ~ radiation balance
селекти'вный ~ selective balance
теплово'й ~ caloric balance, heat balance, thermal balance
углево'дный ~ carbohydrate balance
балансё'р *m* balancer (*in Amphibia larvae*)
балантидио'з balantidiasis, balantidiosis
ба'лка *f* gully, ravine; beam
балл *m* score, point
баллистокардиогра'мма *f* ballistocardiogram
баллистокардиогра'фия *f* ballistocardiography
~ **ско'рости** velocity ballistocardiography
~ **смеще'ния** displacement ballistocardiography
~ **ускоре'ния** acceleration ballistocardiography
ультранизкочасто'тная ~ ultralow-frequency ballistocardiography
баллистоспо'ра *f* ballistospore
бало'бан *m* saker falcon (*Falco cherrug*)
бальза'м *m* balsam
кана'дский ~ Canada balsam
бальзами'н *m bot* balsam *(Impatiens)*
бальзамоно'сный *a* balsamiferous
бальнеоло'гия *f* balneology
бамбу'к *m* bamboo (*Bambusa*)
ба'мия *f* okra, gumbo (*Hibiscus esculentus*)
бана'н *m* banana (*Musa*)
бандику'т *m ther* bandicoot *(Peramelidae)*
банк *m* bank
~ **да'нных** databank
~ **ге'нов** gene bank
~ **после'довательностей** sequence databank
ба'нка *f* jar; cupping glass; bank, bed
ба'нки *f pl* cups, cupping glasses; **ста'вить** ~ apply cupping glasses
банте'нг *m ther* banteng *(Bos javanicus)*
банья'н *m* banyan, banyan tree (*Ficus bengalensis*)
ба'ня *f* bath
водяна'я ~ water bath
баоба'б *m* baobab, monkey-bread tree (*Adansonia digitata*)
бара'н *m* sheep (*Ovis*); ram, buck
го'рный ~ argali (*Ovis ammon*)
сне'жный ~ bighorn (*Ovis canadensis*)
бара'нина *f* mutton
бара'хтаться *v eth* wallow
бара'шек *m* whitecap (*of a wave*); wing nut, thumbscrew
барбари'с *m* berberry, pipperidge tree (*Berberis*)
барви'нок *m* periwinkle (*Vinca*)
барда' *f* stillage, slops
разба'вленная отфильтро'ванная ~ thin stillage
це'льная нефильтро'ванная ~ whole stillge
барогнози'я *f* barognosis
барока'мера *f* altitude chamber, vacuum chamber, altichamber
барореце'птор *m* baroreceptor
барота'ксис *m* batotaxis
баротропи'зм *m* barotropism
барохо'р *m* barochore
баррикади'рование *n eth* phragmosis
барс *m* leopard (Panthera *pardus*)
сне'жный ~ snow leopard *(Uncia uncia)*
барсу'к *m* badger (*Meles meles*)
ба'рхат *m* cork tree (*Phellodendron*); velvet
бархати'стый *a* velutinate, velutinous, velvety
ба'рхатница *ent* brown, satyr butterfly *(Satyridae)*
ба'рхатцы *pl* marigold (*Tagetes*)
барье'р *m* barrier
альвеоля'рно-капилля'рный ~ blood-air barrier, blood-gas barrier
биогеографи'ческий ~ biogeographical barrier
гематоли'кворный ~ blood-cerebrospinal fluid barrier
гематоофтальми'ческий ~ blood-ocular barrier
гематоэнцефали'ческий ~ hematoencephalic barrier, blood-brain barrier
плацента'рный ~ placental barrier, transplacental barrier
приро'дный ~ natural barrier
трансплантацио'нный ~ transplantation barrier, graft immunity
бассе'йн *m* basin, pool
аэрацио'нный ~ aeration basin
водосбо'рный ~ catchment basin, drainage area, drainage basin
о'пытный ~, эксперимента'льный ~ experiment tank

баста'рд *m* bastard
бата'т *m* sweet-potato (*Ipomoea batatas*)
батмотропи'зм *m* bathmotropism
батмотро'пный *a* bathmotropic
бато'метр *m* water bottle
бахрома' *f* fringe, fimbria
бахро'мка *f* fimbria, armilla
бахро'мчатый *a* fimbriate, fimbriated, fringed
бахча' *f* melon field
баци'лла *f* bacillus
бациллоноси'тель *m* bacillicarrier, germ carrier
бацилля'рный *a* bacillary, bacilliform
ба'шня *f* tower
  сило'сная ~ silage tower, silo tower
бег *m* running
  ~ **прыжка'ми** saltation
  рикошети'рующий ~ ricochet
бе'гать *v* run
бе'гающий *a* cursorial
бегемо'т *m* hippopotamus (*Hippopotamus amphibius*)
бе'гство *n* escape, escape behavior
бегуно'к *m* courser (*Cursorius*)
  еги'петский ~ crocodile plover (*Pluvialis aegyptius*)
бедегуа'р *m phytop* bedeguar, bedegar
бе'дренный *a* femoral
бедро' *n* femur, thigh; hip; ham
бе'дствие *n* calamity
  стихи'йное ~ elemental calamity, natural calamity, natural disaster
бежа'ть *v* run
  ~ **и'ноходью** amble
  ~ **ры'сью** trot
без- (*before voiced consonnants*) *a*-, an-, e-, ex-, -less, non-
безазо'тистый *a* anazotic, nitrogen-free
безаперту'рный *a pal* inaperturate, alete
безбелко'вый *a* protein-free; exalbuminous
безболе'зненный *a* painless
безборо'дый *a* beardless
безборо'здный *a* asulcate, anosulcate, acolpate
безво'дный *a* arid, waterless, anhydrous
безволо'сость *f* baldness
безволо'сый *a* hairless, atrichous
безвре'дный *a* uninjurious
безвре'менник *m* meadow saffron (*Colchicum*)
безвы'борный *a* nonselective
безголо'вый *a* acephalous
бездо'мный *a* stray
бездыха'нный a breathless
безжа'берный *a* abranchial, abranchiate
безжгу'тиковый *a* aflagellar
безжи'зненный *a* lifeless, azoic
безжи'лковый *a* ecostate, enervose
беззамко'вый *a zool* ecardinal
беззо'льный *a* ash-free, ashless
беззу'бый *a* edentate
безлепестко'вый *a* apetalous
безле'сность *f* treelessness
безле'сный *a* treeless
безли'стный *a* leafless, aphyllous
безмя'котный *a* nonmedullated, nonmyelinated, amyelinic
безно'гий *a* apodal, apodous, legless
безопа'сность *f* safety
  радиологи'ческая ~ radiological safety
  экологическая ~ environmental safety
безопа'сный *a* safe
безо'стость *f* baldness
безо'стый *a* awnless, beardless, bald, hummel
безро'гий *a* hornless, polled; acerous; hummel
безро'тый *a* astomatous, astomous
безрубцо'вый *a pal* alete
безу'хий *a* earless
безъя'дерный *a* anucleate, acaryote, anuclear, apyrene
безъя'дрышковый *a* anucleolate
безъязычко'вый *a* eligulate
безязы'чный *a* aglossate
безъя'русный *a* unstratified
бе'йра *ther* beira *(Dorcatragus megalotis)*
бека'с *m orn* snipe (*Gallinago*)
бекаси'нник *m* small shot
беко'н *m* bacon
белёк *m* white-coat seal
белена' *f* henbane (*Hyosciamus*)
белё'сый *a* albescent
беле'ющий *a* albicant
бе'лка *f* squirrel (*Sciurus*)
  кра'сная ~ red squirrel *(Tamiasciurus)*
  обыкнове'нная ~ red squirrel *(Sciurus vulgaris)*
  самсо'нистая ~ Samson squirrel
белковоподо'бный *a* albuminoid
белко'вый *a* proteinaceous
белладо'нна *f* deadly nightshade (*Atropa belladonna*)
белова'тый *a* albescent

белогла'зка *orn* white-eye *(Zosterops)*
белоду'шка *f* stone marten (*Martes foina*)
белозо'р *m* grass-of-parnassus (*Parnassia*)
бело'к *m* protein; white (*of eggs*; *of the eye*)
~ **Бенс-Джо'нса** Bence Jones protein
~ **кро'ви** blood protein
~ **моло'чной сы'воротки** whey protein
~ **оболо'чки** coat protein *(of a virus)*
~ **сердцеви'ны** core protein *(of a chromosome)*
~ **слия'ния** *mbiol* fused protein, fusion protein
~ **теплово'го шо'ка** heat shock protein
~ **цитоскеле'та** cytoskeletal protein
аутоимму'нный ~ autoimmune protein
аутологи'чный ~ autologous protein
ви'русный капси'дный ~ viral capsid protein
восстано'вленный ~ reduced protein
гаптенили'рованный ~ hapten-liganded protein
глобуля'рный ~ globular protein
денатури'рованный ~ denatured protein
ДНК-расплета'ющий ~ DNA-unwinding protein
ДНК-свя'зывающий ~ DNA-binding protein
железосодержа'щий ~ iron protein
живо'тный ~ animal protein
запасно'й ~ reserve protein
иммуноге'нный ~ immunologically relevant protein
иску'сственный ~ factitious protein
каналообразу'ющий ~ channel-forming protein, pore-forming protein, cytolytic pore-forming protein
капси'дный ~ core protein
конъюги'рованный ~ conjugated protein
кэппи'рующий ~ capping protein
марке'рный ~ marker protein
метилакцепти'рующий ~ хемота'ксиса methyl-accepting chemotaxis protein
мино'рный ~ minor protein
мультидоме'нный ~; **мультиме'рный** ~ multidomain protein
нати'вный ~ native protein, complete protein
о'бщий ~ crude protein
полноце'нный ~ complete protein
пороформиру'ющий ~ channel-forming protein, cytolytic pore-forming protein
прими'рующий ~ priming protein
просто'й ~ simple protein
процесси'рованный ~ truncated protein
расти'тельный ~ vegetable protein
С-реактивный ~ C-reactive protein
репрессо'рный ~ repressor protein
ретиа'рный ~ retiaric protein
рецепто'рный ~ sensor protein
Са-свя'зывающий ~ Ca-binding protein
сенсо'рный ~ sensor protein
сконструи'рованный ~ designer protein
слива'ющий ~ fusogenic protein
сли'тый ~ *mbiol* fused protein, fusion protein
сократи'тельный ~ contractile protein
спира'ль-дестабилизи'рующий ~ helix-destabilizing protein
структу'рный ~ structural protein
сы'вороточный ~ serum protein
трансактиви'рующий ~ transacting protein
фибрилля'рный ~ fibrous protein
фотооки'сленный ~ photooxidized protein
химе'рный ~ chimeric protein
хитшо'ковый ~ heat shock protein
чужеро'дный ~ foreign protein
яи'чный ~ ovalbumin
якорный ~ anchor protein, anchorage protein
бело'к-носи'тель, бело'к-перено'счик *m* carrier-protein
белокро'вие *n* leukemia, white blood disease
белокры'лка *f* white fly *(Aleurodidae)*
белокры'льник *m* water arum (*Calla*)
бело'к-трансактива'тор *m* transacting protein
белоло'з *m* white willow, common willow (*Salix alba*)
белоу'с *m* mat-grass (*Nardus*)
белоцветко'вый *a* albiflorous
белоцве'тник *m* snowdrop *(Leucojum)*
бельмо' *n* leukoma
беля'нка *f ent* sulfur burrefly *(Pieridae); bot* milk-white russula (*Russula delica*)
~ **ре'пная** turnip butrterfly, cabbage butterfly *(Pieris rapae)*
~ **капу'стная** cabbage white *(Pieris brassicae)*
бензили'рование *n* benzylation
бензоили'рование *n* benzoylation
бергамо'т *m* bergamot, bergamot orange (*Citrus bergamia*)

бе'рег *m* shore, coast, bank (*e.g., exposed shores of all coasts*)
~ **мо'ря** seacoast, seashore
отло'гий ~ low coast
берегову'шка *f* shore fly (*Ephydridae*)
берегоукрепле'ние *n* bank protection
бере'за *f* birch (*Betula*)
~ **борода'вчатая** common birch (*Betula verrucosa*)
~ **бума'жная** spoolwood *(Betula papyrifera)*
~ **карельская** speckled birch, curled birch
~ **ка'рликовая** dwarf birch, Arctic birch (*Betula nana*)
березня'к *m* birch forest
бере'зовик *m* rough boletus, shaggy boletus (*Leccinum scabrum*)
бере'менная *a* pregnant, gravid, heavy
бере'менность *f* pregnancy, gestation, fetation, gravidity
бересkле'т *m* spindle-tree (*Evonymus*)
бе'рест *m* elm (*Ulmus*)
береста' *f* birch bark
берестня'к *m* elm forest
бе'ри-бе'ри *n* beri-beri, beriberi
бе'ркут *m* golden eagle (*Aquila chrysaetus*)
берло'га *f* den, lair, couch, squat
бес- (*before voiceless consonants*) *a*-, an-, e-, non-, -less
бескилево'й *a* ecarinate
бескислоро'дный *a* anoxic, oxygen-free
бескле'точный *a* acellular, cell-free
бесклубо'чковый *a* aglomerular
бесклю'вый *a* beakless, erostrate
бескопы'тный *a* hoofless
беско'рмица *f* fodder shortage
бескро'вный *a* bloodless
бескры'лый *a* apterigial, apterous, wingless
беспа'нцирный *a* armorless, shell-less, nonthecate, nonloricate
беспло'дие *n* sterility, infertility, barrenness, agenesia, agenesis
беспло'дность *f* infertility, barrenness, apogeny
беспло'дный *a* infertile; barren, sterile
беспозвоно'чное *n* invertebrate
беспозвоно'чный *a* invertebrate, backboneless
беспоко'ить *v* vex
беспоко'йство *n* anxiety, agitation, restlessness, uneasiness, unrest
~ **из-за разлу'ки** *eth* separation distress
беспокро'вный *a bot* naked, achlamydeous
беспо'лый *a* agamic, agamous, asexual, sexless
беспоря'дочный *a* disorderly, chaotic, irregular, random; *bot* inordinate
бесприли'стниковый *a* exstipulate
бесприцве'тниковый *a* bractless
беспы'льниковый *a* anantherous, antherless
бессемядо'льный *a* acotyledonous
бессемя'нный *a* seedless, apyrene
бесси'лие *n* impotence, impotency
полово'е ~ impotence
бессозна'тельный *a* unconscious
бессо'нница *f* sleeplessness
бессосу'дистый *a* avascular
бесстe'бельный *a* stemless, acauline, acaulescent, acaulous, acormose
бестычи'нковый *a* anandrous, instaminate
бесхво'стый *a* tailess, acaudate, ecaudate, anurous
бесхлорофи'лльный *a* achlorophyllous
бесцветко'вый *a* flowerless, ananthous
бесцве'тный *a* colourless, achromatic
бесчелюстно'й *a* agnathous, agnathostomatous
бесчерешко'вый *a bot* stipeless, sessile
бесчешу'йный *a* alepidote
бесшпо'рцевый *a* spurless, ecalcarate
бесщелево'й *a pal* alete
бе'та-во'лны *f pl*, **бе'та-ри'тм** *m* beta activity
бета-разнообра'зие *n* beta-diversiy
бе'тель *m* betel pepper (*Piper betle*)
бе'шенство *n* rabies
биатори'новый *a* biatorine *(lichen)*
бивале'нт *m* bivalent
закры'тый ~, **кольцево'й** ~ ring bivalent
нера'вный ~ unequal bivalent
откры'тый ~ rod bivalent
прецентри'ческий ~ precentric bivalent
бивале'нтный *a* bivalent
би'вень *m* tusk, tush
бивольти'нный *a* bivoltine
бига'мия *f* bigamy
бигль *m* beagle
бие'ние *n* beat, beating, throb, throbbing
~ **пу'льса** pulsation
~ **се'рдца** heart throb, throbbing of the heart
бизо'н *m* bison, buffalo (*Bison bison*)
биколлатера'льный *a* bicollateral
билатера'льный *a* bilateral
биливерди'н *m* biliverdin
билируби'н *m* bilirubin

бильгарцио'з *m* bilharziasis, schistosomiasis
бима'стия *anat* bimastism
бина'рный *a* binary, binomial
бинокуля'р *m* binocular
бинокуля'рный *a* binocular
бинт *m* bandage
бинтуро'нг *ther* binturong *(Arctictis)*
био- bio-
биоаку'стика *f* bioacustics
биоана'лиз *m* bioassay
биобла'ст *m* bioblast
биовыщела'чивание *n* bioleaching
биога'з *m* biogas
биогене'з *m* biogenesis
биогенети'ческий *a* biogenetic
биогени'я *f* biogeny
биоге'нный *a* biogenic, biogenous
биогеогра'фия *f* biogegraphy
биогеохи'мия *f* biogeochemistry
биогеоцено'з *m* biogeocenosis, biogeocenose
биогидроло'гия *f* biohydrology
биода'тчик *m* biosensor
биодеграда'ция *f* biodegradation
биодемогра'фия *f* biodemography
биодина'мика *f* biodynamics
биоинжене'рия *f* bioengineering
биокатализа'тор *m* biocatalist
биоката'лиз *m* biocatalysis
биокиберне'тика *f* biological cybernetics, biocybernetics
биокли'мат *m* bioclimate
биоклиматоло'гия *f* bioclimatology
биоли'зис *m* biolysis
биолити'ческий *a* biolytic
био'лог *m* biologist
биологи'ческий *a* biological, biologic
биоло'гия *f* biology
~ **кле'тки** cell biology
~ **разви'тия** developmental biology
~ **челове'ка** human biology
ква'нтовая ~ quantum biology
косми'ческая ~ space biology, bioastronautics
молекуля'рная ~ molecular biology
радиацио'нная ~ radiation biology
биолюминесце'нция *f* bioluminescence
био'м *m ecol* biome
биоманипуля'ция *f* biomanipulation
биома'сса *f* biomass, standing crop
~ **ви'да** species biomass
биомедици'нский *a* biomedical
биометри'ческий *a* biometric
биоме'трия *f* biometrics, biometry
биомеха'ника *f* biomechanics
биомо'ниторинг *m* biomonitoring
бионакопле'ние *n* bioaccumulation
био'ника *f* bionics
бионо'з *m* bionosis
бионо'мия *f* bionomics, bionomy
био'нт *m* biont
биоповрежде'ние *n* biodeterioration
биополиме'р *m* biopolymer
биопрепара'т *m* biopreparation, biological preparation
биопро'ба *f* bioassay, biological assay, biological test
биопромы'шленность *f* bioindustry
биопси'я *f* biopsy
аспирацио'нная ~ suctional biopsy
после'довательно-многокра'тная ~ serial biopsy
пункцио'нная ~ needle biopsy
сери'йная ~ serial biopsy
биоразнообра'зие *n* biodiversity
биореа'ктор *m* bioreactor
биори'тм *m* biorhythm
би'ос *m* bios
биосе'нсор *m* biosensor
биоси'нтез *m* biosynthesis
беззатра'вочный ~ unprimed biosynthesis
биосисте'ма *f* biosystem
биосистема'тика *f* biotaxonomy
биосо'ма *f* biosome
биосоциоло'гия *f* biosociology
биспорангиа'тный *a* bisporangiate
биоста'нция *f* biological research station
биостати'стика *f* biostatistics
биостати'ческий *a* biostatic
биостимуля'тор *m* biostimulant, biological stimulant
биостратигра'фия *f* biostratigraphy
бисубстра'тный *a* bisubstrate
биосфе'ра *f* biosphere
биосфе'рный *a* biospheric
био'та *f* biota
биотелеметри'я *f* biotelemetry
биоте'ст *m* bioassay
биоте'хника *f* biotechnics, biological engineering
биотехноло'гия *f* biotechnology
~ **окружа'ющей среды'** environmental biotechnology
биоти'н *m* biotin
биотинили'рование *n* biotinylation

биоти'п *m* biotype
биоти'ческий *a* biotic
биото'нус *m* biotonus
биото'п *m* biotope, community habitat
маргина'льный ~ marginal habitat
биотро'н *m* biotron
биотэ'нк *m* biotank
биофа'г *m* biophage
биофи'зика *f* biophysics
биофи'лия *f* biophylia
биофи'льтр *m* biofilter, biological filter
биофи'т *m* biophyte
биофотогене'з *m* biophotogenesis
биофото'лиз *m* biophotolysis
биохими'ческий *a* biochemical
биохи'мия *f* biochemistry
биохо'р *m* biochore
биохро'м *m* biochrome
биоцено'з *m* biocenose, biocenosis, biocoenosis, community, biocommunity
~ **гнию'щего наво'за**, ~ **компо'ста** syrphetobios
биоценоло'гия *f* coenobiology, biocenology, biocenotics, biosociology
биоцено'н *m* biocenon
биоци'д *m* biocide
биоци'кл *m* biocycle
биочи'п *m* biochip
биоэколо'гия *f* bioecology
биоэлектри'чество *n* bioelectricity
биоэлектро'ника *f* bioelectronics
биоэнерге'тика *f* bioenergetics
бипинна'рия *f zool* bipinnaria
биполя'рность *f* bipolarity
биполя'рный *a* bipolar
бирюзо'вый *a* turquoise
бирючи'на *f bot* privet (*Ligustrum*)
бисексуа'льный *a* bisexual
бисериа'льный *a* biserial
бисло'й *m* bilayer
би'смарк *m*:
~ **кори'чневый** Bismark brown
би'тва *f eth* battle
бить *v* beat; **его' бьет лихора'дка** he is shivering with fever
~ **гарпуно'м** harpoon
~ **за'дом** kick (*of a horse*)
~ **ко пы'том** stamp, paw
~ **на лету'** shoot on the wing
~ **ры'бу острого'й** spear fish
~ **тюле'ней** seal
~ **хвосто'м** *eth* lash the tail, swish the tail
би'ться *v* throb
битю'г *m* dray horse
бифациа'льный *a* bifacial
бифидобакте'рия *f* bifid bacterium
бифурка'ция *f* bifurcation
бихевиори'зм *m* behaviorism
бихевиори'стский *a* behavioral
бицикли'ческий *a* bicyclic
бластениоспо'ра *f* blasteniospore
бласте'я *f* blastaea, blastea
бласто- blasto-
бластогене'з *m* blastogenesis
бластоде'рма *f emb* blastoderm, germinal membrane
бластоди'ск *m emb* blastodisk
бластозоо'ид *m* blastozooid
бластокине'з *m* blastokinesis
бластоме'р *m emb* blastomere
бластомико'з *m* blastomycosis
~ **Ги'лкриста** Gilchrist blastomycosis
бластонейропо'р *m* blastoneuropore
бластопо'р *m* blastopore
бластоспо'ра *f* blastospore
бластости'ль *m* blastostyle
бластотоми'я *f* blastotomy
бластохо'р *m* blastochore
бластоце'ль *f emb* blastocoele, blastocele
бластоци'ста *f emb* germ bladder, blastocyst
бластоци'т *m* blastocyte
бласттрансформа'ция *f* blast-cell transformation
бла'стула *f emb* blastula, bladder germ, blastodermic vesicle, vesicular germ
бле'дный *a* pale
бле'кнуть *v* wither
блефаропла'ст *m* blepharoplast
бле'яние *n* bleat
бле'ять *v* baa, bleat
близне'ц *m* twin
близнецы' *m pl* twins
двуяйцо'вые ~ biovular twins, dioval twins, dizygotic twins
однояйцо'вые ~ monoval twins, monozygotic twins, identical twins, uniovular twins, enzygotic twins
разнояйцо'вые ~ fraternal twins
близору'кий *a* myopic, nearsighted
близору'кость *f* myopia, shortsightedness, nearsightedness
блок *m* trochlea, pulley, block
~ **При'бнова** Pribnow's box

~ **созревания** maturation block, maturational block
~ **Хо'гнесса** Hogness' box
ано'дный ~ anodal block
не'рвный ~ nerve block
рендомизи'рованный ~ *biom* randomized block
блока'да *f* blockade, block
~ **пучка' Ги'са** bundle-branch block
~ **не'рва** nerve block
~ **се'рдца** heart block
атриовентрикуля'рная ~ antegrade block
внутрижелу'дочковая ~ arborization block, intreventricular block
внутрипредсе'рдная ~ intra-atrial block
интермитти'рующая ~ intermittent block
интравентрикуля'рная ~ arborization block, intraventricular block
интраатриа'льная ~ intra-atrial block
межжелу'дочковая ~ bundle-branch block
не'рвно-мы'шечная ~ neuromuscular block
перемежа'ющаяся ~ intermittent block
синоаурикуля'рная ~ sinoauricular block
части'чная ~ partial block
блоки'ровать *v* block
блокиро'вка *f* block
ано'дная ~ anodal block
блокови'дный *a* trochlear
блокполимериза'ция *f* blockpolymerization
блоксополимериза'ция *f* blockcopolymerization
блоки'ровать *v* block
бло'ттинг *m* blotting
~ **ферме'нтов** enzymoblotting
блоха' *f* flea (*Pulex*)
водяна'я ~ water flea *(cladoceran; Daphnia)*
челове'ческая ~ human flea (*Pulex irritans*)
бло'шка *f*: **земляна'я** ~ flea beetle (*Halticinae*)
блюдцеви'дный *a* patelliform, salverform, acetabuliform
бля'шка *f* plaque, plate
мото'рная ~ motor plate
пе'йерова ~ Peyer's patch, Peyer's gland
боб *m* bean
~ **кака'о** cacao bean
бобови'дный *a* bean-shaped, fabiform
бобо'вый *a* fabaceous, leguminous
бобр *m* beaver (*Castor fiber*)
америка'нский ~ American beaver, Canadian beaver *(Castor canadensis)*
речно'й ~ European beaver *(Castor fiber)*
бобы' *m pl* bean (*Vicia*)
богара' *f* dry farming, bogharic farming
богомол *ent* mantid *(Mantidae)*
~ **обыкнове'нный**, ~ **религио'зный** European mantid, praying mantid *(Mantis religiosa)*
бода'ние *n* horning
бода'ть *v* butt, gore
бода'ться *v* butt, gore
бодли'вость *n* buttness
бодли'вый *a* butting
бо'дрствующий *a* awake, wakeful
бодя'к *m* thistle (*Cirsium*)
бой *m* fight, battle
~ **быко'в** bullfight
петуши'ный ~ cockfight
бо'йня *f* abbatoir, slaughterhouse
бок side, flank
бока'л *m* goblet
глазно'й ~ eyecup
бока'лец *m* scypha
бокалови'дный, бока'льчатый *a* cyathiform, scyphiform, cup-shaped
бокалови'дная кле'тка goblet cell, flask cell
бока'льчик *m* cyathium
боково'й *a* lateral
бокопе'стичный *a* pleurogynous
боконе'рвный *a* side-veined
бокопло'дность *f* pleurocarpy
бокопло'дный *a* pleurocarpic, pleurocarpous, cladocarpous, cladogenous, side-fruiting
бокоходя'щий *a* laterigrade
бокоцве'тный *a* pleuranthous, lateriflorous
бокс *micr* box
боксёр *m* boxer
бокси'рование *n eth* boxing
бди'тельность *f eth* alertness
белковоподо'бный *a* albuminoid
боле'зненный *a* painful
болезнетво'рность *f* pathogenicity
болезнетво'рный *a* pathogen, pathogene, pathogenic
боле'знь *f* disease, sickness, illness
~ **А'йерсы** Ayerza's disease
~ **Альцге'ймера** Alzheimer's disease
~ **Ауе'ски** Aujeszky's disease
~**деформированных листьев клевера** trifolium clubleaf *(caused by a virus)*
~ **Жильбе'ра** Gilbert's disease

~ **иммунных комплексов** immune deposit disease
~ **коша'чьей цара'пины** cat-scratch disease
~ **Минама'ты** Minamata disease
~ **Паркинсо'на** Parkinson's disease
~ **от коша'чьих цара'пин** cat-scratch disease, cat- scratch fever
~ **раневых опухолей растений** wound tumor disease *(caused by a virus)*
~ **Хо'джкина** Hodgkin's disease
~ **Хо'джсона** Hodgson's disease
~ **Чагаса** Chagas disease *(caused by Schizotrypanum cruzi)*
~ **Шарко'** Charcot's disease
аддисо'нова ~ Addison's disease
базе'дова ~ Basedow's disease
борнхо'льмская ~ Bornholm disease
бра'йтова ~ Bright's disease
голландская ~ вязов Dutch elm disease *(caused by Ceratocystis ulmi)*
гомологичная ~ homologous disease, wasting disease
го'рная ~ mountain sickness, high-altitude disease
гречи'шная ~ fagopyrism
желчнока'менная ~ cholelithiasis
инфекцио'нная ~ infectious disease, communicable disease
ишеми'ческая ~ се'рдца myocardial ischemia
кессо'нная ~ caisson disease
морска'я ~ sea sickness
пайерсова ~ винограда Pierce's disease *(caused by a virus)*
попуга'йная ~ *med* parrot fever
профессиональная ~ industrial disease, occupational disease
розе'точная ~ rosette disease
слюннока'менная ~ sialolithiasis
со'нная ~ sleeping sickness, tsetse-fly disease
тяже'лая ~ serious illness
у'ровская ~ Urov disease
цейло'нская ~ Ceylon disease
шо'ковая ~ shock disease

болезнетво'рный *a* disease-producing
болеутоля'ющий *a* analgetic
болиголо'в *m bot* hemlock *(Conium)*
боло'нка *f* Bolognese
боло'тистый *a* marshy, moory, boggy, swampy, paludal, uliginous
боло'тница *f* spike rush *(Eleocharis)*
боло'тный *a* paludal, uliginous
боло'то *n* swamp, marsh, moor, fen, bog, swampland
вереща'тниковое ~ heath bog, heather bog
верхово'е ~ high bog, raised bog, raised moss, raised moor, upland bog, upland swamp, mire
вы'пуклое ~ domed bog
ги'пновое ~ hypnum bog
гря'дово-мочажи'нное ~ hummock-ridge bog
доли'нное ~ valley bog, valley mire
зыбу'чее ~ trembling bog, floating bog
камышо'вое ~ bulrush fen
клю'квенное ~ cranberry bog
ключево'е ~ spring bog, spring fen
кочкова'тое ~ hillock bog, tussock bog
крупноосо'ковое ~ high sedge fen
куста'рниковое ~ shrub bog
куста'рничковое ~ subshrub bog, dwarf shrub bog
лесно'е ~ wooded bog, forest bog, wood swamp
лиша'йниковое ~ lichen bog
ма'нгровое ~ mangrove swamp
мелкоосо'ковое ~ low sedge fen, small sedge moor
мохово'е ~ mossery, moss, moss bog, moss fen, moss moor
надво'дное ~ superaquatic moor
низи'нное ~ lowland moor, lowland swamp, fen, flood-plain marsh, lowland bog
ольхо'вое ~ alder swamp
омброге'нное ~ ombrogenic bog
осо'ковое ~ sedge fen, sedge marsh
перехо'дное ~ transitional swamp, transitory bog, transitional mire
пло'ское ~ flat bog, flat moor
по'йменное ~ valley bog, river marsh
пуши'цевое ~ cotton grass swamp, cotton grass bog
соле'ное ~ salt marsh, salt-water marsh
солончако'вое ~ salt swamp, salt marsh
сфа'гновое ~ peat moss bog, sphagnum bog
торфяно'е ~ peat bog, peat swamp
травяно'е ~ grass bog, marsh
тростнико'вое ~ reed swamp
болотове'дение *n* telmathology, peat land science
болотолюби'вый bog-loving
боль *f* pain

головна'я ~ headache
де'ргающая ~ shooting pain, throbbing pain, lanciniating pain
жгу'чая ~ burning pain
загруди'нная ~ retrosternal pain
зубна'я ~ toothache
мучи'тельная ~ troublesome pain, excruciating pain
мы'шечная ~ myalgia
но'ющая ~ aching pain, dull pain, boring pain
о'страя ~ acute pain, keen pain, stinging pain, sharp pain
си'льная ~ violent pain, throe
сто'йкая ~ persistent ache, steady pain, unwavering pain, lasting pain
тупа'я ~ dull pain
фанто'мная ~ stump neuralgia

больно'й *m* patient
амбулато'рный ~ outpatient, ambulatory patient, ambulant patient, walking patient
безнадёжный ~ fatal case
беспомощный ~ helpless patient
инфекционный ~ contagious patient, infectious patient, infective patient
лежа'чий ~ bed patient
неконтактный ~ uncooperative patient
оперированный ~ operated patient
ослабленный ~ debilitated patient
ходя'чий ~ walking patient, ambulatory patient, ambulant patient

большеберцо'вый *a* tibial
большегла'зый *a* big-eyed
большо'й *a* big, large, great, bulky
больша'я высота' high altitude
~ **па'лец** first finger

бомба'ж *m* bloat, swell
бомбарди'р *m* bombardier beetle (*Brachinus*)
боните'т *m* quality of locality
~ **насажде'ния** quality of stand

бонитиро'вка *f* bonitation, assessment
~ **почв** soil assessment

бор *m* pine forest; millet-grass (*Milium*); drill; boron
~ **беломо'шник** lichen pine forest
~ **зеленомо'шник** pine forest with green mosses

бордо'вый *a* claret, claret-colored, dark red
бореа'льный *a* boreal
боре'ц *m* monkshood, wolfsbane (*Aconitum*)
борза'я *f* borzoi
бормоту'шка *f* booted warbler (*Hippolais caligata*)
бо'ров *m* hog
борови'к *m* edible boletus, cep, penny bun (*Boletus edulis*)
борода' *f* beard
борода'вка *f* wart
борода'вник *bot* dockcress *(Lampsana)*
борово'й *a*:
борова'я дичь upland fowl

борода'вчатый *a* verrucate, verrusose, verrucous, warty, warted
борода'тка, борода'стик *orn* barbet *(Megalaima)*
борода'тый *a* bearded, barbate
борода'ч *m orn* lammergeier, lammergeyer, bearded vulture (*Gypaetus barbatus); bot* bluestem (*Andropogon*)
борода'ч-ягня'тник *m* lammergeier, bearded vulture (*Gypaetus barbatus*)
боро'дка *f* beard; barb (*of a feather*)
боро'дочка *f* barbicel, barbule (*of a feather*)
борозда' *f* furrow, groove, fissure, sulcus; row
~**аммо'нова ро'га** *anat* dentate fissure
~ **дробления** *emb* cleavage furrow
~ **запя'стья** carpal groove
~ **мо'зга** cerebral sulcus
~ **сошника'** vomeral sulcus
вене'чная ~ **се'рдца** coronary sulcus
вено'зная ~ venous groove
гаррисо'нова ~ Harrison's groove
губна'я ~ labial groove
запира'тельная ~ obturator groove
заты'лочно-височ'чная ~ *anat* occipitotemporal sulcus
зри'тельная ~ *anat* calcarine sulcus
латитудина'льная ~ *emb* latitudinal furrow
лё'гочная ~ pulmonary sulcus
медулля'рная ~ medullary groove
межбугорко'вая ~ bicipital groove
меридиона'льная ~ *emb* meridional furrow
не'бная ~ palatine groove, palatinal groove
обоня'тельная ~ olfactory sulcus
окклюзио'нная ~ occlusion groove
око'льная ~ collateral fissure
парапсида'льная ~ parapsidal furrow
плу'жная ~ *agr* plowed furrow
подбуго'рная ~ hypothalamic sulcus
посевна'я ~ *agr* drill furrow

разва'лочная ~, разва'льная ~ *agr* open furrow
реше'тчатая ~ ethmoidal groove
рола'ндова ~ Rolandic fissure
си'львиева ~ groove of Sylvius
слё'зная ~ lacrimal groove
со'нная ~ *anat* carotid groove, carotid sulcus
среди'нная ~ median sulcus
шпо'рная ~ *anat* calcarine fissure, calcarine sulcus
центра'льная ~ central fissure
ягоди'чная ~ gluteal furrow

боро'здка *f* furrow, groove, stria, sulcus, channel, scrobe
~ **ко'жи** skin furrow
~ **оссифика'ции** ossification groove
~ **Ранвье'** Ranvier sulcus, ossification groove
ампулля'рная ~ ampullary sulcus
антенна'льная ~ *ent* antennal scrobe
больша'я ~ major groove (*in DNA*)
зубна'я ~ dental furrow
ма'лая ~ minor groove (*in DNA*)
межпредсе'рдная ~ interatrial groove
меридиона'льная ~ *emb* meridional furrow
полова'я ~ genital furrow
рёберная ~ costal furrow
экваториа'льная ~ equatorial furrow

бороздчатоли'стный *a* furrow-leaved
бороздчатопло'дный *a* aulacocarpous
бороздчатосемя'нный *a* furrow-seeded
бороздчатосте'бельный *a* furrow-stemmed
боро'здчатый *a* grooved, furrowed, sulcate
борони'ть *v agr* harrow
боронова'ние *n agr* harrowing
бо'ртик *m*:
боково'й ~ *ent* connexivum
бо'ртничество *n* wild-honey farming
борть *f* hive of wild bees
борщеви'к *m bot* cow-parsnip (*Heracleum*)
борьба' *f* fighting, struggle, wrestle, contest; control
~ **за существова'ние** struggle for existence
~ **с вреди'телями** pest control
~ **с загрязне'нием** pollution control
~ **с сорняка'ми** weed control
~ **с шу'мами** noise control
~ **с эро'зией** erosion control
~ **ше'ями** *eth* neck-fighting, necking, neck-wrestling
биологи'ческая ~ biological control
конкуре'нтная ~ competitive fighting
хими'ческая ~ chemical control

ботанизи'рка *f* vasculum
ботанизи'ровать *v* botanize
бота'ник *m* botanist
бота'ника *f* botany
прикладна'я ~ applied botany, economic botany
ботани'ческий *a* botanic
ботва' *f* tops, haulm
ботри'дий *m* bothridium
ботриоцефалё'з *m helm* bothriocephalosis
бо'трия *f* bothrium
воронкообра'зная ~ funnelar bothrium
ботули'зм *m* botulism
боченкообра'зный, бочкови'дный *a* dolioliform, barrel-shaped
боя'знь *f* fear
боя'нусов: ~ **о'рган** *zool* organ of Bojanus
боя'рышник *m bot* hawthorn, thorn (*Crataegus*)
боя'рышница *f ent* thorn butterfly, black-veined white *(Aportia crataegi)*
брадидиа'стола *f* bradydiastole
брадикини'н *m* bradykinin
брадипно'э *phys* bradypnoea
брадиси'стола *phys* bradysystole
брадители'я *f* bradytely
бра'жник *m ent* hawk moth, hawk, sphinx, hummingbird moth (*Sphingidae*)
~ **мё'ртвая голова'** dead's head hawk *(Acherontia atropos)*
~ **молоча'йный** spurge hawkmoth *(Celerio euphorbiae)*
браконье'р *m* poacher
браконье'рский *a* illegal
браконье'рство *n* poaching
браконье'рствовать *v* poach
бранхиа'льный *a* branchial
бранхиокардиа'льный *a* branchiocardiac
бранхиомери'я *f* branchiomery
бранхиомико'з *m* branchiomycosis
брать *v* take
~ **барье'р** fence
брахиа'ция *f* brachiation
брахибла'ст *m* brachyblast
брахидактили'я *f* brachydactyly, brachydactylia
брахимейо'з *m* brachymeiosis
брахиодо'нтный *a* brachydont
брахиоля'рия *f* brachiolaria

брахисклереи'да *f* brachysclereid
брахицефа'льный *a* brachycephalic
бра'чный *a* breeding, nuptial
бревно' *n* log
верши'нное ~ top log
комле'вое ~ butt log
бре'гма *f* bregma
бред *m* delirium, delusions
~ **Манья'на** delirium of Magnan, delusions of Magnan
бреди'на *f* goat willow, sallow (*Salix caprea*)
бриологи'ческий *a* bryological
бриоло'гия *f* bryology
бриотерофи'т *m* bryotherophyte
бровь *f* brow, eyebrow, supercilium
броди'льный *a* fermantative, zymotic
броди'ть *v* ferment; roam
бродя'жка *f* telotroch
бродя'чий *a* vagile, vagrant, nomadic; stray (*of a dog*)
бродя'щий *a* fermenting
броже'ние *n* fermentation, fermenting
~ **в пото'ке** bulk fermentation
акти'вное ~ active fermentation
анаэро'бное ~ anaerobic fermentation
ацетонобути'ловое ~ acetone-butanol fermentation
ацетоноэти'ловое ~ acetone-ethanol fermentation
аэро'бное ~ aerobic fermentation
бактериа'льное ~ bacterial fermentation
вя'лое ~ sluggish fermentation
галловоки'слое ~ gallic-acid fermentation
гетероферментати'вное молочноки'слое ~ heterolactic fermentation
глицери'новое ~ glycerol fermentation
глюконовоки'слое ~ gluconic-acid fermentation
гомоферментати'вное молочноки'слое ~ homolactic fermentation
глуби'нное ~ submerged fermentation, subsurface fermentation
гни'лостное ~ putrefactive fermentation
дрожжево'е ~ yeast fermentation
завершё'нное ~ complete fermentation
итаконовоки'слое ~ itaconic-acid fermentation
лимонноки'слое ~ citric-acid fermentation
масляноки'слое ~ butyric fermentation
мо'крое ~ wet fermentation
молочноки'слое ~ lactic fermentation, lactic-acid fermentation
непреры'вное ~ continuous fermentation
низово'е ~ low fermentation, bottom fermentation
откры'тое ~ open fermentation
пове'рхностное ~ surface fermentation
пропионовоки'слое ~ propionic-acid fermentation
самопроизво'льное ~ spontaneous fermentation
слизево'е ~ slime fermentation
сме'шанное ~ mixed fermentation
спиртово'е ~ alcoholic fermentation, ethyl-alcohol fermentation
спонта'нное ~ spontaneous fermentation
уксусноки'слое ~ acetic fermentation
фумаровоки'слое ~ fumaric-acid fermentation
холо'дное ~ cool fermentation
эндоге'нное ~ endogenous fermentation
бромдезоксиуриди'н, БДУ *m* bromdeoxyuridine
бромметили'рование *n* bromomethylation
бромураци'л *m* bromouracil
бромформили'рование *n* bromoformylation
бронено'сец *m* armadillo (*Dasypus*)
бронзо'вка *f* rose chafer (*Cetonia aurata*)
бронх *m* bronchus
верхнедолево'й ~ proximal bronchus
гла'вный ~ extrapulmonary bronchus, main bronchus, primary bronchus
нижнедолево'й ~ distal bronchus
сегмента'рный ~ segmental bronchus
трахеа'льный ~ tracheal bronchus
бронхиа'льный *a* bronchial
бронхио'ла *f* bronchiole
бронхи'т *m* bronchitis
бронхоальвеоля'рный *a anat* bronchovesicular
бронхомото'рный *a* bronchomotor
бронхоскопи'я *f* bronchoscopy
бронхотрахеа'льный *a* bronchotracheal
броса'ться *v* fly at (*of a dog*); attack
бросо'к *m* dart
брусни'ка *f* cowberry, clusterberry (*Vaccinium vitis-idaea*)
брусни'чник *m* cowberry grove
бруцелле'з *m* brucellosis
брыже'ечный *a* mesenterial, mesenteric
брыже'йка *f anat* mesentery, frill
брюшна'я ~ ventral mesentery

спинна'я ~ dorsal mesentery
бры'згальце *n* spiracle
брыка'ться *v* buck, kick, fling
брылы' *pl* hanging lips (*of dogs*)
брысь *i* shoo (*to a cat*)
брю'ква *f* rutabaga, swede, rooted turnip, Russian turnip (*Brassica napus var. napobrassica*)
брю'квенница *f ent* gray-veined white *(Pieris napi)*
брю'хо *n* belly, maw
брюши'на *f* peritoneum, abdominal membrane
брюши'нный *a* peritoneal
брю'шко *n* abdomen, belly, venter
сидя'чее ~ sessile abdomen
стебе'льчатое ~ petiolated abdomen
брю'шно-грудно'й *a* abdominothoracic
брюшно'й *a* ventral, abdominal
бубе'нчик *m* ladybell (*Adenophora*)
бубо'н *m* bubo
буго'р *m* protuberance, tuber, eminence
ло'бный ~ frontal protuberance, frontal eminence
седа'лищный ~ ishial tuberosity
теменно'й ~ parietal protuberance
скулово'й ~ malar eminence
бугорко'вый *a* grumous
бугоро'к *m* protuberance, tubercle, boss, eminence; cone (*e.g., of a tooth*); hill, hillock, button
~ **оплодотворе'ния** entry cone, exudation cone, fertilization cone
~ **ушно'й ра'ковины** Darwinian tubercle, Woolner's tubercle
акросо'мный ~ acrosomal cone
воспринима'ющий ~ entrance cone, receptive spot, entry cone, exudation cone, fertilization cone
гло'точный ~ pharyngeal spine
да'рвинов ~ Darwinian tubercle
кутикуля'рный ~ cuticular hillock
лобко'вый ~ pubic spine
ло'нный ~ pubic spine
не'рвный ~ nerve hillock
подборо'дочный ~ genial tubercle
полово'й ~ genital tubercle, genital eminence
ризо'идный ~ rhizoid protuberance
семенно'й ~ seminal hillock
со'нный ~ carotid tubercle
суставно'й ~ articular tubercle, articular eminence
эктоплацента'рный ~ *ther* ectoplacental cone
буго'рчатый *a* knobby, bosselated, bumpy, torulose, tuberculose, tubercular, tuberculate, nodular, tuberous, warted
бугри'стость *f anat* tuberosity
бугри'стый *a* hilly, hillocky, hummocky; bumpy, knobby, torose, grumous, tuberous
буди'ть *v* wake
бу'дра *f* ground ivy (*Glechoma*)
буера'к *m* gully, ravine
бузина' *f* elder (*Sambucus*)
бу'йвол *m* buffalo (*Bubalus bubalus*)
азиатский ~ water buffalo, Indian buffalo*(Bubalus arnee)*
африканский ~ African fuffalo, Cape buffalo *(Syncerus caffer)*
индийский ~ water buffalo, Indian buffalo*(Bubalus arnee)*
чёрный ~ African fuffalo, Cape buffalo*(Syncerus caffer)*
бу'йный *a* violent; wild; lush, luxuriant, heavy, exuberant (*of vegetation*)
бу'йствовать *v* rage
бук *m* beech (*Fagus*)
бука'шник *m bot* Shepherd's scabious *(Jasione)*
бу'квица *f bot* betony (*Betonica*)
буке'т *m* bouquet, bunch of flowers; flavour (*of tea*, *wine*)
букка'льный *a* buccal
булава' *f* club
нерасчлене'нная ~ inarticulate club
пласти'нчатая ~ lamellate club
була'вка *f* pin
энтомологи'ческая ~ entomological pin, insect pin
булавови'дный *a* clublike, clavate
булавоно'сный *a* club-bearing
булавопло'дный *a* club-fruited
булавостебе'льчатый *a* club-footed
булавоцветко'вый *a* club-flowered
була'ный *a* light-bay
булы'жник *m* cobble-stone
бульбу'ль *m* bulbul (*Pycnonotidae*)
бульдо'г *m* bulldog
бульо'н *m* broth
глицери'новый ~ glycerin broth
глюкозосы'вороточный ~ glucose serum broth

декстрозофосфа'тный ~ dextrose phosphate broth
мясно'й ~ beef-infusion broth
мясопепто'нный ~, **МПБ** meat infusion broth, beef-extract broth, plain broth
нитра'тный ~ nitrate broth
очи'щенный ~ clarified broth
пепто'нный ~ peptone broth
пита'тельный ~ nutrient broth
ры'бный ~ fish broth
са'харный ~ sugar broth
сы'вороточный ~ serum broth
трипсинизи'рованный ~ trypsin broth
яи'чный ~ egg infusion broth
бума'га *f* paper
диазобензилоксиметилцеллюло'зная ~ DBM paper
диазоти'рованная ~ diazotized paper
диазоти'рованная аминобензилоксимети'ловая ~ diazotized aminobensyloxymethyl paper
диаминоэтиламиноэтилцеллюло'зная ~ DEAE paper
индика'торная ~ indicator paper, test paper
ла'кмусовая ~ litmus paper
ли'пкая ~ flypaper
реакти'вная ~ exploration paper, reagent paper
фильтрова'льная ~ bibulous paper, filter paper, absorbent paper
бума'жистый *a bot* papery
буньяви'рус *m* bunyavirus
бур *m* borer
бура' *f* borax
бургоми'стр *m orn* burgomaster, glaucous gull (*Larus hyperboreus*)
бурдо'н *m* burdo, burdon
буреве'стник *m* shearwater, petrel (*Puffinus*)
антаркти'ческий ~ Antarctic petrel *(Thalassoica antarctica)*
гига'нтский ~ Mother Carey's goose *(Macronectes giganteus)*
ка'пский ~ Cape pigeon (*Daption capensis*)
бурело'м *m* windbreak, windfall, breakage
буре'ние *n* boring
бурозу'бка *f ther* common shrew, long-tailed shrew *(Sorex)*
бурси'т *m* bursitis
бурст, бурст-оча'г *m* burst
бурт *m* pile
бурунду'к *m* chipmunk (*Tamias*)
восточноамерика'нский ~ hackee *(Tamias striatus)*
бурча'ние *n* growling, rumbling
бурча'ть *v* growl
бу'рый *a* brown
бурья'н *m* tall grasses, tall dense weeds
бу'стер *m* booster
бу'стер-до'за *f* booster
бу'стер-иммуниза'ция *f* booster immunization
бу'стер-инъе'кция *f* booster shot, overshot
~ **реиммуниза'ции** postrecovery overshot
бу'стер-эффе'кт *m* booster effect
бу'стинг *m imm* booster immunization
бу'сы *pl* beads
стекля'нные ~ glass beads
бутерла'к *m* water-purslane (*Peplis*)
бутили'рование *n* butylation
буто'н *m* flower bud
бутониза'ция *f* flower-bud formation, budding
бутылеобра'зный, бутылкови'дный *a* lageniform, bottle-shaped
буты'лка *f* bottle
бу'фер *m* buffer
ге'левый ~ gel buffer
карбона'тный ~ carbonate buffer
солево'й ~ salt buffer
станда'ртный ~ standart buffer
фосфа'тный ~ phosphate buffer
экстрацио'нный ~ extraction buffer
электро'дный ~ electrode buffer
бу'ханье *n* bump (*of a bittern*)
бу'хать *v* bump (*of a bittern*)
бу'хта *f* bight
бушбо'к *m ther* bush buck *(Tragelaphus scriptus)*
бушме'йстер *m herp* bushmaster *(Lachesis muta)*
бык *m* bull, ox, neat; pier (*of a bridge*)
му'скусный ~ musk-ox, musk sheep (*Ovibos moschatus*)
первобы'тный ~ urus *(Bos primigenius)*
бы'стрый *a* agile, quick, rapid, fast, swift
бы'чий *a* bovine
бычо'к *m* bull calf
бычо'к-кастра'т *m* steer
бэр, биологи'ческий эквивале'нт (*m*) **рентге'на** rem, roentgen-equivalent-man
бюдже'т *m* budget
бюкс *m* weighing bottle
бюльбю'ль *m orn* bulbul *(Pycnonotidae)*

бюре'тка *a lab* burette
~ для титрова'ния titrating burette
градуи'рованная ~ volume burette

В

ваготоми'я *f* vagotomy
ваготони'н *m* vagotonin
ваготони'я *f* vagotony
ваготропный *a* vagotropic
ва'гусный *a* vagal
ва'женка *f ther* female reindeer
вазицентри'ческий *a* vasicentric
вазопресси'н *m phys* vasopressin
ва'йя *f bot* frond
вакуолиза'ция *f* vaculization
вакуолизи'рованный *a* vacuolated, vaculate
вакуо'ль *f* vacuole
  ~ с кле'точным со'ком sap vacuole
  аутофаги'ческая ~ autophagic vacuole
  контракти'льная ~ contractile vacuole, water-expulsion vacuole
  паразитофо'рная ~ parasitophorous vacuole
  периальга'льная ~ perialgal vacuole
  пиноцито'зная ~ pinocytotic vacuole
  пищевари'тельная ~ digestive vacuole, food vacuole
  сократи'тельная ~ contractile vacuole
  фагоцита'рная ~ phagocytic vacuole
  эндоцито'зная ~ endocytic vacuole
вакуоля'рный *a* vacuolar
вакуо'ля *f* vacuole
вакуо'м *m* vacuome
вакци'на *f* vaccine
  ~ с адъюва'нтом adjuvant vaccine
  авианизи'рованная ~ avianized vaccine
  ассоции'рованная ~ mixed vaccine
  аттенюи'рованная ~ attenuated vaccine
  аутоге'нная ~ autogenous vaccine, autologous vaccine, personal vaccine, corresponding vaccine, homologous vaccine
  безаллерге'нная ~ allergen-free vaccine
  вируле'нтная ~ virulent vaccine
  генноинжене'рная ~ engineered vaccine
  жива'я ~ live vaccine
  инактиви'рованная ~ killed vaccine, dead vaccine
  комбини'рованная ~ mixed vaccine
  осла'бленная ~ attenuated vaccine, weakened vaccine
  поливале'нтная ~ polyvalent vaccine, multipartial vaccine, multivalent vaccine, mixed vaccine
  противови'русная ~ virus vaccine
  рекомбина'нтная ~ engineered vaccine
  уби'тая ~ killed vaccine, dead vaccine
  формалинизи'рованная ~ formolated vaccine
вакцина'льный *a* vaccinal
вакцина'тор *m* vaccinator
вакцина'ция *f* vaccination, vaccinization, immunization
  ма'ссовая ~ cohort immunization, mass immunization
  перви'чная ~ primary immunization
  пла'новая ~ planned immunization
вакцини'рованный *a* vaccinated
вакцини'ровать *v* vaccinate
вакци'нный *a* vaccinal
вале'жник *m* deadfallen wood, slash, windfallen trees, windthrown trees
вале'нтность valence
  экологи'ческая ~ ecological valence
валериа'на *f* valerian *(Valeriana)*
вали'дность *f* validity
  вне'шняя ~ external validity
  вну'тренняя ~ internal validity
вали'дный *a* valid
ва'лик *m* torus, torulus
  ногтево'й ~ nail wall, wall of the nail
вали'н *m* valine
ва'лка *f* felling
  ~ ле'са felling, timber cutting
  группова'я ~ multistem felling
валу'й *m* fetid russula, stinking russula (*Russula foetens*)
валу'х *f* bullock
ва'льва *f* valve
ва'льдшнеп *m* woodcock (*Scolopax rusticola*)
валькова'тый *a* terete, teretial, cylindrical
валя'ться *v* wallow
вампи'р *m* vampire, vampire bat (*Desmodontidae*)
  обыкнове'нный ~ vampire bat *(Desmodus)*
ванили'н *m* vaniline
ванили'новый *a* vanillic
ва'нночка *f* pan, dish
  препарова'льная ~ dissecting pan, dissecting dish
вара'кушка *f* blue throat, blue-throated robin, blue-throated warbler (*Luscinia svecica*)
вара'н *m herp* monitor (*Varanus*)
  комо'дский ~ Komodo dragon *(Varanus komodoensis)*

варене'ц *m* boiled fermented milk
вариа'бельность *f* variability
вариа'нса *f biom* variance
  генотипи'ческая ~ genotypic variance
  фенотипи'ческая ~ phenotypic variance
  эпистати'ческая ~ epistatic variance
вариа'нт *m* variant
  ~ **ассоциа'ции** variant of association
вариа'нта *f biom* variant
вариа'ция *f* variation
  внеза'пная ~ brusque variation
  гомологи'ческая ~ homologous variation
  оста'точная ~ remaining variation
  по'чковая ~ bud variation
  сбаланси'рованная ~ balanced variation
  сомати'ческая ~ somatic variation
  соматогенная ~ somatogenic variation
  фа'зовая ~ phase variation
вариете'т *m* variety
варико'зный *a* varicose
вариото'п *m* variotypic epitope
варьи'ровать *v* vary
васкуляриза'ция *f* vascularization
васкуля'рность *f* vascularity
васкуля'рный *a* vascular
ва'та *f* cotton wool
ва'хта *f bot* bogbean, buckbean (*Menyanthes trifoliata*)
введе'ние *n* administration (*e.g., of a* drug); injection; intromission; introduction
  ~ **с пи'щей** dietary administration
  бы'строе ~ slug injection
  дро'бное ~ intermittent administration
  ора'льное ~ oral administration
  парэнтера'льное ~ parenteral adminsitration
  перора'льное ~ oral administration
  повто'рное ~ repeated adminsitration
  преры'вистое ~ intermittent administration
вводи'ть *v* introduce; administer; inject
ввора'чиваться *v* inturn
ВВФ, Всеми'рный фонд (*m*) **охра'ны приро'ды** WWF, World Wildlife Fund
вдаю'щийся *a* incurrent
вдох inspiration
  акти'вный ~ active inspiration
вдува'ние *n* insufflation
вдыха'ние *n* inhalation, inspiration
вдыха'тельный *a* inspiratory
вдыха'ть *v* inhale
вегетализа'ция *f* vegetalization
вегетати'вный *a* vegetative
ве'ер *m* fan; *bot* rhipidium
веерови'дный *a* flabellate, flabelliform
веероне'рвный *a* flabellinerved
веерообра'зный *a* flabellate, rhipidate
везикуля'рный *a* vesicular
ве'йник *m* bluejoint, wood-reed, shorthear (*Calamagrostis*)
вейсмани'зм *m* weismannism
ве'ко *n* eyelid, palpebra, blepharon
ве'ктор vector
  ~ **для клони'рования** *mbiol* cloning vector
  ~ **замеще'ния** replacement vector
  ~ **со вста'вкой** insertional vector, insertion vector
  ~ **экспре'ссии** expression vector
  бакулови'русный ~ baculovirus vector
  инсерцио'нный ~ insertional vector, insertion vector
  клони'рующий ~ cloning vector
  мгнове'нный ~ instantaneous vector
  мобилизу'емый ~ mobilizable vector
  моме'нтный ~ instantaneous vector
  насы'щенный ~ high-titer vector
  перестано'вочный ~ permutation vector
  плазми'дный ~ plasmid vector
  простра'нственный ~ spatial vector
  ретрови'русный ~ retroviral vector
  самозамыка'ющийся ~ self-ligated vector
  трансдуци'рующий ~ transducing vector
  характеристи'ческий ~ characteristic vector
  экспресси'рующий ~ expression vector
  энха'нсер-конститути'вный ~ enhancer-constitutive vector
вела'мен *m* velamen
велиге'р *m zool* veliger
величина' *f* size; value
  ~ **погре'шности** *biom* extent of error
  ~ **сцепле'ния** *gen* linkage value
  дискре'тная случа'йная ~ discrete random variable
  за'данная ~ desired value
  кауза'льная ~ explanatory variable
  коне'чная ~ *biom* finite
  многоме'рная случа'йная ~ multidimensional random variate
  натура'льная ~ full size, full scale, life size
  непреры'вная случа'йная ~ continuous random variable
  норма'льная случа'йная ~ normal random variable

норми'рованная случа'йная ~ standardized variate
преде'льная ~ limit value
приближё'нная ~ rough value
случа'йная ~ variate
теорети'чески рассчи'танная ~ theoretical value
эффекти'вная ~ популя'ции effective population size
ве'люм *m* velum
ве'на *f* vein
~ **желто'чного мешка'** *emb* omphalomesenteric vein
алланто'исная ~ allantoic vein, umbilical vein
бе'дренная ~ femoral vein
брюшна'я ~ abdominal vein
воро'тная ~ portal vein
головна'я ~ cephalic vein
диафрагма'льная ~ phrenic vein
желто'чная ~ vitelline vein
за'дняя кардина'льная ~ postcardinal vein
ма'ленькая ~ veinlet
основна'я ~ basal vein
подключи'чная ~ subclavian vein
подколе'нная ~ popliteal vein
подкрыльцо'вая ~ axillary vein
субкардина'льная ~ subcardinal vein
яре'мная ~ jugular vein
венери'ческий *a* venereal
вене'ц *m* crown
лучи'стый ~ radiating crown
вене'чный *a* coronary, coronal, coronoid
вене'чное кровообраще'ние coronary circulation
~ **отро'сток** coronoid process
венодви'гательный *a* venomotor
вено'зный *a* venous
венопрессо'рный *a* venopressor
вентили'ровать *v* air, ventilate
вентилято'рный *a* ventilatory
вентиля'ция *f phys* ventilation
лё'гочная ~ lung ventilation, pulmonary ventilation
максима'льная ~ maximal ventilation
произво'льная ~ voluntary ventilation
вентра'льный *a* ventral
венцеобра'зный *a* coroniform
ве'нчик *m bot* corolla
губови'дный ~ lip-shaped corolla
двугу'бый ~ bilabiate corolla
зигомо'рфный ~ zygomorphous corolla
колоко'льчатый ~ bell-shaped corolla
раздельнолепе'стный ~ choripetalous corolla
спайнолепе'стный ~ gamopetalous corolla
тру'бчатый ~ tubular corolla
венчикови'дный *a bot* corollaceous
венчиконо'сный *a bot* corolliferous
ве'рба *f bot* violet willow (*Salix daphnioides*)
вербе'йник *m bot* loosestrife (*Lysimachia*)
верблю'д *m* camel (*Camelus*)
двуго'рбый ~ two-humped camel, Bactrian camel (*Camelus bactrianus*)
одного'рбый ~ dromedary, Arabian camel, one-humped camel (*Camelus dromedarius*)
верблю'дица *f* female camel
верблю'дка *f ent* snake fly, raphidia (*Raphidia*); *bot* tick-seed (*Corispermum*)
верблюжо'нок *m* colt
верболо'з *m* pussy willow, silver willow (*Salix acutifolia*)
ве'реск *m* heather (*Calluna*)
верете'нник *m* godwit (*Limosa*)
большо'й ~ stone plover (*Limosa limosa*)
верете'нница *f* slow worm (*Anguis*)
веретено' *n cyt* spindle
~ **деле'ния созрева'ния** maturation spindle
~ **тяготе'ния** attraction spindle
лимби'ческое ~ *phys* limbic spindle
метафа'зное ~ metaphase spindle
мы'шечное ~ muscle spindle
не'рвно-сухожи'льное ~ neurotendinal spindle
я'дерное ~ nucleospindle
веретенови'дный, веретенообра'зный *a* fusiform, spindle-shaped
вереща'тник *m* heath, heathy land
вереща'ть *v* squall
вермици'д *m* vermicide
ве'рность *f*: ~ **ви'да** exclusiveness
ве'рный *a* infallible *(e.g., of experimental results)*
вероя'тность *f* probability
~ **отка'за** rejection probability
апостерио'рная ~ posterior probability
априо'рная ~ prior probability
довери'тельная ~ acceptance probability, confidence probability
усло'вная ~ conditional probability
вертише'йка *f* wryneck (*Jynx torquilla*)
вертлу'г *m* trochanter

верту'н *m*: **сосно'вый** ~ *phytop* pine twisting rust (*caused by Melampsora pinitorqua*)
вертя'чка *f* turn sickness; whirligig beetle *(Gyrinidae)*
верхне- supra-
верхнегубно'й *a* labral
верхнезаты'лочный *a* supraoccipital
верхнечелюстно'й *a* supramaxillary
верхове'нство *n* supremacy
верхово'й *a* surface (*of fermentation*); raised (*of a bog*); saddle (*of a horse*)
верхо'вье *n* upper reaches, river head
верхо'м *adv* astride, on horseback
верхопло'дный *a* acrocarpous
верхушечноцветко'вый *a* acranthed
верху'шечный *a* apical
верху'шка *f* apex, top, tip, head, vertex, umbo
~ **ле'гкого** apex of a lung
~ **ушно'й ра'ковины** tip of the auricle
верши'на *f* vertex (*of an angle*); top, summit (*of a mountain*); head (*of a gully*); eye-end *(of a potato tuber)*
вершкова'ние *n* tipping
вес *m* weight
~ **сыро'й тка'ни** fresh weight
~ **те'ла** body weight
абсолю'тно сухо'й ~ ovendry weight
возду'шносухо'й ~ air-dry weight
живо'й ~ live weight
избы'точный ~ overweight
молекуля'рный ~ molecular weight
объё'мный ~ volume weight
сухо'й ~ dry weight
весё'лка *f bot* common stinkhorn, wood witch (*Phallus impudicus*)
весе'нний *a* spring, vernal
весну'шка *f* freckle
весня'нка *f* stone fly (*Plecoptera*)
вестибуля'рный *a* vestibular
весы' *pl* balance, scales
аналити'ческие ~ analytical balance
пружи'нные ~ compression balance
торсио'нные ~ torsion balance
ветвено'сный *a* ramiferous, ramigerous
ветви'стый *a* branched, branching, branchy, ramose, ramate, cladose, ramulose, ramified, ramulous, boughy, brachiferous, brachigerous
ветви'ться *v* branch, ramify
ветвле'ние *n* branching, ramification
ветвь *f* branch, bough; ramus
~ **вилообра'зного хвоста'** *helm* furcal ramus
~ **второ'го поря'дка** secondary branch
~ **пе'рвого поря'дка** primary branch
~ **фу'рки** furcal ramus
бокова'я ~ lateral branch
гла'вная ~ leading branch
ма'точная ~ scaffold branch
плодонося'щая ~ fruit-bearing branch, fruiting branch
ростова'я ~ vegetative branch
скеле'тная ~ scaffold branch
цветоно'сная ~ flower-bearing branch
ветвя'щийся *a* branching
ве'тер *m* wind
поры'вистый ~ gusty wind
ветерина'р *m* veterinarian, veterinary
ветеринари'я *f* veterinary medicine, veterinary science
ветерина'рный *a* veterinary
ве'тка *f* branch, bough
ветла' *f* white willow, common willow (*Salix alba*)
ве'точка *f* branchlet, twig, sprig
ветрова'л *m* windbreak, windfall
ветрозащи'тный *a* wind-protective
ветроопыле'ние *n* wind-pollinating
ветроопыля'емый *a* anemophilous, wind-pollinated
ветроусто'йчивый *a* wind-resisting, wind-firm
ветчина' *f* ham
вех *m* water-hemlock (*Cicuta*)
вечнозеле'ный *a* evergreen, sempervirent, aiphyllous, indeciduous
ве'шенка *f* oyster mushroom (*Peurotus ostreatus*)
ве'шний *a* spring, vernal
вещество' *n* matter, substance, material
~ **коры'** cortical substance
~, **мобилизу'ющее липи'ды** lipid-mobilizing substance
~, **подавля'ющее рост** growth inhibiting substance
~, **стимули'рующее рост** growth-promoting substance, growth-stimulating substance
~ **утомле'ния** fatigue substance
абиоти'ческое ~ nonliving material
абсорби'рующее ~ absorbing agent
агглютини'рующее ~ agglutinating substance

антимитоти'ческое ~ antimitotic substance
бактерици'дное ~ bactericide, germicidal agent
балла'стное ~ ballast substance
белко'вое ~ proteic substance
бе'лое ~ white matter, white substance
вкусово'е ~ flavoring agent
группоспецифи'ческие вещества' кро'ви *pl* blood group substances
дегидри'рующее вещество dehydrating agent
докле'точное ~ precellular matter
живо'е ~ living material
жиромобилизу'ющее ~ lipid-mobilizing substance
загрязня'ющее ~ pollutant, contaminant
запасно'е ~ storage material
иммy'нное ~ phylactic agent
индуци'рующее ~ inducing substance
компа'ктное ~ ко'сти compact substance of bone
кра'сящее ~ staining agent
лету'чее ~ volatile substance, volatile
липотро'пное ~ lipotropic substance
ма'точное ~ *ent* queen substance
мозгово'е ~ medullary substance
неживо'е ~ nonliving material
неоргани'ческое ~ inorganic matter
органи'ческое ~ organic matter
основно'е ~ ground substance
паху'чее ~ odorous substance
пита'тельное ~ nutrient
пло'тное ~ close substance
пове'рхностно-акти'вное ~ surfactant, surface-active agent, surface-active substance
подсла'щивающее ~ sweetener
посторо'ннее ~ foreign substance
просветля'ющее ~ clarifying agent
рогово'е ~ horny substance
рола'ндово ~ *hist* Rolando's gelatinous substance
ростово'е ~ growth substance
связу'ющее ~ binding agent
се'рое ~ gray matter, gray substance *(of brain)*
стимули'рующее ~ promotoring agent
сто'йкое загрязня'ющее ~ persistent pollutant
студени'стое ~ gelatinous substance
сухо'е ~ dry substance, dry matter
тигро'идное ~ tigroid substance, tigroid body
токси'чное ~ toxic agent
угнета'ющее ~ inhibitory substance
хромотро'пное ~ chromotropic substance
цементи'рующее ~ cement substance
ядови'тое ~ toxic agent
вещество'-носи'тель carrying agent
вживле'ние *n* implantation
вживле'нный *a* implanted
взаимоде'йствие *n* interaction
~ **ге'нов** gene interaction
~ **Т- и В-лимфоци'тов** *imm* T-B interaction
бело'к-белко'вое ~ protein-protein interaction
внутримолекуля'рные взаимоде'йствия *pl* intramolecular interactions
гидрофо'бное ~ hydrophobic interaction
межкле'точное ~ cell-cell interaction
межмолекуля'рные взаимоде'йствия *pl* intermolecular interactions
сбаланси'рованное ~ balanced interaction
взаимоде'йствовать *v* interact
взаимоотноше'ния *pl* interactions
внутривидовы'е ~ intraspecific interactions
межвидовы'е ~ interspecific interactions
взаимопо'мощь *f* aid-giving behavior
взаимосвя'зь *f* interconnection
взаимотормозя'щий *a* interinhibitive
вызрева'ние *n* aging
взба'лтывание *n* agitation
взбира'ться *v* clamber up (*e.g., of* vine); climb up
взвесь *f* suspension
~ **отмы'тых кле'ток** washed cell suspension
бактери'йная ~ bacterial suspension
взви'згивание *n* yelp
взви'згивать *v* yelp
взгляд *m* stare, gaze
отведе'нный ~ *eth* avert stare
при'стальный ~ *eth* stare
вдра'гивать *v* startle
взду'тый *a* ventricose
взду'тие *n* inflation, distention, bloating, swelling; puff (*of a chromosome*)
взлета'ть, взлете'ть *v* take flight, take wing, take to the air
взмах *m* swing, flap; wag
~ **крыла'** wing beat

взмы'ленный *a* foamy, lathery
взро'слый *a* adult
взрыв *m* explosion
  ~ **потребле'ния** consumption explosion
  демографи'ческий ~ demographic explosion, population explosion, population outbreak
  популяцио'нный ~ population explosion
взрыхлё'нный *a* loosened
взъеро'шивать *v* ruffle up
взя'тие *n:* ~ **образцо'в** sampling
взя'ток *m* forage, honey gathering
вибракуля'рия *f* vibraculum
вибри'рующий *a* vibratile
вибри'сса *f* vibrissa
вибри'ссы *f pl* whiskers
вибрио'н *m* vibrio
  вибрион Эль-То'р El Tor vibrio
вибриононоси'тель *m* vibriocarrier
виброкардиограмма *f* vibrocardiogram
вива'рий *m* vivarium
вивипа'рия *f* viviparity, vivipary
вивисе'кция *f* vivisection
вивотокси'н *m* vivotoxin
виго'нь *f* vicuna, vicugna (*Vicugna vicugna*)
вид *m* appearance; kind; view; species
  ~, **бога'тый фо'рмами** demoid species
  ~ **по'чвы** kind of soil
  ~ **сбо'ку** lateral view
  ~ **све'рху** dorsal view
  ~ **сза'ди** rear view
  ~ **сни'зу** ventral view, bottom view
  абориге'нный ~ native species, indigenous species, local species
  аллопатри'ческий ~ allopatric species
  безме'рный ~ nondimensional species
  биологи'ческий ~ biological species
  бу'ферный ~ buffer species
  ве'рный ~ exclusive species, faithful species
  викари'рующий ~ vicarious species, substitute species, geminate species
  внедре'нный ~ introduced species
  внедря'ющийся ~ invader
  вне'шний ~ appearance, external appearance, habit
  восстано'вленный ~ restored species
  второстепе'нный ~ secondary species
  вы'мерший ~ extinct species, deleted species, epiobiotic species
  вырожда'ющийся ~ degenerate species
  деструкти'вный ~ destructive species
  диагности'ческий ~ diagnostic species
  ди'кий ~ wildlife species
  доче'рний ~ daughter species
  дифференциа'льный ~, **дифференци'рующий** ~ differential species
  замеща'ющий ~ vicariad, vicariant
  занесе'нный ~ adventitious species, tramp species
  индика'торный ~ indicator species
  индиффере'нтный ~ indifferent species
  исчеза'ющий ~ endangered species, vanishing species
  исче'знувший ~ extinct species, deleted species, epiobiotic species
  ключево'й ~ key species
  консервато'рный ~ conserving species
  конста'нтный ~ constant species
  конструкти'вный ~ constructive species
  космополи'тный ~ cosmopolitan species
  крипти'ческий ~ cryptic species
  линнеевский ~ linneon
  ма'ссовый ~ dominant
  матери'нский ~ mother species
  ме'стный ~ native species, local species, indigenous species
  мигри'рующий ~ migratory species
  мономорфи'ческий ~, **мономо'рфный** ~ monomorphic species
  монотипи'ческий ~ monotypic species
  морфологи'ческий ~ morphological species, morphospecies
  направля'ющий ~ directive species
  нейтра'льный ~ neutral species
  о'бщий ~ habit view
  обы'чный ~ common species
  перехо'дный ~ bridging species
  пионе'рный ~ pioneer species, pioneering species
  повсеме'стно распространё'нный ~ cosmopolitan species
  подчине'нный ~ subordinate species
  полимо'рфный ~ polymorphic species
  политипи'ческий ~ polytypic species
  полноце'нный ~ good species
  постоя'нный ~ constant species
  предпочита'ющий ~ preferent species, preferential species
  привлека'ющий ~ directive species
  привя'занный ~ selective species
  промысло'вый ~ commercial species
  различи'мый ~ distinct species

разнохозя'йственный ~ heteroecious species
ре'дкий ~ rare species
рели'ктовый ~ deleted species, epiobiotic species
репрезентати'вный ~ representative species
ро'дственный ~ allied species
сбо'рный ~ collective species
селекти'вный ~ selective species
симпатри'ческий ~ sympatric species
синхро'нный ~ synchronic species
скры'тый ~ cryptic species
случа'йный ~ casual species, incidental species
сни'женный альпи'йский ~ dealpine species
сопряжё'нный ~ accessory species, associated species, accompanying species, companion species
сопу'тствующий ~ accessory species, associated species, accompanying species, companion species
средообразу'ющий ~ edificator, environment-forming species
субдомина'нтный ~ subdominant species
тетрапло'идный ~ digenomic species
типово'й ~ type species
тузе'мный ~ native species, indigenous species, local species
угрожа'емый ~ threatened species, endangered species
уязви'мый ~ vulnerable species
физиологи'ческий ~ physiological species
характе'рный ~ characteristic species
чу'ждый ~ alien species, strange species
эндеми'чный ~ endemic species
вид-близне'ц *m* twin species
вид-вселе'нец *m* invading species
вид-двойни'к *m* sibling species
вид-индика'тор *m* indicator species
вид-колони'ст *m* colonist species, invading species
видово'й *a* specific
видообразова'ние *n* speciation
аллопатри'ческое ~ allopatric speciation
центробе'жное ~ centrifugal speciation
«взрывно'е» ~ explosive speciation
географи'ческое ~ geographic speciation, geographical speciation
полусимпатри'ческое ~ semisympatric speciation
симпатри'ческое ~ sympatric speciation
видоспецифи'чный *a* species-specific
вид-перено'счик *m* vector species
визг *m* yelp, scream, shriek, screech
визжа'ть *v* yelp, scream, shriek, squeal
визуализа'ция *f* visualization
ви'ка *f* vetch (*Vicia*)
викариа'нт vicariad, vicariant
высо'тный ~ altitudinal vicariant
географи'ческий ~ geographical vicariant
сезо'нный ~ seasonal vicariant
экологи'ческий ~ ecological vicariant
викариа'т *m* vicariat
викари'рование *n* vicariation
вика'рный *a* vicarious
вику'нья *f* vicuna, vicugna (*Vicugna vicugna*)
ви'лка *f* fork, furcula
ДНК-репликати'вная ~ DNA-replication fork
пры'гательная ~ spring
репликати'вная ~ replicative fork
вилконо'сный *a* dicranophorous
вило'к *m* head of cabbage
ви'лочка *f* furcula
вилт *m* wilt
ви'льчатый *a* bifurcate, bifurcated, furcate, forcipate
виля'ние *n* wagging
~ **хвосто'м** wagging
виля'ть *v* wag
~ **хвосто'м** wag
винно-кра'сный *a* vinaceous, vinous
виногра'д *m* grape, vine (*Vitis*)
~ **белесоватый** mustang grape *(Vitis candicans)*
~ **винный** common grape, European grape *(Vitis vinifera)*
~ **зи'мний** frost grape, chicken grape (*Vitis cordifolia*)
~ **Изабе'лла** fox grape (*Vitis labrusca*)
~ **изю'мный** raisin grape
~ **круглолистный** bullace grape, muscadine grape *(Vitis rotundifolia)*
~ **культу'рный** grape vine (*Vitis vinifera*)
~ **лесной** woodland grape *(Vitis silvestris)*
~ **ле'тний** summer grape (*Vitis aestivalis*)
~ **лисий** chicken grape, frost grape, mountain grape, raccoon grape, river grape, riverside grape, Spanish grape, winter grape *(Vitis vulpina)*
~ **муска'тный** muscadine grape (*Vitis rotundifolia*)

~ **позднеспе'лый** late vine, late grape
~ **сизоватый** native grape *(Vitis hypoglauca)*
~ **скальный** bush grape, rock grape, sand grape *(Vitis rupestris)*
виногра'дарство *n* grape culture, viticulture
виногра'дарь *m* grape grower, vine grower
виногра'дина *f* grape berry
виногра'дник *m* vineyard
винт *m* screw
микрометри'ческий ~ micrometer screw
регулиро'вочный ~ adjusting screw
устано'вочный ~ adjusting screw
винтово'й *a* helical, helicine, spiral
виро'ид *m* viroid
вирио'н *m* virion
виросо'ма *f* virosome
вируле'нтность *f* virulence
вируле'нтный *a* virulent
ви'рус *m* virus, ultramicrobe
~ **гепати'та** hepatitis virus
~ **гри'ппа** influenza virus
~ **иммунодефици'та** immunodeficiency virus
~ **иммунодефици'та челове'ка** human immunodeficiency virus
~ **индуци'рованной лейкеми'и** induced leukemia virus
~ **Кокса'ки** Coxsackie virus
~ **лимфолейко'за** lymphatic leukemia virus
~, **передава'емый гриба'ми** fungus-transmitted virus
~ **полио'мы** polyoma virus
~ **просто'го ге'рпеса** herpes simplex virus
~ **расте'ний** plant virus
~ **сарко'мы Ра'уса** Rous sarcoma virus
~ **Сенда'й** Sendai virus
~ **синци'тия цыпля'т** chicken syncitial virus
~ **Сме'ди** smedivirus
~ **со спира'льной симметри'ей** helical symmetric virus
~ **СПИ'Да** AIDS virus
~ **Эпште'йна-Ба'рра** Epstein-Barr virus
адапти'рованный ~ adapted virus
аденоассоции'рованный ~ adenoassociated virus
аэроге'нный ~ air-borne virus
артропоно'зный ~ arthropod-borne virus
аттенуи'рованный ~ attenuated virus
бактериа'льный ~ bacterial virus
вируле'нтный ~ virulent virus
гибри'дный ~ hybrid virus
гранули'рованный ~ pelleted virus
дефекти'вный ~, дефе'ктный ~ defective virus
зре'лый ~ mature virus
индика'торный ~ indicator virus
инта'ктный ~ intact virus
интерфери'рующий ~ interfering virus
инфекцио'нный ~ infective virus
исключе'нный ~ excluded virus
исхо'дный ~ intact virus
капринизи'рованный ~ caprinized virus
кише'чный ~ enteric virus
компете'нтный ~ competent virus
контамини'рующий ~ contaminating virus
лапинизи'рованный ~ lapinized virus
лате'нтный ~ latent virus
лимфотро'пный ~ lymphotropic virus
маскиро'ванный ~ masked virus
ме'дленный ~ slow virus
ме'ченый ~ labeled virus
многокра'тно пасси'рованный ~ high passage virus
незре'лый ~ immature virus
неизменё'нный ~ intact virus
неинфекцио'нный ~ latent virus
неполноце'нный ~ incomplete virus
неро'дственные ви'русы *pl* viruses-satellites
онкоге'нный ~ oncogenic virus, tumor virus
осла'бленный ~ attenuated virus
пантро'пный ~ pantropic virus
персисти'рующий ~ persistent virus
политро'пный ~ pantropic virus
расти'тельный ~ plant virus
РНК-содержа'щий онкоге'нный ~ RNA-tumor virus
сенсибилизи'рованный ~ sensitized virus
сиро'тский ~ orphan virus
сопу'тствующий ~ satellite virus
субуме'ренный ~ submoderate virus
уме'ренный ~ moderate virus
фильтру'ющийся ~ filterable virus, filter-passer, ultravirus
экзоге'нный ~ exogenous virus
эндоге'нный ~ endogenous virus
энтеротро'пный ~ enterotropic virus
вируле'нтность *f* virulence
вируле'нтный *a* virulent
ви'рус-возбуди'тель causal virus

ви'русный *a* viral
вирусоло'гия *f* virology
~ **живо'тных** animal virology
вирусоноси'тель *m* virus carrier
вирусоноси'тельство *n* virus carrying, virus-carriage
ви'рус-помо'щник *m* helper virus
амфотро'пный ~ amphotropic helper virus
ви'рус-сателли'т *m* virus-satellite
виска'ча *f ther* viscacha
вискози'метр *m* viscosimeter
вислопло'дный *a* crenocarpous
вислокры'лка *f* orl fly (*Sialis*)
вислоу'хий *a* flap-eared, lop-eared
висо'к *m* temple
висо'чный *a* temporal
висцера'льный *a* visceral
висцеромото'рный *a* visceromotor
висцеротро'пный *a* viscerotropic
वися'чий *a* pensile (*e.g., of a nest*), pendulous, pendent
вися'щий *a* hanging, pendent
витали'зм *m* vitalism
витали'ст *m* vitalist
виталисти'ческий *a* vitalistic
вита'льный *a* vital
витами'н *m* vitamin
витаминиза'ция *f* vitaminization
витаминизи'ровать *v* vitaminize
витамино'зный *a* vitaminous
вителли'н *m* vitellin
вителлогене'з *m* vitellogenesis
вителлоге'нный *a* vitellogenous
вителлофа'г *m* vitellophag
витрифика'ция *f* vitrification
вить *v* weave
~ **гнездо'** build *a* nest, nest, nidificate
ви'ться *v* climb (*of plants*), run; curl, wave (*of hair*); hover (*e.g., of insects*); circle
вихля'й *m* houbara bustard (*Clamydotis undulata*)
вицини'зм *m* vicinism
ВИЧ, вирус иммунодефици'та челове'ка human immunodeficiency virus
ви'шня *f* cherry (*Cerasus*)
вклад *m eth* investment
~ **ма'тери** maternal investment
~ **отца'** paternal investment
~ **роди'телей** parental investment
вкли'нивающийся *a* incurrent
включа'ющий *a* including
включе'ние *n* inclusion; incorporation; body; *bioch* insertion
цитоплазмати'ческое ~ cytoplasmic inclusion
я'дерное ~ nuclear inclusion
включё'нный *a* included
вкраплё'нный *a* ingrained
вкус *m* taste; sense of taste
вкусово'й *a* gustatory
вла'га *f* moisture; humor
водяни'стая ~ aqueous humor
досту'пная ~ available moisture
капилля'рная ~ capillary moisture
недосту'пная ~ unavailable moisture
стеклови'дная ~ *anat* vitreum
влага'лище *n anat* vagina; axil, boot (*of a leaf*), sheath; hose (*of a maize ear*)
~ **глазно'го я'блока** bulbar sheath
~ **не'рва** nerve sheath
~ **хо'бота** rostellar sheath
волосяно'е ~ hair sheath
крахмалоно'сное ~ starch sheath
листово'е ~ leaf sheath
ма'ленькое ~ *bot* vaginula
не'рвное ~ nerve sheath
синовиа'льное ~ synovial sheath
сухожи'льное ~ tendon sheath
влага'лищный *a* vaginal
влага'льце *n bot* vaginella
влагоё'мкость *f* moisture capacity
влаголюби'вый *a* hydrophilic, hydrophilous, water-loving
влагооборо'т *m* hydrological cycle, water cycle
вла'жность *f* humidity, moisture
~ **во'здуха** air humidity
~ **завяда'ния** wilting moisture
абсолю'тная ~ absolute humidity
избы'точная ~ excess moisture
максима'льная ~ maximum humidity
относи'тельная ~ relative humidity
регули'руемая ~ controlled humidity
уде'льная ~ specific humidity
вла'жный *a* humid, moist, damp, wet
влече'ние *n* drive, desire
полово'е ~ sexual desire, sexual urge, libido
влия'ние *n* influence, effect
~ **замести'телей** *bioch* effect of groups
~ **сосе'днего соо'бщества** edge effect
бесконе'чно ма'лое ~ infinitesimal effect
вме'рзший *a* imbedded in ice
вмести'лище *n* conceptacle

вмести'тельный *a* capacious
внебрюши'нный *a* extraperitoneal
внедоме'нный *a* extra-domain
внедре'ние *n* introduction *(e.g., of a new species)*; intrusion; intussusception, incorporation
внедря'ть *v* introduce *(e.g., a new species)*, incroporate; implant; promote
внедря'ться *v* intrude
внежа'берный *a* extrabranchial
внежелу'дочковый *a* extraventricular
внезаро'дышевый *a* extraembryonic
внекапсуля'рный *a* extracapsular
внекише'чный *a* extraenteric, extraintestinal
внекле'точный *a* extracellular, exocellular
внели'стный *a* extrafoliaceous
внема'точный *a* extrauterine
внемитохондриа'льный *a* extramitochondrial
вненуцелля'рный *a* extranucellar
внепа'зушный *a* extra-axillary
внеполово'й *a* extragenital
внепо'ровый *a* extraporate
внепо'чечный *a* extrarenal
внесе'ние *n* application
~ **разбры'згиванием** spray application
глубо'кое ~ subsoil application
двухле'нточное ~ split row application
дро'бное ~ split application
ле'нточное ~ band application
ле'нточное ~ удобре'ний band fertilization
разбро'сное ~ broadcast application
ра'зовое ~ single application
ручно'е ~ hand application
внесерде'чный *a* extracardiac, exocardial
внесосу'дистый *a* extravascular
внетычи'ночный *a* extrastaminal
внехромосо'мный *a* extrachromosomal
внецветко'вый *a* extrafloral
внечерепно'й *a* extracranial
внешнесе'тчатый *a* extrareticulate
вне'шний *a* outer, exterior, external, extrinsic, ectal
внея'дерный *a* extranuclear
внима'ние *n* attention
избира'тельное ~ selective attention
вноси'ть *v* apply
~ **удобре'ния** fertilize
вну'тренний *a* inner, interior, internal, inside, intrinsic
железа' вну'тренней секре'ции endocrine gland
вну'тренности *f pl* entrails, insides, viscera
вну'тренностный *a* visceral, splanchnic
внутри- intra-, endo-, end-
внутриареа'льный *a* intraneous
внутриартериа'льный *a* intra-arterial, endarterial
внутрибрюши'нный *a* intraperitoneal
внутрибрюшно'й *a* intra-abdominal
внутриве'нный *a* intravenous
внутривидово'й *a* intraspecific
внутривлага'лищный *a* intravaginal
внутриглазно'й *a* intraocular
внутригрудно'й *a* intrathoracic
внутритру'бный *a* intratubal
внутрижелто'чный *a* intravitelline
внутрижелу'дочковый *a* intraventricular
внутрижелудо'чный *a* intragastric
внутризерни'стый *a* intragranulate
внутрикапсуля'рный *a* intracapsular
внутрикле'точный *a* intracellular
внутриклона'льный *a* intraclonal
внутрико'жный *a* intracutaneous, intradermal
внутрилё'гочный *a* intrapulmonic
внутрили'стный *a* intrafoliaceous
внутрима'точный *a* intrauterine
внутримолекуля'рный *a* intramolecular
внутриморщи'нистый *a pal* intrarugulate
внутримы'шечный *a* intramuscular
внутрине'рвный *a* endoneural
внутриносово'й *a* intranasal
внутрипло'дник *m* endocarp
внутриполостно'й *a* intracavitary
внутрипо'чечный *a* intrarenal
внутрипредсе'рдный *a* intra-atrial
внутрипузы'рный *a* intravesical
внутриродово'й *a* intrageneric
внутриселезё'ночный *a* intrasplenic
внутрисерде'чный *a* endocardiac, endocardial, intracardiac
внутрисе'тчатый *a pal* intrareticulate
внутрисомати'ческий *a* intrasomatic
внутрисосу'дистый *a* intravascular
внутристе'ночный *a* intraparietal
внутристо'лбчатый *a pal* intrabaculate
внутрисуста'вный *a* intra-articular
внутритычи'ночный *a* intrastamineous
внутриушно'й *a* intra-aural, endotic
внутрихрящево'й *a* intracartilaginous
внутричерепно'й *a* intracranial
внутрия'дерный *a* intranuclear
вода' *f* water

атмосфе'рная ~ meteoric water
бидистиллиро'ванная ~ redistilled water, bidistilled water
водопрово'дная ~ tap water
гигроскопи'ческая ~ hygroscopic water
грунто'вая ~ ground water
деионизи'рованная ~ deionized water
дистиллиро'ванная ~ distilled water
досту'пная ~ available water
жаве'левая ~ Javelle solution, Javelle water
же'сткая ~ hard water
ирригацио'нная ~ irrigation water
капилля'рная ~ capillary water
конституцио'нная ~ constitution water
ливнева'я ~ storm water
метаболи'ческая ~ metabolic water
минера'льная ~ mineral water
морска'я ~ sea water
мя'гкая ~ soft water
недосту'пная ~ nonavailable water, unavailable water
неочи'щенная сто'чная ~ raw wastewater
неусвоя'емая ~ nonavailable water, unavailable water
обеззара'женная сто'чная ~ sterilized wastewater
ороси'тельная ~ irrigation water
очи'щенная ~ renovated water
пепто'нная ~ peptone water
питьева'я ~ drinking water, potable water
плё'ночная ~ film water, pellicular water
поливна'я ~ irrigation water
по'ровая ~ intersticial water
пре'сная ~ fresh water
прото'чная ~ running water, flowing water, lotic water
свобо'дная ~ free water
свя'занная ~ fixed water, bound water, combined water
солонова'тая ~ brackish water
стоя'чая ~ stagnant water, lentic water
та'лая ~ melt water
тяжё'лая ~ heavy water
фреати'ческая ~ phreatic water
чи'стая ~ pure water
води'тель *m*: ~ **ри'тма** time-giver
води'тельство *n* leadership
воднохозя'йственный *a* aquicultural
водоё'м *m* water body, body of water, water
водозабо'р *m* water intake
водола'з *m* diver; Newfoundland dog
водолюби'вый *a* water-loving
водонепроница'емый *a* waterproof, water-tight
водоно'сный *a* water- bearing, water-carrying
водообеспе'ченность *f* water availability
водооборо'т *m* water cycle
глоба'льный ~ global water cycle
водопо'й *m* watering; drinking place, watering place
водопотребле'ние *n* water consumption
водопрово'д *m anat* **aqueduct**
~ **сре'днего мо'зга** Sylvian aqueduct
си'львиев ~ Sylvian aqueduct
водоразде'л *m* watershed
водораствори'мый *a* water-soluble
водоросле'вый *a* algal
во'доросль *f* alga
водосбо'р *m bot* columbine *(Aquilegia)*
водосто'йкий, водопро'чный *a* water-stable
водохрани'лище *n* reservoir, dam, man-made lake
во'ды *f pl of* **вода'**
бытовы'е сто'чные ~ *pl* residental waters, domestic waste
ве'шние ~ spring floods, vernal floods
вну'тренние ~ *pl* territorial waters
возвра'тные ~ *pl* return waters
городски'е сто'чные ~ *pl* municipal waters, urbanic waste waters
грунто'вые воды *pl* subterranean waters, underground waters
неочи'щенные сто'чные ~ crude sewage
околопло'дные ~ amniotic fluid
очи'щенные сто'чные ~ sewage effluents
по'чвенные ~ *pl* soil waters
промы'шленные ~ *pl* industrial waters, commercial waters
промы'шленные сто'чные ~ industrial waste
сто'чные ~ sewage water, sewage, waste water, waste effluents
водяни'ка *f* crowberry (*Empetrum*)
водяни'стость *f* aquosity, wateriness
водяни'стый *a* watery
водя'нка *f* edema, hydrops, dropsy
водяно'й *a* aquatic
ВОЗ, Всеми'рная организа'ция (*f*) **здравоохране'ния** WHO, World Health Organisation
возбуди'мость *f* excitability, irritability
~ **мышц** muscular irritability
мы'шечная ~ muscular irritability

не'рвная ~ nervous irritability
рефлекто'рная ~ reflex excitability, reflex irritability
возбуди'тель *m* pathogen, agent, causative agent; exciter *(a nerve);* incitant
~ **боле'зни** agent of disease
~ **инфе'кции** infectious agent, agent of infection
возбужда'емость *f* excitability
возбужда'ть *v* excite
возбужда'ющий *a* excitative, excitatory, excitant
возбужде'ние *n* arousal, excitement, agitation; excitation, incitation, irritation; stimulation, induction
ано'дное ~ anode excitation
возвра'тное ~ recurrent excitation
латера'льное ~ lateral excitation
мозгово'е ~ cerebral excitement
полово'е ~ sexual excitement
психи'ческое ~ psychic excitement
размыка'тельное ~ break excitation
усло'вно-рефлекто'рное ~ conditioned excitation
чрезме'рное ~ superexcitation
эмоциона'льное ~ emotional excitement
возбуждё'нный *a* exited, hot, aglow; activated
возвра'т *m* relapse (*of disease*); return
~ **ме'тки** tag return
промысло'вый ~ yield to the fishery
возвра'тный *a* recurrent
~ **тиф** recurrent fever, relapsing fever
возде'йствие *n* effect, influence, action, agency, impact
дли'тельное ~ long-term exposure
допусти'мое ~ acceptable exposure
мультипликати'вное ~ multiplicative effect
непосре'дственное ~ immediate effect
несмерте'льное ~ survivable exposure
о'строе ~ acute exposure
экологи'ческое ~ ecological impact
возде'лываемый *a* cultivated
возде'лывание *n* cultivation
возде'лывать *v* cultivate, till
~ **зе'млю** till the land
~ **рис** grow rice
~ **сад** garden
во'здух *m* air
атмосфе'рный ~ atmosphere air
вдыха'емый ~ inhaled air, inspired air, respiratory air
вдыха'емый и выдыха'емый ~ breathing air
возвра'тный ~ return air
выдыха'емый ~ expired air, expirate
защемлё'нный ~ entrapped air
надзе'мный ~ ground air
оста'точный ~ residual air, minimal air
по'чвенный ~ soil air, entrapped air
припо'чвенный ~ ground air
резе'рвный ~ compemental air, complementary air
рециркули'рующий ~ return air
воздухонепроница'емый *a* air-tight, air-locked
воздухоно'сный *a* aeriferous
воздухопроница'емый *a* air-permeable
возду'шно-сухо'й *a* air-dry
возду'шный *a* air, aerial
возника'ть *v* emerge, appear
возобновле'ние *n* reproduction
возобновля'ть *v*: renew; ~ **лесонасажде'ние** reforest
во'зраст *m* age; instar
~ **по Бине'-Симо'ну** Binet age
~ **лакта'ции** milking age
~ **насажде'ния** age of stand
~ **нача'ла яйцекла'дки** age at first egg, lying age
~ **отъе'ма** weaning age
~ **полово'го созрева'ния** age of puberty
~ **спе'лости**, ~ **хозя'йственной спе'лости** exploitable age (*of forests*)
абсолю'тный ~ absolute age
анато'мо-физиологи'ческий ~ physiological age
биологи'ческий ~ physiological age
грудно'й ~ infancy
деторо'дный ~ childbearing age
календа'рный ~ chronological age
морфологи'ческий ~ anatomical age
подсо'сный ~ sucking age
ра'нний де'тский ~ infancy
репродукти'вный ~ childbearing age
сре'дний ~ mean age
хронологи'ческий ~ chronological age
экологи'ческий ~ ecological age
возраста'ние *n* increase
возраста'ть *v* increase
вой *m* howl
во'йлок *m* felt; tomentum (*e.g., on leaves*)

во'йлочный *a* felted, tomentose
война' *f* war
экологи'ческая ~ ecological war
вокализа'ция *f* vocalization
вока'льный *a* vocal
вокругротово'й *a* circumoral
вол *m* ox, bullock
во'ла *f* vola *(of diatoms)*
вола' *genetive of* **вол**
волды'рь *m* blister, bleb, wheal
волк *m* wolf (*Canis lupus*)
гри'вистый ~ maned wolf (*Chrysocion*)
земляно'й ~ earth wolf *(Proteles cristatus)*
кра'сный ~ dhole, red dog (*Cuon*)
обыкнове'нный ~ wolf, common wolf, European wolf *(Canis lupus)*
пчели'ный ~ bee-killer wasp (*Philantus*)
волкода'в *m* wolf dog
волна' *f* wave
~ **возбужде'ния** excitation wave
~ **Тра'убе-Ге'ринга** Traube-Hering wave
анакроти'ческая ~ anacrotic wave
блужда'ющая ~ travelling wave
дикроти'ческая ~ dicrotic wave, recoil wave
отражё'нная ~ reflected wave
популяцио'нные во'лны *pl* population waves
пу'льсовая ~ pulse wave
волне'ние *n* trouble, unrest, excitation, excitement; rough water
волнистолепестко'вый *a* wave-petaled
волнистоли'стный *a* wave-leaved
волни'стый *a* wavy, undulate, flexuous
воло'душка *bot* hare's-ear *(Bupleurum)*
волокни'стый *a* fibrous, fibrillar, stringy
волокно' *n* fiber, fibre, filament, string; grain (*of timber*)
~ **Брю'кке** Brucke's fiber
~ **денти'на** dentin fiber, Tomes' fiber
~ **Ко'рфа** Korff's fiber, radial fiber of dentin
~ **ксиле'мы** xylary fiber, xylem fiber
~ **митоти'ческого веретена'** spindle fiber
~ **Пуркинье'** Purkinje's fiber
~ **То'меса** Tomes' fiber
~ **Шарпе'я** Sharpey's fiber, perforating fiber
~ **Шти'ллинга** Stilling's fiber
~ **Э'бнера** Ebner's fiber, tangential fiber of dentin
адренерги'ческое ~ adrenergic fiber
ассоциати'вное ~ association ~ fiber
аффере'нтное ~ afferent fiber
аффи'нное ~ affinity fiber
безмя'котное ~ naked fiber, nonmedullated fiber, gray nerve fiber, Remak's fiber
бе'ргмановское ~ Bergmann's fiber
вневеретё'нное ~ extrafusal fiber
внутриверетё'нное ~ intrafusal fiber
гетероге'нное склере'нхимное ~ extraxylary fiber
дви'гательное ~ motor fiber
динами'ческое фузимото'рное ~ fusimotor dynamic fiber
интрафуза'льное ~ intrafusal fiber
ка'бельное не'рвное ~ ropelike nerve fiber
коллаге'новое ~ collagen fiber, collagenous fiber, collagenic fiber
миели'новое ~ myelin fiber, medullated fiber
мохови'дное ~ *anat* moss fiber
мы'шечное ~ muscle fiber, muscular fiber, myofibril, myofibrilla
мю'ллерово ~ Muller's fiber
мя'котное ~ myelin fiber, medullated fiber
не'рвное ~ nerve fiber, nerve filament
перви'чное ~ embryo fiber
постганглиона'рное ~ postganglionic fiber
преганглиона'рное ~ preganglionic fiber
преколлаге'новое ~ precollagenous fiber
пробода'ющее ~ perforating fiber, Sharpey's fiber
радиа'льное ~ дентина radial fiber of dentin
расти'тельное ~ plant fiber, vegetable fiber
рема'ково ~ Remak's fiber
ретикуля'рное ~ reticular fiber
решё'тчатое ~ **эласти'на** strainerlike fiber of elastin
серде'чное мы'шечное ~ cardiac muscular fiber
стати'ческое фузимото'рное ~ fusimotor static fiber
тангенциа'льное ~ **денти'на** tangential fiber of dentin, Ebner's fiber
холинерги'ческое ~ cholinergic ~ fiber
экстрафуза'льное ~ extrafusal fiber
эффере'нтное ~ efferent fiber
волоко'нце *n* filament
мы'шечное ~ muscular fibril
не'рвное ~ nerve fibril
то'нкое ~ fibril, fibrilla
волоко'нчатый *a* filamentary

во'лос *m* hair; **сгоня'ть** ~ dehair
ко'нский ~ horsehair
остево'й ~ top hair
пухово'й ~ down hair
волоса'тик *m* hairworm (*Gordius*)
волоса'тый *a* hairy
волоси'стость *f* villosity, hairyness, pilosity
волоси'стый *a* villose, villous, pilose, piliferous, hairy
волоски' *pl of* **волосо'к** hairs; *bot* fuzz
волоскови'дный *a* piliform
волосне'ц *m* wild rye *(Elymus)*
волосови'дный *a* capillaceous, capilliform, trichoid, piliform
волосо'к *m* hair
жгу'чий ~ stinging hair, urticating hair
желе'зистый ~ glandular hair
корнево'й ~ *bot* fibril, fibrilla, root hair, root fibril
крапи'вный ~ stinging hair, urticating hair
кро'ющий ~ clothing hair
осяза'тельный ~ tactile hair
слухово'й ~ auditory hair
чувстви'тельный ~ sense hair
волосообра'зный *a* hairlike, trichoid
во'лосы *m pl* hair
волчи'ца *f* she-wolf, bitch wolf
волчо'к *m bot* epicormic branch, sucker, surculus
во'льва *f bot* volva
волье'ра *f* aviary (*for birds*), open-air cage
волюти'н *m* volutin
воню'чка *f* striped skunk (*Mephitis*)
вооруже'ние *n* armament, armature (*of, e.g.,* spines on some structure, etc.)
воркова'ние *n* crooning
воркова'ть *v* coo, curr
воробе'й *m* sparrow (*Passer*)
воробьи'ный *a* passerine
во'рон *m* raven (*Corvus corax*)
воро'на *f* crow (*Corvus*)
се'рая ~ hooded crow (*Corvus cornix*)
чё'рная ~ carrion crow (*Corvus corone*)
воронё'ный *a* blued
воро'ний *a* corvine
воро'нка *f* funnel; *bot* trumpet
~ **Берле'зе** Berlese funnel
~ **Бе'рмана** Baermann funnel
~ **Бю'хнера** Buchner funnel
~ **Ту'лгрена** Tullgren funnel
дели'тельная ~ dropping funnel
ка'пельная ~ dropping funnel
ресни'чная ~ ciliated funnel
воронкови'дный, воронкообра'зный *a* infundibular, infundibuliform, funnelform
вороно'й *a* solid black, black (*of a horse*)
~ **с отли'вом** glossy black
во'роно-ча'лый *a* blue-roan
воро'та *pl* hilum, hilus
~ **инфе'кции, входны'е ~ инфе'кции** entry of infection, infection atrium
~ **о'ргана** hilum, hilus
воротничо'к *m* collar
воро'тный *a* portal
вороши'ть *v* ted (*hay*)
ворс *m* nap, pile
ворси'нка *f* villus
ворсинкови'дный *a* villiform
ворси'нчатость *f* villosity
ворси'нчатый *a* villous, villose, villiferous
ворси'стость *f* hairiness
ворси'стый *a* nappy, hairy, lanate
ворся'нка *f* teasel (*Dipsacus*)
ворча'ние *n* growl, snarl
ворча'ть *v* growl
воск *m* wax
воскови'дный *a* wax-like
воскови'ца *f* cere
восково'й *a* waxen, wax, waxy, cereous; ~ **налё'т** wax bloom
восколи'стный *a* waxy-leaved
воскono'сный *a* ceriferous, wax-producing
воскопло'дный *a* waxy-fruited
воскосемя'нный *a* waxy-seeded
воскоцве'тный *a* waxy-flowered
воспале'ние *n* inflammation
гно'йное ~ purulent inflammation
ро'жистое ~ erysepelatous inflammation
воспалё'нный *a* inflamed
воспали'тельный *a* inflammatory
воспита'ние *n* training, fostering
перекре'стное ~ cross-fostering
восполне'ние *n* recovery
восприи'мчивость *f* receptivity, susceptibility
восприи'мчивый *a* susceptible (*e.g., to diseases)*
воспринима'емый *a* perceptible
воспринима'ть *v* perceive
восприя'тие *n* perception, susception
~ **изображе'ний** picture perception
~ **простра'нства** space perception
~ **це'лостного о'браза** Gestalt perception
зри'тельное ~ visuognosis

зри'тельное ~ простра'нства optokinetic perception
чу'вственное ~ sense perception
воспроизведе'ние *n* reproduction, procreation
воспроизво'дство *n* reproduction
восстана'вливать *v* restore, renew; reclaim; reduce
восстанови'тель reducing agent
восстановле'ние *n* restoration, renewal; regeneration; reduction; recovery
~ **в жи'дкой среде'** liquid holding recovery
~ **на минима'льной среде'** minimal medium recovery
~ **отхо'дов** waste recovery
~ **приро'дных ресу'рсов** restoration of natural resourses
~ **ро'ста** growth recovery
восьми- octo-
восьмиря'дный *a* octostichous
вошь *f* louse *(Pediculus)*
головна'я ~ human head louse *(Pediculs humanus)*
ка'рповая ~ carp louse (*Argulus*)
кни'жная ~ book louse *(Troctes divinatorius)*
лобко'вая ~ crab louse (*Phthirius pubis*)
платяна'я ~ clothes louse, body louse (*Pediculus vestimenti*)
впаде'ние *n* inflow *(e.g., of a river)*
~ **в спя'чку** entrance into hibernation
впа'дина *f* socket, hollow, cavity
вертлу'жная ~ cotyloid cavity, socket of the hip, coxal cavity
глазна'я ~ eyehole, eyepit, eyesocket, orbital chamber, orbital cavity
суста'вная ~ лопатки glenoid cavity
та'зиковая ~ cotyloid chamber, coxal chamber, coxal cavity
впи'тывание *n* imbibition
впи'тывать *v* imbibe
впры'скивание *n* injection, shot
вя'чивание *n* invagination, ingrowth
впя'чиваться *v* invaginate
вражде'бный *a* adverse
враста'ние *n* intussusception, ingrowth
врач *m* physician, doctor
ветерина'рный ~ veterinarian
зубно'й ~ dentist
ле'чащий ~ attending physician
враща'тельный *a* gyral
враще'ние *n* rotation, gyration
опти'ческое ~ optical rotation
отрица'тельное ~ negative rotation
положи'тельное ~ positive rotation
пра'вое ~ dextrorotation
вред *m* damage, harm
вреди'тель *m* pest, crop-eater
вре'дность *f* hazard, harmfulness
~ **произво'дства** occupational hazard
вре'дный *a* detrimental, harmful, noxious, pernicious
вре'мя *n* time
~ **генера'ции** generation time
~ **жи'зни** lifetime
~ **кле'точной генера'ции** cell generation time
~ **конта'кта** contact time
~ **лё'та** flight time
~ **опо'ры** contact time *(in walking)*
~ **осемене'ния** breeding time
~ **перено'са ноги'** stride time
~ **полужи'зни** half lifetime
~ **полураспа'да** half lifetime
~ **полуудале'ния** half-disappearance time
~ **предъявле'ния** presentation time
~ **реа'кции** reaction time
~ **релакса'ции** relaxation time
~ **рефле'кса** reflex time
~ **свё'ртывания кро'ви** blood clotting time, coagulation time
~ **спа'ривания** breeding time, pairing time
ассимиляцио'нное ~ assimilation time
биологи'ческое ~ biological time
действи'тельное ~ жи'зни actual lifetime
есте'ственное ~ жи'зни natural lifetime
запре'тное ~ closed season
сре'днее ~ генера'ции mean generation time
сре'днее ~ жи'зни mean lifetime
врождё'нный *a* congenital, innate, inborn, inherent
вса'сывание *n* suction, sucking; absorption, imbibition, intake, inhalation
вса'сывать *v* absorb, imbibe, inhaust, inhale
«всё и'ли ничего'» all-or-none
вселе'нец *m* colonizer, invader
вселе'ние *n* immigration
всея'дный *a* omnivorous, pantophagous
вска'рмливать *v* suckle, bring up
вскрыва'ть *v* dissect
вскрыва'ться *v* explode *(e.g., of fruit),* burst, break
вскрыва'ющийся *a* explosive *(e.g., of fruit)*

вскры'тие *n* dissection, autopsy
вспа'ивать *v* nurse, rear, rear on milk
вспа'хивать *v* plow up, till
  ~ **под пар** fallow
вспа'шка *f* plowing, tillage
  безотва'льная ~ boardless plowing
  зя'блевая ~ autumn plowing
вспе'ниватель foaming agent
всплеск *m* dash
вспу'гивать, вспугну'ть *v* frighten away; put up (*birds*)
вспы'шка *f* outbreak; flash
вста'вка *mbiol* insert, insertion
вста'вочный *a* inserted, intercalary
встреча'емость *f* occurrence, frequency, frequency of occurrence, incidence
встря'хивание *n* agitation
всхо'д *m* sprout, sprouting, plantling, plantule
всходи'ть *v* sprout, germinate, emerge
всхо'жесть *f* germinative capacity, germinating ability, germinability, germinating power
всхо'жий *a* germinable
вторга'ться *v* invade, intrude
вторже'нец *m* intruder
вторже'ние *n* invasion, intrusion, encroachment, incursion
второуко'сный *a* aftermath
втя'гивание *n* retraction
вход *m* inlet, input
  аффере'нтный ~ aflerent inlet
  обоня'тельный ~ olfactory inlet
  сенсо'рный ~ sensory input, sensory inlet
  синапти'ческий ~ synaptic inlet
  та'зовый ~ superior pelvic strait, pelvic inlet
вхожде'ние n intussusception
вши'вость *f* pediculosis, lousiness, phthyriasis
вши'вый *a* lousy
выбира'ть *v* choose, select; haul (*nets*)
вы'бор *m* sampling
  двойно'й ~ double sampling
  квазислуча'йный ~ quasi-random sampling
  ко'свенный ~ indirect sampling
  многоступе'нчатый ~ multistage sampling
  многофа'зовый ~ multiphase sampling
  не вполне' случа'йный ~ judgement sampling
  непосре'дственный ~ direct sampling
вы'борка *f* sample, extract; sampling; lift (*nets*)
  ~ **по гру'ппам** quota sampling
  вероя'тностная ~ probability sample
  взаимнопроника'ющие вы'борки *pl* interpenetrating samples
  гла'вная ~ master sample
  гнездова'я ~ cluster sample, cluster sampling, nested sampling
  группова'я ~ cluster sample
  лотере'йная ~ lottery sampling
  неслуча'йная ~ nonrandom sample
  объединё'нная ~ combined sample
  однокра'тная ~ single sample, unitary sampling
  па'рные вы'борки *pl* paired samples
  повто'рная ~ duplicated sample
  посло'йная ~ stratified sampling
  произво'льная ~ random sample
  пропорциона'льная ~ quota sample
  пропорциона'льная посло'йная ~ proportional stratified sampling
  проста'я ~ simple sample
  расслоё'нная ~ stratified sample
  репрезентати'вная ~ representative sample, representative extract
  решё'тчатая ~ lattice sampling
  самовзве'шенная ~ self weighting sample
  сбаланси'рованная ~ balanced sample
  системати'ческая ~ systematic sampling
  случа'йная ~ random sample, random sampling
  упоря'доченная ~ ordered sample
  усечё'нная ~ truncated sample
  фикси'рованная ~ fixed sample
выбрако'ванный *a* culled
выбра'сывание *n* ejection
выбра'сывать *v* bolt (*e.g., a flower stalk*)
  ~ **стре'лку** *v bot* bolt
выбра'сывающийся *a* explosive
вы'брос *m* discharge, emission; release
  ~ **в атмосфе'ру** discharge into the air
  антропоге'нный ~ man-made emission
  преде'льно допусти'мый ~, ПДВ maximum permissible discharge
  промы'шленный ~ industrial discharge
  регули'руемый ~ regulated emission
  случа'йный ~ accidental discharge
выбуха'ние *n* extrusion
выва'живание *n* playing (*in fishing*)
выведе'ние *n* rearing; hatching, raising
вы'вих *m* dislocation, luxation, abarticulation

выводи'ть *v* breed, raise, rear; hatch; exterminate; lead out, remove; deduct, deduce, conclude
~ **но'вую поро'ду** breed a new strain, raise *a* new strain
~ **пото'мство** brood
выводи'ться *v* hatch out
вы'водка *f* exercising (*of horses*); *genetive of* **вы'водок**
выводко'вый *a* nidifugous, precocial
выводно'й *a* exhalant (*e.g., of a siphon of mollusks*), excurrent
вы'водок *m* brood, litter, covey, hatch
выводя'щий *a* excurrent, deferent
выгиба'ть *v* arch, bend
~ **спи'ну** arch its back, hump (*of a cat*)
вы'гон *m* pasture, grazing land, pasture land, walk
вы'гонка *f agr* forcing
выгоня'ть *v* force; drive out
вы'гул *m* pasture; enclosure
выгу'ливать *v:*
выгу'ливать соба'ку to walk the dog
выда'вливание *n* expression, expulsion
выда'вливать *v* press out; strip (*e.g., milt of fish*)
выдаю'щийся *a* exserted
выдвижно'й *a* protractile, protrusible
выделе'ние *n* excretion; discharge; isolation, recovery; liberation; abjection (*of spores*)
~ **же'лчи** biliation, bilification
~ **мокро'ты** expectoration
выделе'ния *n pl* discharge, waste; excrements
вы'деленный *a* isolated
выдели'тельный *a* excretory
выделя'ть *v* excrete, secrete; select, choose; isolate; release
~ **смолу'** gum
выде'рживать *v* withstand; expose
вы'держка *f* ripening, seasoning, aging, curing, maturing; exposure
вы'дох *m* exhalation, expiration
вы'дра *f* otter (*Lutra*)
выдыха'емый *a* exhalant
выдыха'ние *n* exhalation
выдыха'тельный *a* expiratory
выдыха'ть *v* expire
выеда'ние *n* consumption, grazing
избира'тельное ~ selective grazing
выезжа'ть *v* break in, train (*a horse*); depart, go out
вы'езженный *a* trained, made (*a horse*)
вы'емка *f* excision, valley, vallecula
аксилля'рная ~ *bot* axillary excision
преаксилля'рная ~ *bot* preaxillary excision
вы'емчатый *a* emarginate, sinuate, sinuous
вы'жженный *a* burnt-over
выжива'емость *f* survival, survival rate, survirorship rate, survival ability
выжива'ние *n* survival
выжива'ть *v* survive
выжига'ние *n* burning out, burning, firing
вы'жимки *f pl* pomace
выздора'вливание *n* recovery, convalescence
выздора'вливать *v* recover
выздоровле'ние *n* recovery, convalescence
вызрева'ние *n* aging, ripening
вызрева'ть *v* ripen
вызыва'ть *v* excite
выка'рмливать *v* nurture
вы'кидыш *m* abortion, miscarriage; slink
вы'кинуть *v* have an abortion, have a miscarriage, slink; throw out, discard
вы'клев m hatching
выкола'шивание *n* earing, ear formation
выкорче'вка *f*, **выкорче'вывание** *n* uprooting, stubbing, grubbing
выкорче'вывать *v* grub, grub out, grub up
вылеза'ть *v* crawl out; fall out (*of hair*)
вы'лет *m* emergence
вы'лов *m* catch
вылупле'ние *n* emergence, hatching; *ent* exclosion
вылупля'ться *v* hatch, emerge, burst
вылу'щивание *n med* enucleation
вымерза'ние *n* frostkilling, winterkilling, frost injury
вы'мерший *a* extinct
вымира'ние *n* extinction
вымира'ть *v* die out
вымыва'ние *n* flushing
вы'мя *n* udder, dug
выноси'ть *v* tolerate; carry away, carry out
вынос'ливость *f* tolerance, endurance, hardiness
~ **к ускоре'ниям** G tolerance
выно'сливый *a* tolerant, hardy
выносящ'ий *a* deferent, efferent
выпаде'ние *n* prolapse; shedding
выпа'лывать *v* weed
выпа'ривание evaporation, vaporization
~ **в ва'кууме** vacuum evaporation

выпа'риватель evaporator
вы'пас *m* pasture, grazing land; grazing
неограни'ченный ~ free grazing
чрезме'рный ~ overgrazing
выпле'вывать *v* spit out
вы'ползок *m* slough, shed snakeskin
вы'полненный *a* plump (*e.g., of a cavity, of a grain*)
вы'пот *m* exudate, effusion
выпотева'ние *n* extravasation
выпотева'ть *v* exude
вы'потрошить *v* eviscerate
выпра'шивание *n eth* begging, begging behavior
~ **пи'щи** food soliciting, food begging
ритуа'льное ~ ritualized begging
выпотева'ние *n* transudation
выпрева'ние *n* damping-off, damping-out
выпрямля'ться *v* erect oneself, stand erect
вы'пуклость *f* protuberance, promimence, convexity; *zool* venter
вы'пуклый *a* convex, gibbous
выпуска'ть *v* cast off (*falcons*); let out
вы'пученный *a* goggle (*of* eyes)
выпь *f* bittern (*Botaurus*)
выпя'чивание *n* evagination, protrusion
ротово'е ~ oral evagination
вы'работка *f* elaboration, development, formation
~ **на'выка** acquisition of skill, habit formation
~ **на'выка нажа'тия на рыча'г** bar-press training
предвари'тельная ~ preliminary training
выраже'ние *n* expression
лицево'е ~ facial expression
нейтра'льное ~ лица' *eth* neutral face, relaxed face
рассла'бленное ~ лица' *eth* neutral face, relaxed face
вы'раженный *a*: **я'рко** ~ strongly pronounced, well marked
выра'щивание *n* raising, growing
кле'точное ~ cage raising
выра'щивать *v* raise, cultivate
выреза'ние *n* excision
вы'резка *f* incisure, notch
~ **вертлу'жной впа'дины** acetabular notch
седа'лищная ~ sciatic notch
яре'мная ~ jugular notch
вырожда'ться *v* degenerate
вырожда'ющийся *a* degenerative
вырожде'ние *n* degeneration, degeneracy, degradation
вы'рожденность *f* degeneracy
~ **генети'ческого к'ода** degeneracy of genetic code
вы'рост *m* process, projection, outgrowth
~ **с опо'рой** strutted process *(in diatoms)*
двугу'бый ~ labiate process *(in diatoms)*
за'мкнутый ~ occluded process *(in diatoms)*
ло'бный ~ frontal process
печё'ночный ~ *zool* hepatic cecum
по'днятый ~ strutted process *(in diatoms)*
хорда'льный ~ notochordal process
щелеви'дный ~ labiate process *(in diatoms)*
вы'рубка *f* cutting, clearing; cut-over land, logged land
вы'садка *f agr* setting, setting-out
выса'живать *v agr* bed out
выса'ливание *n* salting-out
взаи'мное ~ mutual salting-out
дро'бное ~ divided salting-out
выса'сывание *n* exhaustion
выса'сывать *v* exhaust
высвобожда'ть *v* release
высвобожде'ние *n* release
климати'ческое ~ climatic release
вы'сев *m* seeding
высева'ть *v* sow, seed
вы'севки *pl* chaff
выси'живание *n* hatch
выси'живать *v* sit; hatch
~ **я'йца** sit on eggs, cover eggs
выска'бливание *n* excision
высле'живать *v* track
высо'кий *a* tall, high; deep-bodied (*of fish*)
высокоави'дный *a* high-avid
высокотра'вье *n* tall herbage, tall grasses, tall herbaceous vegetation
высота' *f* height, altitude; depth *(of fish)*
~ **в хо'лке** withers height
~ **проле'та** flight altitude
выставля'ть *v* expose
вы'стилка *f* lining
сли'зистая ~ mucous lining
эпителиа'льная ~ epithelial lining
выстре'ливание *n bot* discharge
высту'кивание *n* percussion
вы'ступ *m* projection, lappet
выступа'ющий *a* protrudent, exserted

высу'шивание, высыха'ние *n* exsiccation, insiccation
выта'лкивание *n* expulsion, extrusion
выта'лкивающийся *a* explosive *(e.g., of seeds)*
выта'пливание *n* rendering
выта'птывание *n* trampling
выта'птывать *v* trample
вытесне'ние *n* extrusion, exclusion; displacement
конкуре'нтное ~ competitive replacement
вы'топка *f* rendering
выть *v* howl
вытя'гивание *n* exhaustion
вы'тяжка *f* extract, infusion
по'чвенная ~ soil extract
вы'тянувшийся *a* leggy (*of a plant*)
вы'тянутый *a* oblong, elongated, stretched out
вы'ход *m* outlet, exit, egress, vent; emergence; yield; *geol* outcrop
~ **в стре'лку** *bot* bolting
~ **в тру'бку** stalk-shooting
~ **ди'чи** game yield
~ **фа'га** phage burst
~ **флуоресце'нции** fluorescence yield, yield of fluorescence
ква'нтовый ~ фотоси'нтеза quantum yield of photosynthesis
та'зовый ~ inferior pelvic strait
выходи'ть *v* hatch, emerge (*e.g., from a pupa*); go out
~ **в ко'лос** come into ear
~ **в лист** foliate
~ **в тру'бку** spire, boot
вы'хухоль *m* water mole, desman (*Desmana*)
выщела'чивание *n* leaching
бактериа'льное ~ bacterial leaching
выщела'чивать *v* leach
выщемле'ние *n* excision
вьюно'к *m* bindweed (*Convolvulus*)
вьюро'к *m* finch (*Fringilla*); brambling (*Fringilla montifringilla*)
да'рвинов ~ Darwin's finch *(Geospizinae)*
земляно'й ~ Darwin's finch *(Geospizinae)*
канаре'ечный ~ canary, canary bird (*Serinus canaria*)
вью'щийся *a* climbing, twining, twisting, winding, tendrilar
вя'жущий *a* astringent
вяз *m* elm (*Ulmus*)
вяза'ть *v* bind, tie
~ **в снопы'** sheaf
вя'зель *m* crown vetch (*Coronilla*)
вязи'га *f* dried spinal cord of sturgeon
вя'зка *f* service; binding
~ **снопо'в** tying of sheaves
вя'зкий *a* viscid, viscous; uliginous, miry, marsgy, swampy
вя'зкость *f* viscosity
вну'тренняя ~ intrinsic viscosity
структу'рная ~ structural viscosity
вя'ление *n* sun-curing
вя'лый *a* flaccid, flabby, insipid; sluggish (*of a horse*); soft *(e.g., of a muscle)*
вя'нуть *v* fade, wither, shrivel, wilt
вя'хирь *m* ring dove, wood pigeon (*Columba palumbus*)

Г

га'битус *m* habit, habitus
прямостоя'чий ~ *bot* erect habit
распросте'ртый ~ *bot* prostrate habit
гав, гав-га'в *i* woof-woof, bow-wow
гавиа'л *m* gavial *(Gavialis)*
га'га *f* eider, eider duck (*Somateria*)
обыкнове'нная ~ common eider *(Somateria mollissima)*
очко'вая ~ spectacled eider *(Somateria fischeri)*
сиби'рская ~ Steller's eider *(Polysticta stelleri)*
сте'ллерова ~ Steller's eider *(Polysticta stelleri)*
га'га-гребену'шка *f* king eider *(Somateria spectabilis)*
гага'ра *f* loon, diver (*Gavia*)
гага'рка *f* scoot, noddy, razorbill, razor-billed auk (*Alca torda*)
бескры'лая ~ great auk, garefowl (*Pinguinus impennis*)
ма'лая ~ dovekie, little auk (*Plautus alle*)
гадро'м *m*, **гадро'ма** *f* hadrome
гадроцентри'ческий *a* hadrocentric
гад *m* reptile *or* amphibian
га'ды *m pl* amphibians and reptiles
гадю'ка *f* viper, adder (*Vipera*)
газе'ль *f* gazelle (*Gazella*)
жира'фовая ~ giraffe gazelle *(Litocranius walleri)*
кювье'рова ~ edmi gazelle *(Gazella cuvieri)*
обыкнове'нная ~ mountain gazelle *(Gazella gazella)*
ры'жая ~ red gazelle *(Gazella rufina)*
саха'рская ~ Addra gazelle, dama gazelle *(Gazella dama)*
газе'ль-да'ма *f* dama gazelle *(Gazella dama)*
газе'ль-до'ркас *f* dorcas gazelle *(Gazella dorcas)*
газоанализа'тор *m* gas analyzer
регистри'рующий ~ recording gas analyzer
термохими'ческий ~ thermochemical gas analyzer
термомагни'тный ~ thermomagnetic gas analyzer
газо'н *m* lawn
газообме'н *m* respiratory metabolism
га'ичка *f* chickadee *(Parus)*
средиземномо'рская ~ sombre tit (*Parus lugubris*)
гаймори'т *m* maxillary sinusitis
га'йно *n* dray, drey
галакто- galacto-
галакто'за *f* galactose
га'лка *f* daw (*Corvus monedula*)
галл *m* gall, cecidium
блюдцеобра'зный ~ cup gall
корнево'й ~ root gall
коро'нчатый ~ crown gall *(caused by Dendrophagus globosus)*
многока'мерный ~ polythalamous gall
наплывно'й ~ excrescence gall
однока'мерный ~ monothalamous gall
чё'рный ~ то'поля aspen black gall *(caused by Phoma etheridgei)*
га'ллица *f* gall midge, gall gnat (*Cecidomyiidae*)
галлюцина'ция *f* hallucination
галлюциноге'н *m* hallucinogen
галлюциноге'нный *a* hallucinogenic
галобакте'рия *f* halobacterium, halophilic bacterium
галобио'нт *m* halobiont
галоби'ос *m* halobios
галогено'лиз *m* halogenolysis
галоиди'рование *n* halogenation
галолимни'ческий *a* halolimnic
галомо'рфный *a* halomorphic
гало'п *m* gallop
коро'ткий ~ short gallop
латера'льный ~ rotatory gallop
ле'гкий ~ canter, easy gallop, hand gallop
манежный ~ short gallop
полево'й ~, приба'вленный ~ extended gallop
ре'звый ~ racing gallop
галофи'л *m* halophile
галофи'льный *a* halophilic, halophilous
галофи'т *m* halophyte
галофо'б *m* halophobe
галофо'бный *a* halophobous
га'лстучник *m* ring plover, ringed plover (*Charadrius hiaticula*)
гальванота'ксис *m* galvanotaxis
гальвануто'нус *m* galvanotonus
гальванотропи'зм *m* galvanotropism
гамадри'л *m* hamadryas baboon, sacred baboon *(Papio hamadryas)*

гаме'та *f* gamete
несбаланси'рованная ~ unbalanced gamete
сбаланси'рованная ~ balanced gamete
гамета'нгий *m* gametangium
гаметангиогами'я *f* gametangiogamy
гамети'ческий gametic
гаметога'мия gametogamy
гаметогене'з *m* gametogenesis
гаметого'ний *m* gametogonium
гаметотоки'я gametotoky
гаметофи'лл *m* gametophyll
гаметофи'т *m* gametophyte
гаметофо'р *m* gametophore
гаметоци'ста gametocyst
гаметоци'т *m* gametocyte
гамма-разнообра'зие *n* gamma-diversity
гамо'биум *m* gamobium
гамогене'з *m* gamogenesis
гамогенети'ческий *a* gamogenetic
гамого'ния *f* gamogony
гамоде'м *m* gamodeme
гамо'н *m* gamone
гамо'нт *m* gamont
гамонтога'мия *f* gamontogamy
гамонтоци'ста *f* gamocyst
гамосте'ла *f bot* gamostele
гамостели'ческий *a bot* gamostelic
гамосте'лия *f bot* gamostely
гамотропи'зм *m* gamotropism
гамофа'за *f* gamophase
гамофилли'я *f* gamophylly
га'нглиевый *a* ganglionic, ganglionary
га'нглий *m* ganglion
~ **преддве'рия** vestibular ganglion
~ **тройни'чного не'рва** trigeminal ganglion
автоно'мный ~ autonomic ganglion
база'льный ~ basal ganglion
бараба'нный ~ tympanic ganglion
брыже'ечный ~ mesenteric ganglion
вентрикуля'рный ~ ventricular ganglion
вестибуля'рный ~ vestibular ganglion
висцера'льный visceral ganglion
гассе'ров ~ Gasserian ganglion, semilunar ganglion, solar ganglion
головно'й ~ brain ganglion
грудно'й ~ thoracic ganglion
дополни'тельный ~ accessory ganglion
кароти'дный ~ Bock's ganglion
клинонё'бный ~ Meckel's ganglion
коле'нчатый ~ geniculate ganglion
коллатера'льный ~ collateral ganglion
надгло'точный ~ supraesophageal ganglion
нодо'зный ~ nodose ganglion
осно'вно-нё'бный ~ sphenopalatine ganglion
паравертебра'льный ~ paravertebral ganglion
подчелюстно'й ~ submaxillary ganglion
полулу'нный ~ Gasserian ganglion, semilunar ganglion, solar ganglion
ресни'чный ~ ciliary ganglion
сенсо'рный ~ sensory ganglion
симпати'ческий ~ sympathetic ganglion
слухово'й ~ acoustic ganglion, otic ganglion
спина'льный ~ spinal ganglion
спинномозгово'й ~ spinal ganglion
спира'льный ~ spiral ganglion
спира'льный ~ **ули'тки** auditory ganglion, cochlear ganglion
термина'льный ~ terminal ganglion
узлова'тый ~ nodose ganglion
фронта'льный ~ frontal ganglion
чре'вный ~ coeliac ganglion
цилиа'рный ~ ciliary ganglion
черепно'й ~ cranial ganglion
чре'вный ~ coeliac ganglion
ше'йный ~ cervical ganglion
ганглиобла'ст *m* ganglioblast
ганглио'зный, ганглиона'рный *a* ganglionic, ganglionary, gangliar
ганглиоци'т *m* gangliocyte
гангре'на *f* gangrene
~ **Фурнье'** Fournier's disease
ганобла'ст *m* ganoblast
гано'идный *a* ganoid
ганои'н *m* ganoine
гапло- haplo-
гаплобио'нт *m* haplobiont
гаплодипло'ид *m* haplodiploid
гаплодиплоиди'я *f* haplodiploidy
гаплодипло'идный *a* haplodiploid
гаплодипло'нт *m* haplodiplont
гаплодо'нтный *a* haplodont
гаплозиго'тный *a* haplozygous
гапло'зис *m* haplosis
гапло'ид *m* haploid
гаплоидиза'ция *f* haploidization
гаплоиди'я *f* haploidy
гапло'идный *a* haploid
гаплокони'дия *f* haploconidium

гаплоко'рм *m bot* haplocorm
гапломи'кт *m gen* haplomict
гапло'нт *m* haplont
гаплополипло'ид *m* haplopolyploid
гаплосо'мик *m* haplosomic
гаплосоми'ческий *a* haplosomic
гаплоспо'ра *f* haplospore
гаплостемо'нный *a* haplostemonous
гаплоти'п *m gen* haplotype
~ **«по'лный дом»** extended haplotype
наполови'ну запо'лненный ~ half-null haplotype
гаплофа'за *f* haplophase
гаплофи'т *m* haplophyte
гаплохлами'дный *a* haplochlamydeous
гапте'н *m* hapten, haptene
~ **Фо'рсмана** Forssman's hapten
перекрё'стно реаги'рующие гапте'ны *pl* cross-reacting haptens
гапте'н-аутомодифика'тор *m* autocoupling hapten
гаптенили'рованный *a* haptenated
гаптенили'рование *n* haptenating
гаптоге'н *m* haptogene
гаптоглоби'н *m* haptoglobin
гаптотропи'зм *m* haptotropism
гаптофо'р *m* haptophore
гаре'м *m* harem
гармо'ника *f* harmonic *(in bioacoustics)*
пе'рвая ~ first harmonic
основна'я ~ fundamental harmonic
гармонический *a* harmonic
гарцева'ть *v* prance
га'ршнеп *m* small snipe, half snipe, jack snipe (*Lymnocryptes minima*)
гарь *f* burned-out forest, fire site
гастра'льный *a* gastral, gastric
гастре'я *f* gastraea
гастри'т *m* gastritis
гастроваскуля'рный *a* gastrovascular
гастрокардиа'льный *a* gastrocardiac
гастроге'нный *a* gastrogenic
гастроскопи'ческий *a* gastroscopic
гастроскопи'я *f* gastroscopy
гастроце'ль *f* gastrocoel
га'струла *f* gastrula
гаструля'ция *f* gastrulation
гау'р *ther m* gaur *(Bos gaurus)*
гаусто'рия *f* haustorium
гая'л *ther m* gayal *(Bibos frontalis)*
гваю'ла *f*: ~ **серебри'стая** *bot* guayule *(Parthenium argentatum)*
гвере'ца *ther* guereza *(Colobus)*
гвозди'ка *f* pink (*Dianthus*); *pl* cloves
гедони'ческий *a* hedonic
гейтоногa'мия *f* geitonogamy
гекистоте'рм *m* hekistotherm
гекко'н *m* gecko (*Gekkonidae*)
гексапло'ид *m* hexaploid
гексаплоиди'я *f* hexaploidy
гексапло'идный *a* hexaploid
гексахлора'н *m* hexachlorane
гектоко'тиль *m* hectocotylus
геликоспо'ра *f* helicospore
гелио- helio-
гелио'з *m* heliosis
гелиота'ксис *m* heliotaxis
гелиотропи'зм *m* heliotropism
гелиофи'л *m* heliophile
гелиофи'льный *a* heliophilous
гелиофи'т *m* heliophyte
гелиофо'бный *a* heliophobic, heliophobous
гелофи'т *m* helophyte
гель *m* gel
~ **окиси кремния** silica gel
ага'ровый ~ agar gel, agar-agar gel
агаро'зный ~ agarose gel
адсорбцио'нный ~ adsorbent gel
гранули'рованный ~ granulated gel
концентрирующий ~ stacking gel *(in electrophoresis)*
крахма'льный ~ starch gel
перекрестносши'тый ~ cross-linked gel
полиакрилами'дный ~ polyacrylamide gel
разделя'ющий ~ sieving gel
сепарирующий ~ sieving gel *(in electrophoresis)*
фильтру'ющий ~ sieving gel
гель-электрофоре'з *m* gel electrophoresis
~ **с инве'рсией полюсо'в** field-inversion gel electrophoresis
двуме'рный ~ two-dimensional gel electrophoresis
гельми'нт *m* helminth
гельминто'з *m* helminthiasis, helminthism
гельминто'лог *m* helminthologist
гельминтоло'гия *f* helminthology
гельминтофа'уна *f* helminth fauna
гель-электрофоре'з *m* gel electrophoresis
гем *m* heme, haem
гема- hema-, haema-
гемагглютина'ция *f* hemagglutination
гема'льный *a* hemal

гематаллоскопи'я *f* hematalloscopy
гемати'н *m* hematin
гематобла'ст *m* hematoblast
гематоге'нный *a* hematogenic, hematogenous
гематоиди'н *m* hematoidin
гематокри'т *m* hematocrit
гематоксили'н *m* hematoxylin
~ **Делафи'льда** Delafield's hematoxilin
~ **Рего'** Regaud's hematoxylin
~ **Э'рлиха** Ehrlich's hematoxylin
желе'зный ~ по Гейденга'йну Heidenhain's iron hematoxylin
гематологи'ческий *a* hematologic, hematological
гематоло'гия *f* hematology, hemology
гемато'ма *f* hematoma
гематохро'м *m* hematochrome, haematochrome, hemachrome
гематоэнцефали'ческий *a* hematoencephalic
гематури'я *f* hematuria
геми- hemi-
гемигапло'ид *m* hemihaploid
гемиглоби'н *m* hemiglobin
гемизиго'та *f* hemizygote
гемикарио'н *m* hemikaryon
гемикриптофи'т hemicryptophyte
гемиметаболи'ческий *a* hemimetabolic
гемиметаболи'я *f* hemimetaboly
геми'н *m* hemin
геминиви'рус *m* geminivirus
гемипло'ид *m* hemiploid
гемитропи'зм *m* hemitropism
гемисферэктоми'я *f* hemispherectomy
гемиптериго'ид *m* hemipterygoid
гемитропи'зм *m* hemitropism
гемитро'пный *a* hemitropous
гемицеллюло'за *f* hemicellulose
гемицикли'ческий *a* hemicyclic
гемиэндофи'т *m* hemiendophyte
ге'ммула *f* gemmule
гемо- hemo-, haemo-
гемобла'ст *m* hemoblast
гемогистиобла'ст *m* hemohistioblast
гемоглоби'н *m* hemoglobin, hematoglobulin
восстано'вленный ~ reduced hemoglobin
гемоглобинури'я *f* hemoglobinuria
гемогра'мма *f* hemogram, blood count
гемодина'мика *f* hemodynamics
гемозои'н *m* hemozoin
гемоко'ния *f* hemoconia
гемокульту'ра *f* hemoculure, blood culture
гемо'лиз *m* hemolysis, hematolysis, hematocytolysis, hemoclasia, hemocatheresis
комплементзави'симый ~ complement-mediated hemolysis
иммy'нный ~ hemolysis
гемолиза'т *m* hemolysate
гемолизи'н *m* hemolysin, haemolysin
гемолизи'рование *n* hemolyzation
гемолизи'ровать *v* hemolyze
гемолизи'роваться *v* hemolyze
гемоли'мфа *f* hemolymph
гемолити'ческий *a* hemolytic, hematolytic, hemoclastic
гемо'метр *m* hemometer
~ **Сали'** hemoglobinometer Sahli
гемопоэ'з *m* hemopoiesis, hematopoiesis, hematogenesis, hemogenesis, blood formation, hemapoiesis, hematosis
экстрамедулля'рный ~ extramedullary hemopoiesis
фета'льный ~ fetal hemopoiesis
гемопоэти'н *m* hemopoietin
гемопоэти'ческий *a* hemopoietic, hematopoietic, hemafacient, hematogenic, hematogenous
геморраги'ческий *a* hemorrhagic
гемоста'з *m* hemostasis
гемостати'ческий *a* hemostatic
гемотерапи'я *f* hemotherapy
гемотокси'ческий *a* hemotoxic
гемофили'я *f* hemophilia
гемохро'м *m* hemochrome
гемохромоге'н *m* hemochromogen
гемоце'ль *f* hemocoel
гемоциани'н *m* hemocyanin
гемоци'т *m* blood cell, hemic cell, hemacyte, hemocyte
грануля'рный ~ granular hemocyte
малоподви'жный ~ sessile hemocyte
сесси'льный ~ sessile hemocyte
гемоцитобла'ст *m* hemocytoblast, hemagonium
гемоцито'метр *m* hemocytometer, hemacytometer, hemacytometer chamber, hematimeter
~ **Бю'ркера** Burker's cell
~ **Тома'-Це'йсса** Thoma-Zeiss hemocytometer
ген *m* gene
~**«дома'шнего хозя'йства»** housekeeping gene

~ **иммунного отве'та** immune response gene, Ir gene
~ **иммунной супре'ссии** Is gene
~, **контроли'руемый по'лом** sex-controlled gene
~ **несовмести'мости** incompatibility gene
ге'ны, образу'ющие бло'ки *pl* block genes
~, **ограни'ченный по'лом** sex-limited gene
~ **подавле'ния иммуните'та** Is gene
~ **с несто'йким проявле'нием** inconstant gene
~ **стери'льности** sterility gene
~ **супре'ссии иммунного отве'та** immunosuppression gene
аддити'вный ~ additive gene
бу'ферный ~ buffer gene, buffering gene
внехромосо'мный ~ extrachromosomal gene, extranuclear gene
внея'дерный ~ extrachromosomal gene, extranuclear gene
гипа'рхный ~ hyparchic gene
гипомо'рфный ~ hypomorphic gene
главный ~ key gene, major gene, master gene
голандри'ческий ~ holandric gene
гомологи'чный ~ homologous gene
диагини'ческий ~ diagynic gene
диандри'ческий ~ diandric gene
домина'нтный ~ dominant gene
до'норский ~ donor gene
дополни'тельный ~ supplementary gene
дупликa'тный ~ duplicate gene
индуци'бельный ~ inducible gene
ине'ртный ~ inert gene
иску'сственный ~ artificial gene
комплемента'рный ~ complementary gene
ко'мплексный ~ complex gene
криптоме'рный ~ cryptomeric gene
лаби'льный ~ labile gene
лета'льный ~ lethal gene
ма'лый ~ minor gene
матери'нский ~ maternal gene
межвидово'й ~ interspecific gene
мими'ческий ~ mimic gene
мозаи'чный ~ mosaic gene
молча'щий ~ silent gene
мутаби'льный ~ mutable gene, inconstant gene
незави'симый ~ independent gene
непостоя'нный ~ inconstant gene
нефункциони'рующий ~ silent gene
одноко'пийный ~ single-copy gene
основно'й ~ key gene, major gene, master gene
отцо'вский ~ paternal gene
перекрыва'ющийся ~ overlapping gene
плейотро'пный ~ pleiotropic gene
ползу'щий ~ scrambler gene
полиме'рный ~ polymeric gene
полулета'льный ~ semi-lethal gene
пре'дковый ~ ancestral gene, precursor gene
предра'нний ~ pre-early gene
преры'вистый ~, **расщепленный** ~ split gene
промежу'точный ~ intermediate gene
«пры'гающий» ~ jumping gene
регулято'рный ~ regulatory gene
рецесси'вный ~ recessive gene
синтези'рованный ~ artificial gene
сло'жный ~ compound gene
структу'рный ~ structural gene
субституцио'нный ~ substitution gene
сце'пленный ~ linked gene
сце'пленный с по'лом ~ sex-linked gene
триплика'тный triplicate gene
химе'рный ~ chimeric gene
хромосо'мный ~ chromosomal gene
эквилока'льный ~ equilocal gene
элемента'рный ~ elementary gene
эпистати'ческий ~ epistatic gene

генеалоги'ческий *a* genealogic, genealogical
генеало'гия *f* genealogy
ге'незис *m* genesis, origin
генерализа'ция *f* generalization
генерати'вный *a* generative, genesic
генера'ция *f* generation
антитети'ческие генера'ции *pl* antithetic generations
генери'ровать *v* generate
генери'ровать *v* generate
гене'тика *f* genetics
~ **растений** phytogenetics
~ **живо'тных** animal genetics
~ **поведе'ния** behavioral genetics
~ **челове'ка** human genetics
биохими'ческая ~ biochemical genetics
медици'нская ~ medical genetics, clinical genetics
молекуля'рная ~ molecular genetics
популяцио'нная ~ population genetics
прикладна'я ~ applied genetics

радиацио'нная ~ radiation genetics
генети'ческий *a* genetic, genetical
ген-ингиби'тор *m* inhibiting gene
ген-интенсифика'тор *m* intensifying gene, extension gene
генита'лии *f pl* genitals
генита'льный *a* genital
ген-модифика'тор *m* modifier gene
ген-мута'тор *m* mutator gene
«ген на ген» gene-for-gene;
взаимоде'йствие «ген на ген» gene-for-gene relationship
ге'нный *a* genic
геновариа'ция *f* genovariation
генови'д *m* genospecies
геногеогра'фия *f* gene geography
ген-ограничи'тель *m* restriction gene
геноде'м *m* genodeme
генокли'н *m* genocline
геноко'пия *f* genocopy
гено'м *m* genom, genome
гено'мный *a* genome, genomic
геноме'р *m* genomere
геноне'ма *f* genonema
генони'м *m* genonymum
ген-ограничи'тель *m* restriction gene
ген-опера'тор *m* operator gene
ген-переключа'тель *m* switch gene
геносо'ма *f* genosoma
геноти'п *m* genotype, genetic complex
оста'точный ~ residual genotype
генотипи'рование *n* genetic typing
генотипи'ческий *a* genotypic, genotypical
генотокси'ческий *a* genotoxic
генофе'на *m* genophene
генофо'нд *m* gene pool
ген-переключа'тель *m* switch gene
ген-подави'тель *m* suppressor gene
ген-регуля'тор *m* regulatory gene
ген-секре'тор *m* secretor gene
ген-«сирота'» *m* "orphane" gene
ген-супре'ссор *m* suppressor gene
ген-усили'тель *m* intensifying gene, extension gene
генцианвиоле'т *m* gentian violet
геобио'нт *m* geobiont
геобота'ника *f* phytosociology, plant sociology, geobotany
геоботани'ческий *a* geobotanical
геогельми'нт *m* geohelminth
географи'ческий *a* geographical, geographic
геогра'фия *f* geography
~ **почв** soil geography
~ **расте'ний** phytogeography, plant geography
~ **расти'тельности** geography of vegetation
ботани'ческая ~ botanical geography, plant geography
геодиотропи'зм *m* geodiotropism
геокарпи'ческий *a* geocarpic
геокриптофи'т *m* geocryptophyte
геоксе'н *m* geoxene
геона'стия *f* geonasty
георги'н, георги'на *f* dahlia, georgina (*Dahlia*)
геота'ксис *m* geotaxis
геотропи'зм *m* geotropism
геотропи'ческий *a* geotropic
геофили'я *f* geophily
геофи'т *m* geophyte
геоцентри'ческий *a* geocentric
геоэкоти'п *m* geoecotype
гепаднави'рус *m* hepadnavirus
гепа'рд *m* cheetah, chita (*Acinonyx jubatus*)
гепари'н *m* heparin
гептаме'рный *a* heptamerous
гепати'т *m* hepatitis
гепатоге'нный *a* hepatogenic
гепатози'н *m* hepatosin
гепатоло'гия *f* hepatology
гепатопа'нкреас *m* hepatopancreas, liver-pancreas
гепаторена'льный *a* hepatonephric
гепатоци'т *m* hepatocyte, hepatic cell
гептаме'рный *a* heptamerous
гера'нь *f* crane's-bill (*Geranium*)
герба'рий *m* herbarium
гербици'д *m* herbicide, weed killer, plant killer
гормона'льный ~ hormone herbicide
гербици'дный *a* herbicidal, weed-killing
герену'к *m* giraffe gazelle (*Litocranius walleri*)
герма'рий *m* germarium
гермафроди'т *m* hermaphrodite, bisexual
гермафродити'зм *m* hermaphroditism, androgynism, androgyny
и'стинный ~ true hermaphroditism
гермафроди'тный *a* hermaphroditic, bisexual
гермина'лий *m* germinal
герминати'вный *a* germinative, germinal
гермици'д *m* germicide

геронтоло'гия *f* gerontology
геронтоморфо'з *m* gerontomorphosis
ге'рпес *m* herpes
~ **Ма'нсона** Manson's herpes
герпето'лог *m* herpetologist
герпетоло'гия *f* herpetology
геспери'дий *m* hesperidium
гетеро- hetero-
гетероагглютини'н *m* heteroagglutinin
гетероантиге'н *m* heteroantigen, heterocell antigen
гетероантиге'нный *a* heteroantigenic
гетероантисы'воротка *f* heteroantiserum
антиидиотипи'ческая ~ antiidiotypic heteroantiserum
гетероантите'ло *n* heteroantibody
гетеробази'дия *f* heterobasidium
гетеробласти'ческий *a emb* heteroblastic
гетеробрахиа'льный *a cyt* heterobrachial
гетеровакци'на *f* heterologous vaccine, heterotypic vaccine
гетерогаме'та *f* heterogamete
гетерогаме'тность *f* heterogamety
гетерогаме'тный *a* heterogametic
гетерогами'я *f* heterogamy, anisogamy
гетерога'мный *a* heterogamic, heterogamous
гетерогене'з *m* heterogenesis
гетерогениза'ция *f* heterogenization
гетероге'нность *f* heterogeneity
~ **фитоцено'за** heterogeneity of phytocenosis
идиотипи'ческая ~ idiotypic heterogeneity
иммунохими'ческая ~ immunochemical heterogeneity
гетероге'нный *a* heterogenous
гетерогено'мный *a* heterogenomatic
гетерогено'та *f* heterogenote
гетерогеофи'т *m* heterogeophyte
гетерогибридо'ма *f* heterohybridoma
гетерогони'я *f* heterogony
гетеродиме'р *m* heterodimer
гетеродиме'рный *a* heterodimeric
гетеродина'мный *a* heterodynamous
гетеродо'нтный *a* heterodont
гетеродро'мный *a* heterodromous
гетероду'плекс *m* heteroduplex
гетерозиго'та *f* heterozygote
~ **по инве'рсии** inversion heterozygote
~ **с концево'й избы'точностью** terminal redundancy heterozygote
гетероду'плексная ~ heteroduplex heterozygote
ко'мплексная ~ complex heterozygote
мутацио'нная ~ mutational heterozygote
структу'рная ~ structural heterozygote
гетерозиго'тность *f* heterozygosity, heterozygosis
гетерозиго'тный *a* heterozygous
гетеро'зис *m* heterosis, crossbred vigor
мутацио'нный ~ mutational heterosis
гетеро'зисный *a* heterotic
гетероимму'нный *a* heteroimmune
гетероиммуниза'ция *f* heteroimmunization
гетерокарио'з *m* heterocaryosis
гетерокарио'н *m* heterokaryon
гетерокариотиза'ция *m* heterocaryotization
гетерокариоти'ческий *a* heterocaryotic
гетерока'рпный *a* heterocarpous
гетерокине'з *m* heterokinesis
гетерокли'нный heteroclinous
гетероклити'ческий *a* heteroclytic
гетероконъюга'т *m* heteroconjugate
гетеролецита'льный *a* heterolecithal
гетерологи'чный *a* heterologous
гетероме'р *m* heteromer
гетеромери'я *f* heteromery
гетерометаболи'ческий *a* heterometabolic
гетеромиело'ма *f* heterohybridoma
гетероми'ксис *m* heteromixis
гетероморфи'зм heteromorphism
гетеромо'рфный *a* heteromorphic
гетероморфо'з *m* heteromorphosis
гетероплази'я *f* heteroplasia
гетеропланоби'ос *m* heteroplanobios
гетероселе'кция *f* heteroselection
гетероси'напсис *m gen* heterosynapsis
гетеросома'льный *a* heterosomal
гетероспецифи'чный *a* heterospecific
гетероспорангиати'вный *a* heterosporangic
гетероспо'рия *f* heterospory
гетероспо'ровый *a* heterosporic, heterosporous
гетеростили'я *f* heterostyly
гетероталли'зм *m* heterothallism, heterothally
гетероталло'мный *a* heterothallic
гетеротетрапло'ид *m* heterotetraploid
гетеротипи'ческий *a* heterotypic
гетероото'пный *a* heterotopic
гетеротро'ф *m* heterotroph
гетеротрофи'я *f* heterotrophy
гетеротро'фность *f* heterotrophy
гетеротро'фный *a* heterotrophic, allotrophic

гетероферментати'вный *a* heterofermentative
гетерофертилиза'ция *f* heterofertilization
гетерофилли'я *f* heterophylly
гетерофи'т *m* heterophyte
гетерофи'тный *a* heterophytic
гетерохлами'дный *a* heterochlamydeous
гетерохроматиза'ция *f* heterochromatization
гетерохромати'новый *a* heterochromatic
гетерохромати'ческий *a* heterochromatic
гетерохрома'тин *m* heterochromatin
  структу'рный ~ constitutive heterochromatin
  факультати'вный ~ facultative heterochromatin
гетерохромосо'ма *f* heterochromosome
гетерохро'нность *f* heterochronism, heterochrony
гетероце'льный *a* heterocoelous
гетероцентри'ческий *a* heterocentric
гетероцерка'льный *a* heterocercal
гетероцикли'ческий *a* heterocyclic
гетероци'ста *f* heterocyst
гетероэпито'п *m* heteroepitope
гиали'н *m* hyaline
гиали'новый *a* hyaline, hyalin, hyaloid, vitreous, glassy
гиало- hyalo-
гиаломе'р *m* hyalomere
гиалопла'зма *f* hyaloplasm
гиалосо'ма *f* hyalosome
гиалоспо'ра *f* hyalospore
гиалуронида'за *f* hyaluronidase
гиаци'нт *m* hyacinth (*Hyacinthus*)
  водяно'й ~ water hyacinth (*Eichhornia*)
гибберелли'н *m* gibberellin
гиббо'н *m* gibbon *(Hylobates)*
гиберна'ция *f* hibernation
ги'бкость *f* flexibility
  конформацио'нная ~ conformational flexibility
гибри'д *m* hybrid, cross
  вегетати'вный ~ vegetative hybrid, graft hybrid
  дикариоти'ческий ~ dicaryotic hybrid
  и'стинный ~ true hybrid
  ло'жный ~ false hybrid
  межвидово'й ~ interspecies hybrid, species hybrid
  межлине'йный ~ hybrid between lines
  межродово'й ~ intergeneric hybrid
  неполово'й ~ asexual hybrid
  нерасщепля'ющийся ~ permanent hybrid
  отдалё'нный ~ distant hybrid
  полигено'мный ~ polygenomatic hybrid
  полипло'идный ~ polygenomatic hybrid
  постоя'нный ~ permanent hybrid
  приви'вочный ~ graft hybrid
  произво'дный ~ derivative hybrid
  реципро'кные ~ы *pl* reciprocal hybrids
  сомати'ческий ~ somatic hybrid
  сомати'ческий кле'точный ~ somatic cell hybrid
  структу'рный ~ structural hybrid
  Т-кле'точный ~ T-cell hybrid
  четверно'й ~ quaternary hybrid
  я'дерно-цитоплазмати'ческий ~ nucleocytoplasmic hybrid
гибридиза'ция *f* hybridization
  ~ **кле'ток** cell hybridization
  аллопатри'ческая ~ allopatric hybridization
  вычита'тельная ~ subtractive hybridization
  интрогресси'вная ~ introgressive hybridization
  иску'сственная ~ artificial hybridization
  молекуля'рная ~ molecular hybridization
  напра'вленная ~ directed hybridization, purposeful hybridization
  одностор'нняя ~ unilateral hybridization
  отдалё'нная ~ distant hybridization, remote hybridization
  симпатри'ческая ~ sympatric hybridization
  субтракти'вная ~ subtractive hybridization
гибридизи'ровать *v* hybridize
гибри'дность *f* hybridism, hybridity
гибри'дный *a* hybridous
гибридо'ма *f* hybridoma
  В-кле'точная ~ B-cell hybridoma
  гетерологи'ческая ~ heterohybridoma
  межвидова'я ~ interspecies hybridoma
  перви'чная ~ primary hybridoma
  Т-кле'точная ~ T-cell hybridoma
  трансге'нная ~ transgenic hybridoma
гига'нт *m* giant
giganti'зм *m* giantism, gigantism
гигантоци'т *m* gigantocyte
гига'нтский *a* giant
гигие'на *f* hygiena
  ~ **труда'** hygiene of work, hygiene of labor
  коммуна'льная ~ communal hygiene, municipal hygiene
  ли'чная ~ personal hygiene

гигиени'ст *m* hygienist
гигиени'ческий *a* hygienic
гигиени'чный *a* hygienic
гигро- hygro-
гигрогра'мма *f* hygrogram
гигрокине'з *m* hygrokinesis
гигро'метр *m* hygrometer
гигроморфи'зм *m* hygromorphism
гигроскопи'ческий *a* hygroscopic
гигроскопи'чность *f* hygroscopicity
гигросциафи'т hygrosciaphyte
гигрофи'льный *a* hygrophilous
гигрофи'т *m* hygrophyte
гигроха'зия *f* hygrochasy
гидати'да *f* hydatid
гидато'да *f* hydatode, water gland
гидатофи'т *m* hydatophyte
гидradeни'т *m* hydradenitis
гидрази'д *m* hydrazide
~ **изоникоти'новой кислоты'** isonicotinic acid hydrazide
~ **малеи'новой кислоты'** maleic hydrazide
гидра'нт *m* hydranth
гидрата'за *f* hydratase
гидрата'ция *f* hydration
гидрати'рование *n* hydration
гидри'рование hydrogenation
гидро- hydro-
гидробиоло'гия *f* hydrobiology
гидробио'ника *f* hydrobionics
гидробио'нт *m* hydrobiont, hydrocole
гидрогено'лиз *m* hydrogenolyse
гидрогеофи'т *m* hydrogeophyte
гидро'ид *m* hydroid
гидро'идный *a* hydroid
гидрока'рпный *a* hydrocarpic
гидроконденса'ция *f* hydrocondensation
гидрокортизо'н *m* hydrocortisone
гидрокриптофи'т *m* hydrocryptophyte
гидроксиалкили'рование *n* hydroxyalkylation
гидроксили'рование *n* hydroxylation
гидроксиметили'рование *n* hydroxymethylation
гидрола'за *f* hydrolase, hydrolysing enzyme
гидро'лиз *m* hydrolysis
кисло'тный ~ acid hydrolysis
щелочно'й ~ alkaline hydrolysis
ферментати'вный ~ enzymic hydrolysis
гидролиза'т *m* hydrolysate
гидролизова'ть *v* hydrolyze
гидролити'ческий *a* hydrolytic
гидро'м *m* hydrome
гидромегате'рм *m* hydromegatherm
гидромелиора'ция *f* hydrotechnical amelioration
гидромо'рфный *a* hydromorphic
гидроморфо'з *m* hydromorphosis
гидрона'стия *f* hydronasty
гидрополимериза'ция *f* hydropolymerization
гидрополи'п *m* hydropolyp
гидропо'ника *f* hydroponics
гидрори'за *f* hydrorhiza
гидростати'ческий *a* hydrostatic
гидростеро'н *m* hydrosterone
гидросфе'ра *f* hydrosphere
гидроте'ка *f* hydrotheca
гидротропи'зм *m* hydrotropism
гидротропи'ческий *a* hydrotropic
гидрота'ксис *m* hydrotaxis
гидротропи'ческий *a* hydrotropic
гидрофи'льный *a* hydrophilic, hydrophilous
гидрофи'т *m* hydrophyte, hydrophytic plant
гидрофи'тный *a* hydrophytic
гидрофо'бность *f* hydrophobicity, hydropathy
гидрофо'бный *a* hydrophobic
гидрохино'н *m* hydroquinone
гидрохори'я *f* hydrochory
гидрохо'рный *a* hydrochorous
гидроце'ле *n* hydrocele
гидроэколо'гия *f* hydroecology
гие'на *f* hyena (*Hyaena*)
полоса'тая ~ striped hyena (*Hyena hyaena*)
пятни'стая ~ spotted hyena (*Crocuta crocuta*)
ги'льдия *f* guild
ги'мен *m* hymen
гимена'льный *anat* hymenal
гимениа'льный *a* hymenial
гиме'ний *m* hymenium
гименофо'р *m* hymenophore
Ги'мза-ди'ск *m* Giemsa band, G-band
гимнопла'ст *m* gymnoplast
гин- gyn-
гинандри'ческий *a* gynandrous
гинандро'ид *m* gynandroid
гинандромо'рф *m* gynandromorph
гинандроморфи'зм *m* gynandromorphism
гинандрофо'р *m* gynandrophore
гинантери'ческий *a* gynantherous
гинеко'лог *m* gynecologist, gynaecologist
гинеколо'гия *f* gynecology, gynaecology

гинекофо'р *m* gynaecophore
гине'цей *m* gynoecium
гиноба'зис *m* gynobase
гиногамо'н *m* gynogamone
гиногене'з *m* gynogenesis
гинопо'дий *m* podogynium
гиноспо'ра *f* gynospore
гиносте'гий *m* gynostegium
гиносте'мий *m* gynostemium
гинофо'р *m* gynophore
гио'ид *m* hyoid
гио'идный *a* hyoidean, hyoid
гиости'льный *a* hyostylic
гипа'нтий *m* hypanthium
гипанто'дий *m* hypanthodium
гипа'нтрум *m* hypantrum
гипер- hyper-
гипервариа'бельность *f* hypervariability
гипервариа'бельный *a* hypervariable
гипервентиля'ция *f* hyperventilation
гипервитамино'з *m* hypervitaminosis
гипергаммаглобулинеми'я *f* hypergammaglobulinemia
гипергапло'ид *gen* hyperhaploid
гипергаплоиди'я *f* hyperhaploidy
гипергонади'зм *m* hypergenitalism, hypergonadism
гипердипло'ид *m* hyperdiploid
гипереми'рованный *a* hyperemic
гипереми'я *f* hyperaemia, hyperemia
  реакти'вная ~ reactive hyperemia
гиперестези'я *f* hyperesthesia
гипериммуниза'ция *f* hyperimmunization
гипериннерва'ция *f* hyperinnervation
гиперкине'з *m* hyperkinetic disorder, hyperkinesia
гиперку'б *m* hypercube
  лати'нский ~ *biom* Latin hypercube
гиперметаморфо'з *m* hypermetamorphosis
гипермо'рф *m* hypermorph
гипермо'рфный *a* hypermorphous, hypermorphic
гиперморфо'з *m* hypermorphosis
гипермута'ция *f* hypermutation
гиперосмоти'ческий *a* hyperosmotic
гиперпарази'т *m* hyperparasite
гиперпаразити'зм *m* hyperparasitism
гиперпитуитари'зм *m* hyperpituitarism
гиперплази'я *f* hyperplasia
  лимфоретикуля'рная ~ lymphoreticular hyperplasia
гиперпласти'ческий *a* hyperplastic
гиперпло'ид *m* hyperploid
  части'чный ~ *gen* sectional hyperploid
гиперплоиди'я *f* hyperploidy
гиперпростра'нство *n* hyperspace
гиперрепликa'ция *f* hyperreplication
гиперрефле'ксия *f* hyperreflexia
гипертели'я *f* hypertely
гиперте'нзия *f* hypertension, hypertonia, supertension
гипертерми'я *f* hyperthermia
гипертони'ческий *a* hypertonic
гипертони'чность *f* hypertonicity
гипертони'я *f* hypertension, hypertonia, supertension
гипертрихо'з *m* hypertrichosis, hypertrichiasis
гипертрофи'ческий *a* hypertrophic
гипертрофи'я *f* hypertrophy, hypertrophia
  компенсато'рная ~ compensatory hypertrophy
  физиологи'ческая ~ physiological hypertrophy
гиперфа'гия *f* hyperphagia
гиперфу'нкция *f* hyperactivity
гиперхромати'зм *m* hyperchromatism
гиперчувстви'тельность *f* hypersensitivity
гиперэкспре'ссия *f* hyperexpression
гипестези'я *f* hypaesthesia, hypesthesia
гипно'дия *f* hypnody
гипно'з *m* hypnosis
гипноспо'ра *f* hypnospore
гипноспора'нгий *m* hypnosporangium
гипнотизе'р *m* hypnotist
гипнотизи'ровать *v* hypnotize
гипо- hypo-
гипобаза'льный *a* hypobasal
гипобази'дия *f* hypobasidium
гипобари'ческий *a* hypobaric
гипова'льва *f* hypovalve
гиповентиля'ция *f* hypoventilation
гиповитамино'з *m* hypovitaminosis
гипогапло'ид *gen* hypohaploid
гипогаплоиди'я *f* hypohaploidy
гипогапло'идный *a gen* hypohaploid
гипогене'з *m* hypogenesis
гипоги'нный *a* hypogynous
гипогнати'ческий *a* hypognathous
гиподе'рма *f* hypodermis, hypoderma
гиподерма'льный *a* hypodermal, hypodermic
гиподинами'я *f* hypodynamics
гиподиплоиди'я *f* hypodiploidy

гипоко'тиль *m* hypocotyl, hypocotyl hook
гипоксанти'н *m* hypoxanthine
гипоксида'ция *f* hypoxidation
гипоксидо'з *m* hypoxidosis
гипокси'ческий *a* hypoxic
гипокси'я *f* hypoxia
гипомо'рф *m* hypomorph
гипомо'рфный *a* hypomorphic
гипоморфо'з *m* hypomorphosis
гипоневра'льный *a* hyponeural
гипони'хий *m* hyponychium
гипопета'льный *a* hypopetalous
гипопи'гий *m* hypopyge
гипоплази'я *f* hypoplasia, hypoplasy
гипопласти'ческий *a* hypoplastic
гипопло'ид *m* hypoploid
гипоплоиди'я *f* hypoploidy
  части'чная ~ partial hypoploidy
гипопо'дий *m* hypopodium
гипополипло'ид *gen* hypopolyploid
гипореакти'вность *f* hyporesponsiveness
гипорефле'ксия *f* hyporeflexia
гипоспора'нгий *m* hyposporangium
гипоста'з *m* hypostasis
гипостати'ческий, гипостати'чный *a* hypostatic
гипосто'м *m* hypostome
гипосфе'н *m* hyposphene
гипотала'мус *m* hypothalamus
гипота'ллус *m* hypothallus
гипо'теза *f* hypothesis
  ~ **антиге'нной концентра'ции** antigen-focusing hypothesis
  ~ **Ве'генера** Wegener's hypothesis, continental drift hypothesis
  ~ **деле'ции кло'нов** clonal deletion hypothesis
  ~ **индуци'рованного соотве'тствия** induced fit hypothesis
  ~ **коллинеа'рности после'довательностей** sequence hypothesis
  ~ **материко'вого дре'йфа** continental drift hypothesis, Wegener's hypothesis
  ~ **миниге'нов** minigene hypothesis
  ~ **мише'ни** target hypothesis
  ~ **отта'лкивания** repulsion hypothesis
  ~ **па'рных реце'пторов** two-receptor hypothesis
  ~ **после'довательности** sequence hypothesis
  ~ **промискуите'тной рекомбина'зы** promiscuous recombinase hypothesis
  ~ **се'ти** network hypothesis
  ~ **со'бственных маркё'ров** self marker hypothesis
  ~ **стволово'й кле'тки** stem cell hypothesis
  ~ **фокуси'рующего антиге'на** antigen-focusing hypothesis
  бирецепто'рная ~ two-receptor hypothesis
  индивидуалисти'ческая ~ individualistic hypothesis, principle of species individuality
  монорецепто'рная ~ one-receptor hypothesis
  нулева'я ~ *biom* null hypothesis, zero hypothesis
гипотензи'вный *a* hypotensive
гипоте'ка *f* hypotheca
гипоте'ций *m* hypothecium
гипотони'ческий *a* hypotonic
гипотони'чность *f* hypotonicity
гипотони'чный *a* hypotonic
гипотони'я *f* hypotension, hypotonia
гипофарингеа'льный *a* hypopharyngeal
гипофа'ринкс *m* hypopharynx
гипо'физ *m* hypophysis, pituitary gland, pituitary body
гипофи'за *bot* hypophysis
гипофиза'рный *a* hypophyseal, hypophysial, pituitary
гипофи'лл *m* hypophyll
гипофлео'дный *a bot* hypophleodal
гипофу'нкция *f* hypofunction
гипохи'лий *m* hypochilium
гипохо'рда *f* hypochord
гипоцерка'льный *a* hypocercal
гиппока'мп *m* hippocamp
гипполо'гия *f* hippology
гипселодо'нтный *a* hypsodont
гипсова'ние *n* gypsum treatment
гипсодо'нтный *a* hypsodont
гипсолюби'вый *a* gypsophilous
гипура'лия *f* hypural bone
гирсути'зм *m* hirsutism
гируди'н *m* hirudin
гирча' *f* selinum (*Selinum*)
гистами'н *m* histamine
гистере'зис *m* hysteresis
гистеросо'ма *f* hysterosoma
гистеротели'я *f* hysterotely
гистиди'н *m* histidine
гистиобла'ст *m* histioblast

гистиоци'т *m* histocyte, histiocyte, fixed macrophage, resident macrophage, stable macrophage, tissue-fixed macrophage
гистогене'з *m* histogenesis
гистоге'нный *a* histogenic
гистогра'мма *f* histogram, bar graph
ко'нтурная ~ contour histogram
то'чечная ~ **интерва'лов** joint interval histogram
гистозо'йный *a* histozoic
гисто'лиз *m* histolysis
гистолити'ческий *a* histolytic
гистологи'ческий *a* histological
гистоло'гия *f* histology
гисто'н *m* histon, histone
гистопатоло'гия *f* histopathology
гистоплазмо'з *m* histoplasmosis *(caused by Histoplasma)*
гистосовмести'мость *f* histocompatibility
гистотипи'рование *n* tissue typing
гистото'п *m* histotope
гистотро'фный *a* histotrophic
гистофизиоло'гия *f* histophysiology
гистохими'ческий *a* histochemical
гистохи'мия *f* histochemistry, histological chemistry
гистоци'т *m* histocyte
гитарообра'зный *a* pandurate, pandurated, panduriform
ги'фа *f* hypha
аскоге'нная ~ ascogenous hypha
внедря'ющаяся ~ penetrating hypha
воро'нинская ~ Woronin's hypha
воспринима'ющая ~ receptive hypha
плодонося'щая ~ fruiting hypha
соедини'тельная гифа bridging hypha
спороносящая ~ fruiting hypha
гифа'зма *f* hyphasma
гифомице'т filamentous fungus (*Hyphomycetes*)
гла'дкий *a* smooth; glabrous; *pal* psilate
гладкогребе'нчатый *a* psilolophate
гладкоколо'сый *a* smooth-spiked
гладколи'стный *a* smooth-leaved
гладкопло'дный *a* smooth-fruited
гладкопокро'вный *a* psilotegillate
гладкосемя'нный *a* smooth-seeded
гладкоцветко'вый *a* smooth-flowered
гладкошё'рстный *a* smooth-haired
глаз *m* eye; **вооруже'нным гла'зом** with an aided eye; **невооруже'нным гла'зом** with an unaided eye; **с вы'пученными глаза'ми** goggle-eyed; **глаза'** *pl* **на вы'кате** bulging eyes
аппозицио'нный ~ day eye
вооруже'нный ~ aided eye
воро'ний ~ herb Paris (*Paris quadrifolia*)
дневно'й ~ day eye
инверти'рованный ~ inverse eye
невооруже'нный ~ naked eye, unaided eye
неинверти'рованный ~ converse eye
ночно'й ~ night eye
сидя'чий ~ sessile eye
сло'жный ~ compound eye
стебе'льчатый ~ projecting eye, stalked eye
суперпозицио'нный ~ night eye
телескопи'ческий ~ telescopic eye
теменно'й ~ parietal eye
гла'зки *m pl*: **аню'тины** ~ pansy, heartsease (*Viola tricolor*)
глазни'ца *f* orbit, eye socket
глазни'чно-ло'бный *a* oculofrontal
глазни'чно-носово'й *a* oculonasal
глазни'чный *a* orbital, ophthalmic
глазно'й *a* ocular, ophthalmic
глазодви'гательный *a* ophthalmogyric, optokinetic, oculogyric, oculomotor
глазо'к *m* (*pl* **глазки'**) ocellus, eye; *bot* eyelet; spot, speculum (*e.g., on a wing of birds);* bud, eye (*of a potato tuber*); *bot* oculus; cave (*in cheese*)
зиму'ющий ~ wintering eye
крылово'й ~ pterostigma
спя'щий ~ dormant eye
глазча'тый *a* ocellate, ocellated, oculate
гладкоше'рстный *a* short-haired
гландулоци'т *m* glandulocyte, gland cell, glandular cell
гле'ба *f* gleba
гле'була *f* glebula
глееобразова'ние *n* gleization
глей *m* glei, gley
глиа'льный *a* glial
гли'я *f* glia
глика'н *m* glycan
гликеми'я *f* glycemia
глико- glyco-
гликоге'н *m* glycogen
гликоге'нный *a* glycogenic, glycogenous
гликогено'лиз *m* glycogenolysis
гликогенолити'ческий *a* glycogenolytic
гликозаминоглика'н *m* glycosaminoglycan
гликозиди'рование *n* glycosidation

гликозили'рование *n* glycosilation
гликока'ликс *m* glycocalix
глиоксисо'ма *f* glyoxysome
глико'лиз *m* glycolysis
гликопротеи'д *m* glycoprotein
гликофи'т *m* glycophyte
гли'на *f* clay
  пылева'тая ~ silty clay
гли'нистый *a* argillaceous, clayish
глинозе'м *m* clay
глиокса'ль *m* glyoxal
глиоци'т *m* glia cell
глист *m* helminth
глицеральдеги'д *m* glyceraldehyde
глицери'н *m* glycerol
глици'н *m* glycin, glycine
глици'ния *f* wisteria *(Wisteria)*
глоби'н *m* globin
глобули'н *m* globulin
глода'ть *v* gnaw
гло'жущий *a* browsing
гломеруля'рный *a* glomerular
гло'сса *f* glossa
глосса'рий *ent* glossarium
глоссопо'дий *m* glossopodium
глота'ть *v* swallow, gulp, gorge
гло'тка *f* throat, gullet, gorge, pharynx
глото'к *m* draught
гло'точно-основно'й *a* pharyngobasilar
гло'точный *a* pharyngeal
глубина' *f* depth
  ~ **нарко'за** anesthetic depth
  ~ **промерза'ния** frost level
  ~ **фокуси'рования** focal depth
глубино'мер *m* depth gauge
глупы'ш *m orn* fulmar, Atlantic fulmar, northern fulmar mallie, malliemauk (*Fulmarus glacialis*)
глутами'н *m* glutamine
глутаральдеги'д *m* glutaraldehyde
глуха'рь *m* wood grouse, great grouse, capercaille (*Tetrao urogallus*)
глухота' *f* deafness, surdity
глуши'ть *v* overrun *(of weeds)*; stun *(fish)*
глушь *f* backwoods
  лесна'я ~ backwoods
глюко'за *f* glucose
глюкозиди'рование glucosidation
глюкози'метр glucosimeter
глюконеогене'з gluconeogenesis
глюкосо'ма glucosome
глянцеви'тый *a* vernicose
гляциоло'гия *f* glaciology
гнатио'н *m* gnathion
гнатоба'за *f* gnathobase
гнатопо'д *m* gnathopod
гнатосо'ма *f* gnathosoma
гнать *v* drive (*of cattle*); ride, ride hard (*of a horse*)
гна'ться *v* pursue
гнев *m* anger
гнедо'й *a* bay
гнезда'рь *m* nestling
гнезди'ться *v* nest
гнездо' *n* nest; locule, loculus
  муравьи'ное ~ formicary, anthill
  оси'ное ~ vespiary
  эпителиа'льное ~ epithelial loculus
гнездова'ние *n* nesting
  парасоциа'льное ~ parasocial nesting
  субсоциальное ~ subsocial nesting
гнездо'вье *n* breeding ground, nesting ground, nesting place, rookery, breeding site, nesting site
гнездострое'ние *n* nest building, nidification
гнесиога'мия *f* gnesiogamy
гни'да *f ent* nit
гние'ние *n* rot, rotting, putrefaction, decay, decomposition
гнило'й *a* rotten, putrid
гнило'стный *a* putrescent
гнилу'шка *f* sponk
гниль *f* rot
  ~ **древеси'ны** timber decay *(caused by fungi)*
  ~ **корнево'й ше'йки** crown rot *(caused by Phytophthora parasitica)*
  бактериа'льная ~ bacterial rot
  бе'лая ~ white rot *(caused by fungi)*
  бу'рая ~ brown rot *(caused by fungi)*
  ве'ршинная ~ blossom-end rot rot *(caused by fungi)*
  го'рькая ~ виногра'да grape bitter rot *(caused by Melanconium fuligineum)*
  го'рькая ~ плодо'в bitter rot *(caused by fungi Glomerella cingulata)*
  го'рькая ~ я'блок apple rot *(caused by Gloeosporium fructigenum)*
  заболо'нная ~ sap rot *(caused by fungi)*
  комлева'я ~ butt rot
  корнева'я ~ root rot *(caused by fungi)*
  корнева'я ~ бахче'вых культу'р watermelon root rot *(caused by Monosporascus cannoneallus)*

корнева'я ~ бобо'вых bean root rot *(caused by Fusarium)*
корнева'я ~ кукуру'зы corn root rot *(caused by Phialophora radicicola)*
корнева'я ~ ржи root decay of rye *(caused by Fusarium culmorum)*
корнева'я ~ со'и soybean root rot *(caused by Corynespora cassicola)*
мя'гкая ~ soft rot
пе'пельная ~ charcoal rot *(caused by Sclerotium bataticola)*
плодо'вая ~ fruit rot *(caused by bacteria and fungi)*
ра'нняя ~ карто'феля early blight of potato *(caused by Alternaria solani)*
ро'зовая ~ карто'феля pink rot *(caused by Phytophthora erythroseptica)*
се'рая ~ черни'ки blueberry bunch rot *(caused by Botrytis cinerea)*
сердцеви'нная ~ дере'вьев heart rot *(caused by Fomes)*
сердцеви'нная ~ свё'клы heart rot *(caused by Mycosphaerella tabifica)*
стеблева'я ~ stalk rot rot
суха'я ~ dry rot *(caused by fungi)*
суха'я ~ кукуру'зы corn ear rot *(caused by Diplodia zeae)*
у'гольная ~ charcoal rot *(caused by Sclerotium bataticola)*
чё'рная ~ black rot *(caused by fungi and viruses)*
чё'рная корнева'я ~ black root rot *(caused by fungi)*

гнить *v* rot, decay, putrefy
гноеро'дный *a* suppurative, pyogenic
гнои'ться *v* suppurate
гной *m* pus
гнойни'к *m med* abscess
гнойничо'к *m med* pustule
гно'йно-воспали'тельный *a* pyo-inflammatory
гно'йный *a* purulent, suppurative
гнотобио'нт *m* gnotobiont
гнотобио'тика *f* gnotobiotics
гну *m ther* gnu *(Connochaetes)*
гнус *m* vermin, bloodsucking arthropods
говя'дина *f* beef
го'голь *m* goldeneye, garrot (*Bucephala*)
ма'лый ~ bufflehead, buffle-headed duck (*Bucephala albeola*)
обыкнове'нный ~ garrot (*Bucephala clangula*)

го'гот, гогота'нье *m* cackling, cackle (*of geese*), gabble
гогота'ть *v* cackle (*of geese*), squawk, gabble
год *m* year
~ отрожде'ния мо'лоди brood year
лё'тный ~ swarm year
малово'дный ~ low-water year
многово'дный ~ high-water year
семенно'й ~ seed year
урожа'йный ~ bumper-crop year

годова'лый *a* year-old
годови'к *m* yearling, one-year-old
го'кко *orn* guan *(Cracidae)*
голаркти'ческий *a* holarctic
голендобиоти'ческий *a* holendobiotic
голендофи'т *m* holendophyte
голеносто'пный *a* talocrural
го'лень *f* shank, shin, crus
голо- holo-; gymno-
голобази'дия *f* holobasidium
голобласти'ческий *a* holoblastic
голова' *f* head; **де'сять голо'в скота'** ten head of cattle
~ са'хару loaf of sugar, sugar-loaf
~ сы'ра *a* cheese

голова'стик *m* frog larva, tadpole
головетви'стый *a* naked-twigged
голо'вка *f* head, capitulum; glome, glomus (*a kind of inflorescence*); glans *(of penis)*
~ льна flax boll
~ фа'говой части'цы phage head
колосови'дная ~ *bot* earhead
суста'вная ~ *anat* articular head

головно'й *a* cephalic
головня' *f* smut (*Ustilago*)
~ зла'ков leaf smut *(caused by Ustilago striaeformis)*
~ пы'льников anther smut *(caused by Ustilago violacea)*
~ со'рго whole-head smut *(caused by Sphacelotheca reiliana)*
воню'чая ~ пшени'цы ball smut, bunt smut, stinking smut *(caused by Tilletia caries, Tilletia tritici)*
голо'вчатая ~ со'рго sorghum head smut *(caused by Peronoscleropora sorghi)*
мо'края ~ пшени'цы ball smut, bunt smut, stinking smut *(caused by Tilletia caries, Tilletia tritici)*
покрыва'ющая ~ closed smut *(caused by Ustilago kolleri)*

пузы'рчатая ~ кукуру'зы boil smut *(caused by Ustilago zeae)*
пы'льная ~ про'са kernel smut *(caused by Sphaceolotheca cruenta)*
стеблева'я ~ ржи' flag smut, stripe smut *(caused by Urocystis occulta)*
твё'рдая ~ зерна' covered smut *(caused by Ustilago hordei, Ustilago levis)*
твё'рдая ~ пшени'цы ball smut, bunt smut, stinking smut *(caused by Tilletia caries, Tilletia tritici)*
твё'рдая ~ ячменя' barley smut *(caused by Ustilago hordei, Ustilago nigra)*
головогру'дь *f* cephalothorax
головокруже'ние *n* vertigo, dizziness, giddiness
голо'вчатый *a* cephalate, cephalated, capitate
гологи'нный *a* hologynic
го'лод *m* hunger, starvation, famine
«во'лчий» ~ insatiable appetite, bulimia, hyperorexia
голода'ние *n* starvation, fasting, fast, hunger
белко'вое ~ protein deprivation, protein deficiency
во'дное ~ water deprivation
кислоро'дное ~ oxygen deficiency, air hunger
голода'ть *v* starve, hunger, famish, fast
голода'ющий *a* starving
голожа'берный *a* nudibranchiate
голозё'рный *a* bare-grained
голозо'йный *a* holozoic
голокорешко'вый *a* naked-rooted
голокри'нный *a* holocrine
голокры'лый *a* naked-winged
гололи'стный *a* naked-leaved
голопарази'т *m* holoparasite
голопло'дный *a* gymnocarpous, naked-fruited
голопнейсти'ческий *a* holopneustic
голопти'ческий *a* holoptic
голоростко'вый *a* gymnoblastous
го'лос *m* voice
голосемя'нный *a* gymnospermous
голосово'й *a* vocal
голостебе'льчатый *a* bare-stalked
голоти'п *m* holotype
голофи'т *m* holophyte
голофи'тный *a* holophytic
голоцветко'вый *a* gymnanthous, naked-flowered
голоце'н *m* holocene
голоценоти'ческий *a* holocoenotic
голубево'д *m* pigeon breeder
голуби'ка *f* bog bilberry (*Vaccinium uliginosum*)
голуби'ный *a* columbine
голу'бка *f* dove
голубо'й *a* light-blue, pale-blue, sky-blue
бриллиа'нтовый ~ brilliant blue
крези'ловый ~ cresyl blue
кума'сси бриллиа'нтовый ~ coomassie brilliant blue
метилтимо'ловый ~ methylthymol blue
ни'льский ~ Nile blue
тетразо'лиевый ~ tetrazolium blue
целести'новый ~ celestin blue
э'ванс ~ Evans blue
голубо'к *m diminutive of* **голубь**
ка'пский ~ pintado, pintado petrel, Cape petrel, Cape pigeon (*Daption capenses*)
го'лубь *m* pigeon, dove (*Columba*)
венцено'сный ~ crowned pigeon *(Goura cristata)*
декорати'вный ~ fancy pigeon
одича'вший ~ feral pigeon
полудома'шний ~ dovecote pigeon
почто'вый ~ carrier pigeon, homing pigeon
си'зый ~ rock dove, rock pigeon, wild pigeon (*Columba livia*)
стра'нствующий ~ passenger pigeon (*Ectopistes migratorius*)
голубя'нка *f ent* azure
голубя'тня *f* loft, pigeon-loft, dove-cot, dove-cote
го'лый *a* glabrous, bare, bald; nacked
гомаксо'нный *a* homaxonic
гомео- homeo-
гомеобо'кс *m* homeobox
гомео'зис *m* homeosis
гомео'зисный *a* homeotic
гомеокине'з *m* homeokinesis
гомеоме'рный *a* homoiomerous
гомеомо'рфа *f* homeomorph
гомеопа'тия *f* homeopathy
гомеоре'зис *m* homeorhesis
гомеоси'напсис *m* homeosynapsis
гомеоста'з, гомеоста'зис *m* homeostasis
иммуʼнный ~ immunological homeostasis
популяцио'нный ~ population homeostasis
физиологи'ческий ~ physiological homeostasis
гомеоста'т *m* homeostat

гомеостати'ческий *a* homeostatic
гомеотипи'ческий *a* homeotypic
гомо- homo-
гомобази'дия *f* homobasidium
гомобласти'ческий *a* homoblastic
гомобрахиа'льный *a* homobrachial
гомогаме'та *f* homogamete
гомогаме'тный *a* homogametic
гомога'мия *f* homogamy
гомога'мный *a* homogamous
гомогена'т *m* homogenate
гомогенети'ческий *a* homogenetic
гомогениза'тор *m* homogenizer
~ **Да'унса** Dounce homogenizer
гомогениза'ция *f* homogenization
гомоге'нность *f* homogeneity
гомогенный *a* homogeneous
гомогено'та *f* homogenote
гомого'ния *f bot* homogony
гомоде'рмный *a emb* homodermic
гомодиме'р *m* homodimer
гомодинами'ческий *a* homodynamic
гомодинами'я *f* homodynamy
гомодро'мный *a* homodromous
гомозиго'та *f* homozygote
гомозиго'тность *f* homozygosis, homozygosity
гомозиго'тный *a* homozygous
гомойосмоти'ческий *a* homoiosmotic
гомойоте'рмный *a* homoiothermal, homeothermal
гомокарио'н *m* homokaryon
гомока'рпный *a* homocarpous
гомолецита'льный *a* homolecithal
гомо'лиз *m* homolysis
гомо'лог *m* homologue, homolog
гомологизи'ровать *v* homologize
гомологи'чный *a* homologous
гомоло'гия *f* homology
~ **после'довательностей** *mbiol* sequence homology
оста'точная ~ residual homology
гомоме'рия *f* homomery
гомомо'рфный *a* homomorphic, homomorphous
гомоморфо'з *m* homomorphosis
гомоно'мный *a* homonomous
гомопласти'ческий *a* homoplastic
гомоселе'кция *f* homoselection
гомоси'напсис *m* homosynapsis
гомосома'льный *a gen* homosomal
гомоспецифи'ческий *a* homospecific
гомостили'я *f* homostyly
гомоталли'зм *m* homothallism
гомоталли'чный, гомоталло'мный *a* homothallic
гомотипи'ческий *a* homotypic
гомото'нность *f* homotoneity
гомотранспланта'т *m* homotransplant
гомотранспланта'ция *f* homotransplantation
гомотро'пный *a* homotropic
гомоферментати'вный *a* homofermentative
гомохлами'дный *a* homochlamydeous
гомоцентри'ческий *a* homocentric
гомоцерка'льный *a* homocercal
гомофи'т *m* homophyte
гомофи'тный *a* homophytic
гомохлами'дный *a* homochlamydeous
гомоцентри'ческий *a* homocentric
гомоцерка'льный *a* homocercal
гон *m* rut, rutting, ruttishness
гона'да *f* gonad
мужска'я ~ male gonad
гонадогене'з *m* gonadogenesis
гонадолибери'н *m* gonadotropin-releasing hormone
гонадосомати'ческий *a* gonadosomatic
гонадотропи'н *a* gonadotropin
гипофиза'рный ~ pituitary gonadotropin
хориони'ческий ~ chorion gonadotropic hormone
гонадотро'пный *a* gonadotropic
гона'нгий *m* gonangium
гонапофи'з *m* gonapophysis
гонеокли'н *m* goneoclin
гонидиа'льный *a* gonidial, gonidic
гонидиофо'р *m* gonidiophore
гони'дия *f* gonidium
гонимобла'ст *m* gonimoblast
го'нион *m* gonion
гониоци'ста *f* goniocyst
гоно- gono-
гонобо'бель *m* bog bilberry (*Vaccinium uliginosum*)
гонbacku'кт *m* gonoduct
гономери'я *f* gonomery
гононефрото'м *emb* gononephrotome
гонопо'дий *m* gonopodium
гонорре'я *f* gonorrhea
гоносо'ма *f* gonosome
гоносфе'ра *f* gonosphere
гоноте'ка *f* gonotheca
гонофа'г *m* gonophage
гонофо'р *m* gonophore

гоноце'ль *f* gonocoel
гоноци'т *m* gonocyte
го'нчая *f* course dog, hound
гоня'ть *v* fly, race (*of pigeons*)
~ **на ко'рде** lunge *a* horse
гоня'ться *v* chase
гора'л *m* goral, goat antelope (*Nemorhaedus goral*)
горб *m* hump, hunch, humpback, hunchback
горба'тый *a* humpbacked, humped, crookbacked, gibbous
гордеи'н *m* hordein
го'рец *m* jointweed, knotweed (*Polygonum*)
го'речь *f* bitter
горизо'нт *m* horizon
водоно'сный ~ water-bearing horizon
вы'щелоченный ~ leached horizon
ги'псовый ~ gypsic horizon
гле'евый ~ gley horizon, gleyic horizon
оподзо'ленный ~ podzolized horizon
сто'лбчатый ~ columnar horizon
горихво'стка *f* redstart (*Phoenicurus*)
горицве'т *m* adonis (*Adonis*); marsh gilliflower *(Coronaria flos-cuculi)*
го'ричник *m* hog's-fennel *bot (Peucedanum)*
го'рлица *f* turtle dove (*Streptopelia*)
ко'льчатая ~ collared turtle dove (*Streptopelia decaocto*)
обыкнове'нная ~ turtle dove (*Streptopelia turtur*)
го'рло *n* throat
горлово'й *a* gular, throat
горля'нка *f* bottle gourd, calabash (*Lagenaria vulgaris*)
гормого'ний *m* hormomogon, hormogone, hormogonium
гормо'н *m* hormone
~ **аденогипо'физа** adeno-pituitary hormone
~ **гипо'физа** hypophyseal hormone, pituitary hormone
~ **жё'лтого те'ла яи'чника** luteal hormone
~ **за'дней до'ли гипо'физа** posterior pituitary hormone
~ **катаболи'ческого де'йствия** catabolic hormone
~ **коры' надпо'чечников** adrenal hormone
~ **ли'ньки** molting hormone
~ **поджелу'дочной железы'** pancreatic hormone
~ **ро'ста** growth hormone
~ **щитови'дной железы'** thyroid hormone
~ **яи'чника** ovarian hormone
адипокинети'ческий ~ adipokinetic hormone
адренокортикотро'пный ~, АКТГ adrenocorticotrophic hormone, adrenocorticotropic hormone
андроге'нный ~ androgenic hormone
антидиурети'ческий ~ antidiuretic hormone
гипофиза'рный ~ hypophyseal hormone, pituitary hormone
гонада'льный ~ gonadal hormone, sex hormone
гонадотро'пный ~ *m* gonadotropin, gonadotrophic hormone
же'нский полово'й ~ female sex hormone
кортика'льный ~ cortical hormone
лактоге'нный ~ galactopoietic hormone, lactogenic hormone
лютеинизи'рующий ~, ЛГ luteinizing hormone
лютеотро'пный ~, ЛТГ luteotrophic hormone
маммотро'пный ~ mammotropic hormone
минералокортико'идный ~ mineralocorticoid hormone
мужско'й полово'й ~ male sex hormone
нейроэндокри'нный ~ neuroendocrine hormone
нейтрогипофиза'рный ~ neurohypophyseal hormone
нимфальный ~ nymphal hormone
паратирео'идный ~ parathyroid hormone
паращитови'дный ~ parathyroid hormone
плацента'рный ~ placental hormone
полово'й ~ gonadal hormone, sex hormone
ранево'й ~ wound hormone
расти'тельный ~ plant hormone
полово'й ~ sex hormone, gonadal hormone
соматотро'пный ~ somatotropic hormone
стеро'идный ~ steroid hormone
тими'ческий ~ thymic hormone
тирео'идный ~ thyroid hormone
тиреостимули'рующий ~ thyroid-stimulating hormone
тиреотропный ~ thyrotrophic hormone
фолликулостимули'рующий ~, ФСГ follicle-stimulating hormone
эстроге'нный ~ estrogenic hormone
ювени'льный ~ juvenile hormone
гормона'льный *a* hormonal, hormonic
гормонообразова'ние *n* hormonopoiesis, hormonogenesis

гормонообразу'ющий *a* hormonogenic, hormonopoietic
гормонопоэ'з *m* hormonopoiesis, hormonogenesis
гормоци'ста *f* hormocyst
горноста'й *m* common weasel, greater weasel, short-tailed weasel, ermine stoat (*Mustela erminea*)
городча'тый *a* crenate
горотели'я *f* horotely
горотели'ческий *a* horotelic
горо'х *m* pea (*Pisum*)
горо'шек *m* vetch *(Vicia)*
души'стый ~ lady pea, sweet pea (*Lathyrus odoratus*)
мыши'ный ~ tufted vetch, bird's tare (*Vicia cracca*)
горо'шина *f* pea
горта'нный *a anat* laryngeal
горта'нь *f* larynx
горчи'ца *f* mustard (*Sinapis*)
горшо'к *m* pot
горшо'чек *m diminutive of* **горшок**
перегно'йный ~ muck block
торфоперегно'йный ~ peat-compost block
торфяно'й ~ peat block
го'рький *a* bitter
горькопло'дный *a* bitter-fruited
горя'нка *f bot* barrenwort *(Epimedium)*
горя'чий *a* hot; spirited, bean, fiery, high-mettled (*of a horse*)
горя'чка *f* fever
го'фер *m ther* gopher *(Geomyidae)*
граб *m* hornbeam (*Carpinus*)
граби'нник *m* oriental hornbeam *(Carpinus orientalis)*
грабо'вник *m* hornbeam forest
гравила'т *m bot* avens (*Geum*)
гравиро'вка *f* etching
~ **табака'** tobacco etch
гравита'ция *f* gravitation
градие'нт *m* gradient
~ **давле'ния** pressure gradient
~ **парциа'льного давле'ния** partial pressure gradient
~ **пло'тности** density gradient
~ **по'ля** field gradient
~ **при'знаков** character gradient
~ **регенера'ции** regeneration gradient
~ **среды'** environmental gradient
анима'льный ~ animal gradient
анима'льно-вегетати'вный ~ animal-vegetal gradient
вегетати'вный ~ vegetal gradient
дорсовентра'льный ~ dorsoventral gradient
желто'чный ~ yolk gradient
желу'дочковый ~ ventricular gradient
морфогенети'ческий ~ morphogenetic gradient
осево'й ~ axial gradient
осмоти'ческий ~ osmotic gradient
пере'дне-за'дний ~ anteroposterior gradient
простра'нственный ~ spatial gradient
градоби'тие *n* hail damage
градоло'гия *f ent* gradology
градуи'ровать *v* graduate
гра'дус *m* degree
~ **Бомэ'** Baume degree
~ **Ке'львина** Kelvin degree
~ **Реомю'ра** Reaumur degree
~ **فаренге'йта** Fahrengeit degree
~ **Це'льсия** Celsius degree
гра'дусо-де'нь *m* degree-day
гракл *orn* grackle *(Quiscalus)*
лодкохво'стый ~ boat-tail grackle *(Quiscalus major)*
гра'кса *f* gurry
грамициди'н *m* gramicidin
грамотрица'тельный *a* Gram-negative
грамположи'тельный *a* Gram-positive
гра'на *f bot* granum
грана'т *m* pomegranate (*a fruit of Punica*); granatum *(a kind of fruit)*
грана'тник *m* pomegranate (*Punica*)
грани'ца *f* boundary, limit
~ **кла'сса** *biom* class limit
высо'тная ~ altitudinal limit
грани'цы довери'тельного интерва'ла *pl* confidence limits, fiducial limits
крити'ческая ~ *biom* rejection limit
гра'нула *f* granule, graun; pellet (*e.g., of feed or pesticide*)
~ **зимоге'на Берна'ра** Bernard's granule
~ **полово'го хромати'на** granule of sex chromatin
азурофи'льная ~ azurophilic granule
акросо'мная ~ acrosomal granule
ацидофи'льная ~ acidophilic granule
осмиофи'льная ~ osmiophilic granule
хромаффи'нная ~ chromaffin granule
грануле'ма *f* granuloma

~ **Ба'нга** Bang's granuloma
гранули'рование *n* granulation
гранули'рованный *a* granular; pelleted
гранули'ровать *v* granulate
гранулобла'ст *m* granuloblast
гранулоци'т *m* granular leukocyte, granulocyte
ацидофи'льный ~ eosinophilic granulocyte
нейтрофи'льный ~ neutrophilic granulocyte
гранулоцитопоэ'з *m* granulocytopoiesis
грануля'рный *a* granular, granulate
грануля'ция *f* granulation
пахио'нова ~ Pacchionian body
гра'ус *m* moorfowl (*Lagopus scoticus*)
грач *m* rook (*Corvus frugilegus*)
грачо'вник *m* rookery
гребе'нчатый *a* pectinate, cristate
гре'бень *m* crown (*of a bird*); pecten, crista, crest, ridge; comb
~ **ра'ковины** *anat* turbinated crest
заты'лочный ~ occipital crest, occipital ridge
клинови'дный ~ sphenoid crest
краево'й ~ marginal ridge *(in diatoms)*
ло'бный гребень frontal ridge
надсосцеви'дный ~ supramastoid crest
нё'бный ~ palatine crest
носово'й ~ nasal crest
подвздо'шный ~ *anat* iliac crest
подвисо'чный ~ infratemporal crest
стридуляцио'нный ~ stridulatory file
гребешко'вый *a bot* crested
гребешо'к *m* pecten, comb, cockscomb, crest, crista; scallop (*Pecten*)
слухово'й ~ acoustic crest, ampullary crest
гребнеперегоро'дчатый *a* carinimurate
гре'динг *m* grading
гре'йпфрут *m* grapefruit (*Citrus paradisi*)
гре'на *f* silkworm eggs, grain
гречи'ха *f* buckwheat (*Fagopyrum*)
гриб *m* fungus (*Fungi*), mushroom
базидиа'льный ~ basidium fungus, palisada fungus *(Basidiomycetes)*
берё'зовый ~ shelf fungus *(Polyporus sulphureus)*
возду'шно-во'дный ~ aero-aquatic fungus
головне'вый ~ smut, smut fungus, bunt fungus *(Tilletiaceae) ?????*
голока'рпный ~ holocarpic fungus
голосу'мчатый ~ exoascale fungus *(Exoascales)*
домо'вый ~ house fungus, dry-rot fungus (*Serpula lacrymans*); cellar fungus *(Thelephoraceae, Polyporaceae)*
древоразруша'ющий ~ wood-attacking fungus, wood-rotting fungus
дрожа'лковый ~ trembling fungus *(Tremellales)*
дья'вольский ~ Devil's Boletus (*Boletus satanas*)
же'лчный ~ bitter bolete, bitter boletus (*Tylopilus felleus*)
коро'вий ~ bovinus mushroom, shallow pored boletus (*Suillus bovinus*, *Boletus bovinus*)
мицелиа'льный ~ filamentous fungus *(Hyphomycetes)*
настоя'щий ~ true fungus *(Eumycota)*
ни'зший ~ lower fungus
осе'нний ~ honey fungus, boot-lace fungus (*Armillariella mellea*)
паразити'ческий ~ parasitic fungus
патоге'нный ~ pathogenic fungus
пе'речный ~ peppery boletus (*Suillus piperatus*)
печё'ночный ~ beefsteak fungus, liver fungus *(Fistulina hepatica)*
пласти'нниковый ~, пласти'нчатый ~ gill fungus (*Agaricaceae*)
пле'сневый ~ mold fungus
по'льский ~ bay boletus (*Boletus badius*)
по'чвенный ~ soil fungus, geophilous fungus
ржа'вчинный ~ rust fungus (*Urediniales*)
рога'тиковый ~ coral fungus (*Clavariaceae*)
сапрофи'тный ~ saprophytic fungus
сатани'нский ~ satan's boletus, satanic boletus, Satan's mushroom (*Boletus satanas*)
су'мчатый ~ ascomycete, sac fungus, cup fungus *(Ascomycetes)*
съедо'бный ~ edible fungus
тру'бчатый ~ pore fungus
у'стричный ~ oyster fungus (*Pleurotus ostreatus*)
фитопатоге'нный ~ plant pathogenic fungus
хитри'диевый ~ chytrid
хи'щный ~ predacious fungus
це'зарский ~ caesar's mushroom (*Amanita caesarea*)

«ча'йный ~» tea fungus *(симбиоз дрожжей и молочнокислых бактерий)*
черни'льный ~ ink mushroom *(Coprinus)*
шляпочный ~ mushroom, pileate fungus *(Agaricales)*
энтомопатоге'нный ~ entomopathogenic fungus
эпифи'тный ~ epigenous fungus
ядови'тый ~ poisonous fungus
гриб-бара'н maitake mushroom *(Grifola frondosa)*
гриб-зо'нтик пестрый, гриб-зонтик большой parasol mushroom (*Macrolepiota procera*)
грибко'вый *a* fungic, fungous
грибни'ца *f* mycelium
грибно'й *a* fungal, fungic, fungous
грибо'к *m* fungus, mould
дрожжево'й ~ yeast
кефи'рный ~ kefir grain
ни'тчатый ~ thread fungus
пле'сневый ~ mold, mould
грибни'к *m* mushroomer, mushroom picker
грибни'ца *f myc* mycelium
гри'ва *f* mane, mane hair, crest
гри'вистый *a* jubate, ruffed, crested
гри'зли *m* grizzly bear (*Ursus arctos*)
грима'са *f eth* grimace
~ **с оска'лом** bared-teeth face
грисбо'к *ther* grysbok *(Raphicerus melanotis)*
гриф *m orn* vulture (*Aegyptiidae*)
африка'нский белобо'кий ~ white-backed vulture *(Pseudogyps africanus)*
африка'нский белоголо'вый ~ white-headed vulture *(Trigonoceps occipitalis)*
африка'нский уша'стый ~ lappet-faced vulture *(Torgos tracheliotus)*
короле'вский ~ king vulture *(Sarcoramphus papa)*
па'льмовый ~ palm-nut vulture *(Gypohierax angolensis)*
гриф-инде'йка *m* turkey vulture *(Cathartes aura)*
грифелё'к *m* style
гроздеви'дный *a* bunchlike, botryoidal, racemose, agminate, agminated
гроздь *f* bunch, cluster, truss; raceme
~ **бана'нов** hand of bananas
виногра'да bunch of grapes
ры'хлая ~ loose raceme
грубоше'рстный *f* rough-wool, coarse
груди'на *f* breastbone, sternum
груди'нно-ключи'чный *a* sternoclavicular
груди'нно-ре'берный *a* sternocostal
груди'нный *a* sternal
грудно'й *a* pectoral, thoracic
грудь *f* breast
груз *m* load
генети'ческий ~ genetic burden
груздь *m* milk fungus, milk mushroom, milk-cup (*Lactarius*)
~ **че'рный** sombre lactarid, base toadstool, ugly toadstool, ugly milk-cup (*Lactarius necator*)
гру'минг *m* grooming behavior
грунт *m* ground
вя'зкий ~ sticky ground
гру'ппа *f* group
~ **дере'вьев** block of trees
~ **интерстери'льности** *myc* intersterility group
~ **кла'ссов** *bot* class group
~ **кро'ви** blood group
~ **кро'ви Да'ффи** Duffy's blood group
~ **кро'ви Ке'лла** Kell's blood group
~ **кро'ви Ки'дда** Kidd's blood group
~ **кро'ви Лютера'на** Lutheran's blood group
~ **молоды'х о'собей** juvenile group
~ **ри'ска** risk group, high-risk group
~ **самцо'в** all male group
~ **сцепле'ния** linkage group
алломимети'ческая ~ живо'тных allomimetic group
вне'шняя ~ outgroup
возрастна'я ~ age group
гаптофо'рная ~ haptophore group
монотети'ческая ~ monothetic group
нулева'я ~ zero group
полититети'ческая ~ polythetic group
простети'ческая ~ prosthetic group
разме'рная ~ size group
репродукти'вная ~ reproductive group
самцо'вая ~ all male group
токсофо'рная ~ toxophore group
успе'шно развива'ющаяся ~ successful group
группирова'ние *n* grouping
группирова'ть *v* group
группиро'вка *f* grouping, clustering
пионе'рная ~ pioneer community
гру'ша *f* pear (*Pyrus*)
груша'нка *f* shinleaf (*Pyrola*)
грушеви'дный *a* pyriform

гры'жа *f* hernia
~ **Вельпо'** Velpeau's hernia
~ **Тре'йца** Treitz's hernia
~ **невправи'мая** ~ irreducible hernia
ущемле'нная ~ incarcerated hernia, strangulated hernia
гры'жник *m* herniary (*Herniaria*)
грызе'ние *n* gnawing
грызть *v* gnaw
грызу'н *m* rodent (*Rodentia*)
гряда' *f agr* bed; bank (*of clouds*)
~ **барха'нов** row of sand dunes
го'рная ~ mountain range
гря'дка *f* bed
грязелече'ние *n* pelotherapy, mudtherapy
гуана'ко *n* guanaco (*Lama guanicoe*)
гуани'н *m* guanine
гуа'но *n* guano
гуанотрофи'я guanotrophy
гуанофо'р *m* guanophore
гуаха'ро *orn* guacharo *(Steatornis caripensis)*
губа' *f* lip, labium
~ **бластопо'ра** *emb* blastoporal lip
ве'рхняя ~ labrum; upper lip
за'ячья ~ cleft lip, hare lip
ни'жняя ~ labium; bottom lip, lower lip, under lip
половые губы *pl* vulvar lips
губа'ч *m* Indian bear *(Melursus)*
губерна'кулум *helm* gubernaculum
гу'бка *f zool* sponge (*Spongia*); *bot* tree fungus
~ **бере'зовая** birch polypore (*Piptoporus betulinus*)
~ **дубо'вая** maze-gill (*Daedalea quercina*)
~ **корнева'я** root fomes (*Heterobasidion annosum*)
губно'й *a* labial
губно-нё'бный *a* labiopalatine
губно-подборо'дочный *a* labiomental
губови'дный *a* labelloid
губодесново'й *a* labiogingival
губоно'гие *n pl* chilopodans, chilopods *(Chilopoda)*
губоцве'тный *a* labiate, labiatiflorous
гу'бчатый *a* spongiose, spongious, spongy; cancellate, cancellated, cancellous, fungoid, fungous
гу'бы *f pl* lips
гуде'ть *v* drone
гу'зка *f* rump
гуля'вник *m bot* loesel *(Sisymbrium)*
гуме'нник *m* bean goose (*Anser fabalis*)
гуми'новый *a* humic
гумифика'ция *f* humification
гумифици'рованный *a* humified
гуммо'з *phytop* gumming, gummosis
гумно' *n* threshing floor, barn
гу'мор *m* humor
гумора'льный *a* humoral
гу'мус *m* humus
гу'мусовый *a* humic
гумусообразова'ние *n* humus formation
гурт *m* herd
гуса'к *m* gander, he-goose
гу'сеница *f* caterpillar
волоса'тая ~ woolly bear caterpillar
гусё'нок *m* gosling
густо'й *a* thick, dense; luxuriant, lush (*of vegetation*)
~ **лес** dense forest
~ **густа'я трава'** thick grass
густы'е сли'вки *pl* thick cream
густопсо'вый *a* thick-haired (*of borzoi dogs*)
густота' *f* density
~ **древосто'я** thickness of stand
~ **стоя'ния ви'да** *bot* density
гусы'ня *f* goose
гусь *m* goose
голубо'й ~ blue goose *(Chen caerulescens)*
голубокры'лый ~ blue-winged goose *(Cyanochen cyanopterus)*
еги'петский ~ Egyptian goose *(Alopochen aegyptiaca)*
красноголо'вый ~ ruddy-headed goose *(Chloephaga rubidiceps)*
ма'лый ~ pygmy goose *(Nettapus)*
полула'пчатый ~ magpie goose *(Anseranas semipalmata)*
се'рый ~ gray goose, lag goose (*Anser anser*)
гуттаперчено'с *m* gutta-percha plant
гутта'ция *f* guttation
гу'ща *f* grounds (*of coffee*); **в гу'ще ле'са** in the thick of the forest
кофе'йная ~ coffee grounds
гюрза' *f* Levantine viper (*Vipera lebertina*)

Д

дава'ть *v* give
~ **лека'рство** administer a medicine
~ **плоды'** yield fruit
~ **отро'стки** flush
~ **побе'ги** sprig, flush
~ **по'чки** bud
~ **припло'д** breed
~ **ростки'** tiller
~ **урожа'й** yield, yield a harvest
дави'лка *f* dead fall
дави'ть *v* crush, press, squeeze, trample
давле'ние *n* pressure, tension
~ **большо'й чи'сленности населе'ния** population pressure
~ **в лё'гочной арте'рии** pulmonary pressure
~ **в ма'лом кругу'** pulmonary pressure
~ **диффу'зии** diffusion pressure
~ **набуха'ния** swelling pressure
~ **наполне'ния** filling pressure
~ **насыще'ния** saturation pressure
~ **отбо'ра** selection pressure, pressure of selection
~ **при вдо'хе** inspiratory pressure
~ **при вы'дохе** expiratory pressure
~ **расслабле'ния** *phys* relaxation pressure
~ **растворё'нного вещества'** solute pressure
артериа'льное ~ arterial pressure, arterial tension
биоосмоти'ческое ~ bioosmotic pressure
вено'зное ~ venous pressure, intravenous tension
внутрибрюши'нное ~ abdominal pressure
внутриглазно'е ~ intraocular pressure
внутрисерде'чное ~ intracardiac pressure
внутричерепно'е ~ intracranial pressure
гидростати'ческое ~ hydrostatic pressure
диастоли'ческое ~ diastolic pressure
капилля'рное ~ capillary pressure
коне'чное ~ end pressure
корнево'е ~ root pressure
крити'ческое ~ закры'тия critical closing pressure
кровяно'е ~ blood pressure
мутацио'нное ~ mutation pressure
онкотическое ~ oncotic pressure
осмоти'ческое ~ osmotic pressure, osmolarity
парциа'льное ~ partial pressure
пищево'дное ~ esophageal pressure
пульсово'е ~ pulse pressure
систоли'ческое ~ systolic pressure
со'бственно-внутрисосу'дистое ~ intrinsic blood pressure
ту'ргорное ~ turgor pressure
функциона'льное ~ functional pressure
эволюцио'нное ~ evolution pressure
эффекти'вное кровяно'е ~ effective blood pressure
эффекти'вное осмоти'ческое ~ effective osmotic pressure
дактилогнати'т *m zool* dactylognathite
дактилогра'мма *f* dactylogram
дактилогра'фия *f* dactylography
дактилоскопи'ческий *a* dactyloscopic
дактилоско'пия *f* dactyloscoy
дальнозо'ркий *a* long-sighted
дальнозо'ркость *f* farsightedness, longsightedness, hyperopia, hypermetropia
дама'н *m* hyrax *(Procavia)*
да'нные *pl* data, findings
~ **иссле'дования** findings
~ **лаборато'рного**
~**иссле'дования** laboratory findings
посме'ртные ~ postmortem findings
дансили'рование *n* dansylation
дарвини'зм *m* Darwinism
да'тчик *m* sensor, transducer
дистанцио'нный ~ remote sensor
пьезоэлектри'ческий ~ piezoelectric transducer
фотоэлектри'ческий ~ photoelectric sensor
двенадцатипе'стичный *a* dodecagynous
двенадцатитычи'нковый *a* dodecandrous
дви'гательный *a* motive, motorial
движе'ние *n* movement, motion,locomotion
~ **наме'рения** intention movement
~ **по спира'ли** torsion movement
~ **по ти'пу «флип-флоп»** flip-flop motion
~ **сна'** *bot* sleeping movement
амебо'идное ~ ameboid locomotion, ameboid motion
анафа'зное движение anaphase movement
взаимосвя'занные движе'ния *pl* associated movements

враща'тельное ~ rotatory motion, gyration, circular movement, rolling movement
вырази'тельное ~ expressive movement
глота'тельное ~ swallowing movement
дыха'тельное ~ respiratory movement
индуци'рованное ~ induced movement
кардиопневмати'ческое ~ cardiopneumatic movement
комфо'ртное ~ *eth* comfort movement
морфогенети'ческое ~ morphogenetic movement
нача'льное ~ intention movement
ориенти'рующее ~ orienting movement
парадокса'льное ~ paradoxic movement
пищедобыва'тельное ~ food-procuring movement
произво'льное ~ voluntary movement
просле'живающее ~ глаз pursuit eye movement
ресни'чное ~ ciliary locomotion, ciliary motion
рефлекто'рное ~ reflex movement
саккади'ческое ~ глаз saccadic eye movement
сигна'льное ~ signal movement
скользя'щее ~ gliding motion
содру'жественное ~ associated movement, synkinesis
торсио'нное ~ torsion movement
то'чное ~ skilled movement
уступа'ющее ~ relaxed movement
у'стьичное ~ *bot* stomatal movement
формати'вное ~ formative movement
хореифо'рмное ~ choreiform movement
целенапра'вленное ~ purposive movement

двойно'й *a* double, binary, binate, geminate, geminated
дво'йня *f* twins
двор *m* yard
пти'чий ~ poultry yard, fowl-yard
ско'тный ~ farmyard
дво'рик *m*: ~ **у'стьица** stomatal chamber
дворня'га, дворня'жка *f* mongrel (*a* dog)
двоякоды'шащее *n* dipnoan
двояко- bi-, di-
двояково'гнутый *a* biconcave
двояковы'пуклый *a* biconvex
двоякоды'шащий *a* amphipneustic; dipnoan
двоякозазу'бренный *a* bicrenate
двоякозу'бчатый *a* duplicodentate
двоякоизо'гнутый *a* birecurvate
двоякопи'льчатый *a* doubly serrate
двоякопреломля'ющий *a* birefringent
двоякопе'ристо-надре'зный *a bot* bipinnatifid
двоякопе'ристо-разде'льный *a bot* bipinnatipartite
двоякопе'ристо-рассечё'нный *a bot* bipinnatisect
двоякопе'ристый *a* bipinnate
двоякопи'льчатый *a* biserrate
дву- two-, di-, dupli-, double-
двуаперту'рный *a* ditreme
двуборо'здный *a* disulcate
двубра'тственный *a* diadelphous
двубра'чный *a* bigamous
двубрю'шный *a anat* biventral
двуверши'нный *a* bicuspid
двуветви'стый *a* biramous
двугла'вый *a* bicapitate, bicephalous, bicipital
двугла'вая мы'шца biceps, biceps muscle
двугне'здный *a* bilocular, biloculate, biloculine, dithecal
двуго'рбый *a* two-humped
двугребе'нчатый *a* bipectinate, two-combed, two-crested
двугу'бый *a* bilabiate, bilabial
двудо'льный *a* dicotyledonous; bilobate, bilobated, bilobular
двудо'мность *f* dioecy
двудо'мный *a* diecious, dioecious, diclinous
двужа'берный *a* dibranchiate
двужеле'зистый *a bot* biglandular, two-glanded
двузао'стренный *a* biacuminate
двузу'бчатый, двузу'бый *a* bidental, bidentate
двука'псульный *a* bicapsular
двука'рпельный *a* bicarpellate
двукилево'й *a* bicarinate
двукле'точный *a* bicellular
двукоготко'вый *a* biunguiculate
двукоро'бочный *a* bicapsular
двуко'сточковый *a* dipyrenous
двукра'тно-закру'ченный *a* double-twisted
двукры'лый *a* dipterous
двулепе'стный *a* bipetalous, dipetalous
двуле'тний *a* biennial
двуле'тник *m* biennial
двули'стный *a* bifoliate, diphyllous
двумешко'вый *a* bisaccate, disaccate

двунадре'зный *a* bifid
двунапра'вленный *a* bidirectional
двуно'гий *a* biped
двунадре'зный *a* bifid
двунитево'й *a* double-stranded
двуно'гий *a* biped
двуно'гое *s* biped
двуости'стый *a* biaristate
двупа'лый *a* didactyl
двупа'рный *a* bijugate, bigeminate
двупе'ристо-надре'зный *a* bipinnatifid
двупе'ристо-разде'льный *a* bipinnatipartite
двупе'ристо-рассечё'нный *a* bipinnatisect
двупе'ристый *a bot* bipinnate
двупе'стичный *a* digynous
двупи'льчатый *a* biserrate
двупласти'нчатый *a* bilamellar, bilaminar, bilaminated
двуплё'нчатый *a* bipaleolate
двуплодоли'стиковый *a* bicarpellate
двупо'лый *a* bisexual
двупо'ровый *a* diporate
двупо'чечный *a bot* bigemmate
двуприцве'тниковый *a* bibracteate
двупятни'стый *a* bimaculate, bimaculated
двуразде'льный *a* bipartite, bifid, biseptate
двурассече'нный *a* bisectional
двуоре'шковый dicoccous
двуро'гий *a* bicornute, bicornous, bicorn, two-horned
двуру'кий *a* bimanous
двусемя'нка *f* diachenium
двусемя'нный *a* dispermous, two-seeded
двуси'льный *a* didymous
двускла'дчатый *a* biplicate
двусло'жный *a* binate
двуспира'льный *a* double-helical
двуспо'ровый *a* disporous
двуство'рчатый *a* bivalve, bivalved, bivalvular
двусто'лбчатый *a pal* duplibaculate
двусторо'нний *a* bilateral, bifacial, ambilateral
двуто'чечный *a* bipunctate
двутычи'нковый *a* diandrous
двуу'стка *f* fluke
  коша'чья ~ Siberian liver fluke *(Opisthorchis felineus)*
  пече'ночная ~ liver fluke (*Fasciola hepatica*)
двуутро'бный *a* didelphic
двуу'хий *a* biauriculate
двух- bi-, di-, two-, double-
двухборо'здный *a* bisulcate, bisulcated
двухвале'нтный *a* bivalent
двухверши'нный *a* bicuspid; bimodal
двухви'льчатый *a* two-forked
двухгодови'к *m ich* two-year-old fish
двухжгу'тиковый *a* bicilate, biciliated, biflagellate
двухжи'лковый *a bot* binervate, two-nerved, two-veined
двухзо'нтичный *a bot* biumbellate
двухка'мерный *a* bilocular, dithecal, bicameral, two-celled
двухкла'панный *a* bivalvular
двухкле'точный *a* two-celled
двухколю'чковый *a* bispinose
двухло'пастный *a* bilobate, bilobated, bilobular
двухлучево'й *a* biradial, diactinal
двухме'рный *a* two-dimensional
двухме'сячный *a* bimestrious
двухнитево'й *bioch a* two-stranded
двухоболо'чковый *a* bitunicate
двухосно'вный *a* bibasic
двухполуборо'здный *a pal* diplodemicolpate
двухпо'люсный *a* bipolar
двухря'дный *a* two-rowed, two-ranked, biserial, displostichous
двухсло'йный *a* two-layer, two-layered, bilamellar, bilaminar, bilaminate, bilaminated
двухстади'йный *a* two-stage
двухступе'нчатый *a* two-stage
двухсуста'вной *a* biarticulate, biarticulated
двухфа'зный *a* biphase
двухцветко'вый *a* biflorate, biflorous
двухцве'тный *a* bicolorous, bicolor, dichromatic
двухцепо'чечный *a bioch* double-stranded, two-stranded
двучле'нистый *a* biarticulate, biarticulated
двухэта'пный *a* two-stage
двухъя'дерный *a* binuclear, binucleate
двухъя'русный *a* two-layered, two-storied
двуча'шечный *a* bicalyculate
двушпо'рцевый *a* bicalcarate
двуяйцево'й *a* biovular, biovulate, dizygotic
ДДТ 1,1,1-трихлор-2,2-бис (п-хлорфенил)этан *m* DDT 1,1,1-trichloro-2,2-bis (p-chlorophenyl)ethane
де- de-
деадапта'ция *f* deadaptation

деамина'за *f* deaminase
деваста'ция *f* devastation
дево'н *m* Devonian
де'вственный *a* virgin, primeval (*e.g., forest*)
девяси'л *m* elecampane (*Inula*)
дегельминтиза'ция *f* dehelmintization
дегенерати'вный *a* degenerative
дегенера'ция *f* degeneration, degeneracy
~ **паре'нхимы** parenchymatous degeneration
жирова'я ~ fatty degeneration
ретрогра'дная ~ retrograde degeneration
травмати'ческая ~ traumatic degeneration
транснейро'нная ~ transneuronal degeneration
дегидра'ция *f* dehydration
дегидрогена'за *f* dehydrogenase
дегенери'ровать *v* degenerate
де'готь *m* tar, coal-tar, wood-tar
деграда'ция *f* degeneration, degradation
дезактива'ция *f* decontamination, deactivation
дезинва'зия *f* disinvasion
дезинсе'кция *f* disinsection, disinsectization
дезинтегра'ция *f* disintegration
дезинфе'кция *f* disinfection
дезодора'ция *f* deodoration
дезоксирибонуклеа'за *f* deoxyribonuclease
де'йствие *n* action, effect
~ **излуче'ния** radiation effect
аллергоге'нное ~ allergenic effect, allergenetic effect
биологи'ческое ~ biological effect
биохими'ческое ~ biochemical effect
гербици'дное ~ herbicidal effect
гормона'льное ~ hormonal action, hormonal effect
заде'рживающее ~ inhibitory action
избира'тельное ~ selective effect
иммунодепресси'вное ~ immunodepressive effect
ингиби'рующее ~ inhibiting effect, inhibitory effect, inhibitory action
комбини'рованное ~ combined effect
кооперати'вное ~ cooperative effect
кумуляти'вное ~ cumulative action, cumulative effect
о'бщее ~ systemic effect
оптима'льное ~ optimal effect
оста'точное ~ residual effect
отдалё'нное ~ remote effect
пермисси'вное ~ permissive effect
побо'чное ~ side effect
подавля'ющее ~ inhibitory action
приторма'живающее ~ retarding effect
психи'ческое ~ mental effect
психомото'рное ~ psychomotor action
регули'рующее ~ controlling effect
рефлекто'рное ~ reflex action; reflex effect
синерги'ческое ~ synergic effect
систе'мное ~ systemic action
совме'стное ~ cooperative effect
соче'танное ~ joint effect
специфи'чески динами'ческое ~ specific dynamic action
стимули'рующее ~ promoting effect, stimulating effect
тормозя'щее ~ inhibitory action
эконо'мящее ~ sparing effect
дейтерогене'з *m* deuterogenesis
дейтеротоки'я *f* deuterotoky
дейтероко'н *m* deuterocone
дейтомери'т *m* deutomerite
дейтопла'зма *f* deutoplasm
дейтоцере'брум *m* deutocerebrum
декапита'ция *f* decapitation
декарбоксила'за *f* decarboxylase
декарбоксили'рование *n* decarboxylation
окисли'тельное ~ oxidative decarboxylation
декоди'рующий *a* decoding
декомплементи'ровать *v* decomplementize
декомпре'ссия *f* decompression
взрывна'я ~ explosive decompression
внеза'пная ~ explosive decompression
преры'вистая ~ interrupted decompression
деконъюга'ция *f* deconjugation
декорати'вный *a* ornamental, fancy
декортика'ция *f* decortication
декреме'нт *m* decrement
дексиотро'пный *a* dexiotropic, dextrorse
декстра'льный *a* dextral
декстри'н *m* dextrin, dextrine
декстрове'рсия *f* dextroversion
декстро'за *f* dextrose, dextroglucose
декумба'ция *f* decumbation
~ **я'русов** decumbation of layers
деламина'ция *f* delamination
де'лать *v* make; do
~ **ло'гово** bed
~ **сто'йку** set, standard (*of a dog*)
деле'ние *n* division
~ **кле'тки** cell division, cell fission

~ **на'двое** binary fission
~**созрева'ния** maturation division
второ'е ~ в мейо'зе homeotypic division
гомеотипи'ческое ~ homeotypic division
мейоти'ческое ~ meiotic division
митоти'ческое ~ mitotic division
мно'жественное ~ multiple fission
попере'чное ~ transverse fission
прямо'е ~ amitotic division
редукцио'нное ~ meiosis
эквацио'нное ~ equation division

деле'ция *f* deletion
~ **кло'на** elimination of clone, deletion of clone, clonal deletion
антиге'нная ~ antigenic deletion

де'ло *n*:
запове'дное ~ reserve management and studies
лесно'е ~ forestry

дельфи'н *m* dolphin (*Delphinidae*)

деля'нка *f* plot
контро'льная ~ control plot, check plot
основна'я ~ main plot
расщеплё'нная ~ split plot
семенна'я ~ seed plot
учё'тная ~ record plot

демагнифика'ция *f* demagnification

демерса'льный *a* demersal

демиелиниза'ция *f* demyelinization

демогра'фия *f* demography

демонстра'ция *f eth* advertisement, display
~ **гениталий** genital display
~ **оска'ла** bared-teeth display
~, **отвлека'ющая внима'ние** distraction display
~ **па'ховой о'бласти** inguinal display
~ **угро'зы** threat display
бокова'я ~ broadside display, side display, lateral display
запу'гивающая ~ intimidation display
отвлека'ющая внима'ние ~ distraction display
социа'льная ~ social display
эпидейкти'ческая ~ epideictic display

денатура'ция *f* denaturation
кисло'тная ~ acid denaturation
теплова'я ~ thermal denaturation
щелочна'я ~ alkali denaturation

дендра'рий *m* arboretum

дендри'т *m* dendrite

дендри'тный *a* dendritic

дендро- dendro-

дендробио'нт *m* dendrobiont

дендрогра'мма *f* dendrogram

дендроло'гия *f* dendrology

дендроме'тр *m* dendrometer

дендроме'трия *f* dendrometry

дендрофи'л *m* dendrophil

дендрофи'льный *a* dendrophilous

дендрохроноло'гия *f* dendrochronology

дендрофло'ра *f* dendroflora

денерви'рованный *n* denervated

денитрифика'тор *m* denitrifier

денитрифика'ция *f* denitrification

денти'н *m* dentin, dentine

деонтоло'гия *f* deontology

депигмента'ция *f* depigmentation

депиля'ция *f* depilation

депо' *n* depot
жирово'е ~ fat depot

депо'-адъюва'нт *m* depot-forming adjuvant

деполяриза'ция *f* depolarization
ни'зкая ~ above-threshold depolarization

депресса'нт *m* depressant

депресси'вный *a* depressive

депре'ссия *f* depression, dejection
ано'дная ~ *phys* anode depression
инбре'дная ~ inbreeding depression
не'рвно-мы'шечная ~ neuromuscular depression
распространя'ющаяся ~ spreading depression
распространя'ющаяся ~ по Леао Leao's depression

депре'ссор *m* depressor

деприва'ция *f* deprivation
зри'тельная ~ visual deprivation
монокуля'рная ~ monocular deprivation
пищева'я ~ food deprivation
предме'тная ~ deprivation of pattern vision
сенсо'рная ~ sensory deprivation

депси'д *m* depside

де'пула *f* depula

дератиза'ция *f* deratization

де'рбник *m* merlin, pigeon hawk (*Falco coumbarius*)

дерга'ч *m* corncrake, landrail (*Crex crex*)

де'ргающийся *a* working (*e.g., of cheeks*), twitching, jerking

дереве'й *m* yarrow (*Achillea*)

де'рево *n* tree; wood; arbor
~ **с веерообра'зной кро'ной** fan-trained tree
~ **на опу'шке** border tree

~ **путеше'ственников** traveller's tree *(Ravenala madagascariensis)*
ветвя'щееся ~ branching tree
гвозди'чное ~ clove tree *(Syzigium aromaticum)*
декорати'вное ~ ornamental tree
дупли'стое ~ hollow tree
ды'нное ~ **обыкнове'нное** melon tree *(Carica papaya)*
зави'сшее ~ lodged tree
заподсо'ченное ~ tapped tree
иу'дино ~ Judas tree *(Cercis siliquastrum)*
клеймё'ное ~ marked tree
колба'сное ~ sausage tree *(Kigelia aethiopica)*
корнео'тпрысковое ~ burr wood
кофе'йное ~ coffee tree *(Coffea arabica)*
ли'ственное ~ foliage tree
ло'вчее ~ decoy tree
неветвя'щееся ~ tufted tree
низкоро'слое ~ bushy tree
орехопло'дное ~ nut tree
отде'льно расту'щее ~ single tree
плаку'чее ~ weeping tree
плодо'вое ~ fruit tree
подчинё'нное ~ subordinate tree
раски'дистое ~ shaggy tree
рожко'вое ~ carob tree *(Ceratonia siliqua)*
светолюби'вое ~ light-demanding wood
семенно'е ~ seed tree, mother tree, parent tree
суховерши'нное ~ stag-headed tree
сухосто'йное ~ dead standing tree
теневыно'сливое ~ shade-bearing wood
угнетё'нное ~ suppressed tree
фа'утное ~ defective tree
флагообра'зное ~ flag-shaped tree
формо'ванное ~ shaped tree, clipped tree
хво'йное ~ conifer
хле'бное ~ **цельноли'стное** breadfruit tree *(Artocarpus integrifolia)*

деревяни'стый *a* ligneous
деревя'нный woody, wooden; ligneous
дереза' *f* box-thorn (*Lycium*)
де'рен *m* dogwood (*Cornus*)
держа'ть *v* keep; ~ **соба'к** keep dogs
держиде'рево *n* Christ's-thorn (*Paliurus spina-christi*)
де'рма *f* derm, derma
дерма'льный *a* dermic
дерматоге'н *m* dermatogen
дерматоло'гия *f* dermatology
дермато'м *m* dermatome
дерматомико'з *m* dermatomycosis
дерматофи'т *m* dermatophyte
дермобла'ст *m* dermoblast
дермографи'зм *m* dermographism
дерн *m* sod, turf
дерни'на *f* sod, sod mat, grassy turf, sward
пло'тная ~ heavy sod, dense turf
дерни'стый *a* cespitose, caespitose, swardy, turfy
дернова'ние *n* sodding
дернови'на *f* sod, sod mat, grassy turf, sward
дерно'вый *a* soddy, turfy; cespitose, caespitose
дернообразова'ние *n* sod formation
дернообразу'ющий *a* sod-forming, turf-forming
деря'ба *f* missel, missel thrush, mistle thrush *(Turdus viscivorus)*
десенсибилизация *f* desensitization
перекрё'стная ~ cross desensitization
десквама'ция *f* desquamation
десми'н *m* desmin
десмогнати'ческий *a* desmogmathous
десмо'з *m* desmose
десмо'н *m* desmone
десмосо'ма *f* desmosome
десмотру'бочка *f* desmotubule
десмоци'т *m* desmocyte
десмурги'я *f* desmurgy
десна' *f* gum, gingival
десо'рбция *f* desorption
полева'я ~ field desorption
дестробиля'ция *f* destrobilation
дестру'кция *f* destruction
дете'ктор *m* detector
~ **дви'жущихся грани'ц** moving-edge detector
~ **о'бщего затемне'ния** net dimming detector
~ **дли'тельного контраста** sustained contrast detector
детерге'нт *m* detergent
детермина'нт *m* determinant
детермина'нта *f* determinant
~ **кле'точной пове'рхности** cell-surface epitope
~ **конформацио'нного ти'па** conformation-dependent determinant
~ **непреры'вного ти'па** continuous determinant, sequential determinant

~ **преры'вистого ти'па** discontinuous determinant
~ **секвенциа'льного ти'па** continuous determinant, sequential determinant
аллоантиге'нная ~ allodeterminant
антиге'нная ~ antigenic determinant, epitope, antigenic epitope
антиге'нная ~ моле'кулы-носи'теля carrier determinant
антиге'нная ~ о'стова framework epitope
B-кле'точная антиге'нная ~ B-cell epitope
вариотипи'ческая ~ variotypic epitope, framework epitope
вируснейтрализу'ющая антиге'нная ~ virus neutralizing epitope
видоспецифи'ческая антиге'нная ~ species-specific epitope
гапте'новая ~ haptenic determinant
группоспецифи'ческая антиге'нная ~ group-specific epitope
досту'пная антиге'нная ~ accessible epitope
карка'сная антиге'нная ~ framework epitope
клонотипи'ческая ~ антите'ла clonotypic determinant
ключева'я антиге'нная ~ immunodominant epitope
консервати'вная антиге'нная ~ conserved epitope
конформацио'нно-зави'симая ~ conformational determinant
маскиро'ванная ~ covert determinant, hidden determinant
маскиро'ванная антиге'нная ~ masked epitope
мута'нтная антиге'нная, мути'ровавшая антиге'нная ~ mutant epitope
неза'нятая ~ free epitope
незащищё'нная антиге'нная ~ accessible epitope
неэкрани'рованная антиге'нная ~ accessible epitope
низкоиммуноге'нная ~ bystander epitope
основна'я антиге'нная ~ immunodominant epitope
откры'тая антиге'нная ~ accessible epitope
перекрыва'ющиеся антиге'нные детермина'нты *pl* overlapping epitopes
пла'вающая антиге'нная ~ mobile epitope
пове'рхностная антиге'нная ~ cell-surface epitope
повторя'ющаяся антиге'нная ~ repeating epitope
подви'жная антиге'нная ~ mobile epitope
приро'дная антиге'нная ~ natural epitope
реактоге'нная ~ free epitope
рецепторсвя'зывающая антиге'нная ~ receptor binding epitope
свобо'дная ~ free epitope
скеле'тная антиге'нная ~ framework epitope
скры'тая антиге'нная ~ masked epitope
скры'тая ~ covert determinant, hidden determinant
сло'жная ~ compound determinant
типоспецифи'ческая антиге'нная ~ type-specific epitope
углево'дная ~ carbohydrate epitope
уника'льная антиге'нная ~ unique antigenic determinant, unique epitope
штаммоспецифи'ческая антиге'нная ~ strain-specific epitope
эволюцио'нно стаби'льная антиге'нная ~ conserved epitope

детермина'ция *f* determination
динами'ческая ~ dynamic determination
лаби'льная ~ labile determination
отрица'тельная ~ negative determination
прогресси'вная ~ progressive determination

детё'ныш *m* cub, baby, young
де'тка *f bot* daughter bulb, bulbil, ratoon
детоксика'ция *f* detoxication, detoxification
детоуби'йство *n eth* infanticide, pup killing
детри'т *m* detritus
детритоя'дный *a* detritus-consuming, detrivorous
дефека'ция *f* defecation
дефибрилля'тор *m* defibrillator
дефибрини'рование *n* defibrination
дефинити'вный *a* definitive
дефици'т *m* deficit
~ **вла'жности** saturation deficit
~ **давле'ния диффу'зии** diffusion pressure deficit, turgor deficit
~ **насыще'ния** turgor deficit

дефолиа'нт *m* defoliant
дехроматиза'ция *f* dechromatization
де'ятельность *f* activity, performance

~ **се'рдца** cardiac performance
вы'сшая не'рвная ~ higher nervous activity
не'рвно-психи'ческая ~ neuro-psychic activity
рефлекто'рная ~ reflex activity
джейра'н *m* goitred gazelle, granulated gazelle, Persian gazelle (*Gazella subgutturosa*)
джек *m* houbara bustard (*Clamydotis undulata*)
джу'нгли *pl* jungle
джут *m* jute (*Corchorus*)
дзере'н *m* zeren, black-tailed gazelle (*Procarpa gutturosa*)
монго'льский ~ Mongolian gazelle *(Procapra gutturosa)*
ди- di-
диа- dia-
диагини'ческий *a* diagynic
диа'гноз *m* diagnosis
дифференциа'льный ~ differential diagnosis
диагно'стика *f*
листова'я ~ **пита'ния** foliar diagnostics of nutrition
диагности'ческий *a* diagnostic
диагра'мма *f* diagram
~ **Ве'нна** *phys* Venn diagram
~ **разбро'са** *biom* scatter diagram
~ **рассе'ивания** *biom* scatter diagram
~ **цветка'** floral diagram
мутацио'нная ~ **Да'йхоффа** Dayhoff's mutational diagram
пыльцева'я ~ pollen diagram
диа'да *f* diad
диадро'мный *a* diadromous
диазотрофи'я *f* diazotrophy
диакине'з *m* diakinesis
диа'лиз *m* dialysis
равнове'сный ~ equilibrium dialysis
диандри'ческий *a* diandric
диапа'уза *f* diapause
диапеде'з *m* diapedesis
диартро'з *m* diarthrosis
диаста'з *m* diastase
диасте'ма *f* diastem, diastema
диа'стер *m* diaster
диа'стола *f* diastole
~ **предсе'рдий** auricular diastole
~ **желу'дочков** ventricular diastole
диастоли'ческий *a* diastolic
диате'з *m* diathesis
диатерми'я *f* diathermy
диатоме'я *f* diatom (*Bacillariophyta*)
диатоми'н *m* diatomin
диатоми'т *m* diatomite, diatomaceous earth
диато'мовый *a* diatomaceous
диатропи'зм *m* diatropism, transverse tropism
диатропи'ческий *a* diatropic
диауксі'я *f* diauxia
диафра'гма *f* diaphragm; core wall (*of a dam*); lens stop, stop
диафрагма'льно-ободо'чный *a* phrenicocolic
диафрагма'льно-плевра'льный *a* phrenicopleural
диафототропи'зм *m* diaphototropism
дива'ла *f bot* knavel (*Scleranthus*)
диверге'нтный *a* divergent
диверге'нция *f* divergence
~ **популя'ций** population divergence
диверги'рующий *a* divergent
диверти'кул *m* diverticulum
~ **Ме'ккеля** Meckel's diverticulum
дигамети'ческий *a* digametic
дигапло'ид *m* dihaploid
дигенети'ческий *a* digenetic
диге'нный *a* digenic
дигеномати'ческий *a* digenomatic
дигибри'д *m* dihybrid
дигитоксико'за *f* digitoxicose
дигитокси'н *m* digitoxin
дигитони'н *m* digitonin
дигокси'н *m* digoxin
дигони'ческий *a* digonic
дигонопо'ровый *a* digonoporous
Дие'го-антиге'н *m* Diego antigen
дие'та *f* diet
~ **Ба'линта** Balint's diet
~ **Джоване'тти** Giovanetti's diet
~ **Мейленгра'хта** Meulengracht's diet
безбелко'вая ~ protein-free diet
бессолева'я ~ low-salt diet, salt-free diet
вегетариа'нская ~ vegetarian diet
высокобелко'вая ~ high-ptotein diet
высокалори'йная ~ high-calorie diet
диабети'ческая ~ diabetic diet
закрепля'ющая ~ constipating diet
моло'чная ~ milk diet
овощна'я ~ vegetable diet
послабля'ющая ~ laxative diet

разгру'зочная ~ fasting diet, unloading diet
стро'гая ~ rigid diet, strict diet, meagre diet, spare diet
фрукто'вая ~ fruit diet
щадя'щая ~ sparing diet, bland diet, light diet, smooth diet
диз- dis-
дизентери'я *f* dysentery
аме'бная ~ amoebic dysentery
дизиготи'ческий *a* dizygotic
дизъюнкти'вный *a* disjunctive
дизъю'нкция *f* disjunction
дикарио'н *m* dikaryon, dicaryon
дикариотиза'ция *f* dikaryotization, dicaryotization
дикариофа'за *f* dikaryophase, dicaryophase
дикариофи'т *m* dicaryophyte, dikariophyte
ди'кий *a* wild; tameless
дикобра'з *m* porcupine (*Hystrix*)
дикорасту'щий *a* wild
дикро'та *f* dicrotic wave, recoil wave
дикроти'зм *m* dicrotism
диктиокине'з *m* dictyokinesis
диктиосо'ма *f* dictyosome
диктиоспо'ра *f* dictyospore
дилато'метр *m* dilatometer
дим *m* deme
диме'р *m* dimer
димериза'ция *f* dimerization
диме'рный *a* dimeric
дими'ксис *m* dimixis
димину'ция *f* diminution
~ **хромати'на** chromatin diminution
~ **хромосо'м** chromosome diminution
димити'ческий *a bot* dimitic
диморфи'зм *m* dimorphism
полово'й ~ sex dimorphism, sexual dimorphism
сезо'нный ~ seasonal dimorphism
я'дерный ~ nuclear dimorphism
димо'рфный *a* dimorphic, dimorphous
дина'мика *f* dynamics
~ **популя'ции** population dynamics
~ **чи'сленности** dynamics of abundance, population dynamics
гамильто'нова ~ Hamiltonian dynamics
сезо'нная ~ seasonal distribution
ди'нго *m, f* dingo (*Canis dingo*)
диоксигена'за *f* dioxygenase
диокси'н *m* dioxin
дипланети'зм *m* diplanetism
дипле'врула *f* dipleurula
дипло- diplo-
диплобио'нт *m* diplobiont
дипло'з *m* diplosis
дипло'ид *m* diploid
двойно'й ~ double diploid
части'чный ~ partial diploid
диплоидиза'ция *f* diploidization
диплоиди'я *f* diploidy
дипло'идный *a* diploid
диплокарио'н *m* diplokaryon
диплоко'кк *m* diplococcus
диплоксили'ческий *a* diploxylic
диплоне'ма *f* diplonema
дипло'нт *m* diplont
диплорнави'рус *m* diplornavirus
диплосо'ма *f* diplosoma
диплостемо'нный *a* diplostemonous
диплоте'на *f* diplotene
диптеро'идный *a* dipteroid
диптеро'лог *m* dipterologist
дисахари'д *m* disaccharide
дисбактерио'з *m* dysbacteriosis
дисбала'нс *m* imbalance
иммунорегулято'рный Т-кле'точный ~ immunoregulatory T-cell imbalance
дисге'незис, **дисгенези'я** *m* dysgenesia, dysgenesis
диск *m* disk, disc
~ **А** A-disk, Q-disk
~ **Ме'ркеля** Merkel's disk, tactile disk
~ **Се'кки** Secchi's disk
анизотро'пный ~ A-disk
вста'вочный ~ intercalated disk
заро'дышевый ~ blastodisk, embryonic disk, germ disk, germinal disk
изотро'пный ~ I-disk, isotropic disk, isotropous disk
имагина'льный ~ imaginal disk
попере'чный ~ transverse disk
приса'сывательный ~ suctorial disk
све'тлый перви'чный ~ primary light disk
тёмный перви'чный ~ primary dark disk
я'дерный ~ nuclear disk
ди'ск-электрофоре'з *m* disk electrophoresis
дискли'макс *m* disclimax
дискобласти'ческий *a* discoblastic
дискобла'стула *f* discoblastula
дискова'ние *n* disk tillage
дискови'дный *a* discal, discoid, discoidal
дискомфо'рт *m* discomfort
дископла'зма *f* discoplasm

дискоцветко'вый *a* disciflorous
дискрази'я *f* dyscrasia
дискре'тность *f*:
~ **расти'тельного покро'ва** discontinuity of vegetation
дискримина'тор *m* discriminator
дисморфокариоци'т *m* dysmorphokaryocyte
дисмута'ция *f* dismutation
дисо'мик *m* disomic
диспара'тный *a* disparate
дисперги'рование *n* dispersion
дисперми'я *f* dispermia
диспе'рсия *f* variance; dispersion
~ **вы'борки** sampling variance
~ **оптического вращения, ДОВ** optical rotatory dispersion
~ **оши'бки** error variance
аддити'вная ~ additive variance
генотипи'ческая ~ genotypic variance
о'бщая ~ total variance
оста'точная ~ residual variance
диспе'рсность *f* dispersity
диспире'ма *f* dispireme
дисплази'я *f* dysplasia
~ **Я'ффе-Лихтенста'йна** Jaffe-Lichtenstein disease
дисплоиди'я *f* dysploidy
диссе'кция *f* dissection
диссимиляцио'нный *a* dissimilatory
диссимиля'ция *f* dissimilation
диссогени'я *f* dissogeny
диссоко'нх *m* dissoconch
диссофи'т *m* dissophyte
дистальноборо'здный *a* anacolpate
дистальнопо'ровый *a* anaporate
диста'льный *a* distal
диста'нция *f* distance
~ **бе'гства** flight distance
индивидуа'льная ~ individual distance
патристи'ческая ~ patristic distance
флористи'ческая ~ floristic distance
дистрофика'ция *f* dystrophication
дистрофи'я *f* dystrophia, dystrophy
алимента'рная ~ nutritional dystrophy
дистро'фный *a* dystrophic
дисфу'нкция *f* disfunction
дисхро'нный *a* dischronous
диуре'з *m* diuresis
диурети'ческий *a* diuretic
дифазми'да *f* diphasmide
дифенилами'н *m* diphenylamine
дифилети'ческий *a* diphyletic
дифиодо'нтный *a* diphyodont
дифицерка'льный *a* diphycercal
дифференциа'л *m* differential
селекцио'нный ~ selection differential
дифференциа'льный *a* differential
дифференциа'ция, дифференциро'вка *f* differentiation
~ **осевы'х структу'р** axial differentiation
~ **яи'чников** ovarian differentiation
гистологи'ческая ~ histological differentiation
зави'симая ~ dependent differentiation
кле'точная ~ cell differentiation, cellular differentiation
незави'симая ~ independent differentiation
овариа'льная ~ ovarian differentiation
полова'я ~ sex differentiation
простра'нственная ~ regional differentiation
региона'льная ~ regional differentiation
термина'льная ~ terminal differentiation
функциона'льная ~ functional differentiation, corporative differentiation
эритро'идная ~ erythroid differentiation
диха'зий *m* dichasium, dichasial cyme
дихога'мия *f* dichogamy
дихопти'ческий *a* dichoptic
дихоти'ческий *a* dichotic
дихотоми'ческий *a* dichotomous
дихотоми'я *f* dichotomy
дихофи'з *m* dichophysis
дихрои'зм *m* dichroism
кругово'й ~ circular dichroism
магни'тный кругово'й ~ magnetic circular dichroism
колеба'тельный кругово'й ~ vibrational circular dichroism
дихрои'ческий *a* dichroic
дихромази'я *f* dichromasy
дице'нтрик *m* dicentric
дицикли'чный *a* dicyclic
дичеразведе'ние *n* game breeding
дичь *f* game, wildlife; wild fowl
борова'я ~ upland fowl
водопла'вающая ~ waterfowl
кру'пная ~ big game
ме'лкая ~ small game
перна'тая ~ fowl, wild fowl, game birds, feather game
дичо'к *m* wilding, wildling
длaневи'дный *a* palmate, palmatiform
длина' *f* length

~ **всего' те'ла** total length
~ **дня** day length
~ **те'ла** length of the body
абсолю'тная ~ total length
крити'ческая ~ дня critical day length
длинно- dolicho-
длинноволокни'стый *a* long-stapled
длинноголо'вый *a* dolichocephalic
длиннодне'вный *a bot* long-day
длинноклю'вый *a* long-billed, longirostral
длинноколо'сый *a* long-spiked, dolichostachous
длинноколючковый *a* long-thorned
длиннокры'лый *a* longipennate
длиннолепе'стный *a* long-petaled
длинноли'стный *a* long-leaved
длинноли'цый *a* dolichofacial
длинноно'гий *a* long-legged, leggy
длиннопе'стичный *a bot* long-styled, long-pistillate
длиннопло'дный *a* dolichocarpous
длинноро'гий *a* longicorn
длинноство'льный *a* long-boled, long-stemmed
длинностебе'льчатый *a* long-stalked
длинносто'лбчатый *a* dolichostylous
длинноу'сый *a ent* longicorn
длинноу'хий *a* long-eared
длиннохво'стый *a* long-tailed
длинноцветко'вый *a* long-flowered
длинноше'ий *a* long-necked
длинноше'рстный *a* long-haired, long-wooled, shaggy
ДНК дезоксирибонуклеи'новая кислота' *f* DNA deoxyribonucleic acid
бесполе'зная ~ junk DNA
бессмы'сленная ~ junk DNA
ве'кторная ~ vector DNA
внехромосо'мная ~ extrachromosomal DNA, extranuclear DNA
внея'дерная ~ extranuclear DNA, extrachromosomal DNA
гено'мная ~ genomic DNA
гетероду'плексная ~ heteroduplex DNA
двуни'тевая ~ double-stranded DNA
двуни'тевая лине'йная ~ linear duplex DNA
двухцепо'чечная ~ double-stranded DNA
денатури'рованная ~ denatured DNA
депротеинизи'рованная ~ naked DNA
за'мкнутая кольцева'я ~ closed circular DNA
кинетопла'стная ~ kinetoplast DNA
кольцева'я ~ circular DNA
комплемента'рная ~ complementary DNA (cDNA)
лине'йная ~ linear DNA
ма'тричная ~ template DNA
метили'рованная ~ methylated DNA
ме'ченая по конца'м ~ end-labeled DNA
митохо'ндриальная ~ (мт ДНК) mitochondrial DNA (mt DNA)
«молча'щая» ~ silent DNA
нати'вная ~ native DNA
некоди'рующая ~ noncoding DNA
ник-трансли'рованная ~ nick-translated DNA
однонитева'я ~ single-stranded DNA
одноцепо'чечная ~ single-stranded DNA
плазми'дная ~ plasmid DNA
повторя'ющаяся ~ repetitive DNA
рекомбина'нтная ~ recombinant DNA
ренатури'рованная ~ renatured DNA
самогибридизи'рующаяся ~ foldback DNA
сателли'тная ~ satellite DNA
сверхспира'льная ~ supercoiled DNA, superhelical DNA
спе'йсерная ~ spacer DNA
хлоропла'стная ~ (хл ДНК) chloroplast DNA (chl DNA)
центроме'рная ~ centromeric DNA
цикли'ческая ~ circular DNA, cyclic DNA
цитоплазмати'ческая ~ cytoplasmic DNA
чужеро'дная ~ foreign DNA
«эгоисти'чная» ~ selfish DNA
я'дерная ~ (яДНК) nuclear DNA (nDNA)
я'дрышковая ~ nucleolar DNA
ДНКа'за *f* deoxyribonuclease, DNase
ДНК-гира'за *f* DNA-gyrase
ДНК-зонд *m* DNA probe
ДНК-«пассажи'р» *m* passenger DNA
ДНК с повто'рами repetitive DNA
дно *n* bottom; bed; fundus
~ **доли'ны** valley floor
~ **реки'** river bed
глазно'е ~ eyeground
доба'вка *f* additive
вкусова'я ~ flavor additive
хлебопека'рная ~ baking additive
доба'вочный *a* adventitious, accessory
доберма'н-пи'нчер *m* Doberman pincher
добра'живание *n* after fermentation, end fermentation

доброка'чественный *a* benign, nonmalignant
добыва'ние *n* acquisition
добы'ча *f* prey; take (*in hunting*);
поджида'ть добы'чу lie in wait for the prey
дове'рчивый *a* confiding
дог *m* mastiff
до'гма *f* dogma
центра'льная ~ *mbiol* central dogma
дое'ние *n* milking, milking operation
ручно'е ~ hand milking
дождева'ние *n* sprinkling, sprinkler irrigation, overhead irrigation, spray irrigation
дождеви'к *m myc* puffball (*Lycoperdon*)
дождеме'р *s* udometer
дождь *m* rain
грибно'й ~ rain during sunshine
грозово'й ~ thunderstorm
кисло'тный ~ acid rain
ли'вневый ~ pouring rain
ме'лкий ~, морося'щий ~ drizzling rain, drizzle
обложно'й ~ continuous rain, steady rain, widespread rain, driving rain
проливно'й ~ pouring rain, torrential rain, downpour
си'льный ~ hard rain, heavy rain
до'за *f* dose, dosage
~ ге'на gene dosage
~ облуче'ния irradiation dose
~ 50-%-ной цитопатоге'нности 50% end point cytopathogenic dose
агглютини'рующая ~ agglutinating dose
анафилактоге'нная ~ shocking dose, anaphylaxis-provoking dose
больша'я ~ significant dose
быстроубива'ющая ~ fast-killing dose
допусти'мая ~ permissible dose
защи'тная ~ protective dose
иммуноге'нная ~ immunogenic dose
индика'торная ~ indicator dose
кумуляти'вная ~ cumulative dose
лета'льная ~ lethal dose, fatal dose
лета'льная минима'льная ~ lethal minimal dose
многокра'тная ~ multiple dose
нача'льная ~ initial dose
о'бщая ~ integral dose
однокра'тная ~, однора'зовая ~ single dose
повто'рная ~ антиге'на recall dose, booster dose
подпоро'говая ~ subliminal dose
поро'говая ~ threshold dose
ра'зовая ~ single dose
разреша'ющая ~ shocking dose, challenging dose; anaphylaxis-provoking dose
ревакцини'рующая ~ booster dose, recall dose
сенсибилизи'рующая ~ sensitizing dose
субагглютини'рующая ~ subagglutinating dose
субгемагглютини'рующая ~ subhemagglutinating dose
сублета'льная ~ sublethal dose
сумма'рная ~ total dose, cumulative dose
су'точная ~ daily dose
тка'невая ~ tissue dose
токси'ческая ~ toxic dose
толера'нтная ~ tolerance dose
уда'рная ~ shocking dose, challenging dose, special shock dose; anaphylaxis-provoking dose
уде'льная ~ specific dose
шо'ковая ~ shocking dose; anaphylaxis-provoking dose
эффекти'вная ~ effective dose
доза'ривание *n* after-ripening
дози'метр *m* dosimeter, radiation meter, radiation monitor
дозиметри'я *f* dosimetry
радиацио'нная ~ radiation dosimetry
дози'рование *n* dosing
дози'ровать *v* dose
дозиро'вка *f* dosage
дозо'р *m*: **лесно'й дозор** forester beat
дозрева'ние *n* after-ripening
дои'ть *v* milk
до'йка *f* milking
долг *m* debt
кислоро'дный ~ oxygen debt
долгове'чность *f* longevity
экологи'ческая ~ ecological longevity
долгове'чный *a* longevous
долговре'менный *a* long-term
долгоживу'щий *a* long-lived
долголе'тний *a* longevous
долгоно'жка *f* crane fly (*Tipulidae*)
долгоно'сик *m* weevil, snout beetle (*Curculionidae*)

~ **большо'й сосно'вый** larger pine weevil *(Hylobius abietis)*
~ **свеклови'чный** beet root weevil, sugar-beet weevil *(Bothynoderes punctiventris)*
~ **хло'пковый** boll weevil *(Anthonomus grandis)*
долгосро'чный *a* long-term
доли'на *f* valley, vale, dale, dell
долихоцефали'я *f* dolichocephalism, dolichocephaly
долотови'дный, долотообра'зный *a* dolabriform
до'лька *f* lobule, lacinule
до'льчатый *a* lobed, lobate, laciniate, cleft
до'ля *f* lobe; fraction, part
~ **Брока'** limbic lobe
~ **лё'гкого** lung lobe, lobe of the lung
~ **околоцве'тника** perianth lobe
~ **пе'чени** hepatic lobe
висо'чная ~ temporal lobe *(of the brain)*
за'дняя ~ гипо'физа posterior pituitary
заты'лочная ~ occipital lobe, end lobe *(of the brain)*
зри'тельная ~ *ent* optic lobe
квадра'тная ~ quadrate lobe *(of the liver)*
лимби'ческая ~ limbic lobe *(of the brain)*
ло'бная ~ frontal lobe *(of the brain)*
обоня'тельная ~ olfactory lobe *(of the brain)*
пере'дняя ~ гипо'физа anterior pituitary
ма'ссовая ~ fraction of total mass
теменна'я ~ parietal lobe *(of the brain)*
хвоста'тая ~ Spiegelian lobe *(of the liver)*
дома'шний *a* domestic
доме'н *m* domain
~ **тяжё'лой це'пи** *mbiol* heavy chain domain
внекле'точный ~ *mbiol* outer domain
карбоксиконцево'й цитоплазмати'ческий ~ *mbiol* COOH-cytoplasmic domain
трансмембра'нный ~ *mbiol* transmembrane domain
доместика'ция *f* domestication
до'мик *m* case (*e.g., of caddis worms*), lorica (*e.g., of rotifers*)
домина'нт *m*, **доминанта** *f* dominant
~ **по сте'пени частоты' frequency** dominant
~ **форма'ции formation** dominant
домина'нтность *f* dominance
~ **необы'чных при'знаков** fancier's dominance
альтерни'рующая ~ alternating dominance
непо'лная ~ inclomplete dominance, partial dominance
несто'йкая ~ irregular dominance
по'лная ~ complete dominance
усло'вная ~ conditional dominance
домина'нтный *a* dominant
домина'тор *m* dominator
доминиге'н *m* dominigene
домини'рование *n* dominance, domination
~ **верху'шки** *bot* apical dominance
~ **в соо'бществе** social dominance
запа'здывающее ~ delayed dominance
обрати'мое ~ shifting dominance
ступе'нчатое ~ top dominance
домини'ровать *v* dominate
домини'рующий *a* dominant
до'нник *m* sweet clover (*Melilotus*)
до'нный *a* bottom-living
до'нор *m* donor
~ **электро'нов** electron donor
иммунокомпете'нтный ~ immune donor
до'нце *n* collum, disk, corm, fundus (*of an onion*)
доопыле'ние *n* supplementary pollination, extrapollination
доочи'стка *f* afterpurification
дополни'тельный *a* accessory
дорза'льный *a* see **дорса'льный**
дородово'й *a* antenatal
дорса'льный *a* dorsal
дорсовентра'льный *a* dorsoventral
дот-бло'ттинг *m* dot blotting
дра'ка *f* fight, fighting
драко'н *m* dragon
лету'чий ~ flying dragon (*Draco*)
древеси'на *f* wood, timber
весе'нняя ~ spring wood, early wood
втори'чная ~ secondary wood
делова'я ~ commercial timber, lumber, commercial lumber
кольцепо'ровая ~ ring-porous wood
кососло'йная ~ cross-fibered wood, spiral-grained wood
крупносло'йная ~ wide-ringed wood
ле'тняя ~ summer wood, late wood
мелкосло'йная ~ narrow-ringed wood, fine-grained wood, close-grained wood
свилева'тая ~ curly wood

спе'лая ~ adult wood
стволова'я ~ trunk timber, stem wood
узо'рчатая ~ figured wood
фа'утная ~ defective wood
широкосло'йная ~ wide-ringed wood
древе'сница *f*: ~ **въе'дливая** leopard moth (*Zeuzera pyrina*)
древе'сно-коло'сый woody-spiked
древе'сно-корешко'вый woody-rooted
древе'сный *a* arboreal, arboreous, woody
дре'во *n*: **генеалоги'ческое ~, родосло'вное** ~ genealogical tree
древови'дный *a* treelike, dendriform, dendritic, dendroid, arbuscular, arboreous, arborescent, subarborescent
древогрызу'щий *a* hylotomous
древоразруша'ющий *a* wood-attacking
древосто'й *m* forest stand, stand of trees, stand, stock of trees, timber stand
древото'чец *m* wood-boring animal
дрейф *m* drift
~ **ге'нов** genetic drift
генети'ческий ~ genetic drift
континента'льный ~ continental drift
мейоти'ческий ~ meiotic drift
мутацио'нный ~ mutational drift
дрё'ма *f* campion (*Melandrium*)
дрема'ть *v* doze, slumber
дремо'та *f* drowsiness, sleepiness, somnolence
дрему'чий *a* dense, thick
дрена'ж *m* drainage; *med* drain
~ **Ди'вера** Deaver's drain
~ **Ке'ра** Kehr's drain
дрессирова'ть *v* train, tame
дрессиро'вка *f* training
дримофи'т *m* drymophyte
дроби'на *f* spent grains, spent brewers' grains, grains
дробле'ние *n emb* cleavage; fission, embryonic fission; segmentation; crushing, grinding
асимметри'чное ~ asymmetrical cleavage
билатера'льное ~ bilateral cleavage
вертика'льное ~ vertical cleavage
голобласти'ческое ~ holoblastic cleavage
горизонта'льное ~ horizontal cleavage, latitudinal cleavage
двусторо'ннее ~ bilateral cleavage
дискоида'льное ~ discoidal cleavage
дополни'тельное ~ accessory cleavage
левосторо'ннее ~ leiotropic cleavage, sinistral cleavage
меридиона'льное ~ meridional cleavage
мозаи'чное ~ mosaic cleavage
непо'лное ~ incomplete cleavage
неравноме'рное ~ unequal cleavage
полиэмбриона'льное ~ polyembryonic cleavage
правосторо'ннее ~ dextral cleavage, dexiotropic cleavage
пове'рхностное ~ superficial cleavage
по'лное ~ total cleavage
равноме'рное ~ equal cleavage
радиальное ~ radial cleavage
регуляцио'нное ~ regulatory cleavage
спира'льное ~ spiral cleavage
части'чное ~ incomplete cleavage
экваториа'льное ~ equatorial cleavage
дровосе'к *m* capricorn beetle, longhorn (*Cerambycidae*); woodcutter
дрожа'ть *v* tremble, shake, quiver, quaver; tremble, shiver (*with cold*); tremble, shake (*with fear*); flutter (e.g, *of leaves of aspen*)
дрожжево'й *a* yeasty
дрожжеподо'бный *a* yeastlike
дро'жжи *pl* yeast
~ **верхово'го броже'ния** top fermentation yeast
~ **низово'го броже'ния** bottom fermentation yeast
аспороге'нные ~ asporogenous yeast, nonspore-forming yeast
ви'нные ~ wine yeast (*Saccharomyces*)
деля'щиеся ~ fission yeast
ди'кие ~ wild yeast
зада'точные ~ pitching yeast
зерни'стые ~ granular yeast
кормовы'е ~ feed yeast, fodder yeast (*Candida utilis*)
ксилозоперераба'тывающие ~ xylose-fermenting yeast
культу'рные ~ cultivated yeast
ма'точные ~ stock yeast culture, barm
метано'льные ~ methanolic yeast
метилотро'фные ~ methylotrophic yeast
пе'карские ~ baker's yeast (*Saccharomyces cerevisiae*)
пивны'е ~ beer yeast, brewer's yeast (*Saccharomyces*)
пищевы'е ~ food yeast
почку'ющиеся ~ budding yeast
пруссо'ванные ~ compressed yeast

спиртовы'е ~ distiller's yeast (*Saccharomyces*)
спорообразу'ющие ~ spore-forming yeast, sporulating yeast
флокули'рующие ~ flocculating yeast
хлебопека'рские ~ baker's yeast, bakery yeast (*Saccharomyces cerevisiae*)
хлопьеви'дные ~ flocculating yeast
«чё'рные ~» "black" yeast

дрожь *f* trembling, shivering

дрозд *m* thrush, ouzel, ousel (*Turdus*)
пе'вчий ~ song thrush *(Turdus philomelos)*
че'рный ~ blackbird, merle (*Turdus merula*)

дро'зд-деря'ба *m* mistle thrush, misel thrush, missel bird, missel (*Turdus viscivorus*)

дрозо'фила *f* fruit fly *(Drosophila)*

дро'зд-ряби'нник *m* fieldfare (*Turdus pilaris*)

дрок *m* broom, woodwaxen (*Genista*)

дромаде'р, дромеда'р *m* Arabian camel, dromedary (*Camelus dromedarius*)

дромотропи'зм *m* dromotropism

дромотро'пный *a* dromotropic

дронт *m* dodo (*Didus ineptus*)

дрофа' *f* bustard (*Otis*); great bustard (*Otis tarda*)

дря'блый *a* flabby, flaccid, loose

дря'ква *f* sowbread (*Cyclamen*)

дря'хлость *f* anility

дуб *m* oak (*Quercus*)
~ **про'бковый** cork oak (*Quercus suber*)

дуби'льный *a* tannic

дубле'ние *n* tanning, tannage, tanning process, dressing

дубня'к *m* oakery, oak-forest, oak-wood

дубови'к *m* lurid boletus (*Boletus luridus*)

дубо'к *m* oaklet, oakling

дубоно'с *m* grosbeak, hawfinch (*Coccothraustes*)
обыкнове'нный ~ hawfinch (*Coccotharustes coccothraustes*)

дубра'ва *f* oakery, oak-forest, oak-wood

дуга' *f* arch, arc
~ **ао'рты** aortic arch
~ **лё'гочной арте'рии** pulmonary arc
~ **непа'рной ве'ны** azygous arch
альвеоля'рная ~ alveolar arch
висцера'льная ~ visceral arch
втора'я жа'берная ~ *emb* postaural arch
гио'идная ~ hyoid arch, lingual arch, arch of tongue, parietohemal arch
жа'берная ~ gill arch, branchial arch
ко'ртиева ~ Corti's arch
ладо'нная ~ palmar arch
лобко'вая ~ subpubic arch
ло'нная ~ pubic arch
надбро'вная ~ superciliary arch
нё'бно-язы'чная ~ glossopalatine arch
нейра'льная ~ neural arch
нижнечелюстна'я ~ mandibular arch
подкрыльцо'вая ~ axillary arch
пряма'я спина'льная ~ *phys* direct spinal arc
рё'берная ~ costal arch
рефлекто'рная ~ *phys* reflex arc
скулова'я ~ zygomatic arch, subocular arch, malar arch
спина'льная ~ spinal arch
сухожи'льная ~ tendinous arch
тре'тья жа'берная ~ *emb* thyroid arch
челюстна'я ~ zygomatic arch

дугови'дный *a* arcuate

дугоне'рвный *a bot* curvinervate

дугообра'зный *a* arcuate

ду'дка *f* fistular stem, piping stem

ду'дник *m* angelica (*Angelica*)

ду'дчатый *a bot* fistular

ду'жка *f* arculus
нё'бная ~ palatine arch

дуодена'льный *a* duodenal

ду'пель *m* great snipe (*Gallinago media*)

дуплика'ция *f* duplication
танде'мные дуплика'ции *pl* tandem duplications

дупли'стый *a* hollow

дупло' *n* hollow (*in a tree*), cavity (*in a tooth*)

дупля'нка *f* log house, nest box

дурма'н *m* stramony (*Datura*)

дурни'шник *m* cocklebur (*Xanthium*)

ду'рра *f* durra, dura (*Sorghum vulgare*)

дуст *m* dust

ду'тыш *m* pouter

душ *m* shower, douche
~ **Шарко'** Charcot's douche

души'стый *a* odorant, odoriferous, fragrant

души'ца *f* marjoram (*Origanum majorana*)

Д-ферме'нт *m* D-enzyme, disproportionating enzyme

ды'ня *f* melon (*Cucumis melo*)

ды'рчатый *a* foraminous

ды'хало *n* blowhole, spout (*of a whale*)

ды'хальце *n* breathing opening, spiracle; stigma

дыха'ние *n* respiration, breathing

~ **«всё и'ли ничего'»** all-or-nothing breathing
~ **Био'та** Biot's respiration
~ **Куссма'уля** Kussmaul respiration
~ **максима'льной глубины'** all-or-nothing breathing
~ **плода'** fetal respiration
~ **Чейна-Стокса** tidal respiration, periodic breathing
анаэро'бное ~ anaerobic respiration
атакси'ческое ~ ataxic breathing
аэробно'е ~ aerobic respiration
брюшно'е ~ abdominal respiration
брюшно'е диафрагма'льное ~ ventral breathing
вне'шнее ~ external respiration
вну'треннее ~ internal respiration
внутриутро'бное ~ fetal respiration
грудно'е ~ thoracic beathing, costal respiration
диафрагма'льное ~ abdominal respiration, ventral breathing, diaphragmal respiration
жа'берное ~ gill breathing, branchial respiration
же'сткое rough respiration
затрудне'нное ~ labored respiration
иску'сственное ~ artificial respiration
кле'точное ~ cell respiration
ко'жное ~ dermal respiration
лё'гочное ~ lung respiration, pulmonary respiration
нитра'тное ~ nitrate respiration
основно'е ~ basal respiration
оста'точное ~ residual respiration
периоди'ческое ~ periodic breathing
пове'рхностное ~ shallow respiration
рефлекто'рное ~ reflex respiration
светово'е ~ light respiration
солево'е ~ salt respiration
споко'йное ~ quiet breathing
темново'е ~ dark respiration
тканево'е ~ tissue respiration
трахеа'льное ~ tracheal respiration
тяже'лое ~ heavy breathing
учаще'нное ~ hurried respiration
экзоге'нное ~ exogenic respiration, extra respiration
эндоге'нное ~ endogenic respiration

дыха'тельный *a* respiratory
дыша'ть *v* breath
дю'на *f* dune
закрепле'нная ~ arrested dune, stabilized dune
подви'жная ~ wandering dune, mobile dune

дя'гиль *m* garden angelica (*Archangelica*)
дя'тел *m* woodpecker (*Picidae*)
большо'й пё'стрый ~ greaterspotted woodpecker *(Dendrocopos major)*
зелё'ный ~ green woodpecker *(Picus viridis)*
ма'лый пё'стрый ~ lesserspotted woodpecker *(Dendrocopos minor)*
чё'рный ~ black woodpecker *(Dryocopus martius)*

Е

евге'ника *f* eugenics
е'герь *m* game warden
едини'ца *f* unit
~ **акти'вности** activity unit
~ **биогеографи'ческого райони'рования** biogeographic unit
~ **измере'ния unit**
~ **ка'рты** *gen* map unit
~ **коли'чества теплоты'** thermal unit
~ **освещё'нности** light unit
~ **расти'тельного покро'ва** unit of vegetation
~ **транскри'пции** transcription unit, transcriptional unit
~ **эволю'ции** unit of evolution
антиге'нная ~ antigen unit
бляшкообразу'ющая ~ plaque-forming unit, PFU
бонитиро'вочная ~ evaluation unit
колониеобразу'ющая ~, КОЕ *f*, colony-forming unit, CFU
кормова'я ~ feed unit, fodder unit
крыси'ная ~ rat unit
мыши'ная ~ mouse unit
нейромото'рная ~ neuromotor unit
не'рвная ~ nerve unit
опера'бельная таксономи'ческая ~ operational taxonomic unit
панмикти'ческая ~ panmictic unit
плодоплацента'рная ~ feto-placental unit
произво'льная ~ измере'ния arbitrary unit
системати'ческая ~ systematic unit
сукцессио'нная ~ syngenetic unit
таксономи'ческая ~ taxonomic unit
теплова'я ~ thermal unit
функциона'льная ~ functional unit
функциона'льная ~ гено'ма genome functional unit
хорологи'ческая ~ chorologic unit
ёж *m* hedgehog, urchin (*Erinaceus*)
морско'й ~ sea-urchin (*Echinoidea*)
уша'стый ~ eared hedgehog *(Hemiechinus)*
ежа' *f* cock's-foot (*Dactylis*); *genitive of* **ёж**
ежеви'ка *f* dewberry (*Rubus caesius*)
ежеголо'вник *m* bur-reed (*Sparganium*)
ежему'ха *f* bristle fly, tachinid fly (*Tachinidae*)
ежи'ха *f* female hedgehog
ежо'вник *m* barnyard grass (*Panicum crus-galli*)
езда' *f* riding
верхова'я ~ equitation, riding
е'здить *v* ride
~ **верхо'м** ride (*on horseback*)
ель *f* spruce (*Picea*)
е'льник *m* spruce forest
е'льник-долгомо'шник haircup-moss spruce forest
ё'мкость *f* capacity
~ **вдо'ха** inspiratory capacity
~ **ландша'фта** landscape carrying capacity
~ **обме'на** exchange capacity
~ **па'стбища** grazing capacity
~ **уго'дья** land capacity
жи'зненная ~ vital capacity
жи'зненная ~ ле'гких lung capacity, vital capacity
информацио'нная ~ information capacity
мембра'нная ~ membrane capacity
о'бщая ~ лё'гких total lung capacity
оста'точная ~ лё'гких residual lung capacity
поле'зная ~ net capacity
по'лная ~ лё'гких total lung capacity
функциона'льная ~ functional capacity
функциона'льная оста'точная ~ equilibrum capacity, functional residual capacity *(of lungs)*
емура'нчик *m* thick-tailed three-toed jerboa (*Stylodipus*)
ено'т *m* racoon (*Procyon*)
е'рник *m* dwarf birch, Arctic birch (*Betula nana*); yernik (*stand of Betula nana*)
есте'ственник *m* naturalist
есте'ственный *a* natural
естествозна'ние *n* natural science
естествоиспыта'тель *m* natural scientist
есть *v* eat; **дым ест глаза'** the eyes smart with smoke
ехи'дна *f* anteater (*Trachyglossus*)

Ж

жаба *f* toad (*Bufo*)
водяна'я ~ aquatic toad *(Pseudobufo)*
грудна'я ~ *med* angina pectoris
зелё'ная ~ green toad, variable toad *(Bufo viridis)*
настоя'щая ~ true toad *(Bufonidae)*
обыкнове'нная ~ common toad *(Bufo bufo)*
се'рая ~ common toad *(Bufo bufo)*
жа'ба-повиту'ха *f* midwife toad *(Alytes obstetricans)*
жа'ба-малю'тка *f* oak toad *(Bufo quercicus)*
жа'берный *a* branchial
жа'бра *f* gill
трахе'йная ~ *ent* tracheal gill
жа'воронок *m* lark (*Alaudidae*)
полево'й ~ (*Alauda arvensis*)
жа'жда *f* thirst
жако' *m* gray parrot (*Psittacus erithacus*)
жа'лить *v* sting, bite
жа'ло *n* sting, stinger
жаропонижа'ющий *a* antipyretic
жасми'н *m* jasmine (*Jasminum*)
жа'тва *f* harvest, harvesting
жать *v* mow, reap; press
жва'чка *f* chew, cud
жва'чное *n* ruminant (*Ruminantia*)
жва'чный *a* ruminant
жгут *m* tourniquet, garrot
~ **Ви'нера** Viner's tourniquet
~ **Ки'дде** Kidde's tourniquet
~ **Э'смарха** Esmarch's tourniquet
жгу'тик *m* flagellum
за'дний ~ posterior flagellum
поля'рно располо'женный ~ polar flagellum
попере'чный ~ transverse flagellum
жгутикова'ние *n* flagellation
амфитрихиа'льное ~ amphitrichous flagellation
лофотрихиа'льное ~ lophotrichous flagellation
монотрихиа'льное ~ monotrichous flagellation
перитрихиа'льное ~ peritrichous flagellation
жгу'тиковый *a* flagellar
жгутиконо'сный *a* flagellated, flagellate
жгу'честь *f* pungency
жгу'чий *a* stinging, urticant; pungent; hot; ~ **пе'рец** hot pepper
жева'ние *n* mastication; rumination
жева'тельный *a* masticatory, manducatory
жева'ть *v* chew, masticate
~ **жва'чку** ruminate, chew the cud
желати'н *m* gelatin, gelatine
мясопепто'нный ~, **МПЖ** beef-extract gelatin
пла'зменный ~ plasma-coated gelatin
желатиниза'ция *f* gelatinization
желатинизи'рованный *a* gelated
желати'новый *a* gelatinous
желва'к *m* lump, knob; struma
железа' *f* gland, glandula
~ **бараба'нной по'лости** tympanic gland
~ **вну'тренней секре'ции** endocrine gland, closed gland, ductless gland, internal secretion gland, vascular gland
~ **двенадцатипе'рстной кишки'** duodenal gland, Brunner's gland
~ **дна желу'дка** fundic gland
~ **желобова'того сосо'чка** Ebner's gland, gland of vallate papilla
~ **желу'дка** gastric gland
~ **коллетериа'льная** ~ colleterial gland
~ **кра'йней пло'ти** preputial gland, Tysonian gland
~ **Кра'узе** Krause's gland, accessory lacrimal gland
~ **Ли'ттра** Littre's gland
~ **Монтго'мери** Montgomery's gland, areolar gland
~ **мочеиспуска'тельного кана'ла** urethral gland
~ **нару'жной секре'ции** ectocrine gland
~ **обоня'тельной о'бласти** Bowman's gland, olfactory gland
~ **околососо'чкового кружка'** areolar gland, ~ Montgomery's gland
~ **преддве'рия влага'лища** Bartholin's gland, vulvovaginal gland
~ **проникнове'ния** *helm* penetration gland
~ **семяприё'мника** spermathecal gland
~ **сме'шанной секре'ции** endoexocrine gland, mixed gland
~ **Та'йсона** Tysonian gland, preputial gland
~ **Це'йса** gland of Zeiss
~ **Э'бнера** Ebner's gland, gland of vallate papilla
~ **эндоме'трия** uterine gland

альвеоля'рная ~ alveolar gland, acinous gland, saccular gland
ана'льная ~ anal gland
апокри'нная ~ apocrine gland
аромати'ческая ~ scent gland, stink gland
ацино'зная ~ acinous gland, alveolar gland
бартоли'нова ~ Bartholin's gland, vulvovaginal gland
бе'дренная ~ femoral gland
би'ссусная ~ byssal gland, byssus gland
бо'уменова ~ Bowman's gland, olfactory gland
бру'ннерова ~ Brunner's gland, duodenal gland
бульбоуретра'льная ~ bulbo-urethral gland, Cowper's gland
ви'лочковая ~ thymus gland
вкусова'я ~ gustatory gland
воскова'я ~ wax gland
выдели'тельная ~ excretory gland
га'зовая ~ gas gland
гермафроди'тная ~ ovariotestis
головна'я ~ head gland
головна'я шелкоотдели'тельная ~ *ent* cephalic silk gland
голокри'нная ~ holocrine gland
доба'вочная ~ accessory gland
доба'вочная слё'зная ~ accessory lacrimal gland, Krause's gland
дюфу'рова ~ Dufour's gland
желто'чная ~ yolk gland, vitelline gland
желу'дочная ~ gastric gland
жирова'я ~ adipose gland
зелё'ная ~ green gland *(of crustaceans)*
зо'бная ~ thymus gland
известко'вая ~ calciferous gland
кароти'дная ~ carotid gland
кишечная ~ intestinal gland
клубо'чковая ~ convoluted gland, coil gland, glomerate gland
ко'жная ~ cutaneous gland; hedonic gland *(of reptiles)*
коллетериа'льная ~ colleterium
кольцева'я ~ ring gland, Weismann's gland
ко'пчиковая ~ *orn* preen gland, coccygeal gland, oil gland
ку'перова ~ Cowper's gland, bulbo-urethral gland
либеркю'нова gland of Lieberkuhn, Lieberkuhn's gland, ~ enteric gland
лицева'я ~ face gland
максилля'рная ~ maxillary gland
мерокри'нная ~ merocrine gland
многокле'точная ~ multicellular gland
моло'чная ~ milk gland, lactiferous gland, mammary gland
моноптихиа'льная ~ single-rowed gland
нё'бная ~ palatine gland
нидамента'льная ~ nidamental gland
ножна'я ~ foot gland *(of rotifers)*
однокле'точная ~ *bot* simple gland
околоне'рвная ~ neural gland
околоу'шная ~ parotid gland
околощитови'дная parathyroid gland, parathyroid
паращитови'дная ~ parathyroid gland
па'ховая ~ groin gland
паху'чая ~ scent gland, stink gland
пе'нная ~ *ent* Batelli's gland, froth gland
пере'дняя губна'я ~ rostral gland *(of snakes)*
пилори'ческая ~ pyloric gland
поджелу'дочная ~ pancreas, pancreatic gland
подчелюстна'я ~ mandibular gland
подъязы'чная ~ sublingual gland
полиптихиа'льная ~ many-rowed gland
полова'я ~ genital gland, germ gland, sex gland
по'товая ~ sudoriferous gland, sweat gland, perspiratory gland, sweat gland
предста'тельная ~ prostate, prostate gland
прида'точная ~ accessory gland
прикрепи'тельная ~ adhesive gland
проста'я ~ *zool* simple gland
проторака'льная ~ prothoracic gland
пряди'льная ~ *ent* spinning gland, silk-spinning gland
ректа'льная ~ rectal gland
ресни'чная ~ ciliary gland, Moll's gland
са'льная ~ sebaceous gland, sebiferous gland; oil gland; gland of Zeiss *(of eyelid)*
секрето'рная ~ secretory gland
семенна'я ~ seminal gland
скорлупо'вая ~ shell gland
слё'зная ~ lacrimal gland, tear gland
слизи'стая ~ mucous gland
сло'жная ~ compound gland
сло'жная альвеоля'рная ~ racemose gland
слюнна'я ~ salivary gland
со'бственная ~ желу'дка acid gland
солева'я ~ *zool* salt gland
торака'льная ~ thoracic gland

тру'бчатая ~ tubular gland
фунда'льная ~ acid gland
хвостова'я ~ caudal gland
хоботко'вая ~ rostellar gland
цеме'нтная ~ cement gland
черни'льная ~ ink-bag
ше'йная ~ cervical gland
шёлкоотдели'тельная ~ *ent* silk gland
шишкови'дная ~ pineal body, pineal gland
щитови'дная ~ thyroid, thyroid gland
экзокри'нная ~ exocrine gland, externally secreting gland
экскрето'рная ~ excretory gland
эндокри'нная ~ endocrine gland, ductless gland, closed gland, vascular gland
ядови'тая ~ venom gland
язы'чная ~ lingual gland
яи'чная ~ Blandin's gland

желе'зисто-пи'льчатый *a bot* glandulose-serrate
желе'зисто-опушё'нный *a bot* glandulosous
желе'зистый *a* adenose, adenic, adeniform, adenoid, glandlike, glandiform, glandular; ferrous
желё'зка *f* glandule
желе'зница *f* follicle mite *(Demodex folliculorum)*
~ **бы'чья** cattle follicle mite *(Demodex bovis)*
~ **коша'чья** cat follicle mite *(Demodex cati)*
~ **соба'чья** dog follicle mite *(Demodex canis)*
желе'зо *n* iron
за'кисное ~ ferrous iron
о'кисное ~ ferric iron
желе'зка *f* glandule
желе'зница *f* follicle mite *(Demodex folliculorum)*
железобакте'рия *f* iron bacterium
желна' *f* black woodpecker *(Dryocopus martius)*
же'лоб *m* groove, vallecula, valley
~ **для со'нной арте'рии** carotid groove
горта'нно-трахеа'льный ~ laryngotracheal groove
запира'тельный ~ *anat* obturator groove
яре'мный ~ jugular groove
желобова'тый *a* valleculate
желобо'к *m* groove
~ **не'рвной тру'бки** neural groove
ногтево'й ~ nail groove

желобчатоли'стный *a* canalifolious
желo'бчатый *a* channelled, grooved, furrowed, sulcate, valleculate
желтозё'м *m* yellow earth
желто'к *m* yolk
белко'вый ~ protein yolk
бе'лый ~ white yolk
жирово'й ~ fatty yolk
яи'чный ~ egg yolk
желтоло'зник *m* purple willow (*Salix purpurea*)
желтопло'дный *a* xanthocarpous
желтопу'зик *m* glass-lizard, glass-snake (*Pseudopus apodus*)
желтофио'ль *m* wallflower (*Cheiranthus*)
желто'чник *m* vitelline gland, yolk gland, vitellarium
желто'чный *a* vitelline
желту'ха *f* jaundice, icterus
желту'шный *a* icteric, icteroid, icteritious
же'лтый *a* yellow
акриди'новый ~ acridine yellow
ализари'новый ~ alizarin yellow
бриллиа'нтовый ~ brillinat yellow
метани'ловый ~ metanil yellow
нафто'ловый ~ naphthol yellow
основно'й ~ basic yellow
про'чный ~ fast yellow
суда'н ~ Sudan yellow
тита'новый ~ titan yellow
я'нус ~ Janus yellow
желудеобра'зный *a* balanoid
желудепло'дный *a* balanocarpous
желу'док *m* stomach, ventricle, ventriculus
жева'тельный ~ masticatory stomach
желе'зистый ~ glandular stomach
му'скульный ~ masticatory stomach
мы'шечный ~ *orn* gizzard stomach
желу'дочек *m* ventricle
ле'вый ~ aortic ventricle
ма'лый ~ по Па'влову, па'вловский ~ Pavlov pouch, Pavlov stomach, miniature stomach
желудочковый *a* ventricular
желу'дочно-диафрагма'льный *a* gastrophrenic
желу'дочно-кише'чный *a* gastrointestinal
желу'дочно-ободо'чный *a* gastrocolic
желу'дочный *a* gastric, stomachal, stomachic, ventricular
же'лудь *m* acorn
желчего'нный *a* bile-expelling

желчека'менный *a* cholelithic
желчеобразова'ние *n* bilification, biligenesis
желчеобразу'ющий *a* cholepoietic
желчеотделе'ние *n* biliation, bile secretion
же'лчный *a* biliary
желчь *f* bile, gall
же'мчуг *m* pearls
жемчу'жина *f* pearl
иску'ственно вы'ращенная ~ cultured pearl
же'нский *a* female, feminine
женьше'нь *m* ginseng (*Panax*)
жердня'к *m* polewood, pole stand; small-size pole timber
же'рдочка *f* perch
жердь *f* pole, stake
жере'бая *a* in foal, pregnant (*of a mare*)
жеребе'нок *m* colt, foal
жеребе'ц *m* stallion
жереби'ться *v* foal
жере'бчик *m* colt
жерля'нка *f* fire-bellied toad, fire-belly (*Bombina*)
~ **желтобрю'хая** yellow-bellied toad (*Bombina variegata*)
~ **краснобрюхая** fire-bellied toad, red-bellied toad (*Bombina bombina*)
же'ртва *f* prey, victim
жест *m eth* gesture
прося'щий ~ begging gesture
жестикули'ровать *v* gesture
жестколи'стный *a* stiff-leaved, hard-leaved, sclerophyllous
жечь *v* burn; bite (*e.g., of pepper*)
живе'ц *m* live bait, bait fish
живи'ца *f* galipot, gallipot
жи'вность *f* livestock
живо'й *a* living
живоко'сть *f* larkspur (*Delphinium*)
живоло'вка *f* live trap
живородя'щий *a* love-bearing, viviparous, zoogonous
живорожде'ние *n* vivipary, viviparity, zoogony
живо'т *m* abdomen, belly, venter
втя'нутый ~ sunken abdomen, scaphoid abdomen
о'стрый ~ acute abdomen, burst abdomen, surgical abdomen
отвиса'ющий ~ dropped abdomen
животново'д *m* cattle raiser, aninal husbandman
животново'дство *n* cattle breeding, animal husbandry, livestock raising, livestock farming
мясно'е ~ beef husbandry
живо'тное *n* animal
~, **подве'ргнутое стре'ссу** stressed animal
взро'слое ~ mature animal
восприи'мчивое ~ susceptible animal
вре'дное ~ injurious animal, noxious animal
всея'дное ~ omnivore, omnivorous animal
гипоталами'ческое ~ hypothalamic animal
гнотобио'нтное ~ gnotobiote
годова'лое ~ yearling
гомойоте'рмное ~ homoiotherm, warn-blooded animal
двусторо'нне-симметри'чное ~ bilateral animal *(Bilateria)*
дома'шнее ~ domestic animal
жва'чное ~ ruminant, ruminant animal, cud-chewing animal
иммунизи'рованное ~ immunized animal
имму'нное ~ immune animal
инбре'дное ~ inbred animal
индика'торное ~ sentinel animal, surveillance animal
контро'льное ~ control animal
копы'тное ~ hoofed animal
лаборато'рное ~ laboratory animal, experimental animal
многокле'точное ~ many-celled animal *(Metazoa)*
многопло'дное ~ multiparous animal
молодо'е ~ young animal
нежва'чное ~ nonruminant animal
нежела'тельное ~ nuisance animal
неинфици'рованное ~ clean animal
непарнокопы'тное ~ solid-hoofed animal
обу'ченное ~ trained animal
охо'тничье-промысло'вое ~ game animal
парнокопы'тное ~ cloven-hoofed animal
пасу'щееся ~ grazer
плотоя'дное ~ preying animal, carnivorous animal
подо'пытное ~ experimental animal, test animal
пойкилоте'рмное ~ poikilotherm, cold-blooded animal
полуодома'шненное ~ semidomesticated animal
прикреплё'нное ~ stationary animal

растениея'дное ~ plant-eating animal, plant-eater, vegetarian animal, phytophagous animal
рога'тое ~ horned animal
светя'щееся ~ phosphorescing animal
сельскохозя'йственное ~ farm animal
сенсибилизи'рованное ~ primed animal
спина'льное ~ spinal animal
спленэктоми'рованное ~ splenectomized animal
станда'ртное лаборато'рное ~ conventional animal
стери'льное ~ germ-free animal
талами'ческое ~ thalamic animal
теплокро'вное ~ homoiotherm, cold-blooded animal
травоя'дное ~ herbivorous animal, weed-eater, graminivorous animal
убо'йное ~ slaughter animal
хи'щное ~ preying animal, predator, carnivorous animal
холоднокро'вное ~ poikilotherm, cold-blooded animal
целоми'ческое ~ coelomate animal *(Coelomata)*
четвероно'гое ~ four-footed animal
чле'нистое ~ segmented animal, articulated animal *(Articulata)*

живу'чий *a* tenacious
живу'чка *f* bugle-weed (*Ajuga*)
жига'лка *f* stable fly (*Stomoxys calcitrans*)
жи'дкость *f* liquid, fluid
~ **брюшно'й по'лости** peritoneal fluid
амниоти'ческая ~ amniotic fluid
бордо'сская ~ Bordeaux liquid
внекле'точная ~ extracellular fluid, extracellular water
внесосу'дистая ~ extravascular fluid
внутриглазна'я ~ intraocular fluid
интерстициа'льная ~ interstitial fluid
культура'льная ~ culture fluid, culture liquid
надоса'дочная ~ supernatant fluid
околопло'дная ~ waters
оте'чная ~ transudate
отпу'гивающая ~ repugnatorial fluid
перивисцера'льная ~ perivisceral fluid
перивителли'новая ~ perivitelline fluid
пло'дная ~ fetal fluid
семенна'я ~ seminal fluid
серо'зная ~ serous fluid
синовиа'льная ~ joint oil, joint fluid, synovial fluid, joint water
спинномозгова'я ~ cerebrospinal fluid
тка'невая ~ interstitial fluid, tissue fluid
цереброспина'льная ~ spinal fluid

жизнеде'ятельность *f* life activity, vital activity
жи'зненность *f* vitality, vital power
жи'зненный *a* vital
жизнеобеспе'чение *n* life support, life sustenance
жизнеспосо'бность *f* vitality, viability
жизнеспосо'бный *a* viable
жизнесто'йкий *a* hardy, tough, enduring
жизнь *f* life
внутриутро'бная ~ antenatal life
внеутро'бная ~ extrauterine life, postnatal life

жи'лка *f* veine, nerve, nervure; *bot* rib
ана'льная ~ anal vein
во'гнутая ~ concave vein
вы'пуклая ~ convex vein
гла'вная ~ крыла' principal wing vein
диста'льная ~ distal vein
доба'вочная случа'йная ~ adventitious vein
дополни'тельная ~ accessory vein
дополни'тельная краева'я ~ marginal accessory vein
дополни'тельная продо'льная ~ intercalary vein
заузелко'вая попере'чная ~ postnodal cross vein
кардина'льная ~ cardinal vein
коса'я ~ oblique vein
коса'я свя'зывающая ~ oblique brace vein
медиа'льная попере'чная ~ medial cross vein
медиокубита'льная попере'чная ~ mediocubital cross vein
печё'ночная ~ hepatic vein
плечева'я попере'чная ~ humeral cross vein
попере'чная ~ cross vein
попере'чная ~ сектора sectorial cross vein
посткубита'льная попере'чная ~ postcubital cross vein
предузелко'вая ~ antecubital cross vein
продо'льная ~ longitudinal vein
радиа'льная попере'чная ~ radial cross vein

радиомедиа'льная попере'чная ~ radiomedial cross vein
сериа'льная ~ serial vein
сло'жная ~ compound vein
сре'дняя ~ midvein, midrib
субкоста'льная ~ subcostal vein
жилкова'ние *n* nervation, neuration, venation, veining
дихотоми'ческое ~ dichotomous venation
дланеви'дное ~ palmate venation
паралле'льное ~ parallel venation
пе'ристое ~ pinnate venation
ретикуля'рное ~ dictyodromous venation, netted venation, reticulated venation
се'тчатое ~ dictyodromous venation, netted venation, reticulated venation
жилкова'тый *a bot* veined, venose, nervate, ribbed
жило'й *a* resident; freshwater, non-anadromous (*of fish*)
жи'молость *f* honeysuckle (*Lonicera*)
жир *m* fat
нутряно'й ~ visceral fat, interior fat
ры'бий ~ fish oil, cod-liver oil
жира'ф *m* giraffe (*Giraffa*)
жире'ть *v* fatten
жи'рность *f* fatness, fat content
жирнохво'стый *a* fat-tailed
жи'рный *a* fatty
жирова'ть *v* fatten
жирово'й *a* fatty, adipose
жирообразу'ющий *a* lipogenetic, lipogenous
жиропо'т *m* suint, wool sweat
жирораствори'мый *a* fat-soluble, liposoluble
жирорасщепля'ющий *a* lipolytic, lipoclastic
жиря'к *m orn* oilbird *(Steatornis caripensis)*
жиря'нка *f bot* butterwort (*Pinguicula*)
жить *v* live
жмых *m* oil cake
жнивье' *n* stubble
жом *m* pulp
жордано'н *m* jordanon
жу'желица *f* ground beetle (*Carabus*)
жу'жжальце *n ent* balancer, halter
жужжа'ние *n* buzz, buzzing, hum, drone, droning
жужжа'ть *v* buzz, drone, whirr, hum (*of flies*)
жук *m* beetle (*Coleoptera*)
грибно'й ~ fungus beetle *(Myceptophagidae)*
колора'дский ~ potato beetle, Colorado potato beetle (*Leptinotarsa decemlineata*)
коротконадкры'лый ~ road beetle *(Staphylinidae)*
ма'йский ~ May beetle (*Melolontha*)
мали'нный ~ raspberry beetle, fruitworm beetle *(Byturidae)*
музе'йный ~ museum beetle *(Anthrenus museorum)*
пластинчатоу'сый ~ lamellicorn beetle, leaf-horned beetle *(Lamellicornia)*
хле'бный ~ cereal chafer *(Anisoplia)*
жук-носоро'г *m* rhinoceros beetle (*Dynastes*)
жук-оле'нь *m* stag beetle (*Lucanus cervus*)
жук-щелку'н *m* click beetle (*Elateridae*)
жула'н *m* red-backed shrike (*Lanius collurio*)
жураве'льник *m* heron's-bill (*Erodium*)
жура'вль *m* crane (*Grus*)
бе'лый ~ white crane (*Grus leucogeranus*)
венцено'сный ~ crested crane, crowned crane (*Balearica pavonina*)
дау'рский ~ white-naped crane *(Grus vipio)*
се'рый ~ common crane, gray crane *(Grus grus)*
жура'вль-краса'вка *m* demoiselle crane (*Anthropoides virgo*)
журча'лка *f* flower fly, hover fly (*Syrphidae*)
~ лу'ковая onion bulb fly (*Eumerus strigatus*)
жу'хлый *a* withered
жу'хнуть *v* wither, shrivel
жу'чка *f* scute (*on sturgeons*)

З

забива'ть *v* slaughter; sacrifice
забо'й *m* slaughter
забола'чивание *n* marsh formation, bog formation, paludification, swamping, bogging
заболева'емость *f* morbidity
заболева'ние *n* disease, illness, sickness
ви'русное ~ viral disease
грибко'вое ~ fungus disease
иммунодефици'тное ~ immunodeficient disease
микро'бное ~ germ disease
очаго'вое ~ focal disease
паразита'рное ~ parasitic disease
протозо'йное ~ protozoan disease
профессиона'льное ~ occupational disease, industrial disease
споради'ческое ~ sporadic disease
трансмисси'вное ~ transmissible disease
заболе'ть *v* fall ill, fall sick, contract a disease
заболо'нник *m ent* engraver (*Scolytus*)
за'болонь *f bot* sapwood, alburn, pith sheath
заболо'ченный *a* swampy, boggy
забо'та *f* care, caregiving behavior
~ **обо'их роди'телей** biparental care
~ **о пото'мстве** caregiving behavior
матери'нская ~ maternal care
отцо'вская ~ male parental care, paternal care
роди'тельская ~ parental care
забо'титься *v* take care
забра'сывание *n*: ~ **ног** capping (*of a horse*)
забро'шенный *a* abandoned
заве'рнутый *a* reflected, reflexed
завива'ние *n* curling, twisting; strophotaxis
~ **голо'вок** heading
завива'ющийся *a* twisting, coiling
завиру'шка *f* accentor (*Prunella*)
лесна'я ~ hedge sparrow, dunnock (*Prunella modularis*)
зави'симость dependence, relation
~ **Бро'ди-Бертала'нффи** the Brody-Bertalanffy relationship, the Brody-Bertalanffy relation
~ **«до'за-эффе'кт»** dose-effect relationship, dose-effect relation
гормона'льная ~ hormone dependence:
функциона'льная ~ functional relation
зави'сящий *a:*
~ **от пло'тности** density dependent
завито'й *a* coiled, curled, twisted, involute, circinate, circumflex
завито'к *m* curl, coil; spire, body whorl (*of a shell*), bostryx; vortex; volute (*a kind of inflorescence*)
заво'д *m* factory, plant
бонда'рный ~ barrel factory
броди'льный ~ fermentation plant
бума'жно-целлюло'зный ~ wood-pulp and paper mill
виноку'ренный ~ distillery
гидро'лизный ~ hydrolysis plant
дрожжево'й ~ yeast plant
коже'венный ~ tannery
комбико'рмовый ~ formula-feed plant
ко'нный ~ stud farm
консе'рвный ~ cannery, canning plant, canning factory
копти'льный ~ smoke-house
ледоде'лательный ~ ice-making plant
лесопи'льный ~ sawmill
льнообраба'тывающий ~ flax-mill
маслобо'йный ~ oilery, oil mill
маслоде'льный ~ butter factory
моло'чный ~ dairy plant, dairy factory
мылова'ренный ~ soap factory, soap plant, soap works
пивова'ренный ~ brewery, brewing plant
посо'льный ~ saltery
рыбово'дный ~ fish hatchery, fish farm
рыбоперераба'тывающий ~ fish-processing house
семеочисти'тельный ~ seed cleaning plant
сыроде'льный ~ cheese-making plant
хлопкоочисти'тельный ~ gin house, ginnery
за'водь *f* dead channel, slough, backwater
«завора'живание» *n eth* fascination behavior
завора'живать *v* fascinate
заворо'т *m med* malrotation
~ **кишо'к** volvulus
завыва'ние *n* howl, whimper
завыва'ть *v* howl
завяда'ние *n* wilting
вре'менное ~ temporary wilting
дли'тельное ~ permanent wilting
начина'ющееся ~ incipient wilting
начина'ющееся дли'тельное ~ initial permanent wilting

непроходя'щее ~ ultimate wilting
завяда'ть *v* fade, wither
завя'зывание *n bot* setting (*of fruit*)
~ **плодо'в** fruit inception
завя'зывать *v* set; tie
~ **плоды'** set the fruits
завя'зываться *v bot* set
за'вязь *f bot* ovary
ве'рхняя ~ superior ovary
многогне'здная ~ multilocular ovary
ни'жняя ~ inferior ovary
однокарпе'льная ~ monocarpellary ovary
полуни'жняя ~ half-inferior ovary
одногне'здная ~ unilocular ovary
трёхгнё'здная ~ three-celled ovary
завя'ливание *n* withering
зага'р *m* sun-tan, tan
заглазни'чный *a* postorbital
загла'тывание *n* ingestion
загла'тывать *v* ingest, swallow
загло'хший *a* overgrown, neglected
заглуша'ть *v agr* smother
заглуше'ние *n* smothering
за'гнанный *a* coursed (*a hare*); winded, overridden (*a horse*)
загна'ть *v* bay (*in hunting*); overdrive, override (*a horse*)
загнива'ть *v* begin to rot, taint, putrefy
заго'н *m* pen, enclosure
загоня'ть *v* drive in
загри'вок *m* nape, back of neck
загруди'нный *a* retrosternal
загрязне'ние *n* pollution, contamination
~ **атмосфе'ры** air pollution
~ **среды'** environmental pollution
антропоге'нное ~ anthropogenic pollution
глоба'льное ~ global pollution
теплово'е ~ heat pollution, thermal pollution
шумово'е ~ noise pollution
загрязни'тель *m* pollutant
загрязня'ть *v* pollute, contaminate
зад *m* back; behind, stern
зада'ча *f* task, aim, problem
~ **вы'бора по образцу'** matching-from-sample problem
~ **по различе'нию предметов** object-discrimination problem
~ **сравне'ния с образцо'м** *eth* matching-to-sample task
двухальтернати'вная ~ two-decision problem
заде'лка *f* covering (*of seeds*); embedding, imbedding
заде'лывать *v* bury (*e.g., seeds*); embed
задержа'ть, заде'рживать *v* hold (*e.g., breath*); detain; hamper; delay
заде'ржка *f* retardation, inhibition; delay, arrest; *ent* retinaculum (*in Collembola*)
~ **деполяриза'ции** depolarization delay
~ **перено'са** transfer delay
~ **подкрепле'ния** delay of reward
~ **разви'тия** arrest of development
атриовентрикуля'рная ~ atrio-ventricular delay
синапти'ческая ~ synaptic delay
центра'льная ~ central delay
задерне'ние *n* sod formation, sodding, swarding
задерно'ванный *a* matted
задерно'вывать *v* cover with turf, sward, cover with sward
задне-, post-
заднегру'дь *f* metathorax
заднепрохо'дно-ко'пчиковый *a* anococcygeal
заднепрохо'дный *a* anal
заднеспи'нка *f* metanotum
за'дний *a* posterior
задо'лженность *f* debt
кислоро'дная ~ oxygen debt
зажива'ть *v* cicatrize
заживле'ние *n* cicatrization
зажи'м *m lab* clamp
~ **Бильрота** Bilrot's hemostatic forceps
~ **Бле'лока** Blalock's clamp
~ **Ко'хера** Kocher's clamp
~ **Лю'эра** Luer's forceps
~ **Мо'ра** Mohr clamp
~ **Пеа'на** Pean's forceps
~ **Ха'ртмана** Harmann's forceps
~ **Хо'лстеда** Halstead's forceps
~ **Хо'фмана** Hoffman's clamp
винтово'й ~ Hoffman's clamp
пружи'нный ~ Mohr clamp
зазу'бренный *a* indented, serrated
заика'ние *n* stammering
заиле'ние, **заиливание** *n* silting, silting up, mud accumulation, acumulation of mud
зайчи'ха *f* doe hare
зака'зник *m* preserve
зака'ливание *n* hardening, habituation
зака'лка *f* hardening

закаля'ть *v* harden, increase resistance, habituate
зака'пывать *v* embed
заква'ска *f* leaven, barm; starter culture, starter
заква'шивание *n* addition of starter
закиса'ние *n* souring
закла'дка *f* setting up
закла'дывать *v* install (*e.g., hotbeds*); block, stuff up (*e.g., ears*); harness (*of horses*)
заключа'ть *v* embed, imbed
заключе'ние *n* embedding, imbedding; conclusion; resolution
заколоси'ться *v* came into ear
зако'н *m* law
~ **больши'х чи'сел** *biom* law of large numbers
~ **«всё и'ли ничего'»** all-or-nothing law
~ **Долло'** Dollo's law
~ **изме'нчивости** law of variation
~ **Ли'биха** Liebig's law, law of minimum
~ **ма'лых чи'сел** *biom* law of small numbers
~ **Ме'нделя** Mendel's law
~ **ми'нимума** law of minimum, Liebig's law
~ **незави'симого распределе'ния ге'нов** law of independent assortement
~ **обра'тных квадра'тов** *biom* inverse-square law
~ **относи'тельности** law of relativity
~ **оши'бок** *biom* error law
~ **приро'ды** law of nature
~ **Пуассо'на** Poisson's law
~ **расщепле'ния при'знаков** law of segregation
~ **рекапитуля'ции** law of recapitulation, recapitulation theory, biogenetic law, Haeckel's law
~ **се'рдца** law of the heart
~ **Ста'рлинга** law of the heart
~ **Торнда'йка** *eth* law of effect
~ **ускоре'ния** law of acceleration
~ **эфф'екта** *eth* law of effect
биогенети'ческий ~ biogenetic law, Haeckel's law, recapitulation theory
экспоненциа'льный ~ exponential law
закра'ина *f* flange ice
закрепле'ние *n* fixation, stabilization
~ **дюн** dune fixation
~ **песко'в** sand stabilization, sand fixation
~ **по'чвы** soil fixation
случа'йное ~ *phys* random fixation
закру'ченность *f phytop* twist
закру'чивание *n* coiling, curling, twisting; strophotaxis
закры'тие *n* closure, close, finish (*e.g., of the season*)
закрытопузы'рный *a* physoclistous
за кру'ченный *a* contorted, circinate
заку'кливание *n phytop* pupation
~ **зла'ков** cereal mosaic disease
~ **овса'** oat mosaic
заку'порка *f* occlusion, obstruction
закуста'ривание *n* bushing
залё'жка *f* herd, rookery, ground (*of seals*)
де'тная ~, **ще'нная** ~ breeding colony (*of seals*)
за'лежь *f* fallow soil
залива'ть *v* embed
зали'вка *f* embedding
заложе'ние *n* initiation
заложи'ть *v* establish (e.g., a garden)
залуже'ние *n* meadow formation, grassland establishment
зама'зка *f* covering (*for wounds of plants*)
зама'чивание *n* steeping
зама'чивать *v* steep (*e.g, seeds*)
замедле'ние delay
заме'дленный delayed
замедли'тель *m* inhibitor
замедля'ть delay; slow down
заме'на *f* replacement, substitution
~ **алле'ля** gene substitution
~ **без измене'ния смы'сла кодо'на** synonymous substitution
~ **основа'ний** *gen* base substitution
замени'тель *m* substitute, surrogate; replacer
~ **кро'ви** blood substitute
замеща'ющий *a* vicarious
замеще'ние *n* replacement, substitution, displacement
~ **в кольце** ring substitution
~ **в цепи** chain substitution
~ **тка'ни** replacement of tissue
согласо'ванное ~ *bioch* concerted displacement
замира'ние *n* freezing behavior, shammed death, freezing
замо'к *m* hinge, cardo (*of a shell*)
замора'живание *n* freezing
~ **в бло'ках** block freezing
~ **углекислото'й** carbon freezing
бы'строе ~ quick freezing, rapid freezing

глубо'кое ~ deep freezing
замора'живание-высу'шивание *n* freeze-drying
замора'живание-ска'лывание *n* cryofracture, freeze-fracture
замора'живание-травле'ние *n* freeze-etching
за'морозок *m* light frost, frost, cold spell
за'мша *f* chamois leather, oil leather, suede, shammy leather
замыка'ние *n* closure
усло'вное ~ conditioned closure
перекрё'стное ~ crossed closure
занеду'жить *v* fall ill
зано'сный *a* strange
зао'стренный *a* mucronate, mucronasted, acuminate, pointed
западня' *f* decoy
запа'здывание *n* lagging, delay
запа'здывающий delayed
запа'л *m* broken wind (*of a horse*)
запарши'веть *v* become mangy
запа'с *m* stock
о'бщий ~ total stock
промысло'вый ~ commercial stock
запаса'ние *n* hoarding, storing
~ **ко'рма,** ~ **пи'щи** food storing, food caching
за'пах *m* smell, odor, odour, scent
за'тхлый ~ stale smell
злово'нный ~ smelly odour
запечатле'ние *n* imprinting
~ **ко'рма** food imprinting
~ **ме'ста** locality imprinting
~ **привя'занности** filial imprinting
~ **сти'мула реа'кции сле'дования** filial imprinting
~ **пе'сни** song imprinting
обоня'тельное ~ olfactory imprinting
полово'е ~ sexual imprinting
запле'сневелый *a* moldy
запле'сневеть *v* must
заплыва'ть *v* occlude (*of a wound on a tree*); swim in
запове'дник *m* reserve, reserved area, conservation area, sanctuary, nature reserve
биосфе'рный ~ biosphere reserve
госуда'рственный ~ state reserve
охо'тничий ~ warren
запре'т *m* closure, ban
запру'да *f* dam, dike, weir
запру'живать *v* dam, dike, impound, weir
запряга'ть *v* harness, tackle up, tackle
запря'жка *f* team
~ **ве'ером** fan-wise team
~ **цу'гом** tandem-type team
~ **четве'ркой** four-horse team
запя'стный *a* carpal
запя'стье *n* wrist, carpus
заража'емость *f* susceptibility to infection
заража'ть *v* infect with, infest; inoculate; contaminate
зараже'ние *n* infecting, infection, infestation, contamination
~ **кро'ви** blood poisoning
~ **ме'стности** ground contamination
~ **пле'сневыми гриба'ми** mold contamination
бактериа'льное ~ bacterial contamination
внутриутро'бное ~ transplacental infection
конта'ктное ~ immediate contamination
мно'жественное ~ multiple infection
продукти'вное ~ productive infection
радиоакти'вное ~ radioactive contamination
трансплацента'рное ~ transplacental infection
зараже'нность *f* infection rate
зараже'нный *a* infested
зара'за *f* infection
водяна'я ~ waterweed (*Elodea*)
зарази'ха *f* broom-rape, choke vetch (*Orobanche*)
зара'зный *a* contagious, infectious, infective
зараста'ние *n* overgrowing
зараста'ть *v* occlude (*of a wound on a tree*), overgrow, grow over
заро'дыш *m* embryo, germ
аборти'вный ~ abortive embryo
адвенти'вный ~ *bot* adventive embryo
анимализо'ванный ~ animalized embryo
вегетализо'ванный ~ vegetalized embryo
зре'лый ~ mature embryo
кури'ный ~ chick embryo
многокле'точный ~ multicellular embryo
недора'звитый ~ abortive embryo
прида'точный ~ adventive embryo
сформирова'вшийся ~ mature embryo
шестикрючко'вый ~ *helm* hexacanth embryo
заро'дышевый *a* embyonic, embyonal, germinal
за'росль *f* thicket, brake, bed, vegetation bed, stand

за'ростковый *a* prothallial
за'росток *m* prothallium
заро'сший *a* overgrown, grown over, weedy
~ **ле'сом** forested
~ **тростнико'м** reedy
зару'бка *f* slash
зарубцо'вываться *v* cicatrize
заря' *f* daybreak, down; sunset
заря'нка *f* robin, redbreast (*Erithacus rubecula*)
заса'да *f* ambush
заса'дчик *m* ambuscader, sit-and-wait predator
заса'сывание *n* suction
засе'в *m* sowing, inoculation
засева'ть *v* sow with, plant with; inoculate
засе'ивать *v* sow
заселе'ние *n* stocking; colonization
заселе'нный *a* populated; stocked
заселя'ть *v* populate
заси'женный *a* flyspecked
~ **му'хами** flyspecked, flyblown
засло'нка *f anat* valve
бауги'ниевая ~ Bauhin's valve
засну'ть *v* fall asleep
засоле'ние *n* salting, salinification, salinization, salification
засоре'нность *f* littering, pollution; weedyness
засоре'нный *a* weedy, weeded, contaminated
засоря'ть *v* infest (*with weeds*); contaminate
засо'хший *a* dried; dead (*of leaves*)
засто'йный *a* stagnant
застыва'ние *n* setting
за'суха *f* drought
си'льная ~ severe drought
физиологи'ческая ~ physiological drought
физи'ческая ~ physical drought
засухоусто'йчивость *f* drought resistance, drought tolerance
засухоусто'йчивый *a* drought-resisting, siccostable
засу'шливость *f* aridity
засу'шливый *a* arid, xeric, droughty
засы'пать *v* (*perf.*) fill up, fill in; bury under
засыпа'ть *v* (*imperf.*) fall asleep; die (*of fish*); fill in, fill up; bury under; ~ **овса' ло'шади** give a horse some oats
засыха'ть *v* wither
зата'ивание *n* consealment, hiding behavior
зата'иватель *m* hider
затаить *v*: ~ **дыха'ние** hold (one's) breath
затаи'ться *v* hide
затопля'емый *a* inundable, floodable
затрави'ть *v* hunt
затра'вка *f* primer
затра'вленный *a* coursed, baited *(by dogs, e.g., a hare*)
заты'лок *m* occiput, poll
заты'лочный *a* occipital, nuchal
затя'гиваться *v* skin over
зау'шный *a* postotic
зацвета'ние *n* efflorescence
заце'пка *f* tenaculum, retinaculum, catch
зача'тие *n* conception, conceiving
зача'ток *m* primordium, germ, rudiment, bud
~ **коне'чности** limb rudiment
~ **крыла'** *ent* wing bud
~ **мочето'чника** ureteric bud
~ **о'ргана** organ rudiment
зубно'й ~ *emb* tooth bud
зача'точный *a* primordial, rudimentary
зашива'ть *v* suture
защека' *f ent* postgena
защи'та *f* shielding (*from radiation*), protection
~ **от уда'рных перегру'зок** impact protection
~ **расте'ний** plant protection
биологи'ческая ~ biological shield
иммy'нная ~ immunodefence
защи'тный *a* protective
защища'ть *v* protect
за'яц *m* hare, true hare (*Lepus*)
морско'й ~ bearded seal (*Erignatus barbatus*)
за'яц-беля'к *m* variable hare, mountain hare, polar hare (*Lepus timidus*)
за'яц-руса'к *m* brown hare, European hare (*Lepus europeus*)
за'яц-тола'й tolai hare *(Lepus tolai)*
за'ячий *a* leporine
звезда' *f* star; *cyt* aster
сперматиче'ская ~ sperm-aster
звездообра'зный *a* starlike, stelliform, asteroid
звё'здочка *f* star, blaze (*a white spot on the forehead*)
земляна'я ~ *myc* earthstar *(Geaster)*
зве'здчатый *a* stellar, stellate
зверё'к *m* small beast
зверобо'й *m* Saint-John's-wort (*Hypericum*); hunter, trapper
зверова'ние *n* hunting

зверово'д *m* fur farmer, fur breeder
зверово'дство *n* animal breeding, fur farming
  пушно'е ~ fur farming, fur farm industry
зверolo'в *m* hunter, trapper
зверolo'вство *n* hunting, trapping
зверофе'рма *f* fur farm
зверь *m* beast
  за'гнанный ~ animal at bay
  пушно'й ~ fur-bearing animal, fur-producing animal
зверю'шка *f* small beast
звон *m* ringing
  ~ **в уша'х** ringing in the ears
звоне'ц *m* midge, harlequin fly (*Chironomidae*)
звук *m* sound
  умиротворя'ющие звуки *pl eth* appeasement vocalization
здоро'вый *a* healthy, sound; salubrious; wholesome
  ~ **кли'мат** salubrious climate
  здоро'вая пи'ща wholesome food
здоро'вье *n* health
здравоохране'ние *n* public health
зе'бра *f* zebra (*Equus zebra*)
зе'бу *m* zebu (*Bos indicus*)
зев *m* fauces; *bot* gorge, throat
  льви'ный ~ *bot* snapdragon (*Antirrhinum*)
  ма'точный ~ *anat* mouth of the womb
зева'ть *v* yawn
зево'к *m* yawn
зево'та *f* yawn, yawning
зеленова'тый *a* virescent
зеленоцветко'вость *f* chloranthy
зелену'шка *f orn* greenfinch (*Chloris chloris*)
зеленчу'к *m bot* weasel-snout (*Galeobdolon*)
зелё'ный *a* green
  бриллиа'нтовый ~ brilliant green, diamond green
  бромкрезо'ловый ~ bromocresol green
  лиссами'новый ~ lissamine green
  малахи'товый ~ malachite green
  метиле'новый ~ methylene green
  мети'ловый ~ methyl green
  основно'й ~ basic green, malachite green
  про'чный ~ fast green
зе'лень *f* greenery, verdure; greens
землева'ние *n* earthing
земледе'лие *n* farming, agriculture
  бога'рное ~ dryland cultivation, dry farming, rain-fed farming
  ороша'емое ~ irrigated agriculture, irrigation farming
  подсе'чно-огнево'е ~ burnt-earth farming
  сухо'е ~ dry agriculture
землепо'льзование *n* land use, land utilization
землеро'йка *f* shrew (*Soricidae*)
землетрясение earthquake
землеустро'йство *n* land management, land regulation
земли'стый earthy
земля' *f* earth, land, soil
  бро'шенная ~ abandoned land, idle land
  вы'жженная ~ burnt-over land
  диато'мовая ~ kieselguhr
  забро'шенная ~ abandoned land
  парнико'вая ~ hotbed earth
  па'хотная ~ arable land, plowing land
  пусту'ющая ~ abandoned land, idle land
  цели'нная ~ virgin land
земляни'ка *f* strawberry (*Fragaria vesca*)
земляно'й earthy
земново'дное *n* amphibian (*Amphibia*)
  безногое ~ limbless amphibian *(Gymnophiona)*
  бесхвостое ~ tailless amphibian *(Anura)*
  хвостатое ~ tailed amphibian *(Caudata)*
земново'дный *a* amphibious, amphibian
зе'ркало *n* mirror, speculum
  ~ **воды'** water plane
зе'ркальце *n ent* areolet
зерни'стость *f* granularity, grain, *med* stippling
  мауре'рова ~ *cyt* Maurer's dots
зерни'стый *a* grained, granular, granulous, granulated, acinose
зерно' *n* grain
  алейро'новое ~ aleurone grain, aleurone granule
  вы'полненное ~ plump grain, plump seed
  кефи'рное ~ kefir grain, kefir seed
  кормово'е ~ fodder grain
  кофе'йное ~ coffee grain
  крахма'льное ~ starch grain
  продово'льственное ~ food grain
  пыльцево'е ~ pollen grain
  хлорофи'лловое ~ chlorophyll grain
  щу'плое ~ hollow grain, shrunken grain
  ячме'нное ~ barleycorn
зернобобо'вый *a* leguminous
зерно'вка *f bot* caryopsis; *ent* seed beetle (*Bruchidae*)

фасо'левая ~ bean weevil (*Acanthoscalides oblectus*)
зернов'й *a* cereal
зерновы'е *pl* corn, grain-crops, cereals
кормовы'е ~, фура'жные ~ coarse grain
зернохрани'лище *n* granary, grain barn
зерноя'дный *a* granivorous
зе'рнышко *n* grain, granule, kernel, pip
зиго- zygo-
зигода'ктильный *a* zygodactyl, zygodactyle
зигодо'нтный *a* zygodont
зигоморфи'зм *m* zygomorphy
зигомо'рфный *a* zygomorphic, zygomorphous
зигоне'врия *f* zygoneury
зигоспорока'рпный *a* zygosporocaric
зигосфе'н *m* zygosphene
зиго'та *f* zygote
зигота'ксис *m* zygotaxis
зиготи'ческий *a* zygotic
зиготропи'зм *m* zygotropism
зигофи'т *m* zygophyte
зигофо'р *m* zygophore
зимнезеле'ный *a* wintergreen
зимня'к *m* rough-legged hawk, roughleg (*Boteo lagopus*)
зимова'ть *v* winter
зимо'вка *f* wintering, hibernation
зимоге'н *m* zymogen
зимогра'мма *f* zymogram
зимоло'гия *f* zymology
зиморо'док *m* kingfisher, halcyon (*Alcedo*)
зимосто'йкий *a* winterhardy
зимосто'йкость *f* winter hardiness, winter resistance
зимоусто'йчивость *f* tolerance for winter conditions
злак *m* grass (*Gramineae*); cereal
деревяни'стый ~ woody grass
дернови'нный ~ bunch grass
дерново'й ~ sod grass
дернообразу'ющий ~ cespitous grass, turf forming grass
верхово'й ~ high grass, tall-growing grass
высо'кий ~ high grass
колосово'й ~ spiked grass
кустово'й ~ bunchgrass, bushy grass
ни'зкий ~ low grass
низово'й ~ short grass
плотнодернови'нный ~ firm bunchgrass
плотнокустово'й ~ firm bunchgrass, compact-tussock grass, densely tufted grass
рыхлокустово'й ~ loose-bunch grass, loose-tussock grass
ры'хлый ~ loose bunchgrass
зла'ковый *a* gramineous
злаколи'стый *a* graminifolious
златогла'зка *f* golden-eyed fly *(Chrysopidae)*
златогу'зка *f* brown-tail moth (*Nygmia phaeorrehaea*)
злово'нный *a* fetid, stinking, malodorous
злока'чественность *f* malignancy
злока'чественный *a* malignant, pernicious
змееви'дный *a* serpentiform, anguine
змееше'йка *f* snake bird (*Anhinga*)
змеи'ный *a* serpentine
змеея'д *m* serpent eagle, snake eagle *(Circaetus)*
змея' *f* snake (*Serpentes*)
очко'вая ~ common cobra, spectacled cobra (*Naja naja*)
эскула'пова ~ healing serpent *(Elaphe longissima)*
знак *m* sign
зри'тельный ~ visual sign
слухово'й ~ auditory sign
усло'вный ~ sign
знако'миться *v* acquaint
значе'ние *n* value; significance, meaning, sense
~ **переме'нной** variate value
абсолю'тное ~ *biom* absolute value
действи'тельное ~ true value
жела'тельное ~ desired value
и'стинное ~ true value
крити'ческое ~ critical value
наблюдё'нное ~ *biom* observed value
нача'льное ~ starting value, initial value
оконча'тельное ~ final value
оце'ночное ~ estimated value
поро'говое ~ liminal value, threshold value
предвари'тельное ~ provisional value
предска'занное ~ predicted value
презумпти'вное ~ presumptive value
прогнози'руемое ~ predicted value
проспекти'вное ~ prospective value
скорректи'рованное ~ *biom* adjusted value
сре'днее су'точное ~ daily mean
теорети'ческое ~ theoretical value
ценоти'ческое ~ coenotic value
экстрема'льное ~ extreme value
зоб *m* *med* goiter, struma; *orn* crop, maw
~ **Ри'деля** Riedel's goiter

~ **Хасимо'то** Hashimoto's disease, Hashimoto's goiter
зо'бик *m*: **медо'вый** ~ honey bag, honey stomach
зо'еа *f* zoea
золоту'ха *f* scrofula
зо'льность *f* ash content
зо'на *f* zone, band, belt
~ **брызг** spray zone
~ **Ге'да** Head's zone
~ **гемо'лиза** *imm* active area
~ **гибридиза'ции** hybrid zone
~ **жи'зни** zone of life
~ **за'плеска** splash zone, spray zone
~ **ко'рковой прое'кции** cortical projection area
~ **обще'ственного по'льзования** public access area *(e.g., beaches)*
~ **по'лного подавле'ния** zone of complete inhibition
~ **прибо'я** surf zone
~ **пролифера'ции** proliferation zone
~ **расти'тельности** floral belt, vegetation zone
~ **регенера'ции** regeneration area
~ **риско'ванного земледе'лия** area of risk farming
~ **ро'ста** region of growth, growth zone
~ **снего'в** niveal zone
~ **уме'ренного кли'мата** temperate zone
~ **фотоси'нтеза** photosynthetic zone
~ **эквивале'нтности** *imm* eqivalence zone
адапти'вная ~ adaptive zone
адора'льная ~ мембранелл *prot* adoral spiral zone
аркти'ческая ~ frigid zone
ассоциати'вная ~ association zone, association area
бу'ферная ~ buffer zone
ветрозащи'тная ~ windbreak zone
дви'гательная ~ *anat* motor area
жи'зненная ~ life zone
засу'шливая ~ arid area
защи'тная ~ protective belt
зеле'ная ~ green belt
ко'рковая проекцио'нная ~ cortical projection area
краева'я ~ marginal zone
краева'я ~ Лисса'уера Lissauer's zone
лесна'я ~ forest zone
лесостепна'я ~ forest-steppe zone
лесоту'ндровая ~ tundra forest belt
ло'бная ассоциати'вная ~ frontal association area
мё'ртвая ~ dead zone
перви'чная соматосенсо'рная ~ first somatic sensory area
перехо'дная ~ transitional zone
поглоща'ющая ~ absorbing zone
прибре'жная ~ coastal zone
прикорнева'я ~ rhizosphere
прили'вно-отли'вная ~ tidal zone, intertidal zone
проксима'льная ~ proximal zone
промежу'точная ~ intermediate zone
рефлексоге'нная ~ reflexogenic zone
речева'я ~ language zone
ростова'я ~ growth zone
снегова'я ~ niveal zone
степна'я ~ steppe zone
таё'жная ~ taiga zone
три'ггерная ~ trigger zone
тропи'ческая ~ tropical zone
ту'ндровая ~ tundra zone
уме'ренная ~ temperate zone
уме'ренно тё'плая ~ warm-temperate zone
уме'ренно холо'дная ~ cold-temperate zone
экономи'ческая ~ economic zone, 200-mile exclusive economic zone
зона'льность *f* zonality, zonation
~ **расти'тельности** zonality of vegetation
зона'льный *a* zonal
зонд *m* probe
глазно'й ~ eye probe, lacrimal sound
кле'точный ~ cell probe
коди'рующий ~ *mbiol* coding probe
молекуля'рный ~ molecular probe
мультиэлектро'дный ~ multiple electrode probe
некоди'рующий ~ *mbiol* noncoding probe
«ник»-трансли'рованный ~ *mbiol* nick-translated probe
спи'новый ~ spin probe
флуоресце'нтный ~ fluorescent probe
фотоаффи'нный ~ photoaffinity probe
зо'нтик *m bot* umbel
ло'жный ~ cyme, false umbel
сло'жный ~ compound umbel
зо'нтичек *m* umbellula, umbellule
зо'нтичный *a* umbellate
зоо- zoo-
зообе'нтос *m* zoobenthos
зообиоти'ческий *a* zoobiotic

зоогео'граф *m* zoogegrapher
зоогеографи'ческий *a* zoogeographical
зоогеогра'фия *a* zoogeography, zoological geography
зоогле'я *f* zooglea, zoogloea
зоо'ид *m* zooid
зоологи'ческий *a* zoological
зооло'гия *f* zoology
по'чвенная ~ soil zoology
зооморфо'з *m* zoomorphosis
зооно'з *m* zoonosis, zoonotic disease
зооно'мия *f* zoonomy
зоопалеонтоло'гия *f* paleozoology
зоопа'рк *m* zoological garden, zoo
зоопланкто'н *m* zooplankton
зоопсихоло'гия *f* zoopsychology
зооса'д *m* zoological garden, zoo
зооспо'ра *f* zoospore, swarm spore, swarm cell, swarmer
зооспора'нгий *m* zoosporangium
зоote'хник *m* zootechnician
зоотехни'ческий *a* zootechnic
зоотехни'я *f* zootechny
зоото'мия *f* zootomy
зоофи'льный *a* zoophilous
зоофи'т *m* zoophyte
зоохо'р *m* zoochore
зоохо'рный *a* zoochoric
зооцено'з *m* zoocenosis
зрачко'вый *a* pupillary
зрачо'к *m* pupil
си'льно су'женный ~ pin-hole pupil
щелеви'дный ~ cat's-eye pupil
зре'лость *f* maturity, ripeness
биологи'ческая ~ biological ripeness
полова'я ~ sexual maturity, puberty
преждевре'менная половая ~ early sexual maturity
съё'мная ~ picking maturity
физиологи'ческая ~ physiological maturity, biological ripeness
зре'лый *a* mature, ripe
зре'ние *n* sight, eyesight, vision
бинокуля'рное ~ binocular vision
дневно'е ~ daylight vision, day sight, day vision, photopic vision
мезопи'ческое ~ mesopic vision
мозаи'чное ~ mosaic vision
ночно'е ~ night vision, night sight, scotopic vision
о'строе ~ keen eyesight
простра'нственное ~ spatial vision
скотопи'ческое ~ night sight, night vision, scotopic vision
стереоскопи'ческое ~ stereoscopic vision, solid vision
су'меречное ~ twilight vision, mesopic vision
фотопи'ческое ~ day sight, day vision, photopic vision
цветово'е ~ color vision, chromatic vision
зреть *v* ripen
зри'тельный *a* optic, visual
зуб *m* tooth
~ **му'дрости** wisdom tooth
акродо'нтный ~ acrodont tooth
боково'й ~ lateral tooth
большо'й коренно'й ~ azzle tooth, cheek tooth, molar tooth, multicuspid tooth, tricuspid tooth
вре'менный ~ temporary tooth
глазно'й ~ eye tooth
гло'точный ~ pharyngeal tooth
жева'тельный ~ grinding tooth
коренно'й ~ molar tooth, molar, grinder, grinding tooth, cheek tooth
латера'льный ~ lateral tooth
лофодо'нтный ~ lophodont tooth
лу'нчатый ~ selenodont tooth
ма'лый коренно'й ~ premolar tooth
моло'чный ~ milk tooth, deciduous tooth, baby tooth, primary tooth
плевродо'нтный pleurodont tooth
плотоя'дный ~ carnassial tooth, sectorial tooth
постоя'нный ~ secondary tooth, succedaneous tooth, permanent tooth
секу'щий ~ sectorial tooth
селенодо'нтный ~ selenodont tooth
скла'дчатый ~ lophodont tooth
шата'ющийся ~ loose tooth
эмбриона'льный ~ *ent* egg tooth
ядови'тый ~ fang
яйцево'й ~ egg tooth
зубе'ц *m* tooth; wave (*on ECG*)
эмбриона'льный ~ hatching spike
зубно'й *a* dental
зубови'дный *a* odontoid
зубо'к *m* clove, chive, bulbil
зубр *m* wisent (*Bison bonasus*)
зубро'вка *f* holy-grass (*Hierochloe*)
зу'бчик *m* denticle
зу'бчатость *f* serration, dentation

зубча'тый *a* dentate, indented, serrate, toothed
зу'бчик *m* denticle
  слухово'й ~ auditory tooth, hair tooth
зу'бы *m pl* teeth
  оска'ливать ~ bare teeth, show teeth
зуд *m* pruritis, itch
зу'день *m*: **чесо'точный** ~ *m* itch mite (*Sarcoptes scabiei*)
зуе'к *m* plover (*Charadrius*)
  каспи'йский ~ Caspian plover, oriental sand plover (*Charadrius asiaticus*)
  крикли'вый ~ killdeer, killdeer plover (*Charadrius vociferus*)
  ма'лый ~ little ringed plover *(Charadrius dubius)*
зыбу'н *m* scraw bog
зыбь *f* ripple; swell
  мё'ртвая ~ swell, swell waves
зю'зник *m* water hoarhound (*Lycopus*)
зя'блик *m* chaffinch (*Fringilla coelebs*)

И

и'ва *f* willow, osier (*Salix*)
~ **ко'зья** goat willow *(Salix caprea)*
~ **корзи'ночная** basket willow, osier willow, velvet osier *(Salix viminalis)*
~ **ло'мкая** brittle willow, crack willow *(Salix fragilis)*
~ **плаку'чая** weeping willow, drooping willow, ring willow (*Salix babylonica*)
ива'н-да-ма'рья *f* blue cowwheat (*Melampyrum nemorosum*)
ива'н-ча'й *m* fire weed (*Chamaenerion angustifolium*)
и'ва-шелю'га *f* pussy willow, silver willow (*Salix acutifrons*)
ивня'к *m* willow stand
и'волга *f* oriole (*Oriolus*)
игла' *f* needle; spine; quill (*of a porcupine*)
~ **Деша'на** Dechamp's needle
~ **для подко'жных инъе'кций** hypodermic needle
~ **для посе'ва** inoculating needle
~ **Фра'нка** Francke's spring lancet
~ **шпри'ца** hypodermic needle
инъекцио'нная ~ entry needle
препарова'льная ~ dissecting needle, teasing needle
стекля'нная ~ glass needle
иглови'дный, **иглообра'зный** *a* acerate, acicular, spicular, spiniform, needle-shaped, belonoid
иглодержа'тель *m med* needle-holder
~ **Матье'** Mathieu's needle-holder
иглоука'лывание *n* acupuncture
иго'лка *f* needle; spine *(of a hedgehog)*
иго'лочка *f* acicle, aciculum, spinule
иго'льчатый *a* acerate, spicular
игра' *f* play, play behavior
ид *m gen* id
ида'нт *m gen* idant
идентифика'ция *f* identification
идентифици'ровать *v* identify
иденти'чность *f* identity
иденти'чный *a* identical
идеогра'мма *f* ideogram, ideograph
идеомото'рный *a* ideomotor
идио- idio-
идиоадапта'ция *f* idioadaptation
идиобиоло'гия *f* idiobiology
идиобла'ст *m* idioblast
идиовариа'ция *f* idiovariation
идиовентрикуля'рный *a* idioventricular
идиога'мия *f* idiogamy
идиогра'мма *f* idiogram
идиопепти'д *m* idiopeptide
идиопла'зма *f* idioplasm
идиосинкрази'ческий *a* idiosyncratic
идиосинкрази'я *f* idiosyncrasy
идиосинкрати'ческий *a* idisyncratic
идиосо'ма *f* idiosoma
идиоти'п *m* idiotype
генети'чески детермини'рованный ~ original idiotype
идиотипи'рование *n* idiotype assay
идиоти'пия *f* idiotypy
идиохромати'н *m* idiochromatin
идиохромосо'ма *f* idiochromosome
идти' *v* go; **идти по сле'ду** trail, spoor
иерархи'ческий *a* hierarchic
иера'рхия *f* hierarchy
~ **инсти'нктов** instinct hierarchy
лине'йная ~ linear hierarchy
синтаксономи'ческая ~ syntaxonomical hierarchy
избега'ние *n* escape, avoidance, avoidance behavior, avoidant behavior
обусло'вленное ~ *eth* conditioned avoidance
усло'вно-рефлекто'рное ~ conditioned avoidance
избира'тельность *f* selective ability, selectivity
избира'тельный *a* elective, selective
избы'точность *f* redundancy
концева'я ~ terminal redundancy
изверже'ние *n* ejection, discharge; eruption (*of vulcano*)
известкова'ние *n* lime application, lime treatment
известня'к *m* chalkstone, limestone
и'звесть *f* lime
извива'ться *v* wriggle
изви'лина *f* gyrus, convolution
~ **Бро'ка** Broca's convolution
~ **мо'зга** convolution of the brain
ве'рхняя висо'чная ~ first temporal convolution
зу'бчатая ~ dentate gyrus
ле'нточная ~ fasciolar gyrus
мозгова'я ~ gyrus
надкраева'я ~ supramarginal convolution

ни'жняя височ'ная ~ third temporal convolution
ни'жняя ло'бная ~ inferior frontal gyrus, Broca's convolution
парагиппокампа'льная ~ parahippocampal gyrus
паратермина'льная ~ Zuckerkandl's convolution
подмозо'листая ~ Zuckerkandl's convolution
поясна'я ~ callosal convolution, cingulate convolution, cingulate gyrus, convolution of cingulum
предцентра'льная ~ anterior central convolution
пряма'я ~ straight gyrus
сво'дчатая ~ limbic lobe
сре'дняя височ'ная ~ second temporal convolution
углова'я ~ angular convolution
язы'чная ~ lingual gyrus
изви'листость *f* sinuosity
изви'листый *a* flexuous, tortuous
изви'тость *f* tortuosity
извлече'ние *n* extraction
извраще'ние *n* perversion; inversion
изги'б *m* bend, curvature, flexure, incurvation
~ **сре'днего мо'зга** *emb* cephalic flexure
геотропи'ческий ~ geotropic curvature
мостово'й ~ *anat* pontine flexure
ростово'й ~ growth curvature
сигмови'дный ~ sigmoid flexure
среднемозгово'й ~ cranial flexure, mesencephalic flexure
фототропи'ческий ~ phototropic curvature
ше'йный ~ спинно'го мо'зга *emb* cervical flexure
изгиба'ние *n* inflection
изгиба'ть *v* incurvate
изгиба'ться *v* bend, curve
и'згородь *f* fence
жива'я ~ hedge, plant hedge
колю'чая ~ thorny hedge
издава'ть *v* utter (*a sound*), give vent to (*e.g., croaking*); emit (*a smell*)
изжо'га *f* heartburn
изи'дий *m* isidium
излече'ние *n* recovery
изле'чивать *v* cure
излечи'мость *f* curability
излечи'мый *a* curable, healable
излуча'ть *v* irradiate
излуча'ющий *a* irradiant
излуче'ние *n* radiation
ви'димое ~ visible radiation
диффу'зное ~ diffuse radiation, scattered radiation
жё'сткое ~ hard radiation
индуци'рованное ~ induced radiation
инфракра'сное ~ infrared radiation
ионизи'рующее ~ ionizing radiation
коллими'рованное ~ narrow-beam radiation
митогенети'ческое ~ mitogenetic radiation
наведё'нное ~ induced radiation
напра'вленное ~ beamed radiation
обратнорассе'янное ~ back-scattered radiation
отражё'нное ~ reflected radiation, back-scattered radiation
па'дающее ~ incident radiation
парази'тное ~ spurious radiation, stray radiation
побо'чное ~ spurious radiation, stray radiation
проника'ющее ~ penetrating radiation
прярассе'янное ~ forward-scattered radiation
рассе'янное ~ diffuse radiation, scattered radiation
рентге'новское ~ roentgen radiation, X-ray radiation
теплово'е ~ heat radiation, heat radiation
ультрафиоле'товое ~ ultraviolet radiation
фо'новое ~ background radiation
фотосинтети'чески акти'вное ~ photosynthetic active radiation
электромагни'тное ~ electromagnetic radiation
излу'чина *f* bend *(of a river)*
измельча'ть *v* mince; break
измельче'нный *a* minced
измене'ние *n* change, variation
~ **окружа'ющей среды'** environmental change
~ **пара'метра раздраже'ния** stimulus parameter variation
антропоге'нное ~ man-made change
внутриге'нное ~ intragenic change
внутримолекуля'рное ~ intramolecular change
конформацио'нное ~ conformational change
лока'льное ~ local variation

насле'дственное ~ hereditary change, autogenous variation
посме'ртное ~ postmortem change
посттранскрипцио'нное ~ post-transcriptional change
предсме'ртное ~ submortem change
ступе'нчатое ~ step change
изме'нчивость *f* variability, variation, mutability
~ **при'знака** variability of characteristic
альтернати'вная ~ alternative variability
внутривидова'я ~ intraspecific variation
внутрипопуляцио'нная ~ intrapopulation variation
возрастна'я ~ age variation, age variability
географи'ческая ~ geographic variation
группова'я ~ group variation
дискре'тная ~ discrete variation
индивидуа'льная ~ individual variation
корреляти'вная ~ correlated variability
межвидова'я ~ interspecific variation
напра'вленная ~ directional variation
насле'дственная ~ genetic variation
непреры'вная ~ continuous variation, continuous variability
определё'нная ~ determinate variation
оста'точная ~ residual variability
потенциа'льная ~ potential variability
преры'вистая ~ discontinuous variation, discontinuous variability
скры'тая ~ cryptic variation, cryptic variability
случа'йная ~ chance variation
фенотипи'ческая ~ phenotypic variation
экологи'ческая ~ ecological variation
изме'нчивый *a* variable, mutable
изменя'ться *v* vary, variate, change
измере'ние *n* measurement
~ **до'зы** dosage measurement
непосре'дственное ~ direct measurement
периоди'ческие измере'ния *pl* consecutive measurements
то'чное ~ precision measurement
изнеможе'ние *n* exhaustion
изнуре'ние *n* attrition, inanition
изнуря'ть *v* exhaust
изо- iso-
изоакце'пторный *a mbiol* isoaccepting
изоалле'ль *f* isoallel
изоаллели'зм *m* isoallely
изоаллерге'н *m* isoallergen
изоаллоти'п *m* isoallotype
изоа'нта *f* isoanth
изоантиге'н *m* isoantigen
изоантисы'воротка *f* isoantiserum
изоантите'ло *n* isoantibody
изоби'ловать *v* teem
изоби'лующий *a* teeming
изоби'льный *a* abundant, affluent
изоволюми'ческий, изоволюметри'ческий *a* isovolumic
изогаме'та *f* isogamete
изогами'я *f* isogamy
изога'мный *a* isogamous
изогемагглютина'ция *f* isohemagglutination
изогене'з *m* isogenesis
изогени'я *f* isogeneity
изоге'нный *a* isogenic
изогено'мный *a* isogenomatic
изогнати'ческий *a* isognathous
изо'гнутый *a* bent, curved, incurvate
изогони'я *f* isauxesis
изоде'м isodem
изодеми'ческий *a* isodemic
изодицентри'ческий *a* isodicentric
изодо'нтный *a* isodont
изозиго'тность *f* isozygoty
изозиго'тный *a* isozygotic
изози'м *m* isozym, isozyme
изоиммуниза'ция *f* isoimmunization
изоимму'нный *a* isoimmune
изокарпи'ческий *a* isocarpous
изоко'ртекс *m* isocortex
изоли'рование *n* isolation
изолецита'льный *a* isolecithal
изоли'ровать *v* isolate
изоля'т *m* isolate
географи'ческий ~ geographic isolate
перифери'ческий ~ marginal isolate
изоля'ция *f* isolation
географи'ческая ~ geographic isolation
полова'я ~ sexual isolation
посткопуляти'вная репродукти'вная ~ postmating reproductive isolation
простра'нственная ~ distance isolation, spatial isolation
репродукти'вная ~ reproductive isolation
фенологи'ческая ~ phenological isolation
изоме'р *m* isomere
изомера'за *f* isomerase, isomerizing enzyme
изомериза'ция *f* isomerization
восстанови'тельная ~ reductive isomerization
изомери'я *f* isomery, isomerism

изоме'рный *a* isomeric, isomerous
изометри'ческий *a* isometric
изоме'трия *f* isometry
изоморфи'зм *m* isomorphism
изомо'рфный *a* isomorphic
изопеди'н *m* isopedin
изопло'ид *m* isoploid
изополипло'ид *m* isopolyploid
изополиплоиди'я *f* isopolyploidy
изоспо'ра *f* isospore
изостази'я *f* isostasy
изотахофоре'з *m* isotachophoresis
  капилля'рный ~ capillary isotachophoresis
  препарати'вный ~ preparative isotachophoresis
изоте'рма *f* isotherm
изотерми'ческий *a* isothermal
изотиоциана'т *m* isothiocyanate
изоти'п *m* isotype
изотипи'я *f* isotypy
изотони'ческий, **изотони'чный** *a* isotonic
изотони'я *f* isotonicity
изото'п *m* isotope
  радиоакти'вный ~ radioactive isotope, radioisotope
изотрансплaнта'ция *f* isologous transfer
изотрисоми'я *f* isotrisomy
изотропи'я, изотро'пность *f* isotropy
изофа'г *m* isophage
изофе'на *f* isophene
изофе'нный *a* isophenic, isophenous
изоферме'нт *m* isoenzyme, isozym, isozyme
изофи'та *f* isophyte
изохромати'ческий *a* isochromatic
изохромосо'ма *f* isochromosome
изохро'нный *a* isochronic, isochronous
изоцено'з *m* isocoenosis
изоэлектрофокуси'рование *n* isoelectrofocusing
изре'женный *a* thinned, lighted
изре'живание *n* thinning
изре'занность *f* embayment (*of a shore*)
изуче'ние *n* study, investigation
  ~ **возрастно'го соста'ва популя'ции** growth-rate study
  дистанцио'нное ~ remote sensing
  сравни'тельное ~ comparative study
изъязвле'ние *n* ulceration
изъязвля'ть *v* ulcerate, canker
изъязвля'ться *v* ulcerate
ико'та *f* hiccup, hiccough, singultation
икоти'п *m* icotype
ико'тник *m* hoary madwort (*Berteroa*)
икра' *f* roe, hard roe (*inside fish*), eggs, spawn (*laid*), caviar (*a food item*); calf (*of a leg*)
икри'нка *f* egg (*of fish*)
  ~ **на ста'дии глазка'** eyed egg
ил *m* silt, sludge
  акти'вный ~ acivated sludge
и'листый *a* silty, oozy, uliginous
иллю'зия *f* illusion
илосбо'рник *m* sludge chamber
илоя'д *m* mud-eater
илоя'дный *a* mud-eating, limivorous
ильм *m* elm (*Ulmus*)
имагина'льный *a* imaginal
има'го *n* imago
имби'рь *m* ginger (*Zingiber*)
имита'ция *f* imitation
  ~ **невесо'мости** zero G imitation
иммерсио'нный *a* immersion
имме'рсия *f* immersion
  ма'сляная ~ oil immersion
иммигра'нт *m* immigrant, incomer
иммигра'ция *f* immigration
иммигри'рующий *a* immigrant
иммобилиза'ция *f* immobilization
  ~ **ферме'нтов** enzyme immobilization
иммобилизо'ванный *a* immobilized
иммортализа'ция *f* immortalization
  ~ **кле'ток** cell immortalization
иммуниза'ция *f* immunization, immunifaction
  ~ **конге'нного реципие'нта** congenic immunization
  ~ **неконге'нного реципие'нта** noncongenic immunization
  акти'вная ~ active immunization
  аллоге'нная ~ allogenic immunization
  внутриселезё'ночная ~ intrasplenic immunization
  пасси'вная ~ passive immunization
  региона'льная ~ local immunization
  сери'йная ~ cohort immunization, mass immunization
  целенапра'вленная ~ intentional immunization
иммунизи'ровать *v* immunize
  ~ **изоантиге'ном** isoimmunize
иммуните'т *m* immunity, immune resistance, immunoresistance, immunodefence
  ~ **новоро'жденного** neonatal immunity
  ~ **пло'да** fetal immunity

абсолю'тный ~ complete immunity
акти'вный ~ active immunity
антителонезави'симый ~ antibody-independent immunity
видово'й ~ specific immunity
внутриутро'бный ~ пло'да fetal immunity
вре'менный ~ temporary immunity
врожде'нный ~ inborn immunity, congenital immunity, inherent immunity
гумора'льный ~ humoral immunity, antibody-mediated immunity
депрессио'нный ~ promunity
дли'тельный ~ lasting immunity
есте'ственный ~ native immunity, natural immunity
защи'тный ~ protective immunity
иску'сственный иммунитет acquired immunity, adaptive immunity, artificial immunity
кле'точно-опосре'дованный ~ cell-bound immunity, cell-mediated immunity
кле'точный ~ cell-bound immunity, cellular immunity
конституциона'льный ~ constitutional immunity, preexisting immunity
напряже'нный ~ high-grade immunity, high-level immunity
насле'дственный ~ genetic immunity
неоната'льный ~ neonatal immunity
непродолжи'тельный ~ short-lived, short-term immunity
нестери'льный ~ nonsterile immunity
несто'йкий ~ short-lived immunity, short-term immunity
оста'точный ~ residual immunity
пасси'вный ~ passive immunity
перекре'стный ~ cross immunity
перината'льный ~ perinatal immunity
племенно'й ~ herd immunity
подавля'ющий ~ epistatic immunity
поливале'нтный ~ multivalent immunity, mixed immunity
популяцио'нный ~ herd immunity
предсуществующий ~ preexisting immunity
приобрете'нный ~ acquired immunity, adaptive immunity, artificial immunity
приро'дный ~ native immunity, natural immunity
продолжи'тельный ~ durable immunity, life-long immunity, long-lived immunity
противови'русный ~ virus-induced immunity
расовоспецифи'ческий ~ racial immunity
си'льный ~ high-grade immunity, high-level immunity
сла'бый ~ low-grade immunity
стери'льный ~ sterile immunity
стой'кий ~ durable immunity, life-long immunity, long-lived immunity
транзитор'ный ~ nonsterile immunity
трансплантацио'нный ~ graft immunity
уме'ренный ~ medium immunity
упрежда'ющий ~ preemptive immunity
фета'льный ~ fetal immunity

иммy'нный *a* immune
иммуно- immuno-
иммуноана'лиз *m* immunoassay
~ **с ла'тексным усиле'нием** latex enhanced immunoassay
~ **с подсчё'том части'ц** particle-counting immunoassay
~ **с туше'нием флуоресце'нции** fluorescence quenching immunoassay
коли'чественный ~ immunoquantification
турбидиметри'ческий ~ particle-enhanced turbidimetric immunoassay
флуоресце'нтный ~ fluorescent immunoassay
энзимокапилля'рный ~ capillary enzyme immunoassay
иммунобиоло'гия *f* immunobiology
иммунобла'ст *m* immunoblast
иммунобло'т *m* immunoblot
иммунобло'ттинг *m* immunoblotting
многояче'йковый ~ multitrack immunoblotting
мультитре'ковый ~ multitrack immunoblotting
иммуноге'ль-хроматогра'фия *f* immune gel filtration
радиа'льная ~ radial immune gel filtration
иммуногематоло'гия *f* immunohematology
иммуноге'н *m* immunogen
прими'рующий ~ priming immunogen
сла'бый ~ poor immunogen
иммуногене'тика *f* immunogenetics
иммуноге'нность *f* immunological potency, immunizing power, immunogenicity
иммуноге'нный *a* mmunogenic, immunifacient
иммуногистоло'гия *f* immunohistology
иммуноглобули'н *m* immunoglobulin, Ig

гибри'дный ~ chimeric immunoglobulin
комплементсвя'зывающий ~ complement-fixing immunoglobulin
мембра'нный ~ surface immunoglobulin
пове'рхностный ~ surface immunoglobulin
термолаби'льный ~ heat-labile immunoglobulin
химе'рный ~ chimeric immunoglobulin
иммуногра'нула *f* immunobead
магни'тная ~ magnetic immunobead
иммунодефици'т *m* immunodeficiency, immunodeficiency disorder
вариа'бельный неклассифици'руемый ~ common variable immunodeficiency
комбини'рованный ~ combined immunodeficiency
неклассифици'руемый ~ generalized immunodeficiency
иммунодиффу'зия *f* immunodiflusion
радиа'льная ~ radial immunodiflusion
иммуноимпри'нтинг *m* imprint immunofixation
иммунокомпете'нтность *f* immunocompetence
иммунокомпете'нтный *a* immunocompetent
иммуноконглютини'н *m* immunoconglutinin
иммуноконъюга'т *m* immunoconjugate
иммунолипосо'ма *f* immunoliposome
антимише'невая ~ site-directed immunoliposome
pH-регули'руемая ~ pH-sensitive immunoliposome
сайт-напра'вленная ~ site-directed immunoliposome
терморегули'руемая ~ heat-sensitive immunoliposome
иммуно'лог *m* immunologist
иммунологи'ческий *a* immunological
иммуноло'гия *f* immunology
~ **о'пухолей** cancer immunology
~ **плода'** fetal immunology
~ **разви'тия** developmental immunology
кле'точная ~ cellular immunology
сравни'тельная ~ comparative immunology
трансплантацио'нная ~ transplantation immunology
эксперимента'льная ~ experimental immunology
иммуномедиа'тор *m* immunomediator
быстроде'йствующий ~ fast-acting immunomediator
короткоживу'щий ~ short-lived immunomediator
супрессо'рный ~ immunomediator of suppression
иммуномикроноси'тель *m* immunobead
иммуномодуля'тор *m* immune-response modulating factor, immunomodulating factor
иммунопаразитоло'гия *f* parasite immunology
иммунопатогене'з *m* immunopathogenesis
иммунопотенциа'тор *m* immunopotentiator
иммунопреципита'т *m* immunoprecipitate
иммунопреципита'ция *f* immunoprecipitation
иммунорадиоавтогра'фия *f* immunoautoradiography
иммунорегуля'ция *f* immunoregulation
~ **изоти'пом** isotype-specific immunoregulation
изоти'п-специфи'ческая ~ isotype-specific immunoregulation
иммунореофоре'з *m* immunorheophoresis
иммуноселе'кция *f* immunoselection
иммуносе'нсор *m* immunosensor
иммуносорбе'нт *m* immunosorbent, immunoadsorbent
иммуносо'рбция *f* immunoadsorption
твёрдофа'зная ~ solid state immunoadsorption
иммуностимули'рование *n* immunopotentiation
иммуносупре'ссия *f* immunosuppression
дли'тельная ~ long-term immunosuppression
донор-специфи'ческая ~ donor-specific immunosuppression
избира'тельная ~ split immunosuppression
кле'точно-опосре'дованная ~ cell-mediated immunosuppression
кумуляти'вная ~ cumulative immunosuppression
селекти'вная ~ split immunosuppression
специфи'ческая ~ specific immunosuppression
стаби'льная ~ long-term immunosuppression

ятроге'нная ~ iatrogenic immunosuppression
иммуносупре'ссор *m* immunosuppressor
иммуносфе'ра *f* immunobead
иммунотокси'н *m* immunotoxin
иммунотопогра'фия *f* immunotopography
иммуносфе'ра *f* immunobead
иммуноторможе'ние *n* immunoinhibition
иммунофенотипи'рование *n* immunophenotyping
иммунофизиоло'гия *f* immunophysiology
иммунофикса'ция *f* immunofixation
иммунофильтра'ция *f* immunofiltration
иммунофлуоресце'нция *f* immunofluorescence
непряма'я ~ indirect immunofluorescence
многоцве'тная ~ multicolor immunofluorescence
мультимаркё'рная ~ multicolor immunofluorescence
иммунохи'мия *f* immunochemistry
иммуноци'т *m* immunocyte
коммити'рованный ~ committed immunocyte
иммуноцитоло'гия *f* immunocytology
иммуноцито'ма *f* immunocytoma
иммуноцитоме'трия *f* immunocytometry
иммуноцитохи'мия *f* immunocytochemistry
иммуноэлектрофоре'з *m* immunoelectrophoresis
~ **в промежу'точном ге'ле** intermediate gel immunoelectrophoresis
встре'чный ~ countercurrent immunoelectrophoresis
двуме'рный ~ two-dimensional immunoelectrophoresis
коли'чественный ~ quantitative immunoelectrophoresis
лине'йный ~ line immunoelectrophoresis
перекрё'стно-лине'йный ~ crossed line immunoelectrophoresis
перекрё'стный ~ crossed immunoelectrophoresis
противото'чный ~ countercurrent immunoelectrophoresis
раке'тный ~ по Ла'уреллу rocket immunoelectrophoresis
сли'тный раке'тный ~ fused rocket immunoelectrophoresis
сравни'тельный ~ comparative immunoelectrophoresis
та'ндемный перекрё'стный ~ tandem crossed immunoelectrophoresis, dual second step crossed immunoelectrophoresis
иммуноэмбриоло'гия *f* fetal immunology
импеда'нс *m* impedance
входно'й ~ input impedance
импланта'т *m* implant
импланта'ция *f* implantation
импоте'нция *f* impotence, impotency
импрегни'ровать *v* impregnate
импрегна'ция *f* impregnation
~ **серебро'м** silver impregnation
импри'нтинг *m eth* imprinting
гено'мный ~ genomic imprinting
гормона'льный ~ hormonal imprinting
и'мпульс *m* impulse
афферe'нтный ~ afferent impulse
восходя'щий ~ ascending impulse
дви'гательный ~ motor impulse
дыха'тельный ~ respiratory impulse
кортикоспина'льный ~ corticospinal impulse
не'рвный ~ nervous impulse
нисходя'щий ~ descending impulse
серде'чный ~ cardiac impulse
инактива'тор *m* inactivator
инактива'ция *f* inactivation
~ **комплеме'нта** complement depletion
инбре'дный *a* inbred
инбри'динг *m* inbreeding
инвагина'ция *f* invagination
инвагини'ровать *v* invaginate
инвази'рованный *a* infested
инва'зия *f* invasion, infestation
мно'жественная ~ multiple invasion
инве'рсия *f* inversion
~ **вертика'льных поясо'в** inversion of altitudinal belts, inversion of zonality
~ **температуры** temperature inversion
асимметричная ~ asymmetric inversion
ацентрическая ~ acentric inversion
две смежные инверсии *pl* tandem inversion *(in one chromosome)*
дисцентрическая ~ dyscentric inversion
комплексная ~ compound inversion
конформационная ~ conformational inversion
накладывающиеся инверсии *pl* overlapping inversions
температурная ~ temperature inversion
трансцентрическая ~ transcentric inversion

эуцентрическая ~ eucentric inversion
инверти'рование *n* inverting
инверти'рованный *a* inverted
инверти'ровать *v* invert
инверти'руемый *a* inversive
инверти'рующий *a* inverting
инверти'рующийся *a* inversive
инве'ртный *a* invert
инволю'ция *f* involution
возрастна'я ~ age involution
ингаля'ция *f* inhalation
ингиби'рование *n* inhibition
~ **коне'чным проду'ктом** end-product inhibition
аллоге'нное ~ allogenic inhibition
аллостери'ческое ~ allosteric inhibition
конкуре'нтное ~ selective inhibition
мгнове'нное ~ instantaneous inhibition
неконкуре'нтное ~ noncompetitive inhibition
необрати'мое ~ irreversible inhibition
парциа'льное ~ fractional inhibition
по'лное ~ complete inhibition
постепе'нное ~ progressive inhibition
постоя'нное ~ permanent inhibition
сла'бое ~ low inhibition
я'вственное ~ definite inhibition
ингиби'рующий *a* inhibitory
ингиби'тор *m* inhibitor
аллостери'ческий ~ allosteric inhibitor
избира'тельно де'йствующий ~ selective inhibitor
и'стинный ~ true inhibitor
конкуре'нтный ~ competitive inhibitor
неконкуре'нтный ~ noncompetitive inhibitor
необрати'мо де'йствующий ~ irreversible inhibitor
неспецифи'чный ~ nonspecific inhibitor
обрати'мо де'йствующий ~ reversible inhibitor
свобо'дный ~ free inhibitor
специфи'чный ~ specific inhibitor
эффекти'вный ~ effective inhibitor
инде'йка *f* turkey hen, turkey (*Meleagris gallopavo*)
и'ндекс *m* index
~ **агрега'ции** index of aggregation
~ **ассоциа'ции** index of association
~ **Блэ'кмена** Blackman index
~ **выжива'емости** survival index
~ **гомоге'нности** index of homogeneity
~ **гомото'нности** index of homotoneity
~ **Даля** Dahl index
~ **Дже'глама** Jeglum index
~ **Ивао** Iwao index
~ **избира'тельности** index of selectivity, index of electivity
~ **изоля'ции** isolation index
~ **Каяма** Kayama index
~ **Кертиса** Curtis index
~ **Ке'ршоу** Kershaw index
~ **кише'чной па'лочки** coli index
~ **кли'максовой адапта'ции** *bot* climax adaptation number
~ **Коттэма** Cottam index
~ **Ли'нкольна** Lincoln index
~ **Ллойда** Lloyd index
~ **листово'й пове'рхности** leaf area index
~ **Макинто'ша** McIntosh index
~ **Ма'ргалефа** Margalef index
~ **метафа'зной конъюга'ции** metaphase pairing index
~ **Мора'веца** Moravec index
~ **Мориси'та** Morisita index
~ **Му'ра** Moore index
~ **наполне'ния желу'дка** index of fullness of stomach, stomach filling index
~ **напряже'ние - вре'мя** *phys* tension-time index
~ **нейтрализа'ции ви'руса** neutralization index of virus
~ **непреры'вности** continuum index
~ **Нума'ты** Numata index
~ **Пе'терсена** Petersen index
~ **пищево'го схо'дства** index of food similarity
~ **продукти'вности** productivity index
~ **производи'теля** sire index
~ **Раа'бе** Raabe index
~ **разнообра'зия** diversity index
~ **рекомбина'ции** recombination index
~ **Ре'ньи-Ра'о** Renyi-Rao index
~ **реци'клинга** *imm* recycling index
~ **свя'занности плея'ды** cohesiveness of *a* cluster index
~ **серде'чный** ~ cardiac index
~ **Си'мпсона** Simpson index
~ **систоли'ческой рабо'ты** stroke work index
~ **ску'ченности** crowding index
~ **су'ммы при'знаков** character index
~ **супре'ссии** *imm* suppressor cell index

~ **уде'рживания** retention index *(in chromatography)*
~ **Фрэ'кера-Бри'шла** Fracker-Brischle index
~ **частоты' мито'зов** mitotic index
~ **частоты' мута'ций** mutation index
~ **частоты' реве'рсий** reversion index
~ **черепно'й** ~ cephalic index
~ **Ше'ннона-Уи'вера** Shannon-Weaver index
~ **Ште'ккера** Stocker index
~ **электи'вности** index of electivity
гонадосомати'ческий ~ gonadosomatic index
грудно'й ~ thoracic index
группово'й ~ group index
ква'нтово-хими'ческий ~ quantum chemical index
крестцо'вый ~ sacral index
митоти'ческий ~ mitotic index
паразита'рный ~ parasitic index
сакра'льный ~ sacral index
селекцио'нный ~ selection index
системи'ческий ~ stroke volume index
торака'льный ~ thoracic index
фагоцита'рный ~ phagocytic index
черепно'й ~ cephalic index
индиви'д *m* individual
индивидуализа'ция *f* individualization
индивидуа'льный *a* individual
индивидуа'ция *f* individuation
индиви'дуум *m* individual
инди'го *n* indigo (*Indigofera*)
индика'тор *m* indicator
~ **сме'ны аспе'ктов** index of succession
~ **фенологи'ческих смен** index of succession
фенологи'ческий ~ index of succession
индика'ция *f* indication
и'ндри *m ther* indri *(Indri)*
короткохво'стый ~ indri lemur *(Indri indri)*
инду'зий *m* indusium
инду'ктор *m emb* inductor, inducer
анома'льный ~ abnormal inductor
головной ~ head inductor
заднеголовно'й ~ deuterencephalic inductor
кроссо'верный ~ crossover inducer
мезодерма'льный ~ mesodermalizing agent, spinocaudal agent
нейтра'льный ~ neutralizing agent, dorsalizing agent
переднеголовно'й ~ archencephalic inductor
переднемозгово'й ~ forebrain inductor, archencephalic agent
ту'ловищный ~ trunk inductor, spinocaudal inductor
инду'кция *f* induction
~ **профага** prophage induction
~ **ротовых структур** mouth induction
~ **толерантности** tolerance induction
~ **туловищных структур** spinocaudal induction
автономная ~ autonomous induction
ассимилятивная ~ assimilative induction
гетерогенетическая ~ heterogenetic induction
гомогенетическая ~ homogenetic induction
зиготная ~ zygotic induction
комплементарная ~ complementary induction
косвенная ~ indirect induction
мезодермальная ~ mesodermal induction
нейральная ~ neural induction
отрицательная ~ negative induction
первичная ~ primary induction
переднеголовная ~ archencephalic induction
последовательная ~ successive induction
прямая ~ direct induction
сери'йная ~ sequential induction
соматическая ~ somatic induction
ферментативная ~ enzyme induction
фотосинтетическая ~ photosynthetic induction
эмбриональная ~ embryonic induction
индуци'рованный *a* induced
индуци'руемый *a* inducible
индю'к *m* turkey cock, gobbler
индюшо'нок *m* turkey chick
ине'ртность *n* inanimation
ине'ртный *a* inactive, inanimate
инжене'рия *f* engineering
белко'вая ~ protein engineering
генети'ческая ~ genetic engineering
ге'нная ~ gene engineering
инжи'р *m bot* fig (*Ficus carica*)
инио'н *m* inion
инициа'тор *m* initiator
инициа'ция *f mbiol* initiation

иници'рование *n* initiation
инквили'н *m* inquiline
инкроссбре'д *m* incrossbred
инкроссбри'динг *m* incrossbreeding
инкро'ссинг *m* incrossing
инкруста'ция *f* incrustation
инкусти'рованный *a* incrusted
инкуба'тор *m* incubator
инкубацио'нный *a* incubative, incubatory
инкуба'ция *f* incubation
инкуби'ровать *v* incubate
инкуда'тный *a* incudate
инкумба'ция *f* incumbation
инкурва'ция *f emb* incurvation
иннерва'ция *f* innervation, nervation, neuration, nerve supply
  перекрё'стная ~ cross innervation
иннерви'ровать *v* innervate
иноко'мма *f* inocomma
инокули'рование *n* inoculation
инокули'ровать *v* inoculate
ино'кулюм *m* inoculum
инокуля'т *m* inoculum
инокуля'тор *m* inoculator
  многото'чечный ~ multipoint inoculator
инокуля'ция *f* inoculation
иноро'дный *a* foreign
инотропи'зм *m* inotropism
инотро'пный *a* inotropic
инофра'гма *f* inophragma
и'ноходь *f* amble, rack, pace; **бежа'ть и'ноходью, идти' и'ноходью** amble
инса'йт *m eth* insight
инсекта'рий *m* insectary, insectarium
инсектици'д *m* insecticide
инсера'ция *f* inseration
инсе'рт *m mbiol* insert
инсе'рция *f mbiol* insertion
инсоля'ция *f* insolation
инсти'нкт *m* instinct
  ~ **до'ма** homing
  ~ **размноже'ния** reproductive instinct
  ~ **самосохране'ния** self-preservative instinct
  нересто'вый ~ spawning instinct
  полово'й ~ sex instinct
  ста'дный ~ flocking instinct, herd instinct
  роди'тельский ~ parental instinct
  хи'щнический ~ predatory instinct
изоля'ция *f* insulation
инстинкти'вный *a* instinctive
инсули'н *m* insulin
инсу'льт *m* insult, insultus
инсуля'рный *a* insular
инта'ктный *a* intact
интегра'за *f* integrase
интегра'ция *f* integration
  ~ **фа'га** phage integration
  биологи'ческая ~ biological integration
  не'рвная ~ nervous integration
интегри'рование *n* integration
интегуме'нт *m* integument
интелле'кт *m* intellect, intelligence, mentality
  иску'сственный ~ artificial intelligence
интенси'вность *f* intensity, rate
  ~ **ассимиля'ции** assimilation rate
  ~ **дыха'ния** respiration intensity, respiratory rate
  ~ **круговоро'та** turnover rate
  ~ **обме'на веще'ств** metabolic rate
  ~ **основно'го обме'на** basal metabolic rate
  ~ **отбо'ра** selection intensity
  ~ **транспира'ции** transpiration rate
  ~ **фотоси'нтеза** photosynthetic rate
  ~ **эксплуата'ции** rate of exploitation
  потенциа'льная ~ **фотоси'нтеза** potential rate of photosynthesis
интенсифика'тор *m* intensificator
интер- inter-
интерва'л *m* interval
  ~ **ме'жду про'бами** intertrial interval
  ~ **ме'жду соса'нием** nonsuckling interval
  довери'тельный ~ *biom* confidence interval
  за'мкнутый ~ *biom* closed interval
  кла'ссовый ~ class interval
  межтре'левый ~ *orn* intertrill
  наиме'ньший довери'тельный ~ *biom* shortest confidence interval
  откры'тый ~ *biom* open interval
  совме'стные довери'тельные интерва'лы *pl biom* simultaneous confidence intervals
интербри'динг *m* interbreeding
интерграда'ция *f* intergradation
  непреры'вная ~ *gen* gliding intergradation
интерзона'льный *a* interzonal
интеркаля'рный *a* intercalary
интеркаля'ция *f* intercalation
интеркине'з *m* interkinesis
интерлейки'н *m* interleukin
интерлоба'рный *a anat* interlobar
интерлобуля'рный *a anat* interlobular
интерло'кинг *m* interlocking

интермеди'н *m* intermedin
интернализа'ция *f* internalization
интеркро'сс *m* intercross
интероцепти'вный *a* interoceptive
интероце'птор *m* interoceptor
интерпрета'ция *f* interpretation
интерредупликa'ция *f* interreduplication
интерсе'кс *m* intersex
интерсексуа'льность *f* intersexuality
интерстерильный *a* interincompatible
интерстициа'льный *a* interstitial
интерфа'за *f* interphase
интерфере'нция *f* interference
~ **хиа'зм** chiasma interference
~ **хромати'д** chromatid interference
высо'кая отрица'тельная ~ high negative interference
межхромосо'мная ~ interchromosomal interference
отрица'тельная ~ negative interference
цитологи'ческая ~ cytological interference
интерферо'н *m* interferon
инти'ма *f* intima
инти'на *f* intine
интоксика'ция *f* intoxication
интразона'льный *a* intrazonal
интрамура'льный *a* intramural
интраскульпту'ра *f* intrasculpture
интрогресси'вный *a* introgressive
интрогре'ссия *f* introgression
интроду'кция *f* introduction
интродуци'ровать *v* introduce
интроми'ссия *f* intromission
многокра'тные интроми'ссии *pl* multiple intromissions
одино'чная ~ single intromission
интрон *m* intron
интро'рзный *a bot* introrse
интроспе'кция *f* introspection
интуба'ция *f* intubation
интуссусце'пция *f* intussusception
инцисти'ровавшийся *a* encysted
интэкзи'на *f pal* intexine
интэксти'на *f pal* intextine
инули'н *m* inulin
инфантили'зм *m* infantilism
инфанти'льный *a* infantile
инфа'ркт *m* infarct, infarction
~ **миока'рда** myocardial infarction
инфекти'вность *f* infectivity
инфекцио'нный *a* infectious, contagious
инфе'кция *f* infection
~ **дыха'тельных путе'й** respiratory infection
~, **передава'емая членистоно'гими** arthropod-borne infection
алимента'рная ~ alimentary infection
бытова'я ~ house infection
возду'шная ~ aerial infection, airborne infection
возду'шно-ка'пельная ~ droplet infection
врождё'нная ~ congenital infection
втори'чная ~ consecutive infection
инаппара'нтная ви'русная ~ inapparent virus infection
ка'пельная ~ droplet infection
конта'ктная ~ contact infection
лате'нтная ~ latent infection, silent infection, dormant infection, hidden infection
маскиро'ванная ~ masked infection, occult infection
о'бщая ~ systemic infection
очаго'вая ~ focal infection
пылева'я ~ dust infection
скры'тая ~ masked infection, occult infection
сме'шанная ~ concurrent infection, mixed infection
сопу'тствующая ~ concurrent infection
трансмисси'вная ~ arthropod-borne infection
хрони'ческая ~ persistent infection
эндоге'нная ~ endogenous infection
инфильтра'т *m* infiltrate, infiltration
инфильтра'ция *f* infiltration
инфлюэ'нт *m* influent
информа'ция *f* information
генети'ческая ~ genetic information
информосо'ма *f* informosome
информофе'р *m* informofere
инфрамаргина'льный *a* inframarginal
инфундибуля'рный *a* infundibular
инцисти'рование *n* encystment, encystation, cyst formation
инцисти'роваться *v* encyst
инцу'хт *m bot* inbreeding
инъе'кция *f* injection, shot
инъеци'ровать *v* inject
иониза'ция ionization
ионофо'р *m* ionophore
каналообразу'ющий ~ channel-forming ionophore
иохимби'н *m* yohimbin

ипекакуа'на *f* ipecacuanha (*Psychotria ipecacuanha*)
ирга' *f* Juneberry (*Amelanchier*)
иридоци'т *m* iridocyte
иррадиа'ция *f* irradiation
  динами'ческая ~ dynamic irradiation
  стати'ческая ~ static irradiation
иррита'ция *f* irritation
исключе'ние *n* exclusion, exception
  алле'льное ~ allele exclusion
  взаи'мное ~ mutual exclusion
  конкуре'нтное ~ competitive exclusion
  презиго'тное ~ prezygotic exclusion
  эквацио'нное ~ *cyt* equational exception
ископа'емое *n* fossil
  живо'е ~ living fossil
  руководя'щее ~ index fossil, guide fossil
ископа'емый *a* fossil
искорене'ние *n* eradication, extermination
искривле'ние *n* tortuosity
искривле'нный *a* curved, crooked, irregular
испаря'емость *f* evaporating capacity
испо'д *m* underside (*of insect wings*)
испо'льзование *n* use
  многосторо'ннее ~ multiple use
  обы'чное ~ common use
  практи'ческое ~ common use
  факти'ческое ~ actual use
испо'льзовать *v* use, utilize
испражне'ние *n* defecation, evacuation of the bowels
испражне'ния *n pl* faeces, excrements
испуска'ть *v* emit
  ~ **лучи'** irradiate
испыта'ние *n* examination, test
  ~ **на всхо'жесть** germination test
  ~ **на морозосто'йкость** freezing test
  ~ **по ка'честву пото'мства** progeny test
  биологи'ческое ~ bioassay, biological assay
  лаборато'рное ~ bench test, laboratory test
испы'тывать *v* test
иссека'ть *v* exsect, resect
иссече'ние *n* excision, resection
иссле'дование *n* investigation, examination, study, exploration, research, test
  ~ **в о'пытной устано'вке** pilot plant study
  ~ **кро'ви** blood analysis, blood examination
  ~ **опера'ции** *biom* operational research
  ко'мплексное ~ integrated study
  круглогоди'чное ~ twelve-month study
  макроскопи'ческое ~ gross examination
  микроскопи'ческое ~ microscopic examination
  ориентиро'вочное ~ pilot study
  полево'е ~ field study
  предвари'тельное ~ pilot study
  радиоизото'пное ~ isotopic examination
  ректа'льное ~ rectal examination
  рентгенологи'ческое ~ X-ray examination
исследова'ть *v* research, test, explore, investigate, examine
истече'ние *n* efflux
исто'рия *f* history
  ~ **боле'зни** case record
  есте'ственная ~ natural history
исто'д *m* milkwort (*Polygala*)
исто'чник *m* source
  ~ **изме'нчивости** *biom* source of variation
истоща'ть *v* exhaust, deplete
  ~ **по'чву** impoverish the soil
истоще'ние *n* exhaustion, emaciation; inanition; depletion, impoverishment
  ~ **комплемента** complement depletion
  ~ **не'рвной систе'мы** nervous breakdown
  ~ **озо'нового сло'я** ozone depletion
истоще'нность *f* emaciation
истоще'нный *a* exhausted; depleted, impoverished
истребле'ние *n* extermination
истребля'ть *v* exterminate
исчеза'ть *v* vanish, disappear
исчезнове'ние *n* extinction; disappearance
исче'знувший *a* extinct
исче'рченность *f* striation; banding (*of chromosomes*)
  база'льная ~ basal striation
  попере'чная ~ cross striation
исче'рченный *a* striated
исчисле'ние *n* calculus
  ~ **предика'тов** predicate calculus
  многозна'чное пропозициона'льное ~ many-valued propositional calculus
  пропозициона'льное ~ propositional calculus
  расши'ренное пропозициона'льное ~ extended propositional calculus
  у'зкое ~ предика'тов restricted predicate calculus
  функциональное ~ functional calculus
итера'ция *f* iteration
ито'г *m* total, sum, result
  о'бщий ~ *biom* grand total
ито'граф *m* itograph

ихневмо'н *m ther* Egyptian mongoose (*Herpestes ichneumon*)
ихтиоло'гия *f* ichthyology
ишеми'я *f med* ischemia
~ **миока'рда** myocardial ischemia

Й

йе'ти *m* yeti
йоди'рование *n* iodination
йод *m* iodine
  белко'во-свя'занный ~ protein-bound iodine
йодофи'льный *a* iodophilic

К

каба'н *m* boar; wild boar (*Sus scrofa*)
кабарга' *f* musk deer (*Moschus*)
кабачо'к *m bot* vegetable marrow (*Cucurbita pepo var. giraumontia*)
кавале'р *m ent* swallowtail, swordtail (*Papilio*)
каве'рна *f* cavern
каверно'зный *a* cavernous
кавита'ция *f* cavitation
кадавери'н *m* cadaverin, cadaverine
када'стр *m* cadastre
кады'к *m anat* Adam's apple
каё'мка *f* border, edge, margin
  щё'точная ~ *cyt* brush border
каёмкообра'зный *a* marginiform
каза'рка *f orn* goose *Branta; ent* weevil *Rhynchites*
  белоще'кая ~ barnacle goose (*Branta leucopsis*)
  краснозо'бая ~ red-breasted goose *(Branta ruficollis)*
  чё'рная ~ brant goose, brent goose, rode goose, rood goose *(Branta bernicla)*
казеи'н *m* casein, caseine
казео'зный *a* cheesy
казуа'р *m* cassowary (*Casuarius*)
кайма' *f* edge, border, margin, limbus
  ки'сточковая ~, **щё'точная** ~ *cyt* brush border
кайнозо'йский *a* Cenozoic
ка'йра *f* murre, kiddaw, willock, guillemot (*Uria*)
  длинноклю'вая ~ Atlantic murre, common murre, common guillemot, foolish murre *(Uria aalge)*
  короткоклю'вая ~ thick-billed murre *(Uria lomvia)*
  толстоклю'вая ~ thick-billed murre, thick-billed guillemot, Brunnich's guillemot *(Uria lomvia)*
  тонкоклю'вая ~ scoot, scout, Atlantic murre, common murre, common guillemot, foolish murre (*Uria aalge*)
какаду' *m* cockatoo (*Kacatoeinae*)
  ро'зовый ~ galah *(Eolophus roseicapillus)*
кака'о *n* cocoa-tree (*Theobroma cacao*); cocoa
какогене'з *m* cacogenesis
какогени'ческий *a* cacogenic
какоми'цли *m* cacomistle cat, ringtail *(Bassariscus)*
ка'ктус *m* cactus (*Cactaceae*)
ка'ктусовый *a* cactaceous
кал *m* fecal matter, feces
  перворо'дный ~ meconium
кала'н *m* sea otter, kalan (*Enhydra lutris*)
калга'н *m* tormentil (*Potentilla erecta*)
кале'ндула marigold *(Calendula)*
кале'чение *n* mutilation
кали'диум *m bot* kalidium
ка'лико *phytop* calico
кали'на *f* water elder, arrowwood, white-rod (*Viburnum*)
кали'птра *f bot* calyptra, calypter, moss-cap
калиптроге'н *m* calyptrogen
калло'зный *a* callosal, callose
ка'ллюс *m* callus, tylosis
  корнево'й ~ root callus
  стволово'й ~ wood callus
каллюсогене'з, каллюсообразова'ние *m* callusing, callusogenesis
ка'ловый *a* fecal
калори'йность *f* caloricity
калори'йный *a* calorific, calorigenic
калори'метр *m* calorimeter
калориме'трия calorimetry
  адиабати'ческая дифференциа'льная ~ adiabatic differential calorimetry
  дифференциа'льная скани'рующая ~ differential scanning calorimetry
  скани'рующая ~ scanning calorimetry
кало'рия *f* calorie, calory
калу'жница *f* marsh marigold, gools, kingcup (*Caltha palustris*)
кальви'ль calville
кальцефи'л *m* calciphil
кальцефи'льный *a* calciphilous, calcicole, calcicolous
кальцефи'т *m* calciphyte
кальцефо'б *m* calciphobe, calcifuge
кальцефо'бный *a* calciphobous, calcifugous
кальцитони'н *m* calcitonin
кальцификация *f* calcification
камамбе'р *m* camembert
камбиа'льный *a* cambial
ка'мбий *m* cambium
  про'бковый ~ cork cambium
камедено'сный *a* gummiferous
камедетече'ние *n* gumming, gummosis
ка'медь *f* resin, gum

камени'стый *a* stony, *anat* petrosal
ка'менка *f* wheatear (*Oenante*)
каменноу'гольный *a* carboniferous
камену'шка *f* harlequin duck (*Histrionicus histrionicus*)
ка'мень *m* stone; calculus
зубно'й ~ odontholith, tartar, dental calculus
же'лчный ~ gallstone, cholelith, biliary calculus
ка'мера *f* chamber, cell
~ **А'ббе-Це'йсса** Abbe-Zeiss apparatus
~ **Алги'ра** Algire's chamber
~ **Бо'йдена** Boyden's chamber
~ **гла'за** chamber of the eye
~ **иску'сственного кли'мата** climatic chamber, climatizer
~ **Марбру'ка** Marbrook chamber
бра'чная ~ nuptial chamber
вла'жная ~ moist chamber, dampening chamber, humidified box
выводко'вая ~ brood pouch, brood chamber, brood cavity
гермети'ческая ~ sealed chamber
диффузио'нная ~ diffusion chamber
ку'колочная ~ pupal chamber
культура'льная ~ culture chamber
морози'льная ~ freezing chamber, refrigerating chamber
отсто'йная ~ settling chamber
пере'дняя ~ anterior chamber
семенна'я ~ *bot* seed cavity
смеси'тельная ~ mixing chamber
суши'льная ~ drying chamber
сцинтилляцио'нная ~ scintillation camera
сче'тная ~ counting chamber, counting cell
хроматографи'ческая ~ chromatogram chamber
ца'рская ~ royal cell
камнеша'рка *f* turnstone, red-legged plover *(Arenaria interpres)*
кампилотро'пный *a* campylotropous
камподео'идный *a* campodeiform
камптодро'мный *a* camptodrome
камы'ш *m* bulrush, club-rush (*Schoenoplectus lacustris*)
камы'шница *f* moorhen (*Gallinula*)
камышо'вка *f* millerbird (*Arctocephalus*)
вертля'вая ~ aquatic warbler (*Arctocephalus paludicola*)
кана'ва *f* ditch, trench
кана'вка *f* groove, furrow, channel
ло'вчая ~ pitfall trap
кана'л *m* canal, channel, *anat* duct
~ **Ги'са** His' canal
~ **гу'бчатого вещества'** diploic channel
~ **зри'тельного не'рва** optic canal
~ **кле'точной коммуника'ции** cell communication channel
~ **лицево'го не'рва** Fallopian aqueduct
~ **остео'на** canal of osteon, Haversian canal
~ **ули'тки** cochlear canal
~ **ше'йки ма'тки** cervical canal of the uterus
ана'льный ~ anal canal
атриовентрикуля'рный ~ atrioventricular canal
бе'дренный ~ femoral canal
боково'й выдели'тельный ~ excretory lateral canal
большо'й нё'бный ~ pterygopalatine canal
входно'й ~ on-center channel *(in the eye retina)*
выནося'щий ~ efferent duct
выходно'й ~ off-center channel *(in the eye retina)*
га'версов ~ canal of osteon, Haversian canal
желто'чный ~ omphaloenteric duct, omphalomesenteric duct, umbilical duct, vitelline duct, yolk duct
заднепрохо'дный ~ anal canal
ио'нный ~ ion channel
ка'лиевый ~ potassium channel
ка'льциевый ~ calcium channel
кальцийзави'симый ка'лиевый ~ calcium-dependent potassium channel
камени'стый ~ stone canal *(in echinoderms)*
ке'лликеров ~ Kolliker's canal
кольцево'й ~ ring canal
крылови'дный ~ pterygoid canal
крылонё'бный ~ pterygopalatine canal
лауре'ров ~ *helm* Laurer's canal, canal of Laurer
лицево'й ~ facial canal, fallopian canal, Fallopian aqueduct
ме'дленный ка'льциевый ~ slow calcium channel
мочеполово'й ~ nephrogonoduct
на'триевый ~ sodium channel
нервно-кише'чный ~ *emb* neurenteric canal
нижнечелюстно'й ~ mandibular canal

носонё'бный ~ incisive canal, nasopalatine canal
носослё'зный ~ nasal canal, nasolacrimal duct
отводя'щий ~ efferent duct
па'ховый ~ inguinal canal, abdominal canal
пита'тельный ~ nutrient canal
подглазни'чный ~ infraorbital canal
позвоно'чный ~ medullary canal, spinal canal
полукру'жный ~ semicircular canal
потенциалозави'симый ~ voltage-dependent channel
приводя'щий ~ radiating canal *(of a contractile vacuole)*
пронефри'ческий ~ pronephric duct
резцо'вый ~ incisive canal, nasopalatine canal
родово'й ~ parturient canal
светочувстви'тельный ~ light-sensitive channel
семявынося'щий ~ deferent canal
семяизверга'тельный ~ ejaculatory duct
сенсо'рный ~ sensory canal
слё'зно-носово'й ~ nasolacrimal duct
сосу'дистый ~ vascular channel
стено'нов ~ incisive canal, nasopalatine canal
спиральны'й ~ **ули'тки** ganglionic canal
фалло'пиев ~ facial canal, fallopian canal, Fallopian aqueduct
хорда'льный ~ *emb* notochordal canal
хордомезодерма'льный ~ archenteric canal, chordo-mesodermal canal
шле'ммов ~ Schlemm's canal, ciliary canal

кана'лец *m* canaliculus, ductule
~ **перви'чной по'чки** mesonephric tubule
га'версов ~ Haversian space, tubule
денти'нный ~ dentinal canal
извито'й ~ convoluted tubule, spiral tubule
мочево'й ~ uriniferous tubule
по'чечный ~ kidney tubule, renal tubule
прямо'й ~ straight tubule
прямо'й по'чечный ~ collecting tubule
семенно'й ~ seminiferous tubule, testicular tubule
собира'тельный ~ discharging tubule

кана'льчатый *a* channelled, canaliculate, canalicular, tubular, tubulate

канаре'йка *f* canary, canarybird (*Serinus canaria*)

кана'тик *m* cord
полово'й ~ genital cord
пупо'чный ~ umbilical cord, navel string
семенно'й ~ spermatic cord, testicular cord

кана'тник *m* flowering maple (*Abutilon*)

кандидо'з *m* candidiasis *(caused by Candida albicans)*

каннибали'зм *m* cannibalism

каннибалисти'ческий *a* cannibalistic

ка'нтер *m* canter

канceроге'н *m* carcinogen

канцерогене'з *m* carcinogenesis

канцероге'нный *a* carcinogenic

канцеро'лиз *m* carcinolysis

канцеролити'ческий *a* carcinolytic

каню'к *m* buzzard (*Buteo*)
большеклю'вый ~ insect hawk (*Boteo magnirostris*)
короле'вский ~ ferruginous hawk *(Buteo regalis)*
короткохво'стый ~ short-tailed buzzard *(Buteo brachyurus)*
мохноно'гий ~ roughleg, rough-legged buzzard, rough-legged hawk (*Buteo lagopus*)
обыкнове'нный ~ common buzzard *(Buteo buteo)*
пусты'нный ~ Harris' hawk *(Parabuteo unicinctus)*
ры'бный ~ black-collared hawk *(Busarellus nigricollis)*
полоса'тый пе'гий ~ barred princeps hawk *(Leucopternis princeps)*
чё'рный кра'бовый ~ black hawk *(Buteogallus anthracinus)*
ширококры'лый ~ broad-winged hawk *(Buteo platypterus)*

кап *m for* burl, gnarl, snag

ка'персник *m*, **ка'персы**, **ка'перцы** *pl* caper, caper-bush (*Capparis spinosa*)

капиба'ра *f ther* capybara *(Hydrochoerus hydrochaeris)*

капилли'ций *m* capillitium

капилля'р *m* capillary
~ **выдели'тельной систе'мы** excretory capillary
артериа'льный ~ arterial capillary
вено'зный ~ venous capillary

капилля'рный *a* capillary

капка'н *m* steel trap; **ста'вить** ~ trap

каплу'н *m* capon

ка'пля *f* drop
вися'чая ~ hanging drop

каприо'ль *f* capriole
каприфи'га *f* caprifig
каприфика'ция *f* caprification
каприфо'ль *f* goat's-leaf *(Lonicera caprifolium)*
капси'д *m* capsid
капсида'ция *f* capsidation
капсоме'р *m* capsomere
ка'псула *f* theca
карпофо'р *m* tecaphore
ка'псула *f*
ка'псула *f* capsule, sheath, case, theca
~ **Бо'умена** capsule of glomerulus, Bowman's capsule
~ **Гли'ссона** Glisson's capsule, perivascular fibrous capsule
~ **по'чки** renal capsule
~ **Шумля'нского** capsule of glomerulus, Shumlyansky's capsule
глазна'я ~ eye capsule
жирова'я ~ fatty capsule, adipose capsule
носова'я ~ *emb* nose capsule
обоня'тельная ~ nose capsule
слухова'я ~ ear capsule
суста'вная ~ joint capsule
теко'нова ~ Tenon's capsule
яйцева'я ~ egg capsule
ка'псульный *a* capsular
капу'ста *f* cabbage (*Brassica*)
~ **брюссе'льская** Brussels sprouts (*Brassica oleracea gemmifera*)
~ **за'ячья** stonecrop (*Sedum*)
~ **ква'шеная** sour cabbage, fermented cabbage, sauerkraut
~ **кита'йская** pakchoi *(Brassica chinensis)*
~ **кольра'би** turnip-rooted cabbage, kohlrabi (*Brassica oleracea var. gongiloides*)
~ **полева'я** turnip, field turnip, Swedish turnip *(Brassica campestris)*
~ **саво'йская** savoy cabbage (*Brassica oleracea var. sabauda*)
~ **цветна'я** cauliflower, broccoli (*Brassica oleracea var. botrytis*)
капу'стница *f* cabbage white butterfly (*Pieris rapae*)
капуци'н *m* capuchin *(Cebus)*
кара'бкаться *v* scramble
карава'йка *f* glossy ibis *(Plegadis falcinellus)*
карага'на *f* pea tree, pea shrub (*Caragana*)
карага'ч *m* cork elm (*Ulmus suberosa*)
карака'л *m* caracal, caracal lynx *(Felis caracal)*
кара'ковый *a* dark bay
каракулево'дство *n* karakul sheep raising
кара'куль *m* astrakhan fur, karakul
каракульча' *f* karakulcha
карамо'ра *f* crane fly (*Tipula*)
каранти'н *m* quarantine
карапа'кс *m* carapace
карапу'зик *m* hister beetle, mimic beetle, steel beetle (*Histeridae*)
карбамили'рование, карбамоили'рование *n* carbamoylation, carbamylation
карбамилэтили'рование, карбамоилэтили'рование *n* carbamoylethylation, carbamylethylation
карбоксила'за *f* carboxylase
карбоксили'рование *n* carboxylation
восстанови'тельное ~ reductive carboxylation
карбокси'льный *a* carboxyl, carboxylic
карбоксиметили'рование *n* carboxymethylation
кардамо'н *m* cardamon, cardamom (*Elettaria cardamomum*)
кардиа'льный *a* cardiac
кардина'л *m* cardinal *orn (Cardinalis)*
кардиоге'нный *a* cardiogenic
кардиогра'мма *f* cardiogram
кардио'граф *m* cardiograph
кардиогра'фия *f* cardiography
кардио'лог *m* cardiologist
кардиомензура'тор *m* cardiomensurator
кардиостимули'рующий *a* cardioaccelerator
кардиостимуля'тор *m* cardioaccelerator
ка'рдия *f* cardiac orifice
кари'бу *m* caribou *(Rangifer tarandus)*
ка'риес *m* caries
карио- karyo-, cario-
кариогами'я *f* caryogamy, kariogamy
кариоге'н *m* karyogene
кариогра'мма а karyogram
кариоде'сма *f* karyodesma
карио'ид *m* karyoid
кариокине'з *m* karyokinesis, mitosis
кариокинети'ческий *a* karyokinetic
кариоли'зис *m* karyolysis
кариоли'мфа *f* karyolymph
кариологи'ческий *a* karyological, karyologic
кариоло'гия *f* karyology, caryology

кариоме'р *m* karyomere
кариони'да *f* caryonide
кариопла'зма *f* karyoplasm
кариоплазмати'ческий *a* caryoplasmatic
кариоре'ксис *m* karyorhexis
кариоти'п *m* karyotype
асимметри'чный ~ asymmetric karyotype
произво'дный ~ derived karyotype
симметри'чный ~ symmetric karyotype
фундамента'льный ~ fundamental karyotype
кариотипи'ческий *a* karyotypic
кариофо'р *m* karyophore
ка'рканье *n* croaking, caw
ка'ркать *v* croak
ка'рлик *m* dwarf
ка'рликовость dwarfism, dwarfishness, nanism, stunt; *bot* brachysm; *phytop* dwarf
жё'лтая ~ лу'ка onion yellow dwarf
непятни'стая курча'вая ~ карто'феля potato unmottled curly dwarf
жё'лтая ~ карто'феля potato yellow dwarf
жё'лтая ~ табака' tobacco yellow dwarf
ка'рликовый *a* dwarfish, pygmean, pygmaean, nanous; *bot* brachytic
карма'н *m* pocket, pouch
~ **Ра'тке** Rathke's pouch, Rathke's pouch, hypophyseal recess
гипофиза'рный ~ hypophyseal recess
гло'точный ~ pharyngeal recess
жа'берный ~ branchial pouch, gill pouch, gill pouch
надшишкови'дный ~ suprapineal recess
шишкови'дный ~ pineal recess
карми'н *m* carmine
кароти'дный *a* carotid
кароти'н *m* carotene, carotin
каротиногене'з *m* carotenegenesis
каротино'ид *m* carotenoid
карпелля'рный *a* carpellary
карпого'ний *m* carpogonium
карполо'гия *f* carpology
карпоспе'рм *m* carposperm
карпоспо'ра *f* carpospore
карпоспорангий
карпоспора'нгий *m* carposporangium
карпофи'т *m* carpophyte
карпофо'р *m* carpophore
карст *m* karst
ка'рта *f* map, chart
~ **генети'ческая** ~ genetic map
~ **зача'тков** fate map
~ **изогие'т** isohyetal map
~ **оса'дков** isohyetal map
~ **распределе'ния оса'дков** rainfall map
~ **расти'тельности** vegetation map
~ **секвени'рования** *mbiol* sequence map
~ **сцепле'ния** linkage map
геоботани'ческая ~ vegetation map
комплементацио'нная ~ complementation map
пепти'дная ~ peptide map
по'чвенная ~ soil map
физи'ческая ~ *mbiol* physical map
хромосо'мная ~ chromosome map
карти'на *f* picture, pattern
~ **кро'ви** blood count, hemogram, blood picture
карти'рование *n* mapping
~ **кле'точных пове'рхностей** cell surface mapping
~ **эпито'пов** epitope mapping
геоботани'ческое ~ mapping of vegetation, vegetation mapping
мембра'нное ~ cell surface mapping
многото'чковое ~ multipoint mapping
мультило'кусное ~ multipoint mapping
пепти'дное ~ peptide mapping
расти'тельности vegetation mapping
рестрикцио'нное ~ restriction site mapping
хромосо'мное ~ chromosome mapping
картофелехрани'лище *n* potato storage house
карто'фель *m* potato (*Solanum tuberosum*); potatoes
молодо'й ~ new potatoe
поса'дочный ~ seed potato
продово'льственный ~ food potato, ware potato
сла'дкий ~ sweet potato (*Ipomoea batatas*)
карто'шка *f* potatoes (*colloquial*)
кару'нкула *f* caruncle
карце'рула *f bot* carcerule
карциноло'гия *f* carcinology
карцино'ма *f* carcinoma
ка'рьер *m* career
каса'тик *m* flag (*Iris*)
каска'д *m* cascade
~ **Ха'гемана** Hageman pathway
ка'ста *f* caste
настоя'щая полова'я ~ *ent* first reproductive caste

ненастоя'щая полова'я ~ *ent* second reproductive caste
полова'я ~ reproductive caste
эргато'идная полова'я ~ *ent* third reproductive caste
кастра'т *m* castrate; spado, spade
кастра'ция, **кастри'рование** *f* castration, gelding
паразита'рная ~ parasitic castration
функциона'льная ~ functional castration
кастри'ровать *v* castrate, geld
кат *m bot* khat, kat (*Catha edulis*)
катаболи'зм *m* catabolism
катаболи'т *m* catabolite
катаболи'тный *a* catabolite
катаболи'ческий *a* catabolic
катадро'мный *a* catadromous
катагене'з *m* catagenesis
катадро'мный *a* catadromous
катакро'та *f* decaying limb
катакроти'зм *m* catacrotism
катала'за *f* catalase
катале'псия *f* catalepsy
каталепти'ческий *a* cataleptic
ката'лиз catalysis
кисло'тный ~ acid catalysis
осно'вный ~ base catalysis
ферментати'вный ~ enzyme catalysis
многоступе'нчатый ~ multistep catalysis
полифункциона'льный ~ polyfunctional catalysis
катализа'тор *m* catalyst, catalyzer
заверша'ющий ~ finishing catalyst
подгота'вливающий ~ preparatory catalyst
защища'ющий ~ protecting catalyst
многосторо'нний ~ versatile catalyst
катаплази'я *f* cataplasia
катапле'ксия *f* kataplexy
катара'кта *f* cataract
~ **Коппо'к** Coppock's catarct
розе'тчатая ~ rosula cataract
катаро'бный *a* katharobic
катаро'метр *m* katharometer
катаста'з *m* catastasis
катастро'фа *f* catastrophe, disaster
экологи'ческая ~ eco-catastrophe, ecological catastrophe
кататермо'метр *m* katathermometer
кататони'ческий *a* catatonic
кататони'я *f* catatonia
катафи'лл *m* cataphyll
катафоре'з *m* cataphoresis
катаэлектрото'н *m phys* cataelectrotonus
катего'рия *f* category
собира'тельная ~ collective category
таксономи'ческая ~ taxonomic category
катена'ция *f* catenation
катепси'н *m* cathepsin
кате'тер *m* catheter
катетериза'ция *f* catheterization
катехи'н *m* catechin
катехо'л *m* catechol
катоптри'ческий *a* catoptric
кату'шка *f* orb snail (*Planorbidae*); reel, spool
катэлектрото'н *m* katelectrotonus
каудализа'ция *f* caudalization
кауда'льно *adv* caudad
кауда'льный *a* caudal
ка'удекс *m* caudex
кауза'льный *a* causal
каули'дий *m* caulid
каулифло'рия *f* cauliflory
каулока'рпный *a* caulocarpous
каутериза'ция *f* cauterization, cautery
каучуконо'с *m* rubber plant
кахекси'я *f* cachexia, cachexy
кача'лка *f* shaker
ка'чественный *a* qualitative
ка'чество *n* quality
вкусовы'е ка'чества *pl* palatability, tastiness
втори'чное ~ secondary quality
перви'чное ~ primary quality
качу'рка *f* petrel, mother Carey's chicken (*Hydrobatidae*)
ма'лая ~ storm petrel (*Hydrobates pelagicus*)
ка'ша *f* gruel, cereal
ба'риевая ~ barium meal
овся'ная ~ porridge
ка'шель *m* cough
кашта'н *m* chestnut (*Castanea*)
кашта'новый *a* chestnut
ква'гга *f* quagga (*Equus quagga*)
квадра'т *m* square
лати'нский ~ *biom* Latin square
маги'ческий лати'нский ~ *biom* magic Latin square
реше'тчатый ~ *biom* lattice square
самосопряже'нный лати'нский ~ *biom* self conjugate Latic square

свя'занный лати'нский ~ *biom* tied Latin square
сопряжё'нные лати'нские квадра'ты *pl biom* conjugate Latin squares
сре'дний ~ *biom* mean square
сре'дний ~ **оши'бок** *biom* error mean square
транспони'рованные лати'нские квадра'ты *pl biom* conjugate Latin squares
квадривале'нт *m* quadrivalent, tetravalent
квадривале'нтный *a* quadrivalent
квадрипле'кс *m* quadriplex
квазиупру'гий *a* quasielastic
ква'канье *n* croaking
ква'кать *v* croak, sing (*of frogs*)
ква'ква *f* night heron, black-crowned night heron (*Nycticorax*)
зелё'ная ~ green-backed night heron *(Butorides striatus)*
ква'кша *f* tree frog, tree-toad *(Hylidae)*
ква'ртал *m for* compartment (*of a forest*)
кварте'т *m* quartet
квас *m* kvass (*a fermented drink of rye bread*)
квасцы' *pl* alum, alumen
алюмоаммо'ниевые ~ ammonia alum
алюмока'лиевые ~ potassium alum
железоаммиа'чные ~ iron alum
жжё'ные ~ burnt alum, exsiccated alum
хро'мовые ~ chrome alum
кватерниза'ция *f* quaternization
квохта'ть *v* cackle
кво'чка *f* brood hen, sitting hen
ке'а *m orn* kea *(Nestor notabilis)*
кедр *m* cedar (*Cedrus*)
~ **гимала'йский** Himalayan cedar, deodar cedar (*Cedrus deodara*)
~ **ла'данный** incence cedar (*Libocedrus decurrens*)
~ **лива'нский** cedar of Lebanon *(Cedrus libani)*
~ **сиби'рский** Siberian cedar (*Pinus sibirica*)
кедро'вка *f orn* nutcracker (*Nucifraga*)
кедро'вник *m* cedar forest, cembretum
кейло'н *m* chalone
ке'клик *m* chukar, chuckar, rock partridge *(Alectoris)*
кембри'йский *a* Cambrian
кена'ф *m bot* gambo hemp, mallow hemp, Deccan hemp, hemp mallow, kenaf (*Hibiscus cannabinus*)
кенгуру' *m* kangaroo *(Macropus)*
древе'сный ~ tree kangaroo *(Dendrolagus)*
кенды'рь *m bot* dogbane, Indian hemp, kendyr (*Apocynum*)
керати'н *m* keratin
кератиниза'ция *f* keratinization
керати'т *m* keratinization
керато'идный *a* keratoid
ке'рвель *m* chervil (*Anthriscus cerefolium*)
кефали'н *m* cephalin
киберне'тика *f* cybernetics
ки'ви *m* kiwi (*Apterix*); kiwi fruit
киво'к *m* wag
кився'к *m* snake millipede (*Julida*)
кида'ться *v* attack (*of a dog*)
кизельгу'р *m* kieselguhr
кизи'л *m* Cornelian cherry, dogwood (*Cornus mas*)
кизя'к *m* manure cake, dried dung
кила' *f* hernia; clubroot (*of Cruciferae*)
~ **крестоцве'тных** clubroot, anbury *(caused by Plasmodiophora brassicae)*
килева'тый *a* keeled, carinate, carinated, cariniform
килеви'дный *a* keel-shaped, cariniform, tropeic
килево'й *a* carinal
ки'ллерный *a* killer
ки'ллер-фа'ктор *m* killer factor
киль *m* keel, carina
~ **груди'ны** *orn* lophosteon
кимо'граф *m* kymograph
кинестезио'метр *m* kinesthesiometer
кине'з *m* kinesis
кинезиоло'гия kinesiology
кинестези'я *f* kinesthesia
кинестети'ческий *a* kinesthetic
кинети'да *f* kinetid
кинетокардиогра'мма *f* kinetocardiogram
кинетокардиогра'фия *f* kinetocardiography
кинетоде'сма *f* kinetodesma
кине'та *f* kinety
кине'тика *f* kinetics
~ **анамнести'ческой реа'кции** anamnestic kinetics, boostable kinetics
стациона'рная ~ steady-state kinetics
кинетоде'сма *f* kinetodesma
кинетопла'ст *m* kinetoplast
кинетосо'ма *f* kinetosome
кинетохо'р *m* kinetochore
кинжалови'дный *a* dagger-shaped
киноло'гия *f* cynology
кинопла'зма *f* kinoplasm

киносъё'мка *f* motion-picture filming
цейтра'ферная ~ time-lapse filming
кипари'с *m* cypress (*Cupressus*)
кипре'й *m* willow-herb, willowweed (*Epilobium*)
кирказо'н *m* snakeroot *(Aristolochia)*
киселеобра'зный *a* tremelloid
кис-кис' *i* puss-puss (*calling a cat*)
кисли'ца *f* wood-sorrel, shamrock *(Oxalis acetosella)*
кисловатый *a* sourish
кислоро'д *m* oxygen
кислота' *f* acid
абиети'новая ~ abietic acid, abietinic acid
абсци'зовая ~ abscisic acid
агарици'н овая ~ agaric acid ~, agaricic acid
адени'ловая ~ adenylic acid
аденилоянта'рная ~ adenylosuccinic acid
аденозиндифо'сфорная ~, **АДФ** adenosine diphosphoric acid
аденозинмонофо'сфорная ~, **АМФ** adenylic acid
аденозинтрифо'сфорная ~, **АТФ** adenosine triphosphoric acid
адипи'новая ~ adipic acid ~, adipinic acid
азо'тная ~ nitric acid
аконити'новая ~ aconitic acid
акри'ловая ~ acroleic acid ~, acrylic acid
алифати'ческая ~ aliphatic acid
аллантои'новая ~ allantoic acid
алло'новая ~ allonic acid
аллофа'новая ~ allophanic acid
альги'новая ~ alginic acid
альдобио'новая ~ aldobionic acid
аминоадипи'новая ~ aminoadipic acid
аминобензо'йная ~ aminobenzoic acid
аминоглута'ровая ~ aminoglutaric acid
аминоизома'сляная ~ aminoisobutyric acid
аминоимидазолкарбокси'льная ~ amino-imidazole carboxylic acid
аминокапро'новая ~ aminocaproic acid
аминолевули'новая ~ aminolevulinic acid
аминома'сляная ~ aminobutyric acid
аминооксима'сляная ~ aminohydroxybutyric acid
аминопимели'новая ~ aminopimelic acid
ани'совая ~ anisic acid
антрани'ловая ~ anthranilic acid
арахидо'новая ~ arachidonic acid
арахи'новая ~ arachic acid
арсо'новая ~ arsonic acid
аскорби'новая ~ ascorbic acid, hexuronic acid, cevitamic acid
аспараги'новая ~ asparaginic acid, aspartic acid
асперги'лловая ~ aspergillic acid
ауринтрикарбо'новая ~ aurintricarboxylic acid
ацетилмура'мовая ~ acetylmuramic acid
ацетомоло'чная ~ acetolactic acid
ацетондикарбо'новая ~ acetonedicarboxylic acid
ацетооксима'сляная ~ acetohydroxybutyric acid
ацетоу'ксусная ~ acetoacetic acid, diacetic acid
ацетэру'ковая ~ erucylacetic acid, nervonic acid
барбиту'ровая ~ barbituric acid
беге'новая ~ behenic acid
бензо'йная ~ benzoic acid
виннока'менная ~ tartaric acid
вольфра'мовая ~ tungstic acid
галакто'новая ~ galactonic acid
га'лловая ~ gallic acid
гентизи'новая ~ gentisic acid
гиалуро'новая ~ hyaluronic acid
гиббере'лловая ~ gibberellic acid
гидантои'новая ~ hydantoic acid
гиднока'рповая ~ hydnocarpic acid
гиппу'ровая ~ hippuric acid, urobenzoic acid
гиптаге'новая ~ hiptagenic acid, nitropropionic acid
глико'левая ~ glycolic acid, hydroxyacetic acid
гликохо'левая ~ glycocholic acid
глиокси'ловая ~ glyoxylic acid, ethanal acid
глицери'новая ~ glyceric acid
глицирре'товая ~ glycyrrhetic acid
глутами'новая ~ glutamic acid, glutaminic acid
глута'ровая ~ glutaric acid
глюко'новая ~ gluconic acid
глюкуро'новая ~ glucuronic acid
гепта'новая ~ heptanoic acid, heptylic acid
глутами'новая ~ aminoglutaric acid
гомогентизи'новая ~ homogentisic acid, alcapton
гуани'ловая ~ guanylic acid
гуло'новая ~ gulonic acid

дегидрохо'левая ~ dehydrocholic acid
дегидрошики'мовая ~ dehydroshikimic acid
дезоксирибонуклеи'новая ~, ДНК deoxyribonucleic acid
дезоксиуриди'ловая ~ deoxyuridylic acid
дезоксицитиди'ловая ~ deoxycytidylic acid
дека'новая ~ decanoic acid
диалу'ровая ~ dialuric acid
диаминопимели'новая ~ diaminopimelic acid
дигидрофо'левая ~ dihydrofolic acid
длинноцепо'чная жи'рная ~ long chain fatty acid
додека'новая ~ dodecanoic acid
дьенко'ловая ~ djenkolic acid
жё'лчная ~ bile acid
жи'рная ~ fatty acid
индолилма'сляная ~ indolebutyric acid
индолилу'ксусная ~, ИУК indoleacetic acid
какоди'ловая ~ cacodylic acid
капри'ловая ~ caprylic acid, n-octanoic acid
какоди'ловая ~ dimethylarsinic acid
капри'новая ~ capric acid
капро'евая ~ caproic acid, hexanoic acid
капролеи'новая ~ caproleic acid
капро'новая ~ capronic acid
карбами'новая ~ carbamic acid
карбо'ловая ~ phenylic acid
карми'новая ~ carminic acid
кетоглута'ровая ~ ketoglutaric acid
кинуре'новая ~ kynurenic acid
клофибри'новая ~ clofibric acid
коби'новая ~ cobinic acid
кори'чная ~ cinnamic acid
кофеи'новая ~ caffeic acid
крото'новая ~ crotonic acid
ксанти'ловая ~ xanthylic acid
ксантуре'новая ~ xanthurenic acid
кума'ровая ~ coumaric acid
куми'новая ~ cumic acid ~, cuminic acid
кути'новая ~ cutinic acid
лактами'новая ~ lactamic acid, acetylneuramic acid
ланоцери'новая ~ lanoceric acid
лаури'новая ~ lauric acid
лауролеи'новая ~ lauroleic acid
левопима'ровая ~ levopimaric acid
левули'новая ~ levulinic acid
лизерги'новая ~ lysergic acid
лимо'нная ~ citric acid
лино'левая ~ linoleic acid
линоле'новая ~ linolenic acid
липо'евая ~ lipoic acid
литохо'левая ~ lithocholic acid
малеи'новая ~ maleic acid
мало'новая ~ malonic acid
манно'новая ~ mannonic acid
ма'сляная ~ butyric acid
мевало'новая ~ mevalonic acid
мелило'товая ~ melilotic acid
мелисси'новая ~ melissic acid
меркаптока'рбоновая ~ mercaptan acid
меркапту'ровая ~ mercapturic acid
метиленаминома'сляная ~ methyleneaminobutyric acid
метилянта'рная ~ methylsuccinic acid
мефена'мовая ~ mefenamic acid
мико'ловая ~ mycolic acid
моло'чная ~ lactic acid
мочева'я ~ uric acid
муко'новая ~ muconic acid
муравьи'ная ~ formic acid
мура'мовая ~ muramic acid
мiристи'новая ~ myristic acid
миристоле'новая ~ myristoleic acid
мурихо'левая ~ muricholic acid
мускари'новая ~ muscarinic acid
муци'новая ~ mucic acid
мышьяко'вая ~ arsenic acid
мышьяко'вистая ~ arsenous acid
налиди'ксовая ~ nalidixic acid
нейрами'новая ~ neuraminic acid
никоти'новая ~ nicotinic acid
нитрилтрёху'ксусная ~ nitrilotriacetic acid
нуклеи'новая ~ nucleic acid
оксалу'ровая ~ oxaluric acid
оксиантрини'ловая ~ hydroxyanthranilic acid
оксипировиногра'дная ~ hydroxypyruvic acid
оксиянта'рная ~ malic acid
олеано'левая ~ oleanolic acid
олеи'новая ~ oleic acid
опиа'новая ~ opianic acid
орниту'ровая ~ ornithuric acid
оро'товая ~ orotic acid
орсели'новая ~ orsellinic acid
о'смиевая ~ osmic acid
пальмити'новая ~ palmitic acid

пальмитолеи'новая ~ palmitoleic acid
панто'евая ~ pantoic acid
пантоте'новая ~ pantothenic acid
парааминобензо'йная ~ p-amino-benzoic acid
параба'новая ~ parabanic acid
пекти'новая ~ pectic acid, pectinic acid
пеларго'новая ~ pelargonic acid
пента'новая ~ pentanoic acid
пикри'новая ~ picric acid, nitroxanthic acid
пима'ровая ~ pimaric acid
пимели'новая ~ pimelic acid
пипеколи'новая ~ pipecolic acid
пировиногра'дная ~ pyruvic acid, methylsuccinic acid
пирога'лловая ~ pyrogallic acid
пироглутами'новая ~ pyroglutamic acid
полиадени'ловая ~ polyadenylic acid
префе'новая ~ prephenic acid
пропио'новая ~ propionic acid
протокате'ховая ~ protocatechuic acid
псевдоуриди'ловая ~ pseudouridylic acid
пулвилло'ровая ~ pulvilloric acid
рибонуклеи'новая ~ ribonucleic acid
салици'ловая ~ salycilic acid
са'харная ~ saccharic acid
себаци'новая ~ sebacic acid
сенецио'новая ~ senecioic acid
се'рная ~ sulfuric acid
сиа'ловая ~ sialic acid, acetylneuraminic acid
сина'повая ~ sinapic acid
сини'льная ~ hydrocyanic acid, cyanhydric acid
сли'зевая ~ mucic acid
соля'ная ~ hydrochloric acid
сорби'новая ~ sorbic acid
стеари'новая ~ stearic acid
сульфо'новая ~ sulfonic acid
таурохо'левая ~ taurocholic acid
тейхуро'новая ~ teichuronic acid
тетрагидрофо'левая ~ tetrahydrofolic acid
тимиди'ловая ~ thymidylic acid
тио'ктовая ~ thioctic acid
тиоциа'новая ~ thiocyanic acid
трийодбензо'йная ~, ТИБК triiodobenzoic acid
трикарбо'новая ~ thymidylic acid
тро'повая ~ tropic acid
туберкулостеари'новая ~ tuberculostearic acid
у'гольная ~ carbonic acid
у'ксусная ~ acetic acid, ethanoic acid
ундеце'новая ~ undecenoic acid
уриди'ловая ~ uridylic acid
уро'новая ~ uronic acid
феру'ловая ~ ferulic acid
фити'новая ~ phytic acid
флави'новая ~ flavinic acid
фо'лиевая ~ folic acid
фоли'новая ~ folinic acid
фосфати'дная ~ phosphatidic acid
фосфоглицери'новая ~ phosphoglyceric acid
фосфоглюко'новая ~ phosphogluconic acid
фосфопантоте'новая ~ phosphopantothenic acid
фо'сфорная ~ phosphoric acid
фосфорнова'тистая ~ hypophosphorous acid
фосфоэнолпировиногра'дная ~ phosphoenolpyruvic acid
френози'новая ~ phrenosinic acid
фруктуро'новая ~ fructuronic acid
фта'левая ~ phthalic acid
фумара'мовая ~ fumaramic acid
фума'ровая ~ fumaric acid
фуси'довая ~ fusidic acid
хелидо'новая ~ chelidonic acid
хенодезоксихо'левая ~ chenodeoxycholic acid
хи'нная ~ kinic acid
хиноли'новая ~ quinolinic acid
хлорнова'тистая ~ hypochlorous acid
хола'новая ~ cholanic acid
хо'левая ~ cholic acid
холеи'новая ~ choleic acid
хори'змовая ~ chorismic acid
церебро'новая ~ cerebronic acid, phrenosinic acid
цероти'новая ~ cerotic acid
циа'нисто-водоро'дная ~ cyanhydric acid, hydrocyanic acid
циа'новая ~ cyanic acid
циану'ровая ~ cyanuric acid
цистеи'новая ~ cysteic acid
цитиди'ловая ~ cytidylic acid
цитрама'левая ~ citramalic acid
шики'мовая ~ shikimic acid
щаве'левая ~ oxalic acid
щавелевоу'ксусная ~ oxaloacetic acid
щавелевоянта'рная ~ oxalsuccinic acid

элаиди'новая ~ elaidic acid
эру'ковая ~ erucic acid
этилендиаминтетрау'ксусная ~ ethylenediaminetetraacetic acid
я'блочная ~ malic acid
кисло'тность *f* acidity
акти'вная ~ actual acidity, actual acidity
и'стинная ~ actual acidity, actual acidity
о'бщая ~ total acidity
кисло'тный *a* acid, acidic
кислотоусто'йчивость *f* acid fastness
кислотоусто'йчивый *a* acid-fast
ки'слый *a* sour; acid, acidic
киста' *f* cyst
кистеви'дный *a* botryoid, botryoidal, racemose
кистено'сный *a* racemiferous
кистеобра'зный *a* botryoid, botryoidal, racemose
ки'сточка *f* tuft
кисточкови'дный *a* brush-like
кисть *f* hand; bunch, cluster, acervulus; *bot* raceme, racemation, truss
~ **виногра'да** bunch of grapes
~ **руки'** hand
виногра'дная ~ bunch of grapes
небольша'я ~ racemule
кит *m* whale (*Cetacea*)
кита'йка *f* Chinese apple (*Malus prunifolia*)
киша'щий *a* teeming
кише'ть *v* teem with, swarm with
кише'чник *m* intestine
кише'чный *a* intestinal, enteral, enteric
кишка' *f* intestine, gut
двенадцатипе'рстная ~ duodenum
задняя ~ hind intestine
ободо'чная ~ colon, segmented intestine
перви'чная ~ primitive gut
пере'дняя ~ foregut, fore intestine
подвздо'шная ~ ileum, twisted intestine
пряма'я ~ rectum, straight intestine
слепа'я ~ caecum, cecum, blind gut, blind intestine
сре'дняя ~ midgut
то'лстая ~ large intestine, large gut, large bowel, great gut
то'нкая ~ small intestine, small gut, narrow intestine, little gut
то'щая ~ jejunum, empty intestine
кишне'ц *m bot* coriander (*Coriandrum*)
кла'вус *m* clavus
клади'стика *f* cladictics
кладисти'ческий *a* cladistic
кла'дка *f* clutch, egg mass; egg laying, laying
кладогене'з *m* cladogenesis
кладогра'мма *f* cladogram
кладо'дий *m* cladode, cladophyll
кладомиа'льный *a bot* cladomial
клазматоци'т *m* clasmatocyte
кла'пан *m* valve, flap
~ **лё'гочной арте'рии** pulmonary valve
аорта'льный ~ aortic valve
вдыха'тельный ~ inspiratory valve
выдыха'тельный ~ expiratory valve
выпускно'й ~ exhaust valve
двуство'рчатый ~ bicuspid valve
дыха'тельный ~ respiratory valve
запра'шивающий ~ demand valve
контро'льный ~ check valve
манти'йный ~ mantle flap
митра'льный ~ bicuspid valve, mitral valve
спира'льный ~ spiral valve
трехство'рчатый ~ tricuspid valve, tricuspidal valve
кла'панный *a* valvate, valvular
класс *m* class, category
~ **ассоциа'ций** association class
~ **боните'та** quality class
~ **домини'рования** dominance class
~ **домини'рующих дере'вьев** crown class
~ **местообита'ния** site class
~ **собы'тий** *biom* class of events
~ **содомини'рующих дере'вьев** codominant crown class
~ **частоты' встреча'емости** frequency class
возрастно'й ~ age class
годово'й ~ year class
мода'льный ~ *biom* modal class
откры'тый ~ *biom* open-ended class
рекомбинацио'нный ~ recombination class
классифика'ция *f* classification
~ **по одному' при'знаку** one-way classification
~ **сероти'пов** serological classification
дихотоми'ческая ~ dichotomic classification
иерархи'ческая ~ hierarchial classification
мно'жественная ~ manifold classification
нумери'ческая ~ numerical classification
программи'рованная ~ automatic classification

серологи'ческая ~ serological classification
классифици'ровать *v* categorize
кла'стер *m* cluster
~ **ге'нов** gene cluster
дуплици'рованный ~ duplicated cluster
кла'стер-ана'лиз *m* cluster analysis
клёв *m* biting, bite
клева'ние *n* pecking
ло'жное ~ sham pecking
побо'чное ~ collateral pecking
клева'ть *v* dab, peck (*of birds*); bite (*of fish*)
клева'ть на *v* bite at, nibble at (*of fish*)
кле'вер *m* clover, trefoil (*Trifolium*)
клевери'ще *n* clover sod
клево'к *m* peck, dab
клей *m* adhesive, glue, paste
по'чковый ~ *bot* blastocolla
пчели'ный ~ bee glue, propolis
ры'бий ~ fish glue, isinglass
клейдои'ческий *a* cleidoic
клейкови'на *f* gluten
клейкови'нность *f* gluten content
клейме'ние *n* marking
кле'йстер *m* paste
крахма'льный ~ starch paste, starch adhesive
клейстога'мия *f* cleistogamy
клейстога'мный *a* cleistogamic, cleistogamous
клейстока'рпий *a* cleistocarp
клейстока'рпный *a* cleistocarpous
клейстоте'ций *m* cleistothecium
клека'чка *f* bladdernut (*Staphylea*)
кле'кот *m* squawk, scream
клекота'ть *v* squawk, scream
кле'мма *f* clamp
клён *m* maple (*Acer*)
~ **бе'лый** sycamore maple (*Acer pseudoplatanus*)
~ **саха'рный** sugar maple, sweet maple *(Acer saccharum)*
~ **тата'рский** Tatarian maple *(Acer tataricum)*
клено'вник maple forest
клептобио'з *m* cleptobiosis
клептога'мия *f eth* kleptogamy
клептопаразити'зм *m* kleptoparasitism
клёст *m* crossbill *(Loxia)*
~ **белокры'лый** two-barred crossbill, white-winged crossbill *(Loxia leucoptera)*
клёст-ело'вик *m* common crossbill, red crossbill *(Loxia curvirostra)*
клёст-сосно'вик *m* parrot crossbill *(Loxia pytyopsittacus)*
кле'тка *f* cell; cage
~ **а'цинуса** acinar cell
~ **Бе'тца** Betz cell
~ **HeLa** HeLa cell
~ **Ге'нзена** Hensen's cell, outer limiting cell
~ **гибридо'мы** hybridoma cell
~ **гно'я** pus cell
~ **Го'льджи** Golgi cell, stellate nerve cell
~ **Де'йтерса** Deiters' cell, outer phalangeal cell, prop cell
~ **Кла'удиуса** Claudius' cell, outer supporting cell
~ **клубо'чковой зо'ны** glomerulosa cell
~, **нагру'женная антитела'ми** antibody-armed cell, antibody-coated cell
~ **Ко'рти** Corti's cell
~ **ко'стного мо'зга** marrow cell
~ **кро'ви** blood cell, hemic cell
~ **культура'льного происхожде'ния** culture-origin cell
~ **Ку'пфера** Kupfer cell, stellate endothelial cell
~ **Лангха'нса** Langhans cell
~ **Ле'йдига** Leydig's cell
~ **Марша'на** Marchand's cell, adventitional cell
~ **мостови'дного эпите'лия** pavement cell
~ **невроглии'и** glia cell
~ **па'мяти** *imm* memory cell
~ **Па'нета** Paneth cell, intestinal acidophil cell
~ **пе'чени** hepatic cell
~ **пло'ского эпите'лия** squamous epithelial cell
~, **прилипа'ющая к стеклу'** glass-adherent cell
~, **прими'рованная гапте'ном** hapten-primed cell
~ **Пуркинье'** Purkinje's cell
~ **раздраже'ния** irritation cell, Türk irritation cell
~ **Рамо'н-и-Каха'ла** cell of Ramon-y-Cajal
~ **Ри'дера** Rieder cell
~ **Руже'** Rouget cell, adventitional cell
~ **с включе'ниями** inclusion-bearing cell
~, **свя'зывающая гапте'н** hapten-binding cell

~ **Серто'ли** Sertoli's cell, testicular follicular cell
~ **склере'нхимы** mechanical cell
~ **сли'зистой тка'ни** myxocete
~ **спина'льного га'нглия** spinal ganglion cell
~ **с ресни'чным пла'менем** flame cell
~ **стебелька'** stalk cell
~ **стро'мы** stromal cell
~ **Тю'рка** Türk irritation cell
~ **Фана'на** cell of Fanana, radiate glial cell
~ **Фре'йнда** Freund's cell
~ **цитотрофобла'ста** Langhans cell
~ **Шва'нна** Schwann's cell, sheath cell
~ **Шу'льце** Schultze's cell
агглютини'рованные кле'тки *pl* clumped cells
адвентициа'льная ~ adventitional cell, Marchand's cell, Rouget cell
аллоге'нная ~ allogenic cell, allogenerated cell
амебо'идная ~ amoeboid cell
антеридиа'льная ~ antheridal cell
антиге'н-отвеча'ющая ~ antigen-responsive cell
антиге'н-представля'ющая ~ antigen-presenting cell, presenter cell
антиге'н-презенти'рующая ~ antigen-presenting cell, presenter cell
антиге'н-специфи'ческая ~ antigen-specific cell
антиге'н-чувстви'тельная ~ antigen-sensitive cell
антителообразу'ющая ~ antibody-forming cell, antibody-generating cell, antibody-producing cell
антителопродуци'рующая ~ antibody-forming cell, antibody-generating cell, antibody-producing cell
антителосодержа'щая ~ antibody-containing cell
аргирофи'льная ~ argyrophil cell, enterochromaffin cell, granulobasal cell, pheochrome cell
археспориа'льная ~ archesporial cell
астрогиа'льная ~ spider cell, astrocyte
аутоагресси'вная ~ autoaggressive cell
аутоантиге'нная ~, **аутологичная** ~ self cell, self-restricted cell, self- specific cell
аутореакти'вная ~ self-reactive cell, autologous responding cell
ацидофи'льная ~ acidophil cell, alpha cell
ацидофи'льная кише'чная ~ intestinal acidophil cell, Paneth cell
база'льная ~ basal cell
базальнозерни'стая ~ argyrophil cell, enterochromaffin cell, granulobasal cell, pheochrome cell
базофи'льная ~ basophil cell
бактериа'льная ~ bacterial cell
безъя'дерная ~ acaryocyte, acaryote
бесспа'йковая ~ nonspiking cell
биполя'рная ~ bipolar cell
блужда'ющая ~ migratory cell, wandering cell
бляшкообразу'ющая ~ plaque-forming cell
бокалови'дная ~ caliciform cell, chalice cell, flask cell, goblet cell, bottle cell, beaker cell
бурстобразу'ющая ~ burst-forming cell
вегетати'вная ~ vegetative cell
веретенообра'зная ~ spindle cell
верху'шечная ~ apical cell
вкусова'я ~ taste cell
вну'тренняя волоско'вая ~ Corti's cell
водопроница'емая ~ *bot* hydropot cell
волоско'вая ~ hair cell
волоско'вая сенсо'рная ~ acoustic hair cell, bristle cell
воротничко'вая ~ collar cell, choanocyte
вса'сывающая ~ absorbing cell
вуа'левая ~ veiled cell
вуалеви'дная ~ veiled cell
ганглио'зная не'рвная ~ ganglionic nerve cell, Purkinje's cell
гаптенили'рованная ~ haptenylated cell
гемопоэти'ческая ~ hemopoietic cell
генерати'вная ~ *bot* generative cell
гибри'дная ~ hybrid cell
гига'нтская ~ giant cell
гипофиза'рная ~ pituitary cell
глиа'льная ~ Го'льджи glial Golgi cell, epithelial glial cell
грамотрица'тельная ~ Gram-negative cell
грамположи'тельная ~ Gram-positive cell
грудна'я ~ chest, thorax, thoracic cage
двухъя'дерная ~ binucleate cell
деля'щаяся ~ dividing cell
денти'нная ~ dentinal cell
децидуа'льная ~ decidual cell
диста'льная ретина'льная ~ distal retinula cell
доче'рняя ~ daughter cell

дрожжева'я ~ yeast cell
ЕА-розеткообразу'ющая ~ EA-rosetting cell
ЕАС-розеткообразу'ющая ~ EAC-rosetting cell
Е-розеткообразу'ющая ~ E-rosetting cell
есте'ственная цитотокси'ческая ~ natural cytotoxic cell
желе'зистая ~ gland cell, glandular cell
желе'зистая ~ яи'чка Leydig's cell
желто'чная ~ yolk cell
жгутиконо'сная ~ whip cell
жива'я ~ living cell
жирова'я ~ fat cell, fatty cell, adipose cell
замеща'ющая ~ substituting cell
замыка'ющая ~ *bot* guard cell
заро'дышевая ~ germ cell, germinal cell, formative cell
звё'здчатая ~ star cell, starlike cell, stellate cell
звё'здчатая не'рвная ~ stellate nerve cell, Golgi cell
зимоге'нная ~ zymogenic cell
зре'лая ~ mature cell
иммобилизо'ванная ~ immobilized cell
иммунокомпете'нтная ~ immune cell, immunocompetent cell
иммуноотвечаю'щая ~ responder cell
иммунореакти'вная ~ responder cell
иммунорегулято'рная ~ immunoregulatory cell
инициа'льная ~ initial cell
инсуля'рная ~ islet cell
интерстициа'льная ~ interstitial cell
исхо'дная ~ stem cell
камени'стая ~ stone cell
кана'льцевая ~ canal cell
кластеробразу'ющие кле'тки *pl* cluster-forming cells
ко'лбочковидная зри'тельная ~ cone cell
коллаге'н-синтези'рующая ~ collagen-producing cell
коммити'рованная ~ committed cell
корзи'нчатая ~ basket cell
костеобразу'ющая ~ osteoblast
ко'стная ~ bone cell, osseous cell, osteocyte
кроветво'рная ~ hemopoietic cell
кубови'дная ~ cuboidal cell
лимфати'ческая ~ lymph cell
лютеи'новая ~ lutein cell
матери'нская ~ mother cell, parent cell
мезе'нхимная ~ mesenchyme cell
ме'лкая ~ Рамон-и-Каха'ла small cell of Ramon-y-Cajal
митра'льная ~ mitral cell
многоя'дерная ~ multinucleate cell
мононуклеа'рная ~ mononuclear cell
мужска'я полова'я ~ male cell
мультиполя'рная не'рвная ~ nerve multipolar cell
нару'жная пограни'чная ~ Hensen's cell, outer limiting cell
нару'жная подде'рживающая ~ Claudius' cell, outer supporting cell
нару'жная фала'нговая ~ Deiters' cell, outer phalangeal cell, prop cell
натура'льная цитотокси'ческая ~ natural cytotoxic cell
недифференци'рованная ~ undifferentiated cell
незре'лая ~ immature cell
некоммити'рованная ~ uncommitted cell
неприми'рованная ~ unprimed cell
неразмножа'ющаяся ~ nonproliferating cell
не'рвная ~ nerve cell
несинге'нная ~ histoincompatible cell
нефлуоресци'рующая ~ fluorescence-dull cell
обкла'дочная ~ accessory cell
обоня'тельная ~ olfactory cell, Schultze's cell
образова'тельная ~ formative cell
«обу'ченная» антиге'ном ~ antigen-educated cell, antigen-experienced cell, antigen-exposed cell
«обу'ченная» ~ "educated" cell
одноя'дерная ~ mononuclear cell
опо'рная ~ foot cell
о'пухолевая ~ tumor cell
островна'я ~ islet cell
осяза'тельная ~ touch cell, tactile cell
палочкови'дная зри'тельная ~ rod cell
паренхимато'зная ~ parenchymatous cell
париета'льная ~ parietal cell
перви'чная ~ elementary cell, initial cell
перви'чная кровяна'я ~ primitive blood cell
перви'чная полова'я ~ primordial germ cell
пигме'нтная ~ pigment cell
пирамида'льная ~ pyramidal cell
пита'ющая ~ nurse cell, feeder cell

пло'ская ~ squamous cell
плюрипоте'нтная ~ pluripotential cell
побо'чная ~ *bot* accessory cell
подви'жная ~ mobile cell
подде'рживающая ~ floor cell, supporting cell
подка'рмливающая ~ feeder cell
поко'ящаяся ~ dormant cell, durative cell, quiescent cell, resting cell, noncycling cell, nonproliferating cell
покры'тая гемолизи'ном ~ hemolysin-coated cell
полова'я ~ generative cell, germ cell, germinal cell, sex cell, sexual cell
полулу'нная ~ demilune cell
поля'рная ~ pole cell
прекоммити'рованная ~ precommitted cell
премитоти'ческая ~ premitotic cell
«поту'хшая» ~ fluorescence-dull cell
прекоммити'рованная ~ precommitted cell
прими'рованная иммунореакти'вная ~ primed responder cell
прими'рованная носи'телем ~ carrier-primed cell
приро'дная цитотокси'ческая ~ natural cytotoxic cell
проталлиа'льная ~ prothallial cell
пылева'я ~ dust cell
ра'ковая ~ cancer cell, cancerous cell, malignant cell
расти'тельная ~ plant cell
резе'рвная ~ reserve cell
репопули'рующая ~ repopulating cell
ресни'чная ~ iliated cell
рестрикти'рованная стволова'я ~ restricted stem cell
ретина'льная пигме'нтная ~ accesory pigment cell
рогова'я ~ horn cell
розеткообразу'ющая ~ rosette-forming cell
светочувстви'тельная ~ visual cell
светя'щаяся ~ light-producing cell, photogenic cell
свобо'дная ~ free cell
секрето'рная ~ secretory cell
сенсибилизи'рованная ~ allergized cell, sensitized cell
серпови'дная ~ sickle cell
се'стринская ~ sister cell
симпатикотро'пная ~ sympathicotropic cell
симпатохромаффи'нная ~ sympathochromaffin cell
ситови'дная ~ sieve cell
сли'зисто-белко'вая, сли'зисто-серо'зная ~ trophochrome cell
сомати'ческая ~ body cell, somatic cell
сопровожда'ющая ~ companion cell
спа'йковая ~ spiking cell
сперматоге'нная ~ spermatogenous cell
спороге'нная ~ sporogenous cell
спорообразу'ющая ~ sporogenous cell
стволова'я ~ founder cell, stem cell
стволова'я лимфо'идная ~ common lymphoid stem cell
стволова'я миело'идная ~ common myeloid stem cell
стебелько'вая ~ stalk cell
стимули'рованная ~ triggered cell
стимули'рующая ~ stimulator cell
столбова'я ~ pillar cell
стрека'тельная клетка nettling cell, stinging cell, nematocyst, lasso cell
строма'льная ~ stromal cell
супре'ссорная ~ suppressor cell
«танцу'ющая ~» *eth* jiggle cage
тапета'льная ~ *bot* tapetal cell
термина'льная ~ terminal cell
тотипоте'нтная ~ totipotent cell
трахеи'дная ~ tracheidal cell
ту'чная ~ mastocyte, mast cell, labrocyte
флуоресци'рующая ~ fluorescence-bright cell
фолликуля'рная ~ follicular cell
фолликуля'рная ~ яи'чка testicular follicular cell, Sertoli's cell
хлорофиллоно'сная ~ chlorophyll-containing cell
хромаффи'нная ~ chromaffin cell
хрящева'я ~ cartilaginous cell
цикли'рующая ~ cycling cell
цитотокси'ческая ~ cytocidal cell, cytotoxic cell
чешу'йчатая ~ squamous cell
чувстви'тельная ~ sense cell, sensory cell
шва'нновская ~ Schwann's cell, sheath cell
шипови'дная ~ prickle cell
шипови'дная эпителиа'льная ~ spiny epithelial cell
экзокри'нная ~ exocrine cell

эма'левая ~ enamel cell
эмбриона'льная ~ embryo cell, embryonic cell
энтерохромаффи'нная ~ argyrophil cell, enterochromaffin cell, granulobasal cell, pheochrome cell
эпенди'мная ~ ependymal cell
эпидерма'льная ~ epidermal cell
эпителиа'льная ~ epithelial cell
эффе'кторная ~ effector cell
кле'тка-гибри'д *f* hybrid cell
кле'тка-головоло'мка *f* puzzle box
кле'тка-ки'ллер *f* killer cell, killer, attacker cell
аллоге'нная ~ allogenerated killer cell, allospecific killer
аутологи'чная ~ self killer
есте'ственная ~ natural killer cell
лимфокинактиви'рованная ~ lymphokine-activated killer
приро'дная ~ natural killer cell
натура'льная ~ natural killer cell
нетипи'рованная ~ anomalous killer
кле'тка-корму'шка *f* feeder cell
кле'тка-мише'нь *f* target cell, target
гистосовмести'мая ~ matching target
неме'ченная ~ cold target
«холо'дная» ~ cold target
кле'тка-«ня'ня» *f* nurse cell
кле'тка-основа'тельница *f* founder cell
кле'тка-партнё'р *f*:
миело'мная ~ myeloma partner
кле'тка-«пассажи'р» *f* "passenger" cell
кле'тка-помо'щник *f* helper cell
кле'тка-предше'ственник *f* progenitor cell, progenitor
коммити'рованная ~ committed progenitor
монопоте'нтная ~ monopotent progenitor
олигопоте'нтная ~ oligopotent progenitor
плюрипоте'нтная ~ pluripotent progenitor
примити'вная ~ лимфоци'тов primititve lymphoid progenitor
клетка-продуцент *f*:
~ тяжё'лых цепе'й иммуноглобули'нов heavy-chain producer
~ лё'гких цепе'й иммуноглобули'нов light-chain producer
кле'тка-респо'ндер *f* responder cell
прими'рованная ~ primed responder cell
кле'тка-спу'тник *f bot* companion cell
глиа'льная ~ satellite glial cell
кле'тка-стимуля'тор *f* stimulator cell
кле'тка-суперпродуце'нт *f* high-producer cell
кле'тка-супре'ссор *f* suppressor cell
есте'ственная ~ natural suppressor cell
натура'льная ~ natural suppressor cell
приро'дная ~ natural suppressor cell
кле'тка-уби'йца *f* killer cell
кле'тка-фи'дер *f* feeder cell
кле'тка-хе'лпер *f* helper cell
аутореакти'вная ~ autoimmune helper cell
кле'тка-эффе'ктор *f* attacker cell
кле'точка *f* cellule
кле'точный *a* cellular
клетча'тка *f* cellulose, fibre
подко'жная жирова'я ~ subcutaneous fat
расти'тельная ~ cellulose
сыра'я ~ crude fiber
клешнено'сный *a* cheliferous
клешня' *f* pincer, claw, chela, nipping claw
ло'жная ~ subchela
клещ *m* mite (*Acarina*); tick (*Ixodoidea*)
амба'рный ~ true mite *(Acaridae)*
мучно'й ~ true mite *(Acaridae)*
сы'рный ~ cheese mite *(Tyroglyphus siro)*
клещеви'на *f bot* castor bean, castor-oil plant (*Ricinus*); damage of a hide caused by ticks
клещеви'дный *a* forcipate
клещево'й *a* mite-borne
кле'щик *m* mite
паути'нный ~ spider mite (*Tetrarhynchus*)
кли'зма *f* enema; **ста'вить кли'зму** give an enema
кли'макс *m* climax, climacterium; climax association
вероя'тный ~ potential climax
послепожа'рный ~ fire climax
потенциа'льный ~ potential climax
кли'макс-ко'мплекс *m* climax complex
кли'макс-моза'ика *f* climax pattern
климактери'ческий *a* climacteric
кли'мат *m* climate
ари'дный ~ arid climate
бореа'льный ~ boreal climate
вла'жный ~ damp climate
континента'льный ~ continental climate
морско'й ~ marine climate, maritime climate
мягкий ~ mild climate
перехо'дный ~ transitional climate
полуари'дный ~ semiarid climate
полузасу'шливый ~ semiarid climate

пусты'нный ~ desert climate
средиземномо'рский ~ Mediterranean climate, Etesian climate
сухо'й ~ arid climate
тропи'ческий ~ tropical climate
уме'ренно тё'плый ~ lukewarm climate
уме'ренный ~ temperate climate
холо'дный ~ frigid climate
климати'п *m* climatype
климати'ческий *a* climatic
климатоло'гия *f* climatology
климатро'н *m* climatizer
климогра'мма *f* hytherograph
клини'ческий *a* clinical
клинови'дно-нижнечелюстно'й *a* sphenomandibular
клинови'дно-обратнояйцеви'дный *a* cuneate-obovoid
клинови'дный *a* cuneate, cuneated, cuneiform
клиногеотропи'зм *m* klinogeotropism
клинокине'з *m* klinokinesis
клинокубови'дный *a* cuneocuboid
клинообра'зный *a* cuneate, cuneated, cuneiform
клиноста'т *m* clinostat, klinostat
клинота'ксис *m* klinotaxis
кли'нтух *m* stock dove (*Columba oenas*)
кли'ренс *m* clearance
иммунный ~ immune clearance
по'чечный ~ renal clearance
тка'невый ~ tissue clearance
клите'ллюм *m* clitellum
кли'тор *m* clitoris
клоа'ка *f* cloaca
клоака'льный *a* cloacal
клобучко'вый *a* cucullate, cucullated
клобучо'к *m* cucullus
клок *m* wisp; bit, piece, shred; **разорва'ть в кло'чья** tear to shreds
клок се'на wisp of hay
клон *m* clone, clon, klon, single-cell culture
аутореакти'вный ~ self-reactive clone, autoreactive clone
«запрещённый» ~ forbidden clone
рекомбина'нтный ~ recombinant clone
самоподде'рживающийся ~ self-maintaining clone
клона'льность *f* clonality
клона'льный *a* clonal
клони'рование *n* cloning
клони'руемый *a* clonable
кло'новый *a* clonal
клоноге'нность *f* clonogenicity
клоноге'нный *a* clonogenic, clonable
клонорхо'з *helm* clonorchiasis
клоноте'ка *f* clonotheque
~ **ге'нов** gene clonotheque
клоноти'п *m* clonotype
кло'нус *m* clonus
клоп *m* bug, true bug (*Hemiptera*)
посте'льный ~ bedbug (*Cimex lectularius*)
земляно'й ~ chinch bug, ground bug (*Lygaeidae)*
клоп-солда'тик *m* red soldier bug (*Pyrrhocoris apterus)*
клохта'нье *n* cackle, cluck
клохта'ть *v* cackle, cluck
клочо'к *m anat* flocculus
нару'жный ~ paraflocculus (*a lobe of the cerebellum*)
клубене'к *m* nodule; knob, tubercle
корнево'й ~ root nodule
клу'бень *m* tuber
клубенько'вый *a* tubercular
клубнево'й *a* tuberous
клубнелу'ковица *f* bulbotuber, bulbous tuber, corm
клубнено'сный *a* tuberiferous
клубнеобразова'ние *n* tuberization
клубни'ка *f* strawberry (*Fragaria viridis*)
клубо'к *m* glomerulus, glomus, glomerule; skein
ко'жный ~ cutaneous glomerus
пло'тный ~ *cyt* compact skein
ры'хлый ~ *cyt* loose skein
клубо'чек *m* glomerulus, glomerule; cluster (*a compound fruit*)
~ **по'чечного те'льца** Malpighian glomerulus
мальпи'гиев ~ Malpighian glomerulus, Malpighian body
клубочкообразу'ющий *a bot* club-forming
клу'мба *f* bed, flower bed, garden bed
клу'ша *f* mother hen
клык *m* fang, canine tooth, cuspid tooth, tush
клюв *m* beak, bill, neb
клю'вик *m* rostel, rostrulum
клювикови'дный *a* rostelliform
клювови'дно-акромиа'льный *a* coracoacromial
клювови'дно-ключи'чный *a* coracoclavicular
клювови'дно-плечево'й *a* coracohumeral

клюволи'стный *a* beak-leaved
клювообра'зный *a* beaked
клювопло'дный *a* beak-fruited
клю'ква *f* cranberry (*Oxycoccus*)
~ **боло'тная** European cranberry *(Oxycoccus palustris)*
~ **крупнопло'дная** marsh cranberry, American cranberry *(Oxycoccus macrocarpus)*
клю'квенник *m* cranberry bog
клю'нуть *v* bite; ~ **на прима'нку** bite at the bait
ключ *m* key to
ключи'ца *f* collar bone, clavicle
ключи'чно-грудно'й *a* clavipectoral
ключи'чный *a* clavicular
кля'ча *f* jade, hack
кмин *m bot* cumin (*Cuminum*)
кни'га *f* book
кра'сная ~ Red Data Book
племенна'я ~ herd book
родосло'вная ~ studbook
кни'да *f* cnida
книдобла'ст *m* cnidoblast
книдоци'ль *m* cnidocil
книдоци'ста *f* cnida
кни'жка *f* psalterium, omasum, booklet
ле'гочная ~ *zool* lung book
чешу'йная ~ scale book
княжени'ка *f* arctic bramble (*Rubus arcticus*)
ко- co-
коагули'ровать *v* coagulate, curdle
коагули'рующий *a* coagulative
коагули'рующийся *a* coagulable
коагуля'нт *m* coagulant
коагуля'т *m* coagulate
коагуля'тор *m* coagulator
коагуля'ция *f* coagulation
коадапта'ция *f* coadaptation
коаксиа'льный *a* coaxial
коа'кция *f* coaction
коа'ла *f* coala bear (*Phascolarctos cinereus*)
коали'ция *f eth* coalition
коаперту'рный *pal* coaperturate
коаркта'ция *f* coarctation
коа'ти *m ther* coati, coatimundi *(Nasua)*
коацерва'т *m* coacervate
коацерва'ция *f* coacervation
кобалами'н *m* cobalamin
кобе'ль *m* male dog
ко'бра *f* cobra
древе'сная ~ tree cobra (*Pseudohaye*)
еги'петская ~ banded cobra (*Naja haje*)
индий'ская ~ common cobra (*Naja naja*)
короле'вская ~ king cobra (*Ophiophagus*)
ошей'никовая ~ spitting cobra (*Hemachatus haemachatus*)
пусты'нная ~ desert black snake (*Walterinnesia*)
среднеазиа'тская ~ oxus cobra (*Naja oxiana*)
ко'бчик *m* red-footed falcon, kestrel (*Falco vespertinus*)
кобы'ла *f* mare
верхова'я ~ saddle mare
жерё'бая ~ pregnant mare
скакова'я ~ running mare
кобы'лка *f* grasshopper (*Acrididae*)
кобы'лочка *f* leafhopper (*Cicadellidae*)
ковариа'ция *f* covariation
~ **с запа'здыванием аргуме'нта** lag covariation
ковё'р *m* mat *(of vegetation)*
ковы'ль *m* feather-grass (*Stipa*)
ковыря'ние *n:* ~ **в носу'** nose-picking
коге'зия *f* cohesion
когнити'вный *a* cognitive
когo'рта *f* cohort
когото'к *m* claw, unguis, unguiculus
ко'готь *m* claw; talon
код *m* code
временно'й ~ timing code
вы'рожденный генети'ческий ~ degenerated code
генети'ческий ~ genetic code, hereditary code
генети'ческий ~ **без запяты'х** commaless code
нуклеоти'дный трипле'тный ~ nucleotide triplet code
ко'декс *m* code
Междунаро'дный ~ **зоологи'ческой номенклату'ры** International code of zoological numenclature
коди'рование *n* coding
~ **сигнала** signal coding
ана'логовое кодирование analog coding
дискре'тное кодирование digital coding
коди'рующий *a* coding
кодомина'нтость *f* codominance
кодо'н *m* codon
бессмы'сленный ~ nonsense codon
вырожде'нный кодон degenerated codon

инициа'торный ~ initiation codon, initiator codon, start codon
иниции'рующий ~ initiation codon, initiator codon, start codon
квадрупле'тный ~ quadruplet codon
смыслово'й ~ sense codon
ста'ртовый ~ initiation codon, initiator codon, start codon
фази'рующий ~ phasing codon
КОЕ, колониеобразу'ющая едини'ца *f* CFU, colony-forming unit
ко'жа *f* skin; leather; hide; harp seal (*Phoca groenlandica*)
гуси'ная ~ gooseflesh
полова'я ~ sexual skin
со'бственно ~ inner skin, true skin
сыромя'тная ~ row hide
кожа'н *m* brown bat (*Eptesius*)
кожее'д *m ent* dermestid beetle *(Dermestes)*
ветчи'нный ~ bacon beetle *(Dermestes lardarius)*
музе'йный ~ museum beetle *(Anthrenus museorum)*
ко'жистый *a* lethery, coriaceous, skinny
ко'жица *f* peel, cuticle, skin, rind
ко'жно-му'скульный *a* skin-muscular
ко'жный *a* cutaneous
кожура' *f* rind, peel, skin, hull
коза' *f* goat, she-goat
козёл *m* goat (*Capra*); billy goat, he-goat, buck
~ **отпуще'ния** scapegoat
безоа'ровый ~ bezoar goat, Persian wild goat *(Capra aegagrus)*
борода'тый ~ bezoar goat, Persian wild goat *(Capra aegagrus)*
водяно'й ~ waterbuck (*Kobus*)
козлёнок *m* yeanling, goatling
козли'на *f* goat skin
козли'ный *a* goaty
козлоборо'дник *m* goat's-beard (*Tragopogon*)
козодо'й *m* nighthawk, nightjar, dorhawk (*Caprimulgus*)
козя'вка *f* small insect (*colloquial*); booger *(dry slime in nostrils)*
маврита'нская ~ cadelle (*Tenebrioides mauritanicus*)
ольхо'вая ~ alder leaf beetle *(Agelastica alni)*
коинтегра'т *m* cointegrate
коинтегра'ция *f* cointegration
ко'итус *m* coition, coitus
койо'т *m* coyote *(Canis latrans)*
ко'ка *f* coca bush (*Erythroxylum coca*)
кокаи'н *m* cocaine
коканцероге'н *m* cocarcinogen
коканцерогене'з *m* cocarcinogenesis
кокк *m* coccus
ко'ккер-спание'ль *m* cocker spaniel
ко'кковый *a* coccal
коккого'н *m* coccogone
кокколи'т *m* coccolith
коккосфе'ра *f* coccosphere
ко'кон *m* cocoon
коконопря'д *m* eggar (*Lasiocampidae*)
коко'с *m* coconut palm (*Cocos nucifera*); coconut
коко'шка *f* false chanterelle (*Hygrophoropsis aurantiaca*)
ко'кса *f* coxa
кокса'льный *a* coxal
коксоподи'т *m* coxopodite
кокцидио'з *m* coccidiosis
ко'лба *f* flask
~ **Бу'нзена** Bunsen flask, flat-bottomed flask
~ **Вюрца** Wurtz flask
~ **Кьельда'ля** Kjeldahl flask
~ **Кра'узе** bulb of Krauze
~ **Ру** Roux flask
~ **с притё'ртой про'бкой** glass-stoppered flask
~ **Эрленме'йера** Erlenmeyer flask
~ **Фре'йденрейха** Freudenreich flask
враща'ющаяся ~ spinner flask
встря'хиваемая ~ shake flask, shaken flask
кони'ческая ~ conical flask
круглодо'нная ~ round flask, round bottom flask
ме'рная ~ volumetric flask measuring bottle, graduated flask
перего'нная ~ boiling flask, distilling flask
титрова'льная ~ titration flask
ко'лбочка *f* cone (*in the eye structure*)
~ **сетча'тки гла'за** retinal cone
коле'нный *a* genicular
коле'но *n anat* knee, *bot* stemnode
надсемядо'льное ~ epicotyl
подсемядо'льное ~ hypocotyl, hypocotyl hook, caulicle
коле'нчато-изо'гнутый, коле'нчатый *a* geniculate, genuflexuous

колео'птиль *m* coleoptile
колеори'за *f* coleorhiza
колесо' *n* wheel
~ **акти'вности**, **«бе'личье»** ~ activity wheel, running wheel
колесови'дный *a* rotate
колибактериофа'г *m* colibacteriophage
колибакте'рия *f* colibacillus
коли'бри *m*, *f* hummingbird (*Trochilidae*)
колимови'рус *m* caulimovirus
коли'т *m* colitis
колифа'г *m* coliphage
коли'чественный *a* quantitative
коли'чество *n* quantity
ко'лкий *a* spiny, spiniscent, spinose, spinous, prickly
коллаге'н *m* collagen
коллаге'новый *a* collagenic
ко'ллапс *m* collapse
коллатера'ль *f* collateral
сосу'дистые коллатера'ли *pl* vascular collaterals
коллатера'льный *a* collateral
коллекти'рование n collection
колле'ктор *m* collector
~ **фра'кций** fraction collector
коллекциони'рование *n* collecting
колле'кция *f* collection
~ **культу'р** culture collection
~ **типовы'х культу'р** type culture collection
нау'чная ~ research collection
обзо'рная ~ survey collection
определи'тельная ~ identification collection
системати'ческая ~ taxonomic collection
специа'льная ~ survey collection
колле'нхима *f* collenchime
пласти'нчатая ~ lamellar collenchyme
ры'хлая ~ lacunar collenchyme
уголко'вая ~ angular collecnchyme
колленци'т *m* collencyte
ко'лли *m*, *f* collie
коллима'тор *m* collimator
коллима'ция *f* collimation
коллобла'ст *m* colloblast
колло'дий, **колло'диум** *m* collodion, collodium
колло'ид *m* colloid
коллоида'льный, **колло'идный** *a* colloid, colloidal
коло'дец *m* well
центра'льный ~ *lab imm* center well
осево'й ~ *lab imm* center well
нару'жный перифери'ческий ~ *lab imm* outer well
ко'лок *m* insular grove *(in forest-steppe)*
ко'локол *m* bell, pileus, umbrella
ма'точный ~ *helm* uterine bell
колоколообра'зный *a* campaniform
колоко'льчатый *a* campaniform
колоко'льчик *m* bellflower, harebell (*Campanula*)
колониа'льный *a* colonial
колониеобразу'ющий *a* colony-forming
колониза'ция *f* colonization
колонизу'емый *a* recipient
коло'ния *f* colony; rookery (*of birds*)
~ **бакте'рий** bacterial clump
~ **кроветво'рных кле'ток** hemopoietic colony
~ **пингви'нов** penguin rookery
~ **птиц** rookery
~ **сме'шанного ти'па** mixed colony, multilineage colony
блестя'щая ~ glossy colony
гемопоэти'ческая ~ hemopoietic colony
гетеромо'рфная ~ heteromorphic colony
гла'дкая ~ *micr* smooth colony
глуби'нная ~ subsurface colony
гнездова'я ~ *orn* breeding colony
гомомо'рфная ~ homomorphic colony
и'стинная ~ pure colony
кормоге'нная ~ kormogene colony
кроветво'рная ~ hemopoietic colony
монокаппакле'точная ~ "kappa-only" colony
монолиниеспецифи'ческая ~ single lineage colony
монолямбдакле'точная ~ "lambda-only" colony
муко'идная ~ mucoid colony
неодноро'дная ~ heteromorphic colony
одноро'дная ~ homomorphic colony
олиголине'йная ~ oligolineage colony
пове'рхностная ~ surface colony
сектoриа'льная ~ sectored colony
сли'зистая ~ mucoid colony
сме'шанная ~ mixed colony, multilineage colony
то'чечная ~ pinpoint colony
шерохова'тая ~ rough colony
коло'нка f column

~ **для афи'нной хроматогра'фии** affinity column
~ **с иммобилизо'ванными антитела'ми** antibody affinity column
барбота'жная ~ bubbled column
хроматографи'ческая ~ chromatographic column
кондициони'рованная ~ cured column *(in chromatography)*
гидроксиапати'товая ~ hydroxyapatite column

коло'нна *f* column

колоно'к *m* Siberian weasel (*Mustela sibirica*)

коло'нчатый *a* columnar

колори'метр *m* colorimeter

колориметри'я *f* colorimetry

ко'лос *m* ear, spike
безо'стый ~ bald ear
ветви'стый ~ branched spike
ло'жный ~ false spike
нали'вшийся ~ filled ear
однобо'кий ~ one-sided spike
по'лный ~ filled ear
просто'й ~ simple spike
ры'хлый ~ loose spike, lax spike, loose ear
сжа'тый ~ compressed spike
сло'жный ~ compound spike

колоси'стый *a* spicate, spicigerous

колоси'ться *v* come into ear, form ears, ear, spire

колосови'дный *a* spicate

колоско'вый *a* spicate; **колосковая чешуя'** glume

колосови'к *m bot* early rough boletus (*Leccinum scabrum*) (*early boletus appeared when cereals come into ear)*

колосо'к *m* spikelet (*of angiosperms*), cone (*of horsetail*), spicule
ве'рхний ~ apical spikelet
втори'чный ~ spiculum, spikelet
души'стый ~ vernal grass, sweet vernal-grass (*Anthoxanthum odoratum*)
ни'жний ~ basal spikelet
многоцветко'вый ~ many-flowered spikelet
одноцветко'вый single-flowered spikelet
плоду'щий ~ fertile spikelet
сидя'чий ~ sessile spikelet

колосоно'сный *a* spiciferous

колося'щийся *a* eared

колоше'ние *n* earing, heading, ear emergence

колпа'к *m* cap
стекля'нный ~ bell jar

колпачкови'дный *a* calyptriform, mitriform

колпачо'к *m* calyptra, calypter, cap, moss cap, mitra
нави'нчивающийся ~ screw stopper
противозача'точный ~ contraceptive cap

ко'лпица *f* spoonbill (*Platalea leocordia*)

колхици'н *m* colchicin

колцеми'д *m* colcemid

кольцева'ние *n* ringing, banding
~ **птиц** birdbanding

кольцева'ть *v* ring

кольцево'й *a* annular

кольцева'ние *n* ringing

кольцеви'дный *a* ringed

кольцепо'ровый *a* ring-porous

кольцо' *n* ring, annulus
~ **Кабо'** Cabot's ring, basophilic ring
~ **Манчи'ни** Mancini ring
ана'льное ~ anal ring
базофи'льное ~ Cabot's ring, basophilic ring
бе'дренное ~ femoral ring
ве'дьмино ~ fairy ring
вну'треннее па'ховое ~ abdominal inguinal ring
возрастно'е ~ **на зу'бе** dentine ridge, tooth ridge
вста'вочное ~ intercalary band
годи'чное ~ annual ring, growth ring
доба'вочное ~ supernumerary ring (*on a fish scale*), disturbance ring *(on a shell)*
клонзапира'ющее ~ cloning ring
лимфати'ческое ~, лимфоадено'идное глo'точное ~, лимфо'идное ~ lymphoepithelial ring of pharynx, Pirogoff tonsillar ring
моро'зное ~ frost ring
па'ховое ~ abdominal ring, inguinal ring
поля'рное ~ *zool* polar ring
пупо'чное ~ navel ring
уплотни'тельное ~ sealing ring *(in chromatography)*
хромосо'мное ~ chromosome ring

ко'льчатый *a* ringed, annulate

колюмелля'рный *a* columellar

колючепло'дный *a* acanthocarpous

колючещетини'стый *a* spear-bristled

колю'чий *a* pricky, barbed, acanthous, acanthaceous, acanthoid, lappaceous, spiculate, spinous, spiniscent, spinose, spiny, thorned, thorny, briery

колю'чка *f* spine, bur, burr, barb, acantha, prickle, thorn

верблю'жья ~ camel's-thorn (*Alhagi camelorum*)

ком *m* lump; clod

ко'ма *f* coma

кома'р *m* mosquito (*Culicidae*)

~ **маляри'йный** malaria mosquito, anopheline mosquito*(Anopheles)*

~ **обыкнове'нный** northern house mosquito *(Culex pipiens)*

кома'рик *m* small midge (*colloquial*)

грибно'й ~ fungus midge, fungus gnat (*Mycetophilidae*)

комбико'рм *m* combined feed, formula feed, mixed feed, combined fodder

комбина'ция *f* combination

кольра'би *f* turnip-rooted cabbage, kohlrabi (*Brassica oleracea var. gongiloides*)

кольцева'ть *v* ring

колю'чий *a* spiculate, spiculiferous, spiculigerous

колю'чка *f* spiculum

коме'ль *m* butt

комиссу'ра *f* commissure

комиссура'льный *a* commissural

комкова'тый *a* lumpy; cloddy *(of soil)*

комменса'л *m* commensal

комменсали'зм *m* commensalisms

комменсали'ст *m* commensal

комме'рческий commercial

комми'скуум *m* commiscuum

коммити'рование *n imm* commitment

коммуника'ция *f* communication

иску'сственная ~ artificial communication

комо'к *m* lump, clot; **сверну'ться в** ~ roll oneself into *a* ball

жирово'й ~ **Биша**' fat ball of Bichat, buccal fat pad, sucking pad

комо'лый *a* polled, hornless

компа'ртмент *m* compartment

компартментализа'ция *f* compartmentalization, compartmentation

компенсато'рный *a* compensatory

компенса'ция *f* compensation

~ **доз** *gen* dosage compensation

компенсиру'ющий *a* compensatory

компете'нтность *f* competence, competency

~ **кле'тки** cellular competence

компенте'нция *f* competence, competency

региона'льная ~ *emb* regional competence

компиля'ция *m* compilation

ко'мплекс *m* complex, *phys* combination

~ **антиге'н–антите'ло** antigen-antibody complex

~ **Го'льджи** Golgi apparatus

~ **модифика'торов** modifer complex

~ **соо'бществ** community complex

антиге'нный ~ antigenic complex

габитуа'льный ~ habit complex

ге'нный ~ gene complex

гетерога'мный ~ heterogamic complex

гидрофи'льный иммун'ный ~ hydrophilic immune complex

гла'вный ~ **гистосовмести'мости** major histocompatibility complex

зерни'стый паравезикуля'рный ~ granular paravesicular complex

иммун'ный ~ immune complex

клона'льный ~ clone complex

коадапти'рованный ге'нный ~ coadapted gene complex

мембраноатаку'ющий ~ membrane attack complex

мультиме'рный ~ modifer complex

мультимолекулярный ~ multimolecular complex

пероксида'зо-антипероксида'зный ~ peroxidase-antiperoxidase complex, PAP complex

полиферме'нтный ~ modifer complex

после'довательный ~ successive combination

приро'дный ~ natural complex

приро'дный территориа'льный ~ natural territorial complex

промежу'точный ~ intermediate complex

светособира'ющий ~ light-harvesting complex

синаптонема'льный, синаптоне'мный ~ synaptonemal complex

скотово'дческий ~ cow-calf enterprise

стромозащи'тный антиге'нный ~ stromata-protective antigen complex

транскрипцио'нный ~ transcriptional complex

фермент-субстра'тный ~ enzyme-substrate complex

эди'пов ~ Oedipus complex

хлорофи'лл-белко'вый ~ chlorophyll-protein complex

циркули'рующий иммy'нный ~ circulating immune complex
ко'мплексность *f* complexity
ко'мплексный *a* complex
комплеме'нт *m* complement
мембраносвя'зный ~ membrane-bound complement
о'бщий гемолити'ческий ~ total hemolytic complement
комплемент-отклоня'ющий *a* complement-deviating
комплемента'рность *f* complementarity
комплемента'рный *a* complementary
комплемента'ция *f* complementation
межалле'льная ~ interallelic complementation
межге'нная ~ intergenic complementation
внутриге'нная ~ intragenic complementation
компози'ция *f* composition
флористи'ческая ~ floristic composition
компоне'нт *m* component
~ вариа'нсы component of variance
~ диспе'рсии component of dispersion, component of variance
~ изме'нчивости component of variability
второстепе'нный ~ minor component
неотъе'млемый ~ integral component
обы'чный ~ normal component
основно'й ~ major component
отцо'вский ~ sire component
приро'дный ~ natural constituent
центра'льный ~ ви'руса virus core component
компо'ст *m* compost
компости'ровать *v* compost
компто'н-эффе'кт *m* Compton effect
кона'рия *f* conarium
конверге'нция *f* convergence
конверги'ровать *v* converge
конве'рсия *f* conversion
~ ге'на gene conversion
антиге'нная ~ antigen conversion
лизоге'нная ~ lysogenic conversion
фа'говая ~ phage conversion
конвульси'вный *a* convulsive
конву'льсия *f* convulsion
конгениа'льный *a* connatural
конге'нный *a* congenic
конглютина'ция *f* conglutination
конгре'ссия *f* congression
конденса'т *m* condensate
конденса'тор *m* condenser
конденса'ция *f* condensation
~ хромосо'мы chromosome condensation
алли'льная ~ allylic condensation
бензои'новая ~ benzoin condensation
восстанови'тельная ~ reductive condensation
сложноэфи'рная ~ ester condensation
конденси'ровать, конденси'роваться *v* condense
конде'нсор *m* condenser
~ А'ббе Abbe's condenser
кондициони'роване *n* conditioning
кондициони'ровать *v* condition
ко'ндор *m* condor *(Vultur gryphus)*
коневo'дство *n* horse raising, horse husbandry
конё'к *m* pipit (Anthus)
конема'тка *f* mare
коне'ц *m* end
о'стрый ~ point end *(of an egg)*
пупови'нный ~ stem end (*of a potato*)
коне'чность *f* extremity, limb
грудна'я ~ thoracic limb
за'дняя ~ hind limb
клешнено'сная ~ chelate limb
пере'дняя ~ fore limb
торака'льная ~ thoracic limb
ту'ловищная ~ trunk limb
челюстна'я ~ feeding limb
конидиа'льный *a* conidial
конидиока'рпий *m* conidiocarp
конидиоспо'ра *f* conidiospore
конидиофо'р *m* conidiphore
кони'дия *f* conidium
ме'лкая ~ conidiole
кониофа'г *m* coniophage
кони'ческий *a* fastigiate
конусообра'зный *a* fastigiate
конкатеме'р *m* concatemer
конкатена'т *m* concatenate
конкла'в *m* conclave
конкорда'нтный concordant
конкреме'нт *m* calculus, concretion
конкуре'нт *m* competitor
конкуре'нтный *a* competitive
конкурентоспосо'бность *f* competitive ability
конкуре'нция *f* competition
внутривидова'я ~ intraspecific competition

интерференцио'нная ~ intereference competition, encounter competition
межвидова'я ~ interspecific competition
эксплуатацио'нная ~ exploitation competition, consumptive competition
коннекти'в *m* connective
коннекти'н *m* connectin
коно'ид *m* conoid body
конопля' *f* hemp (*Cannabis sativa*)
конопля'нка *f* linnet (*Carduelis cannabina*)
консерва'нт *m* preservative, preserving agent, conservative
консерва'ция *f* preservation
~ **замора'живанием** freeze preservation
консе'рвы *m pl* canned food
мясны'е ~ canned meat
овощны'е ~ canned vegetables
ры'бные ~ canned fish
консисте'нция *f* consistence, consistency
ко'нский *a* equine
консо'рт *m* consort
консорти'зм *m* consortism
консо'рция *f* consortium
консоциа'ция *f* consociation
консо'ция *f* consocies, consocion
конспецифи'чный *a* conspecific
конста'нта *f* constant
~ **аллометри'ческого ро'ста** allometric constant
~ **ассоциа'ции** association constant
~ **диссоциа'ции** dissociation constant
~ **диффу'зии** diffusion constant
~ **иониза'ции** ionization constant
~ **Михаэ'лиса** Michaelis constant
~ **равнове'сия** equilibrium constant
~ **седимента'ции** sedimentation constant
~ **ско'рости** velocity constant
~ **сродства'** affinity constant
~ **стаби'льности** stability constant
актива'торная ~ activator constant
аллометри'ческая ~ allometric constant
ингиби'торная ~inhibitor constant
субстра'тная ~ substrate constant
конста'нтность *f* constance, constancy
~ **ви'дов** constance of species
конста'нтный *a* constant
конститути'вный *a* constitutive
конститу'ция *f* constitution
констри'ктор *m* constrictor
консульта'ция *f* consultation
генети'ческая ~ genetic consultation, genetic counseling
консуме'нт *m* consumer
консуммато'рный *a* consummatory
контагио'зный *a* contagious
конта'кт *m* contact
зу'бчатый межклеточный ~ denticulate intercellular union
межкле'точный ~ intercellular union
пальцеви'дный межкле'точный ~ digitiform intercellular union
фока'льный ~ focal contact
континента'льный *a* continental
ре'зко ~ extreme continental
конти'нуум *m* continuum
контралатера'льный *a* contralateral
контра'ст *m* contrast
поведе'нческий ~ behavioral contrast
положи'тельный поведе'нческий ~ positive behavioral contrast
фа'зовый ~ phase contrast
флористи'ческий ~ floristic contrast
контрасупре'ссия *f* contrasuppression
контро'ль *m* control, supervision
визуа'льный ~ eye control
генети'ческий ~ genetic control
генотипи'ческий ~ genotypic control
гормона'льный ~ hormonal control
дистанцио'нный ~ remote control
санита'рный ~ sanitary inspection
транскрипцио'нный ~ transcriptional control
трансляцио'нный ~ translational control
цветово'й ~ color-matched control
конту'зия *f* contusion, concussion
ко'нтур *m* outline
нару'жный ~ outline
конура' *f* kennel
ко'нус *m* cone
~ **нараста'ния** apical cone, cone of origin, vegetative cone
~ **нараста'ния побе'га** *bot* apical dome
~ **ро'ста** growth cone
артериа'льный ~ pulmonary cone
кристалли'ческий ~ *ent* crystalline cone
обоня'тельный ~ *ent* sense cone
полово'й ~ *helm* genital cone
хруста'льный ~ crystalline cone
конфигура'ция *f* configuration
конфли'кт *m* conflict
~ **типа «приближение – избегание»** *eth* approach-avoidance conflict
~ **родителей и потомства** *eth* parent-offspring conflict

конформацио'нный *a* conformational
конформа'ция *f* conformation
конфо'рмер *m* conformer
конхиоли'н *m* conchiolin
конхиоло'гия *f* conchology
концентра'тор *m:* **центробе'жный ~** centrifugal concentrator
концентра'ция *f* concentration
  абсолю'тно лета'льная ~ absolute lethal concentration
  безопа'сная ~ safe concentration
  допусти'мая ~ permissible concentration
  исхо'дная ~ initial concentration
  коне'чная ~ final concentration
  крити'ческая ~ critical concentration
  лета'льная ~ lethal concentration
  максима'льная ~ highest concentration
  норма'льная ~ normal concentration
  объё'мная ~ volume concentration
  поро'говая ~ threshold concentration
  преде'льно допусти'мая ~, ПДК maximum allowable concentration, MAC, critical concentration
  смерте'льная ~ lethal concentration
  среднелета'льная ~ average lethal concentration
конце'пция *f* concept
  градие'нтная ~ gradient concept
концы' *pl of* **коне'ц**
  «ли'пкие» концы' *pl* cohesive ends, sticky ends *(of DNA)*
  «тупы'е» концы' *pl* blunt ends, overhang ends *(of DNA)*
конъюга'нт *m* conjugant
конъюга'т *m* conjugate
  антителотокси'новый ~ antibody-toxin conjugate
  мише'нь-эффе'кторный ~ effector-target conjugate
конъюга'ция *f* conjugation, pairing
  ~ по ти'пу «игра' в пятна'шки» touch-and-go conjugation, touch-and-go pairing
  гетероге'нная ~ heterogenic pairing
  избира'тельная ~ selective pairing
  нача'льная ~ initial pairing
  непра'вильная конъюгация illegitimate conjugation, illegitimate pairing
  непреры'вная ~ complete pairing
  петлеви'дная ~ torsion pairing
  по'лная ~ complete pairing
  преры'вистая ~ discontinuous pairing
  торсио'нная ~ torsion pairing
  эктопи'ческая ~ ectopic pairing
конъюги'ровать *v* conjugate
конъюнкти'ва *f* conjunctiva
конъюнктиви'т *m* conjunctivitis
  ~ Парино' Parinaud conjunctivitis
конь *m* horse, stead
  троя'нский ~ the Trojan horse
коню'га *f orn* auklet (*Aethia*)
  больша'я ~ crested auklet *(Aethia cristatella)*
  ма'лая ~ whiskered auklet *(Aethia pygmaea)*
коню'га-кро'шка least auklet *(Aethia pusilla)*
коню'шня *f* horse barn
кооперати'вность *f mbiol* cooperativity
  отрица'тельная ~ negative cooperativity
координа'ция *f* coordination
  ~ расти'тельности coordination of vegetation
копа'лка *f* trowel
копа'тельный *a* fossorial
копа'ть *v* dig
  ~ карто'шку lift, lift potaoes, dig potatoes
копа'ться *v* grub about, dig
копа'ющий *a* fossorial
ко'пийность *f gen* copy number
копна' *f* mow
ко'пра *f* copra, palm nut
копрезента'ция *f imm* copresentation
копроантите'ло *n* coproantibody
копробио'нт *m* coprobiont
копроге'нный *a* coprogenous
копрозо'йный *a* coprozoic
копроли'т *m* coprolite
копрофи'льный *a* coprophilous
копрофи'т *m* coprophyte
копти'льня *f* smokehouse
копти'ть *v* smoke, cure in smoke
ко'пула *f* basihyal bone
копули'ровать *v* copulate; whipgraft
копулиро'вка *f* splice grafting, copulation
  ~ с язычко'м tongue grafting, whip grafting, whip-and-tongue grafting
копуляти'вный *a* copulative, copulatory
копуля'ция *f* copulation
  внебра'чная ~ extra-pair copulation
копче'ние *n* smoking
копче'ный *a* smoked
ко'пчик *m* coccyx, coccygeal bone, tail bone
ко'пчиковый *a* coccygeal
копы'тень *m* wild ginger (*Asarum*)

копы'тка *f* Pallas' sandgrouse (*Syrrhaptes paradoxus*)
  обыкнове'нная ~ Pallas' sandgrouse (*Syrrhaptes paradoxus*)
копы'тный *a* ungulate, hoofed
копы'то *n* hoof; **бить копы'том** stamp, hoof
копытообра'зный *a* hoof-shaped
копы'тце *n* hoof
копьевидногу'бый *bot* spear-lipped
копьевидноли'стный *a* spear-leaved
копьено'сный *a* spear-bearing
копьецо' *n* hastula
кор *m* core
  ~ **фа'га** phage core
кора' *f* bark, cortex, rind
  ~ **головно'го мо'зга** brain cortex, cerebral cortex
  ~ **надпо'чечника** adrenal cortex
  ассоциати'вная ~ association cortex
  дуби'льная ~ tan bark
  земна'я ~ earth crust
  зри'тельная ~ optic cortex, visual cortex
  перви'чная зри'тельная ~ primary visual cortex
  плекте'нхимная ~ fungal cortex
  сенсо'рная ~ sensory cortex
  слухова'я ~ acoustic cortex, auditory cortex
  соматосенсо'рная ~ somatosensory cortex, somesthetic cortex
  чешу'йчатая ~ scaly bark
  шелушаща'яся ~ shell bark
  экстрастриа'рная ~ extrastriate cortex
корако'ид *m* coracoid bone
кора'лл *m* coral
коралли'т *m* corallite
кора'лловый *a* coral
корамо'ра *f* crane fly (*Tipula*)
кораци'дий *m* coracidium
коре'мия *f* coremium
корена'стый *a* cobby
коренни'к *m* shaft horse, wheel horse
коренно'й *a* aboriginal; ~ **зуб** molar
ко'рень *m* root; **пуска'ть корни** root
  ~ **во'лоса** hair root
  ~ **ре'пчатой фо'рмы** napiform root
  агресси'вный ~ aggressive root
  боково'й ~ lateral root, branch root
  возду'шный ~ aerating root, aerial root, pneumatophore root, breathing root, crampon
  втя'гивающий ~ contractile root
  гла'вный ~ main root, tap root
  гроздеви'дные мяси'стые ко'рни *pl* clustered fleshy roots
  доскови'дный ~ tabular root, plank-buttress root
  дыха'тельный ~ breathing root, aerating root, aerial root, pneumatophore
  запаса'ющий ~ storage root
  заро'дышевый ~ *bot* radicle, radicule
  и'рный ~ sedge root *(Acorus calamus)*
  клубневи'дный ~ tuberous root
  мочкова'тый ~ coronal root, fibrous root, racemose root
  мяси'стый ~ fleshy root
  омоложе'нный ~excised root
  опо'рный ~ brace root, buttress root, supporting root
  пита'ющий ~ feeding root
  пове'рхностный ~ shallow root
  подре'занный ~ excised root
  ползу'чий ~ creeping root
  прида'точный ~ additional root, adventitious root, secondary root
  прилипа'ющий ~ adhesive root
  стержнево'й ~ taproot, main root
  узлово'й ~ nodal root
  ходу'льный ~ prop root, stilt root
  чёткови'дный ~ nodose root
  я'корный ~ strut root, anchor root, anchoring root
ко'рень-прице'пка *f* holdfast root
корешко'вый *a* radicular
корешо'к *m* radicle, radicule, rootlet
  ~ **не'рва** nerve root
  ~ **чувстви'тельного не'рва** sensory root
  за'дний ~ *anat* dorsal root
  заро'дышевый ~ *bot* embryo root, primordial root, seminal root
  перви'чный ~ *m bot* radicle, radicule, embryo root, primordial root, seminal root
  пере'дний ~ *anat* ventral root
корзи'нка *f* head, calathidium, calathium (*a kind of inflorescence*)
корзинкови'дный *a* basket-shaped
корзи'ночка *f* basket, corbicula (*of bees*)
кориа'ндр *m* coriander *(Coriandrum sativum)*
ко'риум *m* corium
кори'чневый *a* brown
  би'смарк ~ Bismark brown
  суда'н ~ Sudan brown
кори'чник *m* cinnamon-tree (*Cinnamomum*)
ко'рка *f* crust, cortex
  апельси'нная ~ orange peel

по'чвенная ~ soil crust
хле'бная ~ bread crust
ко'рковый *a* cortical, crustose
~ **лиша'йник** crustose lichen
коркообразова'ние *n* crustification
корм *m* feed, fodder, food, feedstuff, pabulum
гранули'рованный ~ pelleted feed
гру'бый ~coarse fodder
есте'ственный ~ natural forage
жи'дкий ~ slop feed
зеле'ный ~ green-cut feed
излю'бленный ~ favorite food
личи'ночный ~ larval food
мя'гкий ~ soft feed
намо'ченный ~ soaked feed
объё'мистый ~ bulky feed
па'стбищный ~ pasture forage
подно'жный ~ pasture, grass, pasture forage
сухо'й ~ dry food
кормё'жка *f* feeding
корми'дий *m* cormidium
корми'ть *v* feed
~ **гру'дью** suckle
кормле'ние *n* feeding, alimentation
мни'мое ~ sham feeding, fictitious feeding
принуди'тельное ~ forced feeding
ко'рмность *f* food capacity
кормово'й *a* fodder, forage, feeding, pabular; stern
кормодобыва'ние *n* forage, foraging
кормофи'т *m* cormophyte
корму'шка *f* manger, feeder, feedbox, trough
~ **для птиц** *f* birdfeeder
корнеа'льный *a* corneal
корневи'дный *a* rootlike, rhizoid
корневи'ще *n* rhizome, rootstock, creeping rootstock
запаса'ющее ~ storage rhizome
корневищеви'дный *a* rhizomorphous
корневи'щный *a* rhizomatous
корнево'й *a* radical
корнегрызу'щий *a* rhizophagous
корнено'жка *f* rhizopod *(Rhizopoda)*
корнено'сный *a* rhizophorous
корнеобра'зный *a* rhizomorphous
корнеобразу'ющий *a* rhizogenic, rhizogenous
корнео'тпрысковый *a* creeping-rooted
корнепло'д *m* root crop, storage root
корнепло'дный *a* rhizocarpous
корнерасположе'ние *n* root arrangement, rhizotaxis
корнесо'бственный *a* own-rooted, scion-rooted
корнецветко'вый *a* radiciflorous, rhizanthous
корнея'дный *a* rhizophagous, radicivorous
корноу'хий *a* crop-eared
коро'бка *f* box
черепна'я ~ cranium
коро'бочка *f* boll, pod, capsule, fruitcase
семенна'я ~ boll
коро'ва *f* cow (*Bos taurus*)
безро'гая ~ hummel cow, hornless cow
бодли'вая ~ butting cow
до'йная ~ cow in milk, milk cow, milker cow
комо'лая ~ hummel cow, hornless cow
сте'льная ~ cow in calf
малоудо'йная ~ poor yelding cow
моло'чная ~ dairy cow
сте'льная ~ calver
сухосто'йная ~ dry cow
тугодо'йная ~ hard-milking cow
коро'вка *f* small cow (*diminutive*)
бо'жья ~ lady-bird, lady beetle, ladybug (*Coccinellidae*)
коро'вник *m* cow barn
коровобизо'н *m* cattalo, catalo
коровя'к *m* mullein, mullein dock (*Verbascum*); cow manure
короe'д *m* bark beetle (*Ipidae*)
короле'к *m* kinglet (*Regulus*)
желтоголо'вый ~ golden-crested wren (*Regulus regulus*)
красноголо'вый ~ firecrest (Regulus *ingnicapillis*)
коро'на *f* corona
коронави'рус *m* coronavirus
корона'рный *a* coronary; **корона'рное кровообраще'ние** coronary circulation
~ **сосу'д** coronary vessel
коро'нка *f* crown, corona (*of a tooth*)
коро'ста *f* scab
короста'вник *m* corn scabious (*Knautia*)
коросте'ль *m* crake, corn crake, landrail (*Crex*)
коротко- brevi-, brachy-, short-
короткоборо'здный *a pal* brevicolpate
коротковетви'стый *a* short-branched
короткоголо'вость *f* brachycephalism, brachycephaly

короткодне'вный *a* short-day
короткозу'бчатый *a* short-toothed
короткоживу'щий *a* short-lived
короткокисти'стый *a* brachybotryous
коротоклю'вый *a* brevirostrate
коротколоси'стый *a* short-eared
короткокры'лый *a* brachypterous, brevipennate, short-winged
короткolепе'стный *a* short-petaled
коротколи'стный *a* brevifoliate
коротколо'пастный *a* short-lobed
коротконадкры'лый *a* brachelytrous
коротконе'рвный *a bot* short-nerved
коротколи'стный *a* short-leaved
коротконо'гий *a* short-legged, breviped, brachypodous
короткоости'стый *a* short-awned
короткопа'лость *f* brachydactyly
короткопе'стичный *a* short-styled
короткопло'дный *a* short-fruited
короткоро'гий *a* brachycerous
короткостволь'ный *a* short-boled, short-stemmed
короткостебе'льчатый *a* short-stalked
короткоу'сый *a* brachycerous
короткоу'хий *a* short-eared
короткоцветко'вый *a* short-flowering
короткошё'рстный *a* short-wooled
короткошипова'тый *a* short-thorned
короткохво'стый *a* brevicaudate, brachyural, short-tailed
ко'рпус *m* trunk; body (*of a horse*); **придти' к фи'нишу на два ко'рпуса впере'д** win by two lengths
корпу'скула *f* corpuscle
корреляти'вный *a* correlative
корреляцио'нный *a* correlated
корреля'ция *f* correlation
~ **ра'нгов** rank correlation
~ **с запа'здыванием аргуме'нта** lag correlation
внутригруппова'я ~ intraclass correlation
внутрикла'ссовая ~ intraclass correlation
вы'борочная ~ sample correlation
генети'ческая ~ genetic correlation
криволине'йная ~ curvilinear correlation
лине'йная ~ linear correlation
ло'жная корреляция illusory correlation, spurious correlation
максима'льная ~ maximal correlation
мно'жественная корреляция multiple correlation
нелине'йная ~ nonlinear correlation
норма'льная ~ normal correlation
обра'тная ~ invert correlation
о'бщая ~ overall correlation
оста'точная ~ residual correlation
отрица'тельная ~ negative correlation
по'лная корреляция overall correlation, total correlation
прямолине'йная ~ linear correlation
ра'нговая корреляция rank correlation
части'чная корреляция partial correlation
ча'стная ~ partial correlation
фенотипи'ческая ~ phenotypic correlation
кортекс cortex
кортизо'н *m* cortisol
кортикализа'ция *f* corticalization
кортика'льный *a* cortical
кортико'ид *m* corticoid
кортикопета'льный *a* corticipetal
кортикостеро'ид *m* corticosteroid
кортикостеро'идный *a* corticosteroid
кортикостеро'н *m* corticosteron
кортикотропи'н *m* corticotropin
кортикофуга'льный *a* corticifugal
корти'на *f bot* cortina
кортицио'идный *myc* corticioid
корчева'ние *n*, **корче'вка** *f* stubbing, grubbing up, uprooting
ко'рчиться *v* writhe
~ **от бо'ли** writhe with pain
ко'ршун *m* kite (*Milvus*)
че'рный ~ black kite, pariah kite (*Milvus korshun*)
корь *f* measles
корьё' *n* bark
дуби'льное ~ tan bark
коря'га *f* snag
ко'свенный *a* indirect
коси'ть *v* mow, cut; squint
~ **сачко'м** sweep
косма'тый *a* shaggy, hairy, long-haired, jubate
косми'да *f* cosmide
косми'н *m* cosmine
космо'идный *a* cosmoid
космополи'т *m* cosmopolite
косогла'зие *n med* strabismus, squint, heterotropia
диверги'рующее ~ divergent squint
конверги'рующее ~ convergent squint
расходя'щееся ~ divergent squint

содру'жественное ~ concomitant squint
сходя'щееся ~ convergent squint
косого'р *m* hillside, slope
косо'й *a* slant, oblique
косола'пый *a* club-footed, club-stalked
кососло'й *m* cross grain *(in wood)*
кососло'йный *a* cross-fibered
коста'льный *a ent* costal
костеобразу'ющий *a* ossific, osteogenetic, osteoplastic
костё'р *m* brome-grass, quake-grass, chess (*Bromus*); fire, campfire
кости'стый *a* bony
ко'стно-суставно'й *a* osteoarticular
ко'стный *a* osteal, osseous, bony
ко'сточка *f* bonelet, ossicle; kernel, pyrene, stone (*e.g., in a cherry*)
гипура'льная ~ hypural bone
слухова'я ~ ear bone, auditory ossicle
ко'сточковый *a* drupaceous; ~ **плод** stone fruit; **ко'сточковая плодо'вая культу'ра** stone fruit crop
костре'ц *m* rump, sacrum
кость *f* bone
~ **запя'стья** carpal bone
~ **и'нков** Inca bone
~ **предплю'сны** tarsal bone
~ **та'за** pelvic bone
~ **лицево'го че'репа** facial bone
бе'дренная ~ thigh bone
безымя'нная ~ hip bone, innominate bone
берцо'вая ~ cannon bone
бокова'я заты'лочная ~ exoocipital bone *(in fish)*
бокова'я клинови'дная ~ laterosphenoid bone
бокова'я обоня'тельная ~ exethmoid bone
большеберцо'вая ~ shin bone
вене'чная ~ coronary bone, short pastern
ве'рхняя заты'лочная ~ supraoccipital bone
ве'рхняя ушна'я ~ epiotic bone
висо'чная ~ temporal bone
воздухоно'сная ~ pneumatic bone
воро'нья ~ coracoid bone
гемиптериго'идная ~ hemipterygoid bone
глазни'чно-клинови'дная ~ orbitosphenoid bone, sphenolateral bone *(of Anura)*
глоссохиа'льная ~ glossohyal bone
голо'вчатая ~ capitate bone
горохови'дная ~ pisiform bone
гу'бчатая ~ spongy bone
заднеглазни'чная ~ postorbital bone
заднело'бная ~ postfrontal bone
заднеушна'я ~ opisthotic bone
за'дняя крылови'дная ~ metapterygoid bone
заты'лочная ~ occipital bone
зубна'я ~ dental bone
квадра'тная ~ quadrate bone
квадра'тно-скулова'я ~ quadratojugal bone
клинови'дная ~ sphenoid bone, cuneiform bone, pterygoid bone, sphenolateral bone *(of Urodela)*
кожно-сочлено'вная ~ gonial bone
компа'ктная ~ compact bone
копы'тная ~ coffin bone
крестцо'вая ~ sacral bone
крылови'дная ~ pterygoid bone
крыло-клинови'дная ~ alisphenoid
кры'шечная ~ splenial bone
крючкови'дная ~ hamate bone, unciform bone
кубови'дная ~ cuboidal bone, cuboid
ладьеви'дная ~ navicular bone, scaphoid bone, shuttle bone
лобко'вая ~ sharebone, pubic bone
ло'бная ~ frontal bone
локтева'я ~ cubitus
лучева'я ~ spoke bone
малоберцо'вая ~ brooch bone, calf bone, peroneal bone
межмы'шечная ~ intermuscular bone
межтеменна'я ~ interparietal bone, Inca bone
межчелюстна'я ~ intermaxillary bone
надвисо'чная ~ *herp* supratemporal bone
надуглова'я ~ supra-angular bone, *herp* epigonial bone
нару'жная крылови'дная ~ ectopterygoid bone
нару'жная решё'тчатая ~ ectethmoid bone
нё'бная ~ palatine bone
носова'я ~ nasal bone
пласти'нчатая ~ splenial bone
околоушна'я ~ periotic bone *(of fish)*
основна'я ~ basilar bone
основна'я сочлено'вная ~ autarticular bone
основна'я ~ basilar bone

основна'я висо'чная ~ basitemporal bone
основна'я заты'лочная ~ basioccipital bone
основна'я клинови'дная ~ basisphenoid bone
переднело'бная ~ prefrontal bone
переднеушна'я ~ prootic bone
пере'дняя клинови'дная ~ presphenoid bone
пласти'нчатая ~ splenial bone
пло'ская ~ flat bone
пневмати'ческая ~ pneumatic bone
подвздо'шная ~ haunch bone, iliac bone, huckle bone, sidebone
подъязы'чная ~ hyoid bone, lingual bone, tongue bone
покро'вная ~ investing bone, dermal bone
полулу'нная ~ lunate bone
полуптериго'идная ~ hemipterygoid bone
предлобко'вая ~ prepubis
предносова'я ~ prenasal
предсле'зная ~ adlacrimal bone
покро'вная ~ investing bone, membrane bone
предносова'я ~ adnasal bone
предтеменна'я ~ preparietal
предчелюстна'я ~ premaxillary bone
премаксилля'рная ~ incisive bone
приа'пова ~ baculum
пу'товая ~ pastern (*of the horse*)
пя'стная ~ basidigital bone
пя'точная ~ heel bone, bone of heel
резцо'вая ~ incisive bone, premaxillary bone
реше'тчатая ~ ethmoid bone, cribriform bone
седа'лищная ~ ishiadic bone, ishial bone, seat bone
сезамови'дная ~ sesamoid bone
симплекти'ческая ~ *ich* mesotympanic bone
скулова'я ~ jugal bone, cheek bone, malar bone, zygomatic bone
слё'зная ~ lacrimal bone
слоно'вая ~ ivory, elephant ivory
сочлено'вная ~ articular bone
сре'дняя обоня'тельная ~ mesethmoid bone
сто'лбчатая ~ *herp* epipterygoid bone
та'зовая ~ haunch bone, hip bone, innominate bone
тара'нная ~ ankle bone, huckle bone
теменна'я ~ bregma bone, parietal bone
трёхгра'нная ~ triquetrous bone
тру'бчатая ~ tubular bone, cylindrical bone, pipe bone
углова'я ~ angular bone
ушна'я ~ periotic bone
хрящева'я ~ cartilage bone
челюстна'я ~ maxillary bone
чешу'йчатая ~ squamosal bone
щё'чная ~ jugal bone *(of fish)*
язы'чная ~ basihyal bone

костяни'ка *f* stone bramble (*Rubus saxatilis*)
костя'нка *f* stone fruit, drupaceous fruit, drupe
сбо'рная ~ aggregate drupe
костя'нковый *a* drupaceous
костяно'й *a* bony
косу'ля *f* roe deer, roe (*Capreolus capreolus*)
косьба' *f* cutting, grass-cutting, mowing
кося'к *m* school, shoal, swarm, run, stock
косячо'к *m* slant
~ **ага'ра** agar slant
кот *m* cat, tomcat, gib
барха'нный ~ dune cat, sand cat *Felis margarita)*
котё'нок *m* kitten, kit
ко'тик *m* fur seal (*Callorhynus ursinus*)
безгаре'мный саме'ц ~ idle seal
морско'й ~ fur seal (*Callorhynus ursinus*)
се'верный морско'й ~ sea-bear, sea-cat, northern fur, Alaska fur seal *(Callorhinus ursinus)*
ко'тики *m pl* rabbit-foot clover *(Trifolium arvense)*
ко'тик-одноле'ток *m* gray pup, silvery pup
котиледо'н *m* cotyledon
котиледо'нный *a* cotyledonary, cotyledonous
коти'п *m* cotype
коти'ться *v* kitten, bring forth
котлови'на *f* basin
кото'вник коша'чий *m bot* catnip *(Nepeta cataria)*
котрансляцио'нный *a mbiol* co-translational
котра'нспорт *m mbiol* cotransport
кофа'ктор *m* cofactor
ко'фе *m* coffee tree (*Coffea*); coffee
~ **в зё'рнах** coffee beans, coffee in the bean
быстрораствори'мый ~ instant coffee, soluble coffee
мо'лотый ~ ground coffee

раствори'мый ~ instant coffee, soluble coffee
кофеи'н *m* caffeine
коферме'нт *m* coenzyme, cofactor, coferment
кохлеа'рный *a* cochlear
кохроматогра'фия *f* co-chromatography
обратнофа'зная ~ reversed-phase co-chromatography
коча'н *m* cabbagehead, head, head of cabbage
~ **капу'сты** cabbagehead
кочево'й *a* nomadic
кочеды'жник *m bot* female fern, lady fern, spleenwort (*Athyrium*)
кочеры'га *f* cabbage stump
кочеры'жка *f* cabbage stump, cabbage stalk; cob (*of a corn ear*)
ко'чка *f* tussock, hummock
кочкова'тый *a* tussocky, hummocky
кочу'ющий *a* migrant
коша'ра *f* sheep barn
коша'чий *a* cattish, catlike, feline
коша'чья ла'пка *f bot* cat's-foot *(Antennaria)*
коше'ние *n* cutting, mowing; sweeping (*collection of insects from grass with a net*)
кошени'ль *f* cochineal, cochineal insect, coccus *(Coccus cacti)*
ко'шка *f* cat (*Felidae*); crowfoot grapple
дома'шняя ~ cat (*Felis*)
до'хлая ~ dead cat
камышо'вая ~ jungle cat (*Felis chaus*)
лесна'я ~ European wild cat *(Felis sylvestris)*
коэкспре'ссия *f* coexpression
коэффицие'нт *m* coefficient, index, quotient, rate, factor
~ **автокорреля'ции** autocorrelation coefficient
~ **аккомода'ции** accommodation coefficient
~ **аллометри'и** coefficient of allometry
~ **ассоциа'ции** coefficient of association
~ **Бравэ** Bravais's coefficient
~ **вариа'ции** coefficient of variation
~ **воспроизво'дства** coefficient of generation
~ **встреча'емости** frequency index
~ **выжива'емости** survival index
~ **гигроскопи'чности** hygroscopic coefficient
~ **Дайса** Dice's coefficient
~ **диспе'рсности** dispersion ratio
~ **диффу'зии** diffusion coefficient
~ **Жакка'ра** Jaccard's coefficient of community
~ **зави'симости ме'жду величи'нами** coefficient of association
~ **завяда'ния** wilting point, wilting coefficient
~ **заиле'ния** silt factor
~ **изотони'чности** isotonic coefficient
~ **иммигра'ции** immigration coefficient
~ **инбри'динга** inbreeding coefficient
~ **испо'льзования** use factor
~ **испо'льзования кислоро'да** oxygen-utilization coefficient, oxygen utilization quotient
~ **конди'ции** coefficient of condition
~ **корреля'ции** coefficient of correlation, correlation coefficient
~ **кроссинго'вера** correlation coefficient
~ **Ко'ула** Cole's coefficient
~ **Коха** Koch's coefficient
~**кусти'стости** stooling coefficient
~ **лета'льности** lethal index
~ **мно'жественной корреля'ции** coefficient of multiple correlation
~ **моля'рной эксти'нкции** molar extinction coefficient
~ **Мориси'та** Morisita's coefficient
~ **насыще'ния** saturation coefficient, saturation ratio
~ **неопределё'нности** coefficient of nondetermination
~ **обра'тной свя'зи** feedback factor
~ **о'бщности** coefficient of community, similarity coefficient, index of similarity
~ **передвиже'ния** translocation quotient
~ **перемеще'ния** translocation quotient
~ **Пи'рсона** Pearson's coefficient
~ **плодови'тости** coefficient of fertility
~ **поглоще'ния** absorption coefficient, absoptivity, absorption factor
~ **поле'зного де'йствия** efficiency, performance
~ **поро'зности** porosity factor *(of soil)*
~ **поте'рь** loss factor
~ **проница'емости** permeability coefficient
~ **пути'** path coefficient
~ **разбро'са** scattering coefficient
~ **разделе'ния** fractionation factor
~ **ра'нговой корреля'ции** rank correlation coefficient
~ **рассе'яния** scattering coefficient
~ **регре'ссии** regression coefficient

~ **родства'** coefficient of parentage, kinship coefficient
~ **рожда'емости** birth-rate coefficient
~ **ро'ста** growth coefficient
~ **сегрега'ции** segregation coefficient
~ **селе'кции** selection coefficient
~ **совпаде'ния** coefficient of coincidence
~ **согла'сия** coefficient of concordance
~ **согласова'ния** coefficient of agreement
~ **сопряжё'нности** coefficient of contingency
~ **состоя'тельности** coefficient of consistence
~ **Спи'рмена** Spearman's coefficient
~ **схо'дства** quotient of similarity
~ **сцепле'ния** factor of adhesion
~ **терминализа'ции** terminalization coefficient
~ **транспира'ции** transpiration coefficient
~ **увлажне'ния** precipitation-evaporation ratio
~ **упи'танности** coefficient of condition, coefficient of fatness, condition factor
~ **Ха'ккера-Да'йса** Hacker-Dice coefficient
~ **ча'стной корреля'ции** coefficient of partial correlation
~ **экстин'кции** extinction coefficient
~ **Ю'ла** Yule coefficient
~ **коэффициент** respiratory coefficient
азо'тный ~ nitrogen factor
ассимиляцио'нный ~ photosynthetic coefficient, assimilation quotient
биномиа'льный ~ binomial coefficient
гидротерми'ческий ~ hydrothermic coefficient
довери'тельный ~ confidence coefficient
дыха'тельный ~, **ДК** respiratory coefficient, respiratory quotient
изотони'ческий ~ isotonic coefficient
кормово'й ~ food coefficient, feeding ratio
о'бщий ~ overall coefficient
попра'вочный ~ correction factor, evaluation factor
попра'вочный ~ **на во'зраст** age correction factor
путево'й ~ path coefficient
середи'нный ~ **корреля'ции** medial correlation coefficient
сериа'льный ~ **корреля'ции** serial correlation coefficient
температу'рный ~ temperature coefficient
фено'льный ~ phenol coefficient
фотосинтети'ческий ~ photosynthetic coefficient
экономи'ческий ~ economic coefficient

краево'й *a* marginal, craspedodromous, limbic

краечерешко'вый *a* palaceous

край *m* edge, margin, border, limbus
вибрисса'льный ~ facial ridge, vibrissal ridge
запира'тельный ~ *cyt* closure margin
окаймлё'нный ~ bordered border, margined border
рё'берный ~ costal margin

крайт *m* krait (*Bungarus*)

краниа'льный *a* cranial

краниоло'гия *f* craniology

краниометри'я *f* craniometry

кра'нц-структу'ра *f* kranz structure

крапи'ва *f* nettle (*Urtica*)
~ **глуха'я** white dead nettle (*Lamium album*)

крапи'вник *m* wren (*Troglodytidae*)

крапи'вница *f* tortoise-shell buttefly (*Vanessa urticae*); *med* urticaria, hives, nettle rash
~ **Ми'льтона** Milton's disease

кра'пинка *f* fleck, freckle, mottle

кра'пчатость *f* mottling, spottiness; *phytop* mottle
~ **сте'бля карто'феля** potato-stem mottle
~ **табака'** tobacco mottle

краса'вка *f* deadly nightshade (*Atropa belladonna*); demoiselle crane (*Anthropoides virgo*)

краси'льный *a* tinctorial

краси'тель *m* stain, dye
~ **Ги'мза** Giemsa stain
~ **Ле'йшмана** Leishman stain
~ **Май-Грю'нвальда** May-Grunwald stain
~ **по Гра'му** Gram stain
ази'новый ~ azine dye
антрахино'новый ~ anthraquinone dye
вита'льный ~ vital stain, vital dye, intravital stain
дифференциа'льный ~ differential stain
индами'новый ~ indamine dye
индиго'идный ~ indigoid dye
индика'торный ~ indicator dye
индофено'льный ~ indophenol dye
ки'слый ~ acid dye
ксанте'новый ~ xanthene dye
основно'й ~ basic dye

пищево"й ~ food dye
прижи'зненный ~ intravital stain, vital stain
протравно'й ~ mordant dye
трехцве'тный ~ Ма'ллори Mallory's triple stain
трёхцве'тный ~ Флемминга Flemming tricolour stain
флуоресце'нтный ~ fluorescent dye
я'дерный ~ nuclear stain
кра'сить *v* stain
кра'ска *f* paint; dye, stain
необраста'ющая ~ antifouling paint
краснозё'м *m* red soil, red earth
краснокло'п *m* stainer (*Dysdercus*)
~ **бескры'лый** *m* red soldier bug *(Pyrrhocoris apterus)*
красноле'сье *n* conifer forest, coniferous forest, needle-leaved forest, softwood forest, pine forest
краснота' *f* rubor, erythema
краснота'л *m* pussy willow, silver willow (*Salix acutifolia*)
красноте'лка *f* harvest mite (*Trombidiidae*)
красну'ха *f* red-spot disease (*of carps*)
кра'сный *a* red
акриди'новый ~ acridine red
ализари'новый ~ alizarin red
бромфено'ловый ~ bromophenol red
бы'стрый ~ fast red
дуразоло'вый ~ durazol red
ко'нго ~ Congo red, cosmos red, cotton red, direct red
крезоло'вый ~ cresol red
мети'ловый ~ methyl red
нафти'ловый ~ naphthyl red
нейтра'льный ~ neutral red, toluylene red
руте'ний ~ ruthenium red
трипа'новый ~ trypan red
фено'ловый ~ phenol red
я'нус ~ Janus red
краспе'дий *m* craspedium, craspedon
кра'сться *v* slink
кратерообра'зный *a* crateriform
крати'кула *f* craticula *(in diatoms)*
кратковре'менный *a* short-term
краткосро'чный *a* short-term
крахма'л *m* starch
живо'тный ~ animal starch, hepatic starch, liver starch
расщепля'ющий ~ starch-splitting
крахма'листый *a* starchy
крахмалообразу'ющий *a* amylogenic, starch-forming
крахмалосодержа'щий *a* starch-containing
крахма'льный *a* amylaceous
кра'чка *f* tarn, tern, terrick (*Sterna*)
боло'тная ~ tarn (*Chlidonias*)
креати'н *m* creatine
креатини'н *m* creatinine
кренофи'льный *a* crenophilous
крепита'ция *f* crepitation
крепити'рующий *a* crepitant
кресс-сала'т *m* garden cress, cress (*Lepidium sativum*)
крест *m* cross
петро'в ~ *bot* Toothwort (*Lathrea*)
кресте'ц *m* sacral bone, sacrum, rump
крестови'к *m* cross spider (*Aranea diadema*)
кресто'вник squaw-weed *(Senecio)*
крестообра'зный *a* cruciform
крестоцве'тный *a* cruciferous
крестцо'во-ко'пчиковый *a* sacrococcygeal
крестцо'вый *a* sacral
кре'чет *m* gyrfalcon, gerfalcon, Arctic falcon, northern falcon (*Falco gyrfalco*)
крече'тка *f* sociable plover (*Chettusia gregaria*)
крива'я *f* curve
~ **активи'рования све'том** photoactivation curve
~ **Бертала'нфи** Bertalanffy curve, von Bertalanffy curve
~ **возрастно'й инволю'ции** age-involution weight curve
~ **воспроизво'дства** reproduction curve, recruitment curve
~ **встреча'емости** frequency curve
~ **выжива'емости** survivalship curve
~ **Гейдельбе'ргера** antibody-binding curve
~ **де'йствия** action curve
~ **до'за-эффе'кт** dose-effect curve
~ **зави'симости от вре'мени** time curve
~ **мы'шечного сокраще'ния** muscle curve
~ **накопле'ния** accumulation curve
~ **нараста'ющих часто'т** accumulation curve
~ **населе'ния** population curve
~ **обуче'ния** learning curve
~ **одино'чного ци'кла размноже'ния** one-step growth curve, single-step growth curve
~ **поглоще'ния** absorption curve

~ **пополне'ния** reproduction curve, recruitment curve
~ **распределе'ния по частоте'** distribution curve
~ **регре'ссии** regression curve
~ **ро'ста** growth curve
~ **свя'зывания антите'л** antibody-binding curve
~ **сме'ртности** mortality curve
~ **титрова'ния** titration curve
~ **титрова'ния антисы'воротки** antibody-binding curve
~ **часто'т** frequency curve
~ **«число' ви'дов-пло'щадь»** species-area curve
гиперболи'ческая ~ hyperbolic curve
двухверши'нная ~ bimodal curve
интегра'льная ~ accumulation curve
калибро'вочная ~ calibration plot
колоколови'дная ~ bell-shaped curve
логисти'ческая ~ logistic curve
многоуда'рная ~ multihit curve
одноверши'нная ~ unimodal curve
одноступе'нчатая ~ размноже'ния one-step growth curve, single-step growth curve
одноуда'рная ~ one-hit response curve
преципитацио'нная ~ antibody-binding curve
результи'рующая ~ resultant curve
светова'я ~ light curve
сигмо'идная ~ sigmoid curve
теорети'ческая ~ theoretical curve
титрацио'нная ~ titration curve
эксперимента'льная ~ experimental curve
экспоненциа'льная ~ exponential curve

кривизна' *f* curvature
больша'я ~ желу'дка greater curvature of stomach
ма'лая ~ желу'дка lesser curvature of stomach

криво'й *a* curved, crooked
криволе'сье *n* elfin woodland, crooked forest
кри'зис *m* crisis, solution (*of a disease*)
экологи'ческий ~ ecological crisis

крик *m eth* call, cry, screech
~ **бе'дствия** distress call
~ **сепара'ции** separation call
~ **трево'ги** alarm call, cry of alarm
большо'й ~ great call
отры'вистые кри'ки *pl* staccato cries
призы'вный ~ advertisement call
у'тренний ~ morning call

крино'м *m* crinome
крио- cryo-
криобиоло'гия *f* cryobiology
криоге'нный *a* cryogenic
криозащи'та *f* cryoprotection
криоиммуноглобули'н *m* cryoimmunoglobulin
криоконсерва'ция *f* cryopreservation
криопланкто'н *m* cryoplankton
криоповрежде'ние *n* freezing damage, cryoinjury
криопреципита'ция *f* cryoprecipitation
криопроте'ктор *m* cryoprotector
криопроте'кция *f* cryoprotection
криоска'лывание *n* cryofracture
криоско'п *m* cryoscope
криоскопи'я *f* cryoscopy
криоста'т *m* cryostat
криотропи'зм *m* cryotropism
криофикса'ция *f* cryofixation
криофи'л *m* cryophil
криофи'т *m* cryophyte
кри'пплинг *m*: **имму'нный** ~ immunocrippling

кри'пта *f anat* crypt
кише'чная ~ intestinal crypt, Lieberkuhn's crypt
либеркю'нова ~ Lieberkuhn's crypt, intestinal crypt

криптагглютино'ид *imm* cryptagglutinoid
крипти'ческий *a* cryptic
крипто- crypto-
криптобио'з *m* cryptobiosis
криптоге'нный *a* cryptogenous
криптоме'рия *f* cryptomerism, cryptomery
криптомито'з *m* cryptomitosis
криптоспоридио'з *m* cryptosporidiosis
криптосто'ма *f* cryptostome
криптото'п *m* cryptodeterminant
криптофи'т *m* cryptophyte
криптохиме'ра *f* cryptochimaera
крисскро'ссинг *m* crisscrossing
кристаллиза'тор *m* crystallizing dish, crystallizer
кристалли'ческий *a* crystalline
кристалло'ид *m* crystalloid

крите'рий *m* criterion, test
~ **аддити'вности** test of additivity
~ **гомоге'нности** homogeneity test
~ **достове'рности** test of significance
~ **зна'ков** *biom* sign test
~ **зна'чимости** test of significance

~ **лине'йности** test of linearity
~ **норма'льности** test of normality
~ **одноро'дности** test of homogeneity
~ **оплодотворя'емости** fertility criterion
~ **перестано'вки** permutation test
~ **подстано'вки** permutation test
~ **разма'ха** range test
~ **случа'йности** test of randomness
~ **смеще'ния** sippage test
~ **согла'сия** test of fit, test of goodness of fit
~ **сопряже'нности** *bot* association test
~ **среднеквадрати'чной оши'бки** square criterion
~ **с тремя' реше'ниями** three-decision test
~ **Стью'дента** Student's test
~ **хи-квадра'т** chi-squared test
биномиа'льный ~ binomial test
инвариа'нтный ~ homogeneity test
консервати'вный ~ conservative test
несмещё'нный ~ unbiased test
односторо'нний ~ one-sided test
приближё'нный ~ approximate test
середи'нный ~ corner test
симметри'чно ограни'ченный ~ equal-tails test
состоя'тельный ~ consistent test
усло'вный ~ conditional test
эффекти'вный ~ efficient test

крича'ть *v* cry, hoot; chitter (*of gulls*); bray (*of a donkey*)
крова'вый *a* bloody
кровезамени'тель *m* blood substitute
кро'велька *f* aril
кровено'сный *a* circulatory
кровепарази'т *m* blood parasite
кроветворе'ние *n* hemapoiesis, hemopoiesis, hematopoiesis, hemogenesis, hematogenesis, blood formation
кроветво'рный *a* hemopoietic, hematopoietic, hematogenetic, blood-forming, blood-making, blood-producing
кровоизлия'ние *n* blood effusion, blood extravasation, hemorrhage
кровообраще'ние *n* blood circulation, circulation
артериа'льное ~ arterial circulation
вене'чное ~ coronary circulation
иску'сственное ~ artificial circulation, extracorporeal circulation
капилля'рное ~ capillary circulation
коллатера'льное ~ collateral circulation
компенси'рующее ~ compensatory circulation
корона'рное ~ coronary circulation
мозгово'е ~ cerebral circulation
о'бщее ~ general circulation
перекрё'стное ~ cross circulation
перифери'ческое ~ peripheric circulation
печё'ночное ~ hepatic circulation
плацента'рное ~ placental circulation
регули'руемое ~ controlled circulation

кровоподте'к *m* ecchymosis
кровопоте'ря *f* hemorrhage, loss of blood
кровопуска'ние *n* bloodletting
кровоснабже'ние *n* blood supply
кровосо'с *m* bloodsucker
кровосо'ска *f ent* louse fly, spider fly, ked (*Hippobosca*)
кровососу'щий *a* bloodsucking
кровотече'ние *n* bleeding, hemorrhage
артериа'льное ~ arterial hemorrhage, bright bleeding
вено'зное ~ venous bleeding, venous hemorrhage

кровото'к *m* bloodstream, blood current, blood flow
избы'точный ~ surplus blood flow
обра'тный ~ retrograde blood stream, retrograde blood flow
фетоплацента'рный ~ placentofetal flow

кровоточа'щий *a* bleeding
кровоточи'вый *a* hemorrhagic
кровоточи'ть *v* bleed
кровь *f* blood; **карти'на кро'ви** hemogram, blood picture, blood count
артериа'льная ~ bright-red blood
аутологи'чная ~ autologous blood
вено'зная ~ dark blood, dark-red, blood black blood, blue blood
гепаринизи'рованная ~ heparinized blood
катиони'тная ~ cationitic blood
ла'ковая ~ laky blood
несовмести'мая ~ incompatible blood
оксала'тная ~ oxalate blood
оксигени'рованная ~ oxygenated blood
перифери'ческая ~ peripheric blood
консерви'рованная ~ stored blood
совмести'мая ~ compatible blood
це'льная ~ whole blood
цитра'тная ~ citrated blood

кровяни'стый *a* bloody
кровяно'й *a* hematic, sanguineus, bloody
кроко'ди'л *m* crocodile (*Crocodile*)

гребни'стый ~ estuarine crocodile *(Crocodylus porosus)*
ни'льский ~ Nile crocodile *(Crocodylus niloticus)*
крокоди'ловый *a* crocodilian
кро'лик *m* rabbit (*Oryctolagus*)
кроликово'д *m* rabbit breeder, rabbit keeper, warrener
кроликово'дство *n* rabbit husbandry, rabbit raising
крольча'тник *m* rabbit warren, rabbit house, rabbitry
кролье'нок *m* baby rabbit
крольчи'ха *f* doe rabbit
кро'мка *f* edge
кро'на *f* crown, krone
плаку'чая ~ drooping crown
кро'ншнеп *m* curlew (*Numenius*)
сре'дний ~ whimbrel (*Numenius phaeopus*)
кросс *m* cross
кроссбре'д *m* crossbred
кроссбри'динг *m* crossbreeding
кроссинго'вер *m* crossing-over
гониа'льный ~ gonial crossing-over
двойно'й комплемента'рный ~ complementary double crossing-over
двойно'й прогресси'вный ~ progressive double crossing-over
двойно'й регресси'вный ~ regressive double crossing-over
инверсио'нный ~ inversion crossing-over
межалле'льный ~ interallelic crossing-over
непра'вильный ~ illegitimate crossing-over
нера'вный ~ unequal crossing-over
прогресси'вный ~ progressive crossing-over
сомати'ческий ~ somatic crossing-over
крот *m* mole (*Talpidae*)
обыкнове'нный ~ common mole, Eurasian mole, European mole *(Talpa europaea)*
крото'вина *f* mole cast, molehill
кроха'ль *m* merganser, harle, fish duck, sawbill (*Mergus*)
большо'й ~ goosander (*Mergus merganser*)
кроши'ть *v* mince
круг *m* circle; **круги'** (*pl*) **на воде'** ripples on water
~ **рас** circle of races
~ **сы'ра** cheese
артериа'льный ~ arterial ring
артериа'льный ~ большо'го мо'зга circle of Willis
большо'й ~ кровообраще'ния systemic circulation, systemic circuit, greater circulation
вилли'зиев ~ circle of Willis, arterial ring *(of the brain)*
лё'гочный ~ кровообраще'ния lesser circulation, pulmonary circulation
ма'лый ~ кровообраще'ния lesser circulation, pulmonary circulation, pulmonary circuit
круглоголо'вка *f* toad agama, toad-headed agama (*Phrynocephalus*)
круглозу'бчатый *a* crenate
кругполи'стный *a* round-leaved, rotundifolius
круглопло'дный *a* cyclocarpous, round-fruited
круговоро'т *m* turnover
~ **воды'** water circulation
годи'чный ~ annual turnover
кругоресни'чный *a* peritrichous
кружи'ть над *v* fly round
кру'жка *f* mug, jug
~ **Э'смарха** Esmarch's irrigator
кружо'к *m* areola, circle, disk, orbiculus
околососко'вый ~ mammary areola
круп *m zool* croup; *med* croup
крупа' *f* groat, groats; sleet, granular snow, snow pellets, soft hail
гре'чневая ~ buckwheat
ма'нная ~ semolina
овся'ная ~ oatmeal
перло'вая ~ pearl barley, pearled barley
я'чневая ~ pot barley
крупно- macr-, macro-
крупнозла'чник *m* tall grass community
крупноколо'сый *a* long-spiked
крупнопло'дный *a* large-fruited, marcocarpous, megalocarpous
крупноразнотра'вник *m* tall grass community
крупносе'тчатый *a* wide-meshed
крупнотычи'нковый *a* macrandrous
крупночашели'стиковый *a* macrosepalous
крупноя'чеистый *a* wide-meshed
кру'пный *a* large, big; coarse; large-scale
крупору'шка *a* hulling mill, peeling mill
круто'й *a* steep; hard-boiled (*of eggs*)
круши'на *f bot* buckthorn (*Rhamnus*)
крыжо'вник *m* gooseberry (*Grossularia*)

крыла'н *m* fruit bat (*Pteropodidae*)
крыла'тка *f* samara, key fruit, key, winged fruit, winged seed
крыла'тый *a* winged, alate
крыло' *n* wing, ala
~ **но'са** wing of nostril
пере'днее ~ fore wing, front wing
крылови'дный *a* winglike, aliform, alar, alary, pteroid, pterigoid
крылово'й *a* alary
крылоно'гий *a* pteropod, aliped
крылoости'стый *a* pterygospinal
кры'лышко *n orn* bastard wing, alula
кры'ночка *f bot* pyxidium, pyxis
кры'са *f* rat (*Rattus*)
бе'лая ~ albino rat
се'рая ~ common rat, brown rat, Norway rat (*Rattus norvegicus*)
чё'рная ~ black rat, roof rat, house rat, ship rat (*Rattus rattus*)
крысё'нок *m* infant rat
крысё'нок-отъё'мыш *m* weanling rat
кры'ска *f* rat-tailed maggot (*Eristalis*)
крысоло'вка *f* rat-trap
кры'ша *f* roof, tectum
~ **бараба'нной по'лости** roof of tympanum
~ **че'репа** skullcap
кры'шечка *f* lid, opercle, operculum
кры'шка *f* lid, cap
жа'берная ~ gill cover, opercle, operculum
крючкова'тый *a* uncinate, hamate, hooked
крючколи'стный *a* hook-leaved
крючкопло'дный *a* hook-fruited
крючо'к *m* hook, crotchet, hamus, hamulus
~ **с наса'дкой** baited hook
ацетабуля'рный ~ acetabular hook
двупоря'дковые крю'чья *pl ent* biordinal hooks
многопоря'дковые крю'чья *pl ent* multiordinal hooks
однопоря'дковые крю'чья *pl ent* uniordinal hooks
ротово'й ~ oral hook
рыболо'вный крючок fish hook
эмбриона'льный ~ embryonal hook
кря'канье *n* quack
кря'кать *v* quack
кря'ква *f* mallard, mallard duck (*Anas platyrhynchos*)
ксанто'з *m* xanthosis
ксантолейци'т *m* xantholeucite
ксантофи'лл *m* xanthophyll
ксантофо'р *m* xanthophore
ксениобио'з *m* xeniobiosis
ксе'ния *f* xenia
ксено- xeno-
ксеноантиге'н *m* xenoantigen
ксеноантисы'воротка *f* xenoantiserum
ксеноантите'ло *n* xenoantibody
ксенобио'з *m* xeniobiosis
ксенобио'тик *m* xenobiotic
ксеноге'нность *f* xenogenicity
ксеноге'нный *a* xenogenic
ксенодиагно'стика *f* xenodiagnosis
ксенопарази'т *m* xenoparasit
ксеносо'ма *f* xenosome
ксенотрансплантa'т *m* xenograft
ксенотрансплантa'ция *f* xenogenic grafting, xenografting, heterologous transfer, xenogenic transfer
ксеноэпито'п *m* xenoepitope
ксери'ческий *a* xeric
ксеро- xero-
ксеробо'л *m bot* xeriobole, xerobole
ксерогеофи'т *m* xerogeophyte
ксеромо'рф *m* xeromorph
ксероморфи'зм *m* xeromorphy
ксеромо'рфный *a* xeromorphic, xeromorphous
ксероморфо'з *m* xeromorphosis
ксероте'рм *m* xerotherm
ксеротерми'ческий *a* xerothermic
ксерофи'льный *a* xerophilous, siccocolous
ксерофи'т *m* xerophyte
ксерофи'тный *a* xerophytic
ксиле'ма *f* xylem
втори'чная ~ secondary xylem
двухлучева'я ~ diarch xylem
меза'рхная ~ mesarch xylem
перви'чная ~ primary xylem
промежу'точная ~ mesarch xylem
экза'рхная ~ exarch xylem
энда'рхная ~ endarch xylem
ксиленциано'л *m* xylene cyanol
ксилоге'н *m* xylogen
ксилока'рп *m* xylocarp
ксило'л *m* xylene, xylol
ксилосахаро'за *f* xylosucrose
ксилото'мия *f* xylotomy
ксилофа'г *m* xylophage
ктени'дий *m* ctenidium
ктено'идный *a* ctenoid
ктырь *m* assasin fly, hawk fly, robber fly (*Asilidae*)

кубареви'дный *a* turbinate
кубита'льный *a* cubital
кубови'дно-ладьеви'дный *a* cuboideonavicular
кубы'шка *f* yellow pond-lily, yellow water lily, spatterdock (*Nuphar*); *ent* egg pod
кувши'нка *f* water lily (*Nymphaea*)
кувшинови'дный *a* pyxidate
кувши'нчатый *a* urceolate, ascidiform
кувши'нчик *m* urceole
куда'хтанье *n* cackle, clucking, squawk, gabble
куда'хтать *v* cackle, cluck, squawk, gabble
кудла'тый *a* shaggy
кудрева'тый *a* crispate, crisped
кудрявоцветко'вый *a* curl-flowered
кузне'чик *m* grasshopper (*Tettigoniidae*, *Acrididae*)
кукаба'ра *f* kookaburra *(Dacelo)*
куклоро'дный *a* pupigenous, pupiparous
кукова'ть *v* cuckoo
ку'колка *f* pupa
  акти'вная ~ active pupa
  бочонкообра'зная ~ barrel-shaped pupa, coarctate pupa
  откры'тая ~ free pupa
  свобо'дная ~ free pupa
  покры'тая ~ obtected pupa
  педогенети'ческая ~ paedogenetic pupa
куколкообра'зный *a* pupiform
ку'колочный *a* pupal
куко'ль *m* corn cockle (*Agrostemma githago*)
ку-ку' *i* cockoo
кукуру'за *f* maize, corn, Indian corn (*Zea mays*)
куку'шка *f* cuckoo (*Cuculus*)
ку'кша *f* Siberian jay (*Cractes infaustus*)
кула'к *m* fist
  сжа'тый ~ clenched fist
кула'н *m* Asiatic wild ass, onager (*Equus hemmionus*)
кулды'кать *v* gobble
кули'га *f* cloud
кули'к *m* sandpiper, stint (*Calidris*)
кули'к-соро'ка *m* catcher (*Haematopinae*)
кульба'ба *f* fall dandelion (Leontodon)
культива'р *m* cultivar
культива'ция *f* cultivation
культиви'рование *n* culturing, cultivation, culture
  аксе'нное ~ axenic cultivation
  глуби'нное ~ submerged cultivation
  ма'ссовое культивирование large-scale culture
культиви'ровать *v* cultivate
культиви'руемый *a* cultured, cultivated
культиге'н *m* cultigen
культу'ра *f* culture; crop; **в чи'стой культу'ре** in pure culture
  ~ **в вися'чей ка'пле** hanging-drop culture
  ~ **в жи'дкой среде'** liquid culture
  ~ **Ви'тлока - Уа'йта** Whitlock-White culture
  ~ **де'кстеровского ти'па** Dexter-type culture
  ~ **защищё'нного гру'нта** glasshouse culture
  ~ **из одно'й кле'тки** single-cell culture
  ~ **кле'ток** cell culture
  ~ **мазко'м** smear culture
  ~ **на ага'ре** agar culture
  ~ **на по'чвенной вы'тяжке** soil-water culture
  ~ **на предме'тном стекле'** slide culture
  ~ **на ско'шенном ага'ре** slant culture
  ~ **откры'того гру'нта** outdoor culture
  ~ **под откры'тым не'бом** outdoor culture
  ~ **ро'ллерного ти'па** roller-type culture
  ~ **с подпи'ткой** fed batch culture
  ~ **с постоя'нным переме'шиванием** "spinner" culture
  ~**тка'ни** tissue culture
  ага'ровая ~ agar culture
  аксе'нная ~ axenic culture
  альгологи'чески чи'стая ~ unialgal culture
  ароматообразу'ющая ~ aroma-producing culture
  бактериологи'чески чи'стая ~ bacteria-free culture
  бахчё'вые культу'ры *pl* cucurbits crop, melons and gourds
  бескле'точная ~ cell-depleted culture
  бога'рная ~ dryland culture
  бульо'нная ~ broth culture
  во'дная ~ water culture
  встря'хиваемая ~ agitated culture
  вы'гоночная ~ forcing culture
  гидропо'нная ~ hydroponic culture, soilless culture, soilless crop, gravel culture
  глуби'нная ~ submerged culture
  гря'дковая ~ ridge culture
  густа'я ~ crowded culture
  двухле'тняя ~ biennial culture

двухсло'йная ~ two-layer culture
дли'тельная ~ long-term culture
долгоживу'щая ~ long-term culture
желати'новая ~ gelatin culture
загрязнё'нная ~ impure culture
замещё'нная ~ replacement culture
интродуци'рованные культу'ры *pl* alien crops
исхо'дная ~ stock culture
ка'пельная ~ droplet culture
кло'новая ~ clone culture
кормова'я ~ forage crop, feed crop, feeding crop
ко'сточковая плодо'вая ~ stone fruit crop
кровяна'я ~ blood culture
крупномасшта'бная ~ large-scale culture
лаборато'рная ~ laboratory culture
ма'сличная ~ oil-bearing culture, oil crop, oil-producing crop, oil-yielding crop
ма'ссовая ~ mass culture, large-scale culture
междуря'дная ~ interrow crop
микро'бная ~ germ culture
многостади'йная непреры'вная ~ multistage continuous culture
моноксе'нная ~ monoxenic culture
моносло'йная ~ monolayer culture
непреры'вная ~ continuous culture
обогати'тельная ~ enrichment culture
овощна'я ~ vegetable crop
одноле'тняя ~ annual crop
односло'йная ~ monolayer culture
одностади'йная непреры'вная ~ one-stage continuous culture
ози'мая ~ winter crop
оранжере'йная ~ glasshouse culture, greenhouse culture
о'рганная ~ organ culture
орехопло'дная ~ nut-fruited crop, nut crop
ороша'емая ~ irrigated crop
первичная ~ primary culture
периоди'ческая ~ batch culture
пове'рхностная ~ surface culture, stroke culture
погружё'нная ~ submerged culture
подсевна'я ~ undersowing crop
подсе'чная ~ shifting forest crop
покро'вная ~ cover crop
полунепреры'вная ~ semicontinuous culture
по'чвенная ~ soil culture
почвоочища'ющая ~ cleaning crop
предше'ствующая ~ advance crop
продлё'нная ~ extended culture
пропашна'я ~ tilled crop
самосе'вная ~ self-seed crop
севооборо'тная ~ rotating crop
сельскохозя'йственная ~ crop
си'лосная ~ ensilage crop, silage crop
синхронизи'рованная ~ synchronized culture
сме'шанная ~ mixed culture
сме'шанная ~ лимфоци'тов, СКЛ mixed lymphocyte culture
сопу'тствующая ~ companion crop
стациона'рная ~ steady-state culture
субконфлюэ'нтная ~ subconfluent culture
суспензио'нная ~ suspension culture
сыра'я ~ crude culture
техни'ческая ~ technical crop
тка'невая ~ tissue culture
турбидиста'тная ~ turbidistate culture
уплотне'нная ~ companion crop
фура'жная ~ fodder crop, forage crop, hay crop
хемоста'тная ~ chemostat culture
чи'стая ~ axenic culture, pure culture; **в чи'стой культу'ре** in pure culture
ярова'я ~ spring crop

культура'льный *a* cultural
культу'ра-уплотни'тель *f* companion crop
культу'рный *a* cultivated, cultured
культя' *f* stump
умани'ка *f* blackberry (*Rubus fruticosus*)
кумари'н *m* coumarin
кумули'ровать *v* cumulate
кумули'роваться *v* cumulate
кумуляти'вный *a* cumulative
кумуля'ция *f* cumulation
кумы'с *m* koumiss
кунжу'т *m* sesame, beniseed, benne (*Sesamum*)
куни'ца *f* marten (*Martes*)
ку'па *f* cluster, clump, krone, group (*of trees*)
купа'лка *f eth* wallow
купа'льница *f* globe-flower, double buttercup (*Trollius*)
купа'ние *n* bathing
~ **в пыли'** *eth* dust-bathing
купа'ться *v* bathe
~ **в пыли'** *eth* take s dust bath
купе'на *f* Solomon's-seal (*Polygonatum*)
~ **изме'нчивая** sealwort *(Polygonatum commutatum)*

купи'ровать *v* cup off, arrest
ку'пол *m* dome
купре'й *m* kouprey *(Bos sauveli)*
купы'рь *m* wild chervil, beaked chervil, beakchervil (*Anthriscus*)
курага' *f* dried apricots, dried pitted and halved apricots
кура'ре *n* curare
курари'н *m* curarine
курга'нник *m* long-legged buzzard (*Buteo rufinus*)
кургу'зый *a* bob-tailed
курдю'к *m* fat tail (*of sheep*)
курдю'чный *a* fat-rumped (*of sheep*)
ку'рица hen, chicken (*Gallus gallus*)
  куста'рниковая ~ brush turkey *(Aepypodius)*
  со'рная ~ brush turkey *(Talegalla)*
курлы'канье *n* clang (*of cranes*)
курлы'кать *v* clang (*of cranes*)
курово'дство *n* chicken farming, poultry farming
куропа'тка *f* partridge (*Perdix*); ptarmigan (*Lagopus*)
  бе'лая ~ willow ptarmigan *(Lagopus lagopus)*
  ка'менная ~ rock partridge *(Alectoris)*
  се'рая ~ common partridge, European partridge, gray partridge *(Perdix perdix)*
  тундряна'я ~ ptarmigan, rock ptarmigan (*Lagopus mutus*)
куро'рт *m* resort
курти'на *f* clump, bed, grove
курча'вость *f phytop* curl, curly top
  ~ **табака'** tobacco stripe-and-curl disease
курча'вый *a* fuzzy (*of hair*); curled
куря'тник *m* hen house
куса'ть *v* bite, snap
куса'ться *v* bite
ку'скус *m* cuscus *(Phalanger)*
куст *m* bush, shrub
  кокаи'новый ~ coca (*Erythroxylum coca*)
  креозо'товый ~ creosote bush *(Corillea)*
куста'рник *m* bush, shrub, shrubbery, frutex
  декорати'вный ~ ornamental bush
  древови'дный ~ arbuscle
куста'рниковый *a* frutescent, fruticose, shrubby
куста'рничек *m* dwarf shrub, low shrub, undershrub
куста'рничковый *a* fruticulose
кусти'стость *f* tilling capacity, bushiness
кусти'стый *a* bushy, dumetose, dumose, fruticose; ~ **лиша'йник** fruticose lichen
кусти'ться *v* tiller
кути'кула *f* cuticle
  эмбриона'льная ~ embryonic cuticle
кутикуля'рный *a* cuticular
кути'н *m* cutin
кутиниза'ция *f* cutinization
ку'тис *m* cutis
ку'цый *a* bob-tailed, dock-tailed, tailless
ку'ча *f* pile, heap
  муравьи'ная ~ ant hill
куще'ние *n* tillering
кэ'ппинг *m imm* capping
  антителоиндуци'рованный ~ antibody-induced redistribution
кэппи'рующий *a* capping
кэп-са'йт *m* cap site

Л

лаби'льный *a* labile
лабири'нт *m eth* maze; *anat* labyrinth
~ **решё'тчатой ко'сти** ethmoidal labyrinth
~ **с одни'м вхо'дом** one-choice maze
каска'дный ~ elevated maze
ко'стный ~ bony labyrinth
перепо'нчатый ~ membranous labyrinth
Т-обра'зный ~ T maze
лабиринтодо'нтный *a* labyrinthodont
ла'биум *m* under lip, labium
лаборато'рия *f* laboratory
иссле'довательская ~ exploratory laboratory, research laboratory
полева'я ~ field laboratory
лаброци'т *m* labrocyte
ла'брум *m* upper lip, labrum
лава'нда *f* lavender (*Lavandula*)
лавр *m* laurel *(Laurus)*
лаврови'шня *f* laurel cherry (*Laurocerasus officinalis*)
лаг-пери'од *m* lag period
лаг-фа'за *f* lag phase
ла'дан *m* frankincense
ладаноно'сный *a* incense-bearing
ладо'нный *a* palmar, thenal
ладо'нь *f* palm, the heel of hand
ладьеви'дный, ладьеобразный *a* navicular, cymbifrom, scaphoid, boat-shaped
ла'зающий *a* scansorial, scandent, climbing
лазо'ревка *f* nun, blue tit, willowbiter (*Parus caeruleus*)
лай *m* bark, barking, yelp, jap, woof
визгли'вый ~ shrill-bark
ла'йка *f* husky, wolf dog; kid-skin
лайониза'ция *f* lyonization
лака'ть *v* lap up
ла'кмус *m* litmus
лакри'ца, **лакри'чник** *m* licorice, liquorice (*Glycyrrhiza*)
лакта'ция *f* lactation, lactogenesis, galactosis, galactopoiesis
лактоальбуми'н *m* milk albumin
лактоге'н *m* lactogen
лактоге'нный *a* lactogenic, lactigenous
лакто'за *f* lactose
лактофлави'н *m* lactoflavin
лаку'на *f* lacune, lacuna
~ **Га'ушипа** lacuna of Howship
разъе'денная ~ lacuna of Howship
лакуна'рный *a* lacunar
лакфио'ль *f* wallflower (*Cheiranthus*)
ла'ма *f* lama (*Lama*)
ламарки'зм *m* lamarckism, lamarckianism
ламиа'ция *f* lamiation
ла'мия *f* lamies
ла'мпа *f* lamp
~ **лампа дневно'го све'та** day-light lamp
бактерици'дная ~ bactericidal lamp, germicidal lamp, uviol lamp
ландша'фт *m* landscape
ланоли'н *m* lanolin
ла'ндыш *m* lily-of-the-valley (*Convallaria*)
ланце'тник *m* lancelet (*Amphyoxus*)
ланцетови'дный *a* lanceolate
ланцетоли'стный *a* lance-leaved
ланцетообразный *a* lanceolate
лань *f* fallow deer (*Cervus dama*)
ла'па *f* paw, foot, pad; **стук ла'пами** stamping
ла'пка *f ent* tarsus
коша'чья ~ cat's ear, cat's foot, pussy's toes, pussytoe (*Antennaria*)
лапча'тка *f* five-finger, cinquefoil (*Potentilla*)
ла'пчатый *a* palmate
ларвици'д *m* larvicide, larvicidal agent
ла'ска *f* weasel (*Mustela nivalis*)
ласт *m* paddle, flipper, flapper
ла'стовень *m* swallowwort (*Antitoxicum*)
ла'сточка *f* swallow, martin, martinet (*Hirundinidae*)
берегова'я ~ sand martin, bank swallow (*Riparia riparia*)
городска'я ~ martin, martinet, house martin (*Delichon urbica*)
ла'сточка-берегову'шка bank swallow, sand swallow (*Riparia riparia*)
ла'та-тип *m* lata-type
ла'текс *m* latex
латерализа'ция *f* lateralization
латера'льный *a* lateral
латери'т *m* laterite
латерофле'ксия *f* lateroflexion
лаци'ния *f* lacinia
ла'ять *v* bark, yelp, jap
лебеда' *f* orach, pigweed (*Atriplex*)
ле'бедь *m* swan (*Cygnus*)
ма'лый ~, ту'ндровый ~ Bewick's swan *(Cygnus bewickii)*

черноше'йный ~ black-necked swan *(Cygnus melanocoryphus)*
чё'рный ~ black swan *(Cygnus atratus)*
ле'бедь-клику'н *m* whooper swan (*Cygnus cygnus*)
ле'бедь-саме'ц *m* pen swan
ле'бедь-труба'ч *m* trumpeter swan *(Cygnus buccinator)*
ле'бедь-шипу'н *m* swan, mute swan (*Cygnus olor*)
лев *m* lion (*Panthera leo*)
муравьи'ный ~ *ent* ant lion (*Myrmileon*)
лева'тор *m* levator
левко'й *m* stock, gilliflower (*Matthiola*)
левовраща'ющий *a* levorotatory; sinistrorse
левозаве'рнутый *a* levotropic
левозавито'й *a* sinistral
левонапра'вленный *a* votropic
левору'кость *f* left-handedness, mancinism
ле'вый *a* left, sinistral
лега'вая *f* gundog
лё'гкий *a* light
лё'гкое *n* lung
легго'рн *m* Leghorn
лё'гочный *a* pulmonary
лёд *m* ice
бли'нчатый ~ cake ice, lily pan ice, pancake ice, plate ice
па'ковый ~ pack ice
сто'лбчатый ~ columnar ice
торо'систый ~ pack ice
ледни'к *m* glacier
ледни'чник *m* snow fly (*Boreus*)
леднико'вый *a* glacial
ледохо'д *m* ice run, ice drift, drifting of ice
лежа'ть *v* lie
ле'жбище *n* rookery (*of fur seals*), ground
лё'жка *f* lair
лё'жкий *a* long-keeping
лезть *v* climb; fall out (*of hair*)
лейкеми'я *f* leukemia, leukosis
лейкеми'ческий *a* leukemic
лейко- leuko-
лейкоагглютина'ция *f* leukoagglutination
лейкоагглютини'н *m* leukoagglutinating activator
лейковори'н *m* leukovorin
лейколизи'н *m* leukolysin
лейкопени'я *f* leukopenia, leucopenia, leucocytopenia
лейкопла'ст *m* leucoplast, leucoplastid, leucite
лейкопоэ'з *m* leukopoiesis, leucopeiesis, leukocytogenesis
лейкотрие'н *m* leukotriene
лейкоцита'рный *a* leukocytic, leucocytal
лейкоцито'з *m* leukocytosis
лейци'н *m* leucine
лейцинэнкефали'н *m* Leu-enkephalin
лейко'з *m* leukosis, leukemia
лейкоци'т *m* leukocyte, leucocyte, white blood cell
базофи'льный ~ basophil, basophile, basophilic leukocyte
блужда'ющий ~ erratic leukocyte
зерни'стый ~ granular leukocyte
мононуклеа'рный ~ mononuclear leukocyte
нейтрофи'льный ~ neutrophilic leukocyte
осё'длый ~ resident leukocyte
подви'жный ~ motile leukocyte
полимо'рфно-я'дерный ~ polymorphonuclear leukocyte
сегменти'рованный ~ segmented leukocyte
тканево'й ~ resident leukocyte
эозинофи'льный ~ eosinophilic leukocyte
лейкоцита'рный *a* leukocytal, leukocytic
лейкоцито'з *m* leukocytosis
пищевари'тельный ~ digestive leukocytosis
лейшманио'з *m* leishmaniasis
америка'нский ~ American leishmaniasis
висцера'льный ~ visceral leishmaniasis
инди'йский висцера'льный ~ Assam fever, dumdum fever
ко'жный ~ cutaneous leishmaniasis, dermal leishmaniasis
лека'рство *n* drug, medicament, medicine
лекти'н *m* lectin
~ **проро'стков пшени'цы** wheat germ lectin
~ **семя'н клещеви'ны** castor bean lectin
лектоти'п *m* lectotype
ле'мминг *m* lemming (*Lemmus*)
копытный ~ arctic lemming, hoofed lemming, collared lemming *(Dicrostonyx torquatus)*
настоя'щий ~ lemming *(Lemmus)*
норве'жский ~ Norway lemming *(Lemmus lemmus)*
оше'йниковый ~ arctic lemming, hoofed lemming *(Dicrostonyx torquatus)*

сиби'рский ~ Siberian lemming, brown lemming (*Lemmus sibiricus*)
леммоци'т *m* lemmocyte
лемни'ск *m* lemniscus
лему'р *m* lemur *(Lemuridae)*
лён *m* flax, lint (*Linum*)
~ **двухле'тний** pale flax *(Linum bienne)*
~ **долгуне'ц** fiber flax (*Linum usitatissimum proles elongata*)
~ **кудря'ш** crown flax (*Linum usitatissimum proles brevimulticaulis*)
~ **обыкнове'нный** lint *(Linum usitatissimum)*
~ **слаби'тельный** fairy lint *(Linum catharicum)*
куку'шкин ~ haircap-moss (*Polytrichum commune*)
лён-долгуне'ц *m* fiber flax *(Linum usitatissimum elongata)*
лене'ц *m* toadflax *(Thesium)*
лени'вец *m* sloth (*Bradypodidae*)
лён-кудря'ш oil flax, crown flax (*Linum usitatissimum proles brevimulticaulis*)
лено'к *m bot* goldilocks (*Linosyris*); *ich* lenok (*Brachymistax lenok*)
ле'нта *f* band, tape
о'рденская ~ underwing moth (*Catocala*)
ленте'ц *m* tapeworm
широ'кий ~ fish tapeworm, broad tapeworm (*Diphyllobothrium latum*)
лентиви'рус *m* lentivirus
лентообра'зный *a* taeniate, taenioid
ле'нточница *f* underwing moth (*Catocala*)
леопа'рд *m* leopard (*Panthera pardus*)
лепесткови'дный *a* petaliform, petaline, petaloid, petaloideous, petal-shaped
лепестко'вый *a* petaline
лепесто'к *m* petal
боково'й ~ бобо'вых wing petal
жа'берный ~ gill filament
закру'ченный ~ involute petal
ле'пет *m eth* babbling
лепоци'т *m* lepocyte
ле'пра *f* leprosy
~ **Лу'сио-Альвара'до** Lucio-Alvarado leprosy
лепто'м *m* leptome
лептоне'ма *f* leptonema
лептоте'на *f* leptotene
лептоцентри'ческий *a* leptocentric
лептоцефали'да *f* leptocephalid
лес *m* forest, woodland
~ **на корню'** standing timber
байра'чный ~ forest in steppe ravines
берё'зовый ~ birch forest
боло'тистый ~ paludal forest
бореа'льный хво'йный ~ boreal coniferous forest
бу'ковый ~ beech forest
вереща'тниковый ~ heath forest
вечнозелё'ный ~ evergreen forest
вла'жный го'рный ~ cloud forest
вла'жный тропи'ческий ~ tropical rain forest, moist tropical forest
втори'чный ~ secondary forest
вы'борочный ~ selection forest
галере'йный ~ gallery forest, fringe forest, fringing forest, fringe woodland
глухо'й ~ dense forest, thick forest
го'рно-таё'жный ~ mountain taiga forest
го'рный ~ mountain forest
гра'бовый ~ hornbeam forest
де'вственный ~ virgin forest
дождево'й ~ rain forest
дрему'чий ~ dense forest, thick forest
дубо'во-гра'бовый ~ hornbeam-oak forest
ело'вый ~ spruce forest
жестколи'стный ~ sclerophyllous forest, hard-leaved forest
заболо'ченный ~ paludal forest
запове'дный ~ reserved forest, sanctuary forest, protection forest
защи'тный ~ protective forest
зимнезелё'ный ~ winter green forest
клено'вый ~ maple forest
колю'чий ~ thorn forest, thorn woodland
коренно'й ~ primary forest
летнезелё'ный ~ deciduous summer forest
листвени'чный ~ larch forest
ли'ственный ~ broad-leaved forest, deciduous forest, foliage forest
листопа'дный ~ deciduous forest
мелколи'ственный ~ small-leaved forest, parvifoliate forest
муссо'нный ~ monsoon forest
низкостволь'ный ~ coppice wood, shrubwood
па'льмовый ~ palm forest
первобы'тный ~ primeval forest
пи'хтовый ~ fir forest
по'йменный ~ bottomland forest, flood plain forest, inundated forest, glade
полусухо'й тропи'ческий ~ semidry tropical forest

по'рослевый ~ shrubwood, coppice forest
прибре'жный ~ riparian forest, riparian woodland
редкосто'йный ~ light forest
самши'товый ~ box forest
светлохво'йный ~ light coniferous forest
сме'шанный ~ mixed forest
сосно'вый ~ pine forest
строево'й ~ building timber
субальпи'йский ~ orohylile forest
субтропи'ческий вечнозелё'ный ~ subtropical evergreen forest
темнохво'йный ~ dark coniferous forest
тропи'ческий ~ tropical forest
тропи'ческий дождево'й ~ tropical rain forest, moist tropical forest
туга'йный ~ riparian forest
уще'льный ~ gorge forest, ravine forest
хво'йный ~ conifer forest, coniferous forest, needle-leaved forest, softwood forest
ча'стый ~ dense forest
широколи'ственный ~ broad-leaved forest
эвкали'птовый ~ gum-tree forest

леса' *f* fishing line; *pl of* **лес**
ле'са *genetive of* **лес**
леси'стый *a* wooded, woody, forest-clad
лесни'чество *n* forestry
лесни'чий *m* forester
лесно'й *a* sylvan, sylvestral, forest
лесове'дение *n* forest science
лесово'дство *n* forestry, silviculture, sylviculture, forest management
лесово'дческий *a* sylvicultural, silvicultural
лесовосстановле'ние *n* reforestation, reafforestation
лесозаготови'тельный *a* lumbering, logging
лесозаготo'вка *f* lumbering, logging
лесозащи'та *f* forest protection
лесо-лугово'й *a* forest-meadow
лесоматериа'л *m* timber
лесомелиора'ция *f* forest reclamation
лесонасажде'ние *n* afforestation, forestation
защи'тное ~ shelter forest
полезащи'тное ~ field-protective forestation
лесопа'рк *m* recreational forest
лесопова'л *m* tree felling
лесопо'льзование *n* forest management
лесопоса'дка *f* afforestation, forestation
лесосе'ка *f* cutting site, felling site, cutting area, felling area
лесоспла'в *m* logging
лесосте'пь *f* forest-steppe
лесоту'ндра *f* forest-tundra
лесоустро'йство *n* forest management
леспромхо'з *m* integrated logging-lumbering enterprise
лесс *m* loess
ле'стница *f*:
~ **преддве'рия** staircase of vestibule
иерархи'ческая ~ social ladder
лесхо'з *m* forestry, district forestry
лет *m* flight
лета'ль *f* lethal
аутосо'мная ~ autosomal lethal
безусло'вная ~ unconditional lethal
гамети'ческая ~ gametic lethal
домина'нтная ~ dominant lethal
зиготи'ческая ~ zygotic lethal
кле'точная ~ cell lethal
пыльцева'я ~ pollen lethal
рецесси'вная ~ recessive lethal
сбаланси'рованные летали *pl* balanced lethals
усло'вная ~ conditional lethal
лета'льность *f* lethality
гибри'дная ~ hybrid lethality
полиге'нная ~ polygenic lethality
лета'льный *a* lethal
лета'ть *v* fly
лета'ющий *a* flying, volant
ле'тка *f* entrance (*in beehive*)
ле'тний *a* aestival, summer
лето'к *m* entrance, bee-entrance (*of a beehive*)
лету'чка *f* pappus
лецити'н *m* lecithin
лече'ние *n* treatment, therapy, therapeutics, cure
~ **зубо'в** treatment of teeth, dental treatment, dental therapy, dental care
лече'бный *a* therapeutic, curative
лечи'ть *v* treat, cure, heal; **лечить** (*with something*) treat with; ~ **от** treat for
лещи'на *f* hazel-nut, hazel (*Corylus avellana*)
лиа'за *f* lyase
лиа'на *f* liane, liana
ли'вень *m* storm rain
ли'вер *m* liver
лига'за *f* ligase
лигаме'нт *m anat* ligament
термина'льный ~ terminal ligament
лига'нд *m* ligand
лигату'ра *f* ligature, ligation

лиги'рование *n mbiol* ligation
лигни'н *m* lignin
лигнифика'ция *f* lignification
ли'гула *f* ligule, ligula
ли'дер *m* leader
ли'дерство *n* leadership
лиена'льный *a* lienal, splenic
лиза'т *m* lysate
  кле'точный ~ cell lysate
  бескле'точный ~ cell-free lysate
лиза'ть *v* lick
лизиге'нный *a* lysigenicous
лизика'рпный *a* lysicarpous
лизи'ровать *v* lyse
ли'зис *m* lysis
  ~ **извне'** lysis from without
  ~ **эритроци'тов** erythrocyte lysis
  неспецифи'ческий ~ bystander lysis
  преждевре'менный ~ premature lysis
  реакти'вный ~ reactive lysis
  сливно'й ~ confluent lysis
  спонта'нный ~ snap lysis, spontaneous lysis
  циани'дный ~ cyanide lysis
лизогене'з *m* lysogenesis
лизогениза'ция *f* lysogenization
лизоге'нность *f* lysogenicity
лизосо'ма *f* lysosome
лизоци'м *m* lysozym, lysozyme
лизуне'ц *m* lick
ли'лия *f* lily (*Lilium*)
лима'н *m* lagoon
лимаци'д *m* limacide
лимби'ческий *a* limbic
лими'т *biom* limit
лимноло'гия *f* limnology
лимнологи'ческий *a* limnological
лимнофи'т *m* limnophyte
лимо'н *m* lemon (*Citrus limon*)
лимо'нник *m bot* schizadra (*Schizandra*)
лимо'нница *f* brimstone butterfly (*Gonepteryx rhamni*)
лимосфе'ра *bot* limosphere
ли'мфа *f* lymph, lymphatic fluid
  ~ **Ко'ха** Koch's lymph
лимфати'ческий *a* lymphatic
лимфобла'ст *m* lymphoblast
лимфоге'нный *a* lymphogenic
лимфогра'мма *f* lymphogram
лимфо'идный *a* lymphoid
лимфоки'н *m* lymphokine
  В-кле'точный ~ B-cell-derived lymphokine
  иммунорегулято'рный ~ afferent lymphokine
  эффе'кторный ~ efferent lymphokine
лимфоло'гия *f* lymphology
лимфо'ма *f* lymphoma, lymphoid tumor
  ~ **Бе'ркитта** Burkitt's lymphoma, Burkitt's tumor
лимфообразу'ющий *a* lymphogenous, lymphogenic
лимфопени'я *f* lymphocyte depletion
  глубо'кая ~ generalized lymphocyte depletion
  региона'льная ~ focal lymphocyte depletion
  тота'льная ~ generalized lymphocyte depletion
  фока'льная ~ focal lymphocyte depletion
  лимфопоэ'з *m* lymphization
лимфоци'т *m* lymphocyte
  ~ **селезё'нки** splenic lymphocyte
  активи'рованный ~ activated lymphocyte
  анома'льный ~ ill-defined lymphocyte
  аутоагресси'вный ~ self-reactive lymphocyte
  аутореакти'вный ~ self-reactive lymphocyte
  грануля'рный ~ granular lymphocyte
  дифференци'рованный ~ "adult" lymphocyte, end-stage lymphocyte
  зерни'стый ~ granular lymphocyte
  зре'лый ~"adult" lymphocyte, end-stage lymphocyte
  инта'ктный иммунокомпете'нтный ~ initially responsive lymphocyte
  коммити'рованный ~ committed lymphocyte
  конволюцио'нный ~ convoluted lymphocyte
  ма'лый ~ small lymphocyte
  неадгези'вный ~ nonadherent lymphocyte
  неделя'щийся ~ quiescent lymphocyte
  неприми'рованный ~ unprimed lymphocyte
  отжива'ющий ~ convoluted lymphocyte
  пережива'ющий ~ convoluted lymphocyte
  перифери'ческой кро'ви peripheral blood lymphocyte
  поко'ящийся ~ quiescent lymphocyte
  прекоммити'рованный ~ precommitted lymphocyte

промежу'точный ~ medium lymphocyte
эффе'кторный ~ effector lymphocyte
лимфоцитопоэ'з *m* lymphocytopoiesis
лимфоцито'з *m* lymphocytosis
лимфоци'т-эффе'ктор m effector lymphocyte
лимфоге'нный *a* lymphogenic
лимфообразу'ющий *a* lymphogenous
лимфогра'мма *f* lymphogram
лимфо'идный *a* lymphoid
лимфоло'гия *f* lymphology
лимфотокси'н *m* lymphotoxin
лимфоци'т *m* lymphocyte
лимфоци'т-ки'ллер *m* killer lymphocyte
лимфоцито'з *m* lymphocytosis
лимфоцитопоэ'з *m* lymphocytopoiesis
лине'йно-ланце'тный *a* linear-lanceolate
лине'йно-ланцетови'дный *a* lance-linear
лине'йно-мечеви'дный *a* linear-ensate
лине'йно-продолгова'тый *a* linear-oblong
лине'йный *a* linear
ли'нза *f* lens
кутикуля'рная ~ *helm* cuticular lens
иммерсио'нная ~ immersion lens
ма'сляно-иммерсио'нная ~ oil-immersion lens
ли'ния *f* line, strain
~ **кле'ток** cell line
~ **клоноге'нных кле'ток** cloned cell line
~ **поведе'ния** line of behavior
~ **преципита'ции** *imm* line of precipitation, precipitation line
~ **регре'ссии** regression line
~ **Уо'ллеса** Wallace's line
~ **Фа'рро-Вальде'йера** Farro-Waldeyer's line
аксилля'рная ~ axillary furrow
аутосе'ксная ~ autosexing line
бессме'ртная ~ immortalized cell line
бокова'я ~ *zool* lateral line
генеалоги'ческая ~ line of breeding, line
гибри'дная ~ hybrid strain
же'нская ~ distaff side
изоге'нная ~ isogenic strain
иммортализо'ванная ~ клеток immortalized cell line
инбре'дная ~ inbred line, I- line, closed line, inbred strain
кле'точная ~ cell line
клона'льная кле'точная ~ cloned cell line
коизоге'нная ~ coisogenic strain
конге'нная ~ congenic line
конге'нно-резисте'нтная ~ congenic line
кро'вная ~ blood line
манти'йная ~ *zool* pallial line
мужска'я ~ spear side
неинбре'дная ~ noninbred strain
продукти'вная ~ successful line
пункти'рная ~ dotted line
роди'тельская кле'точная ~ parental line
соско'вая ~ mammary line, nipple line
стволова'я ~ *gen* stem line
трансге'нная ~ transgenic line
уо'ллесова ~ Wallace's line
штрихова'я ~ broken line
штрих-пункти'рная ~ dot-and-dash line
чи'стая ~ *gen* pure line
чистопоро'дная ~ purebred line, pure strain
ли'ния-анализа'тор *m* tester strain
ли'нкер *m mbiol* linker
линнео'н *m* linneon
ли'нька *f* molt, moult, molting, ecdysis; mew (*of birds*)
ку'колочная ~ pupal molt
личи'ночная ~ larval molt
постэмбриона'льная ~ postembryonic molt
линя'ть *v* molt, moult, cast the coat; mew (*of birds*)
лиоферме'нт *m* lyo-enzyme
лиофилиза'ция *f* lyophilization, cool dehumidification, freeze drying
лиофи'льность *f* lyophilic property
лиофо'бность *f* lyophobic property
лиоцито'з *m* lyocytosis
ли'па *f* lime, linden, basswood, linn (*Tilia*)
липи'д *m* lipid, lipide
ли'пкий *a* sticky, adhesive
липня'к *m* lime-tree forest
липо- lipo-
липогене'з *m* lipogenesis
липо'ид *m* lipoid
липо'идный *a* lipoid
липока'ик *m* lipocaic
липоксе'ния *f* lipoxeny
липоксе'нный *a* lipoxenous
липо'лиз *m* lipolysis
липолити'ческий *a* lipolytic, fat-splitting
липопротеи'д *m* lipoprotein
~ **высо'кой пло'тности**, **ЛВП** high density lipoprotein
~ **сре'дней пло'тности** intermediate density lipoprotein

~ **ни'зкой пло'тности, ЛНП** low density lipoprotein
липосо'ма *f* liposome
липотро'пный *a* lipotropic
липофи'льный *a* lipophil
липофо'р *m* lipophore
липохро'м *m* lipochrome
липоци'т *m* fat cell, fatty cell
лире'лла *f* lirella
лирови'дный *a* lyrate
лироли'стный *a* lyre-leaved
лирохво'ст *m* lyrebird (*Menura*)
лиса' *f* fox (*Vulpes*)
ли'сий *a* vulpine
лиси'ца *f* fox (*Vulpes*)
степна'я ~ corsac fox *(Vulpes corsac)*
лету'чая ~ flying fox *(Pteropus)*
ры'жая ~ red fox *(Vulpes vulpes)*
чё'рно-бу'рая ~ silver fox *(Vulpes fulvus)*
лиси'чка *f bot* chanterelle (*Cantarellus cibarius*)
лисово'дство *n* fox breeding
лисохво'ст *m* meadow foxtail, foxtail (*Alopecurus*)
лист *m* leaf
буго'рчатый ~ knobbed leaf
влага'лищный ~ sheathing leaf
во'гнутый ~ concave leaf
во'дный ~ water leaf
вы'емчатый ~ sinuate leaf, emarginate leaf, dedalous leaf
глянцеви'тый ~ sleek leaf
два'жды тройчатосло'жный ~ biternate leaf
двоякоперистосло'жный ~ bipinnately compound leaf, pinnately decompound leaf
двупарнопе'ристый ~ bijugate leaf
дельтови'дный ~ deltoid leaf
до'льчатый ~ laciniated leaf, cleft leaf, lobulose leaf
ду'дчатый ~ fistulose leaf, hollow leaf
жело'бчатый ~ channeled leaf
клинови'дный ~ wedge-shaped leaf
копьеви'дный ~ spear-shaped leaf, halberd-shaped leaf, hastate leaf
кро'ющий ~ bract, flag leaf, cataphyllary leaf, jacket leaf, covering leaf
курча'вый ~ curled leaf
мелкопе'ристый ~ moss leaf
морщи'нистый ~ crinkled leaf
мяси'стый ~ fleshy leaf
неопада'ющий ~ persistent leaf
ова'льный ~ oval leaf, ovate leaf
одноле'тний ~ summer leaf
односторо'нний ~ unifacial leaf
остроконе'чный ~ cuspidate leaf
па'льчато-ло'пастный ~ palmatilobate leaf, palmately lobed leaf
пальчатопе'ристый ~ digitinervate leaf, finger-veined leaf
па'льчато-разде'льный ~ palmatipartite leaf, palmately parted leaf, digitipartite leaf
па'льчатый ~ digitate leaf, palmate leaf
пе'ристо-не'рвный ~ pinnately veined leaf, feather-nerved leaf, feather-veined leaf
перистосло'жный ~ pinnately compound leaf, compound pinnate leaf
прикорнево'й ~ root leaf, basal leaf, bottom leaf
прямостоя'щий ~ erect leaf
пятни'стый ~ dotted leaf
рассечё'нный ~ dissected leaf, cut leaf
сердцеви'дный ~ heart-shaped leaf, cordate leaf
складчатый ~ *mbiol* pleated sheet
треуго'льный ~ deltoid leaf
тру'бчатый ~ hollow leaf, boot leaf
чашеобра'зный ~ concave leaf
шерохова'тый ~ scabrous leaf
листва' *f* leafage, foliage
ли'ственница *f* larch (*Larix*)
листвени'чник *m* larch forest
ли'ственный *a* foliar
листобло'шка *f* jumping plant louse (*Psylla*)
листовё'ртка *f* leaftier, leaf-rolling moth, tube-builder (*Tortricidae*)
листови'дный *a* leaf-shaped, foliaceous
листо'вка *f bot* follicle, hose
листово'й *a* foliar
листогры'з, листое'д *m* leaf beetle (*Chrysomelidae*)
листо'к *m* leaflet; folium, layer
висцера'льный ~ *emb* splanchnopleure, visceral layer
заро'дышевый ~ *emb* germ layer, germinal layer, blastophyllum
париета'льный ~ *emb* parietal layer
листоно'гий *a zool* phyllopodous
листоно'сный *a* frondiferous
листообра'зный *a* phylloid
листообразу'ющий *a* phyllogenetic
листообъе'млющий *a* amplexifoliate
листопа'д *m* leaf fall, exfoliation
листопа'дный *a* deciduous, therophyllous

листорасположе'ние *n* phyllotaxis, phyllotaxy, leaf arrangement, leaf position
двуря'дное ~ distichous leaf arrangement
муто'вчатое ~ whorl leaf arrangement, whorl leaf aestivation, verticillate leaf arrangement
очередно'е ~ alternate leaf arrangement, alternating leaf position
перекрё'стно-па'рное ~ opposite arrangement
спира'льное ~ spiral arrangement
супроти'вное ~ opposite leaf aestivation, decussate leaf arrangement
листосложе'ние *n* foliation, ptyxis, vernation, leafing, convolution
свё'рнутое ~, скру'ченное ~rolled-up convolution
листо'чек *m* leaflet, foliole, pinna
лити'ческий *a* lytic
литоде'сма *f* lithodesma
литора'ль *f* littoral
литора'льный *a* littoral; intertidal
литофи'льный *a* lithophilous
литофи'т *m* lithophyte
литоци'ст *m* lithocyst
лихени'зм *m* lichenism
лихеноло'гия *f* lichenology
лихора'дить *v* fever
лихора'дка *f* fever
~ **Бва'мба** Bwamba fever
~ **де'нге** Aden fever, bouquet fever
~ **доли'ны Сан Хоаки'н** San-Joaquin fever, San-Joaquin Valley fever *(caused by Coccidioides immitis)*
~ **Ла'сса** Lassa fever
~ **Оро'я** Oroya fever
~ **Пе'ля-Эбште'йна** Pel-Ebstein disease
~ **Фо'рта Брегг** Fort Bragg fever
геморраги'ческая ~ hemorrhagic fever
жё'лтая ~ yellow fever
киасану'рская ~ Kyasanur Forest disease
клещева'я ~ tick-bite fever, tick fever
мальти'йская ~ Malta fever, Maltese fever
марсе'льская ~ Marseilles fever
моски'тная ~ pappataci fever, phlebotomus fever
ремитти'рующая ~ remittent fever
сенна'я ~ hay fever, pollen fever
япо'нская речна'я ~ inundation fever
лихора'дочный *a* feverish
лицево'й *a* facial
лице'нзия *f* license, permit
~ **на отстре'л ди'чи** game certificate
охо'тничья ~ hunting license
рыболо'вная ~ fishing license
лицо' *n* face
насторо'женное ~ alert face
личи'нка *f* larva
~ **му'хи** maggot
де'зоровская ~ Desor's larva
инвазио'нная ~ infective larva, invasion larva
камподеови'дная ~, камподео'идная ~ campodeiform larva
карабо'идная ~ caraboid larva
педогенети'ческая ~ paedogenetic larva
рабдитови'дная ~ rhabditiform larva
скарабео'идная ~ scarabaeoid larva
сколито'идная ~ scolytoid larva
трилоби'тная ~ trilobite larva
филяриеви'дная ~ filariform larva
хрущеобра'зная ~ scarabaeiform larva
циприсови'дная ~ cypris larva
личинкоя'дный *a* larvivorous, larva-eating
личи'ночный *a* larval
лиша'й *m med* lichen
опоя'сывающий ~ zoster, shingles
стригу'щий ~ ringworm of scalp
лиша'йник *m bot* lichen (*Lichenes*)
~ **ма'нна** manna lichen *(Lecanora esculenta)*
борода'тый ~ barbate lichen
ко'рковый ~ crustose lichen
кусти'стый ~ fruticose lichen
листова'тый ~ foliose lichen
накипно'й ~ crustose lichen
стра'нствующий ~ erratic lichen
лишайникови'дный *a* lichenoid
лише'ние *n* deprivation
лихени'зм *m* lichenism
лихеноло'гия *f* lichenology
лоб *m* forehead, front, frons
ло'бия *f* bonavist (*Dolichos lablab*)
ло'бно-верхнечелюстно'й *a* frontomaxillary
ло'бно-заты'лочный *a* fronto-occipital
ло'бно-носово'й *a* frontonasal
ло'бно-скулово'й *a* frontomalar
ло'бно-теменно'й *a* frontoparietal
ло'бный *a* frontal
лобко'во-бе'дренный *a* pubofemoral
лобко'во-предста'тельный *a* puboprostatic
лобко'во-пузы'рный *a* pubovesical
лобко'вый *a* pubic
лобо'к *m* pubis

лобопо'дия *f* lobopodium
лов *m* haul, fishing, fishery
лови'ть *v* catch
лову'шка *f* trap, decoy; **ста'вить лову'шку** trap
~ **с захло'пывающейся две'ркой** enclosure trap
перено'сная ~ portable trap
светова'я ~ light trap
я'щичная ~ box trap
лову'шка-дави'лка *f* break-back trap, span trap
лову'шка-корра'ль *f* corral trap
лову'шко-день *m* trap-day
лову'шко-ночь *f* trap-night
лову'шко-су'тки *pl* trap-day
ло'говище *n* den, lair, retreat
лог *m* ravine
ло'гово *n* den, lair
лоди'кула *f* lodicule
ло'дочка *f bot* carina, keel
лоды'жечный *a* malleolar
лоды'жка *f* ankle, malleolus
ло'же *n* bed; receptacle; torus; channel (*of river*); gunstock
ногтево'е ~ nail bed
ложно- pseudo-
ложноно'жка *f* pseudopod, pseudopodium
ложноспи'нка *f* pseudonotum
лоза' *f* rod, cane, wither, vine, sarment
виногра'дная ~ grapevine, vine, vine shoot
лозня'к *m* willow bush
локализа'ция *f* localization
лока'льный *a* local
ло'ква *bot* loquat *(Eriobotrya japonica)*
локомото'рный *a* locomotor, locomotory
локомо'ция *f* locomotion
ло'коть *f* elbow
локтево'й *a anat* cubital, ulnar, anconal, anconeal
локтезапя'стный *a anat* ulnocarpal
ло'кус *m gen* locus, gene locus
~ **гомоталли'чности** homothallic locus, HM locus
~ **имму'нного отве'та** immune-related locus
~ **компенси'рованного отве'та** Cr locus, compensative response locus
~ **переключе'ния сигна'ла SMI locus**
маркё'рный ~ marker locus
сло'жный ~ compound locus
ло'мкий *a* fragile
ло'мкость *a* fragility
ломоно'с *m* leatherflower, virgin's bower (*Clematis*)
лонча'к *m* yearling bear
ло'пастность *a* lobation
ло'пастный *a* lobar
ло'пасть *f* lobe, lappet
глазна'я ~ *ent* ocular lobe
ма'ленькая ~ lobule
преана'льная ~ *ent* preanal lobe
протоцеребра'льная ~ *ent* protocerebral lobe
лопа'тка *f* bladebone, shoulder blade, scapula
лопатообра'зный *a* paliform, spatulate
лопа'точка *f* pala, spatula
лопа'точно-ключи'чный *a* scapuloclavicular
лопа'точный *a* scapular
ло'паться *v* burst, break (*of buds*)
ло'пающийся *a* dissilient
лопу'х *m* burdock (*Arctium*)
лордо'з *m* lordosis
ло'ри *m* loris *(Lorisidae)*
то'лстый ~ slow lemur *(Nycticebus)*
то'нкий ~ slender loris *(Loris)*
лосе'нок *m* elk calf
лоску'т *m* flap, graft
ко'жный ~ skin flap, skin graft, cutaneous flap
тканево'й ~ tissue graft
лосня'щийся *a* lustrous
лось *m* elk (*Alces alces*)
лото'к *m orn* nesting hollow
ло'тос *m* lotus (*Nelumbium*)
лофодо'нтный *a* lophodont
лофофо'р *m* lophophore
лофоцерка'льный *a* lophocercal
лох *m* oleaster (*Eleagnus*); male salmon in spawning condition
лоха'нка *f anat* pelvis
по'чечная ~ kidney pelvis
лохма'тый *a* shaggy, pannose
лохмоно'гий *a* booted
лохмоно'гость *f* shank feathering
лоциа'ция *f ecol* lociation
ло'ция *f ecol* locies
ло'шадь *f* horse (*Equus*)
~ **Пржева'льского** Przewalski's wild horse (*Equus caballus*)
бегова'я ~ race horse
брыкли'вая ~ bucking horse
верхова'я ~ saddle horse

взмы'ленная ~ lathery horse
вислоза'дая ~ sloping-croup horse
ворона'я ~ black horse
выездна'я ~ parade horse
вы'езженная ~ trained horse
вью'чная ~ sumpter horse
гнеда'я ~ bay horse
горя'чая ~ spirited horse
коренна'я ~ shaft horse, wheel horse
ломова'я ~ dray horse
норо'вистая ~ skittish horse
пристяжна'я ~ trace horse
прогу'лочная ~ pleasure horse
рабо'чая ~ working horse
скакова'я ~ race horse
упряжна'я ~ coacher, vanner, carriage horse, draught horse, draft horse, driving horse
упря'мая ~ skittish horse
ло'шадь-полукро'вка *f* half-bred horse
лошади'ный *a* equine
лоша'к *m* hinny
лощи'на *f* dell, hollow
луб *m* liber, bast, bast fiber, phloem
луг *m* meadow, grassland
альпи'йский ~ alpine meadow
аркти'ческий ~ arctic meadow
боло'тистый ~ swamp meadow, peat-land meadow
боло'тистый кочкова'тый ~ tussock swamp meadow
ве'йниковый ~ reedgrass meadow
вла'жный ~ wet meadow
есте'ственный ~ native meadow
закочка'ренный ~ tussock swamp meadow
заливно'й ~ water meadow
кочкова'тый ~ tussock meadow
мо'крый ~ wet meadow
мя'тличный ~ bluegrass meadow
низи'нный ~ lowland meadow
низкотра'вный ~ short grass meadow
остепнё'нный ~ steppe meadow
по'йменный ~ bottom meadow, floodplain meadow, inundated meadow
полеви'цевый ~ bentgrass meadow
пу'стошный ~ rough meadow
пусты'нный ~ desert meadow
пыре'йный ~ quackgrass meadow
разнотра'вный ~ forb meadow, grass meadow, herb meadow
солончако'вый ~ salt meadow
суходо'льный ~ dry meadow, upland meadow
лугово'дство *n* meadow management
лугово'й *a* pratal
лужа'йка *f* grassplot, lawn; glade
лузга' *f* peelings, hull, husk, shell
лу'згать *v* husk, hull
лук *m* onion (*Allium*)
~ **на перо'** salad onion, green onion
~ **гадю'чий** grape-hyacinth (*Muscari*)
~ **гуси'ный** gagea (*Gagea*)
~ **зеле'ный** green onion
~ **зи'мний** cibol (*Allium fistulosum*)
~ **медве'жий** ramson, bear's onion (*Allium ursinum*)
~ **ре'пчатый** onions, bulb onion *(Allium cepa)*
~ **песча'ный** cibol (*Allium fistulosum*)
~ **пони'клый** wild nodding onion *(Allium cernuum)*
~ **сала'тный** salad leek *(Allium kurrat)*
~ **сиби'рский** shore onion *(Allium sibiricum)*
лук-бату'н *m* rock onion, spring onion, cibol, Welsh onion, Japanese leek *(Allium fistulosum)*
лук-поре'й *m* leek (*Allium porrum*)
лук-резане'ц *m* chive (*Allium schoenoprasum*)
лук-сево'к *m* seed onion
лук-татарка *m* Japanese leek *(Allium fistulosum)*
лук-шало'т *m* eschalot (*Allium ascalonicum*)
лу'ковица *f* onion; bulb, bulbus
~ **ао'рты** *anat* bulb of aorta, aortic bulb
~ **во'лоса** hair bulb
~ **пищево'да** *helm* bulb of esophagus
вкусова'я ~ taste bud
волосяна'я ~ hair bulb
доче'рняя ~ daughter bulb
неврофибрилля'рная ~ *anat* neurofibrillar bulb
обоня'тельная ~ olfactory bulb
плё'ночная ~ *bot* tunicated bulb
лу'ковичка *f* clove
луковицеобра'зный *a* bulbous, bulbar, bulblike
лук-скорода' *m* chive (*Allium schoenoprasum*)
луксу'рианс *m* luxuriance
лук-тата'рка spring onion (*Allium fistulosum*)

лу'нка *f* cavity, well (*e.g., of a slide*), lunule
 ~ для антиге'на antigen well
 ~ для антисы'воротки antiserum well
 ~ для антите'ла antibody well
 ~ на микротитрова'льном планше'те microtiter well
 контро'льная ~ control well
 фронта'льная ~ *ent* frontal lunule
лу'нчатый *a* selenodont (*of teeth*)
лунь *m* harrier (*Circus*)
 боло'тный ~, камышо'вый ~ marsh harrier, duck harrier (*Circus aeruginosus*)
 полево'й ~ hen harrier, marsh hawk (*Circus cyaneus*)
лу'па *f* magnifier, lens
 препарова'льная ~ dissecting lens, dissection microscope
лупи'н *m* lupine *(Lupinus)*
лу'ток *m* smew (*Mergus albellus*)
луч *m* ray; finger (*of sea stars*)
 ~ гидролока'тора echo-ranging beam
 актини'чные лучи' *pl* actinic rays
 боково'й ~ side lobe *(in hydroacoustics)*
 древеси'нный ~ xylem ray
 жа'берный ~ branchiostegial ray
 жё'сткие лучи' *pl* hard rays
 инфракра'сные лучи' *pl* infrared rays
 ксиле'мный ~ *bot* xylem ray
 лубяно'й ~ *bot* phloem ray
 митогенети'ческие лучи' *pl* mitogenetic rays, M rays
 оста'точный ~ residual rays
 па'дающий ~ incident ray
 плавнико'вый ~ fin-ray
 положи'тельный ~ positive ray
 поля'рный ~ polar ray
 рассе'янные лучи' *pl* stray rays
 рентге'новские лучи' *pl* X-rays, Roentgen rays
 светово'й ~ light beam, light ray
 световы'е ви'димые лучи' *pl* visible light rays
 сердцеви'нный ~ *bot* medullar ray, medullary ray, pith ray, primary ray
 сло'жный ~ *bot* aggregate ray
 тепловы'е лучи' *pl* heat rays
 ультрафиоле'товые лучи' *pl* ultraviolet rays
лучезапя'стный *a anat* radiocarpal
лучелоктево'й *a anat* radioulnar
луще'ние *n* husking, hulling, shelling
 ~ стерни' stubble breaking, skim plowing
лущи'ть *v* husk, shell, hull, dehull, peel; crack (*nuts*)
лы'ко *n* bast
лысу'ха *f* coot, water hen (*Fulica atra*)
льви'нка *f* chameleon fly, armed fly, soldier fly (*Stratiomys*)
льви'ный *a* leonine
льви'ца *f* lioness
льново'дство *n* flax raising
льня'нка *f* toadflax (*Linaria*)
любозна'тельный, **любопы'тный** *a* inquisitive
любопы'тство *n* curiosity
люминисце'нция *f* luminiscence
лю'рик *m* little auk, dovekey, dovekie (*Plautus alle*)
 пё'стрый ~ marbled murrelet (*Brachyramphus marmoratus*)
лютеиниза'ция *f* luteinization
лю'тик buttercup, crowfoot (*Ranunculus*)
лю'тый *a* ferocious
люце'рна *f* alfalfa (*Medicago*)
люцифери'н *m* luceferin, luciferine
ляга'ть, **ляга'ться** *v* kick
лягуша'тник *m bot* frogbit *(Hydrocharis)*
лягушо'нок *m* frogling, froglet
лягу'шка *f* frog (*Ranidae*)
 настоя'щая ~ true frog *(Rana)*
 озе'рная ~ marsh frog, lake frog, laughing frog *(Rana ridibunda)*
 остромо'рдая ~ moor frog (*Rana arvalis*)
 прудо'вая ~, съедобная ~ edible frog (*Rana esculenta*)
 травяна'я ~ common frog, grass frog, brown frog (*Rana temporaria*)
 шпо'рцевая ~ clowed toad, clawed frog (*Xenopus*)
лягушка-бык *m* bullfrog (*Rana catesbeiana*)
лягушка-голиа'ф *m* goliath frog (*Rana goliath*)

M

МАГАТЭ', Междунаро'дное аге'нтство по а'томной эне'ргии IAEA, International Atomic Energy Agency
магнитобиоло'гия *f* magnetobiology
магнитотропи'зм *m* magnetotropism
мадрепо'ровый *a* madreporic
мазо'к *m* smear, smear preparation, film, film preparation, streak preparation
~ **кро'ви** blood smear, blood film, thin blood film
окра'шенный ~ stained smear
све'жий ~ fresh smear
термофикси'рованный ~ heat-fixed smear
то'лстый ~ **кро'ви** thick blood film
цитоспи'новый кле'точный ~ cytospin cell smear
фикси'рованный ~ fast smear
фикси'рованный нагрева'нием ~ heat-fixed smear
мазо'к-отпеча'ток *m* touch smear, impression smear
мазь *f* ointment, liniment
~ **Вилькинсо'на** Wilkinson's ointment
ма'йка *f* oil beetle (*Meloe*)
ма'йна *f orn* myna *(Acridotheres)*
ма'йник *m bot* bead-ruby (*Maianthemum*)
майора'н *m bot* marjoram *(Majorana)*
~ **шипр** pot marjoram, wild marjoram *(Majorana onites)*
~ **садо'вый** sweet marjoram *(Majorana hortensis)*
маи'с *m* maize, corn (*Zea mays*)
мак *m* poppy (*Papaver*)
~ **о'пийный** opium poppy (*Papaver somniferum*)
мака'к *m* macaque *(Macaca)*
мака'к-ре'зус rhesus macaque *(Macaca mulatta)*
макло'к *m* external angle of ilium
ма'ковый *a* papaverous
макро- macro-
макроавторадиогра'фия *f* macroautoradiography
макробла'ст *m* macroblast
макрогаме'та *f* macrogamete
макрогаметогене'з *m* megagametogenesis
макрогаметофи'т *m* megagametophyte
макрогаметоци'т *m* macrogametocyte, megagametocyte
макрога'мия *f* macrogamy
макрогли'я *f* macroglia
макроглобули'н *m* macroglobulin
макрогони'дия *f* macrogonidium
макрокони'дия *f* macroconidium
макроконъюга'нт *m* macroconjugant
макролецита'льный *a* megalecithal
макроли'д *m* macrolide
макроме'р *m* macromere
макромоле'кула *f* macromolecule
макромолекуля'рный *a* macromolecular
макромута'нт *m* macromutant
макромута'ция *f* macromutation
макронуклеоци'т *m* macronucleocyte
макрону'клеус *m* macronucleus
макрону'триент *m* macronutrient
макроорганизм'зм *m* macroorganism
макроскле'ра *f* macrosclere
макросклереи'да *f* macrosclereid
макроскопи'ческий *a* macroscopic, macroscopical
макроскопи'я *f* macroscopy
макросо'ма *f* macrosome
макросоми'я *f* macrosomia
макроспо'ра *f* macrospore
макроспора'нгий *m* macrosporangium
макроспорангиофо'р *m* macrosporangiophore
макроспорофи'лл *m* macrosporophyll
макростилоспо'ра *f* macrostylospore
макрострукту'ра *f* macrostructure
макроте'рм *m* macrotherm, macrothermophyte, megatherm
макрофа'г *m* macrophage
активи'рованный ~ elicited macrophage
акти'вно фагоцити'рующий ~ "angry" macrophage
антигенпредставля'ющий ~ antigen-presenting macrophage
антигенпрезенти'рующий ~ antigen-presenting macrophage
«воинствующий» ~ "angry" macrophage
осё'длый ~ fixed macrophage, resident macrophage, stable macrophage, tissue-fixed macrophage
типи'чный ~ classical macrophage
макрофанерофи'т *m* macrophanerophyte
макрофа'уна *f* macrofauna
макрофибри'лла *f* macrofibril
макрофи'т *m* macrophyte

макрофи'ты *m pl* macrovegetation
макрофло'ра *f* macroflora
макрохе'та *f* macrochaeta
макроци'ста *f* macrocyst
макроци'т *m* macrocyte
макроцита'рный *a* macrocytic
макроэволю'ция *f* macroevolution, macrophylogenesis
макси'лла *f* maxilla
максилля'рный *a* maxillary
максимиза'ция *f* maximization
маку'шка *f* vertex, apex, crop; crown (*of head*)
малаколо'гия *f* malacology
мала'шка *f* soft-winged flower beetle (*Melyridae*)
малёк *m* young fish, fry, hatchling
мали'на *f* raspberry (*Rubus idaeus*)
мали'нник *m* raspberry brake; *ent* raspbery beetle *(Byturidae)*
мали'новый *a* crimson; raspberry
маллеорама'тный *a* malleoramate
мало- oligo-
малоберцо'вый *a* fibular
малозу'бчатый *a* oligodontous
малокле'точный *a* few-celled
малокорнево'й *a* oligorhizous
малокро'вие *n* anemia
малокро'вный *a* anemic
малоли'стный *a* oligophyllous
маломе'рный *a* undersized
малоости'стый *a* few-awned
малопло'дный *a* oligocarpous
малокорнево'й *a* oligorhizous
малотычи'нковый *a* oligandrous, oligostemonous
малоцветко'вый *a* oliganthous, few-flowered
маляри'йный *a* malarial, malarious
маляри'я *f* malaria
  подо'страя ~ dumb ague
  тропи'ческая ~ falciparum malaria
  трёхдне'вная ~ tertian malaria, vivax malaria
  четырёхдне'вная ~ quartan malaria
ма'льва *f* mallow *(Malva)*
мальто'за *f* maltose
мальтузиа'нство *n* Malthusianism
ма'мба *f herp* mamba *(Dendroaspis)*
маммоло'гия *f* mammalogy
маммотропи'зм *m* mammotropism
ма'монт *m* mammoth (*Elephas primigenius*)
маму'ра *f* arctic bramble (*Rubus arcticus*)
мангабе'й *m ther* mangabey, mangaby *(Cercocebus)*
ма'нго *n*, *m* mango, mango tree (*Mangifera*)
манго'льд *m* chard, leaf beet, mangold (*Beta vulgaris var. cicla*)
мангу'ст *m* mongoose, common mongoose *(Herpestes)*
  водяно'й ~ water mongoose *(Atilax)*
  еги'петский ~ Egyptian mongoose, ichneumon *(Herpestes ichneumon)*
мангуста'н *m bot* mangosteen *(Garcinia mangostana)*
мандари'н *m* mandarin, mandarine, tangerine, king orange (*Citrus nobilis*)
мандари'нка *f* mandarin duck (*Aix galericulata*)
манди'була *f* mandible
мандибуля'рный *a* mandibular
мандраго'ра *f bot* mandrake *(Mandragora)*
мандри'лл *m ther* mandrill *(Papio sphinx)*
манже'тка *f* lady's mantle (*Alchemilla*)
манио'к *m* tapioca, manioc, manihot, cassava (*Manihot*)
манипули'рование *n* handling; manipulative behavior
  ~ **камня'ми** stone handling
манипуля'ция *f* manipulation
  ~ **ору'дием** instrumental manipulation
  ~ **предме'тами** object manipulation
  неоруди'йная ~ noninstrumental manipulation
  оруди'йная ~ instrumental manipulation
ма'нна *f* manna
  ~ **лиша'йниковая** manna lichen, manna (*Lecanora esculenta*)
ма'нник *m* manna-grass (*Glyceria*)
манни'т *m* mannitol
мано'к *m* call, decoy
манометри'я *f* manometry
манти'йный *a* pallial
ма'нтия *f* mantle
ману'бриум *m* manubrium
ману'л *m* manul (*Felis manul*)
марабу' *m* marabou stork, adjutant stork, marabou (*Leptoptilus crumeniferus*)
мара'л *m* Siberian stag, maral (*Cervus elaphus*)
маргари'тка *f* daisy, true daisy (*Bellis*)
маргина'льный *a* marginal
маре'на *f bot* madder *(Rubia)*
марикульту'ра *f* mariculture

мари'та *f helm* marita
маркё'р *m* marker
антиселекти'вный ~ contra-selective marker
пове'рхности surface marker
пове'рхностный ~ surface marker
по'здний ~ "eclipsed" marker
селекти'рованный ~ selected marker
селекти'руемый ~ selectable marker
термина'льный ~ "eclipsed" marker
Т-кле'точный ~ T-cell marker
трансге'нный ~ transgenic marker
маркиро'вка *f* marking
мармозе'тка *f* plumed marmoset *(Callithrix)*
марты'шка *f* guenon (*Ceropithecus*)
маршру'т *m* route
~ **переле'та** flyway
марь *f bot* goosefoot (*Chenopodium*); larch peatbog moss forest
ма'ска *f* mask
маскирова'ть *v* mask
маскиро'вка *f* camouflage, masking
обра'тная ~ backward masking
пряма'я ~ forward masking
тенева'я ~ shadow masking, shadow camouflage
маскиро'ванный *a* absconded
маскиру'ющий *a* consealing
маскови'дный *a bot* masked
маскулиниза'ция *f* masculinization
маслё'нок *m bot* yellow-brown boletus, slippery jack (*Suillus luteus*)
масли'на *f* olive (*Olea*)
ма'сличность *f* oil content
ма'сличный *a* oleiferous, oily
масло *n* oil
ара'хисовое ~ peanut oil, arachis oil, groundnut oil
бергамо'тное ~ oil of bergamot, bergamot oil
вазели'новое ~ petrolatum oil
гвозди'чное ~ clove oil
иммерсио'нное ~ immersion oil
ке'дровое ~ cedarwood oil, cedar oil
коко'совое ~ coconut oil
кукуру'зное ~ corn oil, maize oil
ма'ковое ~ poppy-seed oil, poppy oil
пе'рсиковое ~ peach-kernel oil
подсо'лнечное ~ sunflower-seed oil, sunflower oil
сиву'шное ~ fusel oil
сливочное ~ butter
со'евое ~ soya-bean oil, soya oil
эфи'рное ~ essential oil
маслоно'сный *a* oleiferous
маслораствори'мый *a* oil-soluble
масляни'стый *a* oily
ма'сса *f* mass
убо'йная ~ slaughter mass
масса'ж *m* massage
масси'в *m* expense
лесно'й ~ forest area, forested area
ма'ссовый *a* large-scale, mass
масс-фрагментогра'фия *f* mass-fragmentography
ма'стакс *m* mastax
мастоци'т *m* mast cell
мастурба'ция *f eth* masturbation
масть *f* color, paint, coat color
одноцве'тная ~ whole color
масшта'б *m* scale
лаборато'рный ~ laboratory scale
промы'шленный ~ industrial scale
укрупнё'нный ~ large scale
материа'л *m* material; cloth
межкле'точный ~ matrix, intercellular matrix
поса'дочный ~ planting stock
посевно'й ~ inoculum
матери'нский *a* maternal
матери'нство *n* maternity
мате'рия *f* matter
жива'я ~ living matter
нежива'я ~ inanimate matter, nonliving matter
мате'рка *f* pistillate hemp
ма'тка *f* womb, uterus; dam; queen bee
дополни'тельная ~ *ent* complemental queen, substitute queen
эргато'идная ~ *ent* ergatoid queen
ма'товый *a* dull, lusterless
ма'точный *a* uterine; ~ **раство'р** stock solution, mother solution
матра'с *m* matrass, flattened bottle
~ **Ру** Roux flask
ма'трикс *m* matrix, intercellular matrix
~ **митохо'ндрии** mitochondrial matrix
цитоплазмати'ческий ~ matrix
ма'трица *f* matrix; template
~ **фа'кторных коэффицие'нтов** factor matrix
врождё'нная ~ innate template
вы'работанная ~ *eth* acquired template
вы'рожденная ~ singular matrix

едини'чная ~ unit matrix
квадра'тная ~ quadratic matrix, square matrix
ковариацио'нная ~ covariance matrix, dispersion matrix
корреляцио'нная ~ correlation matrix
обра'тная ~ inverse matrix, reciprocal matrix
слухова'я ~ auditory template
социометри'ческая ~ *eth* sociometric matrix
транспони'рованная ~ **коэффицие'нтов** *biom* allocation matrix
устано'вленная ~ acquired template
характеристи'ческая ~ characteristic matrix
ма'тричный *a* template
матрокли'нный *a* matroclinal, matroclinous
мать-и-ма'чеха *f* colt's foot (*Tussilago farfara*)
мах *m* swing
маха'ть *v* wag (*of a* dog's tail); wave
махо'рка *f* rustic tobacco, common tobacco (*Nicotiana rustica*)
махро'вость *f* doubling, doubleness
махро'вый *a* double, double-flowering, pleiopetalous
мацера'ция *f* maceration
мацери'ровать *v* macerate
мачо'к *m* redweed *(Papaver rhoeas)*
маш *m* mung bean (*Phaseolus aureus*)
ме *i* baa, blah
мега- mega-
мегазооспо'ра *f* megazoospore
мегакариоци'т *m* megakaryocyte
мегалобла'ст *m* megaloblast
мегало'па *f zool* megalope
мегалоци'т *m* megalocyte
мегаскле'ра *f zool* megasclere
мегасо'ма *f* macrosome
мегаспо'ра *f* megaspore
мегаспора'нгий *m* megasporangium
мегаспороци'т *m* megasporocyte
мегаспорофи'лл *m* megasporophyll
мегате'рм *m* macrotherm, macrothermophyte, megatherm
мегафа'уна *f* megafauna
мёд *m* honey
ди'кий ~ wild honey
па'девый ~ honeydew honey
со'товый ~ chunk honey, comb honey, hive honey
медве'дица *f* she-bear; tussock moth (*Actiidae*)
медве'дка *f* mole cricket (*Gryllotalpa*)
медве'дь *m* bear, true bear (*Ursus*)
бамбу'ковый ~ cat bear *(Ailuropoda melanoleuca)*
бе'лый ~ polar bear, ice bear, sea bear, white bear (*Ursus maritimus*)
бу'рый ~ brown bear (*Ursus arctos*)
очко'вый ~ spectacled bear *(Tremarctos ornatus)*
пеще'рный ~ cave bear *(Ursus spelaeus)*
медве'жий *a* ursine
медиа'льный *a* medial
медиа'на *f biom* median
медиастина'льный *a* mediastinal
медиовентра'льный *a* medioventral
медиодорса'льный *a* mediodorsal
медиолатера'льный *a* mediolateral
медиа'тор *m* mediator
меди'на *f* medine
медици'на *f* medicine
косми'ческая ~ space medicine
наро'дная ~ folk medicine
суде'бная ~ forensic medicine, legal medicine
медици'нский *a* medical, medicinal
медо'вый *a* melleous
медоно'с *m* honey plant, bee plant, meliferous plant
медоно'сность *f* nectar bearing
медоно'сный *a* melliferous
медосбо'р *m* honey yield
медососу'щий *a* mellisugent
медоя'дный *a* melliphagous, mellisugent, mellivorous
меду'за *f* jelly-fish, medusa
медузо'ид *m* medusoid
медузо'идный *a* medusoid
медулля'рный *a* medullary
медуни'ца *f* lungwort (*Pulmonaria*)
медя'нка *f* smooth snake (*Coronella*)
меж- inter-
межа' *f* balk, boundary
межвидово'й *a* interspecific
междоу'злие *n* internode, merithallus
стеблево'е ~ subcrown internode
междуря'дный *a* interrow
ме'жень *f* low water season
межеу'мок *m* cross
межкле'тник *m* intercellular duct
межкле'точный *a* intercellular

межко'чье *n* space between hummocks
межледнико'вый *a* interglacial
межли'нечный *a* intermoult
межнитево'й *a* intestrand
межсортово'й *a* intervarietal
межцистро'нный *a* intercistronic
мезаксо'н *m* mesaxon
меза'рхный *a* mesarch
мезга' *f* pomace, pulp
мездра' *f* flesh, fleshings, underskin
мездре'ние *n* fleshing
мездри'ть *v* flesh, scrape
мезентериа'льный *a* mesenterial, mesenteric
мезенте'рий *m* mesentery
  латера'льный ~ lateral mesentery
мезе'нхима *f* mesenchyme
мезе'нхимный *a* mesenchymatous
мези'на *f* mesine
мезентериа'льный *a* mesenterial, mesenteric
мезо- meso-
мезобла'ст *m* mesoblast
мезобласте'ма mesoblast
мезобласти'ческий *a* mesoblastic
мезогле'я *f* mesogloea
мезогидрофи'т *m* mesohydrophyte
мезогидрофи'тный *a* mesohydrophytic
мезогнатио'н *m* mesognathion
мезоде'рма *f* mesoderm, somatic mesoblast
мезодермализа'ция *f* mesodermalization, mesodermization
мезодерма'льный *a* mesodermal, mesoblastic
мезозо'йский *a* Mesozoic
мезока'рдий *m* mesocardium
мезока'рпий *m* mesocarp
мезокорако'ид *m* mesocoracoid
мезоко'тиль *m* mesocotyl
мезоло'гия *f* mesology
мезоме'р *m* mesomere
мезомито'з *m* mesomitosis
мезомо'рфия *f* mesomorphy
мезомо'рфный *a* mesomorphic, mesomorphous
мезоне'фрос *m* mesonephros
мезонэкзи'на *f pal* mesonexine
мезопи'ческий *a* mesopic
мезопластро'н *m* mesoplastron
мезоподиа'льный *a bot* mesopodial
мезоптери'гий *m* mesopterygium
мезоптериго'ид *m* mesopterygoid
мезосапро'бный *a* mesosaprobic
мезосо'ма *f* mesosoma
мезоспе'рм *m* mesosperm
мезоспо'рий *m* mesosporium
мезосте'рнум *m anat* mesosternum
мезоте'лий *m* mesothelium
мезоте'рм *m* mesotherm
мезоте'рмный *a* mesothermous, mesothermophilous
мезоте'ций *m* mesothecium
мезото'ракс *m ent* mesothorax
мезотро'пный *a* mesotropic
мезотро'фный *a* mesotrophic
мезофа'уна *f* mesofauna
мезофи'лл *m* mesophyll
  околопучко'вый ~ paraveinal mesophyll
мезофи'льный *a* mesophilous
мезофи'т *m* mesophyte
мезофи'тный *a* mesophytic
мезофлоэ'ма *f* mesophloem
мезофра'гма *f* mesophragma
мезоцентра'льный *a* mesocentrous
мезоцерка'рия *f helm* mesocercaria
мезоцефали'ческий *a* mesocephalic
мезоци'кл *m* mesocycle
мезэкзи'на *f pal* mesexine
мезоэктоде'рма *f* mesectoderm
мезоэнтоде'рма *f* mesendoderm, mesentoblast
мейо'з *m* meiosis, meiotic division
  двухступенчатый ~ two-step meiosis
  нару'шенный колхици'ном ~ C-meiosis
  одноступенчатый ~ one-step meiosis
мейоспо'ра *f* meiospora
мейоспора'нгий *m* meiosporange
мейота'ксия *f* meiotaxy
мейоти'ческий *a* meiotic
мейофилли'я *f bot* meiophylly
мейоци'т *m* meiocyte
мекзи'на *f pal* mexine
меко'ний *m* meconium
мелани'зм *m* melanism
  индустриа'льный ~ industrial melanism
мелани'н *m* melanin
мелано- melano-
меланобла'ст *m* melanoblast
меланогене'з *m* melanogenesis
мелано'з *m phytop* melanose
меланосо'ма *f* melanosome
меланоти'ческий *a* melanotic
меланофо'р *m* melanophore
меланоци'т *m* melanocyte
меланхоли'ческий *a* melancholic
мелиора'ция *f* amelioration, reclamation

мелково'дье *n* shallow water
мелкооло'вчатый *a* capitellate
мелкодо'льчатый *a* lacinulate
мелкозазу'бренный *a* crenulate, crenulated
мелкозу'бчатый *a* denticulate, denticulated
мелкоиго'льчатый *a* echinulate
мелкокуста'рниковый *a* fruticulose
мелколепе'стный *a* micropetalous
мелколи'стный *a* microphyllous
мелколе'сье *n* low forest, undergrowth, understory, underwood
мелконадрезно'й *a* lacinulate
мелкопи'льчатый *a* serrulate, serratulate
мелкопло'дный *a* microcarpous
мелкореше'тчатый *a* cribellate
мелкоча'шечковый *a* microcalycular
мелкошипова'тый *a* aculeolate
мелкоштрихо'ванный *a* striolate
мелово'й *a* chalky; Cretaceous
ме'льница *f* mill
мембра'на *f* membrane
~ **капилля'ра** capillary membrane
альвеоля'рная ~ alveolar membrane
асимметри'чная ~ asymmetric membrane
база'льная ~ basal membrane
вестибуля'рная ~ Reissner's membrane
вне'шняя ~ outer membrane
вну'тренняя ~ inner membrane, internal membrane
двухсло'йная ~ bilayer membrane
депротеинизи'рованная ~ skeleton-free membrane
ионообме'нная ~ ion-exchanger membrane
капилля'рная ~ capillary membrane
кле'точная ~ cell membrane, cellular membrane
микросома'льная ~ microsomal membrane
митохондриа'льная ~ mitochondrial membrane
многосло'йная ~ multilayer membrane
нару'жная ~ outer membrane
нитроцеллюло'зная ~ nitrocellulose membrane
нуклеопо'ровая ~ nucleopore membrane
ортовезикуля'рная ~ orthovesicular membrane
перитрофи'ческая ~ peritrophic membrane
плазмати'ческая ~ plasma membrane
плевроперикардиа'льная ~ pleuropericardial membrane
плевроперитонеа'льная ~ pleuroperitoneal membrane
полиме'рная ~ polymeric membrane
по'ристая ~ porous membrane
постсинапти'ческая ~ postsynaptic membrane
пресинапти'ческая ~ presynaptic membrane
проница'емая ~ permeable membrane
псевдотрахеа'льная ~ *ent* pseudotracheal membrane
ре'йснерова ~ Reissner's membrane
синапти'ческая ~ synaptic membrane
сочлено'вная ~ *ent* articular membrane
ундули'рующая ~ undulating membrane
фотосинтети'ческая ~ photosynthetic membrane
хорионалланто'исная ~ chorioallantoic membrane
цитоплазмати'ческая ~ cytoplasmic membrane
щёчно-гло'точная ~ *emb* buccopharyngeal membrane
элемента'рная ~ elementary membrane, unit membrane
я'дерная ~ nuclear membrane
яйцева'я ~ egg membrane
мембране'лла *f* membranella
мендели'зм *m* Mendelism
менингоэнцефалит meningoencephalitis
амё'бный ~ amoebic meningoencephalitis
мени'ск *m* meniscus
осяза'тельный ~ meniscus of touch
суставно'й ~ crescentic cartilage, semilunar cartilage
мениско'ци'т *m* meniscocyte
менопа'уза *f* menopause
менструа'льный *a* menstrual
менструа'ция *f* menstruation, menses
менота'ксис *m* menotaxis
меоспо'ра *f* meospore
ме'ра *f* measure
~ **изме'нчивости** measure of variability
ме'ры по охра'не приро'дных бога'тств *pl* conservation measures
~ **расстоя'ния Евкли'да** Euclidean distance
мерзлота' *f* frozen ground
ве'чная ~ permafrost
меридиа'льно-слитнополуборо'здный *a* *pal* syndemicolpate
меридиа'н *m* meridian

стоматоге'нный ~ *prot* stomatogenic meridian
мерика'рпий *m* mericarp
ме'рин *m* gelded horse
мерино'с *m* merino
мeристе'ла *f bot* meristele
мeристе'ма *f* meristem
бокова'я ~ flank meristem, lateral meristem
вста'вочная ~ intercalary meristem
интеркаля'рная ~ intercalary meristem
основна'я ~ ground meristem
оста'точная ~ residual meristem
сосу'дистая ~ vascular meristem
меристемати'ческий *a* meristematic
меристи'ческий *a* meristic
меркапта'н *m* mercaptan
мермито'идный *a* mermithoid
меро- mero-
мероблa'ст *m* meroblast
меробласти'ческий *a* meroblastic
мерога'мия *f* merogamy
мерога'струла *f* merogastrula
мерогене'з *m* merogenesis
мерогено'та *f* merogenote
мерого'н *m* merogon
мерого'ния *f* merogony
мерозиго'та *f* merozygote
мерозо'ит *m* merozoite
мерокине'з *m* merokinesis
мерокри'нный *a* merocrine
мероми'ксис *m* meromixis
мероморфо'з *m* meromorphosis
меросперми'я *f* merospermy
мероспо'ра *f* merospore
меротоми'я *f* merotomy
мероцено'н *m* merocoenon
мероце'рк *m* merocercus
мероци'т *m emb* merocyte
мертворожда'емость *f* natimortality
мертворожде'нный *a* deadborn, stillborn
ме'ртвый *a* dead
мерца'ние *n* fibrillation
ме'ссенджер *m* messenger
ме'стность *f* land, country, terrain, ground
боло'тистая ~ moorland, marshland
гори'стая ~ highland country
заболо'ченная ~ marshland, moorland
леси'стая ~ forest country
неро'вная ~ rough country
окружа'ющая ~ surrounding locale
откры'тая ~ exposed district
пересече'нная ~ rugged ground, rough ground, broken ground, bumpy ground
типова'я ~ type locality
ме'стный *a* local, native, aboriginal, indigenous
ме'сто *n* place; site
~ **водопо'я** drinking, place, watering place
~ **выведе'ния мо'лоди** breeding place
~ **ра'змножения** breeding place
де'тское ~ placenta
место'м *m* mestome
местообита'ние *n* habitat, site
месторожде'ние *n* deposit, bed, field
мета- meta-
метабази'дия *f* metabasidium
метабио'з *m* metabiosis
метаболи'зм *m* metabolism
~ **карбо'новых кисло'т по ти'пу толстя'нковых** crassulacean acid metabolism
эндергони'ческий ~ endergonic metabolism
эндоге'нный ~ endogenous metabolism
перви'чный ~ primary metabolism
втори'чный ~ secondary metabolism
метаболи'т *m* metabolite
метаболи'ческий *a* metabolic
метаболиме'тр *m* metabolimeter
метаболи'я *f* metaboly
метага'струла *f* metagastrula
метагене'з *m* metagenesis
метаги'ния *f* metagyny
метадро'мный *a* metadromous
метаконтра'ст *m phys* metacontrast
метаксени'я *f* metaxenia
метаксиле'ма *f* metaxylem
метаме'р *m* metamere, merome, merosome
метамери'я *f* metamerism, merism
метаме'рность *f* merism
метаме'рный *a* metameric
метамиелоци'т *m* metamyelocyte
метамито'з *m* metamitosis
метамо'рфный *a* metamorphic
метаморфо'з *m* metamorphosis
непо'лный ~ incomplete metamorphosis
по'лный ~ complete metamorphosis
метаморфопси'я *f phys* metamorphopsia
метана'уплиус *m* metanauplius
мета'ндрия *f* metandry
метане'фрос *m* metanephros
метано'лиз *m* methanolysis
метаплази'я *f* metaplasia

метапла'зма *f* metaplasm
метапле'вра *f ent* metapleura
метаплевра'льный *a* metapleural
метапнейсти'ческий *a* metapneustic
метапопуля'ция *f* metapopulation
метаптери'гий *m* metapterygium
метасинде'зис *m* metasyndesis
метаскуте'ллум *m ent* metascutellum
метаску'тум *m ent* metascutum
метасо'ма *f* metasoma
метаста'з *m* metastasis
метастати'ческий *a* metastatic
метасте'рнум *m* metasternum
метастигмати'ческий *a* metastigmate
метатала'мус *m anat* metathalamus
метатро'фный *a* metatrophic
метатро'х *m* metatroch
мета'ть *v* spawn, cast forth (*eggs*, *of fish*); lay
~ **икру'** spawn, oviposit
метафа'за *f* metaphase
метафра'гма *f ent* metaphragma
метахромази'я *f* metachromasy
метахромати'ческий *a* metachromatic
метахро'нность *f* metachronism
метацерка'рия *f helm* metacercaria
метгемоглоби'н *m* methemoglobin
метё'лка *f bot* panicle, tassel
однобо'кая ~ one-sided panicle
пони'клая ~, **пони'кшая** ~ drooping panicle
раски'дистая ~ spreading panicle
ры'хлая ~ lax panicle
сжа'тая ~ contracted panicle
ску'ченная ~ crowded panicle
щиткови'дная ~ umbel-like panicle
метелконо'сный *a* panicled
метелкообра'зный *a* coremiform
мете'ль *f* snowstorm
мете'льчатый *a* panicular, paniculate
метили'рование *n* methylation
метилнитрозомочеви'на *f* methylnitrosourea
метиони'н *m* methionine
метионинэнкефали'н *m* Met-enkephalin
ме'тить *v* tag, label
ме'тка *f* tag, label, mark, marker, marking
и'мпульсная ~ pulse label
ольфакто'рная ~ scent marking
цветна'я ~ color mark
метла' *f* broom
ве'дьмина ~ *phytop* witches' broom
ме'тод *m* method, mode, technique, technology
~ **адсо'рбции - элю'ции** adsorption-elution method
~ **ага'ровых бло'ков** agar block method
~ **ага'ровых слоё'в** agar-layer technique
~ **ага'р-фильтра'ции** agar-filtration technique
~ **аналити'ческого диск-электрофоре'за** analytical disk method
~ **антисмысловы'х мРНК** antisense mRNA technology
~ **бактополо'сок** bacto-stripe method
~ **«беспоря'дочной стрельбы'»** *mbiol* shot gun technique
~ **ближа'йшего сосе'да** nearest neighbour method
~ **ближа'йшей о'соби** closest individual method
~ **ближа'йших сосе'дей** nearest-neighbor method
~ **блоки'рования** blocking method
~ **бля'шек** plaque method, method of plaque-forming cells, plaque technique
~ **бля'шек по Е'рне** Jerne's plaque assay, LGH, localized-in-gel hemolysis assay
~ **борьбы'** method of controlling
~ **Бре'я** Bray method
~ **«вверх и вниз»** *biom* up-and-down method
~ **взаи'много усредне'ния** reciprocal averaging
~ **взаи'мной информа'ции** mutual information method
~ **взве'шенных па'рных групп** weighted pair-group method
~ **взве'шивания лимфоузло'в** lymph-node weight assay
~ **взве'шивания селезё'нок** spleen weight assay
~ **внутригруппово'й диспе'рсии** within-group dispersion method
~ **враща'ющихся колб** roll-bottle method
~ **враща'ющихся проби'рок** roll-tube method, "spinner-culture" method
~ **выра'внивания** method of fitting constants
~ **выра'щивания в кюве'тах** shallow pan method
~ **гапте'новой по'мощи** hapten help method

~ **гла'вных компоне'нт** principal component analysis
~ **Грейг-Сми'та** Greig-Smith method
~ **Гу'дола** Goodall method
~ **Да'йса** Dice method
~ **двойно'го кэ'ппинга** cocapping technique
~ **двойно'й иммунодиффу'зии** double immunodiffusion technique
~ **двойны'х антите'л** sandwich method, double-antibody method
~ **Джа'нси** Jancey method
~ **Дже'глама** Jeglum method
~ **диагона'льного электрофоре'за** diagonal electrophoretic technique
~ **ди'сков** *micr* disk technique
~ **диффу'зии в ага'р** agar diffusion method
~ **диффу'зии в гель** gel-diffusion method
~ **заде'ржки в ге'ле** gel retardation assay, band-shift assay
~ **замора'живания** freeze-etch method
~ **определе'ния во'зраста** ageing method, aging method
~ **осажде'ния в желати'не** gelatin sedimentation method
~ **замора'живания - высу'шивания** freezing-drying method
~ **замора'живания-замеще'ния** freeze-substitution technique
~ **замора'живания-отта'ивания** freeze-thaw method
~ **избира'тельного окисле'ния** selective oxidation method
~ **изомо'рфного замеще'ния** isomorphous replacement method
~ **изото'пного разведе'ния** isotope dilution method
~ **изуче'ния на пло'щади в 1 кв. фут** *plsoc* square-foot method
~ **имму'нной адге'зии** immunoadherence technique
~ **имму'нной преципита'ции** immunoprecipitation method
~ **имму'нных отпеча'тков** immunoreplica technique
~ **иммунобло'ттинга** immune transfer method
~ **имму'нного прилипа'ния** immunoadherence technique
ме'тоды иммунологи'ческого ана'лиза *pl* immunoassay technology
~ **иммуноферме'нтный** ~ immunoenzyme method
~ **имму'нных отпеча'тков** immunoreplica technique
~ **иссле'дования в тё'мном по'ле** dark field method
~ **квадра'тов** *plsoc* quadrate method
~ **Ки'нга** *ecol* King method
~ **кле'точной сортиро'вки** cell-sorting technique
~ **«ключе'й»** clue method
~ **колеса'** *plsoc* wheel method, wheel point method
~ **косми'дной «прогу'лки»** "cosmid walking" technique
~ **коше'ния сачко'м** *ent* sweep-net method
~ **культиви'рования в кюве'тах** shallow pan method
~ **Кьельда'ля** Kjeldahl procedure, Kjeldahl method
~ **лиди'рующей грани'цы** "leading-front" technique
~ **ли'зиса кле'ток** killing technique
~ **Ло'ури** Lowry method
~ **максима'льного правдоподо'бия** method of maximum likelihood, method of ml
~ **микроразво'док** microplating method
~ **микроразли'вок** microplating method
~ **моносло'я** monolayer technique
~ **Мо'нте-Ка'рло** Monte Carlo method
~ **мо'стика-изоля'тора** *phys* bridge-insulator method
~ **наиме'ньших квадра'тов** method of least squares
~ **накло'нной иглы'** *bot* inclined point method
~ **негати'вного окра'шивания** negative staining technique
~ **обесцве'чивания** bleaching method
~ **обнажё'нного квадра'та** *plsoc* denuded quadrate method
~ **обогаще'ния** enrichment method
~ **обнаже'нного квадра'та** denuded quadrate method
~ **одно'й вы'борки** one-sample method
~ **«окра'ски при заключе'нии в среду'»** enclosure staining
~ **окра'шивания по Граму** Gram's method, Gram's technique
~ **оплодотворе'ния in vitro** IVF technique
~ **Орло'ци** Orloci method

~ **осредне'ния** averaging method
~ **остано'вленной струи'** stopped flow method
~ **отбе'ливания** bleaching method
~ **«отпеча'тков па'льцев»** "fingerprinting" technique
~ **«отпеча'тка подо'швы»** *mbiol* foot-printing technique
~ **па'рного сравне'ния** paired comparison method
~ **пепти'дных карт** "fingerprinting" technique
~ **переме'нного ра'диуса** *plsoc* variable radius plotless method
~ **пересека'ющихся ли'ний** *plsoc* line-interception method, line-intercept method
~ **«перетасо'вки экзо'нов»** exon-shuffling technique
~ **Пилу'** Pielou method
~ **пласти'нчатых разво'док** plate method, plating technique
~ **плё'нки со съё'мным сло'ем** stripping film method
~ **пове'рхностного культиви'рования** surface culture method
~ **подавле'ния бляшкообразова'ния** plaque inhibition method
~ **подавле'ния эритропоэ'за** erythropoiesis inhibition assay
~ **позити'вного окра'шивания** method of positive staining
~ **пои'мка - ме'чение - повто'рная пои'мка** *ecol* capture-mark-recapture method
~ **по'лного изъя'тия** *ecol* removal method
~ **посло'йного среза'ния** *bot* stratifying clip method
~ **приши'вки** coupling technique technique
~ **проб и оши'бок** trail-and error method
~ **про'бных площаде'й** method of sample plots
~ **проме'ров** *plsoc* distance method
~ **просветле'ния** method of clarification
~ **простра'нственного исключе'ния** spatial exclusion method
~ **противото'чного распределе'ния** countercurrent distribution method
~ **прото'чной иммунофлуоресце'нции** flow immunofluorescence method
~ **равнове'сного диа'лиза** dialysis equilibrium method
~ **разведе'ния** dilution method; breeding method
~ **разведе'ния краси'теля** dye dilution method
~ **разграни'чивающих фаз** interfacial technique
~ **рассе'янной затра'вки** random primer method
~ **рекомбина'нтных ДНК** recombinant DNA technology
~ **реконстру'кции** reconstruction method
~ **Рида-Ме'нча** Reed-Muench method
~ **розеткообразова'ния** rosette method
~ **розе'ток** rosette method
~ **«самоуби'йства»** suiciding technique
~ **сахаро'зного мо'стика** sucrose gap method, sucrose gap technique
~ **свя'зывания** coupling technique technique
~ **свя'зывания краси'теля** dye-binding method
~ **седиментацио'нного равнове'сия** sedimentation equilibrium technique
~ **серебре'ния** silver method
~ **сери'йных разведе'ний** method of serial dilution, serial dilution method, limiting dilution method, double dilution technique, dual dilution technique
~ **с затра'вкой** priming method
~ **сину'зий** synusium method
~ **скачка' давле'ния** pressure jump method
~ **скачка' раствори'теля** solvent-jump method
~ **скользя'щих сре'дних** moving average method
~ **скри'нинга косми'д** "cosmid walking" technique
~ **слоё'в** layer method
~ **случа'йных уко'лов** *plsoc* random point frames method
~ **с ту'шью** Indian-ink method
~ **счё'та пузырько'в** bubble-counting method, bubbling method
~ **твёрдофа'зного иммунoферме'нтного ана'лиза** ELISA technique
~ **температу'рного скачка'** temperature-jump method
~ **теплово'го шо'ка** heat shock technique
~ **удлине'ния затра'вки** primer-extension method
~ **уко'сов** cut method

~ **ультрафильтра'ции** micropore technique
~ **фикса'ции краси'теля** dye-binding method
~ **фикса'ции напряже'ния** *phys* voltage-clamp method
~ **флота'ции** flotation method
~ **флуоресци'рующих антите'л** immunofluorescence method, fluorescent antibody technique
~ **Ха'рвелла** Harvell's hot phenol-water method
~ **Хи'лла** Hill method
~ **холодово'го шо'ка** cold shock technique
~ **«ша'хматного ана'лиза»** "checker board" assay
~ **экзоге'нных коло'ний** exocolonizing assay
~ **электрофоре'за** electrophoretic technique
~ **электрофоре'за по Лэ'ммли** Laemmli electophoretic technique
акрифлави'новый ~ флокуля'ции *micr* acriflavine flocculation technique
биологи'ческий ~ bioassay technique
биологи'ческий ~ борьбы' biological control
весово'й ~ gravimetric method
гапте'н-хе'лперный ~ hapten help method
графи'ческий ~ graphical method
гру'бый ~ crude technique
изото'пный ~ tracer method
иммерсио'нный ~ immersion method
иммуноадгезио'нный ~ immunoadherence technique
иммунологи'ческий ~ окра'шивания зо'лотом immunogold staining
иммунологи'ческий ~ окра'шивания серебро'м immunosilver staining
иммунонефелометри'ческий ~ immunoturbidimetric method
иммунотокси'новый ~ immunotoxin method
иммунофлуоресце'нтный ~ immunofluorescence method
ка'пельно-ча'шечный ~ drop plate method
коли'чественный ~ иммунобло'ттинга quantitative immunoblotting assay
колориметри'ческий ~ colorimetric method
манипуляцио'нно-иссле'довательский ~ manipulation-exploration method
нето'чный ~ crude technique
однофа'кторный ~ *biom* single factor method
описа'тельный ~ descriptive method method
перманганa'тный ~ permanganate method
плазмометри'ческий ~ plasmometric method
полунепреры'вный ~ semicontinuous method
поляризацио'нно-опти'ческий ~ polarization optical method
радиоуглеро'дный ~ radiocarbon method
ра'зностно-дисперсио'нный ~ variate-difference method method
релаксацио'нный ~ relaxation technique
ро'ллерный ~ "spinner-culture" method
серологи'ческий ~ antibody-mediated method
скля'ночный ~ light-and-dark-bottle technique, light-and-dark-bottle method *(of determination of primary production)*
спектрофотометри'ческий ~ spectrophotometric method
твёрдофа'зный ~ solid-phase method
температу'рно-поплавко'вый ~ temperature-float method
то'чечный ~ *plsoc* point contact method, point frame method, steel point method
хитшо'ковый ~ heat shock technique
хлорами'новый ~ chloramine-T method
цитолити'ческий ~ killing technique
ча'шечный ~ plate method, plating technique, cup-plate technique
ча'шечный градие'нтный ~ gradient-plate technique
ча'шечный ~ культиви'рования plate culture method
ча'шечный ~ подсчё'та plate count method
ча'шечный ~ ре'плик replica plating technique

мето'дика *f* technique, methods
~ **непреры'вного культиви'рования** continuous culture technique
~ **после'дующего введе'ния** afterloading technique

метоксили'рование *n* methoxylation
мето'пион *m* metopic point, metopion
метосульфа'т *m* methosulfat

метотрекса'т *m* methylaminopterin
метрате'рм *m helm* metraterm
ме'тула *f* metula
ме'тчик *m* marker
метэ'струс *m* metestrus
мех *m* fur, fell
механи'зм *m* mechanism
~ **изоля'ции** isolation mechanism
~ **обра'тной свя'зи** feedback mechanism
~ **отрица'тельной обра'тной свя'зи** negative feedback mechanism
~ **поддержа'ния по'зы** postural mechanism
~ **положи'тельной обра'тной свя'зи** positive feedback mechanism
врождё'нный пусково'й ~ *phys* innate releasing mechanism
врождё'нный разреша'ющий ~ *eth* innate releasing mechanism
запреща'ющий ~ arresting mechanism
иммунологи'ческий ~ immunologic mechanism
кассе'тный ~ cassete mechanism
кла'панный ~ *eth* valve mechanism
контро'льный ~ control mechanism
оце'нивающий ~ appraisal mechanism
проксима'льный ~ *eth* proximate mechanism
пусково'й ~ trigger, releasing mechanism
разреша'ющий ~ releasing mechanism
регулято'рный ~ control mechanism
меха'ника *f* mechanics
~ **разви'тия** developmental mechanics
механореце'птор *m* mechanoreceptor
мечеви'дный *a* gladiate, ensiform, sword-shaped
мечевидноли'стный *a* sword-leaved
мечели'стный *a* xiphophyllous
ме'чение *n* tagging, marking; labeling, labelling
~ **го'лосом** sound marking
аффи'нное ~ affinity labeling
визуа'льное ~ visual marking
за'паховое ~ olfactory marking, scent marking
и'мпульсное ~ pulse labeling
конкуре'нтное ~ competitive labeling
ольфакто'рное ~ olfactory marking, scent marking
радиоакти'вное ~ radioactive marking
территориа'льное ~ territorial marking
фотоаффи'нное ~ photoaffinity labeling
мечено'сный *a* sword-bearing
ме'ченый *a* marked
мечеобра'зный *a* ensiform, xiphoid
меша'лка *f* stirrer, agitator, mixer
магни'тная ~ magnetic stirrer
периоди'ческая ~ intermittent agitator
мешкови'дный *a* sacciform, saccular
мешко'вый *a bot* saccate
мешкообразный *a* sacciform, saccular, saccate, bursiform
мешо'к *m* bag, sac, pouch
~ **Амбу'** Ambou breathing bag
~ **Ли'стера** Lister bag
возду'шный ~ air sac, *orn* pneumatocyst
горлово'й ~ gular sac
грудно'й воздухоно'сный ~ *orn* thoracic air sac
диа'лизный ~ dialysis sac
желто'чный ~ yolk sac, yolk bag, vitelline sac, vitellicle
заро'дышевый ~ embryo sac
защe'чный ~ cheek pouch
ко'жно-му'скульный ~ skin-muscular sac
конъюнктива'льный ~ conjunctival sac
моноспори'ческий заро'дышевый ~ monosporic embryo sac
пло'дный ~ fetal sac, gestation sac, bag of waters
слё'зный ~ lacrimal sac, tear sac
споровый ~ sporosac
тетраспори'ческий заро'дышевый ~ tetrasporic embryo sac
черни'льный ~ ink sac
эмбриона'льный желто'чный ~ embryonic yolk sac
мешо'чек *m* sac, sack, pouch, saccule, sacculus, sacculus
волосяно'й ~ hair pouch, hair sac
зубно'й ~ dental sac, tooth sac
лимфати'ческий ~ lymph sac
сле'зный ~ dacryocyst
мешо'чница *f* bagworm (*Psychidae*)
мига'ние *n* nictitation, nictation, winking, blinking, eyeblink
усло'вно-рефлекто'рное ~ conditioned eyeblink
мига'ть *v* blink, wink, nictate
мигра'нт *m* migrant
мигра'ция *f* migration
ма'ссовая ~ mass migration
ночна'я ~ nocturnal migration
мигре'нь *f* migraine
мигри'ровать *v* migrate

мигри'рующий *a* migrating, migratory
ми'дия *f* mussel *(Mytilus)*
миели'н *m* myeline
миелиниза'ция *f* myelination
миело- myelo-
миелобла'ст *m* myeloblast
миелоге'нный *a* myelogenous
миело'идный *a* myeloid
миеломе'р *m* myelomere
миелопоэ'з *m* myelopoiesis
миелоци'т *m* myelocyte
мизи'нец *m* little finger, minimus
мико- myco-
микобактери'н *m* mycobacterin
мико'з *m* mycosis
~ **ко'жи** dermatomycosis
миколо'гия *f* mycology
микопла'зма *f* mycoplasm, mycoplasma
микори'за *f* mycorhiza, mycorrhiza
везикуля'рно-арбускуля'рная ~ VA mycorhiza
эктендотрофи'ческая ~ ectendotrophic mycorhiza
эктотро'фная ~ ectotrophic mycorhiza
эндотро'фная ~ endotrophic mycorhiza
микори'зный *a* mycorrhizal
микостеро'л *m* mycosterol
микотоксико'з *m* myctotoxicosis
микотокси'н *m* mycotoxin
микотро'ф *m* mycotrophic plant
микотро'фный *a* mycotrophic
микофаги'я *f* mycophagy
микофло'ра *f* mycoflora, mycobiota
микоцеци'дий *m* mycocecidium
микоци'д *m* mycocide
микро- micro-
микроана'лиз *m* microanalysis, microassay
иммунофлуоресце'нтный ~ microimmunofluorimetry
спот-антиге'нный ~ antigen spot microanalysis
рентге'новский ~ X-ray microanalysis
микроасоциа'ция *f* microassociation
микроавторадио'граф *m* microautoradiograph
микроавторадиогра'фия *f* microautoradiography
микроаэрофи'л *m* microaerophil
микроаэрофи'льный *a* microaerophilic
микро'б *m* microbe, germ
микробиологи'ческий *a* microbiologic, microbiological
микробиоло'гия *f* microbiology
~ **во'здуха** air microbiology
~ **пищевы'х проду'ктов** food microbiology
морска'я ~ marine microbiology
о'бщая ~ general microbiology
пищева'я ~ food microbiology
по'чвенная ~ soil microbiology
сельскохозя'йственная ~ agricultural microbiology
техни'ческая ~ industrial microbiology
микробла'ст *m* microblast
микро'бный *a* microbial, microbic, microbal, microbian, germinal
микробюре'тка *f lab* microburet, microburette
микровесы' *pl lab* microbalance
микроворси'нка *f* microvillus
микрогаме'та *f* microgamete
микрогаметофи'т *m* microgametophyte
микрогаметоци'т *m* microgametocyte
микрогематокри'т *m lab* microhematocrit
микроги'на *f* microgyne
микрогли'я *f* microglia
микрогониди'я *f* microgonidium
микрогоре'лка *f lab* microburner
микрогра'нула *f* microsphere
микродиспенсер *m lab* microdispenser
микродозатор *m lab* microdispenser
микрозо'нд *m lab* microprobe
микрозооспо'ра *f* microzoospore
микроигла' *f lab* microneedle
микрокиносъе'мка *f* microcinematography
микрокли'мат *m* microclimate
микроко'кк *m* micrococcus
микрокони'дия *f* microconidium
микроконъюга'нт *m* microconjugant
микроко'см *m* microcosm
микроку'зница *f lab* microforge
микрокульту'ра *f* microculture
микрокультиви'рование *n* microculturing
микролепе'стный *a* micropetalous
микролу'ч *m* microbeam
микроманипуля'тор *m lab* micromanipulator
микроманипуля'ция *lab* micromanipulation
микроме'р *m* micromere
микроместообита'ние *n* microhabitat
микроме'тод *m* micromethod
~ **Кьельда'ля** micro-Kjeldahl method
~ **определе'ния кислоро'да по Ви'нклеру** micro-Winkler technique

микроме'тр *m* micrometer (*a unit*)
микро'метр *m* micrometer (*a measuring device*)
винтово'й ~ screw micrometer
окуля'рный ~ eyepiece micrometer
микрометри'я *f lab* micrometry
микромиелоци'т *m* micromyelocyte
микромута'ция *f* micromutation
микро'н *m* micron
микроноси'тель *m* microcarrier
микрояче'йка *f lab* microcell
микронуклеа'ция *f* micronucleation
микрону'клеус *m prot* micronucleus
микрону'триент *m* micronutrient
микроокруже'ние *n* microenvironment
гемопоэти'ческое ~, кроветво'рное ~ hemopoietic microenvironment
микрооргани'зм *m* microorganism, microbe
азотфикси'рующий ~ nitrogen-fixing microorganism
алкалифи'льный ~ alkaliphilic microorganism
ацидофи'льный ~ acidophil microorganism
газообразу'ющий ~ acidophil microorganism
галофи'льный ~ halophilic microorganism
гетеротро'фный ~ heterotrophic microorganism
гни'лостный ~ putrefactive microorganism
дрожжеподо'бный ~ yeastlike microorganism
мезофи'льный ~ mesophilic microorganism
мета'новый ~ methanic microorganism
метано'льный ~ methanolic microorganism, methylotrophic microorganism
метилотро'фный ~ methanolic microorganism, methylotrophic microorganism
мицето'мный ~ mycetomal microorganism
по'чвенный ~ soil microorganism
слизеобразу'ющий ~ mucoid microorganism
микропалеонтоло'гия *f* +micropaleontology
микропарази'т *m* microparasite
микро'пиле *n* micropyle
микропиля'рный *a* micropylar
микропипе'тка *f lab* micropipet, micropipette
микропланше'т *m lab* microplate, microtray
микроподви'д *m* microsubspecies
микропрепара'т *m lab* microslide, slide mount
микропроектор *m lab* microprojector
микропси'я *f* micropsia
микроптери'зм *m* micropterism
микропучо'к *m* microbeam
микрорадиография *f lab* microradiography
микрореспиро'метр *m* microrespirometer
микросеквени'рование *n* microsequencing
мелкосемя'нный *a* microspermous
микросе'рия *f* microsere
микросе'точка *f* microreticulum
кле'точная ~ cellular microreticulum
микросжига'ние *n lab* microincineration
микросподогра'фия *f lab* microincineration
микроско'п *m* microscope; **под микроскопом** under *a* microscope, with a microscope, by a microscope
~ **с нагрева'тельным сто'ликом** hot stage microscope
~ **сравне'ния** comparison microscope
~ **с телевизио'нным экра'ном** television microscope
бинокуля'рный ~ binocular microscope
измери'тельный ~ measuring microscope, comparator microscope
инверти'рованный ~ inverted microscope
интерференцио'нный ~ interference microscope
инфракра'сный ~ infrared microscope
ла'зерный ~ laser microscope
операцио'нный ~ surgical microscope, operative microscope
опти'ческий ~ optical microscope, light microscope
поляризацио'нный ~ polarizing microscope
ра'стровый тунне'льный ~ scanning tunneling microscope
ра'стровый электро'нный ~ scanning electron microscope
рентге'новский ~ X-ray microscope
светово'й ~ light microscope
скани'рующий тунне'льный ~ scanning tunneling microscope
скани'рующий электро'нный ~ scanning electron microscope
стереоскопи'ческий ~ stereoscopic microscope
темнопо'льный ~ dark-field microscope
тенево'й ~ shadow microscope

трансмиссио'нный электро'нный ~ transmission electron microscope
ультрафиоле'товый ~ ultraviolet microscope
фазовоконтра'стный ~, **фазоконра'стный ~** phase-contrast microscope, phase microscope
флуоресце'нтный ~ fluorescence microscope
цветово'й контра'стный ~ color-contrast microscope
широкопо'льный ~ wide-field microscope
электро'нный ~ electron microcope

микроскопи'рование *n* microscopy, microscopic examination

микроскопи'ческий *a* microscopic, microscopical

микроскопи'я *f* microscopy
акусти'ческая ~ acoustic microscopy
иммуносорбцио'нная электро'нная ~ immunosorbent electron microscopy, solid phase immunoelectron microscopy
иммунофлуоресце'нтная ~ immunofluorescence microscopy
иммуноэлектро'нная ~ immunoelectron microscopy
и'мпульсная рентге'новская ~ flash X-ray microscopy
интерференцио'нная ~ interference microscopy
конфока'льная ~ confocal microscopy
криоге'нная ~ cryogenic microscopy
криоэлектро'нная ~ cryoelectron microscopy
поляризацио'нная ~ polarization microscopy
прижи'зненная ~ vital microscopy
просве'чивающая электро'нная ~ transmission electronic microscopy, transmission electron microscopy
ра'стровая иммуноэлектро'нная ~ immunoscanning electron microscopy
ра'стровая электро'нная ~, РЭМ raster electronic microscopy, scanning electronic microscopy
светова'я ~ light microscopy
скани'рующая иммуноэлектро'нная ~ immunoscanning electron microscopy
скани'рующая электро'нная ~, РЭМ raster electronic microscopy, scanning electronic microscopy
темнопо'льная ~ dark-field microscopy, dark-ground microscopy
трансмиссио'нная электро'нная ~ transmission electronic microscopy, transmission electron microscopy
ультрафиоле'товая ~ ultraviolet microscopy
фа'зово-контра'стная ~ phase-contrast microscopy
флуоресце'нтная ~ fluorescence microscopy
электро'нная ~ electronic microscopy, electron microscopy

микросни'мок *m* micrograph

микросо'ма *f* microsome

микросоми'я *f* microsomia

микросоо'бщество *n* microcommunity

микроспектроско'п *m lab* microspectroscope

микроспектрометри'я *f* microspectrometry
ультрафиоле'товая ~ ultraviolet microspectrometry
рентге'новская ~ X-ray microspectrometry

микроспектрофото'метр *m* microspectrophotometer

микроспектрофотометри'я *f* microspectrophotometry

микроспо'ра *f* microspore

микроспора'нгий *m* microsporangium

микроспоридио'з *m* microsporidiosis

микроспорогене'з *m* microsporogenesis

микроспорофи'лл *m* microsporophyll

микроспорофо'р *m* microsporophore

микроспороци'т *m* microsporocyte

микросреда' *f* microenvironment

микростилоспо'ра *f* microstylospore

микростро'ма *f* microstroma

микрострукту'ра *f* microstructure

микросфе'ра *f* microsphere
микросфе'ры, сенсибилизи'рованные антиге'нами *pl* antigen-coated microspheres

микросфери'ческий *a* microspheric

микроте'льце *n* microbody

микроте'рм *m* mirotherm

микрото'м *m* microtome
замора'живающий ~ freezing microtome
колё'сный ~ rotary microtome
ротацио'нный ~ rotation microtome
са'нный ~ sledge microtome, sliding microtome

микротру'бочка *f* microtubule
микрофа'уна *f* microfauna
микрофибри'лла *f* microfibril
микрофиля'рия *f* microfilaria
микрофила'мент *m* microfilament
микрофи'льтр *m lab* microstrainer
микрофильтра'ция *f* microfiltration
  ~ **в попере'чном пото'ке** cross-flow microfiltration
микрофи'т *m* microphyte
микрофитоцено'з *m* microphytocoenosis
микрофло'ра *f* microflora, microscopic flora
микрофотогра'фия *f* photomicrography; photomicrograph, micrograph, microphotography
  электро'нная ~ electron micrograph, electronic microphotography
микрофото'метр *m lab* microphotometer
  фотоэлектри'ческий ~ photoelectric microphotometer
микрофотосни'мок *m* photomicrograph
микрофтальми'я *f* microphthalmia
микрофотогра'фия *f* microphotography, photomicrography; photomicrograph, micrograph
  электро'нная ~ electronic microphotography, electron micrograph
микрофотосни'мок *m* photomicrograph
  электро'нный ~ electron photomicrograph
микрохи'мия *f* microchemistry
микрохирурги'я *f* microsurgery, micrurgy, microdissection
микрохромосо'ма *f* microchromosome
микроцентрифу'га *f lab* microfuge
микроце'рк *m* microcercus
микроцефа'льный *a* microcephalic
микроци'ста *f* microcyst
микроци'т *m* microcyte
микроциркуля'ция *f* microcirculation
микроцита'рный *a* microcytic
микроэволю'ция *f* microevolution
микроэлектро'д *m* microelectrode
  двухство'льный ~ double-barreled microelectrode
  ион-селекти'вный ~ ion-selective microelectrode
микроэлеме'нт *m* microelement, micronutrient element, minor element, trace element, minor constituent, trace constituent
  пита'тельный ~ micronutrient
микрурги'я *f* microsurgery, micrurgy, microdissection
микроядро' *n prot* micronucleus
миксо- mixo-
миксобакте'рия *f* myxobacterium
миксобио'з *m* mixobiosis
миксовариа'ция *f* mixovariation
миксови'рус *m* myxovirus
миксопло'ид *m* mixoploid
миксоплоиди'я *f* mixoploidy
миксоптери'гий *m* myxopterygium
миксотрофи'я *f* mixotrophy
миксотро'фный *a* mixotrophic
миксохиме'ра *f* mixochimaera, mixote
милиа'рный *a* miliary
милдью' *f phytop* mildew *(caused by Plasmopara)*
мимети'зм *m* mimesis, mimetism
мимети'ческий *a* mimetic
ми'мик *m gen* mimic
ми'мика *f* facial expression
мимикри'я *f* mimicry
  бе'йтсовская ~ Batesian mimicry
  вока'льная мимикрия vocal mimicry
  идиотипи'ческая ~ idiotypic mimicry
  иммунологи'ческая ~ immunologic mimicry
  молекуля'рная ~ molecular mimicry
ми'на *f ent* mine
  ~ **с пальцеви'дными хода'ми** digitate mine
  ~ **в листе'** leaf mine
минда'лина *f anat* tonsil; *bot* almond
  ~ **мозжечка'** tonsil of cerebellum
  нёбная ~ faucial tonsil, palatine tonsil
  тру'бная ~ eustachian tonsil
минда'ль *m* almond (*Amygdalus*)
  земляно'й ~ earth almond, chufa flat sedge, chufa (*Cyperus esculentus*)
  очи'щенный ~ nibbed almond
мине'р *m* miner
минералокортико'ид *m* mineralocorticoid
миниге'н *m* minigen
минакле'тка *f* minicell
миникольцо' *n mbiol* minicircle
минима'кс *m biom* minimax
минимиза'ция *f* minimization
ми'нимум-ареа'л *m* minimum areal
минисегме'нт *m* minisegment
минисе'ть *f* mininetwork
  иммy'нная ~ immune mininetwork
миобла'ст *m* myoblast
миоге'н *m* myogen

миогене'з *m* myogenesis
миоге'нный *a* myogenic
миогли'я *f* myoglia
миоглоби'н *m* myoglobin
миоглобули'н *m* myoglobulin
миогра'мма *f* myogram
миогра'ф *m* myograph
миогра'фия *f* myography
миодинами'ческий *a* myodynamic
миози'н *m* myosin
миока'рд *m* myocard, myocardium, cardiac muscle
миокине'з *m* myokinesis
миокини'н *m* myokinine
миокло'нус *m* myoclonia, myoclonus
  рефлекто'рный ~ reflex myoclonus
миоле'мма *f* myolemma
миоло'гия *f* myology
мио'ма *f* myoma
миоме'р *m* myomere, muscle segment
миометра'льный *a* myometrial
миоме'трий *m* myometrium
мионевра'льный *a* myoneural
мионевроци'т *m* myoneurocyte
мионе'ма *f* myoneme
миопати'я *f* myopathy, myopathia
миопи'я *f* myopia
миополя'рный *a* myopolar
миосе'пта *f* myoseptum
миоспа'зм *m* myospasm
миоспо'ра *f* miospore
миотакти'ческий *a* myotactic
миото'м *m* myotome, muscle segment
миотони'я *f* myotonia
миофибри'лла *f* myofibrilla, muscular fibril
миофибрилло'мер *m* myofibrillomere
миофиламе'нт *m* myofilament
миофри'ск *m* myophrisk
миоце'ль *f* myocoel, myocoele
миоци'т *m* myocyte
миоэпителиа'льный *a* myoepithelial
мир *m* world; peace
  живо'тный ~ fauna
  расти'тельный ~ flora
мирабе'ль *f* cherry plum, myrobalan plum (*Prunus cerasifera*)
мираци'дий *m* miracidium
мирмеко- myrmeco-
мирмеколо'гия *f* myrmecology
мирмекохо'р *m* myrmecochorous plant
мирмекохори'я *f* myrmecochory
мирмекохо'рный *a* myrmecochorous
мирмекофи'л *m* myrmecophil
мирмекофили'я *f* myrmecophily
мирмекофи'льный *a* myrmecophilous
мирмекофи'т *m* myrmecophyte
мирмекофо'б *m* myrmecophobe
мирмекофо'бный *a* myrmecophobic
мирози'н *m* myrosin
мирт *m* myrtle (*Myrtus*)
миссе'нс-мута'нт *m* missense mutant
миссе'нс-мута'ция *f* missense mutation
миссе'нс-супре'ссор *m* missense-supressor
мито- mito-
митоге'н *m* mitogen
митогенети'ческий *a* mitogenetic
митоге'нность *f* mitogenicity
митоге'нный *a* mitogenetic
мито'з *m* mitosis, mitoschisis
  ~, **нарушенный колхици'ном** C-mitosis
  эндореституцио'нный ~ endorestitutional
  реституцио'нный ~ restitutional mitosis
  восстанови'тельный ~ restitutional mitosis
митопла'ст *m* mitoplast
митосо'ма *f* mitosome
митоти'ческий *a* mitotic
митохондриа'льный *a* mitochondrial
митохо'ндрия *f* mitochondria
митра'льный *a* mitral
мицелиеви'дный *a* mycelioid
мице'лий *m* mycelium
  вегетати'вный ~ vegetative mycelium
  возду'шный ~ aerial mycelium
  втори'чный ~ secondary mycelium
  дикариофи'тный ~ dicaryon mycelium, secondary mycelium
  многоле'тний ~ perennial mycelium
  некле'точный ~ unicellular mycelium
  нечле'нистый ~ unicellular mycelium
  одноклe'точный ~ unicellular mycelium
  перви'чный ~ primary mycelium
  сапротро'фный ~ saprobic mycelium
  септи'рованный ~ septate mycelium
  се'тчатый ~ reticulate mycelium
мице'лла *f* micella
мицелля'рный *a* micellar
мицето'ма *f* mycetome, mycetoma
мицетофа'г *m* mycetophage
мицетоци'т *m* mycetocyte
мише'нь *f* target
  ~ **для кле'тки-ки'ллера** cytotoxic target
  нагру'женная антитела'ми ~ antibody-coated target

сенсибилизи'рованная антитела'ми ~ antibody-coated target
этало'нная ~ reference target
младе'нческий *a* infantile
млеконо'сный *a bot* lactiferous
млекопита'ющее *n* mammal (*Mammalia*)
млекопита'ющий *a* mammalian
мле'чник *m* latex tube, lactifer; sea milkwort (*Glaux*); lactarius, milk mushroom (*Lactarius*)
мле'чный *a* lacteal, lactiferous
мнемо'ника *f* mnemonics
мнемони'ческий *a* mnemonic
мнемота'ксис *m* mnemotaxis
много- multi-, poly-, pluri-
многоборо'здный *a pal* polypericolpate
многоборо'здчатый *a* multisulcate
многобра'тственный *a* polyadelphous, polydelphous
многобуго'рчатый *a* multituberculate
многовале'нтность *a* multivalency
многовале'нтный *a* plurivalent
многоветви'стый *a* polycladous, multiramose
многовитко'вый *a* multispiral
многогла'зый *a* multiocular
многогне'здный *a* multilocular, plurilocular, multicamerate
многодо'льный *a* multilobate
многоды'рчатый *a pal* polyforate
многожгу'тиковый *a* multiflagellate
многожилкова'тый *a bot* multicostate
многожи'лковый *a* multinervate, multicarinate
многозу'бчатый *a* multidentate
многока'мерный *a* plurilocular, polycystic, multicamerate, multichamber
многокаска'дный *a* many-stage
многоки'левый *a* multicarinate
многокле'точный *a* multicellular, many-celled, pluricellular
многокра'тный *a* manifold
многоле'тний *a* perennial
многоле'тник *m* perennial plant, perennial
многоли'стный *a* polyphyllous, multifoliate
многолисто'чковый *a* multifoliolate
многоло'пастный *a* pluripartite
многолучево'й *a* multiradiate
многоме'рный *a* multivariate; multidimensional
многому'жество *n* polyandry
многонадре'зный *a* multifid
многоно'жка *f zool* centipede, myriapod (*Myriapoda*); *bot* polypody (*Polypodium*)
многоо'сный *a* pluriaxial
многоости'стый *a* many-awned
многопа'лость *f* polydactyly, polydactylism
многопа'лый *a* polydactyl
многопа'рный *a* multijugate
многоперегоро'дчатый *a* multiseptate
многопе'стичный *a* polygynous
многопласти'нчатый *a* multilaminate
многопо'люсный *a* multipolar, pluripolar
многоприцве'тниковый *a* many-bracted
многопучко'вый *a bot* multifasciculate
многоразде'льный *a* multifid, multipartite
многоспо'ровый *a* polyporate
многоразде'льный *a* multipartite, multifid, pluripartite
многоресни'чный *a* multiciliate
многоря'дный *a* multiserial, polyserial, plural-row, multiseriate, polyseriate, polystichous, pluriserial
многосемядо'льность *f* pleiocotyledony
многосемядо'льный *a* polycotyledonous
многосемя'нный *a* many-seeded
многосло'жный *a* decompound
многосло'йный *a* multiplayer
многоство'льный *a* deliquescent
многостебе'льный *a* multicaulis
многосторо'нний *a* multilateral
многосторо'нность *f* versatility
биохими'ческая ~ biochemical versatility
многостручко'вый *a* multisiliquous
многоступе'нчатый *a* many-stage
многосуста'вный *a* multarticulate
многото'чечный *a bot* multipunctate
многотычи'нковый *a* polyandrous, polystemonous, multistaminate
многоузлово'й *a* multinodal
многохозя'йственность *f* pleioxeny
многоцветко'вый *a* polyanthous, multiflowered
многоцве'тный *a* multicolored
многоцикли'ческий *a* pleiocyclic
многочле'нистый *a* multiarticulate, many-segmented
многочле'нный *a* manifold
многощу'пальцевый *a* multitentaculate
многоя'дерный *a* multinuclear, multinucleate, polykaric, plurinuclear, plurinucleate, polynucleate
многоя'дный *a* plurivorous
многоязычко'вый *a bot* multitongued

многояйцево'й *a* multiovulate
многоя'дность *f* polyphagy
многоя'дный *a* polyphagous, plurivorous
многояче'истый *a* polybrochate
мно'жественно-разноборо'здный *a pal* multiheterocolpate
мно'жественно-сто'лбчатый *a pal* multibaculate
мно'жественность *f* plurality, multiplicity
~ **зараже'ния** *f* multiplicity of infection
мно'жественный *a* multiple
мобилиза'ция *f* mobilization
моби'льность *f* mobility
моби'льный *a* mobile
мога'р *m* foxtail millet, Italian millet (*Setaria italica*)
моги'льщик *m* burying beetle, carrion beetle (*Necrophorus*)
мода'льность *f* modality
мо'да *f biom* mode
модели'рование *n* modeling, simulation
компью'терное ~ computer simulation
маши'нное ~ computer simulation
моде'ль *f* model
~ **агрессивного поведения** model of aggressive behavior
~ **антиге'на** antigenic model
~ **антиге'нного мо'стика** antigen-bridge model
~ **выпе'тливания-выреза'ния** looping out-excision model
~ **Дели'зи-Перельсо'на** patching model
~ **жи'дкой моза'ики** fluid mosaic model
~ **иску'сственного интелле'кта** artificial intelligence model
~ **когна'тной по'мощи** model of antigen-specific help
~ **латера'льного торможе'ния** lateral inhibition model
~ **Ми'тчисона** antigen-bridge model
~ **По'ртера** Porter's structure
~ **«хи'щник-же'ртва»** predator-prey model
~ **Хо'ллидея** Holliday's model
~ **Хо'нжо-Катао'ки** allelic deletion model
антиге'нная ~ antigenic model
идиопепти'дная ~ "peptidic self" model
математи'ческая ~ mathematical model
фа'кторная ~ *biom* factor model
модифика'тор *m* modificator, modifier
модифика'ция *f* modification
дли'тельная ~ persistent modification
посттранскрипцио'нная ~ post-transcriptional modification, post-translational modification
модуля'рный *a* modular
модуля'тор *m bioch* modifier, modulator
~ **домина'нтности** dominance modifier
~ **кроссинго'вера** crossing-over modifier
~ **мутаби'льности** mutability modifier
пепти'дный ~ peptide modulator
плюс-ми'нус ~ plus-minus modifier
модуля'ция *f* modulation
моё'вка *f* kittiwake (*Rissa*)
можжеве'льник *m* juniper (*Juniperus*)
моза'ик *m gen* mosaic
~ **по по'лу** sex mosaic
гоносо'мный ~ gonosomic mosaic
хромосо'мный ~ chromosome mosaic
моза'ика *f* mosaic, mottle
~ **овса'** oat mosaic
~ **расти'тельности** vegetation pattern
инфекцио'нная ~ infectious variegation
кольцева'я ~ ring mosaic
мозаици'зм *m gen* mosaicism
мозаи'чность *f* mosaic, mosaicism, patchiness; mottle
мозаи'чный *a* mosaic
мозг *m* brain
большо'й ~ great brain
головно'й ~ brain, cerebrum
жё'лтый ко'стный ~ yellow bone marrow
коне'чный ~ end brain, telencephalon
ко'стный ~ marrow, bone marrow, medulla
кра'сный ко'стный ~ red bone marrow
межу'точный ~ thalamencephalon, diencephalon, interbrain
обоня'тельный ~ smell brain
продолгова'тый ~ myelencephalon
промежу'точный ~ thamencephalon, diencephalon, interbrain
спинно'й ~ spinal cord, dorsal cord, spinal marrow, spinal medulla, myelon
сре'дний ~ mesencephalon, midbrain
мозгово'й *a* cerebral, medullary
мозжечко'вый *a* cerbellar
мозжечо'к *m* cerebellum, little brain
мозо'листость *f* callosity
мозо'листый *a* callosal
мозо'ль *f* callosity, callus, corn, horn
мокре'ц *m vet* malanders; biting midge, black gnat (*Ceratopogonidae*)

мокри'ца *f zool* wood louse, pill bug (*Oniscoidea*); *bot* chickweed (*Stellaria media*)
мокро'та *f* sputum, expectoration
мо'крый *a* wet, moist, damp
моле'кула *f* molecule
  катени'рованная ~ catenated molecule
  ДНК-катени'рованная ~ DNA catenated molecule
молекулоноси'тель *m* molecular carrier
молекуля'рный *a* molecular
моллю'ск *m zool* shellfish, mollusk, mollusc, molluscan shellfish, shell, molluscan (*Mollusca*); *med* molluscum
  брюхоно'гий ~ gastropod *(Gastropoda)*
  двуство'рчатый ~ bivalved mollusk, clam, mussel (*Bivalvia*)
  крылоно'гий ~ sea butterfly (*Pteropoda*)
  ра'ковинный ~ shell-bearing mollusk *(Conchifera)*
молоди'ло *bot* houseleek *(Sempervivum)*
моло'дка *f* pullet
молодня'к *m* young, young growth; young animals
молодо'й *a* young, juvenile; adolescent
мо'лодь *f* young
моло'зивный *a* colostric
моло'зиво *n* foremilk, colostrum, protogala, first milk
моло'ки *f pl* milt, milt roe, soft roe
молоко' *n* milk
  жи'дкое ~ watery milk
  ла'кмусовое ~ litmus milk
  матери'нское ~ breast milk
  парно'е ~ new milk
  сбо'рное ~ bulked milk
  сгуще'нное ~ condensed milk
  сухо'е ~ powdered milk
  це'льное ~ whole milk
молотогла'в *m* hammerhead (*Scopus umbretta*)
молото'чек *m anat* malleus, hammer bone
молоча'й *m* spurge (*Euphorbia*)
молочко' *n apic*: **ма'точное** ~ royal jelly
моло'чность *f* milk yield
моло'чный *a* lactic, lactescent
моль *f* moth (*Tinea*, *etc.*); mole
  платяна'я ~ clothes moth (*Tineola biseliella*)
моме'нт *m* moment
  абсолю'тный ~ *biom* absolute moment
  фа'кторный ~ *biom* factorial moment
мона'да *f* monad
мона'ндрия *f* monandry
мона'стер *m cyt* monaster
монаура'льный *a phys* monaural
моне'тный *a:*
  обра'зование «моне'тных сто'лбиков» rouleaux formation
  скле'ивание эри'троцитов в моне'тные сто'лбики nummulation
мо'нитор *m* monitor
мо'ниторинг *m* monitoring
  авиацио'нный ~ aircraft monitoring
мон- mon-
мона'да *f* monad
мона'ндрия *f* monandry
монаура'льный *a* monaural
мона'шенка *f* nun moth (*Lymantria monacha*)
моно- mono-
моновольти'нный *a* monovoltine
монога'мия *f* monogamy
монога'мный *a* monogamous
моногапло'ид *m* monohaploid
моногене'з *m* monogenesis
моногенети'ческий *a* monogenetic, monogenous
моногени'я *f* monogeny
моногено'мный *a* monogenomic
моногибри'д *m* monohybrid
моногиния *f* monogyny
моноги'нный *a* monogynous
моногони'я *f* monogony
монозиго'тный *a* monozygotic, monoovular, unioval
монокли'макс *m* monoclimax
моноклона'льность *f* monoclonality
моноклона'льный *a* monoclonal
монокульту'ра *f* monoculture
монокулярный *a* monocular
мономери'я *f* monomery
мономе'рный *a* monomeric
мономити'ческий *a bot* monomitic
мономорфи'зм *m* monomorphism
мономо'рфный *a* monomorphic, monomorphous
мононуклеа'р *m* mononuclear leukocyte, mononuclear cell
мононуклеа'рный *a* mononuclear
мононуклеоти'д *m* mononucleotide
монооксигена'за *f* monooxygenase
монопло'ид *m* monoploid
монопло'идный *a* monoploid

монопло'нт *m* monoplont
монопо'дий *m* monopodium
моносахари'д *m* monosaccharide
моносло'й *m* monolayer
  адгези'вный ~ adherent monolayer
  седименти'рующий ~ sedimentary monolayer
моносо'ма *f* monosome
моносо'мик *m* monosomic
моносоми'ческий *a* monosomic
моносперми'ческий *a* monospermic, monospermous
моносперми'я *f* monospermy
моноспора'нгий *m* monosporangium
моноспо'ра *f* monospore
моносте'ла, моносте'ль *f bot* monostele
монотели'я *f ent* monothely
монотети'ческий *a* monothetic
моноти'п *m* monotype
монотипи'ческий, монотипи'чный *a* monotypic
моното'пный *a* monotopic
монотриха'льный *a* monotrichous
монотро'пный *a* monotropic
монотропи'я *f* monotropy
монофа'гия *f* monophagy
монофакториа'льный *a* monofactorial, unifactorial
монофилети'ческий *a* monophyletic
моноха'зий *m* monochasium
монохромати'ческий *a* monochromic, monochromatic
моноцено'з *m* monocoenosis
моноцентри'ческий *a* monocentric
моноце'рк *m* monocercus
моноцикли'ческий *a* monocyclic
моноцистро'нный *a* monocistronic
моноци'т *m* monocyte
моноцита'рный *a* monocytic
моноцито'з *m* monocytosis
монти'ровать *v lab* mount
монтиро'вка *f lab* mounting
мопс *m* pug
морга'ть *v* wink, blink
моргани'да *f* Morgan unit
мо'рда *f* snout, face, muzzle, neb; fish trap
  во'гнутая ~ dished face
  заострё'нная ~ pointed face
морж *m* walrus, морж *(Odobaenus rosmarus)*
мори'лка *f* killing jar, killing bottle
морко'вь *f* carrot (*Daucus carota*)
морозосто'йкий *a* frostproof
морозосто'йкость, морозоустойчивость *f* frost hardness, freezing tolerance, frost tolerance, frost resistance
морозоусто'йчивый *a* frostproof
моро'шка *f* cloudberry (*Rubus chamaemorus*)
морско'й *a* marine, maritime, sea
мо'рула *f* mulberry body, morula
мо'рфа *f* morph
морфалла'ксис *m* morphallaxis
морфакти'н *m* morphactin
морфе'ма *f eth* morpheme
морфи'зм *m* morphism
морфи'н *m* morphine
морфо- morpho-
морфови'д *m* morphological species
морфогене'з *m* morphogenesis, morphogeny
морфогенети'ческий *a* morphogenetic
морфогра'фия *f* morphography
морфо'з *m* morphosis
морфокли'макс *m* morphoclimax
морфологи'ческий *a* morphological
морфоло'гия *f* morphology
  функциона'льная ~ functional morphology
  эволюцио'нная ~ evolutionary morphology
морфометри'ческий *a* morphometric
морфометри'я *f* morphometrics
морфопла'зма *f* morphoplasm
морщи'на *f* wrinkle, crinkle, corrugation, crease
морщи'нистость *f* rugosity
морщи'нистый *a* rugose, rugate, puckered
морщи'нка *f* crinkle
моря'нка *f* old-squaw, long-tailed duck (*Clangula hyemalis*)
моски'т *m* moth-fly, gnat (*Phlebotomus*)
моско'вка *f* coalmouse, coal tit (*Parus ater*)
мост *m* bridge
  анафа'зный ~ anaphase bridge
  бивале'нтный ~ bivalent bridge
  ло'жный ~ *mbiol* matrix bridge
  межконтинента'льный ~ intercontinental connection
  хромати'дный ~ chromatid bridge
  хромосо'мный ~ chromosome bridge
мо'стик *m* bridge
  антиге'нный ~ antigen bridge
  дисульфи'дный ~ *mbiol* disulfide bridge
  межкле'точный ~ intercellular bridge, cell bridge

протоплазмати'ческий ~ protoplasmic bridge
скутелля'рный ~ *bot* scutellar bridge
моти'в *m* motive
мотива'ция *f* motivation
побуди'тельная ~ incentive motivation
мотонейро'н *m* motor neuron, motor neurone, motoneuron, motor unit
мото'риум *m* motorium
мотылько'вый *a bot* papilonaceous
мох *m* moss (*Musci*); moor, bog
акрока'рпный ~ top-fruiting moss
бокопло'дный ~ side-fruiting moss
водяно'й ~ water moss *(Fontinalis)*
зелёный ~ true moss *(Bryales)*
ирла'ндский ~ carragheen moss *(Cladonia islandica)*
исла'ндский ~ Iceland moss (*Cetraria*) (*a lichen*)
оле'ний ~ reindeer moss, cup moss (*Cladina rangiferina*) (*a lichen*)
торфяно'й ~ peat moss, bog moss (*Sphagnum*)
мохна'тый *a* tomentose, villous
мохноно'гий *a* feather-legged, plumiped, cootie
мохови'на *f* mossy bog, moss-covered swamp
мохообра'зный *a* muscoid, bryophytic
моча' *f* urine, urina
мочажи'на *f* pool (*in bog*); hollow
мочеви'на *f* urea
мочево'й *a* uric, uretic, urinary, urinous
мочего'нный *a* diuretic
мочеиспуска'ние *n* urination, micturition, miction, emiction
мочеполово'й *a* urinogenital, urogenital, genitourinary
мочето'чник *m* ureter
мочи'льный *a* steeping
мо'чка *f* lobule of the ear, lobe of the ear, flap of the ear; *bot* fibril, fibrilla; retting
~ **ко'рня** root fibril
~ **у'ха** lobule of the ear, lobe of the ear, flap of the ear
мочкова'тый *a* fascicular, fibrillose, fibrous
мо'шка *f* blackfly, buffalo gnat, turkey gnat (*Simuliidae*)
мошо'нка *f* scrotum
мо'щность *f* power, force, vigor; capacity; depth (*of soil*)
~ **годово'го кла'сса** year-class strength
~ **поколе'ния** brood strength, year-class strength
~ **проро'стков** seedling vigor
~ **расте'ний** plant vigor
гибри'дная ~ hybrid vigor
мра'морный *a* marmorate, marble
МСОП, Междунаро'дный сою'з охра'ны приро'ды и приро'дных ресу'рсов IUCN, International Union for the Conservation of Nature and Natural Resources
му *i* moo
мужчи'на *m* male
мужско'й *a* male
мука' *f* flour, meal
бурова'я ~ *ent* wormhole dust, frass
карто'фельная ~ potato starch
костяна'я ~ bone flour
ры'бная мука fish meal, fish flour
сенна'я мука ground hay
муко'за *f* mucosa
муко'ид *m* mucoid
муко'идный *a* mucoid
муколити'ческий *a* mucolytic
мукопепти'д *m* mucopeptide
мукополисахари'д *m* mucopolysaccharide
мукопротеи'д *m* mucoprotein
му'кро *m* mucro
мул *m* mule
мулль *m* mull
мультивале'нт *m* multivalent
мультивольти'нный *a* multivoltine
мультидилю'тер *a lab* multidiluter
мультипаразити'зм *a* multiparasitism
мультиполя'рный *a* multipolar
мультисэ'мплер *a lab* multipipetter
мульча' *f* mulch, mulching
мульчи'рование *n* mulching
мульчи'ровать *v* mulch
мунтжа'к *m ther* muntjac *(Muntiacus)*
мураве'й *m* ant (*Formicidae*)
зо'нтичный ~ fungus-growing ant, gardener ant, leaf-cutting ant, parasol ant *(Atta)*
лесно'й ры'жий ~ common red ant, hill ant, red ant, wood ant, red wood ant *(Formica rufa)*
о'гненный ~ imported fire ant *(Solenopsis)*
мураве'й-бо'чка *m* plerergate
мураве'й-во'ин *m* dinergate, soldier ant

мураве'й-листоре'з *m* fungus-growing ant, gardener ant, leaf-cutting ant, parasol ant *(Atta)*
мураве'й-солда'т *m* soldier ant
мураве'йник *m* ant hill, formicary
мураве'й-рабо'чий *m* ergate
мураве'й-солда'т *m* dinergate
муравле'ние *n* anting
муравье'д *m* anteater (*Myrmecophagidae*)
муравьежу'к *m* ant beetle (*Thanasimus formicarius*)
муреинопла'ст *m* mureinoplast
мурлы'канье *n* purr
коша'чье ~ cat's purr
мурлы'кать *v* purr
мускари'н *m* muscarine
му'скул *m* muscle
дви'гательный ~ motor
замыка'ющий ~ closing muscle, adductor muscle *(in mollusks)*
мускулату'ра *f* musculature
му'скул-фикса'тор *m* fixation muscle
муст *m* must
муста'нг *m* mustang
мутаби'льность *f* mutability
мутаби'льный *a* mutable
мутаге'н *m* mutagen, mutagene
мутагене'з *m* mutagenesis
~ **ме'тодом микроинъе'кции** insertional mutagenesis
инсерцио'нный ~ insertional mutagenesis
неспецифи'ческий ~ random mutagenesis
хими'ческий ~ chemical mutagenesis
сайт-напра'вленный ~ site-directed mutagenesis, site-specific mutagenesis
сайт-специфи'ческий ~ site-directed mutagenesis, site-specific mutagenesis
мутаге'нность *f* mutagenicity
мутаге'нный *a* mutagenic, mutagenous
мута'за *f* mutase
мута'нт *m* mutant
~ **опа'л** opal mutant
~ **о'хра** ochre mutant
~ **с деле'цией** deletion mutant
амбивале'нтный ~ ambivalent mutant
ауксотро'фный ~ auxotrophic mutant, auxotroph mutant
бессмы'сленный ~ nonsense mutant
звё'здчатый ~ star mutant
конститути'вный ~ constitutive mutant
обра'тный ~ reverse mutant
опера'торно-конститути'вный ~ operator-constitutive mutant
перви'чный ~ primary mutant
пласти'дный ~ plastid mutant
«поправи'мый» ~ reparable mutant
прототро'фный ~ prototrophic mutant, prototroph mutant
поведе'нческий ~ behavioral mutant
по'лный ~ full mutant
прозра'чный ~ clear mutant, c-mutant
проса'чивающийся ~ leaky mutant
протека'ющий ~ leaky mutant
спонта'нный ~ spontaneous mutant
супресси'руемый ~ suppressor-sensitive mutant, sus mutant
температу'рный ~ temperature mutant
температурочувстви'тельный ~ temperature-sensitive mutant
усло'вно-лета'льный ~ conditional lethal mutant
фагоусто'йчивый ~ phage-resistant mutant
мута'нт-«самоуби'йца» *m imm* "suicide" mutant
мута'ция *f* mutation
~ **истоще'ния** depletion mutation
~ **коне'чной то'чки** end point mutation
~ **кру'пных отли'чий** large-scale mutation
~ **недоста'точности** loss mutation
~ **опа'л** opal mutation
~ **о'хра** ochre mutation
~ **пласти'д** ~ plastid mutation
~ **со вста'вкой** insertion mutation
~ **со сдви'гом ра'мки** phase-shift mutation, frame-shift mutation
~ **с те'чью** leaky mutation
анафрагми'ческая ~ anaphragmic mutation
бессмы'сленная ~ nonsense mutation
биохими'ческая ~ biochemical mutation
бло'ковая ~ block mutation
блужда'ющая ~ floating mutation
взаимоде'йствующие мута'ции ~ *pl* interacting mutations
внутриге'нная ~ intragenic mutation
гамети'ческая ~ gametic mutation
ге'нная ~ gene mutation
гено'мная ~ genomic mutation, ploidic mutation
гипермо'рфная ~ hypermorphic mutation
гипомо'рфная ~ hypomorphic mutation
гла'вная ~ major mutation

гомологи'чная ~ homologous mutation
зна'ковая ~ phase-shift mutation, frame-shift mutation, sign mutation
инсерцио'нная ~ insertion mutation
ключева'я ~ key mutation
коли'чественная ~ numerical mutation
комплемента'рная ~ complementing mutation
конститути'вная ~ constitutive mutation
лета'льная ~ lethal mutation
ло'жная ~ spurious mutation, false mutation
ма'ссовая ~ mass mutation
ма'лая ~ minor mutation
межге'нная ~ intergenic mutation
модифици'рующая ~ modifying mutation
молча'щая ~ silent mutation
моноло'кусная ~ micromutation
моноса'йтовая ~ single-site mutation
неви'димая ~ invisible mutation
нейтра'льная ~ neutral mutation
нерастека'ющаяся ~ nonleaky mutation
обра'тная ~ back mutation, backward mutation, reverse mutation
паралле'льная ~ parallel mutation
повто'рная ~ recurrent mutation
поля'рная ~ polar mutation
по'чковая ~ bud mutation, sport
прямая ~ direct mutation, forward mutation
регулято'рная ~ control mutation
рецесси'вная ~ recessive mutation
сайт-напра'вленная ~ site-directed mutation
систе'мная ~ systemic mutation
случа'йная ~ chance mutation
смыслова'я ~ sense mutation
сомати'ческая ~ somatic mutation, vegetative mutation
спонта'нная ~ spontaneous mutation
субвита'льная ~ subvital mutation, detrimental mutation
сублимина'льная ~ subliminale mutation
супервита'льная ~ supervital mutation
супрессо'рная ~ suppressor mutation
схисти'ческая ~ schistic mutation
термини'рующая ~ chain-terminating mutation
то'чковая ~ point mutation
хромосо'мная ~ chromosome mutation
мути'ровать *v* mutate
му'тность *n* opacity, turbidity, dullness
муто'вка *f* whorl, verticil
~ **вертициллё'зного ти'па** *myc* verticillated whorl
муто'вчатый *a* whorled, verticillate, multispiral, multicipital
муто'н *m* muton
мутуали'зм *m* mutualism
муфло'н *m* moufflon *(Ovis ammon)*
му'ха *f* fly (*Muscidae*)
~ **це-це'** tsetse, tsetse fly (*Glossina*)
ко'мнатная ~ housefly (*Musca domestica*)
лу'ковая ~ onion fly *(Delia antiqua)*
мясна'я ~ blowfly *(Calliphoridae)*
наво'зная жё'лтая common yellow dung fly *(Scatophaga stercoraria)*
нефтяна'я ~ petroleum fly (*Psilopa petrolei*)
па'дальная ~ blowfly, bluebottle (*Calliphoridae*)
се'рая мясна'я ~ common gray blowfly, flesh fly *(Sarcophaga carnaria)*
си'няя мясна'я ~ bluebottle blowfly *(Calliphora erythrocephala)*
шпа'нская ~ Spanish fly, blister beetle (*Lytta vesicatoria*)
мухоло'вка *f* flytrap; *orn* flycatcher (*Muscicapidae*)
вене'рина ~ Venus flytrap (*Dionaea*)
мухоло'вка-пестру'шка *f* pied flycatcher (*Muscicapa hypoleuca*)
мухомо'р *m* fly agaric (*Amanita muscaria*)
муци'н *m* mucin
муциноге'н *m* mucinogen
му'шка *f*: a small fly (*diminutive*)
ге'ссенская ~ Hessian fly (*Mayetiola destructor*)
зла'ковая ~ eye gnat, chlopoid fly *(Chloropidae)*
плодо'вая ~ fruit fly, pomace fly (*Drosophila*)
шве'дская ~ frit fly *(Oscinella frit)*
шпа'нская ~ Spanish fly, blister beetle (*Lytta vesicatoria*)
мушмула' *f* medlar (*Mespilus*)
мча'ться *v* race
мша'нка *f zool* moss animalcule, bryozoan, moss animalcule (*Bryozoa*); *bot* pearlwort, pearlweed (*Sagina*)
мша'ра *f* mossy swamp
мши'стый *a* mossy, moss-covered
мыча'ние *n* bellow, bellowing, moo, mooing
мыча'ть *v* low, moo

мыша'стый *a* mouse-coloured
мышеви'дные *m pl* mice
мышеви'дный *a* murine, mouselike
мыше'й *m* bristle grass (*Setaria viridis*)
мышело'вка *f* mousetrap
мышехво'стник *bot* mousetail *(Myosurus minimus)*
мы'шечно-ко'жный *a* musculocutaneous
мы'шечный *a* muscular, musculary
мыши'ные *m pl* mice
мыши'ный *a* murine
мы'шка *f* armpit; small mouse, mousey (*diminutive*)
мышле'ние *n* thinking
  абстра'ктное ~ abstract thinking
  конкре'тное ~ concrete thinking
мышкова'ть *v* catch mice (*of a fox*)
мышо'вка *f* birch mouse, bush mouse *(Sicista)*
мы'шца *f* muscle
  ~ **ве'ка** palpebral muscle
  ~ **Ивано'ва** Ivanov's muscle
  ~ **ма'тки** uterine muscle
  бы'страя ~ phasic muscle
  веретенообра'зная ~ fusiform muscle
  гла'дкая ~ nonstriated muscle, organic muscle, smooth muscle, plain muscle, unstriated muscle
  грудна'я ~ pectoral muscle
  двугла'вая ~ biceps muscle, biceps
  двупе'ристая ~ bipennate muscle
  дельтови'дная ~ acromiohumeral muscle
  жева'тельная ~ mastication muscle, masseter
  камбалови'дная ~ soleus muscle
  кольцева'я ~ circular muscle
  коса'я ~ skew muscle
  кругова'я ~ orbicular muscle
  крылови'дная ~ wing muscle
  лета'тельная ~ flight muscle
  ме'дленная ~ postural muscle
  непроизво'льно сокраща'ющаяся ~ involuntary muscle
  отводя'щая ~ abductor, abducent muscle
  папилля'рная ~ papillary muscle
  по'зная ~ postural muscle
  попере'чно-полоса'тая ~ cross-striated muscle, cross-striped muscle, striated muscle, striped muscle, voluntary muscle
  портня'жная ~ tailor's muscle
  поясни'чная ~ psoas muscle
  приводя'щая ~ adductor muscle, adductor
  продо'льная ~ longitudinal muscle
  произво'льно сокраща'ющаяся ~ voluntary muscle
  серде'чная ~ cardiac muscle, myocard, myocardium
  скеле'тная ~ skeletal muscle, somatic muscle
  сосо'чковая ~ papillary muscle
  тони'ческая ~ postural muscle
  трапециеви'дная ~ cowl muscle
  фази'ческая ~ phasic muscle
  фибрилля'рная лета'тельная ~ fibrillar flight muscle
  цилиа'рная ~ ciliary muscle
  червеобра'зная ~ ки'сти fiddle muscle
  широча'йшая ~ спины' broadest muscle of back
мы'шца-антагони'ст *f* antagonistic muscle
мы'шца-сжима'тель *f* compressor
мы'шца-синерги'ст *f* synergistic muscle
мышь *f* mouse (*Mus*)
  длинноу'хая ~ gerbil mouse *(Malacothrix)*
  домо'вая ~ house mouse (*Mus musculus*)
  жи'рная ~ fat mouse *(Steatomys)*
  ле'мминговая ~ bog mouse *(Synaptomys)*
  лесна'я ~ field mouse, wood mouse (*Apodemus sylvaticus*)
  лету'чая ~ bat (*Chiroptera*)
  полева'я ~ striped field mouse (*Apodemus agrarius*)
мышь-малю'тка harvest mouse (*Micromys minutus*)
мы'щелковый *a* condylar
мы'щелок *m* condyle
мя'гкий *a* mild, soft; malacoid
мягкопе'рый *a ich* soft-finned, soft-rayed, malacopterous
мягкопокро'вный *a* malacodermous
мягкоте'лка *f* soldier beetle, leather-winged beetle (*Cantharididae*)
мягкоте'лое *n* mollusk *(Mollusca)*
мягкоте'лый *a* malacoid
мязга' *f* squash
мяки'на *f* chaff, husk
мя'киш *m* crumb (*soft part of bread*)
мя'котный *a* medullated
мя'коть *f* pulp, squash, pap, meat, flesh
  ~ **плода'** fruit pulp
мяси'стость *f* succulence, succulency, fleshiness
мяси'стый *a* fleshy, carnose; succulent
мя'со *n* meat; flesh

мясокомбина'т *m* meat processing plant, slaughtering and meat processing plant
мясцо' *n* caruncle
сле'зное ~ lacrimal caruncle
мя'та *f* mint *(Mentha)*
мя'тлик *m* meadow-grass (*Poa*)
мяу, **мяу-мяу** *i* miaow
мяу'канье *n* mewing, miaow
мяу'кать *v* mew

Н

набе'г *m* irruption
набега'ющий *a* imbricate, imbricated
наблюда'ть *v* observe, watch
наблюде'ние *n* observation, watch; supervision
визуа'льное ~ direct observation, visual observation, sighting
враче'бное ~ medical supervision
набо'р *m* set
~ **для радиоиммуноана'лиза** radioimmunoassay kit, RIA kit
иммунодиагности'ческий ~ immune monitoring kit
хромосо'мный ~ chromosomal complement, chromosomal set
набуха'ние *n* swelling
му'тное ~ *phys* cloudy swelling
полово'е ~ perineal swelling, sexual swelling
набуха'ть *v* swell
наве'ска *f* weighed portion, batch, sample
навига'ция *f* navigation
навикуло'идный *a* naviculoid
наводне'ние *n* inundation, flood, catastrophic flood
наво'з *m* manure, dung
ко'нский ~ horse manure, horse dung
перепре'вший ~ decomposed dung
наво'зник *m ent* dung beetle (*Coprinae*); *bot* ink cap (*Coprinus*); field mushroom (*Agaricus campestris*)
навозохрани'лище *n* dung-yard, manure yard
на'вык *m* skill
дви'гательный ~ motor skill, motor habit
сенсо'рный ~ sensory habit
нагное'ние *n* suppuration, fester
нагнои'ться *v* suppurate, fester
наго'рье *n* highland
нагру'зка *f* load, loading
антропи'ческая ~, **антропоге'нная** ~ anthropogenic loading
бытова'я ~ residential load
весова'я ~ weight load
генети'ческая ~ genetic load
допусти'мая ~ allowable loading, permissible loading
мы'шечная ~ muscular exercise
па'стбищная ~ grazing pressure, stocking level
рекреацио'нная ~ recreational load
стре'ссовая ~ stress load
физиологи'ческая ~ physiological stress
физи'ческая ~ physical exercise
нагу'л *m* feeding period, fattening
нагу'ливать *v* fatten
нагу'ливаться *v* fatten
НАД, никотинами'д-адени'ндинуклеоти'д *m* NAD, nicotinamide adenin dinucleotide
надбро'вный *a* superciliary
надвисо'чный *a* supratemporal
надво'дный *a* above-water
надглазни'чный *a* supraorbital
надгло'точный *a* epipharyngeal, suprapharyngeal
надгорта'нник *m* epiglottis
надгруди'нный *a* suprasternal
наджа'берный *a* suprabranchial
надзе'мный *a* above-ground, overground, epiterranean, supraterranean, supraterraneous, superterraneous
надзо'р *m* supervision, surveillance, watch
ветерина'рный ~ veterinary inspection
иммунологи'ческий ~ immunosurveillance
санита'рный ~ health control
надкле'точный *a* supracellular
надклю'вье *n* culmen
надключи'чный *a* supraclavicular
надколе'нник *m* patella
надкостница *f* periost, periosteum
надкры'лье *n ent* elytron, elytrum, wing cover, wing sheath
надлопа'точный *a* suprascapular
надмолекуля'рный *a* supramolecular
надмы'щелковый *a* supracondylar
надмы'щелок *m* epicondyle
надоеда'ние *n* annoyance
надоеда'ть *v* annoy
надо'й *m* milk yield, yield of milk
надо'стный *a* supraspinal
надпа'зушный *a bot* supra-axillary
надпоро'говый *a* supraliminal
надпо'чечник *m* adrenal gland, adrenal body, suprarenal
надпо'чечный *a* epinephral, suprarenal
надрё'берный *a* supracostal
надре'з *m* incision, incisure, slash, insection, notch
надреза'ние *n* incision

надре'занный *a* incised
НАДФ, никотинами'д-адениндинуклеотидфосфа'т *m* NADP, nicotinamide adenin dinucleotide phosphate
надхвостово'й *a* supracaudal
надчерепно'й *a* supracranial
надчре'вный *a* epygastric
нажа'тие *n* pressing
~ **на рыча'г** *eth* lever-pressing
нажи'вка *f* living bait
наживля'ть *v* bait (*in fishing*)
наза'льный *a* nasal
назва'ние *n* name
видово'е ~ specific name
ме'стное ~ vernacular name
наро'дное ~ popular name, common name, trivial name
отбро'шенное ~ rejected name
приго'дное ~ valid name
примени'мое ~ valid name
при'нятое ~ accepted name
родово'е ~ generic name
назе'мный *a* terrestrial, terraneous
на'зион *m* nasal point
называ'ть *v* name
нака'лывание *n* pinning
нака'пливание *n* accumulation
наки'дка *f* cast net
накла'дывать *v*: put on; superimpose; load
~ **лигату'ру** ligate
~ **повя'зку** apply a bandage
~ **швы** put in stitches, suture
накле'йка *f* label; **накле'йка этике'ток** labeling labelling
накло'нный *a* inclined
накова'льня *f* anvil bone, incus
накома'рник *m* mosquito net
накопле'ние *n* accumulation
~ **о'бщего азо'та** total nitrogen production
~ **сухо'го вещества'** dry weight production
сопряже'нное ~ coupling accumulation
налега'ющий *a* incumbent
налё'т *m* bloom
восково'й ~ waxy bloom, wax coating, bloom *(on fruit)*
мучни'стый ~ mealy bloom, powdery bloom, powdery coating *(on fruit)*
са'жистый ~ sooty bloom
налива'ться *v* ripen (*of fruit*), form (*of grain*)
налито'й *a* laden; ripe
нали'чник *m ent* clypeus
нали'чниковый *a* clypeal
наложе'ние *n* superposition, imposition, application
~ **бандажа'** bandaging
~ **лигату'ры** ligation
~ **шва** suture
~ **ши'ны** splinting
~ **щипцо'в** application of forceps
намо'рдник *m* muzzle
нани'зм *m* nanism, dwarfishness, nanosomia
наноподзо'л *m* nanopodzol
нанофанерофи'т *m* nanophanerophyte
напада'ть *v* attack, strike
нападе'ние *n* attack, offence; infestation
наперстя'нка *f* foxglove (*Digitalis*)
напи'тывать *v* imbue
наплы'в *m* burr, burl, callus, excrescence
корнево'й ~ root swelling
наполне'ние *n* filling
диастоли'ческое ~ diastolic filling
наполни'тель m bulking agent
направи'тельный: направительное те'льце directing body, polar body, polocyte
направле'ние *n* trend, direction
~ **скре'щивания** direction of cross
~ **эволю'ции** evolutionary trend
напряга'ть *v* exert
напряже'ние *n* exertion, effort; voltage
напыле'ние *n* sputtering
напылё'нный мета'ллом metal-shadowed
нарко'з *m* narcosis, anesthesia
наркома'ния *f* narcomania
нарко'тик *m* narcotic
наркоти'ческий *a* narcotic
наро'ст *m* gnarl (*on a tree*), knag, snag; excrescence, wart
нару'жное *n a* medicine for external use
нару'жный *a* external, exterior, outer, ectal
наруше'ние *n* disturbance, disorder, failure, distortion
~ **восприя'тия** imperception
~ **обме'на веще'ств** metabolic disorder
~ **поведе'ния** behavior disorder
~ **ро'ста** growth disorder, growth disturbance
~ **соотноше'ния** ratio distortion
нарци'сс *m* daffodil (*Narcissus*)
~ **ло'жный** daffodil *(Narcissus pseudonarcissus)*
нары'в *m med* abscess
нары'вник *m* blister beetle, oil beetle (*Meloidae*)

наря'д *m* dress, livery
бра'чный ~ nuptial dress, breeding color, spawning dress, spawning livery
демонстрати'вный ~ display costume
насади'ть *v* plant; pin, stick
наса'дка *f* bait
насажде'ние *n for* plantation, stand, planting
~ **с просве'тами** gappy stand, incomplete stand
госпо'дствующее ~ principal stand, main stand
иску'сственное ~ artificial stand
матери'нское ~ nurse-wood
многоя'русное ~ multistoried stand
одново'зрастное ~ even-aged stand, uniform stand
одноя'русное ~ single-storied stand
откры'тое ~ open stand, isolated stand, light stand
пересто'йное ~ overmatured stand
подчине'нное ~ secondary stand
по'лное ~ fully stocked stand, complete stand
разново'зрастное ~ all-aged stand, unevened stand
сме'шанное ~ mixed stand
со'мкнутое ~ closed stand, dense stand
спе'лое ~ mature stand
чи'стое ~ pure stand
наса'живать *v* pin, stick; mount
~ **нажи'вку** bait
насе'дка *f* brood hen, brooding hen, clucking hen, sitting hen, sitter
насеко'мое *n* insect (*Insecta*)
вре'дное ~ injurious insect, noxious insect, destructive insect
первичнобескры'лое ~ apterous insect *(Apterygota)*
поле'зное ~ beneficial insect
скрыточелюстно'е ~ entognathous insect *(Entognatha)*
кровососу'щее ~ blood-sucking insect
хи'щное ~ predacious insect
насеко'мое-вреди'тель *n* injurious insect, noxious insect
насеко'мое-вреди'тель зерна' *n* grain insect
насеко'мое-галлообразова'тель *n* gall-producing insect
насекомоопыля'емый *a* entomophilous
насекомоя'дное *n* insectivore, insect-eater *(Insectivora)*
насекомоя'дный *a* entomophagous, insectivorous, insect-eating
населе'ние *n* population
населя'ть *v* populate
насе'ст *m orn* roost, henroost, perch
насечё'нный *a* indented
насе'чка *f* incisure, insection
~ **Шми'дта-Лантерма'на** Schmidt-Lantermann incisure
наси'женный *a* set, incubated
наси'живать *v* sit
насле'дование *n* inheritance
~ **при'знаков** inheritance of characters
альтернати'вное ~ alternative inheritance
ограни'ченное по'лом ~ sex-limited inheritance
полисо'мное ~ polysomic inheritance
сцеплё'нное с по'лом ~ sex-linked inheritance
насле'довать *v* inherit
насле'дственность *f* heredity, inheritance
~ **коли'чественных при'знаков** quantitative inheritance
аллосо'мная ~ allosomal inheritance
ауксоплазмати'ческая ~ auxoplasmatic inheritance
внехромосо'мная ~ extrachromosomal inheritance
внея'дерная ~ extranuclear inheritance
дискре'тная ~ particulate inheritance
ка'чественная ~ qualitative inheritance
коли'чественная ~ quantitative inheritance
матери'нская ~ maternal inheritance, maternal inheritance
матрокли'нная ~ matroclinal inheritance
мендели'рующая ~ Mendelian inheritance
многофа'кторная ~ multiple inheritance
модификацио'нная ~ modification inheritance
неге'нная ~ nongenic inheritance
однолине'йная ~ unilinear inheritance
патрокли'нная ~ patroclinal inheritance
перекрё'стная ~ criss-cross inheritance
плазмати'ческая ~ plasmatic inheritance
пласти'дная ~ plastid inheritance
полиге'нная ~ polygenic inheritance
сли'тная ~ blending inheritance
сме'шанная ~ blended inheritance
цитоплазмати'ческая ~ cytoplasmic inheritance
насле'дственный *a* hereditary, inheritable
насле'дуемость *f* heritability

насле'дуемый *a* hereditable, heritable, inheritable
на'сморк *m* rhinitis, nasal cold
си'льный ~ heavy cold
насо'с *m* pump; pompe
ио'нный ~ ionic pump
ка'лиевый ~ potassium pump
катио'нный ~ cation pompe, cation pump
метаболи'ческий ~ metabolic pompe
на'триевый ~ sodium pompe, sodium pump
наст *m* smow crust
наста'ивать *v* infuse
насти'ческий *a* nastic
на'стия *f* nastic movement
насто'й *m* infusion
насто'йка *f* tincture
настора'живать *v* set (*of a trap*)
настороже' *adv* on the alert; **быть** ~ be on guard
настор'женность *f* vigilance, alertness
настор'женный *a* shy, alert; set, wary
настор'жка *f* trigger
настоя'щий *a* natural
насту'рция *f* nasturtium (*Tropaeolum*)
насыща'ть *v* imbue
насыще'ние *n* saturation
~ **вла'гой** imbibition
светово'е ~ **фотоси'нтеза** light saturation of photosynthesis
насы'щенность *f* saturation
насы'щенный *a* saturated; packed (*of a cenosis*)
ната'льный *a* natal
ната'скивать *v* train; drag in
нати'вный *a* native
натоща'к adverb on empty stomach
натра'вливать *v* set on
нату'ра *f* nature; **плати'ть нату'рой** pay in kind
натурализа'ция *f* naturalization
натурали'ст *m* naturalist
ю'ный ~ young naturalist
натура'льный *a* natural
натяже'ние *n* tension
пове'рхностное ~ surface tension
науплиа'льный *a* naupliar
на'уплиус *m* nauplius
нахожде'ние *n* occurrence
нахо'хлиться *v* ruffle up, erect the feathers
нача'ло *n* principle
де'йствующее ~ active principle, effective agent
нача'льный *a* initial, incipient
начина'ть *v* initiate
начина'ющийся *a* incipient
наше'ствие *n* irruption
~ **саранчи'** grasshopper plague
начина'ющийся *a* incipient
наше'ствие *n* incursion, invasion, infestation
ная'да *f ent* naiad; *bot* naias *(Najas)*
неадаптациони'зм *m* nonadaptionism
неадапти'вный *a* nonadaptive
неадеква'тность *f* inadequacy
~ **раздраже'ния** stimulus inadequacy
неадеква'тный *a* inadequate
неакти'вный *a* inactive, indolent
неантиге'нный *a* nonantigenic
неаперту'рный *a* nonaperturate
неаркти'ческий *a* nearctic
небелко'вый *a* nonprotein
неблагоприя'тный *a* adverse, inimical, unfavorable
не'бный *a* palatal, palatine
нё'бо *n* palate, roof of mouth
мя'гкое ~ soft palate
твё'рдое ~ hard palate
небонитиро'ванный *a* unculled
невесо'мость *f* weightlessness, zero-gravity state, imponderability
невесо'мый *a* imponderable
неветвя'щийся *a* nonbranching
неви'димый *a* invisible
невозбуждё'нный *a* unexcited
невозобнови'мый *a* unrenewable
нево'льный *a* involuntary
нево'ля *f* captivity
невооруже'нный *a* unarmed, inermous; ~ **глаз** naked eye
невосприи'мчивость *f* insusceptibility, insensibility, immunity
невосприи'мчивый *a* insensible, insensitive, unsusceptible
невоспринима'емый *a* inconceivable
невралги'я *f* neuralgia
невра'льный *a* neural
невриле'мма *f* neurilemma
неври'т *m* neurite
невро- neuro-
невроглия'я *f* neuroglia
невродендро'н *m* neurodendron
невро'з *m* neurosis
невроло'гия *f* neurology, neuroscience

невромери'я *f* neuromery
невромио'н *m* neuromion
невроме'р *m* neuromere
невропатоло'гия *f* neuropathology
невропи'ль *m* neuropil, neuropile
невропо'дий *m* neuropodium
невроце'ль *m* neurocoel
невроци'т *m* neurocyte, nerve cell
  многомо'рфный ~ multiform neurocyte
невсхо'жесть *f* nongermination
невыжива'ние non-survival
  ~ **неприспосо'бленных** non-survival of the non-fit
невыно'сливый *a* nonhardy, nonresistant
невыявля'емый *a* undetectable
негнездово'й *a* nonbreeding
него'дный *a* unfit
негомоге'нный *a* inhomogeneous
недви'жимый *a* immobile
неделя'щийся *a* nondividing
недержа'ние *n* incontinence
недиссоции'рованный *a* undissociated
недифференци'рованный *a* undifferentiated
недиффунди'рующий *a* indiffusible
недо- in-, pre-, un-, under-
недоеда'ющий *a* underfed
недозре'лый *a* unripe, underripe
недока'рмливание *n* underfeeding, undernourishment
недоло'в *m* underfishing
недомога'ние *n* indisposition, fatigue
недоно'шенный *a* premature
недооце'нка *f* underestimation
недоразви'тие *n* aplasia
недора'звитый *a* undeveloped, immature, rudimentary
недореплика'ция *f* underreplication
недоро'д *m* poor harvest, poor crop, harvest failure, crop failure, shortcoming of a crop
недоста'точность *f* incompetence, incompetency, insufficiency, failure, deficiency, inadequacy, deficit
  ~ **пита'ния** malnutrition
  минера'льная ~ mineral deficiency
  пищева'я ~ dietary deficiency, nutritional deficiency
  функциона'льная ~ incompetence
недоста'точный *a* inefficient
недостове'рный *a* unreliable
недосту'пный *a* unavailable, impervious
недотро'га *f* snapweed (*Impatiens*)
не'дра *pl* underground, mineral resources
  ~ **земли'** mineral resources, the bowels of the earth
недроби'вшийся *a* uncleaved
неду'г *m* disease, ailment
нежела'тельный *a* undesirable
нежё'сткий *a* nonrigid
неживо'й *a* nonliving
нежизнеспосо'бный *a* inviable, nonviable, unvital
незабу'дка *f* forget-me-not (*Myosotis*)
незави'симый *a* independent
незавяда'ющий *a* nonwilting
незагрязнё'нный *a* uncontaminated
незакалё'нный *a* nonhardened
незако'нченный *a* imperfect
незамени'мый *a phys* essential
незаме'тный *a* imperceptible, inconspicuous
незаражё'нный *a* noninfected, uncontaminated; noninoculated
незаса'женный *a* unplanted
незасолё'нный *a* nonsaline *(of soil)*
незащище'нный *a* unprotected; outdoor (*of ground*)
нездоро'вый *a* unhealthy
незиготи'ческий *a* azygoid
низкоро'слый *a* undersized
незре'лость *f* immaturity
  функциона'льная ~ functional immaturity
незре'лый *a* unripe, green; immature
неизбира'тельный *a* nonselective
неизоме'рный *a* anisomerous
неиммуноге'нный *a* nonimmunogenic
неиндуци'бельность noninducibility
неинкапсули'рованный *a* unencapsulated
неинфици'рованный *a* uncontaminated
неисправи'мый *a* irreclaimable
неисче'рченный *a* unstriated
нейри'т *m* axon
нейро- neuro-
нейробио'лог *m* neuroscientist
нейробиоло'гия *f* neurobiology, neuroscience
нейробла'ст *m* neuroblast
нейроге'нный *a* neurogenic, neurogenous
нейрогипо'физ *m* neurohypophysis
нейрогу'мор *m* neurohumor
нейрогумора'льный *a* neurohumoral
нейрализа'ция *f* neuralization
нейроиммуноло'гия *f* neuroimmunology
нейроиммунохи'мия *f* neuroimmunochemistry
нейрокри'нный *a* neurocrine
нейро'лиз *m* neurolysis

нейромедиа'тор *m* neurotransmitter
нейромото'рный *a* neuromotor
нейро'н *m* neuron, neurone, neurocyte, nerve unit
  адренэрги'ческий ~ adrenergic neuron
  аффере'нтно-эффере'нтный ~ bipolar neuron
  аффере'нтный ~ afferent neuron
  биполя'рный ~ bipolar neuron
  бы'стрый ~ fast neuron
  возбужда'ющий ~ excitatory neuron
  вста'вочный ~ internuncial neuron
  ГАМК-эрги'ческий ~ GABA-ergic neuron
  дви'гательный ~ motor neuron
  за'лповый ~ bursting neuron
  ко'рково-геникуля'рный ~ corticogeniculate neuron
  коро'ткий адренэрги'ческий ~ short adrenergic neuron
  кортикотекта'льный ~ corticotectal neuron
  постганглиона'рный ~ postganglionic neuron
  регули'рующий ~ adjustor neuron
  рецепто'рный ~ receptor neuron
  ретикулоспина'льный ~ reticulospinal neuron
  ретикуля'рный ~ reticular neuron
  сенсо'рный ~ sensory neuron
  тормозно'й ~ inhibitory neuron
  универса'льный па'узовый ~ omnipause neuron
  униполя'рный ~ unipolar neuron
  центра'льный моноаминсодержа'щий ~ central monoamine neuron
  центробе'жный ~ efferent neurone
  центростреми'тельный ~ afferent neuron
  чувстви'тельный ~ sensory neuron
  эффере'нтный ~ efferent neuron
нейропепти'д *m* neuropeptide
нейропи'ль *m* neuropil, neuropile
нейропла'зма *f* neuroplasm
невропо'дий *m* neuropodium
нейропо'р *m* neuropore
нейросекрето'рный *a* neurosecretory
нейросекре'ция *f* neurosecretion
нейроси'напс *m* neurosynapse
нейротокси'н *m* neurotoxin
нейротропи'зм *m* neurotropism
нейротрофи'ческий *a* neurotrophic
нейрофибри'лла *f* neurofibril, neurofibrilla
нейрофизиоло'гия *f* neurophysiology
нейроце'птор *m* neuroceptor
нейроэпите'лий *m* neuroepithelium
нейроэтоло'гия *f* neuroethology
не'йрула *f* neurula
нейруля'ция *f* neurulation
нейтрализа'тор *m* neutralizer
нейтрализа'ция *f* neutralization
  перекрё'стная ~ cross neutralization
нейтрали'зм *m* neutralism
нейтрализова'ть *v* neutralize
нейтра'льный *a* neutral
нейтропени'я *f* neutropenia
нейтрофи'л *m* neutrophil, neutrophilic leukocyte
  поляризо'ванный ~ polarized neutrophil
  полимо'рфно-я'дерный ~ polymorphonuclear neutrophil
  розеткообразу'ющий ~ rosette-forming neutrophil
  сегментноя'дерный ~ segmented neutrophil
  палочкоя'дерный ~ stab neutrophil, rod neutrophil, banded neutrophil
нейтрофили'я *f* neutrophilia
нейтрофи'льный *a* neutrophilic, neutrophilous
некатоpo'з *m helm* necatoriasis
некле'точный *a* noncellular
неклубенько'вый *a* nonnodulating
некоагули'руемость *f* incoagulability
неколю'чий *a* inermous
неконтроли'руемый *a* uncontrollable
неконъюги'рованный *a* unconjugated
некоопериати'вность *f mbiol* noncooperativity
некра'сящийся *a* achromatophilous, achromophilous
некристаллизу'ющийся *a* uncrystallizable
некробио'з *m* necrobiosis
некробиоти'ческий *a* necrobiotic
некро'з *m* necrosis
некро'лиз necrolysis
некрота'ксис *m* necrotaxis
некроти'ческий *a* necrotic
некрофа'г *m* necrophage
некрофа'гия necrophagy
некрофи'льный *a* necrophilous
некроцито'з necrocytosis
некси'н *m* nexin
  протеолити'ческий ~ protease nexin
некта'р *m* nectar

нектари'н *m* nectarine (*Persica nucipersica*)
некта'рник *m* nectary, nectar gland, honeycup
некта'рный *a* nectarous
нектароно'сный *a* nectariferous
некто'н *m* nekton
нектосо'ма *f* nectosome
нектофо'р *m* nectophore
некультиви'руемый *a* uncultivated *(of plants)*
нелиня'ющий *a* nonmolting
нелистопа'дный *a* nondecidous
немаме'р *m* nemamere
немате'ций *m* nemathecium
нематобла'ст *m* nematoblast
нематоге'нный *a* nematogene
немато'да *f* nematode, threadworm (*Nematoda*)
  карто'фельная ~ golden nematode, potato cyst nrmatode (Heterodera rostochiensis)
  пшени'чная ~ wheat nematode (Anguina tritici)
  свеклови'чная ~ beet cyst nematode, sugarbeet nematode (Heterodera schachtii)
нематодо'з *m hel m* nematodosis
нематодоло'гия *f* nematology
нематоци'ста *f* nematocyst
неметаме'рный *a* nonmetameric
неме'ченный *a* unlabeled, untagged
немигри'рующий *a* nonmigratory
ненадё'жный *a* unreliable
ненапра'вленный *a* nondirectional
ненасле'дуемый *a* nonheritable
ненасы'щенный *a* unsaturated; unfilled *(of biotope)*
ненорма'льный *a* abnormal, anomalous
нео- neo-
неоантиге'н *m* neoantigen
необла'ст *m* neoblast
необнару'живаемый *a* undetectable
необою'дный *a* irreciprocal
необрабо'танный *a* untreated, nontreated, uncultivated, undeveloped
необраста'ющий *a* antifouling
необрати'мость *f* irreversibility, inconvertibility
  ~ **эволю'ции** irreversibility of evolution
необрати'мый *a* irreversible, nonreversible
необъе'зженный *a* unbroken, unhandled, green (*of a horse*)
неоге'я *f* Neogaea
неограни'ченный *a* unrestricted
неодарвини'зм *m* Neo-Darwininsm
неодина'ковость *f* inequality
неоднозна'чность *f* ambiguity
  ~ **антикодо'на** anticodon wobble
  ~ **кодо'на** codon wobble
неодноро'дность *f* heterogeneity
неодноро'дный *a* inhomogeneous, nonhomogeneous; impure
неокаймлё'нный *a bot* immarginate
неока'рпия *f* neocarpy
неокинети'ческий *a* neokinetic
неокисля'ющийся *a* inoxidable
неокра'шенный *a* uncolored
неокра'шивающийся *a* nonstaining
неокультуренный *a* uncultivated *(of soil)*
неоламарки'зм *m* Neo-Lamarckism
неомо'рф *m* neomorph
неомо'рфный *a* neomorphic
неоморфо'з *m* neomorphosis
неомыля'ющийся *a* unsaponifiable
неопада'ющий *a* indeciduous, nondeciduous
неоперё'нный *a* unfledged
неоплази'я *f* neoplasia
неопла'зма *f* neoplasm, newgrowth
неопласти'ческий *a* neoplastic
неоплодотворё'нный *a* unfertilized
неопределё'нный *a* indefinite, indeterminate, neutral
неопредели'мый *a* indefinable
нео'птиль *m* neoptile, neossoptile
неопушё'нный *a bot* impuberal
неоргани'ческий *a* inorganic
неосвоенный *a* uncultivated, undeveloped *(of land)*
неося'заемый *a* impalpable
неотбрако'ванный *a* unculled
неотвеча'емость *f* nonresponsiveness
  иммунологи'ческая ~ unresponsiveness
неотени'ческий *a* neotenic
неотени'я *f* neoteny, neotenia, neoteinia
неоти'п *m* neotype
неотипово'й *a* neotype
неотропи'ческий *a* neotropical
неотсортиро'ванный *a* ungraded
неофили'я *f eth* neophilia
неофи'т *m* neophyte
неофо'бия *f eth* neophobia
неоцентроме'ра *f* neocentromere
неоци'т *m* neocyte
неочи'щенный *a* unpurified, unrefined, impure, uncleaned
неощути'мый *a* impalpable

неоэндеми'зм *m* neo-endemism
непарнокопы'тное *n* odd-toed ungulate *(Perissodactyla)*
непарнокопы'тный *a* odd-toed, solid footed, solidungulate, soliped, perissodactyl
непарнопа'лый *a* odd-toed, inpairidigitate, perissodactyl
непарнопе'ристый *a bot* imparipinnate, odd-pinnate
непа'рный *a* unpaired, impar, azygous; **непа'рный шелкопря'д** gypsy moth (*Ocneria dispar*)
непатоге'нный *a* nonpathogenic
непаху'чий *a* scentless
неперева'ренный *a* undigested
неперевари'мый *a* indigestible
неперекрыва'ние *n* nonoverlapping
неплеменно'й *a* nonbreeding
неплодови'тость *f* infertility, infecundity
неплодоно'сный *a* nonbearing, unfruitful
неплодоро'дие *n* infertility, infecundity
неплодоро'дный *a* unfruitful, infertile, nonfertile, infecund; nonbearing *(of soil)*
неповреждё'нный *a* noninjured, intact, unbroken
неподви'жность *f* immobility
неподви'жный *a* immobile, immovable, nonmobile, nonmotile, motionless, stationary
неподкрепля'емый *a phys* unrewarded
неподстри'женный *a* uncut
неподходя'щий *a* unfit
непокры'тый *a* uncovered
неполнозу'бый *a* edentate
неполнота *f* incompleteness, imperfection
~ **геологической летописи** imperfection of the geological record
неполноце'нный *a* inadequate
непо'лный *a* incomplete, semicomplete, imperfect
неполовозре'лый *a* impuberal
неполя'рный *a* nonpolar
непоражё'нный *a* unaffected
непо'ртящийся *a* nonperishable
непоса'женный *a* unplanted
непостоя'нный *a* inconstant
непра'вильный *a* irregular
непреры'вный *a* sustained
неприе'млемый *a* undesirable
неприме'тный *a* inconspicuous
неприми'рованный *a imm* unprimed
неприру'ченный *a* untaimed, tameless, feral
неприспособи'тельный *a* inadaptive
неприспосо'бленность *f* inadaptability
неприхотли'вый *a* undemanding, unpretentious
непродукти'вный *a* unproductive
непроизводи'тельный *a* unproductive
непроизво'льный *a* involuntary, unintentional, unprompted
непроница'емость *f* impenetrability, impermeability
непроница'емый *a* impervious, impermeable
непротеи'новый *a* nonprotein
непротра'вленный *a* untreated *(of seeds)*
непроходи'мость *f* impassability, obstruction
непроходи'мый *a* impervious
непроце'женный *a* unstrained
непу'ганный *a* tame
нера'венство *n* inequality
неравнобо'кий *a* anisopleural
неравнокры'лый *a* anisopterous
неравноли'стность *f* anisophylly
неравноли'стный *a* anisophyllous
неравноло'пастный *a bot* anisolobous
неравноме'рность *f* inequality
неравноме'рный *a* inhomogeneous, irregular
неравнопле'чий *a* heterobrachial
неравнополя'рный *a* anisopolar
неравностворчатый *a* inequivalve
неравночле'нный *a* anisomerous
неразба'вленный *a* undiluted
неразведё'нный *a* undiluted
неразветвлё'нный *a* unbranched
неразличи'мый *a* invisible
нераскрыва'ющийся *a* indehiscent
нераствори'мый *a* insoluble, indissoluble, nonsoluble
нерастворя'ющийся *a* undissolving
нерастре'скивающийся *a* indehiscent
нерасхожде'ние *n* nondisjunction
~ **хромати'д** chromatid nondisjunction
~ **при мито'зе** mitotic nondisjunction
нерасчленё'нный *a* nonsegmented
нерв *m* nerve
~ **Ге'ринга** cardiac sinus nerve
~ **кароти'дного си'нуса** cardiac sinus nerve
~ **преддве'рия** vestibular nerve
~ **се'рдца** cardiac nerve
~ **преддве'рия** vestibular nerve
~ **Цио'на-Лю'двига** depressor nerve
адренерги'ческий ~ adrenergic nerve
аффере'нтный ~ afferent nerve

бе'дренный ~ femoral nerve
блужда'ющий ~ pneumogastric nerve, vagus nerve
бранхиоме'рный ~ branchiomeric nerve
вазомото'рный ~ vasomotor nerve
вегетати'вный autonomic nerve
вестибуля'рный ~ vestibular nerve
вкусово'й ~ gustatory nerve
возвра'тный ~ recurrent nerve
глазни'чный ~ ophthalmic nerve
глазодви'гательный ~ oculomotor nerve
гло'точный ~ pharyngeal nerve
горта'нный ~ laryngeal nerve
грудно'й ~ thoracic nerve
дви'гательный ~ motor nerve, somatic nerve
жа'берный ~ branchiomeric nerve
жева'тельный ~ masticator nerve
зри'тельный ~ ophthalmic nerve
ко'жный ~ cutaneous nerve
крестцо'вый ~ sacral nerve
лицево'й ~ facial nerve
межко'стный ~ interosseous nerve
межре'берный ~ intercostal nerve
обонятельный ~ olfactory nerve
отводя'щий ~ abducent nerve
перифери'ческий ~ peripheral nerve
подчре'вный ~ hypogastric nerve
подъязы'чный ~ hypoglossal nerve
поясни'чный ~ lumbar nerve
пре'ссорный ~ pressor nerve
седа'лищный ~ sciatic nerve
секрето'рный ~ secretory nerve
семенно'й ~ spermatic nerve
сердечный ~ cardiac nerve
симпати'ческий ~ sympathetic nerve
синокароти'дный ~ carotid sinus nerve
слухово'й ~ acoustic nerve, auditory nerve
сосудодвигательный ~ vasomotor nerve
сосудорасширя'ющий ~ vasodilatator nerve
сосудосу'живающий ~ vasoconstrictor nerve
спинномозгово'й ~ spinal nerve
тормозно'й ~, тормозя'щий ~ inhibitory nerve
тройни'чный ~ trigeminal nerve
трофи'ческий ~ trophic nerve
трохлеа'рный ~ trochlear nerve
ушно'й ~ auricular nerve
холинерги'ческий ~ cholinergic nerve
центробе'жный ~ efferent nerve, exodic nerve
центростреми'тельный ~ afferent nerve, esodic nerve
че'репно-мозгово'й ~ cranial nerve
чувстви'тельный ~ sensory nerve
ше'йный ~ cervical nerve
эффере'нтный ~ efferent nerve
языкогло'точный ~ glossopharyngeal nerve
язы'чный ~ lingual nerve
яре'мный ~ jugular nerve

нерв-депре'ссор *m* depressor nerve
не'рвно-кише'чный *a* neurenteric
не'рвно-мы'шечный *a* neuromuscular, nervimuscular
не'рвно-сосу'дистый *a* neurovascular
не'рвный *a* nerval, neural, nervous
нервомото'рный *a* nervimotor
нерво'н *m* nervone
нервю'ра *f* nervure
нереаги'рование *n* nonresponsiveness
нереаги'рующий *a* nonreactive
нерегенери'рованный *a* unreclaimed
нерегуля'рный *a* inconstant, irregular; ~ **пульс** irregular pulse
нерезисте'нтный *a* nonresistant
неро'вный *a* uneven, rough; erratic (*e.g., of flight*); ~ **пульс** irregular pulse
неро'дственный *a* unrelated
нерожа'вшая *a f* nulliparous
не'рпа *f ther* seal *(Pusa)*
байка'льская ~ Baikal seal *(Pusa sibirica)*
каспи'йская ~ Caspian seal *(Pusa caspia)*
ко'льчатая ~ ringed seal, jar seal, rough seal *(Pusa hispida)*

несбаланси'рованность *f* imbalance
несбра'живаемый *a* unfermentable
несвё'ртываемость *f* incoagulability
несветя'щийся *a* nonluminous
несвя'занный *a* unconjugated
несегменти'рованный *a* inarticulate, nonsegmented
неселекти'вный *a* nonselective
несимметри'чный *a* asymmetrical
неско'шенный *a* uncut
несма'чиваемость *f* nonwettability
несма'чивающий, несма'чивающийся *a* nonwetting
несме'шанный *a* unblended
несме'шиваемый *a* immiscible
несме'шивающийся *a* immiscible

несмещё'нный *a biom* unbiased
несоверше'нный *a* imperfect
несовмести'мость *f* incompatibility, disparity
антиге'нная ~ antigenic disparity
гамети'ческая ~ gametic incompatibility
гибри'дная ~ cross incompatibility
иммуногенети'ческая ~ immunogenetic disparity
иммунологи'ческая ~ immunogenetic disparity
протоплазмати'ческая ~ protoplasmatic incompatibility
экологи'ческая ~ ecological incompatibility
несовмести'мый *a* incompatible
несогласу'ющийся *a* incompatible
несоотве'тствие *n* inadequacy
несоотве'тствующий *a* inadequate
несочлене'нный *a* inarticulate
неспе'лый *a* unripe, green, immature
неспорообразу'ющий *a* asporous
неспосо'бность *f* inability
неспосо'бный *a* inefficient
несре'занный *a* uncut
нестаби'льность *f* instability
нестаби'льный *a* unstable
нестерилизо'ванный *a* unsterilized
нестери'льный *a* unsterile
несто'йкий *a* nonpersistent, nonhardy, unstable
несто'мчивость *n* perseverance *(of a dog)*
нестрелку'ющийся *a* nonbolting *(e.g., of onions)*
несу'шка *f* laying bird, laying fowl, laying hen, hen in lay
несхо'дство *n* inequality
несъедо'бный *a* inedible, uneatable, not edible
нетокси'чный *a* nontoxic
нетопы'рь *m* pipistrelle (*Pipistrellus*)
нетро'нутый *a* pristine, virgin, undisturbed, unaffected
нетрудоспосо'бность *f* disablement, disability, incapacity, invalidity
неувяда'ющий *a* nonwilting
неудовлетвори'тельный *a* insufficient
неукорени'вшийся *a* nonrooted
неукроти'мый *a* uncontrollable, incoercible; **неукроти'мая рво'та** uncontrollable vomiting
неуравнове'шенность *f* imbalance
неурожа'й *m* bad crop, bad harvest, crop failure, poor crop, harvest failure, failure of crop
неусва'иваемый *a* inassimilable, nondigestible
неусвоя'емый *a* inassimilable, nondigestible
неустанови'вшийся *a* immature *(of a feature)*
неусто'йчивость *f* instability
неусто'йчивый *a* inconstant, labile
нефело'метр *m* nephelometer
нефелометри'я *f* nephelometry
неферменти'руемый *a* unfermentable
нефильтро'ванный *a* unstrained
неформалинизи'рованный *a* unformalinized
нефотосинтези'рующий *a* nonphotosynthetic
нефридиа'льный *a* nephridial
нефри'дий *m* nephridium
нефридиопо'р *m* nephridiopore, nephropore
нефри'т *m* nephritis
нефроге'нный *a* nephrogenic
нефроми'ксий *m* nephromixium
нефро'н *m* nephron
нефросто'м *m* nephrostome
нефрото'м *m* nephrotome, nephromere
нефроце'ль *m* nephrocoel
нефроци'т *m* nephrocyte
нецветко'вый *a* nonflowering
нехи'щный *a* nonpredatory
нечё'ткий *a* indistinct
нечё'тный *a* odd
нечувстви'тельность *f* insensitivity, insensibility, insusceptibility
нечувстви'тельный *a* insensible, insensitive, unsusceptible
~ **к бо'ли** indolent
неэффекти'вный *a* inefficient
нея'вный *a* implicit
неядови'тый *a* nontoxic, innocuous, innoxious
нея'ркий *a* inconspicuous *(of color)*
нея'сный *a* indistinct
нея'сыть *f* barred owl (*Strix varia*)
длиннохво'стая ~ Ural owl *(Strix uralensis)*
пё'страя ~ barred owl *(Strix varia)*
рога'тая ~ crested owl *(Lophostrix cristata)*
ниаци'н *m* niacin
нигрози'н *m* nigrosin
нидамента'льный *a* nidamental

нида'ция *f* nidation
нижнебоково'й *a* inferolateral
нижнеза'дний *a* inferoposterior
нижнемедиа'нный *a* inferomedian
нижнепере'дний *a* inferoanterior
нижнесреди'нный *a* inferomedian
ни'жний *a* inferior
низбега'ющий *a* decurrent, peripterous
низи'на *f* lowland, low country
низкоро'слость *f* stunting, dwarfism
низкоро'слый *a* dwarfish, short, shortish, low-growing, undersized, stunted
низкостово'льник *m* shrubwood, coppice wood
низкотра'вье *n* short grass
никоти'н *m* nicotine
никотинами'д *m* nicotinamide
никотинамид-адени'ндинуклеоти'д, НАД *m* nicotinamide adenin dinucleotide, NAD
никотинамид-адени'ндинуклеоти'дфосфа'т, НАДФ *m* nicotinamide adenin dinucleotide phosphate, NADP
никтина'стия а nyctinasty
ни'мфа *f ent* nymph
нимфа'льный *a* nymphal
ниста'гм *m* nystagmus
  диссоции'рованный ~ incongruent nystagmus
  интенцио'нный ~ end-position nystagmus
  клони'ческий ~ jerky nystagmus
  оптокинети'ческий ~ optokinetic nystagmus
  послевраща'тельный ~ postrotational nystagmus
  толчкообра'зный ~ jerky nystagmus
  устано'вочный ~end-position nystagmus
нисходя'щий *a* decurrent
нитеви'дный, нитеобра'зный *a* filamentous, filiform, filose
нитрифика'тор *m* nitrifier
нитрифика'ция *f* nitrification
нитрифици'ровать nitrify
нитроба'ктер *m* nitrobacter
нитрова'ние nitration
  окисли'тельное ~ oxidative nitration
нитрозифика'ция *f* nitrosification
нитро'лиз *m* nitrolysis
нитрофи'л *m* nitrophil
нитрофи'льный *a* nitrophilous
нитрофо'ска *f* nitrophoska
нитча'тка *f* filamentous alga, conferva
ни'тчатый *a* filamentous, filiform, filose
нить *f* filament, fiber, thread, strand
  концева'я ~ terminal filament
  мезентериа'льная ~ mesenteric filament
  тычи'ночная ~ stamen filament
  хромосо'мная ~ chromosome thread
ни'ша *f* niche
  пищева'я ~ food niche
  пусту'ющая ~ vacant niche
  экологи'ческая ~ ecological niche, environmental niche
новообразова'ние *n* neoplasm, newgrowth
  эволюцио'нное ~ innovation
новорожде'нный *a* newborn, neonatal
нога' *f* leg, foot
  брюшна'я ~ abdominal leg
  за'дняя ле'вая ~ near hind leg
  за'дняя пра'вая ~ off hind leg
  копа'тельная ~ burrowing leg
  ложная ~ *ent* abdominal leg
  максилля'рная ~ maxilliped, maxillipede, jawfoot
  пере'дняя ле'вая ~ near fore leg
  пере'дняя пра'вая ~ off fore leg
  плавательная ~ swimming leg
  пры'гательная ~ hopping leg, leaping leg
  хвата'тельная ~ seizing leg, seizing limb
  ходи'льная ~ ambulatory leg, walking leg, gradient leg, walking limb
ногтево'й *a* ungual
ноготки' *m pl* pot-marigold (*Calendula*)
ногото'к *m zool* claw; *bot* unguis
но'готь *m* nail, unguis
ногохво'стка *f* sprintail (*Collembola*)
ногоче'люсть *f* maxilliped, maxillipede, jawfoot
нож *m* knife
  микрото'мный ~ section knife
  окулиро'вочный ~ budding knife
  приви'вочный ~ grafting knife, pruning knife, pruner
  садо'вый ~ garden knife
ножеви'дный, ножеобра'зный *a* cultrate, cultrated, cultriform
но'жка *f* stalk, limb, stem, stipe, peduncle, pedicel, pedicle, crus
  ~ **глазно'го пузыря'** optic pedicle
  лягу'шечьи но'жки *pl* frogs legs
  пла'вательная ~ swimmeret
  чё'рная ~ *phytop* blackleg, black stem
но'жницы *pl* scissors, shears
  ~ **Ку'пера** Cooper's scissors

~ **Ри'хтера** Richter's scissors
изо'гнутые ~ curved scissors
остроконе'чные ~ pointed scissors
садо'вые ~ pruning scissors
тупоконе'чные ~ blunt scissors
хирурги'ческие ~ surgical scissors
ножно'й *a* pedal
ноздрева'тый *a* cavernous (*e.g., cheese*)
ноздря' *f* nostril
нозоло'гия *f* nosology
нозофи'т *m* nosophyte
номади'зм *m* nomadism
номенклату'ра *f* nomenclature
бина'рная ~ binary nomenclature
биномиа'льная ~ binomial nomenclature
номогене'з *m* nomogenesis
номогра'мма *f* nomogram
но'нсенс-мута'нт *m* nonsense mutant
но'нсенс-мута'ция *f* nonsense mutation
но'нсенс-супре'ссор *m* nonsense-suppressor
нора' *f* burrow, hole
норадренали'н *m* noradrenaline
но'ричник *m* figwort *(Scrophularia)*
но'рка *f* burrow; European mink, marshotter *(Mustela lutreola)*
америка'нская ~ American mink *(Mustela vison)*
европе'йская ~ European mink *(Mustela lutreola)*
но'рма *f* norm, standard
~ **адапта'ции** adaptive norm
~ **примене'ния** rate of application
~ **реа'кции** norm of reaction, reaction norm
адапти'вная ~ adaptive norm
альтернати'вная ~ **реа'кции** alternative reaction norm
нормализа'ция *f* normalization
норма'льный *a* standard
нормобла'ст *m* normoblast
нормобласти'ческий *a* normoblastic
нормотензи'вный *a* normotensive
нормотони'ческий *a* normotonic
нормотро'пный *a* normotropic
нормохро'мный *a* normochromic
нормоци'т *m* normocyte
норо'вистость *f* tricky disposition
норо'вистый *a* balky (*of a horse*)
нос *m* nose
горба'тый ~ aquiline nose, hooked nose
зао'стренный ~ pinched nose
курно'сый ~ turned-up nose
орли'ный нос aquiline nose
носа'тый *a* nasute
но'сик *m* beak, rostellum
носи'лки *pl* litter
носи'тель *m* carrier
близкоро'дственный ~ related carrier
гомологи'чный ~ homologous carrier
здоро'вый ~ healthy carrier
иммуноге'нный ~ immunogenic carrier
неконъюги'рованный ~ unconjugated carrier, free carrier
неро'дственный ~ unrelated carrier
несвя'занный ~ free carrier, unconjugated carrier
свобо'дный ~ unconjugated carrier, free carrier
чужеро'дный ~ unrelated carrier
носово'й *a* nasal, rhinal
носогубно'й *a* nasolabial
носоло'бный *a* nasofrontal
носонёбный *a* nasopalatine
носогло'тка *f* nasopharynx, rhinopharynx
носоро'г *m* rhinoceros (*Rhinocerotidae*)
бе'лый ~ white rhinoceros, wide-mouthed rhinoceros *(Ceratotherium simum)*
инди'йский ~ Indian rhinoceros *(Rhinoceros unicornis)*
суматра'нский ~ wooly rhinoceros, Sumatran rhinoceros *(Dicerorhinus sumatrensis)*
чё'рный ~ black rhinoceros *(Diceros bicornis)*
нотоге'я *f* Notogaea
нотоплевра'льный *a* notopleural
нотопо'дий *m* notopodium
но'тум *m ent* notum
ноцицепти'вный *a* nociceptive
ноцице'птор *m* nociceptor
ноче'вка *f* roosting time (*of birds*)
обще'ственная ~ communal roost
ночецве'тный *a* nyctanthous
ночле'г *m orn* roost
ночни'ца *f* common bat (*Myotis*)
ночно'й *a* nocturnal, vespertine
но'-шпа *f med* no-spa
нрав *m* disposition, temperament
послу'шный ~ kind disposition
нуклеа'за *f* nuclease
~ **репара'ции** repair nuclease
~ **рестри'кции** restriction nuclease
«упако'вочная» ~ headful nuclease
нуклеа'ция *f* nucleation
нуклеиниза'ция *f* nucleination

нуклеи'новый *a* nucleic
нуклео- nucleo-
нуклеогисто'н *m* nucleohistone
нуклеоде'сма *f* nucleodesma
нуклеози'д *m* nucleoside
нуклео'ид *m* nucleoid
нуклеокапси'д *m* nucleocapsid
нуклеоля'рный *a* nucleolar
нуклеоми'ксис *m* nucleomixis
нуклеопепти'д *m* nucleopeptide
нуклеопла'зма *f* nucleoplasm, nucleohyaloplasm
нуклеопротеи'д *m* nucleoprotein
нуклеосо'ма *f* nucleosoma
нуклеоти'д *m* nucleotide
нуклеофи'льный *a* nucleophilic
нуклеоцентросо'ма *f* nucleocentrosome
нуллипле'кс nulliplex
нуллисо'мик *m* nullisomic
нумери'ческий *a* numerical
нута'ция *f* nutation
нутриево'дство *n* nutria farming
нутрици'зм *m* nutricism
ну'трия *f* swamp beaver, nutria (*Myocastor coypus*)
нутряно'й *a*: нутряной жир interior fat, visceral fat
нуха'льный *a* nuchal
нуце'ллус *m* nucellus
ныро'к *m* diving duck (*Aythya*)
  красноголо'вый ~ dunbird, pochard (*Aythya ferina*)
ныря'ние *n* diving
ныря'ть *v* dive, plunge, take to the water
ньюфаундле'нд *m* Newfoundland
нюх *m* scent, power of scent
ню'хать *v* sniff, snuff, smell, nose

О

оа'зис *m* oasis
обвё'ртка *f* involucre, envelope, sheath, spathe
листова'я ~ husk (*of a maize ear*)
ча'стная ~ partial involucre
обве'ртывающий *a* wrapping
обвива'ться *v* clasp
обвива'ющий *a* enwrapping
обви'тый *a* obvolute
обгла'дывание *n* gnawing, browsing
обгла'дывать *v* browse
обгрыза'ние *n* browsing
обдира'ние *n* hulling, peeling
обдира'ть *v* strip
~ **ту'шу кита'** flay
~ **кору'** bark, debark
~ **ту'шу** skin
обедне'ние *n* depauperation, degradation, deterioration
обеднё'нный *a* depauperated
обез- (*before voiced vowels*) a-, de-, dis-, im-, un-
обезбо'ливание *n* anesthesia, anaesthesia, anaesthetization, anesthetization
обезво'женный *a* dehydrated, dewatered
обезво'живание *n* dehydration, dewatering
одномоме'нтное ~ nongraded dehydration
обезво'живать *v* dewater, dehydrate
обезвола'шивание *n* unhairing, dehairing, depilation, unwooling
обезвола'шивать *v* dehair
обезгла'вливать *n* decapitate
обездви'живание *n* immobilization
обездви'живать *v* immobilize
обезжи'ренный *a* fat-free, defatted
обезжи'ривание *n* defatting, degreasing
обеззара'живание *n* disinfecting, disinfection, sterilization
обеззара'живать *v* disinfect, decontaminate
обезле'сение, обезлесивание *n* deforestation, disafforestation
обезле'сить *v* deforest
обезья'на *f* ape
обезья'ний *a* simian, apish
обезья'нник *m* monkey house
обелио'н *m anat* obelion
обес- (*before voiceless vowels*) de-, ex-, -less
обескро'вленный *a* bloodless, exsanguinated
обескро'вливание *n* bleeding
обескро'вливать *v* bleed
обессо'ливание *n* desalinization, desalting
обесцве'чивать *v* destain
обеспло'живать *v* sterilize
оби'лие *n* abundance
оби'льный *a* abundant
обита'тель *m* inhabitant, dweller
обкла'дка *f* sheath
~ **сосу'дистых пучко'в** *bot* bundle sheath
обла'ва *f* beat, beating up
охо'тничья ~ enclosing of game
обла'ивать *v* bark at
о'бласть *f* region; field, domain, sphere
~ **безразли'чия** *biom* indifference zone
~ **ге'нов иммy'нного отве'та** immune response region, lr region
~ **Ir-ге'нов** immune response region, lr region
~ **го'лени** crural region
~ **гомоло'гии** region of homology
~ **изгиба** *mbiol* "elbow" region
~ **ма'нтии** mantle zone
~ **неприя'тия гипо'тезы** rejection region
~ **нечувстви'тельности** dead zone
~ **обита'ния** habitat, locality
~ **ограни'ченного оби'лия** area of specific abundance
~ **переключе'ния** switch region
~ **пита'ния** region of alimentation
~ **подпоро'гового возбужде'ния** subliminal field
~ **постоя'нного оби'лия** area of normal abundance
~ **постоя'нной вредоно'сности** area of normal abundance
~ **предпле'чья** antebrachial region
~ **предпочте'ния** *biom* preference region, preference zone
~ **приня'тия** *biom* acceptance zone
~ **приня'тия гипо'тезы** acceptance region
~ **продолже'ния вы'бора** *biom* zone of continued sampling
~ **проме'жности** perineal region
~ **проявле'ния** display area
~ **спа'ривания** region of pairing
~ **«та'лии»** hinge region, "link" region
~ **эпизоди'ческих размноже'ний** area of occasional abundance
~ **эпизоди'ческого оби'лия** area of occasional abundance
австрали'йская ~ Australian region

анима'льная ~ *emb* animal region
антибореа'льная ~ antiboreal region
бореа'льная ~ boreal region
вариа'бельная ~ *imm* variable region, V region
вегетати'вная ~ *emb* vegetal region
височ'чная ~ temporal region
внедо'менная ~ extra-domain region
восто'чная ~ Oriental region
голаркти'ческая ~ Holarctic region
диверги'рующая ~ *gen* divergent region
довери'тельная ~ *biom* confidence region
заты'лочная ~ occipital region
коди'рующая ~ *gen* coding region
консервати'вная ~ *gen* conserved region
консервати'вная ~ **белка'** conserved protein domain
конста'нтная ~ constant region, C region, invariable region
крити'ческая ~ *biom* critical region; rejection region, rejection zone
надчре'вная ~ epigastric region
неаркти'ческая ~ Nearctic region
неотропи'ческая ~ Neotropic region, Austro-Columbian region
нетрансли'руемая ~ *gen* untranslated region
ограни'чивающая ~ flanking region
палеаркти'ческая ~ Palearctic region
па'ховая ~ inguinal region
подключи'чная ~ infraclavicular region
подмы'шечная ~ axillary region
подо'бная ~ similar region
подпоро'говая ~ subliminal field
подрё'берная ~ hypochondriac region
поясни'чная ~ lumbar region
прецеребра'льная ~ *zool* rhynchodaeum *(in nemertines)*
пупо'чная ~ umbilical region
скулова'я ~ zygomatic region
спектра'льная ~ spectral region
теменна'я ~ parietal region
Т-кле'точная ~ *anat* T-cell zone
фаунисти'ческая ~ faunal region
фланки'рующая ~ flanking region
флористи'ческая ~ floral region
шарни'рная ~ hinge region, "link" region
энха'нсерная ~ enhancer region
эфио'пская ~ Ethiopian region

облегче'ние *n* facilitation
гетеросинапти'ческое ~ *phys* heterosynaptic facilitation
социа'льное ~ biosocial facilitation, social facilitation

облепи'ха *f* sea-buckthorn (*Hippophae rhamnoides*)
облесе'ние *n* afforestation
облесе'нный *a* afforested, stocked
облё'т *m* flight
ориентиро'вочный ~ *apic* orientation flight
очисти'тельный ~ *apic* cleansing flight
облета'ть *v* fall off (*of leaves*); fly all over
облете'лый *a* bare (*of plants*)
облига'тный *a* obligatory, obligate
о'блик *m* habit
обли'ственный *a* leafed, leafy, leaf-bearing
облуче'ние *n* irradiation
~ **га'мма-луча'ми** gamma irradiation
~ **инфракра'сными луча'ми** infrared irradiation
~ **рентге'новскими луча'ми** X-irradiation
~ **ультрафиоле'товыми луча'ми** ultraviolet irradiation
дро'бное ~ fractionated radiation, split-dose radiation
комбини'рованное ~ combination radiation
о'бщее ~ general irradiation
однокра'тное ~ single exposure irradiation, single exposure, single exposition
о'строе ~ acute exposition, acute irradiation
хрони'ческое ~ chronic exposure, chronic exposition
фракциони'рованное ~ fractionated radiation, split-dose radiation
облучё'нный *a* irradiated
облущи'ть *v* dehull, hull, husk, shell
облысе'ть *v* get bald, become bald, grow bald
обма'н *m*: ~ **зре'ния** optical illusion
обме'н *m* exchange, interchange; metabolism
~ **веще'ств** metabolism
~ **веще'ств при по'лном поко'е** resting metabolism
~ **в капилля'рах** capillary exchange, transcapillary exchange
~ **информа'цией** information transfer
~ **элеме'нтами фа'уны** faunal interchange
азо'тистый ~ nitrogen metabolism, nitrogenous metabolism
акти'вный ~ active metabolism
белко'вый ~ protein metabolism
во'дный ~ water exchange
водоро'дный ~ hydrogen exchange

двууда'рный ~ *gen* two-hit exchange
жирово'й ~ fat metabolism, adipose metabolism
изото'пный ~ isotopic exchange
ио'нный ~ ion exchange
катио'нный ~ cation exchange
кислоро'дный ~ oxygen exchange
кисло'тно-щелочно'й ~ acid-base metabolism
межмолекуля'рный ~ intermolecular exchange
минера'льный ~ mineral metabolism
основно'й ~ basal metabolism, fasting metabolism, standard metabolism, resting metabolism
промежу'точный ~ intermediate metabolism
са'харный ~ sugar metabolism, saccharometabolism
сегме'нтный ~ segmental interchange
солево'й ~ salt metabolism
темново'й ~ dark metabolism
тиол-дисульфи'дный ~ SH – S-S exchange
транскапилля'рный ~ transcapillary exchange
углево'дный ~ carbohydrate metabolism
хромати'дный ~ chromatid interchange
хромосо'мный ~ chromosome interchange
энергети'ческий ~ energy metabolism

обме'нивать *v* exchange
обмороже'ние *n* frostbite
обморо'женный *a* frost-bitten
о'бморок *m* syncope
о'бморочный *a* syncopal, syncopic
обнаже'ние *n* exposition
~ **корне'й** exposition of roots
~ **при отли'ве** tidal exposition
обнаруже'ние *n* discovery
обнару'живать discover
обновле'ние *n* rejuvenescence, renovation, renewal
обновля'ть *v* renew
обно'жка *f apic* pollen load
обню'хивание *n* sniffing
обню'хивать *v* scent about, sniff, snuff
обогаща'ть *v* enrich
обогаще'ние *n* enrichment
электи'вное ~ selective enrichment
ободо'к *m* annulus
вста'вочный ~ *bot* intercalary annulus
обоепо'лость *f* androgynism, androgyny
обоепо'лый *a* ambisexual, bisexual, hermaphroditic, androgynous, *bot* teleianthous, telianthous, monoecious
оболо'чка *f* coat, coating, envelope, membrane, tunic
~ **Бру'ха** Bruch's membrane
~ **ви'руса** viral envelope
~ **головно'го мо'зга** brain tunic
~ **не'рва** nerve tunic
~ **Не'смита** Nasmyth's membrane
~ **оплодотворе'ния** fertilization membrane
~ **плода'** birth membrane
~ **се'мени** seed coat
адвентициа'льная ~ adventitious membrane
бело'чная ~ scleral coat
блестя'щая ~ zona pellucida
внезаро'дышевая ~ extraembryonic membrane
десцеме'това ~ Descemet's membrane
децидуа'льная ~ *anat* caduca, caduceous membrane, decidual membrane
желто'чная ~ vitelline membrane, yolk membrane
заро'дышевая ~ embryonic membrane
защи'тная ~ protecting coat
кле'точная ~ cell wall, cell enlvelope
миели'новая ~ medullary sheath, myelin sheath
мозгова'я ~ brain tunic, meninx
мы'шечная ~ muscular coat
нару'жная ~ outer coat
отпада'ющая ~ *anat* caduca, decidual membrane, caduceous membrane
паути'нная ~ arachnoid membrane, arachnoid, arachnoid sheath
пигме'нтная ~ pigment coat
пло'дная ~ fetal membrane, fetal sac, gestation sac, birth membrane
плодо'вая ~ fruit coat
ра'дужная ~ iris
семенна'я ~ seed coat
серо'зная ~ serous coat, serous tunic
се'тчатая ~ гла'за nervous tunic of eyeball
сли'зистая ~ mucosa, mucous membrane, mucous coat, jelly coat, mucosal tunic, mucous tunic
сли'зистая ~ но'са Schneiderian membrane
серо'зная ~ serous membrane, serous tunic
синовиа'льная ~ synovial membrane
скорлупо'вая ~ shell membrane
сли'зистая ~ mucosal tunic, mucous tunic

сосу'дистая ~ vascular membrane, vascular tunic
стеклови'дная ~ hyaloid membrane
фа'говая ~ phage membrane
флоэ'мная ~ phloem sheath
хоро'идная ~ choroid coat
шва'нновская ~ Schwann's sheath, neural sheath
я'дерная ~ nuclear enlvelope
обоня'ние *n* sense of smell, olfaction, osmesis, rhinesthesia
обоня'тельный *a* olfactory, olfactic, olfactive, osmetic, rhinal
обоня'ть *v* olfact, smell
оборо'на *f* defense
оборони'тельный *a* defensive
оборо'т *m* rotation; revolution; whorl (*of a shell*)
обору'дование *n* equipment
лаборато'рное ~ labware
стациона'рное ~ fixed equipment
обостре'ние *n* exacerbation
обо'чина *f* roadside, wayside
обоюдоо'стрый *a* ancipitous
обраба'тывать *v* cultivate (*land*), till (*land*), process (*e.g., leather*)
обрабо'тка *f* processing, treatment
камера'льная ~ cameral treatment
теплова'я ~ heat treatment
математи'ческая ~ mathematical treatment
о'браз *m* shape; image
«~ иско'мого» *eth* search image
~ **жи'зни** mode of life, mode of existence, on-goings
«вну'тренний ~» *imm* internal image
зри'тельный ~ *phys* visual object
сидя'чий ~ жи'зни fixed mode of life
образе'ц *m* pattern; sample, specimen, example
контро'льный ~ check sample, control sample
процесси'рованный ~ processed sample
сме'шанный ~ composite sample
образова'ние *n* formation; education
~ **же'лчи** bilification
~ **ко'жи** cutification
~ **ко'рки** crusting
~ **пе'ны** foaming
~ **пе'тель** looping
~ **пигме'нта** chromogenesis
~ **побе'гов** shooting
~ **семя'н** seeding
природоохра'нное ~ environmental education
экологи'ческое ~ ecological education
образо'вывать *v* form
~ **коча'н** loaf
обраста'ние *n* fouling; epibiose
микробиологи'ческое ~ microbiological fouling
обратнодельтови'дный *a* obdeltoid
обратноконусови'дный *a* obconic
обратноланце'тный, обратноланцетови'дный *a* oblanceolate
обратносердцеви'дный *a* obcordate
обратночерепи'тчатый *a* obimbricate
обратнояйцеви'дный *a* obovate, obovoid, inversely egg-shaped
обреза'ние *n* pruning; circumcision
обреза'ть *v* crop, cut off; prune, trim, string *(runners of strawberry)*, dress *(trees)*; circumcise
обре'зка *f* pruning
~ **клю'ва** debeaking
обро'сший *a* overgrown; barnacled
обруба'ть *v* cut off, chop off; dock (*a tail*)
обру'бленный *a* truncated, succise
обру'шивание *n* shelling, hulling, peeling; falling down, collapsing
обру'шивать *v* hull
обры'згивать *v* spray
обсле'дование *n* survey, examination, study
дли'тельное ~ follow-up study
единовре'менное ~ cross-sectional study
ма'ссовое ~ screening study
обтека'емый *a* streamlined
обтека'ние *n* circumwallation
обуча'ть *v* train
обуче'ние *n* learning
~ **в лабири'нте** maze learning
~ **ме'тодом проб и оши'бок** trial and error learning
~ **мно'жественному различе'нию** multiple-discrimination learning
~ **наблюде'нием** observational learning
~ **распознава'нию о'бразов** discrimination learning
ассоциати'вное ~ associative learning
дифференци'рованное ~ discrimination learning
инструмента'льное ~ instrumental learning
облига'тное ~ obligatory learning

перцепти'вное ~ perceptual learning
простра'нственное ~ spatial learning
скры'тое ~ latent learning
стохасти'ческое ~ contingency learning
обхва'т *m* circumference
обшелушивать *v* dehull (*seeds*)
обще'ние *n* intercourse, communication
обще'ственность *f* sociability
обще'ственный *a* social
о'бщество *n* community
о'бщий *a* common
о'бщина *f* community
объезжа'ть *v* train, break in (*a horse*); drive round, go round; drive all over
объе'кт *m* object
замеща'ющий ~ *eth* substitute object
пищево'й ~ food object, food item
объекти'в *m* objective, lens
иммерсио'нный ~ immersion objective
объё'м *m* volume
~ **вдо'ха** inspiratory volume
~ **вы'доха** expiratory volume
~ **движений** range of motions
~ **лё'гких** lung volume
~ **циркули'рующей кро'ви** circulating blood volume
дыха'тельный ~ breathing volume, tidal volume, respiratory volume
дыха'тельный ~ во'здуха respiratory air, tidal air
коне'чно-диастоли'ческий ~ enddiastolic volume
коне'чно-систоли'ческий ~ endsystolic volume
лё'гочный ~ lung volume
мину'тный ~ minute volume
мину'тный ~ се'рдца cardiac output
оста'точный ~ tidal volume, tidal air, residual air, residual volume, stationary air
оста'точный ~ лё'гких pulmonary residual volume, residual air, stationary air
парциа'льный уде'льный ~ partial specific volume
резе'рвный ~ вдо'ха inspiratory reserve volume
резе'рвный ~ во'здуха reserve air, supplemental air
резе'рвный ~ вы'доха expiratory reserve volume
систоли'ческий ~ stroke volume, systolic volume
систоли'ческий ~ се'рдца stroke output
уда'рный ~ stroke volume, systolic volume
уда'рный ~ се'рдца stroke volume, stroke output
уде'льный ~ specific volume
чи'стый уде'рживаемый ~ net retention volume *(in chromatography)*
чувстви'тельный ~ sensitive volume
элюи'рующий ~ elution volume
объе'млющий *a* embracing, clasping, equitant
объё'мометр *m* volumeter
самопи'шущий ~ recording volumeter
объясне'ние *n* explanation
~ **коне'чных причи'н** *eth* ultimate explanation
проксима'льное ~ *eth* proximate explanation
обызвествле'ние *n* calcification
обы'скивание *n* grooming behaviour, grooming
обы'чный *a* usual, ordinary
овалоци'т *m* ovalocyte
овальбуми'н *m* ovalbumin
ова'льно-ланцетови'дный *a* lance-oval, ovate-lanceolate
ова'льно-окру'глый *a* ovate-orbicular
ова'льно-остроконе'чный *a* ovate-acuminate
ова'льно-продолгова'тый *a* ovate-oblong
ова'льно-эллипсоида'льный *a* ovate-ellipsoidal
ова'льный *a* oval, ovate, oviform
оварио'ла *f* ovarole
овариэктоми'ровать *v* spay
овариэктоми'я *f* ovariectomy, spay
овершу'т *m* overshoot
овё'с *m* oat (*Avena*)
ове'чий *a* ovine
ови'зм *m* ovism
овика'пт *m* ovicapt, oocapt
ови'н *m* drying barn
овипа'рия *f* oviparity
овоапога'мия *f* ooapogamy
овобласте'ма *f* ooblastema
овогене'з *m* ovogenesis, oogenesis
овоглобули'н *m* ovoglobulin
ового'ний *m* oogonium
о'вод *m* botfly (*Oestridae*)
бы'чий ~ ox bot, ox botfly (*Hypoderma bovis*)
бы'чий подко'жный ~ bomb fly *(Hypoderma bovis)*

бы'чий полоса'тый ~ heel fly *(Hypoderma lineatum)*
овечий ~ sheep botfly *(Oestrus ovis)*
желу'дочный ~ stomach botfly, horse bot fly *(Gasterophilidae)*
подко'жный ~ warble fly *(Hypodermatidae)*
ово'идный *a* ovoid
овулято'рный *a* ovulatory
овуляцио'нный *a* ovulatory
ОВОС, оце'нка возде'йствия на окружа'ющую среду' environmental impact assessment, EIA
овоско'п *m* eggtester
овоскопи'рование *n* candling
овоце'нтр *m* ovocenter
овоци'т *m* ovocyte
о'вощ *m* vegetable
овощево'д *m* vegetable grower
овощево'дство *n* vegetable farming, vegetable growing
овощно'й *a* vegetable
овсю'г *m* wild oat, poor oat, wheat oat (*Avena fatua*)
овся'ница *f* fescue (*Festuca*)
овся'нка *f* bunting (*Emberiza*); oatmeal; oatmeal porridge
овсяно'й *a* oaten, avenaceous
овули'ровать *v* ovulate
овуляцио'нный *a* ovulatory
овуля'ция *f* ovulation
спонта'нная ~ spontaneous ovulation
овца' *f* sheep (*Ovis*); ewe; **ста'до ове'ц** flock of sheep, herd of sheep
выбрако'ванная ~ cull ewe
кара'кульская ~ karakul sheep
курдю'чная ~ fat-rumped sheep
мерино'совая ~ merino sheep
остри'женная ~ shorn sheep
по'льзовательная ~ flock sheep
суя'гная ~ pregnant sheep
тонкору'нная ~ fine-fleece sheep
я'ловая ~ barren sheep
овцебы'к *m* muck buffalo (*Ovibos moschatus*)
овцево'д *m* sheep breeder
овцево'дство *n* sheep husbandry
овча'рка *f* sheep dog
овча'рня *f* fold
овчи'на *f* sheep pelt, sheepskin
о'гарь *m* ruddy shelduck (*Casarca ferruginea*)
оги'ва *f biom* ogive
оглее'ние *n* gleization, gleyization, gleying
огли'нивание *n* argillification
огне'вка *f* grass moth (*Crambus*)
оголе'нный *a* glabrate, bald; denuded
огоро'д *m* kitchen garden, vegetable garden
ограниче'ние *n* restriction
генети'ческое ~ genetic restriction
ограни'ченный *a* localized, restricted
огре'х *m agr* gap (*e.g., in ploughing*)
огрыза'ться *v* snarl at, snap
огры'зок *v* end, core (*of an apple*)
огу'зок *m* rump
огуре'ц *m* cucumber (*Cucumis sativus*); cucumber, cucumber fruit
бе'шеный ~ squirting cucumber (*Eckballium elaterium*)
парнико'вый ~ frame cucumber
огуре'чник borage (*Borago officinalis*)
одерно'ванный *a* sodded
одерно'вка *f* sodding
одина'ковый *a* uniform; identical
одине'ц *m* sigleton
одино'чный *a* single
одиночнопло'дный *a* single-fruited
одиночноцвету'щий *a* single-flowering
одича'вший *a* feral
одича'ть *v* run wild
одно- uni-, one-, mono-, mon-
одноаперту'рный *a pal* monotreme
одноборо'здный *a pal* monosulcate, monocolpate
однобра'тственный *a* monadelphous
однобра'чный *a* monogamous
однобуго'рчатый *a* unicuspid *(of teeth)*
одноверши'нный *a* unimucronate
одноветви'стый *a* uniramous, uniramose
одногнё'здный *a bot* unilocular, monothalamous, monothecious
одноголо'вчатый *a bot* monocephalous
одногу'бый *a* unilabiate
однодо'льный *a bot* monocotyledonous, unicotyledonous
однодо'мный *a* monoecious, synoecious, synoicous
одножа'берный *a* unibranchiate
одножгу'тиковый *a* uniflagellate
одножи'лковый *a* uninerved
однозернянка *f bot* einkorn (*Triticum monococcum*)
однозу'бчатый *a* unidentate
однока'мерный *a* monolocular, unilocular, unicamerate; *zool* monothalamous

однока'псульный *a* unicapsular
одноклe'точный *a* one-celled, single-celled, unicellular, unicellate, monocelled
одноколо'сый *a bot* monostachyous
однокоро'бочковый *bot* unicapsular
одноко'сточковый *a* monopyrenous
однокры'лый *a bot* monopterous
однолепе'стный *a bot* monopetalous, unipetalous
одноле'тний *a* annual, annotinous
одноле'тник *m* annual
одноли'стный *a bot* monophyllous, unifoliate
одноло'пастный *a* unilobar
одномешко'вый *a pal* monosaccate
однонитево'й *a* single-stranded
однообра'зный *a* uniform
одноосево'й *a* uniaxial
одноо'сный *a* uniaxial
однопа'лость *f* monodactyly
однопа'лый *a* monodactylous, unidactyl
однопа'рный *a* unijugate
однопе'стичный *a bot* monogynous
однопло'дный *a bot* monocarpic, monocarpous, monotocous
одноплодоли'стиковый *a bot* monocarpellary
однопокро'вный *a* monochlamydeous
однопо'лый *a* unisexual
одноприцве'тничковый *a bot* unibracteate
одноро'гий *a* monocerous
одноро'дность homogeneity
  простра'нственная ~ spatial homogeneity
одноро'дный *a* uniform, homogeneous
однорубцо'вый *a* monolete
одноры'льцевый *a bot* monostigmatous
одноря'дный *a* uniserial, monostichous, one-row
односемядо'льный *a bot* monocotyledonous
односемя'нный *a bot* monospermic, monospermous, one-seeded
односло'йный *a* monostratal, one-layered, unilaminate
односпира'льный *a* unispiral
односпо'ровый *a* monosporous
одноство'льный *a* monocormic
односто'лбиковый *a bot* monostylous
одностороʼнний *a* unilateral, unifacial
одноти'пность *f* uniformity of type
однохозя'йный *a* autoecious, autoxenous
одноцветко'вый *a* single-flowered, uniflorous
одноцве'тный *a* monochroic, monochromatic
одноцепо'чечный *a mbiol* single-stranded
одночашели'стиковый *a bot* monosepalous
одночле'никовый *a* one-segmented, uniarticulate
одночле'нный *a bot* monomeric
однощелево'й *a* monolete
одноя'дерный *a* uninucleate, uninuclear, mononuclear
одноя'дность *f* monophagy
однояйцо'вый *a* unioval, monoovular, one-egg; monogerminal
одноя'русный *a* one-layered
одома'шненный *a* domesticated, tame
одома'шнивание *n* domestication
одонтобла'ст *m* odontoblst
одонтокла'ст *m* odontoclast
одонтоло'гия *f* odontology
одонтофо'р *m* odontophore
одревесне'ние *n* lignification
одуванчик *m* dandelion (*Taraxacum officinale*)
ое'ций *m* oecium, ooecium
оживле'ние *n* revival, reanimation, resuscitation
оживля'ть *v* resuscitate, revive
ожида'ние *n* expectation, expectancy
  ~ **вознагражде'ния** *eth* reward expectancy
  математи'ческое ~ mathematical expectation
  усло'вное математи'ческое ~ conditional expectation
ожи'ка *f* wood-rush (*Luzula*)
ожире'ние *n* adiposity, obesity
ожо'г *m* burn, burning, scald, scorch
  ~ **коры'** bark blister
  ~ **ли'стьев** *phytop* leaf scorch *(caused by Leptosphaeria)*
  лучево'й ~ radiation blister
  радиацио'нный ~ radiation burn
  со'лнечный ~ sunblister, sunburn, sunscald, sunscorch, solar burn
оздоровле'ние *n* sanitation
озелене'ние *n* landscaping, setllement gardening
озернё'нность *n* grain content
  ~ **ко'лоса** grain content
озе'рный *a* lacustrine, limnic; limnicolous
о'зеро *n* lake
  сле'зное ~ *anat* lacrimal lake
о'зимь *f* winter crop
ознакомле'ние *n* familiarization
озно'б *m* chill, shiver

озо'н *m* ozone
озони'ровать *v* ozonize
ои'дий *m myc* oidium
ойкопла'ст *m* oikoplast
окаймле'нный *a* fringed, bordered
окамене'лый *a* petrified, fossilized, fossil
ока'пи *m ther* okapi *(Okapia)*
ока'пывать *v* earth
океа'н *m* ocean
океана'рий *m* oceanarium
о'кисел *m* oxide
окисле'ние *n* oxidation
  ферментати'вное ~ enzymic oxidation
  фотохими'ческое ~ photochemical oxidation
окисле'ние – восстановле'ние *n* oxidoreduction
окисли'тель *m* oxidant, oxidizer, oxidizing substance
окисли'тельно-восстанови'тельный *a* redox
окисли'тельный *a* oxidative
окисля'ть *v* oxidize
о'кись *f* oxide
окклюзио'нный *a* occlusal
окклю'зия *f* occlusion
окно' *n* window, *anat* fenestra
  ~ **преддве'рия** oval window
  ~ **ули'тки** round window
  кру'глое ~ *anat* round window
  акусти'ческое ~ acoustic window, sound window *(in bioacoustics)*
  ова'льное ~ *anat* oval window
около- par-, para-, peri-
околовено'зный *a* paravenous, perivenous
околопе'стичный *a bot* perigynous
околопло'дник *m bot* pericarp
околоу'шный *a* parotid
околоцве'тник *m bot* perianth, floral envelope
  двойно'й ~ double perianth
  колесови'дный ~ rotate perianth
  чашечкови'дный ~ calyciform perianth
окольцо'ванный *a* ringed
оконча'ние *n* ending, finishing, termination, terminal
  ~ **дви'гательного не'рва** motor nerve ending
  ~ **чувстви'тельного не'рва** sensory ending
  адренэрги'ческое не'рвное ~ adrenergic nerve ending
  втори'чное ~ *anat* secondary ending
  гроздеви'дное не'рвное ~ grapelike nerve ending
  кольцеспира'льное не'рвное ~ annulospiral nerve ending
  не'рвное ~ nerve terminal, nerve termination, nerve ending
  перви'чное ~ *anat* primary ending
  пласти'нчатое ~ plate ending
  пресинапти'ческое ~ presynaptic terminal
  сте'лющееся ~ *anat* trail ending
  холинэрги'ческое не'рвное ~ cholinergic nerve ending
око'пник *bot* comfrey *(Symphytum)*
око'рка *f* barking, rossing
окостенева'ть *v* ossify
окостене'ние *n* ossification
  ко'жное ~ dermoossification
око'т *m* lambing, fell (*of sheep*)
окочене'лость *n* rigidity
окочене'ние *n* rigidity
  тру'пное ~ postmortem rigidity, cadaveric rigidity
о'кра *f bot* okra *(Abelmoschus esculentus)*
окра'с *m* color (e.g., *of dogs*)
окра'ска *f* color, colour, coloration; stain, staining
  ~ **опере'ния** plumage color
  аллосемати'ческая ~ allosematic coloration, pseudoaposematic coloration
  антикрипти'ческая ~ anticryptic coloration
  апатети'ческая ~ apatetic coloration
  бра'чная ~ breeding color, attire of nuptial
  вита'льная ~ vital staining
  вспы'хивающая ~ flash coloration
  двойна'я ~ double staining
  дифференциа'льная ~ differential staining
  дополни'тельная ~ counterstaining
  защи'тная ~ assimilative staining, cryptic coloration, concealing coloration, allocryptic coloration
  имити'рующая ~ procryptic coloration
  контра'стная ~ contrastive staining, counter staining
  крипти'ческая ~ cryptic coloration, allocryptic coloration, assimilative coloration, concealing coloration
  меня'ющаяся ~ confusing coloration
  опознава'тельная ~ episematic coloration
  отпу'гивающая ~ startle coloration

покрови'тельственная ~ assimilative coloration, protective coloration
предостерега'ющая ~ allosematic coloration
предупрежда'ющая ~ sematic coloration, aposematic coloration
привлека'ющая ~ alluring coloration, pseudoepisematic coloration
прима'нивающая ~ alluring coloration, pseudoepisematic coloration
прокрипти'ческая ~ procryptic coloration
расчленя'юшая ~ disruptive coloration
семати'ческая ~ sematic coloration, aposematic coloration
фо'новая ~ background staining
окра'шивание *n* staining; tinction
~ **мазко'в** smear staining
~ **по Ги'мза** Giemsa staining
~ **по Гра'му** Gram staining
~ **по ме'тоду Ле'ффлера** Loffler's staining
~ **по ме'тоду Ра'йта** Wright's staining
~ **по ме'тоду Ци'ля-Не'льсена** Ziel-Nelsen staining
~ **сре'зов** section staining
~ **фо'на** background staining
вита'льное ~ intravital staining, vital staining
иммуноцитохими'ческое ~ immunocytochemical staining
негати'вное ~ negative staining
непрямо'е ~ indirect staining
неспецифи'ческое ~ nonspecififc staining
прижи'зненное ~ vital stain, intravital staining, intravitam stain
прогресси'вное ~ progressive staining
прямо'е ~ direct staining
суправита'льное ~ supravital staining
окра'шивать *v* stain
окра'шивающийся *a* stainable
окри'кивание *n orn* mobbing
окро'л *m* kindling
о'круг *m* district
биогеографи'ческий ~ biotic district
окру'глый *a* subcircular, orbicular
окси- oxy-
оксигемоглоби'н *m* oxyhemoglobin, oxygenated hemoglobin
оксигена'за *f* oxygenase
оксигена'тор *m* oxygenator
оксигена'ция *f* oxygenation
оксигени'ровать *v* oxygenate
окси'д *m m* oxide
оксида'за *f* oxidase
оксиди'ровать *v* oxidize
оксидосо'ма *f* oxidosome
оксилофи'т *m* oxylophyte
окси'метр *m* oxymeter
оксинитрова'ние oxinitration
оксипа'тия *f* oxypathia
окситропи'зм *m* oxytropism
оксифи'л *m* oxyphil, oxyphile
оксифи'льный *a* oxyphilous
оксифи'т *m* oxyphyte
оксифо'б *m* oxyphobe
оксифо'бный *a* oxyphobous
окситоци'н *m* oxytocin
окто- octo-
октопло'ид *m* octoploid
октопло'идный *a* octoploid
оку'кливание *n* pupation
оку'кливаться, **оку'клиться** *v* pupate
окули'ровать *v* graft
окулиро'вка *f* inoculation, grafting, graftage, budding, cion budding
окули'ст *m* oculist
окульту'ривание *n* cultivation (*of soil*)
~ **ландша'фта** landscape development
окуля'р *m* eyepiece, eyelens, eyeglass
окуля'р-микро'метр *m* eyepiece micrometer
оку'ривание *n* fumigation
оку'танный *a* obvolute
оку'чивание *n* earthing
оку'чивать *v* earth
олеа'ндр *m* oleander (*Nerium*)
оледене'ние *n* glaciation
оленево'дство *n* deer farming
оленё'к *m* mouse deer, chevrotain (*Tragulus*)
оле'ний *a* corvine
оле'нь *m* deer, antlered ruminant (*Cervidae*)
~ **Дави'да** Father David's deer *(Edaphurus davidianus)*
благоро'дный ~ red deer (*Cervus elaphus*)
пятни'стый ~ sika deer (*Cervus nippon*)
се'верный ~ reindeer (*Rangifer tarandus*)
олену'ха *f* doe
олиго- oligo-
олигогали'нный *a* oligohaline
олигоге'н *m* oligogene
олигоге'нный *a* oligogenic
олигогли'я *f* oligoglia
олигодендрогли'я *f* oligodendroglia
олигодендроци'т *m* oligodendrocyte
олигодинами'ческий *a* oligodynamic
олигокинези'я *f* oligokinesia

олиголецита'льный *a* oligolecithal
олигоме'р *m* oligomer
олигомериза'ция *f* oligomerization
олигоме'рный *a* oligomerous
олигонитрофи'льный *a* oligonitrophilic
олигонуклеоти'д *m* oligonucleotide
олигопире'нный *a* oligopyrene
олигопо'дный *a* oligopod
олигосапро'бный *a* oligosaprobic
олигосахари'д *m* oligosaccharide
олиготакси'я *bot* oligotaxy
олиготрофи'я *f* oligotrophy
олиготро'фный *a* oligotroph, oligotrophic
о'луша *f* solan, solan goose, booby, gannet (*Sula*)
ольфактоло'гия *f* olfactology
ольфакто'метр *m* olfactometer
ольха' *f* alder (*Alnus*)
~ **кле'йкая** black alder, ooler (*Alnus glutinosa*)
~ **се'рая** speckled alder (*Alnus incana*)
~ **чё'рная** black alder (*Alnus glutinosa*)
ольша'нник *m* alder forest
оля'пка *f* dipper, water ouzel (*Cinclus*)
омброфи'л *m* ombrophil
омброфо'б *m* ombrophobe
оме'жник *m* water dropwort (*Oenanthe*)
оме'ла *f* mistletoe (*Viscum*)
омертвева'ть *v* necrotize
омертве'ние *n* necrosis
оммати'дий *m* ommatidium
омматофо'р *m* ommatophore
омнипоте'нтный *a* omnipotent
омола'живание, омоложе'ние *n* rejuvenescence
омша'ник *m* apiary house, barn for wintering bees
она'гр *m* onager, Asiatic wild ass (*Equus hemmionus*)
онда'тра *f* musk beaver (*Ondatra zibethica*)
онко- onco-
онкобело'к *m* oncoprotein
онкоге'н *m* oncogene
онкоге'нность *f* oncogenicity
онколо'гия *f* oncology
онкорнави'рус *m* oncornavirus
онкосфе'ра *f helm* oncosphere
онкоти'ческий *a* oncotic
онтогенез *m* ontogeny, ontogenesis
онтогенети'ческий *a* ontogenetic, ontogenetical
онтокли'н *m* ontocline
онтомута'ция *f* ontomutation
онхоцерко'з *m* onchocercosis
ооапога'мия *f* ooapogamy
oобласте'ма *f* ooblastema
оогаме'та *f* oogamete
оога'мия *f* oogamy
оогене'з *m* oogenesis, ovogenesis
ooго'ний *m* oogonium
оокине'з *m* ookinesis
оокине'та *f* ookinete
ооле'мма *f* oolemma
ооcо'ма *f* oosome
ооспо'ра *f* oospore
оостеги'т *m* oostegite
оосфе'ра *f* oosphere
ооте'ка *f* ootheca
ооти'да *f* ootid
ооти'п *m* ootype
оофи'т *m* oophyte
ооце'нтр *m* ovocenter
ооци'ста *f* oocyst, ovocyst
ооци'т *m* oocyte, ovocyte
~ **пе'рвого поря'дка** primary oocyte
~ **второ'го поря'дка** secondary oocyte
опа'вший *a* dead (*of leaves*)
опа'д *m* litter, tree waste
листово'й ~ litter, tree waste
опада'ние *n* cast, fall, falling off
~ **ветве'й** cladoptosis
~ **ли'стев** leaf fall
опада'ть *v* fall (*e.g., of leaves*)
опада'ющий *a* deciduous
опаде'ние *n* fall, falling off, drop, abscission
опа'л *m* scald; opal
со'лнечный ~ sun scald
опарши'веть *v* become mangy
опа'рыш *m* maggot (*used as bait or for fish feeding*)
опа'сность *f* danger, peril, hazard, risk
генети'ческая ~ genetic hazard
опа'сный *a* dangerous, perilous
опаха'ло *n* vane; rhipidium (*a kind of inflorescence*)
опе'нок *m* honey fungus, honey agaric (*Armillariella mellea*)
опера'тор *m* operator
опера'ция *f* operation
операция Винкельма'на-Бе'ргмана Winkelmann-Bergmann's operation
опере'ние *n* feathering, plumage
ювени'льное ~ juvenile plumage

оперённый *a* feathered, feathery, plumose, fledged
опери'вшийся *a* full-fledged
опери'ровать *v* operate, perform operation, do operation, make operation, carry out operation
опери'роваться *v* be operated
опе'ркулум *m* operculum, opercle
оперо'н *m* operon
оперя'ться *v* feather, fledge
описа'ние *n* description
~ **расти'тельного покро'ва** description of plant cover
геоботани'ческое ~ geobotanical description
первонача'льное ~ first description, original description
опистио'н *m* opisthion
опистоне'фрос *m* opisthonephros
описторхо'з *m helm* opisthorchiasis
опистосо'ма *f* opisthosoma
опистоце'льный *a anat* opisthocelous
о'пиум *m* opium
опла'та *f:* ~ **ко'рма** food conversion ratio
оплодотворе'ние *n* fertilization, impregnation, fecundation
двойно'е ~ double fertilization
избира'тельное ~ selective fertilization
иску'сственное ~ artificial fertilization
мно'жественное ~ multiple fecundation
моноспе'рмное ~ monospermic fertilization, monospermic fecundation
перекрё'стное ~ cross-fertilization, cross-fecundation, xenogamy
полиспе'рмное ~ polyspermic fertilization, polyspermic fecundation
оплодотворя'емость *f* conception rate
оплодотворя'ть *v* fertilize, impregnate, fecundate, ingravidate
оповеще'ние *n eth* advertisement
оподзо'ливание *n* podsolization
опо'ек *m* buttermilk skin
опознава'тельный *a* episematic
опо'р: **во весь опор** at full speed, at top speed, at full gallop
опо'ра *f* fulcrum
опорожне'ние *n* ejection
~ **желу'дочка** ventricular ejection
опоро'с *m* farrow, farrowing
опороси'ться *v* farrow
опо'ссум opossum *(Didelphidae)*
опоя'санный *a* belted, zonate
оппози'ция *f* opposition
определе'ние *n* determination, identification; definition
~ **во'зраста** aging, ageing
~ **гру'ппы кро'ви** blood typing
~ **ка'чества воды'** water quality assessment
~ **по'ла** sex determination
~ **чи'сленности популя'ции** population estimation
колориметри'ческое ~ colorimetric estimation
прото'чно-инъекцио'нное ~ flow injection determination
определё'нный *a* strict; definite
непра'вильно ~ misidentified
определи'тель *m* key to
определя'ть *v* identify
опробкове'ние *n* suberization, suberification
опры'скивание *n* spraying, spinkling
~ **мочо'й** *eth* urine spraying, enurination
опры'скиватель *m* sprayer
опры'скивать *v* spray; dabble
опсониза'ция *f* opsonization
опсони'н *m* opsonin
опсономе'трия *f* opsonometry
оптимиза'ция *f* optimization
о'птимум *m* optimum
экологи'ческий ~ ecological optimum
опти'ческий *a* optic
оптогра'мма *f* optogram
оптокинети'ческий *a* optokinetic
опто'метр *m* optometer
оптомото'рный *a* optomotor
опуска'ться *v* descend; alight (*e.g., of birds*)
опустоше'ние *n* devastation
опусты'нивание *n* desertification
опу'хнуть *v* swell
о'пухолевый *a* tumorous
о'пухоль *f* swelling, tumor
~ **Бре'ннера** Brenner tumor
~ **Броди'** Brodie's tumor
~ **Ви'льмса** Wilms tumor
~ **Гра'вица** Grawitz's tumor
~ **карма'на Ра'тке** Rathke's pouch tumor
~ **Ко'дмана** Codman's tumor
~ **Кру'кенберга** Krukenberg's tumor
~ **Пе'нкоста** Pancoast's tumor
~ **Уо'ртина** Warthin's tumor
~ **Шо'па** Shope tumor
~ **Эрдге'йма** Erdheim tumor
~ **Ю'инга** Ewing's tumor

доброка'чественная ~ benign tumor, innocent tumor
злока'чественная ~ malignant tumor
пальпи'руемая ~ palpable tumor
опуше'ние *n* downiness
опушенно-ли'стный *a* downy-leaved
опуше'нность *f* downiness, pubescence
опушё'нный *a* pubescent, tomentose, nappy, downy, lanate
опу'шка *f* margin, forest edge
опуще'ние *n med* ptosis, falling, descent
опыле'ние *n* pollination, fertilization
~ **ве'тром** wind pollination
~ **насеко'мыми** insect pollination
иллегити'мное ~ illegitimate pollination
иску'ственное ~ artificial pollination, controlled pollination
перекрё'стное ~ cross-pollination, open pollination
подво'дное ~ hydrophilous pollination
ро'дственное ~ close pollination
случа'йное ~ random pollination
опы'ливание *n* dusting
опыли'тель *m* pollinator, pollinizer
опыля'емый пти'цами ornithophilous
опыля'ть *v* pollinate, fertilize
о'пыт *m* experiment; experience
~ **в кле'тке** cage experiment
~ **двойны'м слепы'м ме'тодом** double blind experiment
~ **по выведе'нию** rearing experiment
~ **по кормле'нию** feeding experiment
~ **«пульс-че'йз»** *mbiol* pulse chase experiment
~ **с вытесне'нием ме'тки** chasing experiment
жи'зненный ~ experience
контро'льный ~ check experiment, control experiment
многокра'тный ~ multiplied experiment
многофа'кторный ~ multiple-factor experiment
полево'й ~ field experiment
предвари'тельный ~ trial experiment
предше'ствующий ~ *eth* previous experience, early experience
разве'дочный ~ trial experiment
ра'нее приобретё'нный ~ early experience
созна'тельный ~ conscious experience
холосто'й опыт blank experiment
о'пытный *a* experimental; experienced
ора'льно *adv* orally
ора'льный *a* oral
орангута'н *m* orangutan (*Pongo pygmaeus*)
ора'нжевый *a* orange
акриди'новый ~ acridine orange
мети'ловый ~ methyl orange
оранжере'я *f* glass house
орби'та *f* eyepit, eyehole, eyesocket, orbit
орбита'ль *f* orbital
запо'лненная ~ filled orbital
свобо'дная ~ vacant orbital
орбита'льный *a* orbital
запо'лненная ~ filled orbital
свобо'дная ~ vacant orbital
орга'зм *m* orgasm
о'рган *m* organ
~ **вку'са** organ of taste, taste organ, gustatory organ
~ **выделе'ния** eliminative organ
~ **Ге'йера-Гро'ссера** Heyer-Grosser organ
~ **Гра'бера** *ent* Graber's organ
~ **дыха'ния** respiratory organ
~ **Же'не** Gene's organ
~ **зре'ния** visual organ, organ of sight
~ **механи'ческого чу'вства** mechanical sense organ
~ **обоня'ния** olfactory organ, organ of smell
~ **осяза'ния** organ of touch
~ **пищеваре'ния** anchoring organ
~ **прикрепле'ния** anchoring organ, adhesive organ
~ **равнове'сия** organ of equilibrium
~ **размноже'ния** reproductive organ
~ **свече'ния** light organ, luminous organ
~ **слу'ха** acoustic organ, organ of hearing, auditory organ
~ **совокупле'ния** intromittent organ
~ **хими'ческого чу'вства** chemical sense organ
~ **Цукерка'ндля** Zuckerkandl's organ, Zuckerkandl's body, organ of Zuckerkandl
~ **чу'вства** sense organ
би'ддеров ~ Bidder's organ
боя'нусов ~ organ of Bojanus
гребе'нчатый ~ cribriform organ
джон'стонов ~ Johnston's organ
доба'вочный полово'й ~ accessory genital organ
же'нский полово'й ~ female breeding organ
жира'льдесов ~ Giraldes' organ
запаса'ющий ~ *bot* storage organ

звуково'й ~ sound-producing organ, musical organ
ке'беров ~ Keber's organ
коне'чный выдели'тельный ~ excretory end organ
концево'й ~ end organ
копуляти'вный ~ intromittent organ
ко'ртиев ~ Corti's organ, spiral organ
краево'й ~ marginal organ
кроветво'рный ~ blood-forming organ, blood-making organ, hemopoietic organ
кювье'ров ~ Cuverian organ
ле'йдигов ~ Leydig's organ
лимфо'идный ~ lymphoid organ
околома'точный ~ paruterine organ
парапинеа'льный ~ parietal organ
париета'льный ~ parietal organ
перви'чный лимфо'идный ~ primary lymphoid organ
перви'чный полово'й ~ primary reproductive organ
перифери'ческий лимфо'идный ~ peripheral lymphoid organ
пищевари'тельный ~ digestive organ
полово'й ~ generative organ, genital organ, sex organ, organ of generation
постантенна'льный ~ *ent* postantennal organ
прикрепи'тельный ~ anchoring organ, attachment organ
сверля'щий ~ piercing organ
светопреломля'ющий ~ light-refracting organ
совокупи'тельный ~ copulatory organ
сошнико'во-носово'й ~ vomeronasal organ, Jacobson's organ
спира'льный ~ Corti's organ, spiral organ
стридуляцио'нный ~ stridulating organ, rasping organ
субгенуа'льный ~ subgenual organ, supra-tympanal organ
субрадуля'рный ~ subradular organ
термоскопи'ческий ~ thermoscopic organ
тимпана'льный ~ tympanal organ
тру'бчатый преана'льный ~ tubular preanal organ
фагоцита'рный ~ phagocytic organ
хвата'тельный ~ grasping organ
хордотона'льный ~ chordotonal organ, fiddle-stringlike organ
электри'ческий ~ electric organ
эма'левый ~ enamel organ
эндокри'нный ~ endocrine organ
якобсо'нов ~ Jacobson's organ organ, vomeronasal organ
о'рган-мише'нь *m* target organ
органе'лла *f* organelle, organoid, cell organ
организа'тор *m emb* organizer
втори'чный ~ secondary organizer
головно'й ~ head organizer
перви'чный ~ primary organizer
ту'ловищный ~ trunk organizer
я'дрышковый ~ nucleolar organizer
органи'зм *m* organism
аутореакти'вный ~ autologous responder
аэро'бный ~ oxybiotic organism
модуля'рный ~ modular organism
однокле'точный ~ *m* unicell
пищево'й ~ food organism
по'чвенный ~ soil organism
органи'зменный *a* organismal, organismic
органи'ческий *a* organic
органогене'з *m* organofaction, organogenesis, organogeny
органогра'фия *f* organography
органо'ид *m* organoid, organelle, cell organ
органолепти'ческий *a* organoleptic
органоло'гия *f* organology
органотро'пность *f* organotropy
ордина'ция *f* ordination
га'уссова ~ Gaussian ordination
оре'л *m* eagle
ры'бный ~ fishing eagle *(Ichthyophaga)*
степно'й ~ steppe eagle, tawny eagle *(Aquila rapax)*
оре'л-змееяд *m* serpent eagle (*Circaetus*)
оре'л-скоморо'х *m* bateleur (*Terathopius ecaudatus*)
оре'х *m* nut
~ **кэ'шью** cashew nut (*produced by Anacardium occidentale*)
америка'нский ~ Brazil nut, Para nut, cream nut (*Bertholletia excelsa*)
водяно'й ~ water chestnut (*Trapa*)
гре'цкий ~ Circassian walnut, English walnut, European walnut, French walnut, Italian walnut, Persian walnut, Turkish walnut *(Juglans regia)*
земляно'й ~ peanut, earthnut (*Arachis*)
ке'дровый ~ pine nut, cedar nut (*produced by Pinus sibirica*)
лесно'й ~ hazelnut
маньчжу'рский ~ Manchurian walnut *(Juglans mandschurica)*

муска'тный ~ nutmeg (*Myristica*); mace
пусто'й ~ deaf nut
орехови'дный *a* nuciform
оре'ховка *f* nutcracker (*Nucifraga*)
орехопло'дный *a* nuciferous, carycarpous, nut-bearing; **орехопло'дная культу'ра** nut-fruited crop
орехотво'рка *f* gallfly (*Cynipidae*)
оре'шек *m* nutlet, coccus
дуби'льный ~ *phytop* gall, oak gall
черни'льный ~ *phytop* oak apple
оре'шник hazelnut (*Corylus avellana*)
ориента'ция *f* orientation
~ **по небе'сному сво'ду** celestial orientation
ночна'я ~ nocturnal orientation
ориенти'рование *n* orientation
ориентиро'вка *f* orientation
о'рикс *m* oryx *(Oryx gazella)*
орла'н *m* sea eagle (*Haliaeetus*)
белоголо'вый ~ bald ealgle (*Haliaeetus leucocephalus*)
орлан-белохво'ст *m* white-tailed eagle (*Haliaeetus albicilla*)
орле'нок *m* eaglet
орли'ный *a* aquiline; **орлиный нос** aquiline nose
орли'ца *f* hen eagle, eagle-hen
орля'к *m* braken, brake fern, eagle fern, upland fern (*Pteridium aquilinum*)
орнити'н *m* ornithin
орнито'з *m* ornithosis, parrot fever
орнито'лог *m* ornithologist
орнитоло'гия *f* ornithology
орнитофа'уна *f* avifauna
орнитофи'л *m* ornithophilous plant
орнитофи'льный *a* ornithophilous
орогове'ние *n* hornification, cornification, keratinization
орофи'т *m* orophyte
ороше'ние *n* watering, irrigation
~ **дождева'нием** sprinkling
внутрипо'чвенное ~ subsoil irrigation, subsurface irrigation, subterranean irrigation
ка'пельное ~ drop irrigation, drip irrigation
пове'рхностное ~ surface irrigation
подпо'чвенное ~ subsoil irrigation, subsurface irrigation, subterranean irrigation
удобри'тельное ~ fertilizer sprinkling
орсеи'н *m* orcein
орто- ortho-
ортовезикуля'рный *a* orthovesicular
ортогене'з *m* orthogenesis
ортоденти'н *m* orthodentine
ортодро'мный *a* orthodromic
ортокине'з *m* orthokineis
ортоко'мплекс *m* orthocomplex
ортомиксови'рус *m* orthomixovirus
ортоплоиди'я *f* orthoploidy
ортоптеро'идный *a* orthopteroid
ортоселе'кция *f* orthoselection
ортостати'ческий *a* orthostatic
ортотропи'зм *m* orthotropism
ортотопи'ческий *a* orthotopic
ортотро'пный *a* orthotropic, orthotropous
ортотру'бочка *f* orthotubule
ортофи'т *m* orthophyte
ортофори'я *f* orthophoria
ору'жие *n* weapons
биологи'ческое ~ biological weapons
оса' *f* wasp (*Vespa*)
доро'жная ~ spider wasp, spider-hunting wasp, mason wasp *(Pompilidae)*
настоя'щая ~ social wasp *(Vespidae)*
обыкнове'нная ~ common wasp *(Paravespula vulgaris)*
одино'чная ~ potter wasp, solitary wasp, mud wasp *(Eumenidae)*
пилю'льная ~ potter wasp, solitary wasp, mud wasp *(Eumenidae)*
ро'ющая ~ digger wasp, sphecoid wasp *(Sphecidae)*
ры'жая ~ red wasp *(Paravespula rufa)*
складчатокры'лая ~ social wasp *(Vespidae)*
осади'тель *m* precipitator, precipitant
оса'дки *m pl* precipitation
атмосфе'рные ~ *pl* precipitation
оса'док *m* precipitate, sediment, residue
осажда'емость *f* precipitability
осажда'ть *v* precipitate
осажда'ться *v* precipitate
осажде'ние *n* precipitation
оса'нка *f* posture, attitude
освежева'ть *v* skin, dress
освети'тель *m* illuminator
~ **для микроско'па** microscope lamp; substage lamp
освеще'ние *n* lighting, illumination
есте'ственное ~ natural light
непреры'вное ~ continuous light
темнопо'льное ~ dark field illumination, dark ground illumination
освиде'тельствование *n* examination

освобожда'ть *v* free, liberate
освое'ние *n* development
~ **земе'ль** land development
осево'й *a* axial, rachial, rachidial
осе'длый *a* resident, settled, nonmigratory
осёл *m* ass, donkey (*Equus asinus*)
ди'кий ~ *m* onager, Asiatic wild ass (*Equus hemmionus*)
бурида'нов ~ Buridan's ass
осемене'ние *n* insemination; service
гетероспе'рмное ~ heterospermic insemination
гомоспе'рмное ~ homospermic insemination
иску'сственное ~ artificial insemination
осеменя'ть *v* inseminate, spermatize
оси'на *f* aspen (*Populus tremula*)
оси'новый *a* aspen
оси'ный *a* vespine
оска'л *m eth* grin; **демонстра'ция оска'ла** bared-teeth display
приве'тственный ~ greeting grin
оска'ливать *v* bare teeth
~ **зу'бы** bare teeth, show teeth
о'скулюм *m zool* oscule
осмете'рия *f* osmeterium
ослабева'ть *v* grow weak, fail, abate
ослабле'ние *n* attenuation
осле'нок *m* colt, foal
осли'нник *m* evening-primrose (*Oenothera*)
осли'ца *f* she-ass
валаа'мова ~ Balaam's ass
осложне'ние *n* complication
осмиофи'льный *a* osmiophil
осмоме'тр *m* osmometer
осморегуля'ция *f* osmoregulation, osmotic regulation
о'смос *m* osmose, osmosis
капилля'рный ~ capillary osmose
отрица'тельный ~ negative osmose
обра'тный ~ reversed osmose
осмоти'ческий *a* osmotic
осмо'тр *m* examination
осмотропи'зм *m* osmotropism
осмотрофи'я *f* osmotrophy
осмотро'фный *a* osmotrophic
осмофи'льный *a* osmophil
основа'ние *n* base
~ **листа' base** of leaf
~ **околоцве'тника** base of perianth
~ **ру'бчика се'мени** omphalodium
~ **че'репа** base of skull, basilar region
~ *mbiol* **мино'рное** ~ minor base
~ **пури'новое** ~ purine base
основа'тельница *f ent* stem mother
осо'бенность *f* character
о'собь *f* individual, specimen
взро'слая ~ adult
высоотвеча'ющая ~ good responder, high responder
низкоотвеча'ющая ~ low responder, poor responder
за'дняя доче'рняя ~ *prot* opisthe
о'собь-часово'й *f* sentinel
осови'дный *a* vespine
осое'д *m* honey buzzard (*Pernis apivorus*)
осо'ка *f* sedge (*Carex*)
осоко'рь *m* black poplar (*Populus nigra*)
осола'живание *n* malting
осола'живатель *m* malting agent
осо'т *m* sow thistle (*Sonchus*)
о'спа *f* pox
оссеи'н *m* ossein
оста'вленный *a* abandoned
остано'вка *f* arrest, stop
~ **дыха'ния** respiratory standstill, respiratory arrest
~ **се'рдца** cardiac arrest
~ **се'рдца в диа'столе** diastolic arrest
~ **се'рдца в си'столе** systolic arrest
постепе'нная ~ *phys* obsolescence
оста'тки *m pl* remains, residues
пору'бочные ~ *pl* slash
органи'ческие ~ *pl* organic residues
расти'тельные ~ *pl* plant residues
эпителиа'льные ~ epithelial debris
оста'ток rest, remnant, residue, resudual
~ **перего'нки** distillation residue
заро'дышевый ~ embryonal rest
кисло'тный ~ acid residue
ку'бовый ~ distillation residue
эпителиа'льный ~ epithelial remnant
оста'точный *a* residual
остеобла'ст *m* osteoblast
остеогене'з *m* osteogenesis
остеоге'нный *a* osteogenic, ossific, osteoplastic
остеоденти'н *m* osteodentine
остео'ид *m* osteoid
остеокла'ст *m* osteoclast
остеоло'гия *f* osteology
остеомиели'т *m* osteomyelitis
~ **Гарре'** Garre's disease
остео'н *m* osteon

остеопласти'ческий *a* osteoplastic
остеохондра'льный *a* osteochondral
остеоци'т *m* osteocyte, bone corpuscle
остепне'ние *n* steppification, steppization, steppe formation
ости'стость *f* beardedness
ости'стый *a* awned, bearded, barbate
о'стия *f* ostium
о'стов *m* framework, stroma
осторо'жность *f* guard
острие' *n* point, tip
ка'пельное ~ dripping tip, rain tip (*of a* leaf)
остри'женный *past participle* shorn
о'стров *m* island
острова'тый *a* acutate
острово'к *m* island, islet
~ **Лангерга'нса** *anat* Langerhans islet, pancreatic island, islet of Langerhans, island of Langerhans
кровяно'й ~ blood island
панкреати'ческий ~ *anat* pancreatic islet, pancreatic island, island of Langerhans
ре'йлиев ~ *anat* island of Reil
остроконе'чный *a* mucronate, mucronated, pointed, cuspidate, acuminate, cuspidate
остроли'стный *a* acutifoliate
о'стро-пё'стро *n* milk thistle *(Silybum Marianum)*
острота' *f* acuity, acuteness (*e.g., of vision*)
~ **зре'ния** visual acuity, acuity of vision, acuteness of vision
~ **обоня'ния** olfactory acuity
~ **ощуще'ния** sensory acuity
~ **слуха** hearing acuity, auditory acuity, acuity of hearing
повы'шенная ~ **зре'ния** hyperacuity
остроцветко'вый *a* acutiflorous
острошипова'тый *a* oxyacanthous
острощети'нистый *a* exasperate
о'стрый *a* acute, sharp, pointed; hot (*e.g., pepper*), acrid, pungent; strong (*of cheese*)
ость *f bot* awn, arista, spine, beard, barb; *zool* guard hair
~ **клинови'дной ко'сти** alar spine
~ **лопа'тки** scapular spine
бараба'нная ~ tympanic spine
бло'ковая ~ trochlear spine
зубча'тая ~ scabrous awn
пе'ристая ~ plumose awn
подбоpо'дочная ~ mental spine
носова'я ~ nasal spine
подвздо'шная ~ iliac spine
седа'лищная ~ ischial spine, sciatic spine
осциллогра'мма *f* oscillogram
осциллогра'фия *f* oscillography
осцилломе'трия *f* oscillometry
осыпа'ние *n* fall (*of seeds*)
осыпа'ться *v* fall, shatter (*of seeds, leaves*)
осыпа'ющийся *a* deciduous
осыха'ние *n* drying-up
ось *f* axis, rachis
~ **ко'лоса** ear axis
~ **цветка'** floral axis
поля'рная ~ pole axis
попере'чная ~ transapical axis *(of diatoms)*
продо'льная ~ apical axis *(of diatoms)*
центра'льная ~ pervalvar axis *(of diatoms)*
электри'ческая ~ *phys* electric axis, electrical axis
осьмино'г *m* octopus (*Octopoda*)
осяза'ние *n* tactile sense, sense of touch, touch
осяза'тельный *a* tactile, tactual
осяза'ть *v* feel
от *preposition*: **защища'ть от заболева'ний** protect against diseases; **меры защиты от** protective measures against; **сре'дство от головно'й бо'ли** headache remedy; **страда'ть от боле'зни** suffer from an illness; **умере'ть от боле'зни** die of illness; **умере'ть от ра'ны** die from a wound; **умере'ть от я'да** die by poison; **жеребе'нок от...и** foal by...out of
ота'ва *f* aftergrass, aftermath, regrowth
отбива'ть *v* ride (*an animal from the herd*)
отбива'ться *v* stray from
отби'вшийся *a*: ~ **от стада** stray
отби'ться *v* stray from
~ **от ста'да** stray
отбо'р *m* selection, sorting, screening
~ **внутри' де'мов** intrademe selection
~ **по геноти'пу** genotypic selecion
~ **под влия'нием среды'** environmental selection
~ **по по'лным си'бсам** full-sib selection
~ **по полуси'бсам** half-sib selection
~ **по пото'мству** progeny selection
~ **по семена'м** seed selection
~ **по феноти'пу** phenotypic selection
~ **проб** sampling
~ **про'тив домина'нтного алле'ля** selection against a recessive allele

~ про'тив рецесси'вного алле'ля selection against a dominant allele
~ среди' де'мов interdeme selection
~ чи'стых ли'ний pure-breeding selection
апостати'ческий ~ apostatic selection
внутривидово'й ~ intraspecific selection
внутриполово'й ~ intrasexual selection
внутрипопуляцио'нный ~ intrapopulation selection
втори'чный ~ secondary selection
есте'ственный ~ natural selection
гамети'ческий ~ gamete selection
группово'й ~ group selection
дизрупти'вный ~ disruptive selection
заро'дышевый ~ germinal selection
зача'тковый ~ germinal selection
иску'сственный ~ artificial selection
канализи'рующий ~ canalizing selection
кло'новый ~ clone selection, sib selection
лине'йный ~ line selection, linear selection
ма'ссовый ~ mass selection
межвидово'й ~ interspecific selection
межгруппово'й ~ intergroup selection
межлине'йный ~ interline selection
межпопуляцио'нный ~ interpopulation selection
нака'пливающий ~ cumulative selection
напра'вленный ~ directional selection
нормализу'ющий ~ normalizing selection
органи'ческий ~ organic selection
первонача'льный ~ original selection
периоди'ческий ~ periodic selection, recurrent selection
периоди'ческий односторо'нний ~ one-way recurrent selection
периоди'ческий реципро'кный ~ recurrent reciprocal selection
повто'рный ~ вы'борок repeated sampling, sampling on successive occasions
повторя'ющийся ~ recurrent selection
полово'й ~ sexual selection
после'довательный ~ вы'борок sequential sampling
приспособи'тельный ~ adaptive selection
разруша'ющий ~ destructive selection
ра'нний ~ early selection
семе'йный ~ kin selection
системати'ческий ~ проб systematic sampling
случа'йный ~ random selection
случа'йный ~ вы'борок random sampling
стабилизи'рующий ~ stabilizing selection
флуктуи'рующий ~ fluctuating selection
частотозави'симый ~ density-dependent selection, frequence-dependent selection
отбира'ть *v* select
отбрако'вка *f* cull
отбрако'вывать *v* cull
отбро'сы *m pl* waste
отва'л *m* dump
отва'р *m* decoction, broth
ри'совый ~ rice broth
сенно'й ~ hay broth
отведе'ние *n* abduction, lead, derivation; recording
~ сенсо'рных и'мпульсов sensory recording
прямо'е ~ direct lead
то'чечное ~ focal recording
отведе'нный *a* avert (*of stare*); diverted
отверга'ть *v* reject
отве'рнутый *a* revolute
отве'рстие *n* aperture, orifice, foramen, opening, fenestra, ostiole, ostium
~ Маже'нди Magendie's foramen
~ мочето'чника urethral orifice
ана'льное ~ anus, rectal orifice
большо'е заты'лочное ~ foramen magnum
верху'шечное ~ apical foramen
влага'лищное ~ vaginal orifice
вну'треннее слухово'е ~ internal auditory foramen
дыха'тельное ~ breathing opening
заднепрохо'дное ~ anus, rectal orifice
запира'тельное ~ obturator foramen
зри'тельное ~ optic foramen
крестцо'вое ~ sacral foramen
межжелу'дочковое ~ interventricular foramen
надглазни'чное ~ supraorbital foramen
нару'жное слухово'е ~ external auditory foramen
пита'тельное ~ nutrient foramen *(in a bone)*
позвоно'чное ~ vertebral foramen
полово'е ~ genital opening
резцо'вое ~ incisive foramen
ротово'е ~ mouth opening, buccal orifice; actinostome *(of sea anemones, sea stars)*
са'льниковое ~ epiploic foramen
скуловисо'чное ~ zygomaticotemporal foramen
скулоглазни'чное ~ zygomaticoorbital foramen

скулолицево'е ~ zygomaticofacial foramen
шилососцеви'дное ~ stylomastoid foramen
яре'мное ~ jugular foramen
отве'т *m* response
адапти'вный имму'нный ~ adaptive immune response
аллоимму'нный ~ alloresponse
аутоимму'нный ~ antiself response, autoimmune response, autospecific response
безадъюва'нтный имму'нный ~ adjuvant-free response
втори'чный ~ secondary response
втори'чный имму'нный ~ anamnestic response, boosted response, secondary immune response, memory response
вы'званный ~ evoked response
гормона'льный ~ hormonal response
закреплё'нный ~ fixated response
зри'тельный ко'рковый ~ visual cortical response
конкуре'нтный ~ competitive response
непродолжи'тельный имму'нный ~ short-lived immune response
электроко'жный ~ electrodermal response
ответвле'ние *n* offset, offshoot, branch, derivation
отвеча'ть *v* respond
отви'слый *a* pendulous
отводи'ть *v* decoy away (*of a bird using, e.g., sham cries of pain*)
отво'док *m* offset, offshoot, layer, runner, ratoon, sapling, stool, cutting
отводя'щий *a* abducent (*of a muscle*), revehent
отворо'ченный *a* recurvate
отвраще'ние *n* aversion, abhorrence
отгоня'ть *v* repel
отгрыза'ть *v* bite off
отгры'зть *v* bite off
отдалё'нный *a* remote
отда'ча *f* output
отде'л part, portion
брюшно'й ~ abdominal region
отделе'ние *n* abjunction; isolation; department
отделя'емое *n* discharge
отделя'ть *v* separate, remove, abject; ride (*an animal from the herd*)
отду'шка *f* flavour
о'тдых *m* rest
отё'к *m* edema, dropsy, hydrops
~ **Кви'нке** Quincke's edema
отё'л *m* calving; fawning (*of deer*)
отели'ться *v* calve; fawn (*of deer*)
оте'чный *a* edematic, edematous, dropsical
отка'рмливать *v* fatten, engross
откла'дка *f* oviposition
~ **яи'ц** oviposition
откла'дывать *v* lay
~ **икру'** spawn
~ **я'йца** oviposit
отклоне'ние *n* deviation, derivation, deflection
~ **компле'мента** deviation of complement, complement deviation, complement deflection
~ **от регре'ссии** deviation from regression
абсолю'тное ~ absolute deviation
вероя'тное ~ probable deviation
вну'треннее ~ intrinsic deflection
допусти'мое ~ standard deviation
имму'нное ~ immune deviation
квадрати'чное ~ square deviation
максима'льное ~ maximum deviation
норма'льное ~ normal deviation
сре'днее ~ mean deviation, average deviation
среднеквадрати'чное ~ mean-square deviation, root-mean-square deviation
станда'ртное ~ standard deviation
эпистати'ческое ~ epistatic deviation
отклоня'ться *v* deviate
отко'рм *m* fattening, engrossing
откорми'ть *v* give finish
отко'рмленный *a* finished, fat, fattened, fatted
отко'рмочный *a* fattening
открыва'ние *n* opening
~ **у'стьиц** stomatal opening
открыва'ть *v* discover
откры'тие *n* discovery
открытопузы'рный *a* physostomous
откры'тый *a* open; outdoor; **под откры'тым не'бом** outdoor
откуси'ть *v* bite off
отку'сывать *v* bite off
отла'вливать *v* catch, trap, capture
отли'в low tide
отло'в *m* catching, capture, trapping
вы'борочный ~ selective trapping, selective capture
отлови'ть *v* catch, trap
отложе'ние *n* deposit
отложения *n pl* sediments

до'нные ~ bottom sediments
отло'женный *participle* laid
отлуча'ть *v* separate from
~ от ма'тери wean
отме'тина *f* mark; star (*on forehead*)
отмира'ние *n* die-off
отмира'ющий *a* moribund
отмороже'ние *n* frostbite
отморо'женный *a* frost-bitten
отморо'зить *v* injure by frostbite
отнерести'вшийся *a* spent
отнима'ть *v* remove, take away; amputate; subtract
~ но'гу amputate a leg
~ от груди' wean
отнима'ться *v* be paralysed; **у него' отняла'сь нога'** his leg is paralysed; **у него' отнялся' язы'к** he has lost the use of his tongue
отноше'ние *n* relation, relationship; ratio
~ диспе'рсий variance ratio
~ домина'нт-соподчинё'нный *eth* leader-follower relationship
~ домини'рование-подчине'ние *eth* dominance-submission relation
~ зиго'т zygotic ratio
~ основа'ний *mbiol* base ratio
~ правдоподо'бия likelihood ratio
~ ро'ста growth ratio
~ хи'щник-же'ртва prey-predator relationship
вариа'нтное ~ variance ratio
внутривидовы'е отноше'ния *pl* intraspecies relations
корреляцио'нное ~ correlation ratio
межвидовы'е отноше'ния *pl* interspecies relations
ме'нделевское ~ *gen* Mendelian ratio
пищевы'е отноше'ния *pl* food relations
я'дерно-плазмати'ческое ~ karyoplasmatic ratio, nucleocytoplasmic ratio, nucleoplasmic ratio
я'дерно-я'дрышковое ~ nucleonucleolar ratio
отня'тие *n* removal
ото'гнутый *a* deflexed, excurved
отоко'ний *m* otoconium
отоли'т *m* otolith, ear stone
отоларинго'лог *m* otolaryngologist
отоларинголо'гия *f* otolaryngology
отоокципита'льный *a* oto-occipital
отоpo'ченный *a* bordered
отоци'ст *m* otocyst
отпеча'ток *m* impression; replica; scar (*of a muscle on a bivalve shell*)
спо'ровый ~ *bot* spore print
отпочко'вываться *v* bud off
о'тпрыск *m* sarment, spear, sapling, scion, stool layer, runner, sobole; offset, offshoot, offspring
корнево'й ~ ratoon, surculus
отпу'гивать *v* scare; repel
отта'лкивать *v* repel
отта'лкивающий *a* repellent
отпу'гивающий *a* repugnatorial
отра'ва *f* poison
отравле'ние *n* poisoning, intoxication
пищево'е ~ food poisoning
отравля'ющий *a* poisonous
отраста'ние *n* regrowth
отреза'ть *v* cut off; excise; lop (*of branches*)
отре'зать *v* have cut off
отро'сток *m* ramification, appendage, appendix; projection, outgrowth, process; *bot* shoot, sprout, chit, tiller; *zool* tine (*of horns*)
~ Фо'ля Folian process
боево'й ~ bez tine
боково'й ~ offset, offshoot
вене'чный ~ coronoid process
верхнечелюстно'й ~ maxillary process
головно'й ~ head process
зонтикови'дный ~ umbraculum
зубови'дный ~ эпистрофе'я odontoid process
клювови'дный ~ coracoid process
ко'стный ~ bony process
ло'бный ~ brow tine
локтево'й ~ olecranon
мечеви'дный ~ груди'ны xiphoid process, xiphioid appendix
невра'льный ~ neurapophysis
носово'й ~ nasal process
ости'стый ~ spinous process
пере'дний ~ молото'чка Folian process
рогообра'зный ~ hornlike process
слепо'й ~ cecum, caecum
сосцеви'дный ~ mammillary process
сочлено'вный ~ zygapophysis
спинно'й ~ dorsal process
суставно'й ~ articular process
цилиа'рный ~ ciliary process
челюстно'й ~ mandibular process
яре'мный ~ jugular spine

отруба'ть *v* lop off (*e.g., a tail*), cut off
о'труби *pl* bran
отры'гивать *v* belch, regurgitate
отры'жка *f* belching, eructation
отря'д *m* order
  спаса'тельный ~ rescue party
отса'сывание *n* suction, sucking-off, aspiration, exsufflation
отсека'ть *v* lop (*of branches*)
отсла'ивание *n* exfoliation, desquamation
отсле'живание *n* tracking
отслое'ние *n* exfoliation, desquamation
отстава'ние *n* lagging
отсто'й *m* residue
отсто'йник *m* sedimentation basin, sedimentation tank, settling basin, settling tank, sludge tank, clarification tank
  ~ **сто'чных вод** sewage tank
отстре'л *m* shooting, killing
  вы'борочный ~ selective killing
  рациона'льный ~ rational hunting
  регули'руемый ~ regulated hunting
отступа'ние *n* retreat
  ~ **ледника'** glacial retreat
отсу'тствие *n* failure, absence
отсыха'ть *v* wither
отта'ивание *n* thawing
отта'лкивание *n* repulsion
о'ттепель *f* thaw
отто'к *m* outflow, outflux
оттопы'ренный *a* squarrose, bulging out
отторже'ние *n* rejection
  ~ **транспланта'та** graft rejection, transplant rejection
оття'нутый *a* tapered
отха'ркивание *n* expectoration
отхо'д *m* mortality, losses; retreat
отхо'ды *m pl* waste
  ~ **животново'дства** livestock waste
  бытовы'е ~ domestic waste
  лесосе'чные ~ logging waste
  промы'шленные ~ trade waste
отцвести' *v* cease blossoming
отцве'тший *a* deflowered
отце'живать *v* strain off; strip (*sex products from a fish*)
отцо'вский *a* paternal
отцо'вство *n* paternity
отчлене'ние *n* abjunction
отше'льнический *a* eremobic
отщепле'ние *n* removal
отщепля'ть *v* remove
отъё'м *m* weaning
отъе'мыш *m* weanling
офиоплуте'ус *m* ophiopluteus
оха'пка *f* lock (*e.g., of hay*), armful
охва'т *m* amplexus; scope, range
охва'тывающий *a* embracing, amplectant
охлажде'ние *n* chilling
охлажде'нный participle chilled (*e.g., meat*)
охора'шиваться *v* preen
охо'та *f* hunting; rut
  ~ **на пти'цу** fowling
  полова'я ~ rut, ruttishness
  промысло'вая ~ commercial hunting
  руже'йная ~ shooting
  соколи'ная ~ falconry
  спорти'вная ~ sport hunting
охо'титься *v* hunt
охо'тник *m* hunter
охра'на *f* protection, conservation
  ~ **здоро'вья** health protection
  ~ **и рациона'льное испо'льзование земе'ль** land conservation
  ~ **и рациона'льное испо'льзование подзе'мных вод** groundwater conservation
  ~ **матери'нства** maternity protection
  ~ **приро'ды** conservation, nature conservation, nature protection, conservation of nature, wildlife preservation
  ~ **террито'рии** *eth* patrolling
  ~ **травяно'го покро'ва** grass conservation
  ~ **труда'** labor protection
охрящеве'ние *n* cartilagination, cartilagnification, chondrification
оцело'т *m* ocelot *(Felis pardalis)*
оце'нивание *n* estimation
оце'нивать *v* estimate
оце'нка *f* assessment, evaluation, appraisal, estimation, estimate; estimator
  ~ **биома'ссы** estimate of standing crop
  ~ **во'дных ресу'рсов** water resources appraisal
  ~ **воздействия на окружающую среду, ОВОС** environmental impact assessment, EIA
  ~ **максима'льного правдоподо'бия** maximum likelihood estimator
  ~ **на глаз** estimate by eye
  ~ **погре'шности** estimation of error
  ~ **по ме'тоду регре'ссии** regression estimate
  ~ **состоя'ния среды'** environmental assessment

абсолю'тная несмещё'нная ~ absolutely unbiased estimator
асимптоти'чески несмещё'нная ~ asymptotically unbiased estimator
асимптоти'чески эффекти'вная ~ asymptotically efficient estimator
визуа'льная ~ visual estimate
доста'точная ~ sufficient estimator
комбини'рованная регрессио'нная ~ combined regression estimate
лине'йная несмещё'нная ~ linear unbiased estimate
несмещё'нная ~ unbiased estimate, unbiased estimator
неэффекти'вная ~ inefficient estimator
оптима'льная ~ optimum estimator
по'лная ~ omnibus estimate, overall estimate
приблизи'тельная ~ approximate estimate
регрессио'нная ~ regression estimator
состоя'тельная ~ consistent estimator
товарове'дческая ~ trade analysis
эффекти'вная ~ efficient estimator

оцепене'вший *a* torpid

оцепене'ние *n* numbness, stupor, stupefaction, torpor
холодо'вое ~ chill coma

оча'г *m* locus, focus
~ **зараже'ния** infection focus
приро'дный ~ hot spot, natural focus
эпидеми'ческий ~ effective disease area, sick area

очаго'вость nidality
приро'дная ~ natural nidality

очаго'вый *a* focal

очередно'й *a* alternate

очере'т *m* bog rush (*Schoenus*)

очерта'ние *n* outline

оче'сывание *n* bolling (*of flax*)

очи'н *m* calamus

очи'стка *f* purification; disposal
~ **воды'** water purification
~ **лесосе'к** slash disposal
~ **осажде'нием** precipitation purification
~ **от радиакти'вных загрязне'ний** decontamination
~ **сто'чных вод** sewage disposal, purification of sewage, sewage treatment, effluent treatment
биологи'ческая ~ biological treatment
предвари'тельная ~ coarse purification

очи'ток *m* stonecrop (*Sedum*)

очища'ть *v* clean, cleanse; purify, refine; dress; peel, shell; rectify

очище'ние *n* purification
~ **от бакте'рий** bacterial purification

оши'бка *f* error
~ **аппроксима'ции** approximation error
~ **вы'борки** sampling error
~ **выра'внивания** error of fitting
~ **измере'ния** measurement error
~ **копи'рования** *gen* copy error
~ **наблюде'ния** error of observation, observation error
~ **отсчё'та** error of reading
~ **пе'рвого ро'да** error of the first kind, type I error
~ **приближе'ния** approximation error
~ **произво'льной вы'борки** random sampling error
~ **репроду'кции** *gen* copy error
~ **сре'днего** error of mean
~ **ра'зности** error of difference
абсолю'тная ~ absolute error
вероя'тная ~ probable error
допусти'мая ~ admissible error
ко'синусная ~ cosine error
нако'пленная ~ cumulative error
относи'тельная ~ relative error
скорректи'рованная ~ corrected error
случа'йная ~ random error
сре'дняя ~ mean error, average error
сре'дняя квадрати'ческая ~ mean-square error, standard error
сре'дняя квадрати'ческая ~ диспе'рсии standard error of variance
сре'дняя квадрати'ческая ~ оце'нки standard error of estimate
сре'дняя квадрати'ческая ~ ра'зности standard error of difference
сре'дняя квадрати'ческая ~ сре'дней standard error of mean
сре'дняя квадрати'ческая ~ су'ммы standard error of sum
станда'ртная ~ mean-square error, standard error
станда'ртная ~ диспе'рсии standard error of variance
станда'ртная ~ оце'нки standard error of estimate
станда'ртная ~ ра'зности standard error of difference
станда'ртная ~ сре'дней standard error of mean

станда'ртная ~ станда'ртного отклоне'ния standard error of standard deviation
станда'ртная ~ су'ммы standard error of sum
ощени'ться *v* pup (*of dogs*), whelp, cub
ощи'пывание *n* picking (*of a bird*)
ощи'пывать *v* defeather, pick
ощи'пывающий *a* browsing
ощу'пывание *n* palpation
ощуща'ть *v* feel
ощуще'ние *n* feeling, sensation, esthesis
ло'жное ~ false sensation
рефлекто'рное ~ referred sensation, transferred sensation

П

па'ва *f orn* peahen
павиа'н *m* baboon (*Papio*)
~ **ану'бис** anubis baboon, Doguera baboon, olive baboon *(Papio anubis)*
до'геровский ~ anubis baboon, Doguera baboon, olive baboon *(Papio anubis)*
жё'лтый ~ yellow baboon *(Papio cynocephalus)*
плащено'сный ~ hamadryas baboon, sacred baboon *(Papio hamadryas)*
павли'н *m* peacock, peafowl (*Pavo cristatus*)
молодо'й ~ peachick
па'водок *m* flood
па'губный *a* pernicious, fatal
па'далица *f* fallen fruit, fruit droppage, fruit drop
па'даль *f* carrion
па'дать *v* fall, drop
па'дди *m* paddy
падё'ж *m* loss
~ **скота'** loss of cattle
паде'ние *n* drop
па'дуб *m* holly (*Ilex*)
падь *f* honeydew; gully
па'зуха *f anat* sinus, atrium; *bot* axil
~ **бараба'нной по'лости** tympanic sinus
~ **решё'тчатой ко'сти** ethmoid sinus
вернехнечелюстна'я ~ maxillary sinus
заднепрохо'дная ~ rectal sinus
ло'бная ~ frontal sinus
мо'рганиева ~ rectal sinus
основна'я ~ basilar sinus
прида'точная ~ **но'са** accessory sinus of nose
прямокише'чная ~ rectal sinus
па'зушный *a* axillary
па'ка *f ther* spotted cavy *(Cuniculus paca)*
паклеви'дный, паклеобразный *a* towlike, stupeous, stupose
па'кля *f* oacum
пакси'лла *f* paxilla
пал *m* grassland fire, burning, fire
палатоквадра'тный *a* palatoquadrate
палеаркти'ческий *a* palearctic
палео- paleo-
палеобиоло'гия *f* paleobiology
палеобота'ника *f* paleobotany
палеоге'н *m* paleogene
палеогене'з *m* paleogenesis
палеогене'тика *f* paleogenetics
Палеоге'я *f* Palaeogaea
палеозо'йский *a* Paleozoic
палеозооло'гия *f* paleozoology, zoopaleontology
палеокра'ний *m* paleocranium
палеолимнологи'ческий *a* paleolimnological
палеолимноло'гия *f* paleolimnology
палеонто'лог *m* paleontologist
палеонтоло'гия *f* paleontology
палеосинхороло'гия *f* paleosynchorology
палеосинхроноло'гия *f* paleosynchronology
палеосинэколо'гия *f* paleosynecology
палеоце'новый *a* Paleocene
палеоэколо'гия *f* paleoecology
пале'тка *f* pallet, *ent* palette
па'лец *m* finger (*of hand*), toe (*of foot*), dactyl, digit
~ **ки'сти** finger
~ **ноги'** toe
~ **стопы'** toe
безымя'нный ~ ring finger
большо'й ~ thumb, pollex, first finger
за'дний ~ hallux (*of birds*)
сре'дний ~ long finger
указа'тельный ~ forefinger, index finger
палингене'з *m* palingenesis
палингенети'ческий *a* palingenetic
палиндро'м *m mbiol* palindrome
палиногра'мма *f* palynogram
палинологи'ческий *a* palynological
палиноло'гия *f* palynology
палиномо'рфа *f* palynomorpha
палиса'дник *m* front yard, front garden
палиса'дный *a* palisade
паллиати'вный *a* palliative, mitigating
па'ллиум *m* pallium
паломи'но *n* palomino
па'лочка *f* rod
сенна'я ~ hay bacterium (*Bacillus subtilis*)
палочкови'дный *a* baculate, baculiform, bacillar, bacillary, bacilliform
па'лочник *m ent* walking stick (*Phasmodea*)
па'льма *f* palm (*Palmae*)
~ **кате'ху** betel-nut-palm *(Areca catechu)*
аре'ковая ~ betel-nut-palm (*Areca catechu*)
капу'стная ~ cabbage palm (*Euterpe oleracea*)
коко'совая ~ coconut palm, coco palm (*Cocos nucifera*)

са'говая ~ sago palm (*Metoxylon*)
фи'никовая ~ date palm (*Phoenix dactylifera*)
пальма'рный *a* palmar
пальме'лла *f* palmella
пальмеллеви'дный *a* palmelloid
пальме'тта *f* palmetto
пальмити'н *m* palmitin
па'льпа *f* palp, palpus
пальпа'ция *f* palpation
пальпиге'р *m* palpiger
пальпи'ровать *v* palpate
пальпи'руемый *a* palpable
пальпита'ция *f* palpitation
пальпифе'р *m* palpifer, palpus-bearer
пальпо'н *m zool* palpon
пальцеви'дный *a* dactylic, digital
пальцево'й *a* dactylar, digital
пальцеобра'зный *a* dactyloid, digital
пальцеходя'щий *a* digitigrade
пальчатови'дный *a* palmatiform
па'льчато-жилкова'тый *a* palmatinerved
пальчатоло'пастный *a* palmatilobate
пальчатоне'рвный *a bot* palmately veined, actinodrome, actinodromous, digitinervate, palmatinerved
па'льчато-пе'ристый *a* digitipinnate
па'льчато-разде'льный *a bot* digitipartite, palmatipartite
па'льчато-рассечё'нный *a* palmatisected
па'льчатый *a* palmate, digitate, dactylate
па'мпа *f* pampa
пампрода'ктильный *a* pamprodactylous
па'мять *f* memory
~ **на антиге'н** antigen-specific memory
~ **на отдалё'нные собы'тия** remote memory
~ **на «своё»** self-specific memory
антиге'н-специфи'ческая ~ antigen-specific memory
долговре'менная ~ long-term memory, reference memory
зри'тельная ~ eye memory, visual memory
идиотипи'ческая ~ idiotypic memory
иммунологи'ческая ~ immunological memory
кратковре'менная ~ short-term memory, recent memory
логи'ческая ~ logical memory
операти'вная ~ recent memory, short-term memory
подсозна'тельная ~ subconscious memory
рабо'чая ~ working memory
слухова'я ~ aural memory
чу'вственно-о'бразная ~ affect memory
пан- pan-
панагглютина'ция *f* panagglutination
панари'ций *m* panaritium
панаркти'ческий *a* panarctic
панга'мия *f* pangamy
панга'мный *a* pangamic
панге'н *m* pangen
пангене'з, пангене'зис *m* pangenesis
пангидро'з *m* panhydrosis
панголи'н *m* scaly anteater (*Manis*)
па'нда *f* panda *(Ailurinae)*
пандеми'чный *a* pandemic
пандеми'я *f* pandemic
пане'ль *f* panel
~ **моноклона'льных антите'л** panel of monoclonal antibodies
этало'нная стандартизо'ванная ~ reference panel
панзоо'тия *f* panzootic
панкли'макс *m* panclimax
панкреати'н *m* pancreatin
панкреати'т *m* pancreatitis
панкреати'ческий *a* pancreatic
панмикси'я *f* panmixia, panmixis
панми'кстный, панмикти'ческий *a* panmictic
паноисти'ческий *a* panoistic
пантана'ль *f* pantanal
панте'ра *f* panther *(Panthera pardus)*
пантои'л *m* pantoyl
пантокри'н *m* pantocrine
пантотена'т *m* pantothenate
пантропи'ческий *a* pantropical
па'нты *pl* antlers of young Siberian stag
панформа'ция *f* panformation
панформио'н *m* panformion
панфотометри'ческий *a* panphotometric
па'нцирный *a* loricate
па'нцирь *m* armour, shell, carapace, case, mail, testa; lorica (*of a rotifer*), test (*of a diatom*), frustule (*of a diatom*)
папа'йя *f* papaya (*Carica papaya*)
папи'лла *f* papilla
ана'льная ~ *ent* anal papilla
папилло'ма *f* papillary tumor
папилля'рный *a* papillary
папи'рус *m* papyrus *(Cyperus papyrus)*
папoвави'рус *m* papovavirus
па'поротник *m* fern (*Felicidae*)

~ **оле'ний рог** staghorn fern (*Platycerium*)
древови'дный ~ tree fern
па'поротниковый *a* filical
папоротниколи'стный *a* fern-leaved
па'пула *f* papule, papula
пар *m* vapour, steam; *agr* fallow
теку'чий ~ flowing steam
чё'рный ~ black fallow
па'ра *f* couple
пара- para-
па'ра *f* pair
~ **алле'лей** factor pair
~ **нуклеоти'дов** base pair
конъюгацио'нная ~ mating pair
параагглютини'н *m* minor agglutinin, partial agglutinin
парааденови'рус *m* para-adenovirus
парабаза' льный *a* parabasal
парабио'з *m* parabiosis
парабио'нт *m* parabiont
парабиоти'ческий *a* parabiotic
парабла'ст *m* parablast
параварииа'ция *f* paravariation
парага'нглий *m* paraganglion
поясни'чно-аорта'льный ~ aorticolumbar paraganglion
парага'струла *f* paragastrula
парагелиотропи'зм *m* paraheliotropism
парагене'з *m* paragenesis
парагенео'н *m* parageneon
парагло'сса *f ent* paraglossa
параде'ма *f* parademe
параде'рма *f* paraderm
параде'сма *f* paradesmus
паради'димис *m* paradidymis
парадокса'льный *a* paradoxical
парази'т *m* parasite
~ **расте'ний** plant parasite
вну'тренний ~ internal parasite
внутрикле'точный ~ intracellular parasite, cell parasite
внутрико'жный ~ intradermal parasite
внутриэритроцита'рный ~ endoglobular parasite
вре'менный ~ temporary parasite
втори'чный ~ secondary parasite
гетероксе'нный ~ heteroxenous parasite
гнездово'й ~ brood parasite
живо'тный ~ animal parasite
кише'чный ~ intestinal parasite
кле'точный ~ cell parasite
корнево'й ~ root parasite
кровососу'щий ~ blood-sucking parasite
нару'жный ~ epizoic parasite, external parasite
облига'тный ~ holoparasite, obligate parasite
постоя'нный ~ permanent parasite
разнохозя'йный ~ heteroxenous parasite
расти'тельный ~ plant parasite
случа'йный ~ incidental parasite
тка'невый ~ tissue parasite
факультати'вный ~ facultative parasite
паразита'рный *a* parasitic, parasitical, parasitogenic
паразитеми'я *f* parasitemia
паразити'зм *m* parasitism
полово'й ~ *eth* sexual parasitism
паразити'ровать *v* parasitize
паразити'ческий *a* parasitic
паразито'з *m* parasitic disease, parasitosis
паразито'ид *m* parasitoid
паразито'лог *m* parasitologist
паразитоло'гия *f* parasitology
паразитоноси'тель *m* parasite carrier
паразитоноси'тельство *n* parasitosis
паразитофо'рный *a* parasitophorous
парака'ллюс *m* paracallus
парака'рпий *m* paracarpium
парака'рпный *a* paracarpous
паракли'макс paraclimax
эдафи'ческий ~ edaphic paraclimax
пара'кме *n* paracme
параксиа'льный *a* paraxial
парализова'ть *v* paralyze, benumb
парализу'ющий *a* paralyzant
парал'ич *m* palsy, paralysis
~ **Бе'лла** Bell's palsy
~ **Бро'ун-Сека'ра** Brown-Sequard paralysis
~ **Гу'блера** Gubler's paralysis
~ **Дежери'н-Клю'мпке** Dejerine-Klumpke's syndrome
~ **Дюше'нна-Э'рба** Duchenne-Erb paralysis
~ **Клю'мпке** Klumpke's paralysis
~ **Куссма'уля-Ландри'** Kussmaul-Landry paralysis
~ **Це'нкера** Zenker's paralysis
иммунологи'ческий ~ immunological paralysis
рефлекто'рный ~ reflex paralysis
параллели'зм *m* parallelism
параллелотропи'зм *m* parallelotropism

параллельноне'рвный *a* parallelinervate, parallelodrome, parallel-veined
паралле'льный *a* parallel
параллерги'я *f* parallergy
пара'метр *m* parameter
параметри'ческий *a* parametric
парамиксови'рус *m* paramyxovirus
парами'л *m* paramylon, paramylum
парамиотони'я *f* paramyotonia
парамо'рфа *f* paramorph
парамутуали'зм *m* paramutualism
парануклеи'н *m* paranuclein
парану'клеус *m* paranucleus
парапепси'н parapepsin
парапла'зма *f* paraplasm
парапле'гия *f* paraplegia
~ **По'тта** Pott's paraplegia
парапо'дия *f* parapodium, parapodial limb
парапо'физ *m* parapophysis
парапро'кт *m* paraproct
парапсида'льный *a* parapsidal
парапсихоло'гия *f* parapsychology
парaптеро'н *m* parapteron
парарефле'ксия *f* parareflexia
парарозанили'н *m* pararosaniline
парасексуа'льный *a* parasexual
параселекти'вность *f* paraselectivity
парасемати'ческий *a* parasematic
парасимпати'ческий *a* parasympathetic
параси'напсис *m* parasynapsis, parasyndesis
параси'ндез *m* parasynapsis, parasyndesis
парасистоли'я *f* parasystoly
параспо'ра *f* paraspore
парастери'льность *f* parasterility
парасти'ха *f* parastichy
парасфено'ид *m* parasphenoid
парасфено'идный *a* parasphenoid
парате'ций *m* parathecium
паратгормо'н *m* parathormone
парати'п *m* paratype
паратипи'ческий *a* paratypic
парато'идный *a* paratoid
парато'мия *f* paratomy
паратони'ческий *a* paratonic
парато'п *m* paratope
паратрофи'ческий *a* paratrophic
парафациа'льный *a* parafacial
парафи'за *f* paraphysis
парафи'н *m* paraffin
парафи'лл *m* paraphyll
парафототропи'зм *m* paraphototropism
парафро'нс *m ent* parafrons
парахорда'ль *f* parachordal
парахорда'льный *a* parachordal
парацеллюло'за *f* paracellulose
парашю'т *m bot* parachute
парвови'рус *m* parvovirus
паре'з *m* paresis
паре'ние *n* soaring, hovering
парентера'льный *a* parenteral
паре'нхима *f* parenchyma, parenchymatous tissue
краева'я ~ border parenchyma
ло'жная ~ spurious tissue
лубяна'я ~ phloem parenchyma
гу'бчатая ~ lacunose parenchyma, spongy parenchyma
сердцеви'нная ~ medullar parenchyma
сто'лбчатая паренхима columnar parenchyma, palisade parenchyma
паренхимато'зный *a* parenchymatous
паре'нхимный *a* parenchymal
паренхи'мула *f* parenchymula
париета'льный *a* parietal
пари'ть *v* hever, soar
па'рить *v* stew, steam
парк *m* park
национа'льный ~ national park
приро'дный ~ natural park
парло'дий *m* parlodion
пармеза'н *m* Parmesan-cheese, Parmesan
парни'к *m* hotbed
парнокопы'тное *n* even-toed ungulate, even-hoofed mammal, cloven-hoofed mammal, paridigitate mammal *(Artiodactyla)*
парнокопы'тный *a* even-toed, cloven-footed, artiodactyls
парнокопы'тное *n* paridigitate
парнопа'лое *n* paridigitate *(Artiodactyla)*
парнопа'лый *a* artiodactyl
парнопе'ристый *a bot* paripennate
парноцветко'вый *a* geminiflorous
па'рный *a* paired, bijugate, binate, geminate, geminated
паро'йкия *f* paroecia
партени'та *f* parthenita
партеноапога'мия *f* parthenapogamy
партенога'мия *f* parthenogamy, parthenomyxis
партеногене'з *m* parthenogenesis
аборти'вный ~ abortive parthenogenesis
дипло'идный ~ parthenapogamy

иску'сственный ~ artificial parthenogenesis, experimental parthenogenesis
мужско'й ~ patrogenesis
факультати'вный ~ facultative parthenogenesis
есте'ственный ~ natural parthenogenesis
сомати'ческий ~ parthenapogamy
спонта'нный ~ spontaneous parthenogenesis
случа'йный ~ spontaneous parthenogenesis
партеногенети'ческий *a* parthenogenetic
партенокарпи'ческий *a* parthenocarpic
партенока'рпия *f* parthenocarpy
партеноспо'ра *f* parthenospore
партено'т *m* parthenote
партиге'н *m* partigen
партнё'р *m* partner, associate, mate
конге'нный ~ congenic partner
па'рус *m* velum, *bot* standard, banner, vane, flag, vexillum
пере'дний мозгово'й ~ medullary velum
па'русник *m* swallowtail, kite butterfly (*Papilio*)
парце'лла *bot* parcel
парциа'льный *a* partial
парша' *f* favus, scab, mange, tetter
~ **карто'феля** potato scab *(caused by Actinomyces scabies)*
поро'шистая ~ **карто'феля** powdery scab of potato *(caused by Spongospora subterranea)*
чёрная ~ bare patch *(caused by Rhizoctonia solani)*
парши'вый *a* mangy, scabby
парэнтера'льный *a* parenteral
па'сека *f* apiary
паслён *m* nightshade
~ **сладкого'рький** bitter-sweet nightshade (*Solanum dulcamara*)
па'сока *f* xylem exudate
пасса'ж *m* passage, subculture, transfer, subinoculation
~ **на кури'ных эмбрио'нах** egg passage
непреры'вный сери'йный ~ continued serial passage
однокра'тный ~ single passage
очередно'й ~ subclone
повто'рный ~ repeated passage
после'довательный ~ successive passage
сери'йный ~ serial passage
слепо'й ~ blind passage
пасса'нт *m bot* passant
пасси'рование *n* passaging, passage
многокра'тное ~ high passaging
пасси'ровать *v* subculture
паста *f* pap, paste
па'стбище *n* pasture, pasturage, range, grazing land, feeding ground, pasture ground, lea
общи'нное ~ community pasture, common pasture
па'стбищный *a* pascual
пасте'ль *f* pastel *(coloration of mink)*
бурогла'зая ~ brown-eye pastel
пастериза'тор *m* pasteurizer
~ **периоди'ческого де'йствия** batch pasteurizer
~ **непреры'вного де'йствия** continuous pasteurizer, continuous-flow pasteurizer
пастериза'ция *f* pasteurization
мгнове'нная ~ flash pasteurization
момента'льная ~ flash pasteurization
дли'тельная ~ holder pasteurization
пастеризова'ть *v* pasteurize
пастерна'к *m* parsnip (*Pastinaca sativa*)
пасти' *v* herd, shepherd
пасти'сь *v* graze
пасту'х *m* cowherd, cowboy, herdsman
пастушо'к *m* rail (*Rallus*)
пасть *m* chaps, gorge, mouth
пастьба' *f* grazing, pasturage, pasturing
во'льная ~ open herding
заго'нная ~ rotational grazing
пасынкова'ние *n* sucker removal, pruning
пасынкова'ть *v* prune
па'сынок *m* sucker, side shoot
пасю'к *m* common rat, brown rat (*Rattus norvegicus*)
пате'лла *f* patella
пателлярный *a bot* patellar
патиссо'н *m* custard squash, patty pan squash, scalloped squash (*Cucurbita pepo var. patisson*)
пато- patho-
патобиоло'гия *f* pathobiology
патогене'з *m* pathogenesis, pathogenesy, pathogeny, pathogenicity
патогенети'ческий *a* pathogenetic
патоге'нность *f* pathogenicity, pathogenic power
патоге'нный *a* pathogen, pathogene, pathogenic
па'тока *f* syrup, treacle

кукуру'зная ~ Каро' Karo corn syrup
пато'лог *m* pathologist
патологи'ческий *a* pathologic, pathological
патоло'гия *f* pathology
~ расте'ний phytopathology, plant pathology, vegetable pathology
профессиона'льная ~ occupational pathology
ча'стная ~ special pathology
патологоа'натом *m* pathoanatomist
патологоанатоми'ческий *a* pathoanatomical
патологогистологи'ческий *a* pathohistologic
патотокси'н *m* pathotoxin
патоморфоло'гия *f* pathomorphology
патофизиоло'гия *f* pathophysiology
патофоре'з pathophoresis
патрогене'з *m* patrogenesis
патрого'ния *f* patrogony
патрокли'ния *f* patrocliny
патрокли'нность *f* patrocliny
патрокли'нный *a* patroclinous, patriclinous
па'ттерн *m* pattern
~ ме'чения labeling pattern
дви'гательный ~ motor pattern
локомото'рный ~ locomotor pattern
простра'нственно-временно'й ~ spatio-temporal pattern
патуле'ктор *m* patulector
па'уза *f* pause
~ о'тдыха rest pause
компенсато'рная ~ compensatory pause
полимера'зная ~ "polymerase" pausing
пау'к *m* spider *(Araneida)*
~ «чё'рная вдова'» black widow spider *(Latrodectes mactans)*
водяно'й ~ water spider *(Argyroneta aquatica)*
воро'нковый ~ brass spider (*Agelena naevia*), funnel-web spider (*Agelenidae*)
молодо'й ~ spiderling
назе'мный ~ ground spider *(Gnaphosidae)*
паук-бокохо'д *m* crab spider (*Thomisidae*)
паук-волк wolf *m* spider, ground spider (*Lycosidae*)
паук-землеко'п *m* purse-web spider (*Atypus abboti*)
пау'к-кра'б crab spider *(Thomisidae)*
пау'к-крестови'к garden spider *(Araneus diadematus)*
паук-кругопря'д *m* orb-web spider, garden spider (*Argiopidae*)
паукообра'зное *n* araneid, araneidan (*Araneida*)
пау'к-охо'тник fishing spider, raft spider *(Dolomedes)*
паук-птицея'д *m* bird spider (*Avicularia avicularia*)
пау'к-скаку'н jumping spider *(Salticidae)*
паук-ткач че'рный black widow (*Latrodectes mactans*)
паути'на *f* web, cobweb, spiderweb, net
паучо'к *m* spiderling
пах *m* groin
пахине'ма *f* pachynema
пахите'на *f* pachytene
па'ховый *a* inguinal
па'хота *f* tillage, plowing
зя'блевая ~ fall plowing
па'хотный *a* arable, plowable
паху'чий *a* odoriferous, odorous
пацие'нт *m* patient
пацифи'ческий *a* Pacific
па'чули *pl bot* patchouli (*Pogostemon patchouli*)
па'шня *f* plowing land, arable land; plowed field
ПДК, преде'льно допусти'мая концентрация MAC, maximum allowable concentration
пега'нка *f* shelduck, shelgoose (*Tadorna*)
пе'гий *a* pied, piebald
педигри' *f* pedigree
педипа'льпа *f* pedipalp, pedipalpus
педицелля'рия *f* pedicellaria
педога'мия *f* pedogamy, paedogamy
педогене'з *m* pedogenesis, paedogenesis
педоморфи'зм *m* paedomorphism
педомо'рфный *a* paedomorphic
педоморфо'з *m* paedomorphosis, paedomorphism
пека'н *m ther* pekan *(Martes pennanti)*
пекине'с *m* pekingese
пекти'н *m* pectin
пекти'новый *a* pectic
пекто'за *f* pectose
пелика'н *m* pelican (*Pelecanus*)
кудря'вый ~ Dalmatin pelican (*Pelecanus* crispus)
ро'зовый ~ white pelican, European white pelican, great white pelican *(Pelecanus onocrotalus*)
се'рый ~ gray pelican, spot-billed pelican *(Pelecanus philippensis)*

розовоспи'нный ~ pink-baked pelican *(Pelecanus rufescens)*
пелли'кула *f* pellicle, pellicula
пелобио'нт *m* pelobiont
пелоге'нный *a* pelogenous
пелофи'т *m* pelophyte
пе'на *f* foam, froth, scum
пенекли'макс *m* peneclimax
пенетра'нтность *f* penetrance
пенетра'нтный *a* penetrant
пенё'к *m* stump
пенекли'макс peneclimax
пенепле'н peneplain
пенетра'нта *zool* penetrant
пенетра'нтность penetrance
пенетра'нтный penetrant
пе'ние *n* singing, song, pipe
антифона'льное ~ *eth* duetting, antiphonal singing
па'рное ~ pair song
пе'нис *m* penis, phallus
пе'ниться *v* froth
пеницилли'н *m* penicillin
пе'нна *orn* penna
пе'нница *f* spittlebug (*Cercopidae*)
пенообразова'ние *n* froth
пенообразова'тель *m* foaming agent, frother
пеноотдели'тель *m* skimming tank
пе'ночка *f* warbler (*Phylloscopus*)
ды'мчатая ~ willow warbler *(Phylloscopus fulgiventer)*
зелё'ная ~ greenish warbler *(Phylloscopus trochiloides)*
чернобро'вая ~ black-browed warbler *(Phylloscopus cantator)*
пе'ночка-весни'чка *f* willow warbler *(Phylloscopus trochilus)*
пе'ночка-зарни'чка *f* yellow-browed warbler *(Phylloscopus inornatus)*
пе'ночка-тенько'вка *f* brown willow warbler *(Phylloscopus collybitus)*
пентагона'льный *a* pentagonal
пентапло'ид *m* pentaploid
пентапло'идный *a* pentaploid
пентасо'мик *m* pentasomic
пентасоми'ческий *a* pentasomic
пень *m* stump, stub
пенька' *f* hemp
пенькочеса'льный *a* hemp-scutching
пепси'н *m* pepsin
пепсини'ровать *v* pepsinate
пепсиноге'н *m* pepsinogen
пепти'д *m* peptide
активацио'нный ~ activation peptide
аутологи'чный ~utologous peptide
гастри'н-высвобожда'ющий ~, ГВП gastrin-releasing peptide
иммуносупрессо'рный ~ immunosuppressive peptide
концево'й ~ tail peptide
ли'дерный ~ leader peptide
сигна'льный ~ identificaction peptide
термина'льный ~ tail peptide
пептида'за *f* peptidase
пептидили'рование *n* peptidylation
пептидоглика'н *m* peptidoglycane
пепто'н *m* peptone
пептониза'ция *f* peptonization
пептонизи'роваться *v* peptonize
пе'рвенец firstling
перве'рсия *f* perversion
первичноро'тый *a* archaestomatous
перви'чный *a* initial, primary
первобы'тный *a* primeval, primitive, virgin
первого'док *m* yearling
первозда'нный *a* primeval
первоисто'чник *m* protosource
первонача'льный *a* initial, pristine
первоописа'ние *n* first description
первопричи'на *f* first cause, ultimate cause
первоте'лка *f* first calver
первоцвет *m* primrose, paigle *(Primula)*
перга' *f* beebread, ambrosia
пердомина'нт *m* perdominant species, predominant
пере- de-, over-, pro-, sub-, super-, trans-
переалкили'рование *n* transalkylation
переамиди'рование *n* transamidation
переамини'рование *n* transamination
переаминометили'рование *n* transaminomethylation
переацетили'рование *n* transacetylation
переацили'рование *n* transacylation
перебра'живание *n* overfermentation
перебро'д *m* overfermentation
перебро'ска *f* transfer
межбассе'йновая ~ сто'ка interbasin diversion
межрегиона'льная ~ сто'ка interregional water transfer
перева'ривание *n* digestion
перева'р *m:* ~ **Ха'ртли** Hartley's broth
перева'ривать *v* digest
перевари'мость *f* digestibility

перевари'мый *a* digestible
перевё'рнутый *a bot* resupinate
перевива'ть *v* transfer
переви'вка *f* subinoculation, transfer, transplantation
перивителли'новый *a* perivitelline
переви'тый *a* intorted, tangled
перевы'пас *m* excessive grazing
перевя'зка *f* dressing, bandging; ligation; marbled polecat, mottled polecat (*Vormella*)
~ **пупови'ны** umbilical ligation
перевя'зочный *a* dressing
перевя'зывать *v* bandage, dress
перегидри'рование *n* transhydrogenation
перегно'й *m* humus, mold
перего'нка *f lab* distillation
~ **с водяны'м па'ром** steam distillation
суха'я ~ dry distillation
перегоро'дка *f* septum, phragma, wall, interstice; **с прямыми перегородками** *pal* rectimurate
~ **предсе'рдий** atrial septum
~ **с по'рами** *bot* pseudoseptum
втори'чная ~ septulum
междо'льковая ~ interlobular septum
межжелу'дочковая ~ interventricular septum, ventricular septum
межмы'шечная ~ intermuscular septum
межпредсе'рдная ~ interatrial septum
небольша'я ~ septulum
носова'я ~ internasal septum, nasal septum
перфори'рованная ~ *bot* pseudoseptum
предсе'рдно-желу'дочковая ~ atrioventricular septum
прозра'чная ~ pellucid septum *(in the brain)*
перегоро'дчатый *a* septate
перегре'в *m* overheat
перегруппиро'вка *f* redistribution, rearrangement
передава'ть *v* transmit, transfer
переда'тчик *m* transmitter, *phys* transducer
переда'ча *f* transmission, transfer
~ **культу'ры** *eth* cultural transmission
~ **от поколе'ния к поколе'нию** vertical transmission
~ **сигна'ла** signaling
«вертика'льная» ~ vertical transmission
«горизонта'льная» ~horizontal transmission
моносинапти'ческая ~ monosynaptic transfer
насле'дственная ~ vertical transmission
не'рвно-мы'шечная ~ neuromuscular transmission
однонапра'вленная ~ unidirectional transmission
односторо'нняя ~ *gen* uniparental transmission
полисинапти'ческая ~ polysynaptic transfer
синапти'ческая ~ synaptic transmission
трансмембра'нная ~ сигна'ла transmembrane signaling
трансовариа'льная ~ transovarial transmission
передаю'щийся *a* transmissible
передвиже'ние *n* transport
~ **растворё'нных веще'ств** transport of solutes
акти'вное ~ active transport
радиа'льное ~ *bot* radial transport
переде'лка *f* modification
~ **на'выка** habit modification, reversal learning
переднебрю'шие *n* preabdomen *(in spiders)*
переднегру'дь *f* prosternum, *ent* prothorax
переднеспи'нка *f ent* pronotum
пере'дний *a* anterior
пereже'вывать *v* chew, masticate
пережива'ние *n* survival
перезимо'вывание *n* wintering
перезимо'вывать *v* overwinter, winter, hibernate
перезрева'ть *v* overripen
перезре'лость *f* overmaturity, dead-ripe state
перезре'лый *a* overmature, overripe
перейо'н *m* pereion
перейопо'д *m zool* pereiopod, peraeopod
перека'рмливание *n* overfeeding, superalimentation
перека'рмливать *v* overfeed
перека'т *m* riffle, rapid; bar
перекати'-по'ле *n* baby's-breath (*Gypsophila paniculata*)
перека'тывание *n* rolling
~ **яи'ц** egg rolling
перекли'чка *f* roll-call
перекре'ст *m* chiasm, decussation
~ **бло'ковых не'рвов** decussation of trochlear nerves
~ **Ме'йнерта** fountain decussation of Meynert
~ **пе'тель** decussation of fillet

~ **пирами'д** motor decussation, pyramidal decussation
~ **Форе'ля** Forel's decussation
дви'гательный ~ motor decussation, pyramidal decussation
чувстви'тельный ~ decussation of fillet
перекрёстноопыля'ющийся *a* allogamous
перекре'щиваться *v* decussate
перекрыва'ние *n* overlapping
дифференциа'льное ~ differential overlapping
~ **генети'ческого ко'да** genetic code overlapping
перекрыва'ющийся *a* imbricate
перелеза'ние *n* crawling over
переле'ска *f* liverleaf (*Hepatica*)
переле'сок *m* coppice
перелё'т *m* migration, flight, passage
перелё'тный *a* migratory
перелива'ние *n* transfusion
перело'в *m* overfishing
перело'г *m* fallow, fallow land, old field, lea
перело'м *m* break, breaking, fracture; discontinuity; crisis
закры'тый ~ closed fracture
оско'льчатый ~ comminuted fracture
переме'на *f* change
~ **ноги'** change of leg
~ **положе'ния** postural change
переме'нная *f* variable
ве'кторная ~ vector variable
дополни'тельная скользя'щая ~ slack variable
зави'симая ~ dependent variable
канони'ческая ~ canonic variable
ка'чественная ~ qualitative variable
коли'чественная ~ quantitative variable
незави'симая ~ independent variable
ограни'ченная ~ bounded variable
промежу'точная ~ intervening variable
случа'йная ~ aleatory variable, random variable, chance variable
теорети'ческая ~ theoretical variable
функциона'льная ~ functional variable
экзоге'нная ~ exogenous variable
эндоге'нная ~ endogenous variable
переметили'рование *n* transmethylation
переме'шивание *n* agitation
перемеща'ть *v* transfer
перемеще'ние *n* transition, transport, transposition, transference, displacement
пасси'вное ~ passive transport
перемы'чка *f* bridge, isthmus
ко'стная ~ bone bridge
сухожи'льная ~ tendinous intrsection
перенапряже'ние *n* overstrain
перенаселе'ние *n* overpopulation, overcrowding, congestion
перенаселе'нность *f* overpopulation, overcrowding
перенаселё'нный *a* congested
перенаселя'ть *v* overpopulate
перенасыще'ние *n* oversaturation
перено'с *m* transfer, transmission, transport, vection, passage
~ **ге'нов** transgenation, transgenesis
~ **на'выка** *eth* transfer of training
~ **насеко'мыми** insect transmission
~ **насыще'ния** saturation transfer
~ **по пищево'й це'пи** trophic transfer
~ **че'рез плаце'нту** transplacental passage
~ **электро'нов** electron transport
~ **эне'ргии** energy transfer
ве'кторный ~ ге'нов vector-mediated gene transfer
гомоя'дерный ~ поляриза'ции homonuclear polarization transfer
внутримолекуля'рный ~ эне'ргии intramolecular energy transfer
одностороо'нний ~ one-way transfer
прото'нный ~ proton transfer
трипле'т-трипле'тный ~ эне'ргии triplet-triplet energy transfer
«челно'чный» ~ *gen* shuttle transfer
центростреми'тельный ~ somatopetal transport
переноси'ть *v* transmit, transfer
перено'сица *f* bridge of nose
перено'счик *m* transmitter, vector, carrier, transmitting agent, carrying agent, mediator
~ **водоро'да** hydrogen carrier
~ **кислоро'да** oxygen carrier
переописа'ние *n* redescription
переоткла'дывать *v* redeposit
переохлажда'ть *v* overchill
переохлажде'ние *n* supercooling, surfusion
переохлаждё'нный *a* overchilled
перепа'рхивать *v* flit
пе'репел *m orn* quail (*Coturnix coturnix*)
перепеля'тник *m orn* sparrow hawk (*Accipiter nisus*)
пе'репись *f* census
~ **населе'ния** census of population
многокра'тная ~ multiple census

обра'тная ~ inverse census
повто'рная ~ repeated census
пряма'я ~ direct census
сельскохозя'йственная ~ census of agriculture

переполне'ние *n* overstocking

перепо'нка *f* membrane; palama (*in waterfowl*)
бараба'нная ~ eardrum, tympanic membrane, drum membrane
гло'точная ~ pharyngeal membrane
жа'берная ~ branchiostegal membrane
клоа'чная ~ cloacal membrane
ко'ртиева ~ *anat* tectorial membrane
мига'тельная ~ nictitant membrane, haw
пла'вательная ~ web, webbing, flipper
покро'вная ~ *anat* tectorial membrane

перепончатокры'лое *n* hymenopteran *(Hymenoptera)*
жа'лящее ~ aculeate hymenopteran *(Aculeata)*
стебельчатобрю'хое ~ petiolate hymenopteran *(Apocrita)*

перепончатокры'лый *a* hymenopterous

перепо'нчатый *a* webbed, membranous

перепрева'ние *n* firefanging

перепрева'ть *v* firefang

перепривива'ние *n* passaging

перепроизво'дство *n* overproduction

перепро'мысел *m* overhunting

перераспределе'ние *n* redistribution

перераста'ть *v* overgrow

перереза'ть *v* sever, cut

перере'зка *f* section, cutting, dissection

перерожде'ние *n* degeneration, degeneracy
амило'идное ~ amyloid degeneration
ва'ллеровское ~ Wallerian degeneration
гиали'новое ~ vitreous degeneration
жирово'е ~ fatty degeneration
злока'чественное ~ neoplastic transformation
уо'ллерово ~ Wallerian degeneration

переро'сший *a* overgrown

переры'в *m* gap, interruption, break

переса'дка *f* replanting, relocation, transfer; grafting, transplantation
~ **в гомологи'чную о'бласть** homotopic transplantation
~ **в негомологи'чную о'бласть** heterotopic transplantation
~ **тка'ни** tissue transfer
аутопласти'ческая ~ autoplastic transplantation, autologous grafting, autoplastic grafting
гетеропласти'ческая ~ heteroplastic transplantation, heterologous grafting, heteroplastic grafting
гомопласти'ческая ~ homoplastic transplantation, homologous grafting, homoplastic grafting
ксенопласти'ческая ~ xenoplastic transplantation

переса'живание *n* relocation

переса'живать *v* relocate, transfer, transplant, displant

пересе'в *m* reinoculation, subcloning, subculture, subculturing, passage, transfer, replanting; reseeding
~ **шта'ммов** passage of strains

пересева'ть *v* subculture

переселе'нец *m* migrant; colonicus (*of an aphid*)

переселе'ние *n* relocation, emigration

переселя'ть *v* relocate

пересме'шник *m* mocker, mockingbird, thrasher (*Mimus*)

переспе'лый *a* overmature, overripe

перестано'вка *f* rearrangement; *biom* permutation
гено'м-цикли'ческие перестано'вки *pl* genome circular permutations

пересто'й *m* dead-ripe stage

пересто'йный *a for* overripe, overmatured, overmature, declining

перестро'йка *f* rearrangement; *cyt* displacement
асимметри'чная хромосо'мная ~ asymmetric chromosome rearrangement
сомати'ческая ~ *gen* somatic rearrangement
центра'льная ~ *eth* central reset

пересту'пень *m* bryony (*Bryonia*)

перетя'жка *f* strangulation, constriction

переувлажне'ние *n* overwetting

переуплотнё'нный *a* overstocked

переуплотня'ть *v* overstock

переутомле'ние *n* overfatigue, overstrain, overwork

перехва'т *m* isthmus, constriction
~ **Ранвье'** *hist* Ranvier's constriction, annular isthmus
кольцево'й ~ annular isthmus

перехо'д *m* transition, pass

~ **спира'ль-клубо'к** helix-coil transition
аллостери'ческий ~ allosteric transition
безизлуча'тельный ~ radiationless transition *(in photosynthesis)*
излуча'тельный ~ radiative transition *(in photosynthesis)*
конформацио'нный ~ conformational transition, conformation transition
структу'рный ~ transconformation
теплово'й ~ thermal transition
фа'зовый ~ phase transition

перехо'дный *a* transitional

пе'рец *m* pepper (*Piper*)
~ **горо'шком** pepper in grains, pepper in corns, hole pepper
жгу'чий ~ hot pepper, chili pepper (*Capsicum*)
кайе'нский ~ Cayenne pepper, hot pepper (*Capsicum annuum*)
мо'лотый ~ powdered pepper, cracked pepper
овощно'й сла'дкий ~ paprica (*Capsicum frutescens*)
сла'дкий ~ sweet bell red pepper *(Capsicum frutescens var. grossum)*
стручко'вый ~ red pepper (*Capsicum*)
чё'рный ~ black pepper (*Piper nigrum*)

пе'речный *a* peppery
переше'ек *m* isthmus
пери- peri-
периальга'льный *a* perialgal
перибла'ст *m* periblast
перибласти'ческий *a* periblastic
перибла'стула *f* periblastula
перибле'ма *f* periblem
периваскуля'рный *a* perivascular
перивисцера'льный *a* perivisceral
перивителли'новый *a* perivitelline
периганглиона'рный *a* periganglionic
перига'струла *f* perigastrula
перигема'льный *a* perihemal
периги'ний, периги'ниум *m* perigynium
периго'ний *m* perigonium
периде'рма *f* periderm
периде'сма *f* peridesm
периди'димис *m anat* perididymis
пери'дий *m* peridium
перика'мбий *m* pericambium
перика'рд *m* pericardium, pericardial sac, heart sac
перикардиа'льный *a* pericardiac, pericardial
перика'рдий *m* pericardium
перикарио'н *m* perikaryon
перика'рпий *m* pericarp
периклина'льный *a* periclinal
перикольпора'тный *a* pericolporate
перили'мфа *f* perilymph
периме'трий *m* perimetrium
перими'зий *m* perimysium
перинеа'льный *a* perineal
перине'врий *m* perineurium

пери'од *m* period, season, time
~ **внутрикле'точного ро'ста** intracellular growth period
~ **возобновле'ния** reproduction period
~ **вы'паса** grazing season
~ **го'на** rutting season
~ **запре'та** closed season
~ **инкуба'ции** incubation period, incubation time
~ **лакта'ции** lactation period
~ **ли'ньки** molting season
~ **нагу'ла** period of feeding
~ **нараста'ния** rise period
~ **наси'живания** brood time
~ **невозбуди'мости** refractory period
~ **неподви'жности** quiescent period
~ **плодоноше'ния** bearing season, fruiting season
~ **поко'я** dormant season, dormant period, rest period
~ **полово'й акти'вности** sexually active period, sexual season
~ **полувыведе'ния** half-life, effective half-life
~ **полуобме'на** half-time exchange
~ **полувыведе'ния** half-life, effective half-life
~ **полураспа'да** half-life period, half-life
~ **полово'й охо'ты** heat period
~ **предупрежде'ния** warning period
~ **размноже'ния** reproduction period, breeding season
~ **регулиро'вки** regulation period
~ **ро'ста** growth period, growing season
~ **сезо'нной сукце'ссии** biotic season
~ **спа'ривания** mating period, mating season, breeding time
~ **стагна'ции** stagnating period
~ **те'чки** heat period
~ **увяда'ния** wilting period
~ **цвете'ния** blooming period, flowering period
~ **яйцекла'дки** laying period, laying season

безморо'зный ~ frost-free season, frostless season
бра'чный ~ mating season
вегетацио'нный ~ growing season, vegetative season, vegetation season, vegetation period
инкубацио'нный ~ incubation period
кембри'йский ~ Cambrian period
крити'ческий ~ critical period, crucial period
лате'нтный ~ latent period, latent time
леднико'вый ~ glacial period, ice age, glacial age
мелово'й ~ Cretaceous period
нересто'вый ~ spawning period, spawning season
па'стбищный ~ grazing season
послеродово'й ~ puerperal period
препате'нтный ~ prepatent period
рефракте'рный ~ refraction time
светово'й ~ light period
сезо'нный ~ **поко'я** aestivation period
скры'тый ~ latent time
трети'чный ~ Tertiary period
фенологи'ческий ~ biotic season
чувстви'тельный ~ sensitive period
ю'рский ~ Jurassic period
периоди'чность *f* periodicity
су'точная ~ daily periodicity
лу'нная ~ lunar periodicity
периодо'нт *m* periodont
периодонта'льный *a* peridental
перио'ст *m* periost, periosteum
перио'стракум *m* periostracum
перипла'зма *f* periplasm
перипла'ст *m* periplast
перипнейсти'ческий *a ent* peripneustic
перипро'кт *m* periproct
periса'рк *m* perisarc
периспе'рм *m* perisperm
периспо'ра *f* perispore
периспора'нгий *m* perisporangium
периссоплоиди'я *f* perissoplidy
периста'льтика *f* peristalsis
обра'тная ~ reversed peristalsis, antiperistalsis
перистальти'ческий *a* peristaltic
петисто'м *m* peristome
пе'ристо-ло'пастный pinnatilobate
пе'ристо-надре'зный *a* pinnatifid
перистоне'рвный *a bot* feather-veined, penninerved, pinninervate
перистокры'лка *f* ptiliid beetle *(Ptiliidae)*
перисторазде'льный *a* pinnately divided, pinnately parted, pinnatipartite, pinnately decompound
перисторассече'нный *a* pinnately sected, pinnatisect, pinnatisected
перистосло'жный *a* pinnately compound
пе'ристость *f* pinnation
пе'ристый *a* pinnate, plumate; **пе'ристое о'блако** cirrus
перите'ций *m* perithecium
примити'вный ~ prothecium
перитонеа'льный *a* peritoneal
перитре'ма *f* peritrema
перитрихиа'льный *a micr* peritrichous
перитрофи'ческий *a* peritrophic
перитро'ха *f* peritroch
перифери'ческий *a* peripheral, peripheric
перифи'за *f* periphysis
перифито'н *m* periphyton
перифито'нный *a* periphytic
перифо'риум *m bot* periphorium
перихе'ций *m* perichaetium
перихо'ндрий *m* perichondrium
перихорда'льный *a* perichordal
перицентра'льный *a* pericentral, pericentric
перици'кл *m* pericycle
перици'т *m* pericyte
перколя'тор *m* percolator
перку'ссия *f* percussion
перламу'тр *m* mother-of-pearl, nacre
перламу'тровый *a* nacreous
пермисси'вный *a* permissive
пе'рмь *f* Permian period
перна'тые *pl* birds
перна'тый *a* feathered, feathery
перницо'зный *a* pernicious
перо' *n* feather, plume, pen
большо'е рулево'е ~ хвоста' main tail feather
ко'нтурное ~ contour feather
кро'ющее ~ covert feather, covert, tectrixб hard feather
махово'е ~ oar feather, quill feather, flight feather, remix
махово'е ~ пе'рвого поря'дка primary feather
мя'гкое покро'вное ~ soft body feather
нитеви'дное ~ filoplume
оспопривива'тельное ~ vaccination pen
покро'вное ~ covert feather, covert, tetrix
прида'точное ~ aftershaft

пухово'е ~ semiplume, down feather, plumule, prepenna
рулево'е ~ tail feather, rectix
пероксида'за *f* peroxidase
пероксисо'ма *f* peroxisome
перпендикуля'рный *a* perpendicular
персевера'ция *f* perseveration
пе'рсик *m* peach (*Prunus persica*)
персикосли'ва *f* plum peach (*Prunus persica* x *Prunus salicina*)
персисте'нция *f* persistence, persistency
~ **ви'руса** virus persistence
персисти'рование *n* persistence
персисти'рующий *a* persisting, persistent
перстневи'дно-гло'точный *a* cricopharyngeal
перстневи'дный *a* cricoid
перстнетрахеа'льный *a* cricotracheal
перстнечерпалови'дный *a* cricoarytenoid
перстнещитови'дный *a* cricothyroid
перфорато'рий *m* perforatorium
перфори'н *m* perforin
перфори'рованно-покро'вный *a* punctitegillate
перфори'рованный *a* foraminous
перфори'ровать *v* perforate
перфуза'т *m* perfusate
перфу'зия *f* perfusion
пе'рхоть *f* dandruff, dandriff, scurf
перце'пция *f* perception
перчи'нка *f* pepper-corn
песе'ц *m* Arctic fox, blue fox (*Alopex lagopus*)
голубо'й ~ blue fox (*Alopex lagopus*)
пески' *pl of* **песок** sand
барха'нные ~ barchan sands
закреплё'нные ~ fixed sands
незакреплё'нные ~ blown sands
подви'жные ~ blown sands
пескожи'л *m* lug worm (*Arenicola*)
песколюби'вый *a* psammophilous
пескоукрепля'ющий *a* sand-binding
пе'сня *f* song
~ **уха'живания** courtship song
амо'рфная ~ amorphic song
бесструкту'рная ~ amorphic song
пласти'чная ~ plastic song
территориа'льная ~ territorial song
ювени'льная ~ juvenile song
песо'к *m* sand
по'чечный ~ renal sand
песо'чник *m orn* stint, sandpiper (*Calidris*)
пе'ссимум *m* pessimum
пе'стик *m bot* pistil; pestle
~ **из одно'й карпе'ли** apocarpous pistil, simple pistil
~ **с удлинё'нным сто'лбиком** dolichostylous pistil
пестици'д *m* pesticide
пе'стичный *a* pistillate
пестрокры'лка *f* fruit fly, flower fly *(Tephritidae)*
пёстроли'стный *a* poecilophyllous
пё'стрый *a* marked, versicolored
пестря'к *m* checkered beetle (*Cleridae*)
песту'н *m* bear cub (*over a year old*)
песча'ник *m* sandstone
песча'нка *f* gerbil (*Gerbillus*)
когти'стая ~ clawed jird, Mongolian gerbil (*Meriones unguiculatus*)
песча'ный *a* arenaceous, sandy, sabulous
песь *f med* vitiligo
петлеви'дный, петлеобра'зный *a* ansiform, ansate, brochal
петля' *f* loop, ansa; snare; brochus (*of pollen grains*)
~ **Ге'нле** Henle's loop
~ **для пересе'вов** *micr* transfer loop
~ **нефро'на** nephronic loop, Henle's loop
~ **с культу'рой** infected loop
антикодо'новая ~ *mbiol* anticodon loop
пла'тиновая ~ platinum loop
петрофи'т petrophyte
петру'шка *f* parsley (*Petroselinum*)
пету'х *m* cock, rooster
бойцо'вый ~ game cock
петь *v* sing, pipe
печа'тка *f apic* capping
пече'ночный *a* hepatic
пе'чень *f* liver, hepar
печь *f* oven, furnace
му'фельная ~ muffle furnace
пасте'ровская ~ dry-air sterilizer, hot-air sterilizer
пешня' *f* ice-chisel
пеще'ра *f* cave
пеще'рный *a* cavernicolous, cave-dwelling, spelean
пеще'ристый *a* cavernous
пестроли'стность *f* variegation
пестроли'стный *a* variegated-leaved, poecilophyllous
пе'стрый *a* versicoloured, pied, multicolored
пи'во *n* beer
пивова'р *m* brewer

пивоваре'ние *n* brewage, brewing
пивова'рня *f* brewery, brew-house
пиги'дий *m* pygidium
пигме'й *m* pygmy
пигме'нт *m* pigment
пигмента'ция *f* pigmentation
пигме'нтный *a* pigmental, pigmentary
пигости'ль *m* pygostyle
пигохо'рд *m* pygochord
пи'жма *f bot* tansy (*Tanacetum*)
пиза'нг *m bot* plantain banana (*Musa paradisiaca*)
пик *m* spike, peak
~ **поглоще'ния** absorption peak
~ **чи'сленности** population peak
адапти'вный ~ adaptive peak
популяцио'нный ~ population peak
селекти'вный ~ selective peak
уси'ленный ~ overshoot peak, overshoot spike
пикирова'ть *v* thin out
пики'ровать *v* dive
пикиро'вка *f* thinning out; singling; top removal
пикни'да *f* pycnidium, pycnium
пикни'ческий *a* picnic
пикногони'дия *f* pycnogonidium
пикно'з *m* pycnosis
пикнокони'дия *f* pycnoconidium
пикно'метр density bottle, gravity bottle
пикноти'чный *a* pycnotic
пикоднави'рус *m* picodnavirus
пикорнави'рус *m* picornavirus
пику'льник *m* hemp nettle (*Galeopsis*)
пили'дий *m zool* pilidium
пили'льщик *m* sawfly (*Tenthredinidae*)
пилови'дный *a* serratiform, serriform
пилокарпи'н *m* pilocarpine
пило'м *m* pylome
пилори'ческий *a* pyloric
пило'рус *m anat* pylorus
пилоцисти'да *f* pilocystidium
пилоэре'кция *f eth* pilo-erection
пи'льчато-зу'бчатый *a* serrate-dentate
пи'льчатоли'стный *a* serrate-leaved
пи'льчато-есни'тчатый *a* serrate-ciliate
пи'льчато-цветко'вый *a* serrate-flowered
пи'льчатый *a* serrate, saw-toothed, dentate
пингви'н *m* penguin (*Spheniscidae*)
~ **Аде'ли** Adelie penguin (*Pygoscelus adeliae*)
галапаго'сский ~ Galapagos penguin (*Spheniscus mendiculus*)
золотоволо'сый ~ macaroni penguin (*Eudyptes chrysolophus*)
импера'торский ~ emperor penguin (*Aptenodytes fosteri*)
короле'вский ~ king penguin, royal penguin (*Aptenodytes patagonica*)
магелла'нов ~ Magellan penguin (*Spheniscus magellanicus*)
папуа'сский ~ gentoo penguin (*Pygoscelis papua*)
патаго'нский ~ king penguin, royal penguin (*Aptenodytes patagonica*)
очко'вый ~ jackass penguin, spectacled penguin, black-footed penguin (*Spheniscus demersus*)
хохла'тый ~ rock hopper penguin, crested penguin (*Eudyptes cristatus*)
пинну'ла *f zool* pinnule
пиноцито'з *m* pinocytosis
пиноцито'зный *a* pinocytotic
пинце'т *m* forceps, pincers
~ **Пеа'на** Pean's forceps
анатоми'ческий ~ mouse-tooth forceps
изо'гнутый ~ curved forceps
око'нчатый ~ fenestrated forceps
остроконе'чный ~ pointed forceps
пи'нчер *m* pincher
пио'н *m* peony (*Paeonia*)
пионе'р *m* pioneer
пипети'ровать *v* pipet
пипе'тка *f* pipet, pipette, dropping tube, dropper
~ **с то'рмозом** braking pipet
автомати'ческая ~ automatic pipet
высокото'чная ~ precision pipet
градуи'рованная ~ graduated pipet, measuring pipet
измери'тельная ~ graduated pipet, measuring pipet
многокана'льная ~ multichannel pipettor
обра'тная ~ backfill pipet
прецизио'нная ~ precision pipet
пипе'ттор *m lab* pipettor
пирами'да *f* pyramid
~ **биома'ссы** *ecol* biomass pyramid
~ **чи'сел** *ecol* pyramid of numbers, ecological pyramid, Eltonian pyramid, food pyramid
желто'чная ~ yolk pyramid
мальпи'гиева ~ *anat* Malpighian pyramid

экологи'ческая ~ ecological pyramid
пирено'ид *m* pyrenoid
пиренока'рпий *m* pyrenocarp
пиренофо'р *m* pyrenophore
пиридокси'н *m* pyridoxine
пиримиди'н *m* pyrimidine
пирогалло'л *m* pyrogallol
пироге'нный *a* pyrogenic
пирогра'мма *f* pyrogram
пирофи'т *m* pyrophyte
пирофосфа'т *m* pyrophosphate
пирофосфорили'рование n pyrophosphorylation
пироге'нность *f* pyrogenicity
пироге'нный *a* pyrogenic
писк *m* squeak, peep, cheep; whine, whining (*of a mosquito*)
комари'ный ~ whining of mosquitoes
пистилло'дий *m* pistillode
пита'ние *n* feeding; aliment; nutrition; nourishment, diet, food; alimentation (*of a river*); recharge (*of ground water*)
~ **че'рез ли'стья** foliar nutrition
автотро'фное ~ autotrophic nutrition
безвы'борное ~ non-selective feeding
вне'шнее ~ external feeding
гетеротро'фное ~ heterotrophic nutrition
голозо'йное ~ holozoic nutrition, phagotrophy
иску'ственное ~ artificial feeding
корнево'е ~ root nutrition
минера'льное ~ inorganic nutrition, mineral nutrition
недоста'точное ~ subalimentation
некорнево'е ~ foliar nutrition
нерациона'льное ~ improper feeding
сбаланси'рованное ~ balanced feeding
сме'шанное ~ mixed feeding
пита'тельность *f* nutritive value, nutritional value, feeding power
пита'тельный *a* nutritional, nutritious, nutrient, nutritive, nourishing, alimentary
пита'ть *v* feed, nourish
пита'ться *v* feed on
пито'мник *m* nursery, seminary
садо'вый ~ nursery garden
селекцио'нный ~ breeding nursery
пито'н *m* python *(Python)*
питуици'т *m* pituicyte, pituitary cell
пить *v* drink
питье' *n* drinking; drink
питьево'й *a* drinking, potable
пи'хта *f* fir (*Abies*)
пихта'рник *m* fir forest
пи'хтовый *a* firry
пи'ща *f* food, foodstuff, nutriment, nutrition, nourishment, pabulum, diet
бактериа'льная ~ bacterial diet
безбелко'вая ~ protein-free diet
живо'тная ~ animal diet, animal food
здоро'вая ~ wholesome food
расти'тельная ~ vegetable food
пища'ть *v* peep, pip, squeak
пищеваре'ние *n* digestion
внекле'точное ~ extracellular digestion
внутрикле'точное ~ intracellular digestion
желу'дочное ~ gastric digestion
кише'чное ~ intestinal digestion
межкле'точное ~ intracellular digestion
присте'ночное ~ parietal digestion
слю'нное ~ salivary digestion
пищевари'тельный *a* digestive
пищево'д *m* esophagus, oesophagus
пищево'дный *a* esophageal, oesophageal
пищево'й *a* nutritional, alimentary, pabular
пищева'я промы'шленность food industry
пищево'е расте'ние food plant
пищу'ха *f* tree creeper (*Certia*); pika, mouse hare, piping hare, rat hare, whistling hare (*Ochotona*)
пия'вка *f* leech (*Hirudinea*)
медици'нская ~ medicinal leech *(Hirudo medicinalis)*
пла'вательный *a* natatorial
пла'вающий *a* natant, swimming
пла'вни *pl* overflow land, plavni (*flooded emergent vegetation*)
плавни'к *m* fin; flotsam, driftwood
рукообра'зный ~ pseudobrachium
плавничо'к *m* finlet
плаву'нчик *m orn* phalarope (*Phalaropus*)
плагиокли'макс *m* plagioclimax
плагиотропи'зм *m* plagiotropism
плагиотро'пный *a* plagiotropic
пла'зма *f* plasma, plasm
аутологи'чная ~ autologous plasm
заро'дышевая ~ germ plasm, germinal plasm
иммуʼнная ~ immune plasm
плазмаге'н *m* plasmagene
плазмале'мма *f* plasmalemma
плазмалоге'н *m* plasmalogen
плазма'ль *f* plasmal

плазмати'ческий *a* plasmatic
плазмаци'т *m* plasmacyte, plasma cell
плазми'да *f* plasmid
~ **несовмести'мости** incompatibility plasmid
крипти'ческая ~ cryptic plasmid
рекомбина'нтная ~ recombinant plasmid
самораспространя'ющаяся ~ promiscuous plasmid
синтети'ческая ~ synthetic plasmid
челно'чная ~ shuttle plasmid
плазми'н *m* plasmin
плазминоге'н *m* plasminogen
плазмогами'я *f* plasmogamy
плазмоде'сма *f* plasmodesm, plasmodesma
плазмо'дий *m* plasmodium, plasmode
плазмодиока'рпий *m* plasmodiocarp
плазмодиотрофобла'ст *m* plasmodiotrophoblast
плазмо'лиз *m* plasmolysis
во'гнутый ~ concave plasmolysis
вы'пуклый ~ convex plasmolysis
колпачко'вый ~ cap plasmolysis
нача'льный ~ incipient plasmolysis
пограни'чный ~ limiting plasmolysis
стимули'рованный ~ stimulative plasmolysis
су'дорожный ~ cramped plasmolysis, spasmodic plasmolysis
плазмоли'тик *m* plasmolyzing agent
плазмолити'ческий *a* plasmolytic
плазмо'н *m* plasmon, plasmone
плазмо'птиз *m* plasmoptysis
плазмоци'т *m* plasmocyte
плако'да *f* placode
головна'я ~ cephalic placode
ли'нзовая ~ lens placode
обоня'тельная ~ nasal placode, olfactory placode
слухова'я ~ auditory placode
эпибранхиа'льная ~ epibranchial placode
плако'идный *a* placoid
плаку'н-трава' *f* purple loosestrife (*Lythrum salicaria*)
плаку'чий *a bot* weeping, pendant, pendent, drooping
план *m* design, plan
~ **о'пыта по схе'ме лати'нского квадра'та** latin-square design
~ **о'пыта по схе'ме просто'й решё'тки** simple lattice design
~ **о'пыта по схе'ме прямоуго'льника** rectangular lattice design
~ **преобразова'ния приро'ды** plan for the transformation of nature
циркуля'рный ~ **строе'ния** circular plan
плани'рование *n* planning
экологи'ческое ~ ecological design
плани'ровать *v* design
планиро'вка *f* playing out; layout; grading, levelling
~ **са'да** garden layout
беспорядочная ~ random arrangement
ландшафтная ~ informal arrangement
пейзажная ~ informal arrangement
планкто'н *m* plankton
планогаме'та *f* planogamete
планозиго'та *f* planozygote
планосо'ма *f* planosome
планоспо'ра *f* planospore
пла'нта *f ent* planta
планта'ция *f* plantation
виногра'дная ~ vineyard
пла'нула *f* planula, planula larva
планше'т *m* tray
~ **Бе'рнера** Boerner's tray
агглютинацио'нный ~ agglutination tray, sport plate, multicell tray
аналити'ческий ~ assay plate
гемагглютинацио'нный ~ hemagglutinin tray
гемолити'ческий ~ plaquing plate
иммунодиффузио'нный ~ immunodiffusion plate
культура'льный ~ culture plate
многолу'ночный ~ cluster plate, multiwell plate, multicell tray, sport plate
плассо'н *m* plasson
пласти'да *f* plastid
пластидо'м *m* plastidome
пласти'н *m* plastin
пласти'на *f* plate
пласти'нка *f* plate, platelet, lamella, blade, pallet
~ **листа'** leaf blade
ага'ровая ~ agar plate
амбулакра'льная ~ ambulacral plate
ана'льная ~ anal plate
база'льная ~ basal plate
бокова'я ~ lateral plate
вы'емчатая ~ *myc* sinuate gill
га'версова ~ Haversian lamella
гимениа'льная ~ *myc* gill

глазна'я ~ ocular plate
гребна'я ~ paddle plate, swimming plate *(in Ctenophora)*
жа'берная ~ gill plate
желто'чная ~ yolk plate
закрылова'я ~ *ent* prehalter
замыка'ющая ~ closing plate
зерни'стая ~ **Э'бнера** granular layer of Ebner, Ebner's granular plate
интерамбулакра'льная ~ interambulacral plate, ambital plate
кле'точная ~ cell plate
концева'я ~ end plate
кровяна'я ~ blood plate, platelet
крылови'дная ~ *anat* alar plate
ло'бная ~ frontal plate
ло'жная ~ *bot* pseudolamina
мадрепо'ровая ~ madreporic plate
метафа'зная ~ equatorial plate, metaphase plate, nuclear plate
надкрылова'я ~ *ent* pterygode, pterygodum
не'рвная ~ *emb* neural plate, medullary plate
перви'чная ~ primitive plate
поля'рная ~ pole plate
по'ровая ~ pore plate
прехорда'льная ~ prechordal plate
прикреплё'нная ~ *myc* adnexed gill
приро'сшая ~ *myc* adnate gill
продыря'вленная ~ cribriform plate
раздели'тельная ~ septal plate
решё'тчатая ~ cribriform plate
ротова'я ~ oral plate
ситови'дная ~ *bot* sieve plate
сифононо'сная ~ *myc* gill-shaped plate
соединя'ющая ~ connecting plate
счё'тная ~ counting plate
теменна'я ~ *ent* genovertical plate
штрихова'я ~ streak plate, streaked plate
эпимера'льная ~ pleuron *(in crustaceans)*
пластинкообра'зный *a* placoid
пластинчатоклю'вый *a* lamellirostral
пластинчатоу'сый *a* lamellicorn
пласти'нчатый *a* lamellar, lamellate, lamellose, laminated; *myc* gill-bearing
пласти'ческий *a* plastic
пласти'чность *f* flexibility, compliance
адапти'вная ~ adaptive flexibility
генети'ческая ~ genetic flexibility
поведе'нческая ~ behavioral plasticity
фенотипи'ческая ~ phenotypic flexibility
пласти'чный *a* plastic
пластогами'я *f* plastogamy
пластоге'н *m* plastogene
пластоди'м *m* plastodeme
пластоко'нт *m* plastocont
пластоме'р *m* plastomere
пластохинон *m* plastoquinone
пластоциани'н *m* plastocyanin
пластро'н *m zool* plastron
пла'стырь *m* plaster
пла'та *f* fee, charge, payment, toll
~ **нату'рой** payment in kind
попё'нная ~ felling fee, forest tax, stumpage
плата'н *m* plane-tree (*Platanus*)
платибаза'льный *a* platybasal
плати'зма *f* platysma
плау'н *m* club-moss (*Lycopodium*)
плаце'бо *n* placebo
плаце'нта *f* placenta, afterbirth
база'льная ~ basal placenta
бидискоида'льная ~ bidiscoidal placenta
гемохориа'льная ~ hemo-chorial placenta
дискоида'льная ~ discoidal placenta
диффу'зная ~ diffuse placenta
неотпада'ющая ~ nondeciduous placenta
непо'лная ~ semiplacenta
отпада'ющая ~ deciduous placenta
хориоалланто'исная ~ chorio-allantoic placenta
эндотелиохориа'льная ~ endothelio-chorial placenta
эпителиохориа'льная ~ epithelio-chorial placenta
плацента'рный *a* placental
плацента'ция *f* placentation
краева'я ~ marginal placentation
маргина'льная ~ marginal placentation
осева'я ~ axile placentation
париета'льная ~ parietal placentation
присте'нная ~ parietal placentation
свобо'дная центра'льная ~ free central placentation
плач *m* weeping; bleeding, exudation (*of plants*)
~ **расте'ний** bleeding, exudation
пла'шка *f* plate; break-back trap, span trap
пла'щ *m:*
~ **головно'го мо'зга** *m* pallium
пле'ва *f* film, membrane, covering
де'вственная ~ hymen, maidenhead
пле'вел *m* rye-grass (*Lolium*)

пле'вра *f* pleura
плевра'льный *a* pleural
плеврапо'физ *m* pleurapophysis
плеври'т *m* pleuricy, pleuritis
плевровисцера'льный *a* pleurovisceral
плевродо'нтный *a* pleurodont
плевромито'з *m* pleuromitose
плевропеда'льный *a* pleuropedal
плевроспо'ровый *a* pleurosporous
плевроце'нтр *m anat* pleurocentrum
плевроцеребра'льный *a* pleurocerebral
плевроцисти'да *f* pleurocystidium
плезиомо'рфия *f* plesiomorphy
плезиомо'рфный *a* plesiomorphic
плейомери'я *f* pleiomery
плейопета'льный *a bot* pleiopetalous
плейотакси'я *f* pleiotaxis, plelotaxy
плейотропи'я *f* pleiotropy
  биохими'ческая ~ biochemical pleiotropy
  иммунологи'ческая ~ immunologic pleiotropy, immunological pleiotropy
  и'стинная ~ genuine pleiotropy
плейотро'пный *a* pleiotropic
плейофи'льный *a* pleiophyllous
плейоха'зий *m* pleiochasium
плекте'нхима *f* plectenchyma
плектридиа'льный *a micr* plectridial
племенно'й *a* pedigree, throughbred;
  племенна'я кни'га stud book, herd book
плё'нка *f* film, membrane, pellicle, hull, husk, palea
  ~ **прицве'тника** bract husk
  пове'рхностная ~ surface film
  формва'ровая ~ formvar film
  фотографическая ~ photographic film
плёнкообразова'тель *m* filming agent
плё'нчато-ли'стный *a* membrane-leaved
плё'нчато-корешко'вый *a* membrane-rooted
плё'нчатый *a* membranaceous, pellicular, pelliculated, filmy, scarious, hulled, chaffy
плеога'мия *f* pleogamy
плеомо'рф *m* pleomorph
плеоморфи'зм *m* pleomorphism, pleomorphy
плеомо'рфный *a* pleomorphic
плеопо'д *m* pleopod
плеотре'мный *a pal* pleotreme
плероцерко'ид *m* plerocercoid
пле'сень *f* mould, mold, must
  голуба'я ~ blue mold
  зелё'ная ~ green mold *(caused by Trichoderma harzianum)*
  хле'бная ~ bread mold *(caused by Monilia sitophila)*
  чё'рная ~ black mold *(caused by Aspergillus niger)*
плесневе'ние *n* mold attack
плести' *v* spin (*e.g., spiderweb*)
плетеви'дный *a* whiplike, lorate
плетизмогра'мма *f* plethysmogram
плетизмо'граф *m* plethysmograph
плетизмогра'фия *f* plethysmography
  импеда'нсная ~ impedance plethysmography
плетизмота'ллюс *m bot* plethysmothallus
плеть *f* creeping stem, runner, procumbent stem, prostrate stem, vine (*e.g., of cucumbers*); lash
плечево'й *a* humeral, brachial
плечеголовно'й *a* brachiocephalic
плечелоктево'й *a* brachiocubital
плечо' *n* arm, upper arm; shoulder; limb, arm (*of a chromosome*)
плод *m anat* foetus, fetus; *bot* fruit
  ~, **рассе'ивающий семена'** disseminanteous fruit
  ~ **с перика'рпом** pericarpous fruit
  ~ **с семена'ми** seed-bearing fruit
  ~ **шипо'вника** hip, rose hip
  апока'рпный ~ apocarpous fruit
  вскрыва'ющийся ~ dehiscent fruit
  дро'бный ~ schizocarpic fruit, secedenous fruit, separating fruit, mericarp
  затверде'вший ~ hard fruit
  и'стинный ~ true fruit, defined fruit
  ко'сточковый ~ stone fruit, drupaceous fruit
  крыла'тый ~ pterodium
  лизика'рпный ~ lysicarpous fruit
  ло'жный ~ false fruit, accessory fruit, spurious fruit, pseudocarp
  невскрыва'ющийся ~ indehiscent fruit
  недозре'лый ~ immature fruit
  недоно'шенный ~ premature fetus, immature fetus
  не'жный ~ delicate fruit, fragile fruit
  незре'лый ~ crude fruit
  односемя'нный ~ one-seeded fruit, pip fruit
  партенока'рпный ~ parthenocarpic fruit
  просто'й ~ simple fruit
  сбо'рный ~ aggregate fruit
  синка'рпный ~ syncarpous fruit
  скоропо'ртящийся ~ perishable fruit

сло'жный ~ sorosis
со'чный ~ fleshy fruit, soft fruit
сухо'й ~ dried fruit
пло'дик *m* carpel
пло'дник *m* carpel
пло'дный *a* fetal
плодови'тость *f* fertility, fecundity, breeding power, breeding performance
плодови'тый *a* fecund, fertile, prolific, fruitful, polytocous
плодово'д *m* fruit grower
плодово'дство *n* pomiculture
плодо'вый *a* fruit, fruiting
плодожо'рка *f* seed worm (*Laspeyresia*)
я'блонная ~ codling moth (*Laspeyresia pomonella*)
плодоко'рм *m* mast, pannage
плодоли'стик *m bot* carpel, carpophyl, carpophyll
за'мкнутый ~ closed carpel
кондупликa'тный ~ conduplicate carpel
ло'жный ~ pseudosperm
обьединё'нные плодоли'стики *pl* constituent carpels
одино'чный ~ solitary carpel
откры'тый ~ open carpel
пельта'тный ~ peltate carpel
пло'тный ~ solid carpel
полупло'тный ~ semisolid carpel
просто'й ~ simple carpel
редуци'рованный ~ reduced carpel
сло'женный вдоль ~ conduplicate carpel
сло'жный ~ complex carpel
со'мкнутый ~ closed carpel
стери'льный ~ sterile carpel
явнопельта'тный ~ manifest peltate carpel
плодоно'жка *f bot* fruit stalk, fruit stem
плодоно'с *m* carpophore
плодоноси'ть *v* fructify, fruit, bear fruit
плодоно'сный *a* fruit-bearing, fruitful, prolific; benign *(soil)*
плодонося'щий *a* fruit-bearing, fructiferous
плодоноше'ние *n* fruitage, fructification
по'лное ~ heavy bearing
плодообразова'ние *n* fruitification
плодоро'дие *n* fertility, crop-producing power
убыва'ющее ~ diminishing fertility
плодоро'дность *f* fecundity
плодоро'дный *a* fertile, fecund, fruitful, prolific
плодосме'н *m* crop succession
плодохрани'лище *n* fruit house
плодоя'дный *a* frugivorous, carpophagous
пло'идность *f* ploidy
пло'ский *a* flat
плоскоголо'вый *a* flatheaded
плоскогo'рье *n* tableland, highland, upland
плоскокле'точный *a* planocellular
плосколи'стный *a* flat-leaved
плосконе'рвный *a bot* flat-veined
плоскопло'дный *a* flat-fruited
плоскопокро'вный *a* planitegillate
плоскостебе'льчатый *a* flat-stalked
плоскосто'пие *n* platypodia, flatfoot
плоскостручко'вый *a* flat-podded
пло'скость *f* plane
~ **дробле'ния** cleavage plane
~ **основа'ния** *bioch* base plane
~ **ско'ла** cleavage plane
зри'тельная ~ visual plane
попере'чная ~ transapical plane *(in diatoms)*
продо'льная ~ apical plane *(in diatoms)*
сагитта'льная ~ sagittal plane
ство'рковая ~ valvar plane *(in diatoms)*
фока'льная ~ focal plane
фронта'льная ~ frontal plane
плоскоте'лка *f* cucujid beetle *(Cucujidae)*
плоскоцветко'вый *a* flat-flowered
плоти'на *f* dam, dike, dyke, weir
бобро'вая ~ beaver dam
плотноме'р *m lab* density tester
пло'тность *f* density
~ **ко'лоса** head density
~ **популя'ции** population density
~ **посе'ва** *micr* density of plating
~ **поса'дки** stocking rate, stocking density (*e.g., in aquaculture*)
~ **у'стьиц** stomatal density
антиге'нная ~ antigen density
гапте'нная ~ haptenation density, haptenic density
опти'ческая ~ absorbancy, absorbency, optical density
плаву'чая ~ buoyant density
пло'тный *a* firm
плотофи'т *m* plotophyte
плотоя'дный *a* carnivorous, sarcophagous
плоть *f* flesh
кра'йняя ~ foreskin, prepuce
площа'дка *f*: site, ground, plot
~ **Раункие'ра** Raunkiaer's quadrat
про'бная ~ sample area, sample plot
произво'льно взя'тая ~ randomized plot

пло'щадь *f* area
~ **вы'рубки** felling area, cutting area
~ **основа'ний** *plsoc* basal cover, basal area
ассимиляцио'нная ~ assimilating area
водосбо'рная ~ drainage area
кормова'я ~ feeding ground
листова'я ~ leaf area
кормова'я ~ feeding area
уде'льная ~ **листа'** specific leaf area
учё'тная ~ accounting area, discount area
эффекти'вная ~ effective area
плуте'ус *m* pluteus
плюрипоте'нтный *a* pluripotent
плю'ска *f bot* cup, acorn cup, cupule
плюсна' *f* instep, metatarsus, shank, cannon, hind cannon
плюснево'й *a* metatarsal
плющ *m* ivy (*Hedera*)
пневматофо'р *m* pneumatophore; aerating root, aerial root, breathing root, pneumatophore root
пневмогра'мма *f* pneumogram
пневмо'граф *m* pneumograph
пневмогра'фия *f* pneumography
пневмони'я *f* pneumonia
пневмоперитоне'ум *m* pneumoperitoneum
пневмосто'м *m* pneumostome
пневмота'ксис *m* pneumotaxis
пневмото'ракс *m* pneumothorax
побе'г *m* spray, sprig, shoot, runner, sprouting, stool, sobole, spear, burgeon, sapling, scion, tiller, provine, propagule; **дава'ть побе'ги** sprig
~ **корнево'го происхожде'ния** root shoot
вегетати'вный ~ innovation shoot
верши'нный ~ leading shoot
гла'вный ~ leader, leading shoot
диагеотро'пный ~ diageotropic shoot
жирово'й ~ nourishing shoot
ле'тний ~ lammas shoot
плоду'щий ~ fertile shoot
ползу'чий ~ runner
прида'точный ~ adventitious shoot
ростово'й ~ continuance shoot, leading shoot
спя'щий ~ resting shoot
сте'лющийся ~ trail
стволово'й ~ epicormic shoot
удлинё'нный ~ extending shoot, long shoot
укоро'ченный ~ short shoot
цветоно'сный ~ floral shoot

побегопроизводи'тельность *f* shoot-forming capacity
побе'жка *f* running, runway
~ **в колесе'** wheel running
побере'жье *n* coast
поби'тый *a*: ~ **моро'зом** frost-bitten
побужда'ть *v* arouse, initiate, provoke
побужда'ющий *a* provocative
побужде'ние *n* motivation, motive
пова'диться *v* get into the habit of, fall into the habit of
пова'дка *f* habit
поведе'ние *n* behavior, habit
~ **в лабири'нте** maze behavior
~ **приве'тствия** greeting behavior
агонисти'ческое ~ agonistic behavior
агресси'вное ~ aggressive behavior
аллеломимети'ческое ~ allelomimetic behavior
амбивале'нтное ~ ambivalent behavior
апосемати'ческое ~ aposematic behavior
аппете'нтное ~ appetitive behavior
аффилиати'вное ~ affiliative behavior
бра'чное ~ mating behavior, rutting behavior, consort behavior
верба'льное ~ verbal behavior
видоспецифи'ческое ~ species specific behavior
врожде'нное ~ innate behavior
гнездово'е ~ nesting behavior
группово'е ~ team behavior
демонстрацио'нное ~ demonstrative behavior, display behavior, display
заверша'ющее ~ consummatory behavior
защи'тное ~ defensive behavior
зри'тельно-опосре'дованное ~ visually guided behavior
игрово'е ~ play behavior
инстинкти'вное ~ instinctive behavior
иссле'довательское ~ investigative behavior, exploratory behavior
компроми'ссное ~ compromise behavior
консуммато'рное ~ consummatory behavior
концептуа'льное ~ conceptual behavior
кооперати'вное ~ cooperative behavior
манипуляцио'нное ~ manipulative behavior
маркиро'вочное ~ marking behavior
маскиро'ванное ~ cryptic behavior
матери'нское ~ maternal behavior
миграцио'нное ~ migratory behavior

напра'вленное ~ directive behavior
обороні'тельное ~ defensive behavior
опера'нтное ~ operant behavior
пищево'е ~ feeding behavior, eating behavior
подража'тельное ~ imitative behavior
подчине'нное ~ submissive behavior, conformation behavior
поиско'вое ~ searching behavior
полово'е ~ sexual behavior
посткопулято'рное ~ postcopulatory behavior
прекопулято'рное ~ precopulatory behavior
приспособи'тельное ~ adaptive behavior
ра'нговое ~ positional behavior
репродукти'вное ~ reproductive behavior
ритуализи'рованное ~ ritualized behavior, ritualistic behavior
роди'тельское ~ parental behavior
сексуа'льное демонстрацио'нное ~ *eth* sexual display
смеще'нное ~ displacement behavior
социа'льное ~ social behavior
спонта'нное ~ spontaneous behavior
ста'дное ~ schooling habit
ста'йное ~ gregarious behavior, aggregating behavior, aggregative behavior
стереоти'пное ~ stereotyped behavior, fixated behavior
«суеве'рное» ~ superstitious behavior
супру'жеское ~ consort behavior
территориа'льное ~ territorial behavior
транспозицио'нное ~ transposition behavior
умиротворя'ющее ~ appeasement behavior
устано'вочное ~ adjustive behavior
целеустремле'нное ~ purposeful behavior, goal-seeking behavior
цепно'е ~ chain behavior
эмоциона'льное ~ affective behavior

поведе'нческий *a* behavioral
пове'рхностно-акти'вный *a* surface-active
пове'рхностно-се'тчатый *a* suprareticulate
пове'рхностный *a* surficial
пове'рхность *f* surface
~ **разреза** cut surface
~ **слизистой** mucosal surface
абсорбирующая ~ absorption surface
ассимилирующая ~ assimilating surface
всасывающая ~ absprption surface
жева'тельная ~ mensa; grinding surface *(of a tooth)*
листова'я ~ leaf area
лицева'я ~ face
окклюзио'нная ~ occlusal surface *(of a tooth)*
разгиба'тельная ~ extensor surface
ре'жущая incisal surface *(of a tooth)*

повили'ка *f* dodder (*Cuscuta*)
повинова'ться *v* obey
пови'слый *a* drooping
по'вод *m* cause
поводо'к *m* leader, gangling, snood (*of a fish-hook*); dog-lead
поврежда'ть *v* damage
поврежде'ние *n* injury, damage
~ **моро'зом** frost injury
~ **насеко'мыми** insect injury
~ **при нагрева'нии** heat injury
иммунологи'ческое ~ immunospecific injury
оста'точное ~ remaining damage
радиацио'нное ~ radiation damage

повреждё'нный *a* injured, damaged
биологи'чески ~ biologically damaged

повтор *mbiol* repeat
внутригенный ~ intragenic repeat
длинный концевой ~ long terminal repeat
инвертированный ~ inverted repeat
короткий концевой ~ short terminal repeat

повторе'ние *n* recurrence, iteration, repetition, recurrence
повторносло'жный *a* decomposed
повто'рность *f* number of replications, replication
~ **не'реста** periodicity of spawning

повяза'ть *v* service; pair with
повя'зка *f* bandage, dressing
пога'дка *f zool* pellet, casting
пога'нка *f bot* inedible mushroom, nonedible mushroom, toadstool; *orn* grebe, diver (*Podiceps*)
бе'лая ~ destroying angel (*Amanita virosa*)
бле'дная ~ death cup, deadly amanita (*Amanita phalloides*)
больша'я ~ great crested grebe (*Podiceps cristatus*)
ма'лая ~ dabchick (*Podiceps ruficollis*)
черноше'йная ~ black-necked grebe (*Podiceps ruficollis*)

погиба'ть *v* perish, die, succumb

поги'бший *a* dead
погла'дить *v* pat; ~ **соба'ку** pat a dog
поглоти'тель *m* absorbent
поглоща'ть *v* consume; imbibe
поглоща'ющий *a* absorptive, absorbent
поглоще'ние *n* absorption; consumption; imbibition
~ **корня'ми** root absorption
акти'вное ~ active uptake
кутикуля'рное ~ cuticular uptake
поголо'вье *n* livestock, livestock population
~ **скота** stock, livestock
пого'ня *f* chase
погремо'к *m* rattle-box (*Rhinanthus*)
погре'шность *f* error
~ **выра'внивания** error of fitting
незначительная ~ inappreciable error
суще'ственная ~ appreciable error
погружа'ться *v* plunge, sink, sumerge
погруже'ние *n* immersion, submersion, submergence
погруже'нный *a* submersed, submerged
под *preposition* under; **осетри'на** ~ **хре'ном** sturgeon with horseradish; **по'ле** ~ **ро'жью** rye field, field of rye; **ры'ба** ~ **со'усом** fish in sauce; **уча'сток** ~ **огоро'д** a lot for vegetable garden
~ **ве'чер** towards evening
~ **го'ру** downhill
~ **дожде'м** in the rain
~ **кра'сное де'рево** imitation mahogany
~ **микроско'пом** under a microscope, with a microscope, by a microscope
~ **Москво'й** in the environs of Moscow, near Moscow
~ **наблюде'нием врача** under doctors's care
~ **нарко'зом** under anaesthetic
~ **со'усом** with a dressing
под- sub-, infra, hypo-
подавле'ние *n* suppression, repression, inhibition
усло'вное ~ conditioned suppression
подавля'ть *v* inhibit, repress, depress, suppress
подавля'ющий *a* inhibitory, depressory; overwhelming
пода'тливость *f* compliance
подбе'л *m* andromeda (*Andromeda*)
подбере'зовик *m* rough boletus, shaggy boletus (*Leccinum scabrum*)
подбло'ковый *a anat* infratrochlear
подбо'р *m biom* fitting
~ **криво'й** curve fitting
подборо'док *m* chin, mentum
подборо'дочно-подъязы'чный *a* geniohyoid, hyomental
подборо'дочно-язы'чный *a* genioglossal
подборо'дочный *a* genial, mental
подве'сок *m* suspensor, supensorium
подвздо'шно-гребешко'вый *a* iliopectineal
подвздо'шно-поясни'чный *a* iliolumbar
подвздо'шно-седа'лищный *a* iliosciatic
подвздо'шный *a* ileac, ileal, iliac
подви'д *m* subspecies
подвидообразова'ние *n* subspeciation
подви'жность *f* motility, vagility; mobility; agility
амебо'идная ~ ameboid motility
неиндуци'рованная ~ basal motility, spontaneous motility
спонта'нная ~ basal motility, spontaneous motility
электрофорети'ческая ~ electrophoretic mobility
подви'жный *a* motile, movable; versatile; mobile; agile
подвисо'чный *a* infratemporal
подво'дный *a* subaquatic
подво'й *m* stock
вста'вочный ~ intermediate stock
промежу'точный ~ intermediate stock
подвыва'ние *n* howling, whining, whine
подвыва'ть *v* whine
подглазни'чный *a* infraorbital, suborbital
подгло'точник *m* hypopharynx
подглоточный *a* subpharyngeal
подготови'тельный *a* preparatory
подгото'вка *f* preparation
подгруди'нный *a* infrasternal
подгру'док *m* dewlap, jawl, throatiness
подгру'ппа *f* subgroup
поддержа'ние *n* maintenance
подде'рживающий *a* suspensory
подде'рживать *v* support, sustain; maintain
подде'рживающий *a anat* sustentacular
подде'ржка *f* fulcrum
поддиафрагма'льный *a* hypophrenic, subphrenic
подё'нка *f* mayfly (*Ephemeroptera*)
поде'ргивание *n* jerking, twitch
~ **хвосто'м** tail-flick
поде'ций *m* podetium
поджа'берный *a* infrabranchial, subbranchial

поджида'ть *v* wait for
~ **добы'чу** lie in wait for the prey
подзаты'лочный *a* suboccipital
подзе'мный *a* underground, subterranean, belowground
подзо'л *m* podzolic soil, podzol
подзо'на *f* subzone
подка'рмливать *v* dress (*plants*)
подкла'сс *m* subclass
подклешня' *f* subchela
предкли'макс *m* subclimax
подклю'вье *n* mandible (*of birds*)
подключи'чный *a* infraclavicular
подковоно'с *m* horseshoe bat (*Rhinolophus*)
подковообра'зный *a* horseshoe-shaped
подкогтево'й *a* subungual
подко'жный *a* hypodermal, hypodermic, subcutaneous, subdermal
подколенный *a bot* subnodal; *anat* popliteal
подколе'но *bot* subtribe
подколе'нок *m* hock
подко'рковый *a* subcortical, infracortical
подко'рмка *f* top dressing, additional fertilization, supplemental application, extra nutrition
подкрепле'ние *n* reward, reinforcement
негати'вное ~ negative reinforcement
отстав'ленное ~ delayed reward
пищево'е ~ food reinforcement, food reward
положи'тельное ~ reward stimulation, positive reinforcement
части'чное ~ partial reinforcement
подкрестцо'вый *a* subsacral
подку'ривать *v* smoke (*bees*)
подлё'дный *a* subglacial, under the ice
подлеза'ние *n* crawling under
подле'сок *m* undergrowth, understory, underwood, shrub layer, underbrush
подлё'тыш *m* flapper
подло'жка *f* carrier, substrate; support
активи'рованная ~ activated support
гибридизационная ~ immobilizing membrane, transfer membrane
инертная ~ inert support
полимерная ~ resin support
я'корная ~ anchorage
подлопа'точный *a* infrascapular, subscapular
подмаре'нник *m* bedstraw (*Galium*)
подмо'р *m apic* dead bees
подмы'шечный *a* subaxillary
подмы'шка *f* armpit
поднима'ть *v* raise, lift
~ **целину'** open up virgin lands
подногтево'й *a* subungual
подносови'дный *a* hypocrateriform
подня'ть *v* lift, raise
~ **из ло'говища** to break cover
подня'ться *v* rise, go up; come to the surface;
подня'лся ве'тер wind has risen
подо'бие *n* similarity
геометри'ческое ~ geometrical similarity
подо'бласть *f* subregion
подобра'нхия *f* podobranchia
подо'пытный *a* experimental
подо'рлик *m* spotted eagle (*Aquila*)
большо'й ~ spotted eagle (*Aquila clanga*)
ма'лый ~ lesser spotted eagle, small spotted eagle (*Aquila pomarina*)
подоро'жник *m orn* longspur *(Calcarius); bot* plantain (*Plantago*)
лапла'ндский ~ *orn* Lapland bunting (*Calcarius lapponicus*)
подоси'новик *m bot* orange-cup bolete (*Leccinum aurantiacum*)
подо'стный *a* infraspinous
подо'стрый *a* subacute
подоте'ка *f* podotheca
подотря'д *m zool* suborder
подо'шва *f* sole, pelma, planta, bottom
по'лзательная ~ *zool* creeping sole
подо'швенный *a* plantar
подпа'зушный *a bot* subaxillary, infra-axillary
подпа'лина *f* reddish spot (*e.g., on dog's fur*)
подпаути'нный *a* subarachnoidal
подпа'ховый *a* subinguinal
подпе'сня *f orn* subsong, plastic song
подпе'стичный *a* hypogynous
подпи'тка *f* feeding
подпищево'дный *a* subesophageal, infraesophageal
подподборо'док *m* submentum
подподборо'дочный *a ent* submental
подподъязы'чный *a* infrahyoid
подпоро'говый *a* subthreshold, subliminal
подпоря'док *m bot* suborder
подпо'чва *f* undersoil, bottomsoil, subsoil
подпры'гивать *v* jump
подпу'шек *m* hull fuber
подпу'шка *f* down
по'дпушь *f* down
подража'ние *n* imitation
подража'тельный *a* mimic, imitative

подража'ть *v* mimic, imitate
подразделе'ние *n* subdivision
подра'нок *m* wounded game
подрё'берный *a* infracostal
подре'занный *a* succine
подреза'ть *v* poll (*e.g., of a top of trees*), prune, lop, dress, dock
подре'зать *v* have polled
подре'зка *f* pruning
по'дрезь *f* damage by cut (*of a hide*), flesh cut, butcher's cut
подрё'берный *a* subcostal
подре'занный *a* bobbed
подреза'ть *v* clip (*e.g., wings*), prune, trim
подре'зать *v* have clipped, have pruned
подре'зка *f* pruning, lopping
подро'д *m* subgenus
подро'й *m* afterswarm
подро'ст *m* young growth, regrowth, seedling growth, undergrowth
подротово'й *a* suboral
подря'д *adv* running, on the run, in succession
подса'живание *n* beating-up
подса'чивать *v* tap
подсви'нок *m* hilt, gilt, yelt
подсеме'йство *n* subfamily
подсемядо'льный *a* hypocotyledonary
подсеро'зный *a* subserous
подсе'чка *f* incision, tapping; hooking (*in fishing*)
подсе'чно-огнево'й *a* slash-and-burn
подскулово'й *a* subzygomatic
подсне'жник *m* snowdrop (*Galanthus*)
подсозна'ние *n* subconsciousness; the subconscious
подсозна'тельный *a* infraconscious
подсо'лнечник *m* sunflower (*Helianthus annuus*)
подсо'чка *f* tapping
подстерега'ть *v* lie in wait for
подстила'ть *v* litter
подсти'лка *f* litter, ground litter, mulch
  гнездова'я ~ nesting litter
  лесна'я ~ forest litter, carpet of leaves
  листова'я ~ leaf litter
подстрека'ние *n eth* inciting
подстрига'ть *v* shear, crop, trim
подсчё'т *m* count
  ~ **лейкоци'тов по Ши'ллингу** Schilling blood count
  ча'шечный ~ plate count
подтеменно'й *a* subparietal
подузелко'вый *a ent* subnodal
подузлово'й *a bot* subnodal
поду'шечка *f* pulvillus, pulvinulus, cushion; *bot* pustule
  листова'я ~ pulvinus
подушечкови'дный *a* pulvinar, pulvinate
поду'шечник *m* cushion plant, polster plant, cushion plant formation
поду'шка *f* cushion
  когтева'я ~, **па'льцевая** ~ subunguis
  сандерсо'новская ~, **фолликуля'рная** ~ follicular cushion, Sanderson's cushion
подхвостово'й *a* subcaudal
подхо'д *m* approach
  ко'мплексный ~ team approach
подходя'щий *a* appropriate, congenial, suitable, proper
подцве'чивать *v* tint
подча'шечный *a* calycular
подча'шие *n* calycle, outer calyx, epicalyx
подчелюстно'й *a* inframaxillary, submaxillary
подчине'ние *n* submission, subordination
подчинё'нный *a* subordinate, overtopped
подчиня'ть *v* subdue
подчре'вный *a* hypogastric
подше'рсток *m* undercoat, underfur
подщела'чивание *n* alkalinization, alkalization
подщела'чивать *v* alkalify, alkalize
подъё'м *m* ascent
  ~ **зя'би** autumn ploughing
  ~ **ше'рсти ды'бом** ptiloerection
подъязы'чно-надгорта'нный *a* hyoepiglottic
подъязы'чно-щитови'дный *a* hyothyroid
подъязы'чно-язы'чный *a* hyoglossal
подъязы'чный *a* subglossal, sublingual, hypoglossal
подъя'рус *m* sublayer
поеда'емость *f* edibility
поеда'ние *n* ingestion
поеда'ть *v* eat
пожа'р *m* fire
  бе'глый ~ running fire
  верхово'й ~ *for* crown fire
  лесно'й ~ forest fire
  лугово'й ~ grassland fire
  низово'й ~ *for* surface fire, ground fire, creeping fire
  степно'й ~ steppe fire
  торфяно'й ~ peat-bog fire
пожа'рище *n* burn

пожелте'ть *v* turn yellow (*e.g., of leaves*)
пожира'тель *m* predator
пожира'ть *v* devour, gobble
пожу'хлый *a* withered
пожу'хнуть *v* wither
по'за *f* posture, attitude
~ **выпра'шивания** soliciting posture
~ **замира'ния** freezing posture
~ **зата'ивания** freezing posture
~ **кормле'ния** nursing position
~ **нападе'ния** offensive posture
~ **пораже'ния** defeat posture
~ **сна'** sleeping posture
~ **с по'днятой голово'й** head up posture
агресси'вная ~ attitude of combat
бокова'я защи'тная ~ defensive sideways posture
демонстрацио'нная ~ display
защи'тная ~ defensive posture
настoро'женная ~ alert posture
угрожа'ющая ~ threat posture
позвоно'к *m* vertebra, spondyl, spondyle, spondylus
грудно'й ~ thoracic vertebra
ко'пчиковый ~ coccygeal vertebra
крестцо'вый ~ sacral vertebra
ни'жний поясни'чный ~ basilar vertebra
осево'й ~ odontoid vertebra
поясни'чный ~ lumbar vertebra
хвостово'й ~ caudal vertebra
ше'йный ~ cervical vertebra
позвоно'чник *m* spinal column, backbone, vertebrarium
позвоно'чное *n* vertebrate
позвоно'чный *a* vertebral, vertebrate
позднеспе'лость *f* lateness, late ripeness
позднеспе'лый *a* late, late-ripening
поздне'йший *a* latest
по'здний *a* late; serotinal
поздноцвету'щий *a* late-flowering, serotinal
позелене'ние *n* virescence
пози'ция *f* attitude
позы'в *m* desire, tenesmus, urge
пои'лка *f* water dish, drinking bowl, drinker
пои'мка *f* capture
по'иск *m* search
пойкилопло'идный *a* poikiloploid
пойкилоте'рмный *a* poikilothermal
пойкилоци'т *m* poikilocyte
по'йма *f* flood plain, flood bed, bottomland
по'йнтер *m* pointer
показа'ние *n* indication
показа'тель *m* indicator, marker; index; coefficient; factor, parameter
~ **встреча'емость-оби'лие** frequency-abundance index
~ **гибри'дности** hybrid index
~ **жи'рности** fatness index
~ **кро'ви** blood value
~ **лити'ческой акти'вности** lytic value
~ **преломле'ния** index of refraction, refractive index
~ **рожда'емости** birth rate
~ **сме'ртности** death rate, mortality rate
~ **стереоти'пии** index of stereotypy
~ **то'ковых голосо'в** drumming index
~ **частоты'** frequency index
сре'дний ~ mean value
цветно'й ~ color index
цветно'й ~ **кро'ви** globular value
поко'й *m* dormancy, quiescence, rest, resting
глубо'кий ~ true dormancy
зи'мний ~ winter dormancy
ле'тний ~ summer dormancy
поколе'ние *n* generation, year-class
~ **пре'дков** ancestral generation
беспо'лое ~ asexual generation
гермафроди'тное ~ hermaphroditic generation
доче'рнее ~ filial generation
неурожа'йное ~ weak year-class
пе'рвое ~ **гибри'дов** first filial generation *(F_1)*
подраста'ющее ~ rising generation
полово'е ~ sexual generation
после'дующее ~ succeeding generation
роди'тельское ~ parental generation
урожа'йное ~ strong year-class
поко'рность *f eth* submission
покоря'ть *v* subdue
поко'ящийся *a* latent, resting, quiescent, dormant
покрасне'ние *n* reddening, erubescence
покро'в *m* integument, integumentum; covering, cover, coverage, coat; canopy
~ **кры'льев** *ent* pterotheca *(in a pupa)*
покро'вы те'ла *pl* skin coverings
волосяно'й ~ hair-covering, hair-coat covering, hair coat
жирово'й ~ fat covering
ко'жный ~ integument, cutaneous covering
мё'ртвый по'чвенный ~ dead soil covering
напо'чвенный ~ ground cover

нару'жный ~ outer covering
по'чвенный ~ soil cover, soil covering, soil mantle
расти'тельный ~ plant cover, vegetation cover, vegetational cover
рогово'й ~ horny covering
сне'жный ~ snow cover, snow mantle
со'мкнутый расти'тельный ~ closed canopy
травяно'й ~ herbage, sward, grass cover, grass canopy
шё'рстный ~ hair coat
ювени'льный шё'рстный ~ baby coat
покрови'тельственный *a* protective
покро'вный *a* integumentary, tectorial
покровопло'дный *a* chlamydocarpous
покрыва'ло *n* veil, velum, spathe; cortina
вну'треннее ~ inner veil
краево'е ~ marginal veil
о'бщее ~ *bot* universal veil
ча'стное ~ *bot* partial veil
покрыва'льце *n* indusium, spathella
покрыва'ть *v* cover
покры'тие *n* covering, coating; cover degree, cover area, coverage, cover; service
дернoво'е ~ matting, sod matting
проекти'вное ~ coverage, projective cover degree, projective cover
покрытопло'дный *a* angiocarpous
покрытосемя'нный *a* angiospermous
покры'тый *a* covered; coated
~ **мучни'стым налё'том** farinose
~ **налё'том** pruinose
~ **пе'рьями** plumate
~ **ря'бью** dimply (*of water*)
~ **чешу'йками** squamose, lepidote
~ **шелухо'й** ramentaceous
~ **щети'нками** setaceous
пол *m* gender; floor
по'лба *f* spelt, dinkel wheat (*Triticum spelta*)
~ **во'лжская** emmer, emmer wheat (*Triticum dioicum*)
по'ле *n* area, field, region, space, zone
~ **Бро'дмана** Brodmann's area
~ **Брока'** Broca's area
~ **Ве'рнике** Wernicke's area, Wernicke's field
~ **детермина'ции** determination field
~ **зре'ния** visual field, field of vision
~ **зре'ния микро'скопа** microscopic field, field of microscope
~ **Конге'йма** Cohnheim's field, Cohnheim's area, field of myofibrils
~ **о'ргана** organ field
~ **пото'ка** flow field
~ **размеще'ния** *bot* pattern field
адапти'вное ~ adaptive field
анархи'ческое ~ *prot* anarchic field, disorganized field
вегетати'вное ~ vegetative field
внезаро'дышевое ~ extra-embryonic area
двухме'рное ~ heteropolar field
забро'шенное ~ abandoned field, sod field
завалу'ненное ~ boulder field
заро'дышевое ~ embryonic field
засоре'нное ~ foul field
кортика'льное ~ cortical field
ледяно'е ~ ice floe, floe ice
миофибрилля'рное ~ field of myofibrils, Cohnheim's field, Cohnheim's area
морфогенети'ческое ~ morphogenetic field
обоня'тельное ~ *anat* olfactory area
околообоня'тельное ~ parolfactory area
ороша'емое ~ irrigation field
поля' ороше'ния *pl* disposal fields, sewage fields, filter fields
осево'е ~ axial area *(of diatoms)*
парово'е ~ fallow
па'хотное ~ farm field
подмозо'листое ~ subcallosal area
по'ровое ~ pore field
реце'пторное ~ receptive field
сомэстети'ческое ~ *anat* somaesthetic area
трёхме'рное ~ heteroaxial field
поля' фильтра'ции *pl* disposal fields, sewage fields, filter fields
эмбриона'льное ~ individuation field
полеви'ца *f* bent-grass, silk-grass, cloud-grass (*Agrostis*)
полё'вка *f* vole *(Microtus)*
водяна'я ~ water vole, ground vole *(Arvicola terrestris)*
ка'менная ~ high-mountain vole, mountain vole (*Alticola*)
кра'сная ~ northern redbacked vole *(Clethrionomys glareolus)*
лесна'я ~ grey-sided vole *(Clethrionomys glareolus)*
обыкнове'нная ~ common vole, common field vole, field vole *(Microtus arvalis)*
па'шенная ~ field vole, short-tailed vole *(Microtus agrestis)*

подзе'мная ~ pine vole (*Microtus subterraneus*)
пою'щая ~ singing vole *(Microtus miurus)*
ры'жая ~ grey-sided vole *(Clethrionomys glareolus)*
се'рая ~ vole, field vole *(Microtus)*
ска'льная ~ high-mountain vole, mountain vole (*Alticola*)
снегова'я ~ snow vole *(Microtus nivalis)*
сосно'вая ~ pine vole (*Microtus pinetorum*)
узкочерепна'я ~ narrow-sculled vole (*Microtus regalis*)
полё'вка-эконо'мка *f* root vole, tundra vole *(Microtus oeconomus)*
полево'д *m* field husbandman
полево'дство *n* field husbandry, crop agriculture
полево'й *a* campestral
полега'ние *n* beating-down, lodging (*of crops*)
полега'ть *v* lodge, be lodged (*of crops*)
полё'гший *a* beaten-down (*of crops*)
поле'зность *f* usefulness, utility, practicality
поле'зный *a* useful, beneficial; adjuvant; wholesome; **коэффицие'нт поле'зного де'йствия** efficiency, performance
полени'ка *f* arctic bramble (*Rubus arcticus*)
поле'нница *f* pile of wood
полё'т *m* flight; ballooning (*of spiders*); **в полё'те** on wing
бра'чный ~ mating flight, nuptial flight
демонстрацио'нный ~ display flight
ориентиро'вочный ~ orientation flight
паря'щий ~ hovering flight, soar
по'лзание *n* crawling, creeping
гармошкообра'зное ~ concertina locomotion
по'лзающий *a* creeping, repent
ползти' *v* crawl, creep
ползу'чий *a* creeping, crawling; procumbent, decumbent, repent, gadding, trailing
поли- poly-
полиа'да *f* polyade
полиакрила'мид *m* polyacrylamide
полиаденили'рование *n* polyadenilation
полиалкили'рование *n* polyalkylation
полиа'ндрия *f* polyandry
полиартри'т *m* polyarthritis
~ **Понсе'** Poncet's disease
полибла'ст *m* polyblast
поли'в *m* watering, irrigation
поливакци'на *f* mixed vaccine, polyvalent vaccine
поливале'нтный *a* multivalent
поливольти'нный *a* polyvoltine
полига'мия *f* polygamy
полига'мный *a* polygamous
полигапло'ид *m* polyhaploid
полиге'н *m* polygene
полигене'з *m* polygenesis
полигенети'ческий *a* polygenetic
полигени'я *f* polynegy
полиге'нность *f* polygeny
полиге'нный *a* polygenic
полигеномати'ческий *a* polygenomatic
полигено'мный *a* polygenomatic
полигибри'д *m* polyhybrid
полигини'я *f* polygyny
полиги'нный *a* polygynous
полиголо'вки *f pl* polyheads
полиго'н *m* plot, site, ground, area
полидактили'я *f* polydactylism, polydactyly
поликарпи'ческий *a* polycarpic, polycarpous
поликине'та *f prot* polykinety
поликинети'да *f* polykinetid
поликли'макс *m* polyclimax
поликлона'льность *f* polyclonality
поликонденса'ция *f* condensation polymerization
полилецита'льный *a* polylecithal
полили'нкер *m* polylinker
полиме'р *m* polymer
лине'йный ~ linear polymer
многоцепьево'й ~ branched polymer, multichain polymer
неупоря'доченный ~ random polymer
разветвлё'нный ~ branched polymer, multichain polymer
регуля'рный ~ ordered polymer, ternating polymer
субъедини'чный ~ subunit polymer
полимера'за *f* polymerase
полимериза'ция *f* polymerization
~ **на грани'це фаз** interfacial polymerization
полиме'рный *a* polymerous
полимери'я *f* polymery
полими'кстный *a* polymixic
полимито'з *m* polymitosis
полимо'рф *m* polymorph
полиморфи'зм *m* polymorphism

~ **обще'ственных насеко'мых** social polymorphism
алле'льный ~ allelic polymorphism
антиге'нный ~ antigenic polymorphism
генети'ческий ~ genetic polymorphism
географи'ческий ~ geographical polymorphism
мимети'ческий ~ mimetic polymorphism
неусто'йчивый ~ transient polymorphism
подража'тельный ~ mimetic polymorphism
постоя'нный ~ stable polymorphism
сбаланси'рованный ~ balanced polymorphism
скры'тый ~ cryptic polymorphism
средово'й ~ environmental polymorphism
стаби'льный ~ stable polymorphism
полиморфноя'дерный *a* polymorhonuclear
полимо'рфный *a* polymorphic, polymorphous
полинуклеоти'д *m* polynucleotide
полиови'рус *m* poliovirus
полиопла'зма *f* polioplasm
поли'п *m* polyp
полипепти'д *m* polypeptid
мино'рный ~ minor polypeptide
полипи'д *m* polypode
полипло'ид *m* polyploid
внутрия'дерный ~ endonuclear polyploid
дупликацио'нный ~ duplicational polyploid
многоосно'вный ~ polybasic polyploid
части'чный ~ partial polyploid
полиплоиди'я *f* polyploidy
бу'йная ~ rampant polyploidy
несбаланси'рованная ~ unbalanced polyploidy
структу'рно изменё'нная ~ structurally changed polyploidy
полипло'идный *a* polyploid
полипнейсти'ческий *a* polypneustic
полипо'идный *a* polypoid
полиприсоедине'ние *n* polyaddition
полирибосо'ма *f* polyribosome
полирова'ние *n* polishing
~ **ри'са** rice polishing
полисапро'бный *a* polysaprobic
полисахари'д *m* polysaccharide
полисимметри'чный *a* polysymmetric
полисинапти'ческий *a* polysynaptic
полисо'ма *f* polysome
полисома'тия *f* polysomaty
полисо'мик *m* polysomic
полисоми'я *f* polysomy
полисо'мный *a* polysomic
полисперми'я *f* polyspermy
полиспе'рмный *a* polyspermatous, polyspermous
полиспондили'я *f* polyspondily
полисте'ла *f bot* polystele
полистели'я *f* polystely
полисте'ль *f* polystele
полите'нный *a* polytene
политети'ческий *a* polythetic
политипи'ческий, политипи'чный *a* polytypic
политопи'ческий *a* polytopic
политопи'я *f* polytopy
полито'пный *a* polytopic
политре'мный *a* polytreme
политропи'зм *m* polytropism
политро'пный *a* polytropic
политро'фный *a* polytrophic
полифа'г *m* polyphage
полифаги'я *f* polyphagia
полифакториа'льный *a* polyfactorial
полифени'я *f* polypheny
полифилети'ческий *a* polyphyletic
полифили'я *f* polyphylesis
полифиодо'нтный *a* polyphyodont
полихромази'я *f* polychromasy
полихроматофи'л *m* polychromatophil, polychromatophile
полихроматофили'я *f* polychromatophilia
полихроматоци'т *m* polychromatocyte
полихро'мный *a* polycromatic
полицистро'нный *a* polycistronic
поличехо'л *m* polysheath
полиэдро'з *m* polyhedrosis, polyhedral disease
цитоплазмати'ческий ~ cytoplasmic polyhedrosis
я'дерный ~ nuclear polyhedrosis
полиэмбриони'я *f* polyembryony
полиэнерги'дный *a* polyenergid
полиэтерифика'ция *f* polyesterification
поллина'рий *m* pollinarium
полли'ний *m* pollinium, pollen mass, massula
поллино'дий *m* pollinodium
полнодреве'сный *a* nontapering
полнота' *f* stoutness; *for* density;
недоста'точной полноты' understocked
~ **насажде'ния** density of stocking, degree of closeness, degree of density, crop density

полноволо'сый *a* full-fured
поло'ва *f* chaff
полови'на *f* half
~ **центра'льного довери'тельного интерва'ла** *biom* half-width
~ **по'ля до'пуска** *biom* half-width
~ **те'ла** hemisome
полово'дье *n* flood, spring flood
положе'ние *n* posture
половозре'лость *f* sexual maturity
преждевре'менная ~ sexual precocity
половозре'лый *a* pubertal, puberal
полово'й *a* sexual
по'лог *m* canopy, crown layer
разреже'нный ~ open canopy
со'мкнутый ~ closed canopy
поло'гий *a* gently sloping
положе'ние *n* position, situation, location; posture; status
распростё'ртое ~ prostration
по'лоз *m* whip snake, racer (*Coluber*)
узо'рчатый ~ Pallas' coluber *(Elaphe dione)*
полоро'гий *a* cavicorn
полоса' *f* stripe, band, belt
~ **поглоще'ния** absorption band
ветрозащи'тная ~ wind forest strip, wind belt
защи'тная ~ shelterbelt
лесна'я полезащи'тная ~ wind forest strip
лесозащи'тная ~ forest shelterbelt, forest shelter
полезащи'тная лесна'я ~ shelter belt
полоса'тость *f* barring (*of birds*); streak, stripe
полоса'тый *a* striped, banded, streaky
поло'ска *f* stripe, stria, band; line, streak
~ **ганглио'зной пласти'нки** stria of ganglionic layer
~ **Ге'нзена** Hensen's line
~ **Гу'нтера-Шре'гера** Hunter-Schreger band, Hunter-Schreger line
~ **Кра'узе** Krause's line, Krause's membrane
~ **О'уэна** line of Owen
~ **просветле'ния** Hensen's line
~ **Ре'тциуса** line of Retzius, stria of Retzius
~ **Шре'гера** Schreger's band
вну'тренняя тангенциа'льная ~ **Белларже'** Baillarger's internal tangential band
ге'нзеновская ~ Hensen's stripe
заро'дышевая ~ *emb* germ band
обоня'тельная ~ olfactory stria
осева'я ~ *emb* axial plate
перви'чная ~ germinal streak, primitive streak, primitive trace
просветлё'нная ~ band of illumination
по'лость *f* cavity, hollow, antrum, space
~ **га'йморова** antrum of Highmore, maxillary sinus
~ **гло'тки** pharyngeal cavity
~ **грудно'й кле'тки** cavity of chest, thoracic cavity
~ **железы'** glandular cavity
~ **желу'дка** gastric cavity
~ **зу'ба** pulp cavity
~ **ма'тки** uterine cavity
~ **околосерде'чной су'мки** pericardial cavity
~ **перика'рдия** pericardial cavity
~ **се'рдца** cardial cavity, heart cavity
~ **та'за** pelvic cavity
~ **те'ла** body cavity
~ **че'репа** cavity of skull, cranial cavity, head cavity
амниоти'ческая ~ amniotic cavity
бараба'нная ~ tympanic cavity, drum cavity, tympanum
брюши'нная ~ peritoneal cavity
брюшна'я ~ abdominal cavity, abdominal space
букка'льная ~ *prot* buccal cavity
внезаро'дышевая ~ **те'ла** extraembryonic coelom
возду'шная ~ air sinus, air space
втори'чная ~ **те'ла** celom, coelome, deuterocoel
гастра'льная ~ gastric cavity
дыха'тельная ~ respiratory cavity
жа'берная ~ branchial cavity, gill cavity
костномозгова'я ~ medullary space
манти'йная ~ mantle cavity
перви'чная ~ **те'ла** archicoel
перикардиа'льная ~ pericardial cavity
пищеварительная ~ digestive cavity
плевра'льная ~ pleural cavity
подзаро'дышевая ~ subgerminal cavity
прида'точная ~ **но'са** nasal accessory sinus
ротова'я ~ mouth cavity, oral cavity, buccal cavity
субгермина'льная ~ subgerminal cavity
супрааля'рная ~ supra-alar cavity
суста'вная ~ articular cavity, joint cavity

полоци'т *m* polocyte
полу- hemi-, semi-, demi-
полуанатро'пный *a* semianatropous
полубази'дия *f* hemibasidium
полубла'стула *f* hemiblastula
полуборозда' *f* demicolpus
полуборо'здчатый *a* half-grooved
полуве'рхний *a* half-superior
полуви'д *m* semispecies
полувла'жный *a* semihumid
полуво'дный *a* semiaquatic
полугетерога'мия half-heterogamy
полугнё'здный *a bot* semilocular
полуголо'вый *a ent* hemicephalous
полуди'кий *a* half-wild
полуживо'й *a* half-dead
полужи'дкий *a* semifluid, semiliquid
полузасу'шливый *a* semiarid
полузо'нтик *m* cyme
полузо'нтиковый, полузонтичный *a* cymose (*of inflorescence*)
полука'рликовый *a* semidwarf
полукле'тка *f bot* semicell
полуклубнево'й *a* semituberous
полуколи'чественный *a* semiquantitative
полуколоси'стый *a* half-spiked
полуконсервати'вный *a* semiconservative *(of DNA synthesis)*
полукро'вный *a* half-blooded, half-bred
полукрыла'тый *a bot* half-winged
полукру'жный *a* semicircular
полуку'колка *f* semipupa, pseudonymph
полукуста'рник *m* semifrutex, semishrub, subshrub, half-shrub
полукуста'рниковый *a* suffrutescent, suffruticose, suffruticous
полукуста'рничек *m* dwarf semishrub, dwarf subshrub
полулета'ль *f* semilethal
полулета'льный *a* semilethal
полуло'пастный *a bot* half-lobed
полулу'ние *n* demilune
  ~ **Джиану'цци** crescent of Gianuzzi
полулу'нный *a* crescent, crescentic, semilunar
полумахро'вый *a* half-double, semidouble
полумуто'вка *f* verticillaster
полумуто'вчатый *a* half-whorled
полумы'шь *f* woodland jumping mouse (*Napaeozapus*)
полунадпе'стичный *a* hemiepigynous, half-epigynous
полуни'жний *a bot* half-inferior
полуобра'тный *a* half-inverted
полуобращё'нный *a* half-inverted
полуопушё'нный *a* half-hairy
полуости'стый *a bot* half-awned; semispinal *(of a muscle)*
полупарази'т *m* hemiparasite, semiparasite
полупаразити'ческий *a* hemiparasitic
полуперегоро'дчатый *a* half-partitioned
полуперепо'нчатый *a* semimembranous
полупе'ристый *a* half-pinnate
полупогруже'нный *a* emergent (*of aquatic plants*)
полупокро'вный *a* hemichlamydeous, half-coated
полуприкры'тый *a* semirecondite
полуприлега'ющий *a* half-adherent
полупроница'емый *a* semipermeable
полупрямостоя'чий *a* half-upright
полупусты'нный *a* semidesert
полупусты'ня *f* semidesert
полура'са *f* half race
полураскры'тый *a ent* erectopatent
полусава'нна *f* semisavanna
полусапрофи'т *m* hemisaprophyte, semisaprophyte
полусека'ч *m* half-bull (*a fur seal*)
полуси'бсы *m pl* half-sibs
полусимпатри'ческий *a* semisympatric
полусинтети'ческий *a* semisynthetic
полуслия'ние *n* hemifusion
полусо'н *m* drowse
полустеблеобъе'млющий *a* half-stem-embracing
полустери'ль *f gen* semisteril
полустрептости'льный *a* semistreptostylic
полуто'нкий *a* semithin
полутушка'нчик *m* jumping mouse *(Zapus)*
полуфасе'тка *f* demifacet
полуша'рие *n* hemisphere
  ~ **большо'го мо'зга** hemicerebrum
полу'шник *m* quillwort (*Isoetes*)
полуэндофи'т *m* hemiendophyte, hemiepiphyte
полчо'к *m ther* dormouse (*Glis*)
по'лый *a* hollow, caval
полы'нь *f* sage-brush, worm-wood (*Artemisia*)
по'льдер *m* poder
польза *f* advantage, use
по'люс *m* pole
  ~ **хо'лода** pole of cold

анима'льный ~ animal pole
поля' *n pl* fields
~ **ороше'ния** sewage farm
поля'на *f* clearing, glade
поляриза'ция *f* polarization
пресинапти'ческая ~ presynaptic polarization
поляризо'ванный *a* polarized
поляриме'трия *f* polarimetry
поля'рность *f* polarity
поля'рный a polar
поляро'граф *m* polarograph
полярогра'фия *f* polarography
помера'нец *m* bitter orange (*Citrus aurantium*)
по'месь *f* mongrel, bastard, cross
помёт *m* dung, manure, scat, droppings; brood
кури'ный ~ fowl manure, poultry dung
многопло'дный ~ multiple brood
пти'чий ~ chicken manure
помидо'р *m* tomato (*Lycopersicum esculentum*)
помоло'гия *f* pomology
помо'рник *m* jaeger, skua, skua gull *(Stercorarius)*
большо'й ~ great skua, sea hawk (*Stercorarius skua*)
длиннохво'стый ~ long-tailed skua, Buffon's skua *(Stercorarius longicaudus)*
короткохво'стый ~ Arctic skua, Richardson's skua *(Stercorarius parasiticus)*
сре'дний ~ pomarine skua *(Stercorarius pomarinus)*
южнополя'рный ~ south polar skua *(Stercorarius maccormicki)*
помо'щник *m* helper
~ **по ухо'ду за детё'нышами** helper at the den
~ **по ухо'ду за птенца'ми** helper at the nest
по'мощь *f* aid, aid-giving behavior, helping behavior, helping, help, care
амбулато'рная ~ ambulatory care
ассоции'рованная ~ *imm* link-recognition help
гапте'новая ~ hapten help
когна'тная ~ *imm* cognate help
медици'нская ~ medical care, medical aid
пе'рвая ~ first aid
ско'рая ~ ambulance
помутне'ние *n* opacification, opacity
понести' *v* bolt *(of a horse)*
по'ни *m* pony
поника'ние *n* wilting
поника'ть *v* droop, wilt
пони'клый, поникший *a* drooping
понима'ние *n* understanding, comprehension
поно'с *m* diarrhea
поно'ска *f* an object carried by a dog (*e.g, a stick*)
поня'тливость *f* intelligence
поня'тливый *a* sagacious
попеременнопе'ристый *a* alternipinnate
поперечноборо'здчатый *a pal* transcolpate
поплаво'к *m* float
по'ползень *m* nuthatch (*Sitta europaea*)
пополне'ние *n* recruitment, recruit-stock
попуга'й *m* parrot (*Psittacidae*)
попуга'йчик *m* parrakeet (*Psephotus*, *etc.*)
популя'ция *f* population
~ **биологи'ческая ~ по'чвы** soil-biological population
~ **кле'ток-мише'ней** target population
исхо'дная ~ parent population
кло'новая ~ clonal population
ме'нделевская ~ Mendelian population
отде'льная ~ distinct population
по'ра *f* pore, pit
головна'я ~ head pore
ло'жная ~ pseudopore
окаймлё'нная ~ bordered pit
проро'стковая ~ germ pore
пора' *f* season
пораже'ние *n* lesion, damage, injury
грибко'вое ~ mycotic lesion, mycosis
лучево'е ~ radiation damage, radiation injury
непосре'дственное ~ direct injury
радиацио'нное ~ radiation injury
систе'мное ~ system lesion
скры'тое ~ latent injury
поре'з *m* cut
поре'зать *v* cut
поре'заться *v* cut (*oneself*)
по'ристость *f* porosity
по'ристый *a* porous, poriferous, *anat* cancellous, cancellate, cancellated
по'ровый *a* poral, aperturate
поро'зность *f* porosity
поро'г *m* threshold
~ **гальвани'ческой возбуди'мости** rheobase
~ **неприя'тных ощуще'ний** nuisance threshold

~ **раздраже'ния** threshold of stimulation
~ **различе'ния** differential threshold
порога'мия *f* porogamy
поро'да *f* stock, breed, race
го'рная ~ rock
грубоше'рстная ~ rough wool breed (*of sheep*)
джерсе'йская ~ Jersey breed
древе'сная ~ tree species
йоркши'рская ~ Yorkshire breed
мясна'я ~ table breed (*poultry*)
рыси'стая ~ trotting breed (*of horses*)
сму'шковая ~ fur sheep breed
тонкору'нная ~ fine wool breed
яи'чная ~ egg breed (*poultry*)
поро'дистый *a* pure-bred, throughbred, pedigreed
порожда'ть *v* procreate
поро'к *m* fault, defect
~ **разви'тия** developmental defect
~ **се'рдца** heart disease, valvular defect
порообра'зный *a* poriform
поросё'нок *m* piglet
подсо'сный ~ sucking pig, sucker pig
пороси'ться *v* farrow
по'росль *f* coppice, copse, growth, shoots, soboles
корнева'я ~ root sucker, coppice shoots, stool shoots
пнё'вая ~ stump sprouts, stum shoots, stool shoots
поро'ша *f* newly-fallen snow
пороши'ца *f* cytoproct (*of ciliates*)
порошкови'дный *a* powdery
порошкообра'зный *a* powdery
порошо'к *m* powder
ацето'новый ~ acetone powder
портула'к *m* purslane (*Portulaca*)
пору'бка *f* felling, cutting, chopping; illegal cutting
поруче'йник *m* water parsnip (*Sium*)
порфорин *m* channel-forming protein, cytolytic pore-forming protein
порха'ние *n* flitter
порха'ть *v* flitter
порхуно'к *m orn* flapper
по'рча *f* contamination
порфири'н *m* porphyrine
по'рция *f* batch, lot; portion
поря'док *m* order
~ **домини'рования** dominance order
посади'ть *v* plant (*e.g., potatoes*)
~ **ку'рицу на я'йца** set *a* hen on eggs
~ **на стро'гую дие'ту** put on *a* strict diet
~ **на цепь** chain up
~ **пти'цу в кле'тку** cage a bird
~ **соба'ку на цепь** chain *a* dog
поса'дка *f* setting, planting; landing; hanging (*of netting*)
~ **под кол** dibble seeding
~ **под лопа'ту** spade planting
ле'нточная ~ band planting
посе'в *m* sowing; seeding; *micr* plating; inoculation
~ **штрихо'м** streak inoculation
беспокро'вный ~ coverless sowing
двухстро'чный ~ double-row sowing
ле'нточный ~ band sowing, band planting
ози'мый ~ fall seeding
пункти'рный ~ single-grain sowing
разбросно'й ~ broadcast sowing
рядово'й ~ drill sowing, drilled sowing, drill planting
посевно'й *a* sowing
поседе'ние *n* grayness of the hair, canities, poliosis; bloom (*e.g., of chocolate*)
поселе'ние *n* settlement
послабля'ющий *a* aperient
после'д *m* afterbirth
последе'йствие *n* afteraction, aftereffect, residual effect
после'дний *a* ultimate, last
после'довательность *f* sequence, order
согласо'ванная ~ consensus sequence
после'дующий *a* posterior, subsequent
посленересто'вый *a* postspawning
послеродово'й *a* postnatal, postpartum
послете'чка *f* metestrus
посме'ртный *a* postmortal
посо'л *m* salt-curing
поспева'ть *v* ripen, mellow, color
постабдо'мен *m* postabdomen
постадапта'ция *f* postadaptation
постаксиа'льный *a* postaxial
постана'льный *a* postanal
постантенна'льный *a* postantennal
постганглиона'рный *a* postganglionic
постдикроти'ческий *a* postdicrotic
посткли'макс *m* postclimax
постларва'льный *a* postlarval
постмитоти'ческий *a* postmitotic
постната'льный *a* postnatal
постно'тум *m* postnotum, postscutellum
посторбита'льный *a* postorbital

постоя'нная *f* constant
~ **вре'мени паде'ния** *phys* decay constant
временна'я ~ time constant
постоянный *a* constant, permanent
постпетио'люс *m ent* postpetiole
постсинапти'ческий *a* postsynaptic, postjunctional
постскуте'ллум *m* postscutellum, postnotum
посттрансфузио'нный *a* posttransfusion
постфикса'ция *f* postfixation
постфу'рка *f* postfurca
посте'нный *a* parietal
постоя'нство *n* constancy, stability
~ **ви'дов** constance of species
пострада'вший *a* victim
постэмбриона'льный *a* postembryonic
пот *m* sweat, sudor, perspiration
потенциа'л *m* potential
~ **поврежде'ния** lesion potential
~ **последе'йствия** afterpotential
биоти'ческий ~ biotic potential
вы'званный ~ evoked potential
окисли'тельно-восстанови'тельный ~ oxidation-reduction potential, redox potential
следово'й ~ afterpotential
спа'йковый ~ spike potential
поте'нция *f* potency
проспекти'вная ~ prospective potency
поте'ря *f* loss
потеря кро'ви blood loss, loss of blood
потеря трудоспосо'бности disability
поте'ть *v* sweat, perspire
потово'й *a* sudoriferous, sudoral
пото'к *m* flow, stream
ио'нный ~ ion flux
светово'й ~ light flux
пото'метр *m* potometer
пото'мок *m* descendant
пото'мственный *a* hereditary
пото'мство *n* progeny, offspring, brood; posterity
потоотделе'ние *n* sweating, sudation, perspiration, diaphoresis
потра'ва *f* damage (to *crops*)
потрави'ть *v* trample down, damage (*e.g., grass*); poison
потребле'ние *n* consumption, uptake
~ **эне'ргии** energy expenditure
потребля'ть *v* consume
потре'бность *f* requirement, demand
~ **в воде'** water requirement
~ **в гормо'нах** hormonal requirement
~ **в кислоро'де** oxygen requirement
~ **в пи'ще** food requirement, nutritional requirement
биохими'ческая ~ в кислоро'де, БПК biochemical oxygen demand, BOD
пищева'я ~ food requirement
хими'ческая ~ в кислоро'де chemical oxygen demand
потроха' *pl* insides, entrails, viscera
пти'чьи ~ giblets
сви'ные ~ haslet
потроши'ть *v* eviscerate, disembowel
потя'гивание *n* stretching
похо'дка *f* gait, tread
асинерги'ческая ~ asynergic gait
волоча'щаяся ~ dragging gait
гарцу'ющая ~ prancing
напряже'нная ~ stiff gait
неусто'йчивая ~ unsteady gait
шата'ющаяся ~ staggering gait
поцелу'й *m* kiss
початкови'дный *a* earlike
поча'тковый *a* spadiceous
поча'ток *m* spadix
по'чва *f* soil
алли'тная ~ allitic soil
амфоте'рная ~ amphoteric soil
бу'рая ~ brown soil
вулкани'ческая ~ andosol
вы'щелоченная ~ leached soil
галоге'нная ~ halogenous soil
галомо'рфная ~ halomorphic soil
ги'псовая ~ gypsic soil
гидроге'нная ~ hydrogenic soil
гидромо'рфная ~ hydromorphic soil
гле'евая ~ gley soil
глееви'дная ~ gley-like soil
гли'нистая ~ clay soil
го'рная ~ mountain soil
гу'мусовая ~ humic soil, humus soil
дегради'рованная ~ degraded soil
делювиа'льная ~ deluvial soil
известко'вая ~ lime soil, limy soil
известняко'вая ~ limestone soil
изогу'мусовая ~ isohumic soil
и'листая ~ mud soil
иллювиа'льная ~ illuvial soil
интразона'льная ~ intrazonal soil
кальцемо'рфная ~ calcimorphic soil
карбона'тная ~ calcareous soil
кашта'новая ~ chestnut soil

коллювиа'льная ~ colluvial soil
кори'чневая ~ cinnamonic soil
коркообразу'ющая ~ incrusted soil
кра'сная ~ red soil
крупнокомко'вая ~ coarse soil
ксеромо'рфная ~ xeromorphic soil
латеризо'ванная ~ laterized soil
латери'тная ~ laterite soil
ле'гкая ~ light soil
ле'ссовая ~ loess soil
литоге'нная ~ lithogenic soil, lithic soil, lithosolic soil
малогу'мусовая ~ low-humic soil
маломо'щная ~ thin soil
мелкокомкова'тая ~ fine soil
ме'ргелевая ~ marl soil
мю'ллевая ~ mull soil
нано'сная ~ drift soil, transported soil
нера'звитая ~ infantile soil
огле'енная ~ gleyed soil
огли'ненная ~ clayed soil
ожелезе'нная ~ ferruginous soil
окульту'ренная ~ cultivated soil
оподзо'ленная ~ podzolized soil
пусты'ненная ~ desertified soil
осе'вшая ~ consolidated soil
осолоде'лая ~ solodic soil
па'хотная ~ arable soil
перехо'дная ~ intergrade soil, transitional soil
песча'ная ~ sandy soil
пло'тная ~ tight soil, firm soil
погребе'нная ~ buried soil
подзо'листая ~ podzolic soil
по'йменная ~ flood plain soil
покро'вная ~ casing soil
полигона'льная ~ polygonal soil
ре'говая ~ rego soil
рели'ктовая ~ fossil soil
ры'хлая ~ loose soil, free soil
самомульчи'рующаяся ~ self-mulching soil
светлокашта'новая ~ light-chestnut soil
сиалли'тная ~ siallitic soil
слабора'звитая ~ immature soil
старопа'хотная ~ cultivated soil
сугли'нистая ~ loamy soil
темнокашта'новая ~ dark chestnut soil
то'щая ~ lean soil
тяже'лая ~ heavy soil
уплотне'нная ~ compressed soil
ферримо'рфная ~ ferromorphic soil
цели'нная ~ virgin soil
чернозе'мная ~ chernozem soil
эктадинамо'рфная ectadynamorphic soil
эндодинамо'рфная ~ endodynamorphic soil
эо'ловая ~ aeolian soil

по'чвенный *a* edaphic, soil-dwelling, soil-inhabiting, terricolous
почвове'д *m* pedologist
почвове'дение *n* soil science, pedology, edaphology
почвозакрепля'ющий *a* soil-fixing
почвозащи'тный *a* soil-protective
почвообразова'ние *n* soil-formation, pedogenesis
почвообразу'ющий *a* pedogenic, soil-forming, soil-building
почвоутомле'ние *n* soil fatigation
почесу'ха *f* prurigo
поче'сывание *n eth* scratching
~ **головы'** *eth* head scratching
поче'сываться *v* scratch
по'чечка *f* reniculus
по'чечный *a* renal, nephric; gemmatious
по'чка *f anat* kidney, nephros; *bot* bud
блужда'ющая ~ wandering kidney, movable kidney, floating kidney
верху'шечная ~ crown bud, terminal bud
выводко'вая ~ brood bud
зиму'ющая ~ winter bud, hibernacle, hibernaculum
корнева'я ~ radical bud
па'зушная ~ axillary bud, accessory bud
поко'ящаяся ~ resting bud
прида'точная ~ adventitious bud
спя'щая ~ dormant bud, resting bud
почкова'ние *n* bud reproduction, budding, gemmation, pullulation
почковаться *v* pullulate
по'чковый *a* gemmaceous
почкосложе'ние *n* estivation, aestivation
листово'е ~ prefoliation
цвето'чное ~ prefloration, praefloration
почу'ять *v* scent out
поэти'н *m* poietic hormone
по'яс *m* girdle; belt, zone; cingulum
~ **ве'рхних коне'чностей** pectoral arch
~ **за'дних коне'чностей** pelvic arch
~ **ни'жних коне'чностей** pelvic arch
~ **пере'дних коне'чностей** pectoral arch
грудно'й ~ pectoral arch
лесопа'рковый ~ green belt

плечево'й ~ pectoral girdle, shoulder girdle, thoracic girdle
та'зовый ~ pelvic arch, pelvic girdle
тропи'ческий ~ tropical zone
пояско'вый *a zool* clitellar
поясни'ца *f* loin, small of the back
поясни'чно-грудно'й *a* thoracolumbar
поясни'чный *a* lumbar, lumbal
поясни'чная мы'шца psoas muscle
поясно'й *a* zonal
поясо'к *m* cingulum, clitellum
~ **Каспа'ри** *bot* Casparian strip
пра'вило *n* rule
~ **Бе'ргмана** Bergmann's rule, Bergmann's principle
~ **Джо'рдана** Jordan's rule
~ **«одна клетка - один идиотип»** "one cell-one idiotype" rule
правовраща'ющий *a* dextrorotatory
правозавито'й, правозакру'ченный *a* dexiotropic, dextrorse
правору'кий *a* dextromanual, right-handed, dextral
правору'кость *f* dextrality, right-handedness
прайд *m zool* pride
пра'ймер *m mbiol* primer
пра'йминг-феромо'н *m* priming pheromone
пра'йминг-эффе'кт *m* priming effect
прароди'тель *m* progenitor, grandparent
прасоли'н *m* prasoline
пре- pre-
преадапта'ция *f* preadaptation, prospective adaptation
превали'ровать *v* prevail
превраще'ние *n* metamorphosis; transmutation
~ **по'ла** sex reversal
непо'лное ~ hemimetaboly, semimetamorphosis
преганглиона'рный *a* preganglionic
прегетерокине'з *m* preheterokinesis
прегнентриолдио'н *m* pregnenetrioldione
прегра'да *f* barrier
грудобрюшна'я ~ diaphragm
пред- pre-, pro-, ante-, sub-
предвертлу'жный *a* preacetabular
предви'дение *n* prediction
предга'струла *f* pregastrula
предгаструляцио'нный *a* pregastrular
предгио'идный *a* prehyoid
предглазни'чный *a* preorbital
предгло'тка *f helm* prepharynx
предгруди'нный *a* presternal
преддве'рие *n* vestibule
преде'л *m* ceiling, limit
~ **популя'ции** population ceiling
стабилизи'рованный ~ популя'ции stabilized population ceiling
предетермини'рованный *a* predetermined
преджелу'док *m* proventriculus
предзаты'лочный *a* preoccipital
предзубно'й *a* predentary
пре'дки *m pl* ancestry
предкли'макс *m* subclimax
предкли'н *m anat* precuneus
предкли'нье *n anat* precuneus, quader
предку'колка *f* propupa
предлежа'ние *n* presentation
~ **плода'** presentation
ягоди'чное ~ плода' breech presentation
предли'нечный *a* premolt, premoult
предличи'нка *f* prolarva
предло'бный *a* prefrontal
предлокализа'ция *f* prelocalization
преднагру'зка *f* preload
преднересто'вый *a* prespawning
пре'док *m* ancestor, progenitor
предопределе'ние *n* predetermination
предопределя'ть *v* predetermine
предотвраще'ние *n* prevention
предохране'ние *n* protection
предохрани'тельный *a* protective
предохраня'ть *v* protect
предпле'чье *n* forearm
предплю'сна *f anat* basipodium
предпозвоно'чный *a* prevertebral
предполага'емый *a* putative, supposed
предпо'чка *f* pronephros
предпочте'ние *n* preference
предрасположе'ние *n* predisposition
предродово'й *a* antenatal, prenatal
предро'сток *m* proembryo *(of Charophyta);* protonema
предротово'й *a* preoral
предсенсибилиза'ция *f* presensitization, preimmunization
предсенсибилизи'рованный *a* prechallenged
предсе'рдие *n* auricula, atrium
предсказа'ние *n* prediction
предсме'ртный *a* premortal
предсошни'к *m anat* prevomer
предспо'ра *f* prospore
представле'ние *n* presentation

предтеменно'й *a* preparietal
предте'чка *f* proestrus
предузлово'й *a* antenodal
предупреди'тельный, предупрежда'ющий *a* precautionary, preventive
предше'ственник *m* precursor, predecessor, progenytor
предше'ствующий *a* precursory
предщи'т *m ent* prescutum
предъявле'ние *n* presentation
  предъявле'ние ~ stimulus presentation
презента'ция *f* presentation
презенти'рование *n* presentation
презумпти'вный *a* presumptive
преимагина'льный *a* preimaginal
преиммуните'т *m* preimmunity
преиму'щество *n* advantage
  селекти'вное ~ selective advantage
преинкуба'ция preincubation
прекалликреи'н *m* prekallikrein
прекли'макс *m* preclimax
прекраще'ние *n* arrest, cessation
преломле'ние *n* refraction
  двойно'е ~ double refraction
премоля'р premolar
премоля'рный *a* premolar
пре-мРНК pre-mRNA
прената'льный *a* prenatal, antenatal
преоблада'ние *n* predominance
преоблада'ть *v* predominate, prevail
преоблада'ющий *a* predominant, prevailing, prevalent
преобразова'ние *n* transmutation, transduction
  ~ **эне'ргии** energy transduction
препара'т *m* preparation, specimen, mount
  ~ **тка'ни** tissue specimen
  вла'жный ~ wet mount
  гистологи'ческий ~ tissue specimen
  да'вленный ~ squash mount
  лека'рственный ~ medicinal preparation
  нервномы'шечный ~ nerve-muscle preparation
  тота'льный ~ whole mount
препара'т-мазо'к *m* film preparation, streak preparation
препара'т-отпеча'ток *m* impression preparation, contact preparation
препари'рование *n* preparation
препари'ровать *v* prepare, dress
препилори'ческий *a* prepyloric
препоте'нтный *a* prepotent
препоте'нция *f* prepotence, prepotency
препуциа'льный *a* preputial
препу'ций *m* foreskin, prepuce
пре'рия *f* prairie
преры'вистость *f* discontinuity
пресинапти'ческий *a* prejunctional, presynaptic
преску'тум *m ent* prescutum
пресле'дователь *m* pursuer
пресле'дование *n* chase
пресле'довать *v* chase, pursue
пресмыка'ющийся *a* reptile
преспермати'да *f* prespermatid
пресс *m* press
  герба'рный ~ plant press
прессореце'птор *m* pressure receptor
прессо'рный *a* pressory
пресфено'ид *m* presphenoid
преформа'ция *f* preformation
преформи'рованный *a* precasted *(of gel)*
прехорда'льный *a* prechordal
преципити'н *m* precipitin
прибавле'ние *n* addition
приближе'ние *n* approach, approaching
  ба'йесовское ~ *biom* Bayesian approach
прибо'й *m* surf, battering of waves, beating of waves
прибо'р *m* device, apparatus, instrument
  ~ **для встря'хивания** shaking machine
прибре'жный *a* riparian, coastal, inshore
прибыло'й *a zool* young
приведе'ние *n* adduction
приве'нчик *m* paracolla
приве'с *m* weight gain, weight increment, increase in weight
  су'точный ~ daily weight gain
приве'тствие *n eth* greeting
  ~ **с элеме'нтами угро'зы** threat greeting
  подчине'нное ~ submissive greeting
привива'ть *v bot* engraft, imp
  ~ **глазко'м** bud
приви'вка *f* inoculation, vaccination; graft, grafting, imp
  ~ **в боково'й заре'з** side grafting
  ~ **в ко'рень** root grafting, graft on root
  ~ **в кро'ну** top grafting, crown grafting
  ~ **в паз** slotted side grafting
  ~ **в пенё'к** stub grafting
  ~ **в пропи'л** kerf grafting
  ~ **вращe'п** cleft grafting
  ~ **в щель** cap grafting
  ~ **вы'резанным кли'ном** wedge grafting

~ **за кору'** bark grafting
~ **кли'ном** wedge grafting, notch grafting, inlay grafting
~ **ко'лышком** peg grafting
~ **копулиро'вкой с**
~ **листо'м** leaf grafting
~ **мо'стиком** bridge grafting
~ **неодревесне'вшим черенко'м** herbaceous grafting
~ **обра'тным кли'ном** turned saddle type grafting
~ **расщепле'нным черенко'м** split-scion grafting
~ **сближе'нием** side grafting, approach grafting, inarching
~ **седло'м** saddle grafting
~ **сердцеви'ной на ка'мбий** pith-to-cambium grafting
~ **с про'бкой** cork grafting
~ **спя'щим глазко'м** dormant grafting
~ **черенко'м** cutting grafting, veneer grafting
~ **штифтова'нием** nailed grafting
~ **язычко'м** whip and tongue grafting
бокова'я ~ side rind grafting
двойна'я ~ double grafting
дополни'тельная ~ replacing grafting
зу'бчатая ~ toothed grafting
камбиа'льная ~ cambium grafting
кольцеобра'зная ~ annular grafting
концева'я ~ splice grafting
маши'нная ~ machine grafting
юпи'терская ~ Jupiter grafting
приви'тый *a* grafted
при'вкус *m* flavour, off-flavour
привлека'тельность *f* attractability
приводя'щий *a* adducent
приво'й *m* scion, cion, graft, cleft-graft scion
ве'рхний ~ top scion
второ'й ~ top scion
привыка'ние *n* habituation
привы'чка *f* habit
привы'чки *f pl of* **привычка**
пищевы'е ~ dietary habit
привы'чный *a* habitual
привяда'ние *n* wilting
привя'занность *f* attachment behavior, fidelity
приготовле'ние *n* preparation
прида'ток *m* appendage, appendix
прида'точный *a* accessory, adventitious
придо'нный *a* demersal
придоро'жный *a* roadside, wayside, roadside growing
приё'м *m* intake; dose
прие'млемость *n* acceptability
прижа'тый *a* appressed, adpressed, accumbent; ~ **к земле'** *bot* prostrate
прижи'зненный *a* intravital
прижима'ть *v* press to; ~ **у'ши** lay back the ears
при'знак *m* feature, character
втори'чный полово'й ~ secodary sexual character
зарожда'ющийся ~ incipient character
мендели'рующий ~ Mendelian character
меристи'ческий ~ meristic character
отвлека'ющий ~ deflective mark
полово'й ~ sexual character
прогресси'вный ~ advanced character
счё'тный ~ meristic character
таксацио'нный ~ taxational character
призы'в *m* call
прика'пывание *n* heeling-in
прика'пывать *v* heel in
прика'рмливание *n* baiting
прика'рмливать *v* lure
прико'пка *f* heeling-in
прикорнево'й *a* radical
прикрепле'нный *a* attached, sessile, sedentary
прику'с *m* occlusion, bite
прилега'ющий *a* accumbent
прили'в *m* tide, high tide
прилипа'льце *n bot* retinaculum
прили'стник *m* stipule
ло'жный ~ pseudostipula
прили'стниковый *a* stipular, stipuliform
прили'стничек *m* stipel
прило'в *m* incidental catch, by-catch
прима'нивать *v* lure, decoy
прима'нка *f* bait, lure, decoy; взять приманку snap at bait
примене'ние *n* application; administration
однора'зовое ~ single application
применя'ть *v* administer (*e.g., of a drug*), apply
приме'р *m* instance, example; **на примере** with reference to, with special reference to
прими'рование *n imm* priming
примити'вный *a* primitive, protomorphic
примордиа'льный *a* primordial
примо'рдий *m* primordium
примо'рдный *a* primordium

примо'рский *a* seaside, maritime
прини'женность *f* abjection
принима'ть *v* take
принима'ться *v* take root, establish
при'нинг *m* preening
приноси'ть *v* bring; yield; ~ **плоды'** bear fruit, produce fruit, yield fruit
принося'щий *a* afferent
при'нцип *m* principle
~ **Ба'рнета** Burnet's axiom
~ **«всё и'ли ничего'»** all-or-nothing principle
~ **Га'узе** Gause's principle
~ **конкуре'нтного исключе'ния** competitive exclusion principle
приню'хиваться *v* scent about, sniff
приня'вшийся *a* established
приня'тие *n* adoption
~ **реше'ния** decision-making
приобрете'ние *n* acquirement, acquisition
приобрете'нный *a* acquired
приорите'т *m* priority
~ **в до'ступе** *eth* priority of access
припада'ние *n* crouch; ~ **к земле'** squatting
припада'ть *v*: **припадать к земле'** crouch
припа'док *n* fit, attack, bout
припа'й *m* coast ice, collar ice, fast ice
припло'д *m* litter, offspring
припо'днятый *a* suberect
приползa'ть *v* crawl up, creep up
приправля'ть *v* flavour, dress (*food*)
припу'дренный *a* powdered
приро'да *f* nature; **в приро'де** in the field
приро'дный *a* natural
природопо'льзование *n* nature management
прирожде'нный *a* inherent, intrinsic, inborn
приро'ст *m* accretion, amount of growth, increment
~ **живо'й ма'ссы** liveweight gain
годово'й ~ annual gain
сре'дний су'точный ~ average daily gain
приро'сший *a* adnate, adherent
прируча'ть *v* tame
приру'ченный *a* tame, hand-reared, domesticated
приса'сывательный *a* suctorial, adhesive
приседа'ние *n* squat, squatting, crouch
приседа'ть *v* crouch
присемя'нник *m* aril
присоедине'ние *n* addition
альдо'льное ~ aldol addition
присо'ска *f* sucker
брюшна'я ~ ventral sucker
вооруже'нная ~ armed sucker
за'дняя ~ caudal sucker, subanal sucker
ло'жная ~ pseudosucker
ротова'я ~ buccal sucker, mouth sucker, oral sucker
приспособи'тельный *a* adaptive
приспособле'ние *n* adaptation
защи'тное ~ protective adaptation
морфологи'ческое ~ structural adaptation
проспекти'вное ~ prospective adaptation
эволюцио'нное ~ evolutionary adaptation
экобиоти'ческое ~ ecobiotic adaptation
экоклимати'ческое ~ ecoclimatic adaptation
экотопи'ческое ~ ecotopic adaptation
приспособ'ленность *f* fitness
приспособля'емость *f* adaptability, adaptation power
пристава'ние *n* annoyance
присте'нный *a* parietal
присте'ночный *a* parietal
при'ступ *m* fit, attack, seizure, spell, bout
~ **головокруже'ния** dizzy spell
~ **ка'шля** coughing spell, fit of coughing
си'льный ~ severe attack
притворя'ться *v* sham
притворя'шка *m* spider beetle (*Ptinidae*)
прито'к *m* tributary, tributary stream; influx, inflow
венозный ~ venous inflow
притупле'ние *n* obtusion
прицве'тник *m* bract, floral bract
прицве'тниковый *a* bracteal, bract-shaped
прицве'тничек *m* bracteole, bractlet, prophyll, prophyllum
прице'пка *f* tenaculum, hapteron
прищи'пка *f* topping, top removal
прищи'пывание *n* topping, top removal
приэкваториа'льный *a* subequatorial
про- pro-
проакросо'ма *f* proacrosome
проакцелери'н *m* proaccelerin, prothrombin acceleratror
проа'мнион *m* proamnion
про'ба *f* sample, test, trial, assay
~ **Абдерга'льдена** Abderhalden's test
~ **Вальса'льва** Valsalva test
~ **Ве'бера** Weber's test
~ **Вида'ля** Widal's test
~ **Кви'ка** Quick's test

~ **на гемолити'ческие бля'шки** hemolytic plaque assay
~ **Па'уль-Бунне'ля** Paul-Bunnell test
~ **Пирке'** Pirquet's test
~ **Фо'льгарда** Volhard's test
~ **функциона'льная** проба **functional trial**
~ **Шта'нге** Stange's test
агглютинацио'нная ~ agglutination test
аллерги'ческая ~ allergy test
биологи'ческая ~ bioassay, biological assay
броди'льная ~ laboratory test fermentation
сумма'рная ~ gross sample

проба'нд *m* proband
пробива'ться *v* shoot forth (*of grass*), push up, push through
пробира'ться *v* scramble, skulk
проби'рка *f* test tube
~ **для агглютина'ции** agglutination tube
~ **с ага'ром** agar tube
~ **с бульо'ном** bouillon tube, broth tube
~ **с закра'иной** flange-type test tube
~ **с культу'рой** culture tube
~ **с холосты'м о'пытом** dummy tube
адсорбцио'нная ~ adsorption tube
контро'льная ~ control tube
центрифу'жная ~ centrifuge tube

про'бка *f bot* cork; *lab* cork, stopper, plug
~ **Ди'ттриха** Dittrich's plug
ва'тная ~ cotton stopper
желто'чная ~ yolk plug
зави'нчивающаяся ~ screw cap
ко'рковая ~ cork stopper
притё'ртая стекля'нная пробка ground glass stopper
просве'рлённая ~ vented stopper

про'бковый *a* subereous, suberose
прободе'ние *n* perforation
прободё'нный *a* perforated, bored
пробоотбо'рник *m* sampling apparatus, sampling device, sampling instrument
проведе'ние *n phys* conduction
~ **возбужде'ния** conduction of excitement
восходя'щее ~ и'мпульсов ascending current
нисходя'щее ~ и'мпульсов descending current
центробе'жное ~ и'мпульсов descending current
центростреми'тельное ~ и'мпульсов ascending current

прове'рка *f* inspection
вы'борочная ~ *bioch* sampling inspection

провизо'рный *a* provisional
прови'нция *f* province
биогеохими'ческая ~ biogeochemical province
океани'ческая ~ oceanic province

прови'рус *m* provirus
провитами'н *m* vitamin precursor, provitamin
проводи'мость *f* conductance, conductibility
проводи'ть *v* dip (*of histological sections*); conduct; draw; carry out
про'волочник *m* wireworm (*Elateridae*)
прово'рный *a* agile
прово'рство *n* agility
провоци'ровать *v* provoke
провоци'рующий *a* provocative
прога'лина *f* glade, clearing
прогаме'та *f* progamete
прогамета'нгий *m* progametange
прога'мный *a* progamic
прогене'з *m* progenesis
прогенети'ческий *a* progenetic
прогено'та *f* progenote
прогестеро'н *m* progesterone
прогести'н *m* progestin
проглотти'да *f helm* proglottid, proglottis
прогнати'зм *m* prognathism
прогнати'ческий *a* prognathous
прогно'з *m* forecast, prognosis, prediction
долгосро'чный ~ long-term forecast
краткосро'чный ~ short-term forecast
рыбохозя'йственный ~ fishery prognosis

прогнози'рование *n* prognostication, forecasting
прогнози'ровать *v* forecast, prognosticate
прого'нистый *a* slender
прогоня'ть *v* chase away; drive through
прого'рклый *a* rancid
проградa'ция *f* progradation
програ'мма *f* program
откры'тая ~ *eth* open program

программи'рование *n* programming
прогре'сс *m* progress
~ **при отбо'ре** selection progress
эволюцио'нный ~ evolutionary progress

продвиже'ние *n* progression, advancement
продви'нутый *a* advanced
продово'льствие *n* foodstuffus, food, food supplies, provisions

продолгова'тый *a* elongated, oblong, extended, long-drawn, prolate
продолжи'тельность *f* duration, length
~ **дня** day length
~ **жи'зни** life span, longevity, duration of life
~ **предъявле'ния образца'** sample duration
сре'дняя вероя'тная ~ жи'зни life expectancy
продо'льный *a* longitudinal, anteroposterior
продрома'льный *a* prodromic
продува'ние *n* insufflation
~ **по Поли'тцеру** politzerization
проду'кт *m* product
~ **обме'на веще'ств** metabolic product, metabolite
~ **пита'ния** food stuff
~ **фотоси'нтеза** photosynthate
побо'чный ~ by-product
продукти'вность *f* productivity, production, producing capacity, crop-producing power, performance, production performance
~ **па'стбища** pasture performance
биологи'ческая ~ biological productivity
пыльцева'я ~ pollen production
семенна'я ~ seed production
энергети'ческая ~ *ecol* energy production
продукти'вный *a* productive
проду'кты *pl*: **половы'е** ~ genital products
продукцио'нный *a* productional
проду'кция *f* production
промысло'вая ~ commercial production
продуце'нт *m* producer
перви'чный ~ primary producer
продуци'рование *n* producing, production
продуци'ровать *v* produce
продуци'рующий *a* producing
продыря'вленный *a* perforated, perforate, fenestrated, fenestrate
проекти'рование *n* planning, design
природоохра'нное ~ environmental design
прое'кция *f* projection
прожо'рливый *a* voracious
прозе'нхима *f* prosenchyma
прозо'пиль *m* prosopyle
прозоплекте'нхима *f* prosoplectenchyma
прозра'чный *a* pellucid, transparent, translucent
производи'тель *m* breeder, sire; spawner *(ichthyology)*
производи'тельность *f* productivity, production, power, performance
~ **земли'** land capability
производи'тельный *a* productive, prolific
производи'ть *v* make, produce
~ **на свет** bear
~ **о'пыты** carry out experiments
~ **пото'мство** breed, procreate
произво'дство *n* industry
броди'льное fermentation industry
крупномасшта'бное ~ mass production
произраста'ние *n* growth
произраста'ть *v* grow
происхожде'ние *n* origin, ancestry, lineage, parentage, descent
~ **ви'дов** origin of species
~ **жи'зни** origin of life
о'бщее ~ common ancestry
политопи'ческое ~ polytopic origin
эволюцио'нное ~ evolutionary descent
прока'мбий *m* procambium
прокарио'тный *a* procaryotic
прока'рпий *m bot* procarp
прокиса'ние *n* souring
прокле'вывание *n* pipping
проклю'нуться *v* chip
прокоагуля'нт *m* procoagulant
проко'л *m* puncture
проконверти'н *m* serum acceleratror, prothrombinogen
прокорако'ид *m* procoracoid
проко'с *m* swath
проксима'льно-аперту'рный *a pal* proximaloaperturate
проксима'льный *a* proximal
проктоде'ум *m* proctodeum
пролакти'н *m* prolactin
проле'ска *f* squill (*Scilla*)
пролёт *m* flight
сплошно'й ~ unbroken flight
проли'н *m* proline
пролифера'ция *f* proliferation
клона'льная ~ clonal proliferation, clonogenic proliferation
пролифери'рующий *a* proliferous, proligerous
проло'мник *m* rock-jasmine (*Androsace*)
промастиго'тный *a* promastigote
промегалобла'ст *m* promegaloblast
проме'жность *f* perineum, perinaeum
промежу'ток *m* interval, space
межзу'бный ~ interdental space

промеристе'ма *f* promeristem
прометафа'за *f* prometaphase
промиелоци'т *m* promyelocyte
промискуите'т *m* promiscuity
промито'з *m* protomitosis
промице'лий *m* promycelium
проморфоло'гия *f* promorphology
промо'тор *m mbiol* promoter
промуните'т *m* promunity
промыва'ние *n* ablution
про'мысел *m* hunting, fishing
~ **ко'тиков** fur sealing
~ **кра'бов** crabbing
зверобо'йный ~ hunting, trapping
китобо'йный ~ whaling, whale fisheries, whaling industry
охо'тничий ~ hunting, trapping
пушно'й ~ hunting of fur animals
ры'бный ~ fishery
тюле'ний ~ sealing, seal fishery
промысло'вый *a* game, commercial; legal, legal-sized
промы'шленность *f* industry
виноде'льческая ~ wine industry
гидро'лизная ~ hydrolysis industry
крахмалопа'точная ~ starch industry
моло'чная ~ milk industry
пивова'ренная ~ brewing industry
пищева'я ~ food industry, food manufacturing industry
рыбообраба'тывающая ~ fish-processing industry
са'харная ~ sugar industry
торфяна'я ~ peat industry
целлюло'зно-бума'жная ~ pulp and paper industry
прона'тор *m* pronator
прона'ция *f* pronation
проне'фрос *m* pronephros
пронзенноли'стный *a* perforate
прони'мфа *f* pronymph
проница'емость *f* permeability
ио'нная ~ ion permeability
пронормобла'ст *m* pronormoblast
проно'тум *m ent* pronotum
прону'клеус *m* hemikaryon, pronucleus
пропа'лывать *v* weed
пропарази'т *m* proparasite
проперди'н *m* properdin
пропили'рование *n* propylation
пропинобетаи'н *m* homobetaine
пропи'тывание *n* impregnation
пропле'вра *f ent* propleuron
пропнейсти'ческий *a ent* propneustic
пропо'диум *m* propodium
проподосо'ма *f* propodosoma
пропо'лис *m apic* propolis
пропо'лка *f* weeding
проприоге'нный *a* propriogenic
проприореце'птор *m* proprioreceptor
проприореце'пция *f* proprioreception
проприоце'пция *f* proprioception
проптери'гий *m* propterigium
пропу'льсия *f* propulsion
прораста'ние *n* germination, sprouting, pullulation
~ **на свету'** light germination
прораста'ть *v* germinate, sprout, pullulate
прораста'ющий *a* sprouting
прора'щивание *n* sprouting
проре'живание *n* thinning out
проре'зывание *n* teething
~ **зуба** dental eruption
проре'зываться *v* teethe
проро'сток *m* seedling, germ, plantlet, plantling, plantule
прорубь *f* ice hole
проса'чивание *n* infiltration
просве'т *m* lumen
просветле'ние *n* clarification
просвирня'к *m* mallow (*Malva*)
просе'иваемость *f* boltig ability, screening ability
просе'ивание *n* bolting, sifting, screening
просе'ивать *v* bolt
про'сека *f* ride, opening, trace, lane, alley, glade, swath
про'со *n* millet (*Panicum*)
просо'ма *f zool* prosoma; *mbiol* prosome
просо'рус *m* prosorus
проспо'ра *f* prospore
простагланди'н *m* prostaglandin
проста'та *f* prostate gland, prostate
простати'т *m* prostatitis
просте'йшее *n* protozoan, protozoon
простекобакте'рия *f* prostecobacterium
просте'рнум *m ent* prosternum
простира'ться *v* sprawl
просто'миум *m* prostomium
простра'нство *n* space, tract
~ **Ди'ссе** Disse's space
~ **рогови'чно-ра'дужного угла'** space of irido-corneal angle, Fontana's space
вокругсинусо'идное ~ Disse's space

вы'борочное ~ *biom* sample space
ду'гласово ~ pouch of Douglas
ка'жущееся свобо'дное ~ apparent free space
лимфати'ческое ~ lymph space
межвлага'лищное ~ Tenon's space
мё'ртвое ~ respiratory dead space, dead space
многоме'рное ~ multidimensional space
одноме'рное ~ *biom* one-dimensional space
околожелто'чное ~ perivitelline space
перивителли'новое ~ perivitelline space
перилимфати'ческое ~ perilymphatic space
персона'льное ~ personal area, personal field
поведе'нческое ~ behavior space
подпаути'нное ~ arachnoid space
по'ровое ~ pore space *(in soil)*
ре'тциево ~ Retzius' space
те'ноново ~ Tenon's space
физиологи'ческое ~ physiological space
фонта'ново ~ Fontana's space, space of irido-corneal angle
функциона'льное ~ physiological space
простра'ция *f* prostration
простре'л *m* pasque-flower (*Pulsatilla*); lumbago
просту'да *f* chill, cold
просыпа'ться *v* wake
прота'лкивание *n* pressing through
~ **вперё'д** *n* propulsion
прота'ллиум *m* prothallium, prothallus
протами'н *m* protamine
протандри'ческий *a* protandrous, proterandrous
прота'ндрия *f* protandrism, protandry, proterandry
протарго'л *m* protargol
проте'з *m* prosthesis
протеи'д *m* proteid, proteide
протеина'за *f* proteinase
протеинури'я *f* proteinuria
~ **Бенс-Джо'нса** Bence-Jones proteinuria
проте'й *m* olm (*Proteus*)
протео'за *f* proteose
протеозопепто'н *m* proteose-peptone
протео'лиз *m* proteolysis
протеолитиче'ский *a* proteolytic
протерогини'я *f* proterogyny
протероги'нный *a* proterogynous
протерозо'йский *a* Proterozoic
протеросо'ма *f* proterosoma
проте'циум *m* prothecium
противо- anti-
противови'русный *a* antiviral
противогли'стный *a* antivermicular
противолепестко'вый *a* antipetalous
противоопухоле'вый *a* antitumoral, antineoplastic
противопаразита'рный *a* antiparasitic
противопоказа'ние *n* contraindication
противопока'занный *a* contraindicated, counterindicative
противочу'мный *a* antiplague
противоя'дие *n* antidote (for), antipoision, counterpoison
протимоци'т *m* prothymocyte
проти'ст *m* protist
протистоло'гия *m* protistology
прото- proto-
протобази'дия *m* protobasidium
протобио'нт *m* protobiont
протобла'ст *m* protoblast
протоге'м *m* protheme
протоге'н *m* protogene
протогене'з *m* protogenesis
протогини'я *f* protogyny, proterogyny
протоги'нный *a* proterogynous, protogynous
протозиго'та *f* protozygote
протозо'йный *a* protozoan
протозооло'гия *f* protozoology
прото'к *m* duct; channel, passage
~ **Ге'нзена** Hensen's duct
бартоли'нов ~ Bartholin's duct
блужда'ющий ~ aberrant duct
бота'ллов ~ Bottallo's duct, duct of Bottallo
ва'ртонов ~ Wharton's duct
вирсу'нгов ~ Wirsung's duct
во'льфов ~ Wolffian duct
желто'чный ~ *emb* yolk stalk
же'лчный ~ bile duct, biliary duct, gall duct, bile passage, biliary passage
мле'чный ~ lactiferous duct
риви'ниев ~ Rivinian duct
сантори'ниев ~ Santorini's duct
сле'зный ~ tear duct, lacrimal passage
протокарио'н *m* protokaryon
протокне'ма *f* protocneme
протоко'н *m* protocone
протокони'д *m* protoconid
протоко'нх *m* protoconch
протокоопера'ция *f* protocooperation
протоко'рм *m bot* protocorm

протоксиле'ма *f* protoxylem
протом'ер *m* protomer
протомери'т *m* protomerite
протомито'з *m* protomitosis
протоне'ма *f* protonema
протонефри'дий *m* protonephridium
протони'мфа *f* protonymph
протони'рование *n* protonation
протонофо'р *m* protonophore
протоонкоге'н *m* protooncogene
протопати'ческий *a* protopathic
протопла'зма *f* protoplasm
протоплазмати'ческий *a* protoplasmic, protoplasmatic
протопла'ст *m* protoplast
протоподи'т *m zool* protopodite
прото'ракс *m ent* prothorax
протоспо'ра *f* protospore
протосте'ла *f* protostele
прототи'п *m* prototype
прототро'ф *m* prototroph
прототро'фный *a* prototrophic
прототро'х *m* prototroch
протофибри'лла *f* protofibril
протофлоэ'ма *f* protophloem
протохлорофи'лл *m* protochlorophyll
протоцере'брум *m* protocerebrum
протоцерка'льный *a* protocercal
прото'чность *f* flowage
прото'чный *a* lotic (*of water bodies*)
протоэпифи'т *m* protoepiphyte
протра'ва *f* mordant
протра'вливать *v* dress (*seeds*)
протра'ктор *m* protractor
протромби'н *m* prothrombin, proserosyme
протромбина'за *f* prothrombinase
протромбокиназа *f* prothrombin acceleratror
профа'г *m* prophage
  дефе'ктный ~ defective prophage
  зре'лый ~ mature prophage
профа'за *f* prophase
проферме'нт *m* proenzyme
профила'ктика *f* prevention
профилакти'ческий *a* preventive
про'филь *m* profile
  ~ **элю'ции** elution profile
  иммунологи'ческий ~ immunological profile
прохо'д *m* passage; passage pore *(in diatoms)*
  нару'жный слухово'й ~ ear tract
прохожде'ние *n* passing, going through
  ~ **лабири'нта** maze performance
прохорион *m* prochorion
прохромосо'ма *f* prochromosome
процвета'ющий *a* prosperous
процеду'ра *f* procedure
  грязева'я ~ peloid procedure
  лече'бная ~ treatment procedure
проце'льный *a* procelous
процепти'вность *f eth* proceptivity
  полова'я ~ sexual proceptivity
процерко'ид *m* procercoid
проце'сс *m* process
  воспали'тельный ~ inflammatory process
проце'ссинг *m mbiol* processing
процесси'рование *n:*
  ~ **антиге'на** antigen handling
проце'ссор *m lab* processor
  многояче'йковый ка'пельный ~ multitip processor
прощу'пывание *n* palpation
проэмбрио'н *m* proembryo
проэ'струс *m* proestrus
проявле'ние *n* display, manifestation; development *(photography)*
пруд *m* pond
прудови'к *m* pond snail (*Lymnaeidae*)
пруд-охлади'тель *m* cooling pond
пруса'к *m* German cockroach (*Blattella germanica*)
прут *m* twig, switch; withe
пры'гательный *a* saltatory
пры'гать *v* jump, leap, spring, hop
пры'гающий *a* saltigrade, saltatory
пры'гнуть *v* leap, spring
прыжо'к *m* jump, leap, bounce, bound, spring
  смотрово'й ~ stotting
прыщ *m* pustule
пры'щик *m* papula, papule, pimple, pustule
пря'жка *f* clamp, buckle
прямоборо'здный *a* orthocolpate
прямоголо'вчатый *a* upright-headed
пряможилкова'тый *a* rectinerved, rectivenous
прямо'й *a* straight; erect
  пряма'я кишка rectum
прямокише'чно-ма'точный *a* rectouterine
прямокише'чно-пузы'рный *a* rectovesical
прямоклю'вый *a* rectirostral
прямоколо'сый *a* upright-spiked
прямоколю'чковый *a* upright-spined
прямостоя'чий erect, upright
прямоугольник rectangle
  лати'нский ~ *biom* Latin rectangle

прямоуго'льный *a* rectangular
прямоходя'щий *a* orthograde
прямоцветко'вый *a* upright-flowered
псаммо'н *m* psammon
псаммофи'льный *a* psammophilous, ammophilous, ammocolous
псаммофи'т *m* psammophyte
псевдо- pseudo-
псевдоаллели'зм *m* pseudoallelism
псевдоаллеломо'рф *m* spurious allelomorph
псевдоалле'ль *m* pseudoallele
псевдоапога'мия *f* pseudoapogamy
псевдоапосемати'ческий *a* pseudaposematic
псевдоапоспори'я *f* pseudapospory
псевдобивале'нт *m* pseudobivalent
псевдобра'нхия *f* pseudobranch
псевдога'мия *f* pseudogamy
псевдогапло'ид *m* pseudohaploid
псевдоге'н *m* pseudogene
псевдогемафродити'зм *m* pseudohermaphroditism
псевдоглобули'н *m* pseudoglobulin
псевдоде'рма *f* pseudoderm
псевдодомини'рование *n* pseudodominance
псевдокли'макс *m* pseudoclimax
псевдокони'ческий *a ent* pseudocone
псевдоло'кула *f* pseudoloculus *(in diatoms)*
псевдоми'ксис *m* pseudomixis
псевдомитоти'ческий *a* pseudomitotic
псевдомона'да *f* pseudomonade
псевдони'мфа *f* pseudonymph, semipupa
псевдоодноле'тник *m* pseudannual
псевдопаре'нхима *f* pseudoparenchyma
псевдопартеногене'з *m* pseudoparthenogenesis
  гибри'дный ~ hybridogenous pseudoparthenogenesis
псевдопери'дий *m* pseudoperidium
псевдоплазмо'дий *m* pseudoplasmodium
псевдопо'дия *f* pseudopod, pseudopodium
псевдосемати'ческий *a* pseudosematic
псевдоско'лекс *m* pseudoscolex
псевдотрахе'я *f* pseudotrachea
псевдохиа'зма *f* pseudochiasma
псевдоцело'м *m* pseudocoel
псевдоци'ста *f* pseudocyst
псевдоэписемати'ческий *a* pseudepisematic
псевдоэрга'т *m ent* pseudoergate
пситтако'з *m* psittacosis
пси'хика psychics
психологи'ческий *a* psychologic, psychological
психоло'гия *f* psychology
психотро'пный *a* psychotropic
психофизиоло'гия *f* psychophysiology
психрофи'льный *a* psychrophilic
психро'метр *m* psychrometer
психрофи'т *m* psychrophyte
псо'вый *a* canine
птене'ц *m* nestling, chick, fledgling, fledgeling, poult, pullus
  иску'сственный ~ artificial nestling
птенцо'вый *a* altricial
птери'гиум *m* pterygium
птеригопо'дий *m* pterygopodium
птеригофо'р *m* pterygophore
птеридоло'гия *f* pteridography, pteridology
птеридофи'т *m* pteridophyte
птери'лия *f orn* feather tract
птерио'н *m* pterion
птерога'т *m ent* pterogate
птеропле'вра *f ent* pteropleuron
птероплейри'т *m ent* pteropleurite
птерости'гма *f* pterostigma
пти'ца *f* bird (*Aves*); fowl; poultry
  водопла'вающая ~ waterfowl
  выводко'вая ~ nidifugous bird, praecoce
  дома'шняя ~ poultry, fowl, barnyard fowl
  зо'нтичная ~ umbrella bird *(Cephalopterus)*
  ко'мнатная ~ pet bird
  ма'нная ~ decoy bird
  морска'я ~ sea bird
  назе'мная ~ land bird
  ночна'я ~ night bird
  охо'тничье-промысло'вая ~, охо'тничья ~ game bird
  пе'вчая ~ song bird, songster, singing bird
  переле'тная ~ migratory bird, bird of passage
  подсадна'я ~ decoy bird
  птенцо'вая ~ altrice
  ра'йская ~ bird of paradise (*Paradiseidae*)
  хи'щная ~ bird of prey
пти'ца-секрета'рь *f* secretary bird (*Sagittarius serpentarius*)
птицево'д *m* poultry breeder, poultry husbandman
птицево'дство *n* aviculture, poultry farming, poultry husbandry
пти'чий *a* avian, ornithic

пти'чник *m* poultry yard, poultry house, hen-house
пуга'ть *v* scare
пугли'вый *a* flighty (*of a horse*), wild, shy, timid
пу'дель *m* poodle
пудре'тка *f* powder down feather
пузырё'к *m* bubble; follicle, bladder, vesicle; vial
граа'фов ~ Graafian follicle, Graafian vesicle
желто'чный ~ *helm* vitellin follicle
мальпи'гиев ~ Malpighian vesicle
пузырепло'дный *a* bladder-fruited
пузы'рный *a* cystic, vesical
пузырча'тка *f* bladderwort (*Utricularia*); *med* pemphigus
~ **Хе'йли-Хе'йли** Hailey and Hailey disease
пузы'рчатый *a* bladdery, vesicular
пузы'рь *m* bubble; bladder; blister, cyst
жё'лчный ~ gall bladder, bile cyst
заро'дышевый ~ germ bladder
мочево'й ~ urinary bladder
пла'вательный ~ swim bladder, sound, fish sound, swimming bladder, air bladder
по'лиев ~ Polian vesicle
пу'гало *n* scarecrow
огоро'дное ~ scarecrow
пул *m* pool
энергети'ческий ~ energy pool
пульви'лла *f ent* pulvillus, pulvinulus
пульвилля'рный *a ent* pulvillar
пу'льпа *f* pulp
сульфи'тная целлюло'зная ~ sulfite pulp
пульс *m* pulse
~ **перифери'ческих арте'рий** peripheric pulse
перифери'йный ~ peripheric pulse
ча'стый ~ quick pulse
пульса'ция *f* pulsation, pulse, beating, throbbing
пульси'ровать *v* pulse
пульси'рующий *a* pulsative, pulsatory
пуля'рка *f* spayed hen
пу'ма *f* puma (*Felis concolor*)
пункта'т *m* punctate
пу'нкция *f* puncture, tapping, paracentesis
пу'ночка *f* snow bunting (*Plectrophenax nivalis*)
пупа'вка *f* chamomille (*Anthemis*)
пупа'льный *a* pupal
пупа'рий *m* puparium, pseudopupa
пупови'на *f* funiculus, funicle, funis, umbilical cord, navel string
пупо'к *m* navel, umbilicus
пупо'чно-пузы'рный *a* umbilicovesical
пупо'чный *a* omphalic, umbilical
пупы'рчатый *a* pimple-bearing
пупы'рышек *m* pimple, papula
пури'н *m* purine
пу'рпур *m* purple
пу'рпурный *a* purple
бромкрезо'ловый ~ bromocresol purple
пуска'ть *v* put forth, sprout; ~ **ко'рни** root, take root; ~ **ростки** shoot, sprout
пустельга' *f* kestrel (*Falco tinnunculus*)
пустоцве'т *m* barren flower, unfertile flower, sterile flower
пу'стошь *f* waste ground, heath, heathland, barren land
камени'стая ~ stone heathland
травяни'стая ~ grass heathland
пу'стула *f* pustule
пусты'нный *a* desert
пусты'ня *f* desert
куста'рниковая ~ shrub desert
песча'ная ~ sandy desert
полы'нная ~ sagebruch desert
солончако'вая ~ saline desert
пусты'рник *m* motherwort (*Leonurus*)
пусты'рь *m* waste land, barren land, barren
пути'на *f* fishing season
путресци'н *m* putrescine
путь *m* path, pathway, route, track, tract
~ **Го'верса** Gowers tract, column of Gowers
~ **сближе'ния прону'клеусов** copulation path
~ **Фле'ксига** Flechsig's column, Flechsig's tract
~ **Э'мбдена-Мейерго'фа-Па'рнеса** Embden-Meyerhof-Parnes' pathway
~ **ядра' спе'рмия** entrance path, penetration path
альтернати'вный ~ alternative path, alternative pathway
ассоциати'вный ~ *phys* association tract
аффере'нтный ~ afferent pathway
возмо'жный ~ обме'на possible pathway
гла'вный ~ обмена principal pathway
дополни'тельный ~ обме'на additional pathway
за'дний спи'нно-мозжечко'вый ~ Flechsig's tract
заро'дышевый ~ germ track

зри'тельный ~ visual pathway
класси'ческий ~ classical pathway
ко'рково-спинномозгово'й ~ pyramidal tract
лити'ческий ~ lytic pathway
моносинапти'ческий ~ monosynaptic pathway
не'рвный ~ nerve pathway
нисходя'щий дви'гательный ~ descending motor pathway
паралле'льный ~ обме'на parallel pathway
пирами'дный ~ pyramidal tract
половы'е пути' *pl* genital tracts, reproductive tracts
постули'рованный ~ обме'на postulated pathway
преддве'рно-спина'льный ~ vestibulospinal tract
проводя'щий ~ conduction path
пролё'тный ~ flight route, flyway
прямо'й ~ обме'на direct pathway
«реутилизацио'нный» ~ "salvage" pathway
ростра'льный проекцио'нный ~ rostral projection pathway
сигна'льный ~ signaling pathway
синапти'ческий ~ synaptic pathway
стволово'й ~ brainstem pathway
транскле'точный ~ paracellular pathway
цикли'ческий ~ обме'на cyclic pathway
шунто'вый ~ shunt pathway
эффере'нтный ~ efferent pathway
пуф puff
пу'финг *m* puffing
пух *m* down, floss
гага'чий ~ eiderdown
пучегла'зие *n* exophthalmia
пучегла'зый *a* goggle-eyed, pop-eyed
пучкови'дный *a* agminated, agminate
пучко'вый *a* fascilular
пучо'к *m* cluster, bundle, bunch, fascicle, tuft, lock (*of hair*)
~ **Бу'рдаха** wedge-shaped fascicle
~ **Вик д'Ази'ра** bundle of Vicq d'Azyr
~ **Ги'са** bundle of His
~ **Го'верса** Gowers tract
~ **Го'лля** Goll's fascicle
~ **Фле'ксига** Flechsig's tract
клинови'дный ~ wedge-shaped fascicle
коллатера'льный ~ collateral bundle
кругово'й ~ circular bundle
меридиона'льный ~ meridional bundle
сосу'дисто-волокни'стый ~ fibro-vascular bundle
сос'удистый ~ vascular bundle
пушистопло'дный *a* woolly fruited
пуши'стый *a* downy, pubescent, plumulate, lanuginous, fuzzy, tomentous, tomentose, nappy
пуши'ца *f* cotton grass (*Eriophorum*)
пушни'на *f* peltry, peltries, furs
пушно'й *a* fur-bearing
пушо'к *m* pubescence, down, floss, nap
пчела' *f* bee (*Apis*)
медоно'сная ~ honeybee (*Apis mellifera*)
одино'чная ~ solitary bee
рабо'чая ~ worker bee, working bee
сторожева'я ~ guard bee
пчела'-корми'лица *f* nurse bee
пчела'-разве'дчица *f* scout bee
пчела'-сбо'рщица *f* forager
пчелово'дство *n* apiculture, bee breeding, beekeeping, honeycraft
пчелосемья' *f* honey-bee colony, colony of bees
пче'льник *m* apiary
пшени'ца *f* wheat, corn (*Triticum*)
~ **англи'йская** cone wheat, poulard wheat, rivet wheat *(Triticum turgidum)*
~ **безо'стая** awnless wheat, beardless wheat
~ **двузерня'нка** emmer, emmer wheat (*Triticum dicoccum*)
~ **однозерня'нка** einkorn (*Triticum monococcum*)
~ **ка'рликовая** club wheat *(Triticum compactum)*
~ **мя'гкая** soft wheat (*Triticum vulgare*)
~ **ости'стая** awned wheat, bearded wheat
~ **спе'льта** spelt, dinkel wheat, German wheat (*Triticum spelta*)
~ **твё'рдая** durum wheat, hard wheat, flint wheat (*Triticum durum*)
ози'мая ~ winter wheat, autumn wheat, fall wheat
ярова'я ~ spring wheat
пы'жик *m* young reideer
пыли'нка *f* speck of dust; *bot* pollen grain
пы'льник *m bot* anther
кача'ющийся ~ versatile anther
пыльниконо'сный *a bot* antheriferous
пы'льный *a* dusty, pulverulent
пыльца' *f bot* pollen
пыльцено'сный *a bot* polliniferous

пыльцеулови'тель *m* pollen trap
пыре'й *m* wheat-grass, couch-grass, quack-grass, quitch-grass (*Agropyron*)
пытли'вый *a* inquisitive
пы'шный *a* luxuriant, gross, magnificient, exuberant, lush (*of vegetation*); light, fluffy (*e.g., cake*)
пья'вица *f ent:* **пьявица трехлине'йная** three-lined potato beetle (*Lema trilineata*)
пэ'ннинг *m imm* panning
многостади'йный ~ sequential panning
секвенциа'льный ~ sequential panning
пя'деница *f* geometrid, geometrid moth, looper (*Geometridae*)
пя'стный *a* metacarpal
пятачо'к *m* snout (*of a pig*)
пяти- penta-, quinque
пятилепе'стный *a* pentapetalous
пятили'стный *a* pentaphyllous
пятилучево'й *a* pentactinal
пятипа'лость *f* pentadactylism
пятипе'стичный *a* pentagynous
пятипло'дный *a* pentacarpous
пятиразде'льный *a* quinquefid
пятиря'дный *a* quinquefarious, quinqueserial
пятисемя'нка *a* pentachenium
пятитычи'нковый *a* pentandrous
пятиуго'льный *a* pentagonal
пятичле'нный *a* pentamerous, pentactinal
пя'тка *f* heel
пятни'стость *f* maculation, spotting, blotch, mottle, speck; *phytop* spot, spotting
глазко'вая ~ зла'ков cereal eyespot *(caused by Tapesia yallundae)*
кольцева'я ~ ring spot *(caused by fungi)*
охряна'я зона'льная ~ ascochyta-leaf spot *(caused by Ascochyta)*
смоли'стая ~ tar spot *(caused by Rhytisma)*
пятни'стый *a* maculate, maculated, maculiferous, blotched, spotted, flecked, punctated, punctate
пятно' *n* spot, flock, blotch, macula; mark, star (*on a forehead*); plaque
глазча'тое ~ *ent* eyespot, eyelike spot, ocellated spot, orbicular spot, roundspot
заро'дышевое ~ germinal spot, germinal macula
ложноположи'тельное ~ гемолиза pseudoplaque
насе'дное ~ brood-spot
прозра'чное стери'льное ~ clear plaque
слепо'е ~ optic disk
слухово'е ~ saccular macula, utricular macula
стери'льное ~ phage plaque
пя'тнышко *n* mottle, fleck, speck, macula

Р

ра'бда *f* rhabdus
рабди'т *m* rhabdite
рабдови'рус *m* rhabdovirus
рабдо'ид *m* rhabdoid
рабдоли'т *m* rhabdolith
рабдо'м *m* rhabdom
рабдоме'р *m* rhabdomere
рабо'та *f* work
~, **затра'чиваемая на дыха'ние** work of breathing
~ **се'рдца** cardiac work, heart work
аэро'бная ~ aerobic work
мину'тная ~ **се'рдца** minute heart work
систоли'ческая ~ stroke work, systolic work
у'мственная ~ mental work, brain work
работоспосо'бность *f* working capacity, work capacity
рабо'чий *a* working (*e.g., of cattle*); *m* worker
равни'на *f* plain, flat country
аллювиа'льная ~ alluvial flat
песча'ная ~ sand flat
равни'нный *a* lowland
равно- homo-, iso-
равнове'сие *n* equilibrium, balance
~ **в приро'де** balance of nature
~ **обме'на веще'ств** metabolic equilibrium
безразли'чное ~ neutral equilibrium
биологи'ческое ~ biological balance, biotic balance, biological equilibrium
динами'ческое ~ dynamic balance
до'ннановское ~ Donnan equilibrium
ка'жущееся ~ apparent equilibrium
кисло'тно-щелочно'е ~ acid-base balance, acid-base equilibrium
конформацио'нное ~ conformational equilibrium
мутацио'нное ~ mutation equilibrium
относи'тельное ~ relational balance
подви'жное ~ dynamic equilibrium, dynamic balance
сбаланси'рованное ~ balanced equilibrium
равнозу'бый *a* isodont
равнокле'точный *a* isocytic
равнокры'лый *a* homopterous
равнолепе'стный *a* isopetalous, homopetalous
равноло'пастный *a* isocercal
равноме'рность *f* evenness, uniformity
~ **распределе'ния** equitability
равноотстоя'щий *a* equidistant
равнопа'лый *a* isodactylous
равноресни'чный *a* holotrichous
равноспо'ровый *a* homosporic, homosporous, isosporous
равносто'лбчатый *a* homostyled
равносторо'нний *a* equilateral
радиа'льно-жилкова'тый *a* radiate-veined
радиа'льно-симметри'чный *a* radiosymmetrical, radiate
радиа'льный *a* radial
радиа'ция *f* radiation
адапти'вная ~ adaptive radiation
ионизи'рующая ~ ionizing radiation
оста'точная ~ residual radiation
отражё'нная ~ back radiation
проника'ющая ~ penetrating radiation
со'лнечная ~ solar radiation
сумма'рная ~ total radiation
фотосинтети'чески акти'вная ~ photosynthetic active radiation
радика'л *m* radical
свобо'дный ~ free radical
радика'льный *a* radical
радио- radio-
радиоавто'граф *m* radioautograph, autoradiogram, autoradiograph
радиавтогра'фия *f* radioautography
конта'ктная ~ contact autoradiography
электро'нно-микроскопи'ческая ~ electron microscopic autoradiography
радиоактива'ция *f* radioactivation
радиоакти'вность *f* radioactivity
наведе'нная ~ induced radioactivity
уде'льная ~ specific radioactivity
радиоана'лиз *m* radioassay
радиобиологи'ческий *a* radiobiologic, radiobiological
радиобиоло'гия *f* radiobiology
радиогра'фия *f* radiography
радиоизото'п *m* radioisotope, radionuclide
радиоиммуноана'лиз *m* radioimmunoassay, radioimmunodetection
гомоге'нный ~ fluid-phase radioimmunoassay
конкурентносвя'зывающий ~ competitive binding radioimmunoassay
конкуре'нтный ~ competitive radioimmunoassay

радиоиммуноло'гия *f* radioimmunology
радиоиммунопреципита'ция *f* radioimmunoprecipitation
радиоиммунохроматогра'фия *f* radioimmunochromatography
радиоиммуноэлектрофоре'з *m* radioimmunoelectrophoresis
перекрё'стный ~ crossed radioimmunoelectrophoresis
радио'лиз *m* radiolysis
и'мпульсный ~ pulse radiolysis
радиоло'гия *f* radiology
радио'метр *m* radiometer
радиометри'я *f* radiometry
радиомимети'ческий *a* radiomimetic
радиоморфо'з *m* radiomorphose
радиополимериза'ция *f* radiopolymerization
радиопросле'живание *n* radiotracking
радиорезисте'нтность *f* radioimmunity, radioresistance
радиореце'птор *m* radioreceptor
радиоселе'кция *f* radioselection
радиосенсибилиза'ция *f* radiosensitization
радиотелеметри'я *f* radiotelemetry
радиотерапи'я *f* radiotherapy, radiation therapy
радиоуглеро'д *m* radiocarbon
радиоусто'йчивость *f* radioresistance
радиохроматогра'фия *f* radiochromatography
радиочувстви'тельность *f* radiosensibility, radiosensitivity
радиочувстви'тельный *a* radiosensitive
радиоэколо'гия *f* radioecology
ра'дужница *f ent* long-horned leaf beetle, reed beetle *(Donacia)*
ра'дужный *a* iridescent; **ра'дужная оболочка** iris
ра'дула *f zool* radula, lingual ribbon
разбави'тель *m* diluting agent
разбива'ть *v* break; establish;
~ **сад** lay out a garden
разби'вка *f* laying out, marking out
~ **уча'стка** plot layout
разведе'ние *n* breeding, raising, rearing, propagation, culture, cultivation; dilution, thinning
~ **антисы'воротки** antiserum dilution
~ **живо'тных** animal breeding
~ **поро'ды «в себе»** pure breeding
коне'чное ~ end-point dilution
иску'ственное ~ artificial propagation
лине'йное ~ line breeding
межлине'йное ~ interline breeding
после'довательное ~ serial dilution
ро'дственное ~ close breeding
те'сное ~ close breeding
чи'стое ~ pure breeding
чистолине'йное ~ pure-line breeding
разве'систый *a* effuse, patulent, patulous
разветвле'ние *n* ramification, branching, bifurcation, divergence; crotch
~ **це'пи** chain branching
разветвлё'нный *a* branched, branching, brachiate, brachiferous, ramified, brachigerous, ramate, ramose
разветвля'ться *v* ramify
разветвля'ющийся *a* ramiparous
развива'ться *v* develop
разви'лина *f* crotch
разви'лка *f* bifurcation
разви'тие *n* development
~ **без превраще'ния** development without metamorphosis, ametabolous development
~ **заро'дыша** embryonic development
~ **плода'** fetal development
~ **с непо'лным превраще'нием** hemimetabolous development, paurometabolous development
~ **с по'лным превраще'нием** holometabolous development
детермини'рованное ~ determinate development
заде'ржанное ~ impaired development
канализи'рованное ~ canalized development
когнити'вное ~ cognitive development
недетермини'рованное ~ indeterminate development
мозаи'чное ~ mosaic development
поведе'нческое ~ behavioral development
преждевре'менное полово'е ~ precocious development
регуляцио'нное ~ regulative development
стади'йное ~ stage development
устой'чивое ~ sustainable development
эмбриона'льное ~ embryonic development
разводи'ть *v* breed, rear, raise, cultivate, propagate; dilute;
~ **пилу'** set a saw
разво'дка *f* culture
пласти'нчатая ~ *f* platiculture
разга'р *m:*
в разга'ре ле'та in the top of summer

разгреба'ющий *a* rasorial
разгру'зочный *a* unloading;
разгрузочная дие'та fasting diet, unloading diet
раздвое'ние *n* bifurcation
раздво'енный *a* bifid, bifurcate, bifurcated
разделе'ние *n* division, disunion, separation
~ **труда'** division of labour
хроматографи'ческое ~ separation by chromatography
разделе'нный *a* divided
раздели'мый *a* separable
разде'лка *f* butchering, dressing
разде'лывать *v* dress
~ **гря'дки** prepare beds
~ **ту'шу** dress a carcass
раздельнолепе'стный *a* choripetalous, eleutheropetalous
раздельноли'стный *a* eleutherophyllous
раздельнопо'лость *f* gonochorism, dicliny; *zool* dioecy
раздельнопо'лый *a bot* diclinous; *zool*
раздельночашели'стиковый *a* chorisepalous, eleutherosepalous
разде'льный *a* separate; divided, partite; discrete
разделя'ть *v* separate
раздража'ть *v* stimulate, irritate
раздража'ющий *a* irritative, irritating
раздраже'ние *n* irritation; stimulation
~ **мелька'ющим све'том** flicker stimulation
~ **на расстоя'нии** remote stimulation
обусла'вливающее ~ conditioning stimulation
про'бное ~ test stimulation, testing stimulation
прове'рочное ~ testing stimulation
светово'е ~ photic stimulation
раздраже'нный *a* irritated, hot, angry
раздражи'мость *f* irritability
раздражи'мый *a* irritable
раздражи'тель *m* stimulus, irritant, stimulant, excitator
~, **вызыва'ющий испу'г** startle stimulus
~, **вызыва'ющий отрица'тельную реа'кцию** adversive stimuls
адеква'тный ~ homologous stimulus
безусло'вный ~ unconditioned stimulus
врождё'нный сигна'льный ~ innate sign stimulus
вызыва'ющий ~ eliciting stimulus
дифференциро'вочный ~ differentiating excitator
заверша'ющий ~ consummatory stimulus
насыщающий ~ satition stimulus
неадекватный ~, неспецифический ~ heterologous stimulus
понужда'ющий ~ drive stimulus
подпороговый ~ subliminal stimulus, subthreshold stimulus
пороговый ~ threshold stimulus, liminal stimulus
постоянно действующий ~ stationary stimulus
разреша'ющий ~ releasing stimulus
светово'й ~ photic stimulus
сенсибилизи'рующий ~ sensitizing stimulus
сигна'льный ~ sign stimulus
синтети'ческий ~ synthetic excitator
специфи'ческий ~ homologous stimulus
усло'вный ~ conditioned excitatory, conditioned stimulus
цепно'й ~ chain excitator
раздражи'тельный *a* irascible
разлага'ться *v* decay, decompose
разла'пистый *a* widely ramified (*of a coniferous tree*)
разли'в *m* flood; bottling
нефтяно'й разлив oil spill, oil spillage
различа'ть v discriminate
различа'ться *v* differ
различа'ющий a discriminate
различе'ние *n* discrimination
~ **схо'дства** discrimination of sameness
~ **то'ждества** discrimination of sameness
~ **чи'сленности** numerosity discrimination
зри'тельное ~ visual discrimination
разли'чие *n* difference; discrepancy, distinction
достове'рное ~ significant difference
зна'чимое ~ significant difference
ка'чественное ~ qualitative difference
коли'чественное ~ quantitative difference
полово'е ~ sex difference
различи'тельный a discriminating
разли'чный *a* different, distinct
достове'рно ~ significantly different
разложе'ние *n* decomposition, decay, breakdown, dissociation
бактериа'льное ~ bacterial decomposition
разложи'вшийся *a* decomposed, decayed
разлуче'ние *n* separation

разма'х *m* span, spread, range
~ **кры'льев** wing spread
~ **распределе'ния** spread of distribution
разме'р *m* size
това'рный ~ market size, marketable size
размеще'ние *n* distribution, allocation, disposition
размножа'ть *v* propagate
размножа'ться *v* multiply, breed
размноже'ние *n* reproduction, multiplication, propagation, breeding
~ **отво'дками** layering, propagation by layering
~ **черенка'ми** propagation by cuttings
беспо'лое ~ asexual reproduction, asexual propagation, agamogenesis
вегетати'вное ~ vegetative reproduction, vegetative propagation
де'вственное ~ virginal reproduction
иску'сственное ~ artificial propagation
неусто'йчивое ~ versatile reproduction
партеногенети'ческое ~ virginal reproduction
полово'е ~ sexual reproduction, sexual propagation
цикли'ческое ~ cyclic reproduction
размыва'ние *n* washing out, eroding
~ **бе'рега** shore cutting
размягче'ние *n* softening, ramolissement
~ **косте'й** osteomalacia
~ **мо'зга** encephalomalacia
ра'зница *f* difference, distinction
разно- hetero-
разноборо'здный *a* heterocolpate
разнови'дность *f* variety
разноколю'чковый *a* diversispinous
разнокры'лый *a* heteropterous
разнолепе'стный *a* heteropetalous
разноли'стный *a bot* diversifolious, diversifoliate
разноло'пастный *a* diversilobous
разнолучево'й *a* heteroactinal
разнообли'ственный *a bot* diversifoliate
разнообра'зие *n* diversity; diversification
~ **антите'л** antibody diversity
видово'е ~ species diversity
клона'льное ~ clonal diversification
ограни'ченное ~ restricted diversity
разново'зрастный *a* uneven-aged
разноли'стный *a* heterophyllous
разнопо'лый *a* heterosexual
разнопо'ровый *a* heteroaperturate
разноресни'чный *a* heterotrichous
разноспо'ровость *f* anisospory, heterospory
разносто'лбчатость *f* heterostyly
ра'зность *f biom* difference
артериовено'зная ~ arterio-venous difference
крити'ческая ~ critical difference
наиме'ньшая достове'рная ~ least significant difference
наиме'ньшая зна'чимая ~ least significant difference
сре'дняя ~ mean difference
разнотра'вье *n* herbs, forbs
разнотычи'нковый *a* heterandrous
разнохозя'йный *a* heteroecious, metoxenous
разноцветко'вый *a bot* diversiflorous
разноцве'тный *a* many-coloured, versicolored, heterochronous, diversicolorous
разноя'дерность *f* heterocaryosis
разнояче'истый *a* heterobranchiate
ра'зный *a* different, distinct
разобща'ть *v* separate
разобще'ние *n* isolation, disconnection, separation
разо'рванный *a* fragmented, ruptured; disjunctive (*of a geographical range*)
разоре'ние *n* devastation, destruction
разоря'ть *v* destroy
разраба'тывать *v* develop, elaborate, devise, work out; cultivate (*soil*), till
разрабо'тка *f* development
разраста'ние *n* enlargement, spread
разре'женный *a* thinned (*e.g., of forests*), open
разре'з *m* incision, discision, cutting, cut, section; profile
попере'чный ~ cross-section
по'чвенный ~ soil cross-section
разреза'ть *v* dissect
разреше'ние *n* resolution
разруше'ние *n* breakdown, desintegration, dissolution
~ **клеток ультразвуком** ultrasonic cell desintegration
разры'в *m* disruption, break, breakage, breaking, rupture
втори'чные разры'вы *pl* secondary breakage
двухнитево'й ~ double-strand break
запа'здывающие разры'вы *pl* deferred breakage

изолока'льный ~ isolocal break
изохромати'дный ~ isochromatid break
однонитево'й ~ single-strand break
полухромати'дный ~ half-chromatid break
то'чечный ~ *mbiol* breakpoint
хромати'дный ~ chromatid break
хромосо'мный ~ chromosome break
ДНК- ~ DNA break
разрыва'ние *n* breaking, breakage
разря'д *m* discharge
~ **мно'гих не'рвных элеме'нтов** mass discharge
~ **поврежде'ния** injury discharge
дли'тельный ~ *phys* long-lasting discharge
спа'йковый ~ spike discharge
ра'зум *m* mind
разъедине'ние *n* disconnection, disunion, segregation
разъяре'нный *a* goaded
райо'н *m* area, district
райони'рование *n* zoning; regionalization; geographical demarcation
геоботани'ческое ~ geobotanical subdivision
флористи'ческое ~ floristical subdivision
рак *m* cancer, carcinoma; canker (*of plants*); crayfish, crawfish (*Astacura*)
~ **карто'феля** potato canker *(caused by Synchytrium endobioticum)*
~ **коры' ду'ба** oak canker *(caused by Pezicula cinnamomea)*
~ **Крукенбе'рга** Krukenberg's carcinoma
~ **сосны'** pitch canker *(caused by Fusarium subglutinans pini)*
~ **фрукто'вых дере'вьев** fruit trees canker *(caused by Leucostoma)*
многоле'тний ~ расте'ний perennial cancer *(caused by Leucostoma cineta)*
раки'та *f* varnished willow, brittle willow, crack willow, snap willow (*Salix fragilis*)
ра'ковина *f* shell (*of a mollusk*)
носова'я ~ turbinated bone, turbinal bone
ушна'я ~ auricle
ра'ковый *a* cancerous
рак-отше'льник *m* hermit crab (*Paguridae*)
ра'ма *f* frame
рамбулье' *m* Rambouillet
ра'мень *f* ramen
рамице'рк *m helm* ramicercus
ра'мка *f* frame
~ **счи'тывания** reading frame
откры'тая ~ счи'тывания open reading frame
рамно'за *f* rhamnose
рамфоте'ка *f* rhamphotheca
ра'на *f* wound, injury
ко'лотая ~ punctured wound, stab wound
ле'гкая ~ minor wound
огнестре'льная ~ gunshot wound, firearm wound
операцио'нная ~ operative wound
проника'ющая ~ penetrating wound
размозже'нная ~ crushed wound
рва'ная ~ lacerated wound, avulsive wound
ре'заная ~ incised wound, slash wound, cut
ру'бленая ~ chopped wound
сквозна'я ~ perforating wound
ранг *m* rank, social status
зави'симый ~ dependent rank
основно'й ~ basic rank
социа'льный ~ social rank
рандомиза'ция *f* randomization
непо'лная ~ restricted randomization
ране'ние *n* wound, injury
~ **в упо'р** contact wound
проника'ющее ~ penetrating wound
сквозно'е ~ perforating wound
тяже'лое ~ severe wound
ра'неный *m* wounded
раннеспе'лость *f* early ripeness
раннеспе'лый *a* early-ripening, early
ра'нний *a* early, early ripe, early-ripening
рапа' *f* brine
рапс *m* colza, rape (*Brassica napus var. oleifera*)
рапу'нцель *m* corn salad (*Valerianella olitoria*)
ра'са *f* race, strain
биологи'ческая ~ biologic race
географи'ческая ~ geographic race
кариологи'ческая ~ karyological race
климати'ческая ~ climatic race
лока'льная ~ local race
серпенти'новая ~ serpentine race
физиологи'ческая ~ physiologic race
хромосо'мная ~ chromosome race
экологи'ческая ~ ecological race
раски'дистый *a* spreading, shaggy, branchy, patulent, patulous
раски'дываться *v* sprawl
раскрыва'ние *n bot* dehiscence
~ **отве'рстиями** dehiscence by pores

~ **попере'чной тре'щиной** lateral dehiscence
~ **продо'льной тре'щиной** longitudinal dehiscence
~ **по перегоро'дкам** septicidal dehiscence
~ **кла'панами** valvular dehiscence
раскрыва'ться *v* open; break (*e.g., of flower buds*)
раскряже'вка *f* crosscutting, bucking
раскряже'вывать *v* buck, cut, saw across
~ **бревно** buck a log
ра'совый *a* racial
расоспецифи'ческий *a* racial
распа'д *m* breakdown, dissociation, decay, dissolution
кле'точный ~ cell dissolution
окисли'тельный ~ oxidative breakdown
радиоакти'вный ~ radioactive decay
распада'ться *v* decay
распа'хивать *v* plow up
распла'стывание *n* spreading
~ **кле'ток** cell spreading
распло'д *m* brood
распознава'ние *n* recognition
~ **ро'дичей** kin recognition
~ **«своего'»** autorecognition, cognate recognition, self-non-self recognition
аллоге'нное ~ allorecognition
аутогенное ~ autorecognition
двойно'е ~ corecognition
когнати'вное ~ cognate recognition, self-non-self recognition
расположе'ние *n* arrangement, disposition
расправи'лка *f* spreading board
распределе'ние *n* distribution, partitioning
~ **видово'го оби'лия** species-abundance distribution
~ **Га'усса** *biom* Gauss distribution, normal distribution
~ **Ги'брата** *biom* Gibrat distribution
~ **по во'зрасту** age distribution
~ **по вре'мени** temporal distribution
~ **по глуби'нам** bathymetrical distribution
~ **по разме'рам** size distribution
~ **Пуассо'на** *biom* Poisson distribution
~ **хи-квадра'т** *biom* chi-squared distribution
~ **часто'т** *biom* frequency distribution
асимметри'чное ~ *biom* asymmetrical distribution, skew distribution
бета-биномиа'льное ~ *biom* beta-binomial distribution
бимода'льное ~ *biom* bimodal distribution
биполярное ~ bipolar distribution
вертика'льное ~ vertical distribution
геометри'ческое ~ geometric distribution
гипергеометри'ческое ~ *biom* hypergeometric distribution
группово'е ~ clumped distribution
двуме'рное ~ bivariate distribution
двуме'рное норма'льное ~ *biom* binormal distribution
двуме'рное ~ Пуассо'на *biom* bivariate-compounded Poisson distribution
двухверши'нное ~ *biom* bimodal distribution
динами'ческое ~ dispersal distribution
дискре'тное ~ discrete distribution
допусти'мое ~ tolerance distribution
зона'льное ~ zonation
колоколообра'зное ~ *biom* bell-shaped distribution
логарифми'чески норма'льное ~ *biom* logarithmic normal distribution, lognormal distribution, Gibrat distribution
многоверши'нное ~ *biom* multimodal distribution
многоме'рное ~ multidimensional distribution, multivariate distribution
многоме'рное норма'льное ~ multivariate normal distribution
непреры'вное ~ continuous distribution
норма'льное ~ *biom* normal distribution, Gauss distribution
норми'рованное ~ standardized distribution
ограни'ченное ~ restricted distribution distribution
одноме'рное ~ one-dimensional distribution; univariate distribution
островерши'нное ~ *biom* peaked distribution, leptokurtic distribution
отрица'тельное биномиа'льное ~ *biom* negative binomial distribution
плосковерши'нное ~ flat distribution
полуколи'чественное ~ semi-quantitative distribution
преде'льное ~ limiting distribution
преры'вистое ~ discontinuous distribution
простра'нственное ~ spatial distribution
противото'чное ~ countercurrent distribution
прямоуго'льное ~ rectangular distribution

равноме'рное ~ uniform distribution, regularly spaced distribution
сезо'нное ~ seasonal distribution
случа'йное ~ *biom* random distribution
совме'стное ~ joint distribution, mixed distribution
стациона'рное stationary distribution
теорети'ческое ~ theoretical distribution
то'чечное ~ spotted distribution
треуго'льное ~ triangular distribution
унимода'льное ~ unimodal distribution
усечё'нное ~ truncated distribution
усло'вное ~ conditional distribution
экспоненциа'льное ~ *biom* exponential distribution
распредели'тель *m gen* distributor
распростё'ртый *a* prostrate, prone; decumbent
распростране'ние *n* distribution; spreading, spread, propagation
~ **возбужде'ния** spread of activation
~ **по высоте'** altitudinal distribution
амфипацифи'ческое ~ amphi-Pacific distribution
амфитихоокеа'нское ~ amphi-Pacific distribution
биоцентри'ческое ~ bicentric distribution
вертика'льное ~ vertical distribution
всесве'тное ~ cosmopolitanism
распуска'ние *n bot* breaking (*of buds*)
~ **по'чек** breaking of buds
распуска'ться *v* sprout, burst into leaf
распуше'ние *n* ptiloerection (*in birds*)
распусти'вшийся *a* open
расса'да *f* seedlings
расса'дник *m* nursery, seminary
расса'живание *n* resetting
расса'сывание *n* resorption, resolution
расса'сываться *v* resolve
рассе'ивание *n* dissemination, scattering
~ **спор** spore discharge
рассека'ть *v* dissect
расселе'ние *n* dispersal, dissemination
~ **по во'здуху** ballooning behavior (*of spiders*)
рассече'ние *n* dissection
рассече'нный *a* parted, partite, divided
рассе'яние *n* scattering
~ **нейтро'нов** neutron scattering
~ **рентге'новских луче'й** X-ray scattering
~ **све'та** light scattering
малоуглово'е ~ **нейтро'нов** small-angle neutron scattering
малоуглово'е ~ **рентге'новских луче'й** small-angle X-ray scattering
рентге'новское ~ X-ray scattering
расслабле'ние *n* relaxation
~ **мы'шцы в поко'е** resting compliance
диастоли'ческое ~ **се'рдца** compliance of the heart
изометри'ческое ~ isometric relaxation
расслабля'ть *v* relax
расслое'ние *n* stratification
рассоле'ние *n* desalinization
расстоя'ние *n* distance
~ **ме'жду двумя' совоку'пностями** *biom* distance between two populations
обобщё'нное ~ *biom* generalized distance
расстро'йство *n* disorder; distress, disturbance
~ **движе'ния** motor disturbance
раство'р *m* solution
~ **Де'лани** Delaney's solution
~ **Люголя** Lugol's solution, compound iodine solution
~ **нитроси'него тетразо'лия** NBT solution
~ **Ри'нгера** Ringer's solution
бу'ферный ~ buffer solution
исхо'дный ~ mother solution, stock solution
контро'льный ~ reference solution
концентри'рованный ~ strong solution
кре'пкий ~ strong solution
кра'сящий ~ staining solution
культура'льный ~ culture solution
ма'точный ~ stock solution, mother solution
насы'щенный ~ saturated solution
норма'льный ~ normal solution
о'пытный ~ experimental solution
пита'тельный ~ nutrient solution
осветля'ющий ~ clarifying solution
сбаланси'рованный солево'й ~ balanced salt solution
сла'бый ~ weak solution
солево'й ~ saline solution
станда'ртный ~ standard solution
физиологи'ческий ~ physiological solution, normal saline solution
растворе'ние *n* solution, dissolution
раствори'мость *f* solubility
раствори'мый *a* soluble, solvable
раствори'тель *m* solvent

растворя'ть *v* solve
растворя'ться *v* dissolve; resolve
растворя'ющий *a* solvent
расте'ние *n* plant
~ **вла'жных местообита'ний** drought-escaping plant, drought-evading plant
~ **дли'нного дня** long-day plant
~, **закрепля'ющее по'чву** binding plant
~ **коро'ткого дня** short-day plant
~ **норма'льного дня** day-neutral plant
~, **пита'ющее муравьё'в** myrmecotophic plant
~ **подви'жного гра'вия** chalicad plant
~ **с обме'ном по ти'пу толстя'нковых** CAM plant
~, **цепля'ющееся у'сиками** tendril-climber plant
аромати'ческое ~ aromatic plant
бесполе'зное ~ needless plant
волокни'стое ~ fiber plant, textile plant
гу'мусное ~ humus plant
гумусообразу'ющее ~ humus plant
декорати'вное ~ ornamental plant, adornment plant
древе'сное ~ woody plant, ligneous plant
запалё'нное ~ shrunken plant
зернобобо'вое ~ pulse plant, leguminous plant
зиму'ющее ~ overwintering plant
злакови'дное ~ graminifolious plant
иммy'нное ~ disease-resistant plant
ко'мнатное ~ house plant, indoor plant
лесно'е ~ sylvestral plant
матери'нское ~ maternal plant, female plant
мелкосемя'нное ~ smallseeded plant
мотылько'вое ~ papilionaceous ~ plant *(Papilionaceae)*
насекомоопыля'емое ~ entomophilous plant
назе'мное ~ land plant, terrestrial plant, ground plant
нену'жное ~ needless plant
ни'зшее ~ lower plant, inferior plant
овощно'е ~ vegetable
орехопло'дное ~ nut plant
подо'пытное ~ test plant
подушкови'дное ~ cushion plant
полупогруже'нное ~ emergent plant
роди'тельское ~ parent plant
светолюби'вое ~ sun plant, heliophyte, photophilous plant
ска'льное ~ rupicolous plant, saxicolous plant
со'рное ~ undesirable plant, weed
спайнотычи'нковое ~ gamostaminate plant
съедо'бное ~ edible plant
тепли'чное ~ warmhouse plant, hothouse plant, greenhouse plant, glasshouse plant, stove plant
растениево'д *m* crop grower
растениево'дство *n* plant growing, crop growing, plant raising
расте'ние-гидроиндика'тор *n* water-indicating plant
расте'ние-индика'тор *n* indicator plant
расте'ние-пионе'р *n* pioneer plant
расте'ние-поду'шечник *n* cushion plant
расте'ние-поду'шка *n* cushion plant, polster plant
расте'ние-хозя'ин *n* host plant
растениея'дность *f* vegetarianism
растениея'дный *a* phytophagous
расти' *v* grow, vegetate
расти'тельность *f* vegetation
~ **засолё'нных песко'в** halopsammophilous vegetation
~ **теку'чих вод** rheophilic vegetation
антропоге'нная ~ antropogenic vegetation
боло'тная ~ mire vegetation
бу'йная ~ gross vegetation
влаголюби'вая ~ hydrophilous vegetation
во'дная ~ aquatic vegetation
галофи'тная ~ halophytic vegetation
есте'ственная ~ natural vegetation
жестколи'стная ~ coarse vegetation
жи'дкая ~ poor vegetation
кли'максовая ~ climax vegetation
морска'я ~ marine vegetation
нея'русная ~ nonlayered vegetation
низкоро'слая ~ low vegetation
полуесте'ственная ~ seminatural vegetation
придоро'жная ~ roadside vegetation
примо'рская ~ maritime vegetation
пы'шная ~ gross vegetation, lush vegetation
рели'ктовая ~ relic vegetation
рудера'льная ~ ruderal vegetation
со'рная ~ weed vegetation
сте'лющаяся ~ creeping vegetation
степна'я ~ steppe vegetation
травяни'стая ~ herbaceous vegetation

фригано'идная ~ phryganoid vegetation
я'русная ~ layered vegetation
растительноя'дный *a* phytophagous
расти'тельный *a* vegetable, vegetative, vegetal, vegetational
расти'ть *v* bring up
растормa'живание *n* disinhibition, counterinhibition
расторо'пша *f* milk thistle, Saint-Mary thistle (*Silybum Marianum*)
растре'скивание *n* cracking, splitting; *bot* dehiscence
растре'скиваться *v* crack, split; chap (*of skin*); dehisce
растре'скивающийся *a* dissilient; dehiscent
ра'струб *m bot* ochrea, ocrea
раструбови'дный *a* ocreate
расту'щий *a* growing
растяже'ние *n* distension
растяжи'мость *f* distensibility
~ **грудно'й кле'тки** thoracic compliance
расхо'д *m* discharge
расходя'щийся *a* divergent
расхожде'ние *n* divergence, separation, discrepancy, divarication
~ **хромосо'м** chromosome disjunction
анафа'зное ~ anaphase separation
расцвести' *v* burst in blossom
расцвета'ть *v* burst into flower, come into bloom, blossom out
расцве'тка *f* coloration
расцве'тший *a* open
расчисле'ние *n* calculation
обра'тное ~ back calculation
расчи'стка *f* grubbing (*e.g., of a plot*)
расшире'ние *n* dilatation; broadening, widening
расщепле'ние *n* splitting, segregation, disjunction, scission; decomposition, disintegration, breakdown; cleavage
~ **белко'в** proteolysis
~ **жи'ра** lipolysis
~ **по Э'дману** Edman degradation
бромциа'новое ~ cyanogen bromide cleavage
гено'мное ~ genome segragation
гидролити'ческое ~ hydrolytic degradation
гомолити'ческое ~ homolytic cleavage
ме'нделевское ~ Mendelian segregation
окисли'тельное ~ oxidative breakdown, oxidative cleavage
ооплазми'ческое ~ ooplasmic segregation
поляризо'ванное ~ preferential segregation
преференциа'льное ~ preferential segragation
редукцио'нное ~ reductional segregation
сомати'ческое ~ somatic segregation
ферментати'вное ~ enzymatic degradation
хромати'дное ~ chromatid segregation
эквацио'нное ~ equational segregation
эмбриона'льное ~ embryonic segregation
расщепленнозу'бчатый *a* cleft-toothed
расщеплё'нный *a* divaricate, forcipate
расщепля'ть *v* spilt; decompose
рафи'да *f* raphide
рацема'т *m* racemate
рацемиза'ция *f* racemization
рацио'н *m* ration, diet, allowance
~, **бога'тый белка'ми** high-protein diet
безбелко'вый ~ nonprotein diet, protein-depleted diet, protein-free diet
бессолево'й ~ salt-free diet
кормово'й ~ feed allowance
недоста'точный ~ inadequate ration
несбаланси'рованный ~ inadequate ration
основно'й ~ basic diet
обезжи'ренный ~ fat-free diet
отко'рмочный ~ fattening ration
пищево'й ~ diet
полноце'нный ~ adequate ration
сбаланси'рованный ~ balanced diet
су'точный ~ daily ration, daily allowance
рво'та *f* vomiting, emesis
неукроти'мая ~ uncontrollable vomiting
рво'тный *a* vomitory, vomitive, emetic
ре- re-
реабсо'рбция *f* reabsorption
реаге'нт *m*reagent, reactant
реаги'рование *n* reacting, responding
~ **на отбо'р** selection response
реаги'ровать *v* react, respond
реакти'в *m* reagent
~ **Болтона-Хантера** *mbiol* Bolton-Hunter reagent
~ **Клеланда** Cleland's reagent
~ **Не'сслера** Nessler's reagent
~ **Фе'линга** Fehling's reagent
~ **Шиффа** Schiff's reagent
реактива'ция *f* reactivation
~ **кле'ткой хозя'ина** host cell reactivation
~ **профа'гом** prophage reactivation
мно'жественная ~ multiplicity reactivation

перекрё'стная ~ cross reactivation
теплова'я ~ temperature reactivation
реактиви'ровать *v* reactivate
реакти'вность *f* responsiveness, reactivity
иммунологи'ческая ~ immune responsiveness
перекрёстная ~ cross responsiveness
фотопериодическая ~ photoperiodic responsiveness
реа'ктор *m* reactor
реа'кция *f* reaction, response
~ **Абдерга'льдена** Abderhalden's reaction, Abderhalden's test
~ **абсо'рбции** absorption test
~ **агглютина'ции** agglutinating immunoassay, agglutination test
~ **агрега'ции нейтрофи'лов** granulocyte aggregation assay
~ **Адамке'вича** Adamkiewicz's test
~ **«антиге'н–антите'ло»** antigen-antibody reaction
~ **Аско'ли** Ascoli reaction
~ **Ашге'йма-Цо'ндека** Aschheim-Zondek reaction
~ **бе'гства** escape response, fleeing response
~ **биуре'товая** biuret recation, biuret test
~ **бласт-трансформа'ции лимфоци'тов** blast-transformation reaction, lymphocyte transformation assay
~ **Борде'-Жангу'** Bordet-Gengou test
~ **Вассерма'на** Wassermann reaction, Wassermann test
~ **Ве'йля-Фе'ликса** Weil-Felix reaction
~ **взаимоабсо'рбции** *imm* mirror absorption test
~ **вздра'гивания** startle reaction, startle response
~ **Вида'ля** Widal's reaction
~ **вовлече'ния** recruiting response
~ **«всё или ничего'»** all-or-none reaction
~ **«вхолосту'ю»** idling reaction
~ **вы'бора** choosing reaction, choice behaviour
~ **выпада'ния хло'пьев** flocculation test
~ **Гатэ'** Gate's reaction
~ **гемагглютина'ции** hemagglutination reaction
~ **Герксге'ймера** Herxheimer's reaction
~ **гиперчувстви'тельности** hypersensitivity reaction
~ **Гру'бера-Вида'ля** Gruber-Widal reaction
~ **дыха'ния** respiratory response
~ **заде'ржки агглютина'ции** agglutination inhibition test
~ **заме'дленного ти'па** delayed response
~ **избега'ния** avoidance reaction
~ **иммунодиффу'зии** immunodiffusion test
~ **иммунофлуоресце'нции** immunofluorescence test
~ **исключе'ния** exclusion reaction
~ **кле'точной агглютина'ции** cell agglutination assay
~ **кле'точно-опосре'дованного лимфо'лиза** cell-mediated lympholysis assay
~ **кольцепреципита'ции** interfacial precipitin test, "ring" test
~ **Ла'нге** Lange's test, Lange's reaction
~ **латексагглютина'ции** latex fixation test
~ **ло'жной агглютина'ции** paraserum reaction
~ **Манту'** Mantoux reaction, Mantoux test
~ **митоге'нной стимуля'ции** mitogen-driven assay
~ **на бело'к** protein test
~ **на выключе'ние** off-response
~ **нажа'тия на рыча'г** barpressing response
~ **настоpo'женности** vigilance reaction
~ **непрямо'й гемагглютина'ции** indirect hemagglutination test
~ **односторо'ннего избега'ния** one-way avoidance response
~ **окисле'ния** oxidation reaction, oxidizing reaction
~ **«окри'кивания»** mobbing response *(in birds)*
~ **осажде'ния** precipitation reaction
~ **отвраще'ния** averse response
~ **отодвига'ния** shifting reaction
~ **отторже'ния** rejection reaction
~ **перемеще'ния** shifting reaction
~ **Пирке'** Pirquet's reaction, Pirquet's test
~ **повы'шенной чувстви'тельности** hypersensitivity reaction
~ **помутне'ния** turbidity reaction
~ **по ти'пу «все или ничего'»** "all-or-nothing" response
~ **потира'ния** wiping response
~ **по'чвы** soil reaction
~ **присоедине'ния** addition reaction
~ **прыжка'** hopping reaction
~ **распуше'ния** ptiloerection response

~ **розеткообразова'ния** rosette test
~ **ро'ста** growth response
~ **сверхчувстви'тельности** hypersensitive reaction
~ **свя'зывания** coupling reaction
~ **свя'зывания антиге'на** antigen-binding assay
~ **свя'зывания комплеме'нта** complement-fixation reaction, complement fixation test, complement binding assay, complement consumption assay
~ **си'нтеза** synthetic reaction
~ **сле'дования** following response
~ **сме'шанной культу'ры лимфоци'тов** MLC response
~ **спа'ривания** mating reaction
~ **Сто'ссела** Stossel's reaction
~ **схва'тывания добы'чи** prey-catching response
~ **типи'рования** typing reaction
~ **торможе'ния адге'зии лейкоци'тов** leukocyte adherence inhibition assay
~ **торможе'ния агглютина'ции** agglutination inhibition test
~ **торможе'ния гемагглютина'ции** hemagglutination-inhibition reaction
~ **«трансплanta'т про'тив хозя'ина»** graft versus host reaction
~ **трево'ги** alarm reaction
~ **уклоне'ния** evasive response
~, **ула'вливающая эне'ргию** energy-trapping reaction
~ **Фе'льгена** Feulgen reaction
~ **флокуля'ции** flocculation test
~ **Фо'геса-Проска'уэра** Voges-Proskauer reaction
~ **Хи'лла** Hill reaction
акросо'мная ~ *emb* acrosome reaction
аллерги'ческая ~ allergic reaction, allergic response, inordinate response
аллоимму'нная ~ allospecific response
анамнести'ческая ~ anamnestic response, "recollection" reaction, memory response, booster response
аутоиммунная ~ antiself response, autoimmune response, autospecific response, autoresponse
беззатра'вочная ~ unprimed reaction
биуре'товая ~ biuret test
бласттрансформа'ции blastogenic response
вазомото'рная ~ vasomotor reaction
вока'льная ~ vocal response
втори'чная имму'нная ~ secondary immune response
втори'чная иммунологи'ческая ~ secondary immune response
вы'званная ~ evoked response
высокоэнергети'ческая ~ high-energy reaction
голосова'я ~ vocal response
дви'гательная ~ motor reaction, motor response
депре'ссорная ~ depressory reaction
дыха'тельная ~ respiratory response
заде'ржанная ~ delayed reaction, delayed response
заторможё'нная ~ deferred reaction
защи'тная ~ defense reaction
имму'нная ~ immune response, immunological response
иммунологи'ческая ~ immune response, immunological response
ко'жно-гальвани'ческая ~ galvanic skin response
кортика'льная ~ cortical reaction
лизоге'нная ~ lysogenic response
лине'йная ~ linear response
лити'ческая ~ lytic response
мига'тельная ~ evoked response
многоступе'нчатая ~ multistep reaction
мультисубстра'тная ~ multisubstrate reaction
неингиби'руемая ~ uninhibited reaction
непряма'я ~ иммунофлуоресце'нции indirect immunofluorescence test
обра'тная ~ back reaction
о'бщая ~ органи'зма bodily reaction
одноступе'нчатая ~ single-step reaction
окисли'тельная ~ oxidation reaction, oxidizing reaction
окисли'тельно-восстанови'тельная ~ redox reaction
опера'нтная ~ operant response
оптомото'рная ~ optomotor reaction
ориентиро'вочная ~ orientation response
ориентиро'вочно-поиско'вая ~ orientative-trying reaction
ортостати'ческая ~ orthostatic reaction
перви'чная ~ priming reaction
перви'чная имму'нная, перви'чная иммунологи'ческая ~ primary immune response
пероксида'зная ~ peroxidase reaction

пищева'я ~ feeding response
пло'хо адапти'рующаяся ~ maladaptive reaction
побо'чная ~ side reaction
поведе'нческая ~ behavioral response, behavioral reaction
полова'я ~ mating response
полисубстра'тная ~ multisubstrate reaction
полова'я ~ sexual reaction
произво'льная ~ voluntary response
пряма'я ~ forward reaction
пряма'я ~ иммунофлуоресце'нции direct immunofluorescence test
радиохими'ческая ~ radiochemical reaction
серологи'ческая ~ serum test
систе'мная ~ органи'зма bodily reaction
содру'жественная ~ consensual reaction
сосудодви'гательная ~ vasomotor reaction
специфи'ческая ~ настороженности specific vigilance reaction
стати'ческая ~ static reaction
ступе'нчатая ~ consecutive reaction
темнова'я ~ dark reaction
тигмотакти'ческая ~ thigmotactic response
транзито'рная иммуннная ~ short-lived immune response
усло'вная ~ conditioned response
условнорефлекто'рная ~ conditioned reflex reaction
усло'вно-рефлекто'рная эмоциона'льная ~ conditioned emotional response
ферментати'вная ~ enzymatic reaction
фотомиоклони'ческая ~ photomyoclonic response
фотопериоди'ческая ~ photoperiodic response
цветна'я ~ staining reaction, color reaction
цепна'я ~ chain reaction
экзотерми'ческая ~ energy-releasing reaction, energy-yielding reaction, exergonic reaction, exothermic reaction, heat-generating reaction, heat-producing reaction
экзэргони'ческая ~ energy-releasing reaction, energy-yielding reaction, exergonic reaction, exothermic reaction, heat-generating reaction, heat-producing reaction
элимини'рования elimination reaction
эндотерми'ческая ~ endergonic reaction, heat-absorbing reaction
эндэргони'ческая ~ endergonic reaction, heat-absorbing reaction

реанима'ция *f* resuscitation, revivification, revivescence
реаними'ровать *v* resuscitate, revive
реаранжиро'вка *f mbiol* rearrangement
реафферента'ция *f* reafference
ребё'нок *m* child, infant, baby
грудно'й ~ infant, nursing baby, suckling
недоно'шенный ~ premature infant
ре'берный *a* costal
ребри'стый *a* costate
ребро' *n* rib, costa
и'стинное ~ sternal rib, true rib
ло'жное ~ asternal rib, short rib
подви'жное ~ floating rib
ше'йное ~ cervical rib
рёв *m* roar, bellow
ревакцина'ция *f* renewed vaccination, revaccination
реве'нь *m* rhubarb (*Rheum*)
реверси'вность *f* reversibility
реве'рсия *f* reversion
реверта'за *f* revertase
реверта'нт *m* revertant
реве'ть *v* roar, bellow, low, bray (*of an ass*)
реви'зия *f* revision
ревмати'зм *m* rheumatism
ревмати'ческий *a* rheumatic, rheumatismal
ревмато'идный *a* rheumatoid
ревмато'лог *m* rheumatologist
револьве'р *m* nosepiece (*of a microscope*)
реву'н *m* howler monkey (*Alouatta*)
регенера'ция *f* regeneration
во'льфова Wolffian regeneration
восстанови'тельная ~ reparative regeneration
морфаллакти'ческая ~ morphallactic regeneration
физиологи'ческая ~ physiological regeneration
эпимо'рфная ~ epimorphic regeneration
регенери'ровать *v* regenerate
регио'н *m* region
регионализа'ция *f* regionalization
региона'льный *a* regional
регистра'ция *f* recording, record

видеомагнитофо'нная ~ videotape recording
внекле'точная ~ extracellular recording
внутрикле'точная ~ intracellular recording
ре'гма *f bot* regma
регре'сс *m* retrogress
регресси'вный *a* regressive
регресси'ровать *v* retrogress
регре'ссия *f* regression
доче'рняя ~ *gen* filial regression
едини'чная ~ simple regression
криволине'йная ~ curvilinear regression
лине'йная ~ linear regression
логарифми'ческая ~ logarithmic regression
мно'жественная ~ multiple regression
нелине'йная nonlinear regression
проста'я ~ simple regression
прямолине'йная ~ rectilinear regression
ча'стная ~ partial regression
экспоненциа'льная ~ exponential regression
регули'рование *n* management, control, regulation
~ **по'ла** sex control
~ **с обра'тной свя'зью** feedback control
регули'ровать *v* regulate
регуля'рный *a* regular
регуля'тор *m* regulator
~ **ро'ста** growth regulator
регуля'торный *a* regulatory
регуля'ция *f* regulation
~ **мото'рики** motor control
~ **по ти'пу обра'тной свя'зи** feedback regulation
~ **с обра'тной свя'зью** feedback control
~ **с прямо'й свя'зью** feedforward control
аллостери'ческая ~ allosteric regulation
гормона'льная ~ hormonal regulation
гумора'льная ~ humoral control
нейрогормона'льная ~ neurohormonal regulation
не'рвная ~ nervous regulation
не'рвно-гумора'льная ~ neurohumoral control
понижа'ющая ~ down-regulation
рефлекто'рная ~ reflex regulation, reflex control
эмбриона'льная ~ embryonic regulation
эндокри'нная ~ endocrine regulation
реди'с *m* garden-radish (*Raphanus sativus*)
реди'ска *f* garden-radish (*colloquial*)
ре'дия *f* redia
ре'дкий *a* sparse, spaced, thin; rare
редколе'сье *n for* open woodland, light forest
редкосто'йный *a for* open
редукта'за *f* reductase
реду'ктор *m* reducer
кроссо'верный ~ crossover reducer
реду'кция *f* reduction
двойна'я ~ double reduction
редуплика'ция *f* reduplication
редуце'нт *m* reducer
ре'дька *f* radish (*Raphanus*)
режи'м *m* schedule, mode; *med* regimen
~ **подкрепле'ния** reward schedule
~ **рабо'ты** operating conditions, procedure
пищево'й ~ feeding schedule
посте'льный ~ bed rest
ре'жущий *a* sectorial (*of teeth*), cutting
ре'заться *v* come through (*of teeth*)
ре'звость *f* speed (*of a horse*)
ре'звый *a* high-spirited (*of a horse*)
резе'кция *f* resection
резерва'т *m* sanctuary, reserve, reservation, refuge
охо'тничий ~ game reserve
резервуа'р *m* reservoir, tank
окисли'тельный ~ oxidation tank
смеси'тельный ~ agitation tank
резе'ц *m* incisor, incisor tooth, cutting tooth, front tooth, incisive tooth
резеци'ровать *v* resect
резиде'нт *m* resident
резисте'нтность *f* resistance
~ **капилля'ров** capillary resistance
резисте'нтный *a* resistant
ре'зка *f* sectioning
ре'зкий *a* sharp; ~ **за'пах** pungent smell
ре'зко *adv* sharply; ~ **континента'льный** extreme continental
резона'нс *m* resonance
динами'ческий я'дерный магни'тный ~, **динами'ческий ЯМР** dynamic nuclear magnetic resonance
двойно'й электро'нно-я'дерный ~ electron nuclear double resonance
электро'нный спи'новый ~ electron spin resonance
магни'тный ~ magnetic resonance
я'дерный магни'тный ~, **ЯМР** nuclear magnetic resonance

прото'нный магни'тный ~ proton magnetic resonance
прото'нный я'дерный магни'тный ~, **прото'нный ЯМР** proton nuclear magnetic resonance
резона'нсный *a* reverberatory
резона'тор *m* vocal sac, resonator
резо'рбция *f* resorption
результа'т *m* result
~ **отбо'ра** selection response
~ **разреша'ющего о'пыта** challenge result
отдалё'нный ~ long-term result
коне'чный ~ net result
положи'тельный ~ positive take
ре'зус-фа'ктор *m* rhesus
реиммуниза'ция *f* boosting
рекапитуля'ция *f* recapitulation
реклами'рование *n eth* advertisement
~ **террито'рии** territorial advertisement
рекомбина'нт *m* recombinant
рекомбина'нтный *a* recombinant
рекомбина'ция *f* recombination
~ **с переме'ной ма'триц** copying-choice recombination
аддити'вная ~ additive recombination
генети'ческая ~ genetic recombination
генети'ческая ~ **фа'га** phage genetic recombination
гомологи'ческая ~ homologous recombination
доба'вочная ~ additive recombination
межло'кусная внутрихромосо'мная ~ interlocus intrachromosomal recombination
межхромосо'мная ~ interchromosomal recombination
митоти'ческая ~ mitotic recombination
незако'нная ~ illegitimate recombination
сайт-специфи'ческая ~ site-specific recombination
сомати'ческая ~ somatic recombination
субсексуа'льная ~ subsexual recombination
реко'н *m* recon
реконстру'кция *f* reconstruction
пласти'ческая ~ waxed reconstruction
рекреа'ция *f* recreation
рекру'тинг *m imm* recruitment, recruiting response
рекру'тмент *m imm* recruitment, recruiting response
ректа'льный *a* rectal
ректиграда'ция *f* rectigradation
ректифика'ция *f* rectification
рекультива'ция *f* recultivation
рекупера'ция *f* waste recovery
релакса'ция *f* relaxation
дипо'ль-дипо'льная ~ dipole-dipole relaxation
диэлектри'ческая ~ dielectric relaxation
спин-решё'точная ~ spin-lattice relaxation
ультразвукова'я ~ ultrasonic relaxation
я'дерная магни'тная ~ nuclear magnetic relaxation
я'дерная спи'новая ~ nuclear spin relaxation
рели'кт *m* relic, relict
леднико'вый ~ glacial relict
рели'ктовый *a* relic, relict
релье'ф *m* relief
расчленё'нный ~ broken relief
ре'мез *m* penduline titmouse (*Remiz pendulinus*)
реми'ссия *f* remission, remittence
ремитти'рующий *a* remittent
ремонта'нтный *a* remontant, everbearing
ренатура'ция *f* renaturation
рендзи'на *f* rendzina
ренкло'д *m* greenage, gage, Reine Claude (*Prunus italica*)
ренни'н *m* rennin
ренoперикардиа'льный *a* renipericardial
рентгеногра'фия *f* roentgenography
рентгеноло'гия *f* roentgenology
рентгеноскопи'я *f* roentgenoscopy
рентгенотерапи'я *f* roentgen therapy, X-ray therapy
рео- rheo-
реоба'за *f* rheobase
реогра'мма *f* rheogram
реогра'фия *f* rheography
реокисле'ние *n* reoxidation
рео'метр *m* rheometer
реорганиза'ция *f* reorganization
реореце'птор *m* rheoreceptor
реота'ксис *m* rheotaxis
реотропи'зм *m* rheotropism
реотро'пный *a* rheotropic, rheotactic
реофи'л *m* rheophil
реофи'т *m* rheophyte
ре'па *f* turnip (*Brassica rapa*)
репара'ция *f bioch* repair
~ **ДНК** DNA repair
темнова'я ~ dark repair

пострепликати'вная ~ post-replication repair
рекомбинацио'нная ~ recombinational repair
эксцизио'нная ~ excision repair
репе'й *m* bur
репе'йник *m* burdock (*Arctium*)
репелле'нт *m* repellent
репертуа'р *m* repertoire
поведе'нческий ~ behavioral repertoire
ре'плика *f* replica
ге'нная ~ gene replica
углеро'дная ~ carbon replica
реплика'тор *m gen* replicator
реплика'ция *f gen* replication
~ **хромосо'м** chromosome replication
аборти'вная ~ abortive replication
восстанови'тельная ~ repair replication
двунапра'вленная ~ bidirectional replication
дисперси'вная ~ ДНК dispersive DNA replication
конкуре'нтная ~ competitive replication
консервати'вная ~ ДНК conservative DNA replication
полуконсервати'вная ~ ДНК semiconservative DNA replication
репарацио'нная ~ repair replication
реплико'н *m gen* replicon
ре'пница *f* turnip butterfly *(Pieris rapae)*
репови'дный 'napiform
реполимериза'ция *f* repolymerization
реполяриза'ция *f* repolarization
репресси'ровать *v* repress
репре'ссия *f* repression
~ **по ти'пу обра'тной свя'зи** feedback repression
~, **регули'руемая метаболи'тами** metabolite-controlled repression
катаболи'тная ~ catabolite repression
координи'рованная ~ ферме'нтов coordinated enzyme repression
репре'ссор *m* repressor
репродукти'вный *a* reproductive
репроду'кция *f* reproduction
ре'пчатый *a* napaceous; ~ **лук** onions, bulb onion
реси'нтез *m* resynthesis
ресни'тчато-зу'бчатый *a* ciliate-toothed
ресни'тчатый *a* ciliate, ciliated
ресни'ца *f* eyelash, cilium
ресни'чка *f* cilium
ресничконо'сный *a* blepharophorous
ресни'чный *a* ciliary, ciliate, ciliated
респирато'рный *a* respiratory
респиро'метр *m* respirometer
респо'ндер *m* responder
аутологи'чный ~ autologous responder
рестрикта'за *f* restriction enzyme
рестри'кция *f* restriction
генети'ческая ~ genetic restriction
линиеспецифи'ческая ~ lineage restriction
ресу'рс *m* resource
возобновля'емый ~ renewable resource
досту'пный ~ available resources
лесны'е ресу'рсы *pl* forest resources
невозобновля'емый ~ non-renewable resource
пищево'й ~ food resource
приро'дный ~ natural resource
ресуспенди'ровать *v* resuspend
ретиа'рный *a* retiaric
рети'вый *a* frisky (*of a horse*)
ретикулосо'ма *f* reticulosoma
ретикулоци'т *m* reticulocyte
ретикулоцито'з *m* reticulocytosis
ретикулоэндотелиа'льный *a* reticulo-endothelial
ретикулоэндоте'лий *m* reticuloendothelium
рети'кулюм *m* reticulum
ретикуля'рный *a* reticular
рети'на *f* retina
ретина'льный *a* retinal
ретини'т *m* retinitis
ретинобла'ст *m* retinoblast
ретино'л *m* retinol
ретинохориоиди'т *m*: **околососо'чковый ~ Йе'нсена** Jensen's disease
ретра'ктор *m* retractor
ретра'кция *f* retraction
ретри'вер *m* retriever *(a breed of dogs)*
ретрови'рус *m* retrovirus
ретроград'ация *f* retrogradation
ретрогра'дный *a* retrograde
ретрогре'ссия *f* retrogression
ретроморфо'з *m* retromorphosis
ретропозо'н *m* retroposon
ретротранспозо'н *m* retrotransposon
ре'ференс-раство'р *m* reference solution
ре'ференс-шта'мм *m* reference strain
рефле'кс *m* reflex
~ **А'шнера** Aschner's reflex
~ **Бе'йнбриджа** Bainbridge reflex

~ **Би'нга** Bing's reflex
~ **блужда'ющего не'рва** vagal reflex
~ **враще'ния колеса'** instrumental wheel-turning response
~ **выпрямле'ния** righting reflex
~ **выпрямле'ния ту'ловища** body-righting reflex
~ **Ге'ринга-Бре'йера** Hering-Breuer reflex
~ **Гиллена'-Барре'** Guillain-Barre reflex
~ **Ги'ршберга** Hirschberg's reflex
~ **Го'льца** reflex of Holtz
~ **Го'ффмана** monosynaptic Hoffmannn reflex
~ **двугла'вой мы'шцы** biceps reflex
~ **кароти'дного си'нуса** carotid-sinus reflex
~ **конъюнкти'вы** conjunctival reflex
~ **Моро'** Moro's reflex
~ **на ускоре'ние** acceleratory reflex
~ **Оппенге'йма** Oppenheim's reflex
~ **отда'чи** rebound reflex
~ **побе'жки** runway reflex, runway response
~ **положе'ния** postural reflex, attitudinal reflex
~ **прыжка'** jump reflex
~ **растяже'ния** extension reflex, stretch reflex
~ **растяже'ния мы'шцы** myotatic reflex
~ **расшире'ния зрачка'** iris-dilatation reflex
~ **суже'ния зрачка'** iris-contraction reflex
~ **чре'вного сплете'ния** coeliac-plexus reflex
аксона'льный ~ axon reflex
ахи'ллов ~ Achilles jerk, ankle jerk
безусло'вный ~ unconditioned reflex, inborn reflex
болево'й ~ pain reflex
брюшно'й ~ abdominal reflex
вага'льный ~ vagal reflex
вазомото'рный ~ vasomotor reflex
вестибулоокуля'рный ~ vestibulo-ocular reflex
ви'сцеро-висцера'льный ~ viscero-visceral reflex
врождё'нный ~ congenital reflex, inborn reflex
глазно'й пищево'й ~ ocular feeding reflex
глазосерде'чный ~ oculo-cardiac reflex
глота'тельный ~ swallowing reflex
гло'точный ~ pharyngeal reflex
глубо'кий ~ deep reflex
дви'гательный ~ motor reflex
депрессо'рный ~ depressory reflex
заме'дленный ~ delayed reflex, retarded reflex
замыка'тельный ~ switching reflex
запа'здывающий ~ delayed reflex, retarded reflex
зрачко'вый ~ pupillar reflex, pupillary reflex
инструмента'льный усло'вный ~ instrumental conditioning
класси'ческий усло'вный ~ classical conditioning, Pavlovian conditioning
ко'жно-мы'шечный ~ skin-muscle reflex
ко'жный ~ cutaneous reflex, skin reflex
коле'нный ~ knee jerk, patellar reflex
конъюнктива'льный ~ conjunctival reflex
ко'стный ~ bony reflex
лабири'нтный ~ labyrinthine reflex
локтево'й ~ elbow jerk
мига'тельно-челюстно'й ~ jaw-winking reflex
мига'тельный ~ blink reflex
моносинапти'ческий ~ monosynaptic reflex
мы'шечный ~ muscular reflex
надглазни'чный ~ ophthalmic reflex
непосре'дственный усло'вный ~ direct conditioned reflex
непрямо'й ~ indirect reflex
оборони'тельный ~ defensive reflex
оборони'тельный усло'вный ~ defensive conditioning
оптомото'рный ~ optomotor reflex
окулоцефалогири'ческий ~ oculocephalogyric reflex
ориентиро'вочный ~ orientative reflex, investigatory reflex
открыва'ющий ~ unlocking reflex
па'вловский усло'вный ~ classical conditioning, Pavlovian conditioning
перекрё'стный ~ crossed reflex
пиломото'рный ~ pilomotor reflex
пищево'й ~ food reflex
пищедобыва'тельный ~ food-procuring reflex
повы'шенный ~ overactive reflex, reinforced reflex
подборо'дочный ~ chin jerk
подо'швенный ~ sole reflex
подража'тельный ~ imitating reflex

постура'льный ~ static reflex
приобретё'нный ~ acquired reflex
прямо'й ~ direct reflex
психогальвани'ческий ~ psychogalvanic reflex
разгиба'тельно-локтево'й ~ elbow jerk
распространя'ющийся ~ spreading reflex
рво'тный ~ gag reflex
сгиба'тельный ~ bending reflex, flexion reflex
рецесси'вный усло'вный ~ recessive conditioned reflex
синокароти'дный ~ carotid-sinus reflex
скры'тый ~ concealed reflex
следово'й усло'вный ~ trace conditioned reflex
сло'жный ~ compound reflex
соса'тельный ~ sucking reflex
сосудодви'гательный ~ vasomotor reflex
сосудосу'живающий ~ vasopressor reflex
спинномозгово'й ~ arc reflex, spinal reflex
стати'ческий ~ static reflex
сухожи'льный ~ tendon reflex, tendon jerk
тормозно'й ~ inhibitory reflex
тройни'чно-лицево'й ~ trigemino-facialis reflex
усло'вный ~ conditioned reflex, trained reflex, behavior reflex, Pavlov's reflex
усло'вный ~ второ'го поря'дка second-order conditioning
усло'вный ~ дифференциро'вки differential conditioning
усло'вный ~ отвраще'ния aversive conditioning
фази'ческий ~ phasic reflex
хвата'тельный ~ grasp reflex
цепно'й усло'вный ~ chain conditioned reflex
цилиа'рный ~ ciliary reflex
челюстно'й ~ jaw jerk
ше'йный ~ neck reflex
ше'йный ~ выпрямле'ния neck-righting reflex

рефле'ксия *f* reflection, reflexion
рефлексоге'нный *a* reflexogenic
рефлексо'граф *m* reflexograph
рефлексо'метр *m* reflexometer
рефлексотерапи'я *f* reflexotherapy
рефлекто'рный *a* reflectory
рефракте'рный *a* refractory
рефракто'метр *m* refractometer
рефра'кция *f* refraction, refringence
рефу'гий *m* refuge, refugium
гнездово'е ~ nest refuge
рецепти'вность *f* receptivity
полова'я ~ sexual receptivity
рецепти'вный *a* receptive
реце'птор *m* receptor
~ **гормо'на** hormone receptor
~ **кароти'дного си'нуса** carotid-sinus receptor
~ **прикоснове'ния** touch receptor
~ **растяже'ния** stretch receptor
адренерги'ческий ~ adrenergic receptor
высокоаффи'нный ~ high-affinity receptor
диста'нтный, дистанцио'нный ~ distance receptor, distoceptor
зри'тельный ~ sight sensor
инсули'новый ~ insulin receptor
конта'ктный ~ contact receptor
листови'дный ~ leaflike receptor
низкоаффи'нный ~ low-affinity receptor
обоня'тельный ~ smell sensor
мы'шечный ~ muscle receptor
температу'рный ~ thermoreceptor
фагоцита'рный ~ scavenger receptor
холинерги'ческий ~ cholinergic receptor
реце'пторный *a* receptor
рецесси'в *m* recessive
рецесси'вный *a* recessive
рециди'в *m* relapse, recurrence, recrudescence
рецидиви'рующий *a* recurrent
реци'клинг *imm* recycling
постлити'ческий ~ postlytic recycling
реципие'нт *m* recipient
иммунокомпете'нтный ~ immune recipient
реципро'кный *a* reciprocal
речно'й *a* fluvial, riverine, river
речь *f* speech, speaking, talking
реше'ние *n* solution, solving
~ **зада'ч** *eth* problem solving
~ **ме'тодом проб и оши'бок** trial-and-error solution
Ба'йесовское ~ Bayes solution
возмо'жное ~ feasible solution
допусти'мое ~ feasible solution
основно'е ~ basic solution
приближё'нное ~ approximate solution
решё'тка *f* lattice
куби'ческая ~ cubic lattice
трёхме'рная ~ three-dimensional lattice
тройна'я ~ triple lattice

реше'тчатый *a* cancellated, cancellate, cancellous, latticed, cribriform
ржа'вчина *f phytop* rust (*Uredinales*)
бу'рая ~ brown rust *(caused by Puccinia)*
жё'лтая ~ зла'ков stripe rust, yellow rust *(caused by Puccinia glumarum)*
коро'нчатая ~ овса' crown rust of oats *(caused by Puccinia coronifera)*
ржа'ние *n* neighing
ржа'нка *f* plover, dotterel (*Charadrius*)
ржано'й *a rye*
ржана'я мука' rye flour
ржать *v* neigh, nicker
рибо'за *f* ribose
рибози'д *m* riboside
рибокина'за *f* ribokinase
рибонуклеа'за, **РНКа'за** *f* ribonuclease
рибонуклеози'д *m* ribonucleoside
рибонуклеопротеи'д *m* ribonucleoprotein
рибонуклеоти'д *m* ribonucleotide
рибосо'ма *f* ribosome
мембраносвя'занная ~ membrane-bound ribosome
свобо'дная ~ free ribosome
рибофлави'н *m* riboflavin
рибуло'за *f* ribulose
рибулокина'за *f* ribulokinase
ри'га *f* thrashing barn
риги'дность *f* rigidity, stiffness
децеребрацио'нная ~ decerebrate rigidity
ризо- rhizo-
ризо'ид *m* rhizoid
ризо'идный *a* rhizoid
ризокто'ния *f phytop* rhizoctonia
ризомице'лий *m* rhizomycelium
ризомо'рфа *f* rhizomorph
ризопо'дия *f* rhizopodium
ризосфе'ра *f* rhizosphere
ризофо'р *m* rhizophore
рикоше'т *m* ricochet (*a kind of gait*)
ринови'рус *m* rhinovirus
риноспоридио'з *m phytop* rhinosporidiosis *(caused by Rhinosporidium seeberi)*
ринофо'р *m zool* rhinophore
ринхосто'м *m helm* rhynchostome
ринхоце'ль *m* rhynchocoel
рис *m* rice (*Oryza*); rice, paddy (*grain*)
риск *m* risk
ба'йесовский *biom* ~ Bayes risk
генети'ческий ~ genetic risk
экологи'ческий ~ ecological risk
ри'слинг *m* riesling
рисово'д *m* rice grower
рисово'дство *n* rice growing
рису'нок *m* pattern, design, picture; figure, drawing
~ **крыла'** wing pattern
~ **мозговы'х изви'лин** convolutional pattern
иммунологи'ческий антиге'нный ~ immune pattern
капюшо'нный ~ hooded pattern
ритм *m* rhythm
~ **акти'вности** activity rhythm
~ **гало'па** gallop rhythm, triple rhythm
~ **дыха'ния** respiratory rhythm
~ **се'рдца** cardiac rhythm, heart rhythm
атриовентрикуля'рный ~ *phys* atrioventricular rhythm
дневно'й ~ diurnal rhythm
идиовентрикуля'рный ~ idioventricular rhythm
ко'рковый ~ *phys* cortical rhythm
полусу'точный ~ semidian rhythm
сезо'нный ~ seasonal rhythm
серде'чный ~ cardiac rhythm, heart rhythm
си'нусовый ~ sinus rhythm
синхро'нный ~ synchronal rhythm
су'точный ~ day-night rhythm, diurnal rhythm
трёхчле'нный ~ gallop rhythm, triple rhythm
узлово'й ~ *phys* nodal rhythm
ультрадиа'нный ~ ultradian rhythm
циркадиа'нный ~ circadian rhythm
цирка'дный ~ circadian rhythm
эндоге'нный ~ endogenous rhythm
ритми'чность *f* rhythmicity, rhythm
ритмоме'трия *f* rhythmometry
ритмоводи'тель *m* pacemaker, time-giver
ритуа'л *m eth* ritual
~ **запу'гивания** challenge ritual
~ **приве'тствия** greeting ritual
~ **чи'стки опере'ния** cleaning ritual
ритуализа'ция *f eth* ritualization
ритуализи'ровать *v eth* ritualize
ритуалисти'ческий *a eth* ritualistic
ритуа'льный *a eth* ritual
ри'шта *f helm* Guinea worm, dragon worm, medina worm (*Dracunculus medinensis*)
РНК, рибонуклеи'новая кислота' *f* RNA, ribonucleic acid
~ **ми'нус цепь** RNA minus strand
~ **митохо'ндрий** mitochondrial RNA

~ **плюс цепь** RNA plus strand
~ **хлоропла'стов** chloroplast RNA
гетероге'нная я'дерная ~ гяРНК) heterogenous nuclear RNA (hn RNA)
гисто'новая ма'тричная ~ histone messenger RNA
двунитева'я ~ double-stranded RNA (ds RNA)
инфекцио'нная ~ infectious RNA
информацио'нная ~ (иРНК) messenger RNA (mRNA), translational RNA (tRNA)
комплемента'рная ~ complementary RNA
ма'тричная ~ (мРНК) translational RNA (tRNA), messenger RNA (mRNA), informational RNA
митохондриа'льная ~ (мт РНК) mitochondrial RNA (mt RNA)
низкомолекуля'рная я'дерная ~, няРНК small nuclear RNA
по'здняя ~ late RNA
предра'нняя ~ immediate-early RNA [ribonucleic acid], pre-early RNA
ра'нняя ~ early RNA
репликати'вная ~ replication RNA
рибосо'мная ~ (рРНК) ribosomal RNA (rRNA)
самосплайси'рующая ~ self-splicing RNA
тра'нспортная ~ (тРНК) acceptor RNA, transfer RNA (tRNA), soluble RNA (sRNA), adaptor RNA
хлоропластная ~ (хл РНК) chloroplast RNA (chl RNA)
я'дерная ~ (яРНК) nuclear RNA, nucleus RNA (nRNA)
я'дрышковая ~ nucleolar RNA, nucleolus RNA
РНКа'за *f* ribonuclease, RNase
РНК-затра'вка *f* primer RNA
РНК-предше'ственник, пре-РНК precursor RNA
ро'бкий *a* timid
рог *m* horn
~ **ма'тки** uterine horn
аммо'нов ~ horn of Ammon
ветвя'щийся ~ branched horn
вито'й ~ screw horn
закру'ченный ~ crumpled horn
изо'гнутыйе вниз ~ down-turned horn
лирови'дные рога' *pl* lyre horns
прямо'й ~ straight horn
спинно'й ~ dorsal horn *(in spinal cord)*
спира'льный ~ coiled horn
рога'тый *a* cornute, cornuted, horny, horned
догови'дный *a* horn-shaped
догови'чный *a* corneal
догово'й *a* corneous, horny
рого'з *m* cat's-tail (*Typha*)
рогохво'ст *m ent* horntail (*Sirex*)
род *m* genus
~, **бе'дный ви'дами** depauperate genus
монотипи'чный ~ monotypic genus
политипи'чный ~ polytypic genus
родами'н *m* rhodamine
роди'тель *m* parent
~ **при обра'тном скре'щивании** backcross parent
роди'ть *v* bring forth, parturiate, bear, give birth to, labor
роднико'к *m anat* fontanel, fontanelle
родово'й *a* generic; parturient, puerperal
родоксанти'н *m* rhodoxanthin
родопси'н *m* rhodopsin
родосло'вная *f* pedigree, lineage, parentage, genealogy
родофи'лл *m* rhodophyll
ро'дственный *a* related, allied, akin, affined
родство' *n* relation, relationship; alliance, affinity
генетическое ~ genetic relationship
кро'вное ~ consanguity, blood relationship, consanguineous relation
ро'ды *pl* delivery, parturition, labour, labor, lying-in
~ **дво'йней** twin labor
~ **не'сколькими плода'ми** multiple labor
рое'ние *n* swarming
ро'жа *f* erysipelas
рожа'вшая *a* parous
рожа'ть *v* bring forth, parturiate, bear, give birth to
рожда'емость *f* natality, birth rate
рожда'ть *v* bring forth, partiruate, bear, give birth to
рожда'ющий *a* bearing
рожде'ние *n* birth, delivery, bearing
ро'жистый *a* erysipelatous
рожо'к *m* cornicle
рожь *f* rye (*Secale*)
ро'за *f* rose (*Rosa*)
~ **ветро'в** wind rose
роза'рий *m* rosary, rose garden, rosarium
розе'тка *f* rosette

обоня'тельная ~ olfactory rosette
розеткови'дный *a* rosette-like
розеткообразова'ние *n micr* rosetting
розе'точный *a* rosellate, rosette-like
розмари'н *m* rosemary *(Rosmarinus)*
ро'зочка *f* small rose; rosette; leg-base crab meat
рой *m* swarm; brood
пе'рвый ~ topswarm
рои'ться *v* swarm
роке'т-сала'т *m* roquette (*Eruca sativa*)
рокфо'р *m* Roquefort-cheese
рома'шка *f* matricary (*Matricaria*)
ро'нжа *f* Siberian jay (*Cactes infaustus*)
ропа'лий *m* rhopalium
роса' *f* dew
ло'жная мучни'стая ~ downy mildew, false mildew *(caused by Peronosporales)*
ло'жная мучни'стая ~ зла'ков cereal downy mildew *(caused by Peronosclerospora)*
медвя'ная ~ honeydew
мучни'стая ~ ду'ба oak powdery mildew *(caused by Microsphaera alphitoides)*
мучни'стая ~ ячменя' barley powdery mildew *(caused by Erysiphe graminis hordei)*
настоя'щая мучни'стая ~ powdery mildew, true mildew *(caused by Erysiphales)*
роси'стый *a* bedewed
росомаха, **россома'ха** *f* wolverine, glutton (*Gulo gulo*)
рост *m* growth
~ **внедре'нием** *bot* intrusive growth
~ **молодо'го органи'зма** juvenile growth
~ **наложе'нием** *bot* appositional growth
аддити'вный ~ additive growth
аллометри'ческий ~ allometric growth, allomorphic growth, heterogonic growth
бу'рный ~ heavy growth, plentiful growth, riot growth
бы'стрый ~ tachyauxesis
вегетати'вный ~ vegetative growth
внутрикле'точный ~ intracellular growth
вста'вочный ~ intercalation, intercalary growth
втори'чный ~ secondary growth
геометри'ческий ~ geometrical growth, geometric increase
глуби'нный ~ submerged growth
диаукси'ческий ~ diauxic growth
злока'чественный ~ malignant growth
изометри'ческий ~ isometric growth
интеркаля'рный ~ intercalary growth
интерстициа'льный ~ interstitial growth
кле'точный ~ cellular growth
клона'льный ~ кле'ток clonogenic growth
клоноге'нный ~ кле'ток clonogenic growth
компенсато'рный ~ compensatory growth
лимити'рованный ~ limited growth
лине'йный ~ linear growth
меристи'ческий ~ meristic growth, multiplicative growth
нача'льный ~ early growth
непреры'вный ~ continuous growth
несбаланси'рованный ~ disharmonic growth, unbalanced growth
оби'льный ~ heavy growth, plentiful growth, riot growth
ограни'ченный ~ limited growth
относи'тельный ~ relative growth
параболи'ческий ~ parabolic growth
плоскостно'й ~ growth in surface
пове'рхностный ~ surface growth
погружё'нный ~ submerged growth
пы'шный ~ heavy growth, plentiful growth, riot growth
равноме'рный ~ isogonic growth
скользя'щий ~ *bot* sliding growth
ску'дный ~ poor growth
сла'бый ~ poor growth
стациона'рный ~ steady-state growth
экспоненциа'льный ~ exponential growth
эмбриона'льный ~ embryonic growth
ростки' *m pl* of **росто'к**; **дава'ть ростки** tiller
росто'к *m* sprout, plantlet, plantling, plantule, shoot, sprouting, propagule
ро'струм *m* rostrum
«ро'счерк» *m* flourish (*in the end of a bird's* song)
рося'нка *f* sundew (*Drosera*)
рот *m* mouth
ве'рхний ~ superior mouth, upward-pointing mouth *(of fish)*
втори'чный ~ deuterostoma
ни'жний ~ inferior mouth *(of fish)*
полуни'жний ~ subterminal mouth *(of fish)*
рота'ция *f* rotation
ротово'й *a* oral, stomatic
ро'ща *f* grove

ро'ющий *a* fossorial
рубе'ц *m* scar, cicatrice, cicatrix; rumen, plain tripe, regular tripe
кольцеобра'зный ~ *bot* ring scar
листово'й ~ leaf scar
плаце'нтный ~ placental scar
по'чечный ~ *bot* bud scar
ру'бка *f* felling
~ **гла'вного по'льзования** final felling, main felling
~ **осветле'ния** liberation felling
~ **ухо'да** improvement felling, improvement cutting, cleaning cutting
возобнови'тельная ~ regeneration cutting
вы'борочная ~ selective cutting, selection cutting, selective felling, selection felling, selective logging
годи'чная ~ annual felling
кае'мчатая ~ striped felling
лесовосстанови'тельная ~ regeneration felling
постепе'нная ~ successive felling
проходна'я ~ severance felling
санита'рная ~ sanitary felling, sanitary cutting
сплошна'я~, **сплошнолесосе'чная** ~ clear cutting, clear felling
рубцева'ние *n* cicatrization
рубцева'ться *v* cicatrize
рубцо'вый *a* cicatricial
ру'бчик *m* cicatricle
рубчикови'дный *a* cicatriculiform
рудера'льный *a* ruderal
рудиме'нт *m* rudiment, vestige
рудимента'рный *a* rudimentary
рука' *f* hand, arm
руково'дство *n* leadership; handbook, manual
рукокры'лый *a* chiropterous
румя'нец *m* erubescence
румя'ный *a* erubescent (*e.g., of an apple*)
руно' *n* fleece
пло'тное ~ heavy fleece
ру'сло *n* bed (*of a river*)
кровяно'е русло blood channel
рути'н *m* rutin
руче'йник *m* caddis fly (*Trichoptera*), caddisworm (*its larva*)
ручно'й *a* tame; pet
ры'ба *f* fish (*Pisces*)
ры'бий *a* fishy, fish
рыбоя'дный *a* fish-eating, ichthyophagous, piscivorous
рыво'к *m* swing; jerk
ры'же-ча'лый *a* strawberry roan
ры'жик *m* false flax (*Camelina*); orange agaric, saffron milk-cap (*Lactarius deliciosus*)
ры'ло *n* snout (*including that of fish*); muzzle
ры'льце *n bot* stigma
пе'ристое ~ plumose stigma
ры'льцевый *a bot* stigmatic
рыса'к *m* trotter
орло'вский ~ Orlovsky trotter
рысь *f* trot; lynx (*Felis lynx*)
кру'пная ~ spanking trot
ме'рная ~ balanced trot, cadenced trot
разма'шистая ~ extended trot
ре'звая ~ prompt trot, strong trot, two-beat leaped gait
ритми'чная ~ balanced trot, cadenced trot
со'бранная ~, **укороченная** ~ collected trot, short trot
уче'бная ~ sitting trot
эласти'чная ~ elastic trot
ры'сью *adv* at *a* trot; **бежа'ть** ~ trot; **пуска'ть рысью** rise at the trot; **пусти'ть ло'шадь** ~ trot a horse; **кру'пной** ~ at a round trot
рыть *v* dig
рытьё' *n* digging
ры'ться *v* grub about, scratch
ры'хлый *a* loose, spongious, spongiose; mellow (*of soil*)
рыча'ние *n* growl, snarl
рыча'ть *v* growl
рюм *m* horned lark (*Eremophila alpestris*)
ряби'нник *m* fieldfare (*Turdus pilaris*)
рябо'й *a* pitted, pocked; speckled (*e.g., of birds*); piebald
рябо'к *a* sand grouse (*Pterocles*)
белобрю'хий ~ pin-tailed sand grouse (*Pterocles alchata*)
ря'бчик *m orn* hazel grouse (*Tetrastes bonasia*); *bot* fritillary (*Fritillaria*)
ря'вканье *n* bellowing, roaring, bark
ря'вкать *v* bark, bellow, roar
ряд *m* row, series, range
~ **Фурье'** Fourier series
геометри'ческий ~ geometric series
гомологи'ческий ~ homologous series
трофи'ческий ~ trophic series, trophical series
функциона'льный ~ series of function

С

саблеви'дный *a* acinaciform, aciniform
саблезу'бый *a* saber-toothed
са'бельник *m bot* marsh cinquefoil (*Comarum*)
сава'нна *f* savanna, savannah
савра'сый *a* dun, sorrel, roan
сагитта'льный *a* sagittal
са'го *n* sago
саго'вник *m* fern palm (*Cycas revoluta*)
сад *m* garden
ботани'ческий ~ botanical garden
грибно'й ~ fungus garden
детский ~ kindergarten
запу'щенный ~ neglected garden
зи'мний ~ winter garden
зоологи'ческий ~ zoo
плодо'вый ~, **фрукто'вый** ~ orchard
саджа' *f orn* Pallas' sandgrouse (*Syrrhaptes paradoxus*)
сади'ться *v* sit down; alight (*e.g., of birds*), perch (*e.g., of birds*)
~ **на ло'шадь** mount a horse
~ **на насе'ст** perch
са'дка *f* mount, mounting
са'днить *v* smart
садо'вник *m* gardener
садово'д *m* horticulturist, orchardist
садово'дство *n* horticulture, orcharding
декорати'вное ~ ornamental horticulture
садово'дческий *a* horticultural, gardening
садо'вый *a* garden
~ **пито'мник** nursery garden
садо'к *m* cage, holding cage, retainer
сажа'ть *v* plant
~ **картошку** plant potatoes
~ **курицу на яйца** set a hen on eggs
~ **на строгую диету** put on a strict diet
~ **птицу в клетку** cage a bird
~ **собаку на цепь** chain a dog
са'женец *m* sapling, seedling
са'женцы *m pl* saplings; planting stock
сайга' *f ther* saiga (*Saiga tatarica*)
са'йленсинг *m mbiol* silencing
сайт *m* site
~ **инициа'ции** initiation site
~ **инициа'ции транскри'пции** transcription start site
~ **инициа'ции трансля'ции** translational initiation site
~ **интегра'ции** insertion site, integration site
~ **кэпи'рования** *mbiol* cap site
~ **переключе'ния** switch site
~ **рестри'кции** *mbiol* restriction site
~ **сбо'рки** assembly site
~ **свя'зывания** binding site
~ **термина'ции** termination site
~ **термина'ции транскри'пции** consensus end site
~ **узнава'ния** recognition site, R site
~ **фикса'ции комплеме'нта** complement fixation site, complement fixing site
~ **циклиза'ции** cyclization site
иммунодомина'нтный ~ antigenic site, immunogenic site
инсерцио'нный ~ insertion site, integration site
клеточносвя'зывающий ~ узнавания *mbiol* cell-binding recongition site
маскиро'ванный ~ узнава'ния cryptic recognition site
неза'нятый ~ свя'зывания unbound binding site, unbound binding site
свобо'дный ~ свя'зывания unbound binding site, unbound binding site
скры'тый ~ узнава'ния cryptic recognition site
щелеви'дный ~ cleft-type site, groove-type site
сайт-специфи'чный *a* site-specific
саккади'ческий *a* saccadic
са'ккула *f* sacculus
саксау'л *m bot* saxaul (*Haloxylon*)
саксау'льник *m* Haloxylon desert, Haloxylon woodland
саланга'на *f orn* switlet *(Collocalia)*
салама'ндра *f* salamander (*Salamandra*)
сала'т *m* lettuce (*Lactuca*)
коча'нный ~ head lettuce, heading lettuce
сала'т-ромэ'н cos, cos-salad, cos-lettuce (*Lactuca sativa var. romana*)
салитра'ль *f plsoc* salitral
са'ло *n* fat; ice scum, ice slush
салотопле'ние *n* fat rendering, tallow rendering
са'льник *m anat* omentum
большо'й ~ gastrocolic omentum, greater omentum

ма'лый ~ lesser omentum, gastrohepatic omentum
са'льниковый *a anat* omental, epiploic
са'льный *a* sebaceous
сальта'ция *f* saltation
саме'ц *m* male
~, **прекрати'вший полову'ю де'ятельность** idle male *(e.g., in fur seals)*
гинекомо'рфный ~ gynaecaner
ка'рликовый ~ dwarf male
перифери'йный ~ *eth* peripheral male
эргато'идный ~ ergataner
са'мка *f* female, dam
я'ловая ~ spinster
само- auto-, self-
самоассоциа'ция *f* self-association
самобаланси'рующийся *a* self-balancing
самовнуше'ние *n* self-suggestion
самовозгора'ться *v* firefang
самовоспроизво'дство *n* self-reproduction
самовоспроизводя'щийся *a* self-reproducing, self-replicating
самовосстана'вливающийся *a* self-renewing
самовосстановле'ние *n* self-repair
самовса'сывание *n* self-absorption
самодифференциа'ция, самодифференциро'вка self-differentiation
самозагрязне'ние *n ecol* self-pollution
самозаживле'ние *n* self-healing
самозараже'ние *n* autoinvasion
самоизре'живание *n* self-thinning,autothinning
самоинду'кция *f* self-induction
самокале'чение *n* autotomy, self-mutilation
самокомплемента'рный *a* self-complementary
самоконденса'ция *f* self-condensation
самолече'ние *n* autotherapy, self-curing
самомезодермализа'ция *f* automesodermalization
самонейрализа'ция *f* autoneuralization
самонесовмести'мость *f* self-incompatibility, self-repulsion
самообли'зывание *n* self-licking
самообновля'ющийся *a* self-renewing
самообусло'вливание *n* self-conditioning
самообуче'ние *n eth* self-conditioning
самоограни'чивающийся *a* self-limited
самооплодотворе'ние *n* self-fertilization, autogamy, automixis
самопроизво'льный *a* spontaneous
самоопыле'ние *n* self-pollination, selfing, self-fertilization
самоорганиза'ция *f* self-organization
самоо'тжиг *m mbiol* self-annealing
самоочище'ние *n* autolopping, self-pruning, natural pruning *(of tree trunks from branches); ecol* self-purification, natural purification
самоперева'ривание *n* self-digestion, autodigestion
самопи'сец *m* recorder
ле'нточный ~ chart recorder
однокана'льный ~ single-channel recorder
самопоглоще'ние *n* self-absorption
самоприви'вка *f* self-grafting
саморазви'тие *n* self-development
саморазможа'ющийся *a* self-propagating
самораспознава'ние *n* autorecognition
саморегули'рование, саморегуля'ция *n* self-regulation
самореплика'ция *f* self-replication
самоса'д *m* home-grown tobacco
самосбо'рка *f* self-assembly
самосе'в *m* self-seeding, seedling growth; self-sown plants
самосовмести'мость *f* self-compatibility
самосогрева'ние *n* firefanging, self-heating
самосохране'ние *n* self-preservation
самоспла'йсинг self-splicing
самосплайси'рующий *a* self-splicing
самосраста'ние *n* self-grafting
самостери'льность *f* self-incompatibility, self-sterility
самоудвое'ние *n* autoreduplication
самоузнава'ние *n* self-recognition
самоуравнове'шивающийся *a* self-balancing
самоформирова'ние *n* autoshaping
самши'т *m* box, boxtree (*Buxus*)
сана'ция *f* sanitation
сангвини'ческий *a* sanguine
санитари'я *f* sanitation, sanitary science
санита'рный *a* sanitary
сантони'н *m* santonin
сап *m* glanders
сапробио'нт *m* saprobe, saprobiont
сапро'бность *f* saprobity
сапро'бный *a* saprobic
сапроге'нный *a* saprogenic, saprogenous
сапрозо'йный *a* saprozoic
сапропе'левый *a* sapropelic

сапропе'ль *m* sapropel
сапропе'льный *a* sapropelic
сапротро'фный *a* saprotroph
сапрофа'г *m* saprophage
сапрофа'гия *f* saprophagy
сапрофи'т *m* saprophyte
сапрофи'тный *a* saprophytic
сапса'н *m orn* peregrin, peregrine, peregrin falcon, duck hawk (*Falco peregrinus*)
сарана' *f bot* wild lily (*in Siberia*)
саранча' *f* locust (*Acrididae*)
~ **марокка'нская** Maroccan locust *(Stauronotus maroccanus)*
~ **перелё'тная** Asiatic locust, migratory locust *(Locusta migratoria)*
~ **пусты'нная** desert locust *(Schistocerca gregaria)*
сарко- sarco-
саркобла'ст *m* sarcoblast
саркоде'рма *f* sarcoderm
сарко'довый *a* sarcodic
саркока'рпий *m* sarcocarp
сарко'ид *m* sarcoid
~ **Бе'ка** Boeck's sarcoid
сарколе'мма *f* sarcolemma
сарко'ма *f med* sarcoma
саркоме'р *m* sarcomere
саркопла'зма *f* sarcoplasm
саркоци'ста *f* sarcocyst
са'рыч *m* buzzard (*Buteo*)
сателли'т *m* satellite
сателли'тный *a mbiol* satellite
сатио'н *m* sation
са'тия *f plsoc* saties
сафрани'н *m* safranin
са'хар *m* sugar
виногра'дный ~ grape sugar, glucose
крахма'льный ~ starch sugar
моло'чный ~ lactose, milk sugar
о'бщий ~ total sugar
солодо'вый ~ brewing sugar
фрукто'вый ~ fructose
сахари'метр *m* saccharimeter
сахариме'трия *f* saccharimetry
сахари'стый *a* sugary, sacchariferous
сахаро'за *f* saccharose, sucrose
сахаролити'ческий *a* saccharolytic
сахаро'метр *m* saccharometer
сахароно'с *m* sugar plant
сахароно'сный *a* sugar bearing, sacchariferous
сачо'к *m* net, butterfly-net
водяно'й ~ dip net
сбаланси'рованный *a* balanced
сбаланси'рованная дие'та balanced diet, adequate diet
сбега'ющий *a* decurrent
сбе'жистость *f* decrease, diminution (*of a tree trunk*), taperingness
сбе'жистый *a* tapering (*of a tree trunk*)
сбива'ние *n* whipping up
~ **в ста'до** bunching in herd
сбива'ть *v* whip, whisk, beat; shake down, knock off (*e.g., apples*)
~ **белки'** beat egg whites, beat up egg whites
~ **со сле'да** put off the scent, throw off the scent
сбива'ться *v* bunch
сбитость *f*: ~ **те'ла** blockiness
сбить *v* shake down
~ **пти'цу** drop a bird
~ **со сле'да** put off the scent, throw off the scent
сбли'женный *a* connivent, approximate
сбор *m* picking (*of fruit*)
промысло'вый ~ cropping
сбо'рка *f* assemblage
сбо'рный *a* polythalamous, aggregate (*of a fruit*)
сбра'живаемый *n* fermentable
сбра'живание *n* fermentation
сбра'живать *v* ferment
сбра'живающий *a* fermentative
сбра'сывание *n* mew (*e.g., of horns*)
~ **кры'льев** *ent* dealation
~ **ли'стьев** defoliation
сбра'сывать *v* mew (*of horns*), cast
сброс *phys* shunt
~ **кро'ви сле'ва напра'во** left-to-right shunt
~ **кро'ви спра'ва нале'во** right-to-left shunt
сваля'вшийся *a* felted
сведе'ние *n* clearing
~ **ле'са** forest clearing
свежева'ть *v* skin, dress
свежеско'шенный *a* new-mown
свежеснесе'нный *a* new-laid
свежесо'бранный *a* freshly plucked
свё'кла *f* beet (*Beta*)
~ **кра'сная** red beet *(Beta vulgaris)*
~ **кормова'я** fodder beet *(Beta vulgaris* var. *alba)*

~ **листова'я** chard, leaf beet (*Beta vulgaris var. cycla*)
~ **обыкнове'нная** red beet *(Beta vulgaris)*
~ **са'харная** sugar beet *(Beta vulgaris* var. *saccharifera)*
~ **столо'вая** red beet *(Beta vulgaris)*

свеклово'д *m* beet grower
сверли'льщик *m* borer
сверну'вшийся *a* curdy, clotted, coagulated
свё'рнутый *a* rolled, coiled, involute, convolute
сверну'ться *v* fold (*of leaves*), coil up (*of a snake*); curd, curdle, set (*of milk*), clot, coagulate (*of blood*)
~ **клубо'чком** roll up into a ball

свё'ртываемость *f* coagulability
свё'ртывание *n* clotting, congelation, coagulation (*of blood*), curdling, setting (*of milk*)
свё'ртываться *v* curl, coil up (*of a snake*); curd, curdle (*of milk*), set (*of milk*), clot, coagulate (*of blood*)
свё'ртывающийся *a* coagulative
сверх- hyper-, super-
сверхадапта'ция *f* overadaptation
сверхдомини'рование *n* superdominance, overdominance
сверхмута'ция *f* hypermutation
антигензави'симая ~ antigen-selected hypermutation
сомати'ческая ~ somatic hypermutation

сверхоргани'зм *m* superorganism, supraorganism
сверхпарази'т *m* superparasite
сверхпаразити'зм *m* superparasitism
свнрхраздраже'ние *n* overstimulation
сверхрегенера'ция *f* superregeneration
сверхсаме'ц *m* supermale
сверхса'мка *f* superfemale
сверхси'нтез *m* gratuitous synthesis
сверхспира'льность *f* superhelicity
сверхструкту'ра *f* superstructure
сверхчи'сленный *a* supernumerary
сверхчувстви'тельность *f* supersensitivity
сверхчувстви'тельный *a* supersensitive
сверча'ть *v* chirr
сверчо'к *m ent* cricket (*Gryllus*); *orn* warbler (*Locustella*)
~ **домо'вый** house cricket, garbage cricket *(Acheta domesticus)*
~ **полево'й** field cricket *(Gryllus campestris)*

свет *m* light
актини'чный ~ actinic light
ви'димый ~ visible light
да'льний кра'сный ~ far-red light
дневно'й ~ day light
инфракра'сный ~ infrared light
иску'сственный ~ artificial light
коротковолно'вый ~ short-wave light
монохромати'ческий ~ monochromatic light
неактини'чный ~ nonactinic light
отраже'нный ~ reflected light
па'дающий ~ incident light
плоскополяризо'ванный ~ plane-polarized light
поляризо'ванный ~ polarized light
преломлё'нный ~ refracted light
преры'вистый ~ intermittent light, flash light, flashing light
проходя'щий ~ transmittent light
прямо'й ~ direct light
рассе'янный ~ scattered light, diffused light, stray light
ультрафиоле'товый ~ ultraviolet light

све'тло-гнедо'й *a* light-dun
све'тлый *a* light
светля'к *m* firefly, glowworm, lightning beetle, lightning bug (*Lampyridae*)
светобоязли'вый *a* light-avoiding, lucifugous
светобоя'знь *f* photophobism
световыно'сливый *a* light-enduring
светолюби'вый *a* photophilous, luciphile, light-requiring
светонепроница'емый *a* lightproof
светоощуще'ние *n* photoreception
светопоглоща'ющий *a* light-absorbing
светорассе'ивающий *a* light-diffusing
светорассе'яние *n* light scattering
квазиупру'гое ~ quasielastic scattering
перпендикуля'рное ~ perpendicular light scattering
прямо'е ~ forward light scattering

светособира'ющий *a* light-harvesting
светоусто'йчивый *a* light-tight, lightproof
светочувстви'тельный *a* light-sensitive
светя'щийся *a* luminescent, lychnidiate
свеча' *f* candle; *med* suppository
~ **Беркефе'льда** Berkefeld filter
~ **Шамберла'на** Chamberland filter

свилева'тый *a* cross-grained
свиль *m for* knot, gnarl, curly wood

свина'рник *m* swine barn, pig pen
свини'на *f* pork
сви'нка *f med* mumps; *a* small pig
морска'я ~ Guinea pig, restless cavy (*Cavia*)
свиново'д *m* pig breeder
свиново'дство *n* pig husbandry
свино'й *a* porcine
свинома'тка *f* sow
подсо'сная ~ milking sow
супоро'сая ~ brood sow
свиноро'й *m* Bermuda grass (*Cynodon*)
свину'ха *bot* inrolled paxil, brown roll-rim (*Paxillus involutus*)
свину'шка *bot* paxil, paxillus (*Paxillus*)
свинья' *f* pig, swine (*Sus*)
ди'кая ~ wild boar, wild hog, wild pig (*Sus scrofa*)
морска'я ~ porpoise (*Phocoena*)
свире'пый *a* ferocious
свиристе'ль *m* wasbird, waxwing (*Bombycilla*)
свиса'ть *v* hang down, depend (*e.g., of a tail*)
свиса'ющий *a* pendent
свист *m* pipe (*of a bird*), whistle
сви'ч-сайт *m* switch site
свищ *m med* fistula
свия'зь *f orn* wigeon, widgeon (*Anas penelope*)
свободноживу'щий *a* free-living
свободнопе'стичный *a* apocarpous
свод *m anat* vault, fornix, arch
~ **нё'ба** palatine vault, palatomaxillary arch
~ **стопы'** arch of foot
~ **че'репа** vault of skull
черепно'й ~ arch of cranium, vault of skull
своди'ть *v*: **у него' свело' но'гу** he has a cramp in the leg
~ **борода'вки** remove warts
~ **лес** disforest
сво'дчатый *a* fornicate, fornicated
«своё'» *s* self
«угрожающее ~» *imm* harmful self
изменённое ~ *imm* modified self, altered self
сво'йство *n* charactersitic, property, feature
аддити'вное ~ additive property
кинети'ческое ~ kinetic property
самозакрепля'ющееся ~ self-locking property
спектра'льное ~ spectral property
тони'ческое ~ catch property
функциона'льное ~ functional property
эмердже'нтное ~ emergent property
сво'ра *f* pack (*e.g., of dogs*), kennel
свора'чивание *n* folding
свора'чивать *v* swerve, turn, swing
сво'рка *f* lead, leash
свя'занный *a* linked
свя'зка *f anat* ligament, copula, band, cord
голосова'я ~ vocal band, vocal cord, phonatory band
дельтови'дная ~ deltoid ligament
зу'бчатая ~ denticulate ligament
клювови'дно-ключи'чная ~ coracoclavicular ligament
конусови'дная ~ conoid ligament
крестообра'зная ~ crucial ligament
крестцо'во-бугро'вая ~ great sacrosciatic ligament
кру'глая ~ round ligament
крылови'дная ~ alar ligament
па'ховая ~ inguinal ligament, inguinal arch, femoral arch
пупа'ртова ~ Poupart's ligament, inguinal arch, femoral arch
серпови'дная ~ falciform ligament
су'мочная ~ capsular ligament
фибро'зная ~ fibrous band
ци'ннова ~ ligament of Zinn, zonule of Zinn
широ'кая ~ broad ligament
свя'зник *m bot* connective
свя'зочный *a* ligamentary
связу'ющий *a* connective
свя'зывание *n* binding
~ **комплеме'нта** binding of complement, complement binding
конкуре'нтное ~ competitive binding
конкуре'нтное ~ с белко'м competitive protein-binding
коопераи'вное ~ cooperative binding
равнове'сное ~ equilibrium binding
высокоави'дное ~ high-avidity binding
высокоспецифи'ческое ~ high-avidity binding
неспецифи'ческое ~ nonspecific binding
свя'зывать *v* connect, link
связь *f* connection, bond, bonding, tie, linkage, link; affiliation
~ **сти'мула и реа'кции** stimulus-response relation
бра'чная ~ consort relation
водоро'дная ~ hydrogen bond

гликози'дная ~ glycosidic linkage
двойна'я ~ double bond
ковале'нтная ~ covalent bond
лине'йная ~ linear connection
макроэрги'ческая ~ energy- rich bond, high-energy bond
межконтинента'льные свя'зи *pl* intercontinental connections
метаболи'ческая обра'тная ~ metabolic feedback
нековале'нтная ~ noncovalent bond
непосре'дственная ~ direct relation
обра'тная ~ feedback
обра'тная ~ с запа'здыванием delayed feedback
отрица'тельная обратная ~ negative feedback
па'рная ~ pair bonding
пепти'дная ~ peptide bond
положи'тельная обра'тная ~ positive feedback
причи'нная ~ causal relationship, causal relation
пряма'я ~ direct relation; feedforward
свобо'дная ~ available bond
слухова'я обра'тная ~ auditory feedback
сопряжё'нная ~ conjugated bond
социа'льная ~ *eth* social relation
социа'льные свя'зи *pl* social bonds
усло'вная ~ conditioned connection
фосфодиэфи'рная ~ phosphodiester bond
функциона'льная ~ functional linkage

сгиб *m* bend, crease, fold
локтево'й ~ bend of elbow
коле'нный ~ bend of knee

сгиба'ние *n* flexion, bending
сгиба'тель *m* flexor
сгиба'ть *v* fold, flex, bend
сгла'живание *n biom* smoothing
сгни'вший a sphacelated
сгоня'ть *v* herd, drive
~ **весну'шки** remove freckles
~ **во'лос** dehair

сгруппиро'ванный *a* aggregate
сгу'сток *m* clot, curd, coagulate
сгуща'ть, сгуща'ться *v* condense
сгуще'ние *n* condensation, concretion
сдавле'ние *n* compression
сда'вливать *v* compress
сдвиг *m* shift, slip
~ **ра'мки** *mbiol* frame shift
~ **ра'мки счи'тывания** *mbiol* reading frame shift

сдво'енный *a* geminate, geminated
сдира'ть *v* strip off, peel
~ **ко'жу** strip, flay
~ **кору'** bark, rind, peel

сев *m* sowing
севооборо'т *m* rotation, crop rotation, crop succession
полево'й ~ field crop rotation
пятипо'льный ~ five-course rotation

сегме'нт *m* segment
~ **ге'на** gene segment
~ **груди'ны** sternebra
~ **коне'чности** podomere
~ **те'ла** merome, merosome, somite, somatome
антенна'льный ~ antennal segment
ге'нный ~ gene segment
дифференциа'льный ~ *gen* differential segment
изофе'нный ~ isophene segment
инсерцио'нный ~ insertion segment
интерстициа'льный ~ *gen* interstitial segment
конъюги'рующий ~ *gen* pairing segment

сегмента'рный *a* segmental
сегмента'ция *f* segmentation
втори'чная ~ annulation

сегрега'ция *f* segregation
седа'лищный *a* sciatic, ischiatic, ischiadic
седати'вный *a* sedative
седимента'ция *f* sedimentation
седла'ть *v* tackle, saddle (*a horse*)
седло' *n* saddle
туре'цкое ~ *anat* Turkish saddle

сё'длышко *n* ephippium
седми'чник *m bot* starflower (*Trientalis*)
сезамови'дный *a* sesamoid
сезо'н *m* season
закры'тый ~ shutoff
запре'тный ~ closed season
засу'шливый ~ drought season
охо'тничий ~ hunting season

сезо'нный *a* seasonal
сейсмонасти'ческий *a* seismonastic
сейсмона'стия *f* seismonasty
сейсмота'ксис *m* seismotaxis
сека'тор *m* pruner, pruning scissors
сека'ч *m* bull (*of fur seals*)
секвена'тор *m mbiol* sequenator, sequencer
~ **в га'зовой фа'зе** gas phase sequenator

секвени'рование *n mbiol* sequenation, sequencing
~ **белко'в** amino acid sequencing, protein sequencing
~ **ДНК DNA** sequencing
~ **нуклеи'новой кислоты'** nucleotide sequencing
твердофа'зное ~ solid-phase sequencing
хими'ческое ~ по Ма'ксаму-Ги'лберту Maxam-Gilbert chemical sequencing
секвестра'ция *f* sequestration
секодо'нтный *a* secodont
секре'т *m* secretion
~ **са'льных желё'з** sebum
секрети'н *m* secretin
секрети'ровать *v* secrete
секрето'рный *a* secretory
секре'ция *f* secretion
апокри'нная ~ apocrine secretion
ве'кторная ~ vectorial secretion
вне'шняя ~ exocrine secretion, external secretion
вну'тренняя ~ endocrine secretion, internal secretion
голокри'нная ~ holocrine secretion
мерокри'нная ~ merocrine secretion
нейрогумора'льная ~ neurohumoral secretion
экзокри'нная ~ exocrine secretion, external secretion, excretion
эндокри'нная ~ endocrine secretion, internal secretion, incretion
сексду'кция *f* sexduction
сексоло'гия *f* sexology
сексуа'льность *f* sexuality
сексуа'льный *a* sexual
се'ктор *m* sector
пере'дний ~ ду'жки *ent* anterior arculus
за'дний ~ ду'жки *ent* posterior arculus
сектори'альный *a* sectorial
се'лезень *m* drake
селезе'нка *f* spleen, lien
селезе'ночник *m* golden saxifrage (*Chrysosplenium*)
селезе'ночный *a* splenetic, splenic, lineal
селекти'вный *a* selective
селекти'вность *f* selectivity
селекциони'ровать *v* select
селе'кция *f* selection
~ **зерновы'х** cereal breeding
внутрипопуляцио'нная ~ intrapopulation selection
внутрисомати'ческая ~ intrasomatic selection
втори'чная ~ secondary selection
генотипи'ческая ~ genotypic selection
гибридо'мная ~ postfusional selection
динами'ческая ~ dynamic selection
индивидуа'льная ~ individual selection
интрасексуа'льная ~ intrasexual selection
клона'льная ~ clone selection, sib selection
комбини'рованная ~ combined selection
лине'йная ~ line selection, linear selection
ма'ссовая ~ mass selection
нака'пливающая ~ cumulative selection
напра'вленная ~ directional selection, controlled breeding
пеницилли'новая ~ penicillin selection
периоди'ческая реципро'кная recurrent reciprocal selection
периоди'ческая ~ periodic selection, recurrent selection
по'чковая ~ bud selection
семе'йная ~ family selection
сравни'тельная ~ comparative selection
центробе'жная ~ centrifugal selection
центростреми'тельная ~ centripetal selection
селенодо'нтный *a* selenodont
селенотропи'зм *m* selenotropism
се'льва *f* selva
сельдере'й *m* celery, celeriac (*Apium*)
се'льский *a* rural
се'льфинг *m* selfing
семантиза'ция *f eth* semantization
семати'ческий *a* sematic
семе'йство *n* family
~ **ге'нов** gene family
~ **обращё'нных повто'ров** *mbiol* fold-back family, FB family
~ **после'довательностей** *mbiol* sequence family
~ **сце'пленных ге'нов** linked gene family
мультиге'нное ~ multigene family
семена' *n pl* seeds; **на семена** for seed
кондицио'нные ~ certified seeds
семенни'к *m anat* testicle, testis, spermarium, spermary; *bot* seed plant, seed-bearing plant
семеннико'вый *a* orchic, orchitic
семенно'й *a* seminal, spermatic; seed-bearing
семеново'д *m* seed grower
семеново'дство *n* seed raising
семеноно'сный *a bot* seminiferous

семеприё'мник *m* spermatheca, spermatotheca
се'мечко *n* pip
семиари'дный *a* semiarid
семига'мия *f* semigamy
семили'стный *a* heptaphyllous
семио'тика *f eth* semiotics
семиразде'льный *a* septempartite
семиря'дный *a* heptastichous
семичле'нный *a* heptamerous
семья' *f* family
пчели'ная ~ honey-bee colony, colony of bees
се'мя *n* seed; sperm, semen
~ **без эндоспе'рма** exalbuminous seed
безбелко'вое ~ exalbuminous seed
льняно'е ~ linseed, flaxseed
ма'сличное ~ oilseed
о'чень ме'лкое ~ dust seed
поко'ящееся ~ dormant seed
твё'рдое ~ hard seed
цитва'рное ~ wormseed
семявынося'щий *a* seminiferous
семядо'льный *a* cotyledonary, cotyledonous
семядо'ля *f* cotyledon, seed lobe, seed leaf, seminal leaf
семязача'ток *m* ovule
семя'нка *f* achene, cypsela
семяно'жка *f* seed stalk
семяно'сец *m* placenta
семяпо'чка *f* ovule, s seedbud
ортотро'пная ~ orthotropous ovule
семяпрово'д *m* spermiduct, spermoduct, spermaduct, seminal duct, spermatic vessel
семяприе'мник *m* seminal receptacle
се'мячковые *pl* pip fruits
сенберна'р *m* St. Bernard
сени'льный *a* senile
сенни'к *m* barn
се'но *n* hay
све'жее ~ new-mown hay
сенова'л *m* hayloft, loft
сеноко'с *m* grass-cutting, mowing, haymaking, haying; hay field; mowing time
сеноко'сец *m* harvestman (*Phalangida*)
сеноста'вка *f* pika (*Ochotona*)
сенсибилиза'тор *m* sensibilizator, sensibilizer, sensitizer, sensibilizin
сенсибилиза'ция *f* sensibilization, sensitization, immunization
акти'вная ~ antigen sensitization
аллоге'нная ~ allosensitization
пасси'вная ~ antibody sensitization
перекрё'стная ~ crossed
повто'рная ~ resensitization
сенсибилизи'н *m* sensibilizator, sensibilizer, sensibilizin, sensitizer
сенсибилизи'рованный *a* sensitized
сенсибилизи'ровать *v* sensitize
сенси'лла *f* sensillum
сенсити'вность *f* sensitivity
сенситиза'ция *f eth* concaveation
сенсомото'рный *a* sensomotor, sensorimotor
сенсориа'льный *a* sensorial
сенсо'риум *m* sensorium
сенсо'рно-дви'гательный *a* sensomotor, sensorimotor
сенсо'рный *a* sensory
сепара'тор *m* separator
центробе'жный ~ centrifugal separator
дели'тельный ~ phase separator
се'пия *f* sepia ink, sepia
се'псис *m* sepsis
септицеми'я *f* septicemia
се'пта *f* septum
септа'тный *a zool* septate
се'птик-тэ'нк *m* digestion tank, septic tank
се'ра *f* sulfur
ушна'я ~ cerumen, ear wax
серва'л *m ther* serval *(Felis serval)*
се'рвис-пери'од *m* service period
серде'чно-сосу'дистый *a* cardiovascular, vasculocardiac
серде'чный *a* cardiac
серди'ть *v* anger
се'рдце *n* heart
изоли'рованное ~ isolated heart
ле'вое ~ left heart
лё'гочное ~ pulmonary heart
лимфати'ческое ~ lymph heart
пра'вое ~ pulmonary heart, respiratory heart, right heart, venous heart
сердцебие'ние *n* heartbeat
сердцеви'дный *a* cordiform, cordate, heart-shaped
сердцеви'на *f* pith, core, heart, medulla
плекте'нхимная ~ fungal medulla *(of a lichen)*
сердцеви'нный *a* medullary, pithy
сердцеобра'зный *a* cordate, cordiform, heart-shaped
середи'на *f* medium, middle, midst
серё'жка *f* ament, catkin; jow (*e.g., of a cock*), wattle; dewlap (*of a turkey*)

серёжконо'сный *a* amentaceous
серёжкоцве'тный *a bot* amentiferous
серё'жчатый *a* amentaceous
сери'йный *a* serial; seral
сери'н *m* serine
се'рия *f* series; *ecol* sere
  есте'ственная ~ natural sere
  дернообразу'ющая ~ turf sere
  ксери'ческая ~ xerarch sere
  литофи'тная ~ litharch sere
  о'пытная ~ trial series
  экологи'ческая ~ ecological sere
серкли'макс *m* serclimax
се'рна *f* chamois (*Rupicarpa rupicarpa*)
серни'стый *a*: **~ газ** sulfur dioxide
серодиагно'стика *f* serodiagnostic assay, serodiagnosis
серо'за *f* serose
серо'зно-сли'зистый *a* seromucous
серо'зно-фибрино'зный *a* serofibrinous
серо'зный *a* serosal; serous
сероконве'рсия *f* seroconversion
серокульту'ра *f* seroculture
сероло'гия *f* serology, orrhology
серонегати'вный *a* seronegative
сероотрица'тельный *a* seronegative
серопозити'вный *a* seropositive
серположи'тельный *a* seropositive
серорea'кция *f* seroreaction, serotest
серотe'ст *m* seroreaction, sero
сероти'п *m* serotype
серотипи'рование *n* serologic typing, serotyping
серотокси'н *m* serotoxin
серотони'н *m* serotonin
серп *m* sickle; crescent; falx
  ~ мозжечка' falcula
  жё'лтый ~ yellow crescent
  се'рый ~ gray crescent
серпови'дный *a* falcate, falciform, falculate, semilunar, crescent, crescentic, lunar, lunate, sickle-shaped
серполи'стный *a* sickle-leaved
серполо'пастный *a* sickle-lobed
серпопло'дный *a* sickle-fruited
се'рула *f* serule
серумальбуми'н *m* serum albumin
се'рый *a* gray; **~ я'блоках** dapple-gray
сесквидипло'ид *m* sesquidiploid
сесси'льный *a* sessile, sedentary
се'стринский *a* sisterly
сетеви'дный *a* netlike, retiform, plexiform
сетеобра'зный *a* retiform
се'тка *f* net, netting, network, screen; grid; reticulum, rete; honeycomb stomach, fore stomach
  ~ с у'гольной плё'нкой carbon-coated grid
  противокомари'ная ~ mosquito net
сеткообра'зный *a* reticular
сетно'й *a* net
се'точка *f* reticle, reticulum
  кле'точная ~ cellular reticulum
се'ттер *m* setter
се'ттер-гордо'н *m* Gordon setter
сетча'тка *f anat* retina
сетчато-жилкова'тый *a* retinerved
сетчатокры'лый *a* neuropterous
се'тчато-не'рвный *a* net-veined
сетчатопло'дный *a* dictyocarpous
се'тчатый *a* netted, reticulose, reticulate, reticulated, reticular; *bot* dictyodromous (*of venation*)
сеть *f* net, network; reticulum, rete
  ~ семенника' rete of testicle
  ~ социа'льных свя'зей *eth* social network
  агрануля'рная эндоплазмати'ческая ~ agranular endoplasmic reticulum, smooth endoplasmic reticulum
  гидрографи'ческая ~ hydrographic network
  гла'дкая эндоплазмати'ческая ~ agranular endoplasmic reticulum, smooth endoplasmic reticulum
  грануля'рная эндоплазмати'ческая ~ granular endoplasmic reticulum, rough endoplasmic reticulum
  иммунорегулято'рная ~ immunological circuit, immunoregulatory circuit
  капилля'рная ~ capillary network
  кле'точная ~ cellular network
  кра'сная ~ red body
  миопаравезикуля'рная ~ myoparavesicular reticulum
  не'рвная ~ nerve net
  паути'нная ~ mist net *(for catching small birds)*
  пу'шечная ~ cannon net trap, cannon net
  саркоплазмати'ческая ~ sarcoplasmic reticulum
  сосу'дистая ~ vasculature
  хромати'новая ~ chromatin network
  цитоплазмати'ческая ~ cytoplasmic reticulum
  чуде'сная ~ red body, wonderful network

шерохова'тая эндоплазмати'ческая ~ rough endoplasmic reticulum
эндоплазмати'ческая ~ endoplasmic reticulum
сече'ние *n* section
ке'сарево ~ Caesarean section, Cesarean section
се'чка *f* chaff
се'янец *m* seedling
се'ять *v* sow, seed
сжа'тие *n* compression, contraction, constriction
сжа'тый *a* compressed, coarctate
~ кула'к clenched fist
сжига'ние *n* combustion
сжима'тель *m* sphincter
сжима'ть *v* compress, constrict, coarct, squeeze, clench
сжима'ться *v* contract, coarct, shrink, clench
сжима'ющий *a* contractive, contractile, constrictive
си'блинг *m* sibling, sib
си'вуч *m ther* sea king, Steller's sea lion (*Eumetopias jubatus*)
сигмови'дный *a* sigmoid
сигна'л *m* signal, call
~ безопа'сности safety signal
~ для узнава'ния *mbiol* recognition sign
~ опа'сности danger signal
~ приглаше'ния к игре' play-soliciting signal
~ термина'ции termination signal
~ ти'па включе'ние - выключе'ние on/off signal
аллоге'нный хе'лперный ~ allospecific help
аутоге'нный хе'лперный ~ autospecific help
биоэлектри'ческий ~ bioelectric signal
входно'й ~ *phys* incoming signal
голосово'й ~ vocal expression
градуа'льный ~ *eth* graded signal
за'паховый ~ olfactory signal
и'мпульсный ~ pulse signal *(bioacoustics)*
неверба'льный ~ nonverbal signal
обма'нный ~ *eth* deceptive signal
ольфакто'рный ~ olfactory signal
предупрежда'ющий ~ warning signal
рекомбинацио'нный ~ recombination signal
уси'ливающий ~, усили'тельный ~ augmentative signal
усло'вный ~ arbitrary signal
хе'лперный ~ augmentative signal, command signal
частотномодули'рованный ~ frequency-modulated signal *(bioacoustics)*
сигнализа'ция *f* signaling
сидера'т *m* green manure
сидерофи'л *m* siderophile
сидерофи'льный *a* siderophilic, siderophilous
сиде'ть *v* sit
~ на ко'рточках squat
~ на я'йцах cover eggs, sit on eggs
сидячегла'зый *a* sessile-eyed, edriophthalmic
сидя'чий *a* attached, sessile, sedentary
сиза'ль *m* sisal agave (*Agave sisalana*)
сизарь *m* rock pigeon (*Columba livia*)
си'зый *a* blue-gray, glaucous
си'ла *f* force, strength, power, vigor
~ испаре'ния evaporative power
~ отбо'ра selection force
~ поглоще'ния absorbing power
~ притяже'ния attractive force
~ сцепле'ния adhesive force
~ це'ли target strength (*in echo sounding*)
гибри'дная ~ hybrid vigor
дви'жущая ~ motive power
жи'зненная ~ vital force
имму'нная ~ phylactic power
ио'нная ~ ionic strength
мы'шечная ~ muscle strength
подъё'мная ~ elevating power
селекцио'нная ~ selection force
сосу'щая ~ suction power, suction force, suction tension, water absorbing power
силико'з *m* silicosis
сило'с *m* ensilage, silage, silo
силосова'ние *n* ensiling, ensilage, siloing
силосова'ть *v* ensile, ensilage, silage
силосохрани'лище *n* silo
силосу'емость *f* ensilage
силу'рийский *a* Silurian
сильноде'йствующий *a* potent, powerful, stiff
сильномочкова'тый *a* multifasciate
си'льный *a* strong, powerful, robust, intense, severe, heavy, sthenic
сим- sym-
симбио'з *m* symbiose, symbiosis
вре'менный ~, дизъюнкти'вный ~ disjunctive symbiosis

пищево'й ~ nutritive symbiosis
симбио'нт *m* symbiont, symbiote
симбиоти'ческий *a* symbiotic
симбиотро'фный *a* symbiotrophic
си'мвол *m* sign, symbol
симметри'я *f* symmetry
билатера'льная ~ bilateral symmetry
бокова'я ~ lateral symmetry
двусторо'нняя ~ bilateral symmetry
латера'льная ~ lateral symmetry
радиа'льная ~ radial symmetry
спира'льная ~ spiral symmetry
симпатикотони'ческий *a* sympathicotonic
симпатикотони'я *f* sympathicotonia
симпати'ческий *a* sympathetic
симпатобла'ст *m* sympathicoblast, sympathoblast
симпатомимети'ческий *a* sympathomimetic
симпатри'ческий *a* sympatric
симпатри'я *f* sympatry
симпатэктоми'я *f* sympathectomy
симпла'ст *m* symplast
симпле'ктикум *m* symplectic
симподиа'льный *a* sympodial
симпо'дий *m* sympodium
симпо'рт *m* symport
симпто'м *m* sign, symptom
~ **Арга'йла-Робертсо'на** Argyll-Robertson symptom, Argyll-Robertson sign
~ **А'шнера** Aschner's symptom
~ **Брага'ра** Bragard's sign
~ **Ва'ля** Wahl's symptom
~ **Вестфа'ля** Westphal's sign
~ **Ви'нтриха** Wintrich's sign
~ **Ги'бсона** Gibson murmur
~ **Гордона** Gordon's sign, Gordon's symptom
~ **Да'нса** Dance's symptom
~ **Ке'рнига** Kernig's sign
~ **Ки'вуля** Kywul's symptom
~ **Лара'ша** Larasch's symptom
~ **Ласе'га** Lasegue's sign
~ **Ме'йтуса** Meitus symptom
~ **Пи'ка** Pick's symptom
~ **Ро'мберга** Romberg's symptom, Romberg's sign
~ **Та'уссига** Taussig's symptom
серде'чный ~ heart sign
симптома'тика *f* symptomatology, semeiotics, semiotics, semeiology, semiology
симптомоко'мплекс *m* symptom complex, symptom group
~ **Глена'ра** Glenard's disease
симуля'ция *f* simulation
симфи'з *m* symphysis
симфили'я *f* symphily
симфилло'дий *m* symphyllodium
симфиогене'з *m* symphyogenesis
син- syn-
синакме' *n* synacme, synacmy, synanthesis
сина'нтия *f* synanthy
синантро'пный *a* synanthropic
синапомо'рфия *f* synapomorphy
синапосемати'ческий *a* synaposematic
си'напс *m* synapse
~ **на расстоя'нии** distant synaps
возбужда'ющий ~ excitatory synaps
инвагинацио'нный ~ invaginated synaps
не'рвно-мы'шечный ~ myoneural junction
синапси'дный *a anat* synapsid
си'напсис *m* synapsis
синапте'на *f* synaptene
синапти'ческий *a* synaptic
синаптосо'ма *f* synaptosome
синаптоспермі'я *f bot* synaptospermy
синартро'з *m* synarthrosis
сингами'я *f* syngamy
синга'мный *a* syngamic
сингене'з *m* syngenetic succession
сингене'тика *f* syngenetics
сингенети'ческий *a* syngenetic
сингле'тный *a* singlet
синдактили'я *f* syndactyly, syndactylism
си'ндезис *m* syndesis
синдесмо'з *m* syndesmosis, ligamentous joint
синдесмоло'гия *f* syndesmology
синдиплоиди'я *f* syndiploidy
синдро'м *m* syndrome
~ **бе'гства** *eth* flight syndrome
~ **лише'ния ма'тери** maternal deprivation syndrome
~ **Паркинсо'на** Parkinson's syndrome
~ **приобрете'нного иммунодефици'та, СПИД** acquired immunodeficiency syndrome, AIDS
адаптацио'нный ~ adaptation syndrome
пи'кквикский ~ Pickwckian syndrome
синева' *f* blue stain, blue rot (*of wood*)
~ **древеси'ны** blue timber
синеголо'вник *m* sea-holly (*Eryngium*)
синерги'да *f* synergid

синерги'зм *m* synergism
синерги'ст *m* synergist
синерги'ческий *a* synergistic
синерги'чный *a* synergic, synergetic, synergistic
синерги'я *f* synergy, synergism
синере'зис *m* syneresis
синзооспо'ра *f* synzoospore
си'ний *a* blue, dark blue
~ **Э'ванса** Evans blue
альциа'новый ~ alcian blue
бромтимо'ловый ~ bromthymol blue
бромфено'ловый ~ bromphenol blue
изами'новый ~ isamine blue
метиле'новый ~ methylene blue
мети'ловый ~ methyl blue
толуиди'новый ~ toluidine blue
трипа'новый ~ trypan blue
сини'ца *f* tit, titmouse, tomtit (*Parus*)
больша'я ~ oxeye, great tit (*Parus major*)
длиннохво'стая ~ long-tailed titmouse *(Aegithalos caudatus)*
острохо'хлая ~ tufted titmouse *(Parus bicolor)*
тра'урная ~ black tit *(Parus leucomelas)*
хохла'тая ~ crested tit, crested titmouse *(Parus cristatus)*
уса'тая ~ bearded titmouse *(Panurus biarmicus)*
синкарио'н *m* syncaryon, synkaryon
синкариофи'т *m* syncaryopyte
синка'рпий *m* syncarp
синка'рпия *f* syncarpy
синка'рпный *a* syncarpous, symphycarpous
синко'ний *m* synconium
синкрипти'ческий *a* syncryptic
синморфоло'гия *f* synmorphology
синоаурикуля'рный *a* sinuauricular
синовиа'льный *a* synovial
сино'вия *f* synovial fluid, joint fluid, joint oil, synovia
синокароти'дный *a* sinocarotid
сино'ним *m* synonym
синони'мика *f* synonymy
синосто'з *m* synostosis, synosteosis
синрефле'ксия *f* synreflexia
синтаксо'н *m* syntaxon
синтаксономи'ческий *a* syntaxonomical
си'нтез *m* synthesis
~ **в жи'дкой фа'зе** solution-phase synthesis
~ **с затра'вкой** primed synthesis
асимметри'чный ~ asymmetric synthesis
затра'вочный ~ primed synthesis
избы'точный ~ gratuitous synthesis
индуци'рованный ~ induced synthesis
нерепарати'вный ~ **ДНК** unscheduled DNA synthesis
репарати'вный ~ repair synthesis
твёрдофа'зный ~ solid-phase synthesis
чи'стый ~ net synthesis
синтеза'тор *m* synthesizer
синтези'рованный *a* synthesized
синтони'я *f* syntony
синтрофи'зм *m* synthrophism
синтрофобла'ст *m* syntrophoblast, syncytiotrophoblast
сину'зия *f* synusia
си'нус *m* sinus
ана'льный ~ rectal sinus
вене'чный ~ coronary sinus
вено'зный ~ **скле'ры** venous sinus of sclera
заты'лочный ~ occipital sinus
камени'стый ~ petrosal sinus
кароти'дный ~ carotid sinus
корона'рный ~ coronary sinus
мочеполово'й ~ urinogenital sinus, urogenital sinus
прямо'й ~ *anat* tentorial sinus
со'нный ~ carotid sinus
синусно-предсердный *a* sinoatrial, sinuatrial
синхондро'з *m* synchondrosis
синхороло'гия *f* synchorology
синхрониза'ция *f* synchronization
синхро'нный *a* synchronic, synchronous
синхроноло'гия *f* synchronology
синхрониза'ция *f* synchronization
синхро'нный *a* synchronous
синцитиа'льный *a* syncytial
синци'тий *m* syncytium, syncyte
синцитиотрофобла'ст *m* syntrophoblast, syncytiotrophoblast
синьга' *f orn* scoter (*Melanitta*)
синэкзи'на *f pal* synexine
синэколо'гия *f* synecology
генети'ческая ~ genetic synecology
географи'ческая ~ geographic synecology
синэстези'я *f* synaesthesia
синя'к *m* bruise, echymosis; *bot* viper's bugloss (*Echium*)
сип *m orn* griffon (*Gyps*)
белоголо'вый ~ griffon vulture *(Gyps fulvus)*
си'плый *a* husky, hoarse

сипу'ха *f* barn owl, church owl (*Tyto alba*)
сире'невый *a* lilac
сире'нь *f* lilac (*Syringa*)
систе'ма *f* system
~ **вакуо'лей** vacuolar system
~ **воро'тной ве'ны** portal system
~ **врожде'нных де'йствий** *eth* innate action system
~ **гипотала'мус - гипо'физ - яи'чники** hypothalamic-hypophysial-ovarian axis
~ **гипо'физ - надпо'чечники** pituitary-adrenal axis
~ **Ги'са-Тава'ра** His-Tawara system
~ **групп кро'ви** blood-group system
~ **жизнеобеспе'чения** life support system
~ **землепо'льзования** land-use system
~ **инстинкти'вной регуля'ции** instinctive control system
~ **кровообраще'ния** circulatory system
~ **ли'ний серебре'ния** *prot* silverline system
~ **огнево'й расчи'стки по'чвы** burnt-earth farming
~ **покро'вов те'ла** integumentary system
~ **разведе'ния** breeding system
~ **репара'ции** repair system
~ **свё'ртывания** coagulation system
~ **с обра'тной свя'зью** feedback system
~ **хи'щник-же'ртва** prey – predator system
автоматизи'рованная лаборато'рная ~ automated lab system
автоно'мная не'рвная ~ vegetative nervous system, involuntary nervous system
адапти'вная ферме'нтная ~ adaptive enzyme system
адрена'льная ~ adrenal system
амбулакра'льная ~ ambulacral system, water-vascular system
ассоции'рованная ви'русная ~ association virus system
белоксинтези'рующая ~ protein synthesis system
бескле'точная ~ cell-free system
бина'рная ~ binary system
вегетати'вная не'рвная ~ autonomic nervous system, involuntary nervous system, vegetative nervous system
веретё'нная ~ spindle system
воро'тная ~ portal system
воро'тная ~ пе'чени hepatic portal system
втора'я сигна'льная ~ second signaling system
выдели'тельная ~ excretory system
Га'версова ~ Haversian system
генети'ческая ~ genetic system
гумора'льная ~ humoral system
дополни'тельная ольфакто'рная ~ accessory olfactory system
жива'я ~ living system
закры'тая коммуникацио'нная ~ closed communication system
за'мкнутая ~ closed circuit, bounded system
за'мкнутая ~ регули'рования closed-loop system
за'мкнутая кровено'сная ~ closed sirculatory system
за'мкнутая экологи'ческая ~ closed ecological system
зрительная проекционная ~ optical mapping system
иерархическая ~ hierarchical system
иммерсионная ~ immersion system
иммунная ~ immune system
индуцирующая ~ inducing system
искусственная ~ artificial system
искусственная коммуникационная ~ artificial communication system
компартментная ~ compartmental system
корнева'я ~ rootage, root system
корпускуля'рная ~ насле'дственности corpuscular system of heredity
ко'стная ~ skeletal system
кровено'сная ~ circulatory system, blood-vascular system
кроветво'рная ~ hematopoietic system
лимби'ческая ~ limbic system
лимфати'ческая ~ lymphatic system
мочева'я ~ urinary system
мочеполова'я ~ urogenital system
мочкова'тая корнева'я ~ roots in clusters, fibrous root system
мы'шечная ~ muscular system
неза'мкнутая кровено'сная ~ open curculatory system
нейромедиа'торная ~ neurotransmitter system
не'рвная ~ nervous system
не'рвно-мы'шечная ~ neuromuscular system
неспецифи'ческая проекцио'нная ~ nonslecific projection system
откры'тая ~ open circuit

откры'тая коммуникацио'нная ~ open communication system
отцо'вско-матери'нская ~ paternal-maternal system, P-M system
парасимпати'ческая не'рвная ~ parasympathetic system
перифери'ческая не'рвная ~ peripheral system
пищевари'тельная ~ alimentary system, digestive system
подсе'чная ~ **земледе'лия** slash-and-burn system of agriculture
полиге'нная систе'ма ~ polygene system
полова'я ~ reproductive system, genital system
полуза'мкнутая ~ semiclosed circuit
порта'льная ~ portal system
проводя'щая ~ **се'рдца** cardiac conduction system
проприорецепти'вная чувстви'тельная ~ proprioceptive sensory system
прото'чная ~ continuous-flow system
прочностносопряже'нная ~ tightly coopled system
равнове'сная ~ equilibrium system
реаги'рующая ~ reactive system
репродуктивная ~ reproductive system, genital system
ретикулоэндотелиальная ~ reticulo-endothelial system
ретикулярная активирующая ~ recticular activating system
самовоспроизводящая ~, **самореплицирующая** ~ self-replicating system
сбалансированная ~ equilibriun system
сенсорная ~ sensory system
сердечно-сосудистая ~ vascular system, cardiovascular system, heart-vascular system
симпатическая нервная ~ sympathetic nervous system
синтезирующая ~ synthesizing system
собирающая ~ *phys* collecting system
сопряженная ~ coupled system
сосудистая ~ vascular system
таксономическая ~ taxonomic system
травопо'лная ~ **земледе'лия** ley farming system, lay farming system
тройна'я ~ trinary system
ферме'нтная ~ enzyme system
филогенети'ческая ~ phylogenetic system
центра'льная не'рвная ~ central nervous system
система'тика *f* systematics
системати'ческий *a* systematic
систе'мный *a* systemic
си'стола *f* systole
~ **желудочков** ventricular systole
~ **предсердий** atrial systole, auricular systole
механическая ~ mechanical systole
ретроградная ~ retrograde systole
эктопическая ~ ectopic systole
электрическая ~ electrical systole
систоли'ческий *a* systolic
си'тник *m bot* rush (*Juncus*)
ситня'г *m bot* spike-rush (*Eleocharis*)
си'то *n* sieve
ио'нное ~ ionic sieve
молекуля'рное ~ molecular sieve
ситови'дный *a* cribriform, cribrose
ситови'дная тру'бка *bot* sieve tube
ситови'дная ткань sieve tissue
ситообра'зный *a* cribriform, cribrose
ситотропи'зм *m* sitotropism
сифо'н *m* siphon
вводно'й ~ *zool* inhalant siphon
выводно'й ~ *zool* exhalant siphon
сифонога'мия *a* siphonogamy
сифонога'мный *a* siphonogamic, siphonogamous
сифоногли'ф *m* siphonoglyph
сифонозо'оид *m* siphonozooid
скаково'й *a* race; **скакова'я ло'шадь** race horse
скаку'н *m* racer; tiger beetle (*Cicindelidae*)
ска'лить *v:* ~ **зу'бы** *eth* grin
ска'льпель *m* scalpel
ска'нер *m* scanner
инфракра'сный ~ infrared scanner
компью'терный томографи'ческий ~ computed tomographic scanner
ультразвуково'й ~ supersonic scanner, ultrasonic scanner
ультрафиоле'товый ~ ultraviolet scanner
скани'рование *n* scanning
многове'кторное ~ multiple view scanning
многопроекцио'нное ~ multiple view scanning
радиоизото'пное ~ radionuclide scanning
ультразвуково'е ~ supersonic scanning, ultrasonic scanning
скани'рующий *a* scanning

сканогра'мма *f* scintiscan
ска'пус *m ent* scape
скарабе'й *m* scarab (*Scarabeus*)
~ **свяще'нный** sacred scarab *(Scarabaeus sacer)*
скарифика'ция *f* scarification
ска'рмливать *v* feed
скат *m anat* clivus
скато'л *m* skatole
скатоло'гия *f* scatology
скатофаги'я *f* scatophagy
ска'тываться *v ich* migrate down
скафи'дий *m* scaphidium
скафоцери'т *m* scaphocerite
скачкообра'зный *a* saltatory
скачо'к *m* jump, leap, bound, spring
~ **раствори'теля** *mbiol* solvent-jump
~ **электри'ческого по'ля** electric field jump
температу'рный ~ temperature jump, thermocline
сква'жность *f* porosity
сква'шивание *n* souring
сква'шивать *v* let sour
сквер *m* public garden
скворе'ц *m* starling, grackle (*Sturnus vulgaris*)
обыкнове'нный ~ common starling *(Sturnus vulgaris)*
ро'зовый ~ rose-coloured starling *(Pastor roseus)*
скворе'чник *m* starling-house, bird house
скеле'т *m* skeleton; backbone *(of a macromolecule)*
висцера'льный ~ visceral skeleton
вне'шний ~ external skeleton
вну'тренний ~ internal skeleton
гидростати'ческий ~ hydrostatical skeleton
доба'вочный ~ appendicular skeleton
жа'берный ~ gill arch skeleton
осево'й ~ axial skeleton
хрящево'й ~ chondroskeleton
скелети'ровать *v* skeletonize (*the leaf*)
скеле'тный *a* skeletal
скелетоге'нный *a* skeletogenous
скирда' *f* rick, stack
скирдова'ние *n* stacking
скирдова'ть *v* rick, stack
скиса'ние *n* souring
скиса'ть *v* turn sour, go sour
скла'дка *f* fold, plication, plica, duplicature, crease, corrugation, ruga
~ **в моле'куле антите'ла** antibody fold
~ **Ке'ркринга** Kerckring's fold
~ **те'ла** body fold
амниоти'ческая ~ amniotic fold
бахро'мчатая ~ fimbriated fold
во'льфова ~ Wolffian ridge
гло'точно-надгорта'нная ~ pharyngoepiglottic fold
голосова'я ~ vocal fold
желу'дочная ~ gastric fold
ко'жная ~ skin fold
коста'льная ~ costal fold
кругова'я ~ circular fold, Kerckring's fold
носогубна'я ~ nasolabial fold
оперкуля'рная ~ opercular fold
па'ховая ~ inguinal fold
плавнико'вая ~ fin fold
полова'я ~ genital fold, *emb* genital furrow
прямокише'чно-ма'точная ~ rectouterine fold, Douglas' fold
прямокише'чно-пузы'рная ~ rectovesical fold
субкоста'льная ~ subcostal fold
тру'бно-гло'точная ~ salpingopharyngeal fold
тру'бно-нё'бная ~ salpingopalatine fold
хвостова'я ~ tail fold
черпалонадгорта'нная ~ aryepiglottic fold
язы'чно-надгорта'нная ~ glossoepiglottic fold
ягоди'чная ~ gluteal furrow, gluteal fold
скла'дчатость *f* rugosity
скла'дчатый *a* plicate, plaited, creasy, folded, rugous, rugose, rugate
скла'дывать *v* fold; pile, heap, stack
скле'ивание *n* agglutination, coagglutination
~ **эри'троцитов в моне'тные сто'лбики** nummulation
скле'ивать *v* agglutinate
скле'ивающий *a* agglutinant, agglutinative
скле'ра *f anat* sclera, sclerotica, scleral coat
склера'льный *a* sclerotic
склереи'да *f* sclereid
склере'нхима *f* sclerenchima
склери'на *f* sclerine
склери'т *m* sclerite
склеро- sclero-
склеробази'дия *f* sclerobasidium
склеробла'ст *m* scleroblast

склероге'нный *a* sclerogenic, sclerogenous
склероде'рма *f* scleroderm
склеродерма'льный *a* sclerodermatous
склеро'з *m* sclerosis
склеро'идный *a* scleroid
склерока'рпий *m* sclerocarp
склерокорнеа'льный *a* sclerocorneal
склеротиза'ция *f* sclerotization
склеротизи'рованный *a* sclerotized
склероти'ческий *a* sclerotic
склерото'м *m* sclerotome
склерофи'лл *m bot* sclerophyll
склерофилли'я *bot* sclerophylly
склерофи'т *m* sclerophyte
склеро'ций *m bot* clavus, sclerotium
скле'щивание *n* genital lock
склон *m* slope
  круто'й ~ steep slope
склянка *f* jar, vessel
  ~ **с зави'нчивающейся про'бкой** screw-cap jar
  дели'тельная ~ separating flask
  реакти'вная ~ reactive flask
  широкого'рлая ~ wide-mouthed jar, large-mouthed jar
ско'бка *f med* clip
  ~ **Мише'ля** Michel's clip
ско'лекс *m* scolex
сколио'з *m* scoliosis
сколопи'дий *m* scolopidium
скольже'ние *n* slip
скользи'ть *v* slither *(e.g., of snakes),* slip
ско'льзкий *a* slippery
скопа' *f* fish hawk, osprey (*Pandion haliaetus*)
скопле'ние *n* swarm, aggregation, agglomeration, collection
  бра'чное ~ mating aggregation
ско'пула *f* scopula
скорлупа' *f* shell, husk, hull, nutshell, eggshell
скороспе'лость *f* early maturation, early ripeness, earliness
скороспе'лый *a* early, early ripening
ско'рость *f* velocity, speed, rate
  ~ **адсо'рбции** adsorption rate
  ~ **адсо'рбции ви'русов** virus adsorption rate
  ~ **ассоциа'ции** association rate
  ~ **вдо'ха** inspiratory rate
  ~ **восстановле'ния** recovery rate
  ~ **вы'доха** expiratory rate
  ~ **диссоциа'ции** dissociation rate
  ~ **клубо'чковой фильтра'ции** *phys* glomerular filtration rate
  ~ **метаболи'ческого очище'ния** metabolic clearance rate
  ~ **оплодотворе'ния** rate of fertilization
  ~ **продуци'рования** rate of production, rate of of yield
  ~ **реа'кции** speed of response
  ~ **ро'ста** growth rate
  ~ **тече'ния** current speed
  ~ **эволю'ции** rate of evolution, evolutionary rate
  ~ **экспоненциа'льного ро'ста** intrinsic rate
  абсолю'тная ~ ро'ста absolute growth rate
  аддити'вная ~ ро'ста additive growth rate
  броско'вая ~ burst speed
  кре'йсерская ~ cruising speed
  уде'льная ~ ро'ста specific growth rate
скорпио'н *m* scorpion (*Scorpiones*)
скорпио'нница *f* scorpion fly *(Mecoptera)*
ско'сарь *m ent* scarred snout beetle *(Otiorhynchus)*
  люце'рновый ~ alfalfa snout beetle *(Otiorhynchus ligustici)*
скося'чивание *n* schooling
скося'чиваться *v* school
скот *m* cattle, stock, livestock
  выбрако'ванный ~ culled cattle
  дома'шний ~ live stock
  зако'рмленный ~ overfed cattle
  кру'пный рога'тый ~ cattle, bovine animals, heavy beasts
  ме'лкий рога'тый ~ small ruminants, sheep and goats, light beasts
  моло'чный ~ dairy cattle
  нагу'льный ~ fattening cattle
ско'тник *m* cattleman, herdsman
ско'тница *f* cattlewoman, herdswoman
скотово'д *m* cattle breeder
скотово'дство *n* cattle husbandry, cattle breeding, cattle raising, stock raising
  мясно'е ~ beef cattle breeding
скотопи'ческий *a* scotopic
скотопрого'н *m* catle road
скотота'ксис *m* skototaxis
ско'шенный *a* slant, skew, oblique; mown
скрадо'к *m* hide
скре'бень *m helm* thorny-headed worm *(Acanthocephala)*
скребни'ца *f* horse-comb
скрести'сь *v* scratch

скре'щивание *n* crossing, hybridization, mating, connubium
~ **генети'чески бли'зких ви'дов** congruent crossing
~ **инбре'дных ли'ний** inbred line crossing
~ **си'бсов** sibmating, sibs mating
аналити'ческое ~ test crossing
беспоря'дочное ~ pangamic mating
возвра'тное ~ back crossing
группово'е ~ group mating
двойно'е ~ double crossing
двухфа'кторное ~ two-factor crossing
диалле'льное ~ diallel crossing
инконгруэ'нтное ~ incongruent crossing
иску'сственное ~ artificial crossing
конгруэ'нтное ~ congruent crossing
коне'чное ~ final crossing
межвидово'е ~ interspecific crossing, interspecific hybridization
межлине'йное ~ interstrain crossing
межпоро'дное ~ breed crossing
межродово'е ~ bigeneric crossing
межсортово'е ~ intervarietal crossing, varietal hybridization
мно'жественное ~ multiple crossing
обра'тное ~ backcrossing
полиалле'льное ~ polyallel crossing
прове'рочное ~ check crossing
реципро'кное ~ reciprocal crossing
случа'йное ~ random mating
трансдукцио'нное ~ transductional crossing
трёхлине'йное ~ three-way crossing
трёхпоро'дное ~ three-way crossing
трёхфа'кторное ~ three-factor crossing
четырёхпоро'дное ~ four-way crossing
скре'щивать *v* cross, mate
скре'щиваться *v* mate
скри'нинг *m* screening study, screening
скрипи'ца *f bot* fleecy milk-cap (*Lactarius vellereus*)
скрученноволоси'стый *a* streptopilose
скрученноли'стный *a* streptophyllous
скрученнопло'дный *a* streptocarpous
скру'ченно-скла'дчатый *a* contortuplicate
скрученностебе'льчатый *a* streptocaulis
скру'ченный *a* convoluted, convolute, coiled, contorted
скру'чивание *n* twisting, curling
~ **верху'шки** curly top
~ **ли'стьев** leaf roll
скрыва'ться *v* skulk
скрыва'ющий *a* concealing
скры'тник *m ent* brown scavenger beetle, plaster beetle *(Lathridiidae)*
скры'тный *a* secretive
скрытоборо'здный *a* cryptocolpate
скрытопло'дный *a* cryptocarpous
скрытополя'рный *a bot* cryptopolar
скры'тый *a* concealed, cryptic, nonmanifest
скулёж, **скуле'ние** *n* whimper, whine
скули'ть *v* whine, yelp
скулово'й *a* malar, zygomatic, jugal
скульпти'на *f* sculptine
скунс *m* skunk (*Mephitis*)
скута'льный *a* scutal
скуте'ллюм *m* scutellum
ску'тум *m* scutum
ску'ченность *f* congestion
ску'ченный *a* dense, aggregate, congested
ску'чивание *n* huddling, huddling behavior
слаби'тельное *n* laxative, purgative
слаби'тельный *a* laxative, purgative
слабоветви'стый *a* subramose
слабогородча'тый *a* subcrenate
слабокисло'тный *a* subacid
слабора'звитый *a* immature (*of soil*); underdeveloped
сла'бый *a* weak, feeble
сла'вка *f orn* warbler, whitethroat (*Sylvia*)
сла'вка-завиру'шка *f* lesser whitethroat (*Sylvia curruca*)
сла'дкий *a* sweet, luscious
сладколи'стный *a* sweet-leaved
сладкокорнево'й *a* sweet-rooted
сладкопло'дный *a* sweet-fruited
след *m* track, trail, scent; trace; footprint; spoor; vestige; **идти по следу** be on the scent, follow the scent, follow the trail, follow the spoor, follow the tracks; **напасть на** ~ get on the scent, pick the scent;
потеря'ть ~ lose the scent
горя'чий ~ hot scent
ло'жный ~ wrong scent, false scent
ольфакто'рный ~ olfactory trail
све'жий ~ hot scent
сла'бый ~ cold scent
следи'ть *v* watch, track, follow
слежа'вшийся *a* caked (*e.g., fertilizers*)
слеже'ние *n* tracing, tracking
слё'живаемость *f* caking ability
слеза' *f* tear
слё'зный *a* lacrimal, lacrymal, lachrymal
слезови'дный *a* tearlike

слепе'нь *m ent* gadfly, horsefly, greenhead (*Tabanus*)
слепня'к *m ent* plant bug, mirid bug (Miridae)
слепо'й *a* blind, sightless; ~ **на оди'н глаз** blind in one eye
слепота' *f* blindness, loss of sight, ablepsia, ablepsy
слепушо'нка *f* mole-vole (*Ellobius*)
слепы'ш *m* mole rat (*Spalax*)
слёток *m* fledgling
сли'ва *f* plum (*Prunus*)
~ **колю'чая** sloe *(Prunus spinosa)*
~ **ренкло'д** Reine Claude (*Prunus domestica var. cereola*)
сли'вки *pl* cream
слизеви'к *m* slime mold (*Myxomycetes*)
сли'зень *m* slug (*Limacidae*)
слизеобразу'ющий *a* blennogenous
слизеподо'бный *a* mucoid, muculent
сли'зистый *a* mucous, mucilaginous, slimy
слизня'к *m* slug (*Limacidae*)
слизь *f* mucilage, slime, mucus, blenna
слик *m* slick (*of oil on water surface*)
слипа'емость *f* adhesiveness
слипа'ние *n* agglomeration, coalescence
~ **кле'ток** cell-cell adhesion (*in cell culture*)
~ **ма'трикса** *cyt* matrix stickiness
слипа'ться *v* coalesce
слитносте'левый *a bot* gamostelic
слия'ние *n* fusion, coalescence, union
~ **гаме'т** gametic syngamy
центри'ческое ~ centric fusion
слог *m* syllable
слоеви'ще *n* thallome, thallus, frond
слоеви'щный *a* thalline, thallodal, thallodic
сло'жность *f* complexity
~ **фитоцено'за** complexity of phytocenosis
сло'жный *a* compound, composite
слои'стый *a* lamellar, stratose, slaty
слой *m* layer, coat, stratum
~ **бу'ферных кле'ток** *imm* buffer-cell layer
~ **Ве'йля** basal layer of Weil, fibrous stratum of pulp
~ **Ге'ксли** Huxley's layer, Huxsley's sheath, Huxsley's membrane
~ **Ге'нле** Henle's layer, pale epithelial stratum
~ **па'лочек и ко'лбочек** *anat* columnar layer, bacillary layer
~ **Пуркинье'** Purkinje's layer
~ **шипови'дных кле'ток** prickle cell layer, spiny stratum
адгези'вный ~ adherent layer
бле'дный эпителиа'льный ~ pale epithelial stratum
бу'ферный ~ buffer-cell layer
вне'шний ~ enveloping layer
волокни'стый ~ fibrous layer
волокни'стый ~ **пульпы** basal layer of Weil, fibrous stratum of pulp
генерати'вный ~ generative layer
гиали'новый ~ hyaline layer
гонидиа'льный ~ gonidial layer, algal layer
заро'дышевый ~ germ layer, germinal layer
звукорассе'ивающий ~ sonic scattering layer
зерни'стый ~ Томеса granular layer of Tomes
ко'рковый ~ cortical layer
корнеобита'емый ~ root layer, root zone
кортика'льный ~ cortical layer
костено'сный ~ bone bed
краево'й ~ marginal layer
лубяно'й ~ bast layer
мальпи'гиев ~ malpighian layer
мозгово'й ~ medulla
окружа'ющий ~ enveloping layer
отдели'тельный ~ abscess layer, abscission layer, separation layer, abscission zone
пигме'нтный ~ *anat* pigmented layer
пограни'чный ~ boundary layer
подгимениа'льный ~ subhymenium
подстила'ющий ~ sublayer
покро'вный ~ covering layer
рефлекто'рный ~ *ent* reflector layer
рогово'й ~ *anat* horny layer, corneous layer
ростко'вый ~ **эпиде'рмиса** malpighian layer
скелетоге'нный ~ scleratogenous layer
сли'зистый ~ slime layer
сосо'чковый ~ papillary layer
спороно'сный ~ sporiferous layer
трофоге'нный ~ trophogenous layer
чувстви'тельный ~ *anat* sensory layer
слон *m* elephant (*Elephantidae*)
африка'нский ~ African elephant *(Loxodonta africana)*
инди'йский ~ Indian elephant, Asiatic elephant *(Elephas maximus)*
морско'й ~ sea elephant *(Mirounga)*

сло'ник *m* weevil, snout beetle (*Curculionidae*)
слони'ха *f* elephant cow
слух *m* hearing, audition
слухово'й *a* acoustic, acoustical, auditory
случа'йный *a* sporadic
случа'ть *v* cover, pair, service, serve
случа'ться *v* couple, mate, pair; happen, occur
слу'чка *f* coupling, pairing, mating; service, covering
~ **с вазэктоми'рованным самцо'м** sterile service
слу'щивание *n* desquamation
слюна' *f* saliva, spittle, spit, drivel
слю'нки *f pl:* **«куку'шкины ~»** *ent* cuckoo spit, frog spit
слюнно'й *a* salivary, sialine, sialic
слюнообразу'ющий *a* sialogenous
слюноотделе'ние *n* salivation, sialosis, sialism
слюнотече'ние salivation, salivary discharge
сма'чиваемость *f* wettability, water affinity
сма'чиваемый *a* wettable
сме'жный *a* contiguous
сме'на *f* change, rotation
~ **аспе'ктов** aspect change
~ **запечатле'ния** *eth* imprinting reversal
~ **расти'тельного покро'ва** vegetational change
~ **соо'бществ** succession
сезо'нная ~ seasonal change
сменя'емость *f* alternativeness
сменя'ть *v* rotate
смерте'льный *a* mortal, lethal, fatal
сме'ртность *f* mortality, mortality rate, rate of mortality, death rate, loss
~ **при расселе'нии** dispersion loss
зи'мняя ~ winter loss
птенцо'вая ~ nestling loss
сме'ртный *a* mortal
смерть *f* death
клини'ческая ~ clinical death
притво'рная ~ shamming death, thanatosis
смеси'тель *m* mixer
~ **с лопастно'й меша'лкой** paddle mixer
вибрацио'нный ~ vibration mixer
градие'нтный ~ gradient mixer
смесь *f* mixture
сме'шанно-жилкова'тый *a bot* mixed-nerved
сме'шанный *a* mixed
смеше'ние *n* mixing
сме'шивание *n* blending, compounding, mixing
сбаланси'рованное ~ *biom* balanced confounding
смеше'ние *n* blending
сме'шивать *v* mix, compound
сме'шиваться *v* mix, blend; mingle
смеща'ться *v* shift
смеще'ние *n* dislocation, displacement, slip, shift
смещё'нный *a* dislocated, displaced, shifted; *biom* biased
смиря'ть *v* subdue
смола' *f* resin, gum, pitch
~ **для зали'вки** embedding resin
зернё'ная ~ popcorn resin
ионнообме'нная ~ ion-exchange resin
катионообме'нная ~ cation-exchange resin
смолё'вка *f bot* catchfly, campion (*Silene*)
смоли'стый *a* resinous, bituminous
смолоно'сный *a* resiniferous, succiniferous
смоляно'й *a* resinous
~ **ход** resin duct
сморо'дина *f* currant (*Ribes*)
~ **кра'сная** red currant *(Ribes rubrum)*
~ **чё'рная** European black currant *(Ribes nigrum)*
сморчо'к *m myc* morel (*Morchella*)
смо'рщенный *a* corrugated
смо'рщиваться *v* crumple, wrinkle
смотре'ть *v* look
~ **в упор** stare
смо'ченный *a* wetted
сму'шка *f* lamb pelt
смыка'ние *n* closure
~ **крон** crown closure
смышле'нность *f* intelligence
снабжа'ть *v* supply
снабже'ние *n* supply
~ **пита'тельными вещества'ми** nutruent supply
снасть *f* gear, tackle
снеги'рь *m orn* bullfinch (*Pyrrhula pyrrhula*)
снегозадержа'ние *n* snow retention, retention of snow
снегоочи'стка *f* snow clearing
сниже'ние *n* drop, decrease, lowering
~ **сосу'дистого то'нуса** vasodepression
~ **чи'сленности** decrease in the number

снима'ть *v* take in (*the yield*); pick (*e.g., fruit*); skim off; remove, take down
~ **кожуру'** rind
~ **сли'вки** skim the cream
~ **урожа'й** harvest
~ **чешую'** scale
~ **шку'ру** skin
~ **я'блоки** pick apples
сноп *m* sheaf
~ **све'та** beam of light, cone of light
снотво'рное *n* soporific
снотво'рный *a* somniferous
сноше'ния *n pl* intercourse
половы'е ~ sexual intercourse
сну'лый *a* dead, lifeless
сныть *f* ashweed, goutweed (*Aegopodium podagraria*)
соба'ка *f* dog (*Canis familiaris*)
бродя'чая ~ stray dog
ездова'я ~ sledge dog, working dog, draught dog, husky, Siberian husky
енотови'дная ~ racoon dog (*Myctereutes procynoides*)
лега'вая ~ gundog
служе'бная ~ utility dog
сторожева'я ~ guard dog
охо'тничья ~ hunting dog
пасту'шья ~ herd dog, herding dog, herder, sheep dog
подруже'йная ~ gun dog
цепна'я ~ watch dog
собаково'д *m* dog breeder
собаководство *n* dog breeding
ездово'е ~ sledge-dog breeding
соба'чий *a* canine
соба'чка *f* little dog (*colloquial*)
лугова'я ~ prairie dog (*Cynomys*)
хоботковая ~ forest elephant shrew (*Rhynchocyon*)
собира'ть *v* gather, collect, pick, pluck, harvest, aggregate
~ **грибы'** pick mushrooms
собира'ться *v* congregate
соболева'ние *n* sable hunting
соболева'ть *v* hunt sable
соболево'дство *n* sable breeding
соболе'вщик *m* sable hunter
со'боль *m* sable (*Martes zibellina*)
со'бранный *a* aggregate
собы'тие *n biom* event
взаи'мно исключа'ющие собы'тия *pl* disjoint events, excluding events,mutually exclusive events
дополни'тельное ~ complementary event
достове'рное ~ certain event
равновозмо'жные собы'тия *pl* equally possible events
сло'жное ~ compound event
составно'е ~ compound event
факти'ческое ~ actual event
элемента'рное ~ simple event
сова' *f* owl (*Strigiformes*)
бе'лая ~ snowy owl *(Nyctea scandiaca)*
совёнок *m* owlet
со'вка *f* cutworm (*Noctuidae*)
подгрыза'ющая ~ cut worm (*Noctuidae*)
хло'пковая ~ bollworm (*Heliothis*)
совмести'мость *f* compatibility, match
абсолю'тная ~ complete match
непо'лная ~ partial match
по'лная ~ complete match
части'чная ~ partial match
совокупле'ние *n* coition, coitus, copulation
совокупля'ться *v* copulate
совоку'пность *f* aggregate; *biom* population
бесконе'чная ~ infinite population
генера'льная ~ parent population
дихотоми'ческая ~ dichotomous population
исхо'дная ~ parent population
коне'чная ~ finite population
непреры'вная ~ continuous population
норма'льная ~ normal population
совпаде'ние *n* concordance
согласо'ванность *f* concordance
согласо'ванный *a* concordant
согнутозу'бчатый *a* campylodentatum
согнутои'глый *a* campylocanthous
согнутокистево'й *a* campylobotriose
согнутопло'дный *a* campylocarpous
согнутосемя'нный *a* campylospermous
содержа'ние *n* maintenance, management, keeping; content; contents
~ **в заго'нах** penning
~ **в индивидуа'льных заго'нах** single penning
~ **жи'ра** fat content
безвы'гульное ~ close confinement
кле'точное ~ cage management, cage keeping
сто'йловое ~ confinement
содержа'ть *v* keep
содо'ку *f* sodoku, rat-bite disease, rat-bite fever

содру'жественный *a* concomitant;
содру'жественное косогла'зие concomitant strabismus, concomitant squint
соедине'ние *n* junction, connection, connecting, combination, commissure; conjugation; compound; joint, articulation
аромати'ческое ~ aroma compound
лету'чее ~ volatile compound
макроэрги'ческое ~ high-energy compound
ме'ченое ~ tagged compound
не'рвно-мы'шечное ~ neuromuscular junction
органи'ческое ~ organic compound
пло'тное ~ tight junction
промежу'точное ~ intermediate compound
синовиа'льное ~ synovial articulation
цикли'ческое ~ ring compound
шарни'рное swivel joint
щелеви'дное ~ gap junction
соединё'нный *a* connate, linked
соедини'тельный *a* connective, conjunctive
соединя'ть *v* joint, aggregate, compound, connect
соединя'ться *v* coalesce
сожи'тель *m* cohabitant
созна'ние *n* consciousness; sensorium;
потеря'ть ~ faint, swoon
созна'тельный *a* conscious
созрева'ние *n* ripening, maturation
~ **иммунокомпете'нтности** immune maturation
~ **фа'га** phage maturation
внутрити'мусное ~ *imm* intrathymic maturation
непо'лное ~ semimaturation
ра'ннее ~ early maturation
созрева'ть *v* mature, ripen, mellow
созрева'ющий *a* ripening
созре'вший *a* ripe, mature
со'йка *f orn* jay (*Garrulus*)
сок *m* sap, juice, succus
берё'зовый ~ birch sap
вакуоля'рный ~ vacuolar sap
желу'дочный ~ gastric juice
кише'чный ~ intestinal juice
кле'точный ~ cell sap
мле'чный ~ latex; chyle, milky sap
пищевари'тельный digestive juice
тка'невый ~ tissue juice
я'дерный ~ nuclear sap
сокодвиже'ние *n* circulation of sap
со'кол *m* hawk, falcon (*Falco*)
ка'рликовый ~ falconet *(Microhierax)*
сократи'мость *f* contractility
сократи'мый *a* contractile
сократи'тельный *a* contractile
сократи'тельная вакуо'ль contractile vacuole
сокраща'ться *v* contract
сокраща'ющийся *a* contractile, contractive
сокраще'ние *n* contraction
ано'дно-замыка'тельное ~ *phys* anodal closure contraction, anodic closure contraction
ано'дно-размыка'тельное ~ *phys* anodal opening contraction, anodic opening contraction
ауксотони'ческое ~ auxotonic contraction
серде'чное ~ cardiac contraction, heart contraction
като'дно-замыка'тельное ~ *phys* cathodal closure contraction
като'дно-размыка'тельное ~ *phys* cathodal opening contraction
замыка'тельное ~ *phys* closing contraction
идиомото'рное ~ *phys* idiomuscular contraction
изометри'ческое ~ isometric contraction
изотони'ческое ~ isotonic contraction
серде'чное ~ heartbeat, cardiac beat
тетани'ческое ~ *phys* tetanic contraction
тони'ческое ~ tonic contraction
солани'н *m* solanine
солда'т *m ent* soldier
солево'й *a* saline
солевыно'сливость *f* salt tolerance
солевыно'сливый *a* salt-tolerant
солеме'трия *f* salinometry
солено'сный *a* saliferous
солё'ность *f* salinity
соленоци'т *m* solenocyte
солё'ный *a* salty, salinous
солеро'с *m bot* glasswort (*Salicornia*)
солесодержа'щий *a* saliferous
солеусто'йчивость *f* salinity resistance
со'лнечный *a* solar, sunny
солнцелюби'вый *a* sun-loving
солове'й *m* nightingale (*Luscinia*)
со'лод *m* malt
соло'дка *f bot* licorice, liquorice (*Glycyrrhiza*)
соло'ма *f* straw
соло'менный *a* straw
соло'мина *f* haulm, culm, straw

соло'минка *f a* small haulm, culm, straw
солони'на *f* brined meat
солонова'тый *a* saltish; brakish
солонча'к *m* saline
соль *f* salt
сольвата'ция *f* solvation
солюбилиза'ция *f* solubilization
соля'нка *f bot* saltwort (*Salsola*)
со'ма *f* soma
сомати'ческий *a* somatic
сомато- somato-
соматобла'ст *m* somatoblast
соматобластоме'р *m* somatoblastomere
соматога'мия *f* somatogamy
соматоде'рма *f* somatoderm
соматоло'гия *f* somatology
соматомаммотропи'н *m* somatomammotrophin
~ **хорио'на** chorionic somatomammotrophin
соматопла'зма *f* somatoplasm
соматопле'вра *f* somatopleure
соматосенсо'рный *a* somatosensory
соматотропи'зм *m* somatotropism
соматотропи'н *m* somatotropin
хориони'ческий ~ chorionic somatotropin
соматотро'пный *a* somatotropic
соматофи'т *m* somatophyte
соместези'я *f* somesthesia
соми'т *m* somite, somatome, merome, merosome
со'мкнутость *f* density
~ **крон** closing of leaf canopy, crown contact, crown density
~ **лесно'го по'лога** density of canopy
~ **травяно'го покро'ва** density of sward
со'мкнутый *a* closed
сомэстези'я *f* somesthesia
сон *m* sleep
~ **ли'стьев** leaf slumber
~ **цветко'в** flower slumber
бы'стрый ~ desynchronized sleep, fast sleep, rapid eye movement sleep, REM sleep
гипноти'ческий ~ hypnotic sleep, hypnosis
медленноволно'вый ~ slow-wave sleep
парадокса'льный ~ paradoxical sleep
сонли'вость *f* sleepiness, drowse
со'нный *a* sleepy; carotic
соногра'мма *f eth* sonogram
соно'метр *m* sonometer
со'ня *f ther* dormouse
боялы'чная ~ desert dormouse *(Selevinia betpakdalensis)*
лесна'я ~ forest dormouse *(Dryomys)*
садо'вая ~ garden dormouse, golden dormouse *(Eliomys quercinus)*
со'ня-полчо'к *f* edible dormouse, fat dormouse *(Glis glis)*
соо'бщество *n* community, assemblage, association
~ **боло'та** sterrhium
~ **вечнозелё'ных расте'ний** evergreen community
~ **живо'тных** animal community
~ **конкури'рующих органи'змов** competitive association
~ **культу'рных расте'ний** culture plant community
~ **пустыре'й** chersic community
взаи'мно дополня'ющее ~ complementary association
во'дное ~ aquatic community
втори'чное ~ secondary community
дернообразу'ющее ~ lawn community
заверша'ющее ~ terminal community
заключа'ющее ~ terminal community
за'мкнутое ~ closed community
инициа'льное ~ initial community
кли'максовое ~ climax community
многоя'русное ~ pluristratal community
мозаи'чное расти'тельное ~ mosaic community, pattern community
монодомина'нтное расти'тельное ~ monodominant community
несо'мкнутое расти'тельное ~ open community
несто'йкое ~ fragile community
неусто'йчивое ~ seral community
обобщё'нное ~ abstract community
одноя'русное ~ one-layered community, unistratal community
откры'тое ~ sporadophytium
откры'тое расти'тельное ~ open community
перви'чное ~ primary community,virgin community
перехо'дное ~ transitional community
полидомина'нтное расти'тельное ~ polydominant community
полукульту'рное расти'тельное ~ half culture community
расти'тельное ~ plant community

региона'льное ~ regional community
сапрофи'тное ~ saprium
сери'йное ~ serial community
со'рное расти'тельное weed community
теорети'ческое ~ abstract community
усто'йчивое ~ permanent community, continuously stable community
экстрарегиона'льное ~ extraregional community
сооруже'ние *n* construction
сооруже'ния *n pl of* **сооруже'ние**
очистны'е ~ sewage treatment works
водоочистны'е ~ water treatment works
соо'сный *a* coaxial
соотве'тствие *n* accordance, conformity, compliance, correspondence, concordance
соотноше'ние *n* ratio, rate, relation, relationship
~ **возрастны'х групп** age ratio
~ **вы'год и затра'т** *ecol* benefit/cost ratio
~ **ме'жду ве'сом и длино'й** weight-length relationship
~ **поло'в** sex ratio
иммунорегулято'рное ~ immunoregulatory balance
ме'нделевское ~ Mendelian rate
сопе'ние *n* blowing, snuffing, sniffing
сопе'рник *m* rival, adversary
сопе'рничать *v* rival
сопе'рничество *n* rivalry, contest
сопе'ть *v* sniff
сопли'вый *a* snotty
сопло'дие *n* collective fruit, multiple fruit, infructescence
сопля' *f* nose drift, snot
соподчинё'нный *a* superordinate
соприкаса'ющийся *a* contiguous
сопровожда'ющий *a* accompanying; concomitant
сопротивле'ние *n* resistance
~ **диффу'зии** diffusion resistance
~ **дыха'нию** phys respiratory resistance
~ **сдви'гу** shear strength
~ **сосу'дов** vascular resistance
~ **среды'** environmental resistance
~ **транспира'ции** transpiration resistance
~ **у'стьичное** ~ stomatal resistance
перифери'ческое ~ peripheric resistance
эласти'ческое ~ elastic resistance
сопротивля'емость *f* resistance
~ **органи'зма** body resistance
сопряже'ние *n* coupling
сопряже'нность *f* association
~ **при'знаков** contingency
сопу'тствующий *a* accompanying, concomitant, associated, concurrent
соразме'рный *a* commensurable, commensurate
сорби'т *m* sorbitol
со'рбция *f* sorption
сорбо'за *f* sorbose
сорва'ться *v* break loose, break away
~ **с це'пи** break loose (*of a dog*)
со'рго *n* sorghum, broom corn, Kafir corn, Guinea corn (*Sorghum vulgare*)
сорня'к *m* weed
заво'зный ~, зано'сный ~ carried weed
зиму'ющий ~ wintering weed
зло'стный ~ ill weed, persistent weed
каранти'нный ~ quarantine weed
корневи'щный ~ rhizomatous weed, rootstock weed
пожни'вный ~ stubble weed
рудера'льный ~ ruderal weed
сопу'тствующий ~ associating weed
соро'ка *f* magpie, pie, piemag (*Pica*)
сорокарп *m myc* sorocarp
сороконо'жка *f* centipide
сорокопу'т *m* shrike, butcher-bird (*Lanius*)
сорт *m* variety, cultivar, breed
декорати'вный ~ ornamental variety
засухоусто'йчивый ~ drought-resistant variety
зимосто'йкий ~ winter-hardy variety
контро'льный ~ check variety
ме'стный ~ indigenous variety, local variety
мни'мый ~ dummy variety
морозоусто'йчивый ~ frost-resistant variety
ози'мый ~ winter variety
отцо'вский ~ paternal variety
позднеспе'лый ~ full-season variety, late ripening variety
раннеспе'лый ~ early ripening variety, early-season variety
роди'тельский ~ parental variety
селекцио'нный ~ selection variety
скороспе'лый ~ early ripening variety, early-season variety
среднеспе'лый ~ mid-ripening variety
староме'стный ~ old-local variety
столо'вый ~ table variety
урожа'йный ~ productive variety

ярово'й ~ spring variety
со'рт-кло'н *m* clonal variety
со'рт-ли'ния *m* line variety
со'рт-опыли'тель *m* pollinating variety
со'ртер *m* sorter
кле'точный ~ cell sorter
хромосо'мный ~ chromosome sorter
со'ртинг *m* sorting
кле'точный ~ cell sorting
хромосо'мный ~ chromosome sorting
сортирова'ть *v* sort
сортиро'вка sorting
~ **кле'ток** cell sorting
~ **хромосо'м** chromosome sorting
прото'чная ~ flow sorting *(of cells)*
сортиро'вщик *m* sorter
~ **кле'ток** cell sorter
сортово'й *a* varietal
со'рус *m* sorus, fruit dot
соса'льщик *m* fluke (*Trematoda*)
дигенети'ческий ~ fluke, digenetic fluke (*Digenea*)
кровяно'й ~ blood fluke *(Schistosomatidae)*
печё'ночный ~ common liver fluke *(Fasciola hepatica)*
соса'ние *n* sucking, suction
соса'тельный *a* haustellate
соса'ть *v* suck, suckle
сосе'дний *a* contiguous
соска'бливать *v* scrape off
соско'б *m* scraping
соско'вый *a* mammillary
соскреба'ть *v* scrape off, scrape away, rasp off, rasp away
сосна' *f* pine (*Pinus*)
~ **ке'дровая европе'йская** cembra pine (*Pinus cembra*)
~ **ке'дровая сиби'рская** cedar pine, Siberian cedar (*Pinus sibirica*)
~ **обыкновенная** common pine, Scotch pine (*Pinus silvestris*)
сосно'вый *a* piny; ~ **лес** pine forest
сосня'к *m* pine forest
сосня'к-беломо'шник *m* lichen pine forest
сосо'к *m* nipple, papilla, teat, mamilla, mammilla
~ **зри'тельного не'рва** optic papilla
сосо'чек *m* nipple, papilla, caruncle
~ **во'лоса** hair papilla
~ **ры'льца** *bot* papilla of stigma
грибови'дный ~ fungiform papilla
желобова'тый ~ circumvallate papilla
кони'ческий ~ conic papilla
листови'дный ~ foliate papilla
нитеви'дный ~ filiform papilla
осяза'тельный ~ tactile papilla
подъязы'чный ~ sublingual caruncle
чувстви'тельный ~ sense papilla, sensory papilla
сосочкови'дный *a* papilliform, papillate
соста'в *m* composition
~ **насажде'ния** stand structure
~ **соо'бщества** community composition
~ **те'ла** body composition
видово'й ~ species composition
возрастно'й ~ age composition age-class composition
полово'й ~ sex composition
хими'ческий ~ chemical composition
флористи'ческий ~ floristic composition
составно'й *a* compound
состоя'ние *n* state, condition, status
~ **иммунокомпете'нтности** immune state
~ **носи'тельства** carrier state
~ **поко'я** dormant state
~ **половозре'лости** sexually mature state
~ **равнове'сия** equilibrium state
бессозна'тельное ~ unconsciousness
возбуждё'нное ~ excited state
гиперкинети'ческое ~ hyperkinetic state
депресси'вное ~ depression state
динами'ческое ~ dynamic state
иммунореакти'вности immune state
исхо'дное ~ initial state, original state
клони'ческое ~ clonicity
крити'ческое ~ critical state
нати'вное ~ native state
о'бморочное ~ syncopal state, syncopic state, faintness
о'бщее ~ general condition, systemic condition
пальмеллеви'дное ~ palmella condition
патологи'ческое ~ morbid condition
предо'бморочное ~ precollaptoid state, precollapse
равнове'сное ~ equilibrium state
сбаланси'рованное ~ balanced state
свобо'дное ~ free state
со'нное ~ sleepy state, drowsy state
тяже'лое ~ grave condition, poor condition
усто'йчивое ~ stable state, steady state
физиологи'ческое ~ physiological state

центра'льное ~ возбужде'ния central excitatory state
центра'льное мотивацио'нное ~ central motive state
эндотокси'н-толера'нтное ~ endotoxin-tolerant state
сосу'д *m* vessel, jar
~ **для культиви'рования** culture vessel
~ **для окра'шивания** staining dish
~ **для о'пытов** experimental dish
~ **Дьюа'ра** Dewar flask, vacuum flask
вегетацио'нный ~ pot
втори'чный ~ *bot* secondary vessel
градуи'рованный ~ calibrated vessel, graduated vessel
демонстрацио'нный ~ display jar
желто'чный ~ vitelline vessel
ко'жный ~ cutaneous vessel
ко'льчато-спира'льный ~ *bot* spiro-annular vessel
ко'льчатый ~ *bot* annular vessel
корона'рный ~ coronary vessel
кровено'сный ~ blood vessel
ксиле'мы wood vessel
ле'стничный ~ *bot* scalariform vessel, ladder-shaped vessel
лимфати'ческий ~ lymphatic vessel
мальпи'гиев ~ Malpighian tubule, Malpighian tube, Malpighian vessel
ме'рный ~ calibrated vessel, graduated vessel
мле'чный ~ chyliferous vessel; *bot* latex vessel, milk vessel
перифери'ческий ~ peripheric vessel
по'ристый ~ *bot* pitted vessel
пита'тельный ~ nutrient vessel
прободa'ющий кровено'сный ~ perforating blood vessel, Ebner's vessel
се'тчатый ~ *bot* netted vessel, reticulate vessel
спира'льный ~ spiral vessel
сосу'дисто-волокни'стый *a* fibrovascular
сосу'дистость *f* vascularity
сосу'дистый *a anat* vasal, vascular
сосудодви'гательный *a* vasomotor, vasculomotor
сосудообразу'ющий *a* vasifactive, vasoformative
сосудосу'живающий *a* vasohypertonic, vasoexcitatory, vasopressor, vasoconstrictive
сосуно'к *m* sucker, suckling
сосуществова'ние *n* coexistence
сосуществова'ть *v* coexist, symbiose
сосу'щий *a* suctorial
со'товый *a* faveolate, favose
сотрясе'ние *m med* concussion
~ **головно'го мо'зга** concussion of the brain, brain concussion
со'ты *pl apic* honeycomb, comb, bee's cells
~ **с распло'дом** brood comb
~ **запеча'танные ~** capped comb
~ **тру'тневые ~** drone comb
соха'тый *m* elk (*Alces alces*)
сохране'ние *n* conservation; preservation
соцве'тие *n* inflorescence
ботри'ческое ~ indefinite inflorescence, indeterminate inflorescence, racemose inflorescence
верхоцве'тное ~ inflorescence, definite inflorescence, determinate inflorescence
голо'вчатое ~ cephalanthium
неопределё'нное ~ indefinite inflorescence, indeterminate inflorescence, racemose inflorescence
определё'нное ~ cymose inflorescence, definite inflorescence, determinate inflorescence
рацемо'зное соцветие indefinite inflorescence, indeterminate inflorescence, racemose inflorescence
сло'жное ~ compound inflorescence
цимо'зное ~ cymose inflorescence, definite inflorescence, determinate inflorescence
соцве'тие-султа'н *n* tassel-inflorescence
социабилите'т *m* sociability
~ **расте'ний** sociability of plants
социа'ция *f* sociation
социогра'мма *f* sociogram
социу'ла *f* sociule
со'ция *f* socies
сочета'ние *n* combination
сочи'ться *v* bleed (*of trees*); seep, trickle, exude
сочлене'ние *n* articulation, coarticulation, synarthrosis, joint, junction
атлантоаксиа'льное ~ atlantoaxial joint, atloidoaxoid joint
атлантозаты'лочное ~ atlantooccipital joint
крестцо'во-подвздо'шное ~ sacroiliac joint
ма'точно-тру'бное ~ uterotubal junction
рё'берно-груди'нное ~ costo-sternal joint

шо'вное ~ sutural joint
экзо'н-интро'нное сочленение exon-intron junction
сочленё'нный *a* articulated, jointed
сочлено'вный *a* arthrodial, arthrous
сочленя'ть *v* joint
со'чность *f* succulence, succulency
со'чный *a* juicy, sappy, sapful, luscious, succulent
сошни'к *m* vomer, powshare bone
сошнико'вый *a* vomerine
со'я *f* soya, soya bean, soybean (*Glycine*)
спад *m* decrease, fall, drop; abatement (*of* heat, water); recession (*of flood*), decline (*of flood*)
спада'ться *v* collapse
спаде'ние *n* collapse
спазм *m* spasm, cramp
спазмолити'ческий *a* spasmolytic
спайк *m* spike
уси'ленный ~ overshot spike
спа'йка *f* adhesion, commissure
спайнолепе'стный *a* sympetalous, gamopetalous
спайноли'стный *a* gamophyllous
спайнопло'дный *a* gamocarpous
спайносто'лбиковый *a* gamostylous
спайнотычи'нковый *a* gamostaminate
спание'ль *m* spaniel
спа'ренный *a* biconjugate
спа'ржа *f bot* asparagus (*Asparagus*)
спа'ривание *n* coupling, pairing, mating, connubium; tread (*of birds*)
~ **основа'ний** *mbiol* base pairing
~ **подо'бных** assor(ta)tive mating
ассортати'вное ~ assortative mating, assortive mating
беспоря'дочное ~ pangamic mating
близкоро'дственное ~ consanguinous mating
группово'е ~ flock mating
дизассортати'вное ~ disassortative mating
межпоро'дное ~ interbreed mating
неро'дственное ~ outbred mating
случа'йное ~ random mating
хугсти'новское ~ *mbiol* Hoogsteen pairing
чистопоро'дное ~ purebred mating
спа'ривать *v* pair
спа'риваться *v* pair, couple, mate
спасе'ние n rescue
~ **ви'руса** rescue of virus
~ **генети'ческого маркё'ра** marker rescue
спасти'ческий *a* spasmodic
спат *m* spat
спать *v* sleep
спе'йсер *m* spacer
спектр *m* spectrum
~ **де'йствия** action spectrum
~ **жи'зненных форм** biological spectrum
~ **отраже'ния** reflection spectrum
~ **поглоще'ния** absorption spectrum
~ **пропуска'ния** transmission spectrum
~ **рассе'яния** scattering spectrum
~ **флуоресце'нции** fluorescence spectrum
биологи'ческий ~ biological spectrum
дифференциа'льный ~ поглоще'ния difference absorption spectrum
колеба'тельный ~ vibrational spectrum
непреры'вный ~ continuous spectrum
палинологи'ческий ~ palynological spectrum, pollen spectrum
пыльцево'й ~ palynological spectrum, pollen spectrum
ра'зностный ~ difference spectrum
резона'нсный ~ комбинаци'онного рассе'яния resonance Raman spectrum
спо'рово-пыльцево'й ~ sporo-pollen spectrum
терми'ческий пертурбацио'нный ~ thermal perturbation spectrum
ультрафиоле'товый ~ поглоще'ния ultraviolet absorption spectrum
спектрогра'мма *f* spectrogram
звукова'я ~ sound spectrogram
спектро'граф *m* spectrograph
спектрогра'фия *f* spectrography
спектро'метр *m* spectrometer
инфракра'сный ~ спектрометр infrared spectrometer
спектрометри'я *f* spectrometry
~ **комбинацио'нного рассе'ивания** collisional activation spectrometry
релаксацио'нная ~ relaxation spectrometry
спектроско'п *m* spectroscope
спектроскопи'я *f* spectroscopy
а'томно-абсорбцио'нная ~ atomic absorption spectroscopy
беспла'менная а'томно-абсорбцио'нная ~ flameless atomic absorption spectroscopy
высокотемперату'рная ~ high-temperature spectroscopy
инфракра'сная ~ infrared spectroscopy
ла'зерная ~ laser spectroscopy

ла'зерная рама'новская ~ laser Raman spectroscopy
ла'зерного рассе'яния laser light-scattering spectroscopy
опти'ческих бие'ний light beating spectroscopy
резона'нсная ~ комбинацио'нного рассе'яния resonance Raman spectroscopy
рентгеноэлектро'нная ~ X-ray photoelectron spectroscopy
температу'рная пертурбацио'нная ~ thermal perturbation spectroscopy
фотоэлектро'нная ~ photoelectron spectroscopy
фотоэлектро'нная рентге'новская ~ X-ray photoelectron spectroscopy
электро'нная ~ electron spectroscopy
я'дерная магни'тно-резона'нсная ~ nuclear magnetic resonance spectroscopy, NMR spectroscopy

спектрофото'метр *m* spectrophotometer
прото'чный ~ continuous spectrophotometer
двухлучево'й ~ double-beam spectrophotometer
двухволно'вый ~ dual-wavelength spectrophotometer
регистри'рующий ~ recording spectrophotometer
скани'рующий ~ scanning spectrophotometer

спектрофотометри'я *f* spectrophotometry
и'мпульсная ~ flash spectrophotometry

спелеоло'гия *f* speleology

спе'лость *f* maturity, ripeness
воскова'я ~ wax ripeness, yellow ripeness
мё'ртвая ~ dead-ripe stage
моло'чная ~ milky ripeness
моло'чно-воскова'я ~ milky-wax ripeness
по'лная ~ complete ripeness, dead ripeness
техни'ческая ~ industrial ripeness

спе'лый *a* ripe, mature

спе'рма *f* sperm

спермаго'ний *m* spermogonium

спермата'нгий *m* spermatangium

спермате'ка *f* spermatheca, spermatotheca

спермати'да *f* spermatid, spermatoblast, spermoblast

сперматогене'з *m* spermatogenesis, spermateliosis

сперматоге'нный *a* spermatogenetic, spermatogenous

сперматогониа'льный *a* spermatogonial

сперматого'ний *m* spermatogonium

сперматозо'ид *m* spermatozoid, spermatozoon, spermium, sperm

сперматопла'ст *m* spermatoplast

сперматофо'р *m zool* spermatophore

сперматоци'т *m* spermatocyte
~ **I поря'дка** primary spermatocyte
~ **II поря'дка** secondary spermatocyte

спермаце'т *m* spermaceti

сперма'ций *m* spermatium

сперми'зм *m* spermism

спе'рмий *m* spermatozoid, spermatozoon, spermium, sperm
жгу'тиковый ~ flagellate spermatozoid
нежгу'тиковый ~ nonflagellate spermatozoid

спермиогене'з *m* spermiogenesis, spermateliosis

спермоге'нный *a* spermogenous

спермого'ний *m* spermagone, spermagonium

спермоде'рма *f* spermoderm

спермока'рпий *m* spermocarp

спермообра'зный *a* spermatoid

специализа'ция *f* specialization

специализи'рованный *a* specialized

специфи'чность *f* specificity
у'зкая ~ stringent specificity

спе'рма *f* sperm, semen

спермата'нгий *m* spermatangium

спермате'ка *f* spermatheca, spermatotheca

спермати'да *f* spermatid

сперматогене'з *m* spermatogenesis

сперматоге'нный *a* spermatogenic, spermatogenous

сперматого'ний *m* spermatogonium, spermogonium

сперматозо'ид *m* sperm

сперматозо'иды *m pl* sperm
акти'вные ~ *pl* active sperm
подви'жные ~ *pl* active sperm

сперматопла'ст *m* spermatoplast

сперматофо'р *m* spermatophore

сперматоци'т *m* spermatocyte

спермаце'т *m* spermaceti

сперма'ций *m* spermatium

сперми'зм *m* spermism

спе'рмий *m* spermium

сперми'н *m* spermin

спермогене'з *m* spermiogenesis

спермоге'нный *a* spermogenous

спермого'ний *m* spermogonium, spermagone

спермока'рпий *m* spermocarp
специализа'ция *f* specialization
биологи'ческая ~ biological specialization
специализи'рованный *a* specialized
специфика'тор *m* specifier
специфи'ческий *a* specific
специфи'чность *f* specificity
~ **антиге'нной детермина'нты** epitope specificity
~ **сти'мула** stimulus specificity
мутаге'нная ~ mutagene specificity
простра'нственная ~ sterical specificity
субстра'тная substrate specificity
у'зкая ~ stringent specificity
широ'кая ~ broad specificity
СПИД, **синдро'м** (*m*) **приобрете'нного иммун'ного дефици'та** AIDS, acquired immune deficiency syndrome
спи'кула *f* spicule, spicula, spiculum *(of sponges)*
многолучева'я сфери'ческая ~ spheraster
паре'нхимная ~ flesh spicule
спил *m* saw cut
спина' *f* back; **выгиба'ть спи'ну** arch its back (*of a cat*)
во'гнутая ~ *vet* dipped back, hollow back, low back, saddle back
впа'лая ~ *vet* sunken back
кру'глая ~ sway back
о'страя ~ vet acute back, arched back
пло'ская ~ flat back
прови'слая ~ vet dipped back, hollow back, low back, saddle back
седлова'тая ~ vet dipped back, hollow back, low back, saddle back
суту'лая ~ stooping back, round-shouldered back
спина'льный *a* spinal
спи'нка *f* back
~ **языка'** back of the tongue
спинно-брюшно'й *a* back-front
спинно'й *a* dorsal
спинномозгово'й *a* medullispinal, spinal, cerebrospinal
спиралеви'дный *a* helocoid, spiraliform, sporoid, strombuliform
спирализа'ция *f* helix formation
спира'льно-до'льный *a* spirolobous
спиральнопло'дный *a* spirocarpous
спира'ль *f* spiral, spire, coil, helix
~ **второ'го поря'дка** supercoil
~ **ДНК** DNA helix
~ **Ку'ршмана** Curschmann's spiral
вну'тренняя ~ internal coil
двойна'я ~ double helix
молекуля'рная ~ molecular spiral
непо'лная ~ *mbiol* imperfect helix
несоверше'нная ~ *mbiol* imperfect helix
оста'точная ~ remaining coil
паранемати'ческая ~ paranematic coil
плектонеми'ческая ~ braided coil
реляцио'нная ~ relative coil
станда'ртная ~ stereotypic coil
спира'льный *a* spiral, helical, coiled, volute
~ **кла'пан** *anat* spiral valve
спире'ма *f* spirema
спири'лла *f* spirillum
спирогра'мма *f* spirogram
спиро'граф *m* spirograph
спирогра'фия *f* spirography
спиро'метр *m* spirometer
спирометри'я *f* spirometry
спирохе'та *f* spirochaete, spirochete
спирт *m* alcohol, spirit
абсолю'тный ~ dehydrated alcohol
денатури'рованный ~ denaturated alcohol, denatured alcohol
мети'ловый ~ methyl alcohol, pyroligenous alcohol, wood alcohol
одноа'томный ~ monoatomic alcohol, monohydric alcohol
поливини'ловый ~ polyvinyl alcohol
эти'ловый ~ ethyl alcohol
спиртово'й *a* spirit, spirituous
спи'сок *m* list
~ **ви'дов** species list
фаунисти'ческий ~ faunal list
сплав *m* floating (*of timber*); alloy
сплави'на *f* quacking bog, floating bog, floating mat, quagmire
спла'йсинг *m* splicing
спайсингосо'ма *f* spliceosome
сплайси'ровать *v* splice
спланхнокра'ниум *m* splanchnocranium
спланхноло'гия *f* splanchnology
спланхнопле'вра *f* splanchnopeure
спланхното'м *m* splanchnotom
спланхноце'ль *f* splanchnocoel
спленоци'т *m* splenocyte
сплете'ние *n* plexus; tangle
~ **Рема'ка** Remak's plexus
база'льное ~ basilar sinus
лёгочное ~ pulmonary plexus

межмы'шечное ~ intermuscular plexus, myenteric plexus, Auerbach's plexus
не'рвное ~ nerve plexus
не'рвное ~ Ауэрба'ха intermuscular plexus, Auerbach's plexus, myenteric plexus
плечево'е ~ brachial plexus
подсли'зистое ~ submucous plexus, Meissner's plexus
поясни'чное ~ lumbar plexus
серде'чное ~ cardiac plexus
со'лнечное ~ celiac plexus, solar plexus
хоро'идное ~ choroid plexus

сплошно'й *a* solid
сплю'шка *f orn* scops owl (*Otus scops*)
сплю'щенный *a* flattened, oblate
сплю'щивание *n* flattening
сплю'щивать *n* flatten
спонги'н *m* spongin
спонгиобла'ст *m* spongioblast
спонгио'зный *a* spongiose, spongious
спонгиопла'зма *f* spongioplasm
спонгиоци'т *m* spongiocyte
спондиле'з *m* spondylosis
спондили'т *m* spondylitis
спонта'нный *a* spontaneous
спо'ра *f* spore
втори'чная ~ secondary spore
головнё'вая ~ brand spore
зи'мняя ~ winter spore
ле'тняя ~ summer spore
ма'ленькая ~ sporule, sporidiole, sporidiolum
неподви'жная ~ statospore
нитеви'дная ~ scolecospore
обосо'бленная secondary spore
поко'ящаяся resting spore
червеобра'зная ~ scolecospore

споради'ческий *a* sporadic
спора'нгиевый *a* sporangial
спорангиено'сец *m* sporangiophore
спора'нгий *m* sporangium, sporange, spore case, spore sac
многогнё'здный ~ plurilocular sporangium

спорангиока'рпий *m* sporangiocarp
спорангио'ль *f* sporangiole, sporangiolum
спорангиоспо'ра *f* sporangiospore
спориде'сма *f* sporidesm
споридиено'сный *a* sporidiferous
спори'дия *f* sporidium
споробла'ст *m* sporoblast
спо'ровый *a* cryptogamic, cryptogamous
спорогене'з *m* sporogenesis, spore formation
спороге'нный *a* sporogenous
спорого'ний *m* sporogonium
спорогони'я *f* sporogony
спороде'рма *f* sporoderm
спородо'хий *m* sporodochium
спороду'кт *m* sporoduct
спорозо'ит *m* sporozoite
спорока'рпий *m* sporocarp
спороли'стик *m* sporophyll
спороно'сный *a* sporiferous
споро'нт *m* sporont
спорообра'зный *a* sporoid
спорообразова'ние *n* sporification, sporogenesis, spore formation
спорообразу'ющий *a* sporogenous, spore-forming
спoропла'зма *f* sporoplasm
спoропо'чка *f* sporophydium
спороса'к *m* sporosac
спорота'ллиум *m* sporothallium
спороубива'ющий *a* sporocidal
спорофи'дий *m* sporophydium
спорофи'лл *m* sporophyll
спорофи'т *m* sporophyte, synkaryophyte, syncaryophyte
молодо'й ~ sporeling

спорофо'р *m* sporophore
спороци'ста *f* sporocyst
спороци'т *m* sporocyte
спорт *m* sport
споруля'ция *f* sporulation
спорынья' *f* ergot, spur (*Claviceps*)
спо'рыш *m* knotgrass, knotweed, pinkweed, doorweed (*Polygonum aviculare*)
спо'соб *m* method, mode
~ **примене'ния** method of application
~ **разложе'ния** mode of decomposition
~ **Таро'цци** Tarozzi's method
~ **упако'вки ДНК** mode of DNA packing

спосо'бность *f* ability, capacity, power
~ **к адапта'ции** adaptability
~ **к дифференциро'вке** differentiation potency
~ **к разви'тию** developmental potency
~ **к размноже'нию** reproductive power, productive capacity
~ **к ро'сту** growth power, growing power, growth potential
~ **к сохране'нию о'пыта** *eth* retentivity
~ **к формирова'нию представле'ний** *eth* ideational capacity

~ **предви'деть результа'т де'йствий** *eth* means-end capacity
абсорбцио'нная ~ absorbing power, absorbing capacity
азотсвя'зывающая ~ nitrogen-fixing capacity
азотфикси'рующая ~ nitrogen-fixing capacity
влагоуде'рживающая ~ moisture-retaining power
водопоглоща'ющая ~ water-absorbing capacity
восстана'вливающая ~ reducing power
воспроизводи'тельная ~ breeding capacity, reproductive capacity, begetting power
впи'тывающая ~ blotting capacity
вса'сывающая ~ blotting capacity
газообразу'ющая ~ gassing ability, gassing activity (*e.g., of yeast*)
диастати'ческая ~ diastatic power
дискриминацио'нная ~ *biom* discriminatory power
диффузио'нная ~ diffusion capacity
дыха'тельная ~ breathing capacity
железосвя'зывающая ~ sideropexy power
замеща'ющая ~ displacing power
защи'тная ~ protective potency
кислородсвя'зывающая ~ oxygen-binding capacity
комбинацио'нная ~ combining ability
комплементфикси'рующая ~ complement-fixing ability
конкуре'нтная ~ competitive capacity
максима'льная дыха'тельная ~ maximal breathing capacity
обме'нная ~ exchange capacity
оплодотворя'ющая ~ fertilizing capacity
оплодотворя'ющая ~ спе'рмы semen fertilizing capacity
оста'точная железосвя'зывающая ~ unsaturated iron-binding capacity
отража'тельная ~ по'чвы soil albedo
поглоща'ющая ~ absorbing power
по'рослевая ~ rejuvenation capacity
производи'тельная ~ producing capacity
производи'тельная ~ земли' land capability
проника'ющая ~ корне'й root penetration capacity, capability of root penetration
регенерацио'нная ~ regenerative ability
разреша'ющая ~ resolving power, resolution, defining power
растворя'ющая ~ dissolving capacity, solvent capacity
регенерацио'нная ~ regenerative capacity
свя'зывающая ~ coupling capacity
теплотво'рная caloric power, calorific capacity
транспирацио'нная ~ transpiration capacity
тРНК-акце'пторная ~ tRNA acceptor capacity
уде'рживающая ~ retentivity
у'мственные спосо'бности *pl* mental capacity
усва'ивающая ~ uptaking capacity *(of plants)*
функциона'льная ~ functional capacity
целлюлозоразлага'ющая ~ cellulose decomposing capacity

спу'гивать, спугну'ть *v* frighten away, frighten off
спуск *m* discharge; draining; slope
~ **кро'ви** bleeding
спуска'ть *v* lower; cast off, slip (*dogs*)
~ **кровь** bleed
~ **пруд** drain a pond
~ **соба'ку с це'пи** unchain a dog
спу'танный *a* matted (*e.g., grass*), tangled
спу'тник *m* satellite, companion cell
спя'чка *f* dormancy
зи'мняя ~ hibernation
иску'сственная зи'мняя ~ artificial hibernation
спя'щий *a* dormant, latent; *phys* -sleeping
сравни'мый *a* comparable
сравни'тельный *a* comparative
сраста'ние *n* coalescence, concretion
сраста'ться *v* coalesce
сраще'ние *n* symphysis; adnation, coalescence, concrescence; adhesion
~ **косте'й** synosteosis, synostosis
~ **пальцев** syndactyly
среда' *f* environment, milieu, surroundings; medium
~ **Ги'сса** Hiss medium
~ **для зали'вки** embedding medium, mounting medium
~ **для зама'чивания** steeping medium
~ **для накле'ивания** affixative medium
~ **Дорсе'** Dorset's medium
~ **Дьердонне'** Dierdonne's medium

~ **Ле'ффлера** Loeffler's medium
~ **Ме'йтланда** Meitland medium
~ **на по'чвенной вы'тяжке** soil-water medium
~ **Ногу'чи** Noguchi's medium
~ **Ча'пека** Czapek's medium
абиоти'ческая ~ abiotic environment
ага'ровая ~ agar medium
аэро'бная ~ oxic environment
безбелко'вая ~ protein-free medium
бескислоро'дная ~ anoxic environment
бескле'точная ~ acellular medium, cell-free medium
бессы'вороточная ~ serum-free medium, serum-starved medium
биоти'ческая ~ biotic environment
вне'шняя ~ external environment, external medium
вну'тренняя генети'ческая internal genetic environment
вытесня'ющая ~ displacement medium
ге'левая ~ gel medium
генети'ческая ~ genetic background
генотипи'ческая ~ genic environment, genotypic millieu
двухфа'зная ~ biphasic medium
диализи'рованная ~ dialyzed medium
дифференциа'льная ~ differential medium
есте'ственная ~ natural medium
желати'новая ~ gelatin medium
жё'лчная ~ bile medium
жи'дкая ~ fluid medium, liquid medium
замеща'ющая ~ replacement medium
избира'тельная ~ selective medium
инкубацио'нная ~ incubating medium
иску'сственная ~ artificial medium; artificial environment
карто'фельная ~ potato medium
кондициони'рованная ~ conditioned medium
косми'ческая ~ space environment
кровяна'я ~ blood medium
культура'льная ~ culture medium
минима'льная ~ basal medium
недоста'точная ~ deficient medium
неполноце'нная ~ deficient medium
обогати'тельная ~ enrichment medium
обогащё'нная ~ high-growth enhancement medium
окружа'ющая ~ environment, ambient environment, surrounding medium
пепто'нная ~ peptone medium
пита'тельная ~ nutrient medium, nutritional medium, growth medium
пло'тная ~ solid medium
подде'рживающая ~ maintenance medium, supporting medium, cell maintenance medium
по'лная ~ complete medium
полусинтети'ческая ~ semisynthetic medium
полутвё'рдая ~ semisolid medium
просветля'ющая ~ clearing medium
пропи'тывающая ~ infiltrating medium
рабо'чая ~ handling medium
регули'руемая ~ controlled medium
сбра'живаемая ~ fermentation medium, fermentation liquor
ско'шенная сы'вороточная ~ serum slope
случа'йно-неоднородная ~ randomly-layered medium
стабилизи'рующая ~ supporting medium. maintenance medium
суха'я ~ dehydrated medium
сы'вороточная ~ serum-supplemented medium
физи'ческая ~ physical environment
экологи'чески чи'стая ~ salubrious environment
электи'вная ~ selective medium
средиземномо'рский *a* Mediterranean
среди'нный *a* median
среднегру'дь *f ent* mesothorax, medithorax
сре'дней величины' *a* medium-sized
сре'дний *a* middle, medium; average, mean
сре'днее *n* mean, average
~ **арифмети'ческое** arithmetical mean
~ **взве'шенное** weighted mean
геометри'ческое ~ geometrical mean
и'стинное ~ true mean
квадрати'чное ~ quadratic mean
невзве'шенное ~ unweighted mean
о'бщее ~ overall mean, grand average
произво'льно при'нятое ~ assumed mean
скользя'щее ~ moving average, consecutive mean
скорректи'рованное ~ adjusted mean, corrected mean
экстрема'льное ~ extreme mean
средосте'ние *f* mediastinum
сре'дства *n pl of* **сре'дство**
~ **коммуника'ции** *eth* means of communication

сре'дство *n* agent
адреноблоки'рующее ~ adrenoreceptor blocking agent
анестези'рующее ~ anesthetic
антисепти'ческое ~ antiseptic agent
бактерици'дное ~ bactericide, antibacterial agent
вя'жущее ~ astringent
дезактивацио'нное ~ decontamination agent
дезинсекцио'нное ~ disinsectant
дезинфици'рующее ~ disinfectant, nti-infective agent
дисперги'рующее ~ dispersion agent
жаропонижа'ющее ~ antipyretic, febrifuge
кровоостана'вливающее ~ hemostatic, antihemorrhagic
лека'рственное ~ medicinal agent
микостати'ческое ~ antimycotic agent
мо'ющее ~ detergent
отха'ркивающее ~ expectorant
противовоспали'тельное ~ anti-inflammatory agent
противогрибко'вое ~ antimycotic agent
рво'тное ~ emetic
слаби'тельное ~ purgative, cathartic, laxative, aperient
снотво'рное ~ soporific, somnifacient
сосудорасширя'ющее ~ vasodilator
сосудосу'живающее ~ vasoconstrictor, vasoconstrictor agent
спороубива'ющее ~ sporicide
транквилизи'рующее ~ tranquilizing agent
хлори'рующее ~ chlorinating agent
срез *m* section; cut, slice
~, **сде'ланный от руки'** free-hand section
заморо'женный ~ frozen section
лиофилизи'рованный ~ freeze-dried section
макроскопи'ческий ~ gross section
опти'ческий ~ optical section
парафи'новый ~ paraffin section
попере'чный ~ cross section, transverse section
продо'льный ~ longitudinal section
сагитта'льный ~ sagittal section
свободнопла'вающий ~ free-floating section
сери'йный ~ serial section
то'лстый ~ thick section
ультрато'нкий ~ ultrathin section
фронта'льный ~ frontal section
целлоиди'новый ~ celloidin section
среза'ть *v* crop; cut off
сродство' *n* affinity
генетическое ~ genetic affinity
действенное ~ effective affinity
дифференциальное ~ differential affinity
избирательное ~ elective affinity
терминальное ~ terminal affinity
химическое ~ chemical affinity
срок *m* term, time, period; deadline
~ **го'дности** useful life
~ **слу'жбы** life
~ **хране'ния** shelf life
сростноголо'вчатый *a bot* syncephalic
сростнокорешко'вый *a bot* synorrhizous
сростнолепе'стный *a bot* gamopetalous
сростноли'стный *a bot* gamophyllous
сростнопло'дный *a bot* symphycarpous
сростнопы'льниковый *a bot* synantherous, symphyantherous, syngenesious
сростноцветко'вый *a bot* synanthous
сро'сток *m* concretion
сро'сшийся *a* connate, adnate, concrescent
срыва'ть *v* pluck
сры'гивание *n* regurgitation
сса'дина *f* scratch, excoriation, chap
стабилиза'ция *f* stabilization
стабилизи'ровать *v* stabilize
стаби'льность *f* stability
стаби'льный *a* stable
метаболи'чески ~ metabolically stable
ста'вить *v* put, set, place; apply (*e.g., a compress*); set up (*an experiment*)
~ **ба'нки** cup, apply cupping glasses
~ **диа'гноз** diagnose
~ **капка'н** trap, lay a trap
~ **на дрожжа'х** leaven
~ **термо'метр** take temperature
ста'дия *f* phase, stage
~ **вегетати'вного разви'тия** vegetative stage
~ **возбужде'ния** excitement stage
~ **выздоровле'ния** convalescent stage
~ **вы'хода в тру'бку** booting stage, shooting stage, stooling stage
~ **деле'ния** division stage
~ **инкуба'ции** incubative stage
~ **конъюга'ции** coupling phase
~ **ку'колки** pupal stage
~ **плодоноше'ния** fruiting stage

~ **поко'я** resting stage, quiescent stage, dormant stage
~ **прораста'ния** germination stage
~ **разви'тия** developmental stage
~ **размноже'ния** multiplicative stage
~ **ро'ста** growth stage
базидиа'льная ~ *myc* basidial stage
вегетати'вная ~ vegetative stage, vegetative phase
возрастна'я ~ *zool* instar
гонидиа'льная ~ gonidial stage
заро'дышевая ~ embryonal stage
коне'чная ~ terminal stage
перехо'дная ~ transition phase
после'дняя ~ terminal stage
светова'я ~ photostage
су'мчатая ~ ascigerous stage
эмбриона'льная ~ embryonal stage
ста'дный *a* gregarious
ста'до *n* herd, flock, drove, troop, stock, pack, run, school
~ **павиа'нов** troop of baboons
~ **слоно'в** herd of elephants
за'мкнутое ~ closed herd
легко' распада'ющееся ~ open herd
ма'точное ~ brood stock (*of fish*)
несто'йкое ~ open herd
племенно'е ~ pedigree stock, pedigree flock, breed herd, breeding herd, stud
чистопоро'дное ~ purebred herd, purebred flock
эли'тное ~ elite herd
ста'йка *f* pod (*of birds*)
ста'йность *f* aggregating behavior, aggregative behavior, sociability
ста'йный *a* gregarious, schooling
стаз *m* stasis
стазигене'з *m* stasigenesis
стака'н *m* glass
хими'ческий ~ beaker
стамино'дий *m* staminode, staminodium
станда'рт *m* standard
стандартиза'ция *f* standardization
станда'ртный *a* standard
станови'ться *v* become
~ **на дыбы'** rear, prance
ста'нция *f* station
о'пытная ~ experimental station
очистна'я ~ purification plant
санита'рно-эпидемиологи'ческая ~ sanitary epidemiological station
селекцио'нная ~ breeding station
старе'ние *n* aging, ageing, senescence
старе'ть *v* age, grow old
ста'рик *m orn* murrelet (*Synthliboramphus*)
обыкнове'нный ~ ancient murrelet *(Synthliboramphus antiquus)*
хохла'тый ~ Japanese murrelet *(Synthliboramphus wumizusume)*
стари'к *m* old man
ста'рость *f* senility, old age
ста'рый *a* aged, old
ста'рческий *a* senile, gerontal
ста'ти *f pl* points
стати'стика *f* statistics
биологи'ческая ~ biostatistics
вы'борочная ~ sampling statistics
демографи'ческая ~ population statistics
многоме'рная ~ multivariate statistics
непараметри'ческая ~ noparametric statistics
стати'ческий *a* static
статмокине'з *m* stathmokinesis
статобла'ст *m* statoblast
статоли'т *m* statolith
статоци'ст *m* statocyst
статоци'т *m* statocyte
ста'тус *m* status
репродукти'вный ~ reproductive status
стату'т *m* status
~ **ви'да** status of species
~ **в соо'бществе** social status
стаурофи'лл *m* staurophyll
стациона'рный *a* static, stationary, steady-state
ста'ция *f* station
ста'я *f* run; school (*of fish*); pack (*e.g., of wolves*); band, swarm, covey
~ **птиц** flock of birds
~ **рыб** school of fish, shoal of fish
~ **волко'в** pack of wolves
ствол *m* trunk, stem, bole, shaft, stock
~ **ао'рты** aortic trunk
~ **блужда'ющего не'рва** vagal trunk
артериа'льный ~ arterial trunk
брюшно'й не'рвный ~ ventral nerve trunk
не'рвный ~ nerve stem, nerve trunk
пищево'дно-бронхиа'льный ~ broncho-esophageal trunk
продо'льный не'рвный ~ longitudinal nerve trunk
симпати'ческий ~ sympathetic trunk
спинно'й не'рвный ~ dorsal nerve trunk
стволо'вость *f* stemness

стволово'й *a:* **стволова'я кле'тка** stem cell
створа'живание *n* curding
створа'живать, створа'живаться *v* curdle
ство'рка *f* valve; cusp
~ **кла'пана се'рдца** cusp
~ **со швом** raphe valve *(in diatoms)*
бесшо'вная ~ rapheless valve *(in diatoms)*
втори'чная ~ secondary valve
створо'женный *a* curdy
ство'рчатый *a* valval, valvular, valvate
стебелёк *m* stalk, stem, pedicel, pedicle; *ent* podeon
~ **алланто'иса** allantoic stalk
~ **сомита** somite stalk
глазно'й ~ eye stalk, optic stalk
кристалли'ческий ~ *zool* crystalline style
перви'чный ~ **заро'дыша** embryo stem
сте'бель stem, halm, haulm, stalk, caulis
безли'стный ~ scape
гла'вный ~ *bot* main stalk
дигидроуриди'ловый ~ *mbiol* D-stem
ла'зящий ~ climbing stem
лежа'чий ~ decumbent stem, lodged stem
лианови'дный ~ climbing stem
листоподо'бный ~ leaflike stem
неопушё'нный ~ bald stem
ни'тчатый ~ filamentous stem
подзе'мный ~ underground stem, subterranean stem
ползу'чий ~ creeping stem
по'лый ~ hollow stem, tubular stem
поча'тковый ~ ear stem
пригну'тый к земле' ~ decumbent stem, lodged stem
приподнима'ющийся ~ ascending stem
сте'лющийся ~ creeping stem, procumbent stem, prostrate stem
суккуле'нтный ~ succulent stem
цепля'ющийся ~ climbing stem
стебелько'вый *a* scapose
стебельчатобрю'хий *a* petiolate
стебельчатогла'зый *a* stalk-eyed
стебе'льчатый *a* petiolated, pedunculate, stalked
стеблева'ние *n* shooting, stooling stage
стеблеви'дный *a* cauliform, stipiform
стеблево'й *a* cauline
стебleко'рень *m* stem-root
стеблекорнево'й *a* caulirrhizous
стеблеобра'зный *a* cauliform
стеблеобъе'млющий *a* stem-clasping, amplexicaul
стега'льный *a* stegocrotaphic
сте'кинг *m mbiol* stacking
сте'кинг-самоассоциа'ция *f* stacking self-association
стекло' *n* glass
органи'ческое ~ plexiglas
покро'вное ~ cover glass, cover slip
предме'тное ~ slide, glass slide, microscope slide
предме'тное ~ **с лу'нкой** cavity slide, depression slide
часово'е ~ watch glass
стекля'нница *f ent* clearwing moth, clearwinged moth (*Aegeriidae*)
сте'ла *f bot* stele
стели'ться *v* crawl, creep, trail
сте'лющийся *a* trailing, tracing, creeping, decumbent, procumbent, prostrate, repent, spreading
стель *f bot* stele
сте'льная *a* in calf, pregnant, springing (*of a cow*)
сте'мма *f* stemma
сте'нка *f* wall
~ **заро'дышевого мешка'** *bot* embryo sac wall
~ **сосу'да** vascular wall
~ **те'ла** body wall
брюшна'я ~ abdominal wall
кле'точная ~ cell wall
мы'шечная ~ muscular wall
стенкополо'жный *a* parietal
стено- steno-
стеноба'рный *a* stenobaric
стеноба'тный a stenobathic
стенобио'нтный *a* stenoecic
стеногали'нный *a* stenohaline
стеногидри'ческий *a* stenohydric
стено'з *m* coarctation
стено'йкный *a* stenoecic
стенокарди'я *f* stenocardia
стенонота'льный *a ent* stenonotal
стеноолиготе'рмный *a* cold-stenothermic
стеноспермока'рпия *f* stenospermocarpy
стеноте'рмный *a* stenothermal, stenothermic
стеното'пный *a* stenotopic
стенотро'пный *a* stenotropic
стенофа'гный а stenophagous
стенофо'тный *a* stenophotic
стенохо'р *m* stenochor
стенохо'рный *a* stenochoric
сте'пень *f* degree, extent; power

~ **вы'раженности** degree of manifestation
~ **набуха'ния** degree of swelling
~ **насыще'ния** saturation rate
~ **ози'мости** degree of winter character
~ **оплодотворе'ния** rate of fertilization, conception rate
~ **пло'тности** degree of density
~ **покры'тия** cover degree
~ **приближе'ния** degree of approximation
~ **разбавле'ния** dilution rate
~ **свобо'ды** degree of freedom
~ **со'мкнутости** *plsoc* degree of closeness, degree of density
~ **специфи'чности** degree of specificity
~ **то'чности** accuracy rate, degree of accuracy

степь *f* steppe
высокого'рная ~ high-mountain steppe
го'рная ~ mountain steppe
дернови'нно-зла'ковая ~ bunchgrass steppe
ковы'льная ~ feather-grass steppe, stipa steppe
колючекуста'рниковая ~ thorn steppe
куста'рниковая ~ shrub steppe, shrubby steppe
лугова'я ~ meadow steppe
опусты'ненная ~ desert steppe
полупусты'нная ~ semodesert steppe
пусты'нная ~ desert steppe
разнотра'вная ~ forb steppe, grassland steppe
разнотра'вно-дернови'нно-зла'ковая ~ herb-bunchgrass steppe
са'зовая ~ saz steppe

стервя'тник *m* scavenger vulture (*Neophron percnopterus*)
бу'рый ~ hooded vulture *(Neophron monachus)*

стереобла'стула *f* stereoblastula
стереоизоме'р *m* stereoisomer
стереокине'з *m* stereokinesis
стерео'м *m* stereome
стереопла'зма *f* stereoplasm
стереоскани'рование *n* multiple view scanning
стереоспецифи'чность *f* stereospecificity
стереоспонди'льный *a* stereospondylous
стереота'ксис *m* stereotaxis
стереоти'п *m* stereotype
~ **по'иска добы'чи** prey scheme
стереотропи'зм *m* stereotropism
стереотипи'я *f* stereotypy
дви'гательная motor stereotypy
ритми'ческая rhythmical stereotypy
стереоци'лия *f* stereocilia
сте'ржень *m* quill (*of a feather*), shaft, scape, stem
~ **пера'** shaft of feather, stem of feather
~ **во'лоса** hair shaft, stem of feather
коле'нчатый ~ *bot* articulated rachis
стери'гма *f* sterigma
стерилиза'тор *m* sterilizer
бараба'нный ~ rotary drum sterilizer
парово'й ~ steam sterilizer
стерилиза'ция *f* sterilization; castration
~ **в пла'мени** flame sterilization
~ **като'дными луча'ми** cathode-ray sterilization
~ **кипяче'нием** boiling sterilization
~ **па'ром** steam sterilization, steaming
~ **сухи'м жа'ром** dry-hot sterilization
~ **фильтрацией** filter sterilization
дробная ~ discontinuous sterilization, fractional sterilization, intermittent sterilization
стерилизова'ть *v* sterilize
~ **па'ром** steam
стери'льность *f* sterility, apogeny
~ **при расщепле'нии** segregational sterility
~ **пыльцы'** pollen sterility
гамети'ческая ~ gametic sterility, haplontic sterility
генерацио'нная ~ generational sterility
ге'нная ~ genic sterility
гибри'дная ~ hybrid sterility
же'нская ~ apogyny, female sterility
зиготическая ~ diplontic sterility
мужская ~ male sterility
хромосомная ~ chromosomal sterility
стери'льный *a* sterile
стери'н *m* sterol
стерна'льный *a* sternal
стерне'ллум *m* sternellum
стерни'т *m ent* sternite
сте'рнум *m ent* sternum
стерня' *f* stubble
стеро'ид *m* steroid
анаболический ~ anabolic steriod
конъюгированный ~ conjugated steroid
нейтральный ~ neutral steriod
свободный ~ free steroid
связанный ~ conjugated steroid
стеро'идный *a* steroid

стероидогене'з *m* steroidogenesis
стеро'л *m* sterol
сте'ррула *f* sterrula
стерх *m orn* white crane (*Grus leucogeranus*)
стетоско'п *m* stethoscope
сти'гма *f ent* stigma
стиле'т *m* stylet
стилогони'дия *f* stilogonidium
стилокони'ческий *a* styloconic
стилопо'дий *m* stylopodium
стилоспо'ра *f* stylospore
стилоспо'ровый *a* stylosporous
сти'лтон *m* Stilton cheese, Stilton
сти'мул *m* stimulus, stimulant
~ **взаи'много обще'ния** social stimulus
~, **вызыва'ющий пе'ние** *orn* sing stimulus
безусло'вный ~ unconditioned stimulus
вы'званный ~ provoked stimulus
индуци'рующий ~ inductive stimuls
мотиви'рующий ~ motivating stimuls, priming stimulus
поро'говый ~ threshold stimulus, liminal stimulus
сверхнорма'льный supernormal stimulus
усло'вный ~ conditioned stimulus
стимули'рование *n* stimulation
вне'шнее external stumulation
вну'треннее internal stimulation
стимули'ровать *v* stimulate
стимули'рующий *a* stimulant, stimulatory
сти'мул-мише'нь *m eth* target stimulus
стимуля'тор *m* stimulant, stimulator, stimulating agent
~ **ро'ста** growth factor
биоге'нный ~ biogenic stimulator
и'мпульсный ~ pulse stumulator
полупроводнико'вый ~ transistorized stimulator
электро'нный ~ electronic stimulator
стимуля'ция *f* stimulation
стихи'дий *m bot* stichidium, stichid
стихобази'дия *f* stichobasidium
стла'ник *m* elfin wood, elfin woodland, elfin wood formation
стог *m* stack, rick, mow
~ **се'на** hayrick, haystack
стогова'ние *n* stacking
стогова'ть *v* stack, mow
сто'йка *f* set, point (*of a dog*); upright posture *(of ground squirrels)*; rack, stand; **де'лать стойку** point, come to a point, make a point, set, standard
вертика'льная ~ upright posture
вертика'льная защи'тная ~ upright defensive posture
вертика'льная ~ нападе'ния upright offensive posture
сто'йкий *a* hardy, persistent, nonperishable
сто'йкость *f* endurance, resistance, hardiness
сток *m* flow, discharge; runoff; drain, gutter
~ **с поле'й** agricultural discharge
бытово'й ~ natural stream flow, natural flow
пове'рхностный ~ surface flow
подпо'чвенный ~ subsoil flow
столб *m* column, shaft
~ **Берте'на** Bertin’s column, renal column
~ **Кла'рка** Clarke's column
~ **сво'да** column of fornix
ана'льный ~ anal column, rectal column, Morgagni’s column
грудно'й ~ Clarke's column
заднепрохо'дный ~ anal column, rectal column, Morgagni’s column
за'дний ~ dorsicolumn
позвоно'чный ~ backbone, spinal column, dorsal spine, spine
по'чечный ~ renal column, Bertin’s column
сто'лбик *m* stile, style, columella; *pal* spine
сто'лбиковый *a* columellar
столбня'к *m* tetanus
сто'лбур *m phytop* big bud
сто'лбчатый *a* columnar, paxillate
сто'лик *m* stage (*of a microscope*)
~ **для зали'вки** casting table
нагрева'тельный ~ warm stage
неподви'жный ~ fixed stage
предме'тный ~ stage, objective table
столкнове'ние *n eth* encounter
столо'н *m* stolon, offshoot, creeping stem, procumbent stem, prostrate stem
стомати'т *m* stomatitis
стоматоло'гия *f* stomatology
стоматомико'з *m* stomatomycosis
стомоде'ум *m* stomodaeum, stomodeum
стон *m* moan
стона'ть *v* moan
стоп-отве'т *m* stop-response
стопоходя'щий *a* plantigrade
сторожево'й *a* guard
~ **пёс** watchdog
сторона' *f* side

наве'тренная ~ windward side, weather side
подве'тренная ~ lee side
стоя'ние *n* standing
свобо'дное ~ stand-at-ease
стоя'ть *v* stand; standard (*of a dog*); **дом стои'т на берегу' реки'** the house is situatedon the bank of the river; **стои'т хоро'шая пого'да** the weather keeps fine; **стоят си'льные моро'зы** there are severe frosts; **весь ме'сяц стояла те'плая пого'да** the warm weather lasted all month
~ **ды'бом** stand on end (*of hair*)
~ **на тя'ге** hunt woodcocks at mating season
стоя'чий *a* standing; static, lentic, stagnant (*of water*)
стоя'щий *a* standing
страте'гия *f* strategy
~ **кормодобыва'ния** foraging strategy
~ **расте'ний** plant strategy
альтернати'вная ~ alternative strategy
альтернати'вная репродукти'вная ~ alternative mating strategy
кормодобыва'тельная ~ foraging strategy
репродукти'вная ~ mating strategy, reproductive strategy
фуражиро'вочная ~ foraging strategy
эволюцио'нно-стаби'льная ~ evolutionary stable strategy
стратигра'фия *f* stratigraphy
стратифика'ция *f* stratification
температу'рная ~, **терми'ческая** ~ thermal stratification, thermal lamination
стратифици'ровать *v* stratify
стра'ус *m* ostrich (*Struthio*)
страх *m* fear
стрека'тельный *a* stinging, urticant
стрекоза' *f* dragonfly (*Anisoptera*); damselfly (*Zygoptera*)
равнокры'лая ~ damselfly (*Zygoptera*)
разнокры'лая ~ dragonfly (*Anisoptera*)
стрекота'ние *n* chirm, chirr, chirrup, stridulation
стрекота'ть *v* chirm, chirr, chirp, whirr, stridulate
стре'лка *f* flower stalk (*e.g., of onions*), spire, scape, spear; frog (*of a hoof*); **вы'ход в стре'лку** bolting
~ **копы'та** frog
стрелкова'ние *n* bolting
стрелкова'ться *v* bolt
стрелкови'дный *a* scapiform, scapoid
стрелконо'сный *a* scape-bearing
стрелкоцве'тный *a* scapiflorous
стрелови'дный *a* sagittate, hastate, halberd-shaped
стрелоли'ст *m* arrowhead (*Sagittaria*)
стре'мечко *n* stirrup, stapes
стремле'ние *n* tendency
~ **к ста'дности** flocking tendency
стремни'на *f* riffle
стрено'живать *v* hobble, spancel
стре'пет *m* little bustard (*Otis tetrax*)
стрепсине'ма *f* strepsinema
стрепсите'на *f* strepsitene
стресс *m* stress
антропоге'нный ~ anthropogenic stress
экологи'ческий ~ environmental stress
эмоциона'льный ~ mental stress
стре'ссор *m phys* stressor
стриа'рный *a* striatal
стридуляцио'нный *a* stridulating
стридуля'ция *f* stridulation
стриж *m* swift (*Apodidae*)
стри'жка *f* shearing
стрик *m phytop* streak
~ **карто'феля** leaf-drop streak
тепли'чный ~ glasshouse streak
стрикту'ра *f* stricture
стричь *v* shear (*e.g., sheep*)
строби'ла *f* strobile, strobila
стробилоце'рк *m* strobilocercus
стробиля'ция *f* strobilation
стро'гий *a* strict
строе'ние *n* structure, constitution
кле'точное ~ cellular texture
стро'ма *f* stroma
строчо'к *m* brain mushroom (*Gyromitra esculenta*)
структу'ра *f* structure, constitution, texture
~ **внима'ния** *eth* attention structure
~ **«кле'верного листа'»** *mbiol* cloverleaf structure
~ **популя'ции** structure of population
~ **ти'па «сте'бель-петля'»** stem-and-loop structure
~ **фитоцено'за** structure of phytocenosis
«~ шпи'льки» hairpin-like structure
антиге'нная ~ antigenic structure
возрастна'я ~ age structure
волокни'стая ~ fibrous structure
втори'чная ~ secondary structure
кле'точная ~ cellular structure
компа'ктная ~ tight structure

конформацио'нная ~ conformational structure
межкле'точная ~ intercellular structure
многодо'менная ~ multidomain structure
нати'вная ~ native structure
объё'мная ~ space structure, three-dimentional structure
перви'чная ~ primary structure
петледо'менная ~ loop-domain structure
пласти'нчатая ~ sheet structure
полная ~ complete structure
пространственная ~ space structure, three-dimensional structure
рыхлая ~ loose sructure
сотовая ~ honeycomb structure
специфически изогнутая ~, **специфически свёрнутая** ~ specifically fiolded structure
спиральная ~ helical structure
субмикроскопическая ~ submicroscopic structure
третичная ~ tertiary structure, three-dimensional structure, space structure
трёхмерная ~ three-dimensional structure, space structure
упоря'доченная ~ *mbiol* assembly
четверти'чная ~ quaternary structure
чё'ткая ~ well-defined structure
шпилькообра'зная ~ hairpin-like structure
яче'истая ~ honeycomb structure
структури'рованность *f* structuredness
стру'ма *f* struma
~ **Ри'деля** Riedel's disease
струна' *f* string
спинна'я ~ chorda, cord; *emb* primitive backbone
слухова'я ~ auditory string
сухожи'льная ~ tendinous cord
струнови'дный *a* stringlike
струноли'стный *a* string-leaved
струне'ц *m helm* ascaris (*Ascaris*)
струп *m* scab, crust
стручкови'дный *a* silicular, siliquiform
стручкообра'зный *a* silicular, siliquiform
стручо'к *m bot* silique, pod
александри'йской ~ cassia pod (*Cassia fistula*)
нерастре'скивающийся ~ indehiscent pod
растре'скивающийся ~ dehiscent pod
царегра'дский ~ carob (*Ceratonia siliqua*), carob bean
стручо'чек *m bot* silicle
струя' *f* jet, stream, flow, spurt, sprout
бобро'вая ~ castoreum
непреры'вная ~ *mbiol* continuous flow
остано'вленная ~ *mbiol* stopped flow
сту'день *m* jelly
ва'ртонов ~ gelatin of Wharton, Wharton's jelly, Wharton's gel
стук *m* beating
~ **в грудь** *eth* chest beating
~ **ла'пами** *eth* drumming with paws, stumping
бараба'нный ~ *eth* drumming
стул *m* stool, feces, excrements; chair
ступня' *f* foot
сту'пор *m* stupor
стык *m* junction
экзо'н-интро'нный ~ exon-intron junction
стя'гивать *v* nstrict
стягивающий *a* constrictive
суб- sub-
субальпи'йский *a* subalpine
субапика'льный *a* subapical
субарахноида'льный *a* subarachnoidal
субаркти'ческий *a* subarctic
субассоциа'ция *f* subassociation
субвита'льный *a* subvital
субге'н *m* subgene
субгиме'ниум subhymenium
субгуми'дный *a* subhumid
субдерма'льный *a* subdermal
субдо'мен *m mbiol* subdomain
субдомина'нт *m* subdominant
субдура'льный *a* subdural
субери'н *m* suberin
субзона'льный *a* subzonal
субима'го *n* subimago
субинфлуэ'нт *m* subinfluent
субита'нный *a* subitan, subitaneous
субклеточный *a* subcellular
субко'ста *f* subcosta
субкло'н *m* subclone
субклони'рование *n* subcloning
субкортика'льный *a prot* subcortical
субко'ста *f ent* subcosta
субкути'кула *f* subcuticle, subcuticula
субку'тис *m* subcutis
сублета'льный *a* sublethal
сублима'ция *f* sublimation
сублими'ровать *v* sublimate
субме'нтум *m ent* submentum
субметацентри'ческий *a* submetacentric
субневра'льный *a* subneural

субни'мфа *f* subnymph
субордина'ция *f* subordination
су'борь *f* subor (*forest on poor soils*)
 сло'жная ~ suramen
субплевра'льный *a* subpleural
субполя'рный *a* subpolar
субпопуля'ция *f* subpopulation; subset
 кле'точная ~ cell subset
субсе'рия *f* subsere
субсеро'зный *a* subserous
субсоциа'ция *f* subsociation
субста'нция *f* substance
 ~ **Ни'сcля** chromophilic substance
 гонадотро'пная ~ gonadotropic substance
 се'тчатая ~ reticulated substance
 хромофи'льная ~ chromophilic substance
субстра'т *m* substrate
 акце'пторный ~ acceptor substrate
 иниции'рующий ~ sparkling substrate
 термини'рующий ~ termiating substrate
субстра'т-восстанови'тель *m* reducing substrate
субструкту'ра *f* substructure
субтро'пики *m pl* subtropics
субтропи'ческий *a* subtropical
субумбре'лла *f* subumbrella
субхромати'да *f* subchromatid
субформа'ция *f* subformation
субфосси'льный *a* subfossil
субфрагме'нт *m* subportion
субшта'мм *m* substrain
субъедини'ца *f* subunit
 каталити'ческая ~ catalytic subunit
 регулято'рная ~ regulatiry subunit
 рибосома'льная субъединица ribosome subunit
субэпидерма'льный *a* subepidermal
субэпителиа'льный *a* subepithelial
суво'йка *f* vorticel (*Vorticella*)
сугли'нок *m* loam
 пылева'тый ~ silty loam
сугру'док *m for* sudubrava
суда'н *m* Sudan
 ~ **кори'чневый** Sudan brown
суданофи'льный sudanophil, sudanophilic
су'дорога *f* convulsion, seizure, spasm, cramp
 клони'ческая ~ clonic spasm, clonic convulsion
 тони'ческая ~ tonic spasm
су'дорожный *a* convulsive, spasmodic
судубра'ва *f* sudubrava
суже'ние *n* constriction, contraction, narrowing, stricture
 ~ **сосу'дов** vasoconstriction
су'женно-боро'здчатый *a pal* constricticolpate
су'женный *a* narrowed
су'живать, су'живаться *v* coarct
сук *m* bough, branch; snag
су'ка *f* bitch, doggess
суккуле'нт *m* succulent plant
суккуле'нтность *f* succulence, succulency
сукро'льная *f* pregnant (*of rabbits*)
сукцессио'нный *a* seral
сукце'ссия *f* succession
 автоге'нная ~ autogenic succession
 аллоге'нная ~ allogenic succession
 антропоге'нная ~ anthropogenic succession
 биоге'нная ~, биоти'ческая ~ biotic succession
 векова'я ~ secular succession
 втори'чная ~ secondary succession
 годова'я ~ annual succession
 есте'ственная ~ natural succession
 климатоге'нная ~ climatic succession, climatogenic succession
 несоверше'нная ~ imperfect succession
 о'бщая ~ secular succession
 пироге'нная ~ pyrium, pyrogenic succession
 природная ~ natural succession
 растительная ~ plant succession
 регрессивная ~ retrogressive succsession
 сингенетическая ~ syngenetic succession
 совершенная ~ perfect succession
 ча'стная ~ local succession
 эдафоге'нная ~ edaphogenic succession
 экзодинами'ческая ~, экзоэкогенети'ческая ~ exodynamic succession
 эндодинами'ческая ~, эндоэкогенети'ческая ~ endodynamic succession
сукцинили'рование *n* succinylation
султа'н *m* tassel, anthurus, coma
сульфами'рование *n* sulfamation
сульфа'т *m* sulfate
сульфати'рование *n* sulfation
сульфгемоглоби'н *m* sulfhemoglobin
сульфини'рование *n* sulfination
сульфи'рование *n* sulfonation
сульфоалкили'рование *n* sulfoalkylation

сульфокисле'ние *n* sulfoxidation
сульфолипи'д *m* sulfolipid
сульфометили'рование *n* sulfomethylation
сульфони'рование *n* sulfonation
сульфофика'ция *f* sulfofication
сульфуриза'ция *f* sulfurization
сумасше'ствие *n* madness, insanity, mental disease
су'меречный *a* crepuscular
су'меречное состоя'ние twilight state
су'мка *f* bag, pouch; asc, ascus; bursa, sac
~ **ци'рруса** cirrus sac
околосерде'чная ~ pericardium, heart sac, pericardial sac
пасту'шья ~ *bot* shepherd's purse (*Capsella bursa-pastoris*)
са'льниковая ~ omental bursa
совокупи'тельная ~ *ent* copulatory pouch
фабри'циева ~ bursa of Fabricius, bursal sac
сумкообра'зный *a* bursiform
сумма'ция *f* summation
~ **раздраже'ний** summation of stimuli
~ **сти'мулов** stimulus summation
простра'нственная ~ spatial summation
сумми'рование *n* summation
су'мочка *f* bursicle
су'мчатый *a zool* marsupial; *bot* ascigerous; pouched
суперантиге'н *m* superantigen
суперге'н *m* supergene
суперинфе'кция *f* superinfection
суперовуля'ция *f* superovulation
суперсе'кс *m* supersex
суперсеме'йство *n* superfamily
суперспира'ль *f* supercoil
суперфекунда'ция *f* superfecundation
суперфета'ция *f* superfetation, superimpregnation
супесча'ный *a* sabulous
су'песь *f* loamy sand, sandy loam, sabulous clay
супина'тор *m* supinator
супина'ция *f* supination
супоро'сая *a* in-pig, pregnant (*of a sow*)
супра- supra-
суправита'льный *a* supravital
супралитора'льный *a* supralittoral
супраспира'льный *a* supraspinal
супрастигма'льный *a* suprastigmal
супре'ссия *f* suppression
рецесси'вная ~ recessive suppression
супре'ссор *m* suppressor
~ **бессмы'сленных мута'ций** nonsense-suppressor
~ **кроссинго'вера** crossing-over suppressor
~ **миссе'нс-мута'ций** missense-suppressor
внеге'нный ~ intergenic suppressor
внутриге'нный ~ intragenic suppressor
супроти'вный *a* opposite, decussate
сура'мень *a* suramen (*coniferous forest on relatively rich soils*)
суре'пка *f bot* winter-cress (*Barbarea vulgaris*)
суро'к *m ther* marmot *(Marmota)*
обыкнове'нный ~, степно'й ~ bobak marmot, Siberian marmot *(Marmota bobak)*
сурфакта'нт *m* surfactant
сурчи'на *f* marmot's burrow
суса'к *m bot* flowering-rush (*Butomus umbellatus*)
су'слик *m ther* ground squirrel, spermophile (*Citellus*)
су'сло *n* wort
~ **повы'шенной концентра'ции** heavy wort
бродя'щее ~ fermenting wort
пита'тельное ~ nutrient wort
су'сло-ага'р *m* wort agar
суспе'нзия *f* suspemsion, slurry
суспе'нзор *m* suspensor
суста'в *m anat* joint, articulation
блокови'дный ~ hinge joint, trochlear joint, ginglymoid joint
вене'чный ~ coffin joint
винтово'й ~ screw joint, cochleate joint, spiral joint
враща'тельный ~ pivot joint
голеносто'пный ~ talocrural articulation, mortis joint, ankle joint
двухка'мерный ~ bilocular joint
двухосево'й ~ biaxial joint
заплю'сневой ~ ankle joint
и'стинный ~ articulation, true joint
коле'нный ~ knee joint
колесови'дный ~ pivot joint
комбини'рованный ~ complex joint
ло'жный ~ false joint
локтево'й ~ elbow joint
лучезапя'стный ~ wrist joint
лучелоктево'й ~ radioulnar joint
межфала'нговый ~ knuckle, interphalangeal joint
многоосево'й ~ multiaxial joint

мы'щелковый ~ condylar articulation
надколе'нно-бе'дренный ~ patellofemoral joint
неподви'жный ~ immovable joint, synarthrosis
орехови'дный ~ cup-and-ball joint, fall-and-socket, enarthrosis, enarthrodial joint
плечево'й ~ shoulder joint, glenohumeral joint
пло'ский ~ plane joint, gliding joint
подтара'нный ~ subtalar joint
просто'й ~ simple joint
пу'товый ~ fetlock, fetlock joint
разбо'лтанный ~ flail joint
седлови'дный ~ saddlejoint, saddle articulation
синовиа'льный ~ synovial joint, synovial articulation
скака'тельный ~ hock, ankle joint
сло'жный ~ composite joint, compound joint
тазобе'дренный ~ coxofemoral joint, hip joint, thigh joint
цилиндри'ческий ~ pivot joint, trochoid joint, rotatory joint, trochoid articulation
чашеобра'зный ~ cup-and-ball joint, enarthrosis
шарни'рный ~ trochlear joint, hinge joint, ginglymoid joint
шарови'дный ~ ball-and-socket joint, spheroid joint, spheroidal joint, spheroid articulation, enarthrosis
эллипсо'идный ~ ellipsoidal joint
суста'вчатый *a* articulate, articulated
сухо- xero-
суховерши'нность *f* top drying, top-dry, stag-headedness, dieback
суховерши'нный *a* stag-headed, stag-horned
сухоголо'вчатый *a* xerocephalous
суходо'л *m* upland
суходо'льный *a* upland
сухожи'лие *n anat* tendon, sinew
~ **глазно'й мы'шцы** eye string
~ **разгиба'теля** extensor tendon
~ **сгиба'теля** flexor tendon
ахи'ллово ~ Achilles tendon, calcaneal tendon, heel tendon
перстнепищево'дное ~ cricoesophageal tendon
пя'точное ~ Achilles tendon, calcaneal tendon, heel tendon
сухожи'льный *a* tendinous
сухой *a* dry, arid
сухи'е фру'кты dried fruit
сухопло'дный *a* xerocarpous
сухопу'тный *a* terrestrial
сухосто'й *m* dead wood, dead-tree
сухосто'йный *a* dead (*of trees*)
су'хость *f* dryness
физиологи'ческая ~ physiological dryness
физи'ческая ~ почв physical drought
сухоцветко'вый *a* xeroflorous
сучкова'тый *a* gnarled, knaggy
сучо'к *m* twig; snag, knot, knag (*in wood*)
сушени'ца *f bot* cudweed (*Gnaphalium*)
суши'лка *f* drier
лиофи'льная ~ freeze drier
су'шка *f* drying
~ **распыле'нием** spray-drying
лиофи'льная ~ freeze drying
сушь *f* drought
существова'ние *n* existence
созна'тельное ~ sentient existence
существова'ть *v* exist
суя'гная *a* in-lamb, pregnant, heavy (*of ewe*)
сфа'гновый *a* sphagnous
сфагнофи'льный *a* sphagnophilous
сфацелия *f* sphacelia
сфено'ид *m* sphenoid
сфеноида'льный *a* sphenoid, sphenoidal
сфе'ра *f* sphere
лучи'стая ~ astrocenter, astrosphere
сфере'нхима *f* spherenchyma
сфероме'р *m* spheromere
сферопла'ст *m* spheroplast
бактериа'льный ~ bacterial spheroplast
сфероци'т *m* spherocyte
сфигмогра'мма *f* sphygmogram
сфигмо'граф *m* sphygmograph
сфигмогра'фия *f* sphygmography
сфигмомано'метр *m* sphygmomanometer
сфи'нктер *m anat* sphincter
сфеноида'льный *a* sphenoidal, sphenoid
сфра'гис *m ent* sphragis
схва'тка *f eth* scramble
схва'тки *f pl* pains
родовы'е ~ labour pains, birth pains
схва'тывать *v* grab, snap
схе'ма *f* scheme; circuit
~ **обра'тной свя'зи** feedback circuit
~ **рандомизи'рованных бло'ков** randomized block design
схизога'мия *f* schizogamy
схизогене'з *m* schozogenesis

схизогнати'ческий *a* schizognathous
схизогони'я *f prot* schizogony
схизока'рпий *m* schizocarp
схизо'нт *m prot* schizont
схизостели'я *f bot* schizostely
схизофи'т *m bot* schizophyte
схизоце'ль *f* schizocoel
схо'дство *n* similarity, likeness, resemblance, affinity, congeniality
покрови'тельственное ~ cryptic resemblance, protective resemblance
сходя'щийся *a* connivent
сцепле'ние *n* linkage, connection
~ **ге'нов** genetic linkage
~ **с по'лом** sex linkage
ло'жное ~ *gen* false linkage
межхромати'дное ~ interchromatidal linkage
межхромосо'мное ~ interchromosomal linkage
непо'лное ~ partial linkage
части'чное ~ *gen* partial linkage
сцеплё'нный *a* linked
сцепля'ть *v* link, couple
сцинк *herp* skink *(Scincus)*
сцинтигра'мма *f* scintiscan
сцинтигра'фия *f* scintillation scanning, scintigraphy
сцинтилля'ция *f* scintillation
сциофи'т *m* sciad, sciaphyte, sciophyte
сци'фа *f bot* scypha
сцифисто'ма *f* scyphistoma
сче'тчик *m* counter
~ **га'мма-излуче'ния** gamma counter
~ **ка'пель** drop counter
~ **коло'ний** colony counter
~ **фо'рменных элеме'нтов кро'ви** blood cell counter
~ **части'ц** particle counter
сцинтилляцио'нный счётчик scintillator, scintillation counter, liquid scintillation counter
торцо'вый ~ end-window counter
сшива'ние *n mbiol* ligation
перекрё'стное ~ *mbiol* cross-linking
сшива'ющий *a* ligating
сши'вка *f mbiol* cross-link
~ **тка'ней** tissue bridging
ДНК - белко'вая ~ DNA-protein cross-link
межни'тевая попере'чная ~ interstrand cross-link
съеда'ть *v* eat, eat away
съедо'бность *f* edibility, eatability
съедо'бный *a* edible, eatable, esculent
съём *m* picking (*of fruit*)
съё'мка *f* survey; filming; skinning
возду'шная ~ aerial survey
глазоме'рная ~ sketch survey
маршру'тная ~ route survey, reconnaissance survey
сы'воротка *f* serum, serosity; whey
~ **жерё'бой кобы'лы** pregnant mare serum
~ **новоро'жденных теля'т** newborn calf serum
агглютини'рующая ~ agglutinating serum, clumping serum
аллерге'нная ~ allergenic serum
аллергоспецифи'ческая ~ reaginic serum
аллоге'нная ~ allogenic serum
аллоимму'нная ~ alloimmune serum
антиаллотипи'ческая ~ antiallotype serum
антиидиотипи'ческая ~ anti-idiotypic serum
антитимоцита'рная ~ antithymic cell serum
антитокси'ческая ~ antitoxin serum
антитромбоцита'рная ~ antiplatelet serum
атопи'ческая ~ atopic serum
аутологи'ческая ~ autologous serum
гиперимму'нная ~ hyperimmune serum
декомплементи'рованная ~ complement-deficient serum
имму'нная ~ immune serum, antibody-containing serum, protective serum
контро'льная ~ reference serum
кровяна'я ~ blood serum
моло'чная ~ whey
неаллерге'нная ~ anallergic serum
неимму'нная ~ preimmune serum
нейтрализу'ющая ~ neutralizing serum
основна'я ~ key serum
поливале'нтная ~ multipartial serum, polyvalent serum
преципити'рующая ~ precipitating serum
противозмеи'ная ~ anti-ophidic serum
реаги'нная ~ reaginic serum
сенсибилизи'рующая ~ allergenic serum
стандартизи'рованная ~ reference serum
станда'ртная ~ test serum
фета'льная теля'чья ~ fetal calf serum
сы'вороточный *a* serumal
сыпь *f* rash, eruption

сыпня'к *m* exanthematous fever
сыр *m* cheese
  сыр бри Brie cheese
  о'стрый сыр strong cheese
  ли'мбургский сыр limburger
  эда'мский сыр Edam cheese
сыроде'лие *n* cheese industry, cheese manufacture
сырое'жка *f myc* russula (*Russula*)
сыро'й *a* damp, humid, wet, moist; crude, raw; green (*of wood*)
сы'рость *f* moisture
сырье' *n* raw material, resources
сы'тый *a* satiated
сыть *f bot* cypress-grass (*Cyperus*)
сыч *m orn* owl
  воробьи'ный ~ pygmy owl, gnome owl (*Glaudium passerinum*)
  домо'вый ~ little owl (*Athene noctua*)
  кро'личий ~ burrowing owl *(Athene cunicularia)*
  лесно'й ~ forest little owl *(Athene blewitti)*
  мохноногий ~ boreal owl, Tengmalm's owl (*Aegolius funereus*)
сычу'г *m anat* read, read stomach, rede, rennet stomach, rennet bag, abomasum, fourth stomach
сэкзи'на *f pal* sexine
сяжо'к *m* antenna, feeler

T

таба'к *m* tobacco (*Nicotiana*)
  па'пушевый ~ bundled tobacco
  подвё'рточный ~ binder tobacco
табаково'д *m* tobacco grower
табаково'дство *n* tobacco growing
табли'ца *f* table
  ~ **ассоциа'ций** *ecol* association table
  ~ **выжива'ния** life table
  ~ **сопряжё'нности при'знаков** contingency table
  ~ **часто'т** *biom* frequency table
  генеалоги'ческая ~ generation table
  четырёхпо'льная ~ fourfold table
  фитоценологи'ческая ~ phytosociological table
табу'н *m* herd
тавро' *n* burn
тавтони'м *m* tautonym
тавтоними'я *f* tautonymy
та'гма *f* tagma
тагматиза'ция *f* tagmosis
тае'жный *a* taiga
таз *m anat* pelvis
  большо'й ~ large pelvis
  ма'лый ~ small pelvis
та'зик *m ent* haunch, coxa
та'зиковый *a* coxal
та'зовый *a* pelvic
тайга' *f* taiga
та'йминг *m* timing
тайнобра'чный *a* cryptogamic, cryptogamous
та'кса *f* dachshund
такса'ция *f* valuation, taxation, assessment
  ~ **насажде'ния** estimate of standing crop
таксидерми'ст *m* taxidermist
таксидерми'я *f* taxidermy
та'ксис *m* taxis
  пищево'й ~ trophotaxis
таксо'н *m* taxon
  вышестоя'щий ~ superior taxon
  номина'льный ~ nominal taxon
  номинати'вный ~ nominotypical taxon
  подчинё'нный ~ subordinate taxon
таксономи'ческий *a* taxonomic
таксоно'мия *f* taxonomy
  нумери'ческая ~ numerical taxonomy
таксоцено'з *m* taxocene
такти'льный *a* tactile
такы'р *m* takyr
тала'мус *m anat* thalamus
талло'м *m* thallome, thallus, frond
  псевдопаренхимато'зный ~ pseudoparenchymatous thallus
  слои'стый ~ stratose thallus
талло'мный *a* thalline, thalloid, thallose
талломообра'зный *a* thalloid
таллофи'т *m* thallophyte
та'льник *m* osier-bed
тампо'н *m* tampon
танато'з *m* thanatosis
танатоло'гия *f* thanatology
танатоцено'з *m* thanatocoenosis
та'нец *m* danse
  виля'ющий ~ wagtail danse (*of bees*)
  кругово'й ~ round danse (*of bees*)
танни'н *m* tannin
тапета'льный *a* tapetal
тапе'тум *m* tapetum
тапи'р *m ther* tapir (*Tapirus*)
тарака'н *m* cockroach, roach, black beetle (*Blattoidea*)
  ры'жий ~ German cockroach (*Blatella germanica*)
  се'рый ~ grey cockroach, lobster cockroach*(Nauphoeta cinerea)*
  чёрный ~ oriental cockroach, common cockroach, Asiatic cockroach (*Blatta orientalis*)
тара'нтул *m* wolf spider (*Lycosa*)
тарза'льный *a* tarsal
тарпа'н *m* wild horse (*Equus caballus*)
тата'рник *m* Scotch thistle (*Onopordon*)
таутомериза'ция *f* tautomerization
тахигене'з *m* tachygenesis
тахикарди'я *f* tachycardia
тахители'ческий *a* tachytelic
тахители'я *f* tachytely
творо'г *m* curd-cheese, curd
творо'жистый *a* cheesy, caseous
теа'тр *m*: **анатоми'ческий** ~ dissecting room
те'гмен *m* tegmen, tegumen
те'гула *f ent* tegula
тегуля'рный *a ent* tegular
тегуме'нт *m zool* tegument
те'за и антите'за *f* key couplet
те'ка *f* theca
текодо'нтный *a* thecodont
тексту'ра *f* texture
тексту'рный *a* textural

тектоло'гия *f* tectology
тектириа'льный *a* tectorial
текто'риум *m* tectorium
теку'честь *f* fluidity
~ **мембра'н** membrane fluidity
теку'чий *a* running (*water*); ripe (*fish*)
телега'мный *a* telegamic
телего'ния *f* telegony
телейтопу'стула *f* teleutosorus, teliosorus, telium
телейтоспо'ра *f* teleutospore, teliospore
телейтоста'дия *f* teleutostage, teliostage
телеметри'ческий *a* telemetric
телеметри'я *f* telemetry, remote sensing
биологи'ческая ~ biotelemetry
теле'нок *m* calf
телеомо'рф *m* teleomorph
телеоно'мия *f* teleonomy
телеоргани'ческий *a* teleorganic
телепа'тия *f* telepathy
телереце'птор *m* teleceptor
телескопи'ческий *a* telescopiform
телиоспо'ра *f* teliospore
телипла'зма *f* thelyplasm
телито'кия *f* thelitoky
тели'ться *v* calf; fawn (*e.g., of a roe deer*)
те'лка *f* heifer
те'ло *n* body
~ **ма'тки** body of womb, uterine body
~ **позвонка'** vertebral body, neurocentrum
~ **Цукерка'ндля** Zuckerkandl's body, aortic body
ацето'новое ~ acetone body
база'льное ~ basal body
вере'вчатое ~ restiform body
во'льфово ~ Wolffian body
га'йморово ~ Highmore's body
грибови'дное ~ mushroom body
жёлтое ~ yellow body
жирово'е ~ fat body, adipose body
имму'нное ~ immune body
иноро'дное ~ foreign body
кле'точное ~ *bot* cell body
коле'нчатое ~ geniculate body
ко'ховские тела' *pl* Koch's ~s bodies
межпо'чечное ~ interrenal body
«опо'рное ~» *prot* stem body
парабаза'льное ~ parabasal body
пеще'ристое ~ cavernous body, spongy body
плодо'вое ~ fruit body, carposome
полоса'тое ~ striate body
ресни'чное ~ ciliary body
сосо'чковое ~ papillary body
сосцеви'дное ~ mammillary body
стеклови'дное ~ vitreous body
хромати'новое ~ chromatinic body
шти'довское ~ *prot* Stieda body
цилиа'рное ~ ciliary body
телобла'ст *m* teloblast
телокине'з *m* telokinesis
телоле'мма *f* telolemma
тело'м *m bot* telome
телолецита'льный *a* telolecital
теломе'р *m* telomere
теломити'ческий *a* telomitic
телора'бдион *m* telorhabdion
телоре'з *m* water-soldier (*Stratiotes*)
телоси'напсис *m* telosynapsis, telosyndesis
телосложе'ние *n* constitution, build, habitus, stature
астени'ческое ~ asthenic constitution
телосо'мик *m* telosomic
телосо'ма *f* telosoma
телота'ксис *m* telotaxis
телотро'х *m* telotroch
телофа'за *f* telophase
телофра'гма *f* telophragma
телоцентри'ческий *a* telocentric
те'лум *m* telum
тельсо'н *m zool* telson
те'льце *n* corpuscle, body
~ **Бабе'ш-Э'рнста** Babes-Ernst body
~ **Ба'рра** Barr's body
~ **ви'лочковой железы'** Hassal's corpuscle, thymus corpuscle
~ **Га'мны-Фа'вра** Gamna-Favre body
~ **Ге'ндерсона-Па'терсона** Henderson-Paterson body
~ **Гасса'ля** Hassal's body, Hassal's corpuscle, thymus corpuscle
~ **Го'льджи-Маццо'ни** Golgi-Mazzoni body, Golgi-Mazzoni corpuscle, bulblike corpuscle
~ **Жо'лли** Jolly's body
~ **Ме'йснера** Meissner's body
~ **Ме'лиса** Mehlis' gland
~ **Не'гри** Negri body, Negri corpuscle
~ **Ни'сcля** Nissl body
~ **Ра'ссела** Russell's body
~ **Фа'тер-Пачи'ни** Vater-Pacini corpuscle, corpuscle of Vater-Pacini, lamellated corpuscle
~ **Хе'рринга** Herring's body

~ **Шти'да** Stieda body
ацидофи'льное ~ acidophilic body, Russell's body
белко'вое ~ *bot* protein body
воспринима'ющее ~ receptive body
кароти'дное ~ carotid body
концево'е ~ terminal corpuscle
концево'е не'рвное ~ end corpuscle
ко'пчиковое ~ coccygeal body
луковицеобра'зное ~ Golgi-Mazzoni body, bulblike corpuscle
мальпи'гиево ~ Malpighian body, Malpighian corpuscle
метапласти'ческое ~ metaplastic body
моло'зивное ~ colostrum corpuscle
моло'чное ~ milk corpuscle
мю'ллерово ~ Mullerian body
направи'тельное ~ directing body, polar body, polocyte
осяза'тельное ~ tactile body, touch body, Meissner's corpuscle, tactile corpuscle, touch corpuscle
пачи'ниево ~ Pacinian body
пласти'нчатое ~ lamellated corpuscle, Vater-Pacini corpuscle
полово'е ~ genital corpuscle
поля'рное ~ polar body, directing body, polocyte
по'чечное ~ renal corpuscle
рогови'чное ~ corneal corpuscle
тидема'ново ~ Tiedemann's vesicle
«толка'ющее ~» *cyt* stem body
ультимобранхиа'льное ~ ultimobranchial body
центра'льное ~ central body
чувстви'тельное ~ Мейсснера Meissner's corpuscle, tactile corpuscle, touch corpuscle
эпителиа'льное ~ epithelial body

теля'тина *f* veal
теля'тник *m* calf barn, calf pen, calf house
теменно'й *a* parietal
тёмноокра'шенный *a* phaeic, phaeochrous, dark-colored
темноспонди'льный *a* temnospondylous
темп *m* rate, pace
~ **ро'ста** rate of growth
темпера'мент *m* temperament
температу'ра *f* temperature
~ **выра'щивания** incubation temperature
~ **инкуба'ции** incubation temperature
~ **разви'тия** growth temperature
~ **среды'** ambient temperature
~ **те'ла** body temperature, somatic temperature
база'льная ~ **те'ла** basal body temperature
кардина'льная ~ cardinal temperature
ко'мнатная ~ room temperature
комфо'ртная ~ comfort temperature
крити'ческая ~ critical temperature
лета'льная ~ lethal temperature
минима'льная эффекти'вная ~ minimum effective temperature
нача'льная ~ роста base growth temperature
неусто'йчивая ~ erratic temperature
поро'говая ~ threshold temperature
ректа'льная ~ rectal temperature
эффекти'вная ~ effective temperature

темпери'рование *n* tempering
те'мя *n* vertex, crown
те'нар *m anat* thenar
тенде'нция *f* tendency
основна'я ~ *biom* central tendency
теневыно'сливость *f* shade tolerance
теневыно'сливый *a* shade-tolerant, shade-enduring, umbraticole, umbraticolous
тенелюби'вый *a* shade-requiring, umbraticole, umbraticolous, sciophilous
тени'дия *f ent* taenidium, tenidium
тени'стый *a* umbrous, umbragenous, shady
тенореце'птор *m* tenoreceptor
те'нсор *m* tensor
тень *f* shade; shadow
«тень» эритроци'та phantom corpuscle, ghost corpuscle, shadow corpuscle
тео'рия *f* theory
~ **боковы'х цепе'й** *imm* lateral-chain theory, side-chain theory
~ **Ве'генера** displacement theory
~ **ге'нных ре'плик** gene replica theory
~ **голода'ния ге'нов** theory of gene starvation
~ **«замка' и ключа'»** lock and key theory
~ **заморо'женного слу'чая** frozen accident theory
~ **запрещё'нных кло'нов** *imm* forbidden-clone theory
~ **заро'дышевой пла'змы** theory of germ plasm
~ **заро'дышевых листко'в** germ layer theory
~ **игр** *biom* theory of games
~ **изоля'ции** separation theory
~ **информа'ции** information theory

~ **катастро'ф** cataclysmic theory, catastrophism
~ **конве'рсии** *gen* conversion theory
~ **континента'льного дре'йфа** displacement theory
~ **оди'н ген - оди'н ферме'нт** one gene-one enzyme theory
~ **оптима'льного фуражи'рования** *eth* optimal foraging theory
~ **оши'бок** *biom* error theory
~ **полово'го бала'нса** balance theory of sex
~ **попада'ний** *biom* target theory
~ **преформи'зма** preformation theory, incasement theory
~ **прису'тствия - отсу'тствия** presence-and-absence theory
~ **скачко'в** saltation theory
~ **«скользя'щих ни'тей»** *phys* sliding filament theory
~ **смеща'ющегося равнове'сия** shifting balance theory
~ **уда'ч** *biom* target theory
~ **хиазмати'пии** chiasmatype theory
аллогенети'ческая ~ allogenetic theory
сальтацио'нная ~ saltation theory
следова'я ~ **па'мяти** trace theory of memory
эволюцио'нная ~ theory of evolution
тепли'ца *f* glasshouse, hothouse, greenhouse
теплово'дный *a* warmwater
теплово'й *a* thermal, thermic
теплокро'вный *a* homothermal, warm-blooded, homothermous, homoiothermal, homoiothermic, homeothermal
теплолюби'вый *a* thermophilic, thermophilous
теплообразова'ние *n* thermogenesis
теплообме'н *m* thermoexchange
теплоотда'ча *f* thermolysis
теплоусто'йчивость *f* thermal resistance, thermostability
теплоусто'йчивый *a* thermoresistant, thermotolerant
терапе'вт *m* therapeutist, therapist
терапевти'ческий *a* therapeutic
терапи'я *f med* therapy, therapeutics, therapia, therapeia, therapeusis
замести'тельная ~ replacement therapy, substitution therapy
лучева'я ~ radiation therapy
ра'диевая ~ radium therapy
тератоге'н *m* teratogen
тератогене'з *m* teratogenesis
тератоге'нный *a* teratogenic
тератоло'гия *f* teratology
терато'ма *f* teratoid tumor
терги'т *m* tergite, tergum
тергостерна'льный *a* tergosternal
териоло'гия *f* theriology, mammalogy
тё'рка *f zool* radula
терминализа'ция *f* terminalization
термина'льный *a* terminal
термина'тор *m mbiol* terminator
~ **транскри'пции** transcription terminator
термина'ция *f mbiol* termination
преждевре'менная ~ premature termination
терминоло'гия *f* terminology
терми'т *m ent* termite (*Isoptera*)
терми'тник *m* termitarium
терми'ческий *a* thermal, thermic
термо- thermo-
термогене'з *m* thermogenesis
термоге'нный *a* thermogenic
термогра'мма *f* thermogram
термо'граф *m* recording thermometer, thermograph
термоиндуци'руемый *a* thermoinducible
термоклейстога'мия *f bot* thermocleistogamy
термокли'н *m* thermocline
термолаби'льный *a* thermolabile
термо'лиз *m* thermolysis
термолюминесце'нция *f* thermoluminescence
термо'метр *m* thermometer
максима'льный ~ maximum thermometer
минима'льный ~ minimum thermometer
опроки'дывающийся ~ reversing thermometer
термоме'трия *f* thermometry
термона'стия *f* thermonasty
термопа'ра *f* thermocouple
радиацио'нная ~ radiation thermocouple
термопериоди'зм *m* thermoperiodicity, thermoperiodism
терморегуля'ция *f* thermoregulation
терморезисте'нтный *a* thermoresistant, thermotolerant, thermoduric
термореце'птор *m* thermoreceptor
термоскопи'ческий *a* thermoscopic
термостаби'льность *f* thermostability
термоста'т *m* thermostat
термосто'йкий *a* thermoduric

термосто'йкость *f* thermostability
термота'ксис *m* thermotaxis
термотакси'ческий *a* thermotactic
термото'к *m* thermocurrent
термотропи'зм *m* thermotropism
термофи'льный *a* thermophilic, thermophilous
термофи'т *m* thermophyte
термофикси'рованный *a* heat-fixed
термофи'л *m* thermophil
термоэлеме'нт *m* thermopile
терн, **терно'вник** *m* blackthorn (*Prunus spinosa*)
терофи'т *m* therophyte
терносли'ва *f* damson (*Prunus insititia*)
те'рпкость *f* acerbity
терра'риум *m* terrarium, terrarium cage
терра'са *f* terrace
терраси'рование *n* terracing
терраси'рованный *a* terraced
территориали'зм *m* territorial behavior
территориа'льность *f* territorialism, territoriality
территориа'льный *a* territorial
террито'рия *f* territory, area
~ **домини'рования** dominance territory
бра'чная ~ mating territory
группова'я ~ group territory
колонизи'руемая ~ recipient territory
кормова'я ~ feeding territory
неперекрыва'ющиеся территории *pl* nonoverlapping territories
приро'дная охраня'емая ~ conservation area, reserved area
терье'р *m* terrier
тест *m* test
~ **Го'варда-Рапопо'рта** Howard-Rapoport test
~ **закры'того поля** closed field test
~ **иммуу'нного фагоцито'за** immunophagocytosis test
~ **латексагглютина'ции** latex fixation test
~ **на возбужде'ние** provocative test
~ **на защи'тные антитела** protection test
~ **на различе'ние** *eth* discrimination test
~ **на реакти'вность** provocative test
~ **на толера'нтность** tolerance test
~ **на цис-транс-положе'ние** cis-trans test
~ **обрати'мости фа'кторов** factor-reversal test
~ **откры'того по'ля** open field test
~ **перекрё'стной иммуниза'ции** cross-immunization test
~ **перекрё'стной интерфере'нции** cross-interference test
~ **на расщепле'ние** segregational test
~ **с гетерокарио'ном** heterokaryon test
~ **соотноше'ния рекомбина'нтов** ratio test
гру'бый ~ rough test
зна'ковый ~ *biom* sign test
иммунофагоцита'рный ~ immunophagocytosis test
комплементацио'нный ~ complementation test
конвенциона'льный ~ routine test
ла'текс-иммобилизационный ~ latex fixation test
лу'ковый ~ allium test
предвари'тельный ~ rough test
серодиагности'ческий ~ serodiagnostic test
станда'ртный ~ routine test
тестикуля'рный *a* testicular
тести'рование *n* testing
тест-набо'р *m* test-kit
тестостеро'н *m* testosteron
тетани'ческий *a* tetaniform
тетани'я *f* tetania
те'танус *m* tetanus
те'терев *m* heath cock, black grouse (*Lyrulus tetrix*)
кавка'зский ~ Caucasian grouse *(Lyrurus molokoseiwiczi)*
лугово'й ~ prairie grouse, prairie chicken *(Tympanuchus)*
острохво'стый ~ sharp-tailed grouse *(Tetrao phasianellus)*
те'терев-коса'ч *m* black grouse *(Lyrurus tetrix)*
тё'тка *f* aunt
тетра- tetra-
тетрагона'льный *a* tetragonal
тетра'да *f* tetrad, quartet
~ **Фалло'** Fallot's tetrad, tetrad of Fallot, Fallot's tetralogy, tetralogy of Fallot
тетразо'лий *m* tetrazolium
нитроси'ний ~ nitro blue tetrazolium
тетрало'гия *f* tetralogy
~ **Фалло'** Fallot's tetrad
тетраме'р *m* tetramer
тетрамо'рфный *a* tetramorphic
тетрапло'ид *m* tetraploid

двойно'й ~ double tetraploid
тетраплоиди'я *f* tetraplody
тетраполя'рный *a* tetrapolar
тетраселенодо'нтный *a* tetraselenodont
тетрасо'мик *m* tetrasomic
тетрасоми'ческий *a* tetrasomic
тетраспо'ра *f* tetraspore
тетраспора'нгий *a* tetrasporangium
тетрахори'ческий *a biom* tetrachoric
тетраци'т *m* tetracyte
техноло'гия *f* technology
безотхо'дная ~ non-waste technology
малоотхо'дная ~ low-waste technology
техносфе'ра *f* technosphere
тече'ние *n* current, flow
ламина'рное ~ laminar flow, streamline flow
отжи'мное ~ offshore current
турбуле'нтное ~ turbulent flow, turbulent stream
те'чка *f* estruation, estrus, heat, rut, ruttishness
тиами'н *m* thiamine
тибиотарза'льный *a ent* tibiotarsal
тигмокине'з *m* thigmokinesis
тигмоморфо'з *m* thigmomorphosis
тигмота'ксис *m* thigmotaxis
тигмотропи'зм *m* thigmotropism
тигр *m* tiger (*Panthera tigris*)
тигри'ца *f* tigress
тигро'ид *m* tigroid
тигро'идный *a* tigroid
тик *m med* tic; *bot* teak (*Tectona grandis*)
тимиди'н *m* thymidine
тими'н *m* thymine
тимози'н *m* thymosin
тимо'л *m* thymol
тимоци'т *m* thymocyte
большо'й ~ large thymocyte
зре'лый ~ mature thymocyte
кортика'льный ~ cortical thymocyte
ма'лый ~ small thymocyte
промежу'точный ~ common thymocyte
ра'нний ~ early thymocyte
созрева'ющий кортика'льный ~ late cortical thymocyte
субка'псульный ~ subcapsular thymocyte
тимоцитопоэ'з *m* thymocytopoiesis
тимпана'льный *a* tympanal
ти'мус *m anat* thymus
тимусзави'симый *a* thimic-dependent
тимуснезави'симый *a* thimic-independent
тимья'н *m bot* thyme (*Thymus*)
ти'на *f* scum, sludge, blanket algae
тиндализа'ция *f* tyndalization
тинктория'льный *a* tinctorial
тинкту'ра *f* tincture
тио- thio-
тиоацили'рование *n* thioacylation
тиобакте'рия *f* thiobacterium
тиометили'рование *n* thiomethylation
тиомочеви'на *f* thiourea
тиоциани'рование *n* thiocyanation, thiocyanogenation
тиофи'л *m* thiophil
тиофи'льный *a* thiophilic
тип *m* type, phylum
~ **местообита'ния** site type
~ **расти'тельности** vegetation type, vegetational type
~ **реа'кции** reaction type
~ **совмести'мости** compatibility type
~ **соо'бщества** community type
~ **спарива'ния** mating type
~ **телосложе'ния** constitutional type
брюшно'й ~ **дыха'ния** abdominal breathing
ди'кий ~ wild type
комплемента'рный ~ **спа'ривания** *prot* complementary mating type
номенклату'рный ~ name-bearing type
рё'берный ~ **дыха'ния** costal breathing
типи'рование *n* typing
аллоге'нное ~ allotyping
обраще'нное ~ reverse typing
прямо'е ~ forward typing
серологи'ческое ~ serologic typing
тка'невое ~ tissue typing
типи'руемый *a* typeable
типи'чный *a* typical
типогене'з *m* typogenesis
типоло'гия *f* typology
типу'н *m vet* pip
тиреотропи'н *m* thyrotropin
тиреотро'пный *a* thyroid-stimulating
тирку'шка *f* pratincole *(Glareola)*
тирози'н *m* tyrosine
тирокси'н *m* thyroxin, thyroxine
тирсо'идный *a bot* thyrsoid
тис, **тисс** *m bot* yew (*Taxus*)
тите'стер *m* tea-tester
титр *m* titer
~ **кише'чной па'лочки** coli titer
~ **фа'га** phage titer
инфекцио'нный ~ infectious titer

титрова'ние *n* titration
~ ме'тодом «ша'хматной доски'» checkerboard titration, chessboard titration
обме'нное ~ exchange titration
обра'тное ~ back titration
окисли'тельное ~ oxidative titration
титрова'ть *v* titrate
тиф *m med* typhus
брюшно'й ~ enteric fever, abdominal typhoid, typhoid fever
возвра'тный ~ relapsing fever, recurrent fever
тифлозо'ль *m* typhlosole
тихоокеанский *a* Pacific
тихохо'дка *f* water bear (*Tardigrada*)
тка'невый *a* histic, tissular
тканеспецифи'чный *a* tissue-specific
ткань *f* tissue
~ головно'го мо'зга brain tissue, cerebral tissue
~ же'лтого те'ла luteal tissue
~ моло'чной железы' mammary tissue
~ пло'да *anat* fetal tissue
~, спосо'бная к сокраще'нию contractile tissue
ассимиляцио'нная ~ assimilative tissue
бу'рая жирова'я ~ brown adipose tissue
водоно'сная ~ *bot* water-storage tissue
волокни'стая ~ fibrous tissue
грануляцио'нная ~ granulation tissue
гу'бчатая ~ cancellous tissue
железистая ~ glandular tissue
жирова'я ~ adipose tissue, fatty tissue
запаса'ющая ~ storage tissue
заро'дышевая ~ embryonal tissue
зре'лая ~ mature tissue
измельчё'нная ~ minced tissue
интерстициа'льная ~ interstitial tissue
исхо'дная ~ parent tissue
каверно'зная ~ cavernous tissue
каллусови'дная ~ calluslike tissue
коллаге'новая ~ gelatinous tissue
ко'стная ~ bone tissue, bony tissue, osseous tissue
кроветво'рная ~ hematopoietic tissue, blood-forming tissue
лимфати'ческая ~ lymphatic tissue
лимфо'идная ~ lymphoid tissue
механи'ческая ~ strengthening tissue
мы'шечная ~ muscular tissue
не'рвная ~ nerve tissue
несекрети'рующая ~ nonsecretory tissue
опо'рная ~ supporting tissue
палиса'дная ~ palisade tissue
паренхимато'зная ~ parenchymatous tissue
перви'чная ~ primary tissue
пло'тная соедини'тельная ~ dense connective tissue
подко'жная соедини'тельная ~ subcutaneous connective tissue
постоя'нная ~ permanent tissue
проводя'щая ~ conductive tissue, *bot* vascular tissue
промежу'точная ~ interstitial tissue
радиочувстви'тельная ~ radiosensitive tissue
ра'ковая ~ cancer tissue
расти'тельная ~ plant tissue
ретикуля'рная ~ reticular tissue
рогова'я ~ horny tissue
роди'тельская ~ parent tissue
рубцо'вая ~ cicatrical tissue
ры'хлая соедини'тельная ~ areolar tissue, loose connective tissue
се'тчатая ~ reticular tissue
ситови'дная ~ sieve tissue
сли'зистая ~ mucous tissue
соедини'тельная ~ connective tissue, bindweb
сомати'ческая ~ somatic tissue
сосу'дистая ~ vascular tissue
студени'стая соедини'тельная ~ mucous tissue
фибро'зная ~ fibrous tissue
хромаффи'нная ~ chromaffin tissue
хрящева'я ~ cartilaginous tissue
чужеро'дная ~ foreign tissue, heterologous tissue
эласти'ческая соедини'тельная ~ elastic tissue
эмбриона'льная ~ embryonal tissue
эпителиа'льная ~ epithelial tissue
тка'чик *m* weaver, weaverbird (*Ploceidae*)
воскоклю'вый ~ waxbill (*Estrida astrid*)
Т-лимфоци'т *m* T lymphocyte
тля *f* plant louse, aphid (*Aphidoidea*)
тмин *m* caraway (*Carum carvi*)
това'р *m* goods, wares, mershandise
товары *m pl*: **продово'льственные ~** foodstuffs
ток *m* current; threshing floor
~ де'йствия *phys* action current
~ поврежде'ния *phys* injury current

~ **кро'ви** blood flow, blood stream, circulation
~ **поврежде'ния** *phys* injury current
восходя'щий ~ ascending current, centripetal current
входя'щий ~ inward current
выходя'щий ~ outward current
демаркацио'нный ~ demarcation current
ио'нный ~ ion current
мембра'нный ~ membrane current
стациона'рный ~ steady-state current
транспирацио'нный ~ transpiration stream
токова'ние *n eth* lekking, display
токови'ще *n* lek, lekking ground
токоферо'л *m* tocopherol
токсальбуми'н *m* toxalbumin
токсикоге'нный *a* toxigenic
токсико'з *m* toxicosis
токсикологи'ческий *a* toxicological
токсиколо'гия *f* toxicology
токсикометри'я toxicometry
токси'н *m* toxin
~ **уста'лости** fatigue toxin
и'стинный ~ extracellular toxin, true toxin
токсиноге'нный *a* toxicogenic, toxigenic
токси'чность *f* toxicity
токси'чный *a* toxic, toxiferous
токсо'ид *m* toxoid
токсоплазмо'з *m* toxoplasmosis *(caused by Toxoplasma gondii)*
току'ющий *a* lekking
толера'нтность *f* tolerance
~ **к «своему'»** internal tolerance
аллоге'нная ~ allotolerancy
есте'ственная ~ natural tolerance
иммунологи'ческая ~ immune tolerance, immunological tolerance, anergy
клона'льная ~ clonal anergy
перекрё'стная ~ cross tolerance
перенесё'нная ~ adoptive tolerance, infectious tolerance
приобретё'нная ~ acquired tolerance
толокно' *n* oatmeal
толокня'нка *f* bearberry (*Arctostaphylos uva-ursi*)
толо'чь *v* triturate
толстоголо'вчатый *a* thick-headed
толстогу'бый *a* thick-lipped
толстоко'жий *a* pachydermate, pachydermous, pachydermatous
толстоли'стный *a* pachyphyllous
толстоне'рвный *a bot* thick-veined
толстопло'дный *a* thick-fruited
толстопокро'вный *a* crassitegillate
толстосте'нный *a bot* pachypleurous, pachypleurous, thick-walled
толстоэкзи'нный *a pal* crassiexinous
то'лстый *a* thick, gross, stout; **то'лстая кишка'** large intestine
толчо'к *m* beat; push, jerk, shove, thrust
толчки' *(pl)* **та'зом** *eth* thrusting
верху'шечный ~ apical beat
то'лща *f* mass, strata, inside; **в толще льда'** within ice, within the bulk of ice
~ **воды'** free water
~ **льда'** the bulk of ice
нема'я ~ *geol* barren strata
продукти'вная ~ *geol* productive series
тома'т *m* tomato (*Lycopersicum esculentum*)
томле'ние *n* curing (*of tobacco*)
томогра'фия *f* tomography, imaging
~ **движе'ния** motion imaging
~ **пото'ка** flow imaging
импеда'нсная ~ impedance imaging
трансмиссио'нная ~ transmission tomography
эмиссио'нная ~ emission tomography
я'дерно-магни'тно-резона'нсная ~ NMR imaging
тон *m*: tone
~ **се'рдца** cardiac sound, heart sound
основно'й ~ fundamental tone
чи'стый ~ pure tone
тона'льно-модули'рованный *a* tone-modulated *(bioacoustics)*
тонизи'рующий *a* tonic
тони'чность *f* tonicity
то'нкий *a* thin, tender
тонковолокни'стый *a* fine-filamented
тонкопа'лый *a* leptodactylous
тонкопокро'вный *a* tenuitegillate
тонкосте'нный *a* thin-walled
тонкохво'стый *a* leptocercal
тоноклони'ческий *a* tonoclonic
тонопла'ст *f* tonoplast
тонота'ксис *m* tonotaxis
тонофибри'лла *f* tonofibril
то'нус *m* tonus, tonicity, tension, tone
~ **блужда'ющего не'рва** vagal tone
~ **экстракардиа'льных не'рвов** extracardiac nerve tone
база'льный ~ basal tone
втори'чный ~ secondary tone

мы'шечный ~ muscular tension, muscle tone
пласти'ческий ~ catch tone, viscous tone
тка'невый ~ tissue tone
по'здний ~ late tone
сосу'дистый ~ vascular tone
топина'мбур *m bot* Jerusalem artichoke, earth apple, girasol, girasole (*Helianthus tuberosus*)
то'пкий *a* swampy, boggy
топля'к *m* sunken log, snag
топо- topo-
топогене'з *m* topogenesis
топоде'м *m* topodeme
топокли'н *m* topocline
то'поль *m bot* poplar, cottonwood (*Populus*)
~ **бе'лый** abele (*Populus alba*)
~ **пирамида'льный** Bolle's poplar (*Populus pyramidalis*)
топо'рик *m orn* puffin (*Lunda*)
топота'ксис *m* topotaxis
топотропи'зм *m* topotropism
топоти'п *m* topotype
топотипи'ческий, топотипи'чный *a* topotypical
топофотота'ксис *m* topophototaxis
топохи'мия *f* topochemistry
топта'ть *v* tread (*of male birds*)
топь *f* mire, swamp, swampland, slough, quagmire
тор *m* torus
торака'льный *a* thoracic
то'ракс *m* thorax
торе'ц *m* butt end
тори'ца *f bot* spurry (*Spergula*)
торможе'ние *n* inhibition
~ **а'льфа-ри'тма** alpha inhibition
~ **подкрепле'нием** inhibition by reinforcement
~ **по ти'пу обра'тной свя'зи** feedback inhibition, retro inhibition
~ **ро'ста** growth inhibition
автоге'нное ~ autogenic inhibition
бесконкуре'нтное ~ uncompetitive inhibition
вага'льное ~ vagal inhibition
вне'шнее ~ external inhibition
вну'треннее ~ internal inhibition
гипоталами'ческое ~ hypothalamic inhibition
дифференциа'льное ~ differential inhibition
дифференциро'вочное ~ differentiating inhibition
запа'здывающее ~ delayed inhibition
конкуре'нтное ~ competitive inhibition
конта'ктное ~ contact inhibition
лате'нтное ~ latent inhibition
латера'льное ~ lateral inhibition
перви'чное ~ primary inhibition
пресинапти'ческое ~ presynaptic inhibition
прямо'е ~ direct inhibition
реакти'вное ~ reactive inhibition
рефлекто'рное ~ reflex inhibition
самоконта'ктное ~ self-contact inhibition
се'ченовское ~ Setschenow's inhibition
ступе'нчатое ~ stepwise inhibition
трофи'ческое ~ trophic inhibition
угаса'тельное ~ extinctive inhibition
усло'вное ~ conditioned inhibition
усло'вно-рефлекторное ~ conditioned inhibition
центра'льное ~ central inhibition
тормози'ть *v phys* inhibit, extinguish
тормозя'щий *a* inhibitory
торна'рия *f* tornaria
то'рсия *f* torsion
торф *m* peat
торфобрике'т *m* peat briquette
торфодобы'ча *f* peat extraction, peat cutting, peat digging
торфообразова'ние *n* peat formation
торфообразу'ющий *a* peat-forming
торфя'ник *m* peatbog
торфяни'стый *a* peaty
тотипоте'нтность *f* totipotency
тотипоте'нтный *a* totipotent
точи'льщик *m* borer; *ent* death-watch beetle, dust beetle, furniture beetle *(Anobium)*
то'чка *f* dot, point; honing
~ **вы'бора** choice point *(in a maze)*
~ **замерза'ния** freezing point
~ **насыще'ния** saturation point
~ **овуля'ции** ovulation point
~ **плавле'ния** melting point, fusion point
~ **равнове'сия** equilibrium point
~ **ро'ста** growing point, apical point
~ **ро'ста ко'рня** root growing point
~ **ро'ста побе'га** shoot growing point
~ **росы'** dew point
~ **усто'йчивого завяда'ния** permanent wilting point
~ **эффекти'вности** *biom* efficient point

акце'пторная ~ спла'йсинга branching point
вегетати'вная ~ ро'ста vegetative point
дви'гательная ~ motor point
диспара'тная ~ disparate point
изоио'нная ~ isoionic point
изоэлектри'ческая ~ isoelectric point
компенсацио'нная ~ compensation point
крити'ческая ~ critical point
отправна'я ~ starting point
пере'дняя узлова'я ~ anterior nodal point
светова'я компенсацио'нная ~ light compensation point
эвтекти'ческая ~ eutectic point

трабе'кула *f anat* trabecula

трава' *f* grass, herb
со'рная ~ weed

трави'ть *v* exterminate, poison; bay, run, hunt, bait; pay out (*ropes*)
~ мыше'й exterminate mice
~ соба'ками set the dogs on

травле'ние *n* etching
глубо'кое ~ deep etching
ио'нное ~ ion beam etching

тра'вля *f* baiting

тра'вма *f* trauma, injury, wound

травмати'н *m* traumatin

травмати'ческий *a* traumatic

травматропи'зм *m* traumatropism

тра'вник *m* redshank (*Tringa totanus*)

травосе'яние *n* grass cultivation, grass culture, grass growing

травосме'сь *f* mixed grass crop

травосто'й *m* herbage, sward, grass stand, grass canopy

травоя'дный *a* herbivorous, vegetarian, phytophagous, phytovorous

тра'вы *pl of* **трава**
аромати'ческие ~ aromatic herbs
кормовы'е ~ forage herbs
лека'рственные ~ medicinal herbs
пря'ные ~ flavoring herbs

травяни'стый *a* herbaceous, herblike, grassy

травяно'й *a* grassy, herbaceous, herbal

тракт *m* tract
желу'дочно-кише'чный ~ gastrointestinal tract, gastroenteric tract
зри'тельный ~ optic tract, visual tract
мочеполово'й ~ urogenital tract
пищевари'тельный ~ digestive tract, alimentary canal, alimentary tract, alimentary tube

трал *m* trawl
~ А'йзекса-Ки'дда Isaacs-Kidd trawl
~ Си'гзби Sigsbee trawl

тра'ма *f* trama

транзи'торный *a* transient

транзи'ция *f* transition mutation, transition

транквилиза'тор *m* tranquilizer

транс- trans-

трансапика'льный *a* transapical

трансве'кция *f* transvection

трансверса'льный *a* transverse

трансве'рсия *f* transversion, transversion mutation

трансге'н *m* transgene

трансге'нный *a* transgenic

трансгено'з *m* transgenation, transgenesis

трансгетерозиго'та *f* transheterozygote

трансгре'ссия *f* transgression

трансдетермина'ция *f* transdetermination

трансдукта'нт *m* transductant
усто'йчивый ~ complete transductant

трансду'кция *f* transduction
аборти'вная ~ abortive transduction
неограни'ченная ~ unrestricted transduction
неспецифи'ческая ~ general transduction, generalized transduction
о'бщая ~ general transduction, generalized transduction
ограни'ченная ~ restricted transduction, specialized transduction
сигна'льная ~ signal transduction
специфи'ческая ~ restricted transduction, specialized transduction

трансе'кта *f* transect
ле'нточная трансекта belt transect, streep transect
лине'йная ~ line transect

транскапилля'рный *a* transcapillary

трансконформа'ция *f* transconformation

транскри'пт *m mbiol* transcript

транскрипта'за *f* transcriptase
обра'тная ~ reverse transcriptase

транскри'пция *f* transcription
обра'тная ~ reverse transcription

транслока'ция *f* translocation
анэуцентри'ческая ~ aneucentric translocation
ацентри'ческая ~ acentric translocation
взаи'мная ~ mutual translocation
реципро'кная ~ reciprocal translocation
цикли'ческая ~ cyclic translocation

части'чная ~ partial translocation
трансля'ция *f mbiol* translation
выявленная гибридом ~ hybrid-released translation
генетическая ~ genetic translation
остановленная гибридом ~ hybrid-arrested translation
трансмембра'нный *a* transmembrane
трансметили'рование *n* transmethylation
трансмисси'вный *a* transmissible
трансмура'льный *a* transmural
трансовариа'льный *a* transovarian
транспoложе'ние *n* transconfiguration
транспира'ция *f* transpiration
абсолю'тная ~ absolute transpiration
кутикуля'рная ~ cuticular transpiration
относи'тельная ~ relative transpiration
у'стьичная ~ stomatal transpiration
транспиро'метр *m* transpirometer
транспланта'т *m* transplantat, graft
аутопласти'ческий ~ autoplastic graft
гетеропласти'ческий ~ heteroplastic graft
гетерото'пный ~ heterotopic graft
гомопласти'ческий ~ homoplastic graft
синге'нный ~ syngenic graft
транспланта'ция *f* transplantation, grafting, transfer
~ **в гомологи'чную область** homotopic transplantation
~ **в негомологи'чную область** heterotopic transplantation
аутопласти'ческая ~ autoplastic transplantation
гетеропласти'ческая ~ heteroplastic transplantation
гомопласти'ческая ~ homoplastic transplantation
ксенопласти'ческая ~ xenoplastic transplantation
транспланти'рованный *a* grafted
транспози'ция *f* transposition
транспозо'н *m gen* transposon
тра'нспорт *m* transport
аксона'льный ~ axonal transport
акти'вный ~ active transport
перви'чный электро'нный ~ primary electron transport
прото'н-сопряжё'нный ~ proton-coupled transport
ретрогра'дный ~ retrograde transport
электро'нный ~ electron transport
электроосмоти'ческий ~ electroosmotic transport
энергозави'симый ~ energy-dependent transport
транспорте'р *m* transporter
трансреплика'ция *f gen* transreplication
транссуда'т *m* transudate
транссуда'ция *f* transudation
трансфе'кция *f* transfection
трансформа'нт *m* transformant
трансформа'ция *f* transformation
аллоге'нная ~ allogenic transformation
трансформи'зм *m* transformism
трансфосфорили'рование *n* transphosphorylation
трансэпителиа'льный *a* transepithelial
тра'урница *f* mourning-cloak butterfly (*Vanessa antiopa*)
трахеа'льный *a* tracheal, tracheate
трахеа'ция *f* tracheation
трахеи'да *f bot* tracheid
вазицентри'ческая ~ vasicentric tracheid
лучева'я ~ ray tracheid
тя'жевая ~ strand tracheid
трахео'ла *f* tracheole
трахеомико'з *m phytop* tracheomycosis
трахеотоми'я *f* tracheotomy
трахе'я *f* trachea, windpipe, air tube, breathing tube
тра'ченный *a*: ~ **мо'лью** moth-eaten
трево'га *f* alarm, anxiety, uneasuness, trouble
трелева'ть *v* skid logs
~ **бре'вна** skid logs, drag logs
трелё'вка *f* skidding, snaking, snigging
трель *f eth* warble, trill
трemматоло'гия *f* thremmatology
тре'мор *m* tremor, trembling, shaking, trepidation
тренд *m biom* trend
лине'йный ~ linear trend
нелине'йный ~ nonlinear trend
трено'жник *m* anat tripus
треони'н *m* threonine
трепана'ция *f* trepanation, trephination
трепа'нг *m zool* trepang (*Holothuria*)
трепа'ть *v* scutch, beat (*e.g., flax*); **его трепа'ла лихора'дка** he was feverish, he was shivering with fever
~ **пеньку'** beat hemp
трепета'ние *n* flutter, palpitation
~ **желу'дочков** ventricular flutter
~ **предсе'рдия** atrial flutter, auricular flutter

~ **се'рдца** palpitation of the heart
треск *m* snap
тре'скаться *v* chap, crack, split
трети'чный *a* tertiary
треуго'льник *m* triangle
~ **А'льсберга** Alsberg's triangle
~ **Ве'бера** Weber's triangle
~ **Гри'нфельта** Grynfeltt's triangle
~ **Лаббе'** Labbe's triangle
~ **Лангенбе'ка** Langenbeck's triangle
~ **Ле'сгафта** Lesshaft's triangle
~ **Ле'ссера** Lesser's triangle
~ **Траутма'нна** Trautmann's triangle
~ **Уо'рда** Ward's triangle
~ **Фарабе'фа** Farabeuf's triangle
~ **Эйнтхо'вена** Einthoven's triangle
~ **Эло'та** Elaut's triangle
пети'тов ~ Petit's triangle
треугольнопло'дный *a* trigonocarpous
треугольносемя'нный *a* trigonospermous
трёх- tri-
трёхборо'здный *a* tricolpate
трёхбра'тственный *a bot* triadelphous
трёхбуго'рчатый *a* tricuspid, tricuspidate
трёхверши'нный *a* tricrotic
трёхветви'стый *a* tricrural, trifurcate
трёхви'льчатый *a* trifurcate
трёхгла'вый *a* tricephalous, *anat* tricipital
трёхгнё'здный *a* trilocular
трёхгра'нный *a bot* triquetrous
трёхгу'бый *a* trilabiate
трёхдо'льный, трёхдо'льчатый *a* trilobate, three-lobed
трёхжи'лковый *a* triple-nerved, triple-ribbed, trinervate
трёхзу'бчатый *a* tricuspid, tricuspidate; triconodont *(tooth)*
трёхка'мерный *a* three-celled, trilocular
трёхкилево'й *a* tricarinate
трёхкле'точный *a* three-celled
трёхко'льчатый *a* triannulate
трёхкры'лый *a* tripterous
трёхлепе'стный *a bot* tripetalous
трёхле'тний *a* triennial
трёхли'стный *a* trifoliate
трёхлопастно'й, трёхло'пастный *a* three-lobed, trilobate
трёхлучево'й *a* triactinal, triradial, triradiate, trilete
трёхнадре'зный *a* trifid
трёхне'рвный *a* triple-nerved, triple-ribbed, trinervate
трёхо'стный *a* triaristate
трёхпа'лый *a* tridactyl, tridactylous
трёхпа'рный *a bot* trijugate
трёхперегоро'дчатый *a* triseptate
трёх-пятиразде'льный *a bot* triquinate
трёхразде'льный *a* tripartite, trifid
трёхрассечё'нный *a* trisected
трёхрасщеплё'нный *a* trischistoclastic
трёхря'дный *a* tristichous, triplostichous, triserial, *bot* trifarious
трёхсемя'нный *a* trispermous
трёхсемяпо'чковый *a* triovulate
трёхсло'йный *a emb* tripoblastic
трёхспо'ровый *a* trisporous
трёхство'рчатый *a* tricuspid, tricuspidate, trivalve
трёхсто'лбиковый *a bot* tristylous
трёхтычи'нковый *a* triandrous
трёхцветко'вый *a* trianthous, triflorous
трёхцве'тный *a* tricolor
трёхчле'нистый *a* triarticulate
трёхчле'нный *a* trimerous, three-merous
трёхширокопо'ровый *a* trilatiporate
трёхщети'нковый *a* trisetose
треща'лка *f ent*: **двенадцатито'чечная ~** spotted asparagus beetle (*Crioceris duodecimpunctata*)
треща'ть *v* stridulate
тре'щина *f* crack, split; fissure (*e.g., of soil*)
триа'да *f* triad
~ **Бецо'льда** Bezold's triad
~ **Гетчинсо'на** Hutchinson's triad
~ **Картаге'нера** Kartagener's triad, Kartagener's syndrome
триа'совый *a* Triassic
три'ба *f* tribe
тривале'нт *m* trivalent
трига'мный *a bot* trigamous
тригемини'я *f phys* tripling
триге'нный *a* trigenic
тригетерозиго'та *f* triheterozygote
триждыперистонадре'зный *a* tripinnatifid
триждыперисторассечё'нный *a* tripinnatisect
триждыпе'ристый *a* ternatopinnate, tripinnate
триждытро'йчатый *a* triternate
триконодо'нтный *a* triconodont
трикорнави'рус *m* tricornavirus
трикроти'ческий *a* tricrotic
трилофодо'нтный *a* trilophodont
триме'рный *a* trimerous

тримити'ческий *a bot* trimitic
тримо'рфный *a* trimorphous
триномина'льный *a* trinominal
трипафлави'н *m* trypaflavine
трипле'кс *m* triplex
трипле'т *m* triplet
  иниции'рующий ~ initiation triplet
  термини'рующий ~ termination triplet
трипло'ид *m* triploid
триплоиди'я *f* triploidy
трипло'идный *a* triploid
трипси'н *m* trypsin
трипсино'лиз *m* tryptic cleavage
трипти'ческий *a* tryptic
трипто'за *f* tryptose
триптофа'н *m* tryptophane
трисахари'д *m* trisaccharide
трисо'мик *m* trisomic
трисоми'я *f* trisomy
трито'н *m* newt (*Triturus*)
тритони'мфа *f zool* tritonymph
тритоцере'брум *m ent* tritocerebrum
тритуберкуля'рный *a* tritubercular
триунгули'н *m* triungulin, triungulinid
трихи'дий *m bot* trichidium
трихи'лиум *bot* trichilium
трихи'на *helm* trichina *(Trichinella)*
трихи'т *zool* trichite
трихобла'ст *m* trichoblast
трихоги'на *f* trichogyne
трихо'м *m* trichome
трихотоми'ческий trichotomic, trichotomous
трихоци'ста *f* trichocyst
трихрои'чный thichroic
трихро'мный trichromatic
три'цепс *m* triceps
троглобио'нт *m* troglobiont
тройни'чный *a* trifacial *(nerve),* trigeminal
тройчатосло'жный *a* tricompound
тро'йчатый *a* ternate
тромб *m* thromb, thrombus
тромбастени'я *f* thrombasthenia
  ~ **Гла'нцмана** Glanzmann's disease
тромби'н *m* thrombin
тромбиноге'н *m* thrombinogen
тромбоге'н *m* thrombogen
тромбокина'за *f* thromboplastin
  тка'невая ~ tissue thromboplastin
тромбопласти'н *m* thromboplastin
  тканево'й ~ tissue thromboplastin
тромбоци'т *m* thrombocyte, blood platelet
тромбоцитопоэ'з *m* thrombocytopoiesis

тропа' *f* trail, path, run
  уче'бная ~ nature trail
тропибаза'льный *a* tropibasic
тропи'зм *m* tropism
  отрица'тельный ~ negative tropism
  положи'тельный ~ positive tropism
тропи'н *m* tropine
тропи'нка *f* path
тропи'ческий *a* tropical
тропота'ксис *m* tropotaxis
тропофи'т *m* tropophyte
тропофи'тный *a* tropophytic
тростни'к *m* cane, reed (*Phragmites*)
  са'харный ~ sugar cane (*Saccharum officinarum*)
тростнико'вый *a* arundinaceous
трофалла'ксис *m* trophallaxis
трофи'ческий *a* trophic
трофобио'з *m* trophobiosis
трофобиоти'ческий *a* trophobiotic
трофобла'ст *m* trophoblast
  вневорси'нчатый ~ extravillous trophoblast
  синцитиа'льный ~ syncytial trophoblast
трофоге'нный *a* trophogenic, trophogenous
трофоде'рма *f* trophoderma
трофодина'мика *f* trophic dynamics
трофодинами'ческий *a* trophic-dynamic
трофозо'ит *m* trophozoite, trophozooid
трофолити'ческий *a* tropholytic
трофоло'гия *f* trophology
трофо'нт *m* trophont
трофопла'зма *f* trophoplasm
трофоспо'нгий *m* trophospongium
трофота'ксис *m* trophotaxis
трофотропи'зм *m* trophotropism
трофоци'т *m* trophocyte
троханти'н *m ent* trochantin
трохофо'ра *f* trochophore, trochophore larva
труба' *f* tube, tuba
  евста'хиева ~ Eustachain tube, auditory tube
  фалло'пиева ~ Fallopian tube, uterine tube
труба'ч *m* whelk (*Buccinum*)
труби'ть *v* trumpet
тру'бка *f* tube; **вы'ход в тру'бку** shooting, stalk-shooting
  броди'льная ~ fermentation tube
  брю'шная ~ *ent* ventral tube
  дыха'тельная ~ air tube, breathing tube
  конъюгацио'нная ~ *micr* conjugation tube
  лё'гочная ~ *orn* lung tube

не'рвная ~ nerve tube
пыльцева'я ~ *bot* fertilization tube, pollen tube
ситови'дная ~ *bot* sieve tube
собира'тельная ~ *anat* receiving tube
трубковё'рт *m ent* roller; leaf-rolling weevils (*Attelabidae*)
трубкови'дный *a* tubulate, tubulated
трубкозу'б *m* ant bear (*Orycteropus*)
трубкообра'зный *a* tubulate, tabulated, tubate
тру'бный *a* tubal
тру'бочка *f* tubule
~ **Белли'ни** Bellini's tube
~ **Пфлю'гера** Pfluger's tubule
~ **Ре'йни** Rainey's tube
заро'дышевая ~ *bot* germ tube
паути'нная ~ spinnerule
проро'стковая ~ *bot* germ tube
ростко'вая ~ *bot* germ tube
яйцева'я ~ *ent* egg tube, ovarian tube
тру'бочник *m* sludge worm (*Tubifex*)
трубчатоцветко'вый *a* tubuliflorous
тру'бчатый *a* tubular, tubulose
трудоспосо'бность *f* working capacity
труп *m* cadaver, corpse
тру'пный *a* cadaveric, cadaverous
трупое'д *m* necrophage
тру'тень *m apic* drone
трутови'к *m myc* bracket fungus; tinder fungus (*Fomes*); sponk (*Polyporus*)
трухля'вый *a* rotten, easily falling to pieces
трэйдо'ф *ecol* trade-off
трю'фель *m myc* truffle (*Tuber*)
тряси'на *f* quag, slough, mire, swamp, scraw bog, swampland, quagmire
трясогу'зка *f orn* wagtail (*Motacilla*)
бе'лая ~ white wagtail *(Motacilla alba)*
жё'лтая ~ yellow wagtail *(Motacilla flava)*
трясу'нка *f bot* quacking grass (*Briza*)
туберкуле'з *m med* tuberculosis
~ **ле'гких** pulmonary tuberculosis, tuberculosis of the lungs, consumption, phthisis
туберкули'н *m* tuberculin
тубули'н *m* tubulin
тубуля'рный *a* tubular
тубуля'ция *f* tubulation
туво'дный *a* nonmigratory (*of fish*)
туга'й *m* tugai (*bottomland forest in Central Asia*)
тугоро'слый *a* slow-growing
тузе'мный *a* aboriginal, indigenous
тузлу'к *m* brine
тук *m* fertilizer
тука'н *m orn* toucan *(Rhamphastos)*
ту'лес *m* gray plover, pigeon plover (*Squatarola squatarola*)
ту'ловище *n* trunk, body
туляреми'я *f* tularemia, deer-fly fever, rabbit disease
тунг *bot* tung *(Aleurites)*
ту'ндра *f* tundra
аркти'ческая ~ barren-grounds tundra
бугри'стая ~ hilly tundra
камени'стая ~ stony tundra
кочка'рная ~ hummocky tundra, tussock tundra
лиша'йниковая ~ lichen tundra
осо'ково-пуши'цевая ~ sedge cotton-grass tundra
травяни'стая ~ grass tundra
ту'ндровый *a* tundra
тундряно'й *a only in:* **тундряна'я куропа'тка** rock ptarmigan (*Lagopus mutus*)
туни'ка *f* tunic
тунне'ль *m anat* tunnel
~ **Нэ'еля** Nuel's space, adjoining tunnel, outer tunnel
вну'тренний ~ inner tunnel, Corti's tunnel
ко'ртиев ~ Corti's tunnel, inner tunnel
ту'пик *m* pal, puffin (*Fratercula arctica*)
атланти'ческий ~ Atlantic puffin *(Fratercula arctica)*
тихоокеа'нский ~ horned puffin *(Fratercula corniculata)*
тупо'й *a* blunt
тупоконе'чный *a* blunt, obtuse
тупокры'лый *a* amblyopterous
туполи'стный *a* amblyophyllous
тупопло'дный *a* amblyocarpous
тур *m* auroch, wild ox (*Bos primigenius*)
восточнокавка'зский ~, дагеста'нский ~ East Caucasian tur *(Capra cylindricornis)*
западнокавка'зский ~, куба'нский ~ Caucasian tur, West Caucasian tur *(Capra caucasica)*
тура'ч *m* francolin (*Francolinus*)
турбиди'метр *m* turbidimeter
турбидиметри'я *f* turbidimetry
турбоспина'льный *a* torticone
тургесце'нтный *a* turgid
тургесце'нция *f* turgescence

ту'ргор *m* turgor, turgor pressure, turgescence
турио'н *m bot* turio, turion
турма'н *m orn* tumbler pigeon, tumbler
турпа'н *m orn* scoter (*Melanitta*)
турне'пс *m* turnip (*Brassica rapa*)
ту'хлый *a* tainted (*of meat*); rotten, bad, addle; **тухлое яйцо'** addle egg, rotten egg
ту'чность *f* adiposity
ту'чный *a* gross
ту'ша *f* carcass
ту'шка *f* carcass
тушка'нчик *m* jerboa (*Dipodidae*)
ту'я *f* arborvitae (*Thuja*)
Т-ферме'нт *m* T-enzyme
ты'ква *f* squash, gourd (*Cucurbita pepo*); pepo
~ **буты'лочная** calabash, gourd, bottle gourd, calabash gourd
~ **моча'лочная** dishcloth gourd, sponge gourd (*Luffa cylindrica*)
ты'ква-горля'нка *f* crookneck squash (*Cucurbita pepo var. condensa*)
ты'квенные *n pl* gourds (*Cucurbitaceae*)
ты'квина *f* pepo
ты'льный *a* rear, back
тысячели'стник *m* yarrow (*Achillea*)
тычи'нка *f bot* stamen; raker (*of fish gills*)
двуси'льная ~ didynamous stamen
тычинконо'сный *a* staminiferous, staminate, stamen-bearing
тычи'ночный *a* staminal
тю'вик *m* Levant sparrow hawk (*Accipiter brevipes*)
тюле'нь *m* seal (*Phoca*)
тюльпа'н *m* tulip *(Tulipa)*
тя'вканье *n* yapping
тя'вкнуть *v* yelp
тя'га *f* roding (*of woodcocks*); draft; стоять на тяге hunt woodcocks at mating season
гужева'я ~ animal draft
тягу'чий *a* tough
тяж *m* strand, bundle, cord, band
~ **Би'льрота** Billroth's strand, red pulp cord, Billroth's cord
заро'дышевый ~ germinal cord
лубяно'й ~ bast strand
мы'шечный тяж muscular band
не'рвный ~ nerve cord
перви'чный полово'й ~ primary sex cord
прокамбиа'льный ~ protocambial strand
протоксиле'мный ~ protoxylem strand
сухожи'льный тяж tendinous band
тяжелово'з *m* heavy vanner
тяже'лый *a* heavy; grave (*of condition*), serious (*of illness*), severe, acute
тяну'ть *v* pull, haul
~ **за поводо'к** strain at the leash

У

уабаи'н *m* ouabain
убе'жище *n* retreat, shelter, refuge, refugium, covert
убива'ть *v* kill
убикви'ст *m* ubiquist
убира'ть *v* take in (*the yield*); remove
~ **урожа'й** harvest, reap, gather in, take in
убихино'н *m* ubiquinone
убо'й *m* killing; slaughter
убо'рка *f* harvesting, reaping, gathering in, cutting; cleaning up
убы'ток *m* loss
увеличе'ние *n* magnification, magnification power, magnifying power; increase; augmentation, enhancement, enlargement
уве'нчанный *a* coronate, coronated
увё'ртливый *a* slippery, evasive
уве'чье *n* mutilation
увлажне'ние *n* humidification
увлажня'ть *v* humidify
увяда'ние *n* withering, wilt, wilting
~ **цветко'в** *phytop* blossom wilt
фузарио'зное ~ хлопча'тника *phytop* cotton wilt *(caused by Fusarium oxysporum vasinfectum)*
увяда'ть *v* wither, fade, wilt
увя'дший *a* withered, faded, wilted, sphacelated
угаса'ние *n* extinction, decline
~ **реа'кции** response decrement
уга'сший *a* quenched (*e.g., a reflex*)
углево'д *m* carbohydrate
углеводоро'д *m* hydrocarbon
углеро'д *m* carbon
о'бщий органи'ческий ~ total organic carbon
углова'тый *a* angular, angulate
углово'й *a* angular
углозу'б *m* Asiatic salamander, oriental salamander *(Hynobiidae)*
сиби'рский ~ Siberian salamander *(Salamandrella)*
углоне'рвный *a bot* angulinerved
углубле'ние *n* depression, hollow, recess
угнета'ть *v* inhibit, depress
угнета'ющий *a* inhibitory
угнете'ние *n* inhibition
угнете'нный *a* oppressed, suppressed, stunted, inhibited
угова'ривание *n* coaxing (*of a horse*)
уго'дье *n* land
воднболо'тные уго'дья *pl* wetlands
лесно'е ~ forest land
па'стбищное ~ pasture, grazing land, rangeland, lea, ley
рыболо'вное ~ fishing area, fishing ground
у'гол *m* angle
~ **ветвле'ния** branch angle
~ **гелиотропи'и** heliotropic angle
~ **груди'ны** sternal angle
~ **зре'ния** visual angle
~ **накло'на** tilt angle
~ **ни'жней че'люсти** gonial angle
~ **поворо'та** *mbiol* tilt angle
~ **серде'чно-диафрагма'льный** ~ cardiophrenic angle
акромиа'льный ~ acromial angle
лицево'й ~ gnathic angle
о'стрый ~ acute angle, sharp angle
плечево'й ~ humeral angle
прямо'й ~ right angle
ре'берный ~ costal angle
си'львиев ~ Sylvian angle
тупо'й ~ obtuse angle, blunt angle
челюстно'й ~ angle of jaw, angle of mandible
че'репно-лицево'й ~ craniofacial angle
уголо'к *m* corner
живо'й ~ nature corner, pet's corner
у'голь *m* coal
активи'рованный ~ activated carbon
бу'рый ~ brown coal, lignite
древе'сный ~ charcoal
ка'менный ~ coal
у'горь *m med* comedo; *zool* eel (*Anguilla*)
угри'ца *f* paste eel (*Anguillulidae*); follicle mite (*Demodex folliculorum*)
угро'за *f* threat
~ **откры'той па'стью** *eth* gape threat
удале'ние *n* excision, extirpation
удаление кожуры' skinning
удаля'ть *v* remove
удалять кожуру' skin
уда'р *m* stroke, beat, blow
теплово'й ~ heat stroke
удвое'ние *n* doubling
хромосо'м ~ chromosome doubling
сомати'ческое ~ somatic doubling
удлине'ние *n* elongation

~ **междоу'злия** elongation of internode
~ **коло'сьев** elongation of spikes
удобре'ние *n* fertilizer; fertilization
азо'тное ~ nitrogenous fertilizer
заде'ланное ~ buried fertilizer
слежа'вшееся ~ caked fertilizer
ста'ртовое ~ starter fertilizer
удобря'ть *v* fertilize
удовлетворе'ние *n* satisfaction
~ **поведе'нческого стремле'ния** drive satisfaction
~ **биологи'ческой потре'бности** need satisfaction
удовлетворительность *f* satisfactoriness
~ **пита'ния** nutritional sufficiency
удо'д *m orn* hoopoe (*Upupa epops*)
удуша'ть *v* suffocate
уду'шье *n* suffocation, asphyxia
уж *m herp* grass snake (*Natrix*)
ужа'ление *n* sting
уза' *f apic* propolis, bee glue
узда' *f* bridle
узде'чка *f anat* retinaculum, frenulum, habena, habenula; bridle
~ **языка'** frenulum of the tongue
у'зел *m* node, knot, knob; ganglion
~ **Ашо'фа-Тава'ра** Aschoff-Tawara node
~ **Ки'са-Фля'ка** Keith-Flack node
~ **куще'ния** tillering node
кише'чный лимфати'ческий ~ mesenteric lymph node, mesenteric lymphatic node
лимфати'ческий ~ lymph node, lymphatic node, absorbent gland, conglobate gland, globate gland
многолаку'нный ~ multilacunar node
не'рвный ~ nerve knot
однолаку'нный ~ unilacunar node
перви'чный ~ primitive node
предсе'рдно-желу'дочковый ~ Aschoff-Tawara node, atrioventricular node
регионa'льный лимфати'ческий ~ local lymph node, local lymphatic node
рема'ков ~ Remak's ganglion
семядо'льный ~ cotyledonary node
си'нусно-предсе'рдный ~ sinoauricular node, Keith-Flack node
си'нусный ~ sinoauricular node, Keith-Flack node
ше'йный ~ cervical node
узелко'вый *a* nodular
узело'к *m* nodule
ареоли'рованный ло'жный ~ areolate pseudonodulus *(in diatoms)*
блестя'щий ло'жный ~ luminate pseudonodulus *(in diatoms)*
ге'нзеновский ~ Hensen's node
коне'чный ~ terminal nodule *(in diatoms)*
перви'чный ~ *emb* primitive knot
центра'льный ~ central nodule *(in diatoms)*
эмбриона'льный ~ embryonal knot
узкоголо'вчатый *a bot* stenocephalous
узкоголо'вый *a anat* stenocephalous
узковетви'стый *a* stenocladous
узкокисти'стый *a* stenobotryous
узкоколоси'стый *a* stenostachyous
узколу'ковичный *a* stenobulbous
узкопло'дный *a* stenocarpous
узколепе'стный *a* narrow-petaled, stenopetalous
узколи'стный *a* angustifoliate, narrow-leaved, stenophyllous
узколо'пастный *a* stenolobate
узкоперегоро'дчатый *a pal* angustiseptate
узкоро'тый *a* narrow-mouthed, stenostomatous
узкосло'йный *for a* narrow-ringed, narrow-zoned
узкоте'лка *f* cylindrical bark beetle (*Colydiidae*)
узкоцветко'вый *a* narrow-flowered
узлова'тый *a* gnarled, knotty, nodose, nodulous, nodular, nodulated, knobby, knotted, torose
узнава'ние *n* recognition
~ **по ти'пу «подо'бный - подо'бное»** *imm* like-like recognition
узо'р *m* pattern, desingn
ука'чивание *n* motion sickness, motion disease, sea sickness; rocking, lulling
укла'дка *f* folding
~ **ДНК** DNA-folding
~ **субъедини'ц** subunit folding
трёхме'рная ~ three-dimensional folding
укорене'ние *n* rooting
укореня'емость *f* rooting ability
укореня'ться *v* root, take root; establish
укореня'ющийся *a* rooting
пове'рхностно ~ shallow rooting, superficially rooting
уко'с *m* hay crop
укра'шенный *a* adorned
укро'п *m bot* dill (*Anethum*)

укрыва'ть *v* cover up
~ **соло'мой** straw
укры'тие *n* cover; hide; shelter
уку'с *m* bite, snap; sting (*of an insect*)
игрово'й ~ *eth* play-biting, pseudobiting
укусить *v* bite
укуси'ть за *v* bite at, bite in
ула'р *m orn* snow partridge, snow cock (*Tetraogallus himalayensis*)
у'лей *m apic* hive, beehive
ули'т *m orn* yellowlegs, yellowshanks (*Tringa*)
большо'й ~ greenshank (*Tringa nebularia*)
желтоно'гий ~ lesser yellowlegs *(Tringa flavipes)*
пе'пельный ~ gray-rumped sandpiper (*Heteroscelus brevipes*)
пё'стрый ~ greater yellowlegs, winter yellowlegs *(Tringa melanoleuca)*
ули'тка *f* snail (*Gastropoda*); cochlea; bostryx; helicoid cyme
виногра'дная ~ edible snail, French snail, grapevine snail (*Helix pomatia*)
улиткообра'зный *a* helical, helicine
уло'в *m* take, catch, landing, haul
~ **на едини'цу уси'лия** catch per unit effort, catch per unit of effort
~ **на уси'лие** catch per unit effort, catch per unit of effort
промысло'вый ~ commercial catch
устойчивый ~ sustainable catch
уло'вистость *f* catching rate, catching efficiency
ультра- ultra-
ультрави'рус *m* ultravirus
ультразву'к *m* ultrasound
ультразвуково'й *a* ultrasonic
ультракриото'мия *f* ultracryotomy
ультрамикровесы' *pl* ultramicrobalance
ультрамикроско'п *m* ultramicroscope
ультрамикрото'м *m* ultramicrotom
ультрамикрото'мия *f* ultramicrotomy
ультрамикроэлектро'д *m* ultramicroelectrode
ультрапастериза'ция *f* ultrapasterization
ультраструкту'ра *f* ultrastructure
ультраусто'йчивость *f* ultrastability
ультрафи'льтр *m* ultrafilter
ультрафильтра'ция *f* ultrafiltration
ультрафиоле'товый *a* ultraviolet
ультрацентрифу'га *f* ultracentrifuge
углова'я ~ angle-head ultracentrifuge
препарати'вная ~ preparative ultracentrifuge
ультрацентрифуги'рование *n* ultracentrifugation
уме'ние *n* skill, ability
уменьше'ние *n* narrowing
уме'ренный *a* moderate; temperate
~ **кли'мат** temperate climate
умеренные широ'ты *pl* temperate latitudes
умерщвле'ние *n* killing
умира'ть *v* die
умиротворе'ние *n* appeasement
у'мственный *a* mental
умыва'ние *n* washing
~ **мочо'й** *eth* urine washing
унаво'женный *a* dunged
унаво'живать *v* manure, dung
ундулипо'дия *prot* undulipodium
ундули'рующий *a* undulatory, undulating
ундуля'ция *f* undulation
унивале'нт *m* univalent
за'мкнутый ~ closed univalent
ло'жный ~ false univalent, open univalent
пермане'нтный ~ permanent univalent
спонта'нный ~ spontaneous univalent
унивольти'нный *a* univoltine
униже'ние *n* abjection
униполя'рный *a* unipolar
уничтоже'ние *n* extermination
у'нкус *m* uncus
упако'вка *f* packing
упериза'ция *f* ultrapasterization
упи'танность *f* fatness, finish, fleshing
вы'сшая ~ best finish
ни'зкая ~ poor finish
сре'дняя ~ medium finish
упи'танный *a* finished, fattened, fatted, nourished, well-fed, fleshed, in flesh
пло'хо ~ thin-fleshed
уплотне'ние *n* packing (*of a cenosis*)
уплоще'нный *a* flat, flattened, applanate
упо'р *m*: **смотре'ть в** ~ stare at
употребле'ние *n* use
нару'жное ~ external use
управле'ние *n* management
дистанцио'нное ~ remote control
упря'жка *f* team
ве'ерная ~ fan formation (*of dogs*)
у'пряжь *f* harness, tackle, gear
упря'мый *a* balky, headstrong
уравне'ние *n* equation

~ **для несмещё'нной оце'нки** *biom* unbiased estimating equation
~ **Михаэ'лиса-Ме'нтен** Michaelis-Menten equation
~ **ро'ста** growth equation
~ **Ха'рди-Ва'йнберга** Hardy-Weinberg equation
аллометри'ческое ~ allometric equation
логи'ческое ~ ро'ста logistic equation of growth
характеристи'ческое ~ characteristic equation

урага'н *m* hurricane
ураци'л *m* uracil
урединоспо'ра *f* uredinospore
уреди'ниум *m* uredinium
уредопу'стула *f* uredopustule
уредоспо'ра *f* uredospore, brand spore
уредоста'дия *f* uredostage
урё'ма *f for* urema
уреотели'ческий *a* ureotelic
уре'тра *f* urethra
уретри'т *m* urethritis
уриа'л *m ther* urial *(Ovis ammon vignei)*
уриди'н *m* uridine
урикотели'ческий *a* uricotelic
уркарио'т *m* urkaryote
у'рна *f* urn
у'ровень *m* level
~ **восстановле'ния** replacement level
~ **грунто'вых вод** water table
~ **зна'чимости** *biom* level of significance, significance level
~ **органи'зменный** ~ organismic level
~ **па'стбищной нагру'зки** stocking level
~ **стабилизи'рованный** ~ stabilized level
~ **чи'сленности популя'ции** population level
возмеща'ющий ~ replacement level
довери'тельный ~ *biom* confidence level
допусти'мый ~ permissible level; *biom* tolerance level
крити'ческий ~ *biom* critical level
трофи'ческий ~ trophic level
энергети'ческий ~ energy level

уро'д *m* freak, monster
уроде'ум *m anat* urodaeum
уро'дливость *f* teratism, deformity, monstrosity, malformation
уро'дливый *a* misshapen, malformed, teratic
уро'дство *n* monstrosity, malformation, teratism, deformity
врождё'нное ~ congenital malformation
урожа'й *m* crop, yield, harvest
небыва'лый ~ bumper crop
урожа'йность *f* yielding ability, crop capacity, level of crop yield, crop-producing power
уроло'гия *f* urology
урон *m* loss
уропи'гий *m* uropigium
уропо'д *m* uropod
уросо'ма *f* urosome
урости'ль *m* urostyle
урохо'рд *m* urochord
урохро'м *m* urochrome
уро'чище *n* tract, area, site
урча'ние *n* rumble, rumbling, growling
урча'ть *v* rumble, grumble, growl
урю'к *m* dried unpitted apricot
ус *m zool* moustache, mustache; whisker; barb; feeler; antenna; wattle (*of fish*); *bot* tendril, runner, sarment
уса'дка *f* shrinkage loss
уса'ч *m* capricorn beetle (*Cerambycidae*); barbel (*Barbus*)
усва'ивать *v* assimilate
усва'ивающий *a* assimilative, assimilatory
усвое'ние *n* assimilation
усече'нный *a* truncate, truncated
у'сик *m* antenna, feeler, cirrus, cirrhus; barb, barbel, barbule (of a fish); *bot* tendril
~ **листово'го происхожде'ния** leaf tendril
булавови'дный ~ clavate antenna, club-shaped antenna
голо'вчатый ~ capitate antenna, knobbed antenna
гребневи'дный ~, гребе'нчатый ~ comb-like antenna, pectinate antenna
зу'бчатый ~ dentate antenna
коле'нчатый ~ geniculate antenna, crooked antenna; jointed tendril
листово'й ~ leaf tendril
нитеви'дный ~ filiform antenna, thread-like antenna
пилови'дный ~, пи'льчатый ~ saw-like antenna, serrate antenna
пласти'нчатый ~ lamellate antenna
раздво'енный ~ dichotomous antenna
суста'вчатый ~jointed tendril
узлова'тый ~ nodose antenna; jointed tendril
чёткови'дный ~ moniliform antenna, necklace antenna

чешу'йчатый ~ imbricate antenna
щетинкови'дный ~ bristle-like antenna, setaceous antenna
усикови'дный *a* cirriform, antennatus
усиле'ние *n* enhancement, amplification, increase
иммунологи'ческое ~ immunological enhancement
усили'тель *m* amplifier
~ **биоэлектри'ческих сигна'лов** bioelectrical amplifier
~ **переме'нного то'ка** alternating current amplifier
~ **с обра'тной свя'зью** feedback amplifier
~ **электри'ческих и'мпульсов** electric pulse amplifier
дифференциа'льный ~ differential amplifier
электро'нно-лучево'й ~ electron-beam amplifier
усло'вия *pl* conditions
~ **вне'шней среды'** ambient conditions, environmental conditions
~ **кислоро'дного голода'ния** anoxic conditions
~ **культиви'рования** culture conditions
~ **освеще'ния** light conditions
~ **существова'ния** living conditions
благоприя'тные ~ favorable conditions
вне'шние ~ ambient conditions, environmental conditions
избира'тельные ~ selective conditions
климати'ческие ~ climatic conditions
контроли'руемые ~ controlled conditions
нача'льные ~ initial conditions, starting conditions
неблагоприя'тные ~ adverse conditions, unfavorable conditions
неразреша'ющие ~ nonpermissive conditions
окружа'ющие ~ ambient conditions, environmental conditions
пермисси'вные ~ permissive conditions
пограни'чные ~ boundary conditions
постоя'нные ~ stationary conditions
разреша'ющие ~ permissive conditions
стациона'рные ~ stationary conditions
успе'х *m* success
~ **размноже'ния** breeding success
~ **спа'ривания** mating success
уста'лость *f* fatigue
устано'вка *f* installation, plant; adjustment
гру'бая ~ coarse adjustment (*of a microscope*)
о'пытная ~ pilot plant
то'чная ~ fine adjustment (*of a microscope*)
усто'йчивость *f* stability; resistance
~ **гибри'дов** hybrid resistance
~ **к антибио'тику** antibiotic resistance
бактериа'льная ~ bacterial resistance
вну'тренняя ~ inherent resistance
мно'жественная ~ multiple resistance
мутаге'нная ~ mutagene stability
насле'дственная ~ heritable stability
усто'йчивый *a* resistant; sustained, sustainable
устраша'ющий *a* offensive
устраше'ние *n* intimidation
у'стрица *f* oyster (*Ostrea*)
у'стье *n* orifice; mouth (*of a river*); *anat* ostium, opening; *zool* shell opening
у'стьице *n* stoma
водяно'е ~ water stoma
у'стьичный *a* stomatal
усу'шка *f* shrinkage loss
усы' *pl* moustache, mustache; *zool* whiskers
утё'нок *m* duckling
уте'чка *f* leak, leakage; *gen* wastage
у'тка *f* duck (*Anatidae*)
древе'сная ~ wood duck, tree duck (*Dendrocygna*)
летя'щая на юг ~ southing duck
подсадна'я ~ decoy-duck
се'рая ~ gadwall *(Anas strepera)*
у'тка-широконо'ска *f* broadbill (*Spatula clypeata*)
утконо'с *m* duckbill, platypus (*Ornithorhynchus*)
утолще'ние *n* node
утолще'нный *a* incrassate, incrassated
утомле'ние *n* fatigue
утомля'емость *f* fatiguability
утомля'ть *v* fatigue
у'точка *f diminutive of* **утка**
уха' *f* fish soup
у'ха *genitive of* **у'хо**
уха'живание *n eth* courtship, courtship behavior
совме'стное ~ communal courtship
у'ханье *n:* ~ **совы'** the screech of the owl
у'хать *v* hoot (*of owls*)
у'хо *n* ear
вну'треннее ~ internal ear
нару'жное ~ external ear, outer ear

сре'днее ~ middle ear
медве'жье ~ *bot* common mullein, mullein dock, shepherd's club (*Verbascum thapsus*)
уховё'ртка *f* earwig (*Forficula*)
ухови'дный *a* earlike
ухо'д *m* care, management
~ **за больны'ми** nursing
уходи'ть *v* go off, get away, escape
~ **в нору'** take the earth
ухудше'ние *n* degradation, deterioration; aggravation
уча'сток *m* plot, area, ground, range, site, region
~ **земли'** ground, plot
~ **инициа'ции** initiation site
~ **обита'ния** home range
~ **отведе'ния** *phys* recording site
~ **размноже'ния** increase plot, propagation plot
~ **распознава'ния** recognition site, R site
~ **сбо'рки** assembly site
~ **термина'ции** *mbiol* termination site
акти'вный ~ active site
акце'пторный ~ acceptor site
аллостери'ческий ~ allosteric site
вариа'бельный ~ **антите'ла** antibody variable region
встро'енный акце'пторный ~ built-in acceptor site
гаптенсвя'зывающий ~ hapten-binding site
гибри'дный о'пытный ~ composite plot
гипервариа'бельный ~ *imm* hypervariable region, complementarity-determining region
декоди'рующий ~ decoding site
дополни'тельный ~ additional site
запре'тный ~ sanctuary
защищё'нный концево'й ~ *mbiol* cryptic end site
иммунокомплекссвя'зывающий ~ immune complex binding site
индивидуа'льный ~ home range
ине'ртный ~ **хромосо'мы** inert chromosome region
испо'льзуемый ~ action area
каталити'ческий ~ catalytic site
комплементсвя'зывающий ~ complement fixation site, complement fixing site
лесно'й ~ wood lot, forest division
межцистро'нный ~ intercistronic region
о'пытный ~ plot, experimental plot, trial plot
перекрыва'ющийся ~ overlapping site
постоя'нный ~ **антите'ла** antibody constant region
приуса'дебный ~ home grounds
рекомбини'рованный ~ exchanged region
рецепто'рный ~ receptor site
рыболо'вный ~ fishing ground
свя'зывающий ~ binding site
эктопи'ческий ~ ectopic site
эффе'кторный ~ *mbiol* effector site
учё'т *m* inventory, census
~ **птиц** bird count, bird census
~ **чи'сленности** census
абсолю'тный ~ **чи'сленности** true census
уша'стый *a zool* eared
у'ши *pl of* **у'хо**
стоя'чие ~ stand up-ears, straight ears
ушиб *m* hurt, contusion
ушкова'тый *a bot* eared
ушкови'дный *a* auriculate, auriform, earlike
ушно'й *a* aural, otic
ущемле'ние *n* strangulation, incarceration, entrapment
ущемле'нный *a* strangulated, incarcerated
уще'рб *m* damage
уязви'мость *f* vulnerability
уязви'мый *a* vulnerable

Ф

фа'брика *f* factory
таба'чная ~ tobacco house
фаг *m* phage
вегетати'вный ~ vegetative phage
вируле'нтный ~ strong phage, lytic phage, virulent phage
внекле'точный ~ free phage
внутрикле'точный ~ intracellular phage
гетероимму'нный ~ heteroimmune phage
дефе'ктный ~ defective phage
ДНК-содержа'щий ~ DNA phage
зави'симо-вируле'нтный ~ dependent-virulent phage
зре'лый ~ free phage, mature phage
крипти'ческий ~ cryptic phage
лате'нтный ~ latent phage, symbiotic phage
лизоге'нный ~ lysogenic phage, carried phage
лити'ческий ~ lytic phage, strong phage, virulent phage
ме'лкий ~ minute phage
непо'лный ~ incomplete phage
ни'тчатый ~ filamentous phage
поко'ящийся ~ resting phage
ра'нний ~ early phage
РНК-содержа'щий ~ RNA-phage
роди'тельский ~ parent phage
симбиоти'ческий ~ latent phage, symbiotic phage
сла'бый ~ weak phage
типирующий ~ typing phage
Т-чё'тный ~ T-even phage
трансдуци'рующий ~ transducing phage
уме'ренный ~ temperate phage
фаголизосо'ма *f* phagolysosome
фагосо'ма *f* phagosome
фаготипи'рование *n* phage-typing, bacteriophage typing, phagotyping
фаготро'ф *m* phagotroph
фаготрофи'я *f* phagotrophy
фаготро'фный *a* phagotrophic
фагоусто'йчивость *f* phage resistance
фагоци'т *m* phagocyte
фагоцита'рный *a* phagocytal, phagocytic
фагоцити'ровать *v* phagocytize, phagocyte
фагоцито'з *m* phagocytosis
фа'г-помо'щник *m* helper phage
фа'за *f* stage, phase, period,
~ **ассимиля'ции** assimilating stage
~ **бутониза'ции** bud stage, bud-formation period
~ **восково'й спе'лости** wax stage, wax-ripeness stage
~ **вымё'тывания** bot heading stage, paniculation stage
~ **дыха'ния** respiratory phase
~ **завя'зывания плодо'в** fruit-setting period
~ **«закры'тия»** *imm* "shutdown" phase
~ **изгна'ния** *phys* ejection phase
~ **колоше'ния** heading stage, earing stage, paniculation stage
~ **кри'зиса** *ent* critical phase
~ **куще'ния** tillering period
~ **изгна'ния** *phys* ejection stage, sphygmic phase
~ **изометри'ческая** ~ *phys* isometric phase
~ **изометри'ческого напряже'ния** *phys* isometric phase
~ **колоше'ния** paniculation stage
~ **куще'ния** tillering stage
~ **логарифми'ческого ро'ста** log phase, logarithmic phase
~ **моло'чной спе'лости** milk stage
~ **опорожне'ния** *phys* ejection phase, sphygmic phase
~ **плодоноше'ния** fruiting stage
~ **поко'я** resting stage, non-cycling stage
~ **появле'ния всхо'дов** seedling stage
~ **распознава'ния** *imm* recognition phase
~ **расслабле'ния** *phys* relaxation phase
~ **распуска'ния листьев** leafing stage
~ **ро'ста** growth phase
~ **сенсибилиза'ции** *imm* sensitization phase
~ **созрева'ния плодо'в** fruit ripening stage
~ **стеблева'ния** shoot stage, shooting stage, stooling stage, boot stage, booting stage
~ **сцепле'ния** coupling phase
~ **тестообра'зной спе'лости** dough stage
~ **торможе'ния** inhibitory phase
~ **трубкова'ния** shoot stage, shooting stage, stooling stage, boot stage, booting stage
~ **цвете'ния** flowering stage, flowering phase, flowering period, blooming stage
~ **экспоненциа'льного ро'ста** exponential phase
восходя'щая ~ rising phase

генерати'вная ~ reproductive period
деструкти'вная ~ destructive phase
индукцио'нная ~ induction phase
крити'ческая ~ critical phase
лате'нтная ~ latent phase
логарифми'ческая ~ log phase, logarithmic phase
нисходя'щая ~ coupling phase
парадокса'льная ~ paradoxical phase
продрома'льная ~ prodromal phase
ростова'я ~ growth phase
скры'тая ~ latent phase
стациона'рная ~ *micr* steady stage
стациона'рная ~ ро'ста stationary growth phase
твё'рдая ~ solid phase
травяни'стая ~ herbaceous stage
фенологи'ческая ~ phenoogical stage
э'клипсная ~ eclipse phase
экспоненциа'льная ~ exponential phase
эрупти'вная ~ *ent* eruptive phase
эффе'кторная ~ *imm* effector phase

фаза'н *m* pheasant (*Phasianus*)
~ **а'ргус** Argus pheasant (*Argusianus argus*)
зелё'ный ~ green pheasant *(Phasianus versicolor)*
золото'й ~ golden pheasant *(Chrysolophus pictus)*
обыкнове'нный ~ common pheasant, English pheasant *(Phasianus colchicus)*
уша'стый ~ eared pheasant *(Crossoptilon)*

фази'рование *n* phasing

фазми'да *f* phasmid

файл *m* file

фа'ктор *m* factor, agent
~ **Бе'ренса** Behrens factor
~ **вла'жности** hydric factor
~ **вне'шней среды'** environmental factor
~ **вре'мени** time factor
~ **высвобожде'ния** *mbiol* release factor
~ **защищё'нности** safety factor
~ **инду'кции миелопоэ'за** myelopoietic factor
~ **инициа'ции** initiating factor
~ **Ло'уренса** transfer factor
~ **«нейтра'льности»** *gen* neutral factor
~ **ослабле'ния** diluting factor
~ **отбо'ра** selective factor
~ **переда'чи усто'йчивости** resistance transfer factor
~ **перено'са** transfer factor
~ **плеча'** *gen* humerus factor
~ **распла'стывания макрофа'гов** macrophage spreading factor
~ **реакти'вности** reactive factor, R factor
~ **ро'ста** growth factor
~ **свё'ртывания кро'ви** clotting factor, blood-coagulation factor, coagulation factor
~ **си'гма** sigma factor
~**слипа'ния** clumping factor
~ **соотноше'ния поло'в** sex-ratio factor
~**сопряже'ния** coupling factor
~ **среды'** environmental factor
~ **Стю'арта** Stuart factor
~ **супре'ссии** suppressor factor
~ **термина'ции** termination factor
~ **термина'ции транскри'пции** rho factor
~ **усиле'ния** extension factor
~ **Фле'тчера** Fletcher's factor
~ **фон Виллебра'нда** von Willebrand factor
~**элонга'ции** elongation factor
~ **эффекти'вности** efficiency factor
абиоти'ческий ~ abiotic factor
аддити'вный ~ additive factor
анафилактоге'нный ~ anaphylactogenic factor
антиген-специфи'ческий хе'лперный ~ antigen-specific helper factor
антиген-специфи'ческий ~ супре'ссии antigen-specific suppressor factor
антропоге'нный ~ anthropogenic factor
аутокри'нный ~ роста autocrine growth factor
афа'зный лета'льный ~ aphasic lethal
безусловный лета'льный ~ unconditional lethal
биоти'ческий ~ biotic factor
веду'щий ~ major factor
вне'шний ~ external factor, external agent
вну'тренний ~ internal factor, intrinsic factor
возрастно'й ~ age factor
вре'дный ~ noxious agent
вспомога'тельный ~ accessory factor
второстепе'нный ~ minor factor
гамети'ческий ~ gametic factor
гамети'ческий лета'льный ~ gametic lethal, haplophasic lethal
генети'ческий ~ genetic factor
главный ~ primary factor
гормона'льный ~ hormonal factor
дополни'тельный ~ accessory factor
заде'рживающий ~ arresting factor

зиготи'ческий лета'льный ~ zygotic lethal
иммуномодули'рующий ~ immunomodulating factor
иниции'рующий ~ trigger factor
климати'ческий climatic factor
ключево'й ~ key factor
колониестимули'рующий ~, КСФ colony-enhancing factor, colony-stimulating factor
комплемента'рный ~ complementary factor
кормово'й ~ food factor
кумуляти'вный ~ cumulative factor
лета'льный ~ lethal factor
лимити'рующий ~ limiting factor
ме'нделевский ~ Mendelian factor
многофа'зный лета'льный ~ pluriphasic lethal
монофа'зный лета'льный ~ monophasic lethal
мотивацио'нный ~ motivational factor
мутаге'нный ~ mutagenic agent
нару'жный ~ external factor
насле'дственный ~ hereditary factor
ограни'чивающий ~ limiting factor
определя'ющий ~ master factor
первичный ~ primary factor
пищево'й ~ food factor
плазмати'ческий ~ plasma factor
повреждa'ющий ~ disturbing factor
по'чвенный ~ edaphic factor
причи'нный ~ causative factor
провоци'рующий ~ initiating agent
психологи'ческий ~ psychological factor
разветвля'ющий ~ branching factor
расслабля'ющий ~ relaxing factor
регули'рующий ~ control factor
ростово'й ~ growth factor
супре'ссорный ~ suppressor factor
сы'вороточный ~ serum factor
температу'рный ~ thermal factor
термостаби'льный ~ heat-stable factor
тормозя'щий ~ inhibitive factor
тра'нсферный ~ transfer factor
трансформи'рующий ~ transforming factor
трофи'ческий ~ trophic factor
усло'вный лета'льный ~ conditional lethal
формообразу'ющий ~ formative factor
хе'лперный ~ helper factor
хемотакси'ческий ~ chemotactic factor
цитопати'ческий ~ cytocidal factor, cytopathic factor
цитопатоге'нный ~ cytopathogenic agent
эдафи'ческий ~ edaphic factor
экологи'ческий ~ ecological factor
этиологи'ческий ~ etiological factor, causative agent

фала'нга *f* phalange, phalanx
фа'ллос, фа'ллус *m* phallus, penis
фанерога'мный *a* phanerogamous
фармаколо'гия *f* pharmacology
фармакопе'йный *a* pharmacopeial
фармакопе'я *f* pharmacopeia, pharmacopoeia
фармаце'вт *m* pharmacist, pharmaceutist, apothecary
фармацевти'ческий *a* pharmaceutic, pharmaceutical
фарма'ция *f* pharmaceutics
фарнохино'н *m* farnoquinone
фасе'тка *f* facet
фасо'ль *f* kidney bean (*Phaseolus*)
фасциа'льный *a* fascial
фасциа'ция *f* fasciation
фа'сция *f anat* fascia
фа'уна *f* fauna
~ **беспозвоно'чных** invertebrate fauna
~ **позвоно'чных** vertebrate fauna
ископа'емая ~ fossil fauna
по'чвенная ~ soil fauna
фаунисти'ческий *a* faunistic
фауну'ла *f* faunula
фа'ция *f* facies
фациа'ция *f* faciation
федерио'н *m* federion
фека'лии *f pl* feces
фека'лия *f* fecal pellet
фека'льный *a* fecal, faecal
фелле'ма *f* phellem
феллоге'н *m* phellogen
феллоде'рма *f* phelloderm
фелля'ция *f eth* fellation
феминиза'ция *f* feminization
фен *m* phene
фенилалани'н *m* phenylalanine
феногене'тика *f* phenogenetics
фенодевиа'нт *m* phenodeviant
феноди'м *m* phenodeme
фенокли'н *m* phenocline
феноко'нтур *m* phenocontour
феноко'пия *f* phenocopy
фено'л *m* phenol

фенологи'ческий *a* phenological
феноло'гия *f* phenology
~ **фитоцено'зов** phenology of phytocenoses, synphenology
фено'м *m* phenome
фено'мен *m* phenomenon
~ **Артю'са** Arthus phenomenon
~ **А'шнера** Aschner's phenomenon
~ **Бе'лла** Bell's phenomenon
~ **Бе'ккера** Becker's phenomenon
~ **Борде'-Жангу'** Bordet-Gengou phenomenon
~ **Бро'уинга** Browing phenomenon
~ **Ван А'ллена** Van Allen phenomenon
~ **Венкенба'ха** Wenckenbach phenomenon
~ **Вестфа'ля-Пи'льца** Westphal-Piltz phenomenon
~ **Га'мбургера** Hamburger's phenomenon
~ **Го'льдблатта** Goldblatt phenomenon
~ **Го'фмана** Hoffmann's phenomenon
~ **Грассе'-Го'сселя** Grasset-Gaussel phenomenon
~ **Дебре'** Debre phenomenon
~ **Дени'-Лекле'фа** Denys-Leclef phenomenon
~ **До'ната-Ландште'йнера** Donath-Landsteiner phenomenon
~ **Дю'кворта** Duckworth's phenomenon
~ **Ко'ха** Koch's phenomenon
~ **Ку'шинга** Cushing's phenomenon
~ **Не'йссера-Ве'ксберга** Neisser-Wechsberg phenomenon
~ **Не'йссера-Де'ринга** Neisser-Doering phenomenon
~ **Пу'ла** Pool's phenomenon
~ **Пуркинье'-Самсо'на** Purkinje-Samson's image
~ **Рейно'** Raynaud's phenomenon
~ **Ри'ттера-Ролле'** Ritter-Rollet phenomenon
~ **Ру'ста** Rust's phenomenon
~ **Санаре'лли** Sanarelli phenomenon
~ **Тво'рта** Twort phenomenon
~ **Тинда'ля** Tyndall phenomenon
~ **Фо'рстера** Forster's phenomenon
~ **Фридре'йха** Friedreich's phenomenon
~ **Ха'нта** Hunt's phenomenon
~ **Ха'пке** Hapke's phenomenon
~ **Ха'та** Hata's phenomenon
~ **Хохзи'нгера** Hochsinger's phenomenon
~ **Шелло'нг-Стрисо'уэра** Schellong-Strisower phenomenon
~ **Э'рбена** Erben's phenomenon
~ **ассоциати'вного узнава'ния** *imm* associative recognition phenomenon
~ **«буты'лочного го'рлышка»** *ecol* bottleneck phenomenon
~ **возбужде'ния и сокраще'ния** excitation and contraction phenomenon
~ **возрастно'й инволю'ции** age-involution phenomenon
~ **деспецифика'ции** *imm* "downgrading" phenomenon
~ **заселе'ния** "seeding out" phenomenon
~ **«зо'ны»** *imm* zone phenomenon
~ **«инвази'вной гибридо'мы»** "fluid tumor" phenomenon
~ **интерфере'нции** interference phenomenon
~ **когна'тного распознава'ния** cognate phenomenon
~ **Ландште'йнера** *imm* "faulty perspective" phenomenon
~ **образова'ния «ша'почки»** *imm* capping phenomenon
~ **оседа'ния** "seeding out" phenomenon
~ **«оши'бочного прогно'за»** *imm* "faulty perspective" phenomenon
~ **преиму'щественного пра'ва** *imm* "preemption" phenomenon
~ **«прозо'ны»** prozone phenomenon
~ **распознава'ния «своего'»** cognate phenomenon
~ **«си'няя дуга'»** *phys* blue-arcs phenomenon
иммунологи'ческий ~ immunological phenomenon
ле'стничный ~ staircase phenomenon
неспецифи'ческий ~ bystander phenomenon
перекрё'стный диафрагма'льный ~ crossed phrenic phenomenon
фено'н *m* phenon
феноти'п *m* phenotype
~ **«по'лный дом»** "full house" phenotype
антигенный ~ antigenic phenotype, B phenotype
дефинити'вный ~ differentiated phenotype, mature phenotype
запо'лненный ~ "full house" phenotype
зре'лый ~ differentiated phenotype, mature phenotype
В-кле'точный ~ B phenotype
незре'лый ~ immature phenotype

неотвеча'ющий ~ tolerant phenotype
нестаби'льный ~ fluctuating phenotype
«нулево'й» ~ null phenotype
пове'рхностный ~ *imm* membrane phenotype
промежу'точный ~ transitional phenotype
Т-кле'точный ~ *imm* T phenotype
трансформи'рованный ~ transformed phenotype
фенотипи'ческий *a* phenotypic
фенолфталеи'н *m* phenolphthalein
фено'н *m* phenon
фенофа'за *f* phenological stage
фе'нхель *m* fennel (*Foeniculum*)
феогифомико'з *m micr* phaeohyphomycosis
фео'дий *m* phaeodium
феофити'н *m* pheophytin
ферме'нт *m* enzyme
~ **для иссле'дования структу'ры** *mbiol* structure probing enzyme
~ **для секвени'рования** *mbiol* sequencing enzyme
~, **задаю'щий ско'рость** pacemaker enzyme
~ **перено'са** transferring enzyme
~, **разлага'ющий клетча'тку** cellulolytic enzyme
~ **«разры'ва-сши'вки»** nick-closing enzyme
~ **спла'йсинга** splicing enzyme
адапти'вный ~ adaptive enzyme
акти'вный ~ active enzyme
аллостери'ческий ~ allosteric enzyme
амилолити'ческий ~ amylolytic enzyme, starch-converting enzyme
бактериа'льный ~ bacterial enzyme
ветвя'щий ~ branching enzyme
внекле'точный ~ extracellular enzyme
внутрикле'точный ~ intracellular enzyme
гликолити'ческий ~ glycolytic enzyme
деветвя'щий ~ debranching enzyme
диспропорциони'рующий ~ disproportionating enzyme, D-enzyme
ДНК-раскру'чивающий ~ DNA untwisting enzyme
иммобилизо'ванный ~ immobilized enzyme
индуци'руемый ~ inducible enzyme
ключево'й ~ key enzyme
конденси'рующий ~ condensing enzyme
конститути'вный ~ constitutive enzyme
лаби'льный ~ labile enzyme
липолити'ческий ~ lipolytic enzyme
малат-конденси'рующий ~ malate condensing enzyme
митохондриа'льный ~ mitochondrial enzyme
нати'вный ~ native enzyme
неакти'вный ~ inactive enzyme
окисли'тельно-восстанови'тельный redox enzyme
оса'харивающий ~ saccharifying enzyme
отщепля'ющий ~ cleaving enzyme
пантоатактиви'рующий ~ pantoate activating enzyme
пектолити'ческий ~ pectolytic enzyme
пищевари'тельный ~ digestive enzyme
присоединя'ющий ~ coupling enzyme
протеолити'ческий ~ protein-degrading enzyme, proteolytic enzyme
раскру'чивающий ~ untwisting enzyme, unwinding enzyme
репарати'вный ~ repair enzyme
рестрикцио'нный ~ restriction enzyme
«ста'рый жё'лтый ~» old yellow enzyme
фотореактиви'рующий photoreactivating enzyme
ферментати'вный *a* enzymic, enzymatic, zymotic
фермента'ция *f* fermentation, fermenting, zymosis
диа'лизная ~ dialysis fermentation
защищё'нная ~ protected fermentation
непреры'вная ~ continuous fermentation
полунепре'рывная ~ semicontinuous fermentation
твердофа'зная ~ solid state fermentation, solid-substrate fermentation
ферменте'р *m* fermenter
ферменти'рование *n* fermentation, fermenting
ферменти'рующий *a* fermenting, fermentative
ферме'нт-маркё'р *m* marker enzyme
феромо'н *m* pheromone
матери'нский ~ maternal pheromone
мотиви'рующий ~ priming pheromone
сигна'льный ~ releasing pheromone, signalling pheromone
фертилизи'н *m bioch* fertilizin
ферти'льность *f* fertility
фиали'да *f* phialide
фиа'лка *f* violet (*Viola*)

~ **узамба'рская** African violet *(Saintpaulia ionantha)*
фибри'лла *f* fibril, fibrilla
кле'точная ~ cellular fibril
фибрилля'рный *a* fibrillar
фибрилля'ция *f* fibrillation
фибри'н *m* fibrin
фибриноге'н *m* fibrinogen
фибрино'лиз *m* fibrinolysis
фибринообразу'ющий *a* fibrinogenous
фибробла'ст *m* fibroblast
фиброи'н *m* fibroin
фибро'ма *f* fibroid tumor
фибронекти'н *m* fibronectin
фиброци'т *m* fibrocyte
фи'була *f* fibula *(in diatoms)*
фигля'р *m* bateleur (*Terathopius ecaudatus*)
фи'дер *m* feeder
фи'динг *m* feeding
физиогени'я *f* physiogeny
физиогно'мика *f* physiognomics
физиогра'фия *f* physiography
физиологи'ческий *a* physiological
физиоло'гия *f* physiology
~ **кле'тки** cell physiology
~ **о'рганов чу'вств** sensory physiology
о'бщая ~ general physiology
расте'ний plant physiology, phytophysiology
патологи'ческая ~ pathologic physiology, physiopathology
сравни'тельная ~ comparative physiology
физи'ческий *a* physical
фико- phyco-
фикобили'н *m* phycobilin
фикобилисо'ма *f* phycobilisome
фикобио'нт *m* phycobiont
фикоценоло'гия *f* phycocenology
фикса'тор *m* fixer; fixing fluid; preservative
~ **Буэ'на** Bouin's fluid
~ **Карнуа'** Carnoy's fluid
~ **Рего'** Regaud's fluid
о'смиевый ~ osmium fixative
фикса'ция *f* fixation, fixing
~ **азо'та** nitrogen fixation
~ **в темноте'** dark fixation
~ **в формали'не** formol-fixation
~ **комплеме'нта** complement fixation, alexin fixation
~ **ме'тодом замора'живания – высу'шивания** fixation by freezing and drying
~ **ме'тодом перфу'зии** fixation by perfusion
~ **пара'ми** vapor fixation
анаэро'бная ~ азо'та anaerobic nitrogen fixation
темнова'я ~ dark fixation
фикси'рованный *a* fixed
фикси'ровать *v* fix
филакобио'з *m* phylacobiosis
филаме'нт *m* filament
филети'ческий *a* phyletic
фи'лин *m* eagle owl (*Bubo bubo*)
филло'дий *m* phyllode, phyllody
филлозоо'ид *m* phyllozooid
филлокла'дий *m* phylloclade
филлоксанти'н *m* phylloxanthin
филлоксе'ра *f ent* phylloxera *(Phylloxera)*
филло'м *m* phyllome
филлосифони'ческий *a bot* phyllosiphonic
филлота'ксис *m* phyllotaxis, phyllotaxy
филлохино'н *m* phylloquinone
филогене'з *m* phylogenesis
филогенети'ческий *a* phylogenetic
филогени'я *f* phylogeny
фило'зный *a* filose
филопо'дия *f prot* filopodium
филоценогене'з *m* phylocoenogenesis
фильтр *m lab* filter; *anat* philtrum
~ **Беркефе'льда** Berkefeld filter
~ **Зе'йтца** Seitz filter, Seitz pad
~ **Китаса'то** Kitasato filter
~ **Шамберла'на** Chamberland filter
бактериа'льный ~ bacteria-excluding filter, bacterial filter, germ-tight filter
беззо'льный ~ ashless filter
биологи'ческий ~ biological filter
возду'шный ~ air filter
градоко'лловый ~ gradocoll filter
мембра'нный ~ membrane filter
обо'йменный ~ cartridge filter
обра'тный ~ inverted filter
миллипо'ровый millipore filter
полосово'й ~ band filter *(in bioacoustics)*
по'ристый ~ bank filter
самоочища'ющийся ~ self-cleaning filter
скла'дчатый ~ plaited filter, fold filter
стекля'нный ~ filterglass
стерилизу'ющий ~ sterilizing filter
стру'йный ~ jet filter
узкополо'сный ~ narrow band filter *(in bioacoustics)*

широкополо'сный ~ wide band filter *(in bioacoustics)*
фильтра'т *m* filtrate
фильтра'тор *m* filterer, filter feeder, filter-feeding organism
фильтра'ция *f lab* filtration
многоступе'нчатая ~ multiple filtration
одноступе'нчатая ~ single filtration
фильтрова'ние *n lab* filtration, filtering
фильтрова'ть *v* filtrate, filter
фильтр-свеча' *f lab:* **бактериологи'ческая** ~ bacteriological candle
фильтру'ющийся *a* filterable
фи'мбрия *f* fimbria
фингерли'нг *m* fingerling
фи'ник *m* date palm (*Phoenix dactylifera*); date
фиоле'товый *a* violet
генциа'новый ~ gentian violet
крези'ловый ~ cresyl violet
кристалли'ческий ~ crystal violet
метиле'новый ~ methylene violet
мети'ловый ~ methyl violet
фиссу'ра *f* fissure
язы'чная ~ lingual fissure
фиста'шка *f* pistache, pistachio (*Pistacia vera*)
фи'стула *f* fistula
фити'н *m* phytin
фито- phyto-
фитоагглютини'н *m* phytoagglutinin
фитоактино'метр *m* phytoactinometer
фитоаэро'н *m* phytoaeron
фитобиоти'ческий *a* phytobiotic
фитогемагглютини'н *m* phytohemagglutinin
фитоге'нный *a* phytogenous
фитогормо'н *m* phytohormone
фитогеографи'ческий *a* phytogeographical
фитогеогра'фия *f* phytogeography
фитогра'мма *f* phytograph
фитогра'фия *f* phytography
фитоиндика'тор *m* phytometer
фиотокини'н *m* phytokinin
фитоклиматоло'гия *f* phytoclimatology
фито'л *m* phytol
фито'ма *f* phytoma
фитоме'р *m* phytomer, phyton
фитоно'мия *f* phytonomy
фитопатоло'гия *f* phytopathology, plant pathology
фитосероло'гия *f* phytoserology
фитосоциоло'гия *f* phytosociology
фитостери'н *m* phytosterol
фитотокси'н *m* phytotoxin
фитотокси'чный *a* phytotoxic
фитотро'н *m* phytotron
фитоценоло'гия *f* phytocenology
фитофа'г *m* plant feeder, vegetable feeder, phytophage
фитофа'гия *f* phytophagy
фитофи'л *m* phytophil
фитофи'льный *a* phytophilous
фитохро'м *m* phytochrome
фитоцено'з *m* phytocenosis, phytocoenosis
фитоценоло'гия *f* phytocenology, phytosociology
фитоцено'н *m* phytocoenon
фитоценоти'п *m* phytocoenotype
фитоэдафо'н *m* phytoedaphon
фифи' *m orn* wood sandpiper (*Tringa glareola*)
флавази'н *m* flavazin
флави'н *m* flavin
флаг *m bot* standard
фламби'рование *n* flaming
фламби'ровать *v* flame
флами'нго *m* flamingo *(Phoenicopterus)*
фла'нец *m* flange
фланки'рующий *a mbiol* flanking
флебогра'фия *f* phlebography
флебомано'метр *m* phlebomanometer
флебогра'мма *f* phlebogram
фле'ксия *f* flexion
фле'ксор *m* flexor
фле'мен *m eth* flehmen
флокуля'ция *f* flocculation
фло'ра *f* flora
~ **макрофи'тов** macrophytic flora
бактериа'льная ~ bacterial flora, microbial flora
дикорасту'щая ~ wild flora, wild-growing flora
кише'чная ~ intestinal flora
культу'рная ~ cultivated flora
ме'стная ~ local flora, indigenous flora, native flora
по'чвенная ~ soil flora
флориге'н *m* florigen
флори'стика *f* floristics
флористи'ческий *a* floristic
флота'ция *f* flotation
флоэ'ма *f bot* phloem
нару'жная ~ external phloem
вну'тренняя ~ internal phloem

перви'чная ~ primary phloem
втори'чная ~ secondary phloem
флуктуа'ция *f* fluctuation
годи'чная ~ annual fluctuation
случа'йная ~ random fluctuation
флуоресцеи'н *m* fluorescein
флуоресце'нтный *a* fluorescent
флуоресце'нция *f* fluorescence
со'бственная ~ intrinsic fluorescence
флуоресци'рующий *a* fluorescent
флуоро'метр *m* fluorometer
флуорохро'м *m* fluorochrome
флюс *m med* gumboil
фобота'ксис *m* phobotaxis
фока'льный *a* focal
фокстерье'р *m* fox terrier
жесткоше'рстный ~ wire-haired fox terrier
фо'кус *m* focus
фолидо'з *m* pholidosis
фолли'кул *m* follicle
антра'льный ~ antrum-containing follicle
лимфати'ческий ~ lymph follicle, lymphatic follicle
овариа'льный ~ ovarian follicle
перви'чный ~ primordial follicle
перви'чный лимфо'идный ~ primary lymphoid follicle
примордиа'льный ~ primordial follicle
семенно'й ~ testicular follicle
яи'чниковый ~ ovarian follicle
фолликули'н *m* folliculine
фолликулогене'з *m* folliculogenesis
фолликулообра'зный *a* follicular
фолликуля'рный *a* follicular
фомопсио'з *m* phomopsiosis *(caused by Phomopsis)*
фон *m* background
генети'ческий ~ genetic background
фона'рь *m* lantern, lamp
аристотелев ~ *zool* Aristotle's lantern
фонд *m* resources
лесно'й ~ forest resources
охо'тничий ~ game resources
семенно'й ~ seed stock
фоне'ма *f* phoneme
фонореце'птор *m* phonoreceptor
фонота'ксис *m* phonotaxis
форе'зия *f* phoresia, phoresis, phoresy
фо'рма *f* form
~ **релье'фа** land form
~ **ро'ста** growth form
аберра'нтная ~ aberrant form
бацилля'рная ~ bacillary form
безъя'корная ~ anchor-minus form
биологи'ческая ~ biologic form
бли'зкие фо'рмы *pl* related forms, interconnected forms
бродя'чая ~ free-ranging form
ваги'льная ~ free-ranging form
вы'мершая ~ dead form
гла'дкая форма *micr* smooth form
декорати'вная ~ ornamental form
денатури'рованная ~ denatured form
жи'зненная форма life form
кустова'я ~ bush form
лека'рственная форма dosage form
опти'чески акти'вная ~ optically active form
палочкови'дная ~ *micr* rod form
перехо'дная ~ transitional form
пре'дковая ~ ancestral form
репликати'вная ~ replicative form
ро'дственные фо'рмы *pl* related forms, interconnected forms
руководя'щая форма index form, guide form
скры'тая ~ latent form
физиологи'ческая ~ physiologic form
фильтру'ющаяся ~ filterable form
шерохова'тая форма *micr* rough form
ювени'льная ~ juvenile form
формали'н *m* formalin, formaldehydeб formol
формальдеги'д *m* formaldehyde, formalin
форма'нт *m* formant *(in bioacoustics)*
форма'ция *f* formation
закры'тая ~ closed formation
за'мкнутая ~ closed formation
лугова'я ~ meadow formation
ортовезикуля'рная ~ orthovesicular formation
паравезикуля'рная ~ paravesicular formation
расти'тельная ~ plant formation, phytium
ретикуля'рная ~ reticular formation
сетеви'дная ~ reticular formation
формика'рий *m* formicary, formicarium
формили'рование *n* formylation
формимиди'рование *n* formimidation
формио'н *m* formion
формирова'ние *n* formation; training (*of trees*)
~ **кро'ны** training of crown

~ **на'выка** shaping
~ **поведе'ния** behavior shaping
формирова'ть *v* form; train
формо'лиз *m* formolysis
формообразова'ние *n* morphogenesis, morphogeny
фо'рмула *f* formula
~ **кро'ви** blood count
~ **пересче'та** conversion formula
~ **позвоно'чника** vertebral formula
~ **Ха'рди-Ва'йнберга** Hardy-Weinberg rule
~ **цветка'** flower formula, floral formula
зубна'я ~ dental formula
лейкоцита'рная ~ differential blood count, white blood count
структу'рная ~ constitutional formula
форципа'тный *a* forcipate
фоссилиза'ция *f* fossilization
фосфа'т *m* phosphate
фосфа'таза *f* phosphatase
фосфоаргини'н *m* phosphoarginine
фосфоенолпирува'т *m* phosphoenolpyruvate
фосфолипи'д *m* phospholipid
фосфопротеи'д *m* phosphoprotein
фосфопротеи'н *m* phosphoprotein
фосфоресце'нция *f* phosphorescence
фосфоресци'рующий *a* phosphorescent
фосфорили'рование *n* phosphorylation
~ **бактериа'льного ти'па** bacterial-type phosphorylation
митохондриа'льное ~ mitochondrial phosphorylation
окисли'тельное ~ oxidative phosphorylation
субстра'тное ~ substrate-linked phosphorylation
фотосинтети'ческое ~ photosynthetic phosphorylation
цикли'ческое ~ cyclic photosynthetic phosphorylation
фосфорили'ровать *v* phosphorylate
фосфоритова'ние *n* phosphorite application
фосфоро'лиз *m* phosphorolysis
фосфoферме'нт *m* phosphoenzyme
фото- photo-
фотоавтотро'ф *m* photoautotroph
фотоассимиля'ция *f* photoassimilation
фотобиоло'гия *f* photobiology
фотовосстановле'ние *n* photoreduction
фотовысвобожде'ние *n* photorelease
фотоге'нный *a* photogenic
фотогра'фия *f* photography
фотодеграда'ция *f* photodegradation
фотодиме'р *m* photodimer
фотодимериза'ция *f* photodimerization
фотодина'мика *f* photodynamics
фотодиссоциа'ция *f* photodissociation, photolysis
фотодыха'ние *n* photorespiration
фотоинактива'ция *f* photoinactivation
фотоингиби'рование *n* photoinhibition
фотоинду'кция *f* photoinduction
фотокине'з *m* photokinesis
фотокинети'ческий *a* photokinetic
фотолаби'льный *a* photolabile
фото'лиз *m* photolysis
антителоопосре'дованный ~ antibody-targeted photolysis
ла'зерный и'мпульсный ~ laser flash photolysis
фотолюминесце'нция *f* photoluminescence
фото'метр *m* photometer
фотометри'я *f* photometry
абсорбцио'нная ~ absorption photometry
пла'менная ~ flame photometry
фотоме'чение *n* photolabeling
фотоморфогене'з *m* photomorphogenesis
фотона'стия *f* photonasty
фотообразова'ние *n* photoproduction
фотообрати'мость *f* photoreversibility
фотоокисле'ние *n* photooxidation
фотопери'од *m* photoperiod
ко'нтурный ~ skeleton photoperiod
фотопериоди'зм *m* photoperiodism
фотопериоди'ческий *a* photoperiodic
фотопериоди'чность *f* photoperiodicity
~ **полово'го ци'кла** sexual photoperiodicity
фотопи'я *f* photopia
фотоплагиотропи'я *f* photoplagiotropy
фотоплетизмогра'фия *f* photoplethysmography
фотоприсоедине'ние *n* photoaddition
фотопроте'кция *f* photoprotection
фотопси'я *f* photopsia
фотореа'кция *f* photoreaction
фотореактива'ция *f* photoreactivation
фоторегуля'ция *f* photocontrol
фотореду'кция *f* photoreduction
фоторепара'ция *f* photorepair
фотореце'птор *m* photoreceptor
фотореце'пция *f* photoreception
фотосенсибилиза'ция *f* photosensitization
фотоси'нтез *m* photosynthesis

и'стинный ~ brutto photosynthesis, gross photosynthesis, real photosynthesis, true photosynthesis
кратковре'менный ~ short-term photosynthesis
наблюда'емый ~ apparent photosynthesis, net photosynthesis
сумма'рный ~ brutto photosynthesis, gross photosynthesis, real photosynthesis, total photosynthesis
стре'ссовый ~ stressed photosynthesis
фотосисте'ма *f* photosystem
фотостаби'льный *a* photostable
фотота'ксис *m* phototaxis
фототропи'зм *m* phototropism
фототро'ф *m* phototroph
фототрофи'я *f* phototrophy
фототро'фный *a* phototrophic
фотофа'за *f* photophase
фотофикса'ция *f* photofixation
~ **азо'та** photofixation of nitrogen
фотофо'бный *a* photophobic, phengophobe
фотофо'р *m* photophore
фотофосфорили'рование photophosphorylation, photosynthetic phosphorylation
нецикли'ческое ~ noncyclic photosynthetic phosphorylation
фотохи'мия *f* photochemistry
фотоэлеме'нт *m* photocell
фрагме'нт *m* fragment
~ **Клё'нова** Klenow fragment
~ **Оказа'ки** Okazaki fragment
~**рестри'кции** restriction fragment
мультиге'нный ~ multigen fragment
полиге'нный ~ multigen fragment
рестрикцио'нный restriction fragment
фрагмента'ция *f* fragmentation
фрагмо- phragmo-
фрагмобази'дия *f* phragmobasidium
фрагмопла'ст *m* phragmoplast
фрагмосо'ма *f* phragmosome
фрагмоспо'ра *f* septate spore, phragmospore
фра'за *f* phrase *(in bioacoustics)*
фракциони'рование *n* fractionation
~ **по Ко'эну** Cohn fractionation, low temperature ethanol fractionation
~ **по пло'тности** density fractionation
~ **соля'ми** salt fractionation
вытесни'тельное ~ displacement fractionation
низкотемперату'рное ~ этано'лом Cohn fractionation, low temperature ethanol fractionation
про'бное ~ pilot fractionation
фра'кция *f* fraction
надоса'дочная ~ supernatant fraction
оса'дочная ~ pellet fraction
субкле'точная ~ subcellular fraction
франколи'н *m orn* francolin *(Francolinus)*
фра'трия *f* phratrie, phratry
фреатофи'т *m* phreatophyte
фрега'т *m* frigate bird, man-of-war bird (*Fregata*)
фрезерова'ние *n* rotary tillage
френоло'гия *f* phrenology
фригано'идный *a* phryganoid
фриги'дность *f* frigidity
фрима'ртин *m* freemartin
фрукт *m* fruit
фрукто'за *f* fructose
фру'кты *m pl* fruit
фтордезоксиуриди'н *m* fluorodeoxyuridine
фтори'рование *n* fluorination
фторураци'л *m* fluorouracil
фторуриди'н *m* fluorouridine
фрустра'ция *f* frustration
фукоксанти'н *m* fucoxanthin
фукси'н *m* fuchsin
~ **Ци'ля** carbolic fuchsin
карбо'ловый ~ carbolic fuchsin
ки'слый ~ acid fuchsin
основно'й ~ basic fuchsin, diamond fuchsin
фуксинофи'льный *a* fuchsinophil
фумига'нт *m* fumigant
фунгици'д *m* fungicide
фунду'к *m* filbert (*Corylus maxima*); filbert nut, Lambert nut
функциона'льный *a* functional
функциони'ровать *v* function
фу'нкция *f* function
~ **гомологи'чная** ~ homologous function
~ **иммунорегулято'рная** ~ immunoregulatory function
~ **лине'йная** ~ linear function
~ **логарифми'ческая** ~ logarithmic function
~ **логисти'ческая** ~ logistic function
~ **метаболи'ческая** ~ metabolic function
~ **овариа'льная** ~ ovarian function
~ **правдоподо'бия** *biom* likelyhood function

~ **распределе'ния** *biom* distribution function
~ **яи'чников** ovarian function
когнити'вная ~ cognitive function
кумуляти'вная ~ распределе'ния *biom* cumulative distribution function
ма'тричная ~ template function
ограни'ченная ~ *biom* bounded function
показа'тельная ~ exponential function
характеристи'ческая ~ *biom* characteristic function
экспоненциа'льная ~ exponential function

фура'ж *m* forage, forage fodder
фуражи'ровать *v* forage
фуражиро'вка *f* forage, foraging
фу'рка *f zool* furca
фуру'нкул *m med* furuncle, furunculus, boil
фы'ркать *v* snort, sniff

X

хазмога'мия *f* chasmogamy
хазмофи'т *m* chasmophyte, chasmophilous plant
хазмохомофи'т *m* chasmochomophyte
хала'за *f* chalaza
халазога'мия *f* chalazogamy
хамелео'н *m* chameleon (*Chamaeleontidae*)
хамефи'т *m* chamaephyte
хара'ктер *m* character, mode, pattern
~ **акти'вности** pattern of activity
~ **дробле'ния** cleavage pattern
~ **исче'рченности** banding pattern
~ **поведе'ния** behavior pattern
~ **разви'тия** developmental character
~ **распростране'ния** distributional pattern
характери'стика *f* characteristic
харза' *f* yellow-throated marten (*Martes flavigula*)
хароми'да *f* charomide
ха'тка *f* lodge (*of beavers*)
бобро'вая ~ beaver's lodge
хво'йный *a* coniferous
хво'рост *m* brushwood, dry twigs; pastry straws
хворости'на *f* long branch, long switch
хво'рость *f* illness
хворь *f* illness
хвост *m* tail, cauda; scut (*of hares*), stern (*of a dog*)
~ **поджелу'дочной железы'** tail of pancreas
вилочкообра'зно разветвлё'нный ~ forked tail
высо'ко поста'вленный ~ high-set tail
кри'во поста'вленный ~ wry tail
хвоста'тый *a* caudate, caudiferous, tailed, urodelous
хвостови'дный *a* tail-shaped
хвостово'й *a* caudal, cercal
хвощ *m* horse-tail, scouring rush (*Equisetum*)
хво'я *f* needles, acerose leaves
хела'т *m* chelate
хелати'ровать *v* chelate
хели'рование *n* chelation
хелице'ра *f* chelicera
хемилюминесце'нция *f* chemiluminescence
хемиосмоти'ческий *a* chemosmotic
хемо- chemo-
хемоавтотро'ф *m* chemoautotroph
хемоавтотрофи'я *f* chemoautotrophy
хемоавторо'фный *a* chemoautotrophic
хемоаттракта'нт *m* chemoattractant
хемодифференциа'ция *f* chemodifferentiation
хемокине'з *m* chemokinesis
хемокоммуника'ция *f* chemocommunication
хемолюминесце'нция *f* chemoluminescence
хемоморфо'з *m* chemomorphose
хемона'стия *f* chemonasty
хеморезисте'нтность *f* chemoresistance
хеморецe'птор *m* chemoreceptor
хеморецe'пция *f* chemoreception
хемоси'нтез *m* chemosynthesis
хемосинтети'ческий *a* chemosynthetic
хемосо'рбция *f* chemisorption
хемота'ксис *m* chemotaxis
хемотакси'ческий *a* chemotactic
хемотаксоно'мия *f* chemotaxonomy
хемотропи'зм *m* chemotropism
хемотро'пный *a* chemotropic
хемотрофи'я *f* chemotrophy
хемотро'фный *a* chemotrophic
херсофи'т *m* chersophyte
хетотакси'я *f* chaetotaxy
хиа'зма *f* chiasma
диагона'льная ~ diagonal chiasma, disparate chiasma, progressive chiasma
дистанцио'нная ~ distance chiasma
интерстициа'льная ~ interstitial chiasma
компенси'рующая ~ compensating chiasma
латера'льная ~ lateral chiasma
мно'жественная ~ multiple chiasma
некомпенсированная ~ diagonal chiasma, disparate chiasma, progressive chiasma
несоверше'нная ~ imperfect chiasma
перевё'рнутая ~ reversed chiasma
термина'льная terminal chiasma
тройна'я triple chiasma
четверна'я ~ quadruple chiasma
хиазмати'пия *f* chiasmatypy
по'лная ~ total chiasmatypy
части'чная ~ partial chiasmatypy
хиа'змы *pl of* **хиазма** chiasmata
комплемента'рные ~ complementary chiasmata, digressive chiasmata
реципро'кные ~ reciprocal chiasmata, regressive chiasmata
хиастобази'дия *f* chiastobasidium
хиле'зный *a* chylous

хи'лус *m* chyle
хи'лый *a* weak
химе'ра *f* chimaera, chimera
  гаплохлами'дная ~ haplochlamydeous chimaera
  дихлами'дная ~ dichlamydeous chimaera
  кле'точная ~ cytochimera
  мериклина'льная ~ mericlinal chimaera
  периклина'льная ~ periclinal chimaera
  пласти'дная ~ plastid chimaera, plastide chimaera
  поликлина'льная ~ polyclinal chimaera
  полихлами'дная ~ polychlamydeous chimaera
  приви'вочная ~ graft chimera
  радиацио'нная ~ radiation chimaera
  сектори'альная ~ sectorial chimaera
  хромосо'мная ~ chromosomal chimera
химиза'ция *f* chemicalization
химиотерапи'я *f* chemotherapy
хими'ческий *a* chemical
хи'мия *f* chemistry
  ~ **обме'на веще'ств** metabolic chemistry
  биологи'ческая ~ biological chemistry
  медици'нская ~ clinical chemistry
  физиологи'ческая ~ physiological chemistry
хи'мус *m* chyme
хи'мусный *a* chymous
хини'н *m* quinine
хионофо'бный *a* chionophobous
хира'льный *a* chiral
хиропра'ктика *a* chiropractic
хироптерофи'л *m* chiropterophilous plant
хиропти'ческий *a* chiroptical
хиру'рг *m* surgeon, chirurgeon
хирурги'я *f* surgery, chirurgery
хирурги'ческий *a* surgical
хити'н *m* chitin
хитина'за *f* chitinase
хитиниза'ция *f* chitinization
хитинизи'рованный *a* chitinized
хити'новый *a* chitinous
хитобио'за *f* chitobiose
хитоза'н *m* chitosan
хитри'диевый *a* chytrid
хитри'дий *m* chytridium
хитридиомице'т *m* chytrid *(Chytridiomycetes)*
  моноцентри'ческий ~ monocentric chytrid
хищне'ц *ent* assassin bug *(Reduviidae)*
хи'щник *m* predator, carnivore, beast of prey, bird of prey
хи'щничество *n* predation, carnivorism, predatism
хи'щность *f* predatoriness, rapaciousness, rapacity
хи'щный *a* predatory, predacious, rapacious, raptorial, carvivorous, ravenous, vulturious
  хи'щное расте'ние carnivorous plant
хламидоспо'ра *f* chlamydospore
хлеб *m* bread
  ~ **на корню'** standing corn
  бе'лый ~, пшени'чный ~ white bread
  ржано'й ~, черный ~ rye bread, black bread
  че'рствый ~ stale bread
хлеба' *m pl* grain, crops, cereals
  ози'мые ~ winter grain crops
  яровы'е ~ spring grain crops
хле'бный *a* cereal; corny; **хле'бное де'рево** breadfruit tree (*Artocarpus*)
  ~ **злак** cereal
  ~ **квас** bread kvas
хлеборо'б *m* cereal grower
хлеборо'дный *a* corny; fertile
хлев *m* cattle shed, cowshed, cow barn
  свино'й ~ hogcote
хло'пать *v* flap *(of wings)*
хлопково'дство *n* cotton growing, cotton raising
хло'пковый *a* cottony
хло'пок *m* cotton
хло'пок-сыре'ц *m* raw cotton
хлопу'шка *f* catchfly, campion (*Silene*)
хлопча'тник *m* cotton, cotton plant (*Gossypium*)
  ~ **америка'нский** American long-staple cotton (*Gossypium hirsutum*)
  ~ **а'фро-азиа'тский** Asiatic cotton (*Gossypium herbaceum*)
  ~ **длинноволокни'стый** long-staple cotton, bony cotton
  ~ **инди'йский** common cotton (*Gossipium neglectum*)
  ~ **куста'рниковый** common cotton (*Gossipium neglectum*)
  ~ **мексика'нский** American long-staple cotton (*Gossipium hirsutum*)
  ~ **обыкнове'нный** American long-staple cotton (*Gossipium hirsutum*)
  ~ **перуа'нский** Peruvian cotton (*Gossipium peruvianum*)

~ **травяни'стый** Asiatic cotton *(Gossypium herbaceum)*
хлопьеви'дный *a* flakelike
хлора'тор *m* chlorinator
хлорацетили'рование *n* chloracetylation
хлордекарбоксили'рование *n* chlordecarboxylation
хлоре'нхима *f* chlorenchima
хлорино'лиз *m* chlorinolysis
хлори'рование *n* chlorination
хлорметили'рование *n* chloromethylation
хлоро- chloro-
хлоро'биум-хлорофи'лл *m* chlorobium chlorophyll
хлоро'з *m phytop* chlorosis, yellow disease
хлоро'зный *a* chlorotic
хлорокруори'н *m* chlorocruorine
хлоро'лиз *m* chlorinolysis
хлоропла'ст *m* chloroplast, chloroplastid, green plastid
хлоросо'ма *f* chlorosome
хлороти'ческий, хлороти'чный *a* chlorotic
хлорофи'лл *m* chlorophyll
хлорофилли'д *m* chlorophyllide
хлорофилли'н *m* chlorophyllin
хлорофи'лловый *a* chlorophyllous
хлорофиллоно'сный *a* chlorophyll-bearing
хлорофо'рм *m* chloroform
хлорформили'рование *n* chloroformylation
хлорофо'с *m* chlorophos
хлортиомочеви'на *f* chlorothiourea
хлорэтили'рование *n* chloroethylation
хлорцинкио'д *m* chlor zinc iodine
хлыст *m* full-length log; *ent* virga
хмелегра'б *m* hop-hornbeam (*Ostria*)
хмель *m* hop (*Humulus*)
хмелево'д *m* hop grower
хмелево'дство *n* hop growing, hop production
хоа'на *f* choana
хоаноци'т *m* choanocyte, collar cell
хо'бот *m* trunk, proboscis; introvert (*of sipunculans*)
хоботкови'дный *a* rostelliform
хобото'к *m* rostel, proboscis
ход *m* tread, bearing (*of a horse*); mine, burrow; canal, duct; gallery; run (*of fish*)
~ **Лу'шки** Luschka's duct
вилообра'зный ~ forked gallery
звездообра'зный ~ stellate gallery
личи'нковый ~ larval gallery
лучи'стый ~ stellate gallery
ма'точный ~ egg gallery, mother gallery
нересто'вый ~ spawning run
ру'нный ~ main part of *a* run (*of fish*), mass migration
смоляно'й ~ *bot* gum canal, gum duct, resin duct
ходи'льный *a* ambulatory, ambulatorial
ходи'ть *v* go, walk
ходу'лочник *m* stilt, stilt piper (*Himantopus himantopus*)
ходьба' *f* walking
ходя'чий *a* walking
хожде'ние *n* walking; **при хожде'нии, во вре'мя хожде'ния** in walking
хозя'ин *m* host
вста'вочный ~ insert host
второ'й промежу'точный ~ supplementary host
дефинити'вный ~ definitive host, final host
дополни'тельный ~ accessory host, supplementary host
запасно'й ~ reserve host
облига'тный ~ obligatory host
оконча'тельный ~ definitive host, final host
основно'й ~ principal host
пе'рвый промежу'точный ~ initial host
промежу'точный ~ alternate host, bridging host, intermediate host, paratenic host, secondary host
резе'рвный ~ reserve host
случа'йный ~ accidental host, optional host, permissive host
тре'тий промежу'точный ~ insert host
факультати'вный ~ accidental host, optional host, permissive host
хозя'йство *n* management, enterprise
животново'дческое ~ livestock enterprise
зерново'е ~ grain husbandry
лесно'е ~ forest management, forestry
многоотраслево'е се'льское ~ mixed agriculture
моло'чное ~ dairy enterprise
охо'тничье ~ game husbandry
се'льское ~ agriculture, farming
хозя'йствование *n* management
холеглоби'н *m* choleglobin
холестано'л *m* cholestanol
холестери'н *m* cholesterin
холестеро'л *m* cholesterol
холетели'н *m* choletelin

холецисти'т *m* cholecystitis, gallbladder inflammation
холи'н *m* choline
холинэрги'ческий *a* cholinergic
хо'лка *f* withers
холоднокро'вный *a* cold-blooded
холодосто'йкость, холодоусто'йчивость *f* cold resistance, cold tolerance, cold hardiness
холодоусто'йчивый *a* frigostable
холостя'к *m* bachelor
холоферме'нт *m* holoenzyme
холоэнзи'м *m* holoenzyme
хо'минг *m* homing
хо'минг-эффе'кт *m imm* homing-effect
хомя'к *m* hamster (*Cricetus*)
хондриокине'з *m* chondriokinesis
хондриоко'нт *m* chondriocont
хондрио'м *m* chondriome
хондриоме'р *m* chondriomere
хондриоми'т *m* chondriomite
хондриосо'ма *f* chondriosome
хондриосфе'ра *f* chondriosphere
хондробла'ст *m* chondroblast
хондроге'н *m* chondrogene
хондрогене'з *m* chondrogenesis
хондроге'нный *a* chondrogenic
хондро'идный *a* chondroid
хондроити'н *m* chondroitin
хондрокла'ст *m* chondroclast
хондроци'т *m* chondrocyte
хо'рда *f* chorda
хордомезоде'рма *f* chordomesoderm
хордотона'льный *a* chordotonal
хорё'к *m ther* polecat, fitchew (*Mustela putorius*)
хориоалланто'ис *m* chorioallantois
хориомаммотропи'н *m* choriomammotropin
хорио'н *m* chorion
  безворси'нчатый ~ smooth chorion
  ворси'нчатый ~ shaggy chorion
  гла'дкий ~ smooth chorion
хориони'ческий *a* chorial
хориопета'льный *a* choriopetal
хорога'мия *f* chorogamy
хоро'идный *a* choroid
хорологи'ческий *a* chorologic
хороло'гия *f* chorology
хорь *m ther* polecat, fitchew (*Mustela putorius*)
хохла'тка *f* fumitory, birthwort (*Corydalis*)
хохла'тый *a* comose, comate, tufted, pappose
хохоло'к *m* coma, crest, cop, pappus, floccus; *orn* cirrus, hood; tuft
хохоту'н *m*: **черноголо'вый** ~ great black-headed gull (*Larus ichthyaetus*)
хране'ние *n* storing, storage; **срок хране'ния** shelf life
храп *m*, **храпе'ние** *n* snoring
храпе'ть *v* snore; snort (*of animals*)
хребе'т *m* chine
  спинно'й ~ spine
хрен *m* horse radish (*Armoracia*)
хризали'да *f* chrysalis
хризогони'дий *m* chrysogonidium
хризофи'лл *m* chrysophyll
хрип *m* rale
хрипе'ть *v* wheeze
хромази'я *f* chromasie
хромат- chromat-
хромати'да *f* chromatid
  кольцева'я ~ ring chromatid
  се'стринская ~ sister chromatid
хромати'дный *a* chromatid
хромати'н *m* chromatin
  полово'й ~ sex chromatin
  элиминацио'нный ~ elimination chromatin
хромати'новый *a* chromatinic
хромати'ческий *a* chromatic
хроматобла'ст *m* chromatoblast
хроматогра'мма *f* chromatogram
  ~ **на коло'нке** column chromatogram
  ~ **на бума'ге** paper chromatogram
  ~ **на то'нком сло'е** thin-layer chromatogram
  двухме'рная ~ **на бума'ге** two-dimensional paper chromatogram
хроматогра'фия *f* chromatography
  ~ **в пото'ке** stream chromatography
  ~ **высо'кого разреше'ния** high-performance chromatography
  ~ **га'зовой фа'зы над жи'дкостью** head-space gas chromatography
  ~ **на активи'рованной сефаро'зе** activated sepharose column chromatography
  ~ **на бума'ге** paper chromatography
  ~ **на коло'нках** column chromatography
  ~ **на коло'нке с гидроксиапати'том** chromatography on hydroxyapatite column
  ~ **на молекуля'рных си'тах** molecular sieve chromatography
  ~ **при постоя'нной вла'жности** constant-humidity chromatography

~ **с обращё'нной фа'зой** reversed-phase chromatography
~ **со ступе'нчатым вытесне'нием** step-exclusion chromatography
~ **с перено'сом заря'да** charge-transfer chromatography
абсорбцио'нная ~ absorption chromatography
адсорбцио'нная ~ adsorption chromatography
анионообме'нная ~ anion-exchange chromatography
аффи'нная ~ affinity chromatography
восходя'щая ~ ascending chromatography
вытесни'тельная ~ displacement chromatography, exclusion chromatography
газоадсорбцио'нная ~ gas-solid chromatography
га'зовая ~ gas chromatography
газожи'дкостная ~ gas-liquid chromatography
газожи'дкостная распредели'тельная ~ gas-liquid partition chromatography
гидрофо'бная ~ hydrophobic chromatography
двуме'рная ~ two-dimensional chromatography
жи'дкостная ~ liquid chromatography
жи'дкостная ~ бы'строго разреше'ния fast performance liquid chromatography
жи'дкостная ~ высо'кого давле'ния high-pressure liquid chromatography
жи'дкостная ~ высо'кого разреше'ния high-performance liquid chromatography, high-resolution liquid chromatography
жи'дкостная коло'ночная ~ liquid column chromatography
жи'дкостная ~ ни'зкого давле'ния low-pressure liquid chromatography
жи'дкостная распредели'тельная ~ liquid partition chromatography
жи'дкостная ~ с обращё'нной фа'зой liquid chromatography on a reversed-phase support
иммуноаффи'нная ~ immunoaffinity chromatography
иммунообме'нная ~ immunoexchange chromatography
ионоисключа'ющая ~ ion-exclusion chromatography
ионообме'нная ~ ion-exchange chromatography
ка'пельная противото'чная ~ countercurrent droplet chromatography
колоночная ~ column chromatography
кругова'я ~ circular chromatography
кругова'я ~ на бума'ге circular paper chromatography
непреры'вная ~ continuous chromatography
нисходя'щая ~ descending chromatography
одноме'рная ~ one-dimensional chromatography
пе'нная противото'чная ~ countercurrent foam chromatography
пиролити'ческая га'зовая ~ pyrolysis gas chromatography
препарати'вная ~ preparative chromatography
простра'нственно-эксклюзио'нная ~ steric exclusion chromatography
противото'чная ~ countercurrent chromatography
распредели'тельная ~ partition chromatography
распредели'тельная ~ с обращё'нной фа'зой reversed-phase partition chromatography
рециркуляцио'нная ~ recycling chromatography
тонкосло'йная ~ thin-layer chromatography
центробе'жная элюцио'нная ~ centrifugal elution chromatography
эксклюзио'нная ~ ion-exclusion chromatography
электромиграцио'нная ~ electromigration chromatography
электрофорети'ческая ~ electrophoretic chromatography

хромато'лиз *m* chromatolysis
хроматопла'зма *f* chromatoplasm
хроматотропи'н *m* chromatotropin
хроматофи'льный *a* chromatophilous
хроматофокуси'рование *n* chromatofocusing
хроматофо'р *m* chromatophore
хрома'ть *v* limp
хроми'диум *m cyt* chromidium
хромо- chromo-
хромобакте'рия *f* chromobacterium
хромобла'ст *m* chromoblast
хромобластомико'з *m* chromoblastomycosis

хромоге'нный *a* chromogenic
хромозо'ма see **хромосо'ма**
хромоме'р *m* chromomere
  центроме'рный ~ centromeric chromomere
хромоне'ма *f* chromonema
хромопла'ст *m* chromoplast
хромосо'ма *f* chromosome
  ~, **ограни'ченная по'лом** sex-limited chromosome
  ~ **ти'па «ла'мповой щё'тки»** lamp-brush chromosome, lateral-loop chromosome
  акроцентрическая ~ acrocentric chromosome
  ателомити'ческая ~ atelomitic chromosome
  ацентри'ческая ~ acentric chromosome
  гетеротро'пная ~ heterotropic chromosome
  гетероцентри'ческая ~ heterocentric chromosome
  гига'нтская ~ giant chromosome
  гига'нтская ~ **слюнны'х желё'з** *ent* salivary gland chromosome
  голокинети'ческая ~ holocentric chromosome, holokinetic chromosome
  голоцентри'ческая ~ holocentric chromosome, holokinetic chromosome
  гомологи'чная ~ homologous chromosome
  дикинети'ческая ~ dicentric chromosome, dikinetic chromosome
  дицентри'ческая ~ dicentric chromosome, dikinetic chromosome
  доба'вочная ~ accessory chromosome, odd chromosome
  доче'рняя ~ daughter chromosome
  изодицентри'ческая ~ isodicentric chromosome
  кольцева'я ~ circular chromosome, ring chromosome, closed chromosome
  конденси'рованная ~ folded chromosome
  лимити'рованная ~ limited chromosome
  мегаме'рная ~ megameric chromosome
  медиоцентри'ческая ~ mediocentric chromosome
  метафа'зная ~ metaphase chromosome
  метацентри'ческая ~ metacentric chromosome, mediocentric chromosome
  неоцентри'ческая ~ neocentric chromosome
  плевромити'ческая ~ pleuromitic chromosome
  по'здно реплици'рующая ~ late replicating chromosome
  полиме'рная ~ polymeric chromosome
  полите'нная ~ polytene chromosome
  полицентри'ческая ~ polycentric chromosome
  полова'я ~ sex chromosome
  поляризо'ванная ~ polarized chromosome
  разветвлё'нная ~ branched chromosome
  рекомбина'нтная ~ recombinant chromosome
  сбо'рная ~ compound chromosome
  сла'бо конъюги'рующая ~ loosely paired chromosome
  сло'жная ~ compound chromosome
  сцеплённые хромосо'мы *pl* attached chromosomes
  телоцентрическая ~ telocentric chromosome
хромоти'п *m* chromotype
хромотро'пный *a* chromotropic
хромофи'льный *a* chromophilic
хромофо'бный *a* chromophobic
хромофо'р *m* chromophore
хромоце'нтр *m* chromocenter
хромоци'т *m* chromocyte
хронакси'я *f* chronaxia
хрони'ческий *a* chronic
хронобиоло'гия *f* chronobiology
хронотропи'зм *m* chronotropism
хронотро'пный *a* chronotropic
хру'пкий *a* fragile, breakable
хру'пкость *f* fragility
  осмоти'ческая ~ osmotic fragility
хруста'лик *m* lens
  ~ **гла'за** crystalline lens
хрущ *m ent* chafer (*Melolonthinae*)
  ма'йский ~ May beetle (*Melolontha*)
хруща'к *m ent* flour beetle (*Tribolium*)
хрю'канье *n* grunting
хрю'кать *v* grunt
хряк *m* hog
хрящ *m anat* cartilage, gristle; *geol* grit
  ~ **ве'ка** palpebral cartilage
  ~ **крыла' но'са** alar cartilage, nasal alar cartilage
  ~ **перегоро'дки но'са** septal cartilage
  ~ **слухово'й трубы'** tubal cartilage
  ~ **ушно'й ра'ковины** conchal cartilage
  бараба'нный ~ tympanic cartilage

вре'менный ~ precursory cartilage
гиали'новый ~ hyaline cartilage
доба'вочный ~ accessory cartilage, sesamoid cartilage
доба'вочный носово'й ~ epactile cartilage
жа'берный ~ branchial cartilage
жё'лтый ~ yellow cartilage
клинови'дный ~ Morgagni's cartilage
кольцеви'дный ~ innominate cartilage
межпозвонко'вый ~ intervertebral cartilage
ме'ккелев ~ Meckel's cartilage
неокостенева'ющий ~ permanent cartilage
перстневи'дный ~ cricoid cartilage, innominate cartilage
поля'рный ~ polar cartilage
постоя'нный ~ permanent cartilage
рё'берный ~ costal cartilage
ретикуля'рный ~ reticular cartilage
рожкови'дный ~ Santorini's cartilage
сантори'ниев ~ Santorini's cartilage
сесамови'дный ~ sesamoid cartilage, accessory cartilage
сошнико'во-носово'й ~ Jacobson's cartilage
суставно'й ~ arthroidal cartilage, articular cartilage, investing cartilage
черпалови'дный ~ arytenoid cartilage
щитови'дный ~ shield-like cartilage
эмбриона'льный ~ premordial cartilage
Y-обра'зный ~ triradiate cartilage

хрящеватоли'стный *a* chondrophyllous
хрящева'тый *a* cartilaginous, gristly
хрящеви'дный *a* cartilaginoid, chondroid
хрящево'й *a* cartilaginous, chondral
хрящеобразу'ющий *a* chondrogenic
хрящеподо'бный *a* chondroid
худе'ть *v* grow lean, grow thin, lose flesh
худоба' *f* leanness, thinness
худо'й *a* ill-conditioned, ill-nourished

Ц

ца'пля *f* heron (*Ardeidae*)
бе'лая ~ egret (*Egretta*)
белобрю'хая ~ great white-bellied heron *(Ardea imperialis)*
ма'лая бе'лая ~ little egret, little heron *(Egretta garzetta)*
ночна'я ~ black-crowned night heron *(Nycticorax nycticorax)*
ры'жая ~ purple heron (*Ardea purpurea*)
се'рая ~ common heron, gray heron *(Ardea cinerea)*
цара'панье *n* scratching
цара'пать *v* scratch
цара'пина *f* scratch
цари'ца *f ent* queen
ца'рство *n* kingdom
~ **живо'тных** animal kingdom
~ **растений** plant kingdom
цвести' *v* flower, blow, be in flower, bear bloom
цвет *m* color, colour
дополни'тельный ~ complementary color
цвете'ние *n bot* flowering, bloom, blossom, blossoming, blow, florification, florescence, flowerage
цве'тень *m* pollen
цветколи'стный *a* anthophyllous
цветконо'сный *a* anthophorous, floriferous
цветни'к *m* flower garden
цве'тность *f* color index
~ **воды'** color of water
цветово'дство *n* floriculture, flower gardening
цветое'д *m ent*: **я'блонный** ~ apple blossom weevil (*Anthonomus pomorum*)
цвето'к *m* flower
~ **компа'ктного соцве'тия** floret, floscule
~, **не име'ющий тычи'нок** instaminate flower
~, **опыля'емый ба'бочками** moth flower
~, **опыля'емый му'хами** fly flower
~, **раскрыва'ющийся но'чью** nyctigamous flower
~ **с дли'нными тычи'нками** long-stamined flower
~ **с разли'чными сто'лбиками** heterostyled flower
актиномо'рфный ~ actinomorphic flower
амфифи'льный ~ amphiphilous flower
безлепестко'вый ~ apetalous flower
беспло'дный ~ barren flower, unfertile flower
беспокро'вный ~ achlamydeous flower
беспо'лый ~ neutral flower
верху'шечный ~ apical flower, terminal flower
гипоги'нный ~ hypogynous flower
го'лый ~ naked flower
двупо'лый ~ bisexual flower, hermaphrodite flower, monoclinous flower
двучле'нный ~ dimerous flower
едини'чный ~ single flower
же'нский ~ pistillate flower
закры'тый ~ cleistogamous flower
зигомо'рфный ~ irregular flower
клейстога'мный ~ cleistogamous flower
коне'чный ~ apical flower, terminal flower
краево'й ~ ray floret, marginal flower
махро'вый ~ double flower, polypetalous flower
надпе'стичный ~ epigynous flower
немахро'вый ~ single flower
непра'вильный ~ irregular flower
обоепо'лый ~ bisexual flower, monoclinous flower, hermaphrodite flower
однопо'лый ~ unisexual flower, diclinous flower, imperfect flower
па'зушный ~ axillary flower
пентацикли'ческий ~ pentacyclic flower
пе'стичный ~ carpellary flower, pistillate flower
подпе'стичный ~ hypogynous flower
по'лный ~ complete flower
пра'вильный ~ actinomorphic flower
просто'й ~ single flower
пятикругово'й ~ pentacyclic flower
пятичле'нный ~ pentamerous flower
сло'жный ~ composite flower
спайнолепе'стный ~ sympetalous flower
трёхме'рный ~ trimerous flower
трёхчле'нный ~ trimerous flower
тру'бчатый ~ disk flower
тычи'ночный ~ staminate flower
четырёхчле'нный ~ tetramerous flower
энтомофи'льный ~ entomophilous flower
эпигини'чный ~ epigynous flower
язычко'вый ~ ray flower, semifloret, semifloscule
цветок-лову'шка *m* trap-flower, prehensile flower

цветоло'же *n* receptacle, floral receptacle, floral disk, thalamus, torus
цветоно'жка *f* flower stalk, anthophore
цветоно'с *m* flower stalk
цветоно'сный *a* floriferous
цветоощуще'ние *n* color perception
цветоразличе'ние *n* color perception
цветорасположе'ние *n* flower disposition
цвето'чек *m* flowerlet
цветочувстви'тельность *f* color sensitivity
цвету'щий *a* flowering, blooming, florescent, floriferous, flowered
це'вка *f orn* tarsometatarsus
целенапра'вленность *f* purposiveness
целина' *f* virgin land
  по'днятая ~ newly plowed virgin soil
цели'нный *a* virgin; **цели'нная земля'** virgin land
целлоиди'н *m* celloidin
целлюло'за *f* cellulose
целлюляриза'ция *f* cellulation
целобла'ст *m* coeloblast
целобла'стула *f emb* coeloblastula
целога'струла *f emb* coelogastrula
цело'м *m* celom, coelom, coelome
целоми'ческий *a* coelomic
целомоду'кт *m* coelomoduct
целомопо'р *m* coelomopore
целомоци'т *m* coelomocyte
цельнокра'йний *a* smooth-edged
цельноли'стный *a* integrifolious
це'льный *a* entire
цеме'нт *m* cement
  ко'стный ~ bone cement
  межфибрилля'рный ~ interfibrillar cement
цементобла'ст *m* cementoblast
цена' *f* value, price, cost
  ~ **деле'ния** scale division, value of a division, graduation mark
цене'нхима *f* coenenchym, coenenchyma
це'нность *f* value
  адапти'вная ~ adaptive value
  биологи'ческая ~ biological value
  кормова'я ~ feeding value
  пита'тельная ~ food value, nutritive value
  пищева'я ~ food value, nutritive value
  селекти'вная ~ selective value
  энергети'ческая ~ caloric value, calorific value, energy value
цено'бий *m* coenobium
ценобла'ст *m* coenoblast
ценови'д *m* cenospecies, coenospecies
ценогаме'та *f* coenogamete
ценогене'з *m* cenogenesis, coenogenesis
ценогенети'ческий *a* cenogenetic
цено'з *m* cenosis, coenosis
ценозиго'та *f* coenozygote
ценокли'н *m* coenocline
ценоло'гия *f* cenology
цено'н *m* coenon
ценоса'рк *m* coenosarc
цено'ций *m* coenoecium
центр *m* center
  ~ **Брока'** Broca's center
  ~ **вдо'ха** inspiratory center
  ~ **Ве'рнике** Wernicke's center
  ~ **возникнове'ния** center of origin
  ~ **глота'ния** swallowing center, deglutition center
  ~, **координи'рующий ми'мику** facial center
  ~ **наказа'ния** punishment center
  ~ **происхожде'ния** center of origin
  ~ **распростране'ния** center of distribution, center of dispersal
  ~ **расселе'ния** center of dispersal
  ~**рефле'кса** reflex center
  ~ **ре'чи** speech center
  ~ **свя'зывания** binding site
  ~ **терморегуля'ции** heat-regulating center
  ~ **торможе'ния** center of inhibition
  ~ **тя'жести** center of gravity
  ~ **удово'льствия** pleasure center
  ~ **эре'кции** erection center
  актива'торный ~ activator site
  акти'вный ~ active site
  аминоаци'льный ~ рибосо'мы aminoacyl ribosome site
  вазомото'рный ~ vasomotor center
  вкусово'й ~ gustatory center
  дыха'тельный ~ respiratory center, inspiratory center
  инфекцио'нный ~ фа'га phage infective center
  кле'точный ~ cytocentrum
  медьсвя'зывающий ~ copper site
  не'рвный ~ nerve center
  организацио'нный ~ *emb* organization center
  пептиди'л-тРНК-свя'зывающий ~ рибосо'мы peptidyl ribosome site, peptidyl-tRNA binding site

пептиди'льный ~ рибосо'мы peptidyl ribosome site, peptidyl-tRNA binding site
реакцио'нный ~ reaction site, reactive center
сосудодви'гательный ~ vasomotor center
специфи'чный ~ specificity site
сухожи'льный ~ диафра'гмы central tendon of diaphragm, cordiform tendon

центра'льный *a* central, centric

центрио'ль *f* centriole

центрифу'га *f lab* centrifuge
~ **без перегоро'док в ро'торе** hollow-bowl centrifuge
~ **коро'ткого ра'диуса** short-arm centrifuge
~ **непреры'вного де'йствия** continuous centrifuge
~ **периоди'ческого де'йствия** batch centrifuge
~ **с вибрацио'нной вы'грузкой** oscillating-screen centrifuge
~ **с вибри'рующим ро'тором** oscillating-basket centrifuge
~ **с инерцио'нной вы'грузкой** nozzle-discharge centrifuge, valve-discharge centrifuge
~ **со съё'мом оса'дка ножо'м** knife-discharge centrifuge
~ **со шне'ковой вы'грузкой** conveyor centrifuge, helical-conveyor centrifuge
~ **с пневматической вы'грузкой** pneumatic discharge centrifuge
~ **с подве'ской на коло'нках** center-slung centrifuge
~ **с ручно'й вы'грузкой** hand-discharge centrifuge
~ **с се'тчатым бараба'ном** basket centrifuge
~ **с угловы'м расположе'нием проби'рок** angle centrifuge
аффинацио'нная ~ affination centrifuge
ве'нтильная ~ valve centrifuge
вибрацио'нная ~ vibrating centrifuge
высокоскоростна'я ~ high-speed centrifuge
горизонта'льная ~ horizontally mounted centrifuge
и'мпульсная ~ batch centrifuge
ио'нная ~ ionic centrifuge
испари'тельная ~ evaporative centrifuge
корзи'ночная ~ basket centrifuge, basket centrifugal drier
многока'мерная ~ multichamber centrifuge
многосекцио'нная ~ multichamber centrifuge
насто'льная центрифуга bench centrifuge
обезво'живающая ~ centrifugal drier
осади'тельная ~ sedimentation centrifuge
отсто'йная ~ settling centrifuge
подвесна'я ~ suspended centrifuge
подвесна'я ~ с ве'рхней опо'рой top-suspended centrifuge
проби'рочная ~ test-tube centrifuge
прото'чно-спира'льная ~ flow-trough coil centrifuge
препаративная ~ sedimentation centrifuge
рефрижера'торная ~ cold centrifuge
ручна'я ~ hand centrifuge
саморазгружа'ющаяся ~ self-discharging centrifuge, self-opening centrifuge
фильтру'ющая ~ filtering centrifuge

центрифугальный *a* centrifugal

центрифуги'рование *n lab* centrifugation
~ **в градие'нте пло'тности** density-gradient centrifugation
~ **с непреры'вным прото'ком** continuous-flow centrifugation
высокоскоростно'е ~ high-speed centrifugation
дифференциа'льное ~ differential centrifugation
зона'льное ~ zonal centrifugation, banding centrifugation
низкоскоростно'е ~ low-speed centrifugation
отсто'йное ~ settling centrifugation
противото'чное ~ counterflow centrifugation
равнове'сное ~ equilibrium centrifugation
скоростно'е ~ high-speed centrifugation

центрифуги'ровать *v lab* centrifuge, spin

центрифу'жный *a* centrifugal

центробе'жный *a* centrifugal, axifugal, efferent

центродесмо'з *m* centrodesmose

центролецита'льный *a* centrolecithal

центроме'ра *f* centromere
втори'чная ~ secondary centromere
диффу'зная ~ diffuse centromere
локализо'ванная ~ localized centromere
мно'жественная ~ multiple centromere

полулокализо'ванная ~ semilocalized centromere
центропла'зма *f* centroplasma
центропла'ст *m* centroplast
центросо'ма *f* centrosome, cytocentrum
центростреми'тельный *a* axipetal, centripetal, afferent
центросфе'ра *f* centrosphere
цену'р *m helm* coenurus
цеоли'т *m* zeolite
це'пень *m* tapeworm *(Taenia)*
цепеобра'зный *a* cateniform
це'пкий *a* tenent, tenacious; prehensile
цепля'ющийся *a* clinging
цепо'чка *f* chain
цепочкови'дный *a* catenate
цепочконо'сный *a* catenifеrous, catenigerous
цепь *f* chain; circuit (*e.g., electric*); **на цепи'** on *a* chain; **посади'ть на** ~ chain up; **сажа'ть на** ~ chain up; **сорва'ться с це'пи** break loose; **спусти'ть с це'пи** unchain
~ **нападе'ния** *ecol* chain of attack
~ **нейро'нов** neuronic chain
~ **озе'р** series of lakes
~ **перено'са** transfer chain
~ **перено'са электро'нов** electron transport chain
~ **пита'ния** food chain
~ **пищевы'х органи'змов** chain of food organisms
~ **потреби'телей** chain of transformers
~ **реа'кций** reaction chain
~ **тра'нспорта электро'нов** electron transport chain
бокова'я ~ side chain
го'рная ~ mountain range
двойна'я ~ *mbiol* double chain
дыха'тельная ~ respiratory chain
за'мкнутая ~ closed circuit
инвариа'нтная ~ invariable chain
лё'гкая ~ light chain
одино'чная ~ *mbiol* single chain
окисли'тельно-восстанови'тельная ~ redox chain
пищева'я ~ food chain
регулято'рная ~ regulatory circuit
рефлекто'рная ~ reflex chain
трофи'ческая ~ trophic chain
тяжё'лая ~ heavy chain
церами'д *m* ceramide
цервика'льный *a* cervical
церебра'льный *a* cerebral
цереброваскуля'рный *a* cerebrovascular
церебровисцера'льный *a* cerbrovisceral
церебрози'д *m* cerebroside
цереброспина'льный *a* cerebrospinal
церемо'ния *f eth* ceremony
~ **сме'ны партнё'ров** relieving ceremony *(e.g., at a nest)*
бра'чная ~ courtship display
це'рка *f* circus
церкарие'ум *m helm* cercariaeum
церка'рия *f helm* cercaria, cercaria larva
~ **с раздво'енным хвосто'м** lophocercous cercaria
вилохво'стая ~ furcocercous cercaria
гастеросто'мная ~ gasterostome cercaria
церкоци'ста *f helm* cercocystis
церулоплазми'н *m* ceruloplasmin
цеса'рка *f orn* guinea fowl (*Numida meleagris*)
цефализа'ция *f* cephalization
цефа'лий *m* cephalium
цефали'н *m* cephalin, kephalin
цефало'метр *m* cephalometer
цефаломе'трия *f* cephalometry
цефалоте'ка *f* cephalotheca
цефалото'ракс *m* cephalothorax
цех *m* shop
броди'льный ~ fermentation plant
цианокобалами'н *m* cyanocobalamin
цика'дка *f* leafhopper (*Cicadellidae*)
цикл *m* cycle
~ **бо'дрствование-со'н** wakefulness-sleep cycle
~ **Кре'бса** Krebs cycle, tricarbonic acid cycle, citric acid cycle
~ **разви'тия** development cycle
~ **реплика'ции** round of replication
~ **трикарбо'новых кисло'т** tricarbonic acid cycle, citric acid cycle, Krebs cycle
~ **чередова'ния све'та и темноты'** light-dark cycle
годи'чный ~ circannian cycle
дыха'тельный ~ respiratory cycle
жи'зненный ~ life cycle
кле'точный ~ cell cycle
за'мкнутый ~ closed cycle
лити'ческий ~ lytic cycle
овариа'льный ~ ovarian cycle
овулято'рный ~ ovulatory cycle
околосу'точный ~ circadian cycle
пищево'й ~ food cycle
полово'й ~ sexual cycle

популяцио'нный ~ population cycle
репродукти'вный ~ reproductive cycle, genesial cycle
ростово'й ~ growth cycle
серде'чный ~ cardiac cycle, pulse cycle
фотопериоди'ческий ~ photoperiodic cycle
циркадиа'нный ~, цирка'дный ~ circadian cycle
эстра'льный ~ estrous cycle
цикла'за *f* cyclase
циклиза'ция *f bioch* ring formation, cyclization
цикли'ческий *a* cyclic
цикли'чность *f* cyclicism
цикло'з *m* cyclosis
циклоалкили'рование cycloalkylation
циклодегидрата'ция *f* cyclodehydration
циклодегидри'рование *n* cyclodehydrogenation
циклодимериза'ция *f* cyclodimerization
цикло'з *m* cyclosis
цикло'идный *a* cycloid
циклоконденса'ция *f* cyclocondensation
цикломорфо'з *m* cyclomorphosis
циклоолигомериза'ция *f* cyclooligomerization
циклополимериза'ция *f* cyclopolymerization
циклоприсоедине'ние *n* cycloaddition
цико'рий *m* chickory (*Cichorium*)
~ обыкнове'нный witloof chicory *(Cichorium intybus)*
цилиату'ра *f* ciliature
цили'ндр *m* cylinder
~ Бенс Джо'нса Bence Jones cylinder
кровяно'й ~ blood cast
ме'рный ~ graduated cylinder
осево'й ~ axis cylinder
осево'й ~ Рема'ка Remak's axis cylinder
сосу'дистый ~ *bot* vascular cylinder
цимо'зный *a* cymose
цинга' *f* scurvy, scorbutus
цирка'дный *a* circadian
циркуля'ция *f* circulation; overturn (*of lake water*)
циркумбореа'льный *a* circumboreal
циркумора'льный *a* circumbuccal
циркумполя'рный *a* circumpolar
цирро'з *m* cirrhosis
ци'ррус *m* cirrus
ци'ста *f* cyst
цистеи'н *m* cysteine
цисте'рна *f* cistern
~ грудно'го прото'ка chyle cistern
мозжечко'во-мозгова'я ~ cerebellomedullary cistern
цистернэпифи'т *m* cisternepiphyte
цисти'да *f myc* cystidium
цистидио'ла *f* cystidiole
цисти'н *m* cystine
цисти'т *m* cystitis
цистице'рк *m* cysticercus
цистицерко'з *m* cystcercosis
цистицерко'ид *m* cysticercoid
ци'стный *a* cystic
цисто- cysto-
цистока'рпий *m* cystocarp
цистоку'чка *f myc* cystosorus
цистоли'т *m* cystolith
цистока'рпий *m* cystocarp
цистоспо'ра *f* cystospore
цистро'н *m* cistron
структу'рный ~ structural cistron
цитва'рный *a* santonic, wormseed
цито- cyto-
цитоархитекто'ника *f* cytoarchitecture
цитобиоло'гия *f* cytobiology
цитобла'ст *m* cytoblast
цитога'мия *f* cytogamy
цитоге'н *m* cytogene
цитогене'з *m* cytogenesis, cytogeny
цитогене'тика *f* cytogenetics
цитогенети'ческий *a* cytogenetic
цито'да *f* cytode
цитодиагно'стика *f* cytodiagnosis
цитози'н *m* cytosine
цитозо'ль *m* cytosol
цитоиммунотерапи'я *f* cytoimmunotherapy
цитоки'н *m* cytokine
иммунорегулято'рный ~ immunoregulatory cytokine
цитокине'з *m* cytokinesis, cytocinesis, cytodieresis
цито'лиз *m* cytolysis
цитолизи'н *m* cytolysin
цитолизи'роваться *v* cytolize
цитоли'мфа *f* cytolymph
цитолити'ческий *a* cytolytic
цитоло'гия *f* cytology
цито'м *m* cytome
цитомега'лия *f* cytomegalic inclusion disease
цитомегалови'рус *m* cytomegalovirus
цитоме'р *m* cytomere

цито'метр *m* cytometer
цитометри'я *f* cytometry
~ **в пото'ке** flow cytometry
прото'чная ~ flow cytometry
цитомито'ма *f* cytomitome
цитоморфо'з *m* cytomorphosis
цито'н *m* cyton
цитопатоло'гия *f* cytopathology
цитопла'зма *f* cytoplasm
цитоплазмати'ческий, цитопла'зменный *a* cytoplasmic
цитопоэ'з *m* cytopoiesis
цитопоэти'ческий *a* cytopoietic
цитопро'кт *m prot* cytoproct
цитоскеле'т *m* cytoskeleton
цитосо'ма *f* cytosome
цитостати'ческий *a* cytostatic
цитосто'м *m* cytostome, cell mouth
цитосфе'ра *f* cytosphere
цитота'ксис *m* cytotaxis
цитотаксоно'мия *f* cytotaxonomy
цитоти'п *m* cytotype
цитотокси'ческий *a* cytotoxic
цитотокси'чность *f* cytotoxicity
аллоантигензави'симая ~ alloantigen-specific cytotoxicity, allospecific cytotoxicity
антителозави'симая ~ antibody-dependent cytotoxicity
кле'точно-опосре'дованная ~ cell-mediated cytotoxicity
конта'ктная антителоопосре'дованная ~ contact-dependent antibody-mediated cytotoxicity
цитотропи'зм *m* cytotropism
цитотро'пный *a* cytotropic
цитотрофобла'ст *m* cytotrophoblast
цитофизиоло'гия *f* cytophysiology
цитофи'льный *a* cytophilous
цитофотометри'я *f* cytophotometry
цитохими'ческий *a* cytochemical
цитохи'мия *f* cytochemistry
цитохро'м *m* cytochrome
цитоце'нтр *m* cytocentrum
цитри'н *m* citrin, citrine
цитро'н *m* citron (*Citrus medica*)
цитрулли'н *m* citrulline
ци'трусовый *a* citrous
цифе'лла *f bot* cyphella
цифона'утес *m zool* cyphonautes
цо'кор *m zool* zokor (*Myospalax*)
цука'т *m* candied fruit, candied peel
цыпле'нок *m* chick, poult

Ч

ча'га *f myc* "birch fungus" (*Inonotus obliquus*)
чаево'дство *n* tea growing
чай *m* tea-plant (*Thea*); tea
байхо'вый ~ baikhovi tea
желчего'нный ~ cholagogue tea, cholagogue species
зеле'ный ~ green tea
жи'дкий ~ weak tea
кирпи'чный ~ tile tea, brick tea
кре'пкий ~ strong tea
ли'повый ~ limeleaf tea
мочего'нный ~ diuretic tea, diuretic species
пли'точный ~ brick tea, tile tea
потого'нный ~ sudorific tea, sudorific species
сла'бый ~ weak tea
ча'йка *f* gull, mew (*Laridae*)
бе'лая ~ ivory gull (*Pagophila eburnea*)
больша'я морска'я ~ great black-backed gull *(Larus marinus)*
больша'я поля'рная ~ glaucous gull *(Larus hyperboreus)*
вилохво'стая ~ Sabine's gull *(Xema sabini)*
ма'лая ~ little gull *(Larus minutus)*
обыкнове'нная ~ black-headed gull *(Larus ridibundus)*
озё'рная ~ black-headed gull*(Larus ridibundus)*
поля'рная ~ Iceland gull *(Larus glaucoides)*
речна'я ~ black-headed gull *(Larus ridibundus)*
се'рая ~ gray gull *(Larus modestus)*
серебри'стая ~ herring gull (*Larus argentatus*)
сероголо'вая ~ gray-headed gull *(Larus cirrocephalus)*
серокры'лая ~ glaucous gull, glaucous-winged gull (*Larus glaucescens*)
си'зая ~ common gull, short-billed gull (*Larus canus*)
тё'мная ~ dusky gull *(Larus fuliginosus)*
ча'лый *a* flecked, roan *(of a horse)*
чан *m* tank
броди'льный ~ fermentation tank
части'ца *f* particle
бляшкообразу'ющая ~ plaque-forming particle
внутримембра'нная ~ intramembrane particle
фа'говая ~ phage particle
цитоплазмати'ческая ~ cytoplasmic particle
частота' *f* frequency, rate
~ **а'льфа-ри'тма** alpha frequency
~ **гармо'ники** harmonic frequency *(in bioacoustics)*
~ **геноти'па** genotype frequency
~ **дыха'ния** breathing rate, respiration rate
~ **кла'сса** class frequency
~ **мелька'ний** flicker frequency
~ **мута'ций** mutation rate
~ **отбо'ра проб** sampling frequency
~ **появле'ния аберра'ций** aberration rate
~ **пу'льса** pulse rate
~ **реа'кций нажи'ма на рыча'г** lever-pressing rate
~ **рекомбина'ций** recombination frequency
~ **сердцебие'ний** heart rate
~ **слия'ния мелька'ний** flicker fusion frequency
~ **уби'йства** kill rate
абсолю'тная ~ *biom* absolute frequency
домина'нтная ~ dominant frequency *(in bioacoustics)*
звукова'я ~ acoustical frequency
крити'ческая ~ **мелька'ний** critical flicker frequency
нако'пленная ~ *biom* cumulative frequency
ожида'емая ~ expected frequency
основна'я ~ fundamental frequency *(in bioacoustics)*
относи'тельная ~ relative frequency
составля'ющая ~ component frequency *(in bioacoustics)*
теорети'ческая ~ *biom* theoretical frequency
часту'ха *f bot* water-plantain (*Alisma*)
ча'стый *a* dense (*forest*), thick; rapid, quick (*pulse*)
часть *f* part, portion, fraction
часы' *pl* clock, watch
биологи'ческие ~ biological clock
молекуля'рные ~ molecular clock
ча'хлый *a* stunted, poor
ча'ша *f* bowl:
~ **центрифу'ги** rotating bowl

выпари'тельная ~ evaporation bowl
чашеви'дный *a* calycine, calyciform, cupulate
чашели'стик *m* sepal, calyx lobe
чашеобра'зный *a* cotyloid
ча'шечка *f* cup, calycle, calyx, bell
  зу'бчатая ~ toothed calyx
  концева'я не'рвная ~ terminal calyx of nerve
  надколе'нная ~ kneepan, kneecap
  опада'ющая ~ deciduous calyx
  раздельноли'стная ~ segmented calyx
  сростноли'стная ~ calyx with united segments
чашечкови'дный *a* calycine, calyciform
ча'шечковый *a* calycine
чашечкообра'зный *a* calycine, calyciform
чашечкоцветко'вый *a* calyciflorous
ча'шка *f* dish
  ~ **для культиви'рования** culture dish
  ~ **Пе'три** Petri dish
  ~ **Сте'ндера** *hist* Stender dish
  коле'нная ~, **надколе'нная** ~ kneecap, kneepan, patella
ча'ща *f* thicket, midwood
чегло'к *m orn* hobby falcon (*Falco subbuteo*)
  обыкнове'нный ~ hobby falcon (*Falco subbuteo*)
че'дер *m* Cheddar
чек *m* check basin
  ри'совый ~ rice bay
чека'н *m orn* chat, bushchat, stomechat (*Saxicola*)
чека'нка *f* topping
че'лка *f* forelock (*of a horse*)
челове'к *m* man (*Homo sapiens*)
человекообра'зный *a* anthropoid
челюстно'й *a* gnathic, maxillary
че'люсть *f* jaw; chap
  ве'рхняя ~ upper jaw, maxilla
  ни'жняя ~ lower jaw, mandible
чемери'ца *f bot* false hellebore (*Veratrum*)
чепу'ра *f orn* small heron
чепу'ра-нужда' *f orn* little egret *(Egretta garzetta)*
червеобра'зный *a* vermicular, vermiculate, lumbriciform
черве'ц *m* mealybug (*Pseudococcidae*)
  ла'ковый ~ Indian lac insect (*Laccifer lacca*)
черви'вый *a* grubby
червово'дня *f* cocconery, silkworm nursery
червото'чина *f* pith fleck, wormhole
червь *m* worm
  бродя'чий многощети'нковый ~ free-living polychaete *(Errantia)*
  земляно'й ~ earthworm (*Lumbricidae*)
  ко'льчатый ~ annelid worm, segmented worm *(Annelida)*
  колючеголо'вый ~ proboscis worm, spiny headed worm, thorny-headed worm *(Acanthocephala)*
  корабе'льный ~ shipworm (*Teredo*)
  кру'глый ~ threadworm, roundworm, nematode worm (*Nematoda*)
  ле'нточный ~ tapeworm (*Cestoda*)
  многощети'нковый ~ polychaete, bristle-bearing worm, bristle worm, polychaete worm *(Polychaeta)*
  мучно'й ~ mealworm (*Tenebrio molitor*)
  пло'ский ~ flatworm (*Plathelminthes*)
  ра'тный ~ *ent* armyworm
  шелкови'чный ~ *ent* silkworm (*Bombyx mori*)
череда' *f bot* bur-marygold (*Bidens*)
чередова'ние *n* alteration, rotation
  ~ **поколе'ний** alteration of generations, sequence of generations, digenesis
чередова'ть *v* rotate
чере'муха *f bot* bird cherry (*Padus*)
черемша' *f bot* ramson (*Allium ursinum*)
черенкова'ние *n* grafting
  ~ **листо'м** propagation by leaf
черенкова'ть *v* graft
черено'к *m* graft, cutting; handle, haft
  морско'й ~ razor clam (*Ensis*)
  побе'говый ~ shoot cutting
  стеблево'й ~ stem cutting
че'реп *m* skull
  висцера'льный ~ visceral cranium
  лицево'й ~ visceral cranium
  мозгово'й ~ cerebral cranium
  перепо'нчатый ~ membranous cranium
  удлине'нный ~ hypsicephalic skull
  укоро'ченный ~ brachycephalic skull, brachicephalous skull
черепно'й *a* cranial
черепа'ха *f* turtle, tortoise (*Testudines*)
  боло'тная ~ fresh-water turtle *(Emys)*
  гре'ческая ~ spur-thighed tortoise *(Tesudo graeca)*
  длинноше'яя ~ chicken tortoise *(Deirochelys reticulata)*

зелё'ная ~ edible turtle, green turtle *(Chelonia mydas)*
змеинoше'йная ~ long-necked turtle *(Chelodina)*
кавка'зская ~ spur-thighed tortoise *(Tesudo graeca)*
морска'я ~ marine turtle *(Cheloniidae)*
назе'мная ~ turtle *(Testudo)*
пресново'дная ~ fresh-water turtle, marsh turtle, pond turtle *(Emydidae)*
слоно'вая ~ Galapagos tortoise *(Geochelone elephantopus)*
средиземномо'рская ~ spur-thighed tortoise *(Tesudo graeca)*
среднеазиа'тская ~ steppe tortoise, Central Asian tortoise, Horsfield's tortoise *(Agrionemys horsfieldi)*
степна'я ~ steppe tortoise, Central Asian tortoise, Horsfield's tortoise *(Agrionemys horsfieldi)*
супова'я ~ edible turtle, green turtle *(Chelonia mydas)*
сухопу'тная ~ land tortoise *(Testudinidae)*
черепа'шка *f* shield-backed bug, tortoise beetle (*Eurygaster integriceps*)
вре'дная ~ tortoise beetle (*Eurygaster integriceps*)
черепи'тчато-располо'женный *a bot* imbricate, imbricated
черепицеобра'зный *a* imbricate, imbricated
черепно'й *a* cranial
чересполо'сица *f* strip holding of land
черешко'вый *a* petiolar
чере'шня *f bot* sweet cherry (*Cerasus avium*)
черешо'к *m bot* leafstalk, petiole
черешча'тый *a* petiolate
че'рнеть *f orn* scaup, bluebill (*Aythya*)
морска'я ~ greater scaup duck (*Aythya marila*)
хохла'тая ~ tufted duck (*Aythya fuligula*)
черни'ка *f* bilberry, whortleberry (*Vaccinium myrtillus*)
черни'чник *m* bilberry heath
черноголо'вка *f* self-head (*Prunella*)
чернозё'м *m* chernozem, chernozem soil
высокогу'мусный ~ black chernozem
мо'щный ~ deep chernozem
чернозо'бик *m orn* dunlin (*Calidrus alpina*)
черноле'сье *n* broad-leaved forest, leafy forest
черносли'в *m* dried plums
чернота'л *m* bay-leaf willow, laurel willow, bay willow (*Salix pentandra*)
черноте'лка *f ent* darkling beetle (*Tenebrionidae*)
черну'шка *f* fennel-flower, love-in-a-mist (*Nigella*); sombre lactarid, base toadstool, ugly toadstool, ugly milk-cup (*Lactarius necator*)
чё'рный *a* black
ализари'новый ~ alizarin black
нафто'ловый ~ naphthol black
суда'н ~ Sudan black
эриохро'м ~ eriochrome black
я'нус ~ Janus black
черпалови'дный *a anat* arytenoid
чё'рствый stale
~ **хлеб** stale bread
черта' *f* feature, trait; line
чертополо'х *m bot* thistle (*Carduus*)
чеса'ться *v* scratch (*oneself*), itch
чесно'к *m bot* garlic (*Allium sativum*)
чесно'чница *f* spadefoot (*Pelobatidae*); garlic-mustard (*Alliaria*)
чесо'тка *f med* scabies, mange
чесуча' *f* tussore, tussah, wild silk
четвере'ньки *pl*: **на четвере'ньках** on all fours
четвероно'гий *a* quadruped, quadrupedal, four-footed, tetrapod
четвероно'гое *n* quadruped
четвероpу'кий *a* quadrumanous
четверохо'лмие *n* quadrigeminal bodies
четверти'чный *a* Quaternary
четкови'дный *a* moniliform, necklace-shaped, beaded, granose
четырех- quadri-, tetra-
четырёхдо'льчатый *a* quadrilobate
четырёхкры'лый *a bot* quadrialate
четырёхле'тний *a* quadrennial
четырёхли'стный *a bot* quadriphyllous, quadrifoliate
четырёхло'пастный *a* quadrilobate
четырёхлучево'й *a* quadriradiare
четырехпа'лый *a* tetradactyl
четырёхпа'рный *a bot* quadrijugate
четырёхразде'льный *a* quadrifid, quadripartite
четырёхро'гий *a bot* quadricorn
четырёхство'рчатый *a* quadrivalve, quadrivalvular
четырёхцветко'вый *a bot* quadriflorous

четырехря'дный *a* tetrastichous, quadrifarious

че'хлик *m* calyptra, calypter, cap

корнево'й ~ root cap, pilorhiza

чехликообра'зный *a* calyptriform

чечеви'ца *f* lentil (*Lens*)

чечевицеобра'зный *a* lenticulate, lentiform

чечеви'чка *f* lenticel

чече'тка *f* redpoll (*Carduelis*)

чешуеви'дный *a* lepidoid

чешуено'сный *a* scale-bearing

чешу'йка *f* scale, squame, squamule; ramentum; scute; perula (*of a bud*)

кро'ющая ~ *bot* bract scale, cover scale

рогова'я ~ horny scale

чешу'йчатый *a* scaly, scaled, squamate, squamose, squamous

чешуя' *f* scale, scales; *bot* chaff

ве'рхняя колоско'вая ~ palea

вы'емчатая колоско'вая ~ emarginate glume

заострё'нная колоско'вая ~ awn pointed glume

колоско'вая ~ gluma, glume

ладьеви'дная ни'жняя цветко'вая ~ keeled lemma

лу'ковичная ~ *bot* bulbous scale

нару'жная цветко'вая ~ exterior palea

ни'жняя цветко'вая ~ lemma

плё'нчатая ~ *bot* papery scale

пуста'я колоско'вая ~ empty glume

семенна'я ~ *bot* fruit scale, seed scale

тре'тья колоско'вая ~ sterile lemma

цветко'вая ~ lemma, floral glume, flower glume

чи'бис *m orn* pewit, lapwing (*Vanellus vanellus*)

чиж *m orn* siskin (*Spinus*)

чи'рей *m med* boil, furuncle

чири'к *i* chirp

чири'канье *n* chirp, chirrup

чири'кать *v* chirp, chipper, twitter

чиро'к *m orn* teal (*Anas*)

чиро'к-свистуно'к *m* European teal, green-winged teal *(Anas crecca)*

чиро'к-трескуно'к *m* garganey teal, summer teal *(Anas querquedula)*

чи'сленность *f* number

~ **ви'да** abundance of species

~ **кла'сса** class frequency

число' *n* number

~ **зацепле'ний** *mbiol* linking number

~ **импульсов в минуту, имп/мин** count per minute

~ **ко'пий** *gen* copy number

~ **Рейно'льдса** Reynolds' number

~ **хромосо'м** chromosome number

ассимиляцио'нное ~ assimilation number

гапло'идное ~ haploid number

дипло'идное ~ diploid number

ио'дное iodine number

кисло'тное ~ acid number

ко'довое ~ *gen* coding number

мода'льное ~ *gen* modal number

основно'е ~ *gen* basic number

факти'ческое ~ actual count

чи'стик *m orn* guillemot, kiddaw, tystie (*Cepphus*)

атланти'ческий ~ sea pigeon, black guillemot, short-billed guillemot (*Cepphus grylle*)

обыкнове'нный ~ black guillemot, short-billed guillemot *(Cepphus grylle)*

очко'вый ~ sooty guillemot, spectacled guillemot *(Cepphus carbo)*

тихоокеа'нский ~ pigeon guillemot *(Cepphus columba)*

чи'стить *v* clean, cleanse; dress (*a horse*)

~ **пе'рья** preen, prink

чи'стка *f* grooming

~ **гениталий** genital grooming

агресси'вная ~ aggressive grooming

взаи'мная ~ mutual grooming, social grooming

чистокро'вный *a* pure-blooded, thoroughbred

чистопоро'дность *f* genuineness of strain

чистопоро'дный *a* purebred, thoroughbred

чистоте'л *m* celandine (*Chelidonium*)

чи'стый *a* pure; clean; **чистая вода'** pure water; ~ **во'здух** pure air

чиха'ть *v* sneeze

член *m* limb, member

~ **околоцве'тника** perianth lobe

мужско'й полово'й ~ penis, virile member

полово'й ~ penis

чле'ник *m* segment

членистоно'гое arthropod *(Arthropoda)*

кровососу'щее ~ bloodsucking arthropod

чле'нистый *a* articulate, articulated, jointed

чо'мга *f* great crested grebe (*Podiceps cristatus*)

чуба'рый *a* mottled

чубу'к *m* stalk, grape stalk

чувстви'тельность *f* sensitivity, susceptibility, sensitiveness, sensibility
чувстви'тельный *a* sensitive
чу'вство *n* sense, feeling, sensation
~ **вре'мени** time sense
~ **давле'ния** baresthesia, pressure sense
~ **движе'ния** motion sense
~ **положе'ния** posture sense
~ **простра'нства** space sense
~ **равнове'сия** static sense
мы'шечное ~ kinesthetic sense, muscular sense
статокинети'ческое ~ static sense
стереогности'ческое ~ stereognostic sense
чу'вствовать *v* feel
чу'ждый *a* foreign, alien, strange
чужеро'дный *a* foreign, alien
чума' *f* plague, pest
~ **водяна'я** water-weed, water thyme (*Elodea*)
~ **птиц** fowl pest
~ **свине'й** swine fever
чуми'за *f* green foxtail, bristle grass (*Setaria viridis*)
чутьё' *n* scent, scenting
ве'рхнее ~ air scenting
ни'жнее ~ ground scenting
чу'фа *f* chufa, earth almond (*Cyperus esculentus*)
чу'чело *n* stuffed animal
~ **пти'цы** stuffed bird
чу'ять *v* scent, sniff, snuff

Ш

шаг *m* walk *(a kind of gait)*; tread; step
бы'стрый ~ promt step
лё'гкий ~ easy step
ме'дленный ~ slow gait
пружи'нистый ~ springy step
ро'вный ~ promt step
шака'л *m* jackal (*Canis aureus*)
шала'шник *m* bowerbird (*Ptilonorhynchidae*)
шалфе'й *m bot* sage (*Salvia*)
шампиньо'н *m* true mushroom
~ **двуспо'ровый**, ~ **культиви'руемый** champignon (*Agaricus bisporus*)
~ **лугово'й** field mushroom (*Agaricus campestris*)
~ **обыкнове'нный** field mushroom (*Agaricus campestris*)
шампиньо'нница *f* mushroom house
шар *m* ball
заро'дышевый ~ *helm* egg ball, germ ball
шара'хаться *v* dash aside; shy (*of a horse*)
шара'хнуться *v* bolt, shy
ша'рик *m* globule
моло'чный ~ milk globule
шаро'вка *f mal* fingernail clam (*Sphaerium*)
шафра'н *m bot* saffron (*Crocus*)
ша'шень *m mal* shipworm (*Teredo*)
ше'йка *f* neck,collet, collum; cervix
~ **зу'ба** neck of tooth
~ **ма'тки** neck of womb, uterine neck
~**спе'рмия** sperm neck
корнева'я ~ root collar
ра'ковая ~ crayfish tail
ра'ковые шейки *pl bot* snakeweed, serpent grass (*Polygonum bistorta*)
ше'йкер *m lab* shaker
планше'тный ~ plate shaker
ротацио'нный ~ gyratory shaker, rotary shaker
ше'йный *a* cervical
шелесте'ть *v* rustle
шёлк *m* silk
шелкови'стый *a* silky, sericate
шелкови'ца *f* mulberry (*Morus*)
шелково'д *m* silkworm breeder
шелково'дство *n* silk culture, silkworm husbandry
шелкопря'д *m* silkworm moth
непа'рный ~ gypsy moth (*Ocneria dispar*)
ту'товый ~ silkworm (*Bombyx mori*)
шё'лк-сыре'ц *m* raw-silk
шелла'к *m* shellac
шелуди'вый *a* scabby
шелуха' *f* husk, hull, pod, cod, shuck
шелуше'ние *n* desquamation
шелуши'ть *v* husk, hull, shell, dehull, peel
шелуши'ться *v* peel
шелю'га *f bot* violet willow (*Salix daphnoides*)
шерохова'тый *a* scabrous, rough, asperous
шерсти'стый *a* woolly, lanate, lanose, pannose, ulotrichous
шерсть *f* wool, hair
шерша'вый *a* rough, scabrous, asperous
шея *f* neck
шизога'мия *f* schizogamy
шизогене'з *m* schizogenesis
шизогенети'ческий *a* schizogenetic
шизогнати'ческий *a* schizognathous
шизогони'я *f prot* schizogony
шизо'нт *m prot* schizont
шизостели'я *f* schizostely
шизофи'т *m* schizophyte
шилови'дный *a* subulate, styloid, awl-shaped
шилоклю'вка *f orn* avocet (*Recurvirostra avosetta*)
шилоли'стный *a* awl-leaved
шилохво'сть *f orn* pintail (*Anas acuta*)
ши'льце *n bot* pip (*a* sprout *of* grasses)
шимпанзе' *m* chimpanzee (*Pan troglodytes*)
шип *m* spine, thorn, spur, barb, bur, burr, prickle; spiny sturgeon (*Acipenser nudiventris*)
подви'жный *ent* ~ mobile spine
складно'й ~ erectile spine
ядови'тый ~ venomous spine
шипе'ть *v* hiss (*of snakes*)
ши'пик *m* spinelet, spinule
дендри'тный ~ dendritic spine
шипова'тый *a* aculeate, echinate, echinated, acanthaceous, prickly, thorny, thorned, spiny, spinous
шипови'дный *a* aculeiform
шипо'вник *m* brier, rose (*Rosa*); hip (*of briers*)
шипу'н *m orn* swan, mute swan (*Cygnus olor*)
широколи'стный *a* broad-leaved
широконо'ска *f orn* shoveler (*Anas clypeata*)
шистосомато'з *m* schistosomiasis
~ **Ма'нсона** Manson's schistosomiasis, Manson's disease

шистосомо'з *m* schistosomiasis
ши'шка *f* cone, strobila, strobile; pile
  ворсова'льная ~ fuller's teasel (*Dipsacus*)
  геморроида'льная ~ pile
шишкова'тый *a* torose, bosselated
шишкови'дный *a* cone-shaped, pineal, strobiloid; **шишкови'дная железа** pineal body, pituitary gland
шишконо'сный *a* cone-bearing
шишкоя'года *f* galberry
шкала' *f* scale
  ~ **жи'зненности** scale of vitality
  ~ **измере'ния** scale of measurement
  ~ **Ке'львина** Kelvin scale
  ~ **отноше'ний** ratio scale
  ~ **проекти'вного покры'тия** scale of coverage
  ~ **Реомю'ра** Reaumur scale
  ~ **социабилите'та** scale of sociability
  ~ **Փаренге'йта** Fahrenheit scale
  ~ **Це'льсия** Celsius scale
  интерва'льная ~ interval scale
  колориметри'ческая ~ color scale
  логарифми'ческая ~ logarithmic scale
  поря'дковая ~ ordinal scale
  цветова'я ~ color scale
шкаф *m* case, cabinet
  вытяжно'й ~ hood, exhaust hood
  суши'льный ~ oven, drying oven, dry-air sterilizer, hot-air sterilizer
шку'ра *f* hide, skin, pelt
  свищева'тая ~ grubby skin
шку'рка *f* fur, fell, pelt; sand-paper, emery paper
шлапа'к *m* mixed gait
шлем *m* helmet, hood
  сухожи'льный ~ epicranial aponeurosis
шлемови'дный *a* galeate, galeated, cassideous, hood-shaped
шля'пка *f* cap, pileus
шля'почный *a* pileate
шмележу'к *m* bumble flower beetle (*Euphoria*)
шмель *m* bumblebee (*Bombus*)
  ~ **полево'й** common carder bee *(Bombus agrorum)*
шмель-куку'шка *m* parasitic bumblebee *(Psithyrus)*
шмель-пло'тник carpenter bee *(Xylocopa)*
шнитт-лук *m* chive (*Allium schoenoprasum*)
шов *m* seam, suture, raphe; **швы** *pl*; **накладывать швы** put in stitches
  ~ **Альбе'рта** Albert's suture
  ~ **Гели'** Gely's suture
  ~ **Го'улда** Gould's suture
  ~ **Жобе'ра де Ламба'ля** Jobert de Lamballe's suture
  ~ **Ко'ннела** Connell's suture
  ~ **Ку'шинга** Cushing's suture
  ~ **Ламбе'ра** Lembert suture
  ~ **Панко'аста** Pancoast's suture
  ~ **Паре'** Pare's suture
  ~ **Ри'хтера** Richter's suture
  ~ **Хольсте'да** Halstead's suture
  ~ **Че'рни** Czerny's suture
  вене'чный coronary suture
  гармони'ческий harmonic suture
  гула'рный gular suture
  зу'бчатый ~ *anat* schindylesis, dentate suture, serrated suture
  ко'стный ~ bone suture
  ли'ночный ~ *ent* molting suture
  ло'бный ~ frontal suture
  медиа'льный ~ median suture
  попере'чный transverse suture
  эпикраниа'льный ~ epicranial suture
шо'вный *a* sutural
шок *m* shock
  анафилакти'ческий ~ anaphylactic shock, serum shock, allergic shock
  гипотони'ческий ~ hypotonic shock
  осмоти'ческий ~ osmotic shock
  температу'рный ~ heat shock, thermal shock
  теплово'й ~ heat shock
шо'рох *m* murmur (*of leaves*), rustle
шпа'нка *f ent* blister beetle *(Epicauta)*
  ~ **Фабри'ция** ash-gray blister beetle *(Epicauta fabricii)*
шпа'тель *m* spatula
шпина'т *m* spinach, spinage
шпо'ра *f* spur, calcar
шпорцеви'ный *a* spurlike
шприц *m* syringe
  ~ **однора'зового примене'ния** disposable syringe, single-use syringe
  градуи'рованный ~ dose syringe
  однора'зовый ~ disposable syringe, single-use syringe
  фильтру'ющий ~ Флэ'та-Ла'ндина Flath-Lundin filter syringe
шрам *m* scar
шримс *m zool* shrimp (*Crangonidae*)
шта'бель *m* pile

~ **брё'вен** pile of logs
штамм *m* strain
вакци'нный ~ vaccinating strain, vaccine strain, vaccine seed strain
вируле'нтный ~ virulent strain
гомолизоге'нный ~ homolysogenic strain
ди'кий ~ wild strain
индика'торный ~ indicator strain
и'стинно-лизоге'нный ~ true lysogenic strain
исхо'дный ~ wild strain, stock strain
исхо'дный вакци'нный ~ master strain
кле'точный ~ cell strain
крипти'ческий ~ cryptic strain
лизоге'нный ~ lysogenic strain
монолизоге'нный ~ monolysogenic strain
неразреша'ющий ~ nonpermissive strain
пермисси'вный ~ permissive strain
полилизоге'нный ~ polylysogenic strain
псевдолизоге'нный ~ pseudolysogenic strain
роди'тельский ~ parent strain
спра'вочный ~ reference strain
супре'ссорный ~ suppressor strain
типи'руемый ~ typeable strain
чи'стый ~ pure strain
этало'нный ~ reference strain
штамм-«уби'йца» killer strain
штати'в *m* rack, support, stand, support stand
~ **для бюре'ток** burette support
~ **для су'шки** drying rack
~ **для пипе'ток** pipette rack, pipette stand
~ **для проби'рок** tube rack, test tube support
ште'мпель-пипе'тка *f* piston pipette
штрих *m* streak
шуга' *f* cream ice, ice scum
шум *m* noise, murmur, hum, whir
вено'зный ~ venous hum
внутрисерде'чный ~ cardiac murmur
промы'шленный ~ industrial noise
темново'й ~ dark noise
электри'ческий ~ electrical noise
шуме'ть *v* whir
шунт *m* shunt, bypass
артериовено'зный ~ arteriovenous shunt
вено'зно-артериа'льный ~ venous-arterial shunt
шунти'рование *n* shunting
шунти'ровать *v* shunt
шурша'ть *v* rustle
шю'ттель-аппара'т *m* shaker
~ **с кругов'ым враще'нием** gyratory shaker, rotary shaker
вибрацио'нный возвра'тно-поступа'тельный ~ lateral shaker, reciprocal shaker

Щ

щебета'ние *n* chirp, chirrup
щебета'ть *v* chirp, chirrup
щего'л *m* goldfinch (*Carduelis carduelis*)
щека' *f* cheek
щелево'й *a* rimate, rimose
щё'лканье *n* snap
щелку'н *m* click beetle, snapping beetle (*Elateridae*)
щель *f* slit, cleft, fissure, scissure
  ана'льная ~ anal cleft
  бараба'нно-сосцеви'дная ~ auricular fissure
  генита'льная genital slit
  гла'зерова ~ Glaserian fissure
  глазна'я ~ palpebral fissure
  голосова'я ~ glottis
  жа'берная ~ *emb* branchial fissure, branchial grove, visceral cleft; gill slit, gill opening, branchial aperture
  коне'чная ~ terminal fissure *(in diatoms)*
  крылонё'бная ~ pterygopalatine fissure
  полова'я ~ vulvar cleft
  синапти'ческая ~ synaptic cleft
  срамна'я ~ vulvar cleft
  у'стьичная ~ *bot* stomatic cleft
  хоро'идная ~ choroid fissure
щени'ться *v* pup
щё'нка *f* birth
щё'нная *f* pregnant (*of a dog*)
щено'к *m* whelp (*e.g., of fox, wolf*); cub, pup, puppy
щепа' *f* chips
ще'пка *f* chip
щети'на *f* bristle, bristle hair
щети'нистый *a* bristly, hirsute, strigose, setaceous
щети'нка *f* bristle, seta, setule, setula, cheta, chaeta
  акростиха'льная ~ acrostichal bristle, inner dorsocentral bristle
  ве'рхняя ло'бная ~ ascending frontal bristle
  гипоплевра'льная ~ hypopleural bristle
  глазко'вая ~ ocellar bristle
  ло'бная ~ frontal bristle
  мезоплевра'льная ~ mesopleural bristle
  метаплевра'льная ~ metapleural bristle, trichostichal bristle
  надкрылова'я ~ supra-alar bristle
  орбита'льная ~ orbital bristle
  осяза'тельная ~ sense bristle
  плечева'я ~ humeral bristle
  предкрылова'я ~ prealar bristle
  фронтоорбита'льная ~ fronto-orbital bristle
щетинконо'сный *a* setiferous, setigerous, chaetiferous
щети'нник *m* bristle grass (*Setaria*)
щё'тка *f* brush; pollen brush
щёткообра'зный *a* scopate, scopiform
щё'точка *f* brush; pollen brush; strigil; *ent* scopa
щё'чно-гло'точный *a* buccopharyngeal
щё'чный *a* buccal, cheek
щипа'ть *v* pinch, nip; smart, sting, bite; graze, nibble, browse, pick (*grass*); ~ **паклю** pick oakum
щипцы' *pl* forceps
  пулевы'е ~ bullet forceps
щири'ца *f bot* tassel flower, redroot (*Amaranthus*)
щит *m* shield
  головно'й ~ head shield
щи'тень *m* tadpole shrimp (*Notostraca*)
щиткови'дный *a bot* corymbose, corymbiform
щитко'вый *a* scutal
щитови'дный *a* peltate, peltated, scutate, scutellate, scutiform; *anat* thyroid
щитовка *f* coccid (*Coccoidea*)
щито'вник *m* wood fern (*Dryopteris*)
щито'к *m* scute, scutum, scutellum; shield; pelta; *bot* cyme, corymb
  загри'вковый ~ nuchal shield (*of turtles*)
  заро'дышевый ~ embryonic shield
щитоли'стный *a* scutifolious
щитомо'рдник *m herp* copperhead (*Agkistrodon*)
щитоно'ска *f* tortoise beetle (*Cassididae*)
щитоно'сный *a* peltigerous
щу'пальце *n* feeler, tentacle; horn (*of snails*)
щупальцеви'дный *a* tentaculiform
щу'пик *m* palp, palpus, tentacle
  губно'й ~ labial palpus
  лабиа'льный ~ labial palpus
  максилля'рный ~ maxillary palpus, maxillary tentacle
щу'плый *a* underdeveloped, hollow (*of grain*)
щу'риться *v* narrow eyes
щу'рка *f orn* bee eater (*Meropidae*)

Э

эбули'зм *m* ebulism
эвагини'ровать *v* evaginate
эвапотранспира'ция *f* evapotranspiration
эвапотранспиро'метр *m* evapotranspirometer
эвкли'макс *m* euclimax
эвока'тор *m emb* evocator
эвока'ция *f emb* evocation
эволюти'вный *a* evolutive
эволюциони'ровать *v* evolve
эволюцио'нный *a* evolutionary
эволю'ция *f* evolution
~ **взры'вами** explosive evolution
~ **домини'рования** evolution of dominance
~ **кло'на** clonal evolution
~ **органи'ческого ми'ра** organic evolution
адапти'вная ~ adaptive evolution
докле'точная ~ precellular evolution
ква'нтовая ~ quantum evolution
клона'льная ~ clonal evolution
конверге'нтная ~ convergent evolution
мозаи'чная ~ mosaic evolution
молекуля'рная ~ molecular evolution
напра'вленная ~ directional evolution, rectilinear evolution
неадапти'вная ~ nonadaptive evolution
неопределе'нная ~ indeterminate evolution
непреры'вная ~ continuous evolution
неприспособи'тельная ~ nonadaptive evolution
обра'тная ~ reverse evolution
ортогенети'ческая ~ orthogenetic evolution
паралле'льная ~ parallel evolution
после'довательная ~ consequentional evolution
поступа'тельная ~ rectilinear evolution
преры'вистая ~ discontinuous evolution
приспособи'тельная ~ adaptive evolution
програ'ммная ~ program evolution, programme evolution
прогресси'вная ~ progressive evolution
регресси'вная ~ regressive evolution
ретрогресси'вная ~ retrogressive evolution
се'тчатая ~ reticulate evolution
скачкообра'зная ~ saltatory evolution
случа'йная ~ accidental evolution
спонта'нная ~ spontaneous evolution
филети'ческая ~ phyletic evolution
эрупти'вная ~ eruptive evolution
эври- eury-
эвриба'рный *a* eurybaric
эвриба'тный *a* eurybathic
эвривале'нт *m* euryvalent
эвригали'нный *a* euryhaline
эврио'йкный *a* euryoecic
эвриойки'я *a* euryoky
эврипласти'ческий *a* euryplastic
эврите'рмный *a* eurythermal, eurythermic
эврито'пный *a* eurytopic
эвритро'пный *a* eurytropic
эврихо'р *m* eurychore
эвте'ника *f* euthenics
эвтрофика'ция *f* eutrophication
эвтрофи'рование *n* eutrophication
эвтрофи'я *f* eutrophy
эвтро'фный *a* eutrophic
эгре'тка *f* egret plume, aigrette, egret
эдато'п *m* edatope
эдафи'ческий *a* edaphic
эдафо'н *m* edaphon
эдафофи'т *m* edaphophyte
эдельве'йс *m* edelweiss (*Leontopodium alpinum*)
эдести'н *m* edestin
эдифика'тор *m* edificator
эдомина'нт *m* edominant
эзери'н *m* eserine
эзотери'ческий *a* esoteric
эйтана'зия *f* euthanasia
эйфори'ческий *a* euphoric
эйфори'я *f* euphoria
эка'да *f* ecad, oecad
эква'тор equator
~ **яйца'** egg equator
экваториа'льный *a* equatorial
эквацио'нный *a* equational
эквивале'нт *m* equivalent
~ **вла'жности** moisture equivalent
белко'вый ~ protein equivalent
биологи'ческий ~ рентге'на, бэр roentgen-equivalent-man, mammal-roentgen equivalent, rem
вентиляцио'нный ~ *phys* ventilation equivalent
кормово'й ~ fodder equivalent
осмоти'ческий ~ osmotic equivalent
пищево'й ~ food equivalent

теплово'й ~ heat equivalent
эквивале'нтность *f* equivalence
эквивале'нтный *a* equivalent
эквиденситоме'трия *f* equidensitometry
эквимоля'рный *a* equimolar
эквипотенциа'льность *f* equipotentiality
ЭКГ, электрокардиогра'мма *f* electrocardiogram, ECG
экдеми'чный *a* ecdemic
экдизо'н *m* ecdysone
экза'рхный *a bot* exarch
экзе'ма *f med* eczema, tetter
экземпля'р *m* specimen
~ **в спирту'** spirit specimen
герба'рный ~ herbarium specimen
спиртово'й ~ spirit specimen
экзи'на *f pal* exine
экзинти'на *f pal* exintine
экзи'т *m* exite
экзо- exo-
экзоантиге'н *m* exoantigen
экзобиоло'гия *f* exobiology
экзова'т *m* exovate
экзога'мия *f* exogamy
экзога'струла *f* exogastrula
экзогаструля'ция *f* exogastrulation
экзоге'нный *a* exogenous, exogenic, exogenetic
экзогено'та *f* exogenote
экзоде'рма *f bot* exoderm
экзока'рпий *f* exocarp
экзокони'ческий *a ent* exocone
экзокони'дия *f* exoconidium
экзокри'нный *a* exocrine, exocrinous
экзокути'кула *f* exocuticle
экзо'н *m* exon
экзонуклеа'за *f* exonuclease
экзопери'дий *m* exoperidium
экзопла'зма *f bot* exoplasm
экзоподи'т *m* exopodite
экзоскеле'т *m* external skeleton
экзо'смос *m* exosmosis
экзоспо'ра *f* exospore
экзоспо'рий *m* exosporium
экзосто'з *m* exostosis
экзосто'м *m* exostome
экзоте'ка *f* exotheca
экзоте'ций *m* exothecium
экзотокси'н *m* exotoxin
экзотропи'зм *m* exotropism
экзоферме'нт *m* exoenzyme
экзофори'я *f* exophoria
экзохорио'н *m* exochorion
экзоцело'м *m* exocoelom
экзоцито'з *m* exocytosis
экзу'вий *m* exuvium
экзэкзи'на *f* exoexine
эклампси'я *f med* eclampsia
экобиоти'ческий *a* ecobiotic
экови'д *m* ecospecies
экогене'з *m* ecogenesis
экогенети'ческий *a* ecogenetic
экоде'м *m* ecodeme
экокли'мат *m* ecoclimate
экокли'н *m* ecocline
экологи'ческий *a* ecological
эколо'гия *f* ecology
~ **живо'тных** animal ecology, zooecology
~ **па'стбищ** range ecology
~ **расте'ний** plant ecology
~ **челове'ка** human ecology
глоба'льная ~ global ecology
глуби'нная ~ deep ecology
поведе'нческая ~ behavioral ecology
прикладна'я ~ applied ecology
сельскохозя'йственная ~ agricultural ecology
экопротероа'ндрия *f* ecoproteroandry
экопротероги'ния *f* ecoproterogyny
экосисте'ма *f* ecosystem, ecological system
есте'ственная ~ natural ecosystem
ненару'шенная ~ undisturbed ecosystem
экосфе'ра *f* ecosphere
экоти'п *m* ecotype
экото'н *m* ecotone
экото'п *m* ecotope
экофе'н *m* ecophene, oecophene
экофеноти'п *m* ecophenotype
экоци'д *m* ecocide
эксгибициони'зм *m* exhibitionism
эксконъюга'нт *m* exconjugant
экскреме'нты *m pl* excrements, faeces, egesta
экскрети'рование *n* egestion
экскрети'ровать *v* egest
экскре'ты *m pl* egesta
экскрето'рный *a* excretory, emunctory
экскре'ция *f* excretion
экскурва'ция *f* excurvation
экскурви'рованный *a* excurved
экску'рсия *f* excursion
~ **грудно'й кле'тки** chest excursion, respiratory excursion

дыха'тельная ~ chest excursion, respiratory excursion
эксперти'за *f* assessment, examination
санита'рная ~ sanitary examination
токсикологи'ческая ~ toxicological examination
экологи'ческая ~ environment impact assessment
эксо'ма *f* ecsoma
экспа'нсия *f* expansion
клона'льная ~ clonal expansion
экспериме'нт *m* experiment
~ **по ти'пу «рок-н-ролл»** "rock and roll" experiment
эксперимента'льный *a* experimental
эксперименти'рование *n* experimentation, experiment
экспира'ция *f* expiration
экспланта'т *m* explant
экспланта'ция *f* explantation
эксплуата'ция *f* exploitation
экспози'ция *f* exposition, exposure
экспре'сс-ана'лиз *m* quick analysis, rapid analysis, snap analysis
экспресси'вность *f* expressivity
экспресси'рующий *gen* expressing
экспре'ссия *f* expression
~ **ге'нов** *gen* gene expression
~ **идиоти'па** idiotypic expression
предра'нняя ~ ге'нов *gen* gene immediate-early expression
экспре'сс-лаборато'рия *f* express-laboratory
экссуда'т *m* exudate, exudation
экссуда'ция *f* exudation
экстe'нсор *m* extensor
экстерореце'птор *m* exteroreceptor
экстерореце'пция *f* exteroreception, exteroception
экстерье'р *m* exterior appearance, exterior
эксти'на *f* extine
экстирпа'ция *f* extirpation, excursion, ablation
экстра- extra-
экстравагина'льный *a* extravaginal
экстраге'нный *a* extragenic
экстраги'рование *n* extraction
экстраги'ровать *v* extract
экстразона'льный *a* extrazonal
экстраклина'льный *a* extraclinal
экстрако'пия *f* extracopy
экстра'кт *m* extract
бескле'точный ~ cell-free extract
во'дный ~ aqueous extract
густо'й ~ thick extract
дрожжево'й ~ yeast extract
жи'дкий ~ liquid extract
кукуру'зный ~ corn-steep extract
мясно'й ~ beef extract, meat extract
о'бщий ~ crude extract
солодо'вый ~ malt extract
сумма'рный ~ crude extract
экстрапирамида'льный *a* extrapyramidal
экстраполя'ция *f* extrapolation
экстраси'стола *f* extrasystole
атриовентрикуля'рная ~ atrio-ventriclar extrasystole, junctional extrasystole
возвра'тная ~ return extrasystole
вста'вочная ~ interpolated extrasystole
желу'дочковая ~ infranodal extrasystole, ventricular extrasystole
интерполи'рованная ~ interpolated extrasystole
предсе'рдная ~ auricular extrasystole
предсе'рдно-желу'дочковая ~ atrio-ventricular extrasystole, junctional extrasystole
ра'нняя желу'дочковая ~ premature ventricular extrasystole
ра'нняя предсе'рдная ~ premature auricular extrasystole
экстрастеля'рный *a bot* extrastelar
экстрафлора'льный *a* extrafloral
экстрафуза'льный *a* extrafusal
экстре'мум *m* extremum
усло'вный ~ *biom* constraint extremum, conditional extremum
эксумбре'лла *f zool* exumbrella
эксфлагелля'ция *f* exflagellation
эксфолиа'ция *f* exfoliation
эксцентри'чный excentric
эксце'сс *biom* excess
эксци'зия *f* excision
эксцисти'рование *n* excystation
эктази'я *f* ectasia
экто- ecto-
эктобла'ст *m* ectoblast, primitive ectoderm
эктогене'з *m* ectogenesis
эктоге'нный *a* ectogenous
эктогли'я *f* ectoglia
эктоде'рма *f* ectoderm
нейра'льная ~ *emb* neural ectoderm
перви'чная ~ primitive ectoderm
ротова'я ~ *emb* oral ectoderm
эктодерма'льный *a* ectodermic, ectodermal

эктолецита'льный *a* ectolecithal
эктоме'р *m* ectomere
эктоневра'льный *a* ectoneural
эктонэкзи'на *f pal* ectonexine
эктопарази'т *m* ectoparasite, epizoic parasite, external parasite
эктопи'ческий *a* ectopic
эктопи'я *f* ectopia
эктопла'зма *f* ectoplasm
эктопла'ст *m* ectoplast
эктоптериго'ид *m* ectopterygoid bone, ectopterygoid
эктосо'ма *f* ectosome
эктоспо'ра *f* ectospore
эктосто'з *m* ectostosis
эктосфе'ра *f* ectosphere
эктосэкзи'на *f pal* ectosexine
эктоте'ка *f* ectotheca
эктотро'ф *m* ectotroph
эктотро'фный *a* ectotrophic
эктоферме'нт *m* ectoenzyme
эктофи'т *m* ectophyte
эктофлео'дный *a* ectophloeodic
эктохо'н *m* ectochone
эктоци'ста *f* ectocyst
эктэкзи'на *f pal* ectexine
элайобласт *m* elaeoblast
элайопла'ст *m* elaioplast
элайосо'ма *f* elaiosome
эласти'н *m* elastin
эласти'чный *a* elastic
элате'ра *f bot* elater
электи'вный *a* elective
электро- electro-
электробио'логия *f* electrobiology
электробло'ттер *m* electroblotter
электробло'ттинг *m* electroblotting
электровосстановле'ние *n* electroreduction
электро'д *m* electrode
~ **сравне'ния** reference electrode
ага'р-ага'ровый ~ agar-agar electrode
акти'вный ~ active electrode, exciting electrode, exploring electrode
база'льный ~ basal electrode
вживлё'нный ~ implanted electrode, indwelling electrode, permanently implanted electrode, chronically implanted electrode
вживля'емый ~ implantable electrode
внутрикле'точный ~ intracellular electrode
водоро'дный ~ hydrogen electrode
глуби'нный ~ depth electrode
внекле'точный ~ extracellular electrode
иго'льчатый ~ needle electrode
импланти'рованный ~ implanted electrode, indwelling electrode, permanently implanted electrode, chronically implanted electrode
импланти'руемый ~ implantable electrode
индиффере'нтный ~ silent electrode
ион-селекти'вный ~ ion-selective electrode
кислоро'дный ~ oxygen electrode
отводя'щий ~ pickup electrode, recording electrode
пасси'вный ~ silent electrode
полярографи'ческий ~ polarographic electrode
стекля'нный ~ glass electrode
стержнево'й ~ rod electrode
стимули'рующий ~ stimulating electrode
ша'риковый ~ ball-point electrode
электродеканта'ция *f* electrodecantation
электродиа'лиз *m* electrodialysis
электродиализа'тор *m* electrodialyzer
электроиммуноана'лиз *m* electroimmunoassay
электроиммунофикса'ция *f* electroimmunofixation
электрокардиогра'мма, ЭКГ *f* electrocardiogram, ECG
электрокардио'граф *m* electrocardiograph
электрокардиографи'ческий *a* electrocardiographic
электрокардиогра'фия *f* electrocardiography
электрокардиоско'п *m* electrocardioscope
электрокаутериза'ция *f* electrocauterization
электрокимогра'мма *f* electrokymogram
электрокимогра'фия *f* electrokymography
электрокортикогра'мма *f* electrocorticogram
электро'лиз *m* electrolysis
электролока'ция *f* electrolocation
электромиогра'мма *f* electromyogram
интегри'рованная ~ integrated electromyogram
пове'рхностная ~ surface electromyogram
электромиогра'фия *f* electromyography
электромо'рфа *f* electromorph
электромотонейро'н *m* electromotoneurone
электро'н *m* electron
возбуждё'нный ~ excited electron
вы'битый ~ expelled electron

одино'чный ~ single electron
электронопло'тный *a* electron-dense
электронопрозра'чный *a* electron-light
электроо'смос *m* electroosmosis
электропора'ция *f* electroporation
электрореофоре'з *m* electrorheophoresis
электрота'ксис *m* electrotaxis
электротропи'зм *m* electrotropism
электрослия'ние *n* electrofusion
авидинопосре'дованное ~ avidin-mediated electrofusion
электрото'н *m* electrotonus
электротони'ческий *a* electrotonic
электрофизиоло'гия *f* electrophysiology
электрофокуси'рование *n* electrofocusing
~ **в ге'ле** gel electrofocusing
каска'дное ~ cascade electrofocusing
электрофорегра'мма *f* electrophoregram
электрофоре'з *m* electrophoresis
~ **в агаро'зном ге'ле** agarose gel electrophoresis
~ **в ге'левом бло'ке** slab gel electrophoresis
~ **в крахма'льном бло'ке** starch-slab electrophoresis
~ **в крахма'льном ге'ле** starch-gel electrophoresis
~ **в пульси'рующем по'ле** pulsed-field electrophoresis
~ **в свобо'дном пото'ке** free-flow electrophoresis
~ **на бума'ге** paper electrophoresis
~ **на ге'левых ди'сках** disk gel electrophoresis
~ **на коло'нке** column electrophoresis
~ **на крахма'ле** starch electrophoresis
~ **на поло'сках бума'ги** paper-strip electrophoresis
~ **по Гизе'лиусу** free-flow electrophoresis
~ **с дви'жущейся грани'цей** boundary electrophoresis, moving-boundary electrophoresis
афи'нный ~ affinity electrophoresis
встре'чный ~ counter electrophoresis
двуме'рный ~ two-dimensional electrophoresis
двуме'рный высоково'льтный ~ two-dimensional high-voltage electrophoresis
зона'льный ~ zone electrophoresis
конвекцио'нный ~ convection electrophoresis
микропрепарати'вный ~ micropreparative electrophoresis
мультифа'зный зо'нный ~ multiphasic zone electrophoresis
непреры'вный ~ continuous electrophoresis
одноме'рный ~ one-dimensional electrophoresis
препарати'вный ~ preparative electrophoresis
противото'чный ~ anticonvectant electrophoresis
свобо'дный ~ free electrophoresis
тонкосло'йный ~ thin-layer electrophoresis
фокуси'рующий ~ focusing electrophoresis
электрофорети'ческий *a* electrophoretic
электрошо'к *m* electroshock, electroconvulsive shock
электрошокотрансфе'кция *f* electrotransfection
электроэлю'ция *f* electroelution
электроэнцефалогра'мма *f* electroencephalogram
электроэнцефалогра'фия *f* electroencephalography
элеме'нт *m* element
~ **обра'тной свя'зи** fold-back element, FB element
воспринима'ющий ~ detecting element, sensing element
гла'вный ~ major element
ка'ппа-делети'рующий ~ kappa deleting element
контроли'рующий ~ controlling element
основно'й ~ major element
репродукти'вный ~ reproductive element
чувстви'тельный ~ sensing element, detecting element
фо'рменный ~ кро'ви blood corpuscle
элемент-активатор *m* activator element, Ac element
элементарный *a* elementary
элемент-разобщитель *m* dissociator element
элеоци'т *m* elaeocyte
элимина'ция *f* elimination
кле'точная ~ cellular elimination
митоти'ческая ~ mitotic elimination
сомати'ческая ~ somatic elimination
элимини'ровать *v* eliminate
элита *f* elite

элитный *a* elite
эли'ситор *m* elicitor
эли'та *f* elite, superstrain
эли'тра *f* elytron, elytrum, wing cover
э'ллипс *m* ellipse
  довери'тельный ~ *biom* confidence ellipse
эллиптоци'т *m* elliptocyte
элюант *m* eluant
элюат *m* eluate
элювиальный *a* eluvial
элюент *m* eluent
элюирование elution
элюировать *v* eluate
элю'ция *f* elution
  нисходя'щая ~ downward flow elution
  электрофорети'ческая ~ electrophoretic elution
эма'ль *f* enamel
эмбо'л *m* embolus
эмболи'я *f* embolism
эмбрио- embryo-
эмбриобла'ст *m* embryoblast
эмбриогене'з *m* embryogenesis, embryogeny
эмбриого'ний *m* embryogonium
эмбриолета'льный *a* embryolethal
эмбриологи'ческий *a* embryologic, embryological
эмбриоло'гия *f* embryology
эмбрио'н *m* embryo
  кури'ный ~ chick embryo
эмбриона'льный *a* embryonal, embryonic
эмбриони'я *f* embryony
  нуцелля'рная ~ nucellar embryony
эмбриоте'га *f* embryotega
эмбриотокси'чность *f* embryotoxicity
эмбриотокси'чный *a* embryotoxic
эмбриотрофи'я *f* embryotrophy
эмбриофо'р *m* embryophore
эмерге'нтность *f*: ~ **фитоцено'за** emergence of phytocoenosis
эмерже'нтный *a* emergent
эмигра'ция *f* emigration; outwandering
э'ммер *m* emmer (*Triticum dicoccum*)
эмофи'т *m* emophyte
эмоциона'льность *f* emotionality
эмо'ция *f* emotion
эмпири'ческий *a* empirical
эмпо'диум *m* empodium
э'му *m orn* emu (*Dromiceius*)
эмульга'тор *m* emulsifying agent
эмульги'рование *n* emulsification
эмульги'ровать *v* emulsify
эмульсифика'тор *m* emulsifier
эму'льсия *f* emulsion
эмфизе'ма *f* emphysema
эмфиземато'зный *a* emphysematous
энантиоме'р *m* enantiomer
эна'ция *m bot* enation
энвайронментали'зм *m* environmentalism
энвайронментали'стский *a* environmentalistic
эндартерии'т *m* endarteritis, endoarteritis
  ге'йбнеровский ~ Heubner's disease
эндеми'зм *m* endemism
энде'мик *m* endemic
эндеми'ческий *a* endemic, endemial
эндеми'чность *f* endemicity
энди'вий *m* endive (*Cichorium endivia*)
энди'т *m* endite
эндо- endo-
эндоа'ск *m bot* endoascus
эндобаза'льный *a* endobasal
эндобазидиа'льный *a* endobasidial
эндобази'дия *f* endobasidium
эндобиоти'ческий *a* endobiotic
эндобла'ст *m* endoblast
эндобластоме'р *m* endoblastomere
эндобронхиа'льный *a* endobronchial
эндога'мия *f* endogamy
эндога'мный *a* endogamic, endogamous
эндогене'з *m* endogenesis
эндогени'я *f* endogeny
эндоге'нный *a* endogenic, endogenous
эндогено'та *f* endogenote
эндогони'дий *m* endogonidium
эндогони'ческий *a* endogonic
эндоде'рма *f* endoderm; *bot* endodermis
эндодерма'льный *a* endodermal, enderonic
эндодуплика'ция *f* endoduplication
эндозоохо'р *m* endozoochore
эндока'рд *m* endocardium
эндокарди'т *m* endocarditis
эндока'рпий *m* endocarp
эндока'рпный *a* endocarpous
эндокони'дия *f* endoconidium
эндокра'ниум *m* endocranium
эндокри'нный *a* endocrine, endocrinous
эндокриноло'гия *f* endocrinology, hormonology
эндоксилофи'т *m* endoxylophyte
эндокути'кула *f* endocuticle, endocuticula
эндолизосо'ма *f* endolysosome
эндоли'мфа *f* endolymph
эндолимфати'ческий *a* endolymphatic

эндолитофи'т *m* endolithophyte
эндоме'р *m* endomere
эндоме'трий *m* endometrium
эндоми'зий *m* endomysium
эндоми'ксис *m* endomixis
эндомито'з *m* endomitosis
эндомо'рфия *f* endomorphy
эндомо'рфный *a* endomorphic
эндоне'врий *m* endoneurium
эндонуклеа'за *f* endonuclease
~ **рестри'кции** restriction endonuclease
эндопарази'т *m* endoparasite
живо'тный ~ enterozoon
эндопаразити'ческий *a* endoparasitic
эндоперидий *m* endoperidium
эндопла'зма *f* endoplasm
эндоплазмати'ческий *a* endoplasmic
эндоподи'т *m* endopodite
эндоредупликация endoreduplication
эндополиплоидиза'ция *f* endopolyploidization
эндополиплоиди'я *f* endopolyploidy
эндорфи'н *m* endorphine
эндосапрофити'зм *m* endosaprophytism
эндосимбио'нт *m* endosymbiont
эндоскелет *m* internal skeleton
эндосклеро'ций *m* endosclerotium
эндо'смос *m* endosmos
эндосо'ма *f* endosome
эндоспе'рм *m* endosperm
кле'точный ~ cellular endosperm
рогови'дный ~ horny endosperm
румини'рованный ~ ruminate endosperm
стеклови'дный ~ glassy endosperm, vitreous endosperm
я'дерный ~ nuclear endosperm
эндоспо'ра *f* endospore
эндо'ст *m* endosteum
эндостеа'льный *a* endosteal
эндости'ль *m* endostyle
эндосто'м *m* endostome
эндоте'лий *m* endothelium
эндотелиоци'т *m* endotheliocyte
эндотермический *a* endothermic, endergonic
эндотермия *f* endothermy
эндотециа'льный *a* endothecial
эндоте'ций *m* endothecium
эндотокси'н *m* endotoxin
эндотокси'ческий *a* endotoxic
эндотро'фный *a* endotrophic
эндоферме'нт *m* endoenzyme
эндофи'т *m* endophyte
эндофити'зм *m* endophytism
эндофи'тный *a* endophytic
эндохромоце'нтр *m* endochromocenter
эндоце'ль *f* endocoel
эндоци'кл *m* endocycle
эндоцито'з *m* endocytosis
опосре'дованный реце'пторами ~ receptor-mediated endocytosis
эндэкзи'на *f pal* endexine
энерги'да *f* energid
эне'ргия *f* energy
~ **актива'ции** activation energy
~ **броже'ния** fermentation energy
~ **возбужде'ния** excitation energy
~ **Ги'ббса** Gibbs energy
~ **поддержа'ния** maintenance energy
~ **прораста'ния** germinating power, germinating force, germinative energy, emergence rate
~ **ро'ста** growing capacity
~ **хими'ческой свя'зи** chemical bound energy
вну'тренняя ~ internal energy
досту'пная ~ available energy
и'стинная метаболи'ческая ~ true metabolizable energy
и'стинная обме'нная ~ true metabolizable energy
кинети'ческая ~ motive energy
конформацио'нная ~ conformational energy
лучи'стая ~ radiant energy
метаболи'ческая ~ metabolizable energy
обме'нная ~ metabolizable energy
поглощё'нная ~ absorbed energy
поглощё'нная светова'я ~ absorbed light energy
потенциа'льная ~ potential energy
продукти'вная ~ productive energy
свобо'дная ~ free energy
свя'занная ~ bound energy
со'лнечная ~ solar energy
чи'стая ~ net energy
я'дерная ~ nuclear energy
энергозави'симый *a* energy-dependent
энергопотребля'ющий *a* energy-requiring
энзи'м *m* enzyme
энзимати'ческий *a* enzymic, enzymatic
энзимо'лиз *m* enzymolysis
энзимоло'гия *f* enzymology
энзооти'ческий *a* enzootic
энкефали'н *m* enkephalin, enkephaline

энкла'в *m* enclave
эноци'т *m* oenocyte
энтальпи'я *f* enthalpy
энтобла'ст *m* entoblast
энтери'т *m* enteritis
энтеробла'ст *m* enteroblast
энterови'рус *m* enterovirus
энтерокина'за *f* enterokinase
энтероко'кк *m* enterococcus
энтеромико'з *m* enteromycosis
энтеропро'кт *m* enteroproct
энтеротокси'н *m* enterotoxin
энтероце'льный *a* enterocoelous
энтобла'ст *m* entoblast
энтоде'рма endoderm
  внезаро'дышевая ~ extraembryonic endoderm
  эмбриона'льная ~ entoblast
энтодерма'льный *a* endodermal, enderonic
энтомезобла'ст *m* endomesoblast
энтомезоде'рма *f* endomesoderm
энтомо- entomo-
энтомога'мия *f* entomogamy
энтомоло'гия *f* entomology
  прикладна'я ~ economic entomology
энтомофа'г *m* entomophage
энтомофа'уна *f* entomofauna
энтомофили'я *f* entomophily
энтомофи'льный *a* entomophilous
энуклеа'ция *f* enucleation
энуклеи'ровать *v* enucleate
энфито'тия *f* enphytoty
энха'нсер *m gen* enhancer
энхиле'ма *f* enchylema
энхондра'льный *a* enchondral
энцефализа'ция *f* encephalization
энцефали'т *m* encephalitis
  ~ **Штрю'мпелля-Лейхтенште'рна** Strumpell-Leichtenstern disease
  ви'русный ~ viral encephalitis
  клещево'й ~ tick-borne encephalitis, vernal encephalitis
  лошади'ный ~ equine encephalitis, equine encephalomyelitis
  очаго'вый ~ focal encephalitis
энцефалогра'мма *f* encephalogram
энцефалогра'фия *f* encephalography
эози'н *m* eosin, eosine
эозинобла'ст *m* eosinoblast
эозинофи'л *m* eosinophil granulocyte, eosinophil, eosinophil cell, eosinocyte
эозинофили'я *f* eosinophilia
эозинофилобла'ст *m* eosinophiloblast
эозинофи'льный *a* eosinophilic, eosinophilous
эозиноци'т *m* eosinocyte
эокли'макс *m* eoclimax
эолофи'льный *a* eolophilous
эосе'рия *f* eosere
эпенди'ма *f* ependyma
эпенди'мный *a* ependymal
эпи- epi-
эпибази'дия *f* epibasidium
эпибио'з *m* supercrescence
эпибио'нтный *a* epizoic, supercrescent
эпибио'нт *m* epibiont, epicole, epizoan, epizoon
эпибла'ст *m* primitive ectoderm
эпибле'ма *f* epiblema
эпиболи'ческий *a* epibolic
эпиболи'я *f* epiboly
эпига'мный *a* epigamic
эпигене'з *m* epigenesis
эпигене'тика *f* epigenetics
эпигени'чный *a* epigynous
эпигеноти'п *m* epigenotype
эпигини'чный *a* epigynous
эпиги'ния *f* epigyny
эпигнати'ческий *a* epignathous
эпиго'ний *m* epigonium
эпидемиоло'гия *f* epidemiology
эпидеми'ческий *a* epidemic
эпиде'мия *f* epidemic, epidemy
эпиде'рма *f* epidermis
эпидерма'льный *a* epidermal, ecderonic
эпиде'рмис *m* epidermis, epidermic tissue, ecderon
эпидиди'мис *m* epididymis
эпидура'льный *a* epidural
эпизооти'ческий *a* epizootic
эпизоо'тия *f* epizooty, epizootic
эпизоотоло'гия *f* epizootology
эпизоохо'р *m* epizoochore
эпика'нтус *m* epicanthal fold, palpebronasal fold, epicanthus
эпика'рд *m* epicardium
эпика'рпий *m* epicarp
эпико'тиль *m* epicotyl
эпикри'з *m* epicrisis
эпикути'кула *f* epicuticle, epicuticula
эпилити'ческий *a* epilithic
эпилитофи'т *m* epilithophyte
эпиля'ция *f* epilation, depilation
эпимастиго'тный *a* epimastigote

эпиме'р *m* epimer
эпимера'за *f* epimerase
эпимери'т *m* epimerite
эпими'зий *m* epimysium
эпимиока'рд *m* epimyocardium
эпиморфо'з *m* epimorphosis
эпина'стия *f* epinasty
эпинефри'н *m* epinephrine
эпипарази'т *m* epiparasite
эпипаразити'зм *m* epiparasitism
эпипери'дий *m* epiperidium
эпипла'зма *f* epiplasm
эпипле'вра *f* epipleura
эпипо'дий *m bot* epipodium
эпиподи'т *m* epipodite
эпипо'дия *f zool* epipodium
эпипро'кт *m* epiproct
эпипротеи'н *m* epiprotein
эпиптериго'ид *m* epipterygoid, epipterygoid bone
эписо'ма *f* episome
эписпе'рмий *m bot* episperm
эписта'з *m* epistasis, epistasy
эписта'ция *f gen* epistacy
эпистати'ческий *a* epistatic
эпистерни'т *m* episternum
эпистернум *m* episternum
эписто'м *m* epistome
эпистрофе'й *m* epistropheus
эпитала'мус *m* epithalamus
эпите'ка *f* epitheca *(of diatoms)*
эпителиа'льный *a* epithelial
эпите'лий *m* epithelium, epithelial tissue
  двухря'дный ~ two-rowed epithelium
  желе'зистый ~ glandular epithelium
  желто'чный ~ yolk epithelium
  заро'дышевый ~ germ epithelium, germinal epithelium
  куби'ческий ~ cuboidal epithelium, cubical epithelium
  мерца'тельный ~ ciliary epithelium
  многоря'дный ~ multirowed epithelium
  многосло'йный ~ laminated epithelium, stratified epithelium
  перехо'дный ~ transitional epithelium
  пло'ский ~ pavement epithelium
  призмати'ческий ~ columnar epithelium, cylindrical epithelium
  просто'й ~ simple epithelium
  ресни'чный ~ ciliated epithelium
  слу'щенный ~ cast-off epithelium
  сто'лбчатый ~ columnar epithelium, cylindrical epithelium
  фолликуля'рный ~ follicular epithelium
  цилиндри'ческий ~ columnar epithelium, cylindrical epithelium
  чешу'йчатый ~ scaly epithelium, squamous epithelium
эпите'ма *f* epithem
эпите'ций *m* epithecium
эпитимпа'нум *m* epitympanum
эпито'кный *a* epitokous
эпито'п *m* epitope
  ~ **кле'точной пове'рхности** cell-surface epitope
  В-кле'точный ~ B-cell epitope
  группово'й ~ group-specific epitope
  иммунодетермина'нтный ~ immunodominant epitope
  мино'рный ~ bystander epitope
  углево'дный ~ carbohydrate epitope
  уника'льный ~ unique epitope
эпитро'пный *a* epitropic
эпифа'ринкс *m ent* epipharynx, epiglottis
эпифа'уна *f* epifauna
эпифи'з *m* epiphysis, pineal organ
эпифиза'рный *a* epiphyseal
эпифилакси'я *f* epiphylaxis
эпифи'лл *m* epiphyll
эпифи'т *m* epiphyte
  водозапаса'ющий ~ tank epiphyte
эпифити'зм *m* epiphytism
эпифи'тный *a* epiphytic
эпифито'ид *m* epiphytoid
эпифито'н *m* epiphyton
эпифито'тия *f* epiphytoty, epiphytotics
эпифлоэ'ма *f* epiphloem
эпифра'гма *f* epiphragm
эпихорда'льный *a* epichordal
эпици'т *m* epicyte
эпони'хий *m* eponychium
эргасти'ческий *a* ergastic
эргастопла'зма *f* ergastoplasm
эргастоплазмати'ческий *a* ergastic
эргатандромо'рф *m ent* ergatandromorph
эргато'идный *a ent* ergatoid
эрго'граф *m* ergograph
эргокальциферо'л *m* ergocalciferol
эрго'метр *m* ergometer
эргометри'н *m* ergometrine
эргонови'н *m* ergonovine
эргоно'мика *f* ergonomics
эргостеро'л *m* ergosterol

эрготами'н *m* ergotamine
эрготи'зм *m* ergotism
эрготи'н *m* ergotin
эрде'ль-террье'р *m* Airedale terrier
эре'кция *f* erection
эремофи'т *m* eremophyte
эризими'н *m* erisimin
эриохо'р *m bot* eriochore
эритри'н *m* erythrin
эритробла'ст *m* erythroblast
эритробласти'ческий *a* erythroblastic
эритро'за *f* erythrose
эритрози'н *m* erythrosin
эритрокруори'н *m* erythrocruorin
эритромици'н *m* erythromycin
эритро'н *m* erythron
эритропоэ'з *m* erythropoiesis, erythrogenesis
эритропоэти'н *m* erythropoietin
эритропоэти'ческий *a* erythropoietic
эритроре'ксис *m* erythrorhexis
эритрофа'г *m* erythrophage
эритрофагоцито'з *m* erythrophagocytosis
эритрофи'лл *m* erythrophyll
эритрофо'р *m* erythrophore
эритроци'т *m* erythrocyte
  авидинизи'рованный ~ avidine-coated erythrocyte
  нагру'женный антиге'ном ~ antigen-coated erythrocyte, antigen-sensitized erythrocyte
  сенсибилизи'рованный антиге'ном ~ antigen-coated erythrocyte, antigen-sensitized erythrocyte
  серпови'дный ~ sickle erythrocyte, drepanocyte
  таннизи'рованный ~ tannic-acid-treated erythrocyte
  тринитрофенили'рованный ~ trinitrophenylated erythrocyte
эритроцита'рный *a* erythrocytic
эритроцито'з *m* erythrocytosis
эритроцито'лиз *m* erythrocytolysis
эритроцито'метр *m* erythrocytometer
эритроцитопоэ'з *m* erythropoiesis, erythrogenesis
эритроцитопоэти'ческий *a* erythropoietic
эроге'нный *a* erogenic
эро'зия *f* erosion
  ~ **желу'дка** gastric erosion
  ~ **по'чвы** soil erosion
  ~ **ше'йки ма'тки** cervical erosion
  ветрова'я ~ **по'чвы** soil eolation
эруко'идный, эруцифо'рмный *a* eruciform
эспарцет *m bot* sainfoin, holy clover, esparcet (*Onobrychis*)
эстезиоло'гия *f* esthesiology
эстезиоме'тр *m* esthesiometer
эстера'за *f* esterase
эстерифика'ция *f* esterification
эстраго'н *m bot* tarragon, estragon (*Artemisia dracunculus*)
эстрадио'л *m* oestradiol
эстра'льный *a* estrous, oestrous
эстрио'л *m* oestriol
эстроге'н *m* estrogen, oestrogen
эстро'н *m* estrone, oestron
э'струс *m* estruation, estrus
эстуа'рный *a* estuarine
эта'лий *m* aethalium
этано'л *m* ethyl alcohol
этано'лиз *m* ethanolysis
эта'п *m* stage
этерифика'ция *f* etherification
эти'дий *m* ethidium
  бро'мистый ~ ethidium bromide
этикети'рование *n* labelling, labeling
этикети'ровать *v* label
этике'тка *f* label
этиоли'рование *n* etiolation
этиоли'рованный *a* etiolated
этиоло'гия *f* etiology, aetiology
этиоля'ция *f* etiolation
этиопла'ст *m* etioplast
этиотро'пный *a* etiotropic
этнобота'ника *f* ethnobotany
этногене'з *m* ethnogeny
этногра'фия *f* ethnography
этноло'гия *f* ethnology
этогра'мма *f* ethogram
этоло'гия *f* ethology
этоме'рный *a* ethomerous
этоме'трия *f* ethometry
эторфи'н *m* etorphin
этоэколо'гия *f* behavioral ecology
эу- eu-
эуапога'мия *f* euapogamy
эугали'нный *a* euhaline
эугетеро'зис *m* euheterosis
  сбаланси'рованный ~ balanced euheterosis
  мутацио'нный ~ mutational euheterosis
эуинти'на *f pal* euintine
эука'пния *f* eucapnia
эукарио'т *m* eucaryote, eukaryote

эукариоти'ческий, эукарио'тный *a* eucaryotic, eukaryotic
эукони'ческий *a ent* eucone
эумито'з *m* eumitosis
эупире'нный *a* eupyrene
эупло'ид *m* euploid
эуплоиди'я *f* euploidy
эупло'идный *a* euploid
эупно'э *n* eupnoea
эуселекти'вность *f* euselectivity
эуси'стола *f* eusystole
эусоциа'льность *f eth* eusociality
эуспорангиа'тный *a* eusporangiate
эусте'ла *f bot* eustele
эутана'зия *f* euthanasia
эутерофи'т *m* eutherophyte
эуфоти'ческий *a* euphotic
эухроматиза'ция *f* euchromatization
эухромати'н *m* euchromatin
эухромосо'ма *f* euchromosome
эуцентри'ческий *m* eucentric
эуцефали'ческий *a* eucephalic
эфа'пс *m* ephapse
эфармо'н *m* epharmone
эфармо'ния *f* epharmony, epharmosis
эфеме'р *m* ephemer
эфеме'рный *a* ephemeral, ephemerous
эфиппиа'льный *a* ephippial
эфи'ппий, эфи'ппиум *m* ephippium
эфир *m* ether, ester
  сло'жный ~ ester
эфи'ра *f* ephira
эффе'кт *m* effect
  ~ **Бо'лдуина** Baldwin effect
  ~ **Бо'ра** *phys* Bohr effect
  ~ **Брю'са** *eth* Bruce effect
  ~ **включе'ния** on effect
  ~ **вы'борки** sampling effect
  ~ **выключе'ния** off effect
  ~ **Да'ниша** Danysz effect
  ~ **до'зы** dosage effect
  ~ **Ко'ка** Kok's effect
  ~ **Крэ'бтри** Crabtree effect
  ~ **Ку'лиджа** *eth* Coolidge effect
  ~ **Ку'стера** Custer's effect
  ~ **ме'стной концентра'ции** local concentration effect
  ~ **носи'теля** *imm* carrier effect
  ~ **основа'теля** *gen* founder effect
  ~ **Пасте'ра** Pasteur effect
  ~ **перенасе'ления** crowding effect
  ~ **перено'са** transfer effect
  ~ **прикрыва'ния** *gen* covering effect
  ~ **прими'рования** *imm* priming effect
  ~ **про'бы** sampling effect
  ~ **распростране'ния** spreading effect
  ~ **сосе'дней гру'ппы** neighboring group effect
  ~ **сумми'рования** additive effect
  ~ **усиле'ния** enhancement effect
  ~ **Фе'нна** *phys* Fenn's effect
  аддити'вный ~ additive effect
  благоприя'тный ~ favorable effect
  бу'ферный ~ buffer effect
  вирусостати'ческий ~ virostatic effect
  выра'внивающий ~ equalizing effect
  геоэлектри'ческий ~ geoelectric effect
  депрессо'рный ~ depressor effect
  дли'тельный ~ long-term effect, prolonged effect
  заде'рживающий retarding effect
  заме'дленный ~ delayed effect
  защи'тный ~ protective effect
  изото'пный ~ isotopic effect, isotopic discrimination
  инотро'пный ~ inotropic effect
  кислоро'дный ~ oxygen effect
  коне'чный ~ final effect
  консерви'рующий ~ conserving effect
  краткосро'чный ~ short-term effect
  матери'нский ~ maternal effect
  неблагоприя'тный ~ adverse effect, harmful effect
  необрати'мый ~ irreversible effect
  непостоя'нный ~ variable effect
  нивели'рования leveling effect
  обра'тный ~ converse effect
  парадокса'льный ~ paradoxical effect
  парнико'вый ~ greenhouse effect, hothouse effect
  плейотро'пный ~ *gen* pleiotropic effect
  побо'чный ~ side-effect
  подпоро'говый ~ subthreshold effect
  положе'ния position effect
  поро'говый ~ threshold effect
  постоя'нный ~ permanent effect
  пострадиацио'нный ~ radiation effect
  ревакцинацио'нный ~ booster effect
  ре'зко вы'раженный ~ pronounced effect
  сенсибилизи'рующий ~ sensitizing effect
  совоку'пный ~ net effect
  фа'кторный ~ factorial effect
  части'чный ~ partial effect

широ'ко распространё'нный ~ widespread effect
экото'нный эффект ecotone effect
эффекти'вность *f* efficiency, efficacy, effectiveness
~ **вакци'ны** immunological efficacy
~ **кормле'ния** feed efficiency
~ **пита'ния** feed efficiency
~ **посе'ва** *micr* plating efficiency
~ **раздражи'теля** effectiveness of stimulus
~ **счё'та** counting efficiency
~ **трансформа'ции** transformation efficiency
~ **фотоси'нтеза** photosynthetic efficiency
~ **эвапотранспира'ции** evapotranspiration effectiveness
асимптоти'ческая ~ asymptotic efficiency
относи'тельная биологи'ческая ~, ОБЭ relative biological effectiveness
экологи'ческая ~ ecological efficiency
эффекти'вный *a* efficient, effective
эффе'ктор *m* effector
аллостери'ческий ~ allosteric effector
иммy'нный ~ immune effector
эфферента'ция *f* efference
эффере'нтный *a* efferent; exodic *(a nerve fiber)*
эхиноко'кк *m helm* echinococcus *(Echinococcus)*
эхинококко'з *m* echinococcosis
эхиноплуте'ус *m zool* echinopluteus
э'хо *n* echo
фото'нное ~ photon echo
эхоло'т *m* echo sounder
эце'зис *m* ecesis, oecesis
эци'дий *m* aecidium
эцидиокли'мат *m* ecidioclimate
эцидиоспо'ра *f* aecidiospore
эякуля'ция *f* ejaculation

Ю

ювени'льный *a* juvenile
юга'льный *a* jugal
югуля'рный *a* jugular
ЮНЕП, Програ'мма (*f*) **ООН по окружа'ющей среде'** UNEP, United Nations Environment Programme
юнна'т *m* young naturalist
юро'к *m* brambling (*Fringilla montifringilla*)
ю'рский *a* Jurassic
юфть *f* Russian leather, yuft

Я

я'блоко *n* apple; **в я'блоках** dappled
ада'мово ~ Adam's apple
глазно'е ~ eyeball, eyebulb, eyeglobe, bulb of eye, optic bulb
я'блоня *f* apple-tree (*Malus*)
~ **ди'кая,** ~ **лесная** codlin, codling, crab tree, crab apple, wild apple (*Malus silvestris*)
я'блоня-кита'йка *f* Chinese apple (*Malus prunifolia*)
явнобра'чный *a* phanerogamous, phanerogamic
я'вный *a* distinct, evident, obvious
я'вственный *a* distinct, clear
я'вор *m bot* sycamore, sycamore maple (*Acer pseudoplatanus*)
ягдта'ш *m* game bag
я'гель *m* reindeer moss (*Cladina*) (*a lichen*)
ягне'ние *n* lambing
ягнё'нок *m* lamb, eanling, yeanling
ягни'ться *v* lamb, yean, ean, kid
ягня'тник *m* lammergeier, lammergeyer (*Gypaetus barbatus*)
я'года *f* berry, bacca
бессемя'нная ~ seedless berry
ло'жная ~ false berry
я'годица *f* buttock
ягоди'чный *a* gluteal, pygal, natal
я'годник *m* berry bush, berry field
я'годный *a* baccate; rich in berries
ягодови'дный *a* berry-shaped, bacciform
ягодоно'сный *a* bacciferous
ягодообра'зный *a* berry-shaped, bacciform
ягуа'р *m* jaguar (*Panthera onca*)
яд *m* poison, venom, toxin
грибно'й ~ muchroom poison
змеи'ный ~ snake venom
кише'чный ~ digestive poison
конта'ктный ~ contact poison
митоти'ческий ~ mitotic poison
пчели'ный ~ bee venom, apitoxin
ры'бный ~ ichthyotoxin
систе'мный ~ systemic poison
тру'пный ~ ptomaine
я'дерный *a* nuclear
ядови'тость *f* toxicity
ядови'тый *a* poisonous, venomous, toxic, venenous; poison-bearing, toxiferous
ядохимика'т *m* toxic chemical
я'дрица *f* unground buckwheat
ядро' *n* nucleus; kernel (*of a nut*); meat (*of a nut*)
~ **Бе'хтерева** Bechterev's nucleus
~ **Бу'рдаха** Burdach's nucleus
~ **Го'лля** Goll's nucleus
~ **Даркше'вича** Darkshevich's nucleus
~ **Де'йтера** Deiter's nucleus
~ **дробле'ния** cleavage nucleus, segmentation nucleus
~ **зиго'ты** fusion nucleus
~ **Кла'рка** Clarke's nucleus
~ **кле'тки** cell nucleus
~ **окостене'ния** bony nucleus, ossific nucleus
~ **популя'ции** population nucleus
~ **пыльцево'й тру'бки** tube nucleus, vegetative nucleus
~ **Рамо'н-и-Каха'ла** nucleus of Ramon-y-Cajal
~ **спе'рмия** sperm nucleus
~ **Шва'льбе** Schwalbe's nucleus
~ **Э'дингера** Edinger's nucleus
большо'е монетови'дное ~ cartwheel nucleus
вегетати'вное ~ vegetative nucleus
ве'рхнее стати'ческое ~ Bechterev's nucleus
втори'чное ~ secondary nucleus
генерати'вное ~ generative nucleus
интерстициа'льное ~ nucleus of Ramon-y-Cajal
доба'вочное ~ accessory nucleus
доче'рнее ~ daughter nucleus
за'днее ~ Clarke's nucleus
заднекомиссура'льное ~ Darkshevich's nucleus
изохромоцентри'ческое ~ isochromocentric nucleus
интерфа'зное ~ interphase nucleus, resting nucleus
кинети'ческое ~ kinetic nucleus
клинови'дное ~ Burdach's nucleus
комплемента'рное ~ complementary nucleus
кра'сное ~ red nucleus
латера'льное стати'ческое ~ Deiter's nucleus
медиа'льное стати'ческое ~ Schwalbe's nucleus

окологлазодви'гательное ~ Edinger's nucleus
олигосома'льное ~ oligosomal nucleus
паравентрикуля'рное ~ paraventricular nucleus
перви'чное ~ primary nucleus
поко'ящееся ~ interphase nucleus, resting nucleus
полово'й кле'тки germ nucleus
поля'рное ~ polar nucleus
рабо'чее ~ energic nucleus, metabolic nucleus
реституцио'нное ~ restitution nucleus
сперматоге'нное ~ spermatogenous nucleus
то'нкое ~ Goll's nucleus
хвоста'тое ~ caudate nucleus
я'дрышко *n* nucleolus
я'зва *f* ulcer, canker
~ **желу'дка** gastric ulcer
пробода'ющая ~ perforating ulcer
сиби'рская ~ anthrax
я'звенный *a* ulcerous
язы'к *m* tongue
обло'женный ~ coated tongue, furred tongue
раздво'енный ~ forked tongue
язычко'вый *a bot* ligulate, ligulated
язы'чный *a anat* lingual
язычо'к *m anat* lingula, uvula; *bot* ligula
яи'чник *m* ovary; germarium (*of rotifers*)
крыла'тый ~ *helm* wing-shaped ovary
яи'чниковый *a* ovarian
яйцеви'дный *a* oval, ovate, ovoid, ooid, oviform, egg-shaped
яйцево'д *m* oviduct
яйцеживородя'щий *a* ovoviviparous
яйцеживорожде'ние *n* ovoviviparity
яйцекла'д *m* ovipositor
яйцекладу'щий *a* oviparous, ootocous, egg-laying
яйцекле'тка *f* egg cell, ovicell, ovigerm
яйцеобра'зный *a* egg-shaped
яйцо' *n* egg
~ **в яи'чнике** ovarian egg
зре'лое ~ ripe egg
изолецита'льное ~ isolecithal egg, homolecithal egg
инвазио'нное ~ infective egg
клейдои'ческое ~ cleidoic egg
ле'тнее ~ subitan egg
макролецита'льное ~ megalecithal egg
меробласти'ческое ~ meroblastic egg
мозаи'чное ~ mosaic egg
овариа'льное ~ ovarian egg
олиголецита'льное ~ oligolecithal egg, meiolecithal egg
оплодотворё'нное ~ fertilized egg
регуляцио'нное ~ plastic egg, regulative egg
субита'нное ~ subitan egg
телолецита'льное ~ telolecithal egg
ту'хлое яйцо addle egg, bad egg, rotten egg
эктолецита'льное ~ ectolecithal egg
яйцено'скость *f* egg producing ability, egg production, egg-laying rate
як *m ther* yak (*Bos mutus*)
я'ловая *a* barren, dry (*of a cow*)
я'ловеть *v* be barren
я'ма *f* pit, pitfall
ло'вчая яма pitfall trap
си'лосная яма silo
я'мка *f* fossa, foveola, pit, depression, scrobiculus
~ **стеклови'дного те'ла** hyaloid fossa, patellar fossa
антенна'льная ~ antennal fossa
бе'дренная ~ crural fossa, femoral fossa
вене'чная ~ coronoid fossa
ве'ртельная ~ trochanteric fossa
висо'чная ~ temporal fossa
гипофиза'рная ~ hypophyseous fossa of sphenoid, pituitary fossa, sellar fossa
за'дняя черепна'я ~ postcranial fossa
крылонё'бная ~ pterygopalatine fossa
локтева'я ~ cubital fossa, olecranon fossa
носова'я ~ nasal fossa
обоня'тельная ~ nasal pit, olfactory pit, nasal fossa
окаймлё'нная ~ *cyt* coated pit
перви'чная ~ *emb* primitive pit
пере'дняя черепна'я ~ anterior cranial fossa, precranial fossa
подвздо'шная ~ iliac fossa
подвисо'чная ~ infratemporal fossa
подкрыльце'вая ~ armpit
подло'жечная ~ epigastric fossa
подмы'шечная ~ armpit
подъязы'чная ~ sublingual fossa
ротова'я ~ oral fossa
скулова'я ~ zygomatic fossa
соба'чья ~ *anat* canine fossa
сре'дняя черепна'я ~ mesocranial fossa
суста'вная ~ glenoid fossa

яре'мная ~ jugular fossa
ЯМР, ядерный магнитный резонанс *m* NMR, nuclear magnetic resonance
ЯМР-томография *f* magnetic resonance imaging, MR-imaging, MRI
ямс *m bot* yam (*Dioscorea*)
я'мчатый *a* pitted, scrobiculate
янтароно'сный *a* succiniferous
янта'рь *m* amber
яр *m* high bank, steep bank
яре'мный *a* jugular
яровиза'ция *f* vernalization, yarovization
ярово'й *a* spring
я'рость *f* rage, fury
я'рочка *f* female-lamb
я'рус *m* layer, storey; long-line (*a fishing gear*)
~ **погружё'нных расте'ний** submerged layer
куста'рниковый ~ shrub layer
мохово'й ~ moss layer
травяни'стый ~ herb layer
я'русность *f* layerage, stratification
яру'тка *f bot* penny-cress (*Thlaspi arvense*)
я'сень *m bot* ash (*Fraxinus*)
я'сли *pl* manger, crib, hay basket; day nursery
ясно'тка *f bot* dead nettle (*Lamium*)
я'стреб *m orn* hawk (*Accipiter*)
большо'й ~ goshawk (*Accipiter gentilis*)
двухцве'тный ~ bicoloured sparrow-hawk *(Accipiter bicolor)*
оше'йниковый ~ collared sparrow-hawk *(Accipiter cirrocephalus)*
чё'рно-бе'лый ~ black sparrow-hawk *(Accipiter melanoleucus)*
я'стреб-перепеля'тник *m* sparrow hawk (*Accipiter nisus*)
ятроге'нный *a* iatrogenic
яче'йка *f* faveolus, cell, cellule
~ **Ге'рша** Hersch's cell
~ **для микродиа'лиза** Craig-type dialysis cell
генети'ческая ~ па'мяти genetical storage cell
дискоида'льная ~ discal cell
закры'тая ~ *ent* closed cell
маргина'льная ~ *ent* marginal cell
откры'тая ~ *ent* open cell
решё'тчатая ~ *anat* ethmoidal air cell
сосцеви'дная ~ *anat* mastoid air cell
тру'тневая ~ *apic* drone cell
яче'истый *a* faveolate, favose, alveolate
ячме'нь *m bot* barley (*Hordeum*); *med* hordeolum, sty
~ **безо'стый** awnless barley *(Hordeum sativum* var. *inermis)*
~ **голозё'рный** hull-less barley, naked barley (*Hordeum nudum*)
~ **двухря'дный** two-rowed barley*(Hordeum distichum)*
~ **за'ячий** wall barley, way barley *(Hordeum leporinum)*
~ **ка'рликовый** little barley *(Hordeum pusillum)*
~ **непроро'сший** dormant barley
~**обыкнове'нный** common barley, six-rowed barley *(Hordeum vulgare)*
~ **осоло'женный** malted barley
~ **пивова'ренный** brewing barley
~ **шелуше'ный** hulled barley
я'щерица *f* lizard (*Lacertilia*)
живородя'щая ~ viviparous lizard, common lizard, scaly lizard (*Lacerta vivipara*)
зелё'ная ~ green lizard, emerald lizard (*Lacerta viridis*)
лугова'я ~ meadow lizard *(Lacerta praticola)*
прово'рная ~, пры'ткая ~ sand lizard (*Lacerta agilis*)
я'щурка *f* racerunner (*Eremias*)

Некоторые фамилии

Бейеринк М.В. Beyerinck M.W.
Бейтс Г.У. Bates H.W.
Бенеден Э. Ван Beneden E. van
Бернар К. Bernard C.
Бигон М. Begon M.
Биша М.Ф. Bichat M.F.X.
Боталло Л. Bottallo L.
Боумен У. Bowman W.
Буссенго Ж.Б. Boussigault J.-B.
Бэтсон У. Bateson W.
Бюффон Ж.Л.Л. Buffon G.-L. L.
Ваксман З. Waksman S.
Варминг Й. Warming J.E.
Вебер Э.Г. Weber E.H.
Вейсман А. Weismann A.F.
Вирхов Р. Virchow R.
Гаймор Н. Highmor N.
Гальвани Л. Galvani L.
Гальтон Ф. Galton F.
Гарстанг У. Garstang W.
Гегенбаур К. Gegenbaur K.
Гейденгайн Р.П.Г. Heidenhain R.P.H.
Геккель Э. Haeckel E.
Генле Ф.Г.Я. Henle F.G.J.
Гертвиг О. Hertwig O.
Геснер К. Gesner C.
Гольджи К. Golgi C.
Гольдшмидт Р. Goldschmidt R.
Грааф Р. де Graaf R. de
Грам Х.К.И. Gram H.C.J.
Дарвин Ч.Р. Darwin C.R.
Дезор Э. Desor E.
Декандоль О.П. Candolle A.P. de
Долло Л. Dollo L.
Дорн А. Dohrn A.
Дриш Х. Driesch H.A.E.
Кеннон У.Б. Cannon W.B.
Кох Р. Koch R.
Купфер К.В. фон Kupffer K.W.
Кювье Ж.Л.К.Ф.Д. Cuvier G.
Ламарк Ж.Б.П.А. де М. Lamarck J.B.P.A.de Monet
Лангерганс П. Langerhans P.
Левенгук А. ван Leeuwenhoek A. van
Линней К. Linnaeus C.
Лоренц К. Lorentz K.
Лотка А.Д. Lotka A.J.
Майр Э.У. Mayr E.
Мальпиги М. Malpighi M.
Мендель Г.И. Mendel G.J.
Морган Т.Х. Morgan T.H.
Ниссль Ф. Nissl F.
Одум Ю. Odum E.P.
Осборн Г.Ф. Osborn G.F.
Панет Й. Paneth J.
Пачини Ф. Pacini F.
Полинг Л.К. Pauling L.K.
Ранвье Л.А. Ranvier **L.A.**
Раункиер К. Raunkiaer C.
Рей Д. Ray J.
Реомюр Р.А. Reaumur R.-A.F.
Ру В. Roux W.
Сваммердам Я. Swammerdam J.
Симпсон Д.Г. Simpson G.G.
Соссюр Н.Т. де Saussure N.T. de
Уоллес А.Р. Wallace A.R.
Фаллопий Г. Fallopio G.
Хатчинсон Д. Hutchinson G.
Холдейн Д.Б.С. Haldane J.B.S.
Шванн Т. Schwann T.
Шелфорд В.Э. Shelford V.E.
Шлейден М.Я. Schleiden M.J.
Шпеман Х. Spemann H.
Эймер Т.Г.Г. Eimer T.G.H.
Элтон Ч.С. Elton C.S.

АНГЛО-РУССКИЙ

ENGLISH-RUSSIAN

A

abandoned *a* заброшенный
abarticular *a* внесуставный
abarticulation, abarthrosis *s anat* диартроз, истинный сустав
abatement *s* устранение, уничтожение, уменьшение, ослабление, смягчение
pollution ~ очистка
abaxial *a* абаксиальный, направленный от оси
abbreviated *a* сокращённый, укороченный
abbreviation *s* аббревиация, сокращение
abdomen *s anat* брюшной отдел, живот; *ent* брюшко
abdominal *a* абдоминальный, брюшной
abdominothoracic брюшно-грудной
abducent *a anat* отводящий
abduct *v* отводить
abduction *s* абдукция, отведение
abductor *s anat* абдуктор, отводящая мышца
abenteric *a* внекишечный
aberrant *a* аберрантный, отклоняющийся от нормы
aberration *s* аберрация, отклонение от нормы
abhorrence *s* отвращение; ненависть
ability *s* способность, умение
combining ~ сочетаемость
competitive ~ конкурентоспособность
germinating ~ всхожесть
regenerative ~ регенерационная способность
rooting ~ укореняемость
abirritation *s* успокоение
abject *a eth* приниженный
abjection *s* выделение *(e.g, of spores); eth* приниженность, унижение
abjoint *v* отграничивать перегородками, отделять *(e.g., spores)*
abjunction *s* отделение, отчленение
ablactation *s* прививка сближением, аблактировка; аблактация
ablastous *a* абластический
ablate *v* ампутировать
ablation *s* ампутация; экстирпация
ablaze *a* возбуждённый
able-bodied крепкий, здоровый
abloom *a* цветущий
abludent *a* отклоняющийся
abluent s очищающее средство, дезинфицирующее средство; *a* очищающий, дезинфицирующий
ablution *s* отмывка; промывание
abmigration *s* абмиграция
abnormal *a* аномальный; ненормальный
abnormality *s* нарушение, дефект; аномалия
abolition *s* устранение, уничтожение
abomasal *a* сычужный
abomasum *s anat* сычуг
aboral *a* аборальный
aboriginal *s* абориген, коренной житель; туземец; *a* аборигенный, исконный, коренной; первобытный; местный
abortion *s* аборт, выкидыш; остановка развития; недоразвитие органа
abortive *a* абортивный, недоразвитый; *bot* стерильный
abound *v* изобиловать
above-ground *a* надземный
abranchial, abranchiate *a* безжаберный
abrasion *s* абразия; выравнивание поверхности, сглаживание поверхности
abscess *s* нарыв, абсцесс
abscise *v* опадать *(e.g., of leaves)*
abscission *s* ампутация; *бот.* опадение, сбрасывание *(e.g., of leaves)*
absconded *a* скрытый, маскирующийся, маскированный *(e.g., of character)*
absenteism *s eth* абсентеизм
absinth, absinthe, absinthium *s* полынь горькая *(Artemisia absinthium)*
absolute *a* чистый, беспримесный; абсолютный *(e.g, alcohol);* полный, безусловный, совершенный
absorb *v* абсорбировать, поглощать, впитывать, всасывать
absorbability *s* абсорбционная способность, гигроскопичность
absorbancy *s* оптическая плотность
absorbefacient *a* вызывающий абсорбцию, вызывающий поглощение, вызывающий впитывание, вызывающий всасывание
absorbency *s* оптическая плотность; усвояемость пищи

absorbent *s* абсорбент, поглотитель; абсорбирующий, поглощающий, впитывающий, всасывающий
absorptiometer *s* абсорбциометр
absorption *s* абсорбция, поглощение, впитывание, всасывание
dermal ~ всасывание через кожу
root ~ поглощение корнями
absorptive *a* абсорбирующий, поглощающий, впитывающий, всасывающий
absorptivity *s* абсорбирующая способность, поглощающая способность
abstract *v* экстрагировать, отделять, извлекать
abstraction *s* экстрагирование, отделение, извлечение
abundance *s* обилие, изобилие, большое количество; численность
seasonal ~ сезонная динамика числености
abundant *a* обильный; широко распространённый; часто встречающийся
abutment s межа
abuse *s* неправильное употребление; злоупотребление
acalycine, acalycinous *a bot* асепальный, не имеющий чашечки
acalyculate *a bot* не имеющий подчашия
acantha *s* игла, колючка, шип; остистый отросток позванка
acanthaceous *a* имеющий колючки; шиповатый
acanthion *s* передняя носовая ость
acanthocarpous *a* колючеплодный
acanthocyte s акантоцит, шиповатая клетка
acanthoid *a* шипообразный, иглообразный; колючий; покрытый иглами, покрытый колючками
acanthophorous *a* имеющий колючки; шиповатый
acanthor *s helm* акантор
acanthous *a* колючий, покрытый иглами, покрытый колючками
acapnia s акапния
acariasis s акариаз, акариоз
acaricide *s* акарицид; *a* акарицидный
acarology *s* акарология
acarophilous *a* акарофильный
acarophily *s* акарофилия
acarpel(l)ous *a* беспестичный
acarpous *a* акарпный, без плодов
acaryocyte *s* акариоцит, безъядерная клетка
acaryote *s* акариоцит, безъядерная клетка; *a* безъядерный
acaudate *a* бесхвостый
acaulescent *a* бесстебельный
acaulosia *s* недоразвитие стебля
acauline, acaulous *a* бесстебельный
accelerant *s* ускоритель; катализатор
acceleration *s* акселерация, ускорение
acceleratror *s* ускоритель, катализатор, акцелератор
prothrombin ~ проакцелерин, протромбокиназа
serum ~ проконвертин
accelerin *s* акцелерин
accelerogram акселерограмма
accentor *s orn* завирушка *(Prunella)*
accentuator *s* акцентуатор
accept *v* воспринимать; акцептировать; связывать
acceptability *s* приемлемость
relative ~ предпочтение; сравнительный выбор
acceptor s акцептор
accessory *a* добавочный, дополнительный, придаточный, второстепенный, побочный
accidental *s* случайность; случайный элемент; случайный, временный; *bot* добавочный, поздно заложившийся; второстепенный, побочный
acclimatation *s* акклиматизация
acclimate *v* акклимировать, акклимироваться
acclimation *s* акклимация
acclimatization *s* акклиматизация
acclimatize *v* акклиматизировать. акклиматизироваться
accommodation *s* приспособление; *phys* аккомодация
accouchement *s* роды, разрешение от бремени
accrescence *s* непрерывный рост
accrescent *a* разрастающийся
accrete *v* срастаться; обрастать; прикрепляться

accretion *s* рост; приращение, прирост; обрастание; срастание; результат роста
accumbent *a bot* прижатый, прилегающий
accumulation *s* аккумулирование, накопление, накапливание
accuracy *s* точность
acellular *a* бесклеточный
acentric *a* ацентрический *(e.g., chromosome)*
acephalocyst, acephalocystis *s helm* ацефалоциста
acephalous *a* ацефальный, безголовый; *bot* неголовчатый
acerate *a* игловидный; игольчатый; заострённый
acerb *a* терпкий
acerous *a* игловидный; игольчатый; заострённый; безрогий; не имеющий щупалец
acervate *a* сгруппированный, скученный
acervuline *a* скученный, сгруппированный *(e.g., glands)*, беспорядочно сложенный
acervulus *s myc* ложе; пучок; *бот.* кисть; мозговой песок
acescence s окисление; скисание, закисание
acetabuliform *a* блюдцевидный
acetabulum *s helm* брюшная присоска; *ent* тазиковая впадина
acetalation, acetalization *s* ацетализация, ацеталирование
acetoacetylation *s* ацетоацетилирование
acetobacter *s* ацетобактер
acetoethylation *s* ацетоэтилирование
acetolysis *s* ацетолиз
acetonation *s* ацетонирование
acetonuria *s* ацетонурия
acetoxylation *s* ацетоксилирование
acetoxymethylation *s* ацетоксиметилирование
acetylation *s* ацетилирование
acetylcholine *s* ацетилхолин
acetylization *s* ацетилирование
achaetous *a* не имеющий щетинок
achene, achenocarp *s* семянка
achiral *a мол. биол.* Ахиральный
achlamydate *a mal* не имеющий мантии
achlamydeous *a* беспокровный
achlorophyllous *a* бесхлорофилльный
achlorosis *s* ахлороз, отсутствие хлорофилла
achromat *s* ахроматический объектив, ахромат
achromatic *a* ахроматический, неокрашенный, бесцветный; неокрашивающийся
achromatin *s* ахроматин
achromatolysis *s* ахроматолиз, разрушение ахроматина
achromatophilous *a* некрасящийся, трудно окрашивающийся
achromic *a* ахроматический, неокрашенный, бесцветный; неокрашивающийся
achromophilous *a* некрасящийся, трудно окрашивающийся
achromoreticulocyte *s* ахроморетикулоцит
achromous *a* неокрашенный, непигментированный, бесцветный, ахроматический, неокрашивающийся
acicle *s* ацикула, иголочка, шипик
acicular, aciculiform *a* игольчатый, игловидный, шипиковый
aciculum *s* ацикула, иголочка, шипик
acid *s* кислота; *a* кислый; кислотный
acidate *v* подкислять
acid-fast *a* кислотоустойчивый
acidic *a* кислый; кислотный
acidification *s* подкисление
acidity *s* кислотность
acidize *v* подкислять
acidobiontic *a* живущий в кислой среде
acidofuge *s* ацидотрофный организм
acidolysis *s* ацидолиз, кислотный гидролиз
acidophile *s* ацидофильный организм; *a* ацидофильный
acidophilic *a* ацидофильный
acidophyte *s* ацидофит
acidosis *s* ацидоз
acidotrophic *a* ацидотрофный
acidotus *a* остроконечный
acidproof, acid-resistant *a* кислотостойкий, кислотоустойчивый
acid-treated *a* обработанный кислотой
acidulation *s* подкисление; закисление
acidulous *a* кисловатый; подкислённый
acinacifolious саблевиднолистный, саблелистный
acinaciform *a bot* саблевидный

aciniform *a* гроздевидный; многосемянный
acinose *a* зернистый
acme *s* апогей, акме; зрелый возраст
acne *s med* угри
acoelous *a* ацельный *(of a vertebra)*
acolpate *a pal* безбороздный
acormose *a* бесствольный, бесстебельный
acorn *s* жёлудь
acorncup *s bot* плюска
acotytledon *s* бессемядольное растение
acotyledonous *a bot* бессемядольный
acoustical *a* слуховой
acquaint *v eth* знакомиться
acquired *a* приобретённый, неврождённый
acquirement *s* приобретение; приобретённые навыки
acquisition *s* приобретение; добывание *(e.g., of food)*
~ **of skill** *eth* выработка навыка
acquisitive *a* восприимчивый
acranial *a* акраниальный, бесчерепной
acranthed *a* верхушечноцветовый
acraspedote *a helm* акраспедотный
acrid *a* острый; едкий; раздражающий
acroblast *s* акробласт
acrobryous *a* с верхушечным ростом
acrocarpous *a* акрокарпный, верхоплодный
acrocentric *a* акроцентрический
acrodromal *a bot* остробежный
acrogenous *a* акрогеный
acromial *a* акромиальный
acropetal *a* акропетальный
acrophyte *s* акрофит
acrophytia *s* сообщество акрофитов
acroscopic *a* направленный к верхушке
acrosome *s* акросома, идиосома
acrospore *s* акроспора
acrosporous *a* акроспоровый
acrosyndese, acrosyndesis *s* акросиндез
actin *s* актин
actinobiology *s* актинобиология
actinocarpous *a* актинокарпный
actinodrome, actinodromous *a bot* пальчатонервный
actinogram *s* актинограмма
actinograph *s* актинограф
actinoid *a* актиноидный, лучеобразный
actinology *s* актинобиология
actinomere *s* актиномер, радиальный сегмент
actinometry *s* актинометрия
actinomorphic, actinomorphous *a* актиноморфный; радиально-симметричный, правильный
actinost *s* птеригиофор
actinostome *s zool* ротовое отверстие
actinotrichium *s* актинотрихия, кератотрихия
action *s* действие; воздействие; эффект; работа, деятельность
activation ~ авторитмика
antidromic ~ антидромное возбуждение
autorythmic ~ авторитмика
catalytic ~ каталитическая активация
complement ~ активация комплемента
control ~ регуляция
psychomotor ~ психомоторная реакция, психомоторное действие
reflex ~ рефлекторное действие
stimulative ~ стимуляция
synaptic ~ синаптическое возбуждение
activator *s* активатор
leukoagglutinating ~ лейкоагглютинин
active *a* активный
activity *s* активность, деятельность; жизнедеятельность; **to display** ~ проявлять активность
alpha ~ альфа-ритм, альфа-волны
beta ~ бета-ритм, бета-волны
gamma ~ *phys* гамма-ритм, гамма-волны
helper ~ хелперная активность, активность хелперов
life ~ жизнедеятельность
vital ~ жизнедеятельность
actomyosin *s* актомиозин
actophilous *a* актофильный
actualism *s* актуализм
acuity *s* острота *(e.g., of vision)*
auleate *a* имеющий жало, имеющий шипы, покрытый шипами, остроконечный
aculeiform *a* шиловидный, шиповидный
aculaeolate *a* мелкошиповатый
acumetry *s* аудиометрия, акуметрия
acuminate *a* заострённый, остроконечный
acuminulate *a* тонкозаострённый

acusporide *a pal* мелкодырчатый, мелкопродырявленный
acutate *a* островатый
acute *a* острый
acuteness *s* острота *(e.g., of vision)*
acutiflorous *a* остроцветковый
acutifoliate *a* остролистный
acyclic *a* ациклический; *bot* спиральный
acyesis *s* женская стерильность
acylaminoalkylation *s* амидоалкилирование, ациламиноалкилирование
acylaminomethylation *s* амидометилирование, ациламинометилирование
acylation *s* ацилирование
acyloxylation *s* ацилоксилирование
adactyl *a* не имеющий пальцев
adamantobast *s* адамантобласт
adambulacral *a* адамбулакральный
adaptability *s* адаптивность, приспособляемость
adaptation *s* адаптация, приспособление
prospective ~ проспективное приспособление, преадаптация
adaptiogenesis *s* адаптиогенез
adaptive *a* адаптивный, припособительный
adaptor *s* адаптор; *a* адапторный
tube ~ держатель пробирок; штатив
adaxial *a* адаксиальный
add *v* добавлять
adder *s* ядовитая змея; обыкновенная гадюка *(Vipera berus)*
addition *s* добавление, прибавление, присоединение, добавка
~ **of starter** *micr* заквашивание
additional *a* добавочный, дополнительный, придаточный
additive *a* добавляющий, аддитивный; *s* добавка
addle *a* тухлый, гнилой
adducent *a* приводящий
adduction *s* аддукция, приведение
adductor *s* аддуктор, приводящая мышца
adeciduous *a* неопадающий
adelphogamy *s* адельфогамия
adelphomorphous *a* неопределённой формы, неопределённого строения
adelphoharasite *s* адельфопаразит
adelphotaxis *s* адельфотаксис
adenoblast *s* аденобласт
adenocarcinoma *s* аденокарцинома, железистый рак
adenocarpous *a* аденокарпный, железистоплодный
adenocaulis *a* железистостебельчатый
adenocyte *s* аденоцит, клетка железы
adenogenic *a* железистого происхождения
adenohypophysis *s* аденогипофиз, передняя доля гипофиза
adenoid *s* аденоид; *a* аденоидный
adenophore *s* аденофор
adenophyllous *a* железтолистный
adenopterous *a* жезистокрылый
adenose *a* железистый
adenostemonous *a* железисто-тычиночный
adenotrichous *a bot* железисто-волосистый
adenous *a* железистый
adenovirus *s* аденовирус
adermin *s* адермин
adherence *s* плотное соединение; сцепление; прилипание, адгезия
adherent сросшийся; приросший; прикреплённый к субстрату; вязкий, клейкий
adhesion *s* адгезия, прилипание; сцепление; спайка, рубец
cell-cell ~ межклеточная адгезия; слипание клеток
adhesiveness *s* адгезивность, слипаемость, адгезионная способность
adiactinic *a* адиактинический
adiaphoria *s* рефрактерный период, рефрактерная фаза, рефрактерность
adipocere *s* трупный жировоск
adipocyte *s* адипоцит
adipoleukocyte *s* адиполейкоцит
adipose *a* жирный, жировой
adiposis, adiposity *s* ожирение, тучность
adjacent *a* прилегающий, смежный, соседний, сопредельный
adjustor *s* орган регуляции
adjustment *s* установка, пригонка; приспособленность; *biom* корректировка, поправка
adjutant *s* марабу *(Leptoptilus)*
adjuvant *s* адъювант, стимулятор; синергист; *a* адъювантный, вспомогательный, помогающий, полезный

depot-forming ~ депо-адъювант
adjuvanticity *s* способность стимулировать, адъювантные свойства
administration *s* введение, назначение
admiral *s ent* адмирал *(Pyrameis atalanta)*
admix *v* примешивать
admixture *s* примесь
adnasal *a* прилегающий к носу
adnate *a* сросшийся, сращённый. приросший
adnation *s* сращение
adnexa *s pl* придатки; экстраэмбриональные структуры
adolescaria *s helm* адолескария
adolescence *s* подростковый возраст, пубертаный период, пубертатный возрат
adolescent *s* подросток; *a* юный, молодой
adoption *s* уствоение, принятие
adoptive *a* восприимчивый
adoral *a* адоральный
adorned *a* украшенный
adpressed *a* прижатый
adrenal *s* надпоченик
adrenergic *a* адренергический
adrenocortical *a* адренокортикальный
adrenocorticotropin *s* адренокортикотропин, адренокортикотропный гормон гипофиза, АКТГ
adsorbate *s* адсорбируемое вещество
adsorbent *s* адсорбент, адсорбирующее вещество
adsorption *s* адсорбция
adsternal *a* прилежащий к грудине
adterminal *a* направленный к периферии
adult *s* взрослая особь; *a* взрослый; развитой
adulterant *s* примесь, подмесь
adulteration *s* примешивание, подмешивание
adulthood *s* взрослое состояние
aduncate *a* крючковатый
adust *a* обугленный, выжженный, опалённый солнцем
advanced *a* продвинутый, прогрессивный
advantage *s* преимущество
adventitious *a* придаточный, добавочный, побочный, случайный, адвентициальный
adventive *a* адвентивный
adversary *s eth* соперник
adverse *a* неблагоприятный, вредный, враждебный; *bot* обращенный
adversity *s* неблагоприятная обстановка
advertisement *s* демонстрация *(coloration); eth* оповещение, рекламирование
adynamia *s* адинамия
AE-cellulose *s* аминоэтилцеллюлоза
aecidiospore *s* эцидиоспора
aecidium *s* эцидий
aeciospore *s* эцидиоспора
aecium *s* эцидий
aegis *s* защита
aeolation *s* ветровая эрозия
aeon *s* эон
aeration *s* аэрация; насыщение кислородом
aerenchyma *s* аэренхима
aerial *a* воздушный
aerie *s* гнездо хищной птицы
aeriferous *a* воздухоносный
aeriform *a* газообразный; воздушный
aerobe *s* аэроб, аэробный организм
aerobic *a* аэробный
aerobiology *s* аэробиология
aerobiosis *s* аэробиоз
aerobiotic *a* аэробный
aerochorous распространяемый ветром, анемохорный
aerocyst *s* воздушная циста
aerogen *a* газообразующий
aerogenesis *s* газообразование
aerogenic *micr* аэрогенный, газообразующий
aeromorphosis *s* аэроморфоз
aerophilic, aerophilous *a* аэрофильный
aerophytobiota *s* флора аэрофитов
aerophyte *s* аэрофит
aeroplankton *s* аэропланктон
aeroplethysmograph *s* воздушный плетизмограф
aeroscope *s* аэроскоп
aerosol *s* аэрозоль
aerostat *s ent* воздушный мешок*; orn* воздушная полость
aerotank *s* аэротэнк
aerotaxis *s* аэротаксис
aerotropic *a* аэротропный
aerotropism *s* аэротропизм

aestacyte *s* эстоцит, чувствительная клетка
aestisilvae *s pl* летнезелёные леса
aestival *a* летний
aestivaria *s pl* отделы летников *(in botanical gardens)*
aestivation *s* летняя спячка; почкосложение
aethalium *s* эталий
aetiology *s* этиология
afebrile *a* бестемпературный, безлихорадочный
afetal *a* не содержащий плода; бесплодный
affect *s* аффект; повреждение, поражение; *v* поражать
affection *s* поражение, болезнь, заболевание; влияние
afferent *a* афферентный, центростремительный, приносящий
affiliation *s eth* связь, объединение, аффилиция, установление связи
affined *a* сходный, родственный
affinity *s* родство, сродство, подобие, близость
genetic ~ генетическое сродство, генетическая связь
affixion *s* прикрепление, связанность
affluence *s* наплыв, приток; изобилие
affluent *a* изобильный
afflux *s* приток, прилив *(of blood);* скопление *(of fluid)*
afforestation *s* облесение, лесонасаждение
aflagellar *a* безжгутиковый
afloat *a* плавающий
afteraction *s* последействие
afterbirth *s* плацента, послед, детское место
afterbrain *s* задний мозг
afterbreast *s* загрудинное пространство; *ent* заднегрудь
aftercrop *s* второй урожай
afterdepolarization *s* следовая деполяризация
afterdischarge *s* разряд последействия, следовой разряд
aftereffect *s* последействие
aftergrass *s* отава
aftergrowth *s* отрастание
afterimage *s* следовой образ
aftermath *s* отава; последствия; *a* второукосный
afternystagmus *s phys* постнистагм
afterpotential *s* потенциал последействия,следовой потенциал
after-regeneration of wood *s* послесводочное восстановление леса
afterripening *s* дозревание
aftershaft *s orn* придаточное перо
afterswarm *s apic* подрой
afunction *s* нефункционирование, прекращение функций
afunctional *a* нефункционирующий, недействующий, нефункциональный
agad *s* береговое растение
agalactia, agalaxia *s* агалактия
agama *s herp* агама *(Agama)*
toad ~ круглоголовка *(Phrynocephalus)*
agameon *s* агамеон
agamic агамный, бесполый; криптогамный, тайнобрачный
agamobium *s* агамобий
agamodeme *s* агамодем
agamogenesis *s* агамогенез, бесполое размножение
agamogenetic *a* агамогенетический
agamogony *s* агамогония
agamospermy *s* агамоспермия
agamospirura *s* агамоспирура
agamospore *s* агамоспора, бесполая спора
agamous *a* агамный, бесполый; криптогамный, тайнобрачный
agar, agar-agar *s* агар
beef-extract ~ мясопептонный агар, МПА
iron ~, iron-sulfite ~ железосульфитный агар, среда Вильсона – Блера
meat infusion ~ мясопептонный агар, МПА
agaric *s* съедобный гриб; *pl myc* пластинниковые *(Agaricaceae)*
fly ~ мухомор красный*(Amanita muscaria)*
honey ~ опёнок настоящий *(Armillaria mellea)*
orange ~ рыжик *(Lactarius deliciosus)*
agaricole, agaricolous *a* живущий на грибах
agarose *s* агароза
agar-producing *a* агарообразующий

agave *s* агава *(Agave)*
sisal ~ агава сизалевая, сизаль *(Agave sisal)*
age *s* возраст; определять возраст; период; эпоха; старость; стареть; изменяться во времени; совершеннолетие; поколение
~ **of life** продолжительность жизни
aged старый
age-grade, age-group *s* возрастная группа
ageing *s see* **aging**
agency *s* действие, воздействие; сила, фактор
agenesia, agenesis *s* агенез; бесплодие
agent *s* агент; возбудитель; фактор; способ, средство
~ **of disease** болезнетворный агент, возбудитель болезни
~ **of infection** возбудитель инфекционного заболевания
absorbing ~ абсорбент, абсорбирующее вещество
adrenoreceptor blocking ~ адреноблокирующее средство, адреноблокатор
antidoting ~ противоядие
antifungal ~ противогрибковый препарат
binding ~ связующее вещество
bulking ~ наполнитель
carrying ~ переносчик; вещество-носитель
catalytic ~ катализатор
causative ~ возбудитель; этиологический фактор
clarifying ~ просветляющее вещество
cooling ~ охладитель
coupling ~ сопрягающий агент; *bioch* фактор сопряжения
dehydrating ~ дегидрирующее вещество
detergent ~ детергент
diluting ~ разбавитель
dispersion ~ диспергатор, диспергирующее средство
effective ~ действующее начало
emulsifying ~ эмульгатор
fertilizer ~ удобрение
filming ~ плёнкообразователь
fixing ~ фиксирующее средство
flocculating ~ флокулирующий агент, флокулент
foaming ~ пенообразователь, вспениватель
fungicidal ~ фунгицид
gelling ~ агент желатинизации
germicidal ~ бактерицидное вещество
immune-respose modulating ~ иммуномодулирующий фактор, иммуномодулятор
immunosuppressive ~ иммунодепрессант, иммуносупрессор
infectious ~ возбудитель инфекции, инфекционный агент
inhibiting ~ угнетающий агент, ингибирующий агент, ингибитор
larvicidal ~ ларвицид
mutagenic ~ мутагенный агент, мутагенный фактор
nitrifying ~ нитрификатор
pathogenic ~ возбудитель заболевания, патогенный агент
phylactic ~ иммунное вещество
plasmolyzing ~ плазмолизирующее вещество,плазмолитик
pollution ~ фактор, вызывающий загрязнение внешней среды, загрязнитель
polyploiding ~ вещество, вызывающее полиплоидию
preserving ~ консервант
promotoring ~ стимулирующее вещество, стимулятор
reducing ~ восстановитель; редуцирующая субстанция
staining ~ красящее вещество, краситель, краска
stimulating ~ стимулятор
surface-active ~ поверхностно-активное вещество
toxic ~ токсичное вещество. ядовитое вещество
transmitting ~ переносчик
ageotropism *s* отрицательный геотропизм
agglomeration *s* агломерация, скопление, слипание
agglutinant *s* агглютинирующее вещество; *a* агглютинирующий, склеивающий
agglutinate *s* агглютинировть, склеивать

agglutination *s* агглютинация,склеивание; образование групп, образование скоплений
agglutinative *a* агглютинирующий, склеивающий
agglutinin *s* агглютинин
major ~ макроагглютинин
minor ~, partial ~ парaагглютинин
agglutinogram *s* агглютинограмма
agglutinoscopy *s* агглютиноскопия
agglutometer *s* агглютометр
aggregate *s* агрегат; совокупность; собирать; соединять; *a* скученный; сгруппированный; собранный; *bot* сборный
aggravation *s* ухудшение, обострение
aggregation *s* агрегация, объединение; скопление; группа; группировка; *экол.* ассоциация
aggression *s* агрессия, нападение
aggressive *a* агрессивный
agile *a* быстрый, подвижный, проворный
agility *s* подвижность, проворство
aging *s* созревание, вызревание; старение; износ; определение возраста
agitation *s* взбалтывание; перемешивание; встряхивание
agitator *s* мешалка
aglandular *a* лишённый желез
aglet *s bot* серёжка
aglomerular *a* бесклубочковый
aglossate *a zool* безъязычный
aglow *a* возбуждённый
aglucone *s* аглюкон
aglycone *s* агликон
agmatoploidy *s* агматоплоидия
agminate, agminated *a* гроздевидный; пучковидный
agnate *s* родственник по отцу; *a* родственный по отцу
agnathostomatous, agnathous *a* бесчелюстной
agnation *s* агнация, родство по отцу
agnogenic *a* неясного происхождения; неясной этиологии
agnosia *s* агнозия
agonist *s* агонист, мышца-агонист
agonistic *a* агонистический
agonize агонизировать
agony *s* агония; боль, мучение, страдание
agouti *s ther* агути *(Dasyprocta)*
agranulocyte *s* агранулоцит, незернистый лейкоцит
agriculture *s* земледелие, сельское хозяйство, агрокультура, сельскохозяйственная культура
agrium *s* сообщество культивируемых растений
agrobiology *s* агробиология
agrocenology *s* агроценология
agrocenosis *s* агроценоз
agroecotype *s* агроэкотип
agrology *s* агрология, агропочвоведение, прикладное почвоведение
agronomics, agronomy *s* агрономия
agrophilous *a* привязанный к посевам
agrophytocenosis *s* агрофитоценоз
agrostology *s* агростология
agrotype *s* агротип, раса культурного растения
ague *s* малярия; озноб; дрожь; лихорадка
aguish *a* малярийный; перемежающийся
ai *s ther* трехпальный ленивец, ай *(Bradypus tridactylus)*
AIDS, acquired immunodeficiency syndrome *s* СПИД, синдром приобретённого иммунодефицита
aigrette *s bot* пушок, хохолок; *orn* большая белая цапля *(Egretta alba);* эгретка
aiguille *s* остроконечная вершина, игла, игловидный вырост
ail болеть; чувствовать недомогание; беспокоить; причинять страдания
ailment *s* недуг, нездоровье, недомогание
aiphyllous *a* вечнозелёный
aiphytia *s* айфития
air *s* воздух; воздушный; проветривать, вентилировать; **to take to the** ~ взлететь
soil ~ почвенный воздух; газообразная фаза почвы
air-borne *a* переносимый по воздуху; воздушный
air-cooled *a* с воздушным охлаждением
air-dried *a* высушенный на воздухе
air-dry *a* воздушно-сухой
air-locked *a* воздухонепроницаемый, герметичный
air-permeable *a* воздухопроницаемый, пропускающий воздух

air-tight *a* воздухонепроницаемый, герметичный
airway *s* дыхательный путь, воздухоносный путь
aitchbone *s* крестцовая кость
akin *a* родственный; похожий, сходный, близкий
akinesia, akinesis *s* акинез, неподвижность
akinete *s* акинета
akinetic *a* акинетический
ala *s* крыло
alar *a* крыловидный; *bot* пазушный
alarm *s* смятение, страх, тревога; сигнал тревоги
alary *a* крыловой; крыловидный
alate *a* крылатый; окрылённый
albatross *s* альбатрос *(Diomedea)*
albedo *s* альбедо, отражательная способность
albescent *a* беловатый; белёсый
albicant *a* белеющий
albication *s* побеление, приобретение белой окраски
albiflorous *a* белоцветковый
albinism *s* альбинизм
albino *s* альбинос
albumen *s* белок
albumin *s* альбумин
  egg ~ яичный альбумин, овальбумин
  imperfect ~ неполноценный белок
  milk ~ лактоальбумин, альбумин молока
  serum ~ сывороточный альбумин, серумальбумин
albuminiferous *a* содержащий белок; образуюший белок
albuminiparous *a* образующий белок
albuminoid *s* альбуминоид; белковоподобный
albuminolysis *s* протеолиз, гидролиз белка, расщепление белка
albuminous *a* белковый; альбуминовый
alburn, alburnum *s* заболонь
alcalosis *s* алкалоз
alcohol *s* спирт; этиловый спирт
  amino ~ аминоспирт
  ethyl ~ этиловый спирт, этанол
  graded ~s батарея спиртов
  methyl ~ метиловый спирт, метанол
alcoholic *a* спиртовой
alcoholize *v* обрабатывать спиртом
alcohol-soluble *a* спирторастворимый
alcoholometer *s* спиртометр
alcoholysis *s* алкоголиз, спиртовой гидролиз
alder *s* ольха *(Alnus)*
alderfly *s ent* вислокрылка *(Megaloptera)*
aldolization *s* альдольное присоединение
aldosterone *s* альдостерон, электрокортин, альдокортин
alecithal *a* алецитальный, без желтка *(egg)*
alembic *s* дистиллятор, перегонный куб
alepidote *a* бесчешуйный
alert *s* тревога; сигнал тревоги; настороженный; живой, проворный
alertness *s* настроженность, бдительность; живость, проворство
alete *a* безрубцовый, бесщелевой, безапертурный
aletocyte *s* блуждающая клетка
aletophyte *s* придорожное растение
aleurone *s* алейрон
aleuronic *a* алейроновый
aleurospore *s* алеуроспора
alexin *s* алексин, комплемент
alfalfa *s* люцерна *(Medicago)*
alien *s* интродуцированный вид, экзот; чужой; *a* чужеродный; пришлый
aliform *a* крыловидный
alight *v* садиться, опускаться
aliment *s* пища, питание
alimental *a* пищевой
alimentary *a* алиментарный, пищевой; пищеварительный
alimentation *s* питание, кормление
aliquot *s* аликвотная проба; *a* аликвотный
alisphenoid *s* алисфеноид, крылоклиновидная кость
alive *a* живой; изобилующий, кишащий; действующий, существующий
alkalescence *s* щёлочность; образование щёлочи; ощелачивание
alkali *s* щёлочь; щелочной металл; щелочной катион; зола
alkalify подщелачивать
alkaline *a* щелочной
alkalinity *s* щелочность
alkalinization *s* подщелачивание
alkali-proof, alkali-resistant щёлочеупорный, щёлочестойкий
alkali-soluble *a* щёлочерастворимый

alkalitrophic *a* алкалитропный, базитропный, базифильный
alkalization *s* подщелачивание
alkalize *v* подщелачивать
alkaloid *s* алкалоид
cadaveric ~ птомаин
alkalosis *s* алкалоз
alkylation *s* алкилирование
allantoid *a* аллантоидный, сосискообразный, колбасовидный
allantois *s* аллантоис
allautogamia *s* аллаутогамия
allay *v* успокаивать; ослаблять боль
alldouble *a* полностью махровый
allele *s* аллель
allelic *a* аллельный
allelism *s* аллелизм
allelobrachial *a* аллелобрахиальный
allelogenesis *s* аллелогения
allelomorph *s* аллеломорф
spurious ~ ложный аллеломорф, псевдоаллеломорф
allelomorphic аллеломорфный
allelomorphism *s* аллеломорфизм
allelopathy *s* аллелопатия
allelosomal *a* аллелосомный, аллелосомальный
allelotaxy *s* аллелотаксис
allelotype *s* аллелотип
allergen *s* аллерген
allergenic *a* аллергенный
allergenicity *s* аллергенность
allergic *a* аллергический
allergist *s* аллерголог
allergization *s* сенсибилизация аллергеном, аллергизация
allergology *s* аллергология
allergosis *s* аллергическая болезнь, аллергоз
allergy *s* аллергия
pollen ~ поллиноз, пыльцевая аллергия
alleviate *v* облегчать, смягчать
alliaceous *a* луковый
alliance *s* родственная связь, родство
allied *a* родственный; близкий; связанный, соединённый; смежный
alloantigen *s* аллоантиген, аллотипический антиген
alloantigenic *a* аллоантигенный
allocarpy *s* аллокарпия
allocation *s* размещение; распределение
allochroic *a* аллохроический
allochronic *a* аллохронный
allochthonic, allochthonous *a* аллохтонный; внешний
allocortex *s* аллокортекс
allocycly *s* аллоциклия
allocytotoxicity *s* аллогенная цитотоксичность
allodeterminant *s* аллоантигенная детерминанта, аллоэпитоп
allodiploid *s* аллодиплоид
allodiplomonosomic *s* аллодипломоносомик
allodromy *s* аллоритмия, аллоритмическая экстрасистолия
alloepitope *s* аллоэпитоп, аллоантигенная детерминанта
allogamous аллогамный, перекрёстноопыляющийся
allogamy *s* аллогамия, перекрёстное опыление
allogene *s* аллоген, рецессивный ген
allogenetic *a* аллохтонный; внешний
allogenic *a* аллогенный
allogenization *s* аллогенизация
allograft *s* аллотрансплантант
allografting *s* аллотрансплантация
allogrooming *s eth* аллогруминг
allohaploid *s* аллогаплоид
alloheteroploid *s* аллогетероплоид
alloimmune *a* аллоиммунный
alloimmunity *s* аллоиммунитет
alloimmunize иммунизировать аллоантигеном
alloiobiogenesis *s* чередование поколений
alloiophylly *s phytop* аллойофилия
allokinesis *s* рефлекторное движение; пассивное движение
allometric *a* аллометрический
allometry *s* аллометрия
allomixis *s* алломиксис
allomonodiploid *s* алломонодиплоид
allomonoheteroploid *s* алломоногетероплоид
allomorphism *s* алломорфизм
allomorphosis *s* алломорфоз
alloparalectotype *s* аллопаралектотип
allopathy *s* аллопатия
allopatric *a* аллопатрический
allopatry *s* аллопатрия
allopelagic *a* разноглубинный
alloplasm *s* аллоплазма

alloplasmatic *a* аллоплазматический
alloploidion *s* аллоплоидион
alloploidy *s* аллоплодия
allopolyploid *s* аллополиплоид
allopolyploidy *s* аллополиплоидия
allopreening *s eth* взаимная чистка оперения
allorecognition *s* аллогенное распознавание
alloresponse *s* аллоиммунный ответ
all-or-none *adv phys* «всё или ничего»
allorhythmia *s* аллоритмия, аллоритмическая экстрасистолия
allorhythmic *a* аллоритмический, нерегулярный
allosematic *a* аллосематический
allosensitization *s* аллогенная сенсибилизация, аллосенсибилизация
alloserum *s* аллогенная антисыворотка
allosomal *a* аллосомный
allosome *s* аллосома; половая хромосома
allosteric *a* аллостерический
allostery *s* аллостерия
allostose *s* перепончатая кость
allosyndesis *s* аллосиндез, аллосиндезис
allotetraploid *s* аллотетраплоид
allotetraploidy *s* аллотетраплоидия
allotment *s* участок, делянка
allotolerancy *s* аллогенная толерантность, аллотолерантность
allotope *s* аллотоп, аллотипическая детерминанта
allotopia *s* дистопия, атопия, аллотопия
allotransplantation *s* аллотрансплантация
allotriploid *s* аллотриплоид; *a* аллотриплоидный
allotrophic *a* аллотрофный
allotype *s* аллотип
allotyping *s* аллотипирование, аллогенное типирование
allotypy *s* аллотипия, аллотипический феномен
allovariant *s* аллотипический вариант
allowable *a* допустимый
allowance *s biom* поправка; порция, рацион
allozyme *s* аллозим
allseed *s* многосемянное растение
allure *v* приманивать, подманивать
all-year *a* круглогодичный
allylation *s* аллирование
alm *s* высокогорный луг
almond *s anat* миндалина; миндаль *(Amygdalus)*
earth ~ сыть съедобная, чуфа, земляной миндаль *(Cyperus esculentus)*
alone *a* одиночный
alopecia *s* облысение
alpaca *s* альпака, пака *(Lama pacos)*
alpestrine *a* субальпийский
alpha-diversity *s* альфа-разнообразие
alpine *a* альпийский
alter *v* менять, изменять, переделывать
alteration *s* изменение, перемена; *imm* сдвиг
alternans *s phys* альтернанс
alternate *a* чередующийся; *bot* очередной
alternation *s* альтернация, чередование, смена
~ **of generations** чередование поколений, смена поколений
electric(al) ~ электрическая альтернация
alternativenes *s* сменяемость
alternipinnate *a* попеременноперистый
altherbosum *s* сообщество высоких трав
altichamber *s* барокамера
alticolate *a* высотный, обитающий на высотах
altitude *s* высота; отметка уровня; ~ **above sea level** высота над уровнем моря
altrice *s* птенцовая птица
altricial *a orn* незрелорождающийся; птенцовый
altruism *s eth* альтруизм
altruistic *a* альтруистический
***Alu*-family** *s* Alu-семейство
***Alu*-repeat** *s Alu*-повтор
***Alu*-sequence** *s Alu*-последовательность
alula *s* крылышко
alumen *s* квасцы
alveary *s* наружный слуховой проход
alveococcosis *s* альвеококкоз, альвеолярный эхинококкоз
alveolar *a* альвеолярный
alveolate *a* ячеистый
alveole *s* альвеола, ячейка, пузырёк
alveolodental *a* альвеоло-зубной
alveololabial *a* альвеоло-губной
alveololingual *a* альвеоло-язычный
alveolopalatal *a* альвеоло-нёбный

alysogenic *a* ализогенный
amastigote *s* амастиготная форма; *a* безжгутиковый
amazon *s* муравей-амазонка *(Polyergus)*; *orn* амазон *(Amazona)*
ambergrease, ambergris *s* амбра
ambient *a* окружающий
ambiguous *a* нечёткий, сомнительный, неясный
ambilateral *a* двусторонний
ambisexual *a* амбисексуальный, обоеполый; однодомный
amble *s* иноходь; *v* бежать иноходью
ambler *s* иноходец
ambling *s* иноходь
amblychromatic *a* амблихроматический, слабоокрашиваемый
amblyocarpous *a* тупоплодный
amblyophyllous *a* туполистный
amblyopterous *a* тупокрылый
amboceptor *s* амбоцептор
ambrosia *s* перга; амброзия *(Ambrosia)*
ambulacral *a* амбулакральный
ambulacrum *s* амбулакральная ножка; амбулакр, амбулакральная бороздка; *ent* ходильная нога
ambulatorial, ambulatory *a zool* ходильный
ambuscade *v* нападать из засады
ambuscader *s* засадчик
ambush *s* засада; нападать из засады; находиться в засаде
ameba *s* амёба
ameiosis *s* амейоз
amelification *s* амелификация
amelioration *s* улучшение; мелиорация
ameloblast *s* амелобласт, адамантобласт
amelogenesis *s* амелогенез
amelogenic *a* эмалеобразующий
amenable *a* подверженный; восприимчивый; поддающийся
amensalism *s* аменсализм
ament *s bot* серёжка
amentaceous *a* серёжчатый; серёжконосный
amentiferous *a* серёжкоцветный
ameristic *a* несегментированный
amerospora *s* амероспора
ametabolic *a* аметаболический
ametoecious *a* однохозяйный
ametria *s* отсутствие матки
ametrous *a* не имеющий матки
amicrobic *a* безмикробный, стерильный
amicronucleate *a* амикронуклеарный, безмикронуклеусный
amidation *s* амидирование
amidoalkylation *s* амидоалкилирование, ациламиноалкилирование
amidolysis *s* амидолиз
amidomethylation *s* амидометилирование, ациламинометилирование
amination *s* аминирование
aminoacylation *s* аминоацилирование
aminoacyl-tRNA *s* аминоацил-тРНК
aminoethylation *s* аминоэтилирование
aminolysis *s* аминолиз
amitosis *s* амитоз
amitotic *a* амитотический
amixia *s* амиксия
amixis *s* амиксис
ammocolous *a* псаммофильный, песколюбивый; обитающий на песке
ammonification *s* аммонификация
ammoniotelic *a* аммониотелический
ammophilous *a* псаммофильный, песколюбивый; обитающий на песке
ammoxidation *s* аммоксидирование
amniochorial, amniochorionic *a* относящийся к амниону и хориону
amnion, amnios *s* амнион
amniotic *a* амниотический
amoeba *s* амёба
amoebiasis *s* амебиаз
  hepatic ~ амебиаз печени, амёбный абсцесс печени
  intestinal ~ кишечный амебиаз, амёбная дизентерия
amoebiform *a* амёбоидный; амёбовидный
amoebocyte *s* амёбоцит
amoeboid *a* амёбоидный; амёбовидный
amorph *s* аморф, аморфный ген
amorphism *s* аморфное состояние
amorphous *a* аморфный
amount *s* количество
  ~ **of growth** прирост
ampelography *s* ампелография
amphiapomict *s* амфиапомикт
amphiarthrosis *s* амфиартроз
amphiaster *s cyt* амфиастер, стадия звезды
amphiastral *a* амфиастральный
amphiatlantic *a* амфиатлантический

amphibian *s* амфибия, земноводное *(Amphibia); a* амфибионтный, земноводный
limbless ~ безногое земноводное, червяга *(Gymnophiona)*
amphibious *a* амфибионтный, земноводный
amphiblastic *a* амфибластический
amphiblastula *s* амфибластула
amphibolic *a* амфиболический
amphiboreal *a* амфибореальный
amphicarpous *a* амфикарпный
amphichromatism *s* амфихроматизм
amphichromy *s* амфихроматизм
amphicoelous *a* амфицельный, двояковогнутый
amphicondylous *a* амфикондильный
amphicone *s* амфикон
amphicryptophyte *s* амфикриптофит
amphicyte *s* амфицит, мантийная клетка, мантийный глиоцит
amphid *s* амфида
amphidelphic *a* амфидельфический
amphidiploid *s* амфидиплоид; *a* амфидиплоидный
amphidiploidy *s* амфидиплоидия
amphidisk *s* амфидиск
amphigamy *s* амфигамия
amphigastrium *s* амфигастрий
amphigenesis *s* амфигония
amphigenetic, amphigonic *a* амфигенетический
amphigonium *s* архегоний
amphigony *s* амфигония
amphihaploid *s* амфигаплоид
amphikaryon *s* амфикарион*;* амфинуклеус
amphimict *s* амфимикт
amphimictic *a* амфимиктический
amphimixis *s* амфимиксис
amhinucleolus *s* амфинуклеола, двойное ядрышко
amphinucleus *s* амфикарион*;* амфинуклеус
amphiont *s* амфионт
amphipathic *a* амфипатический
amphiphyte *s* амфифит
amphiploid *s* амфиплоид; *a* амфиплоидный
amphiploidy *s* амфиплоидия
amphipneustic *a* двоякодышащий
amphispermium *s* амфиспермий
amphisporangiatic *a* амфиспорангиатный
amphispore *s* амфиспора
amphistomous *a* амфистомный
amphistyly *s* амфистилия
amphitene *s cyt* амфитена, стадия букета
amphithallic *a* амфиталличный
amphithecium *s* амфитеций
amphitoky *s* амфитокия
amphitrichous *a* амфитрихиальный
amphitrocha *s* амфитроха
amphitropous *a* амфитропный
amphivasal *a* амфивазальный
amphochromatophil *a* амфофильный
amphogenic *a* амфогенный
amphophilic, amphophilous *a* амфофильный
amphoteric *a* амфотерный; двусторонний; двойственный
amphoterotoky *s see* **amphitoky**
amphotony *s* амфотония
ample *a* обильный
amplectant *a* охватывающий, объемлющий
amplexicaul *a* стеблеобъемлющий
amplexifoliate *a* листообъемлющий
amplexus *s* охват
ampliate *a bot* расширенный
amplification *s* амплификация; *gen* амплификация, экстракопирование; увеличение численности
clone ~ амплификация клона, экспансия клона, размножение клона
amplifier *s* усилитель; амплификатор
cytotoxic ~ цитотоксическая клетка-амплификатор
T~ промоторная клетка, T-амплификатор
amplitude *s* амплитуда колебания, размах колебания; радиус действия
ampoule, ampule *s* ампула
ampulla *s anat* ампула, пузырёк
amputation *s* ампутация; отсечение; удаление
amycelious *a* не имеющий мицелия
amyelination *s* демиелинизация
amyelinic *a* безмякотный
amygdala *s bot* миндалина*; anat* миндалина
amygdaline *a* миндалевидный
amylaceous *a* крахмальный; содержащий крахмал, крахмалистый

amylobacterium *s* амилобактерия
amyloclastic *a* амилолитический, расщепляющий крахмал
amylogenesis *s* образование крахмала
amylogenic *a* крахмалообразующий
amyloid *a* амилоидный; крахмальный; крахмалоподобный
amylolysis *s* расщепление крахмала
amylolytic *a* амилолитический, расщепляющий крахмал
amyloplast *s* амилопласт
amylum *s* крахмал
anabiont *s* анабионт
anabiosis *s* анабиоз
anabolic *a* анаболический
anabolism *s* ассимиляция, анаболизм
anabolite *s* анаболит
anaboly *s* анаболия, надставка
anacanthartic *a* вызывающий рвоту, рвотный
anachoresis *s* анахорез
anacolpate *a* дистальнобороздный
anaconda *s* анаконда *(Eunectes)*
anacrotic *a* анакротический
anadromous *a* анадромный
anaemia *s* анемия
anaerobe *s* анаэроб
anaerobic *a* анаэробный
anaerobion *s* анаэроб
anaerobiosis *s* анаэробиоз
anaerogenic *a micr* негазопродуцирующий
anaerophyte *s* анаэробное растение
anaesthesia, anaesthetization *s* анестезия, обезболивание, наркоз
anaesthetize *v* анестезировать
anagenesis *s* анагенез; регенерация ткани
anahormone *s* анагормон, дериват гормона
anal *a* анальный, заднепроходный
analepic *s* стимулирующее средство; *a* стимулирующий, укрепляющий
analgetic *a* болеутоляющий, обезболивающий
analogous *a* аналогичный
analogue *s* аналог
analogy *s* аналогия
analysis *s* анализ
  basal ~ исследование основного обмена
  blot(ting) ~ блот-анализ
  dot-blot ~ дот-блот анализ
  restriction ~ рестрикционный анализ; рестрикционное картирование
  sequence ~ секвенирование
  tracer ~ авторадиография, радиоавтография
  tree-ring ~ исследование стволов по годовым кольцам
analytic, analytical *a* аналитический
analyzer *s* анализатор
  fluorescence-activated cell ~ флуоресцентный сортер клеток
  mass ~ масс-спектрометр
  reaction rate ~ анализатор скорости реакции, рефлектометр
  recording gas ~ регистрирующий газоанализатор
  spectrum ~ спектроанализатор
anamorpha *s* анаморфа
anamorphosis *s* анаморфоз
anandrous *a* бестычинковый
anantherous *a* беспыльниковый
ananthous *a* бесцветковый; не цветущий
anaphase *s* анафаза
anaphylactic *a* анафилактический
anaphylactogen *s* анафилактоген
anaphylatoxin *s* анафилатоксин
anaphylaxis *s* анафилаксия, анафилактическая реакция
  acute ~ анафилактический шок
  systemic ~ общая анафилактическая реакция
anaphysis *s* анафиза
anaplasia *s* анаплазия, дедифференцировка
anaplast *s* лейкопласт
anapolysic *a cyt* с неотделяющимися сегментами
anapophysis *s* анапофиз
anaporate *a* дистальнопоровый
anastomose *v* анастомозировать
anastomosing *a* анастомозирующий; *bot* сетчатонервный
anastomosis *s* анастомоз
anastral *a* анастральный
anatomical *a* анатомический
anatomist *s* анатом
anatomy *s* анатомия
  histologic ~ гистология
  special ~ частная анатомия
anatoxin *s* анатоксин
anatropic *a* анатропный, обратный
anavaccine *s* анавакцина, убитая вакцина

anaxial *a* анаксиальный, асимметричный
anazotic *a* безазотистый
anbury *s phytop* кила крестоцветных *(caused by Plasmodiophora brassicae)*
ancestor *s* предок
ancestral *a* анцестральный, наследственный; родовой; атавистический
ancestry *s* происхождение; предки
anchorage *s* якорная подложка *(in cell culture)*
anchor-minus *a* безъякорный
ancient *a* древний; ископаемый
ancillary *a* вспомогательный; дополнительный
ancipitous *a* двухвершинный; двузубый; *bot* обоюдоострый
anconal *a* локтевой
anconeus *s* локтевая мышца
ancophilous *a* анкофильный
ancylostomiasis *s helm* анкилостомоз
androchorous *a* распространяемый человеком
androclinium *s* андроклиний, пыльниковая ямка
androconium *s* андроконий
androcyte *s* андроцит
androecium *s* андроцей
androecy *s* андроэция
androgametangium *s* антеридий
androgamone *s* андрогамон
androgen *s* андроген, андрогенный гормон, мужской половой гормон
androgenesis *s* андрогенез
androgenetic *a* андрогенетический
androgenic, androgenous *a* андрогенный
androgonidium *s* андрогонидий
androgonium *s* андрогоний
androgyne *s* гермафродит, обоеполый организм
androgynism *s* гермафродитизм, обоеполость
androgynophore *s* андрогинофор
androgynous *a* гермафродитный, обоеполый
androgyny *s* гермафродитизм, обоеполость
andromerogony *s* андромерогония
andropetalous *a* андропетальный
androphilous *a* андрофильный
androphore *s* андрофор, тычиночная трубка
androphyll *s* андрофилл, микроспорофилл
androsome *s* андросома
androsporangium *s* андроспорангий
androspore *s* андроспора, микроспора
anelectrotonus *s* анэлектротон
anemia *s* анемия, малокровие
anemic *a* анемичный, малокровный
anemochore *s* анемохор
anemochorous *a* анемохорный
anemoentomophilous *a* анемоэнтомофильный
anemogamous *a* анемофильный, ветроопыляемый
anemone *s zool* актиния; *bot* ветреница *(Anemone)*
anemophilous *a* анемофильный, ветроопыляемый
anemophobe *s* анемофоб
anemoplankton *m* воздушный планктон, аэропланктон
anemosporic *a* анемоспорический
anemotaxis *s* анемотаксис
anemotropism *s* анемотропизм
aner *s* насекомое-самец, муравей-самец
anergy *s* иммунологическая толерантность, анергия
anesthesia *s* анестезия, обезболивание, наркоз
anesthetization *s* анестезирование
anestrus *s* анэструс
aneugamy *s* анэугамия
aneuploid *s* анэуплоид, анэуплоидный организм; *a* анэуплоидный
aneuploidy *s* анэуплоидия
aneurin, aneurine *s* аневрин, витамин B
anfractuose *a bot* извилистый, спиральный
angenesis *s* регенерация тканей
anger *s* гнев; вызывать гнев; сердить, раздражать
angienchyma *s* сосудистая ткань
angioblast *s* ангиобласт
angiocarpous *a* ангиокарпный, покрытоплодный
angiokinesis *s* сосудодвигательная реакция
angiology *s* ангиология
angioscope *s* капилляроскоп
angiosperm *s* покрытосемянное
angiospermous *a* покрытосемянный
angiosporous *a* ангиоспорический

angiotomy *s* ангиотомия, вскрытие сосуда
angle *s* угол; *v* удить
  leaf ~ пазуха листа
anglehead *s herp* лесной дракон, гоноцефал *(Gonocephalus)*
angling *s* ужение
angry *a* раздражённый; яростный; воспаленный, покрасневший
anguine *a* змеевидный
anguish *s* боль, страдание
angular *a* угловой; угловатый; коленчатый
angulate *a* угловатый
angulinerved *a бот.* углонервный
angustifoliate *a* узколистный
angustiseptate *a pal* узкоперегородчатый
anhistic, **anhistous** *a* бесструктурный
anholocyclic *a* анголоциклический
anhormonia *s* гормональная недостаточность
anhydrobiosis *s* ангидробиоз
anhydrous *a* безводный
anil *s* индиго; *bot* индиго *(Indigofera)*
anility *s* старость, дряхлость
animal *s* животное
  alpha ~ особь альфа, доминант
  anamniotic ~ анамний *(Anamnia)*
  amniotic ~ амниот *(Amniota)*
  beta ~ особь бета
  bovine ~s крупный рогатый скот
  cheiropterous ~, chiropterous ~ рукокрылое *(Chiroptera)*
  chordate ~ хордовое *(Chordata)*
  flower ~ коралловый полип *(Anthozoa)*
  fur-bearing ~, fur-producing ~ пушной зверь
  game ~ охотничье-промысловое животное, дичь
  gnawing ~ грызун *(Rodentia)*
  jointed-footed ~, jointed-legged ~, jointed-limbed ~ членистоногое *(Arthropoda)*
  stuffed ~ чучело
  wing-handed ~ рукокрылое *(Chiroptera)*
  wood-boring ~ древоточец
  young ~ молодое животное; детёныш
animal-borne *a* передающийся животными
animalcule *s* микроскопическое животное
animalization *s* анимализация
anise *s* анис обыкновенный *(Anisum vulgare)*
anisocarpous *a* анизокарпный
anisocytosis *s* анизоцитоз
anisodactylous *a* анизодактильный
anisodont *a* анизодонтный
anisogamete *s* анизогамета
anisogamous *a* анизогамный
anisogamy *s* анизогамия
anisogenomatic *a* анизогеномный
anisogenous *a* анизогенный
anisogeny *s* анизогения
anisolobous *a бот.* неравнолопастный
anisomerous *a* неизомерный, неравночленный; разнозернистый
anisometric *a* анизометрический
anisomery *s* анизомерия
anisophyllous *a* неравнолистный
anisophylly *s* анизофиллия, неравнолистность
anisopleural *a* неравнобокий
anisoploid *s* анизоплоид; анизоплоидный
anisoploidy *s* анизоплоидия
anisopolar *a bot* анизополярный, неравнополярный
anisopolyploidy *s* анизополиплоидия
anisopterous *a bot* неравнокрылый
anisoreflexia *s* анизорефлексия
anisorhythmia *s* аритмия
anisospore *s* анизоспора
anisospory *s* анизоспория, разноспоровость
anisotonic *a* анизотонический
anisotrisomy *s* анизотрисомия
anisotropic *a* анизотропный
anisotropy *s* анизотропия
anitrogenous *a* безазотистый
ankle *s* лодыжка; голеностопный сустав; таранная кость
ankylostomiasis *s helm* анкилостомоз
anlage *s* зачаток; закладка органа
annealing *s mbiol* отжиг, гибридизация
annectent *a* связующий, промежуточный
annidation *s* аннидация
annotinous *a* однолетний; годичный, годовой
annoy *v* беспокоить, надоедать, докучать, раздражать
annoyance *s* надоедание, приставание; раздражение

annual *s* однолетник; *a* однолетний; ежегодный; годовой; годичный
annuation *s* ежегодные экологические учёты
annular *a* кольцевой; кольцевидный
annulate *a* кольцевой; кольчатый
annulation *s* образование колец; вторичная сегментация
annulus *s* кольцо, колечко; *myc* вольва, покрывальце, влагалище
anoestrus *s* анэструс
anogenital *a* анально-половой
anogyny *s* женская стерильность
anomalous *a* аномальный, ненормальный
anomaly *s* аномалия
anorectal *a* аноректальный
anormogenesis *s* атипичное развитие
anorthogenesis *s* анортогенез, эволюция зигзагом
anorthoploid *s* анортоплоид, анэуплоид
anorthospiral *s* анортоспираль
anosmia *s* аносмия
anosulcate *a pal* безбороздный
anovulation *s* ановуляция
anoxaemia, anoxemia *s* аноксемия
anoxia *s* аноксия
anoxic *a* бескислородный
anoxybiont *s* анаэроб, анаэробный организм
anoxybiosis *s* анаэробиоз
anoxybiotic *a* анаэробный
ansa *s anat* петля
ansate *a* петлеобразный, петлевидный; ушковидный
ansated *a bot* с ушком
anserine *s* ансерин; *a* гусиный
ansiform *a* петлеобразный, петлевидный; ушковидный
ant *s* муравей *(Formicidae)*
white ~ термит *(Isoptera)*
antagonism *s* антагонизм; соперничество
antagonist *s* антагонист
metabolic ~ антиметаболит
antagonistic *a* антагонистический
antalkaline *a* нейтрализующий щёлочь
antambulacral *a see* **abactinal**
Antarctogaea *s* Антарктогея
antbird *s orn* муравьеловка, муравьянка *(Formicariidae)*
anteater *s* муравьед *(Myrmecophagidae);* ехидна *(Tachyglossidae)*
porcupine ~ ехидна *(Tachyglossus)*
scaly ~ ящер, панголин *(Manis)*
anteclimax *s* антеклимакс
antedisplacement *s* смещение вперёд
antehypophysis *s* аденогипофиз
antelocation *s* переднее положение, положение спереди, расположение спереди
antelope *f* антилопа
saiga ~ сайгак *(Saiga tatarica)*
antenatal *a* антенатальный, пренатальный, предродовой, дородовой, внутриутробный
antenna *s* антенна, усик, сяжок; *bot* клювик
antennal *a* антеннальный
antennatus *a* усиковидный
antennular *a* антеннулярный
antenodal *a anat* предузловой
antepenultimate *a* третий от конца
anteposition *s* смещение кпереди
anteriad *adv* кпереди
anterior *a* передний
antesporophyll *s* антеспорофилл
anthela *s* сложная метёлка
anthelix *s anat* противозавиток
anthelminthic *a* противоглистный, антигельминтный
anther *s bot* пыльник
antheridial *a* антеридиальный
antheridiophore *s* антеридиеносец
antheridium *s* антеридий
antheriferous *a* пыльниконосный
antherless *a* беспыльниковый
antherozoid, antherozooid *s* антерозоид
anthesis *s* цветение; пыление
anthill *s* муравейник
anthocarpous *a* антокарпный
anthoclinium *s* антоклиний, общее цветоложе
anthodium *s bot* антодий; общая обвёртка; корзинка
anthogenesis *s* антогенез
antholysis *s* антолиз, разрушение цветка
anthophilous *a* антофильный
anthophore *s* антофор, цветоножка
anthophorous *a* цветконосный
anthophyllous *a* цветколистный
anthophyte *s* антофит, сперматофит
anthotaxy *s* антотаксия
anthracnose *s phytop* антракноз, пятнистость растений *(caused by Colletotrichum)*

anthropic *a* обусловленный влиянием человека
anthropocenology *s* антропоценология
anthropocentrism *s* антропоцентризм
anthropochore *s* антропохор
anthropochorous *a* антропохорный
anthropogenesis *s* антропогенез
anthropogenic, anthropogenous *a* антропогенный
anthropoid *a* человекообразный
anthropology *s* антропология
anthropometric *a* антропометрический
anthropometry *s* антропометрия
anthropomorphism *s* антропоморфизм
anthropomorphous *a* антропоморфный
anthropophagy *s* каннибализм, людоедство
anthropophilous *a* антропофильный
anthropophyte *s* антропофит
anthroposphere *s* антропосфера
anthropotomy *s* анатомия человека
anthropozoonosis *s* антропозооноз
anthurus *s bot* султан
antiae *s pl* антии
antiallergenic, antiallergic *a* противоаллергенный, антиаллергенный
antiantibody *s* антиантитело
antiaris *s* анчар, упас *(Antiaris)*
antibacterial *a* антибактериальный
antibiogram *s* антибиограмма
antibiont *s* антибионт
antibiosis *s* антибиоз
antibiotic *s* антибиотик; *a* антибиотический
antiblastic *a* препятствующий росту
antibody *s* антитело
  anaphylactic ~ анафилактическое антитело; реагин
  antihemagglutination ~ гемагглютинрующее антитело, гемагглютинин
  anti-Rh ~ антирезусное антитело, резус-изогемагглютинин
  hemolytic ~ гемолитическое антитело, гемолизин
  leukoagglutinating ~ лейкоагглютинирующее антитело, лейкоагглютинин
  reaginic ~ реагиновое антитело, реагин
antibody-blocking *a* антителоблокирующий, экранирующий антителами
antibody-dependent *a* антителозависимый
anticatalyst *s* антикатализатор
anticholinergic *a* антихолинергический
anticipation *s* антиципация; преждевременность
anticoagulant *s* антикоагулянт
anticoagulative *a* антикоагулирующий
anticodon *s* антикодон
anticomplementary *a* антикомплементарный
anticontagious *a* противоинфекционный
antidigestive *a* тормозящий пищеварение
antidote *s* противоядие
antidromic *a* антидромный, противоположно направленный
antienzyme, antiferment *s* антифермент
antifoam *s* антивспениватель
antifungal *a* противогрибковый
antigen *s* антиген
  neuroblastoma-associated ~ нейробластомный антиген, антигенный маркёр нейобластомы
  nominal ~ псевдоантиген
  skin ~ антиген кожи, дермальный антиген, дермаген
  test ~ тест-антиген, контрольный антиген
antigenic *a* антигенный
antigenicity *s* антигенность, антигенные свойства
antigeny *s* антигения, половой диморфизм
antiglobulin *s* антиглобулин
antihemolytic *a* препятствующий гемолизу, антигемолитический
antihormone *s* антигормон
anti-idiotype *s* антиидиотип
antilobium *s anat* козелок
antimere *s* антимер
antimetabolite *s* антиметаболит
antimicrobial *a* антимикробный
antimitotic *a* антимитотический
antimorph *s* антиморф, антиморфный ген
antimutagene *s* антимутаген
antimycotic *s* фунгицид, противогрибковое вещество; *a* противогрибковый

anting *s* энтинг, «муравление»
antioxidant *s* антиокислитель, антиоксидант
antipaludian *a* противомалярийный
antiparasitic *s* противопаразитарное средство; *a* противопаразитарный
antipathic *a* антагонистический
antipathy *s* антагонизм
antiperistalsis *s* антиперистальтика
antipetalous *a* антипетальный, противолепестковый
antiplague *a* противочумный
antipodal *a* противоположный
antipoison *s* противоядие
antiport *s* антипорт
antiporter *s* антипортер
antiputrescent *a* противогнилостный
antipyretic *s* жаропонижающее средство; *a* жаропонижающий
antisense *a mbiol* антисмысловой
antisensitizer *s* десенсибилизирующее средство
antiseptic *s* антисептик, антисептическое средство; *a* антисептический
antiserum *s* антисыворотка, иммунная сыворотка
antispasmodic *a* антиспазматический
antitermination *s* антитерминация
antithenar *s* возвышение мизинца, гипотенар
antitoxic *a* антитоксический
antitoxin *s* антитоксин
antitragus *s anat* противокозелок
antitrope *s* антитело
antitropic *a* противоположно направленный
antivenin *s* антивенин
antivermicular *a* противоглистный, антигельминтный
antiviral *a* антивирусный, противовирусный
antivitamin *s* антивитамин
antler *s* олений рог; отросток оленьего рога
antler-rubbing *s* чистка рогов
antrorse *a* направленный вперёд
antrum *s anat* полость; пазуха; пещера
anucleate *a* безъядерный
anucleolate *a* безъядрышковый
anuran *s* бесхвостое земноводное *(Anura)*
anurous *a* бесхвостый
anus *s* анус, анальное отверстие, заднепроходное отверстие
anvil *s anat* наковальня
anxiety *s* беспокойство, тревога
aorta *s* аорта
aortic *a* аортальный
apalogamic *a* апалогамический
ape *s* обезьяна
  anthropoid ~, great ~ человекообразное, высший примат *(Anthropoidea);* человекообразная обезьяна *(Pongidae)*
  holy ~ лангур, обыкновенный тонкотел *(Presbytis)*
  red ~ орангутан *(Pongo pygmaeus)*
aperiodicity *s* непериодичность
aperturate *a* апертурный, поровый
aperture *s* отверстие; устье *(of a shell); bot* апертура, пора, проростковое устройство; окно *(of diatoms);* апертура микроскопа
  branchial ~ жаберная щель
  cloacal ~ клоакальное отверстие
apetalous *a* апетальный, безлепестковый
apex *s* апекс, верхушка, вершина
apheliotropism *s* отрицательный гелиотропизм
aphercotropism *s* отклонение от препятствия
aphid *s* тля *(Aphidina)*
aphleby *s* афлебия
aphoria *s* женское бесплодие
aphotic *a* афотический
aphototropism *s* отрицательный фототропизм
aphylactic *a* лишенный иммунитета
aphylaxis *s* афилаксия, иммунодефицит, иммунологическая недостаточность
aphyllous *a* безлистный
aphylly *s* афиллия, отсутствие листьев
apiary *s* пасека, пчельник
apical *a* апикальный, верхушечный
apiculate *a* короткоостроконечный
apiculiform *a* остроконечный
apiculture *s* пчеловодство
apiculus *s* маленькое остриё; *myc* носик оогония
apilary *a bot* апилярный, не имеющий волосков
apileate *a myc* не имеющий шляпки
apish *a* обезьяний
apitoxin *s* апитоксин, пчелиный яд

apivorous *a* питающийся пчёлами
aplacental *a* аплацентарный, не имеющий плаценты
aplanogametangium *s* апланогаметангий
aplanogamete *s* апланогамета, неподвижная гамета
aplanospore *s* апланоспора
aplasia *s* аплазия, недоразвитие, отсутствие развития
aplastic *a* апластический; бесструктурный, бесформенный
aploperistomatous *a* аплоперистоматный
aplostemonous *a* аплостемонный
apnoea *s* апноэ, остановка дыхания; удушье, асфиксия
apobasidium *s* апобазидия
apobiosis *s* смерть; некроз, омертвение
apocarp *s* апокарпий
apocarpous *a* апокарпный, свободнопестичный
apocarpy *s* апокарпия
apochlorosis *s* ахлороз, отсутствие хлорофилла
apochlorotic *a* бесхлорофилльный
apochromatic *a* апохроматический
apocrine *a* апокринный, апокриновый
apocyte *s* апоцит, многоядерная клетка, многоядерный плазмодий
apodal *a* безногий
apodema, apodeme *s* аподема
apodous *a* безногий
apoenzyme *s* апофермент
apogamic, apogamous *a* апогамный
apogamy *s* апогамия
apogeny *s* апогения, бесплодность, стерильность
apogeotropism *s* отрицательный геотропизм
apogyny *s* апогиния, женская стерильность
apolegamy *s* аполегамия
apomeiosis *s* апомейоз
apomict *s* апомикт
apomictic *a* апомиктический
apomixic *s* апомиксис
aponeurosis *s anat* апоневроз
  epicranial ~ сухожильный шлем
apopetalous *a bot* безлепестный
apophyllous *a* безлистный
apophysis *s zool* апофиз, отросток, вырост; *bot* апофиза
apophyte *s* апофит
apoplasmatic *a* апоплазматический
apoplastidous *a* утративший пластиды
apoptosis *s* апоптоз
apoptotic *a* апоптический
apo-repressor *s* апорепрессор
aporogamy *s* апорогамия
aposematic *a* апосематический, отпугивающий
aposeme *s* отпугивающая окраска; отпугивающий признак
apospermy *s* апоспермия, вегетативный апомиксис
aposporogony *s* апоспорогония
apospory *s* апоспория
apostasis *s* кризис болезни; абсцесс, гнойник; *bot* расхождение, раздвижка
apostaxis *s* избыточная экссудация
apostrophe *s* апострофия
apothecium *s* апотеций
apotropous *a* апотропный
apparatus *s* аппарат
  Abbe-Zeiss ~ камера Аббе-Цейсса
  basal metabolism ~ метаболиметр
  bronchial ~ бронхиальное дерево
  canalicular ~ *cyt* аппарат Гольджи, комплекс Гольджи
  ciliary ~ ресничное тело
  diffusion ~ диффузор, диффузионный аппарат
  digestive ~ пищеварительный аппарат, пищеварительная система
  dispensing ~ дозатор, дозирующее устройство
  egg ~ *bot* яйцеклетка с синергидами
  genitourinary ~ мочеполовой аппарат, мочеполовая система
  Golgi ~ *cyt* аппарат Гольджи, комплекс Гольджи
  hatching ~ инкубатор
  microtiter ~ микротитратор
  lymphatic ~ лимфатическая система
  osmotic pressure ~ осмометр
  radiation-measuring ~ дозиметр
  respiratory ~ дыхательная система, органы дыхания
  reticular ~ of Golgi *cyt* аппарат Гольджи, комплекс Гольджи
  shaking ~ прибор для встряхивания
  urogenital ~ мочеполовой аппарат, мочеполовая система
appearance *s* внешний вид; появление; проявление

external ~ экстерьер, внешний вид
gross ~ *anat* макроскопическая картина
appeasement *s eth* умиротворение
appendage *s anat* отросток; придаток; *ent* конечность; *bot* семенной придаток
atrial ~, auricular ~ ушко предсердия
cecal ~ червеобразный отросток, аппендикс; слепой вырост кишечника
appendiculate *a* имеющий отростки *или* придатки
appendix *s anat* отросток; придаток; червеобразный отросток, аппендикс
appetence *s eth* аппетенция; инстинктивное стремление; влечение
appetent *a eth* аппетентный, поисковый
appetite *s* аппетит; инстинктивная потребность; склонность
sexual ~ половое влечение
applanate *a* уплощённый
applanation *s* уплощение
apple *s* яблоня *(Malus);* глазное яблоко
Adam's ~ *anat* адамово яблоко, кадык
earth ~ груша земляная, топинамбур *(Helianthus tuberosus)*
oak ~ *phytop* галл дубовой орехотворки, чернильный орешек
paradise ~ парадизка *(Malus domestica var. paradisiaca)*
appliance *s* приспособление, прибор; применение
application *s* применение; употребление; аппликация, прикладывание, наложение; внесение *(e.g., of fertilizer)*
applicator *s* аппликатор
apply *v* применять; употреблять
apposition *s* аппозиция, присоединение
appraisal *s* оценка, экспертиза; бонитировка
appreciable *a* заметный, ощутимый; поддающийся оценке
apprehension *s* тревога; опасение
appressed *a* прижатый; сдавленный
appressorium *s* апрессорий
approach *s* подход; приближение; доступ
team ~ комплексный подход
approach-avoidance *s eth* приближение - избегание
approaching *s eth* приближение
appropriate *a* подходящий, соответствующий; адекватный
approximate *a* приближённый, приблизительный, почти соответствующий; близко расположенный, сближенный
approximation *s* аппроксимация, приближение
successive ~ *биом.* последовательная аппроксимация, последовательное приближение
apricot *s* абрикос *(Prunus armeniaca)*
apsilate *a* негладкий, скульптурированный
apterium *s orn* аптерий
apteroid *a* с зачаточными крыльями
apterous *a* бескрылый
apterygial *a* бескрылый; без плавников
aptitude *s* пригодность, соответствие; склонность, предрасположенность
apyrene *a emb* безъядерный; *bot* бессемянный
aquarium *s* аквариум
aquatic *a* водный; водяной
aquation *s* гидратирование
aquatosere *s* фитоценоз на влажной почве
aqueduct *s anat* канал, проток; водопровод
aqueous *a* водный; водянистый
aquiculture *s* аквакультура
aquiferous *a* водоносный
aquiline *a* орлиный
aquolysis *s* гидролиз
aquosity *s* водянистость, водянистая консистенция
arable *a* пахотный
arachnoid *s* паутинная оболочка; *a* паутинный; напоминающий паутину; паукообразный
arbitrary *a* произвольный, условный
arbor *s* дерево; древовидная структура
arboreal *a* древесный; обитающий на деревьях
arboreous *a* древесный; древовидный
arborescent *a* древовидный
arboretum *s* дендрарий
arboricolous *a* обитающий на деревьях
arborization *s* древовидное ветвление
arborvitae *s* туя *(Thuja)*
arbovirus *s* арбовирус
arbuscle *s* древовидный кустарник; карликовое дерево; ветвящаяся гаустория

arbuscular *a* древовидный
arc *s* дуга; изгиб
arch *s* дуга, свод
  intestinal ~ *helm* кишечная арка
  nasal ~ спинка носа
  neural ~ нейральная дуга
  palatine ~ нёбная дужка
archaebacteria *s pl* архебактерии
archaeocyte *s* археоцит
archaeostomatous *a* первичноротый
archallaxis *s* архаллаксис
archebiosis *s* архебиоз, возникновение жизни
arched *a* куполообразный, сводчатый, выгнутый
archegone *s* архегоний
archegonial *a* архегониальный
archegoniate *s* архегониальное растение
archegoniophore *s* архегонионосец
archegonium *s* архегоний
archencephalon *s emb* передний мозговой пузырь
archenteron *s* архентерон, первичная кишка
arheocyte *s* археоцит
archeophyte *s* археофит
archespore *s* археспора
archesporium *s* археспорий
archetype *s* архетип
archicarp *s* архикарп
archichlamydeous *a* архихламидный
archicoel *s* первичная полость тела
archigenesis *s* абиогенез
archipallium *s anat* старый плащ
archinephron *s emb* мезонефрос, первичная почка, Вольфово тело
architomy *s* архитомия, архетомия
arciform *a* аркообразный, дугообразный
arcocentrum *s* аркоцентрум
arctalpine *a* арктоальпийский
arctic *a* арктический
arctic-alpine *a* арктоальпийский
Arctogaea *s* Арктогея
arcuation *s* изгиб, искривление
arculus *s ent* аркулус, дужка
arcuate *a* дугообразный, дуговидный
ardella *s* арделла
area *s* область; район; зона; ареал, область обитания; площадь, пространство, участок
  action ~ *eth* используемый участок
  adjacent ~s *pl* соприкасающиеся ареалы
  axial ~ осевое поле *(of diatoms)*
  barren ~ территория, лишённая растительного покрова; эродированная территория
  breeding ~ территория размножения; территория гнездования, гнездовая часть ареала
  catchment ~ водосбор, область водосбора
  central ~ центральное поле *(of diatoms)*
  core ~ *eth* наиболее используемая часть участка обитания; экологический центр участка обитания
  cortical projection ~ зона корковой проекции, корковая проекционная зона
  cortical speech ~ речевая область коры
  dominance ~ территория доминирования
  drainage ~ водосборный бассейн
  hibernating ~ территория зимовки; зимовочная часть ареала; зимовка
  lateral ~ боковое поле *(of diatoms)*
  leaf ~ листовая площадь, листовая поверхность
  minimal ~ минимум-ареал
  natural ~ нетронутая местность
  personal ~ *eth* персональное пространство
  recipient ~ колонизируемая территория
  regeneration ~ зона регенерации; временно заповедный участок
  reserved ~ заповедник
  sample ~ пробная площадка
  subcallosal ~ подмозолистое поле, обонятельное поле Брока
  water ~ акватория
  wilderness ~ нетронутая местность
  wintering ~ место зимовки
areactivity *s* ареактивность
areal *s* ареал
areflexia *s* арефлексия, отсутствие рефлексов
areg *s* песчаная пустыня
arenaceous *a* песчаный
arenicolous *a* псаммофильный, обитающий в песке
areography *s* исследование территории

areola *s anat* ареола; *bot* ареола, площадка
cellular ~ клеточный дворик
mammary ~ околососковый кружок
areolar *a* ареолярный
areolate *a* ареолированный, разделённый на ареолы
areolation *s* ареола; разделение на ареолы
areole *s anat* ареола; *bot* ареола, площадка
areolet *s ent* зеркальце
areometer *s* ареометр
argali *s* архар, аргали, горный баран *(Ovis ammon)*
argentaffin *a* аргирофильный, аргентофильный
argentophilic *a* аргентофильный, аргирофильный
argus *s orn* фазан аргус *(Argussianus)*
argyrophil(ous) *a* аргирофильный, аргентофильный
arid *a* аридный, сухой, засушливый; безводный; бесплодный
aridization *s* аридизация
aridity *s* аридность, засушливость
aril *s bot* ариллус, кровелька, присемянник
arillode *s bot* ложный ариллус
arista *s bot* ость
aristogenics *s* евгеника
arm *s* рука; передняя конечность, передняя лапа; ветвь; отросток; вырост; луч, рука *(of echinoderms);* щупальце, рука *(of cephalopods);* плечо *(of a chromosome)*
lower ~ предплечье
upper ~ плечо
armadillo *s* броненосец *(Dasypus)*
armature *s* защитная структура; панцирь
armilla *s bot* бахромка, оторочка
armor *s* панцирь
armorless *a* беспанцирный
armpit *s* подмышечная ямка
armyworms *s pl* «походные черви», «ратные черви»
arolium *s ent* аролий, пульвилла, подушечка лапки
aroma *s* аромат
aromatic *a* ароматический; соединение ароматического ряда
aromorph *s* ароморфа
aromorphosis *s* ароморфоз
arousal *s* возбуждение; раздражение; активация; пробуждение
arouse *v* активизировать; побуждать; возбуждать
arrangement *s* расположение
informal ~ пейзажная планировка, ландшафтная планировка
arrect *a* прямостоящий, прямой
arrest *s* задержка, остановка; прекращение
arrhenogenic *a* арреногенический
arrhenotoky *s* арренотокия
arrhythmia *s* аритмия
arrhythmic *a* аритмический
artefact *s* артефакт
arterial *a* артериальный
arterialization *s* артериализация
arteriogram *s* артериограмма, рентгенограмма артерии
arteriograph *s* артериограф
arteriography *s* артериография
arteriole *s* артериола
arteriotomy *s* вскрытие артерии
artery *s* артерия
arthral *a* суставной
arthroconidium *s* артроконидия
arthrodia *s* шаровидный сустав, артродия
arthrodial *a* сочленовный
arthropod-borne *a* передающийся членистоногими, артропонозный
arthropod *s* членистоногое *(Arthropoda)*
arthrospore *s* артроспора
arthrosterigma *s* артростеригма, перегородчатая стеригма
arthrous *a* суставной; сочленовный
artichoke *s* артишок *(Cynara scolymus)*
articulate, articulated *a* членистый; сочленённый, суставчатый
articulation *s* сочленение, сустав
synovial ~ синовиальное соединение, синовиальный сустав
artifact *s* артефакт
artificial *a* искусственный
artiodactyl *a* парнопалый; парнокопытный
artiodactyle *s* парнокопытное, парнопалое *(Artiodactyla)*
arundinaceous *a* тростниковый
arylalkylation *s* арилалкилирование
arylation *s* арилирование

aryepiglottic, **arytenoepiglottic** *a* черпаловидно-надгортанный
arytenoid *s* черпаловидный хрящ; *a* черпаловидный
asaccharolytic *a* несахаролитический
ascariasis *s* аскаридоз
ascaricid *a* аскарицидный
ascarid *s* аскарида *(Ascaris)*
ascending *a* восходящий
ascidiform *a bot* кувшинчатый
ascigerous *a bot* сумчатый
ascite *s* асцит
ascocarpus *s* аскокарпий
ascogenous *a* аскогенный
ascogonial *a* аскогониальный; аскогенный
ascogonium *s* аскогоний
ascoma *s* аскома, дисковидный аскокарпий
ascophore *s* аскофор
ascospore *s* аскоспора, сумкоспора
ascostroma *s* аскострома
ascus *s bot* аск, сумка
ascyphous *a* не имеющий сциф
aseity *s* самозарождение
asepsis *s* обеззараживание
aseptate *a zool* без перегородок; асептатный
aseptic *a* асептический
asexual *a* бесполый; вегетативный
ash *s* ясень *(Fraxinus);* зола
ash-free *a* беззольный
asiphonate *a zool* не имеющий сифона
A-site *s* акцепторный участок
asp *s* гадюка *(Vipera; Cerastes);* кобра *(Naja)*
asparagus *s* спаржа *(Asparagus)*
aspect *s* внешний вид; аспект; физиономичность
aspecting *a* аспективный
aspection *s* аспектирование
aspen *s* осина *(Populus tremula)*
aspergillosis *s* аспергиллёз
asperity *s* шершавость, шероховатость
asperous *a* шершавый, шероховатый
asphyxia *s* асфиксия
asphyxiation *s* асфиксия; выпревание
aspidate *a pal* со щитками
aspirate *v* аспирировать, отсасывать
aspiration *s* аспирация, отсасывание, вдыхание
asporogenic, asporogenous *a* аспорогенный
asporous *a* не имеющий спор, неспорообразующий
ass *s* осёл *(Equus asinus)*
assart *s* участок леса, расчищенный под пашню; расчистка леса под пашню
assay *s* испытание, проверка; количественный анализ; проба; образец для анализа
biological ~ биологическая проба, биопроба
blank ~ слепой опыт
EA-rosette ~ анализ EA-розеткообразования, EA-POK анализ
enzyme-linked immunosorbent ~ твёрдофазный иммуноферментный анализ, ЭЛАЙЗА, ELISA, ИФА
enzyme-linked immunospot ~ иммуноферментный спот-анализ
hemolytic complement ~ титрование гемолитического комплемента
hemolytic plaque ~ проба на гемолитические бляшки
idiotype ~ определение идиотипа, идиотипирование
immune slot-blot ~ иммуноанализ методом слот-блоттинга
immunoradiometric ~ иммунорадиометрический анализ, количественный радиоиммуноанализ
Jerne plaque ~ метод бляшек по Ерне, реакция локального гемолиза в геле
latex-based ~ латексный анализ
LGH, localized-in-gel hemolysis ~ метод бляшек по Ерне, реакция локального гемолиза в геле
long-term lymphocyte cytotoxicity ~ долговременный лимфоцитотоксический тест
mass-fragmentographic ~ масс-фрагментографический анализ
microculture infectivity ~ микроанализ инфекционности культуры
microcytotoxicity ~ микроцитотоксический тест
microlymphocytotoxicity ~ лимфоцитотоксический микротест, микролимфоцитотоксический тест
mixed colony ~ анализ смешанного колониеобразования, КОЕ-микст анализ, метод смешанных колоний

serodiagnostic ~ серодиагностика; серодиагностический анализ
serologic ~ серологический анализ
short-term lymphocyte cytotoxicity ~ короткий лимфоцитотоксический тест
assemblage *s* сообщество; скопление; группа; сборка в структуру
~ **of rootlets** корневая система
assembly *mbiol* упорядоченная структура
assessment *s* оценка; бонитировка
assidenous *a bot* сидячий, сидящий
assident *a* сопровождающий, сопутствующий
assimilate *s* ассимилят, продукт фотосинтеза; *v* ассимилировать
assimilation *s* ассимиляция, усвоение
stomatal ~ устьичная ассимиляция, устьичное поглощение
assimilative, assimilatory *a* ассимилирующий, усваивающий
assize *s* таксация
associate *s* партнёр; член сообщества; *v* ассоциировать, соединять, связывать; *a bot* ассоциативный, сопутствующий
association *s* ассоциация, сообщество; *bot* сопряжённость; *zool* сизигий
climax ~ климаксовая ассоциация, климакс; коренная ассоциация, узловая ассоциация
elementary ~ элементарная ассоциация, фация
fundamental ~ корневая ассоциация, узловая ассоциация
plant ~ растительная ассоциация, растительное сообщество
sperm-egg ~ соединение спермия с яйцом
subordinate ~ ассоция, серийная ассоциация
twin ~s ассоциации-близнецы
associes *s* ассоция, серийная ассоциация
associon *s* ассоцион
associule *s* ассоциула
assort сортировать; группировать; классифицировать
assortment *s* сортировка; выбор
variable-diversity-joining ~ реаранжировка V-D-J сегментов, V-D-J реаранжировка
assuage успокаивать; смягчать *(pain)*; утолять *(hunger)*
astatic *a* астатический
astelic *a* не имеющий стели
astely *s* астелия
aster *s* астра *(Aster)*; *cyt* звезда
asterigmate *a* астеригмный
asteroid *a* звездообразный
astogeny *s* астогения
astomatous, astomous *a* безротый
astragalocalcanean *a* таранно-пяточный
astragalus *s* таранная кость
astringent *s* вяжущее средство; *a* вяжущий
astroblast *s* астробласт
astrobiology *s* астробиология
astrobotany *s* астроботаника
astrocenter *s cyt* астросфера, лучистая сфера
astrocyte *s* астроцит, макроглиоцит, астроглиальная клетка
protoplasmic ~ плазматический астроглиоцит, плазматический астроцит, плазматическая астроглиальная клетка
astrosphere *s cyt* астросфера, лучистая сфера
asulcate *a pal* асулькатный, безбороздный
asymmetrical *a* асимметричный, несимметричный
asymmetry *s* асимметрия
asymphytous *a* асимфитный, несросшийся
asymptomatic *a* бессимптомный
asynapsis *s* асинапсис, асиндез
asynchronism асинхронизм
asynchronous *a* асинхронный
asyngamic *a* асингамный
asystole *s* асистолия
ataxia *s* атаксия
atavism *s* атавизм
atavistic *a* атавистический
atelia *s* ателия; неполное развитие
ateliosis *s* ателиоз
atelomitic *a* ателомитический
athalamous *a* аталамический, не имеющий таламуса
athrocytosis *s* атроцитоз
atlantooccipital *a* атлантозатылочный
atlas *s* атлант, первый шейный позвонок
atmosphere *s* атмосфера
atocia *s* бесплодность самок
atocous *a* не дающий потомства, бесплодный; атокный

atomization *s* распыливание, распыление, мелкодисперсное разбрызгивание
atomizer *s* пульверизатор, распылитель; опрыскиватель
atonia *s* атония, потеря тонуса, слабость, вялость
atopy *s* атопия, атопическое заболевание
atoxic *a* неядовитый
ATP adenosine triphosphate *s* АТФ аденозинтрифосфат
atreme *a палин.* безапертурный
atresia *s* атрезия
atretic *a* атретический
atrial *a* атриальный; предсердный
atrichous *a* не имеющий жгутиков; безволосый
atriopore *s* атриопор
atrioventricular *a* атриовентрикулярный, предсердно-желудочковый
atrium *s* полость; пазуха; синус; предсердие; атриум, преддверие; входные ворота инфекции
  infection ~ входные ворота инфекции
atrophy *s* атрофия
atropic *a* смещённый, перемещённый
atropous атропный, прямой
attach *v* прикреплять; прикрепляться
attached *a* прикреплённый, сидячий
attachment *s* прикрепление; привязанность
attack *s* нападение; атака; приступ; припадок; *v* нападать; атаковать; разрушать, разъедать
attain *v* достигать
attempt *s* попытка; *v* пытаться
attend *v* заботиться, ухаживать
attendance *s* наблюдение, уход
attention *s* внимание
attenuation *s* ослабление, истощение; уменьшение; разжижение, разбавление
attenuator *s* аттенюатор, разбавитель
attenuatous *bot* утончённый
attitude *s* поза, осанка; позиция, положение
attractant *s* аттрактант
attractivity *s* привлекательность
attribute *s* свойство, характерная черта
attrition *s* истощение, изнурение
atypical *a* атипичный, нетипичный
aubergine *s* баклажан *(Solanum melongena)*
audibility *s* слышимость
audible *a* слышимый
audition *s* слух
auditory *a* слуховой
augmentation *s* увеличение, прирост, приращение
  ~ of antibody response усиление антителогенеза, стимуляция антителогенеза
augmentor *s* стимулятор
auk *s* птица семейства чистиковых; *pl* чистиковые *(Alcidae)*
  little ~ люрик, малая гагарка *(Alle alle)*
  parakeet ~ белобрюшка *(Cyclorrhynchus psittacula)*
  rhinoceros ~ тупик-носорог *(Cerorhinca monocerata)*
auklet *s* конюга *(Aethia)*
  ancient ~ обыкновенный старик *(Synthliboramphus antiquus)*
  Cassin's ~ алеутский пыжик *(Ptychoramphus aleuticus)*
  Japanese ~ хохлатый старик *(Synthliboramphus wumizusume)*
  parakeet ~, paraquet ~белобрюшка *(Cyclorrhynchus psittacula)*
  rhinoceros ~ тупик-носорог *(Cerorhinca monocerata)*
aulacocarpous *a* бороздчатоплодный
aulix *s* подбугорная борозда, гипоталамическая борозда
aulophyte *s bot* аулофит
aulostomatous *a* с трубчатым ртом
aunt *s eth* тётка
aura *s orn* гриф-индейка *(Cathartes aura)*
aural *a* ушной; слуховой
auricle *s* ушная раковина; ушко предсердия
auricula *s* ушко предсердия
auricular *a* ушной; предсердный
auricularia *s* аурикулярия
auriculate *a* ухообразный; ушковидный; ушковатый
auriform *a* ухообразный; ушковидный
auroch *s* тур *(Bos primigenius);* зубр *(Bison bonasus)*
austral *a* южный
autallogamia *s* аллаутогамия
autarcesis *s* врождённый иммунитет
autecology *s* аутэкология

autoaggressiveness *s* аутоагрессивность
autoallergy *s* аутоаллергия, эндоаллергия
autoanalyzer *s* автоматический анализатор
autoantagonism *s* аутоантагонизм
autoantibody *s* аутоантитело
autoantigen *s* аутоантиген, аутологичный антиген
autoantigenic *a* аутоантигенный
autoantigenicity *s* аутоантигенные свойства, аутоантигенность
autobasidium *s* аутобазидия
autobivalent *s* аутобивалент
autocarpy *s* аутокарпия
autocatalysis *s* автокатализ
autochore *s* автохор, автохорное растение
autochthon *s* автохтонный вид; автохтонный организм
autochthonous *a* автохтонный, местный, коренной
autocinesis *s* автокинез, произвольное движение
autoclave *s* автоклав; автоклавировать
autoclaving *s* автоклавирование
autocytolysis *s* аутоцитолиз, самопереваривание клеток, саморазрушение клеток
autocytotoxin *s* эндогенный цитотоксин
autodecomposition *s* сaмораспад
autodefense *s* резистентность
autodeme *s* аутодем
autodermalia *s pl zool* кожные лучи
autodigestion *s* самопереваривание
autoecious *a* однохозяйный
autoecism *s* однохозяинность
autoepitope *s* аутоантигенная детерминанта, аутоэтопитоп
autofluorescence *s* автофлуоресценция
autogamic, autogamous *a* автогамный, аутогамный
autogamy *s* автогамия, аутогамия, самооплодотворение
autogenesis *s* автогенез, аутогенез
autogenetic *a* автогенетический
autogenic *a* аутогенный, эндогенный
autogenomatic *a* аутогеномный
autogenic, autogenous *a* аутогенный, эндогенный
autogeny *s* автогенез, аутогенез
autograft *s* аутотрансплантат
autografting *s see* **autotransplantation**
autogrooming *s eth* автогруминг, чистка собственного тела
autohemagglutination *s* аутогемагглютинация
autohemagglutinin *s* аутогемагглютинин, аутогемагглютинирующее аутоантитело
autohemolysin *s* аутогемолизин, гемолитическое аутоантитело
autoheteroploid *s* аутогетероплоид
autoimmune *a* аутоиммуный
autoimmunity *s* аутоиммунитет
autoimmunization *s* иммунизация аутоантигеном; аутоаллергизация
autoimmunize иммунизировать аутоантигеном
autoimmunogenicity *s* аутоиммуногенность
autoinfection *s* самозаражение, аутоинфекция
autinvasion *s* аутоинвазия
autokinesia, autokinesis *s* автокинез, произвольное движение
autologous *a* аутологический, собственный
autolopping *s* самоочищение от сучьев
autolysate *s* автолизат
autolysis *s* автолиз
autolytic *a* автолитический
automaticity *s* автоматия
automatism *s* автоматизм
automesodermalization *s* самомезодермализация
automixis *s* аутомиксис, самооплодотворение
automutagen *s* аутомутаген
autoneuralization *s* самонейрализация
autonomic *a* автономный; *phys* спонтанный; вегетативный
autonomous *a* автономный
autoorientation *s* аутоориентация
autoorthotropous *a* пряморастущий
autooxidation *s* самоокисление
autoparasite *s* гиперпаразит, паразит второго порядка; автопаразит
autoparasitism *s* автопаразитизм; гиперпаразитизм, паразитизм второго порядка
autoparthenogenesis *s* автопартеногенез
autophagosome *s* аутофагосома, аутофагическая вакуоль
autophagy *s* аутофагия

autophene *s* аутофен
autophilous *a* самоопыляющийся
autophyte *s* автотрофное растение
autophytic *a* автотрофный
autoplast *s* аутотрансплантат, аутогенный трансплантат
autopodium *s* кисть; стопа
autopolyploid *s* аутополиплоид
autopolyploidy *s* аутополиплоидия
autopreening *s orn* чистка собственного оперения
autopsy *s* аутопсия, вскрытие
autoradiogram, autoradiograph *s* авторадиограмма, радиоавтограф
autoradiography *s* авторадиография, радиоавтография
autorecognition *s* распознавание «своего», аутогенное распознавание, самораспознавание
autoreduplication *s* авторедупликация, самоудвоение
autoresponse *s* аутоиммунный ответ, аутоиммунная реакция
autosampler *s lab* автоматический пробоотборник; автоматическая пипетка
autoscoliotropous *a* изгибающийся при росте
autosensitization *s* аутосенсибилизация
autoshaping *s eth* самоформирование
autoskeleton *s* внутренний скелет
autosomal *a* аутосомный
autosome *s* аутосома, неполовая хромосома
autospore *s* автоспора
autostimulation *s* самостимуляция; самораздражение
autosynapsis, autosyndesis *s* аутосинапс, аутосиндез
autothinning *s* самоизреживание
autotilly *s* аутотиллия
autotolerance *s* аутотолерантность, толерантность к «своему»
autotomy *s* аутотомия, самокалечение
autotransfusion *s* аутотрансфузия, аутогемотерапия
autotransplant *s* аутотрансплантат, аутогенный трансплантат
autotransplantation *s* аутотрансплантация; *bot* самопрививка
autotroph *s* автотроф. автотрофный организм
autotrohic *a* автотрофный
autotrophy *s* автотрофность
autotropism *s* автотропизм
autovaccine *s* аутовакцина
autoxenous *a* однохозяйный
autoxidation *s* самоокисление
auxanography *s* ауксанография
auxanometer *s* ауксанометр
auxillary *a* вспомогательный, добавочный
auxoautotroph *s* ауксоавтотроф
auxochrome *s* ауксохром
auxocyte *s* ауксоцит
auxoheterotroph *s* ауксогетеротроф
auxospireme *s* ауксоспирема
auxospore *s* ауксоспора
auxotroph *s* ауксотроф
auxotrophic *a* ауксотрофный
availability *s* доступность, наличие; пригодность
available *a* доступный; пригодный, полезный; действительный
avalvular *a* не имеющий клапанов
avascular *a anat* бессосудистый
avenous *bot* не имеющий жилок
avens *s bot* гравилат *(Geum)*
average *s biom* средняя величина, среднее арифметическое; *a* средний
weighted ~ средневзвешенное значение
averaging *s* усреднение
aversion *s* аверсия, отвращение
food ~ пищевая аверсия, отвращение к пище
avian *a* птичий
aviary *s* вольер, вольера, помещение для птиц, птичий двор, птичник
avicennietum *s* сообщество мангровых деревьев
avicularium *s* авикулярия
aviculture *s* птицеводство
avidity *s imm* авидность
avifauna *s* орнитофауна
avirulence *s* авирулентность
avirulent *a* авирулентный
avitaminosis *s* авитаминоз
avocado *s* авокадо *(Persea gratissima)*
avocet *s orn* шилоклювка *(Recurvirostra)*
avoid *v* избегать
avoidance *s* избегание
awake *a* бодрствующий

awareness-inducing *a* психостимулирующий
awl-leaved *a* шилолистный
awl-shaped *a* шиловидный
awn *s* ость
awned *a* остистый
awnless *a* безостый
awnlet *s* зачаток ости
axenic *a* аксенический, безбактериальный
axenization *s* аксенизация
axial *a* осевой
axifugal *a* центробежный
axil *s bot* пазуха
axile *a bot* пазушный; осевой
axilla *s* подмышечная ямка
axillary *a bot* пазушный; подмышечный, подкрыльцовый
axiom *s* аксиома
  Burnet's ~ принцип Барнета
axipetal *a* центростремительный, афферентный
axis *s* ось; эпистрофей, осевой позвонок; *bot* ось, осевой орган
  brain ~ ствол мозга
  cerebrospinal ~ головной и спинной мозг
axo-axonal *a* аксо-аксонный
axo-dendritic *a* аксо-дендритный
axogamy *s* аксогамия
axon *s* аксон, нейрит
axoneme *s* аксонема
axoplasm *s* аксоплазма
axopodium *s* аксоподия
axo-somatic *a* аксо-соматический
axostyle *s* аксостиль
aye-aye *s ther* ай-ай, мадагаскарская руконожка *(Daubentonia madagascariensis)*
azocoupling *s* азосочетание
azoic *a* безжизненный
azotization *s* азотизация
azotobacter *s* азотобактер
azure *s* азур *(stain); ent* голубянка
azurophilia *s* азурофилия
azurophilic *a* азурофильный
azygoid *a* незиготический
azygomorphous *a* азигоморфный, асимметричный
azygospore *s* партеноспора
azygous *a* непарный

B

baa *v* блеять
babbling *s eth* лепет
babesiosis *s vet* бабезиоз
baboon *s ther* павиан *(Papio)*
  hamadryas ~ гамадрил, плащеносный павиан *(Papio hamadryas)*
  sacred ~ гамадрил, плащеносный павиан *(Papio hamadryas)*
  yellow ~ бабуин, жёлтый павиан *(Papio cynocephalus)*
baby *s* детёныш; ребенок
bacca *s* ягода
baccate *a* ягодный
bacciferous *a* ягодоносный
bacciform *a* ягодовидный, ягодообразный
bachelor *s* холостяк
bacillary *a* бациллярный, палочковидный
bacillicarrier *s* бациллоноситель
bacillicidal *a* бактерицидный
bacilliform *a* бациллярный, палочковидный
bacillosis *s* бактериальная инфекция
bacillus *s* бацилла, палочка
back *s* спина, спинка, тыльная часть; *a* задний
  ~ **of neck** загривок
  ~ **of the tongue** спинка языка
backbone *s* позвоночник, позвоночный столб; остов, каркас, скелет *(of a macromolecule)*
  polypeptide ~ полипептидный остов
  primitive ~ *emb* спинная струна
backboned *a* позвоночный
backboneless *a* беспозвоночный
backcrossing *s* обратное скрещивание
backflow *s* обратный ток, противоток
back-front *a* спинно-брюшной
background *s* фон
  genetic ~ генетическая среда
backpollination *s* обратное опыление
backwardness *s* отсталость
backwoods *s* лесная глушь
bacterial *a* бактериальный
bactericidal *a* бактерицидный
bactericide *s* бактерицид, бактерицидное вещество
bacteriogenous *a* бактериогенный
bacteriological *a* бактериологический
bacteriology *s* бактериология
bacteriolysis *s* бактериолиз
bacteriophage *s* бактериофаг
bacteriophagy *s* бактериофагия
bacteriorhiza *s* бактериориза
bacterioscopy *s* бактериоскопия
bacteriosis *s* бактериоз, бактериальное заболевание
bacteriosolvent *s* бактериолизин, бактериолитическое вещество
bacteriostasis *s* бактериостаз
bacteriostat *s* бактериостат
bacteriostatic *a* бактериостатический
bacteriotropin *s* опсонин
bacterium *s* бактерия
  hay ~ сенная палочка
  rod ~ *micr* палочка
  spiral ~ спирилла
  test ~ тест-бактерия
bacteroid *s* бактероид; *a* бактероидный
baculate, baculiform *a bot* палочковидный; *pal* столбчатый
baculovirus *s* бакуловирус
baculum *s* желобовая кость, приапова кость; *pal* столбик, столбчатый вырост
badge *s:* **wing** ~ *orn* эполет
badger *s* барсук *(Meles meles);* американский барсук *(Taxidea taxus);* обыкновенный длинноносый бандикут, сумчатая носуха *(Perameles nasuta);* прострел раскрытый, сон-трава *(Pulsatilla patens)*
bag *s zool* сумка, мешок; полость; вымя
  ~ **of waters** плодный мешок, плодный пузырь
  farding ~ рубец
  game ~ ягдташ
  honey ~ медовой зобик
  honeycomb ~ сетка
  rennet ~ сычуг
bagworm *s ent* мешочница *(Psychidae)*
bait *s* приманка; насадка, наживка; *v* наживлять; прикармливать ловить на приманку; травля; травить *(by dogs)*
baited *a* затравленный *(by dogs)*
balance *s* равновесие, баланс; весы; *v* уравновешивать; взвешивать
  immunoregulatory ~ иммунорегуляторное соотношение

balancer *s* балансёр *(of amphibians); ent* жужжальце
balanocarpous *a* желудеплодный
balanoid *a* желудеобразный
balantidiasis, balantidiosis *s* балантидиоз
bald *a* оголённый, обнажённый, лишённый покрова, лысый; безостый
baldness *s* безволосость, безостость
baleful *a* гибельный, губительный; вредный, злобный, злой
balk *s* огрех
balky *a* упрямый, норовистый *(of a horse)*
ball *s* шар; ком, комок; семенная коробочка; шаровидный плод
 ~ **of knee** надколенник, коленная чашка, надколенная чашка
 button ~ платан западный *(Platanus occidentalis)*
 cotton ~ коробочка хлопчатника
 deer ~ олений трюфель *(Elaphomyces granulatus)*
 fatty ~ **of Bichat** *anat* жировое тело щеки, жировое тело Биша
 smut ~**s** споры головни
 touch ~ подушечка пальца, мякоть пальца
ballistocardiography *s phys* баллистокардиография
ballistospore *s* баллистоспора
ballooning *s* полёт *(of spiders)*
ballooning-out *s* выпячивание
balsam *s* бальзам; *bot* недотрога, бальзамин *(Impatiens)*
 Canada ~ канадский бальзам
balsamiferous *a* бальзамоносный; смолоносный
balsawood *s bot* бальза заячья *(Ochroma lagopus)*
bamboo *s* бамбук *(Bambuseae)*
bamboo-partridge *s* бамбуковая куропатка *(Bambusicola)*
ban *s* запрет, запрещение; *v* налагать запрет, запрещать; **to be under a** ~ находиться под запретом, быть запрещённым; **to lift the** ~ снимать запрет
banana *s* банан *(Musa)*
 cooking ~**, plantain** ~ пизанг *(Musa paradisiaca)*
band *s* полоса, зона; тяж, связка; диск *(in a chromosome);* кольцо *(in bird banding);* стая, стадо, гурт, отара; *v* кольцевать, окольцовывать
 adhesive ~ клеевой пояс *(on a tree)*
 contraction ~ полоса сокращения
 error ~ область ошибок
 intercalary ~ вставочный ободок, вставочное кольцо *(of a diatom)*
 moderator ~ мышечная перекладина
 muscular ~ мышечный тяж
 tendinous ~ сухожильный тяж
 bandage *s* повязка
banded *a* ленточный
bandicoot *s* бандикут; *pl* сумчатый барсук, бандикут *(Peramelidae)*
banding *s* кольцевание *(e.g., of birds);* бэндинг, исчерченность *(of a chromosome)*; *phytop* фасциация
bane *s* яд; паслён *(Solanum)*
baneberry *s* воронец колосистый *(Actaea spicata)*
baneful *a* гибельный, губительный; вредный
bank *s* банк; берег; банка, отмель; насыпь
 leveed ~ обвалованный берег
banner *s bot* парус, флаг
banteng *s ther* бантенг *(Bos javanicus)*
banyan *s* баньян, смоковница бенгальская *(Ficus bengalensis)*
baobab *s* баобаб *(Adansonia digitata)*
barb *s* колючка, шип; ус; ость; бородка *(of a feather)*
barbastelle *s ther* широкоушка, широкоух *(Barbastella);* европейская обыкновенная широкоушка, курносый ушан *(Barbastella barbastellus)*
barbate *a bot* остистый; *zool* бородатый
barbed *a* колючий
barbellate *a bot* мелкобородчатый
barberry *s* барбарис *(Berberis)*
barbet *s orn* бородатка, бородастик *(Megalaima)*
barbicel *s* бородочка *(пера)*
barbital *s* диэтилбарбитуровая кислота, веронал
barbtail *s orn* шетинкохвостый горшечник *(Margarornis)*
barbule *s bot* бородка; бородочка *(of a feather);* усик *(of a fish)*
bare *a* голый, обнажённый; лишённый растительности
bareback *a* неосёдланный

bare-grained *a* голозёрный
bare-ground *s ecol* плешина
bare-skinned *a* редковолосый; безволосый
bare-stalked *a* голостебельчатый
bark *s* лай, рявканье; лаять, рявкать; вторичная кора; дубильная кора; хина, хинная корка; *v* образовывать кору, наращивать кору; покрываться корой; сдирать кору
birch ~ береста
bark-blister *s* солнечный ожог коры
barking *s* лай; окорка, сдирание коры; *a* лающий
barley *s* ячмень *(Hordeum)*
barm *s* маточные дрожжи; закваска
barmy *a* дрожжёвой
barn *s* амбар; гумно; сарай; хлев
barnyard-grass *s* ежовник обыкновенный, куриное просо *(Echinochloa crus-galli)*
barochore *s* барохор
barognosis *s phys* барогнозия
baroreceptor *s* барорецептор
barotaxis *s* баротаксис
barotropism *s* баротропизм
barpressing *s eth* нажатие на рычаг
barrage *s* плотина, запруда; *myc* барраж
barred *a* поперечно-полосатый
barrel *s* полая часть *(of a feather);* туловище *(e.g. of a horse); anat* барабанная полость
honey ~ *ent* «медовая бочка»
barrel-shaped *a bot* бочковидный
barren *s* пустырь, пустошь; песчаная равнина, покрытая кустарником; *a* бесплодный; неплодородный, тощий *(soil);* бессемянный; не приносящий плодов, бесплодный
barrenness *s* бесплодие; бесплодность; неплодородность
barrenwort *s bot* горянка *(Epimedium)*
barrier *s* барьер, преграда, препятствие
barring *s orn* полосатость
basad *adv* к основанию
basal *a* основной; базальный; нижний
basal-nerved *a bot* базально-нервный
bask *v* греться на солнце
base *s* основание
~ of leaf пазуха листа, основание листа
blackened stem ~ *phytop* чёрная ножка капусты *(caused by Olpidium brassicae)*
nutritive ~ *ecol* кормовая база
purine ~ пуриновое основание
base-catalysis *s* основной катализ
base-pairing *s mbiol* спаривание оснований
baseplate *s* базальная пластинка
basialveolar *a* базиальвеолярный
basic *a* основной, главный; щелочной, основной
basiconic *a* базиконический
basicranial *a anat* базикраниальный
basidial *a* базидиальный
basidiocarp *s* базидиокарп
basidiole *s* базидиола
basidiophore *s* базидиофор, базидионосец
basidiospore *s* базидиоспора
basidium *s* базидия
basifixed *a* прикреплённый основанием
basigamy *s* халазогамия
basigene *s* базиген
basigenic *a bot* образующий основание
basigynium *s* карпофор
basihyal *s* копула, язычная кость
basikaryotype *s* основной кариотип
basilad *adv* к основанию
basilar *a* низовой, базилярный
basilateral *a* базилатеральный
basilemma *s* базальная мембрана, основная мембрана
basilisk *s herp* василиск *(Basiliscus)*
basin *s* бассейн, резервуар, водоём; пруд; *бот.* вдавленность
sedimentation ~, settling ~ отстойник
basipetal *a* базипетальный
basipodite *s zool* базиподит
basipodium *s* запястье; предплюсна
basisphenoid *s* базисфеноид, основная клиновидная кость
basket *s* корзина; корзиночка *(in bees)*
pollen ~ корзиночка *(in bees)*
basket-shaped *a bot* корзинковидный
basocyte, basophil, basophile *s* базофил, базофильный лейкоцит; базофильная клетка
basophilia *s* базофилия
basophilic *a* базофильный
basophiloblast *s* базофилобласт
basophilous *a* базофильный
bass *s* луб, лыко
bassarisk *s ther* кошачий хорёк, какомицли *(Bassariscus astutus)*

basswood *s* липа *(Tilia)*
bast *s* луб; флоэма
bastard *s* бастард, гибрид, помесь; *bot* отпрыск
bat *s* летучая мышь; *pl* рукокрылые *(Chiroptera)*
brown ~ кожан *(Eptesicus)*
common ~ ночница *(Myotis)*
disk-winged ~ американский присосконог *(Thyroptera)*
epaulet(ted) ~ эполетовый крылан *(Epomophorus)*
fisherman ~ большой рыболов *(Noctilio leporinus)*
free-tailed ~ складчатогуб *(Tadarida)*
gray ~ серая ночница *(Myotis grisescens)*
great ~ рыжая вечерница *(Nyctalis noctula)*
horseshoe(-nosed) ~ подковонос *(Rhinolophus)*
javelin ~ большой копьенос *(Phyllostomus hastatus)*
leaf-nosed ~ обыкновенный листонос *(Hipposideros)*
little brown ~ ночница *(Myotis);* малая бурая ночница *(Myotis lucifugus)*
long-eared ~ ушан *(Plecotus)*
long-winged ~ длиннокрыл *(Miniopterus)*
mouse-eared ~ ночница *(Myotis);* большая ночница *(Myotis myotis)*
night ~ летучая мышь
northern ~ северный кожанок *(Eptesicus nilssoni)*
sac-tailed ~ белый мешкокрыл, белый футлярохвост *(Diclidurus)*
vampire ~ обыкновенный вампир, десмод *(Desmodus)*
batata *s bot* ипомея батат *(Ipomoea batatas)*
batch *s* группа; партия; серия; доза
~ **of eggs** кладка яиц
batcher *s lab* дозатор, дозирующее устройство
bate *s eth* ярость, гнев, бешенство; взмахивать крыльями *(of a falcon)*
bateleur *s* орёл-скоморох, фигляр *(Terathopius ecaudatus)*
bath *s* баня
dust ~ пылевая ванна
water ~ водяная баня
bathing *s* купание
bathmotropism *s* батмотропизм
bathyphytia *s pl* фитоценозы низменностей
batrachian *s* бесхвостая амфибия *(Anura)*
batten *v* откармливаться, жиреть; тучнеть *(of soil)*
battle *s eth* бой, битва, драка
battue *s* облава
bay *s* лай; бухта, залив; *a* гнедой; *v* лаять; загонять, травить, преследовать *(зверя)*
dark ~ тёмно-гнедой; караковый
light ~ светло-гнедой
bay-brown *a* гнедо-саврасый
bay-roan *a* гнедо-чалый
beach *s* побережье, отмель
bead *s* гранула, шарик, сфера
beaded *a bot* чётковидный
beadruby *s bot* майник *(Majanthemum)*
bead *s* бусина
beagle *s* бигль
beak *s* клюв; хоботок; клювообразный выступ; зубец; шип; *bot* носик *(of a silique)*
glume's ~ зубец на колосковой чешуе
sucking ~ сосательный хоботок
beakchervil *s bot* купырь *(Anthriscus)*
beaked *a* клювообразный
beaker *s* химический стакан
beak-fruited *a* клювоплодный
beakful *s* содержимое клюва
beak-leaved *a* клюволистный
beakless *a bot* бесклювый
beam *s* ствол рога; пучок света; луч
incident ~ падающий пучок света
beambird *s orn* серая мухоловка *(Muscicapa striata)*
beamed, beamy *a* с ветвистыми рогами
beaming *s* излучение; облучение
bean *s* боб *(a fruit);* боб, бобы *(a plant)*
cacao ~ боб какао
carob ~ царьградский стручок, рожок
castor ~ клещевина обыкновенная *(Ricinus communis)*
coffee ~ кофейное зерно
English ~, field ~ конские бобы *(Vicia faba)*
mung ~ маш *(Phaseolus aureus)*
soya ~ соя культурная *(Soja hispida)*
bean-shaped *a* бобовидный

bear *s* медведь *(Ursus); ent* медведица *(Arctiidae); v* рождать, производить, родить
ant ~ трубкозуб *(Orycteropus); pl* муравьеды *(Myrmecophagidae)*
cat ~ большая панда, бамбуковый медведь *(Ailuropoda melanoleuca)*
grizzly ~ гризли *(Ursus arctos)*
honey ~ кинкажу, пото *(Potos flavus)*
Indian ~ губач *(Melursus)*
koala ~ коала *(Phascolarctos cinereus)*
sea ~ морской котик *(Callorhinus ursinus);* белый медведь (*Ursus maritimus)*
skunk ~ росомаха *(Gulo)*
sloth ~ губач *(Melursus)*
bearberry *s* толокнянка обыкновенная *(Arctostaphylos uva-ursi)*
beard *s zool* бородка, пучок волосков, пучок перьев; *bot* ость
bearded *a zool* бородатый; *bot* остистый; снабжённый колючками
beardedness *s* остистость
bearding *s*:
~ **of roots** мочковатость корней
beardless *a zool* безбородый; *bot* безостый; *ent* не имеющий волосков
bearing *s* плодоношение; стадия плодоношения; способность плодоношения; плодоносяший; рождение, произведение на свет; *a* рождающий, производящий на свет; **to come into** ~ вступать в пору размножения
heavy ~ полное плодоношение
beast *s* зверь; животное
~ **of prey** хищник
heavy ~s крупный рогатый скот
light ~s мелкий скот
beat *s* толчок, удар; биение, пульсация; систола
apical ~ *phys* верхушечный толчок
ectopic ~ *phys* эктопическая систола
premature ~ *phys* экстрасистола
beating *s* биение, пульсация; взмах *(of wings)*
chest ~ *eth* стук в грудь
beating-down *s* полегание
beating-up *s* подсаживание, пополнение насаждения, досаживание
beautification *s* озеленение
beaver *s* бобр *(Castor)*
musk ~ ондатра *(Ondatra zibethicus)*
sea ~ калан, морская выдра *(Enhydra)*
swamp ~ нутрия *(Myocastor coypus)*
becoming *s* становление, появление, возникновение
bed *s* грядка, клумба; заросль, куртина; *бот.* ложе; лёжка зверя; русло *(e.g., of a river);* дно *(e.g., of a sea);* **to ~ out** сажать, высаживать
biological filter ~ биологический фильтр
flower ~ клумба
heavy ~ густая заросль
nail ~ ногтевое ложе
nursery ~ рассадочная грядка
vascular ~ сосудистое русло
bedbug *s* клоп постельный *(Cimex lectularius)*
bedegar, bedeguar *s* бедегуар, волосатый галл
bedew *v* покрываться росой
bedewed *a* росистый
bedim *v* затемнять, этиолировать
bedstraw *s* подмаренник *(Galium)*
bee *s* пчела *(Apis)*
carder ~ пчела-смолёвщица *(Anthidium);* шмель *(Bombus)*
carpenter ~ пчела-плотник, шмель-плотник *(Xylocopa)*
cuckoo ~ пчела-кукушка *(Nomada);* шмель-кукушка; паразитный шмель *(Psithyrus)*
female ~ пчелиная матка
male ~ трутень
queen ~ матка
bee-eater *s orn* щурка *(Merops)*
beebread *s apic* перга
beech *s* бук *(Fagus)*
beechnut *s* буковый орех
bee-entrance *s apic* леток
beef *s* говядина
bee-family *s apic* пчелиная семья
bee-farming *s* пчеловодство
bee-fly *s* жужжало *(Bombylidae)*
beekeeping *s* пчеловодство
beehive *s* улей; рой пчёл
beestings *s* молозиво
beeswax *s* пчелиный воск
beet *s* свёкла *(Beta)*
leaf ~ свёкла листовая, мангольд *(Beta cicla)*
beetle *s* жук *(Coleoptera)*

alder leaf ~ козявка ольховая *(Agelastica alni)*
alfalfa snout ~ скосарь люцерновый *(Otiorrhynchus ligustici)*
ant ~ муравьежук *(Thanasimus formicarius)*
antelope ~ оленёк *(Dorcus parallelopipedus)*
asparagus ~ трещалка спаржевая *(Crioceris asparagi)*
bacon ~ кожеед ветчинный *(Dermestes lardarius)*
bark ~ короед *(Ipidae)*
barley leaf ~ пьявица красногрудая *(Lema melanopus)*
beaver ~ норолюб *(Leptinidae);* бобровая блоха *(Platypsyllidae)*
beet tortoise ~ щитоноска свекловичная *(Cassida nebulosa)*
black ~ таракан *(Blattodea)*
black pine ~ корнежил чёрный *(Hylastes ater)*
blister ~ нарывник *(Meloidae)*
bombardier ~ бомбардир *(Brachynus)*
bone ~ костоед *(Corynetes)*
broad-horned flour ~ хрущак рогатый *(Gnathocerus cornutus)*
brown scavenger ~ скрытник *(Lathridiidae)*
brown willow ~ листоед ивовый *(Galerucella lineola)*
bumble flower ~ шмележук *(Euphoria inda)*
buprestid ~ златка *(Buprestidae)*
burying ~ могильщик *(Necrophorus)*
cabbage stem flea ~ блошка рапсовая *(Psylliodes chrysocephala)*
cambium ~ заболонник *(Scolytinae)*
capricorn ~ усач, дровосек *(Cerambycidae)*
cellar ~ медляк-вещатель *(Blaps mortisaga)*
cereal leaf ~ пьявица красногрудая *(Lema melanopus)*
checkered ~ пестряк *(Cleridae)*
click ~ щелкун *(Elateridae)*
clock ~ навозник обыкновенный *(Geotrupes stercorarius)*
common bean ~ зерновка фасолевая *(Acanthoscelides obtectus)*
common stag ~ жук-олень *(Lucanus cervus)*
darkling ~ чернотелка *(Tenebrionidae)*
death-watch ~ точильщик *(Anobium)*
dryopid ~ прицепыш *(Dryopidae)*
dumble ~ навозник обыкновенный *(Geotrupes stercorarius)*
dung ~ навозник *(Coprinae)*
dust ~ точильщики *(Anobiidae)*
engraver ~ короед *(Ipidae)*
European ground ~ красотел *(Calosoma)*
feather-winged ~ перистокрылка, перокрылка *(Ptiliidae)*
flea ~ земляная блошка *(Halticinae)*
flour ~ хрущак малый мучной *(Tribolium confusum);* хрущак большой мучной *(Tenebrio molitor)*
fruitworm ~ малинный жук, малинник *(Byturidae)*
fungus ~ плеснеед *(Endomychidae);* грибоед, грибной жук *(Myceptophagidae)*
furniture ~ точильщик *(Anobiidae)*
golden spider ~ притворяшка шелковистый *(Niptus hololeucus)*
Goliath ~ голиаф *(Goliathus)*
ground ~ жужелица *(Carabus)*
hairy fungus ~ грибоед *(Mycetophagidae)*
Hercules ~ геркулес *(Dynastes hercules)*
hister ~ карапузик *(Histeridae)*
lady ~ божья коровка *(Coccinellidae)*
leaf ~ листоед, листогрыз *(Chrysomelidae)*
leather-winged ~ мягкотелка *(Cantharididae)*
lighting ~ светляк *(Lampyridae)*
long-horned ~ усач, дровосек *(Cerambycidae)*
malachiid ~ малашка *(Malachiidae)*
May ~ майский хрущ *(Melolontha)*
mimic ~ карапузик *(Histeridae)*
museum ~ кожеед музейный, музейный жук *(Anthrenus museorum)*
oat leaf ~ пьявица красногрудая *(Lema melanopus)*
oil ~ майка *(Meloe);* нарывник *(Meloidae)*
onion lily ~ трещалка луковая *(Lilioceris mardigera)*
pear blight ~ короед грушевый *(Anisandrus pyri)*

pinch ~ рогач *(Lucanidae)*
potato ~ колорадский жук *(Leptinotarsa decemlineata)*
raspberry ~ малинник, малинный жук *(Byturidae)*
rhinoceros ~ жук-носорог *(Dynastes)*
road ~ стафилинида, коротконадкрылый жук *(Staphylinidae)*
sacred ~ of Egyptians скарабей священный *(Scarabaeus sacer)*
scarred snout ~ скосарь *(Otiorhynhus)*
seed ~ зерновка *(Bruchidae)*
sexton ~ могильщик *(Necrophorus)*
shield ~ листоед, листогрыз *(Chrysomelidae)*
snapping ~ щелкун *(Elateridae)*
snout ~ долгоносик, слоник *(Curculionidae)*
soft-winged flower ~ малашка *(Melyridae)*
soldier ~ мягкотелка *(Cantharididae)*
spider ~ притворяшка *(Ptinidae)*
stag ~ рогач *(Lucanidae)*
steel ~s карапузики *(Histeridae)*
three-lined potato ~ пьявица трёхлинейная *(Lema trilineata)*
tiger ~ скакун *(Cicindela)*
timber ~ сверлильщик *(Lymexylonidae)*
tortoise ~ щитоноска *(Cassida)*
white-marked spider ~ притворяшка-вор *(Ptinus fur)*

begetter *s* самец-производитель
beggar-ticks *s bot* череда *(Bidens)*
begging *s eth* выпрашивание
begoon s баклажан *(Solanum melongena)*
behavior *s eth* поведение; *bioch* свойство
aggregating ~, aggregative ~ стайное поведение, стайность
aid-giving ~ помощь; взаимопомощь
attachment ~ привязанность
avoidance ~, avoidant ~ избегание
ballooning ~ расселение по воздуху *(of spiders)*
begging ~ выпрашивание
caregiving ~ забота, забота о потомстве
choice(-point) ~ реакция выбора
consort ~ брачное поведение, супружеское поведение; брачная связь
courtship ~ ухаживание
escape ~ бегство
expressive ~ выразительные движения
fascination ~ *eth* «завораживание»
freezing ~ замирание
grooming ~ груминг, обыскивание
helping ~ помощь
hiding ~ затаивание
huddling ~ скучивание
lekking ~ токование
manipulative ~ манипуляционное поведение, манипулирование
molecular ~ молекулярные единицы поведения
morphogenetic ~ морфогенетическая реакция
play ~ игровое поведение, игра
territorial ~ территориальное поведение, территориализм

behavioral *a* поведенческий, бихевиористский
behaviorism *s* бихевиоризм
beira *s ther* бейра *(Dorcatragus megalotis)*
beisa *s* антилопа бейза, восточно-африканский орикс *(Oryx gazella beisa)*
belch *s* отрыжка; *v* отрыгивать
belemnoid *a* остроконечный; шиповидный
bell *s bot* чашечка; *zool* колокол
belladonna *s* белладонна *(Atropa belladonna)*
belled *a* расширенный, имеющий раструб
bellflower *s* колокольчик *(Campanula)*
bellow *s* мычание; рёв; *v* мычать; реветь
belly *s* живот; брюшко; первый желудок *(of ruminants)*
belly-button *s* пупок
belonoid *a* иглообразный
belt *s* пояс, зона, район; полоса
adhesion ~ фокальный контакт; адгезионная пластинка
green ~ пояс зелёных насаждений; зелёный пояс
marginal tree ~ опушечное насаждение
wind ~ ветрозащитное насаждение, ветрозащитная полоса

belted *a* опоясанный
bend *s* изгиб, сгиб, излучина *(of a river)*
head ~ головной изгиб

bending *s* сгибание, изгибание, пригибание; изгиб; искривление; кривизна; изогнутость

beneficial *a* благотворный, полезный; целебный
benign *a* мягкий *(climate);* плодоносный *(soil);* доброкачественный *(tumor);* лёгкий *(disease)*
beniseed, benne *s bot* кунжут, сезам *(Sesamum indicum)*
bent *a* изогнутый, кривой; склон холма; полевица *(Agrostis)*
benumb *v* притуплять чувства; вызывать оцепенение; парализовать
benzoylation *s* бензоилирование
benzylation *s* бензилирование
bergamot *s bot* бергамот *(Citrus bergamia)*
berry *s* ягода, мясистый плод; икринка
grape ~ виноградина
June ~ ирга колосистая *(Amelanchier spicata)*
berrying *s* сбор ягод
berrylike *a* ягодовидный, ягодообразный
berrypecker *s orn* фруктоед *(Melanocharis)*
berry-producing *a* образующий ягоды
berry-shaped *a* ягодовидный, ягодообразный
beta-diversity *s* бета-разнообразие
betel *s bot* бетель *(Piper betle)*
betel-nut-palm *s* пальма катеху *(Areca catechu)*
betony *s bot* буквица *(Betonica);* чистец *(Stachys)*
betweenbrain *s* промежуточный мозг
bevy *s* стая; стадо
bezoar *s* безоаровый козёл *(Capra aegagrus)*
bharal *s ther* голубой баран *(Pseudois);* голубой баран, нахур, куку-яман *(Pseudois nayaur)*
biacuminate *a* двузаострённый
biaristate *a* двуостистый
biarticulate, biarticulated *a* двучленистый; двухсуставной
bias *s biom* смещение; систематическая ошибка оценки; склонность; *v* оказывать влияние; склонять
variable ~ переменная систематическая ошибка
biased *a biom* смещённый
biatorine *a* биаториновый *(lichen)*
biauriculate *a bot* двуухий
bibasic *a* двухосновный
bible-bag *s anat* книжка *(of a ruminant)*
bibliotheque *s*:
gene ~ клонотека генов, генотека
bibracteate *a* двуприцветниковый
bibulous *a* впитывающий влагу, гигроскопичный
bicalcarate *a bot* двушпорцевый
bicalyculate *a* двучашечный
bicameral *a* двухкамерный
bicapitate *a* двуглавый
bicapsular *a* двукапсульный, двукоробочный
bicarinate *a bot* двукилевой
bicarpellate *a* двуплодолистиковый, двукарпельный
bicellular *a* двуклеточный
bicephalous *a* двуглавый
biceps *s* двуглавая мышца
bicilate, biciliated *a* двухжгутиковый
bicipital *a* двуглавый
bicollateral *a* биколлатеральный
bicolor, bicolorous *a* двухцветный
biconcave *a* двояковогнутый
biconjugate *a* спаренный
biconvex *a* двояковыпуклый
bicorn, bicornous *a* двурогий
bicrenate *a* двоякозазубренный; удвоенно-зазубренный
bicuspid *s* малый коренной зуб, премоляр; *a* двухвершинный; двустворчатый; двустворчатый клапан
bicyclic *a* бициклический
bidental, bidentate *a zool* двузубый; *bot* двузубчатый; удвоенно-зубчатый
bidirectional *a* двунаправленный
biennial *s* двулетник; двулетний
bifacial *a* бифациальный, двусторонний
biferous *a* дважды цветущий, дважды плодоносящий, ремонтантный
bifid *a* двунадрезный; раздвоенный, двураздельный
biflabellate *a bot* дважды веерный
biflagellate *a* двухжгутиковый
biflex *a* дважды изогнутый
biflorate, biflorous *a* двухцветковый
bifoliate *a* двулистный
bifurcate, bifurcated *a* раздвоенный; вильчатый
bifurcation *s* бифуркация, раздвоение, разветвление, развилка
bigamous *a bot* двубрачный
bigamy *s* бигамия

bigeminal *a* дважды повторённый
bigeminate *a* двупарный
bigemmate *a bot* двупочечный
bigener *s* межродовой гибрид
big-eyed *a* большеглазый
bighorn *s* снежный баран, толсторог, чубук *(Ovis canadensis)*
bight *s* бухта; излучина *(of a river)*
biglandular *a bot* двужелезистый
bijugate *a* двупарный
bilabiate *a* двугубый
bilamellar, bilaminar, bilaminated *a* двупластинчатый; двухслойный
bilateral *a* билатеральный, двусторонний
bilateralism *s* билатеральная симметрия
bilayer *s* двойной слой, бислой; двуслойный
bilby *s* кроличий бандикут, билби *(Macrotis)*
bile *s* жёлчь
bile-expelling *a* желчегонный
bilharziasis *s* бильгарциоз, шистосомоз
biliary *a* жёлчный
biliation *s* желчеотделение
bilification, **biligenesis** *s* желчеобразование
bilimbi *s* огуречное дерево *(Averrhoa bilimbi)*
bill *s* клюв; законодательный акт
  dew ~ серёжка *(of a gobbler)*
bilobate, bilobated, bilobular *a* двухлопастный, двудольный
bilocular *a* двугнёздный; двухкамерный
bimaculate, bimaculated *a* двупятнистый
bimanous *a* двурукий
bimarginate *a bot* дважды окаймлённый, дважды оторoченный
bimastism *s anat* бимастия
bimestrious *a* двухмесячный
binary *a* бинарный, двойной; сдвоенный
binate *a* двойной, парный, растущий парами; двусложный
binaural *a* бинауральный
binding *s* связывание, фиксация; связь
  ~ **of complement** связывание комплемента, фиксация комплемента
  hydrogen ~ водородная связь
bindweb *s* невроглия; соединительная ткань
bindweed *s* вьюнок *(Convolvulus)*
bine *s* побег; плеть; стебель ползучего растения
binervate *a bot* двухжилковый
binocular *s* бинокуляр; бинокулярный
binominal *a* бинарный, двойной; сдвоенный
binominalism *s* бинарная номенклатура
binovular *a* двуяйцовый
binturong *s ther* бинтуронг *(Arctictis)*
binuclear, binucleate *a* двухъядерный
bioaccumulation *s* бионакопление
bioacustics *s* биоакустика
bioassay *s* биологическая проба, биопроба, биоанализ, биотест
bioastronautics *s* космическая биология
bioblast *s* биобласт
biocatalist *s* биокатализатор, фермент
biocatalysis *s* биокатализ
biocenology *s* биоценология
biocenose, biocenosis *s* биоценоз
biocenotics *s* биоценология
biochemical *a* биохимический
biochemistry *s* биохимия
biochip *s* биочип
biochore *s* биохор
biocide *s* биоцид
bioclimate *s* биоклимат
bioclimatology *s* биоклиматология
biocommunity *s* биоценоз
biocybernetics *s* биокибернетика
biocycle *s* биоцикл, жизненный цикл
biocytoculture *s* культура клеток
biodegradation *s* биодеградация, биологическое разложение
biodemography *s* биодемография
biodestruction *s* биологическое разложение
biodeterioration *s* биоповреждение
biodiversity *s* биоразнообразие, биологическое разнообразие
biodynamics *s* биодинамика
bioecology *s* биоэкология
bioelectricity *s* биоэлектричество
bioelectronics *s* биоэлектроника
bioenergetics *s* биоэнергетика
bioengineering *s* биоинженерия
biofilter *s* биофильтр
biofuel *s* биологическое топливо
biogas *s* биогаз
biogenesis *s* биогенез
biogenetic *a* биогенетический
biogenic, biogenous *a* биогенный
biogeny *s* биогения

biogeocenose, biogeocenosis *s* биогеоценоз
biogeochemistry *s* биогеохимия
biogeography *s* биогеография
biohazard *s* биологическая опасность
bioindustry *s* биопромышленность
bioinstrumentation *s* медико-биологическая аппаратура
bioleaching *s* биовыщелачивание
biologic, biological *a* биологический
biologist *s* биолог
biology *s* биология
germ-free ~ гнотобиология
bioluminescence *s* биолюминесценция, биологическое свечение
biolysis *s* биолизис
biolytic *a* биолитический
biomanipulation *s* биоманипуляция
biomass *s* биомасса
biome *s ecol* биом
biomechanics *s* биомеханика
biomedical *a* биомедицинский
biometric *a* биометрический
biometrics, biometry *s* биометрия
biomonitoring *s* биомониторинг
bionics *s* бионика
bionomics, bionomy *s* биономия, экология
bionosis *s* бионоз
biont *s* бионт
biophage *s* биофаг
biophilia *s* биофилия, инстинкт самосохранения
biophotogenesis *s* биофотогенез
biophotolysis *s* биофотолиз
biophysics *s* биофизика
biophyte *s* биофит
biopolymer *s* биополимер
biopotential *s* биоэлектрический потенциал
biopreparation *s* биопрепарат
biopsy *s* биопсия
bioradiology *s* радиобиология
bioreactor *s* биореактор, ферментёр
biorhythm *s* биоритм, биологический ритм
bios *s* биос, жизнь
bioscrubbling *s* биологическая промывка, биологическая очистка
biosensor *s* биосенсор, биодатчик
biosociology *s* биоценология; биосоциология
biosome *s* биосома
biosphere *s* биосфера
biospheric *a* биосферный
biostatic биостатический, тормозящий рост и развитие
biostatistics *s* биостатистика
biostimulant *s* биостимулятор
biostratigraphy *s* биостратиграфия
biosynthesis *s* биосинтез
unprimed ~ беззатравочный биосинтез
biosystem *s* биосистема
biota *s* биота
biotank *s* биотэнк
biotaxonomy *s* биосистематика, таксономия
biotechnology *s* биотехнология
biotelemetry *s* биотелеметрия
biotic *a* биотический
biotinylation *s* биотинилирование, обработка биотином
biotomy *s* анатомирование, препарирование
biotonus *s* биотонус
biotope *s* биотоп
biotransformation *s* биологическое преобразование
biotron *s* биотрон, климатрон, камера искусственного климата
biotype *s* биотип
biovular, biovulate *a* двуяйцевой; с двумя семяпочками
bipaleolate *a* двуплёнчатый
bipalmate *a бот.* дихотомически ветвящийся; дважды пальчатый
biparasite *s* гиперпаразит
biparasitic *a* гиперпаразитический
bipartite *a* двураздельный
bipartition *s* раздвоение
bipectinate *a* двугребенчатый
biped *s* двуногое; *a* двуногий
bipedalism *s* двуногое хождение, бипедализм
bipetalous *a* двулепестный
biphase *a* двухфазный
bipinnaria *s* бипиннария
bipinnate *a bot* двоякоперистый, двуперистый, удвоенно-перистый
bipinnatifid *a bot* двоякоперисто-надрезный, двуперисто-надрезный
bipinnatipartite *a bot* двоякоперисто-раздельный, двуперисто-раздельный

bipinnatisect *a bot* двоякоперисто-рассечённый, дваперисто-рассечённый
biplicate *a* двускладчатый, сложенный вдвое
bipolar *a* биполярный; двухполюсный
bipolarity *s* биполярность
bipunctate *a* двуточечный
biradial *a* двухлучевой
biramous *a* двуветвистый
birch *s* берёза *(Betula)*
  dwarf (Arctic) ~ берёза карликовая, ерник *(Betula nana)*
bird *s* птица
  ~ of Jove орёл
  ~ of Juno павлин
  ~ of prey хищная птица
  frigate ~ фрегат *(Fregata)*
  gallinaceous ~s куриные *(Galliformes)*
  hen ~ самка птицы
  incubator ~s большеног, сорная курица *(Megapodidae)*
  indigo ~ красноногая вдовушка *(Vidua chalybeata)*
  laying ~ несушка
  man-o'-war ~ фрегат *(Fregata)*
  missel ~ дрозд-деряба *(Turdus viscivorus)*
  mutton ~ тонкоклювый буревестник *(Puffinus tenuirostris)*
  pilot ~ птица-попутчик *(Pycnoptilus floccosus)*
  pintado ~ капский голубок *(Daption capensis)*
  secretary ~ птица-секретарь *(Sagittarius serpentarius)*
  silver ~ серебристая мухоловка *(Empidornis semipartitus)*
  skunk ~ рисовый трупиал, рисовая птица, боболинк *(Dolichonyx oryzivorus)*
  snake ~ змеешейка *(Anhinga)*
  stuffed ~ чучело птицы
  tropic ~ фаэтон *(Phaethon)*
  whidah ~, widow ~ вдовушка *(Vidua)*
  wounded ~ подранок
  yellow-billed tropic ~ белохвостый фаэтон *(Phaethon lepturus)*
birdbanding *s* кольцевание птиц
birdcage *s* клетка для птиц
birdfeeder *s* кормушка для птиц
birdhouse *s* скворечник
bird's-nest *s phytop* ведьмина метла
birecurvate *a* двоякоизогнутый
birefringent *a* двоякопреломляющий
birth *s* рождение; **at** ~ при рождении, в момент рождения
  live ~ живорождение, вивипария
  virgin ~ партеногенез
birthrate *s* рождаемость; коэффициент рождаемости
bisaccate *a pal* двумешковый
bisectional *a* двурассечённый
biseptate *a* двураздельный
biserial *a* бисериальный, двухрядный
biserrate *a* двоякопильчатый, двупильчатый
bisexual *s* гермафродит; *a* гермафродитный, двуполый, бисексуальный
bishop *s orn* бархатный ткач *(Euplectella)*
Bismark brown *s* бисмарк коричневый *(histological stain)*
bison *s* зубр *(Bison);* (американский) бизон *(Bison bison)*
  American бизон *(Bison bison)*
  European ~ зубр *(Bison bonasus)*
  Indian ~ гаур *(Bos gaurus)*
  plains ~, wood ~ бизон *(Bison bison)*
bispinose *a* двухколючковый
bisporangiate *a* биспорангиатный
bisporic *a* биспорический
bistipulate *a* с двумя прилистниками
bistort *s bot* горец змеиный, раковые шейки *(Polygonum bistorta)*
bisubstrate *a* бисубстратный
bisulcate, bisulcated *a* двухбороздный
bitch *s* самка; сука
bite *s* укус; след укуса; кусок *(of food);* прикус; клёв *(of fish)*; *v* кусать, кусаться; жалить; разъедать *(of acids);* клевать *(of fish)*
biternate *a bot* двоякотройчатый
biting *s* клев *(of fish)*; *a* острый, едкий
bitter *s* горечь; *a* горький
bitter-fruited *a* горькоплодный
bittern *s* рассол, маточный раствор; выпь *(Botaurus);* волчок *(Ixobrychus)*
bituminous *a* смолистый
bitunicate *a myc* двухоболочковый
biumbellate *a bot* двухзонтичный
biunguiculate *a* двукоготковый
bivalent *s gen* бивалент; бивалентный; двухвалентный
bivalve, bivalved *a* двустворчатый

bivalvular *a* двухклапанный
biventer *s anat* двубрюшная мышца
biventral *a anat* двубрюшный
bivittate *a* двуполосатый
bivoltine *a* бивольтинный
biwa *s bot* японская мушмула, локва *(Eriobotrya japonica)*
black *a* черный; вороной; *s hist* краситель чёрного цвета
blackarm *s phytop* гоммоз хлопчатника *(caused by Xanthomonas malvacearum)*
black-beetle *s* таракан *(Blattidae)*
blackberry *s* ежевика *(Rubus)*
blackbird *s* чёрный дрозд *(Turdus merula);* трупиал *(Icteridae)*
black-brown *a* караковый
blackcap *s orn* черноголовая славка, славка-черноголовка *(Sylvia atricapilla);* черноголовый чекан *(Saxicola torquata)*
blackcock *s* самец тетерева *(Lyrurus tetrix)*
blackfly *s* мошка *(Simuliidae)*
black-grain *s phytop* спорынья зерновых *(caused by Claviceps paspali)*
blackhead *s par* гистомоноз *(caused by Histomonas meleagridis)*
blackhull *s* пшеница черноколоска *(Triticum durum var. caerulescens)*
blackleg s phytop чёрная ножка (caused by fungi and bacteria)
black-maned *a* черногривый
blackspot *s phytop* чёрная пятнистость, альтернариоз *(caused by Alternaria)*
~ **of crucifers** чёрная пятнистость крестоцветных *(caused by Alternaria brassicae)*
blackthorn *s* терновник, тёрн *(Prunus spinosa)*
bladder *s* пузырь; полость; мочевой пузырь; циста
bladder-fruited *a* пузыреплодный
bladdery *a* пузырчатый; пустой, полый
blade *s ent* лациния; лезвие; *bot* пластинка; былинка
distichous ~s *bot* симметричные пластинки
bladebone *s anat* лопатка
bladed *a* облиственный
bland *a* мягкий *(e.g., climate);* успокаивающий
blank *s* контроль, контрольная проба; слепой опыт
blanket *s* защитный слой; отложение; покров
blast *s* порыв, струя
rice ~ *phytop* пирикуляриоз риса *(caused by Piricularia oryzae)*
blastaea, blastea *s* бластея
blastema *s* росток; зародыш; слоевище лишайника
blasteniospore *s* бластениоспора
blastochore *s* бластохор
blastocoel *s emb* бластоцель
blastocolla *s* почковый клей
blastocyst *s emb* бластоциста, зародышевый пузырь
blastocyte *s* бластоцит
blastoderm *s emb* бластодерма
blastodisk *s emb* бластодиск, зародышевый диск
blastogenesis *s* бластогенез, почкование
blastokinesis *s* бластокинез
blastomere *s emb* бластомер
blastoneuropore *s* бластонейропор
blastophyllum *s emb* зародышевый листок
blastopore *s emb* бластопор
blastosphere *s emb* бластула
blastospore *s* бластоспора
blastostyle *s* бластостиль
blastotomy *s* бластотомия
blastozoite *s* бластозоид
blastozooid *s* бластозооид
blastula *s emb* бластула
blastulation *s emb* образование бластулы
blaze *s* «звёздочка» *(on the face of an animal);* метка, клеймо *(on a tree)*
bleach *s* отбеливание; отбеливающее вещество; обесцвечивание; *v* обесцвечивать, обесцвечиваться; отбеливать
bleached-out *a* этиолированный
bleat *s* блеяние; мычание; блеять; *v* мычать
blebby *a* пузырчатый, везикулёзный
bleed *v* кровоточить; истекать кровью; спускать кровь *(of a slaughtered animal);* обескровливать; сочиться, просачиваться; истекать
bleeding *s* кровотечение; обескровливание, спуск крови; *bot* плач

blend *s* смесь; *v* смешивать
blender *s* смеситель
blending *s* смешивание, смешение
blendling *s* межрасовый гибрид
blenna *s* слизь
blennogenous *a* продуцирующий слизь, слизеобразующий
blepharon *s* веко
blepharophorous *a bot* ресничконосный
blepharoplast *s* блефаропласт
blesbok *s ther* лиророгий бубал *(Damaliscus);* блесбок, беломордый бубал *(Damaliscus dorcas phillipsi)*
bley *s см.* **bleak**
blight *s*:
celery early ~ ранний ожог, церкоспороз сельдерея *(caused by Cercospora apii)*
celery late ~ белая пятнистость, септориоз сельдерея *(caused by Septoria apii)*
crown ~ завядание кроны деревьев *(caused by Valsa nivea)*
fire ~ бактериальный ожог плодовых деревьев *(caused by Erwinia amylovora)*
gray ~ серая пятнистость листьев чая *(caused by Pestalotia longiseta)*
head ~ **of winter wheat** фузариоз колосьев озимой пшеницы *(caused by Fusarium)*
horsehair ~ «нитчатая болезнь» древесных пород *(caused by Marasmius)*
late ~ **of potato** фитофтороз картофеля *(caused by Phytophthora infestans)*
leaf ~ пятнистость листьев *(caused by Helminthosporium)*
potato ~ ранняя гниль картофеля *(caused by Alternaria solani);* фитофтороз картофеля *(caused by Phytophthora infestans)*
rice ~ пирикуляриоз риса *(caused by Pyricularia oryzae)*
seedling ~ **of cereals** фузариоз злаков *(caused by Fusarium)*
southern corn leaf ~ глазковая пятнистость листьев кукурузы *(caused by Cochliobolus heterostrophus)*
thread ~ «нитчатая болезнь» древесных пород *(caused by Marasmius)*
tomato winter ~ зимняя болезнь томата *(caused by a virus)*
yellow leaf ~ мозаичная болезнь хлопчатника *(caused by a virus)*
blighting *a phytop* приносящий вред
blind *a* слепой; *s* диафрагма, бленда
blindgut *s* слепая кишка
blindness *s* слепота
blindworm *s herp* слепозмейка *(Typhlops);* ломкая веретенница, медяница *(Anguis fragilis)*
blink *s* мигание; *pl бот.* монция *(Montia)*; *v* мигать
blister *s* волдырь; водяной пузырь; пузырчатое заболевание растений *(caused by Taphrina);* бугорок перламутра; головнёвый гриб *(Ustilaginaceae)*
bark ~ ожог коры
leaf ~ пузырчатость листьев груши *(caused by Taphrina bullata)*
white ~ «белая ржавчина» крестоцветных *(caused by Albugo candida)*
blistered пузырчатый, покрытый пузырями
bloating *s* вздутие
block *s* блок; блокада; *v* блокировать; тормозить; препятствовать
blockage *s* блокада
blockcopolymerization *s* блоксополимеризация
blockiness *s* сбитость тела
blockpolymerization *s* блокполимеризация
blood *s* кровь; род, происхождение; *bot* сок
blooded *a* чистокровный
blood-forming *a* кроветворный
blood-inhabiting *a* живущий в крови
bloodless *a* бескровный; безжизненный; вялый
blood-making, blood-producing *a* кроветворный
blood-pool *s* кровяное депо
bloodstream *s* кровообращение; кровоток; кровяное русло
bloodsucker *s* кровососущее животное, кровосос; *herp* обыкновенный калот *(Calotes calotes); herp* калот-кровосос *(Calotes versicolor)*
bloodsucking *a* кровососущий

blood-vascular *a* сосудистый
bloodwood *s* кампешевое дерево, сандаловое дерево *(Haematoxylon campechianum)*
bloody *a* кровяной; кровянистый; кровавый
bloom *s* цвет, цветение; цвести; цветок; восковой налёт; пушок *(on fruits);* **in ~** в цвету; **to come into ~** расцветать, начинать цвести
  full ~ фаза полного цветения
  wax ~ восковой налёт
bloomed *a:* **closely ~** загущённоцветковый
bloomer *s* цветущее растение
blooming *a* цветущий
blossom *s* цвет, цветение; *v* цвести; распускаться; **in ~** в цвету; **to bear ~** цвести; **to burst in ~** расцвести
  full ~ стадия полного цветения
  initial ~ начало цветения
  partial ~ неполное цветение
blossoming *s* цветение; **to cease ~** отцвести
blossom-time *s* время цветения
blotch *s phytop* пятно; пятнистость; нарост; бородавка
  sooty ~ сажистость яблок *(caused by Gloeodes pomigena)*
  winter wheat leaf ~ септориоз пшеницы, бурая пятнистость пшеницы *(caused by Septoria nodorum)*
blotched, blotchy *a* пятнистый; бугорчатый
blotting *s* блоттинг
  dot ~ дот-блоттинг
  northern ~ нозерн-блоттинг
  southern ~ саузерн-блоттинг
  western ~ вестерн-блоттинг
blow *s* цвет, цветение; удар; кладка яиц *(by flies);* фонтан кита; *v* дуть; выдувать; цвести; класть яйца *(of flies)*
blowball *s* соплодие одуванчика *(Taraxacum officinale)*
blower *s* цветущее растение
blowfly *s* мясная муха, падальная муха *(Calliphoridae)*
blowing *s zool* дыхание, сопение, выдувание; фонтанирование *(of whales);* откладывание яиц *(by flies);* цветение
  soil ~ выдувание почвы
blown а загнанный *(of a horse)*
blown-in *a* принесённый ветром
blubber *s* медуза; подкожное сало; китовый жир, ворвань
blue *a* голубой, синий; *s* голубой краситель, синий краситель; *ent* голубянка *(Lycaenidae)*
  Evans ~ эванс голубой, синий Эванса
  fast ~ быстрый голубой, дианизидин
  mordant ~ хромазурол B
blueberry *s* голубика, черника *(Vaccinium)*
bluebill *s orn* чернеть *(Aythya)*
bluebottle *s* муха падальная синяя *(Calliphora erythrocephala);* василёк синий *(Centaurea cyanus)*
bluebuck *s ther* нильгау *(Boselaphus tragocamelus)*
bued *a* воронёный
blue-dun *a* буланый
blue-gray *a* сизый
bluejoint *s bot* вейник *(Calamagrostis)*
blue-roan вороно-чалый
blue-speckled *a orn* крапчато-голубой
bluestart *s orn* синехвостка *(Luscinia cyanurus)*
bluestem *s bot* бородач *(Andropogon)*
bluetail *s orn* синехвостка *(Luscinia cyanurus)*
bluethroat *s orn* варакушка *(Luscinia svecica)*
bluff *s* плащеносная ящерица *(Chlamydosaurus)*
blunt *a* тупой, тупоконечный; дефосфорилировать, «затуплять концы»
blunt-pointed *a* тупой, тупоконечный
boa *s* удав; обыкновенный удав *(Boa constrictor)*
boar *s* боров; кабан; хряк; самец морской свинки
  wild ~ кабан *(Sus scrofa)*
board *s* доска; совет
  ~ of doctors консилиум
  spreading ~ *ent* расправилка
boatshaped *a* ладьевидный
bobac *s* байбак, обыкновенный сурок *(Marmota baibacina)*
bobtail *s* купированный хвост
BOD, biochemical oxygen demand БПК, биохимическое потребление кислорода

body *s* тело; туловище; корпус *(e.g., of a horse);* тельце; организм; ствол, стебель; большое количество, масса
~ of interest изучаемый объект
~ of nail ногтевая пластинка
~ of water водоём
adrenal ~ надпочечник
anococcygeal ~ заднепроходно-копчиковая связка
anti-intermediary ~ антиамбоцептор, антиантитело
aortic ~ тело Цукеркандля, пояснично-аортальный параганглий
Barr's ~ тельце Барра, гранула полового хроматина
basal ~ базальное тело, базальная гранула, кинетосома
cell ~ *bot* клеточное тело, протопласт
central ~ центральное тельце; центросома
coccygeal ~ копчиковое тельце, копчиковый гломус
conoid ~ шишковидная железа, эпифиз
conoidal ~ *prot* коноид
dentate ~ зубчатое ядро
end ~ *imm* конечное звено
epithelial ~ паращитовидная железа, околощитовидная железа, эпителиальное тельце
Golgi ~ тельце Гольджи, диктиосома
Golgi-Mazzoni ~ тельце Гольджи-Мацони, луковицеобразное тельце
Highmore's ~ гайморово тело; средостение яичка
immune ~ иммунное тело, антитело
infundibular ~ нейрогипофиз, задняя доля гипофиза
intercarotid ~ сонный гломус, каротидный гломус
intermediate ~ амбоцептор
Koch's ~s *pl* коховские тела
Jolly's ~ тельце Жоли, базофильное зерно
Malpighian ~ мальпигиево тельце, лимфатический фолликул; мальпигиев клубочек
mulberry ~ морула
paired ~ies межпочечные тела
pineal ~ шишковидная железа, эпифиз
pituitary ~ гипофиз
quadrigeminal ~ies четверохолмие
receptive ~ воспринимающее тельце; рецептор
red ~ красная сеть
restiform ~ верёвчатое тело, нижняя мозжечковая ножка
sense ~ орган чувств
stem ~ сyt «толкающее тельце», межзональная область; *prot* «опорное тело»
Stieda ~ *prot* тельце Штида, штидовское тело
stored neurosecretory ~ тельце Херринга
suprarenal ~ надпочечник
thyroid ~ щитовидная железа
turbinated ~ носовая раковина
water ~ водоём
Wolffian ~ вольфово тело, мезонефрос

bog *s* болото, топь; трясина
meadow ~ заболоченный луг
quacking ~ топь
scraw ~ трясина, зыбун

bog-bird *s* китоглав, королевская цапля *(Balaeniceps rex)*

bogging *s* образование болота, заболачивание

boggy *a* болотистый, топкий; заболоченный

bog-loving *a* болотолюбивый

boil *s* кипение; кипящая жидкость; выпаривание; фурункул, нарыв; *v* кипеть, кипятить, выпаривать; **to ~ dry** выпаривать досуха
Aleppo ~, Bagdad ~, Biskra ~, Oriental ~ кожный лейшманиоз

bole *s* ствол, штамб, пень
annulated ~ маслёнок *(Suillus luteus)*
bitter ~ желчный гриб *(Tylopilus felleus)*
blood-red ~ сатанинский гриб *(Boletus satanas)*
orange-cap ~ подосиновик *(Leccidium aurantiacum)*
rough ~, shaggy ~ подберёзовик *(Leccidium scabrum)*
yellow ~ маслёнок *(Suillus luteus)*

boll *s* семенная коробочка

bolling *s* очёсывание *(of flax)*

bollworm *s* совка хлопковая, коробочный червь *(Heliothis zea)*
cotton ~ совка хлопковая, коробочный червь *(Heliothis zea)*

pink ~ розовый коробочный червь хлопчатника *(Pectinophora gossypiella)*
bolt *v bot* выбрасывать стрелку; понести *(of a horse);* просеивать; убегать
bolter *s* стрелкующееся растение; растение, выбросившее цветонос; норовистая лошадь
boltering *s* цветушность
bolting *s bot* выход в стрелку; просеивание
bolus *s* шаровидная масса; пищевой комок
bond *s* связь; сцепление
disulfide ~ дисульфидный мостик
bonding *s* связь
pair ~ парная связь
social **~s** *pl* социальные связи
bone *s* кость; *pl* костяк, скелет
anvil ~ наковальня
basihyal ~копула, язычная кость
basisphenoid ~ базисфеноид, основная клиновидная кость
breast ~ грудина
coccygeal ~ копчик
collar ~ ключица
convolute ~ носовая раковина
coracoid ~ коракоид, воронья кость
dumb-bell ~ предсошник
ear ~ слуховая косточка
ectopterygoid ~ эктоптеригоид, наружная крыловидная кость
epihyal ~ верхний сегмент гиоидной дуги
epipterygoid ~ *herp* эпиптеригоид, столбчатая кость
ethmoturbinal ~ обонятельная раковина
facial ~ кость лицевого черепа
finger ~ фаланга пальца кисти
hammer ~ молоточек
hypural ~ гипуралия, гипуральная косточка
mastoid ~ сосцевидный отросток
maxilloturbinal ~ нижняя носовая раковина
nail ~ *zool* концевая фаланга пальца; *anat* ногтевая фаланга
parasphenoid ~ парасфеноид
periotic ~ ушная кость; околоушная кость *(of fish);* пирамида височной кости
petrosal ~ пирамида височной кости
plowshare ~ сошник
procoracoid ~ прокоракоид
sacral ~ крестец, крестцовая кость
sphenoid ~ клиновидная кость; сфеноид
sphenoturbinal ~ раковина клиновидной кости
tail ~ копчик
thigh ~ бедро, бедренная кость
tooth ~ дентин
turbinal ~ носовая раковина
bone-dry *a* абсолютно сухой
bone-forming *a* костеобразующий
bonelet *s* косточка
bone-seeking *a* остеотропный
bonitation *s* бонитировка, качественная оценка
bontebock, bontebuck *s ther* бонтбок, беломордый бубал *(Damaliscus dorcas)*
bonxie *s* большой поморник *(Catharacta skua)*
bony *a* костный, костяной; костистый; сухопарый, поджарый
booby *s orn* олуша *(Sulidae)*
booger s «козявка» *(dry slime in nostrils)*
book *s* книга
lung ~ *zool* лёгочная книжка
booklet *s anat* книжка
booster *s imm* усилитель; бустер; активатор; бустер-доза, ревакцинирующая доза; повторная иммунизация, реиммунизация, ревакцинация
boosting *s imm* повторная иммунизация, реиммунизация, ревакцинация
boot *s bot* обвёртка; влагалище *(of a leaf);* выходить в трубку
booted *a* мохноногий, лохмоногий
booting *s bot* трубкование, выход в трубку
borax *s* бура
bordeaux *s* бордосская жидкость
border *s* граница; край
brush ~ *cyt* щеточная каёмка, кисточковая кайма
bordered *a* окаймлённый, отороченный
borderline *s* краевая линия; физиол. пограничное состояние
boreal *a* бореальный, северный
bored *s bot* прободённый, проколотый
borer *s* сверлильщик, точильщик; бур
boring *s* бурение; сверление сверлящий

borzoi *s* борзая
borrowing-lending *s phys* «заимствование крови»
boscage *s* подлесок; кустарник
boss *s* бугорок, шишка, выпуклость
bosselated *a* бугорчатый, шишковатый
bostryx *s bot* улитка, завиток
bot *s* личинка овода
botanic *a* ботанический
botanist *s* ботаник
botanize *v* ботанизировать; собирать гербарий
botanolite *s* окаменевшее растение
botany *s* ботаника
botfly *s* овод; *pl* носоглоточные оводы *(Oestridae)*
bothridium *s* ботридий
bothriocephalosis *s helm* ботриоцефалёз
bothrium *s* ботрия
botryoid, botryoidal *a* кистевидный, кистеобразный, гроздевидный
bott *s* личинка овода
bottle *s* склянка, бутылка; баллон
  density ~ пикнометр
  Dewar ~ сосуд Дьюара
  dropping ~ капельница; капельный дозатор
  gravity ~ пикнометр
  killing ~ морилка
  roller ~ вращающийся флакон, роллерный флакон, роллер-флакон
  washing ~ промывалка
  water ~ батометр
  weighing ~ сосуд для взвешивания, бюкс
bottle-shaped *a* бутылковидный
bottom *s* основание; подошва; подстилающая порода; дно; низ, нижняя часть
bottomland *s* пойма; низина
bough *s bot* ветвь, скелетная ветвь
boughy *a* ветвистый, развесистый
bouillon *s* бульон
boulder *s* валун
bounce *s* прыжок
bound *s* граница, предел; прыжок, скачок; связанный
boundary *s* граница; пограничный слой, контур
  epitope ~ *imm* эпитопное микроокружение
  forest ~ контур леса
  permeative ~ граница проницаемости
bouquet *s cyt* букет
bout *s* приступ, припадок
bovine бычий; коровий
bow *s* дуга, изгиб; *v* подчиняться
bow-cooing *s eth* поклоны с воркованием
bowel *s* кишка; *pl* кишечник, пищеварительный тракт; *v* потрошить; вынимать внутренности
bowerbird *s* шалашник *(Ptilonorhynchidae)*
box *s* ящик, садок, бокс; самшит *(Buxus);* эвкалипт *(Eucalyptus);* подсачивать *(trees)*
  bird ~ искусственное гнездовье
  dry ~ *lab* вытяжной шкаф; перчаточный бокс
  Hogness' ~ *mbiol* ТАТА-бокс
  humidified ~ влажная камера
  nest ~ дуплянка
  Pribnow's ~ *mbiol* ТАТА-бокс
  puzzle ~ *eth* клетка-головоломка
  Skinner ~ ящик Скинера
  TATA ~ *mbiol* ТАТА-бокс
boxing *s eth* боксирование
boxtree *s* самшит *(Buxus)*
brachelytrous *a* коротконадкрылый
brachial *a* плечевой
brachiate *a* разветвлённый
brachiation *s* брахиация
brachiferous, brachigerous *a* ветвистый, разветвлённый
brachiocephalic *a* плечеголовной
brachiocubital *a* плечелоктевой
brachiolaria *s* брахиолярия
brachium *s* плечо; рука
brachyblast *s* брахибласт, укороченный побег
brachybotryous *a* короткокистистый
brachycephalic *a* брахицефальный
brachycerous *a* короткоусый; короткорогий
brachydactyly *s* брахидактилия, короткопалость
brachymeiosis *s* cyt одноступенчатый мейоз, укороченный мейоз
brachyplast *s* брахибласт, укороченный побег
brachypodous *a* коротконогий
brachypterous *a* короткокрылый
brachysclereid *s* брахисклереида
brachysm *s bot* карликовость

brachytic *a bot* укороченный, карликовый
brachyural *a* короткохвостый
bracken *s bot* чаща; заросли кустарника; орляк обыкновенный *(Pteridium aquilinum)*
brackish *a* солоноватый
bract *s bot* прицветник; кроющий лист
bract-bearing *a bot* прицветниконосный
bracteal *a bot* прицветниковый
bracteate *a bot* имеющий прицветники
bracteole *s bot* прицветничек
bractless *a bot* бесприцветниковый
bractlet *s bot* прицветничек
bractlet-shaped *a bot* прицветничковидный
bradycarpic *a bot* удлинённоплодный
bradydiastole *s phys* брадидиастола, удлинённая диастола
bradygenesis *s* замедленное развитие
bradykinetic *a* медленно движущийся
bradypnoea *s phys* брадипноэ, редкое дыхание
bradysystole *s phys* брадисистола,удлинённая систола
bradytely *s* брадителия
brain *s* головной мозг; *pl* интелект, ум
  abdominal ~ чревное сплетение, солнечное сплетение
  visceral ~ висцеральная область мозга
brainpan *s* черепная коробка, череп
brainpower *s* умственные способности
brainstem *s* стволовая часть мозга
braird *s* первый росток; *v bot* всходить
brake *s bot* чаща; заросли кустарника; орляк обыкновенный *(Pteridium aquilinum)*
bramble *s* сложный плод; ежевика несская *(Rubus nessensis)*
brambling *s orn* юрок, вьюрок *(Fringilla montifringilla)*
bran *s* отруби
branch *s* ветвь, ветка; ветвиться; линия родства; ответвление
  axillary ~ пазушный побег
  epicormic ~ волчок, пасынок, водяной побег
  scaffold ~ маточная ветвь, скелетная ветвь; сук
  side ~ боковое ответвление
branched *a* ветвистый; разветвлённый
branchial *a* бранхиальный, жаберный
branchiate, **branchiferous** *a* имеющий жабры
branchiform *a* жаброобразный
branching *s* ветвление; *a* ветвящийся; ветвистый; разветвление; разветвлённый
branchiocardiac *a* бранхиокардиальный
branchiomere *s* жаберный сегмент
branchiomery *s* бранхиомерия
branchiopallial *a* мантийно-жаберный
branchlet *s* веточка, небольшой побег
branchpoint *s* точка ветвления
branchy *a* разветвленный; ветвистый; раскидистый
brand *s phytop* ржавчинный гриб; «ожог»
brant *s orn* чёрная казарка *(Branta bernicla)*
brash *s* изжога
bray *v* реветь, кричать *(of an ass)*
bread *s* хлеб
  bee ~ перга
breadstuff *s* хлебные злаки
break *s* разрыв; *v* разрывать; выезжать *(a horse)*; **to ~ a bond** разрывать связь; **to ~ down** расщеплять, делить; лопаться, раскрываться *(of buds)*; **to ~ cover** поднять из логовища
breakable *a* хрупкий, ломкий
breakage *s* разрывание; разрыв; *cyt* образование разрывов, образование разломов; бурелом
breakdown *s* разрушение, распад, разложение, расщепление
breaking *s* разрывание; разрыв; распускание *(of buds);* прерывание *(of seed dormancy)*
breakpoint *s mbiol* точечный разрыв
breast *s* грудь; молочная железа, млечная железа
breastbone *s anat* грудина; *ent* лопаточка
breath *s* дыхание; вдох
  apneustic ~ *phys* апноэ
  shallow ~ поверхностное дыхание
breathe *v* дышать; вдыхать; выдыхать; жить, существовать
breathful *a* дышащий, живой
breathing *s* дыхание; *a* дыхательный
breathless *a* бездыханный; запыхавшийся; задыхающийся
breech *s* ягодицы

breed *s* порода; потомство, поколение; сорт; *v* размножаться; давать приплод; вынашивать; высиживать, выводить *(chicks);* воспитывать, обучать; **to ~ true** размножаться в чистоте
breeder *s* производитель; животновод
dog ~ собаковод
fur ~ зверовод
pig ~ свиновод
poultry ~ птицевод
silkworm ~ шелковод
breeding *s* размножение; разведение; выведение; улучшение породы, улучшение сорта; выкармливание; воспитание; воспитание; селекция; вязка *(in dog breeding)*
affinity ~ близкородственное скрещивание
animal ~ звероводство; разведение животных
cattle ~ скотоводство
cross ~ скрещивание
dog ~ собаководство
horse ~ коневодство
pig ~ свиноводство
varietal plant ~ сортовыведение
bregma *s anat* брегма
brephic *a* ювенильный
brevicaudate *a* короткохвостый
brevicolpate *a pal* короткобороздный
breviflexor *s anat* короткий сгибатель
brevifoliate *a* коротколистный
breviped *a* коротконогий
brevipennate *a* короткокрылый
brevirostrate *a* коротколювый
bridge *s* мост; мостик; перемычка
~ of nose переносица
arteriolovenular ~ капиллярная сеть
disulfide ~ *mbiol* дисульфидная связь; дисульфидный мостик
jugal ~ скуловая дуга
bridle *s anat* уздечка; ent уздечка, сцепочный вырост
briery *a* колючий
bright *s* минно-чехликовая моль, моль-чехлоноска *(Incurvariidae)*
brimstone *s* крушинница, лимонница обыкновенная (*Gonepteryx rhamni)*
brindled *a* полосатый; пятнистый
bring *v* приносить; **to ~ forth** рождать, производить; котиться; **to ~ up** выращивать, растить; вскармливать
briskness *s* резвость
bristle *s* щетина; *ent* щетинка
sense ~ осязательный волосок; осязательная щетинка
bristle-bill *s orn* щетинкоклювый бюльбюль *(Bleda)*
bristle-bird *s orn* щетинкоклювка *(Dasyornis)*
bristled *a* имеющий щетинки; щетинистый
prickly ~ шиповато-щетинистый
bristly *a* щетинистый
brittle *a* ломкий, хрупкий; крошащийся
broadbill *s orn* утка-широконоска *(Spatula clypeata)*
broad-leaved *a* широколистный
broad-ringed *a* широкослойный *(of timber)*
brochal *a pal* крупносетчатый, петлевидный
brochus *s pal* петля
brock *s* барсук *(Meles meles)*
brocket *s* двухгодовалый олень; мазама *(Mazama)*
broiler *s* бройлер
broken *a* шероховатый, неровный; сломанный, разбитый; нарушенный, с перерывами *(of sleep);* подорванный, ослабленный *(of health);* вспаханный *(of soil)*
bromdeoxyuridine *s* бромдезоксиуридин, БДУ
brome *s bot* костёр *(Bromus)*
bromoformylation *s* бромформилирование
bromomethylation *s* бромметилирование
bromouracil *s* бромурацил
bronchiole *s anat* бронхиола
bronchomotor *a anat* бронхомоторный
bronchotracheal *a anat* бронхотрахеальный
bronchovesicular *a anat* бронхоальвеолярный
bronchus *s anat* бронх
brood *s* помёт, потомство, выводок; стая; рой; самка-производительница; порода; *apic* детка, черва, расплод; *v* выводить потомство; сидеть на яйцах, высиживать
drone ~ *apic* трутневый расплод
brood-bud *s* соредия
brood-cell *s* гонидия

brood-hen *s* наседка
brooding *s* насиживание
brood-patch *s orn* наседное пятно
broom *s bot* ракитник *(Cytisus)*
witches' ~ *фтп* ведьмина метла *(caused by Crinipellis perniciosa)*
broth *s* бульон; жидкая среда
Hartley's ~ перевар Хартли
hay ~ сенной отвар, сенной настой
brotion, brotium *s* антропогенная сукцессия
brotochore *s* антропохорный организм
brow *s* бровь
brown *a* коричневый, бурый; *s* коричневый краситель
brownbul *s orn* земляной бурый бюльбюль *(Phyllastrephus terrestris)*
browntail *s ent* златогузка *(Euproctis chrysorrhoea)*
browse *s* молодой побег, отпрыск; *v* обрезать побеги; глодать, обгладывать; ощипывать *(e.g., leaves)*
browsing *a* гложущий; ощипывающий
bruise *s* повреждение, желвак, кровоподтёк
brush *s ent* щёточка; пушистый хвост; кустарник; заросль; *bot* бородка; лисохвост луговой *(Alopecurus pratensis); v* зарастать кустарником; обсаживать кустарником
pollen ~ щётка, щёточка *(of a bee)*
brush-like *a bot* кисточковидный
brushrat *s* лесной хомяк, древесная крыса *(Neotoma)*
brushrunner *s orn* жаворонковый печник *(Coryphistera)*
brushwood *s* заросль; кустарник; молодая поросль, подлесок; хворост, валежник
bryocole *s* обитатель мха
bryological *a* бриологический
bryology *s* бриология
bryophytic *a* мохообразный
bryotherophyte *s* бриотерофит, моховой терофит
buccal *a anat* буккальный, щёчный, ротовой
buccinator *s* щёчная мышца
buccogingival *a anat* щёчно-десневой
buccolabial *a anat* щёчно-губной
buck *s* самец *(e.g., of deer, antelope, hare);* козёл; баран; хряк; *v eth* становиться на дыбы; брыкаться
antlered ~ рогатый самец *(оленя)*
blazed ~ блесбок, беломордый бубал *(Damaliscus dorcas)*
blue ~ нильгау *(Boselaphus tragocamelus)*
Bohor reed ~ болотный козёл, обыкновенный редунка *(Redunca redunca)*
bush ~ лесная антилопа *(Tragelaphus);* бушбок *(Tragelaphus scriptus)*
buckeye *s* конский каштан *(Aesculus)*
bucking *s eth* вставание на дыбы
buckle *s myc* пряжка
buckthorn *s bot* жостер *(Rhamnus);* крушина *(Frangula)*
buckwheat *s* гречиха *(Fagopyrum)*
bud *s bot* почка, глазок; давать почки; пускать ростки; бутон; *zool* зачаток; *v* прививать глазком; **to ~ off** отпочковывать(ся)
big ~ *phytop* столбур *(of black currant)*
brood ~ луковичка; выводковая почка
flower ~ цветочная почка, бутон
imaginal ~ *ent* имагинальный диск
limb ~ *emb* почка конечности, зачаток конечности
regeneration ~ регенерационная бластема
seed ~ семяпочка
taste ~ *anat* вкусовой сосочек
bud-bearing *a* почконосный
budbreak *s* распускание почек
budding *s* почкование; окулирование; бутонизация
scion ~ окулировка
top ~ прививка в крону
budgerigar *s* волнистый попугайчик *(Melopsittacus unbulatus)*
budget *s* бюджет
nutrient ~ рацион
budlet *s* почечка
buff *a* палевый
buffalo *s* буйвол, азиатский буйвол *(Bubalus);* африканский буйвол *(Syncerus);* бизон *(Bison bison)*
musk ~ овцебык, мускусный бык *(Ovibos moschatus)*
buffalo-weaver *s orn* буйволовый ткач *(Bubalornis)*

buffer *s* буфер, буферный раствор; буферный ген; *ecol* защитная зона
buffing *s vet* засекание
bufflehead *s orn* малый гоголь *(Bucephala albeola)*
buff-tip *s ent* лунка серебристая, хохлатка-буцефал *(Phalera bucephala)*
bug *s* клоп *(Hemiptera);* мелкое насекомое; мелкое членистоногое
assassin ~ хищнец *(Reduviidae)*
bark ~ подкорник *(Aradidae)*
cannibal ~ хищнец *(Reduviidae)*
caspid ~ слепняк *(Miridae)*
damsel ~ клоп-охотник *(Nabidae)*
lace ~ кружевница *(Tingitidae)*
leaf ~ слепняк *(Miridae)*
leaf-footed ~ краевик (Coreidae)
mirid ~ слепняк *(Miridae)*
pill ~ мокрица *(Armadillidiidae)*
pinch ~ рогач *(Lucanidae)*
pirate ~ хищнец *(Reduviidae)*
plant ~ слепняк *(Miridae)*
red soldier ~ клоп-солдатик, красноклоп бескрылый *(Pyrrhocoris apterus)*
shield-backed ~ клоп-черепашка, щитник-черепашка *(Scutelleridae)*
snapping ~ щелкун *(Elateridae)*
true ~ клоп *(Hemiptera)*
bugle *s bot* живучка ползучая *(Ajuga reptans)*
build *s* строение, конструкция; телосложение; *v* создавать; вить *(a nest);* **to ~ back** востанавливать *(e.g., abundance of a species)*
building *s* строительство
nest ~ гнездостроение
built-in *a* встроенный, вмонтированный; вживлённый, имплантированный
bulb *s* луковица; *anat* продолговатый мозг; луковицеобразное расширение; луковицеобразное утолщение
daughter ~ *bot* дочерняя луковица, детка; зубок
dental ~ зубной сосочек
drip ~ капельница
glass ~ стеклянная колба; стеклянная ампула
Krause's terminal ~ *anat* терминальная колба Краузе
bulbar *a* луковицеобразный
bulbiferous *a* луковиценосный, с луковицами
bulbil *s bot* воздушная луковица, выводковая почка, детка
bulbillate *a* луковичкообразный
bulbing *s* формирование луковицы
bulblet *s bot* воздушная луковица, выводковая почка, детка
bulblike *a* луковицеобразный
bulbotuber *s* клубнелуковица
bulbous *a* луковицеобразный
bulbul *s* бюльбюль, бульбуль *(Pycnonotidae)*
bulbus *s* луковица
bulge *s* вздутие, выступ; *mbiol* петля; скула
bulging *a* разбухший; выпуклый
bulging-out *a* выпячивающийся
bulk *s* основной объём, основная масса
~ **of food** основная масса пищи
bull *s* бык; буйвол; самец *(of a large animal)*; секач; *a* бычий
blue ~ нильгау *(Boselaphus tragocamelus)*
sacred ~ зебу *(Bos indicus)*
bulla *s* пузырь; вздутие
bullace *s* слива домашняя *(Prunus domestica)*
bullate *a* пузырчатый; сморщенный
bulldog *s* бульдог
bullfinch *s* снегирь *(Pyrrhula)*; снегиревая овсянка *(Loxigilla)*
bullfrog *s* лягушка-бык *(Rana catesbiana)*
bullock *s* валух; вол
bully *a eth* драчливый
bulrush *s* камыш *(Scirpus)*
bumblebee *s* шмель *(Bombus)*
bump *s* нарост, шишка; буханье, крик выпи; *v* бухать, кричать *(of bittern)*
bumpy *a* бугорчатый
bunch *s* пучок; гроздь; дерновина; куст; группа; стадо; *v* сбиваться в стадо; сгонять в стадо
bunch-forming *a* кустовой; кустообразующий
bunchgrass *s* кустовой злак
bunching *s*: ~ **in herd** сбивание в стадо
bunchlike *a* гроздевидный, пирамидальнометельчатый
bunchy *a* кустовой
bundle *s* пучок; тяж

~ **of Vicq d'Azyr** пучок Вик д'Азира, сосковидно-бугорный путь
circular ~s anat круговые пучки, волокна Мюллера
collateral ~ коллатеральный пучок
meridional ~s anat меридиональные пучки, волокна Брюкке
Monakow's ~ монаков путь, красноядерноспинальный путь
bunk *s bot* цикорий *(Cichorium)*
bunt *s phytop* твёрдая головня пшеницы, мокрая головня пшеницы, вонючая головня пшеницы *(caused by Tilletia caries, Tilletia tritici)*
bunting *s orn* овсянка *(Emberiza)*
rustic ~ овсянка-ремез *(Emberiza rustica)*
snow ~ пуночка *(Plectrophenax nivalis)*
yellow ~ обыкновенная овсянка *(Emberiza citrinella)*
bunyavirus *s* буньявирус
buoyancy *s* плавучесть
bur *s* шип, колючка; колючее растение
burden *s*: груз
genetic ~ генетический груз
burdock *s* лопух *(Arctium);* дурнишник *(Xanthium)*
burdon *s* бурдон
burette *s* бюретка
burgeon *s* почка; побег; *v* распускаться, пускать почки; прорастать
burgomaster *s orn* бургомистр, большая полярная чайка *(Larus hyperboreus)*
burl *s bot* кап, наплыв
bur-marigold *s bot* череда *(Bidens)*
burn *s* гарь, пожарище; выжигать; ожог
burnet *s ent* пестрянка *(Zygaenidae); bot* кровохлёбка *(Sanguisorba)*
burning *s* выжигание, сжигание, огневая очистка, пал; *phytop* ожог
burnt-over *a* выжженный
burnut *s bot* якорцы *(Tribulus)*
burr *s* шип, колючка; колючее растение; розетка рога
bur-reed *s* ежеголовник *(Sparganium)*
burrow *s* нора, норка, ход; *v* рыть нору; жить в норе, прятаться в норе
marmot's ~ сурчина
ventilating ~ *ent* вентиляционная дыра *(in burrows of bark beetles)*
burrower *s* роющее животное
bursa *s anat* бурса, сумка
bursacyte *s* В-лимфоцит
bursicle *s bot* сумочка
bursiform *a* сумкообразный, мешкообразный
burst *v* внезапно появляться; *ent* вылупляться; *bot* прорастать; раскрываться; лопаться; растрескиваться; бурст, бурст-очаг; **to ~ into leaf** зазеленеть; **to ~ into bloom, to ~ into flower** *v* расцветать
phage ~ выход фага
burster *s*: **egg** ~ *ent* эмбриональный зубец
bury *v* зарывать; закапывать
bush *s* куст; кустарник; *v* обсаживать кустарником; густо разрастаться
bushbaby *s ther* галаго *(Galago)*
bushbuck *s ther* лесная антилопа *(Tragelaphus);* бушбок, антилопа гуиб, пестрая лесная антилопа *(Tragelaphus scriptus)*
bushchat *s orn* чекан *(Saxicola)*
bush-cricket *s ent* настоящий кузнечик, кузнечиковое *(Tettigoniidae)*
bushiness *s* кустистость
bushing *s* закустаривание
bush-lark *s* кустарниковый жаворонок *(Mirafra)*
bushmaster *s herp* бушмейстер, сурукуку *(Lachesis muta)*
bushtit *s* кустарниковая синица *(Psaltriparus)*
bushy *a* кустистый; пушистый, лохматый; покрытый кустарником
bustard *s* дрофа *(Otis)*
great ~ дрофа *(Otis tarda)*
houbara ~ джек, вихляй, дрофа-красотка *(Chlamydotis undulata)*
little ~ стрепет *(Otis tetrax)*
butcherbird *s* флейтовая птица *(Cracticus)*
butt *s* комель; удар рогами; *v* бодать
buttercup *s* лютик *(Ranunculus)*
double ~ купальница европейская *(Trollius europaeus)*
fig-root чистяк весенний *(Ficaria verna)*
butterfly *s* дневная бабочка, мотылёк
apollo ~ аполлон *(Parnassius appollo)*
brimstone ~ лимонница обыкновенная, крушинница *(Gonepteryx rhamni)*

cabbage ~ репница, белянка репная *(Pieris rapae)*
cabbage white ~ капустница, белянка капустная *(Pieris brassicae)*
gossamer-winged ~ голубянка *(Lycaenidae)*
leaf ~ листовидка, каллима *(Kallima)*
milkweed ~данаида *(Danais)*
monarch ~ монарх *(Danaus plexippus)*
mourning-clock ~ траурница *(Vanessa antiopa)*
peacock ~ павлиний глаз дневной *(Vanessa io)*
satyr ~ бархатница, сатирида *(Satyridae)*
sulfur ~ белянка *(Pieridae)*
swallowtail ~ парусник *(Papilionidae)*
thistle ~ ванесса *(Vanessa)*
thorn ~ боярышница *(Aporia crataegi)*
turnip ~ репница, белянка репная *(Pieris rapae)*

butterwort *s bot* жирянка *(Pinguicula)*
buttness *s* бодливость
buttock *s anat* ягодица; бедренная кость
button *s* бугорок
butylation *s* бутилирование
buzz *s* жужжание; *v* жужжать
buzzard *s* сарыч, канюк *(Buteo);* обыкновенный канюк, сарыч *(Buteo buteo)*
common ~ обыкновенный канюк, сарыч *(Buteo buteo)*
honey ~ осоед *(Pernis);* длиннохвостый осоед *(Henicopernis)*
long-legged ~ курганник *(Buteo rufinus)*
rough-legged ~ зимняк, мохноногий канюк *(Buteo lagopus)*
turkey ~ гриф-индейка *(Cathartes aura)*
upland ~ мохноногий курганник *(Buteo hemilasius)*

buzzing *s* жужжание; комариный писк
by-catch *s* прилов
by-effect *s* побочное действие, побочный эффект
by-product *s* побочный продукт; побочный результат

C

cabinet *s* шкаф
drying ~ сушильный шкаф
temperature ~ термостат
cabretta *s* кабретта
cache *s eth* запас пищи; *v* припрятывать пищу
cachexia, cachexy *s* кахексия
caching *s*: **food** ~ запасание корма; устраивание кладовых
cackle *s* кудахтанье, клохтанье, гоготанье; *v* кудахтать, клохтать, гоготать
cackler *s* кудахтающая птица, клохчущая птица
cacogenesis *s* какогенез
cacorhythmic *a* аритмичный
cactaceous *a* кактусовый
cactus *s* кактус *(Cactus)*
cadastre *s* кадастр
cadaver *s* труп
cadaveric *a* трупный
cadavericole *a* живущий на трупах
cadaverous *a* трупный
caduca *s anat* отпадающая оболочка, децидуальная оболочка
caducous *a* отпадающий, временный; *bot* рано опадающий
caecilian *s herp* червяга *(Caeciliidae)*
caecum *s* слепая кишка; слепой отросток
caenogenesis *s* ценогенез
caespitose *a* дернистый, дерновый, растущий пучками
caffeine *s* кофеин
cage *s* клетка, садок; *v* сажать в клетку; содержать в клетке
bony ~ костный остов
holding ~ садок *(для насекомых)*
rearing ~ инсектарий, садок для насекомых
terrarium ~ террариум
caiman *s* очковый кайман *(Caiman)*
cake *s* сгусток (*крови); жмых
manure ~ кизяк
caked *a* отвердевший, затвердевший; сгущённый; осевший; свернувшийся
calabash *s* тыква бутылочная, горлянка *(Lagenaria vulgaris)*
calamary *s* кальмар *(Teuthoidea)*
giant ~ гигантский кальмар, архитевтис *(Architeuthidae)*
calamity *s* бедствие
calamus *s* полая часть пера; пальма ротанговая *(Calamus rotang)*
calathidium, calathium *s bot* корзинка
calcaneus *s* пяточная кость
calcar *s анат.* шпора; *anat* ~ **avis** птичья шпора
calcarate *a* имеющий шпоры
calcicole, calcicolous *a* кальцефильный
calciferous *a* содержащий кальций
calcification *s* кальцификация, обызвествление, накопление солей кальция
calcifuge *s* кальцефоб
calcifugous *a* кальцефобный
calcipexy *s* фиксация кальция в организме
calciphil *s* кальцефил
calciphilous *a* кальцефильный
calciphobe *s* кальцефоб
calciphobous *a* кальцефобный
calciphyte *s* кальцефит, меловое растение, известковое растение
calciprivic *a* лишённый кальция
calculation *s* вычисление; расчёт, подсчёт
back ~ обратное расчисление
calculus *s* исчисление; камень, конкремент
calf *s* телёнок; детёныш *(e.g., of an elephant or whale);* икра *(of a leg);* **in** ~ стельная *(a cow)*
sea ~ обыкновенный тюлень, ларга *(Phoca vitulina)*
calicle *s zool* гидротека; *бот.* эпикаликс, подчашие
calico *s phytop* мозаичная болезнь; калико
call *s* крик, призыв, сигнал; манок
let-go ~ призыв к сближению
warning ~ звук, предупреждающий об опасности
callosal *a* мозолистый; *bot* каллозный
callose *a* имеющий каллюс, имеющий наплыв
callosity *s* мозолистость; затверделость, утолщённость
callous *a* имеющий каллюс, имеющий наплыв

callus *s* каллюс, мозоль, мозолистое утолщение; *bot* каллюс, наплыв; бугорок
callusing, callusogenesis *s* каллюсогенез, каллюсооразование
caloradiance *s* тепловое излучение
caloric *a* теплотворный
caloricity *s* калорийность; теплотворная способность
calorie *s* калория
calorific *a* калорийный; теплообразующий
calorification *s* выделение тепла
calorigenic *a* калорийный; теплообразующий
calorimetry *s* калориметрия
calorstat *s* термостат
calory *s* калория
caltrop *s* четырёхлучевая спикула; *bot* рогульник, чилим, водяной орех *(Trapa);* кальстремия *(Kallstroemia);* якорцы *(Tribulus)*
calvaria *s* свод черепа
calve *v* отелиться, родить детёныша
calver *s* стельная корова
calville *s* кальвиль
calving *s* отёл
calyciflorous *a* чашечкоцветковый
calyciform *a* чашечкообразный, чашечковидный, чашевидный
calycine *a* чашечкообразный, чашечковидный, чашевидный; *бот.* относящийся к чашечке, чашечковый
calycle *s zool* гидротека; *bot* эпикаликс, подчашие
calycular *a bot* подчашечный
calyculate,calyculated *a bot* с подчашием
calyculus *s zool* гидротека
calypter *s* корневой чехлик; *ent* закрыловая пластинка
calyptra *s* калиптра, колпачок, чехлик *(in mosses);* корневой чехлик
calyptrate *a ent* с закрыловой пластинкой; *bot* с калиптрой
calyptriform *a bot* колпачковидный; чехликообразный
calyptrogen *s* калиптроген
calyx *s* чашечка; *ent* чашевидная часть
 outer ~ подчашие
cambial *a* камбиальный
cambiogenetic *a* образующий камбиальный слой
cambium *s* камбий
 cork ~ феллоген, пробковый камбий
camel *s* верблюд *(Camelus)*
 Arabian ~ одногорбый верблюд, дромадер, дромедар *(Camelus dromedarius)*
 Bactrian ~ двугорбый верблюд, бактриан *(Camelus bactrianus)*
camel's-thorn *s bot* верблюжья колючка *(Alhagi camelorum)*
camera *s* камера; полость
camnium *s* антропогенная сукцессия
camomile *s* пупавка *(Anthemis);* ромашка *(Matricaria);* первоцвет *(Primula)*
camouflage *s* маскировка
campaniform *a* колоколобразный, колокольчатый
campion *s bot* смолёвка *(Silene);* волдырник *(Cucubalus);* смолка *(Lychnis)*
campodeiform *a* камподеоидный
camptocarpous *a* согнутоплодный
camptodrome *a* камптодромный, дуговидный
camptotropous *a* кампилотропный
campylobotryose *a* согнутокистевой
campylocanthous *a* согнутоиглый
campylodentatum *a* согнутозубчатый
campylospermous *a* согнутосемянный
campylotropous *a* кампилотропный
canal *s* канал
 alimentary ~ пищеварительный тракт
 anal ~ заднепроходный канал, анальный канал, анальная часть прямой кишки
 arterial ~ артериальный проток, боталлов проток
 ciliary ~ шлеммов канал, венозный синус склеры
 Corti's ~ внутренний туннель, кортиев туннель
 Cotunnius' ~ эндолимфатический проток, вестибулярный водопровод
 dentinal ~ дентинный каналец
 ear ~ наружный слуховой проход
 entodermal ~ *emb* кишечная трубка
 external auditory ~ наружный слуховой проход
 gum ~ *bot* смоляной ход
 His' ~ щитовидно-язычный проток, канал Гиса
 Hoyer's ~ артериовенозный анастомоз

intestinal ~ кишечник
medullary ~ позвоночный канал; костно-мозговая полость
parturient ~ родовой канал, родовые пути
Schlemm's ~ шлеммов канал, венозный синус склеры
Van Hoorne's ~ грудной проток
vestibular ~ лестница преддверия, вестибулярная лестница
vitelline ~ желточный проток
canalicular *a* имеющий небольшие каналы
canaliculus *s* каналец
canalifolious *a* желобчатолистный
canalization *s* васкуляризация; канализация; *phys* формирование рефлекторной дуги
canary, canarybird *s* канареечный вьюрок *(Serinus);* канарейка, канарский канареечный вьюрок *(Serinus canaria)*
cancellate, cancellated *a анат.* решётчатый, сетчатый, губчатый
cancellation *s anat* сетчатое строение, губчатое строение; *phys.* взаимопогашение
cancellous *a anat* решётчатый, сетчатый, губчатый
cancer *s* рак
cancerogenic *a* канцерогенный
cancerology *s* онкология
cancerous *a* раковый, злокачественный
cancrology *s* онкология
candidiasis *s* кандидоз *(caused by Candida albicans)*
candle *s* свеча
cane *s* камыш *(Scirpus);* тростник *(Phragmites);* сахарный тростник *(Saccharum);* лоза; лозы, стебли кустарников
fruiting ~ плодоносящий побег
rattan ~ ротанг *(Peddirohr)*
canids *s pl* собачьи, псовые *(Canidae)*
canine *s* клык; *pl* собачьи,псовые (Canidae*); a* собачий, псовый
canker *s* язва; рак растений; некроз плодовых деревьев; червоточина; *v* изъявлять, разъедать; заражать; губить; разлагать; пожирать
cannibal *s* каннибал
cannibalism *s* каннибализм
cannon *s* плюсна; пясть
fore ~ пясть
hind ~ плюсна
cannula *s* канюля
canopy *s* покров; листовой полог, древесный полог
grass ~ травяной покров, полог трав; травостой
canter *s* кентер, лёгкий галоп
canthus *s* угол глазной щели
cantle *s* долька
cap *s* верхушка; шляпка *(of a mushroom);* чехлик; колпачок, калиптра; *mbiol* кэп
apical ~ *cyt* апикальная шапочка
bishop's ~ луковица двенадцатиперстной кишки
death ~ бледная поганка *(Amanita phalloides)*
duodenal ~ луковица двенадцатиперстной кишки
head ~ of the sperm головной чехлик спермия
ink ~ чернильный гриб *(Coprinus)*
knee ~ коленная чашка
root ~ корневой чехлик
skull ~ свод черепа, крыша черепа
capacious *a* вместительный
capability *s* способность
capacity *s* способность; объём, ёмкость; мощность, производительность
aerobic ~ for work аэробная работа, максимальная аэробная производительность
allergenic ~ аллергенность
blotting ~ всасываемость, впитываемость; впитывающая способность, всасывающая способность
body ~ объём тела
breathing ~ дыхательная способность; дыхательный объем
buffer ~ буферность
calorific ~ теплотворная способность; теплоёмкость
capillary moisture ~ капиллярная влагоёмкость *(of soil)*
carrying ~ переносимый объём; производительность, кормовая продуктивность
cellulose decomposing ~ целлюлозоразлагающая способность
chest ~ объём грудной клетки

coupling ~ связывающая ёмкость; связывающая способность
daily ~ суточная производительность
egg-laying ~ яйценоскость
equilibrum ~ функциональная остаточная ёмкость *(of lungs)*
evaporating ~ испаряемость
exchange ~ ёмкость обмена; обменная способность
filter ~ производительность фильтра
food ~ кормность
fruit-bearing ~ плодоносность
functional ~ функциональная способность; функциональная ёмкость
germinating ~ всхожесть
grazing ~ ёмкость пастбища
growing ~ энергия роста
heat ~ теплоёмкость
hereditary ~ препотентность
immunizing ~ иммуногенность
impregnation ~ оплодотворяемость
keeping ~ лёжкость
layering ~ кустистость
maximal breathing ~ максимальная дыхательная способность, максимальная вентиляция лёгких
membrane ~ электроёмкость мембран, мембранная ёмкость
milking ~ удойность
moisture ~влагоёмкость
net ~ полезная ёмкость; полезная производительность
producing ~ продуктивность, производительная способность
residual lung ~ остаточная ёмкость легких
respiratory ~ дыхательный объём
shoot-forming ~ побегопроизводительность
soil water ~ влагоёмкость почвы
tilling ~ кустистость
total lung ~ общая ёмкость лёгких, полная ёмкость лёгких
water ~ влагоёмкость *(of soil)*
work(ing) ~ работоспособность

caper, caperbush *s* каперсы *(Capparis)*
capercaillie *s* глухарь *(Tetrao);* глухарь *(Tetrao urogallus)*
capillaceous *a* волосной, капиллярный; волосовидный
capillarimeter *s* капиллyриметр
capillarity *s* капиллярность
capillaroscope *s* капилляроскоп
capillaroscopy *s* капилляроскопия
capillary *s* капилляр; *a* капиллярный
capilliform *a* волосовидный
capillitium *s bot* капиллиций
capitate *a* имеющий головку, головчатый; в форме головки; собранный в головку
capitatum *s* головчатая кость
capitellate *a bot* мелкоголовчатый
capitellum *s* головка плечевой кости
capitulum *s bot* головка
capping *s* кэппирование; *imm* образование «шапочки», кэппинг(-феномен); забрасывание ног *(in a horse); apic* печатка, крышечка
capreolate *a бот.* имеющий усики
capricorn *s ent* усач, дровосек
caprification *s* капрификация
caprifig *s* каприфига
capsicum *s* перец однолетний *(Capsicum annuum)*
capsid *s* капсид
capsidation *s* капсидация
capsomere *s* капсомер
capsular *a* капсульный
capsule *s anat* капсула; сумка; коробочка; оболочка; кокон; *bot* урна
end ~ окончание чувствительного нерва
seed ~ семенная коробочка
suprarenal ~ надпочечник
Tenon's ~ влагалище глазного яблока, тенонова капсула
capsuliferous *a anat* с капсулой; *bot* с коробочкой
captive *a* находящийся в неволе
captivity *s* неволя
capture *s* поимка; отлов; добыча; *v* отлавливать
captured *a* пойманный
capuchin *s ther* капуцин *(Cebus)*
capybara *s ther* капибара, водосвинка *(Hydrochoerus hydrochaeris)*
caracal *s ther* каракал, степная рысь *(Felis caracal)*
carapace *s* панцирь, щит, щиток
caraway *s* тмин *(Carum)*
carbamoylation *s* карбамилирование, карбамоилирование

carbamoylethylation *s* карбамилэтилирование, карбамоилэтилирование
carbamylation *s* карбамилирование, карбамоилирование
carbamylethylation *s* карбамилэтилирование, карбамоилэтилирование
carbobromination *s* бромформилирование
carbohydrate *s* углевод
carbon *s* углерод
activated ~ активированный уголь
carbonaceous *a* богатый углеродом; содержащий углерод; углеродный; углеродистый
carbonic *a* содержащий углерод
carboxyl *a* карбоксильный
carboxylation *s* карбоксилирование
reductive ~ восстановительное карбоксилирование
carboxylic *a* карбоксильный
carboxymethylation *s* карбоксиметилирование
carcass *s* туша; тушка; скелет; тело
carcerule *s bot* карцерулла
carcinogen *s* канцероген, канцерогенное вещество, онкогенное вещество
carcinogenesis *s* канцерогенез, онкогенез
carcinogenic *a* канцерогенный, онкогенный
carcinology *s* карцинология
carcinolysis *s* канцеролиз
carcinolytic *a* канцеролитический, онколитический, разрушаюший раковые клетки
carcinoma *s* карцинома, рак
carcinous *a* раковый
card *s* карточка
record ~ карточка для регистрации
cardamom, cardamon *s* кардамон *(Eletraria cardamomum)*
cardiac *a* сердечный; кардиальный
cardinal *s orn* кардинал *(Cardinalis);* кардиналовая овсянка *(Paroaria)*
cardioaccelerator *s* кардиостимулятор; кардиостимулирующий
cardiogenic *a* кардиогенный, сердечного происхождения
cardiogram *s* кардиограмма
cardiograph *s* кардиограф
cardiography *s* кардиография
cardiokinetic *a* возбуждающий сердце
cardiomensurator *s* кардиомензуратор
cardiomuscular *a* относящийся к миокарду
cardionector *s* проводящая система сердца
cardiopneumatic *a* сердечно-лёгочный
cardioscope *s* электрокардиоскоп
cardiotachometer *s* кардиотахометр, сердечный тахометр
cardiovascular *a* сердечно-сосудистый
cardo *s* замок *(of bivalves);ent* кардо, основной членик максиллы
care *s* уход, забота; **to take** ~ заботиться
career *s* карьер
caress *s* ласка; *v* ласкать; гладить
caribou *s* карибу, северный олень *(Rangifer tarandus)*
caries *s* кариес
carina *s anat* киль; *bot* лодочка, киль
carinal *a anat* килевой
carinate, carinated *a anat* имеющий киль; *bot* ладьевидный, килеватый
cariniform *a bot* ладьевидный, килеватый
carinimurate *a bot* гребнеперегородчатый
carmine *s* кармин
carnation *s* гвоздика *(Dianthus)*
carnivore *s* плотоядное животное, хищник; *pl* хищные *(Carnivora);* насекомоядное растение
fin-footed ~ ластоногое *(Pinnipedia)*
carnivorism *s* хищничество
carnivorous *a* плотоядный, хищный
carnose *a* мясистый
carotene *s* каротин, провитамин A
carotenogenesis *s* каротиногенез
carotenoid *s* каротиноид
carotid *s* сонная артерия; *a* каротидный
carotin *s* каротин, провитамин A
carpal *s* запястный сустав; *a* запястный, кистевой
carpel *s bot* карпель, плодолистик; плодник; плодик
carpet *s* ковёр, покрытие
~ **of leaves** лесная подстика
carpodia *s pl* карподии, недоразвитые плодолистики
carpogenous *a* образующий плоды
carpogonium *s* карпогоний
carpology *s* карпология

carpophagous *a* плодоядный, питающийся плодами
carpophore *s* карпофор, плодонос, разросшееся цветоложе
carpophyte *s* карпофит
carposome *s бот.* плодовое тело
carposperm *s* карпосперм, оплодотворённая оосфера
carposporangium *s* карпоспорангий
carpospore *s* карпоспора
carpus *s* запястье
carriage *s* носительство; осанка, посадка
carrier *s* носитель, переносчик; *gen* носитель рецессивного аллеля
bacilli ~ бациллоноситель
germ ~ бациллоноситель
molecule ~ молекулоноситель
parasite ~ паразитоноситель
virus ~ вирусоноситель
carrion *s* падаль; мертвечина; *a* гнилой; гниющий, разлагающийся
carrot *s* морковь *(Daucus carota)*
carry *v* нести
~ **out** проводить
carrying *s* носительство
virus ~ вирусоносительство
carry-over *s* вынос
cartilage *s* хрящ
costal ~ рёберный хрящ; хрящевая часть ребра
cartilagination, cartilaginification *s* образование хряща, охрящевение
cartilaginoid *a* хрящевидный
cartilaginous *a* хрящевой
cartridge *s*:
cloning ~ обойма клонирования
caruncle *s anat* мясцо; сосочек; бородавчатое возвышение; *orn* гребень; *bot* карункула, присеменник
lacrimal ~ слёзное мясцо
Morgagni's ~ средняя доля предстательной железы
sublingual ~ подъязычный сосочек
caryocarpous *a* орехоплодный
caryogamy *s* кариогамия
caryokinesis *s* кариокинез
caryology *s* кариология
caryonide *s* карионида
caryoplasmous *a* кариоплазматический
caryopsis *s bot* зерновка
case *s* коробка; капсула, оболочка; чехол; *энт.* домик
brood ~ выводковая камера; зародышевая камера
egg ~ защитная оболочка яйца
spore ~ спорангий
wing ~ надкрылье, элитра
casebearer *s* чехлоноска; чехликовая моль *(Coleophoridae)*
caseous *a* творожистый
caseweed *s* сумочник пастуший, пастушья сумка *(Capsella bursa-pastoris)*
cashew *s bot* анакардия западная *(Anacardium occidentale)*
cassava *s* маниок *(Manihot)*
cassette *s* кассета; *a* кассетный
cassideous *a* шлемовидный
cassowary *s orn* казуар *(Casuarius)*
cast *s* сбрасывание, опадание; *v* сбрасывать, ронять, опадать; *pl* выбросы, экскременты; терять *(teeth); v* менять *(horns);* выкинуть, родить раньше времени; **to ~ forth** метать икру; **to ~ its coat** линять; **to ~ off** спускать *(dogs)*; выпускать *(falcons)*; **to ~ out** вымётывать икру
earthworm ~s выбросы дождевых червей
mole ~ кротовина
pine-leaf ~ *phytop* шютте, болезнь сеянцев сосны *(caused by Lophodermium pinastri)*
caste *s ent* каста
casting *s orn* погадка
casting-off *s* отторжение
castling *a* недоношенный
castor *s* бобр *(Castor fiber)*; американский бобр *(Castor canadensis);* бобровая струя
castoreum *s* бобровая струя
castrate *s* кастрат; кастрировать; кастрированный
castration *s* кастрация
cat *s* кот; кошка; домашняя кошка *(Felis domesticus)*
black ~ илька, пекан *(Martes pennanti)*
cacomistle ~ какомицли *(Bassariscus)*
civet ~ африканская цивета *(Civettictis civetta)*
fossa ~ фосса *(Cryptoprocta)*
hydrophobic ~ пятнистый скунс *(Spilogale putorius)*

jungle ~ хаус, камышовая кошка, болотная рысь *(Felis chaus)*
manul ~ манул *(Felis manul)*
native ~ пятнистая сумчатая куница *(Dasyurus)*
Pallas' ~, steppe ~ манул *(Felis manul)*
catabolic *a* катаболический
catabolism *s* катаболизм, диссимиляция
catabolite *s* катаболит; *a* катаболитный
catacrotism *s phys* катакротизм
catadromous *a* катадромный, направленный книзу
catagenesis *s* катагенез, ретрогрессивная эволюция
catalepsy *s* каталепсия
catalo *s см.* **cattalo**
catalysis *s* катализ
catalyst, catalyzer *s* катализатор
catamount, catamountain *s* пума *(Felis concolor);* рыжая рысь *(Lynx rufus)*
cataphoresis *s* катафорез
cataphyll *s* катафилл, чешуевидный лист
cataplasia *s* катаплазия, анаплазия
catastasis *s* катастаз
catastrophism *s* теория катастроф
catatonia *s* кататония
catatonic *a* кататонический
catawba *s* виноград лабруска, катауба *(Vitis labrusca)*
catch *s* улов; ловить; *ent* зацепка вилочки; самосев
allowable ~ допустимый улов
incidental ~ прилов
catchability *s* улавливаемость
catcher *s* кулик-сорока *(Haematopinae)*
catchfly *s* растение с липким секретом; смолёвка *(Silene)*
catchment *s* водосбор
categorize *v* классифицировать
category *s* категория; класс; группа
catelectrotonus *s phys* катэлектротон
catena *s* цепь, связь, ряд
catenate *a* расположенный цепочками; цепочковидный
catenation *s* образование цепочек
cateniferous, catenigerous *a* несущий цепочки, цепочконосный
catenular, catenulate *a* расположенный мелкими цепочками
catenuliform в виде мелких цепочек
caterpillar *s* гусеница; ложногусеница
catgut *s* кетгут
catheter *s* катетер
catheterization *s* катетеризация
catkin *s bot* серёжка
cat-o'-mountain *s* пума *(Felis concolor);* рыжая рысь *(Lynx rufus)*
catoptric *a* катоптрический, отражательный
cat's-eye *s bot* вероника дубравная *(Veronica chamaedris); мол.* натика *(Natica)*
cat's-foot *s bot* кошачья лапка *(Antennaria)*
cat's-tail *s bot* рогоз *(Typha)*
cattalo *s* коровобизон
cattish *a* кошачий
cattle *s* скот; крупный рогатый скот
cauda *s* хвост; хвостообразный придаток
caudad *adv* каудально
caudal *a* каудальный, хвостовой; задний
caudalization *s* каудализация
caudate *a* хвостатый; с хвостообразным придатком
caudex *s bot* каудекс
caudicle *s* хвостик; *bot* ножка поллиния
caudiferous *a* хвостатый; с хвостообразным придатком
caulicle *s bot* подсемядольное колено
caulid *s* каулидий
cauliflorous *a* способный к каулифлории
cauliflory *s* каулифлория
cauliflower *s* капуста цветная *(Brassica oleracea var. botrytis)*
cauliform *a* стеблеобразный, стеблевидный
cauligenous *a* возникающий на стеблях
caulimovirus *s* колимовирус
cauline *a* стеблевой; растущий на стебле
caulis *s* стебель *(in herbs)*
caulocarpous *a* каулокарпный
caulorrhizous *a* стеблекорневой
causal *a* причинный, каузальный
causation *s* причинность; причинная связь, зависимость
cause *s* причина, основание, повод, фактор; *v* быть причиной, служить поводом; вызывать; причинять
first ~ первопричина
internal ~ эндогенный фактор
ultimate ~ первопричина
cause-and-effect *s* причина и следствие; причинно-следственный
cauterization, cautery *s* каутеризация

cava *s* полая вена
cave *s* пещера; полость
cave-dwelling *a* пещерный
cavernicolous пещерный
cavernosum *s* пещеристое тело
cavicorn *a* полорогий
cavitary *a* полостной
cavitation *s* кавитация, образование полости
cavity *s анат.* полость, впадина; лунка *(of a slide)*
arachnoid ~ подпаутинное пространство
cerebral ~ желудочек мозга
cleavage ~ *emb* бластоцель
cranial ~ полость черепа
orbital ~ глазная впадина, орбита
pharyngonasal ~ носоглотка
sap ~ вакуоль
segmentation ~ *emb* бластоцель
stomatal ~ *bot* дворик устьица
cavy *s* морская свинка *(Cavia)*
giant water ~ водосвинка *(Hydrochoerus)*
restless ~ морская свинка *(Cavia porcellus)*
rock ~ горная свинка, моко *(Kerodon rupestris)*
spotted ~ пака *(Cuniculus paca)*
water ~ водосвинка *(Hydrochoerus)*
caw *s* карканье
cay *s* коралловый риф; песчаная отмель, банка
cecidium *s phytop* галл
cecum *s* слепая кишка; слепой отросток
hepatic ~ *zool* печёночный вырост
cedar *s* кедр *(Cedrus);* можжевельник *(Juniperus);* цедрела *(Cedrela);* туя *(Thuja);* торрейя *(Torreya);* кипарис *(Cupressus)*
ceiling *s* предельный уровень; максимальное количество
population ~ максимальная численность популяции, предел популяции
stabilize d population ~ стабилизированный предел популяции
celandine *s bot* чистотел *(Chelidonium);* василистник полигамный *(Thalictrum polygamum)*
celeriac, celery *s* сельдерей *(Apium)*
cell *s* клетка; ячейка; камера
~ **of Bergmann's fiber** клетка бергмановского волокна, глиоцит с султаном
~ **of Fanana** клетка Фанана, лучистый глиоцит
accessory ~ *ent* дополнительная ячейка; обкладочная клетка; *bot* побочная клетка *(of stomata); imm. pl* А-клетки
adipose ~ липоцит, жировая клетка
adventitional ~ адвентициальная клетка, клетка Руже, клетка Маршана, перицит
air ~ лёгочная альвеола; воздушная полость, воздухоносная полость
amoebocytic ~ амёбоцит
attacker ~ клетка-эффектор; клетка-киллер
atypical cardiac muscle ~ атипическая сердечная мышечная клетка, атипический сердечный миоцит, волокно Пуркинье
Baer's ~ яйцеклетка
basal ~ базальная клетка; базальная ячейка
basophil ~ базофильная клетка, бета-клетка
bee's ~s соты
beta ~ базофильная клетка, бета-клетка
Betz ~ клетка Бетца, гигантопирамидальный нейрон
B-lineage ~ В-клетка
blood ~ клетка крови, гемоцит
bone ~ костная клетка, остеоцит
cameloid ~ овалоцит, эллиптоцит
cartilaginous ~ хрящевая клетка, хондроцит
centroacinar ~ центроацинозный эпителиоцит, цетроацинозная клетка
chromatophore ~ хроматофор
cleavage ~ бластомер
collagen-producing ~ фибробласт, коллаген-синтезирующая клетка
collar ~, collared flagellate ~ воротничковая клетка, хоаноцит
companion ~ *bot* клетка-спутник
ConA-induced suppressor ~ ConA-супрессор
cone ~ колбочка сетчатки, колбочковидная зрительная клетка

contrasuppressor T ~ контрасупрессорный Т-лимфоцит, Т-клетка-контрасупрессор
Coombs'positive red ~ эритроцит, нагруженный изогемагглютинином
counting ~ счётная камера
couple ~ зигота
Craig-type dialysis ~ ячейка для микродиализа, диализатор Крейга
delayer-type T ~ Т-лимфоцит-дилэйер
demilune ~s полулунные клетки; полулуния Джиануцци
dentinal ~ дентинная клетка, одонтобласт
dust ~s пылевые клетки *(in lungs)*
egg ~ яйцеклетка
enamel ~ эмалевая клетка, адамантобласт, амелобласт
eosinophil ~ эозинофил, эозинофильный лейкоцит
ependymal ~ эпендимоцит, эпендимная клетка
epithelial ~ эпителиальная клетка, эпителиоцит
epithelial glial ~ глиальная клетка Гольджи, эпителиальный глиоцит
E-rosetting ~ Е-розеткообразующая клетка, Е-РОК
ethmoidal air ~ *anat* решётчатая ячейка
fat, fatty ~ жировая клетка, липоцит
feeder ~ питающая клетка, подкармливающая клетка, клетка-кормушка, клетка-фидер
flow-through ~ проточная кювета *(in chromatography)*
founder ~ стволовая клетка; клетка-основательница
ganglionic nerve ~ ганглиозная нервная клетка, ганглиозный невроцит
generative ~ половая клетка; гамета; *bot* генеративная клетка
genetical storage ~ генетическая ячейка памяти
gland, glandular ~ железистая клетка, гландулоцит
glia ~ клетка невроглии, глиоцит
glial Golgi ~ глиальная клетка Гольджи, эпителиальный глиоцит
Golgi ~ клетка Гольджи, звездчатая нервная клетка, звездчатый невроцит
granule ~ нервная клетка-зерно, невроцит-зерно
helper ~ клетка-хелпер, клетка-помощник
hemic ~ клетка крови, гемоцит
hepatic ~ гепатоцит, клетка печени
Hersch's ~ ячейка Герша
high-producer ~ клетка-суперпродуцент
horizontal spindle-shaped nerve ~ горизонтальный веретенообразный невроцит
Hortega ~ клетка Гортега, спинальный дендроглиоцит
host ~ клетка-хозяин; клетка хозяина
hybrid ~ гибридная клетка, клетка-гибрид
immune ~, immunocompetent ~иммунокомпетентная клетка, иммуноцит
immunologically committed lymphoid ~ иммунокомпетентный коммитированный лимфоцит
intestinal acidophil ~ клетка Панета, ацидофильная кишечная клетка, ацидофильный энтероцит
islet ~ островная клетка, инсулярная клетка, инсулоцит
juvenile ~ метамиелоцит, юный гранулоцит
juvenile B ~ ювенильный В-лимфоцит, нестимулированный В-лимфоцит
killer ~ клетка-киллер
Kupffer ~ клетка Купфера, звездчатый эндотелиоцит
liber ~s склеренхима
lymph ~ лимфатическая клетка, лимфоцит
major mastoid air ~ *anat* сосцевидная пещера
male ~ мужская половая клетка, спермий
Marchand's ~ адвентициальная клетка, клетка Руже, клетка Маршана, перицит
mast ~ мастоцит, тучная клетка
mini ~ мини-клетка
mossy ~ плазматический астроцит
natural killer ~ естественная клетка-киллер, природная клетка-киллер, натуральная клетка-киллер, NK-клетка

natural suppressor ~ естественная клетка-супрессор, природная клетка-супрессор, натуральная клетка-супрессор, NS-клетка
nerve ~ нервная клетка, невроцит
nerve multipolar ~ мультиполярная нервная клетка, мультиполярный невроцит
NK-sensitive ~ NC-чувствительная клетка-мишень
nurse ~ клетка-«няня»
osseous ~ остеоцит, костная клетка
osteogenetic ~ остеобласт
pairing ~ гамета
Paneth ~ клетка Панета, ацидофильная кишечная клетка, ацидофильный энтероцит
parietal ~ париетальная клетка; париетальный гландулоцит
"passenger" ~ клетка-«пассажир»
pericapillary ~ клетка адвентиции капилляра, перицит
pheochrome ~ аргирофильная клетка, базальнозернистая клетка, энтерохромаффинная клетка, феохромоцит
pituitary ~ гипофизарная клетка, питуицит
plasma ~ плазматическая клетка, плазмацит
pre-B ~ В-клетка-предшественник, пре-В-клетка
prekiller ~ предшественник клетки-киллера
prickle ~ шиповидная клетка; шиповатый эпидермоцит
primed lymphoid ~ примированный лимфоцит
primed responder ~ примированная иммунореактивная клетка, примированная клетка-респондер
primitive sperm ~ первичная мужская половая клетка, сперматогоний
primordial germ ~ первичная половая клетка, гоноцит
progenitor ~ клетка-предшественник
promotor ~ промоторная клетка, Т-амплификатор
Purkinje's ~ клетка Пуркинье, ганглиозная нервная клетка, ганглиозный невроцит
queen ~ маточная пчелиная ячейка, маточник, нуклеус
quiescent ~ клетка в фазе G_0, покоящаяся клетка
radiate glial ~ клетка Фанана, лучистый глиоцит
red blood ~ красное кровяное тельце, эритроцит
responder ~ иммунореактивная клетка, иммуноотвечающая клетка, клетка-респондер
Rieder ~ клетка Ридера, полиморфный лимфоцит
rod ~ палочковидная зрительная клетка, палочка сетчатки
rod nuclear ~ палочкоядерный лейкоцит
Rouget ~ адвентициальная клетка, клетка Руже, клетка Маршана, перицит
royal ~ *ent* центральная камера, царская камера, брачная камера
satellite glial ~ глиальная клетка-спутник, глиоцит-сателлит
scavenger ~ макрофаг
Schwann's ~ клетка Шванна, шванновская клетка. леммоцит
secondary B ~ В-клетка памяти; вторичная В-клетка-стимулянт
segmentation ~ бластомер
Sertoli's ~ клетка Сертоли, фолликулярная клетка яичка, фолликулоцит яичка
sessile phagocytic ~ оседлый макрофаг, блуждаюшая клетка в покое
shadow ~ гемолизированный эритроцит, «тень» эритроцита
sheath ~ клетка Шванна, шванновская клетка, леммоцит
sickle ~ серповидная клетка; дрепаноцит, серповидный эритроцит
skein ~ ретикулоцит
small ~ of Ramon-y-Cajal мелкая клетка Рамон-и-Кахала, малый ганглионарный невроцит
sperm ~ сперматозоид; сперматогенная клетка *(of mosses)*
spider ~ астроглиальная клетка, астроглиоцит
stab ~, staff ~ нейтрофильный палочкоядерный гранулоцит, палочкоядерный лейкоцит

starlike ~, stellate ~ звездчатая клетка
stellate endothelial ~ клетка Купфера, звездчатый эндотелиоцит
stellate nerve ~ клетка Гольджи, звездчатая нервная клетка, звездчатый невроцит
stimulator ~ клетка-стимулятор, стимулирующая клетка
suppressor ~ клетка-супрессор, супрессорная клетка
suppressor-enriched T ~s обогащённая супрессорами фракция Т-клеток
swarm ~ зооспора
switch T ~ лимфоцит-«переключатель»
target ~ клетка-мишень
testicular follicular ~ клетка Сертоли, фолликулярная клетка яичка, фолликулоцит яичка
T-proliferative ~ промоторная клетка, Т-клетка-амплификатор
T suppressor-cytotoxic ~ цитотоксическая Т-клетка-супрессор
vasoformative ~ ангиобласт
virgin B ~ В-клетка-«девственница»
white blood ~ лейкоцит
white branched epidermal ~ белый отросчатый эпидермоцит
cell-free *a* бесклеточный
cellicolous *a* внутриклеточный
cellifugal *a* движущийся от клетки
cellipetal *a* движущийся к клетке
cellular *a* клеточный; клеточного строения; пористый; сетчатый
cellulate *a* состоящий из клеток; ячеистый, сотообразный
cellulation *s* целлюляризация
cellule *s* клеточка; ячейка
cellulicidal *a* цитоцидный, разрушающий клетки
cellulose *s* клетчатка, целлюлоза
aminoethyl ~ аминоэтилцеллюлоза
carboxymethyl ~, CM ~ карбоксиметилцеллюлоза
diethylaminoethyl ~ диэтиламиноэтилцеллюлоза
P- ~ фосфоцеллюлоза, Р-целлюлоза
cellulotoxic *a* цитотоксический
cembretum *s* кедровник
cement *s* цемент, цементирующее вещество
cementification *s* образование зубного цемента
cementoblast *s* цементобласт
cenanthous *a bot* пустоцветный
cenogenesis *s* ценогенез
cenogenetic *a* ценогенетический
cenology *s* ценология
cenosis *s* ценоз, сообщество
cenosite *s* факультативный эктопаразит
cenospecies *s* ценовид
censuring *s* учёт численности
census *s* учёт численности
center *s* центр; центральная точка; точка окостенения
biological ~ биостанция
Broca's ~ моторная речевая зона, центр Брока, зона Брока
nerve ~ нервный центр; нервный узел, ганглий
centipede *s зоол.* многоножка
central *a* центральный
centric *a* центральный
centrifugal *a* центробежный
centrifugation *s* центрифугирование
centrifuge *s* центрифуга; *v* центрифугировать
centriole *s* центриоль
centripetal *a* центростремительный; афферентный
centrodesmose *s* центродесмоз
centrogenous *a* идущий из центра
centrolecithal *a* центролецитальный
centromere *s* центромера, кинетохор
centroplasm *s* центроплазма
centroplast *s* центропласт
centrosome *s* центросома
centrosphere *s* центросфера
centrum *s* тело *(of a vertebra)*
cep, cepe *s* белый гриб *(Boletus edulis)*
century-plant *s* алоэ, столетник *(Aloe)*
cephalad *adv* краниально
cephalanthium *s* головчатое соцветие
cephalate, cephalated *a* головчатый
cephalic *a* головной; верхний
cephalin *s* кефалин
cephalium *s* цефалий
cephalization *s* цефализация
cephalometer *s* краниометр, цефалометр
cephalometry *s* краниометрия, цефалометрия
cephalostyle *s* передний конец хорды
cephalotheca *s* цефалотека

cephalothoracic *a* головогрудный
cephalothorax *s zool* головогрудь, головогрудной отдел
cerambycid *s* усач, жук-дровосек *(Cerambycidae)*
ceramide *s* церамид
ceratium *s* стручок без перегородки
ceratuba *s ent* трубчатая восковая железа
cercaria *s helm* церкария
cercariaeum *s helm* церкариеум
cercidium *s* мицелий
cercocystis *s гельм.* церкоциста
cercus *s* церка, хвостовой придаток; хвостовой отросток
cere *s orn* восковица
cereal *s* хлеб, злак; *a* хлебный, зерновой
cerebellar *a* мозжечковый
cerebellopontine *a* мосто-мозжечковый
cerebellum *s* мозжечок
cerebral *a* церебральный, мозговой
cerebrifugal *a* идущий от мозга к периферии; эфферентный, центробежный
cerebripetal *a* идущий к мозгу; афферентный, центростремительный
cerebroganglion *s* церебральный ганглий
cerebromedullary *a* цереброспинальный, спинномозговой
cerebrophysiology *s* физиология головного мозга
cerebrospinal *a* цереброспинальный, спинномозговой
cerebrovascular *a* цереброваскулярный
cerebrovisceral *a* церебровисцеральный
cerebrum *s* головной мозг
ceremony *s* церемония
 relieving ~ *eth* церемония смены партнёров
cereous *a* восковой; воскоподобный
ceriferous *a* восконосный
ceroris *s* дисковидная восковая железа
certified *a* апробированный, кондиционный, проверенный
cerumen *s* ушная сера
cervical *a* шейный; шейный нерв; шейная артерия; затылочный; цервикальный
cerviciplex *s* шейное сплетение
cerficofacial *a* шейно-лицевой
cervicomuscular *a* относящийся к мышцам шеи
cervicovaginal *a* шеечно-влагалищный
cervine *a* олений
cervix *s* шея; шейка
cesarean: **cesarean section** *s* кесарево сечение
cespitose *a* дернистый, дерновый; растущий пучками
cestode *s* ленточный червь *(Cestoidea)*
 caryophyllid ~ гвоздичник, кариофиллида *(Caryophyllaeidae)*
 polyzootic ~ настоящая цестода *(Cestoda)*
CFU, colony-forming unit *f* колониеобразующая единица, КОЕ
chaeta *s* щетинка
chaetiferous *a* несущий щетинки, щетинконосный
chaetodont *a* щетинисто-зубчатый
chaetognath, chaetognathan щетинкочелюстное *(Chaetognatha)*
chaetolobate *a* щетинисто-лопастный
chaetophyllous *a* щетинисто-листный
chaetotaxy *s* хетотаксия
chafer *s* хрущ; майский хрущ *(Melolontha)*
 cereal ~ кузька, хлебный жук *(Anisoplia)*
 flower ~ бронзовка *(Cetoniinae)*
 garden ~ хрущик садовый, кузька садовый *(Phyllopertha horticola)*
chaff *s bot* чешуя, маленькая чешуйка, дегенерировавший прицветник; полова, мякина
chaffinch *s orn* зяблик *(Fringilla)*
chaffy *a* плёнчатый
chain *s* цепь; цепочка
 J ~, joining ~, joint ~ J-пептид, соединительный пептид
chalaza *s* халаза, градинка
chalazogamy *s* халазогамия
chalice *s bot* чашечка
chalk *s* мел
chalk-loving *a* известелюбивый
chalkstone *s* известняк
chalky *a* меловой; известковый
challenge *s* контрольное заражение; антигенный стимул
 allergen ~ стимуляция аллергеном, иммунизация аллергеном, аллергизация; провокация
 antigen ~ антигенная стимуляция
 laboratory ~ контрольное лабораторное заражение

chalone *s* кейлон
chamaephyte *s* хамефит
chamber *s* камера; полость
chameleon *s* хамелеон *(Chamaelionidae)*
champ *v* чавкать
champignon *s* шампиньон
change *s* изменение; замена; переход; перемена; *v* изменять, изменяться; заменять
~ **of dominance** обратимое доминирование
~ **of variable** замена переменной
fatty ~ жировое перерождение
hereditary ~ геновариация, наследственное изменение
intergenic ~ межгенная перестройка
malignant ~ малигнизация, злокачественное перерождение
mode ~ изменение вида действия
postural ~ перемена положения
changeability *s* изменчивость
changeable *a* изменчивый, непостоянный; неустойчивый
changer *s* переключатель
channel *s* канал; канавка; *cyt* канал; *ent* штрих; вдавленная линия; бороздка; ход
~ **of infection** путь распространения инфекции
dead ~ старица, заводь
lymphatic ~ лимфатический проток
channeled *a* бороздчатый; желобчатый; канальчатый
chanter *s* лесная завирушка *(Prunella modularis)*
chanterelle *s myc* лисичка настоящая *(Cantharellus cibarius)*
chap *s* челюсть; нижняя челюсть; щека; *pl* пасть
character *s* признак; особенность; свойство; характер; качество; природа
developmental ~ характер развития
killer ~ *prot* свойство «убийцы»
characteristic *s* характеристика; особенность; свойство
chard *s* листовой черешок; свёкла листовая, мангольд *(Beta vulgaris var. cycla)*
charge *s* заряд
charomide *s mbiol* харомида
chart *s* таблица; диаграмма; схема; карта; карточка; *v* картировать, наносить на карту
chase *s* преследование, погоня; *mbiol* чейз, слежение за меткой; *v* гоняться, преследовать; охота; охотиться
chasmochomophyte *s* хазмохомофит
chasmogamy *s* хазмогамия
chasmophyte *s* хазмофит
chat *s orn* чекан *(Saxicola);* каменка *(Oenanthe);* чекановая трясогузка *(Ephthianura)*
chatter *s* треск; стук; щёлкание; *v* стучать; щёлкать
tooth ~ *eth* стук зубами
chaus *s* хаус, камышовая кошка, болотная рысь *(Felis chaus)*
cheathproboscis *s helm* хоботковое влагалище
check *s* контроль, проверка; задержка; препятствие; *a* контрольный, проверочный; *v* контролировать, проверять; приостанавливать, задерживать; препятствовать; ~ **of inoculation** проверка эффективности заражения; **to receive a** ~ иметь задержку *(в росте)*
diagnostic ~ диагностическое испытание
experimental ~ экспериментальная проверка, проверка на опыте
check-cross *s* контрольное скрещивание
checkerboarded *a* решётчатый
cheek *s* щека
cheekbone *s* скуловая кость
cheep *s orn* писк
cheeper *s* птенец
cheesy *a* творожистый, казеозный
cheetah *s* гепард *(Acinonyx jubatus)*
cheilocystidium *s myc* хейлоцистида
cheirophyllous *a* дланелистный
chela *s* клешня
chelate *s* хелат, внутрикомплексное соединение, комплексон
chelation *s* хелирование, образование хелата
chelicera *s* хелицера
chelidon *s* локтевая ямка
cheliferous *a* клешненосный
cheliped *s* клешненосная конечность
chemical *s* химическое вещество; *a* химический

fine ~s *pl* чистые химикаты
chemicalization *s* химизация
chemiluminescence *s* хемилюминесценция
chemosmotic *a* хемиосмотический
chemisorption *s* хемосорбция
chemistry *s* химия
biological ~ биологическая химия, биохимия
histological ~ гистохимия
chemoattractant *s* хемоаттрактант
chemoautotroph *s* хемоавтотроф
chemoautotrophy *s* хемоавтотрофия
chemocommunication *s* хемокоммуникация
chemodifferentiation *s* хемодифференциация
chemokinesis *s* хемокинез
chemoluminescence *s* хемолюминесценция
chemomorphose *s* хемоморфоз
chemonasty *s* хемонастия
chemoreception *s* хеморецепция
chemoreceptor *s* хеморецептор
chemoresistance *s* хеморезистентность
chemosynthesis *s* хемосинтез
chemosynthetic *a* хемосинетический
chemotactic *a* хемотаксический
chemotaxis *s* хемотаксис
chemotaxonomy *s* хемотаксономия
chemotrophic хемотрофный
chemotrophy *s* хемотрофия
chemotropic *a* хемотропный
chemotropism *s* хемотропизм
cheradium *s* сообщество голого песка
cherimoya *s bot* черимойя *(Anona cherimolia)*
chermes *s ent* хермес *(Adelges)*
chernozem *s* чернозём
cherry *s* вишня *(Cerasus);* слива *(Prunus);* черёмуха обыкновенная *(Prunus padus)*
bird ~ черёмуха обыкновенная *(Prunus padus)*
bladder ~ пёсья вишня *(Physalis alkekengi)*
cluster ~ черёмуха обыкновенная *(Prunus padus)*
Indian ~ ирга канадская *(Amelanchier canadensis)*
sour ~ вишня обыкновенная *(Cerasus vulgaris)*
chersic *a* обитающий на пустырях
chersium *s* фитоценоз пустырей
chersophyte *s* херсофит
chervil *s bot* бутень *(Chaerophyllum)*
wild ~ купырь лесной *(Anthriscus silvestris)*
chess *s бот.* костёр *(Bromus)*
chest *s* грудная клетка
chestnut *s* каштан *(Castanea); a* каштановый; гнедой
chestnut-roan *a* рыже-чалый
chevrotain *s ther* оленёк *(Tragulidae)*
chew *s* жвачка; *v* жевать, пережёвывать
chiasma *s* хиазма
chiasmatypy *s* хиазматипия
chiastobasidium *s* хиастобазидия
chick *s* цыплёнок; птенец
turkey ~ индюшонок
chickadee *s orn* гаичка; синица *(Parus)*
chickaree *s* бурундуковая белка, красная белка *(Tamiasciurus)*
chicken *s* курица *(Gallus gallus); цып*лёнок
meadow ~s *pl* пастушковые *(Rallidae)*
chicory цикорий *(Cichorium)*
chiffchaff *s orn* пеночка-теньковка *(Phylloscopus collybitus)*
chigger *s* клещ *Trombicula*
chigoe *s* блоха песчаная *(Tunga penetrans)*
chili *s* перец стручковый *(Capsicum)*
chilling *s* охлаждение
chimaera, chimaera *s gen* химера
chimonophilous развивающийся зимой; растущий зимой
chimpanzee *s* шимпанзе *(Pan)*; обыкновенный шимпанзе *(Pan troglodytes)*
chin *s* подбородок
chinchilla *s* шиншилла *(Chinchilla)*
mountain ~ горная вискача *(Lagidium viscacia)*
chine *s* хребет
chingoe *s* блоха песчаная *(Tunga penetrans)*
chinked *a* трещиноватый, щелевидный, щелеватый
chionium *s* фитоценоз снежного поля
chionophilous *a* хионофильный
chionophobous *a* хионофобный
chionophytia *s* фитоценоз снега
chip *v* проклюнуться

chipmunk *s* бурундук *(Tamias)*
chiral *a* хиральный
chirm *s* стрекотание; птичий щебет; *v* стрекотать; щебетать
chironomid *s* комар-дергун, звонец, хирономида *(Chironomidae)*
chiropterophilous опыляемый летучими мышами
chirp *s* чириканье, щебетание; стрекотание; писк; *v* чирикать, щебетать; стрекотать; пищать
chirr *s* стрекотание; *v* стрекотать
chirrup *s* чириканье, щебетание; стрекотание; писк; *v* чирикать, щебетать; стрекотать; пищать
chit *s* отросток, росток, побег; *v* пускать ростки
chita *s* гепард *(Acinonyx jubatus)*
chitin *s* хитин
chitinized *a* хитинизированный
chitin-lined *a* выстланный хитином
chitinous *a* хитиновый
chiton *s mal* хитон *(Polyplacophora)*
chitter *v* чирикать; кричать *(of gulls)*
chive *s* лук-скорода, шнитт-лук *(Allium schoenoprasum)*; лук сибирский *(Allium sibiricum)*
chlamydate *a* имеющий покров
chlamydocarpous *a* покровоплодный
chlamydospore *s* хламидоспора
chloracetylation *s* хлорацетилирование
chloranthy *s* зеленоцветковость
chlordecarboxylation *s* хлордекарбоксилирование
chlorenchyma *s* хлоренхима, хлорофиллоносная ткань
chlorination *s* хлорирование
chlorinator *s* хлоратор
chlorination *s* хлорирование
chlorinity *s* содержание хлора
chlorinolysis *s* хлор(ин)олиз
chlorocarbonylation *s* хлорформилирование
chloroethylation *s* хлорэтилирование
chloroformylation *s* хлорформилирование
chloromethylation *s* хлорметилирование
chlorophyll *s* хлорофилл
chlorobium ~ хлоробиум-хлорофилл
chlorophyll-bearing *a* хлорофиллоносный
chlorophyllide *s* хлорофиллид
chlorophyllin *s* хлорофиллин
chlorophyllous *a* хлорофилловый
chloroplast *s* хлоропласт
chlorosis *s phytop* хлороз
chlorosome *s* хлоросома
chlorothiourea *s* хлортиомочевина
chlorotic *a* хлоротический, хлоротичный, хлорозный
choana *s* хоана, внутренняя ноздря
choanocyte *s zool* хоаноцит, воротничковая клетка
choice *s* выбор; отбор; *a* отборный
choke *s phytop* чехловидная болезнь злаков *(caused by Epichloe typhina)*
cholagogue *s* желчегонное средство
cholecystis *s* жёлчный пузырь
choleic *a* жёлчный
choleopoiesis *s* жёлчеобразование
cholepoietic *a* желчеобразующий
choleresis *s* желчеобразование; желчеотделение
cholesterol *s* холестерин
cholinergic *a* холинэргический
chondral *a* хрящевой
chondrification *s* образование хряща; охрящевение
chondriokinesis *s* хондриокинез
chondriome *s* хондриом
chondriomere *s* хондриомер
chondriosome *s* хондриосома, митохондрия
chondroblast *s* хондробласт
chondroclast *s* хондрокласт
chondrocranium *s* хрящевой череп
chondrocyte *s* хондроцит, хрящевая клетка
chondrogenesis *s* хондрогенез, образование хряща
chondrogenic *a* хондрогенный, хрящеобразующий
chondroid *a* хондроидный, хрящеподобный
chondrophyllous *a* хрящеватолистный
chondroskeleton *s* хрящевой скелет
chondrosteous *a* имеющий хрящевой скелет
chorda *s* спинная струна, хорда; связка; канатик; тяж
chordomesoderm *s* хордомезодерма
chordotonal *a* хордотональный
chorial *a* относящийся к хориону, хорионический

chorheterosis *s myc* гетерокариотический гетерозис
chorioallantois *s* хориоаллантоис
choriomammotropin *s* хориомаммотропин, лактоген, хорионический соматомаммотропин
chorion *s* хорион
chorionic *a* относящийся к хориону
chorioretinal *a* хориоретинальный
choripetalous *a* раздельнолепестный
choriphyllous *a* раздельнолистный
chorisepalous *a* раздельночашелистиковый
chorogamy *s gen* хорогамия
choroid *s* сосудистая оболочка глаза
chorology *s* хорология
choronomic *a* внешний
chorotype *s* местный тип
chough *s* клушица *(Pyrrhocorax pyrrhocorax)*
chromaffinity *s* окрашиваемость
chromasie *s* хромазия
chromatic *a* хроматический
chromatid *s* хроматида; *a* хроматидный
chromatin *s* хроматин; хроматиновый
basic ~ базихроматин
chromatinic *a* хроматиновый
chromatoblast *s* хроматобласт
chromatofocusing *s* хроматофокусирование
chromatofuge *s* хроматографическая центрифуга
chromatogram *s* хроматограмма
chromatographer *s* прибор для хроматографии
chromatography *s* хроматография
~ on hydroxyapatite column хроматография на колонке с гидроксиапатитом
gel ~ гель-хроматография
chromatolysis *s* хроматолиз
chromatophilous *a* хроматофильный, окрышивающийся
chromatophore *s* хроматофор
chromatoplasm *s* хроматоплазма
chromidium *s cyt* хромидиум, хромидиальная сеть
chromobacterium *s* хромобактерия, пигментная бактерия
chromoblast *s* хромобласт
chromoblastomycosis *s* хромобластомикоз
chromocenter *s* хромоцентр
chromocyte *s* хромоцит, пигментная клетка
chromogenic *a* хромогенный, пигментообразующий
chromomere *s* хромомер
centromeric ~ центромерный хромомер, кинетомер
chromonema *s* хромонема
chromopexic *a* пигментофиксирующий
chromophile *a* хромофильный
chromophobe *s* неокрашивающийся элемент; хромофобная клетка
chromophobic *a* хромофобный, неокрашивающийся
chromophore *s* хромофор
chromoplast *s* хромопласт
chromosome *s* хромосома
chromotype *s* хромотип, хромосомный набор
chronaxia *s* хронаксия
chronic *a* хронический
chronobiology *s* хронобиология
chronotropic *a* хронотропный
chronotropism *s* хронотропизм
chrysalid, chrysalis *s ent* куколка
chrysocarpous *a* золотистоплодный
chrysogonidium *s* хризогонидий
chuckle *v* кудахтать; крякать; гоготать; *a* большой; неуклюжий
chufa *s* сыть съедобная, земляной миндаль, чуфа *(Cyperus esculentus)*
chumiza *s* просо итальянское *(Setaria italica)*
chunky *a* коренастый; толстый, широкий *(e.g., of a fruit)*
chyle *s* млечный сок, хилус; лимфа
chylosis *s* образование млечного сока
chyme *s* химус
chymous *a* химусный
chytridium *s* хитридий
cibol *s* лук зимний, лук песчаный, татарка *(Allium fistulosum)*
cicada *s* цикада *(Cicadidae)*
cicatrice *s* рубец; шрам
cicatricial *a* рубцовый
cicatricle *s* рубчик; зародышевый диск
cicatriculiform *a* рубчиковидный
cicatrization *s* рубцевание
cicatrize *v* заживать, зарубцовываться, рубцеваться; заживлять
ciliary *a* ресничный; мерцательный

ciliate, ciliated *a* реснитчатый, ресничный; имеющий реснички
ciliate *s* инфузория *(Ciliophora)*
ciliate-toothed *a* реснитчато-зубчатый
ciliature *s* цилиатура, ресничный покров
ciliospore *s* зооспора
cilium *s* ресничка; жгутик
cincinnus *s bot* завиток
cinclodes *s orn* водяной печник, трясохвостка *(Cinclodes)*
cingular *a* поясной
cingulum *s anat* пояс; поясок; корневая шейка; поясок; поясковый ободок *(e.g., of diatoms)*
cinnamon *s* коричное дерево *(Cinnamomum)*
cinnamonwood *s* сассафрас лекарственный *(Sassafras officinale)*
cinquefoil *s бот.* лапчатка *(Potentilla)*
cion *s* привой
circadian *a* циркадный, циркадианный
circahoralian *a* околочасовой
circannian *a* годичный
circinate *a* свёрнутый в кольцо, завитой, закрученный
circle *s* круг
circling *s* движение по кругу
circuit *s* схема; цепь
  closed ~ замкнутая цепь; замкнутая система
  immunological ~, immunoregulatory ~ иммунорегуляторный цикл, иммунорегуляторная сеть
  open ~ открытая система
  regulatory ~ регуляторная цепь
  semiclosed ~ полузамкнутая система
circulation *s* циркуляция; кровообращение
  water ~ круговорот воды
circulatory *a* циркуляторный; кровеносный
circumadjacent *a* непосредственно окружающий
circumanal *a* окружающий анус, циркуманальный
circumboreal *a* циркумбореальный
circumbuccal *a* окружающий рот, циркуморальный
circumduction *s* циркумдукция, круговое движение
circumesophageal *a* расположенный вокруг пищевода
circumference *s* обхват
circumferential *a* периферический
circumflex *a* огибающий; оборачивающийся вокруг; завитой
circumgenital *a* окружающий половое отверстие
circumhoralian *a* околочасовой
circumnuclear *a* расположенный вокруг ядра
circumnutation *s bot* круговое вращение
circumoral *a* окружающий рот, циркуморальный
circumpolar *a* циркумполярный
circumvallation *s prot* обтекание
cirrate *a* имеющий усики; *orn* имеющий хохолок
cirrhosis *s* цирроз
cirrhus *s* усик; *orn* хохолок; циррус
cirriform усиковидный
cirrous *a* имеющий усики; *orn* имеющий хохолок
cirrus *s* усик; *orn* хохолок; циррус
cis-position *s* цис-положение
cistern *s* цистерна; полость
cisternepiphyte *s* цистернэпифит
cistron *s* цистрон
citrate *s* цитрат; соль лимонной кислоты
citric лимоннокислый
citrous *a* цитрусовый
civet *s мам.* цивета, виверра *(Viverridae)*
  true ~ цивета, виверра *(Viverra)*
clade *s* монофилетический таксон
cladistics *s* кладистика
cladocarpous *a* бокоплодный
cladode *s* кладодий
cladogenesis *s* кладогенез
cladogenous *a* бокоплодный
cladogram *s* кладограмма
cladomial *a* кладомиальный *(a lichen)*
cladophyll *s* кладодий
cladoptosis *s* ежегодное опадение веточек
cladose *a* ветвистый; с ветвями
cladotype *s* реликт, реликтовая форма
clam *s* двустворчатый моллюск *(Bivalvia)*
clamber *v* взбираться
clammy *a* клейкий, липкий
clamp *s helm* присасывательный клапан; *myc* пряжка; зажим
  patch ~ *phys* фиксация потенциала, пэтч-кламп

voltage ~ *phys* фиксация напряжения, вольт-кламп
clan *s* микроассоциация
clandestine *a* тайный, скрытый
clang *s orn* курлыканье; курлыкать
claret-colored *a* бордовый
clarificant *s hist* просветляющее средство
clarification *s hist* просветление; очистка; осветление
clarifier *s* осветлитель
centrifugal ~ осветляющая центрифуга
clasmatocyte *s* клазматоцит, оседлый макрофаг, гистиоцит
clasp *s* зажим; *энт.* лопасть; обвиваться
clasper *s bot* усик
class *s* класс; группа
year ~ возрастная группа, годовой класс
classification *s* классификация
classify *v* классифицировать
clava *s ent* булава
clavate *a* булавовидный
clavicle *s* ключица
clavicular *a* ключичный
clavicularium *s herp* передний парный щиток
clavus *s ent* клавус; склероций спорыньи, «рожок» спорыньи
claw *s* коготь, коготок; *бот.* ноготок; лапа с когтями; клешня; **on the** ~ *orn* живая
nipping ~ клешня
clawless *a* бескоготковый
clay *s* глина
clay-coloured *a* светло-бурый
clean *v* чистить; *a* чистый
clear *a* чистый, прозрачный, светлый, ясный; *v* очищать
clearance *s* очищение, выведение; клиренс
clear-cut *a* сухой *(appearance)*
cleared *a hist* просветлённый
clearer *s hist* средство для просветления
clearing *s* очистка, расчистка; поляна; вырубка; *hist* просветление
bush ~ очистка от кустарника, борьба с закустаренностью
forest ~ сведение леса
snow ~ снегоочистка
clearwing *s ent* стеклянница *(Sesiidae)*
cleavage *s emb* дробление; расщепление; деградация; распад
enzymic ~ ферментативный гидролиз, энзиматический гидролиз
equal ~ равномерное дробление
homolytic ~ гомолиз, гомолитическое расщепление
hydrolytic ~ гидролиз
peptic ~ пептический гидролиз, гидролиз пепсином
phosphoroclastic ~ фосфоролиз
protein ~, proteolytic ~ протеолиз, протеолитическое расщепление
ring ~ *bioch* дециклизация, размыкание цикла
thermal ~ термолиз, термическое расщепление
tryptic ~ триптический гидролиз, гидролиз трипсином, трипсинолиз
cleft *s* щель; углубление, щель, расселина; *bot* дольчатый
middle-ear ~ барабанная полость
natal ~ ягодичная борозда
cleft-beaked *a bot* расщеплённоклювистый
cleft-footed *a* парнокопытный
cleft-leaved *a* расщеплённолистный
cleft-toothed *a* расщеплённозубчатый
cleg *s* слепень *(Tabanus)*
cleidoic *a* клейдоический, заключённый в оболочку *(egg)*
cleistocarp клейстокарпий
cleistocarpous *a* клейстокарпный
cleistogamic, cleistogamous *a* клейстогамный
cleistogamy *s* клейстогамия
cleistogene *s* растение с клейстогамными цветками
cleistothecium *s* клейстотеций
cleptobiosis *s* клептобиоз
climact(er)ic *a* климактерический
climagraph *s* климаграмма
climate *s* климат
plant ~ фитоклимат
climatic *a* климатический
climatizer *s* камера искусственного климата, климатрон
climatograph *s* климаграмма
climatology *s* климатология
climatype *s* климатип
climax *s* климакс, климаксовое растительное сообщество
climb виться, обвиваться; стелиться

climber *s* вьющееся растение; стелющееся растение
leaf ~ *herp* листолаз *(Phyllobates)*
tendril ~ растение, цепляющееся усиками
climograph *s* климаграмма
clinandrium *s* андроклиний
clinanthium *s* антоклиний
cline *s* клин, градиент признаков; резкая сукцессия; катастрофическая сукцессия
clinedeme *s* клинодем
clinosporangium *s* пикнида
clinostat *s* клиностат
clip *v* стричь; обрезать
clisere *s* климатическая серия
clistase *s* следы ископаемых ценозов
clitellar *a* поясковый
clitellates *s* поясковые черви *(Clitellata)*
clitellum *s* клителлюм, поясок
clitoris *s* клитор
cloaca *s* клоака
cloacal клоакальный
clock *s* часы
clod *s* комок; сгусток
cloddy *a* комковатый *(soil)*
clogging *s* засорение; заиливание
clonable *a* клонируемый; клоногенный
clonal *a* клоновый; клональный
clonality *s* клональные свойства, клональность
clone *s* клон
cloning *s* клонирование
clonogenicity *s* клоногенность
clonorchiasis *s helm* клонорхоз
clonotheque *s* клонотека
gene ~ клонотека генов, генотека
clonotype *s* клонотип
clonus *s* клонус
clop *s* стук *(of hooves)*
closebreeding *s* близкородственное разведение; инбридинг
close-clustered *a* пучковатый
closed *a* закрытый, замкнутый, сомкнутый
close-grained *a* мелкозернистый
close-meshed *a* мелкоячейный
closewing *s ent* огнёвка *(Crambinae)*
closing *s* закрытие, замыкание
~ **of leaf canopy** сомкнутость крон
closterospore *s* многоядерная фрагмоспора
closure *s* закрытие, смыкание; запрет
crown ~ смыкание крон
ring ~ *biom* циклизация, замыкание цикла
stomatal ~ закрытость устьиц
clot *s* сгусток, тромб; свёртываться, коагулировать
clottage *s* тромбоз
clotting *s* свёртывание, коагуляция
cloud *s* рой; кулига; стая
cloudberry *s* морошка приземистая *(Rubus chamaemorus)*
cloudburst *s* ливень, ливневый дождь
cloudy *a* мутный; пасмурный
clove *s* луковичка; долька, зубок
cloven-footed, cloven-hoofed *a* парнокопытный
clover *s* клевер *(Trifolium)*
cloverleaf *s* клеверный лист; *mbiol* структура «клеверного листа»
cloverworm *s* вредитель клевера
club *s энт.* булава
club-bearing *a* булавоносный
club-flowered *a* булавоцветковый
club-footed *a* косолапый; булавостебельчатый
club-forming *a bot* клубочкообразующий
club-fruited *a* булавоплодный
clublike *a* булавовидный
club-moss *s* плаун *(Lycopodium)*
clubroot *s* кила крестоцветных *(caused by Plasmodiophora brassicae)*
club-rush *s* камыш *(Scirpus)*
club-stalked *a* косолапый; булавостебельчатый
clubtop *s myc* булавница *(Clavaria)*
cluck *s* кудахтанье; клохтанье; *v* кудахтать; клохтать
clump *s* группа, скопление; куртина
bacterial ~ колония бактерий
blast ~ скопление бластных клеток, бласт-очаг
clumped *a* скученный
cluneal ягодичный
clunis *s* ягодица
cluster *s* кисть, гроздь, пучок, клубочек; группа, купа; кластер; рой пчёл; *v* расти гроздьями, расти пучками
blossom ~ соцветие
flower ~ рацемозное соцветие, ботрическое соцветие
loose ~ рыхлая гроздь

macrophage-lymphocyte ~ макрофаг-лимфоцитарный кластер
cluster-flowered *a* кистецветковый
clustering *s* образование скоплений; образование кластеров
~ **of receptors** образование рецепторных кластеров
clutch *s* кладка; выводок; высиживать; гнездо с яйцами; лапа с выпущенными когтями; **in the** ~ в когтях
clypeate *a ent* имеющий наличник; наличникообразный; круглый; щитообразный
clypeola, clypeole *s* округлый спорофилл
clypeus *s ent* щит, щиток; наличник
cnida *s* книда, нематоциста, книдоциста
cnidoblast *s* книдобласт
cnidocil *s* книдоциль
coacervate *s* коацерват
coacervation *s* коацервация
coaction *s* коакция, совместное действие
coadaptation *s* коадаптация
coagglutination *s* склеивание
coagulable *a* коагулирующийся, способный к свёртыванию
coagulant *s* коагулянт
coagulate *s* коагулят, сгусток; *v* коагулировать, свёртываться
coagulation *s* коагуляция, свёртывание
coagulative *a* коагулирующий, свёртывающийся
coagulator *s* коагулятор, агент коагуляции
coagulin *s* преципитирующее антитело, преципитин
coalesce *v* соединяться; слипаться; срастаться
coalescence *s* срастание, сращение; слипание; *imm* слияние
coalition *s eth* коалиция
coalmouse *s orn* московка *(Parus ater)*
coaperturate *a pal* коапертурный
coarct *v* сжимать, сжиматься; суживать, суживаться
coarctate *a* сжатый; тесно связанный; плотно сросшийся
coarctation *s* коарктация, сужение; стеноз
coarse *a* грубый; крупный
coarse-fibered *a* грубоволокнистый
coarse-grained *a* грубозернистый
coarse-wooled *a* грубошёрстный
coarticulation *s* сочленение
coat *s* оболочка, плева; покров, слой; покрывать; мех, шёрстный покров
buffy ~ лейкоцитная плёнка
jelly ~, mucous ~ слизистая оболочка
scleral ~ белочная оболочка, склера
seed ~ семенная оболочка, оболочка семени, плева
coati, coatimundi *s ther* носуха, коати *(Nasua)*
coating *s* оболочка; покров
coaxial *a* соосный, коаксиальный
cob *s* глыба, ком; лебедь-самец; чайка *(Larus)*; крупный орех; стержень кукурузного початка
cobalamin *s* кобаламин, витамин B_{12}
cobbing *s* образование початков
cobby *a* низкорослый; коренастый
cobble-stone *s* булыжник
cobra *s* кобра *(Naja)*
cobweb *s* паутина
coca *s* кокаиновый куст *(Erythroxylum coca)*
cocarcinogen *s* коканцероген
cocarcinogenesis *s* коканцерогенез
coccal *a micr* кокковый
coccidiosis *s prot* кокцидиоз
coccogenic, coccogenous *a* вызванный кокками
coccogone *s* коккогон
coccolith *s* кокколит
coccosphere *s* коккосфера
coccus *s bot* орешек; *micr* кокк
coccygeal *a* копчиковый
coccyx *s* копчик
cochineal *s ent* кошениль *(Coccus cacti); pl* карминоносные червецы *(Margarodinae);* кармин
cochlea *s anat* улитка; раковина улитки
cochlear *a* улитковый, кохлеарный
co-chromatography *s* кохроматография
cock *s* петух; самец птиц
~ **of the wood** глухарь (*Tetrao urogallus*)
heath ~ тетерев *(Lyrurus tetrix)*
snow ~ *orn* гималайский улар *(Tetraogallus)*
water ~ *orn* рогатая камышница *(Gallicrex cinerea)*
cockatoo *s* какаду
cockchafer *s* майский хрущ западный *(Melolontha melolontha)*

cockle *s mal* сердцевидка *(Cardium); bot* тысячеголов *(Vaccaria);* плевел *(Lolium)*
cockroach *s* таракан *(Blattodea)*
cockscomb *s* гребень петуха; *bot* целозия гребенчатая, петуший гребешок *(Celosia cristata)*
coconut *s* кокосовый орех
double ~ сейшельская пальма *(Lodoicea maldivica)*
cocoon *s* кокон
coctolabile *a* термолабильный
coctostabile *a* термостабильный
cod *s* стручок; оболочка; шелуха
code *s* кодекс
International code of zoological numenclature Международный кодекс зоологической номенклатуры
coding *s* кодирование; *a* кодирующий
codlin, codling *s* дикая яблоня *(Malus sylvestris)*
codominance *s* кодоминантность
codon *s* кодон
amber ~ терминирующий кодон UAG, терминатор
nonsense ~ бессмысленный кодон, нонсенс-кодон
ochre ~ терминирующий кодон UAA
stop ~, termination ~, terminator ~ терминирующий кодон UAA, терминатор
coefficient *s* коэффициент
coeliac брюшной
coeloblast *s* целобласт
coeloblastula *s* целобластула
coelogastrula *s* целогаструла
coelomate *a* вторичнополостной
coelom, coelome *s* целом, вторичная полость тела
coelomesoblast *s* мезодермальная стенка целома
coelomic *a* целомический
coelomocyte *s* целомоцит
coelomoduct *s* целомодукт
coelomopore *s* целомопор
coelomostome *s* наружное отверстие целомодукта
coelospermous *a bot* сводчатосемянный, «выдолбленный»
coelozoic *a* обитающий в полости тела
coen *s* ценоз, сообщество
coenanthium *s* общее цветоложе; гинантодий, аскон
coenenchym, coenenchyma *s* цененхима
coenobiology *s* биоценология
coenobium *s* ценобий
coenoblast *s* ценобласт
coenocline *s* ценоклин
coenocoel *s zool* общая полость
coenoecium *s* ценоций
coenogamete *s* ценогамета, многоядерная гамета
coenogenesis *s* ценогенез
coenon *s* ценон
coenosarc *s* ценосарк
coenosis *s* ценоз, сообщество
coenospecies *s* ценовид
coenozygote *s* ценозигота
coenurus *s helm* ценур
coenzyme *s* кофермент
coexistence *s* сосуществование
coexpression *s imm* коэкспрессия
cofactor *s* кофактор; кофермент
coferment *s* кофермент
coffee *s* кофейное дерево *(Coffea)*
cognition *s* знание; познание; приобретение знания; познавательная способность
cohabitant *s* сожитель
cohabitation *s* совместное обитание, сообитание, сожительстао
coherence *s* связь, сцепление
coherent *a* соединённый, связанный, сцеплённый
cohesion *s* когезия, сцепление
cohobation *s* повторная дистилляция
cohort *s* когорта
coil *s cyt* спираль; *bot* завиток
coiled *a* скрученный, извитый
coimmunization *s* совместная иммунизация, одновременная иммунизация
coinjection *s* совместная инъекция, одновременная инъекция
cointegrate *s* коинтеграт
cointegration *s* коинтеграция
coition, coitus *s* коитус, совокупление
colation *s* процеживание; фильтрация
colature *s* фильтрат
colcemid *s* колцемид
cold *a* холодный
cold-blooded *a* холоднокровный
cold-killed *a* погибший от холода

cold-loving *a* холодолюбивый
cold-stenothermic *a* стенолиготермный
coleophyllous *a* влагалищнолистный
coleoptile *s* колеоптиль
coleorhiza *s* колеориза, корневое влагалище
colibacillus *s* кишечная бактерия *(Escherichia coli)*
colibacteriophage *s* колибактериофаг
colinearity *s* колинеарность
coliphage *s* колифаг
colitis *s* колит
collagen *s* коллаген
collapse *s* коллапс, депрессия; спадение; *v* спадаться; сплющиваться
collar *s* воротничок; *bot* шейка
head ~ *helm* головной воротничок
root ~ корневая шейка
tail ~ воротничок фагового отростка
collateral *s* коллатераль; коллатеральный
axon ~ парааксон
collect собирать
collection *s* коллекция; сбор материала, коллектирование
collector *s* коллектор; собиратель
~ **of nitrogen** азотособиратель
fraction ~ коллектор фракций
collembolan *s ent* ногохвостка, коллембола *(Collembola)*
collenchyme *s* колленхима
collencyte *s* колленцит
collet *s bot* шейка
colleter *s bot* клеевая ворсинка, волосок, выделяющий клей
colleterium *s ent* коллетериальная железа
colley *s orn* дрозд *(Turdus)*
collimation *s* коллимация
collimator *s* коллиматор
collineality *s* колинеальность
collinous *a* растущий на холмах
collision *s* столкновение
sperm-egg ~ встреча спермия с яйцом
colloblast *s* коллобласт
colloid *s* коллоид; *a* коллоидный
colloidal *a* коллоидный
collum *s* шейка; донце *(of an onion)*
colobus *s ther* толстотел, гвереца, колобус *(Colobus)*
colon *s* ободочная кишка
colonial *a* колониальный
colonicus *s ent* переселенец
colonization *s* колонизация, заселение, расселение
colonizer *s* вселенец
colony *s* колония; поселение
ant ~ муравейник
breeding ~ гнездовая колония; детная залёжка *(of seals)*
honey-bee ~ пчелиная семья
colony-forming колониеобразующий
color *s* цвет; окраска; масть
~ **of water** цветность воды
fur ~ цвет меха
coloration *s* окраска, расцветка; окрашивание
color-blindness *s* цветовая слепота
colorimeter *s* колориметр
colorimetry *s* колориметрия
colorless *a* бесцветный
colostric *a* молозивный
colostrum *s* молозиво
colpoid *a pal* бороздновидный
colt *s* жеребёнок; ослёнок; верблюжонок; детёныш зебры
coltsfoot *s* мать-и-мачеха обыкновенная *(Tussilago farfara);* копытень европейский *(Asarum europaeum);* белокопытник *(Petasites)*
columbine *a* голубиный; *s bot* водосбор *(Aquilegia)*
columella *s* столбик; ось коробочки *(in mosses);* центральная часть тычинки
columellar *a* колюмеллярный, столбиковый
column *s anat* столб, колонна; позвоночник; спинная струна, хорда; *bot* колонка, гиностемий; нервный пучок; колонка
~ **of Gowers** путь Говерса, передний спинномозжечковый путь
Burdach's ~ клиновидный пучок, пучок Бурдаха
enamel ~ *anat* эмалевая призма
Flechsig's ~ путь Флексига, задний спинномозжечковый путь
slender ~ тонкий пучок, пучок Голля
spinal ~, vertebral ~ позвоночник, позвоночный столб
columnar *a* столбчатый; колончатый; удлинённый
columnar-celled *a* цилиндроклеточный

coma *s бот.* султан, пучок волосков, верхушечный пучок, хохолок; кома; оцепенение
chill ~ холодовое оцепенение
seed ~ семенной хохолок
comate *a* с хохолком; хохлатый
comb *s* соты; гребень, гребешок; хохолок *(e.g., in birds)*
pollen ~ *apic* щёточка для сбора пыльцы
combination *s* комбинация, сочетание, соединение; *phys* комплекс
successive ~ последовательный комплекс
comb-shaped *a* гребенчатый; гребневидный
combustion *s* горение; сжигание; озоление
coming-up *s* всходы
commensal *s* комменсал, комменсалист
commensalism *s* комменсализм
commensurable, commensurate *a* соразмерный
commercial *a* промысловый; промышленный; коммерческий
comminute *v* толочь; измельчать; растирать в порошок
commiscuum *s* коммискуум
Commission *s*:
Survival Service ~ Комиссия спасения исчезающих видов при Международном союзе охраны природы
commissure *s* комиссура, соединение, спайка, место сочленения
commitment *s imm* коммитирование
common *a* общий
communication *s* общение; коммуникация
community *s* сообщество
animal ~ сообщество животных, зооценоз
biotic ~ биоценоз
closed ~ замкнутое сообщество; сомкнутый фитоценоз
comose *a* с хохолком, хохлатый
comosporous *a* с волосками на спорах
comparator *s* компаратор
comparative *a* сравнительный
comparium *s* коммискуум
compartment *s* компартмент
compartmentalization, compartmentation *s* компартментализация, пространственное разделение
compatibility *s* совместимость; ~ **for stock and scion** совместимость подвоя и привоя
compensation *s* компенсация
compensatory *a* компенсирующий, компенсаторный
competence, competency *s* компентенция; компентентность
competition *s* конкуренция, борьба за существование
interspecific ~ межвидовая конкуренция, межвидовая борьба
intraspecific ~ внутривидовая конкуренция, внутривидовая борьба
competitive конкурентный
competitor *s* соперник, конкурент
complement *s* комплемент, алексин; хромосомный набор
complementarity *s* комплементарность
complementary *a* комплементарный, дополнительный
complementation *s* комплементация
complement-deviating *a* комплемент-отклоняющий
complex *s* комплекс; *a* комплексный, сложный
genetic ~ генотип
granular paravesicular ~ зернистый паравезикулярный комплекс, эргастоплазма
habit ~ габитуальный комплекс, комплекс повадок, поведение
intercalated-type ~ интеркалярное встраивание
complexity *s* комплексность; сложность
idiotypic ~ идиотипическая вариабельность, идиотипическое разнообразие, множественность идиотипических детерминант
compliance *s* пластичность, податливость; растяжимость, эластичность
complication *s* осложнение
component *s* компонент, составная часть
"late phase" ~ *imm* фаза замедленной гиперчувствительности
population ~s *pl* составные элементы популяции

trace ~ микрокомпонент
composite *a* сложный, составной
composition *s* состав
gross ~ макросостав
compost *s* компост
composting *s* компостирование
compound *s* состав, смесь; *a* составной, сложный; химическое соединение; *v* составлять, смешивать; соединять
addition ~ продукт присоединения, аддукт
storage ~ запасное вещество
compounding *s* смешивание, составление смеси
compress *v* сжимать, сдавливать
compression *s* сжатие
compressor *s* мышца-сжиматель
conarium *s zool* конария; эпифиз
conation *s* волевое усилие; активное действие
concatemer *s mbiol* конкатемер
concatenate *s* конкатенат; цепочкообразный; образующий цепочку
concatenation *s* соединение в виде цепочки
concave *a* вогнутый; впалый
concaveation *s eth* сенситизация
concavo-concave *a* двояковогнутый
concavo-convex *a* вогнутовыпуклый
concealed *a* скрытый, латентный
concealment *s eth* затаивание
concealing *a* скрывающий, маскирующий
conceive *v* оплодотворяться, осеменяться; забеременеть
conceiving *s* зачатие
concentration *s* концентрация
idiotope ~ идиотопная плотность; концентрация идиотипических детерминант
concentrator *s* концентратор
centrifugal ~ концентрирующая центрифуга, центробежный концентратор
concept *s* концепция; представление, понятие; направление
~ **of symmetry** принцип симметрии
body space ~ предметно-пространственное представление
gradient ~ градиентная концепция, теория градиентов
conceptacle *s bot* вместилище; стручок без перегородок
conception *s* концепция; представление, понятие; оплодотворение, осеменение; зачатие, зарождение
conceptus *s ther* оплодотворённое яйцо
conch *s* раковина; ушная раковина
conchology *s* конхиология
conclave *s* конклав, изолированная камера
concomitant сопутствующий, сопровождающий
concordance *s* соответствие, совпадение, согласованность
concordant *a* конкордантный, согласованный
concrescence *s* сращение
concrescent *a* сросшийся
concretion *s* срастание; сгущение; конкремент, сросток
concussion *s* сотрясение, контузия
condensate *s* конденсат
condensation *s* конденсация; сгущение
Bose ~ Бозе-конденсация
reductive ~ гидроконденсация, восстановительная конденсация
condense конденсировать, конденсироваться; сгущать, сгущаться
condenser *s* конденсор; конденсатор
condition *s* условие; состояние; кондиционировать
outdoor ~s *pl* на открытом воздухе
conditional условный; обусловленный
conditioning *s* кондиционирование; установление требуемого состава; выработка условного рефлекса
operant ~ оперантное обучение; оперантное обуславливание
condor *s orn* кондор *(Vultur)*
conductance, conductibility *s* проводимость
conduction *s phys* проведение
conductive *a* проводящий
conductivity *s* проводимость
conduplicate *a* сложенный вдоль
condylar *a* мыщелковый
condyle *s* мыщелок
cone *s* конус; колбочка; бугорок *(of a tooth); bot* шишка; колосок; *mal* колпачковидная раковина
~ **of kidney** пирамида почки

pulmonary ~ *anat* артериальный конус, воронка
sense ~ *ent* обонятельный конус, базиконическая сенсилла
cone-bearing шишконосный
cone-fruited *a* конусоплодный
cone-shaped *a* конусообразный; шишковидный
coney *s* кролик *(Oryctolagus cuniculus);* скалистый даман, пустынный даман *(Procavia)*
confiding *a* доверчивый
configuration *s* конфигурация, форма
bridge-fragment ~ конфигурация мост - фрагмент
cis ~ цис-конфигурация
trans ~ транс-конфигурация
confined *a* ограниченный в распространении
confinement *s* ограничение; стойловое содержание; роды
close ~ безвыгульное содержание
conflict *s* конфликт
confoederata *s* сообщество сосуществующих видов
conformation *s* конформация; структура, форма; экстерьер
conformed *a* сходный по форме; однообразный
conformer *s* конформер
confounding *s:* **balanced** ~ *biom* сбалансированное смешивание
congealing *s* замерзание
congelation *s* замораживание; коагуляция, свёртывание
congener *s* представитель того же рода
congenering *a* принадлежащий к одному роду
congenetic *a* имеющий общее происхождение
congenic *a* конгенный
congenial *a* конгениальный, родственный, близкий; врождённый, прирождённый; благоприятный, подходящий
congeniality *s* сходство; родство
congenital *a* врождённый, прирождённый
congested *a* скученный, перенаселённый; застойный; напитавшийся кровью
congestion *s* перенаселение; скученность; гиперемия; прилив крови
conglobate *a* шаровидно-скученный; шарообразный, сферический
conglomerate *a* собранный в клубок
conglutination *s* конглютинация
congregate *v* собираться
congression *s* конгрессия
conidial *a* конидиальный
conidiiferous *a* образующий конидии
conidiocarp *s* конидиокарпий, пикнида
conidiole *s* мелкая конидия
conidiophore *s* конидиофор
conidiospore *s* конидиоспора
conidium *s* конидия
conifer *s* хвойное дерево
coniferous *a* хвойный
coniophage *s* кониофаг
conjugant *s* конъюгант
conjugate *s imm* конъюгат; конъюгировать; соединяться; сливаться
bead-antibody ~ антитело, иммобилизованное на грануле
cell-bridging antibody ~ антителоформированный «мостиковый» цитоконъюгат, гетероконъюгат
conjugation *s* конъюгация; соединение; слияние
touch-and-go ~ конъюгация по типу «игра в пятнашки»
conjunctiva *s* конъюнктива
connate *a* соединённый; сросшийся
connate *a* врождённый; родственный; реликтовый
connatural *a* природный; конгениальный, родственный, близкий
connect *v* связывать, соединять
connection *s* связь, соединение, сцепление
biocoenotic ~ биоценотический коннекс, взаимозависимость компонентов биоценоза
central ~**s** центродесмоз
functional ~ функциональная зависимость
intercontinental ~**s** *pl* межконтинентальные связи, межконтинентальные мосты
connective *a* соединительный, связующий; *bot* связник; *anat* коннектив
connexivum *s ent* боковой бортик

connivent *a* сближенный; сходящийся
connubium *s* спаривание; скрещивание
conodrymium *s* фитоценоз вечнозелёных растений
conoid *s* коноид
conophorium *s* фитоценоз хвойного леса
consanguineous *a* единокровный, кровный
consanguinity *s* кровное родство
conscious *a* находящийся в сознании; сознательный
consciousness *s* сознание
  subliminal ~ подсознание
conservancy *s* охрана природы; рациональное природопользование; служба охраны природы
conservation *s* сохранение; охрана природы
  fish ~ рыбоохрана
  resource ~ рациональное использование природных ресурсов
conservative *s* консервант, консервирующее средство
conservatory *s* теплица, оранжерея
consistence, **consistency** *s* консистенция
consistent *a* плотный; твёрдый
consocial *a* входящий в ту же ассоциацию
consociation *s* консоциация
consocies *s* консоция
consocietum *s* сообщество
consocion *s* консоция
consociule *s* микросообщество ранга консоции
consort *s* консорт, участник консорции
consortism *s* консортизм
consortium *s* консорция; лишайник
consortship *s eth* супружеская связь, брачная связь
conspecific *a* конспецифичный, принадлежащий к тому же виду
constance, constancy *s* константность, постоянство, стойкость, неизменность
constant *s* константа, постоянная величина; *a* константный, постоянный
constituent компонент, составная часть
  ash ~ зольный элемент
  natural ~ природный компонент
constitution *s* строение, структура; конституция, телосложение
constitutive *a* конститутивный; основной
constrict *v* стягивать, сжимать
constricticolpate *a pal* суженно-бороздчатый
constriction *s* сужение; перетяжка, перехват
constrictive *a* сжимающий, стягивающий
constrictor *s* констриктор, сжимающая мышца; обыкновенный удав *(Boa constrictor)*
consume *v* потреблять, поглощать
consumer *s* консумент, потребитель
  detritus ~ детритофаг, детритоядный организм
consummatory *a eth* консумматорный, завершающий
consumption *s* потребление, поглощение
  overall ~ общее потребление, общий расход
consumptive *a* потребляющий; истощающий
contabescence *s* атрофия тычинок
contact *s* контакт
  crown ~ смыкание лесного полога, сомкнутость крон
  focal ~ фокальный контакт; адгезионная пластинка
contagion *s* заразное начало; инфекционное заболевание; распространение инфекции, передача инфекции
contagious *a* контагиозный, заразный, инфекционный
contaminant *s* загрязнитель, загрязняющее вещество
contaminate *s* загрязнять; заражать; портит
contamination *s* загрязнение; примесь; заражение
  aerial ~ воздушное загрязнение; воздушная инфекция
  airborne ~ обсеменение микрофлорой воздуха
  radioactive ~ радиоактивное загрязнение
content *s* содержание; объём, вместимость, ёмкость; *pl* содержимое
  ash ~ зольность, содержание золы
  caloric ~ калорийность
  fat ~ жирность, содержание жира
  gluten ~ клейковинность
  grain ~ озернённость колоса
  moisture ~ влагосодержание, влажность

oil ~ масличность
protein ~ белковистость
sugar ~ содержание сахаров, сахаристость
water ~ влагосодержание, влажность
contest *s* соперничество, борьба
contiguous *a* смежный, соприкасающийся; соседний
contiguum *s* совокупность смежных местообитаний
contingency *s* сопряжённость признаков
continuity *s* непрерывность, неразрывность, целостность
contorted *a* скрученный; закрученный; искривлённый
contortuplicate *a* скрученно-складчатый
contract *v* сокращаться, сжиматься; заразиться
contractile *a* сокращающийся, сжимающийся; сократительный
contraction *s* сокращение, сужение, сжатие
myotatic ~ сухожильный рефлекс
rigor ~ контрактура
contractive *a* сокращающийся, сжимающийся; способный сокращаться, способный сжиматься
contractor *s* контрактор, стягивающая мышца
contraindication *s* противопоказание
contrainterleukin *s* антиинтерлейкин
contralateral *a* контралатеральный, противоположный
contrast *s* контраст; **to stand in marked** ~ резко отличаться
contrastimulant *s* антагонист стимулирующего действия
contrasuppression *s imm* контрасупрессия
control *s* управление; регуляция; контроль; орган управления; борьба, контроль; *v* управлять; регулировать; контролировать
environmental ~ охрана окружающей среды
hormonal ~ гормональная регуляция, гормональный контроль
remote ~ телеуправление, дистанционное управление; дистанционный контроль
sex ~ регулирование пола
convalescence *s* выздоровление
converge *v* конвергировать, сходиться
convergence *s* конвергенция, сходимость; совмещени
conversion *s gen.* конверсия; превращение, трансформирование
convex *a* выпуклый
convulsion *s* судорога, конвульсия
convolute *a* свёрнутый, скрученный
convolution *s* извилина; свёрнутость; изогнутость; *bot* листосложение
convulsion *s* судорога, конвульсия
convulsive *a* судорожный, конвул ьсивный
cony *s мам.* кролик *(Oryctolagus cuniculus);* пищуха *(Ochotona)*
rock ~ **ies** *pl* пищуховые, сеноставки *(Ochotonidae)*
coo *s* воркование; *v* ворковать
coolant *s* охладитель, хладагент
cooler *s* холодильник; холодная камера
cooligomerization *s* соолигомеризация
cooling *s* охлаждение
coolroom *s* холодильная камера
coomassie brilliant blue *s* кумасси бриллиантовый голубой
coon *s* енот-полоскун *(Procyon lotor)*
coop *s* клетка; *v* сажать в клетку
cooperativity *s mbiol* кооперативность
coordination *s* координация
coot *s orn.*лысуха *(Fulica)*
cootie *s orn* мохноногий
cop *s orn* хохолок; *bot* початок
copious *a* плодовитый; распределённый плотно; обильный
copiousness *s* плодовитость; изобилие
copolymer *s* сополимер
copolymerization *s* сополимеризация
copperhead *s* мокассиновый щитомордник, медноголовый щитомордник *(Agkistrodon contortrix);* великолепная австралийская медноголовая змея *(Austrelaps superbus);* золотистый суслик *(Citellus lateralis)*
coppice *s* молодой лесок, поросль
copra *s* копра
copresentation *s imm* копрезентация, сопредставление
coproantibody *s* копрогенное антитело, копроантитело
coproduct *s* побочный продукт
coprogenous *a* копрогенный

coprolite *s* копролит
coprophagous *a* питающийся экскрементами
coprophilous *a* копрофильный
coprophyte *s* копрофит
coprozoic *a* копрозойный, обитающий в экскрементах
coprozoon *s* копробионт
copse *s* молодой лесок, поросль
copula *s anat* связка; иммунное тело
copulate *v* копулировать; совокупляться
copulation *s* копуляция; совокупление; вязка; *bot* копулировка
copulative *a* соединительный, связующий; копулятивный
copulatory *a* копулятивный
copurification *s* совместная очистка
coracidium *s helm* корацидий
coracoid *s anat* клювовидный отросток
corallite *s zool* кораллит
corbicula *s apic* корзиночка
corbie *s* ворон *(Corvus corax)*
corcule *s bot* зародыш
cord *s* канатик; тяж; связка; пуповина; хорда, спинная струна
  Ferrein's ~ голосовая складка
  germinal ~ зародышевый шнур, зародышевый тяж
  heel ~ пяточное сухожилие
  tendinous ~ сухожильная струна
  umbilical ~ пуповина, пупочный канатик
  vocal ~ голосовая связка
cordate *a* сердцевидный, сердцеобразный
cordiform *a* сердцевидный, сердцеобразный
cordon *s:*
  cuticular cephalic ~ *гельм.* кутикулярный головной шнурок
  horseshoe-shaped ~ *гельм.* подковообразный придаток
core *s* сердцевина; внутренность; ядро; стержень; почвенный монолит, керн, колонка
  phage ~ кор фага, ядро фага, внутренний стержень фага
corecognition *s imm* двойное распознавание
coremiform *a* метёлкообразный
coremium *s* коремия
coriaceous *a* кожистый, жёсткий
coriander *s* кориандр, кишнец посевной *(Coriandrum sativum)*
corium *s* кориум, дерма, кожица
cork *s bot* вторичная кора, вторичный луб; пробка
corky *a* пробковый
corm *s* клубнелуковица
cormidium *s* кормидий
cormoid *a* похожий на побег
cormophyte *s* кормофит, листостебельное растение
cormorant *s* баклан *(Phalacrocorax)*
corncob *s* сердцевина кукурузного початка
corncockle *s bot* куколь *(Agrostemma)*
corncrake *s orn* коростель *(Crex crex)*
cornea *s* роговая оболочка глаза, роговица
corneal *a* корнеальный, роговичный
cornel *s* дёрен *(Cornus);* зёрнышко
corneous *a* роговой; роговидный
cornflower *s* василёк *(Centaurea)*
cornicle *s* рожок, роговидный вырост
cornification *s* ороговение
cornu *s* роговидное образование
cornuted, cornuted *a* рогатый
corny *a* хлебный, зерновой; хлебородный
corolla *s bot* венчик
corollaceous *a* венчиковидный
corollate, corollated, corolliferous *a* имеющий венчик, венчиконосный
corona *s bot* корона, венец; коронка
  ~ **radiata** *emb* лучистый венец
coronal, coronary *a* коронарный, венечный
coronate, coronated *a bot* увенчанный, имеющий корону
coronavirus *s* коронавирус
coronoid *a* венечный
corpse *s* труп
corpus *s* тело; *бот.* корпус
corpuscle *s* тельце; частица; корпускула
  blood ~ форменный элемент крови
  bone ~ костная клетка, остеоцит; остеобласт
  bridge ~ десмосома
  cartilage ~ хондроцит, хрящевая клетка
  ghost ~ «тень» эритроцита
  lymph ~ лимфоцит, лимфатическая клетка

Malpighian ~ мальпигиево тельце, лимфатический фолликул селезёнки
milk ~ молочное тельце, молочный шарик
phantom ~ «тень» эритроцита
red blood ~ эритроцит
renal ~ почечное тельце
reticulated ~ ретикулоцит
shadow ~ «тень» эритроцита
taste ~ вкусовая почка
terminal ~ нервное окончание, концевое тельце
white blood ~ лейкоцит
Zimmermann's ~ тромбоцит, кровяная пластинка
correction *s* поправка
end ~ поправка на крайнее значение
correlated *a* корреляционный, взаимосвязанный
correlation *s* корреляция, связь, зависимость, соотношение
invert ~ обратная корреляция, обратное соотношение
correlative *a* коррелятивный, соответствующий, сходный
corrugated *a* сморщенный
corrugation *s* складка, морщина
corrupt *v* гнить, разлагаться
corruption *s* гниение, порча
cortex *s* кора; корка; кортекс; кортикальный слой; *бот.* первичная кора; кортекс; кортикальный слой
cell ~ кортикальный слой клетки
entorhinal ~ энторинальная область коры
renal ~ корковое вещество почки
visual ~ зрительная кора, зрительная зона коры
cortical *a* корковый, кортикальный
corticalization *s* кортикализация
corticated *a* покрытый корой
corticifugal *a* кортикофугальный, идущий от коры
corticioid *a myc* кортициоидный
corticipetal *a* кортикопетальный, идущий к коре
corticoid *s* кортикоид
corticole, corticolous *a bot* обитающий на коре; растущий на коре
corticosteroid *s* кортикостероид; кортикостероидный
cortina *s myc* покрывало
cortisol *s* кортизон
corvine *a* вороний
corymb *s bot* щиток
corymbiform *a* щитковидный
corymbose *a* щитообразный; щитковидный
coryphad *s* альпийско-луговое растение
coryphium *s* фитоценоз альпийских лугов
cosmet *s* узкокрылая моль *(Momphidae)*
cosmide *s* космида
cosmoid *a* космоидный
cosmopolite *s* космополит, космополитный организм
cost *s* цена, стоимость; затраты
reproductive ~ репродуктивная стоимость
costa *s* ребро; средняя жилка; костальная жилка
costal *a* рёберный; *ent* костальный
costate *a* ребристый; жилковатый
costicartilage *s* рёберный хрящ; хрящевая часть ребра
costiform *a* реброобразный
costmary *s bot* пижма *(Tanacetum)*
costochondral *a* рёберно-хрящевой
costophrenic *a* рёберно-диафрагмальный
costoscapular *a* рёберно-лопаточный
costosternal *a* рёберно-грудинный
costovertebral *a* рёберно-позвоночный
costume *s* костюм
display ~ *eth* демонстративный наряд
coterminous *a* имеющий сходное распределение; граничащий
cotransduction *s* совместная трансдукция
co-translational *a mbiol* котрансляционный
cotransport *s mbiol* котранспорт
cotton *s* хлопчатник *(Gossypium);* вата
cottonwood *s* тополь *(Populus)*
cottony *a* хлопковый; пушистый, мягкий
cotyle *s* вертлужная впадина
cotyledon *s* семядоля; котиледон
cotyledonary, cotyledonous *a* семядольный; имеющий семядолю; котиледонный; несущий котиледон
cotyloid *a* чашеобразный
cotype *s* котип
couch *s* логово, берлога, нора; *v* лежать; притаиться; проращивать

couch-grass *s* пырей ползучий *(Agropyron repens)*
cougar *s* пума, ягуар, *(Felis concolor)*
cough *s* кашляющий звук
counseling *s*: **genetic** ~ генетическая консультация
count *s* счёт, отсчёт; определение количества; учёт; считать; количество; импульс
aerial ~ учёт численности с самолёта
background ~ показатель фона
blood ~ счёт кровяных телец
colony ~ количество колоний
plate ~ чашечный подсчёт
viable ~ количество жизнеспособных микроорганизмов
counter *s* счётчик
Burker's cell ~ гемоцитометр Бюркера
counteraction *s* противодействие
counterclockwise *adv* против часовой стрелки
counterelectrophoresis *s* противоточный электрофорез
counterflow *s* противоток
counterimmunoelectrophoresis *s* встречный иммуноэлектрофорез, противоточный иммуноэлектрофорез, электросинерез
counterindication *s* противопоказание
counterinhibition *s* растормаживание
counter-ion *s* противоион
counterpoison *s* противоядие
counterstain *s* контрастное окрашивание
couple *s* пара; *v* соединяться, спариваться
couplet *s*: **key** ~ теза и антитеза
coupling *s* спаривание; *mbiol* сопряжение, связь; *gen* сцепление; поясница, поясничный отдел; **long in the** ~ с длинной поясницей
course *v* преследовать; гнаться по пятам; гнаться за дичью; бегать
coursed *a* загнанный, затравленный
courser *s orn* бегунок *(Cursorius)*
courtship *s eth* ухаживание
covalent *a* ковалентный
covariance *s* коварианса
covariation *s* ковариация, сопряжённая изменчивость
cover *s* крышка; покрытие; покров; покровное стекло; *v* покрывать; случать; сидеть *(on eggs);* заделывать
basal ~ *plsoc* площадь оснований
gill ~ жаберная крышка
projective ~ проективное покрытие
seed ~ семенная кожура
tall herbaceous ~ высокотравье
wing ~ надкрылье, элитра
coverage *s* покрытие, покров
periosteal ~ надкостница
covering *s* покров; *anat* оболочка, плева; покрытие; случка, спаривание; мазь, замазка
face ~ оброслость морды
serosal ~ серозная оболочка
coverslip *s* покровное стекло
covert *s* убежище, логовище; оперение; кроющие перья
covey *s* выводок, стая
cow *s* корова; самка *(of a large mammal)*
cowbane *s bot* вех ядовитый *(Cicuta virosa)*
cowbird *s* воловья птица, коровий трупиал *(Molothrus)*
cowled *a* имеющий капюшон, имеющий колпачок; капюшонообразный
cowry *s mal* каури, ципрея *(Cypraea)*
cowshed *s* хлев
coxa *s ent* кокса, тазик; тазобедренный сустав
coxal *a ent* коксальный, относящийся к тазику, тазиковый
coxofemoral *a* тазобедренный
Coxsackieviris *s* вирус Коксаки
coyote *s* койот, луговой волк *(Canis latrans)*
coypu *s* нутрия *(Myocastor coypus)*
crab *s* краб; яблоня *(*Malus*)*
crake *s orn* коростель *(Crex);* погоныш *(Porzana)*
cramp *s* судорога; спазм
crampon *s* воздушный корень
cranberry *s* клюква *(Oxycoccus)*
crane *s* журавль *(Grus)*; серый журавль *(Grus grus)*
white ~ стерх, белый журавль *(Grus leucogeranus)*
crane's-bill *s* герань *(Geranium)*
cranial *a* краниальный, черепной
craniate *a* имеющий череп
craniocerebral *a* черепно-мозговой
craniofacial *a* черепно-лицевой
craniology *s* краниология
craniometry *s* краниометрия

cranium *s* череп, черепная коробка
craspedodromous *a bot* краевой
craspedote *a* имеющий парус *(о медузах)*
craspedum *s zool* парус; *bot* краспедий
crassiexinous *a* толстоэкзинный
crassimarginate *a bot* с утолщённым краем
crassitegillate *a bot* толстопокровный
crassula *s bot* очиток *(Sedum)*
crateriform *a* кратерообразный; чашеобразный
craticula *s* кратикула *(in diatoms)*
craw *s orn* зоб
crawl *v* ползти; виться; стелиться; очень медленный шаг
crawler *s* ползучее растение; ползающее животное
crease *s* сгиб; складка
creasy *a* морщинистый; складчатый
creep *v* ползти; виться, стелиться
creeper *s* ползучее растение; ползучий побег; стелющийся побег; корневищный побег; ползающее насекомое; пресмыкающееся; *orn* пищуха *(Certhia);* древолаз *(Dendrocolaptidae)*
creeping *a* ползающий; ползучий, вьющийся; стелющийся
creeping-rooted *a* корнеотпрысковый
cremocarpous *a bot* вислоплодный
crenate *a* зазубренный; *bot* округлозубчатый, городчатый
crenation *s bot* округлая зазубрина
crenophilous *a* кренофильный
crenulate, crenulated *a bot* мелкозубчатый, мелкозазубренный, мелкогородчатый
crepitation *s* крепитация, сухой звук
crescentic *a* в форме полумесяца, полулунный, серповидный
crest *s* гребень, гребешок; хохолок; грива; загривок; выступ
  neural ~ нервный валик, ганглионарная пластинка
crested с гребнем; с хохолком; гривистый, с гривой; *bot* гребешковый
cribellate *a* мелкорешётчатый
cribriform, cribrose *a* ситообразный
cribrum *s* крибрум, сито *(in diatoms)*
cricket *s* сверчок *(Gryllidae)*
  African mole ~ медведка африканская *(Gryllotalpa africana)*
  camel [cave] ~s ложнокузнечиковые *(Rhaphidophoridae)*
  greenhouse stone ~ кузнечик оранжерейный *(Tachycines asynamorus)*
  mole ~ медведка *(Gryllotalpa)*
cricoid *s* перстневидный хрящ
crimp *s* извитость шерсти
crimson *a* малиновый
crinite *a* волосатый; с ворсинками
crinkle *s* морщина, морщинка; *phytop* складчатость; морщинистость
crinome *s* кринoм
crisis *s* кризис; криз
crispate *a* кудреватый
crispation *s* лёгкое конвульсивное движение
crisped *a* кудреватый
crisp-leaved *a* курчаволистный
crisp-winged *a bot* курчавокрылый
crisscrossing *s* крисскроссинг, переменное скрещивание
crissum *s orn* околоклоачная область
crista *s* гребень, гребешок; криста
cristate *a* гребенчатый, с гребешком; *bot* гребешковый
criterion *s* критерий
croak *s* карканье; каркать; кваканье; квакать
crochet *s* крючок; крючковидная щетинка
crocodile *s* крокодил *(Crocodylus)*
crocus *s bot* шафран *(Crocus)*
cron *s* крон
crookbacked *a* горбатый
crooked *a* кривой, изогнутый, искривлённый
crooning *s* воркование
crop *s* сельскохозяйственная культура; урожай; хлеб на корню; приплод; верхушка, макушка; зоб; *v* собирать урожай; сажать, сеять; подстригать, обрезать, срезать; объедать, щипать; **in** ~ в стадии плодоношения
  alien ~ интродуцированная культура, экзот
  mixed grass ~ травосмесь
  moss ~ пушица *(Eriophorum)*
  standing ~ биомасса; запас леса на корню
crop-eared *a* корноухий
cropper *s* зобастый голубь

crosier *s* плоская спиральная раковина; свёрнутый молодой лист папоротника; крючок аскогенной гифы
cross *s* кросс, скрещивание; скрещивать; помесь, гибрид
cross-allergenicity *s* перекрёстная аллергенность
crossbill *s* клёст *(Loxia)*
crossbred *s* кроссбред
crossbreeding *s* кроссбридинг, метизация
crossbridge *s* *mbiol* поперечный мостик
cross-checking *s* проверка скрещиванием
cross-fecundation, cross-fertilization *s* перекрёстное оплодотворение
cross-fibred *a* косослойный
cross-fostering *s* перекрёстное воспитание
cross-grained *a* с извилистыми волокнами; свилеватый
crossing *s* скрещивание
crossing-over *s* кроссинговер
cross-link *s* *mbiol* поперечная сшивка
cross-linked *a* поперечно связанный, поперечно сшитый
cross-linking *s* *mbiol* перекрёстное сшивание
cross-neutralization *s* перекрёстная нейтрализация
cross-pollination *s* перекрёстное опыление
cross-protection *s* перекрёстный иммунитет
cross-reaction *s* перекрёстная реакция
cross-responsiveness *s* перекрёстная реактивность, перекрёстная отвечаемость
cross-section *s* поперечный разрез; поперечное сечение
cross-testing *s* перекрёстное тестирование
crotch *s* развилина, разветвление; промежность
crouch *s* *eth* поза «сгорбливания»; поза «скорчившись»; приседание; *v* приседать; припадать к земле; сжиматься
croup *s* *anat* круп
crow *s* ворона; ворон *(Corvus)*; *pl* врановые *(Corvidae)*; *v* кукарекать
hooded ~ серая ворона *(Corvus cornix)*
crowded *a* скученный, густорасположенный, частый
crowding *s* загущение; скучивание
crown *s* верхушка; венец; крона, верхушка дерева; венчик, розетка; *orn* гребень; макушка, темя; крона прямой кишки, анальное отверстие
~ **of head** макушка, темя
~ **of stem** верхушка стебля
decumbent ~ распростёртая розетка листьев
radiating ~ *emb* лучистый венец
crown-bearing *a* венценосный
crucible *s* тигель
sintered filter ~ тигель для фильтрования, фильтр-тигель, воронка Бюхнера
cruciferous *a* крестоцветный
cruciform *a* крестообразный
crumple *v* сморщиваться; скручиваться
crumpled *a* морщинистый; сморщенный; скрученный
crural *a* бедренный; относящийся к ноге; относящийся к ножке
crus *s* голень
terminal ~ **of nerve** концевая нервная ножка
crush *v* дробить; толочь; измельчать
crust *s* корка; покрываться коркой; панцирь; *herp* щит
crustose *a* корковый; разрастающийся в виде корки
crusty *a* покрытый коркой; твёрдый, жёсткий
cry *s* крик
cryesthesia способность ощущать холод
crymium *s* сообщество тундры
crymnion *s* криопланктон, арктический планктон, полярный планктон
crymophil *s* обитатель тундры
crymophylactic *a* холодоустойчивый
crymophium *s* сообщество тундры
cryobiology *s* криобиология
cryofixation *s* криофиксация
cryofracture *s* криоскалывание, замораживание є скалывание
cryoimmunoglobulin *s* криоиммуноглобулин
cryoinjury *s* криоповреждение
cryophil *s* криофил
cryophyte *s* криофит, криофильное растение

cryoplankton *s* криопланктон, арктический планктон, полярный планктон
cryoprecipitation *s* холодовая преципитация, криопреципитация
cryopreservation *s* криоконсервация
cryoprotectant *s* криопротектор
cryoprotection *s* криозащита, криопротекция
cryoprotector *s* криопротектор
cryoscopy *s* криоскопия
cryostat *s* криостат
cryotropism *s* криотропизм
crypsis *s* криптическая окраска, защитная окраска
crypt *s anat* крипта, углубление
cryptagglutinoid *s imm* криптагглютиноид
cryptic *a* криптический, скрытый, латентный
cryptobiosis *s* криптобиоз
cryptocarp *s* цистокарпий
cryptocarpous *a* скрытоплодный
cryptochimaera *s* криптохимера
cryptocolpate *a* скрытобороздный
cryptocystis *s helm* криптоциста
cryptodeterminant *s* маскированная детерминанта, скрытая детерминанта, латентная детерминанта, криптотоп
cryptogam *s* криптогамное растение, споровое растение
cryptogamic, cryptogamous *a* тайнобрачный, споровый, нецветковый
cryptogenous *a* криптогенный, неизвестного происхождения
cryptomere *s* криптомерный ген
cryptomerism, cryptomery *s* криптомерия
cryptomitosis *s* криптомитоз
cryptophyte *s* криптофит
cryptopolar *a bot* скрытополярный
cryptosporidiosis *s* криптоспоридиоз
cryptostome *s* криптостома, скрытое устьице
cryptozoic *a* живущий скрытно
crystalline *a* кристаллический; прозрачный
crystallizer *s* кристаллизатор
crystalloid *s* кристаллоид
cteinophyte *s* паразитное растение, разрушающее хозяина
ctenidium *s* ктенидий; гребневидная структура
ctetology *s* учение о приобретённых признаках
ctetosome *s* дополнительная хромосома
cub *s* детёныш; щенок
cubital *a* локтевой; *ent* кубитальный
cubitus *s* локтевая кость; *ent* кубитальная жилка
cuboid *s* кубовидная кость
cuckoo *s* кукушка *(Cuculus);* куковать
cucullate, cucullated *a* клобучковый; имеющий клобучок
cucullus *s* клобучок
cucumber *s* огурец *(Cucucmis sativus)*
cucurbit *s* тыква *(Cucurbita)*
cud *s* жвачка; жевать
cuirass *s* твёрдый покров, панцирь, щит
cul-de-sac *s anat* слепой мешок, замкнутое пространство
culebra *s herp* кулебра *(Clelia rustica)*
cull *s* отбор; браковка, отбраковка; *v* отбирать; браковать, отбраковывать
culled *a* выбракованный; отсортированный
culm *s* соломина, стебель злаков; образовывать соломину
culmiferous *a* образующий соломину
cultigen *s* культиген
cultivar *s* культивар, сорт культурного растения
cultivate *v* культивировать, разводить, выращивать
cultivated *a* культурный; обрабатываемый, культивируемый, возделываемый *(soil)*
cultivation *s* культивация, культивирование, разведение, выращивание
  plant ~ растениеводство
cultrate, cultrated, cultriform ножеобразный, ножевидный
cultural *a* культуральный
culture *s* культура; разведение, выращивание; **in pure culture** в чистой культуре
  artificial ~ искусственное разведение
  blood ~ гемокультура
  intensive ~ интенсивное выращивание
  micromonoclonal ~ микропланшетный вариант гибридомы

monoclonal ~ антителопродуцирующая гибридома
single-cell ~ клон; культура, выделенная из одной клетки
"spinner" ~ культура с постоянным перемешиванием, спин-культура
cultured *a* культивируемый; культурный
cumulant *s biom* семиинвариант, полуинвариант
cumulate *a* кумулировать, накапливать, накапливаться
cumulation *s* кумуляция, накопление
cumulative *a* кумулятивный
cuneate, cuneated *a* клиновидный, клинообразный
cuneate-obovoid *a* клиновидно-обратнояйцевидный
cuneiform *a* клиновидный, клинообразный
cup *s* чашеобразная структура; *бот.* плюска; чашечка; *v* принимать чашеобразную форму
acorn ~ плюска
cluster ~ *myc* эцидий
flower ~ *bot* чашечка
gemma ~ почечная корзинка
optic ~ глазной бокал
cup-shaped *a* чашевидный; *bot* бокальчатый
cupulate *a* чашевидный
cupule *s zool* присоска; *bot* плюска
cupuliferous *a bot* плюсконосный
curdle *v* застыть, оцепенеть; свертываться, коагулировать
curiosity *s eth* любопытство
curl *s bot* завиток; *phytop* курчавость
curled *a* курчавый; закрученный; завитой
curlew *s orn* кроншнеп *(Numenius)*
stone ~ авдотка *(Burhinus)*
curl-flowered *a* кудрявоцветковый
curr *v* ворковать; мурлыкать; урчать
currant *s* смородина *(Ribes);* коринка
current *s* ток
blood ~ кровоток
brain action ~ биоток головного мозга
cursorial *a* бегающий
curvature *s* кривизна; искривление; изгиб
curve *s* кривая; *v* изгибаться
curved *a* изогнутый; искривлённый; кривой
curvicaulis *a* изогнутостебельчатый
curvidentate *a* изогнутозубчатый
curviflorus *a* изогнутоцветковый
curvifoliate *a* изогнутолистный
curvinervate *a* дугонервный
cuscus *s ther* кускус *(Phalanger)*
cushat *s orn* вяхирь *(Columba palumbus)*
cushion *s anat* подушка; подушечка; листовая подушечка
endocardial ~ эндокардиальный вырост
cusp *s* бугорок зуба; створка клапана
cuspid *s* клык
cuspidate *a* остроконечный
cut *s* разрез; порез; *mbiol* разрыв; канал; выемка; погон, фракция; *v* резать, разрезать; разделять на фракции; укос; косить; кастрировать; **to ~ out** вырезать, удалять
saw ~ спил
cutaneous *a* кожный
cut-grass *s* трава с режущими краями
cuticle *s* кутикула, кожица
embryonic ~ *ent* эмбриональная кутикула, «рубашечка»
primary ~ эпикутикула, эпидермис
secondary ~ дерма
serosal ~ серозная оболочка, сероза
cuticolous *a* живущий под кожей, подкожный
cuticular *a* кутикулярный
cuticularization, cutification *s* образование кутикулы
cutis *s* кутис
cutter *s* косилка, жатка
section ~ микротом
cutting *s* резание; срезание; кошение, косьба; *for* вырубка; режущий; *bot* черенок, отводок
clear ~ вырубка
grass ~, hay ~ сенокос
section ~ приготовление срезов
cuttlefish *s* каракатица *(Sepia)*
cutworm *s* совка *(Noctuidae)*
cuvette *s* кювета
cyanella *s* цианелла, симбиотическая цианобактерия
cyanocobalamin *s* цианокобаламин, витамин B_{12}
cyanotic *a* синюшный
cyathiform *a bot* бокаловидный
cyathium *s bot* циаций, бокальчик

cyclarthrosis *s* цилиндрический сустав
cycle *s* цикл; период пульсации; кругооборот, круговорот; *v* повторяться циклически; проходить цикл развития
nitrogen ~ круговорот азота
water ~ водооборот, влагооборот
cyclic *a* циклический
cyclicism *s* цикличность
cyclization *s* циклизация
cycloaddition *s* циклоприсоединение
cycloalkylation *s* циклоалкилирование
cyclocarpous *a* круглоплодный
cyclocondensation *s* циклоконденсация
cyclodehydration *s* циклодегидратация
cyclodehydrogenation *s* дегидроциклизация, циклодегидрирование
cyclodimerization *s* циклодимеризация
cyclomorphosis *s* цикломорфоз
cyclooligomerization *s* циклоолигомеризация
cyclopolymerization *s* циклополимеризация
cyclopterous *a bot* круглокрылый
cyclosis *s* циклоз
cylinder *s* цилиндр
cylindrical-stalked *a* вальковато-стебельчатый
cymbiform *a* ладьеобразный, ладьевидный
cymbipholious *a* ладьелистный
cymbocarpous *a* ладьеплодный
cyme *s bot* полузонтик; ложный зонтик
fan-shaped ~ опахало
helicoid ~ *bot* улитка
cymose *a bot* цимозный, полузонтичный; симподиально ветвящийся
cynopodous *a* имеющий невтяжные когти
cyophoria *s* беременность
cyotrophy *s* питание плода
cyphella *s bot* цифелла
cyphonautes *s zool* цифонаутес
cypress *s* кипарис *(Cupressus);* болотный кипарис *(Taxodium);* кипарисовик *(Chamaecyparis);* каллитрис *(Callitris);* кохия *(Kochia)*
cypsela *s bot* семянка
cyriodoche *s* полная сукцессия
cyst *s* циста; *anat* пузырь; киста
hydatid ~ гидатида
cysteine *s* цистеин
cystic *a* цистный; пузырный
cysticercoid *s* цистицеркоид
cystcercosis *s* цистицеркоз
cysticercus *s* цистицерк
cysticolous *a* живущий в цисте
cystidium *s myc* цистида
cystitis *s* цистит
cystocarp *s* цистокарпий
cystogenous *a* образующий цисту
cystolith *s* цистолит
cystophore *s* гифа, несущая цисты
cystosorus *s myc* цистокучка
cystospore *s* цистоспора; карпоспора; циста с зооспорами
cytarme *s* уплощение бластомеров
cytaster *s cyt* звезда, клеточная звезда
cytoadherence *s* клеточная адгезия
cytoarchitecture *s* цитоархитектоника
cytobiology *s* цитобиология, биология клетки
cytocentrum *s* цитоцентр, центросома, клеточный центр
cytochemical *a* цитохимический
cytochemistry *s* цитохимия
cytocinesis *s* цитокинез, клеточное деление
cytode *s* цитода, монера
cytoderm *s* оболочка клетки, клеточная стенка
cytodiagnosis *s* цитодиагностика
cytodieresis *s* цитокинез, клеточное деление
cytogamy *s* цитогамия
cytogene *s* цитоген, плазмоген
cytogenesis *s* цитогенез
cytogenetic *a* цитогенетический
cytogenetics *s* цитогенетика
cytogeny *s* цитогенез
cytoimmunotherapy *s* цитоиммунотерапия
cytokine *s* цитокин
cytokinesis *s* цитокинез, клеточное деление
cytolemma *s* плазмалемма
cytological *a* цитологический
cytology *s* цитология
cytolymph *s* цитолимфа
cytolysis *s* цитолиз
cytolytic *a* цитолитический
cytome *s* цитом
cytomegalovirus *s* цитомегаловирус

cytomere *s* цитомер, пластомер, хондриомер
cytometry *s* цитометрия
cytomorphosis *s* цитоморфоз
cyton *s* цитон
cytopathology *s* цитопатология, клеточная патология
cytophilous *a* цитофильный
cytophotometry *s* цитофотометрия
cytophysiology *s* физиология клетки, цитофизиология
cytoplasm *s* цитоплазма
cytoplasmic *a* цитоплазматический
cytopoiesis *s* цитопоэз
cytoproct *s prot* цитопрокт
cytoskeleton *s* цитоскелет
cytosol *s* цитозоль
cytosome *s* цитосома, клеточное тело
cytosphere *s* цитосфера
cytostatic *a* цитостатический
cytostome *s* цитостом, клеточный рот
cytotaxis *s* цитотаксис
cytotaxonomy *s* цитотаксономия
cytotoxic *a* цитотоксический
cytotoxicity *s* цитотоксичность
  self-restricted ~ аутоцитотоксичность
cytotrophoblast *s* цитотрофобласт
cytotropic *a* цитотропный
cytotropism *s* цитотропизм
cytotype *s* цитотип

D

dab *s* клевок; *v* клевать
dachshound *s* такса
dacryocyst *s* слёзный мешочек
dacryoideous *a* слезовидный
dacryon *s* дакрион
dactyl *s* палец
dactylar *a* пальцевой
dactylic *a* пальцевидный
dactylognathite *s* дактилогнатит
dactyloid *a* пальцеобразный
dactylozooid *s zool* дактилозоид
daddy-long-leg *s ent* долгоножка *(Tipulidae)*
daffodil *s* нарцисс ложный *(Narcissus pseudonarcissus)*
dagger-pointed *a* длинноколючковый
dagger-shaped *a* кинжаловидный
dairy *a* молочный
daisy *s* маргаритка *(Bellis)*
dam *s* мать, матка, самка; плотина
damage *s* вред, повреждение; повреждать
  antibody-mediated ~ антителоопосредованный лизис
  freezing ~ криоповреждение
  hail ~ градобитие
damaged *a* повреждённый
daman *s ther* даман, жиряк *(Procavia)*
dammar *s* смола
damp *a* влажный, гумидный; смачивать, увлажнять
damping-off, damping-out *s фтп* выпревание; полегание; чёрная ножка
damselfly *s* равнокрылая стрекоза *(Zygoptera)*
  black-winged ~ красотка *(Calopteryx)*
damson *s* тернослива *(Prunus insititia)*
dance *s eth* танец
dandelion *s* одуванчик лекарственный *(Taraxacum officinale)*
danger *s* опасность
dangerous *a* опасный
dansylation *s* дансилирование
dap *s* зарубка
dappled *a* в яблоках *(color)*
  ~ **gray** серый в яблоках
dark *a* тёмный
darkness *s* темнота
dark-repair *s* темновое восстановление, темновая репарация
darnel *s* плевел *(Lolium)*
dart *s* жало; быстрое короткое движение, бросок; *mal* «любовная стрела»
Darwinism *s* дарвинизм
dash *s* всплеск
dassy *s ther* даман; скалистый даман (*Procavia*)
dasyphyllous *a* пушистолистный; густолистный
databank *s* банк данных
database *s* база данных
data-processing *s* обработка данных
date *s* финик; срок; период
dauber *s* роющая оса *(Sphecidae)*
dauby *a* клейкий, липкий
dauermodification *s* длительная модификация
daw *s* галка *(Corvus monedula)*
day *s* день; сутки
  trap ~ ловушкосутки
dayfly *s* подёнка *(Ephemeroptera)*
deacetylation *s* деацетилирование
deactivation *s* дезактивация
dead *a* мёртвый; снулый *(fish);* опавший *(leaf);* засохший *(leaf);* сухостойный *(tree);* бесплодный *(soil)*
deadaptation *s* деадаптация
deadborn *a* мертворождённый
dead-nettle *s bot* яснотка *(Lamium)*
dead-ripe *a* созревший
dead-tree *s* сухостой
deaeration *s* деаэрация, удаление воздуха
deafening *s* оглушение; звуконепроницаемый
deaf *a* глухой
deafferentation *s* деафферентация
deafness *s* глухота
deal *s* хвойный лес
dealate *a ent* сбросивший крылья
dealation *s ent* сбрасывание крыльев
dealbate *a bot* покрытый белым налётом
dealkylation *s* деалкилирование
  reductive ~ гидродеалкилирование, восстановительное деалкилирование
deallergization *s* десенсибилизация; антианафилаксия
dealpine *a* сниженный
deamidation *s* деамидирование
deaminization *s* дезаминирование

deaquation *s* обезвоживание, дегидратация
death *s* смерть
thymineless ~ *gen* бестиминовая гибель
deathwatch *s* точильщик пёстрый *(Xestobium rufovillosum);* вошь книжная, книгоед *(Troctes divinatorius);* сеноед домовый *(Atropos pulsatoria)*
debark *v* обдирать кору
debilitate *v* ослаблять, расслаблять
debilitation *s* ослабление, обессиливание, слабость
debris *s* осколки, обломки, остатки
debt *s* долг
oxygen ~ кислородная задолженность, кислородный долг
decalcification *s* декальцификация
decandrous *a* с десятью тычинками
decant *v* декантировать
decapitation *s бот.* декапитация, удаление верхушки; декапитация, обезглавливание; обезглавливать
decapsidation *s* декапсидация
decarboxylation *s* декарбоксилирование
decay *s* распад, гниение, разложение; *phytop* гниль; затухание; *v* распадаться, гнить, разлагаться
decaying *a* разлагающийся, гниющий
decease *s* смерть, кончина; умереть, скончаться
decemfid *a* расчленённый на десять сегментов
decerebration *s* децеребрация, удаление головного мозга
dechromatization *s* дехроматизация
decidua *s* отпадающая плацента
deciduous *a* опадающий; осыпающийся; лиственный, листопадный; молочный *(a tooth)*
decision-making *s eth* принятие решения
declinate *a* загнутый, отклонённый вниз
declining *a* начинающий засыхать; перестойный *(forest)*
declive *s anat* скат
decoction *s* отвар, декокт
decolorant *s* обесцвечивающее средство; отбеливатель
decoloration, decolorization *s* обесцвечивание
decolorize *v* обесцвечивать
decompensation *s* декомпенсация
decomplementize *v* удалять комплемент, декомплементировать
decomposed *a* разложившийся, сгнивший
decomposer *s* редуцент, деструктор, биоредуктор
decomposition *s* разложение, гниение; деструкция
decompound *a bot* многораздельный; многосложный; повторносложный
pinnately ~ перистораздельный
decompression *s* декомпрессия, снижение давления воздуха
decondensation *s* деконденсация, разрыхление
decondensed *a* деконденсированный
deconjugation *s* деконъюгация
decontaminant *s* дезактивационное средство
decontamination *s* деконтаминация; дезактивация; очистка; обеззараживание
decorative *a* декоративный
decorticate *v* окоривать; бескорный; шелушить
decortication *s* декортикация
decoy *s* приманка; манок; приманивать; западня, ловушка; *v* заманивать в ловушку
decoy-duck *s* подсадная утка
decrease *s* уменьшение, убывание, убыль; понижение; спад; сбежистость; *v* уменьшаться, убывать; спадать
decrement *s* уменьшение, убывание, убыль; *phys* декремент
response ~ угасание реакции
decubitus *s* лежачее положение
decumbent *a* стелющийся, ползучий
decurrent *a* сбегающий, нисходящий
decussate *v* перекрещиваться, пересекаться; *a* перекрёстно-парный
decussation *s* перекрёст; перекрещивание
decyclization *s* дециклизация
dedentition *s* выпадение зубов
dedetermination *s* дедетерминация
dedicaryotization *s* дедикариотизация
dedifferentiation *s* дедифференциация, дедифференцировка
dediploidization *s* дедиплоидизация

DE, digestible energy *s* переваримая энергия
deep *s* впадина; пучина; *a* глубокий
deep-bodied *a* высокий *(e.g., of fish)*
deer *s* олень (*Cervidae*); красный зверь
barking ~ мунтжак *(Muntiacus)*
fallow ~ лань *(Cervus dama)*
Father David's ~ олень Давида, милу *(Edaphurus davidianus)*
moose ~ лось *(Alces alces)*
musk ~ кабарга *(Moschus)*; сибирская кабарга *(Moschus moschiferus)*
rib-faced ~ мунтжак *(Muntiacus)*
roe ~ косуля *(Capreolus capreolus)*
defatigation *s* переутомление, перенапряжение, крайняя усталость
defatted *a* обезжиренный
defatting *s* обезжиривание
defecation *s* дефекация
defect *s* порок, дефект, недостаток
defective *a* несовершенный; недостаточный; повреждённый, дефектный, нарушенный
defense *s* оборона, защита
body ~s *pl* защитные силы организма
defensive *a* оборонительный, защитный
defensor *s* защитное приспособление
deferent *a* выносящий, выводящий; отводящий; эфферентный, центробежный
defertilization *s bot* стерилизация
defibrillation *s* дефибрилляция
defibrillator *s* дефибриллятор
defibrination *s* дефибринирование
deficiency *s* дефицит; недостаток, порок; недостаточность; нехватка
cellular immune ~ клеточный иммунодефицит
vitamin ~ авитаминоз
deficit *s* дефицит, нехватка, недостаточность
define *v* давать определение, определять
definite *a* определённый, точный; *bot* цимозный
definition *s* определение
definitive *a* дефинитивный, окончательный, безусловный
deflate *v* выпускать, выкачивать; спадаться *(of a lung)*
deflection *s* отклонение
complement ~ фиксация комплемента, связывание комплемента; отклонение комплемента
deflexed *a* повёрнутый резко вниз, отогнутый
deflocculation *s* дефлокуляция, распад хлопьев
deflorated, deflowered *a* отцветший
defoliant *s* дефолиант
defoliate *v* удалять листья; обезлиственный
deforestation *s* обезлесение, вырубка леса, сведение лесов
deform *v* деформировать, уродовать
deformation *s* деформация
deformed *a* деформированный, изуродованный
deformity *s* деформация; уродство
defunctionalization *s* прекращение функции
degeneracy *s* дегенерация, перерождение, вырождение; деградация; *mbiol* вырожденность
degeneration *s* дегенерация, перерождение, вырождение; деградация
fatty ~ жировая дегенерация, жировое перерождение
degenerative *a* дегенеративный, вырождающийся
degerm *v* дезинфицировать, стерилизовать
deglutition *s* глотание, проглатывание
deglutitive, degluttory *a* глотательный
degradation *s* деградация, ухудшение, вырождение; деградация, разрушение, расщепление, распад
hydrolytic ~ гидролиз, гидролитическое расщепление
degrade *v* деградировать; вырождаться
degranulation *s* утрата зернистости
degreasing *s* обезжиривание
degree *s* степень; градус; качество; степень родства
~ **of closeness** *bot* степень сомкнутости; полнота насаждения
~ **of density** степень плотности; *bot* степень сомкнутости; полнота насаждения
~ **of tapering** сбежистость
degree-day *s* градусодень
degrowth *s* уменьшение массы

dehardening *s* потеря холодостойкости
dehelmintization *s* дегельминтизация
dehematize *v* обескровливать
dehiscence *s bot* раскрывание, растрескивание
dehiscent *a bot* открывающийся растрескиванием
dehorning *s* удаление рогов
dehull *v* лущить
dehumidification *s* обезвоживание; высушивание, сушка
 cool ~ лиофилизация, лиофильное высушивание
dehydratation *s* дегидратация, обезвоживание
dehydration *s* дегидратация, обезвоживание; дегидрирование
dehydrocholesterol *s* дегидрохолестерин
dehydrocorticosterone *s* дегидрокортикостерон
dehydroepiandrosterone *s* дегидроэпиандростерон
dehydrogenation *s* дегидрогенизация
deionization *s* деионизация
deintoxication *s* дезинтоксикация
deionization *s* деионизация
deionizer *s* деионизатор
deirid *s helm* шейная папилла
dejection *s* испражнения, каловые массы; дефекация
dejecture *s* испражнения, каловые массы
delamination *s* расслоение, расслаивание; *emb* деламинация
delay *s* задержка, запаздывание, отставание, замедление; промедление; *v* задерживать, замедлять; медлить
delayed замедленный; запаздывающий
delayer *s*: **T** ~ Т-лимфоцит-дилэйер
deleterious *a* вредный, пагубный, губительный, разрушительный
deletion *s* делеция
delicacy *s* чувствительность
delimit *v* разграничивать
delimitation *s* разграничение
delipidization *s* делипидирование
deliquescent *a* многоствольный; распадающийся; растворяющийся
delivery *s* роды, родоразрешение
dell *s* лесистая долина, лощина
demagnification *s gen* демагнификация
demand *s* потребность
 biochemical oxygen ~ биохимическая потребность в кислороде; окисляемость
demander *s* **light** ~ светолюбивое растение; светолюбивое животное
demarcation *s* разграничение, демаркация
 geographical ~ районирование
deme *s* дим, дем
demelcoline *s* колцемид
demersal *a* демерсальный
demethylation *s* деметилирование
demicircular *a anat* полукружный
demicolporate *a pal* полубороздно-поровый
demicolpus *s pal* полуборозда
demifacet *s* полуфасетка; рёберная ямка
demilune *s anat* полулуние
demisheath *s ent* створка яйцеклада
demography *s* демография
demoiselle *s* журавль-красавка *(Anthropoides virgo); pl* равнокрылые стрекозы *(Zygoptera)*
demyelinization *s* демиелинизация
den *s* логово; берлога
denatant *a* плывущий по течению; мигрирующий по течению
denaturation *s* денатурация
dendrite *s* дендрит
dendritic *a* древовидный; ветвящийся; дендритный
dendrobiont *s* дендробионт, обитатель деревьев
dendrochronology *s* дендрохронология
dendrocolous *a* обитающий на деревьях
dendroflora *s* дендрофлора
dendrogliocyte *s* дендроглиоцит, дендроглиальная клетка
 spinal ~ спинальный дендроглиоцит, спинальная дендроглиальная клетка, клетка Гортега
dendrogram *s* дендрограмма
dendrology *s* дендрология
dendrometer *s* дендроматр
dendrometry *s* дендрометрия
dendron *s* дендрит
dendrophil *s* дендрофил
dendrophilous *a* дендрофильный
denervated *a* денервированный
denervation *s* денервация
denitrification *s* денитрификация
denitrifier *s* денитрификатор

denitrosation *s* денитролизование
denizen *s* натурализовавшийся организм; натурализованный организм
denomination *s* название, обозначение
dens *s* зуб; зубец; зубовидный отросток
dense *a* плотный, густой; сомкнутый; дремучий
dense-flowering *a* скученно-цветковый
densitometer *s* денситометр
densitometry *s* денситометрия
density *s* плотность; густота, полнота; сомкнутость
haptenation ~, haptenic степень гаптенилирования, гаптенная плотность
planting ~ густота посадки
dental *a* зубной
dentate *a* зубчатый, пильчатый
dentate-crenate *a* зубчато-городчатый
dentelation *s* зубчатость
denticle *s* зубчик
denticulate, denticulated *a* мелкозубчатый, имеющий зубчики
dentigerous *a* имеющий зубы
dentin, dentine *s* дентин
dentinogenesis *s* дентиногенез
dentition *s* прорезывание зубов; расположение зубов; зубной аппарат
denture *s* зубной ряд
denucleated *a* энуклеированный, с удалённым ядром
denudation *s* денудация
~ **of ova** удаление оболочки яйца
denuded *a* оголённый, обнажённый; денудированный
deodar *s* кедр гималайский (Cedrus deodara)
deoxyadenosine *s* дезоксиаденозин
deoxycholate *s* дезоксихолат
deoxycytidine *s* дезоксицитидин
deoxyepinephrine *s* дезоксиэпинефрин, эпинин
deoxygenation *s* дезоксигенирование
deoxyguanosine *s* дезоксигуанозин
deoxyribonucleohistone *s* дезоксирибонуклеогистон
deoxyribonucleoproteid *s* дезоксирибонуклеопротеид
deoxyribonucleoside *s* дезоксирибонуклеозид
deoxyribonucleotide *s* дезоксирибонуклеотид
deoxyribose *s* дезоксирибоза
deoxysugar *s* дезоксисахар
deoxythymidine *s* дезокситимидин
daparture *s biom* отклонение
depauperated *a* обеднённый; угнетённый
dependence *s* зависимость
dependent *a* зависящий, обусловленный
dephosphorylation *s* дефосфорилирование
depigmentation *s* депигментация
depletion *s* опустошение; истощение
complement ~ истощение комплемента, инактивация комплемента
depolarization *s* деполяризация
depollination *s* удаление пыльцы
depolymerization *s* деполимеризация
deposit *s* осадок, отложение, налёт; отлагать, осаждать, давать осадок
~ **of moisture** запасы влаги
deposition *s* осаждение, отложение, напластование; нанос
retrogressive ~ удаление наносов; размыв
silt ~ осаждение взвеси; кольматаж
depot *s* депо
depredator *s* вредитель
depress *v* подавлять, угнетать
depressant *s* депрессант, тормозящее средство
immunologic ~ иммунодепрессант
depressed *a* сплющенный, сдавленный, вдавленный; вогнутый; подавленный, угнетённый
depression *s* депрессия; подавление; углубление; впадина; низина
facial ~ *ent* антеннальное углубление, усиковая впадина, усиковая ямка
infrasternal ~ подложечная ямка
neuromuscular ~ нервно-мышечная депрессия, «торможение Ввведенского»
precordial ~ подложечная ямка
presutural ~ *ent* предшовное углубление
selective ~ подавление отбором
depressive *a* депрессивный, подавляющий, угнетающий
depressor *s anat* депрессор
depressurization *s* сброс давления; разгерметизация
deprivation *s eth* депривация

deproteinize *v* извлекать белки
deprotonation *s* депротонирование
depside *s* депсид
depth *s* глубина; высота *(e.g., of fish);* толща, толщина; мощность *(of soil)*
working ~ пределы корнеобитаемого слоя
depula *s* депула, инвагинирующая бластула
derepress *v* дерепрессировать, вновь активировать
derepression *s* дерепрессия, повторная активация
derivation *s* источник, происхождение; отклонение; ответвление; *phys* отведение
derivative *s* дериват, производное; *a* производный
derm(a) *s* дерма, кутис, собственно кожа
dermal *a* дермальный, кожный
dermatitis *s* дерматит
dermatogen *s* дерматоген, эмбриональный эпидермис
dermatome *s* дерматом
dermatophyte *s* дерматофит
dermic *a* дермальный, кожный
dermoblast *s* дермобласт
dermographism *s* дермографизм
dermoossification *s* кожное окостенение
dermophyte *s* дерматофит
desalinization, **desalting** *s* обессоливание
descend *v* происходить; передаваться по наследству; снижаться
descendant, descendent *s* потомок; *a* происходящий, ведущий происхождение
descendental *a* основанный на фактах; обусловленный опытом; натуралистический
decendible *a* передаваемый по наследству
descending *a* нисходящий, направленный вниз
descent *s* происхождение; родословная; снижение; понижение; опущение *(органа)*
description *s* описание
first ~ первоописание, первоначальное описание
desegmentation *s* слияние сегментов, утрата сегментации
desensitization *s* десенсибилизация, антианафилаксия; потеря чувствительности
desert *s* пустыня; пустынный
deserticolous *a* обитающий в пустыне
desertification *s* опустынивание
desexualize *v* кастрировать
desiccant *s* высушивающее средство; *a* высушивающий
desiccation *s* высыхание; высушивание; потеря влаги
design *s* схема; план, проект; узор; *v* проектировать, планировать
desintegration *s* дезинтеграция, разрушение
desire *s* желание, влечение
desman *s* выхухоль *(Desmana)*
desmocyte *s* десмоцит
desmoenzyme *s* десмофермент
desmognathous *a* десмогнатический
desmose *s* десмоз
desmosome *s* десмосома
desmotubule *s* десмотрубочка
desorption *s* десорбция
desquamation *s* десквамация, шелушение, слущивание, отслоение
destabilization *s* дестабилизация
destain *v* удалять краситель, обесцвечивать
destrobilation *s helm* дестробиляция, отторжение члеников
destruction *s* разрушение; повреждение
immune ~ иммунное повреждение; иммунный лизис
destructive *a* деструктивный, губительный, разрушающий
desulfitation *s* десульфитация
desulfonation *s* десульфирование
desulfuration, desulfurization *s* десульфуризация
desynapsis *s* десинапс
detachment *s* отделение; отслаивание; разрыв
detector *s* детектор
gas ~ газоопределитель
detergent *s* детергент, очищающее средство; моющее средство
deterioration *s* ухудшение; порча
determinant *s* детерминант, определитель; *imm* антигенная детерминанта, эпитоп

antigenic ~ антигенная детерминанта, эпитоп
characteristic ~ *biom* характеристический определитель
neoantigenic ~ неодетерминанта; опухолеспецифический антиген
unique antigenic ~ уникальная антигенная детерминанта, уникальный эпитоп, унитоп
determination *s* детерминация, определение, установление
detoxification *s* детоксикация
detrimental *a* вредный
detritus *s* детрит
detritus-consuming, detrivorous *a* детритоядный
deuterencephalon *s emb* задний мозговой пузырь
deuterium *s* дейтерий
deuterium-labeled *a* меченный дейтерием
deuterium-loaded *a* насыщенный дейтерием
deuterocerebrum *s* дейтоцеребрум
deuterocoel *s* дейтероцель, целом, вторичная полость тела
deuterocone *s* дейтерокон
deuterogenesis *s* дейтерогенез
deuterostoma *s emb* вторичный рот
deuterotoky *s* дейтеротокия, амфитокия
deutocerebrum *s* дейтоцеребрум
deutomerite *s* дейтомерит
deutonephros *s* мезонефрос
deutonymph *s* дейтонимфа
deutoplasm *s* дейтоплазма, желток яйцеклетки
devastate *v* опустошать
devastation *s* опустошение, разорение *(e.g., of nests);* девастация
development *s* развитие; освоение, разработка
deviant *s* отклоняющаяся форма
deviate отклоняться
deviation *s* отклонение
~ **of complement** фиксация комплемента, связывание комплемента; отклонение комплемента
absolute ~ абсолютное отклонение, модуль отклонения
complement ~ фиксация комплемента, связывание комплемента; отклонение комплемента
normal equivalent ~ *биом.* нормальный квантиль
devolution *s* ретрогрессивное развитие
Devonian *s* девон, девонский период; девонский
devour *v* пожирать
dew *s* роса
dewatering *s* обезвоживание; обезвоживающий
dewclaw *s* рудиментарный палец
dew-cup *s* манжетка луговая *(Alchemilla pratensis)*
dewlap *s* подгрудок *(in cattle);* orn серёжка
dexiotropic *a* дексиотропный, правозавитой, правозакрученный, направленный по часовой стрелке
dextral *a* правый
dextroglucose *s* декстроза, D-глюкоза
dextrorotation *s* правое вращение
dextrorotatory *a* правовращающий
dextrorse *a* правозавитой, правозакрученный
dextrose *s* декстроза, D-глюкоза
dextroversion *s* декстроверсия
dhole *s* красный волк *(Cuon alpinus)*
diabetes *s* диабет
diacetylation *s* диацетилирование
diachenium *s* двусемянка
diacmic *a* с двумя максимумами развития
diacoel *s* третий желудочек мозга
diactinal *a* двухлучевой
diadelphous *a bot* двухбратственный
diadromous *a bot* диадромный, веерообразный
diageotropic *a* диагеотропный
diageotropism *s* диагеотропизм, поперечный геотропизм
diagnosis *s* диагноз; диагностика
foliar ~ листовой анализ
diagnostic *a* диагностический, отличительный
diagnostics *s* диагностика
diagram *s* диаграмма
diagynic *a* диагинический
diaheliotropism *s* диафототропизм, диагелиотропизм
diakinesis *s* диакинез

dialect *s eth* диалект
dialectic, dialectics *s* диалектика
natural ~s диалектика природы
dialkylation *s* диалкилирование
dialycarpellary *a* раздельноплодолистковый
dialypetalous *a* раздельнолепестковый
dialyphyllous *a* раздельнолистный
dialysepalous *a* раздельночашелистиковый
dialysis *s* диализ
dialystaminous *a* раздельнотычиночный
dialyzable *a* диализуемый
diandric *a* диандрический
diandrous *a* двутычинковый
diantheral *a* двупыльниковый
dianthovirus *s* диантовирус
diapause *s* диапауза
diapedesis *s* диапедез
diaphanous *a* прозрачный, просвечивающий
diaphototropism *s* диафототропизм, диагелиотропизм
diaphragm *s* диафрагма
diarch *a* диархный
diarrhea *s* понос, диарея
diarthrosis *s* диартроз, истинный сустав
diaschistic *a* диасхистический
diaspore *s* диаспора
diastase *s* диастаза, амилаза
diastasis *s phys* диастазис
diastema *s* диастема
diaster *s cyt* диастер, двойная звезда
diastole *s* диастола
diastolic *a* диастолический
diathermy *s* диатермия
diathesis *s* диатез
diatomaceous *a* диатомовый
diatom *s* диатомея, диатомовая водоросль *(Diatomeae)*
diatropic *a* диатропический
diatropism *s* диатропизм
diauxia *s micr* диауксия
diaxon, diaxone *s* биполярная нервная клетка, биполярный невроцит
diaxon, diaxonic *a* двуосный, двухосный
diazocoupling *s* (ди)азосочетание
diazotrophy *s* диазотрофия, азотфиксация
dibasic *a gen* двуосновный
dibranchiate *a* двужаберный
dicaryon *s* дикарион
dicaryophase *s* дикариофаза
dicaryophyte *s* дикариофит
dicaryotic *a* имеющий дикарион
dicaryotization *s* дикариотизация
dicentric *s* дицентрик, дицентрическая хромосома; дицентрический, имеющий две центромеры
dichasium *s* дихазий
dichlamydeous *a* с двойным околоцветником
dichloracetylation *s* дихлорацетилирование
dichogamy *s* дихогамия
dichophysis *s* дихофиз
dichoptic *a энт.* дихоптический, с несоприкасающимися глазами
dichotomous *a* дихотомический
dichotomy *s* дихотомия, дихотомическое ветвление
differential ~ сегрегация зародыша
dichroic *a* дихроический; дихроматический
dichroism *s* дихроизм
dichromasy *s* дихромазия
dichromatism *s* двухцветность
dichromatopsia *s* дихромазия
diclinous *a bot* разнополый, раздельнополый
dicoccous *a bot* двуорешковый
dicolpate *a pal* двубороздный
dicolporate *a pal* двубороздно-поровый
dicot, dicotyledon *s* двудольное растение
dicotyledonous *a* двудольный
dicranophorous *a bot* вилконосный
dicronotrichous *a bot* вильчато-волосистый
dicrotism *s* дикротизм, двуударность пульса
dictionary *s* словарь
DNA ~ of sense words *mbiol* ДНК-словарь смысловых слов
dictyocarpous *a bot* сетчатоплодный
dictyodromous *a bot* сетчатый
dictyokinesis *s* диктиокинез
dictyosome *s* диктиосома
dictyosporangium *s* сетчатый спорангий
dictyospore *s* диктиоспора
dicyclic *a* двуцикличный; двулетний
didacty *a* двупалый
didelphic *a anat* имеющий двойную матку; *zool* двуутробный

dideoxyribonucleoside *s* дидезоксирибонуклеозид
didiploid *s* дидиплоид, амфидиплоид
didymospore *s* двуклеточная спора
didymous *a* соединённый в пары; *bot* двусильный
die *v* умирать, погибать; увядать, засыхать; **to ~ out** вымирать
dieback *s phytop* суховершинность
diecious *a zool* раздельнополый; *bot* двудомный
diencephalon *s* межуточный мозг, промежуточный мозг
dientomophily *s* диэнтомофилия
die-off *s* вымирание; отмирание
diet *s* диета; рацион; питание, пища
dieting *s* подкормка
differ *v* различаться, отличаться
difference *s* различие, разница; отличительный признак; *biom* разность
different *a* различный, разный; несходный, отличающийся
differential *s* дифференциал; *a* отличительный, дифференциальный
differentiation *s* дифференциация, дифференцировка; развитие
  independent ~ независимая дифференцировка, самодифференцировка
  programmed ~ запрограммированный жизненный цикл; путь дифференцировки, дифферон
differentiative *a* дифференцирующий
diffluence *s* растекание
diffraction *s* дифракция
  X-ray ~ дифракция рентгеновских лучей, рентгенодифракция
diffuse *a* диффузный; разбросанный; распространённый; широко ветвящийся; широко расстилающийся
diffusion *s* диффузия
diffusive *a* диффузионный
dig *v* копать, рыть
digametic *a* дигаметический, гетерогаметный
digastric *a anat* двубрюшный
dig-dig *s ther* дик-дик, дикдик *(Madoqua)*
digenesis *s* смена поколений, чередование поколений
digenic *a* дигенный, обусловленный двумя генами
digestible *a* удобоваримый, усвояемый, перевариваемый
digestion *s* усвоение, переваривание; пищеварение; расщепление
  acid ~ кислотный гидролиз
digestive *a* пищеварительный; способствующий пищеварению
digit *s* палец; цифра, разряд
digital *a* пальцевой; пальцевидный, пальцеобразный
digitate *a* имеющий пальцы; *bot* пальчатый
digitigrade *a* пальцеходящий
digitinervate *a bot* пальчато-нервный
digitipartite *a bot* пальчато-раздельный
digitipinnate *a bot* пальчато-перистый
digitule *s* небольшой пальцеобразный вырост
digonic *a* дигонический
digonoporous *a* дигонопоровый
digynous *a* двупестичный; с двумя плодолистиками
dihaploid *s* дигаплоид; дигаплоидный
diheterozygote *s* дигетерозигота, дигибрид
diheterozygous *a* дигетерозиготный
dihybrid *s* дигибрид, дигетерозигота
dikaryon *s* дикарион
dikaryophase *s* дикариофаза
dikaryophyte *s* дикариофит
dikaryotic *a* имеющий дикарион
dikaryotization *s* дикариотизация
dik-dik *s ther* дикдик *(Madoqua)*
dikkop *s orn* авдотка *(Burhinus)*
dilatation *s* расширение
dilate *v* расширять, расширяться
dilation *s* расширение
dilatometer *s* дилатометр
dilator *s* мышца-расширитель; расширитель
dill *s* укроп огородный *(Anethum graveolens)*
diluent *s* разбавитель; *a* разбавляющий
diluter *s* разбавитель
dilution *s* разбавление; разведение
  antibody ~ титр антител; разведение антисыворотки, титр антисыворотки
  antiserum ~ разведение антисыворотки, титр антисыворотки
dimegaly *s* диморфизм по размерам тела
dimension *s* размер; размерность; объём; протяжение

dimer *s* димер
dimeric *a* двучленный; димерный
dimerization *s* димеризация
dimidiate *a* уменьшенный наполовину
diminution *s* уменьшение, убывание; сбежистость; диминуция
dimixis *s* димиксис
dimorphic *a* диморфный
dimorphism *s* диморфизм
dimorphous *a* диморфный
dinergate *s* муравей-солдат, муравей-воин
dineuronic *a* с двойной иннервацией
dingo *s* динго *(Canis dingo)*
dinucleotide *s* динуклеотид
  flavine adenine ~ флавинаденин-динуклеотид, ФАД
  nicotinamide-adenine ~ дифосфопиридин-нуклеотид, никотинамид-аденин-динуклеотид, НАД
dioecious *a zool* раздельнополый; *bot* двудомный
dioecy *s zool* раздельнополость; *bot* двудомность
dionychous *a zool* имеющий два коготка
diorate *a* двупоровый
diorchic *a* имеющий два семенника
dip *s* погружение; раствор для окрашивания; *v* погружать; *hist* проводить
dipetalous *a* двулепестный
diphasic *a* двухфазный
diphasmide *s* дифазмида
diphosphate *s*:
  adenosine ~ аденозиндифосфат, АДФ
  cytidine ~ цитидиндифосфат, ЦДФ
  deoxyadenosine ~ дезоксиаденозиндифосфат, дАДФ
  deoxyguanosine ~ дезоксигуанозиндифосфат, аГДФ
  deoxyuridine ~ дезоксиуридиндифосфат, дУДФ
  guanosine ~ гуазиндифосфат, ГДФ
  inosine ~ инозиндифосфат, ИДФ
  thymidine ~ тимидиндифосфат, ТДФ
  uridine ~ уридиндифосфат, УДФ
diphyletic *a* дифилетический, имеющий двух предков
diphyllous *a* двулистный
diphyodont *a* дифиодонтный
diplanetism *s* дипланетизм
diplecolobous *a* скадчато-дольный
dipleurula *s* диплеврула
diplobiont *s* диплобионт
diplobivalent *s* диплобивалент
diplochlamydeous *a* с двойным околоцветником
diplochromosome *s* диплохромосома, удвоенная хромосома
diplococcus *s* диплококк
diplocyst *s* диплоциста
diplodemicolpate *a pal* двухполубороздный
diplohaplont *s* диплогаплонт
diploid *s* диплоид; *a* диплоидный
  double ~ аутотетраплоид; двойной диплоид
  multiple ~ аллополиплоид
diploidization *s* диплоидизация
diploidy *s* диплоидия
diplokaryon *s* диплокарион
diplokaryotic *a* тетраплоидный
diplonema *s* диплонема
diploneural *a* имеющий два нерва
diploperistomous *a* имеющий двойной перистом
diplophase *s* диплофаза, диплоидная фаза
diplornavirus *s* диплорнавирус
diplosis *s* диплоз
diplosome *s* диплосома
diplostemonous *a* диплостемонный, удвоеннотычинковый
diplostichous *a* двухрядный
diplotegia *s bot* диплотегий, нижняя коробочка
diplotene *s* диплотена
diploxylic *a* диплоксилический
dipnoan *s* двоякодышащее; *a* двоякодышащий
diporate *a* двупоровый
dipper *s* сачок; *orn* оляпка *(Cinclus);* моллюск *Spisula*
dipteran *s* двукрылое насекомое *(Diptera); a* двукрылый
dipteroid *a* диптероидный
dipterologist *s* диптеролог
dipterous *a* двукрылый
dipyrenous *a* двукосточковый
direction *s* направление
disaccate *a pal* двумешковый
disaccharide *s* дисахарид
disappearance *s* исчезновение

disassimilation *s* диссимиляция, распад вещества
disaster *s* бедствие
disbalance *s* дисбаланс; нарушение равновесия; *v* нарушать равновесие
disbolism *s* нарушение обмена веществ
disbrain *v* удалять головной мозг
disbud *v bot* удалять почки
disc *s* диск
discal *a* дискальный, дисковидный
discernible *a* видимый, различимый
discernment *s* способность различать
discharge *s* разряд; опорожнение; выброс; выделение; *bot* выстреливание; рассеивание; *v* опоражнивать; выбрасывать; выделять
trichocyst ~ выстреливание трихоцисты
dischronous *a* дисхронный
disciflorous *a* дискоцветковый
discission *s* рассечение
disclimax *s* дисклимакс
discoblastic *a* дискобластический
discoblastula *s* дискобластула
discocarp *s* апотеций
discoid, discoidal *a* дисковидный
discoloration *s* обесцвечивание, потеря окраски
discomfort *s* дискомфорт
disconnection *s* разъединение, разобщение
discontinuity *s* прерывистость
discoplacenta *s* дискоидальная плацента
discoplasm *s* дископлазма, строма эритроцита
discordance *s* несоответствие, несовпадение
discordant *a* дискордантный, несоответствующий, несовпадающий
discover *v* открывать; обнаруживать
discovery *s* открытие; обнаружение
discrepancy *s* несоответствие, расхождение; различие
discrete *a* раздельный, неслившийся
discriminandum *s phys* различаемый стимул
discriminate *v* различать, отличать; различающий, отличающий
discriminating *a* различительный, отличительный
discrimination *s* различение
discriminator *s* дискриминатор
disease *s* болезнь, заболевание
barley stripe ~ полосатость листьев ячменя *(caused by Helminthosporium gramineum)*
black shank ~ of tobacco «чёрная ножка» табака *(caused by Phytophthora parasitica var. nicotianae)*
cereal mosaic ~ закукливание злаков
Chagas ~ болезнь Чагаса, американский трипаносомоз *(caused by Schizotrypanum cruzi)*
cytomegalic inclusion ~ *phytop* цитомегалия
Dutch elm ~ голландская болезнь вязов *(caused by Ceratocystis ulmi)*
foot and mouth ~ ящур
graft-versus-host ~ реакция «трансплантат против хозяина», РТПХ
homologous ~ гомологичная болезнь, синдром карликовости, рант-синдром, синдром атрофии, вастинг-синдром
hypersensitivity ~ состояние гиперчувствительности
immune ~ имунопатологическое заболевание, иммунопатология
immune deposit ~ болезнь иммунных комплексов, иммунокомплексное заболевание
immunodeficient ~ иммунодефицитное заболевание; иммунопарез
late blight ~ of potato фитофтороз картофеля *(caused by Phytophthora infestans)*
leaf spot ~ пятнистость листьев огурцов *(caused by Cercospora melonis)*
molecular ~s *pl* молекулярные болезни
needle-shedding ~ опадение хвои на всходах
oat pupation ~ закукливание овса
parasitic ~ паразитарное заболевание, паразитоз
polyhedral ~ полиэдроз
potato wart ~ рак картофеля *(caused by Synchytrium endobioticum)*
tobacco stripe-and-curl ~ курчавость табака
yellow ~ хлороз
zoonotic ~ зооноз
disease-producing *a* болезнетворный
disfunction *s* дисфункция, нарушение функции, расстройство функции
dish *s* ванночка; чашка; планшет

crystallizing ~ кристаллизатор
dissecting препаровальная ванночка
water ~ поилка
dishabituation *s* снятие эффекта привыкания
disinfect *v* дезинфицировать
disinfectant *s* дезинфицирующее средство; *a* дезинфицирующий
disinfection *s* дезинфекция
disinfest *v* дезинсектировать; дератизировать
disinfestant *s* инсектицид; ратицид
disinfestation *s* дезинсекция; дератизация
disinhibition *s* растормаживание
disintegration *s* дезинтеграция, распад
disinvasion *s* дезинвазия
disjoining *s* разделение; расщепление
disjunction *s* разъединение, разобщение; дизъюнкция, разрыв; *anat* вывих
chromosome ~ расхождение хромосом
disjunctive *a* дизъюнктивный, разорванный
disk *s* диск; донце луковицы; подпестичный диск
A- ~ диск A, анизотропный диск
floral ~ цветоложе
germinal ~ зародышевый диск, бластодиск; рубчик яйца
H- ~, **Hensen's** ~ полоска H, полоска просветления
M- ~ полоска M, мезофрагма
Merkel's ~ диск Меркеля, осязательный мениск
nuclear ~ ядерный диск
optic ~ слепое пятно; сосочек зрительного нерва
pronotal ~ *ent* щиток переднеспинки
T- ~ полоска T, телофрагма
tactile ~ диск Меркеля, осязательный мениск
dislocate *v* сместиться
dislocation *s* смещение, перемещение, вывих
dismutation *s bioch* диспропорционирование, дисмутация
disomic *s* дисомик; *a* дисомный
disorder *s* расстройство, нарушение
autoimmune ~ аутоиммунное нарушение; аутоиммунный статус
congenital immunodeficiency ~ врождённый иммунодефицит, врождённое иммунодефицитное состояние
hyperkinetic ~ гиперкинез
immunodeficiency ~ иммунодефицит, иммунодефицитное состояние, иммунологическая недостаточность
selective immunodeficiency ~ изолированный иммунодефицит, изолированное иммунодефицитное состояние
disparate *a* непарный; несравнимый
disparity *s* несовместимость
dispenser *s lab* пипетка-дозатор
Terasaki's ~ многоканальная пипетка, диспенсер Терасаки
dispermic *a* диспермический
dispermous *a* двусемянный
dispermy *s* диспермия
dispersion *s* дисперсия, рассеяние, разброс; расселение; распространение; диспергирование
dispersity *s* дисперсность
dispireme *s* диспирема
displacement *s* смещение; замещение, вытеснение; ненормальное положение; *cyt* перестройка
displant *v* высаживать; пересаживать; вырывать с корнем
display *s* проявление; демонстрация, демонстрационное поведение; демонстрационная поза; демонстрационная окраска; *orn* токование; *phys* развёртка
courtship ~ *eth* брачная церемония
disporous *a* двуспоровый
disposal *s* удаление; очистка
disposition *s* размещение, расположение; характер, нрав
disproportionation *s bioch* диспропорционирование, дисмутация
disruptate *a* разорванный
disruption *s* разрыв
dissect *v* разрезать, рассекать; вскрывать, препарировать, анатомировать
dissection *s* диссекция, рассечение; вскрытие, препарирование, анатомирование
dissemination *s* рассеивание, расселение
dissilient *a* лопающийся, растрескивающийся
dissimilation *s* диссимиляция
dissociant *s micr* диссоциант

dissociation *s* диссоциация, распад, разложение
dissoconch *s* диссоконх
dissogeny *s* диссогения
dissolution *s* растворение; расплавление; разрушение, распад
dissophyte *s* диссофит
dissospermous *a* удвоенно-семянный
distal *a* дистальный
distaloaperturate *a pal* дистально-апертурный
distance *s* расстояние
distensibility *s* растяжимость
distension *s* растяжение
distichophyllous *a* двухрядно-облиственный
distichostachous *a* двухрядно-колосистый
distichous *a bot* двухрядный
distillation *s* дистилляция, перегонка, отгонка
distinct *a* явный, явственный, чёткий; различный, разный; отличный, отличающийся, непохожий
distinction *s* различение, распознавание; опознавание; разница, различие, отличие; отличительный признак, характерная черта
distoceptor *s* дистанционный рецептор
distorting *s* деформирование
distortion *s* нарушение
distress *s* расстройство, нарушение
separation ~ беспокойство из-за разлуки
distribution *s* распределение; размещение; распространение
distributor *s gen* распределитель, район
district *s* округ
disturb *v* нарушать; беспокоить
disturbance *s* нарушение, расстройство
disulcate *a pal* двубороздный
disulfide *s* дисульфид; дисульфидный, бисульфидный
disunion *s* разделение; разъединение
disymmetrical *a* дисимметричный
ditch *s* канава; ров; траншея
dithecal *a bot* двухкамерный, двухгнёздный; с двумя отделениями
ditreme *a pal* двуапертурный
ditrochous с двучленным вертлугом
diuresis *s* диурез
diuretic *a* диуретический, мочегонный
diurnal *a* дневной; *ent* однодневный
diurnation *s* дневная спячка; суточная миграция
divaricate *a* дивергентный, с большим расхождением; двураздельный; расщеплённый, вильчатый
divarication *s* расхождение
dive *s* погружение; ныряние; пикирование; *v* погружаться; нырять; пикировать
diver *s* гагара *(Gavia); ent* совка Bellura; ныряльщик
magpie ~, pied ~ *orn* луток *(Mergus albellus)*
divergence *s* дивергенция, отклонение; расхождение; разветвление
divergent *a* дивергентный; отклоняющийся; расходящийся
diversicolorous *a* разноцветный
diversification *s* разнообразие; многообразие
diversiflorous *a* разноцветковый
diversifoliate *a* разнооблиственный
diversifolious *a* разнолистный
diversilobous *a bot* разнолопастный
diversispinous *a* разноколючковый
diversity *s* разнообразие; многообразие; различие, отличие; разновидность
junctional ~ множественность J-сегментов
N-region ~ N-концевая вариабельность
diverticulum *s* дивертикул, слепой вырост, ответвление
divided *a* разделённый; раздельный; рассечённый
pinnately ~ перистораздельный
divider *s* делящаяся особь; делящаяся клетка
divisible *a* делящийся
division *s* деление; раздел, отдел
~ **of labor** разделенипе труда
amitotic ~ прямое деление ядра, амитоз
cell ~ деление клетки, цитокинез
indirect cell ~ митотическое деление клетки, митоз
meiotic ~ мейотическое деление, мейоз; редукционное деление
mitotic ~ митотическое деление, митоз
dizoic *a* содержащий два спорозоита
dizygotic *a* дизиготический, двуяйцевой *(twins)*

DNA, deoxyribonucleic acid *s* ДНК, дезоксирибонуклеиновая кислота
passenger ~ ДНК-пассажир
DNA-binding *s* связывание ДНК; *a* ДНК-связывающий
DNA-probe *s mbiol* зонд, ДНК-проба
DNAse *s* ДНКаза
Dobson, dobsonfly *s ent* вислокрылка *(Megaloptera)*
dock *s* щавель *(Rumex);* обрубленный хвост; *v* обрубать хвост, подрезать хвост
dodder *s* повилика *(Cuscuta)*
dodecagynous *a* двенадцатипестичный
dodecandrous *a* двенадцатитычиночный
dodecylsulfate *s* додецилсульфат, лаурилсульфат
dodo *s orn* дронт *(Didus ineptus)*
doe *s* самка *(of deer, hare, rabbit, rat, mouse, polecat)*
dog *s* собака *(Canis familiaris);* кобель, самец *(of dogs, otter)*
maned ~ гривистый волк *(Chrysocyon)*
mongrel [mutt] ~ дворняжка
prairie ~ луговая собачка *(Cynomys)*
red ~ красный волк *(Cuon)*
tracker ~ ищейка
white-tailed prairie ~ белохвостая луговая собачка *(Cynomis leucurus)*
wild ~ красный волк (Cuon)
doggess *s* сука, самка собаки
doghouse *s* конура
dogma *s* догма
dogwood *s* дёрен, кизил *(Cornus)*
dolabriform *a* топорообразный; долотообразный
doldrums *s* экваториальная зона затишья
dolichocarpous *a* длинноплодный
dolichocephalic *a* долихоцефальный; длинноголовый
dolichofacial *a* длиннолицый
dolichostachious *a* длинноколосый
dolichostylous *a* длинностолбчатый
dolioform *a* бочонкообразный
dolphinarium *s* дельфинарий
domain *s* домен, область
VH- ~ VH-область, VH-домен, вариабельная область тяжёлой цепи
VL- ~ VL-область, VL-домен, вариабельная область лёгкой цепи
dome *s* купол
apical ~ *bot* конус нарастания побега
sense ~ *ent* колоколовидная сенсилла
domestic *a* домашний
domesticated *a* одомашненный
domestication *s* доместикация, одомашнивание
domin *s* нерезко доминирующий организм
dominance *s* доминантность; доминирование, преобладание
dominant *s* доминант; доминантный; массовый вид; доминирующая форма; доминирующий
double ~s *pl* доминантные комплементарные факторы
dominate *v* доминировать, преобладать
domination *s* доминирование, преобладание
dominator *s* доминатор
dominigene *s* доминиген
dominule *s* доминант в микросообществе
donkey *s* осёл *(Equus asinus)*
donor *s* донор
immune ~ иммунокомпетентный донор; сенсибилизированная донорская клетка
doodlebug *s ent* муравьиный лев *(Myrmeleon)*
doorweed *s* горец птичий, спорыш *(Polygonum aviculare)*
dopamine *s* допамин, дофамин
dopaquinone *s* допахинон
dorbeetle *s* землерой, навозник настоящий *(Geotrupes)*
dorhawk *s* козодой *(Caprimulgus);* обыкновенный козодой *(Caprimulgus europaeus)*
dormancy *s zool* спячка; *bot* покой
dormant *a* покоящийся
dormin *s* дормин, абсцизовая кислота
dormouse *s ther* соня *(Gliridae)*
desert ~ селевиния, боялычная соня *(Selevinia betpakdalensis)*
edible ~, fat ~ соня-полчок *(Glis glis)*
hazel ~ орешниковая соня, мушловка *(Muscardinus avellanarius)*
dorsad *adv* дорсально
dorsal *a* дорсальный, дорзальный, спинной
dorsicolumn *s anat* задний столб
dorsicornu *s anat* задний рог
dorsolumbar *a* поясничный

dorsoventral *a* дорсовентральный, спиннобрюшной
dorsum *s* спина
dosage *s* дозирование, дозировка; норма; доза
dose *s* доза; *v* дозировать
dosing *s* дозирование; *бот.* подкормка
dot *s* точка, пятнышко
  fruit ~ *bot* сорус
  Maurer's ~s цитол. маурерова зернистость
dot-blotting *s* дот-блоттинг
dotted *a* точечный
dotterel *s orn* зуёк *(Charadrius); pl* ржанковые *(Charadriidae)*
doty *a for* фаутный
double *a* двойной; махровый; петля *(e.g., of a hare);* убегать по своим следам
double-dish *s* чашка Петри
double-flowering *a* махровый
double-helical *a* двуспиральный
double-stranded *a mbiol* двунитевой, двухцепочечный
doublet *s* дублет
double-twisted *a* двукратно-закрученный
double-walled *a* с двойной оболочкой
doubling *s* удвоение
dough *s* фаза восковой спелости
doughnuts *s* «бублики»
dove *s* голубь; горлица
  stock ~ клинтух *(Columba oenas)*
  turtle ~ обыкновенная горлица *(Streptopelia turtur)*
  wood ~ лесная горлица *(Turtur)*
dovekey, dovekie *s* малая гагарка, люрик *(Plautus alle)*
down *s* пух; подпушь, подпушка; *bot* пушок
downiness *s* опушение; опушённость
down-regulation *s* понижающая регуляция, даун-регуляция
downstream *adv mbiol* справа, ниже; вниз по течению
downy *a* пушистый, покрытый пухом; *bot* опушённый
downy-awned *a* пушисто-остистый
downy-flowered *a* пушисто-цветковый
downy-fruited *a* пушисто-плодный
downy-leaved *a* опушённо-листный
downy-pistiled *a* пушисто-пестичный
downy-toothed *a* опушённо-зубчатый
draco, **dragon** *s herp* летучий дракон *(Draco)*
dragonfly *s* стрекоза; *pl* разнокрылые стрекозы *(Anisoptera)*
  club-tailed ~ дедка *(Gomphus)*
  large ~ коромысло (*Aeschnidae)*
drainage *s* дренаж, дренирование
drake *s* селезень
draught *s* глоток
draw *s* молодой побег; вытяжка
drawtube *s* выдвижной тубус
dray *s* гайно, беличье гнездо
dredge *s* драга; дночерпатель; драгировать
drepanocyte *s* серповидный эритроцит
dress *v* обрезать *(trees);* препарировать; свежевать, разделывать *(carcass);* подкармливать *(plants); протравливать (seeds);* перевязывать *(wound)*
  spawning ~ брачный наряд
dressing *s* подкормка *(of plants);* протравливание *(of seeds);* перевязывание
  flax ~ чесание льна
  wound ~ обработка раны, перевязка раны
drey *s* гайно, беличье гнездо
drier *s* сушилка
  centrifugal ~ обезвоживающая центрифуга, центробежная сушилка
drift *s* дрейф, снос, пассивное перемещение; медленное течение, дрейфовое течение; нанос; ледниковые отложения; *a* дрейфовый, сносимый; *v* дрейфовать; **~ to genetic equilibrium** дрейф к установлению генетического равновесия
drill *s ther* дрил *(Papio leucophaeus);* сверлящий брюхоногий моллюск
drimophilous *a* солелюбивый, галофильный
drimophilus *s* солелюбивый организм, галофил
drink *s* питьё; *v* пить
drinker *s* поилка
  self-filling ~ автопоилка
drinking *s* питьё
  cell ~ пиноцитоз
drive *s* побуждение, стимул; загонять
dromedary *s* дромедар, одногорбый верблюд *(Camelus dromedarius)*

dromotropic *a phys* дромотропный
dromotropism *s* дромотропизм
drone *s* трутень; *v* жужжать, гудеть
droning *s* жужжание
droop *v* увядать, поникать; опускать *(e.g., leaves)*
drooping *a* повислый, поникший
drop *s* капля; снижение, падение; падалица; *v* капать, стекать каплями; падать, опадать; выступать каплями; снижаться, телиться, жеребиться, ягниться; родиться; испражняться
fruit ~ падалица
dropper *s* пипетка
droppings *s* помёт, экскременты животных
dropsy *s* водянка
drought *s* засуха; засушливость
physical ~ физическая засуха, физическая сухость почв
physiological ~ физиологическая засуха, физиологическая сухость почв
drought-resistance *s* засухоустойчивость
drought-resistant *a* засухоустойчивый
droughty *a* засушливый; сухой; безводный
drove *s* стадо
drowse *s* дремота, полусон; сонливость; дремать
drug *s* (лекарственный) препарат; медикамент; наркотик
drum *v eth* хлопать крыльями
drumming *s eth* барабанный стук; ~ **with paws** стук лапами
drupaceous *a bot* костянковый
drupe *s bot* костянка
drupelet *s bot* мелкая костянка
dry *v* сушить; высушивать; *a* сухой; высохший; обезвоженный; засышливый; сухой; яловая
drying *s* сушка; высушивание; высыхание; усыхание
freeze ~ лиофилизация, лиофильная сушка
top ~ суховершинность
drymophyte *s* дримофит
dryness *s* сухость
dualism *s* дуализм
dualistic *a* дуалистический, двойственный
dualization *s* раздвоение
duck *s* утка
buffle-headed ~ малый гоголь *(Bucephala albeola)*
diving ~ чернеть*(Aythya)*
eider ~ гага *(Somateria);* обыкновенная гага *(Somateria mollissima)*
fish ~ крохаль *(Mergus)*
greater scaup ~ морская чернеть *(Aythya marila)*
green ~ утёнок
harlequin ~ каменушка *(Histrionicus histrionicus)*
long-tailed ~ морянка *(Clangula hyemalis)*
mallard ~ кряква *(Anas platyrhynchus)*
mandarin ~ мандаринка *(Aix galericulata)*
mountain ~ каменушка *(Histrionicus histrionicus);* австралийская пеганка *(Tadorna tadornoides)*
ring-necked ~ ошейниковая чернеть *(Aythya collaris)*
rufous-crested ~ красноносый нырок *(Netta rufina)*
tufted ~ хохлатая чернеть *(Aythya fuligula)*
wild ~ дикая утка; кряква *(Anas platyrhynchus)*
duct *s anat* проток, канал
efferent ~ отводящий канал, отводящий каналец, выносящий канал, выносящий каналец; выводной проток
guttural ~ евстахиева труба
omphaloenteric, omphalomesenteric ~ *emb* пупочно-брыжеечный проток; желточный канал
pronephric ~ проток предпочки; пронефрический канал
renal ~ мочеточник
thyroglossal ~ щитовидно-глоточный проход
umbilical ~ *emb* пупочно-брыжеечный проток; желточный канал
vitelline ~ *emb* пупочно-брыжеечный проток; желточный канал
yolk ~ *emb* пупочно-брыжеечный проток; желточный канал
ductance *phys* проницаемость, проводимость
ductless *a* не имеющий протока
ductule *s* каналец, маленький проток
duetting *s* антифональное пение *(bioacoustics)*

dug *s* сосок, вымя
dugong *s ther* дюгонь *(Dugong dugong)*
duiker *s* дукер *(Cephalophus)*
dull *a* матовый; медлительный, вялый
dulosis *s* рабовладельчество у муравьёв
dumeticolous *a* обитающий в кустарниках
dumetose *a* заросший кустарником, поросший кустарником; кустистый
dumetosous *a* кустарниковый
dumetum *s* заросль кустарников
dummy *s* макет; модель; имитация; модельный; контрольный опыт; *a* поддельный; подставной
dumose, dumouse *a* заросший кустарником, поросший кустарником; кустистый
dump *s* свалка
dun *a* коричневато-серый
dung *s* помёт; навоз; экскременты; органическое удобрение; *v* унаваживать, удобрять навозом
dunlin *s orn* чернозобик *(Calidris alpina)*
dunnock *s orn* лесная завирушка *(Prunella modularis)*
duodenal *a* дуоденальный
duodenum *s* двенадцатиперстная кишка
duplex *s mbiol* дуплекс; двойной; диплоидный
duplibaculate *a pal* двустолбчатый
duplicate *s* дублет
duplication *s* дупликация, удвоение
duplicidentate *s* зайцеобразное *(Lagomorpha)*
duplicity *s* двойственность
duplicodentate *a* двоякозубчатый
duration *s* продолжительность
durian *s bot* дуриан *(Durio zibethinus)*
dusk *s* сумерки
dust *s* пыль; дуст, пылевидный препарат
  blossom ~ пыльца
  fungicidal ~ фунгицидный дуст, пылевидный фунгицид
  insecticidal ~ инсектицидный дуст
  wormhole ~ *ent* буровая мука
dust-bathing *s eth* купание в пыли
dusting *s* опыление *(e.g., with pesticides)*
  aerial ~ авиаопыление
dustywing *s ent* пыльнокрыл *(Coniopterygidae)*
dwarf *s* карлик; карликовое животное; карликовое растение; *a* карликовый, низкорослый; *phytop* карликовость; *v* мешать росту; останавливать развитие; задерживаться в развитии
dwarfed *a* отставший в росте
dwarfish *a* карликовый, низкорослый; недоразвитый
dwarfism *s* нанизм, карликовость, низкорослость
dwell *v* жить, обитать
dweller *s* обитатель
diad *s* диада
dye *s* краситель, красящее вещество
  azo ~ азокраситель
dynamics *s* динамика
  trophic ~ трофодинамика
dynamoplastic *a* динамопластический
dynein *s* динеин
dyscrasia *s* дискразия, нарушение гомеостаза
dysentery *s* дизентерия
dysfunction *s* дисфункция, нарушение функции, расстройство функции
dysgalactia *s* расстройство секреции молока
dysgenesia, dysgenesis *s* дисгенезис
dysgenic *a* дисгенический; вырождающийся; способствующий вырождению
dysphotic *a* дисфотический, плохо освещённый
dysplasia *s* дисплазия
dysploidy *s* дисплоидия
dystrophia *s* дистрофия
dystrophic *a* дистрофный
dystrophication *s* дистрофикация
dystrophy *s* дистрофия
dzeren *s ther* дзерен *(Procapra)*

E

eagle *s* орёл; орёл *(Aquila)*
bald ~ белоголовый орлан *(Haliaeetus leucocephalus)*
golden ~ беркут *(Aquila chrysaetos)*
greater spotted ~ большой подорлик *(Aquila clanga)*
harpy ~ гарпия *(Harpia harpyja)*
harrier ~ змееяд *(Circaetus)*
imperial ~ могильник *(Aquila heliaca)*
Pallas' sea ~ орлан-долгохвост *(Haliaeetus leucophrys)*
sea ~ *орлан (Heliaeetus)*
serpent ~, snake ~ змееяд *(Circaetus)*
spotted ~ большой подорлик *(Aquila clanga)*
white-bellied sea ~, white-breasted sea ~ белобрюхий орлан *(Haliaeetus leucogaster)*
white-tailed ~ орлан-белохвост *(Haliaeetus albicilla)*
eagle-hen *s* орлица
eagle-owl *s* филин *(Bubo)*
eaglet *s* орлёнок
ean *v* ягниться
eanling *s* ягнёнок
ear *s* ухо; слух; колос; початок; **to come into** ~ выходить в колос, колоситься, вступать в стадию колошения
eardrum *s* барабанная перепонка
eared *a zool* ушастый; *bot* имеющий ушки; ушковатый; колосящийся
earflap *s* ушная раковина
earhead *s bot* колосовидная головка
earing *s* колошение, выколашивание
earlap *s* мочка уха
earless *a* безухий
earlike *a zool* уховидный, ушковидный; *bot* початковидный
earliness *s* раннеспелость, скороспелость
early *a* ранний; преждевременный; раннеспелый, скороспелый
early-ripe *a* раннеспелый, скороспелый
earn *v* выть
earth *s* земля, почва, грунт; суша; *v* зарывать в землю, закапывать, окучивать, окапывать; **go to ~, take to** ~ скрыться в нору
earth-ball *s* трюфель *(Tuber)*
earthnut *s* земляной орех *(Arachis hypogaea);* трюфель *(Tuber);* земляной каштан *(Conopodium majus);* сыть съедобная, чуфа, земляной миндаль *(Cyperus esculentus)*
earthquake *s* землетрясение
earthstar *s myc* земляная звёздочка *(Geaster)*
earthwolf *s* земляной волк *(Proteles cristatus)*
earthworm *s* дождевой червь, земляной червь *(Lumbricus)*
earthy *a* земляной; землистый
earwax *s* ушная сера
earwig *s ent* уховёртка *(Dermaptera)*
easy-to-root *a* легко укореняющийся
eat *v* есть, поедать; **to ~ away** съедать, пожирать
eatable *a* съедобный
eatability *s* съедобность, пригодность в пищу
eater *s*:
bee ~ *orn* щурка *(Meropidae)*
honey ~ *orn* медосос *(Meliphagidae)*
eating *s* поедание; съедание
ebb *s* ущерб; упадок, плохое состояние; отлив
ebracteate *a* не имеющий прицветников
ebracteolate *a* не имеющий прицветничков
ebulism *s* эбулизм
ecad *s* экада
ecalcarate *a* бесшпорцевый
ecardinal *a zool* не имеющий замка, беззамковый
ecarinate *a bot* бескилевой
ecaudate *a* бесхвостый
eccritic *a* усиливающий экскрецию
ecdemic *a* экдемичный, неэндемичный, пришлый, завезённый
ecderon *s* эпидермис
ecderonic *a* эпидермальный
ecdysis *s* линька
ecdysone *s* экдизон, гормон линьки
ecesis *s* эцезис
ECG, electrocardiogram *s* ЭКГ, электрокардиограмма
echidna *s* ехидна *(Tachyglossus)*
echinate, echinated *a* шиповатый; имеющий шипы, колючий
echinococcosis *s* эхинококкоз
echinocyte *s* пойкилоцит

echinopaedium *s* диплеврула
echinopluteus *s* эхиноплутеус
echinosis *s* пойкилоцитоз
echinulate *a* мелкоигольчатый
echo *s* эхо
ecidioclimate *s* эцидиоклимат
ecize *v* колонизировать; укореняться
eclipse *s bioch* эклипсная фаза, скрытый период; эклипс, послебрачное оперение
eclosion *s ent* вылупление
ecobiotic *a* экобиотический
eco-catastrophe *s* экологическая катастрофа
ecocide *s* экоцид
ecoclimate *s* экоклимат
ecocline *s* экоклин, градиент экосистемы
ecodeme *s* экодем
ecogenesis *s* экогенез
ecogenetic *a* экогенетический
ecological *a* экологический
ecology *s* экология
economic *a* прикладной, практический
ecophene *s* экофен
ecophenotype *s* экофенотип
ecoproteroandry *s* экопротероандрия
ecoproterogyny *s* экопротерогиния
ecospecies *s* эковид
ecosphere *s* экосфера
ecostate *a zool* не имеющий рёбер; *bot* безжилковый, не имеющий жилок
ecosystem *s* экосистема
ecotone *s* экотон
ecotope *s* экотоп, местообитание
ecotype *s* экотип
ecsoma *s* эксома
ectad *adv* кнаружи
ectal *a* наружный, внешний
ectasia *s* эктазия, расширение, растяжение
ectethmoid *s* наружная решетчатая кость
ectexine *s pal* эктэкзина
ectoblast *s* эктобласт, эпибласт
ectoblastomere *s* эктодермальный бластомер
ectochone *s* эктохон
ectocondyle *a* наружный мыщелок, латеральный мыщелок
ectocyst *s* эктоциста
ectocytic *a* внеклеточный
ectoderm *s* эктодерма
 primitive ~ первичная эктодерма, эктобласт, эпибласт
ectodermal, ectodermic эктодермальный
ectoenzyme *s* эктофермент
ectogenesis *s* эктогенез
ectogenous *a* эктогенный, экзогенный
ectoglia *s* эктоглия
ectolecithal *a* эктолецитальный
ectomere *s* эктомер
ectoneural *a* эктоневральный
ectonexine *s pal* эктонэкзина
ectoparasite *s* эктопаразит, наружный паразит
ectophagous *a* объедающий, обгрызающий
ectophloeodic *a* эктофлеодный
ectophyte *s* эктофит, эктопаразитическое растение
ectopia *s* эктопия
ectopic *a* эктопический
ectoplasm *s* эктоплазма
ectoplast *s* эктопласт
ectopterygoid *s* эктоптеригоид
ectosexine *s pal* эктосэкзина
ectospore *s* эктоспора
ectosteal *a* внекостный
ectostosis *s* эктостоз
ectothermic *a* холоднокровный
ectotoxin *s* экзотоксин
ectotheca *s* эктотека
ectotrochanter *s anat* большой вертел
ectotroph *s* эктотроф, эктотрофный организм
ectotrophic *a* эктотрофный
ectozoon *s* эктопаразитическое животное
ectromelia *s* эктромелия
eczema *s* экзема
edacious *a* прожорливый
edaphic *a* эдафический, почвенный
edaphology *s* почвоведение
edaphon *s* эдафон
edaphophyte *s* эдафофит
edatope *s* эдатоп, почвенный биотип
edelweiss *s* эдельвейс *(Leontopodium alpinum)*
edema *s* отёк
edentate *a* беззубый; не имеющий зубцов
edentate *s ther* неполнозубое *(Edentata)*
edge *s* край; кант; грань; кайма; *bot* ребро
 forest ~ опушка леса
 water's ~ урез воды

edgebone *s* крестцовая кость, крестец
edibility *s* съедобность, пригодность в пищу
edible *a* съедобный, годный в пищу
edificator *s* эдификатор
edriophthalmic *a* сидячеглазый
educated *imm* «обученный», коммитированный; дрессированный
education *s* обучение; воспитание; развитие
lymphocyte ~ «обучение» лимфоцитов, коммитирование лимфоцитов
educe *v* выявлять, развивать
eduction *s* выявление; удаление; очищение; извлечение
eelworm *s* нематода
gall ~ угрица, галловая нематода *(Anguinidae)*
effect *s* действие, влияние, воздействие; эффект, следствие, результат; *v* производить, осуществлять
booster ~ анамнестическая реакция, вторичный иммунный ответ, ревакцинационный эффект, бустер-эффект
complement-mediated bacteriolytic ~ комплементзависимый бактериолиз
homing ~ *phys* хоминг-эффект, эффект дома
immediate ~ непосредственное воздействие
immunoenhancing ~ иммунологическое усиление, иммунологическая стимуляция
nondelayed ~ немедленное воздействие
oxygen ~ кислородный эффект, O_2-эффект
priming ~ *imm* эффект примирования, прайминг-эффект
remote ~ отдалённое действие, следствие
residual ~ последействие, остаточное действие
retarding ~ задерживающий эффект, притормаживающее действие
sampling ~ эффект пробы, эффект выборки, значение выборки
stimulating ~ стимулирующее действие, стимуляция
subthreshold ~ подпороговый эффект
tau ~ *phys* тау-эффект
effective *a* действительный; эффективный, действенный; действующий, имеющий силу
effectiveness *s* эффективность
effector *s* эффектор
cytotoxic ~ цитотоксическая клетка-эффектор
efference *s* эфферентация
efferent *a* эфферентный, выносящий, центробежный
effervescence *s* вскипание; образование пены; возбуждение, волнение
effete *a* истощённый, слабый; бесплодный; пустой *(anther)*
efficacy *s* эффективность
efficiency *s* эффективность; производительность; продуктивность; отдача; коэффициент полезного действия; действенность; проявление
biological ~ биологическая продуктивность
breeding ~ оплодотворяемость
cardiac ~ производительность сердца; полезная работа сердца
catching ~ уловистость
coating ~ эффективность покрытия, эффективность напыления; коэффициент сорбции, процент иммобилизации
cytotoxic ~ цитотоксический индекс, показатель литической активности
energy ~ энергетический выход
feed ~ эффективность питания; эффективность кормления
fishing ~ уловистость
physical ~ физическая работоспособность
quantum ~ квантовый выход
spleen seeding ~ f-коэффициент, f-число
efficient *a* действенный, эффективный
efflorescence *s* зацветание, расцветание; сезон расцветания; заплесневение; выкристаллизовывание
effluent *s* истечение; вытекающая жидкость; *pl* сточные воды
effluvium *s* испарение
efflux *s* истечение, вытекание, отток
effort *s* усилие; напряжение
effuse *a bot* развесистый; размётанный; поверхностно-разбросанный
effusion *s* выпот; истечение

eft *s* тритон, саламандра *(Salamandridae)*
egest *v* экскретировать, выбрасывать, выводить
egesta *s pl* экскреты; экскременты
egestion *s* экскретирование, выведение наружу, выделение; дефекация
egg *s* яйцо; икринка; яйцеклетка; *pl* яйца; икра; **to cover ~s, to sit on ~s** сидеть на яйцах, высиживать яйца
  nest ~ подкладыш
  silkworm ~s грена
eggar *s ent* коконопряд *(Lasiocampidae)*
egg-laying *a* яйцекладущий; икромечущий
eggplant *s* баклажан *(Solanum melongena)*
egg-shaped *a* яйцеобразный, яйцевидный
  inversely ~ *bot* обратнояйцевидный
eggshell *s* скорлупа, оболочка яйца
eglandular *a* не имеющий желёз
egret *s* цапля; эгретка; *bot* пушок, хохолок
  little ~ малая белая цапля, чепура-нужда *(Egretta garzetta)*
eider *s* гага *(Somateria)*
  king ~ гага-гребенушка *(Somateria spectabilis)*
eigenvalue *s* собственное число
eigenvector *s* собственный вектор
eiloid *a* спиралевидный
einkorn *s* пшеница однозернянка *(Triticum monococcum)*
eisodic *a* афферентный, центростремительный
ejaculation *s* эякуляция
ejection *s* выбрасывание, извержение, изгнание, опорожнение
  milk ~ молокоотдача
elaeoblast *s* элайобласт
elaeocyte *s* элеоцит
elaeodochon *s* копчиковая железа
elaioplast *s* элайопласт
elaiosome *s* элайосома
elaiosphere *s* капля масла
eland *s ther* канна *(Tragelaphus oryx)*
elastic *a* эластичный; упругий
elater *s* прыгательный придаток; *bot* элатера
elbow *s* локоть
elder *s* бузина *(Sambucus)*
elecampane *s* девясил высокий *(Inula helenium)*
elective *a* элективный, избирательный
electroblotter *s* электроблоттер
electroblotting *s* электроблоттинг
electrocardiogram, ECG *s* электрокардиограмма, ЭКГ
electrocardiograph *s* электрокардиограф
electrocardiography *s* электрокардиография
electrocardioscope *s* электрокардиоскоп
electrocorticogram *s* электрокортикограмма
electrocortin *s* электрокортикостерон
electrode *s* электрод
electrodecantation *s* электродекантация
electrodialysis *s* электродиализ
electrodialyzer *s* электродиализатор, ионный диализатор
electroeluter *s* устройство для электроэлюции
electroelution *s* электроэлюция
electroencephalogram *s* электроэнцефалограмма
electroencephalography *s* электроэнцефалография
electrofocusing *s* электрофокусирование
  zone convection ~ изоэлектрофокусирование
electrofusion *s* электрослияние, слияние методом электрошока
electroimmunoassay *s* электроиммуноанализ
electroimmunofixation *s* электроиммунофиксация, электрофорез с иммунофиксацией
electrokymogram *s* электрокимограмма
electrokymography *s* электрокимография
electrolocation *s* электролокация, ориентация с помощью электрических органов
electrolysis *s* электролиз
electromorph *s* электроморфа
electromotoneurone *s* электромотонейрон
electromyogram *s* электромиограмма
electromyography *s* электромиография
electron *s* электрон
electron-dense *a* электроноплотный
electron-light *a* электронопрозрачный, прозрачный для электронов
electroosmosis *s* электроосмос
electrophoregram *s* электрофореграмма
electrophoresis *s* электрофорез
  disk ~ диск-электрофорез

micro-gel ~ микрогель-электрофорез
electrophoretic *a* электрофоретический
electrophysiology *s* электрофизиология
electroporation *s* электропорация
electroreduction *s* электровосстановление
electrorheophoresis *s* электрореофорез
electroshock *s* электрошок
electrotaxis *s* электротаксис
electrotonic *a* электротонический
electrotonus *s* электротон
electrotransfection *s* трансфекция электрошоком, электрошокотрансфекция
electrotropism *s* электротропизм
element *s* элемент
dissociator ~ элемент-разобщитель, Ds-элемент
FB ~ элемент обратной связи, FB-элемент
fold-back ~ элемент обратной связи, FB-элемент
major mineral ~s макроэлементы
trace ~ микроэлемент; следовой элемент
elementary *a* элементарный; простейший; первоначальный; первичный
elephant *s* слон *(Elephantidae)*
elephant's-ear *s* бегония *(*Begonia*)*
eleutherodactyl *a* со свободным задним пальцем
eleutheropetalous *a* раздельнолепестный
eleutherophyllous *a* раздельнолистный
eleutherosepalous *a* раздельночашелистиковый
elevated *a* приподнятый; *bot* надземный
elevation *s* поднятие, возвышение; высота
eligulate *a bot* безъязычковый, без язычка
eliminate *v* исключать, устранять, элиминировать
elimination *s* устранение, удаление, элиминация
~ **of clone** делеция клона
biological ~ биологическая очистка
immune ~ иммунный клиренс
selective ~ элиминация при отборе
elite *s* элита; *a* элитный
elk *s* лось *(Alces alces);* олень *(Cervus)*
ellipse *s* эллипс
ellipsometry *s* эллипсометрия
elliptocyte *s* эллиптоцит
elm *s* вяз, ильм, берест, карагач *(Ulmus)*
elongated *a* удлинённый
elongation *s* удлинение, вытягивание, растяжение; рост в длину, удлинение; *bioch* элонгация
stem ~ выход в трубку
eluant *s* элюант
eluate *s* элюат; *v* элюировать, извлекать из адсорбента, вымывать
eluent *s* элюент
elution *s* элюция, элюирование
elutriation *s* отмучивание, сцеживание, отстаивание
eluvial *a* элювиальный
elytron, elytrum *s ent* элитра, надкрылье, хитинизированное переднее крыло
emaciation *s* истощение, исхудание
emailloblast *s* энамелобласт, адамантобласт, амелобласт
emarginate *a* выемчатый
emasculate *v* кастрировать
emasculation *s* кастрация
embed *v* заделывать, заливать; вкапывать, закапывать, укреплять
embedding *s* заделка, заливка; имплантация
embowel *v* потрошить
embracing *a* объемлющий, охватывающий *(leaf)*
half-stem ~ полустеблеобъемлющий
embryo *s* эмбрион, зародыш
hexacanth ~ *helm* онкосфера, шестикрючковый зародыш
embryoblast *s* эмбриобласт
embryogenesis, embryogeny *s* эмбриогенез, развитие зародыша
embryogonium *s* эмбриогоний
embryolethal *a* эмбриолетальный
embryologic, embryological *a* эмбриологический
embryology *s* эмбриология
embryonal *a* эмбриональный, зародышевый
embryonate *a* оплодотворённый; содержащий эмбрион
embryonic *a* эмбриональный, зародышевый
embryony *s* эмбриония
embryophore *s* эмбриофор
embryotoxic *a* эмбриотоксичный

embryotrophy *s* эмбриотрофия, питание эмбриона
emerge *v* появляться; возникать; вылупляться
emergence *s* появление; возникновение; вылупление; вылет; *bot* вырост, придаток эпидермиса; **to come into ~** колоситься, вступать в фазу колошения
~ of seedlings прорастание; появление всходов
~ of terrestrial vertebrates выход позвоночных на сушу
ear ~ колошение
root ~ проклёвывание корешка
seedling ~ появление всходов
emergent *a* полупогружённый, эмергентный
emetic *a* рвотный
emiction *s* мочеиспускание
emigration *s* эмиграция, переселение; диапедез
eminence *s anat* возвышение, выступ, бугорок
antithenar ~ возвышение мизинца, гипотенар
articular ~ суставной бугорок
Doyère's ~ моторная бляшка, нервно-мышечный синапс
genital ~ половой бугорок
hypothenar ~ возвышение мизинца, гипотенар
thenar ~ возвышение большого пальца, тенар
emission *s* выделение; эмиссия, испускание, излучение; распространение; выброс
emit *v* выделять; излучать; издавать
emmer *s* эммер, пшеница двузернянка *(Triticum dicoccum)*
emophyte *s* эмофит, погружённое растение
emotion *s* эмоция, чувство; волнение
emotionality *s* эмоциональность
empirical *a* эмпирический
empodium *s* эмподиум
empty *a* пустой, вещественно-неопределённый; неясный; необитаемый
empty-gutted *a* с пустым кишечником
emu *s* эму *(Dromiceius)*
emulsification *s* эмульгирование
emulsifier *s* эмульсификатор
emulsify *v* эмульгировать
emulsion *s* эмульсия
emulsive *a* образующий эмульсию
emunctory *s* экскреторный орган; *a* экскреторный
enantiomer *s* энантиомер
enarthrosis *s* ореховидный сустав
enation *s bot* энаций
encage *v* сажать в клетку
encapsulated *a* инкапсулированный, имеющий капсулу
encapsulation *s* инкапсуляция
encephalic *a* церебральный, мозговой
encephalitis *s* энцефалит
encephalization *s* энцефализация
encephalocoel *s* полость мозга; желудочки мозга; переднее расширение невроцеля
encephalogram *s* энцефалограмма
encephalography *s* энцефалография
encephalon *s* головной мозг
enchondral *a* энхондральный, внутрихрящевой
enchylema *s* энхилема, клеточный сок, цитолимфа
enclave *s* энклав
enclosure *s* оболочка; ограждённое пространство; огороженное место; ограждать
encode *v* кодировать
encoding *s* кодирование
encounter *s eth* столкновение
encroachment *s* вторжение; постепенный захват; выход за пределы
encrustation *s* обрастание; образование корки
encystation, encystment *s* инцистирование
end *s* конец; предел
dead ~ тупик
endanger *v* подвергать опасности
endangered *a* находящийся под угрозой
endarterial *a* внутриартериальный
endarterium *s* внутренняя оболочка артерии
endaural *a* внутриушной
endbrain *s* конечный мозг
endemial *a* эндемический
endemic *s* эндемик; *a* эндемический
endemicity *s* эндемичность
endemism *s* эндемизм

endergonic *a* эндотермический
endermic *a* внутрикожный
enderonic *a* энтодермальный, эндодермальный
endexine *s pal* эндэкзина
ending *s* окончание
endite *s* эндит
endoascus *s bot* эндоаск, внутренняя сумка
endobasal *a* эндобазальный
endobasidial *a* эндобазидиальный
endobasidium *s* эндобазидия
endobiophyte *s* эндопаразитическое растение
endobiotic *a* эндобиотический
endoblast *s* эндобласт, гипобласт, целобласт
endoblastomere *s* эндобластомер, эндодермальный бластомер
endobronchial *a* эндобронхиальный
endocardiac *a* внутрисердечный
endocardial *a* внутрисердечный
endocardium *s* эндокард
endocarp *s* эндокарпий, внутриплодник
endocarpoid *a* внутриплодниковый
endocarpous *a* эндокарпный, внутриплодный
endocavitary *a* внутриполостной
endocoel *s* эндоцель
endoconidium *s* эндоконидия
endocranial *a* внутричерепной
endocranium *s* эндокраниум
endocrine *a* эндокринный
endocrinology *s* эндокринология
endocrinous *a* эндокринный
endocuticle, endocuticula *s* эндокутикула
endocycle *s* эндоцикл
endocytosis *s* эндоцитоз
endoderm *s* энтодерма, эндодерма
endodermal *a* энтодермальный, эндодермальный
endodermis *s bot* эндодерма
endoduplication *s* эндодупликация
endoenzyme *s* эндофермент, внутриклеточный фермент
endogamic, endogamous *a* эндогамный
endogamy *s* эндогамия
endogastric *a* внутрижелудочный
endogenesis *s* эндогенез
endogenic *a* эндогенный
endogenote *s* эндогенота
endogenous *a* эндогенный
endogeny *s* эндогения
endoglobular *a* внутриэритроцитарный
endogonic *a* эндогонический
endogonidium *s* эндогонидий
endolamella *s pal* эндэкзина
endolithophyte *s* эндолитофит
endolymph *s* эндолимфа
endolymphatic *a* эндолимфатический
endolysosome *s* эндолизосома
endomere *s* эндомер
endomesoblast *s* энтомезобласт
endomesoderm *s* энтомезодерма
endometrial *a* внутриматочный
endometrium *s* эндометрий
endomitosis *s* эндомитоз
endomixis *s* эндомиксис
endomorphic *a* эндоморфный
endomorphy *s* эндоморфия
endomysium *s* эндомизий
endoneurium *s* эндоневрий
endonexine *s pal* эндоэкзина
endonuclear *a* внутриядерный
endonuclease *s* эндонуклеаза
  restriction ~ эндонуклеаза рестрикции, рестриктаза
endoparasite *s* эндопаразит, внутренний паразит
endoparasitic *a* эндопаразитический
endoperidium *s* эндоперидий
endophloic *a* обитающий в коре
endophyllous *a* живущий внутри листа; скрытолистный
endophyte *s* эндофит
endophytic *a* эндофитный
endophytism *s* эндофитизм
endoplasm *s* эндоплазма
endoplasmic *a* эндоплазматический
endopleura *s bot* внутренняя оболочка семени
endopodite *s* эндоподит
endopolyploidization *s* эндополиплоидизация
endopolyploidy *s* эндополиплоидия
endoreduplication *s* эндоредупликация
endorphine *s* эндорфин
endorrhizous *a* скрытокорешковый
endosaprophytism *s* эндосапрофитизм
endosclerotium *s* эндосклероций
endosexine *s pal* эндсэкзина
endoskeleton *s* внутренний скелет
endosmos *s* эндосмос
endosome *s* эндосома, кариосома

endosperm *s* эндосперм
endospore *s* эндоспора
endospore-forming *a* образующий эндоспоры
endosporous *a* имеющий эндоспоры
endosteal *a* эндостеальный, внутрикостный
endosteum *s* эндост
endostome *s* эндостом
endostracum *s mal* внутренний слой
endostyle *s* эндостиль
endosymbiont *s* эндосимбионт
endothecial *a* эндотециальный
endothecium *s* эндотеций
endothelial *a* эндотелиальный
endotheliocyte *s* эндотелиальная клетка, эндотелиоцит
endothelium *s* эндотелий
endothermic *a* эндотермический
endothermy *s* эндотермия
endotoxic *a* эндотоксический
endotoxin *s* эндотоксин
endotrophic *a* эндотрофный
endovenous *a* внутривенный
endoxylophyte *s* эндоксилофит
endozoochore *s* эндозоохор
endurance *s* выносливость, сопротивляемость, стойкость
energid *s* энергида
energy *s* энергия
 fixed total ~ общее количество запасённой энергии
 net ~ нетто-энергия, чистая энергия
energy-dependent *a* энергозависимый
energy-requirement *s* потребность в энергии
energy-requiring *a* энергопотребляющий
enervose без жилок, безжилковый
engineering *s* техника, технология
 biological ~ биотехника
englobe *v* фагоцитировать
englobement *s* фагоцитоз
engorge *v* жадно есть, глотать; наливаться кровью
engorgement *s* прожорливость; насасывание кровью; прилив крови
engraft *v* прививать
engraftment *s* прививка; черенок
engraver *s энт.* заболонник *(Scolytus)*
 pine ~ короед сосновый *(Ips pini)*
engulfment *s* поглощение *(e.g., of particles);* заглатывание
enhance *v* усиливать; увеличивать
enhancement *s* усиление; увеличение
enhancer *s* ген-усилитель; энхансер
enhydrous *a* содержащий воду
enisle *v* обособлять, изолировать
enkephalin, enkephaline *s* энкефалин
 leucine ~ лейцинэнкефалин, Leu-энкефалин
 methionine ~ метионинэнкефалин, Met-энкефалин
enlargement *s* расширение; развитие, рост; разрастание, распространение
 cell ~ рост клеток
 lumbar ~ поясничное утолщение
enorganic *a* присущий организму
enphytoty *s* энфитотия
enrich *v* обогащать; улучшать; **to ~ soil** удобрять почву
enrichment *s* обогащение
ensate, **ensiform** *a* мечевидный, мечеобразный
ensilage *s* силосование; силосовать; силосованный корм, силос
ensisternum *s anat* мечевидный отросток
entad *adv* по направлению вовнутрь
ental *a* внутренний
entangle *v* запутываться; поймать в ловушку
entanglement *s bot* сплетение
enteral, enteric *a* кишечный
enteritis *s* энтерит
enteroblast *s* энтеробласт
enterococcus *s* энтерококк
enterocoel *s* энтероцель
enteroproct *s* энтеропрокт
enterotoxin *s* энтеротоксин
enterovirus *s* энтеровирус, кишечный вирус
enterozoon *s* животный эндопаразит
enthalpy *s* энтальпия, теплосодержание
entire *a* цельный, сплошной, полный, целый; некастрированный; чистый, беспримесный; *s* жеребец; некастрированный самец
entire-leaved *a* цельнолистный
entoblast *s* энтобласт, эмбриональная энтодерма
entocondyle *s* внутренний мыщелок
entoectad *adv* кнаружи
entomofauna *s* энтомофауна, фауна насекомых
entomogamy *s* энтомогамия

entomology *s* энтомология
entomophage *s* энтомофаг, насекомоядное животное
entomophagous *a* насекомоядный
entomophilous *a* энтомофильный,опыляемый насекомыми
entophyte *s* растительный паразит
entopic *a* в нормальном положении
entotic *a* внутриушной
entozoan *s* животное-эндопаразит
entozoic *a* обитающий внутри животного
entozoon *s* животное-эндопаразит
entrails *s pl* внутренности, кишки
entrance *s*:
  beehive ~ леток улья
entrap *v* поймать в ловушку; заманивать в ловушку
entropy *s* энтропия
entry *s* вход, вступление
  ~ **of infection** входные ворота инфекции
  sperm ~ проникновение спермия
enucleate *v* удалять ядро, энуклеировать; *a* безъядерный
enucleation *s* удаление ядра, энуклеация
enurination *s eth* опрыскивание мочой
envelope *s* оболочка; *bot* обвёртка; покров; плёнка
  biological ~ биосфера
envenom *v* поражать ядом; испускать яд
environment *s* окружающая среда
  difficult ~s *pl* неблагоприятные условия среды
environmental *a* относящийся к окружающей среде
environmentalism *s* энвайронментализм
environmentalistic *a* энвайронменталистский
enwrapping *a* обвивающий
enzootic *a* энзоотический
enzygotic *a* однояйцовый, монозиготный
enzymatic *a* ферментативный
enzyme *s* фермент, энзим
  activating ~ аминоацил-тРНК-синтетаза
  D- ~ декстрин-трансглюкозилаза, Д-фермент, диспропорционирующий фермент
  disproportionating ~ *см.* **D-enzyme**
  hydrolyzing ~ гидролаза
  isomerizing ~ изомераза
  malate condensing ~ малат-синтаза, малат-конденсирующий фермент, глиоксилат-трансацетаза
  malic ~ малатдегидрогеназа, №яблочный¤ фермент
  marker ~ фермент-маркёр
  nick-closing ~ ДНК-топоизомераза
  old yellow ~ дегидрогеназа восстановленного НАДФ, №старый жёлтый фермент¤
  P- ~ а-глюканфосфорилаза, гликогенфосфорилаза, Р-фермент
  PR- ~ фосфатаза фосфорилазы, PR-фермент
  redox ~ оксиредуктаза, окислительно-восстановительный фермент
  restriction ~ рестрикционный фермент, рестриктаза
  T- ~ олигоглюкан-ветвящая глюкозилтрансфераза, Т-фермент
  transferring ~s ферменты переноса, трансферазы
enzymoblotting *s* блоттинг ферментов
enzymoimmunoassay *s* иммуноферментный анализ, энзимоиммуноанализ
  phase change ~ обменно-фазовый иммуноферментный анализ
enzymology *s* ферментология, энзимология
enzymolysis *s* ферментативное разложение, энзимолиз
eoclimax *s* эоклимакс
eolation *s* выветривание
  soil ~ ветровая эрозия почвы
eolophilous *a* эолофильный, распространяемый ветром
eosere *s* эосерия
eosin, eosine *s* эозин
eosinoblast *s* эозинобласт
eosinocyte, eosinophil *s* эозинофильный гранулоцит, эозинофильный лейкоцит, эозинофил
eosinophilia *s* эозинофилия
eosinophilic *a* эозинофильный
eosinophiloblast *s* эозинофилобласт
eosinophilous *a* эозинофильный
epacme *s* эпакмэ
epactal *a* избыточный
epapillate *a* без сосочков
epapophysis *s* верхний остистый отросток

eparterial *a* расположенный над артерией
epelliculosate *a* без кожицы
ependyma *s* эпендима
ependymal *a* эпендимный
ephapse *s* эфапс
epharmone *s* эфармон
epharmony, **epharmosis** *s* эфармония
ephebic *a* взрослый; зрелый
ephemer *s* эфемер, эфемерный организм
ephemeral *a* эфемерный, однодневный
ephemerous *a* эфемерный, однодневный
ephippium *s anat* турецкое седло; эфиппий, сёдлышко
ephyra *s* эфира
epiarticular *a* надсуставной
epibasidium *s* эпибазидия
epibenthos *s* эпибентос
epibiont *s* эпибионт
epibioses *s pl* обрастания
epibiotic *a* реликтовый; выживающий
epiblema *s* эпиблема
epibody *s* эпиантитело
epibolic *a* эпиболический
epiboly *s* эпиболия
epicalyx *s bot* наружная чашечка, подчашие
epicardium *s* эпикард
epicarp *s* эпикарпий, внеплодник
epichordal *a* эпихордальный
epicingulum *s* поясок эпивальвы *(in diatoms)*
epicole *s* эпибионт
epicondyle *s* надмыщелок
epicotyl *s bot* эпикотиль, надсемядольное колено
epicuticle, epicuticula *s* эпикутикула
epicyte *s* эпицит, подоцит
epidemic *s* эпидемия; эпидемический
epidemiology *s* эпидемиология
epidemy *s* эпидемия
epidermal *a* эпидермальный
epidermis *s* эпидермис; эпидерма
epididymis *s* эпидидимис, придаток яичка
epidural *a* эпидуральный
epifauna *s* эпифауна
epiflora *s* эпифитная флора
epigamic *a* эпигамный
epigastric *a* надчревный
epigastrium *s* надчревная область; *ent* стерниты среднегруди и заднегруди
epigastroid *s* надлобковый хрящ
epigeal *a* наземный
epigenesis *s* эпигенез
epigenetics *s* эпигенетика, динамика развития
epigenotype *s* эпигенотип
epiglottis *s* надгортанник; *энт.* эпифаринкс
epignathous *a* эпигнатический
epigonium *s* эпигоний
epigynous *a* эпигиничный, надпестичный
epigyny *s* эпигиния
epilia *s* сообщество эпифитов
epilithic *a* эпилитический, прикреплённый к камню
epilithophite *s* эпилитофит
epimastigote *s* эпимастиготная форма; *a* эпимастиготный
epimer *s* эпимер
epimerite *s* эпимерит
epimorphosis *s* эпиморфоз
epimyocardium *s* эпимиокард
epimysium *s* эпимизий
epinasty *s* эпинастия
epinephral *a* надпочечный
epinephros *s* надпочечник
epiorganism *s* организм высшего порядка
epiparasite *s* эпипаразит
epiparasitism *s* эпипаразитизм
epiperidium *s* эпиперидий, общий перидий
epipetalous *a* налепестный
epipharyngeal *a* надглоточный; носоглоточный
epipharynx *s* носоглотка; *ent* эпифаринкс
epiphenomenon *s* сопутствующее явление
epiphloem *s* эпифлоэма
epiphragm *s* эпифрагма
epiphylaxis *s* эпифилаксия
epiphyll *s* эпифилл
epiphyseal *a* эпифизарный
epiphysis *s* эпифиз, шишковидное тело
epiphyte *s* эпифит
epiphytic *a* эпифитный
epiphytotics, epiphytoty *s* эпифитотия
epiplasm *s* эпиплазма
epiplastron *s herp* передний парный щиток
epipleura *s* эпиплевра
epiploic *a* сальниковый

epiploon *s* большой сальник
epipodite *s* эпиподит
epipodium *s bot* эпиподий; *zool* эпиподия
epiproct *s* эпипрокт, супраанальная пластинка
epiprotein *s* эпипротеин, регуляторный белок
epipterygoid *s* эпиптеригоид, столбчатая кость
epipubis *s* надлобковый хрящ
epirhizous *a* растущий на корнях
episematic *a* опознавательный
episeme *s* опознавательный признак; опознавательная окраска
episepalous *a* начашелистиковый
episome *s* эписома
episperm *s bot* эписпермий
epispore *s* эриспорий; экзоспора, экзогенная спора
epistacy *s gen* эпистация
epistasis, epistasy *s* эпистаз
epistatic *a* эпистатический
episternal *a* надгрудинный
episternite *s* эпистернит
episternum *s* эпистернум, межключичная кость; рукоятка грудины; *ent* эпистернум
epistome *s* эпистом
epistropheus *s* эпистрофей
epitendineum *s* перитендиний, соединительнотканная оболочка сухожилий
epiterranean *a* надземный
epithalamus *s* эпиталамус, надбугорье
epithalline *a* растущий на талломе
epitheca *s* эпитека, большая створка *(in diatoms)*
epithecium *s* эпитеций
epithelial *a* эпителиальный
epithelium *s* эпителий, эпителиальная ткань
epithem *s* эпитема
epitokous *a* эпитокный
epitope *s* антигенная детерминанта, эпитоп
  antigenic ~ антигенная детерминанта, эпитоп
  B-cell ~ В-клеточная антигенная детерминанта, В-клеточный эпитоп
  bystander ~ низкоиммуногенная детерминанта, минорный эпитоп
  carbohydrate ~ углеводная детерминанта, углеводный эпитоп
  cell-surface ~ детерминанта клеточной поверхности, эпитоп клеточной поверхности, поверхностная антигенная детерминанта
  framework ~ антигенная детерминанта остова, скелетная антигенная детерминанта, каркасная антигенная детерминанта;вариотипическая детерминанта, вариотип
  Goodpasture's ~ антиген гломерулярной базальной мембраны, антиген Гудпасчура
  group-specific ~ группоспецифическая антигенная детерминанта, групповой эпитоп
  immunodominant ~ основная антигенная детерминанта, ключевая антигенная детерминанта, иммунодетерминантный эпитоп
  unique ~ уникальная антигенная детерминанта, уникальный эпитоп, унитоп
  variotypic ~ вариотипическая детерминанта, вариотоп
epitympanum *s* надбарабанное углубление, эпитимпанум
epixylous *a* растущий на древесине
epizoan *s* эпибионт; животное-эктопаразит
epizoic *a* эпибионтный
epizoite *s* эпибионт; животное-эктопаразит
epizoochore *s* эпизоохор
epizoon *s* эпибионт; животное-эктопаразит
epizootic *s* эпизоотия; эпизоотический
epizootiology, epizootology *s* эпизоотология
epizooty *s* эпизоотия
eponychium *s* эпонихий
equation *s* уравнение
equational *a* эквационный
equator *s* экватор
equatorial *a* экваториальный
equidensitometry *s* эквиденситометрия
equidistant *a* равноотстоящий
equilateral *a* равносторонний
equilibrium *s* равновесие; равновесное состояние
equimolar *a* эквимолярный

equine *s* лошадь *(Equus caballus);* лошадиный, конский
equipment *s* аппаратура; оборудование; оснащение
control ~ контрольно-измерительные приборы
fixed ~ стационарная аппаратура; стационарное оборудование
recording ~ регистрирующая аппаратура
equipotentiality *s* эквипотенциальность
equitability *s* равномерность распределения
equitant *a* вкладной, объемлющий *(leaf)*
equivalence *s* эквивалентность
equivalent *s* эквивалент; *a* эквивалентный
equivocal *a* неясный, неопределённый, неопределяемый; двойственный
eradication *s* истребление, уничтожение; вырывание с корнем, искоренение
erect *a* прямостоящий, прямой, вертикальный
erectile *a* способный выпрямляться
~ **spine** складной шип
erection *s* эрекция; выпрямление
erectopatent *a ent* полураскрытый; *bot* частино стелющийся, частично поднимающийся над субстратом
eremic *a* пустынный
eremium *s* сообщество пустынь
eremobic *a* отшельнический, существующий одиночно
eremophilous *a* пустынелюбивый
eremophyte *s* эремофит, растение пустынь
eremus *s* сообщество пустынь
erethic *a* вызывающий повышенную возбудимость; возбудимый, раздражимый
ergastic *s* эргастический, эргастоплазматический
ergastoplasm *s* эргастоплазма
ergatandromorph *s ent* эргатандроморф
ergatandrous *a ent* имеющий эргатоидных самцов
ergataner *s ent* эргатоидный самец
ergate *s* муравей-рабочий
ergatogyne *s ent* эргатоидная самка
ergatogynous *a* имеющий эргатоидных самок
ergatoid *a ent* эргатоидный
ergocalciferol *s* эргокальциферол, витамин D
ergograph *s* эргограф
ergometer *s* эргометр
ergonomics *s* эргономика
ergosterol *s* эргостерол, провитамин D1
ergot *s phytop* спорынья *(caused by Claviceps purpurea)*
erianthous *a* шерстисто-цветковый; пушисто-цветковый
ericetum *s* пустошь
eriocephalous *a* шерстисто-головчатый
eriochore *s bot* эриохор
eriocladous *a* шерстисто-ветвистый
eriocomous *a* имеющий курчавую шерсть; шерстисто-опушённый
eriolepis *a* шерстисто-чешуйчатый
eriophyllous *a* шерстистолистный, с опушёнными листьями
eriospermous *a* шерстистосемянный
eriostachyous *a* шерстисто-колосный
ermel *s* горностаевая моль *(Yponomeutidae)*
ermine *s* горностай *(Mustela)*
erogenic *a* эрогенный
erosion *s* эрозия
erostrate *a* бесклювый
errant *a zool* бродячий; подвижный
errantia *s pl* подвижные организмы
error *s* ошибка, отклонение, погрешность
~ **of fitting** ошибка выравнивания, погрешность выравнивания
erubescence *s* покраснение, румянец
eruciform *a* эруциформный, эрукоидный, гусеницевидный
eructation *s* отрыжка
eruption *s* вспышка эпидемии; прорезывание зуба; сыпь
erythroblast *s* эритробласт
erythrocarpous *a* красноплодный
erythrocyte *s* эритроцит
trinitrophenylated ~ тринитрофенилированный эритроцит, ТНФ-эритроцит
erythrocytic *a* эритроцитарный
erythrocytoblast *s* эритробласт
erythrocytolysis *s* эритроцитолиз, гемолиз
erythrocytometer *s* эритроцитометр, счётная камера для эритроцитов
erythrocytosis *s* эритроцитоз

erythrogenesis *s* эритропоэз, эритроцитопоэз
erythrogonium *s* гемоцитобласт, клетка-предшественница эритроцитов
erythrolysin *s* гемолизин
erythrolysis *s* гемолиз
erythromycin *s* эритромицин
erythron *s* эритрон
erythrophage *s* эритрофаг
erythrophagocytosis *s* эритрофагоцитоз, фагоцитоз эритроцитов
erythrophore *s* эритрофор
erythrophyll *s* эритрофилл
erythropoiesis *s* эритропоэз, эритроцитопоэз
erythropoietic *a* эритропоэтический, эритроцитопоэтический
erythropoietin *s* эритропоэтин
erythropyknosis *s* пикноз эритроцитов
erythrorhexis *s* эритрорексис, разрыв эритроцита
escape *s* утечка, истечение; уход; избегание; бегство; выделение; одичавшее культурное растение; *v* просачиваться
tolerance ~ падение толерантности
escharotic *s* едкое средство; *a* едкий
esculent *a* съедобный, пригодный в пищу
escutcheon *s ent* среднещиток
eseptate *a* не имеющий перегородок
esophageal *a* пищеводный
esophagus *s* пищевод
esoteric *a* эзотерический
espathate *a* не имеющий обвёртки
esquamate *a* не имеющий чешуи; не имеющий чешуек
essence *s* субстанция; сущность; отличительный признак
essential *a* существенный, основной; *bioch* незаменимый
establish *v* основывать, создавать; адаптироваться; bot укореняться, приниматься; **to ~ new varieties** внедрять новые сорта
established *a* устоявшийся; адаптированный; укоренившийся, принявшийся
establishment *s* основание, создание, установление; приживаемость, закрепление самосева; укоренение
ester *s* сложный эфир
esterification *s* эстерификация, образование сложного эфира
esthacyte *s* чувствительная клетка
esthesia *s* восприятие, ощущение, чувство; чувствительность, восприимчивость
esthesiology *s* эстезиология
esthesiometer *s* эстезиометр
esthesioneuron *s* чувствительный нейрон
esthesiophysiology *s* физиология органов чувств
esthesis *s* восприятие, ощущение, чувство; чувствительность, восприимчивость
estimate *s* оценка; *v* оценивать, находить оценку
~ **by eye** оценка на глаз
~ **of standing crop** оценка биомассы; таксация насаждения
estimation *s* расчёт, вычисление; оценивание; оценка
estimator *s* статистическая оценка; формула оценки; лицо, производящее оценку
estipulate *a* не имеющий прилистников
estriate *a bot* без полос
estrous *a* эстральный
estruation, estrus *s* эструс, течка
constant ~ персистирующий эструс, постоянный эструс, непрерывная течка
estuarial, estuarian *a* эстуарный
estuary *s* эстуарий
etaerio *s* сборная костянка
etch *s* гравировать; разъедать
tobacco ~ *phytop* гравировка табака
etching *s* травление
ethanol *s* этанол, этиловый спирт
ethanolysis *s* этанолиз
ether *s* простой эфир
etherification *s* этерификация
ether-soluble *a* растворимый в эфире
ethidium *s* этидий
ethmoid *s* решётчатая кость
ethmopalatine *a* решётчато-нёбный
ethnobotany *s* этноботаника
ethnogeny *s* этногенез
ethnography *s* этнография
ethnology *s* этнология
ethogram *s* этнограмма
ethology *s* этология
ethometry *s* этометрия

etho-mimicry *s* поведенческая мимикрия
etiolation *s* этиолирование, этиоляция
etiology *s* этиология
etioplast *s* этиопласт
euapogamy *s* эуапогамия
eucapnia *s* нормокапния, эукапния
eucaryote *s* эукариот, эукариотный организм
eucaryotic *a* эукариотный
eucentric *a* эуцентрический
eucephalous *a ent* эуцефалический
euchromatin *s* эухроматин
euchromatization *s* эухроматизация
euchromosome *s* эухромосома, аутосома
euclimax *s* эвклимакс
eucone *a ent* эуконический
eudipleural *a* двусторонне-симметричный
eudominant *s* настоящий доминант
eugamic *a* половозрелый
eugenics *s* евгеника
eugonic *a* буйно растущий *(of microorganisms)*
euhaline *a* эугалинный, истинно солоноводный
euheterosis *s* эугетерозис
euhybrid *s* межрасовый гибрид; межвидовой гибрид
euintine *s pal* эуинтина
eukaryote *s* эукариот
eukaryotic *a* эукариотический
eumitosis *s* эумитоз
eumorphic, eumorphous *a* хорошо сформированный, нормально сформированный
eupeptic *a* легко переваривающий
euphotic *a* эуфотический
euplastic *a* образующий ткань
euploid *s* эуплоид; *a* эуплоидный
euploidy *s* эуплоидия
eupnoea *s* эупноэ, нормальное дыхание
eupyrene *a* эупиренный
europhytia *s pl* сообщества гнилых листьев
eurybaric *a* эврибарный
eurychor *s* эврихор
eurychoric *a* с широким ареалом
eurycladous *a* широковетвистый
euryhaline *a* эвригалинный
euryoecic *a* эвриойкный
euryoky *s* эвриойкия
eurypalynous *a* с полиморфной пыльцой
euryphagous *a* всеядный
euryplastic *a* эврипластический
eurythermal, eurythermic *a* эвритермный
eurytopic *a* эвритопный
eurytropic *a* эвритропный
euryvalent *s* эвривалент
euselectivity *s* эуселективность
eusociality *s eth* эусоциальность
eusporangiate *a* эуспорангиатный
eustele *s* эустела
eusystole *s* эусистола, нормальная систола
eutely *s* эутелия
euthanasia *s* эутаназия, безболезненное умерщвление
euthenics *s* эвтеника
eutherians *s pl* плацентарные млекопитающие, высшие звери *(Eutheria)*
eutherophyte *s* эутерофит, однолетник
euthycomous *a* лиссотрихический, гладковолосый
eutrophic *a* эвтрофный, евтрофный
eutrophication эвтрофикация, евтрофикация
eutrophy *s* эвтрофия, евтрофия
eutropic *a* поворачивающийся за солнцем
evaginate *v* эвагинировать, выпячиваться
evagination *s* выпячивание
evaluation *s* оценка; бонитировка
  land ~ бонитировка земель
  sire ~ оценка производителей
evalvate *a anat* не имеющий клапанов
evanescent *a* быстро исчезающий
evaporation *s* испарение; выпаривание
evaporator *s* испаритель; выпариватель
evaporimeter *s* эвапориметр
evapotranspiration *s* эвапотранспирация
evapotranspirometer *s* эвапотранспирометр
even-aged *a* одновозрастный
evenness *s* равномерность распределения
evenomation *s* обезвреживание яда
event *s biom* событие
even-toed *a* парнопалый; парнокопытный
everbearer *s* ремонтантное растение
everbearing *a* ремонтантный
everblooming *a* цветущий круглый год
everglade *s* эверглейд

evergreen *s* вечнозелёное растение; вечнозелёный
evertebrate *a* беспозвоночный
evil-smelling *a* вонючий, дурнопахнущий, зловонный
evocation *s emb* эвокация
evocator *s emb* эвокатор
evoke *v* вызывать, индуцировать
evolute *a* развернувшийся, распустившийся
evolution *s* эволюция, развитие, постепенное изменение; выделение *(e.g., of oxygen)*
evolutionary *a* эволюционный
evolutive *a* эволютивный
evolve *v* развивать, развиваться, эволюционировать; выделять *(e.g., oxygen);* издавать *(odor)*
evolvement *s* развитие, процесс развития
ewe *s* овца; *v* ягниться
exacerbation *s* обострение; раздражение
exalbuminous *a* безбелковый
examination *s* исследование, изучение; осмотр, обследование
example *s* экземпляр; образец
exarch *a bot* экзархный
exasperate *v* усиливать *(pain);* раздражать *(e.g., skin);* шершавый, шероховатый; острощетинистый
excentric *a* эксцентричный; *myc* односторонний
exception *s* исключение; *gen* исключительная особь
excess *s* избыток, излишек; *biom* эксцесс
exchange *s* обмен; *v* обменивать
  air ~ воздухообмен
  cation ~ катионный обмен, катионообмен
  gas(eous) ~ газообмен
  half-time ~ период полуобмена
  heat ~ теплообмен
  respiratory ~ дыхательный газообмен
exchanger *s* антипортер
exciple *s bot* покров апотеция
excision *s* отрезание, вырезание; экстирпация; *mbiol* эксцизия; выщемление; выемка
excitability *s* возбудимость; возбуждаемость; раздражимость
excitant *s* возбуждающее средство, стимулирующее средство; *a* возбуждающий; раздражающий
excitation *s* возбуждение, возбуждённое состояние, волнение; раздражение
excitative *a* возбуждающий
excitator *s* раздражитель
excitatory *a* возбуждающий
excite *v* возбуждать; побуждать; вызывать
excitement *s* возбуждение; волнение
exciter *s* возбудитель; стимулятор
excitometabolic *a* усиливающий обмен веществ
excitomotor, excitomotory *a* вызывающий двигательную активность
excitor *s* стимулятор
excitonutrient *a* возбуждающий аппетит
exclave *s* экстразональная растительность; добавочная железа
exclude *v* исключать
exclusion *s* исключение; вытеснение; *энт.* вылупление
exclusiveness *s* исключительная принадлежность
exconjugant *s* экскоиъюгант
excrement(s) *s* выделения, экскременты
excrescence *s bot* наплыв, нарост, кап; опухоль
excrete *v* выделять, извергать
excretion *s* экскреция, выделение; выделения, экскременты
excretory *a* экскреторный, выделительный
excurrent *a* выводящий, выводной, отводящий; *bot* сбегающий
excursion *s* экскурсия
excurvation *s* экскурвация
excurved *a* экскурвированный; отогнутый; направленный кнаружи
excystation *s* эксцистирование, выход из цисты
exembryonous *a bot* не имеющий зародыша
exendospermous *a* не имеющий эндосперма
exercise *s* нагрузка; упражнения
exert *v* напрягать; оказывать давление, влиять
exertion *s* напряжение, усилие
exflagellation *s* эксфлагелляция
exfoliation *s* шелушение, отслоение; опадение листьев, опадение почечных чешуй

exhalant *a* выдыхаемый; выводной
exhalation *s* выдыхание; выдох
exhale *v* выдыхать; испаряться
exhaust *v* истощать, изнурять; высасывать; опустошать
exhausted *a* истощенный; *imm* адсорбированный, истощенный
exhaustion *s* истощение, изнеможение; вытягивание, высасывание; опустошение
exhaustive *a* истощающий; исчерпывающий
exhibit *v* показывать, обнаруживать; проявляться; **to ~ dependencies** зависеть
exindusiate *a* с голыми спорангиями
exine *s bot* экзина
exintine *s bot* экзинтина
existence *s* существование; жизнь; наличие
existential *a* жизненный; наличный, существующий
exite *s zool* экзит
exoantigen *s* экзоантиген
exobiology *s* экзобиология
exocardial *a* внесердечный
exocarp *s* экзокарпий, внеплодник
exocellular *a* внеклеточный
exochorion *s* экзохорион
exocoelom *s* экзоцелом
exocone *a ent* экзоконический
exoconidium *s* экзоконидия
exocrine, exocrinous *a* экзокринный
exocuticle *s* экзокутикула
exocytosis *s* экзоцитоз
exoderm *s zool* эктодерма; *bot.*экзодерма
exodic *a* эфферентный, центробежный *(of nerve fibers)*
exoenzyme *s* экзофермент
exoexine *s* экзэкзина
exogamy *s* экзогамия, неродственное оплодотворение
exogastrula *s* экзогаструла
exogastrulation *s* экзогаструляция
exogenetic, exogenic *a* экзогенный
exogenote *s* экзогенота
exogenous *a* экзогенный
exon *s* экзон
exonuclease *s* экзонуклеаза
exoperidium *s* экзоперидий
exoplasm *s zool* эктоплазма; *bot* экзоплазма
exopodite *s* экзоподит
exoskeleton *s* наружный скелет
exosmosis *s* экзосмос
exospore *s* экзоспора, экзогенная спора
exosporium *s* экзоспорий
exostome *s bot* экзостом
exostosis *s* экзостоз
exotheca *s zool* экзотека
exothecium *s* экзотеций
exotoxin *s* экзотоксин
exotropism *s* экзотропизм
exovate *s* экзоват
expand *v* увеличиваться в объёме; расширяться; *bot* распускаться
expansion *s* экспансия, расселение; увеличение в объёме; расширение; *bot* распускание
  clonal ~ клональная экспансия, размножение клона
  foliage ~ распускание листьев
  range ~ расширение границ ареала
expectancy *s* ожидание, вероятность
  life ~ средняя вероятная продолжительность жизни
expectation *s biom* ожидание
expenditure *s* расход, потребление
experience *s* опыт, жизненный опыт
experiment *s* эксперимент, опыт; экспериментирование
experimental *a* экспериментальный
experimentation *s* экспериментирование
expirate *s* выдыхаемый воздух
expiration *s* экспирация, выдох
expiratory *a* выдыхательный
expire *v* выдыхать; умирать
explanate *a* уплощённый; распростёртый
explanation *s* объяснение
explant *s* эксплантат
explantation *s* эксплантация
explication *s* развёртывание *(of petals)*; распускание *(of flowers)*
exploitation *s* эксплуатация, использование
  forest ~ лесопользование
explode *v* распускаться *(of flowers);* вскрываться *(of fruit);* давать вспышку развития
exploration *s* исследование
explore *v* исследовать, изучать; зондировать
explorer *s* исследователь; зонд
explosion *s* взрыв

explosive *a* распускающийся *(of flowers);* вскрывающийся *(of fruit);* скачкообразный *(of speciation);* выбрасывающийся, выталкивающийся *(e.g., of spores)*
explure of flower *s* распускание цветка
expose *v* выставлять; выдерживать; подвергать воздействию
exposed *a* открытый; незащищённый; *bioch* доступный
exposition *s* экспозиция; подвергание воздействию; облучение; положение, расположение; обнажение
exposure *s* экспозиция; подвергание воздействию; облучение
antigen ~ антигенная стимуляция
environmental ~ воздействие факторов окружающей среды
expression *s* выражение *(facial); gen* экспрессия; выдавливание
facial ~ лицевое выражение, мимика
sex ~ проявление пола
vocal ~ голосовой звук, голосовой сигнал
expressing *a gen* экспрессирующий
express-laboratory *s* экспресс-лаборатория
expressivity *s* экспрессивность
expulsion *s* выталкивание, изгнание; выдавливание
exsanguination *s* обескровливание
exsect *v* иссекать
exserted *a* выдающийся, выступающий
exsertile *a herp* способный к выпячиванию
exserted *a* выступающий, выдающийся
exsiccant *s* высушивающее средство; *a* высушивающий; высыхающий, засыхающий
exsiccatae *s* высушенные образцы, гербарий
exsiccation *s* высушивание; осушение; обезвоживание; высыхание
exstipulate *a* не имеющий прилистников
exsufflation *s* аспирация, отсасывание
extant *a* ныне живущий
extend *v* простираться; распространяться; разгибать, распрямлять
extension *s* распространение; протяжённость; растяжение, распрямление; разгибание, экстензия
extensor *s* экстензор, мышца-разгибатель
extent *s* размер, величина, степень
~ **of error** величина погрешности
lateral ~ of roots горизонтальное простирание корней
exterior *s* экстерьер; наружный, внешний
extermination *s* искоренение, уничтожение, истребление
exterminative *a* искореняющий, уничтожающий, истребляющий
external *a* наружный, внешний
externalia *s pl* наружные половые органы
exteroception *s* экстерорецепция
exteroceptive *a* воспринятый извне
exteroreception *s* экстерорецепция
exteroreceptor *s* экстерорецептор
extesticulate *v* кастрировать
extinct *a* вымерший; исчезнувший
extinction *s* угасание; вымирание; исчезновение
extine *s* экстина, экзоспорий
extinguish *v* гасить, тормозить
extra-axillary *a* внепазушный
extrabranchial *a* внежаберный
extracapsular *a* внекапсулярный
extracardiac *a* внесердечный, экстракардиальный
extracellular *a* внеклеточный
extrachromosomal *a* внехромосомный
extraclinal *a* экстраклинальный, не входящий в клин
extracopy *s gen* экстракопия
extracranial *a* внечерепной
extract *s* экстракт, вытяжка; выборка; *v* экстрагировать, извлекать
rennet ~ сычужная закваска
extraction *s* экстрагирование, извлечение
extraembryonic *a* внезародышевый
extraenteric *a* внекишечный
extrafloral *a* экстрафлоральный, внецветковый
extrafoliaceous *a* внелистный
extragenic *a* экстрагенный
extragenital *a* внеполовой
extraintestinal *a* внекишечный
extramitochondrial *a* внемитохондриальный
extraneous *a* инородный, посторонний; периферический, маргинальный
extranucellar *a* вненуцеллярный
extranuclear *a* внеядерный

extranutrition *s* подкормка растений
extraovate *s* экзоват
extraperitoneal *a* внебрюшинный
extrapolation *s* экстраполяция
extrapollination *s* доопыление
extraporate *a* внепоровый
extrapyramidal *a* экстрапирамидальный
extrarenal *a* внепочечный
extrareticulate *a* внешнесетчатый
extrastaminal *a* внетычиночный
extrastelar *a* экстрастелярный
extrasystole *s* экстрасистола, добавочная систола
extrauterine *a* внематочный
extravaginal *a* экстравагинальный
extravasation *s* транссудация, выпотевание
extravascular *a* внесосудистый
extraventricular *a* внежелудочковый
extremity *s* край; конец; конечность
extremum *s* экстремум
extrinsic *a* внешний; неприсущий, несвойственный
extrorse *a* обращённый наружу
extrusion *s* выталкивание, вытеснение, изгнание; выбухание
exuberance *s anat* выступ, бугор; плодовитость
exuberant *a* пышно растущий, буйный; плодовитый
exudate *s* эксудат, выделение
  root ~ выделение корней
  xylem ~ *bot* пасока
exudation *s* эксудация, выделение; плач растений
  fruit ~ выпот плодов
exumbrella *s zool* эксумбрелла
exuvium *s* экзувий
eye *s* глаз; глазок; *бот.* глазок; отметина, «яблоко» у лошади; **by** ~ визуально; **strong ~s** хорошее зрение
eyeball *s* глазное яблоко
eyeblink *s* мигание
eyebrow *s* бровь
eyebulb *s* глазное яблоко
eyecup *s* глазной бокал, глазная чаша
eye-end *s* вершина *(of a potato tuber)*
eyeglass *s* окуляр
eyeglobe *s* глазное яблоко
eyeground *s* глазное дно
eyehole *s* глазная впадина, орбита, глазница
eyelash *s* ресница
eyelens *s* окуляр
eyelet *s bot* глазок
eyelid *s* веко
eyepiece *s* окуляр
eyepit *s* глазная впадина, орбита, глазница
eye-ring *s* кольцо вокруг глаз
eyeshot *s* поле зрения
eyesight *s* зрение
eyesocket *s* глазная впадина, орбита, глазница
eyespot *s* глазное пятно; стигма; глазок; глазчатое пятно; *phytop* глазковая пятнистость
eyestalk *s* глазной стебелёк
eyetooth *s* глазной зуб, верхний клык
eyewater *s* слёзы
eyrie *s* гнездо хищной птицы

F

fabaceous *a* бобовый; бобовидный
fabiform *a* бобовидный
face *s* лицо; морда; лицевая поверхность; выражение лица; гримаса; вид спереди
alert ~ *eth* настороженное лицо
bared-teeth ~ *eth* гримаса с оскалом
dished ~ вогнутая морда; вогнутое рыло
fracture ~ поверхность скола
lip-smacking ~ *eth* чмоканье губами
relaxed open-mouth ~ *eth* игровой оскал
facet *s ent* фасетка, простой глазок
facial *a* лицевой
faciation *s* фациация
facies *s* фация
facilitation *s eth* облегчение
facing away *s eth* поза «голова в сторону»
factitious *a* искусственный
factor *s* фактор; причина, движущая сила; коэффициент; показатель; *gen* фактор; гормон; витамин
anaphylactogenic ~ анафилактогенный фактор, анафилактоген
animal protein ~ цианокобаламин, витамин B_{12}
antiberiberi ~ тиамин, аневрин, витамин B_1
C- ~ К-фактор
chemotactic ~ хемотаксический фактор, хемотаксин
citrovorum ~ цитроворум-фактор
colony-enhancing ~, colony-stimulating ~ колониестимулирующий фактор, КСФ
condition ~ коэффициент упитанности; упитанность
decay-accelerating ~ фактор распада; стимулятор гемолиза, DAF-фактор
differentiation-inducing ~ индуктор дифференцировки
differentiation-promoting ~ кофактор дифференцировки
diluting ~ фактор ослабления, ген-ослабитель
disease-producing ~ возбудитель болезни
etiological ~ этиологический фактор; возбудитель болезни
extension ~ фактор усиления, ген-усилитель
Fletcher's ~ фактор Флетчера, прокалликреин
form ~ *bot* видовое число
growth hormone releasing ~ соматолиберин, фактор, стимулирующий выделение гормона роста
growth hormone release inhibiting ~ соматостатин, фактор, угнетающий выделение гормона роста
growth-promoting ~ стимулятор роста
H ~ *gen* фактор плеча, H-фактор
hereditary ~ наследственный фактор, ген
humerus ~ *gen* фактор плеча, H-фактор
I ~ фактор-индуктор, I-фактор
immunomodulating ~ иммуномодулятор, иммуномодулирующий фактор
inducer ~ фактор-индуктор, I-фактор
killer ~ киллер-фактор, фактор-«убийца»
lactogenic ~ лактогенный гормон, пролактин
lethal ~ летальный фактор, летальный ген, леталь
linked ~s *pl* сцепленные гены
myelopoietic ~ фактор индукции миелопоэза, миелопоэтин
N ~, neutral ~ фактор «нейтральности», N-фактор
quality ~ показатель качества
R ~ фактор реактивности, R-фактор; фактор передачи устойчивости, R-фактор
reactive ~ фактор реактивности, R-фактор
recessive ~ рецессивный ген
release ~ *mbiol* фактор высвобождения, рилизинг-фактор
resistance transfer ~ фактор передачи устойчивости, R-фактор
Rh ~, rhesus ~резус-фактор
rho ~ фактор терминации транскрипции
SF ~ фактор стерильности самок, SF-фактор

stringent ~ *mbiol* стринжент-фактор
Stuart ~ фактор Стюарта
Thy-1-inducing ~ *imm* индуктор Thy-1 антигена
factorial *s biom* факториал; *a* факториальный; генный
factor-pair *s* пара генов, пара аллелей
facultative *a* факультативный, необязательный
fade *v* увядать, вянуть; обесцвечивать, обесцвечиваться
fade-proof *a* светопрочный
fading *s* выцветание; увядание; увядающий
blossom ~ отцветание; опадение цветков
faecal *a* фекальный
faeces *s pl* фекалии
fail *v* ослабевать, терять силы; не уродиться
fail-safe *a* надёжный, безотказный, работающий без ошибок *(e.g., equipment)*
failure *s* недостаточность; отсутствие; нарушение; расстройство; разрыв; неудача, неблагоприятный исход
~ **of compensation** декомпенсация
cardiac ~ сердечная недостаточность
fair *a* обильный *(crop);* ясный *(weather)*
fair-flesh *a* хорошо упитанный
fairy-ring *s myc* ведьмино кольцо
falcate, falciform *a* серповидный
falcon *s* сокол *(Falco)*
Arctic ~ кречет *(Falco gyrfalco)*
hobby ~ чеглок *(Falco subbuteo)*
peregrin(e) ~ сапсан *(Falco peregrinus)*
red-footed ~ кобчик *(Falco vespertinus)*
saker ~ балобан *(Falco cherrug)*
sooty ~ серебристый чеглок *(Falco concolor)*
falconet *s* карликовый сокол *(Microhierax)*
falconry *s* соколиная охота
falcula *s* серп мозжечка
falculate *a* серповидный
fall *s* опадение, осыпание; падать, опадать, осыпаться; уменьшение, ослабление; выпадение *(e.g., of teeth);* осыпание *(of seeds);* опадение *(e.g., of leaves);* рождение; рождаться; приплод, выводок; выпадение; осень
blossom ~ отцветание, опадение цветков
dead ~ опадный самолов, ловушка давящего типа
leaf ~ листопад
fall-and-socket *s* ореховидный сустав
fall-born *a* родившийся осенью
fall-off *s* линька, потеря шерсти
fallow *s* залежь, заброшенная пашня, перелог; пар
fallow-deer *s* лань *(Cervus dama)*
fall-planting *s* осенняя посадка
false-negative *a* ложноотрицательный
falx *s* серп, серповидная складка; *zool* хелицера
familiarization *s eth* ознакомление
family *s* семейство; семья
FB [fold-back] ~ *mbiol* B-семейство, семейство обращённых повторов
fan *s* веерная крона
fancy *a* декоративный
fang *s* клык; ядовитый зуб; корень зуба; *zool* хелицера
fanged *a* имеющий клыки
fan-leaved *a* вееролистный
fan-shaped *a* веероводный, веерообразный
fare *s* пища; корм
farina *s* цветочная пыльца; крахмал; белый налёт
farinose *a* мучнистый; покрытый налётом
farm *s* ферма, хозяйство; питомник
sewage ~ поля орошения
farming *s* выращивание
deer ~ оленеводство
fruit ~ плодоводство
fur ~ пушное звероводство
livestock ~ животноводство
seed ~ семеноводство
slash-and-burn ~ подсечно-огневое земледелие
farrow, farrowing *s* опорос; пороситься
fascia *s* фасция; поперечная полоса *(in diatoms)*
fascial *a* фасциальный
fasciation *s* фасциация, образование пучков
fascicle *s* пучок; *bot* гроздь
fascicular *a* пучковый; мочковатый
fasciculate *a* собранный в пучки

fasciculation *s* расположение пучками; *phys* фасцикуляция
fascinate *v* завораживать, гипнотизировать
fascioliasis, fasciolosis *s helm* фасциолёз
fast *a* быстрый; прочный; *s* голодание; *v* голодать
fast-growing *a* быстрорастущий
fastidious *a* прихотливый
fastigiate *a bot* конический, конусообразный; прямопараллельный
fastigium *s* вершина, верхушка
fasting *s* голодание
fastness *s* устойчивость; резистентность
  acid ~ кислотоустойчивость
  light ~ светопрочность, светостойкость
fat *s* жир, сало; *a* жирный, сальный; маслянистый; откормленный, упитанный, тучный; *v* откармливать
  depot ~ жировые отложения
  goose ~ гусиное сало
fatal *a* губительный, пагубный; смертельный
fat-body *s ent* жировое тело
fate *s* метаболический путь
  metabolic ~ метаболический путь
fat-extracted, fat-free *a* обезжиренный
fatiguability *s* утомляемость
fatigue *s* усталость, утомление; утомлять, изнурять
  afferent ~ *phys* афферентное понижение
fatiscent *a* распадающийся, растрескивающийся
fatness *s* жирность; ожирение
fat-soluble *a* жирорастворимый
fat-splitting *a* расщепляющий жир, липолитический
fat-tailed *a* жирнохвостый
fatten *v* нагуливаться; жировать
fattening *s* нагул, откорм
fatty *a* жирный; маслянистый; ожиревший; тучный; жировой
fauces *s* зев, глотка; устье раковины; *bot* зев
fault *s* порок, дефект
fauna *s* фауна, животный мир
  helminth ~ гельминтофауна
faunistic *a* фаунистический
faunology *s* зоогеография
faunula *s* фаунула
faveolate *a* сотовый, ячеистый
faveolus *s* альвеола; ячейка
favorable *a* благоприятный
favose *a* сотовый, ячеистый; покрытый паршой
favus *s phytop* парша *(caused by Trichophyton schoenlein)*
fawn *s* молодой олень; *v* телиться; вилять хвостом *(of a dog),* ласкаться
fawning *s* отёл
F-distribution *s biom* F-распределение
fear *s* страх, боязнь
feather *s* перо; *bot* летучка; *v* оперяться
  powder down ~ пудретка
feathered *a* оперённый; оперившийся
feathering *s* оперение
  fast ~ быстрая оперяемость
  tight ~ плотно прилегающее оперение
feather-legged *a orn* мохноногий; со щётками *(of a horse)*
featherless *a* лишённый оперения
feather-veined *a bot* перисто-нервный
feathery *a* оперённый; опушённый
feature *s* признак, характерная черта
fecal *a* фекальный
feces *s* осадок; фекалии
fecund *a* плодородный; плодовитый
fecundate *v* оплодотворять; делать плодородным
fecundation *s* оплодотворение
fecundity *s* плодородность; плодовитость
federation *s* федерация
federion *s* федерион
feeble *a* слабый, хилый
feed *s* корм; *v* кормить
feedback *s* обратная связь
feedbox *s* кормушка
feeder *s* кормушка; подслой, фидер
  plant ~, vegetable ~ фитофаг, растительноядное животное
feeding *s* кормление, питание, кормёжка; фидинг, подпитка
  artificial ~ искусственное питание; искусственное вскармливание
  courtship ~ кормление в брачной демонстрации
  culture ~ подпитка культуры, подкормка культуры
  window ~ выедание окошек *(in leaves)*
feedstuff *s* корм, кормовой продукт
feel *v* чувствовать, ощущать; осязать; **to ~ out** выпускать щупальце
feeler *s* щупальце; щупик; усик, антенна

feeling *s* чувство, ощущение; осязание
fel *s* жёлчь
feline *a* кошачий
fell *s* шкура, шкурка, мех; рубка, валка
fellation *s eth* фелляция
fell-field *s* каменистая пустыня
felling *s for* рубка, валка
felt *s* войлок
  leaf ~ *phytop* войлочная болезнь *(caused by Anthina citri)*
felted *a* войлочный, сильно опушённый
feltwork *s* фиброзная сеть
female *s* самка, матка; женский
feminine *a* женский
feminization *s* феминизация
feminonucleus *s* женский пронуклеус
femoral *a* бедренный
femur *s* бедро; бедренная кость
fen *s* болото; низинное болото
fencing *s* огораживание; остолбление
fenestra *s anat* окно, отверстие; *ent* прозрачное пятно
fenestrate, fenestrated *a* с многочисленными отверстиями
fenestration *s* фенестрация
fennec *s ther* фенек *(Vulpes zerda)*
feral *a* дикий, неприрученный; одичавший
ferment *s* фермент; *v* ферментировать, сбраживать; бродить
fermentation *s* ферментация, брожение; сбраживание
  continuous ~ непрерывное брожение; непрерывная ферментация
  end ~ дображивание
  laboratory test ~ бродильная проба
  over ~ чрезмерная ферментация, перебрprincipal
  semicontinuous ~ полунепрерывная ферментация, полунепрерывное выращивание
  solid state [solid-substrate] ~ твёрдофазная ферментация, твёрдофазное выращивание, твёрдофазное культивирование
  vacuum ~ вакуум-брожение
fermentative *a* ферментирующий, бродильный, сбраживающий
fermenter *s* ферментёр, бродильный аппарат, бродильный чан
fern *s* папоротник *(Felicidae)*
fern-leaved *a* папоротниколистный
ferocious *a* свирепый, лютый
ferret *s* африканский хорёк *(Mustela furo)*
ferriheme *s* гемин
fertile *a* плодородный, плодовитый
fertility *s* фертильность, плодовитость; плодородие; продуктивность
fertilization *s* оплодотворение; опыление; удобрение
fertilize *v* оплодотворять; опылять; удобрять
fertilizer *s* удобрение
fescue *s bot* овсяница *(Festuca)*
fester *s* нагноенние; *v* нагноиться
fetal *a* плодный
fetation *s* развитие плода; беременность
fetid *a* зловонный, вонючий
fetlock *s* путовый сустав
fetus *s* плод
fever *s* лихорадка
  African tick ~ африканский клещевой риккетсиоз, африканская клещевая лихорадка
  Assam ~ индийский висцеральный лейшманиоз
  black ~ пятнистая лихорадка Скалистых гор, американский клещевой риккетсиоз
  blackwater ~ малярийная гемоглобинурия
  camp ~ сыпной тиф
  Cyprus ~ бруцеллез
  dumdum ~ индийский висцеральный лейшманиоз
  hay ~ поллиноз
  inundation ~ цуцугамуши, японская речная лихорадка
  parrot ~ орнитоз, пситтакоз
  pollen ~ поллиноз
  tick-borne relapsing ~ клещевой возвратный тиф
feverish *a* лихорадочный
few-awned *a* малоостистый
few-celled *a* малоклеточный
few-flowered *a* малоцветковый
fiber *s* волокно; нить; клетчатка
  astral ~ астральная нить
  bast ~ луб
  Brucke's ~ волокно Брюкке, меридиональный пучок
  chromosomal spindle ~ хромосомная нить веретена
  crude ~ сырая клетчатка

hull ~ подпушек
inhibitory ~ тормозящий нерв сердца
Purkinje's ~ волокно Пуркинье, атипический сердечный миоцит

fibration *s* образование волокна; волокнистое строение
fibre *s* волокно; нить; клетчатка
fibril, fibrilla *s* фибрилла, тонкое волоконце; *бот.* мочка *(корня);* корневой волосок
cellular ~ клеточная фибрилла
muscular ~ миофибрилла, мышечное волоконце
nerve ~ нервное волоконце
root ~ *bot* мочка корня; корневой волосок

fibrillar *a* фибриллярный, волокнистый
fibrillation *s* фибрилляция, мерцание; образование волокон
fibrillose *a bot* мочковатый
fibrinogenous *a* фибринообразующий
fibrinolysis *s* фибринолиз
fibroblast *s* фибробласт
fibrocartilage *s* волокнистый хрящ, соединительно-тканный хрящ
fibrocyte *s* фиброцит, десмоцит
fibroplate *s* суставной диск
fibroreticulate *a* сетчато-волокнистый
fibrous *a* волокнистый; *bot* мочковатый
fibrovascular *a* сосудисто-волокнистый
fibula *s* малоберцовая кость; фибула *(in diatoms)*
fibular *a* малоберцовый
fickle *a* неустойчивый, ненадёжный
fiddler *s* манящий краб *(Uca); orn* перевозчик *(Tringa hypoleucos)*
fidelity *s* привязанность
fidgets *s pl* беспокойное состояние
field *s* поле; участок; область
abandoned ~ заброшенное поле, залежь
idle ~ залежь
personal ~ *eth* персональное пространство
sod ~ заброшенное поле, залежь

fieldfare *s orn* рябинник *(Turdus pilaris)*
fig *s* фикус *(Ficus),* фиговое дерево, инжир *(Ficus carica)*
fighting *s* борьба; драка
filaceous *a* состоящий из нитей; волокнистый; мочковатый
filament *s* филамент, нить; волоконце; волокно; тычиночная нить; гифа
bacterial ~ нитчатая форма бактерии
intermediate ~ промежуточный филамент
nerve ~ нервное волокно

filamentary *a* волокончатый
filamentation *s* образование нитей
filamentous *a* нитеобразный, нитевидный, нитчатый, волокнистый, с волокнами
filariasis *s helm* филяриоз
filariform *a* нитевидный
filbert *s* лещина понтийская *(Corylus pontica*); ломбардский орех, фундук *(Corylus maxima)*
file *s* напильник; файл
stridulatory ~ *ent* стридуляционный гребень

filial *a* дочерний; сыновний
filiation *s* отношение родства, происхождение
filical *a* папоротниковый
filicauline *a* тонкостебельный
filiferous *a* несущий нити, нитеносный; гифоносный
filiform *a* нитеобразный, нитевидный; гифоносный
filling *s* наполнение
film *s* плёнка; налёт; мазок; *v* покрываться плёнкой
filming *s* киносъёмка
filmy *a* покрытый плёнкой, плёнчатый
filoplume *s* нитевидное перо
filopodium *s* филоподия, нитевидная псевдоподия
filose *a* нитевидный; филозный, имеющий филоподии *(об амёбах)*
filter *s* фильтр; фильтровать
centrifugal ~ фильтрующая центрифуга

filterable *a* фильтрующийся
filterer, **filter-feeder** *s zool* фильтратор
filterglass *s* стеклянный фильтр
filtering *s* фильтрование
filter-passer *s* фильтрующийся вирус
filtrate *s* фильтрат; *v* фильтровать
filtration *s* фильтрация, фильтрование
gel ~ гель-фильтрация, гель-хроматография
thin-layer gel ~ тонкослойная гель-хроматография

fimbria *s* бахромка; *микр.* фимбрия
fimbriate, fimbriated *a* бахромчатый
filmbrillate *a* мелкобахромчатый
fimicolous *a* обитающий на навозе
fin *s* плавник
finch *s* вьюрок *(Fringilla);* мелкая птица
  Darwin's ~ дарвинов вьюрок, земляной вьюрок *(Geospizinae)*
  thistle ~ чиж *(Spinus spinus)*
find находить, обнаруживать
finding *s* обнаружение; открытие; *pl* данные
  ~s of investigation результаты исследования
  analysis ~s данные анализа
fine *a* тонкий; мелкозернистый, мелкий; высококачественный, чистый
fine-fibrous, fine-filamented тонковолокнистый
fine-grained *a* мелкозернистый; тонкоструктурный
fine-leaved *a* мелколистный
fine-meshed *a* мелкоячейный
fine-textured *a* тонкотекстурный
finger *s* палец *(of hand);* указатель, стрелка
  little ~ мизинец
fingered *a bot* дланевидный, пальчатый
fingernail *s* ноготь пальца руки
fingerprinting *s mbiol* «фингерпринтинг», метод «отпечатков пальцев»
finger-veined *a* пальчатонервный
finish *s* конец, окончание; упитанность; *v* кончать, заканчивать
finished *a* упитанный, откормленный
finite *s biom* конечная величина
finlet *s* плавничок
fir *s* пихта *(Abies);* ель *(Pices); бот.* лжетсуга *(Pseudotsuga)*
  spruce ~ ель обыкновенная *(Picea excelsa)*
fire *s* огонь, пожар; пал
  grassland ~ луговой пожар, пал
fire-belly *s* жерлянка *(Bombina)*
firecrest *s orn* красноголовый королёк *(Regulus ignicapillus)*
firefanging *s* перепревание; самовозгорание
firefly *s ent* светляк *(Lampyridae)*
firetop, fireweed *s* кипрей узколистный, иван-чай *(Epilobium angustifolium)*
fireworm *s* гусеница листовёртки
firm *a* плотный, твёрдый
firry *a* пихтовый; еловый
firstling *s* первенец
fish *s* рыба *(Pisces); v* ловить рыбу
fish-eating *a* рыбоядный
fisher *s ther* илька, пекан, рыболов *(Martes pennanti)*
fissilingual *a* с раздвоенным языком *(напр. о ящерице)*
fission *s* деление; расщепление, дробление, сегментация
fissiparous *a* размножающийся делением
fissural *a* относящийся к трещине; щелевой
fissure *s* щель; борозда; извилина
  oral ~ ротовое отверстие, рот
fissured *a* щелевой; щелевидный
fistula *s* фистула, свищ
fistular *a bot* полый, дудчатый
fit *s* соответствие; приступ, припадок
  best ~ *biom* наилучше подобранный, наиболее соответствующий
  semiaverage ~ *biom* выравненный полусредними
fitchew *s* обыкновенный хорёк, лесной хорёк, чёрный хорёк *(Mustela putorius)*
fitness *s* соответствие, приспособленность; пригодность
fits *s pl* судороги, конвульсии
fittest *a* наиболее приспособленный
fitting *s biom* подбор
five-finger *s bot* лапчатка *(Potentilla)*
fix *v* фиксировать; закреплять; укреплять; сгущать; густеть; твердеть
fixation *s* фиксация; закрепление; укрепление; усвоение; сгущение; *phys* закрепление
  alexin ~ связывание комплемента, фиксация комплемента
  complement ~ связывание комплемента, фиксация комплемента
  macrophage ~ торможение миграции макрофагов
  nitrogen ~ азотфиксация, фиксация азота
fixative *s hist* фиксатор
fixed *a* фиксированный; нелетучий, плохо растворяющийся; (химически) связанный; неподвижный, стационарный; закреплённый
fixer *s* фиксатор

nitrogen ~ азотфиксатор
fixing *s см.* **fixation**
fixity *s* неподвижность
flabby *a* вялый, дряблый, слабый
flabellate, flabelliform *a* веерообразный, веероводный
flabellinerved *a bot* веерорнервный
flabellum *s* веерообразный орган; веерообразная структура
flaccid *a* слабый, вялый, дряблый; *бот.* вялый, увядший, поникший
flag *s* лист удлинённой формы; *бот.* касатик *(Iris);* рогоз широколистный *(Typha latifolia);* дёрн; *v* чахнуть; поникать
flagellant *a* жгутиковый
flagellate, flagellated *a* жгутиковый; жгутиконосный
flagellation *s micr* жгутикование
flagelliform *a* жгутиковидный
flagellosis *s* флагеллоз
flagellula *s* флагеллиспора, зооспора, спора со жгутиками
flagellum *s* жгутик; стелющийся побег, плеть, ус
flake *s* тонкий слой; расслаиваться; *pl* хлопья
flakelike *a* хлопьевидный
flame *s* пламя; *v* фламбировать, стерилизовать пламенем
flaming *s* фламбирование, стерилизация пламенем
flamingo *s* обыкновенный фламинго *(Phoenicopterus)*
flange *s* выступ
flank *s* бок, боковая сторона
flanking *a mbiol* фланкирующий
flap *s* клапан; длинное ухо; взмах крыльев; *v* махать крыльями, хлопать крыльями
~ **of the ear** мочка уха
flap-eared *a* вислоухий
flapper *s* лист; утёнок, подлётыш, порхунок
flask *s* колба, склянка; разливать по колбам
drip ~ капельница
spinner ~ вращающаяся колба; центрифужная пробирка
flat *a* плоский; плотно прилегающий *(e.g., fur);* *s* равнина, низина, низменность
flat-flowered *a* плоскоцветковый
flatfoot *s* плоскостопие
flat-fruited *a* плоскоплодный
flatheaded *a* плоскоголовый
flat-leaved *a* плосколистный
flat-podded *a* плоскостручковый
flat-stalked *a* плоскостебельчатый
flattened *a* уплощённый; сплющенный
flattening *s* уплощение
flat-veined *a bot.* плосконервный
flatworm *s* плоский червь *(Plathelminthes)*
flavin *s* рибофлавин, витамин B_2
flavor *s* привкус
flaw *s*порок, дефект; порыв ветра; порывистый ветер; *v* ломать, повреждать, портить
flax *s* лён *(Linum)*
crown ~ лён-кудряш
fiber ~ лён-долгунец *(Linum usitatissimum elongata)*
oil ~ лён-кудряш
pale ~ лён двухлетний *(Linum bienne)*
spurge ~ волчеягодник обыкновенный, волчье лыко *(Daphne mezereum)*
wild ~ льнянка обыкновенная *(Linaria vulgaris)*
flaxweed *s*льнянка обыкновенная *(Linaria vulgaris)*
flea *s* блоха *(Aphaniptera);* мелкое насекомое
blue ~ блошка крестоцветная *(Phyllotreta cruciferae)*
flea-bitten *a* чубарый *(of a horse)*
fleam *s* ланцет
fleck *s* пятно, пятнышко, крапинка; пятнистость
pith ~ червоточина
flecked *a* пятнистый; чалый *(of a horse)*
fledge опериться; выкармливать птенцов; выстилать перьями
fledged *a* оперённый
fledgeling, fledgling *s* оперившийся птенец
fleece *s* руно, рунная шерсть
fleecy *a* покрытый шерстью, шерстистый
flehmen *s eth* флемен
flesh *s* мясо; тело, плоть; мякоть
flesh-eating *a* плотоядный
flesh-flowered *a* мясисто-цветковый
flesh-headed *a* мясисто-головчатый
fleshiness *s* мясистость; мясистый нарост

flesh-leaved *a* мясисто-листный
fleshy *a* мясистый
flexen *a* светло-желтый, соломенный
flexibacteria *s pl* флексибактерии, извитые бактерии
flexibility *s* пластичность, гибкость
flexible *a* пластичный; гибкий
flexion *s* флексия, сгибание
flexor *s* флексор, мышца-сгибатель
flexuous *a* извилистый; волнистый; витой
flexure *s* изгиб; искривление; кривизна; выгиб
flick *s* лёгкий отрывистый удар; щелчок; резкое движение
flight *s* полёт; лёт; перелёт *(птиц); pl* маховые перья; *v* летать
courtship ~ воздушные демонстрации
cleansing ~ *apic* очистительный облёт
orientation ~ *apic* ориентировочный полёт; ориентирововчный облёт
flightless *a* бескрылый
flighty *a* пугливый
fling *v* брыкаться; кидаться, бросаться
flinty *a* каменистый *(soil);* стекловидный, кремнистый
flipper *s* плавник; плавательная перепонка; ласт
flit *v* перелетать с места на место; порхать; перепархивать
flitter *s* порхание; махание крыльями; *v* порхать; махать крыльями
floatation *s* флотация
floccose *a* хлопьевидно опушённый
flocculation *s* флокуляция
flocculent *a* хлопьевидный; пушистый, покрытый пушком
flocculus *s* клочок, пучок
floccus *s* кисточка *(of the tail);* пуховое перо; *bot* хохолок; *myc* группа нитей, гиф
flock *s* стая *(of birds);* стадо; клочок шерсти; пучок волос
species ~ группа близких видов
floor *s anat* дно, основание
forest ~ поверхность почвы в лесу; лесная подстилка
mouth ~ дно полости рта
flora *s* флора
microscopic ~ микрофлора
floral *a* относящийся к флоре; относящийся к цветку
florescence *s* цветение; время цветения
floret *s* цветок компактного соцветия
ray ~ краевой цветок
floricole, floricolous *a* живущий на цветках
floriculture *s* цветоводство
floriferous *a* цветоносный
florification *s* цветение
florigen *s* флориген, гормон цветения
florigenic *a* стимулирующий цветение
floristic *a* флористический
floristics *s* флористика
floscule *s* цветок компактного соцветия
floss *s* пух, пушок
flotation *s* флотация
flourish *s* «росчерк» *(in the end of bird's song)*
floury *a* мучнистый
flow *s* течение, ток; проток
blood ~ ток крови, кровоток
energy ~ поток энергии
gene ~ поток генов, дрейф генов
honey ~ *apic* взяток, медосбор
saliva ~ слюноотделение
transpiration ~ транспирационный ток
urinary ~ мочеотделение
flower *s* цветок; цветущее растение; цвести; цветение; **to be in** ~ цвести; **to come into** ~ вступать в фазу цветения
unfertile ~ бесплодный цветок, пустоцвет
flowerage *s* цветение
flowerbed *s* клумба
flowered *a* цветущий, в цвету, покрытый цветками
flowerer *s* цветущее растение
floweret *s* цветочек
flowerful *a* изобилующий цветками, полный цветков
flowering *s* цветение; цветущий
flowerless *a* бесцветковый; споровый
fluctuation *s* флуктуация
seasonal ~ s сезонные колебания
vegetational ~ смена растительного покрова, сукцессия
fluff *s* пух, пушок; опушение
fluff up *v* распушить
fluid *s* жидкость; жидкий; жидкостный
cellular ~ клеточный сок
fixing ~ фиксатор
joint ~ синовиальная жидкость, синовия

lymphatic ~ лимфа
molting ~ *ent* секрет линочной железы
Scarpa's ~ эндолимфа
synovial ~ синовиальная жидкость, синовия
testicular ~ сперма, семя
fluidifying *s* разжижение
fluidity *s* текучесть
fluke *s helm* трематода, сосальщик *(Trematoda); pl* хвостовой плавник *(of a whale)*
blood ~ шистосоматида, кровяной сосальщик *(Schistosomatidae)*
brain ~ диплосоматида *(Diplosomatidae)*
fluorescence *s* флуоресценция
fluorescent *a* флуоресцирующий; флуоресцентный
fluorination *s* фторирование
fluorochrome *s* флуорохром
fluorodeoxyuridine *s* фтордезоксиуридин
fluoroimmunoassay *s* иммунофлуоресцентный анализ
fluorometer *s* флуорометр
fluorouracil *s* фторурацил
fluorouridine *s* фторуридин
flush *s* промывка; отпрыск, отросток, побег; прилив крови; приступ; *v* давать отростки, давать побеги; способствовать росту; вспугивать *(birds);* промывать
~ of flowering волна цветения; разгар цветения
first ~ of grass первая зелень
flutter *s* трепетание, дрожание; вибрация; *v* трепетать, дрожать; **the ~ of the wings** трепетание крыльев
fluvial *a* речной
flux *s* истечение; обильное отделение; приток, поступление; *v* истекать, течь, вытекать
fly *s* муха *(Muscidae);* двукрылое; крылатое насекомое
adder ~ies неравнокрылые стрекозы *(Anisoptera)*
ant ~ крылатый муравей; журчалка муравьиная *(Microdon)*
apple-seed chalcid ~ семяед яблонный большой *(Torymus druparum)*
armed ~ львинка *(Stratiomys)*
ascalaphus ~ аскалаф *(Ascalaphus)*
assasin ~ ктырь *(Asilidae)*
bee ~ жужжало *(Bombyliidae)*
black ~ мошка *(Similiidae)*
braulid ~ пчелиная вошь *(Braulidae)*
bristle ~ тахина, ежемуха *(Tachinidae)*
caddis ~ ручейник *(Trichoptera)*
chameleon ~ львинка обыкновенная *(Stratiomys chamaeleon)*
chloropid ~ злаковая мушка *(Chloropidae)*
Chinese-candle ~ фонарница китайская *(Laternaria candelaria)*
circular-seamed ~ круглошовное двукрылое *(Cyclorrhapha)*
common fruit ~ дрозофила фруктовая, дрозофила малая, дрозофила обыкновенная *(Drosophila melanogaster)*
crane ~ долгоножка *(Tipulidae)*
dance ~ies толкунчики *(Empididae)*
demoiselle ~ равнокрылая стрекоза *(Zygoptera)*
drone ~ ильница-пчеловидка *(Eristalis tenax)*
flesh ~ серая мясная муха *(Sarcophagidae)*
forest ~ кровососка лошадиная *(Hippobosca equina)*
frit ~ мушка шведская *(Oscinella frit); pl* злаковые мушки *(Chloropidae)*
fruit ~ дрозофила, плодовая мушка *(Drosophila)*
green ~ тля *(Aphididae)*
green-bottle ~ зелёная падальница *(Lucilia)*
harlequin ~ звонец *(Chironomidae)*
hawk ~ ктырь *(Asilidae)*
Hessian ~ гессенская мушка *(Mayetiola destructor)*
horn ~ жигалка коровья малая *(Haematobia irritans)*
horse bot ~ желудочный овод *(Gasterophilidae)*
lantern ~ фонарница *(Fulgoridae)*
large fruit ~ пестрокрылка *(Tephritidae)*
leaf-miner ~ минирующая мушка *(Agromyzidae)*
long-horned ~ длинноусое двукрылое *(Nematocera)*
louse ~ кровососка *(Hippoboscidae)*
May ~ подёнка *(Ephemeroptera)*

megalopteran ~ вислокрылка *(Megaloptera)*
onion bulb ~ журчалка луковая *(Eumerus strigatus)*
robber ~ ктырь *(Asilidae)*
sand ~ мошка *(Simuliidae)*
scorpion ~ скорпионовая муха, скорпионница *(Mecoptera)*
shore ~ береговушка *(Ephydridae)*
short-horned ~ короткоусое двукрылое *(Brachycera)*
snake ~ верблюдка *(Raphidioptera)*
soldier ~ львинка *(Stratiomyidae)*
Spanish ~ шпанка ясеневая, шпанская мушка *(Lytta vesicatoria)*
spider ~ кровососка *(Hippoboscidae)*
stable ~ жигалка осенняя *(Stomoxys calcitrans)*
stone ~ веснянка *(Plecoptera)*
straight-seamed ~ прямошовная муха *(Orthorrhapha)*
syrphid ~ журчалка *(Syrphidae)*
tachina ~, tachinid ~ тахина, ежемуха *(Tachiniidae)*
throat ~ овод носовой олений *(Gastrophilus nasalis)*
tsetse ~ муха це-це *(Glossina)*

flycatcher *s orn* мухоловка *(Muscicapidae)*
flytrap *s* мухоловка, ловушка для мух
flyway *s* пролётный путь *(перелётных птиц)*
foal *s* жеребёнок, ослёнок; жеребиться
foam *s* пена; *v* пениться
foaming *s* пенообразование
foamy *a* пенистый, пенящийся
focal *a* фокальный, очаговый
focalisation *s* локализация, ограничение распространения
focus *s* фокус, средоточие, центр, очаг *(болезненного процесса)*
focusing *s* фокусирование
isoelectric ~ изоэлектрическое фокусирование, изоэлектрофокусирование
fodder *s* корм; грубый корм; кормовое растение
foetus *s* плод
fog *s* отава, подрост; туман; *bot* бухарник *(Holcus)*
fogging *s* туманообразование; мелкокапельное опрыскивание
foil *s* крышка чашки Петри; след зверя
folacin *s* фолиевая кислота
fold *s* складка, сгиб; загон *(for cattle);* овчарня; *v* складывать, сгибать, загибать
folded *a* складчатый
folding *s* образование складок, укладка, сворачивание
three-dimensional ~ *mbiol* трёхмерная укладка, трёхмерная конформация
foliaceous *a* лиственный; имеющий листву, облиственный; листовидный, листообразный
foliage *s* листья; листва
foliar *a* лиственный
foliate *a* листовидный; *v* выходить в лист; покрываться листьями
foliation *s* листосложение, листовое почкосложение
foliicolous живущий на листьях
foliole *s* листочек; *zool* листообразный орган
foliose *a* густо облиственный
folium *s anat* листок
follicle *s* фолликул, пузырёк; мешочек; сумка; *bot* листовка; кокон
aggregated lymphatic ~s обобщённые лимфатические фолликуы, бляшки Пейера
dental ~ зубной фолликул, зубной мешочек
hair ~ волосяной фолликул, волосяной мешочек
follicular *a* фолликулярный, фолликулообразный
folliculogenesis *s* фолликулогенез
follower *s* «ведомый»; следящее устройство, следящая система
follow-up *s* контроль, учёт; последующее наблюдение
fontanel, fontanelle *s anat* родничок
fonticulus *s anat* родничок; яремная вырезка
food *s* пища; питание; корм
larval ~ *apic* личиночный корм, молочко
food-begging *s* выпрашивание пищи
foodstuff *s* пища, корм, пищевой продукт; кормовой продукт
foody *a* съедобный, пищевой; питательный
foot *s* ступня; нога; лапа

club ~ *phytop* кила *(caused by Plasmodiophora brassicae)*
mouth ~ ногочелюсть
footprint *s* след
footstalk *s ent* стипес, стволик максиллы; стебель; черешок, цветоножка
forage *s eth* фуражировка, кормодобывание; фуражировать, добывать корм; *apic* взяток
forager *s* пчела-сборщица
foraging *s eth* фуражировка, кормодобывание
foram *s* фораминифера *(Foraminifera)*
foramen *s* отверстие
foraminifer, foraminiferan *s* фораминифера *(Foraminifera)*
foraminiferal *a* относящийся к фораминиферам
foraminous *a* дырчатый, имеющий отверстия, перфорированный
forbs *s* разнотравье
force *s* сила; *v* выгонять *(plants)*
germinating ~ энергия прорастания
stacking ~ стэкинг-взаимодействие
forceps *s anat* форцепс; *ent* клещи; пинцет
forcer *s* выгоночное растение
forcing *s* выгонка
forcipate *a* расщеплённый; вильчатый; клещевидный
fore *s* передняя нога; передняя лапа
forearm *s* предплечье
forebear *s* предок; предшественник
forebody *s* передняя часть тела
forebrain *s* передний мозг
forecast *s* прогноз; *v* прогнозировать
forefinger *s* указательный палец
foregut *s* передняя кишка; *emb* головная кишка
forehead *s* лоб
foreign *a* инородный, чужеродный, посторонний, привнесённый
foreignness *s imm* степень чужеродности
forekidney *s emb* пронефрос, предпочка, головная почка
foreleg *s* передняя нога; передняя лапа
forelock *s* чёлка *(of a hoprse)*
foremilk *s* молозиво, колострум
foremolar *s* премолярный зуб, премоляр
foreskin *s* крайняя плоть, препуций
forespore *s* предспора
forest *s* лес
alder ~ ольшаник
aspen ~ осинник
bilberry pine ~ сосняк-черничник
bilberry scrub ~ ельник-черничник
birch ~ берёзовый лес, березняк
bog moss pine ~ сфагновый сосняк
boreal coniferous ~ тайга, бореальный хвойный лес
broad-leaved ~ широколиственный лес, чернолесье
burned-out ~ гарь
cedar ~ кедровник
conifer, coniferous ~ хвойный лес, краснолесье
cowberry spruce ~ ельник-брусничник
croocked ~ криволесье
deer ~ олений заповедник
dense ~ глухой лес
fir ~ пихтарник, пихтовый лес
grass pine ~ травяно-болотный сосняк
grass spruce ~ травяной ельник
haircap-moss spruce ~ ельник-долгомошник
herbaceous-spruce ~ травяной ельник
hornbeam ~ грабовник, грабовый лес
hornbeam-oak ~ грабовый дубняк, дубово-грабовый лес
"islands" ~ лесные острова
larch ~ лиственничник, лиственничный лес
lichen pine ~ сосняк-беломошник, лишайниковый сосняк
light ~ редкостойный лес, редколесье
lime-tree ~ липняк
low ~ мелколесье, низколесье
maple ~ кленовник, кленовый лес
mountain rain ~ влажный тропический горный лес, горная гилея
needle-leaved ~ хвойный лес, краснолесье
oak ~ дубрава
overaged ~ перестойное лесонасаждение
palm ~ пальмовник, пальмовая роща, пальмовый лес
pine ~ сосновый лес, сосняк, бор
pine grass ~ травяной бор
polytric pine ~ сосняк-долгомошник
recreational ~ лесопарк
riparian ~ прибрежный лес, тугай, урема
savanna ~ саванновое редколесье

scrub ~ кустарниковое редколесье
sedge-grass scrub ~ осоково-злаковый ельник
shelter ~ защитное лесонасаждение
small ~ *см.* **low forest**
softwood ~ хвойный лес, краснолесье
spruce ~ еловый лес, ельник
tapestry ~ ковровое редколесье
thorn ~ колючий лес; колючее редколесье
wood sorrel spruce ~ ельник-кисличник
forestation *s* облесение, лесонасаждение
forest-clad *a* лесистый
forester *s* лесничий; обитатель леса
forest-meadow *a* лесо-луговой
forestomach *s* кардиальный отдел желудка; *orn* железистый желудок
forest-range *s* лесное пастбище
forestry *s* лесоводство; лесное хозяйство; лесничество
farm ~ агролесоводство
forest-steppe *s* лесостепь
forest-tundra *s* лесотундра
foretooth *s* передний зуб
foretop *s* чёлка *(of a hoprse)*
fork *s* разветвление, ответвление; раздвоение; *bot* вилочка; *mbiol* вилка
forking *s* вильчатое ветвление, дихотомическое ветвление
form *s* форма; разновидность; внешний вид; очертание; формировать
ancestral ~ предок, предковая форма
biologic ~ биологическая форма; физиологическая раса
growth ~ форма роста, характер роста
interconnected ~s *pl* близкие формы, родственные формы
L- ~ *micr* L-форма
land ~ форма рельефа; ландшафт
smooth ~ *micr* S-форма, гладкая форма
formaldehyde, formalin *s* формалин, формальдегид
formant *s* формант *(in bioacoustics)*
formation *s* формация; ассоциация; формирование, образование, создание; строение, структура
antibody ~ образование антител
blood ~ кровообразование, кроветворение, гемопоэз
bog ~ заболачивание
bone ~ остеогенез, костеобразование
complex ~ *bioch* комплексообразование
cushion plant ~ подушечник
ear ~ выколашивание
elfin wood ~ стланик
helix ~ спирализация
leaf-tube выход в трубку
meadow ~ залужение; луговая формация
new ~ новообразование
ring ~ *bioch* циклизация, образование цикла
rouleaux ~ образование «монетных столбиков»
sod ~ задернение
soil ~ почвообразование
spore ~ спорообразование, спорогенез
steppe ~ остепнение
swamp ~ заболачивание
thorn cushion plant ~ колючеподушечник
formative *a* способный расти, способный развиваться
former *s*:
acid ~ кислотообразующая бактерия
gradient ~ градиентный смеситель
spore ~ спорообразующий микроорганизм
formicary *s* муравейник
formimidation *s* формимидирование
formion *s* формион
formless *a* бесформенный
formol *s* формалин
formolation *s* обработка формалином
formol-fixation *s* фиксация в формалине
formolysis *s* формолиз
formula *s* формула; рецепт
antigenic ~ номенклатура антигенов
formylation *s* формилирование
fornicate, fornicated *a* выгнутый, сводчатый, изогнутый
fornicomissure *s anat* спайка свода, лира Давида
fornix *s* форникс; свод головного мозга; *bot* дуговидная чешуйка
fortification *s* усиление, укрепление, обогащение
vitamin ~ витаминизация
forward *v* ускорять; помогать, способствовать; *a* ранний
forwardness *s* раннее развитие
fossa *s* ямка, впадина; канал

hypophyseous ~ of sphenoid *anat* гипофизарная ямка, турецкое седло
oral ~ ротовая ямка, стомодеум
pituitary ~, sellar ~ *anat* гипофизарная ямка, турецкое седло
fossil *s* ископаемое; *a* ископаемый
fossilization *s* фоссилизация
fossores *s pl* роющие осы *(Sphecidae)*
fossorial *a* роющий, копающий
foster *v* кормить, выкармливать; выхаживать; ухаживать
foul *a* загрязнённый; засорённый
fouling *s* загрязнение, засорение; обрастание
foumart *s* чёрный хорёк *(Mustela putorius)*
four-footed *a* четвероногий
fovea *s* ямка, впадина
foveate *a* вдавленный; ямчатый
foveola *s* ямка, ямочка
foveolate, foveolated *a* мелкоямчатый
foveo-reticulate *a* ячеисто-сетчатый
fovilla *s* содержимое пыльцевого зерна
fowl *s* птица; пернатая дичь; домашняя птица
common scrub ~ джунглевая курица *(Megapodius freycinet)*
guinea ~ цесарка *(Numida)*
jungle ~ кустарниковая курица, дикая курица *(Gallus);* банкивская курица *(Gallus gallus);* джунглевая курица *(Megapodius freycinet)*
migratory ~ перелётная птица
scrub ~ джунглевая курица *(Megapodius freycinet)*
spur ~ южноазиатская куропатка *(Galloperdix)*
fox *s* лисица *(Vulpes);* серая лисица *(Urocyon);* песец *(Alopex);* гиеновидная собака *(Lycaon)*
Arctic ~ песец *(Alopex)*
black ~ илька, пекан *(Martes pennanti)*
blue ~ голубой песец, песец в летнем меху
corsac ~, dog ~ корсак, степная лисица *(Vulpes corsac)*
fennec ~ фенек *(Fennecus zerda)*
large flying ~ большая летучая лисица, калонг *(Pteropus vampyrus)*
maned ~ гривистый волк *(Canis jubatus)*
polar ~ песец *(Alopex)*
Samson ~ лисица «самсон»
white ~ песец *(Alopex)*
fraction *s* фракция
gel ~ гель-фракция
fractionation *s* фракционирование
salt ~ фракционирование солями; высаливание
fractoserial *a bot* прерывисто-рядный
fracture *s* разрыв; трещина; излом; перелом
fragile *a* ломкий, хрупкий
fragility *s* ломкость, хрупкость
fragment *s* фрагмент
blunt-end ~s фрагменты с «тупыми» концами
Fab ~ антисвязывающий домен, Fab-фрагмент, Fab-область
Fc ~ *imm* Fc-фрагмент, Fc-область
fragmentation *s* фрагментация, деление на фрагменты
fragmentimurate *a bot* прерывисто-перегородчатый
fragmentography *s*: **mass** ~ масс-фрагментография
fragrance *s* аромат, благоухание
fragrant *a* ароматный, благоухающий
frail *a* хилый, слабый; хрупкий
frame *s* остов; скелет; каркас; телосложение; рама, рамка; кадр
bony ~ костный остов, скелет
frameshift *s* сдвиг рамки считывания
framework *s* остов; каркас; решётка; сеть; *bot* ситечко
francolin *s orn* турач, франколин *(Francolinus)*
frank *a* явный, выраженный
frankincense *s* ладан
frass *s ent* буровая мука; экскременты насекомых
freckle *s* пятнышко, крапинка
free *a* свободный, неприкреплённый; вольный
free-living *a* свободноживущий
freemartin *s* фримартин
free-nitrogenous *a* безазотистый
free-ranging *a* находящийся на свободе
free-veined *a bot* свободнонервный
freeze-drying *s* лиофилизация, сушка сублимацией, замораживание - высушивание
freeze-etching *s* замораживание - травление

freeze-fracture *s* криоскалывание, замораживание є скалывание
freeze-substitution *s* криозамещение, холодовое замещение
freezing *s* замораживание; вымораживание; замерзание; *eth* замирание
frenate *a ent* имеющий зацепку
frenulum *s ent* зацепка, уздечка
frenum *s anat* уздечка
frequency *s* частота; повторяемость; концентрация
~ **of occurrence** встречаемость
~ **of spawning** повторность нереста
length ~ частотное распределение длин, размерный состав
fresh *a* свежий; пресный
freshening *s* опреснение
fret *s* раздражение, волнение; ссадина; разъедание; *v* разъедать, подтачивать
fretter *s*: **vine** ~ *энт.* филлоксера *(Phylloxera vastatrix)*
friable *a* рыхлый; ломкий; хрупкий
friagem *s* вторжение холодного воздуха в тропики
frigatebird *s orn* фрегат *(Fregata)*
frigostable *a* холодоустойчивый
frill *s anat* брыжейка
fringe *s* бахрома; ворсинки
fringed *a* бахромчатый; окаймлённый
fringe-flowered *a* реснитчато-цветковый
fringe-fruited *a* реснитчато-плодный
fringe-leaved *a* реснитчато-листный
fringe-petaled *a* бахромчато-лепестный
frisk *s* прыжок; *v* скакать, резвиться *(of a horse)*
frog *s* лягушка; стрелка копыта
ant ~ украшенная узкоротая квакша *(Microhyla ornata)*
arrow-poison ~ древолаз *(Dendrobates)*
balloon ~ шарообразный узкорот *(Uperodon)*
banana ~ бананница *(Afrixalus)*
banded rubber ~ краснополосый узкорот *(Phrynomerus bifasciatus)*
barking ~ лающая жаба *(Hylactophryne augusti)*
blacksmith tree ~ квакша-кузнец *(Hyla faber)*
bone-headed ~ панцирноголов *(Trachycephalus)*
tree ~ квакша, древесница *(Hyla);* африканская древесница *(Leptopelis)*
frogbit *s bot* лягушатник, водокрас *(Hydrocharis)*
frogeye *s phytop* кольцевая пятнистость
frog-fly *s ent* цикадка *(Cicadellidae)*
froghopper *s ent* пенница, слюнявица *(Cercopidae)*
frog's-bit *s bot* лягушатник, водокрас *(Hydrocharis)*
frogspawn *s* лягушачья икра
frond *s* вайя, лист папоротника; лист пальмы; таллом, слоевище
frondesce *v* выпускать листья; покрываться листьями
frondescence *s* облиствение
frondiferous *a* листоносный
frons, front *s* лоб; передний зуб, резец
frontal *a* лобный; передний; фронтальный
frontlet *s* пятно на лбу животного
frontomalar *a* лобно-скуловой
frontomaxillary *a* лобно-верхнечелюстной
frontonasal *a* лобно-носовой
fronto-occipital *a* лобно-затылочный
frontoparietal *a* лобно-теменной
frontotemporal *a* лобно-височный
front-tooth *s* передний зуб; резец
frost *s* мороз; морозобойный; заморозок; мерзлота *(in soil);* замораживание; замерзание; *v* замораживать; замерзать; побивать морозом
frost-bitten *a* побитый морозом
frost-hardy *a* морозостойкий
frost-killing *s* вымерзание
frostproof, frost-resistant *a* морозостойкий, морозоустойчивый
frost-shattered *a* разрушенный морозом; морозобойный
froth *s* пена; вспенивание; пенообразование; *v* пениться
frother *s* пенообразователь
fructescent *a* начинающий плодоносить
fructiferous *a* плодоносящий
fructification *s* плодоношение
fructify *a* плодоносить
frugivorous *a* плодоядный
fruit *s* плод; фрукты; *v* плодоносить; **to bear** ~ плодоносить; **to set the ~s** завязывать плоды
aggregate-accessory ~ многокостянка

collective ~ соплодие
drupaceous ~ косточковый плод, костянка
key ~ крылатка
multiple ~ соплодие
samara ~ крылатка
schizocarpic ~, secedenous ~ схизокарпий, дробный плод
separating ~ схизокарпий, дробный плод
stone ~ косточковый плод, костянка
windfall ~ падалица
winged ~ крылатка

fruitage *s* плодоношение; фрукты, плоды
fruit-bearing *a* плодоносный, плодоносящий
fruitcase *s bot* коробочка
fruited *a* с плодами
closely ~ густоплодный
prickly ~ колючеплодный
woolly ~ пушистоплодный

fruitful *a* плодородный; плодоносный; плодовитый
fruiticolous *a* поражающий плоды
fruitification *s* плодообразование; плодоношение
fruitless *a* бесплодный, неплодоносящий
fruit-set *s* завязывание плода
fruitstalk, fruitstem *s* плодоножка
fruitworm *s* гусеница-вредитель плодов
frustration *s* фрустрация
frustule *s* панцирь *(of diatoms)*
frutescent *a* кустарниковый
frutex *s* кустарник
fruticose *a* кустарниковый, кустистый
fruticulose *a* кустарничковый, мелкокустарниковый
fry *s* мальки, молодь
fuchsin *s* фуксин
carbolic ~ карбол-фуксин, карболовый фуксин, фуксин Циля

fuchsinophil *a* фуксинофильный, окрашивающийся фуксином
fuciphagous, fucivorous *a* питающийся водорослями
fugacious *a* быстро увядающий; быстро опадающий
fulcrate *a bot* имеющий опору
fulcrum *s* опора, ствол, стебель; *zool* опора, поддержка; спорофор *(in lichens)*
full-blooded *a* чистокровный; полнокровный
full-blown *a* полностью распустившийся
full-eared *a* с налитым колосом
full-fledged *a* оперившийся
full-grown *a* вполне развившийся
fullmouth *s* взрослое животное
full-term *a* доношенный
fulmar *s orn* глупыш *(Fulmarus)*
giant ~ гигантский буревестник *(Macronectes giganteus)*

fumigant *s* фумигант
fumigation *s* окуривание
function *s* функция; *v* функционировать
antixenic ~ реакция на инородное тело

functional *a* функциональный
Fund *s:* **World Wildlife** ~ Всемирный фонд дикой природы
fundatrix *s* тля - основательница колонии
fundus *s anat* дно; *bot* донце
fungaceous *a* грибовидный
fungal *a* грибной; грибковый
fungate *v* грибовидно разрастаться
fungic *a* грибной; грибковый
fungicide *s* фунгицид
fungicolous *a* поражающий грибы
fungistasis *s* фунгистатическое действие
fungoid *a* напоминающий гриб; губчатый, ноздреватый
fungous *a* грибной; грибковый; губчатый, ноздреватый
fungus *s* гриб; грибок
basidium ~ базидиомицет, базидиальный гриб *(Basidiomycetes)*
cup ~ сумчатый гриб *(Ascomycetes)*
filamentous ~ мицелиальный гриб; гифомицет *(Hyphomycetes)*
honey ~ настоящий опёнок *(Armillaria mellea)*
Irish potato famine ~ фитофтора картофеля *(Phytophthora infestans)*
oyster ~ вешенка обыкновенная *(Pleurotus ostreatus)*
palisada ~ базидиомицет, базидиальный гриб *(Basidiomycetes)*
ray ~ лучистый гриб, актиномицет
saddle ~ строчок *(Helvella)*
shelf ~ берёзовый гриб, чага *(Polyporus sulphureus)*
shoestring ~ настоящий опёнок *(Armillaria mellea)*

slime ~ миксомицет *(Myxomycetes)*
tinder ~ трутовик настоящий *(Fomes fomentarius)*
funicle *s* пуповина; семенной канатик; семяножка; *ent* жгутик
funiculose *a* семяножковый
funiculus *s* пуповина; семенной канатик; семяножка; *ent* жгутик
funis *s* пуповина
funnel *s* воронка
funnelform *a* воронкообразный, воронковидный
fur *s* мех; шкурка; меховой; шерсть; пушок *(on fruit); pl* пушнина
fur-bearing *a* пушной
furca *s ent* фурка; *bot* вилка, развилина
furcate *a* вильчатый; разветветвлённый
furcation *s* разветветвление, раздвоение
furcellate *a bot* мелковильчатый
furcula *s rn* вилочка
furfur *s* чешуйка
furfuraceous *a* покрытый чешуйками
furfurellous *a* мелкочешуйчатый
furrow *s* борозда, бороздка
axillary ~ аксиллярная линия
visceral ~ *emb* жаберный карман
furrowed *a* бороздчатый
furrow-leaved *a* бороздчатолистный
furrow-seeded *a* бороздчатосемянный
furrow-stemmed *a* бороздчатостебельный
fusiform *a* веретеновидный, веретенообразный
fusion *s* слияние
fusocellular *a bot* веретеноклеточный
fust *s* запах плесени
fuzz *s bot* волоски, бородка; пух; пушок
fuzzball *s myc* дождевик *(Lycoperdon bovista)*
fuzzy *a* пушистый, опушённый, покрытый пушком

G

GABA-ergic *a* ГАМК-эргический
gabble *v* кудахтать, гоготать; *s* кудахтанье, гоготанье
gadding *a* ползучий
gadfly *s* слепень *(Tabanus)*
gadwall *s* серая утка *(Anas strepera)*
gag *s* ген капсидных белков
gage *s* измерительный прибор; датчик
gaggle *v* гоготать; *s* гоготание *(of geese)*
gain *m* прирост, увеличение; *v* увеличиваться; прибавлять; ~ **in weight** прибавлять в весе
antigen ~ антигенная конверсия
production ~ увеличение продуктивности
gait *s* походка; аллюр
two-beat leaped ~ резвая рысь
galactic *a* молочный
galactoblast *s* молозивное тельце
galactogenic *a* вызывающий образование молока
galactophagous *a* питающийся молоком
galactophore *s* млечный проток
galactophorous *a* образующий молоко
galactopoiesis *s* образование молока
galactosis *s* секреция молока, лактация
galah *s* розовый какаду *(Eolophus roseicapillus)*
galberry *s bot* шишкоягода
galea *s ent* галеа; *bot* шлем, шлемовидная часть
galeate, galeated *bot* шлемовидный
gall *s* жёлчь; *phytop* галл
crown ~ корончатый галл; корневой рак *(caused by Dendrophagus globosus)*
root ~ корневой галл, корневой клубенёк
gallbladder *s* жёлчный пузырь
gallery *s ent* ход
gallfly *s* орехотворка *(Cynipidae)*; ходообразующее насекомое; *pl* галлицы *(Cecidomyidae)*
gall-inducing *a* вызывающий образование галлов
galling *s* галлообразование
~ **of roots** *s* галлообразование на корнях
gallipot *s* живица, еловая смола
gallivorous *a* поедающий галлы
gallnut *s phytop* орехоподобный галл, «дубильный орешек»
gallop *s* галоп; скакать галопом
cross-legged ~ контргалоп
racing ~ резвый галоп, карьер
galvanocontractility *s* электросократимость
galvanotaxis *s* гальванотаксис
galvanotonus *s* гальванотонус, электротон
galvanotropism *s* гальванотропизм
game *s* дичь; *eth* игра; ~ **bird** охотничье-промысловая птица
black ~ тетерев *(Lyrurus tetrix)*
fur ~ пушные звери
ground ~ зайцы; кролики
gamekeeper *s* егерь
gametangium *s* гаметангий
gamete *s* гамета, половая клетка
gametic *a* гаметический
gametocyst *s* гаметоциста
gametocyte *s* гаметоцит
gametogamy *s* гаметогамия
gametogenesis *s* гаметогенез
gametonucleus *s* ядро гаметы
gametophore *s* гаметофор
gametophyll *s* гаметофилл
gametophyte *s* гаметофит
gametotoky *s* гаметотокия, дейтеротокия, амфитокия
gamic *a* оплодотворённый; половой
gamobium *s* гамобиум
gamocarpellarous *a* со слитными плодолистиками, слитнокарпельный
gamocarpous *a* спайноплодный
gamocyst *s* гаметоциста; гамонтоциста
gamodeme *s* гамодем
gamogenesis *s* гамогенез, гамогония, половое размножение
gamogenetic *a* гамогенетический
gamogony *s* гамогенез, гамогония, половое размножение
gamone *s* гамон
gamont *s prot* гамонт, гаметоцит
gamontogamy *s* гамонтогамия
gamopetalous *a* спайнолепестковый
gamophase *s* гамофаза
gamophyllous *a* спайнолистный, сростнолистный
gamophylly *s bot* гамофиллия
gamosepalous *a* спайночашелистиковый

gamostaminate *a* спайнотычинковый
gamostele *s* гамостела
gamostelic *a* гамостелический, слитностелевый
gamostely *s* гамостелия
gamostylous *a* спайностолбиковый
gamotropism *s* гамотропизм
gander *s* гусак
gangliar *a* ганглионарный
gangliate, gangliated *a* имеющий ганглий
ganglioblast *s* ганглиобласт
gangliocyte *s* ганглиоцит
ganglion *s* ганглий, нервный узел
  auditory ~ спиральный ганглий улитки, кортиев узел
  Bock's ~ каротидный ганглий, сонный узел
  cochlear ~ спиральный ганглий улитки, кортиев узел
  Darkshevich's ~ заднекомиссуральное ядро, ядро Даркшевича
  Remak's ~ ремаков узел
ganglionary *a* ганглионарный
ganglioneuron *s* нейрон ганглия
ganglionic *a* ганглионарный
gannet *s* олуша *(Sula)*
ganoblast *s* ганобласт, амелобласт, адамантобласт
ganoid *a* ганоидный
gap *s* разрыв, перерыв, дизъюнкция; разрыв, пробел, гэп; прогалина
  branch ~ *bot* прорыв ветвления
  interlabial ~ межгубная щель, межгубной промежуток
  leaf ~ лакуна, листовой прорыв
  synaptic ~ синаптическая щель
gape *s* зияние; *v* широко открывать рот; зевать
gaping *s* раскрывание клюва
gapped *a mbiol* содержащий разрывы, содержащий пробелы
garden *s* сад
  flower ~ цветник
  nursery ~ питомник, рассадник
  zoological ~ зоопарк
gardener *s* садовник; *orn* садовник *(Amblyornis)*
gardening *s* садоводство
  settlement ~ озеленение
garganey *s* чирок-трескунок *(Anas querquedula)*
garlic *s* чеснок *(Allium sativum)*
garrot *s* закрутка, жгут; *orn* гоголь обыкновенный *(Bucephala clangula)*
gas *s* газ; *v* выделять газ; насыщать газом
  manure ~ биогаз
gaseous *a* газовый; газообразный
gasometry *s* газометрия; газометрический анализ
gaster *s* живот; брюшко
gastraea *s* гастрея
gastraeum *s* брюшная сторона тела
gastral *a* гастральный, желудочный; брюшной
gastralia *s pl* гастральные нити *(in Coelenterata);* брюшные рёбра *(in reptiles)*
gastric *a* гастральный, желудочный; расположенный в области желудка
gastrocardiac *a* гастрокардиальный, желудочно-сердечный
gastrocoel *s* гастроцель, гастральная полость
gastrocolic *a* желудочно-ободочный
gastroenteric *a* желудочно-кишечный
gastroesophagal *a* желудочно-пищеводный
gastrogenic *a* гастрогенный, желудочного происхождения
gastrointestinal *a* желудочно-кишечный
gastropulmonary *a* желудочно-лёгочный
gastrovascular *a* гастроваскулярный
gastrula *s* гаструла
gastrulation *s* гаструляция
gather *s* сбор, урожай; *v* снимать,собирать, убирать; рвать *(flowers)*; собирать *(fruit)*
gatherer *s* собиратель
  nitrogen ~ азотонакопитель, азотособиратель
gathering *s* собирание
  honig ~ *apic* взяток
gating *s* воротный механизм
gauge *s* измерительный прибор; *v* измерять, мерить
  depth ~ глубиномер
  pressure ~ манометр
  rain ~ дождемер, плювиометр
  weather ~ барометр
  wind ~ анемометр
gauging *s* измерение
gaunt *a* измождённый, исхудалый
gaur *s ther* гаур *(Bos gaurus)*
gayal *s ther* гаял *(Bibos frontalis)*

gazelle *s ther* газель *(Gazella)*
Addra ~ *см.* **dama gazelle**
black-tailed ~ дзерен *(Procapra)*
Clarke's ~ дибатаг *(Ammodorcas clarkei)*
dama ~ газель-дама; сахарская газель *(Gazella dama)*
dorcas ~ газель-доркас *(Gazella dorcas)*
giraffe ~ геренук, жирафовая газель *(Litocranius walleri)*
goitred ~ джейран *(Gazella subgutturosa)*
granulated ~ *см.* **goitred gazelle**
Mongolian ~ монгольский дзерен, зобастая антилопа *(Procapra gutturosa)*
gecko *s herp* геккон *(Gekko)*
geitonogamy *s* гейтоногамия
gel *s* гель
polyacrylamide ~ полиакриламидный гель, ПААГ
silica ~ гель окиси кремния, силикагель
gelated *a* желатинизирующий; студневидный
gelatigenous *a* образующий студнеобразное вещество
gelatin, gelatine *s* желатин; студень
~ **of Wharton** вартонов студень
beef-extract ~ мясопептонный желатин, МПЖ
wort ~ сусло-желатин
gelatinization *s* желатинизация; остудневание, превращение в студень
gelatinous *a* желатиновый; желатинозный; студенистый
gelation *s* превращение в гель, желатинирование
geld *v* кастрировать; *a* кастрированный
gelding *s* кастрирование; кастрированное животное
gemellus *s* близнецовая мышца
geminate, geminated *a* сдвоенный, двойной, парный
gemini *s pl* близнецы; хромосомы бивалента; бивалент в мейозе
geminicolpate *a pal* сдвоенно-бороздный, парнобороздный
geminiflorous *a* парноцветковый
geminivirus *s* геминивирус
gemma *s* листовая почка; зародышевая почка; *zool* почка
gemmaceous *a* почковый
gemmate *a* размножающийся почкованием; имеющий почки
gemmation *s* почкование, размножение почкованием; образование почек
gemmiferous *a* образующий почки; несущий почки
gemmule *s* геммула; маленькая почка, почечка
gemsa *s ther* серна *(Rupicapra rupicapra)*
gemsbok *s ther* сернобык *(Oryx gazella)*
genacerores *s pl ent* циркумгенитальные железы
gena *s* щека
genal *a* щёчный, относящийся к щекам
gender *s* пол
gene *s* ген
alpha ~ ген иммуноглобулина A; ген a-рецептора, a-ген
beta ~ ген b-рецептора, b-ген
delta ~ ген иммуноглобулина D; ген d-рецептора, d-ген
equilocal **~s** *pl* эквилокальные гены, аллели
gamma ~ ген иммуноглобулина G; ген c-рецептора, c-ген
Ir ~ ген иммунного ответа, IR-ген
jumping ~ «прыгающий» ген, транспозон
key ~ главный ген, основной ген; олигоген
major ~, master ~ главный ген, основной ген; олигоген
minor ~ малый ген, полиген
para-allelic **~s** *pl* параллели, псевдоаллели
semi-allelic **~s** *pl* семиаллели, псевдоаллели
twin **~s** *pl* параллели, псевдоаллели
genealogy *s* генеалогия
genepool *s* генофонд
generalization *s* генерализация; обобщение; распространение
generate *v* генерировать, производить, порождать
generation *s* генерация, поколение; род; потомство; размножение
antithetic **~s** *pl* антитетические генерации
rising ~ потомство, подрастающее поколение
spontaneous ~ абиогенез, самопроизвольное зарождение

generative *a* генеративный
generic *a* родовой
genesic *a* генеративный
genesis *s* генез, возникновение, происхождение
gene-spread *s* поток генов
genetic, genetical *a* генетический
genetics *s* генетика
genial *a* подбородочный
genic *a* генный
genicular *a* коленный
geniculate *a* коленчатый
genioglossal *a* подбородочно-язычный
geniohyoid *a* подбородочно-подъязычный
genital *a* генитальный, половой
genitals *s pl* гениталии, половые органы, половой аппарат
genitourinary *a* мочеполовой
genocline *s* геноклин
genocopy *s* генокопия
genofond *s* генофонд
genome *s* геном; *a* геномный
genomere *s* геномер
genomic *a* геномный
genonema *s* генонема, хромонема
genonymum *s* геноним, родовое название
genophene *s* генофен
genospecies *s* геновид
genotoxic *a* генотоксический
genotype *s* генотип, идиотип
genotypic, genotypical *a* генотипический
genovariation *s* геновариация, генная мутация
gentle *a* послушный, смирный
genual *a* коленный
genucalcaneal *a* коленно-пяточный
genuflexuous *a bot* коленчато-изогнутый
genuineness *s* подлинность, чистопородность, чистолинейность
~ **of strain** чистопородность
genus *s* род
genyantrum *s* верхнечелюстная пазуха, гайморова пазуха
geobiocoenosis *s* биогеоценоз
geobiont *s* геобионт, обитатель почвы
geobios *s* совокупность обитателей почвы
geobotanical *a* геоботанический
geobotany *s* геоботаника
geocarpic *a* геокарпический
geocentric *a* геоцентрический
geocryptophyte *s* геокриптофит
geodiotropism *s* геодиотропизм
geoecotype *s* геоэкотип
geogenous *a* обитающий на почве, обитающий в почве
geography *s* география
gene ~ геногеография
zoological ~ зоогеография
geohelminth *s* геогельминт
geometer, geometrid *s ent* пяденица *(Geometridae)*
geonasty *s* геонастия
geonemy *s* хорология
geophagous *a* питающийся почвой
geophilic, geophilous *a* обитающий на почве, обитающий в почве, растущий в почве
geophily *s* геофилия
geophyte *s* геофит, наземное растение
geophytia *s* сообщество геофитов
geoponic, geoponical *a* сельскохозяйственный; земледельческий
georgina *s* георгин, георгина *(Dahlia)*
geosere *s* геологическая серия
geotaxis *s* геотаксис
geotropism *s* геотропизм
geotype *s* геоэкотип
geoxene *s* геоксен
gerbicide *s* гербицид
gerbil, gerbille *s ther* песчанка; карликовая песчанка *(Gerbillus)*
gerenuk *s ther* геренук, жирафовая газель *(Litocranius walleri)*
germ *s* зародыш, семя; проросток; зачаток; почка; микроорганизм; инфицирующее начало; *v* зарождаться
disease-producing [pathogenic] ~ патогенный микроорганизм
tooth ~ зубной зачаток
germarium *s* яичник *(of rotifers); ent* гермарий
germ-free *a* стерильный, безмикробный, безбактериальный, аксенический
germicidal *a* бактерицидный
germicide *s* гермицид, бактерицидный препарат
germinability *s* всхожесть
germinable *a* всхожий

germinal *s* герминалий, место прорастания; *a* герминативный, зародышевый, зачаточный
germinant *s* проросток
germinate *v* развиваться, прорастать, давать почки; порождать
germination *s* прорастание; образование поски; начало роста; зарождение; развитие; всхожесть
germinative *a* герминативный, зародышевый, зачаточный
germless *a* беззародышевый
gestation беременность
gesture *s* жест, выразительное движение; *v* жестикулировать
appeasement ~ умиротворяющее выразительное движение
begging просящий жест
contact initiating ~s *pl* движения, инициирующие социальные контакты
get *v* доставать, добывать; рожать, производить на свет; *s* потомство
ghoral *s ther* горал *(Nemorhaedus goral)*
ghost *s* «тень»
cell ~ «тень» клетки, разрушенная клетка
giantism *s* гигантизм
giardiasis *s* лямблиоз
gib *s* кот
gibbon *s* гиббон *(Hylobates)*
gibbous *a* горбатый; выпуклый
Giemsa *s* краситель Гимза
gigantism *s* гигантизм
gigantocyte *s* гигантоцит, мегалоцит, макроцит
gill *s* жабра; *orn* бородка, серёжки; *myc* гимениальная пластинка, спороносный слой; *v* потрошить рыбу, чистить рыбу
gill-bearing *a* имеющий жабры; *myc* пластинчатый
gill-breathing *a* дышащий жабрами, жабродышащий
gill-slit *s* жаберная щель
gilt *s* подсвинок; молодая свинья
ginger *s bot* имбирь лекарственный *(Zingiber officinale)*
gingiva *s* десна
gingival *a* десневой
ginglymus *s* блоковидный сустав
ginseng *s* женьшень *(Panax)*
giraffe *s* жираф *(Giraffa camelopardalis)*
girdle *s anat* пояс; поясок *(in diatoms); v* кольцевать
girth *s* обхват
gizzard *s orn* мускульный желудок
glabella *s* глабелла, надпереносье
glabrate *a* безволосый; гладкий; *bot* неопушённый, оголённый
glacial *a* ледниковый; ледовый; кристаллизованный
glaciation *s* оледенение; континентальное оледенение
glade *s* прогалина, просека
gladiate *a bot* мечевидный
gland *s* железа
~ of Lieberkuhn либеркюнова железа, кишечная крипта
adrenal ~ надпочечник
aggregate ~, agminated ~ *hist* пейерова бляшка
anterior pituitary ~ аденогипофиз, передняя доля гипофиза
arteriococcygeal ~ копчиковое тельце, копчиковый гломус
budlike ~ *hist* железистая почка
enteric ~ либеркюнова железа, кишечная крипта
genital ~, germ ~ половая железа, гонада
groin ~ паховая железа, паховой лимфатический узел
lymph, lymphatic ~ лимфатический узел
Mehlis' ~ *helm* тельце Мелиса
mushroom ~ *ent* грибовидное тело
nectar ~ *bot* нектарник
pineal ~ эпифиз мозга, шишковидная железа
pituitary ~ гипофиз
plexiform ~ осевой орган *(in echinoderms)*
posterior pituitary ~ нейрогипофиз, задняя доля гипофиза
prepituitary ~ аденогипофиз, передняя доля гипофиза
prostate ~ предстательная железа, простата
renal ~ надпочечник
seminal ~ семенная железа; семенник
sex ~ половая железа, гонада
solitary ~ солитарный фолликул
suprarenal ~ надпочечник

thymus ~ тимус, вилочковая железа, зобная железа
vitelline ~ желточная железа, желточник
water ~ *bot* гидатода
yolk ~ желточная железа, желточник
glandular *a* железистый
glandule *s* железка
glandulocyte *s* гландулоцит, железистая клетка
glandulose-serrate *a bot* железисто-пильчатый
glandulosous *a bot* железисто-опушённый
glans *s* жёлудь; головка
glareal *a* растущий на гравии
glass *s* стекло; склянка; стакан; лупа
bell ~ стеклянный колпак
cover ~ покровное стекло
dropping ~ капельница
graduated ~, measuring ~ мензурка; мерный стакан
test ~ пробирка
glasshouse *s* теплица
glass-lizard, glass-snake *s herp* желтопузик *(Pseudopus apodus)*
glasswing *s ent* стеклянница *(Aegeridae)*
glasswort *s bot* солянка *(Salsola);* солерос *(Salicornia)*
glassy *a* стекловидный; гиалиновый, прозрачный
glaucogonidium *s* сине-зелёная гонидия
glaucous *a* сизый; серовато-зелёный; покрытый беловатым налётом
gleba *s* глеба
glei *s* глей
gleization *s* оглеение, глееобразование
glenoid *a* суставной
gleocystidium *s* глеоцистида
gley *s* глей
glia *s* глия
gliacyte *s* глиоцит
glial *a* глиальный, глиоцитарный
glide *v* скользить
global *a* всемирный, глобальный
globe *s* шар; Земля
eye ~ глазное яблоко
globeflower *s bot* купальница *(Trollius);* купальница европейская *(Trollius europaeus)*
globocellular *a* круглоклеточный
globose *a* шаровидный, шарообразный, сферический
globular *a* шаровидный, шарообразный, сферический; содержащий шарики
globule *s* шарик
terminal ~ *hist* концевая пуговка
globulicidal *a* гемолитический
globulous *a* шариковидный, шарообразный, сферический
glochidiate *a bot* имеющий волоски с крючочками
glochidium *s* глохидий
gloeocarpous *a* глеоплодный
gloeocystidium *s* глеоцистида
glome *s bot* головка
glomerular *a* гломерулярный, клубочковый
glomerule *s hist* клубочек
glomerulus *s hist* клубочек
glomerus *s hist* клубок
cutaneous ~ кожный клубок, орган Гейера - Гроссера
glomus *s hist* клубок; *bot* головка
carotid ~ каротидное тельце
glossa *s* язык; *ent* глосса, язычок
glossal *a* язычный
glossarium *s ent* глоссарий
glossate *a ent* имеющий язычок
glossoepigglottidean *a* язычно-надгортанный
glossohyal *a* язычно-подъязычный
glossopharyngeal *a* языкоглоточный
glossopodium *s* глоссоподий
glottis *s* голосовая щель
glowworm *s* личинка светляка *(Lampyris noctiluca)*
gluconeogenesis *s* глюконеогенез
glucose-free *a* не содержащий глюкозу
glucosidation *s* глюкозидирование
glucosimeter *s* глюкозиметр
glucosome *s* глюкосома
glue *s* клей
gluma *s bot* колосковая чешуя; плёнка
glumaceous *a bot* состоящий из колосковых чешуй; плёнчатый
glume *s bot* колосковая чешуя; плёнка
glumiferous *a bot* плёнконосный
glumiflorous *a* плёнчатоцветный
glutaraldehyde *s* глутаральдегид
gluteal *a* ягодичный
gluten *s* клейковина
gluteus *s* ягодичная мышца

glutton *s* росомаха *(Gulo gulo)*
glycan *s* гликан, полисахарид
glycemia *s* гликемия
glyceraldehyde *s* глицериновый альдегид, глицеральдегид
glycerol *s* глицерин
glycocalix *s* гликокаликс
glycogen *s* гликоген, животный крахмал
glycogenesis *s* образование сахара
glycogenic *a* гликогенный
glycogenolysis *s* гликогенолиз
glycogenolytic *a* гликогенолитический
glycogenous *a* гликогенный
glycogeny *s* образование сахара
glycolysis *s* гликолиз
glycophyte *s* гликофит
glycoprival *a* безуглеводный
glycoprotein *s* гликопротеид
glycosaminoglycan *s* гликозаминогликан, ГАГ
glycosidation *s* гликозидирование
glycosilation *s* гликозилирование
glyoxysome *s* глиоксисома
glyptocarpous *a* бороздчатоплодный
glyptospermous *a* бороздчатосемянный
gnarl *s* узел; кап; нарост, свиль
gnarled *a* сучковатый; искривлённый; узловатый
gnash *v* щёлкать зубами *(e.g., of a dog)*
gnat *s* комар; кровососущее насекомое
  black ~ мокрец *(Ceratopogoniidae)*
  eye ~ злаковая мушка *(Chloropidae)*
  fungus ~ грибной комарик *(Fungivoridae)*
  gall ~ галлица *(Cecidomyidae)*
  resin ~ галлица смоляная *(Retinodiplosis resinicola)*
gnathic *a* челюстной
gnathion *s* гнатион
gnathobase *s* гнатобаза
gnathopod *s* гнатопод
gnathosoma *s* гнатосома
gnathostomatous *a* челюстноротый
gnathotheca *s orn* роговой покров подклювья
gnaw *v* грызть, глодать
gnawer *s* грызун
gnawing *s* грызение; обгладывание
gnesiogamy *s* гнесиогамия, внутривидовое оплодотворение
gnotobiont *s* гнотобионт
gnotobiotics *s* гнотобиотика
gnu *s ther* гну *(Connochaetes)*
goa *s ther* тибетский дзерен *(Procapra picticaudata)*
goanna *s* варан *(Varanus)*
goat *s* козёл; настоящий козёл *(Capra)*
goatling *s* козлёнок
goatsucker *s* козодой *(Caprimulgiformes)*
goaty *a* козлиный
gobble *v* жадно есть; пожирать; кулдыкать
gobbler *s* индюк
godwit *s orn* веретенник *(Limosa)*
goggle *a* выпученный
goiter *s* желвак; зоб
goldcrest *s orn* желтоголовый королёк *(Regulus regulus)*
goldeneye *s orn* гоголь *(Bucephala)*
goldenrod *s bot* золотарник *(Solidago)*
goldfinch *s* щегол *(Carduelis carduelis)*; американский чиж *(Spinus tristis)*
goldhamster *s* сирийский хомячок, золотистый хомячок *(Mesocricetus auratus)*
goldtail *s ent* златогузка *(Euproctis chrysorrhoea)*
gonad *s* гонада, половая железа
  female ~ яичник, женская гонада
  indifferent ~ недифференцированная половая железа
  male ~ семенник, мужская гонада
gonadogenesis *s* гонадогенез
gonadosomatic *a* гонадосоматический
gonadotrophin *s* гонадотропин, гонадотропный гормон
gonadotropic *a* гонадотропный
gonadotropin *s* гонадотропин, гонадотропный гормон
  chorionic ~ хорионический гонадотропный гормон, хорионгонадотропин
gonangium *s* гонангий
gonapophysis *s* гонапофиз
goneoclin *s* гонеоклин
gongylidia *s* вздутия гиф
gonidial, gonidic *a bot* гонидиальный
gonidioferous *a bot* образующий гонидин
gonidiophore *s* гонидофор
gonidium *s bot* гонидия
goniocarpous *a bot* угловато-плодный
goniocladous *a bot* угловато-ветвистый
goniophyllous *a bot* угловато-листный
goniospermous *a bot* угловато-семянный

gonochorism *s zool* раздельнополость; *bot* двудомность
gonochoristic *zool* раздельнополый; *bot* двудомный
gonocoel *s* гоноцель, полость гонады
gonocyte *s* гоноцит
gonoduct *s* гонодукт, половой проток
gonogenesis *s* гаметогенез
gonomery *s* гономерия
gononephrotome *s emb* гононефротом
gonophage *s* гонофаг
gonophore *s* гонофор
gonosome *s* гоносома, половая хромосома
gonotheca *s* гонотека
gonotyl *s* половая присоска
goosander *s orn* большой крохаль *(Mergus merganser)*
goose *s* гусь; гусь *(Anser)*
 bean ~ гуменник *(Anser fabalis)*
 emperor ~ белошей *(Philacte canagica)*
 lesser white-fronted ~ пискулька *(Anser arythropus)*
 pygmy ~ малый гусь, блестящий чирок *(Nettapus)*
 solan ~ олуша *(Sula)*
 swan ~ сухонос *(Anser cygnoides)*
gooseberry *s* крыжовник *(Grossularia)*
goosebumps, goose-flesh *s* гусиная кожа
gopher *s ther* гофер *(Geomyidae)*
goral *s ther* горал *(Nemorhaedus goral)*
gore *v* бодать; протыкать рогами
gorge *s* глотка; пасть; глотать; *orn* зоб; *bot* зев
gorgeret *s* зазубренное жало
gorget *s* ожерелье; *orn* пятно на горле
goshawk *s* ястреб; ястреб *(Accipiter);* большой ястреб, тетеревятник, ястреб-тетеревятник *(Accipiter gentilis)*
gosling *s* гусёнок; *bot* серёжка
gossypimine *s* сафранин О, основной красный
gourd *s* тыква бутылочная, горлянка *(Lagenaria siceraria)*
gout *s* подагра
goutweed *s* сныть обыкновенная *(Aegopodium podagraria)*
gowan *s* маргаритка *(Bellis perennis)*
grab *s* хватательное движение; захват; захватывающее приспособление; *v* захватывать; схватывать
grackle *s orn* гракл *(Quiscalus)*
gradation *s* градация
grade *s* качество, сорт; улучшенная порода; *v* улучшать породу скрещиванием
 evolutionary ~ эволюционная ступень
gradient *s* градиент
 character ~ градиент признаков, клин
grading *s* грединг
gradology *s* градология
graduate *v* градуировать, калибровать; постепенно изменяться
graft *s* трансплантат; привой, привитая часть; *v* пересаживать ткань; черенковать, окулировать; **to ~ by approach** прививать сближением, аблактировать; **to ~ on root** прививать в корень
 allogenic ~ аллотрансплантат
 autoplastic ~ аутопластический трансплантат, аутотрансплантат
 heteroplastic ~ гетеропластический трансплантат, гетеротрансплантат, ксенотрансплантат
 syngenic ~ изотрансплантат, сингенный трансплантат
 tissue ~ тканевой лоскут
grafted *a* пересаженный, трансплантированный; привитой
grafting *s* трансплантация, пересадка; прививка
 approach ~ прививка сближением, аблактировка
 side ~ прививка в боковой зарез; прививка сближением, аблактировка
 splice ~ копулировка; концевая прививка
grain *s* зерно, хлебный злак; зёрнышко, гранула, зерно; зернистость; строение, структура; грена; волокно
 black ~ of corn *phytop* спорынья *(caused by Claviceps purpurea)*
 distiller's ~ барда
 spiked ~s колосовые злаки
 standing ~ хлеб на корню
grained *a* зернистый; гранулированный
gramineous *a* злаковый; травянистый
graminifolious *a* злаколистый
graminivorous *a* травоядный
graminoid *s* злак; *a* злаковидный
Gram-negative *a* грамотрицательный
Gram-positive *a* грамположительный

granadilla *s* страстоцвет, пассифлора *(Passiflora)*
granatum *s bot* гранат
granivore *s* зерноядное животное, семеноядное животное, зерноядная птица, семеноядная птица
granivorous *a* зерноядный, семеноядный
granose *a* чётковидный, буссовидный
granular *a* зернистый, гранулярный, гранулированный
granulate *v* гранулировать; *a* зернистый, гранулярный, гранулированный
granulation *s* грануляция, гранулирование; зернистость
granule *s* гранула, зёрнышко
~ **of sex chromatin** гранула полового хроматина, тельце Барра
basal ~ базальное зерно, кинетосома
granuloblast гранулобласт
granulocyte *s* гранулоцит, зернистый лейкоцит
eosinophilic ~ эозинофил, ацидофильный гранулоцит
neutrophilic ~ нейтрофильный гранулоцит, нейтрофил
granulocytopoiesis *s* гранулоцитопоэз
granum *s* грана
grape *s* виноград *(Vitis)*
grapefruit *s* грейпфрут *(Citrus paradisi)*
grapevine *s* виноградная лоза
graph *s* диаграмма, кривая
bar ~ гистограмма
grasp *v* схватывать; захватывать; зажимать; хвататься; *s* способность быстрого восприятия, понимание, схватывание
grass *s* трава; злак; дёрн; пастбище; *pl* злаковые травы; *v* пастись; зарастать травой; засевать травой
permanent ~es *pl* многолетние злаки, многолетние травы
short ~ короткотравье, низкотравье; *pl* низовые злаки
sod ~es дерновые растения, дерновые злаки
tall ~es бурьян; высокотравье
grassbird *s* длиннохвостая камышовка *(Megalurus)*; большехвостая славка *(Sphenoeacus)*
grassed *a* покрытый травой
grass-feeding *a* травоядный
grasshopper *s* саранча, кобылка *(Acrididae)*; *pl* прыгунчики *(Tetrigidae)*
acridoid ~ саранчёвое *(Acridoidea)*
Egyptian ~ саранча египетская, кобылка египетская *(Anacridium aegyptium)*
Maroccan ~ саранча марокканская *(Stauronotus maroccanus)*
grassland *s* лугопастбищное угодье; сенокосное угодье; луг; пастбище
grass-of-Parnassus *s bot* белозор *(Parnassia)*
grassy *a* травяной; травянистый
gravid *a* беременная; супоросая; суягная; жерёбая; стельная; икряная
gravidity *s* беременность
gravimetry *s* гравиметрия
gravitation *s* гравитация
gray *a* серый
graywing *s orn* серокрылый турач *(Francolinus africanus)*
graze *s* пастьба; *v* пастись, щипать траву; обгрызать, обкусывать, общипывать
grazer *s* пасущееся животное
grazing *s* выпас, пастьба; выедание; пастбище, выгон; обгрызание, обкусывание, общипывание; *a* травоядный; пастбищный
grease *s* жир; жиропот
grebe *s orn* поганка *(Podiceps)*
great-crested ~ чомга, большая поганка *(Podiceps cristatus)*
green *a* зелёный; покрытый зеленью; незрелый, неспелый; *s* растительность; листва
leaf ~ хлорофилл
greenery *s* растительность, зелень
greenfly *s* тля *(Aphidodea)*
greengage *s* слива ренклод *(Prunus domestica ssp. italica)*
greenhead *s* слепень *(Tabanidae)*
greenhouse *s* теплица, оранжерея; вегетационный домик
greenland *s* лугопастбищное угодье; сенокосное угодье; луг; пастбище
greenshank *s orn* большой улит *(Tringa nebularia)*
greenweed *s bot* дрок *(Genista)*
greeting *s eth* приветствие; демонстрация приветствия
nose ~ назоназальный контакт

gregariae *s pl* виды, расселённые группами
gregarious *a* стайный, стадный, образующий скопление; *bot* скученный, собранный в группы, сгруппированный
greylag *s* серый гусь *(Anser anser)*
grid *s* сетка; решётка; сеточка, объектодержатель
grievous *a* вредный
griffon *s orn* сип, гриф *(Gyps);* белоголовый сип *(Gyps fulvus)*
grimace *s eth* гримаса
grin *s* оскал; *v* скалить зубы
  greeting ~ приветственный оскал; улыбка
grinder *s* коренной зуб
  tissue ~ гомогенизатор ткани
grinding *s* размалывание, растирание, размельчение
grison *s ther* гризон, гурон *(Galictis)*
gristle *s* хрящ
gristly *a* хрящевой; хрящеватый
grizzly *s* гризли *(Ursus horribilis)*
groin *s* пах, паховая область
grooming *s eth* чистка, обыскивание
groove *s* борозда, бороздка; желобок; углубление
  branchial ~ *emb* жаберная щель
  carotid ~ жёлоб для сонной артерии; сонная борозда
  nail ~ ногтевой желобок
  neural ~ желобок нервной трубки
  oral ~ *prot* ротовая впадина
grooved *a* бороздчатый; желобчатый; ямчатый
groove-fruited *a* бороздчатоплодный
groove-nerved *a* бороздчатонервный
grosbeak *s orn* обыкновенный дубонос *(Coccothraustes coccothraustes);* щур *(Pinicola);* толстонос *(Pheucticus)*
  golden-winged ~ золотокрылый вьюрок *(Rhynchostruthus socotranus)*
  scarlet ~ чечевица *(Erythrina erythrina)*
gross *a* макроскопический; толстый, тучный; буйный, пышный *(vegetation);* грубый *(feed)*
ground *s* земля, почва, грунт; лежбище
  breeding ~ место размножения; гнездовье; нерестилище, место нереста; щённая залёжка *(of seals)*
  berrying ~ ягодник
  collecting ~ место сбора материала
  experimental ~ опытное поле, опытный участок
  feeding ~ кормовая площадь; пастбище
  lekking ~ *orn* токовище, место тока
  nesting ~ *orn* гнездовье
  pasture ~ пастбище
  practice ~ опытное поле
  wintering ~ место зимовки
groundhog *s* лесной сурок *(Marmota monax)*
groundhopper *s ent* прыгунчик *(Tetrigidae)*
groundnut *s* земляной орех, арахис *(Arachis hypogaea);* апиос клубневая *(Apios tuberosa)*
groundwater *s* грунтовые воды, подземные воды
group *s* группа; *v* группировать; классифицировать
  breath ~ дыхательный цикл
  Rh blood ~ резус-группа крови
  supertypic ~ супертипирующая панель
  toxophore ~ токсофорная группа
  year ~ годовой класс
grouping *s* группирование; классифицирование
  leukocyte ~ типирование лейкоцитов
grouse *s* шотландская куропатка, граус *(Lagopus scoticus);* тетерев *(Lyrurus);* тетерев(-косач) *(Lyrurus tetrix)*
  black ~ тетерев-косач *(Lyrurus tetrix)*
  great ~ глухарь *(Tetrao urogallus)*
  hazel ~ рябчик *(Tetrastes bonasia)*
  Pallas sand ~ саджа, обыкновенная копытка *(Syrrhaptes paradoxus)*
  ruffed ~ воротничковый рябчик *(Bonasa umbellus)*
  Siberian [sickle-winged] ~ дикуша *(Falcipennis falcipennis)*
  spruce ~ дикуша *(Falcipennis)*; канадская дикуша *(Falcipennis canadensis)*
  white [willow] ~ белая куропатка *(Lagopus lagopus)*
  wood ~ глухарь *(Tetrao urogallus)*
grove *s* роща, куртина
grow *v* расти, произрастать; **to ~ into the soil** пускать корни; **to ~ out** прорастать; **to ~ over** зарастать
growing *s* выращивание; культивирование; *a* растущий;

способствующий росту, стимулирующий рост
growl *s* рычание; ворчание; *v* рычать; ворчать, урчать, бурчать
growling *s* урчание, бурчание
grown-up *a* взрослый
growth *s* рост; развитие; произрастание; выращивание; культивирование; растительность; поросль; новообразование; ~ **in surface** *bot* плоскостной рост
full ~ полное развитие
latent ~ латентная фаза роста
malignant ~ злокачественный рост; злокачественная опухоль
new ~ новообразование; опухоль
one-step ~ одиночный цикл размножения
second ~ естественное возобновление; вторичный лес, производный лес, порослевой лес
seedling ~ самосев; подсад
young ~ подрост, молодняк
growth-inhibiting *a* угнетающий рост
growth-promoting *a* ускоряющий рост
growth-regulating *a* регулирующий рост
growth-retarding *a* замедляющий рост
grub *s* червовидная личинка; *v* выкапывать; выкорчёвывать; **to ~ about** копаться, рыться; **to ~ out, to ~ up** выкапывать; выкорчёвывать
grub-ridden *a* червивый *(of mushrooms)*
grumble *v* урчать
grumous *a* бугорковый, бугристый
grumulus *s ent* полярный орган, хвостовая группа клеток
grunt *s* ворчание; хрюканье; *v* ворчать; хрюкать
guacharo *s orn* гуахаро *(Steatornis caripensis)*
guan *s orn* кракс, гокко *(Cracidae);* пенелопа *(Penelopa)*
guanaco *s ther* гуанако *(Lama guanicoe)*
guano *s* гуано
guanotrophy *s* гуанотрофия
guard *s* осторожность, осмотрительность; оборонительное положение; *v* защищать, охранять; **to be on (one's)** ~ быть настороже, быть начеку
guarding:
eth ~ **of female** вождение и охрана самки самцом
eth **mate** ~ охрана партнёра
guava *s bot* гуайява *(Psidium guajava)*
guayule *s bot* гваюла серебристая *(Parthenium argentatum)*
gubernaculum *s helm* губернакулум
guenon *s* мартышка; мартышка *(Cercopithecus)*
guereza *s ther* толстотел, гвереца *(Colobus)*
guild *s* гильдия
guillemot *s orn* чистик *(Cepphus);* кайра *(Uria)*
guira *s ther* гиара *(Euryzygomatomys spinosus)*
guineafowl *s* цесарка *(Numida)*
gulf *s* залив
gull *s* чайка *(Larus);* вилохвостая чайка *(Xema)*
glaucous ~ большая полярная чайка, бургомистр *(Larus hyperboreus)*
great black-headed ~ черноголовый хохотун *(Larus ichthyaetus)*
lesser black-backed ~ клуша *(Larus fuscus)*
skua ~ поморник *(Stercorarius)*
gullet *s* пищевод; глотка
gulp *s* глоток; глотание; глотательные движения; *v* глотать, проглатывать
gum *s* растительный клей, камедь, гумми, смола; десна; *v* выделять камедь, выделять смолу
gummiferous *a* камеденосный
gumming, gummosis *s phytop* гоммоз, гуммоз, камедетечение
gustatory *a* вкусовой
gut *s* кишка
guttate, guttated *a* покрытый каплями, покрытый пятнышками, покрытый крапинками
guttation *s* гуттация
guttiflorous *a* крапчатоцветковый
gutting *s* потрошение
gymnanthous *a* голоцветковый
gymnoblastous *a* голоростковый
gymnocarpous *a* голоплодный
gymnospermous *a* голосемянный
gymnostom(at)ous *a bot* не имеющий перистома
gynaecaner *s ent* гинекоморфный самец
gynaeceum *s* гинецей

gynaecoid *s* яйцекладущий муравей-рабочий
gynaecophore *s* гинекофор
gynander *s* гинандроморф
gynandrism *s* гермафродитизм
gynandroid *s* гинандроид
gynandromorph *s* гинандроморф
gynandromorphism *s* гинандроморфизм
gynandrophore *s* гинандрофор
gynandrous *a bot* гинандрический
gynantherous *a bot* гинантерический
gynobase *s* гинобазис, сёдлышко
gynochore *s* гинохорный организм
gynoecium *s* гинецей
gynogamone *s* гиногамон
gynophore *s bot* гинофор; *zool* женский гонофор
gynopodium *s* пестичная ножка
gynospore *s* гиноспора, женская спора, мегаспора
gynostegium *s* гиностегий
gynostemium *s* гиностемий, колонка
gypsophila *s* гипсолюбивое растение, известелюбивое растение; *bot* качим, гипсолюбка *(Gypsophila)*
gypsophilous *a* гипсолюбивый, известелюбивый
gypsophilia *s pl* фитоценозы на известняках
gypsywort *s bot* зюзник европейский *(Lycopus europaeus)*
gyral *a* вращательный
gyration *s* вращение; вращательное движение; циркуляция
gyrfalcon *s* кречет *(Falco rusticolus)*
gyrose *a* извитый, извилистый
gyrus *s* мозговая извилина
  uncinate ~ крючок

H

habena *s anat* уздечка
habenula *s anat* уздечка; поводок
habit *s* облик, внешний облик, телосложение, габитус; характер развития; привычка; свойство; поведение; повадки
~ **of body** телосложение
~ **of life** образ жизни
dietary ~ пищевые привычки
feeding ~ особенность питания
gregarious ~s стадный образ жизни
growth ~ форма роста
heterotrichous ~ *bot* гетеротрихальная структура, разнонитчатая структура
motor ~ двигательный навык
schooling ~ стадное поведение
sensory ~ сенсорное умение, сенсорный навык
habitable *a* обитаемый
habitat *s* место обитания, местообитание, место распространения; естественная среда
communal ~, community ~ место обитания сообщества
marginal ~ маргинальный биотоп, маргинальное место обитания
habitation *s* обитание; поселение
habitual *a* привычный; обычный; врождённый
habituation *s* привыкание
habitude *s* привычка; склонность; привычное состояние
habitus *s* габитус, внешний облик
hackle *s orn* оперение шеи; шерсть на загривке; перо на насадке
hadrocentric *a bot* гадроцентрический
hadrome *s* гадром, гадрома
haem *s* гем
haggard *a* дикий, неприрученный; свирепый
hair *s* волосы; волос, волосок; щетинка; шерсть; *pl* волосяной покров
absorbing ~ *bot* трихом
bristle ~ щетина
down ~ пуховой волос; пух; подшёрсток
guard ~ ость
hog ~ свиная щетина
horse ~ конский волос
sheep ~ овечья шерсть
soft fine ~ пух
top ~ остевой волос, ость
haircap *s* кукушкин лён *(Polytrichum commune)*
hair-covering *s* волосяной покров
hairiness *s* волосистость; опушённость
hairless *a* безволосый; неопушённый
hairnet *s мик.* капиллиций
hairpin *s mbiol* «шпилька»
hairworm *s helm* волосатик *(Gordius)*
hairy *a* волосатый; волосистый
halberd-shaped *a* стреловидный
halcyon *s* зимородок *(Alcedo)*
half-adherent *a* полуприлегающий
half-awned *a* полуостистый
half-blooded, half-bred *a* полукровный, нечистокровный, смешанного происхождения
half-bull *s* полусекач *(fur seal)*
half-coated *a* полупокровный
half-double *a* полумахровый
half-epigynous *a* полунадпестичный
half-grooved *a* полубороздчатый
half-grown *a* недоразвитый
half-hairy *a* полуопушённый
half-heterogamy *s* полугетерогамия, неполная гетерогамия
half-inferior *a bot* полунижний
half-inverted *a* полуобратный, полуобращённый
half-life *s* период полураспада; период полувыведения
half-lobed *a bot* полулопастный
half-mutant *s* половинный мутант
half-partitioned *a* полуперегородчатый
half-pinnate *a* полуперистый
half-shrub *s* полукустарник
half-sibs *s pl* полусибсы
half-spiked *a* полуколосистый
half-stem-embracing *a* полустеблеобъемлющий
half-superior *a* полуверхний
half-upright *a* полупрямостоячий
half-whorled *a* полумутовчатый
half-width *s biom* половина центрального доверительного интервала, половина поля допуска
half-wild *a* полудикий
half-winged *a bot* полукрылатый
halibios *s* галобиос

hallux *s* первый палец ноги; большой палец стопы; *orn* задний палец
halm *s* стебель; ботва
halo of inhibition *s* зона подавления
halobacterium *s* галобактерия
halobiont *s* галобионт
halobios *s* галобиос
halodrymium *s* сообщество мангровых зарослей
halogenation *s* галоидирование
halogenolysis *s* галогенолиз
halomorphic *a* галоморфный
halophilic, **halophilous** *a* галофильный
halophobe *s* галофоб
halophobous *a* галофобный
halophyte *s* галофит
halter *s ent* жужжальце
ham *s* подколенная область, бедро, ляжка
hamate, hamiform *a* крючкообразный, крючковидный
hamirostrate *a* имеющий крючковидный клюв
hammer *s anat* молоточек; *orn* овсянка *(Emberiza)*
hamper *v* препятствовать; задерживать *(e.g., growth)*
hamster *s* хомяк, хомячок *(Cricetidae)*
hamulus *s* крючочек
hamus *s* крючок, крючковидный отросток
hand *s* рука, кисть руки; лапа, передняя нога; стрелка
~ **of bananas** гроздь бананов
handedness *s* правое *или* левое направление; направление по *или* против часовой стрелки
handle *s* ручка, рукоятка; конец плода у плодоножки; *v* возделывать, выращивать *(plants);* обрабатывать *(soil);* убирать *(crop);* приручать *(animals);* сортировать
handling *s* манипулирование
antigen ~ процессирование антигена
hand-reared *a* выращенный в неволе; приручённый, ручной
haplobiont *s* гаплобионт
haplocaulescent *a bot* имеющий одну ось
haplochlamydeous *a bot* гаплохламидный
haploconidium *s* гаплоконидия
haplocorm *s bot* гаплокорм
haplodiploid *s* гаплодиплоид; *a* гаплодиплоидный
haplodiploidy *s* гаплодиплоидия
haplodiplont *s* гаплодиплонт
haplodont *a* гаплодонтный
haploid *s* гаплоид; *a* гаплоидный
haploidization *s* гаплоидизация
haploidy *s* гаплоидия
haplokinety *s* гаплокинета
haplomict *s gen* гапломикт
haplont *s* гаплонт
haploperistomous *a bot* с простым перистомом
haplophase *s* гаплофаза
haplophyte *s* гаплофит, гаметофит
haplopolyploid *s* гаплополиплоид
haplosis *s gen* гаплозис
haplosomic *s* гаплосомик, моносомик; *a* гаплосомический
haplospore *s* гаплоспора, одиночная спора
haplostemonous *a* гаплостемонный
haplotype *s gen* гаплотип
haplozygous *a* гаплозиготный
hapten *s* гаптен
autocoupling ~ гаптен-аутомодификатор
cross-reacting ~s *pl* перекрёстно реагирующие гаптены
Forssman's ~ гаптен Форсмана
haptenated *a* содержащий пришитый гаптен, гаптенилированный
haptenating *s* связывание с гаптеном, гаптенилирование
haptene *s* гаптен
hapteron *s bot* прицепка
haptic *a* тактильный, осязательный
haptogene *s* гаптоген
haptoglobin *s* гаптоглобин
haptophore *s* гаптофор, гаптофорная область, гаптофорный сайт
haptor *s helm* фиксаторный аппарат
haptotropism *s* гаптотропизм, тигмотропизм
harassement *s*:
sexual ~ *eth* вмешательство в спаривание
hardening *s* закаливание; затвердевание
hardiness *s* выносливость, стойкость
cold ~ холодоустойчивость
frost ~ морозостойкость
winter ~ зимостойкость
hard-leaved *a* жестколистный
hardness *s* твёрдость; стекловидность *(of grain)*

hard-seeded *a* твердосемянный
hard-shelled *a* имеющий твёрдую оболочку, имеющий твёрдую раковину
hardy *a* выносливый, стойкий
hare *s* заяц *(Lepus);* американский жесткошёрстный кролик *(Sylvilagus)*
  European ~ заяц-русак *(Lepus europaeus)*
  tolai ~ заяц-толай *(Lepus tolai)*
  true ~ заяц *(Lepus)*
harem *s* гарем
harle *s orn* крохаль *(Mergus)*
harm *s* вред, ущерб; причинять вред
harmful *a* вредный, опасный
harmless *a* безвредный, безопасный
harmonic *s* обертон; гармоника *(bioacoustics); a* гармонический
harmonical *a* гармонический
harness *s* упряжь, сбруя
harrier *s* лунь *(Circus);* полевой лунь *(Circus cyaneus)*
hart *s* взрослый самец оленя
hartebeest *s* бубал, коровья антилопа *(Alcelaphus buselaphus)*
hartshorn *s* олений рог; отросток оленьего рога
harvest *s* сбор урожая, жатва; *v* собирать, получать *(напр. клетки из культуры)*
  bumper ~ небывалый урожай
  community dry weight ~ чистая продуктивность сообщества
  hay ~ сенокос
harvester *s* харвестер, сборщик клеток, коллектор клеток
  cell ~ харвестер, сборщик клеток, коллектор клеток
  Skatron's cell ~ скатроновский харвестер клеток
harvesting *s* уборка урожая, жатва
  selective ~ выборочная уборка
harvestmen *s pl zool* сенокосцы *(Phalangida)*
hastate *a bot* стреловидный
hastula *s bot* копьецо
hatch *s* высиживание, выведение в инкубаторе; выводок; *v* вылупляться, выводиться; высиживать; выводить в инкубаторе
hatcher *s* инкубатор
hatchery *s* питомник
hatching *s* вылупление, выклев; выведение, инкубирование; *ent* расплод
hat-shaped *a* шляпкообразный
haulm *s* стебель; ботва
haunch *s* бедро; задняя нога; *ent* тазик; *v* сидеть *(of a dog)*
haunt *s* логовище
haustellate *a* сосательный
haustellum *s ent* сосательный хоботок
haustorium *s* гаустория
havoc *s*:
  immunologic ~ иммунологический конфликт, иммуноконфликт
haw *s orn* мигательная перепонка; боярышник *(Crataegus)*
hawfinch *s orn* обыкновенный дубонос *(Coccothraustes coccothraustes);* черноголовый дубонос *(Euphona)*
hawk *s* небольшая хищная птица; сокол *(Falco)*
  American sparrow ~ воробьиная пустельга *(Falco sparverius)*
  barred honey ~ красногрудый осоед *(Pernis celebensis)*
  bat ~ широкоротый коршун *(Machaerhamphus alcinus)*
  bicoloured ~ двухцветный ястреб *(Accipiter bicolor)*
  bird ~ ястреб *(Accipitrinae)*
  crane ~ журавлиный ястреб *(Geranospiza caerulescens)*
  crested honey ~ хохлатый осоед *(Pernis ptilorhynchus)*
  dead's head ~ *ent* мёртвая голова *(Acherontia atropos)*
  double-toothed ~ двузубый коршун *(Harpagus bidentatus)*
  duck ~ болотный лунь *(Circus aeruginosus);* сапсан *(Falco peregrinus)*
  eyed ~ бражник глазчатый *(Smerinthus ocellatus)*
  fish ~ скопа *(Pandion haliaëtus)*
  harrier ~ лунёвый ястреб *(Polyboroides)*
  honey ~ осоед *(Pernis);* осоед *(Pernis apivorus)*
  Levant sparrow ~ европейский тювик *(Accipiter brevipes)*
  marsh ~ полевой лунь *(Circus cyaneus)*
  mosquito ~ малый козодой *(Chordeiles minor); pl* стрекозы *(Odonata)*
  night ~ малый козодой *(Chordeiles minor)*
  pigeon ~ дербник *(Falco columbarius)*

rough-legged ~ мохноногий канюк, зимняк *(Buteo lagopus)*
sea ~ большой поморник *(Stercorarius skua)*
sparrow ~ ястреб-перепелятник *(Accipiter nisus);* воробьиная пустельга *(Falco sparverius)*
hawkmoth *s ent* бражник *(Sphingidae)*
hawksbill *s herp* бисса *(Eretmochelys imbricata)*
hawthorn *s* боярышник *(Crataegus)*
hay *s* сено
ground ~ сенная мука
haylage *s* сенаж
haymaker *s ther* пищуха, сеноставка *(Ochotonidae)*
hazard *s* вредный фактор; риск; опасность
occupational ~ вредность производства
hazel *s* лещина *(Corylus)*
hazelhen *s orn* рябчик *(Tetrastes bonasia)*
hazelwort *s bot* копытень *(Asarum)*
Hb, hemoglobin *s* гемоглобин
head *s* голова; головка; верхушка; *bot* корзинка; крона *(of a tree)*; колос; метёлка *(of grass);* кочан *(of cabbage);* напор, давление; *v* формировать крону
~ **of femur** маклок
avert ~ *eth* отворачивание головы в сторону
cabbage ~ кочан капусты
head-bobbing *s* покачивание головой
head-flowered *a* головчато-цветковый
headgut *s emb* головная кишка
heading *s* колошение, выколашивание; завивание головок; образование кочанов
headquarters *s* местообитание, место жизнедеятельности
head up - tail up *s orn* приподнимание головы и хвоста
heal *v* вылечивать, излечивать, исцелять
health *s* здоровье
hear *v* слышать
hearing *s* слух
heart *s* сердце; сердцевина, ядро
bleeding ~ *bot* дицентра великолепная *(Dicentra spectabilis)*
left ~ левая половина сердца, левое сердце
pulmonary ~ лёгочное сердце; правая половина сердца, правое сердце
respiratory ~, right ~, venous ~ правая половина сердца, правое сердце
heartbeat *s* сердцебиение, сокращение сердца
heart-shaped *a* сердцевидный
heartwood *s* ядровая древесина
heat *s* течка, период течки, период половой охоты; теплота; **on** ~ в период течки, в период половой охоты; **to come in** ~ приходить в половую охоту
~ **of food** калорийность пищи, калорийность кормов
blood ~ нормальная температура тела
heath *s bot* эрика *(Erica);* верещатник
heathberry *s bot* водяника чёрная *(Empetrum nigrum)*
heather *s* вереск *(Calluna)*
heathland *s* пустошь
heat-killed *a* убитый нагреванием
heat-resistant *a* теплоустойчивый
heat-shock *s* тепловой шок
heavy *a* тяжёлый; обильный, буйный *(of vegetation);* беременная, суягная *(of sheep)*
heavy-eared *a* тяжелоколосый
hevecalyx *a* пушисточашечный
hebecarpous *a* пушистоплодный
hectocotylus *s zool* гектокотиль, гектокотилизированное щупальце
hederaceous *a* плющеподобный
hedge, hedgegrow *s* живая изгородь
hedgehog *s* ёж *(Erinaceus)*
he-duck *s* селезень
hedycarpous *a* сладкоплодный
heel *s* пятка; *orn* шпора
~ **of hand** ладонь
heelstring *s* пяточное сухожилие, ахиллово сухожилие
heft *s* черенок; стебель
he-goat *s* козёл
he-goose *s* гусак
he-horse *s* жеребец
heifer *s* тёлка
height *s* высота; рост
hekistotherm *s* гекистотерм
helad *s* болотное растение
helical, helicine *a* улиткообразный; спиральный, винтовой
helicospore *s* геликоспора
heliophile *s* гелиофил, светолюбивый организм

heliophilous *a* гелиофильный, светолюбивый
heliophobic, heliophobous *a* гелиофобный, светобоязливый
heliophyte *s* гелиофит, светолюбивое растение
heliotaxis *s* гелиотаксис
heliotropism *s* гелиотропизм
helix *s* спираль; завиток ушной раковины
alpha ~ *mbiol* альфа-спираль
helix – hinge – helix *s mbiol* спираль – шарнир – спираль
helix-turn-helix *s mbiol* спираль – петля - спираль
hellbender *s zool* скрытожаберник *(Cryptobranchus)*
hellgrammite *s ent* коридал *(Corydalus);* вислокрылка *(Megaloptera)*
helmet *s ent* галеа; *bot* шлем
helminth s гельминт, паразитический червь
soil-transmitted ~ геогельминт
helminthoid *a* червеобразный
helminthologist *s* гельминтолог
helminthology *s* гельминтология
helminthosis *s* гельминтоз
helobiont *s* обитатель болот
helobious *a* болотный, обитающий на болотах
helohylium, helophium *s* сообщество заболоченного леса
helophyte *s* гелофит, болотное растение
help *s* помощь; **to donate ~ to** *imm* осуществлять передачу хелперного сигнала
hapten ~ гаптеновая помощь, потенциирование гаптеном
Ia-specific ~ Ia-специфический хелперный фактор
helper *s* вирус-помощник; клетка-помощник, клетка-хелпер; *eth* помощник; ~ **at the den** помощник по уходу за детёнышами; ~ **at the nest** помощник по уходу за птенцами
helping *s eth* помощь
hemachrome *s* гематохром
hemacyte *s* гемоцит, клетка крови
hemacytometer *s* гемоцитометр
hemaden *s* железа внутренней секреции, эндокринная железа
hemadenology *s* эндокринология
hemafacient *a* кроветворный, гемопоэтический
hemagglutination *s* гемагглютинация, агглютинация эритроцитов
hemagonium *s* гемоцитобласт
hemal *a* кровяной, гемальный
hemapoiesis *s* гемопоэз, кроветворение
hemapoietic *a* кроветворный
hematal *a* относящийся к крови, кровяной
hematalloscopy *s* гematаллоскопия
hemateikon *s* микроскопическая картина крови
hematic *a* кровяной; содержащий кровь; действующий на кровь
hematid *s* эритроцит
hematimeter *s* гемоцитометр
hematobic *a* обитающий в крови, живущий в крови
hematobium *s* обитатель кровяного русла
hematoblast *s* гематобласт; тромбоцит, кровяная пластинка
hematochrome *s* гематохром
hematocrit *s* гематокрит
hematocryal *a* холоднокровный
hematocytolysis *s* гемолиз
hematoencephalic *a* гематоэнцефалический
hematogene *s* гемоцитобласт
hematogenesis *s* гемопоэз, кроветворение
hematogenic, hematogenous *a* гематогенный, кроветворный, гемопоэтическ
hematoglobulin *s* гемоглобин
hematologic(al) *a* гематологический
hematology *s* гематология
hematolysis *s* гемолиз
hematolytic *a* гемолитический
hematomancy *s* гематологическое исследование
hematophagous *a* питающийся кровью
hematopoiesis *s* гемопоэз, кроветворение
hematopoietic *a* кроветворный, гемопоэтический
hematosis *s* гемопоэз, кроветворение
hematothermal *a* теплокровный
hematoxylin *s* гематоксилин
hematozoon *s* кровепаразит
heme *s* гем
hemeranthic, **hemeranthous** *a* цветущий днём

hemerecology *s* экология культурной территории
hemerophilous *a* легко культивируемый
hemerophyte *s* культивируемое растение
hemibasidium *s* полубазидия
hemiblastula *s* полубластула
hemibranch *s* полужабра
hemic *a* кровяной; содержащий кровь; действующий на кровь
hemicellulose *s* гемицеллюлоза
hemicephalous *a ent* полуголовый
hemicerebrum *s* полушарие большого мозга
hemichlamydeous *a bot* полупокровный
hemichordate *a* полухордовый; *s* полухордовое *(Hemichordata)*
hemicryptophyte *s* гемикриптофит
hemicyclic *a* гемициклический
hemiendophyte *s* гемиэндофит, полуэндофит
hemiepigynous *a* полунадпестичный
hemiepiphyte *s* гемиэндофит, полуэндофит
hemifusion *s* полуслияние
hemiglobin *s* гемиглобин, метгемоглобин
hemihaploid *s* гемигаплоид
hemikaryon *s* гемикарион, пронуклеус
hemilateral *a* относящийся к одной стороне
hemimetabolic *a* гемиметаболический
hemimetaboly *s* гемиметаболия, неполное превращение
hemiparasite *s* полупаразит
hemiparasitic *a* полупаразитический
hemiploid *s* гемиплоид
hemipode *s orn* трёхперстка *(Turnix)*
hemipterygoid *s* гемиптеригоид
hemisaprophyte *s* полусапрофит
hemisome *s* половина тела
hemisphere *s* полушарие
hemispherectomy *s* гемисферэктомия
hemitropism *s* гемитропизм
hemizygote *s* гемизигота
hemlock *s bot* болиголов *(Conium)*; тсуга, гемлок *(Tsuga)*
hemoblast *s* гемобласт
hemocatheresis *s* гемолиз, эритроцитолиз
hemochrome *s* гемохром
hemoclasia *s* гемолиз, эритроцитолиз
hemoclastic *a* кроверазрушающий, гемолитический
hemocoel *s* гемоцель
hemoconia *s* гемокония
hemoculture *s* гемокультура
hemocyanin *s* гемоцианин
hemocyte *s* гемоцит, клетка крови
hemocytoblast *s* гемоцитобласт, первичная клетка крови
hemocytogenesis *s* образование кровяных клеток
hemocytometer *s* гемоцитометр
hemocytozoon *s* паразит клеток крови
hemodynamics *s* гемодинамика, динамика кровообращения
hemogenesis *s* гемопоэз, кроветворение
hemoglobin *s* гемоглобин
hemogram *s* гемограмма, формула крови
hemohistioblast *s* гемогистиобласт
hemology *s* гематология
hemolymph *s* гемолимфа
  immune ~ *ent* иммуногемолимфа
hemolysate *s* гемолизат
hemolysis *s* гемолиз
hemolytic *a* гемолитический
hemolyze *v* гемолизировать
hemometer *s* гемометр
hemoplastic *a* кроветворный
hemopoiesis *s* гемопоэз, кроветворение
  extramedullary ~ экстрамедуллярный гемопоэз, внекостномозговое кроветворение
  fetal ~ фетальный гемопоэз, внутриутробное кроветворение плода
hemopoietic *a* кроветворный, гемопоэтический
hemopoietin *s* гемопоэтин
hemorrhage *s* кровотечение
hemorrhagic *a* относящийся к кровотечению, геморрагический
hemostasis *s* гемостаз
hemostatic *a* гемостатический, кровоостанавливающий
hemotoxic *a* гемотоксический
hemotrophe *s* гемотрофное вещество
hemozoin *s* малярийный пигмент, гемозоин
hemozoon *s* кровепаразит
hemp *s* конопля *(Cannabis)*
  Bahama ~ агава сизалевая, сизаль *(Agave sisalana)*
  Deccan ~ кенаф, гибискус коноплёвый *(Hibiscus cannabinus)*
  sisal ~ агава сизалевая, сизаль *(Agave sisalana)*

hen *s* курица; самка птицы
clocking ~ наседка
clucking ~ наседка; арама *(Aramus guarauna)*
common scrub ~ джунглевая курица *(Megapodius freycinet)*
fool ~ канадская дикуша *(Falcipennis canadensis)*
gray ~ тетёрка, самка тетерева *(Lyrurus tetrix)*
hazel ~ рябчик *(Tetrastes bonasia)*
heath ~ тетёрка, самка тетерева *(Lyrurus tetrix)*
laying hen несушка
moor ~ камышница *(Gallinula chloropus)*
mud ~s *pl* пастушковые *(Rallidae)*
pea ~ пава
prairie ~ луговой тетерев *(Tympanuchus)*
purple swamp ~ султанка, султанская курица *(Porphyrio porphyrio)*
scrub ~ большеног *(Megapodius)*
turkey ~ индейка *(Meleagris gallopavo)*
water ~ лысуха *(Fulica)*
henbane *s* белена *(Hyoscyamus)*
henbit *s bot* яснотка *(Lamium)*
henna-red *a* красновато-коричневый
hepadnavirus *s* гепаднавирус
hepar *s* печень
heparin *s* гепарин
hepatic *a* печёночный
hepatitis *s* гепатит
hepatocolic *a* печёночно-ободочный
hepatocystic *a* печёночно-пузырный
hepatocyte *s* гепатоцит, печёночная клетка
hepatoenteric *a* печёночно-кишечный
hepatogastric *a* печёночно-желудочный
hepatogenic *a* гепатогенный, печёночного происхождения
hepatology *s* гепатология
hepatonephric *a* гепаторенальный, печёночно-почечный
hepatopancreas *s* гепатопанкреас
heptagynous *a* семипестичный
heptamerous *a* гептамерный, семичленный
heptandrous *a* семитычинковый
heptaphyllous *a* семилистный
heptastichous *a* семирядный
herb *s* трава, травянистое растение; *pl* разнотравье; *pl* ароматические травы; пряные травы
carpenter's ~ живучка ползучая *(Ajuga reptans)*
great willow ~ кипрей узколистный, иван-чай *(Epilobium angustifolium)*
miscellanous ~s разнотравье
rich ~s буйнотравье
willow ~ кипрей узколистный, иван-чай *(Epilobium angustifolium)*
herbaceous *a* травянистый
herbage *s* травы, травяной покров, травостой
kar ~ растения в горных лощинах
tall ~ высокотравье
herbal *a* травяной
herbarium *s* гербарий
herbicide *s* гербицид
herbivore *s* травоядное животное
herbivorous *a* травоядный
herblike *a* травянистый
herbosa *s pl* травянистые фитоценозы
hercogamy *s* невозможность самооплодотворения
herd *s* стадо; гурт; залёжка *(of seals);* ходить стадом
horse ~ табун
herder *s* пастушья собака
hereditable *a* наследуемый
hereditary *a* наследственный
hereditation *s* влияние наследственности
heredity *s* наследственность
heredoimmunity *s* наследственный иммунитет
heritability *s* наследуемость
heritable *a* наследуемый
herkogamy *s* невозможность самооплодотворения
hermaphrodite *s* гермафродит
hermaphroditic *a* гермафродитный, двуполый
hermaphroditism *s* гермафродитизм, двуполость
hernia *s* грыжа
heron *s* цапля *(Ardeidae)*
black-crowned night ~ кваква, ночная цапля *(Nycticorax nycticorax)*
heronry *s* гнездовье цапель
heron's-bill *s bot* аистник цикутный *(Erodium cicutarium)*
herpesvirus *s* вирус герпеса

herpetologist *s* герпетолог
herpetology *s* герпетология
he-sheep *s* баран
hesperidium *s bot* гесперидий, померанец
heteractinal *a* разнолучевой
heterandrous *a* разнотычинковый
heteraxial *a* разноосный
heteroagglutinin *s* гетероагглютинин
heteroantibody *s* гетероантитело, гетерологичное антитело
heteroantigen *s* гетероантиген, гетерологический антиген
heteroantigenic *a* гетероантигенный
heteroantiserum *s* гетероантисыворотка, гетерологичная иммунная сыворотка
heteroaperturate *a* разнопоровый
heterobasidium *s* гетеробазидия
heteroblastic *a emb* гетеробластический
heterobrachial *a* гетеробрахиальный, неравноплечий
heterobrochate *a* разноячеистый
heterocarpous *a* гетерокарпный, разноплодный
heterocaryon *s* гетерокарион
heterocaryosis *s* гетерокариоз
heterocellular *a* разноклеточный
heterocentric *a* гетероцентрический
heterocercal *a* гетероцеркальный
heterochlamydeous *a* разнороднопокровный, гетерохламидный
heterochromatic *a* гетерохроматиновый, гетерохроматический
heterochromatin *s* гетерохроматин
heterochromatization *s* гетерохроматизация
heterochromosome *s* гетерохромосома, аллосома
heterochromous *a* разноцветный
heterochronic *a* гетерохронный
heterochronism, heterochrony *s* гетерохронность
heterochrosis *s* ненормальная окраска
heteroclinous *a* гетероклинный
heterocoelous *a* гетероцельный
heterocolpate *a* разнобороздный
heteroconjugate *s* гетероконъюгат
heterocrania *s* асимметрия черепа
heterocyclic *a* гетероциклический
heterocyst *s* гетероциста
heterodimer *s* гетеродимер
heterodimeric *a* гетеродимерный
heterodont *a* гетеродонтный
heterodromous *a* гетеродромный
heteroduplex *s* гетеродуплекс
heterodynamous *a* гетеродинамный
heteroecious *a par* разнохозяйный
heteroepitope *s* гетероантигенная детерминанта, гетероэпитоп
heterofertilization *s* гетерофертилизация
heterogamete *s* гетерогамета
heterogametic *a* гетерогаметный
heterogamety *s* гетерогаметность
heterogamic, **heterogamous** *a* гетерогамный
heterogamy *s* гетерогамия; гетерогония
heterogeneity *s* гетерогенность, разнородность, неоднородность
  idiotypic ~ идиотипическая гетерогенность, идиотипическое разнообразие
  immunochemical ~ иммунохимическая гетерогенность, антигенное разнообразие
  restricted ~ ограниченное разнообразие; рестриктированный полиморфизм; ограниченный поликлональный характер
heterogenesis *s* гетерогенез, чередование поколений
heterogenization *s* гетерогенизация
heterogenomatic *a* гетерогеномный
heterogenote *s* гетерогенота
heterogenous *a* гетерогенный, разнородный, неоднородный
heterogeophyte *s* гетерогеофит
heterogony *s* гетерогония
heterohybridoma *s* гетерологическая гибридома, гетерогибридома; гетеромиелома
heteroimmune *a* гетероиммунный
heteroimmunization *s* гетероиммунизация
heteroimmunize *v* иммунизировать гетероантигеном
heterokaryon *s* гетерокарион
heterokaryosis *s* гетерокариоз
heterokaryotic *a* гетерокариотический
heterokinesis *s* гетерокинез
heterolecithal *a* гетеролецитальный
heterologous *a* разнородный, гетерологичный
heteromer *s* гетеромер
heteromery *s* гетеромерия

heterometabolic *a* гетерометаболический
heteromixis *s* гетеромиксис
heteromorphic *a* гетероморфный
heteromorphism *s* гетероморфизм
heteromorphosis *s* гетероморфоз
heteronerved *a bot* разнонервный, разножилковый
heteropetalous *a* разнолепестный
heterophagous *a* разноядный
heterophyllous *a* разнолистный
heterophyte *s* гетерофит
heterophytic *a* гетерофитный
heteroplanobios *s* гетеропланобиос
heteroplasia *s* гетероплазия
heteropodous *a* разночерешковый
heteropolar *a bot* разнополярный
heteropterous *a* разнокрылый
heteropycnosis *s* гетеропикноз
heterorugate *a bot* разноморщинный
heteroselection *s* гетероселекция
heterosexual *a bot* разнополый
heterosis *s* гетерозис
heterosomal *a* гетеросомальный
heterosome *s* гетерохромосома, аллосома
heterospecific *a* гетероспецифичный
heterospermous *a* разносемянный
heterosporangic *a* гетероспорангиативный
heterosporic, heterosporous *a* гетероспоровый, разноспоровый
heterospory *s* гетероспория, разноспоровость
heterostemonous *a* разнотычиночный
heterostyled *a* разностолбиковый
heterostyly *s* гетеростилия, разностолбчатость
heterosynapsis *s gen* гетеросинапсис
heterotetraploid *s* гетеротетраплоид
heterothallic *a* гетероталломный
heterothallism, heterothally *s* гетероталлизм
heterotic *a* гетерозисный
heterotomy *s bot* неравная рассечённость
heterotopic *a* гетеротопный
heterotransplant *s* ксенотрансплантат
heterotransplantation *s* ксенотрансплантация
heterotrichous *a prot* разноресничный
heterotroph *s* гетеротроф
heterotrophic *a* гетеротрофный
heterotrophy *s* гетеротрофия; гетеротрофность
heterotypic *a* гетеротипический
heteroxenia *s* гетероксения
heteroxenous *a par* имеющий несколько хозяев
heterozygosis, **heterozygosity** *s* гетерозиготность
heterozygote *s* гетерозигота
heterozygous *a* гетерозиготный
heuristic *a* эвристический
hexacanth *s helm* шестикрючный зародыш, онкосфера
hexacyclic *a bot* с шестью кругами
hexapetalous *a* шестилепестный
hexaploid *s* гексаплоид; гексаплоидный
hexaploidy *s* гексаплоидия
hexapod *s* насекомое; *a* шестиногий
hexasteminous *a* шеститычиночный
hexicology *s* биономия; экология
hiatus *s* гиатус, разрыв
hibernacle, hibernaculum *s* место зимней спячки; зимующая почка
hibernal *a* зимний
hibernate *v* находиться в зимней спячке; перезимовывать
hibernation *s* гибернация, зимняя спячка; зимовка
hidden *a* скрытый, латентный
hide *s* шкура; кожа; скрадок, укрытие
  oesophageal-intestinal ~ *helm* пищеводно-кишечная заслонка
hider *s eth* затаиватель
hiding *s eth* затаивание
hidrosis *s* потоотделение
hierarchy *s* иерархия
high-altitude *a* высокогорный
high-energy *a* богатый энергией; макроэргический; высокоэнергетический
high-fat *a* с высоким содержанием жира
highmoor *s* верховое болото
high-mountain *a* высокогорный
high-productive *a* высокопродуктивный
high-voltage *a* высоковольтный
highway *s* главный путь
hilar *a anat* относящийся к воротам
hiliferous *a* рубчиконосный
hill *s* бугорок, возвышение; холм
  ant ~ муравейник
  axon ~ *cyt* аксональный холм; *bot* конус нарастания
hill-growing *a* растущий на холмах
hillock *s* холмик, бугорок

axon ~ *cyt* аксональный холм, аксонный холмик
hilt *s* подсвинок
hilum *s* хилус, ворота органа; выемка, вырезка; *bot* рубчик; ядро *(of a starch grain)*
hilus *s* хилус, ворота органа
hind *s* самка благородного оленя
hindbrain *s* задний мозг
hindgut *s* задняя кишка
hindhead *s* затылок
hind-kidney *s* метанефрос, постоянная почка, тазовая почка
hindleg *s* задняя нога, задняя конечость
hindquarter *s* задняя часть туловища
hindrance *s* препятствие
steric ~ *imm* стерическое препятствие, стерическое несоответствие
hindwing *s* заднее крыло
hinge *s zool* кардо; *mal* замок раковины; *imm* шарнир
hinny *s* лошак
hip *s* бедро; плод шиповника
hipbone *s* подвздошная кость
hippo *s* гиппопотам, бегемот *(Hippopotamus amphibius)*
hippocampus *s anat* гиппокамп, аммонов рог, извилина морского конька
hippopotamus *s* гиппопотам, бегемот *(Hippopotamus amphibius)*
hirsute *a* щетинистый; косматый; волосатый
hispid *a* покрытый щетинками
hiss *v* шипеть
histic *a* тканевый
histioblast *s* гистиобласт
histiocyte *s* гистиоцит, осёдлый макрофаг
histoblast *s* гистобласт, слой ткани; *ent* имагинальный диск
histochemical *a* гистохимический
histochemistry *s* гистохимия
histocompatibility *s* гистосовместимость, тканевая совместимость
histocyte *s* гистоцит, осёдлый макрофаг
histogenesis *s* гистогенез
histogenic *a* гистогенный
histogram *s* гистограмма
histoincompatibility *s* гистонесовместимость, тканевая несовместимость
histological *a* гистологический
histology *s* гистология
histolysis *s* гистолиз
histolytic *a* гистолитический, разрушающий ткань
histon, histone *s* гистон
histopathology *s* гистопатология
histophysiology *s* гистофизиология
histoplasmosis *s* гистоплазмоз *(caused by Histoplasma)*
history *s* история
life ~ жизненный цикл, цикл развития
natural ~ естественная история, естествознание
histotome *s* микротом
histotope *s imm* гистотоп
histotrophic *a* гистотрофный
histozoic *a* гистозойный; обитающий в тканях
hive *s* улей; рой пчёл; *v* роиться
hoarding *s* запасание
hoary *a* покрытый белым пушком
hoatzin *s orn* гоацин *(Opisthocomus hoazin)*
hobble *v* хромать
hobby *s orn* чеглок *(Falco subbuteo)*
hock *s* скакательный сустав
Hoechst *s* хёхст *(stain)*
hog *s* свинья; свинья, кабан *(Sus); pl* свиные *(Suidae);* боров; молодая овца
earth ~ трубкозуб *(Orycteropus)*
European wild ~ кабан, дикая свинья *(Sus scrofa)*
horned ~ бабирусса *(Babirousa babirusa)*
musk ~ пекари *(Tayassu tajacu)*
wart ~ бородавочник *(Phacochoerus aethiopicus)*
wild ~ кабан, дикая свинья *(Sus scrofa)*
hogling *s* поросёнок; ягнёнок
holarctic *a* голарктический
hold *s* логовище; берлога
holder *s* штатив
chemotactic ~ хемотаксическая камера
holdfast *s helm* фиксаторный аппарат
hole *s* отверстие; нора; берлога; глазок *(in cheese)*
blow ~ дыхало *(of a whale)*
insect ~ червоточина
shot ~ *phytop* листовая пятнистость с пустотами
worm ~ *phytop* червоточина
holendobiotic *a* голендобиотический
holendophyte *s* голендофит

hollow *s* полость, пустота; полый, пустой; *anat* выемка, впадина; дупло; мочажина
~ of the sacrum крестцовая впадина
nesting ~ углубление в гнезде, лоток
hollow-leaved *a* вогнутолистный
hollow-stalked, hollow-stemmed *a* полостебельный
holly *s bot* падуб, остролист *(Ilex)*
hollyhock *s bot* алтей розовый, шток-роза *(Althaea rosea)*
holobasidium *s* голобазидия
holobiont *s* голофит
holoblastic *a* голобластический
holocarpous *a* цельноплодный
holocoenotic *a* голоценотический, биогеоценотический
holocrine *a* голокринный
holocyclic *a* полноциклический
holoenzyme *s* холофермент, холоэнзим
hologamy *s* гологамия, макрогамия
hologynic *a* гологинный
holoparasite *s* голопаразит, облигатный паразит
holopetalous *a* цельнолепестный
holophyte *s* голофит
holophytic *a* голофитный
holopneustic *a* голопнейстический
holoptic *a* голоптический
holosericeous *a* покрытый мягкими волосками
holosteous *a* имеющий костный скелет
holostyly *s* протостилия
holotetanus *s* общий тетанус, сплошной тетанус, слитный тетанус
holothurian *s zool* голотурия, морской огурец *(Holothuria)*
holotrichous *a prot* равноресничный
holotype *s* голотип
holozoic *a* голозойный
holt *s* нора; роща; лесистый холм
homalocarpous *a* плоскоплодный
homalophyllous *a* плосколистный
homaxonic *a* гомаксонный, равноосный
home *s* убежище; обиталище; местожительство; проживание; область распространения, родина; *v* возвращаться домой; прилетать домой
homebred, homegrown *a* доморощенный, местный
homeobox *s mbiol* гомеобокс
homeokinesis *s* гомеокинез, митоз
homeomorph *s pl* гомеоморфа
homeopathy *s* гомеопатия
homeorhesis *s* гомеорезис
homeosis *s* гомеозис, гетероморфоз
homeostasis *s* гомеостаз
homeostat *s* гомеостат
homeostatic *a* гомеостатический
homeosynapsis *s gen* гомеосинапсис
homeothermal *a* гомойотермный; теплокровный
homeotic *a* гомеозисный
homeotypic *a* гомеотипический
homer *s mbiol* упаковочный вектор
homing *s* чувство дома, инстинкт дома, хоминг
homing-effect *s imm* эффект дома, хоминг-эффект
homobasidium *s* гомобазидия
homobetaine *s* пропинобетаин
homobium *s* симбиоз водоросли и гриба
homoblastic *a* гомобластический
homobrachial *a* гомобрахиальный
homocarpous *a* гомокарпный, одинаковоплодный
homocaryon *s* гомокарион
homocentric *a* гомоцентрический
homocercal *a* гомоцеркальный
homochlamydeous *a* гомохламидный, однороднопокровный
homochromous *a* одноцветный
homodermic *a emb* гомодермный
homodimer *s* гомодимер
homodromous *a* гомодромный
homodynamic *a* гомодинамический
homodynamy *s* гомодинамия, гомономная метамерия
homoeandrous *a* имеющий одинаковые тычинки
homoeosis *s* гомоеозис
homogamete *s* гомогамета
homogametic гомогаметный
homogamous гомогамный
homogamy *s* гомогамия, инбридинг, родственное разведение
homogenate *s* гомогенат
homogeneity *s* гомогенность, однородность
homogeneous *a* гомогенный, однородный, общего происхождения; гомопластический
homogenetic *a* относящийся к гомогенезу; гомогенетический
homogenization *s* гомогенизация

homogenizer *s* гомогенизатор
homogenote *s* гомогенота
homogeny *s* общее происхождение; общность генотипа
homogony *s bot* гомогония
homoiomerous *a* гомеомерный
homoiosmotic *a* гомойосмотический
homoiotherm теплокровное животное, гомойотермное животное
homoiothermal *a* гомойотермный, теплокровный
homolecithal *a* гомолецитальный
homologize гомологизировать
homologous *a* гомологичный
homolog, homologue *s* гомолог
homology *s* гомология
homolysis *s* гомолиз
homomery *s* гомомерия
homomorphic *a* гомоморфный
homomorphosis *s* гомоморфоз
homomorphous *a* гомоморфный
homonomous *a* гомономный
homonomy *s* гомономная метамерия
homoosis *s* гомоеозис
homopetalous *a* равнолепестный
homophyadic образующий однотипные побеги
homophyllous *a* однообразно облиствленный
homophyte *s* гомофит
homophytic *a* гомофитный
homoplastic *a* гомопластический
homopteran *s ent* равнокрылое хоботное *(Homoptera)*
homopterous *a ent* равнокрылый
homoselection *s* гомоселекция
homosomal *a gen* гомосомальный
homospecific *a* гомоспецифический
homosporic, homosporous *a* равноспоровый
homostyled *a bot* равностолбчатый
homostyly *s* гомостилия
homosynapsis *s gen* гомосинапсис
homothallic *a* гомоталломный
homothallism *s* гомоталлизм
homothermal, homothermous *a* гомойотермный, теплокровный
homotransplant *s* аллотрансплантат, гомотрансплантат
homotransplantation *s* аллотрансплантация, гомотрансплантация
homotropous *a* гомотропный
homotypic *a* однотипный, гомотипический
homozygosis, **homozygosity** *s* гомозиготность
homozygote *s* гомозигота
homozygous *a* гомозиготный
homunculus *s* гомункул, «человечек»
honey *s* мёд; *bot* донник белый *(Melilotus albus)*
  agaric ~ опёнок настоящий *(Armillaria mellea)*
  chunk [comb] ~ сотовый мёд
  honeydew ~ падевый мёд
  wild ~ дикий мёд
honeybee *s* пчела медоносная *(Apis mellifera)*
honeycomb *s* соты; сотовый; *anat* сетка
honeycup *s* нектарник
honeydew *s* медвяная роса; падь
honeyguide *s orn* медоуказчик *(Indicator)*
honey-laden *a* со взятком
honeysuckle *s* жимолость *(Lonicera)*
honey-sweet *a* медвяный
honk *s* крик диких гусей; хрюканье
hood *s orn* хохолок; *bot* клобучок, шлем; капюшон
  laminar flow ~ ламинарный бокс, ламинарный шкаф, ламинар
hoodie *s* серая ворона *(Corvus cornix)*
hood-shaped *a* шлемовидный
hoody *s* серая ворона *(Corvus cornix)*
hoof *s* копыто, копытце
hoofed *a* копытный
hoofless *a* бескопытный
hoof-shaped *a* копытообразный
hook *s* крючок; бодаться
  frenulum ~ *ent* зацепка
  hypocotyl ~ гипокотиль, подсемядольное колено
  plumular ~ *bot* почечка проростка
hooked *a* крючковатый, кривой; имеющий крючки
hook-fruited *a* крючкоплодный
hook-leaved *a* крючколистный
hookworm *s helm* анкилостома *(Ancylostoma)*
hoopoe *s* удод *(Upupa epops)*
hoot *s* уханье, крик совы; *v* кричать, ухать
hop *s* хмель обыкновенный *(Humulus lupulus); v* прыжок; прыгать

hopper *s* прыгающее беспозвоночное; *ent* толстоголовка *(Hesperiidae)*
horiodimorphism *s* сезонный диморфизм
horizon *s* горизонт
hormocyst *s* гормоциста
hormogon, hormogone, hormogonium *s bot* гормогоний
hormonal *a* гормональный, гормонный
hormone *s* гормон
adipokinetic ~ адипокинетический гормон, адипокинин
adrenocorticotrophic ~, adrenocorticotropic ~ адренокортикотропный гормон, АКТГ
androgenic ~ андроген, андрогенный гормон
chorion gonadotrophic ~, chorion gonadotropic ~ хорионический гонадотропин, хорионгонадотропин
chromatophorotropic ~ меланоцитстимулирующий гормон, мелатонин
corticotropin releasing ~ кортиколиберин
follicle-stimulating ~ фоллитропин, фолликулостимулирующий гормон, ФСГ
galactopoietic ~ пролактин, лактогенный гормон, лютеотропный гормон
lactogenic ~ пролактин, лактогенный гормон, лютеотропный гормон
luteal ~ прогестерон, гормон жёлтого тела яичника
luteinizing ~ лютропин, лютеинизирующий гормон, ЛГ
luteinizing hormone releasing ~ люлиберин
luteotrophic ~ лютеотропный гормон, ЛТГ
male sex ~ мужской половой гормон, андростерон
molting ~ гормон линьки, экдизон
orchidic ~ тестостерон
pancreatic ~ гормон поджелудочной железы; инсулин
parathyroid ~ паратгормон, паращитовидный [паратиреоидный] гормон
poietic ~ поэтин
somatotropic ~ соматотропный гормон, соматотропин, СТГ
testicular ~ тестостерон
thymic ~ тимический гормон, тимический фактор
thyrotropin-releasing ~ тиролиберин
hormonic *a* гормональный, гормонный
hormonogenesis *s* гормонопоэз, гормонообразование
hormonogenic *a* гормонообразующий
hormonology *s* эндокринология
h ormonopoiesis *s* гормонопоэз, гормонообразование
hormonopoietic *a* гормонообразующий
horn *s* рог; роговидный отросток; шип *(in fish);* щупальце *(in snails);* ость; *v* бодать
iliac ~ подвздошная ость
lyre ~s лировидные рога
sacral ~ *abat* крестцовый рожок
hornbeam *s* граб *(Carpinus)*
hornbill *s* птица-носорог *(Bucerotidae)*
hornedness *s* рогатость
hornet *s* крупная оса; шершень *(Vespa);* шершень обыкновенный *(Vespa crabro)*
horn-flowered *a* рогоцветковый
hornification *s* ороговение
horning *s* бодание
horn-leaved *a* роголистный
hornless *a* безрогий; безостый
horn-shaped *a* роговидный
horn-stalked *a* рогостебельчатый
horntail *s ent* рогохвост *(Sirex)*
hornworm *s* гусеница с рогом
horny *a* рогатый; роговой*;* жёсткий, ороговевший
horodimorhism *s* сезонный диморфизм
horotelic *a* горотелический
horotely *s* горотелия
horse *s* лошадь *(Equus caballus)*
gelded ~ мерин
horned ~ гну *(Connochaetes)*
river ~ гиппопотам, бегемот *(Hippopotamus amphibius)*
shaft ~ коренная лошадь, коренник
horsefly *s* слепень *(Tabanus)*
horseradish *s* хрен обыкновенный *(Armoracia rusticana)*
horseshoe-shaped *a* подковообразный
horsetail *s bot* хвощ *(Equisetum);myc* навозник *(Coprinus)*
horticolate *a* обитающий в садах
horticulture *s* садоводство

hose *s bot* фолликула, листовка; влагалище
host *s* хозяин; носитель; реципиент
host-specificity *s* хозяинная специфичность
hot *a* жаркий; возбуждённый, раздражённый
hot-blooded *a* теплокровный
hotspot *s mbiol* горячая точка
houbara *s orn* джек, вихляй *(Chlamydotis undulata)*
hound *s* собака; гончая
basset ~ бассет
house *s* помещение; жилище животного; нора; берлога; гнездо; помещение для животного; клетка, вольер, вольера; домик *(e.g., of a caddis worm)*
log ~ дуплянка
monkey ~ обезьянник
rearing ~ инсектарий, садок для насекомых
housefly *s* муха комнатная *(Musca domestica)*
hover *v* парить; *s* омут
hower-fly *s* журчалка *(Syrphidae)*
howl *s* вой, завывание; *v* выть, завывать
howler *s ther* ревун *(Alouatta)*
huddling *s* скучивание
hull *s* пустой стручок; кожура; плёнка; скорлупа; шелуха; *v* очищать от скорлупы,очищать от шелухи
hulled *a* плёнчатый
hum *s* жужжание; *v* жужжать
human *a* человеческий
humanoid *a* человекообразный
humblebee *s* шмель *(Bombus)*
humeral *a* плечевой
humerus *s* плечевая кость
humic *a* гуминовый; гумусовый
humicolous *a* растущий на перегное
humicular *a* сапрофитный
humid *a* влажный, сырой
humidification *s* увлажнение
humidify увлажнять
humidity *s* влажность, сырость; степень влажности
humification *s* гумификация
humified *a* гумифицированный
humify *v* увлажнять
hummel *a* безрогий; безостый
hummingbird *s* колибри *(Trochilidae)*
hummocky *a* кочковатый
humor *s* гумор, жидкость тела
aqueous ~ водянистая влага *(глаза)*
vitreous ~ **of the ear** эндолимфа
humoral *a* гуморальный
hump *s* горб; выгибать спину *(e.g., of a cat)*
humped *a* горбатый
humus *s* гумус, перегной
hunch *s* горб
hunger *s* голод; голодание; *v* голодать
hunt *s* ловля; охота; *v* ловить; охотиться
hunter *s* охотник
hunting *s* охота
hurricane *s* ураган
husbandry *s* сельское хозяйство; земледелие
animal ~ животноводство
cattle ~ скотоводство
forest ~ лесоводство
horse ~ коневодство
pig ~ свиноводство
sheep ~ овцеводство
husk *s* шелуха; лузга; листовая обвёртка; плёнка; мякинная оболочка; кожица; скорлупа; *v* шелушить; лущить; снимать кожицу
husky *s* лайка
hyaena *s см.* **hyena**
hyaline, **hyaloid** *s* гиалин, гиалиновое вещество; *a* гиалиновый, стекловидный, прозрачный
hyalomere *s* гиаломер
hyaloplasm *s* гиалоплазма
hyalosome *s* гиалосома
hyalospore *s* гиалоспора
hyaluronidase *s* гиалуронидаза
hybrid *s* гибрид; *a* гибридный
inversion ~ гетерозигота по инверсии
myeloma ~ гибридома
hybridism, hybridity *s* гибридность
hybridization *s* гибридизация; скрещивание; *mbiol* отжиг
DNA ~ ДНК-гибридизация
dot ~ дот-гибридизация
hybridize гибридизировать
hybridoma *s* гибридома
hybridous гибридный
hydatid *s* гидатида
hydatode *s* гидатода
hydatophyte *s* гидатофит
hydration *s* гидратация, гидратирование
hydrazide *s* гидразид

hydroacoustics *s* гидроакустика
hydrobiology *s* гидробиология
hydrobionics *s* гидробионика
hydrocarbon *s* углеводород
hydrochory *s* гидрохория, распространение водой
hydrocoel *s* гидроцель
hydrocole *s* гидробионт
hydrocondensation *s* гидроконденсация
hydrocortisone *s* гидрокортизон, кортизол
hydrocryptophyte *s* гидрокриптофит
hydroecology *s* гидроэкология
hydroextractor *s* отжимная центрифуга
hydrogenation *s* гидрирование
hydrogenolyse *s* гидрогенолиз
hydrogeophyte *s* гидрогеофит
hydroid *s* гидроид; *a* гидроидный
hydrolysis *s* гидролиз
hydrolytic *a* гидролитический
hydrolyzate *s* гидролизат
hydrolyze *v* гидролизовать
hydrome *s bot* гидром
hydromegatherm *s* гидромегатерм
hydrometer *s* ареометр
hydromorphic *a* гидроморфный
hydromorphosis *s* гидроморфоз
hydronasty *s* гидронастия
hydropathy *s* гидрофобность
hydrophilic, hydrophilous *a* гидрофильный, влаголюбивый
hydrophobic *a* гидрофобный
hydrophobicity *s* гидрофобность
hydrophoric *a* водоносный
hydrophyte *s* гидрофит, водное растение
hydrophytic *a* гидрофитный, водный
hydrophyton *s* сообщество гидрофитов
hydropolymerization *s* гидрополимеризация
hydropolyp *s* гидрополип, гидроидный полип
hydroponics *s* гидропоника
hydroquinone *s* гидрохинон
hydrorhiza *s* гидрориза
hydrosere *s* водная серия
hydrosphere *s* гидросфера
hydrostatic *a* гидростатический
hydrosterone *s* гидростерон, гидротестостерон, андростан
hydrostome *s* рот гидроидного полипа
hydrotaxis *s* гидротаксис
hydrotestosterone *s* гидростерон, гидротестостерон, андростан
hydrotheca *s* гидротека
hydrotropic *a* гидротропический
hydrotropism *s* гидротропизм
hydrous *a* водный; водянистый
hydroxyalkylation *s* гидроксиалкилирование
hydroxydeoxycorticosterone *s* оксидезоксикортикостерон
hydroxylation *s* гидроксилирование
hydroxymethylation *s* гидроксиметилирование
hyena *s* гиена *(Hyaena)*
hygrochasy *s bot* гигрохазия
hygrogram *s* гигрограмма
hygrokinesis *s* гигрокинез
hygrometer *s* гигрометр
hygromorphism *s* гигроморфизм
hygrophilous *a* гигрофильный, влаголюбивый
hygrophyte *s* гигрофит
hygrosciaphyte *s* гигросциафит
hygroscopic *a* гигроскопический
hylad *s* лесное растение
hylile *a* лесной
hylion, **hylium** *s* климакс леса
hylocolous *a* обитающий в лесу
hylodad *s* растение редколесья
hylophagous *a* поедающий древесину
hylophilous *a* обитающий в лесу
hylophyte *s* лесное растение; мезофит
hylotomous *a* древогрызущий
hymen *s* гимен, девственная плева
hymenal *мик.* гимениальный; *анат.* гименальный
hymeniferous *a* имеющий гимений
hymenium *s bot* гимений
hymenophore *s* гименофор
hymenopteran *s ent* перепончатокрылое *(Hymenoptera)*
hymenopterous *a* перепончатокрылый
hyobranchial *a* гиобранхиальный
hyoepiglottic *a* подъязычно-надгртанный
hyoglossal *a* подъязычно-язычный
hyoid *s* гиоид
hyoidean *a* гиоидный
hyomental *a* подбородочно-подъязычный
hyostylic *a* гиостильный
hyothyroid *a* подъязычно-щитовидный
hypaesthesia *s* гипестезия
hypantrum *s anat* гипантрум

hyparterial *a* находящийся под артерией
hyperactivity *s* повышенная активность; гиперфункция
hyperaemia *s* гиперемия
hyperapophysis *s* заднебоковой отросток позвонка
hyperchromatism *s* гиперхроматизм
hypercube *s* гиперкуб
hyperdiploid *s* гипердиплоид
hyperemia *s* гиперемия
hyperemic *a* гиперемированный
hyperesthesia *s* гиперестезия
hyperexcitability *s* чрезмерная возбудимость
hyperexpression *s* сверхсинтез; сверхпродукция; гиперэкспрессия
hyperfeminization *s* сверхфеминизация
hyperfunction *s* гиперфункция
hypergammaglobulinemia *s* гипергаммаглобулинемия
hypergenitalism, **hypergonadism** *s* гипергонадизм
hyperhaploid *s gen* гипергаплоид, сверхгаплоид
hyperhaploidy *s* гипергаплоидия
hyperimmunity *s* напряжённый иммунитет
hyperimmunization *s* гипериммунизация, интенсивная иммунизация
hyperinnervation *s* гипериннервация, избыточная иннервация
hyperkinesia *s* гиперкинез
hypermasculinization *s* сверхмаскулинизация
hypermetabolism *s* повышенный обмен веществ
hypermetamorphosis *s* гиперметаморфоз
hypermethylation *s* повышенный уровень метилирования
hypermorph *s* гиперморф, гиперморфный ген
hypermorphic *a* гиперморфный
hypermorphosis *s* гиперморфоз, анаболия
hypermotility *s* повышенная подвижность
hypermutation *s* гипермутация, сверхмутация
hyperopia *s* дальнозоркость
hyperosmotic *a* гиперосмотический
hyperparasite *s* гиперпаразит, сверхпаразит
hyperparasitism *s* гиперпаразитизм
hyperpelvic *a anat* расположенный над тазом
hyperphagia *s* гиперфагия
hyperpituitarism *s* гиперпитуитаризм
hyperplasia *s* гиперплазия
hyperplastic *a* гиперпластический
hyperploid *s* гиперплоид
hyperreflexia *s* гиперрефлексия
hyperreplication *s* гиперрепликация
hypersecretion *s* повышенная секреция
hypersensitivity *s* повышенная чувствительность, гиперчувствительность
hypertely *s* гипертелия
hyperthermia *s* гипертермия
hypertonic *a* гипертонический
hypertonicity *s* гипертоничность
hypertoxicity *s* повышенная токсичность
hypertrophic *a* гипертрофический
hypertrophy *s* гипертрофия
hypervariability *s* сверхизменчивость, гипервариабельность
hypervariable *a* сверхизменчивый, гипервариабельный
hypervariability *s* сверхизменчивость, гипервариабельность
hyperventilation *s* гипервентиляция
hypervitaminosis *s* гипервитаминоз
hypesthesia *s* гипестезия
hypha *s myc* гифа
hyphal *a* относящийся к гифам
hyphasma *s* гифазма, рыхлый мицелий
hypnocyst *s* гипноциста, покоящаяся циста
hypnody *s* гипнодия
hypnosporangium *s* гипноспорангий
hypnospore *s* гипноспора, покоящаяся спора
hypnote *s* организм в состоянии спячки
hypoactivity *s* пониженная активность
hypobaric *a* гипобарический
hypobasal *a* гипобазальный
hypobasidium *s* гипобазидия
hypocarpic *a* подплодный
hypochilium *s* гипохилий
hypochondrium *s* подрёберная область
hypochord *s* гипохорда
hypocotyl *s* гипокотиль, подсемядольное колено
hypocotyledonary *a* подсемядольный
hypocrateriform *a* блюдцевидный, подносовидный

hypoderma *s* гиподерма
hypodermal, **hypodermic** *a* гиподермальный, подкожный
hypodermis *s* гиподерма
hypodiploidy *s* гиподиплоидия
hypodispersion *s* равномерное распределение
hypofunction *s* гипофункция
hypogaeic *a* живущий под землёй
hypogastric *a* подчревный
hypogeal *a* живущий под землёй
hypogenesis *s* гипогенез
hypoglossal *a* подъязычный
hypognathous *a* гипогнатический
hypogynous *a* гипогинный, подпестичный
hypohaploid *s gen* гипогаплоид; *a* гипогаплоидный
hypohaploidy *s* гипогаплоидия
hypomorph *s* гипоморф, гипоморфный ген
hypomorphic *a* гипоморфный
hypomorphosis *s* гипоморфоз
hyponeural *a* гипоневральный
hyponome *s* воронка *(in cephalopods)*
hyponutrition *s* недостаточное питание
hyponychium *s* гипонихий, ногтевое ложе
hypopetalous *a* гипопетальный
hypophamine *s* вазопрессин
hypopharyngeal *a* гипофарингеальный, подглоточный
hypopharynx *s* гипофаринкс, подглоточник
hypophloeodal *a bot* гипофлеодный, растущий под корой
hypophrenic *a* поддиафрагмальный
hypophyll *s* гипофилл
hypophyllous *a* расположенный под листом; растущий под листом
hypophyseal, hypophysial *a* гипофизарный
hypophysis *s anat* гипофиз; *bot* гипофиза
hypoplasia *s* гипоплазия
hypoplastic *a* гипопластический
hypoplasy *s* гипоплазия
hypoploid *s* гипоплоид
hypoploidy *s* гипоплоидия
hypopodium *s* гипоподий
hypopolyploid *s gen* гипополиплоид
hypopyge *s* гипопигий
hyporeflexia *s* гипорефлексия
hyporesponsiveness *s* низкая реактивность, гипореактивность; *imm* низкая отвечаемость
hyposecretion *s* пониженная секреция
hyposensitivity *s* пониженная чувствительность
hyposphene *s herp* гипосфен
hyposporangium *s* гипоспорангий
hypostasis *s gen* гипостаз
hypostatic *a* гипостатичный, гипостатический
hypostome *s* гипостом
hypotarsus *s orn* цевка
hypotaxia *s* нарушение координации
hypotension *s* гипотония
hypothalamus *s* гипоталамус
hypothallus *s* гипоталлус, подслоевище
hypothecium *s* гипотеций, субгимениальный слой
hypothesis *s* гипотеза
hypotonia *s* гипотония
hypotonic *a* гипотонический, гипотоничный
hypotonicity *s* гипотоничность
hypotoxicity *s* пониженная ядовитость
hypovalve *s* гиповальва *(in diatoms)*
hypoventilation *s* гиповентиляция
hypovitaminosis *s* гиповитаминоз
hypoxanthine *s* гипоксантин
hypoxia *s* гипоксия, кислородная недостаточность
hypoxic *a* гипоксический
hypoxidation *s* гипоксидация
hypoxidosis *s* гипоксидоз
hypsodont *a* гипсодонтный, гипселодонтный
hyrax *s ther* даман, жиряк *(Procavia)*
hysteresis *s* гистерезис
hysterosoma *s zool* гистеросома
hysterotely *s* гистеротелия
hystogram *s* гистограмма
hytherograph *s* климограмма

I

ianthinous *a* фиолетовый
ibex *s* горный козёл, альпийский козёл *(Capra ibex);* бородатый козёл, безоаровый козёл *(Capra aegargus)*
ibis *s orn* ибис *(Ibis)*
glossy ~ каравайка *(Plegadis falcinellus)*
ice *s* лёд; покрывать льдом; замораживать; замерзать; *a* ледяной; ледовый
clear ~ гололедица
ichneumon *s* египетский мангуст, фараонова крыса *(Herpestes ichneumon); ent* наездник *(Ichneumonidae)*
ichthyophagous *a* рыбоядный
icicle *s* сосулька
icotype *s* икотип
icterus *s* желтуха
ictus *s* вспышка болезни; приступ болезни; удар пульса
id *s gen* ид
idant *s gen* идант
ideation *s* мышление
identical *a* идентичный, одинаковый, тождественный
identifiable *a* определимый, поддающийся определению, поддающийся распознаванию
identification *s* идентификация, определение, распознавание
identify *v* идентифицировать, определять; выяснять систематическое положение
identity *s* идентичность, тождественность; родство; подлинность
ideogram, ideograph *s* идеограмма
ideomotor *a* идеомоторный
idioadaptation *s* идиоадаптация
idiobiology *s* идиобиология
idioblast *s* идиобласт
idiochromatin *s* идиохроматин
idiochromosome *s* идиохромосома, половая хромосома
idioecology *s* аутэкология
idiogamy *s* идиогамия, самоопыление
idiogram *s* идиограмма
idiopeptide *s* идиопептид
idioplasm *s* идиоплазма, зародышевая плазма
idiosoma *s* идиосома
idiosphaerosome *s* акросома, перфораторий
idiosphaerotheca *s* акробласт
idiotype *s* идиотип
self ~ аутоидиотип, набор аутоидиотипических детерминант
idiotypy *s* идиотипия, идиотипическое разнообразие, вариабельность идиотипов
idiovariation *s* идиовариация, мутация
idiovascular *a* собственно сосудистый; относящийся к сосудам
idioventricular *a phys* идиовентрикулярный, собственно желудочковый
ignition *s* прокаливание, озоление
ileac, ileal *a* подвздошный
ileum *s* подвздошная кишка
iliac *a* подвздошный
iliolumbar *a* подвздошно-поясничный
iliosciatic *a* подвздошно-седалищный
ilium *s* подвздошная кость
ill *a* больной
ill-conditioned *a* худой, плохо упитанный
illness *s* болезнь
ill-nourished *a* худой, плохо упитанный
illuminance *s* освещённость
illumination *s* освещение; освещённость
illuminator *s* осветитель
illusion *s* иллюзия
optical ~ обман зрения
image *s* образ
Purkinje-Samson's ~ феномен Пуркинье є Самсона
imaginal *a* имагинальный
imaging *s* получение изображения, визуализация; томография
NMR ~ ядерно-магнитно-резонансная томография, ЯМР-томография
imago *s* имаго, взрослая стадия
imbalance *s* неуравновешенность, несбалансированность, дисбаланс, отсутствие равновесия
imbed *v* заделывать, заливать; вкапывать, укреплять, закапывать
imbedding *s* заделка, заливка
imbibe *v* впитывать, всасывать, поглощать

imbibition *s* впитывание, всасывание, поглощение; насыщение влагой
imbricate, imbricated *a bot* черепитчато-расположенный; черепицеобразный; набегающий
imbue *v* насыщать, напитывать, пропитывать; цитол. окрашивать
imitate *v* имитировать, подражать
imitation *s* имитация, подражание
imitative *a* подражательный
immarginate *a bot* неокаймлённый
immature *a* незрелый, неспелый; недоразвитый; слаборазвитый *(soil);* неустановившийся *(character)*
immaturity *s* незрелость
immersion *s* иммерсия, погружение
immigrant *s* иммигрант; иммигрирующий, вселяющийся
immigration *s* иммиграция, вселение; *emb* иммиграция
immiscible *a* несмешивающийся; несмешиваемый
immobile *a* неподвижный, недвижимый
immobility *s* неподвижность
immobilization *s* обездвиживание, лишение подвижности; иммобилизация, фиксация
immortalization *s* иммортализация
immortelle *s bot* сухоцвет *(Xeranthemum);* бессмертник *(Helichrysum)*
immovable *a* неподвижный; стационарный
immune *a* иммунный
immunifacient *a* иммуногенный, вызывающий иммунитет
immunifaction *s* иммунизация
immunity *s* иммунитет, невосприимчивость
immunization *s* иммунизация; вакцинация; сенсибилизация
allogenic ~ аллогенная иммунизация, аллоиммунизация
booster ~ бустинг, бустер-иммунизация
cohort ~ серийная иммунизация; массовая вакцинация
intentional ~ целенаправленная иммунизация, «прицельная» иммунизация, получение целевой антисыворотки
mass ~ серийная иммунизация; массовая вакцинация
primary ~ первичная вакцинация; сенсибилизация
secondary ~ бустинг, бустер-иммунизация
self ~ аутоиммунизация
supplemental ~ бустинг, бустер-иммунизация
immunizator *s* иммуноген
immunize иммунизировать
immunoadsorbent *s* иммуносорбент
immunoadsorption *s* иммуносорбция
immunoarchitecture *s* иммуноархитектоника, иммуногистологическая структура
immunoassay *s* иммунологический анализ
agglutinating ~ реакция агглютинации
capillary enzyme ~ иммуноферментный анализ в капиллярах, энзимокапиллярный иммуноанализ
dot-blot ~ дот-блот иммуноанализ, дот-иммуноблоттинг
immunodiffusion thin layer ~ тонкослойная иммуногель-хроматография
isotopic ~ радиоиммуноанализ
latex enhanced ~ латекс-иммуноанализ, иммуноанализ с латексным усилением
particle-enhanced turbidimetric ~ турбидиметрический иммуноанализ
immunoautoradiography *s* иммунорадиоавтография
immunobead *s* иммуногранула, иммуномикроноситель, иммуносфера
magnetic ~ магнитная иммуногранула, магнитный иммуномикроноситель
immunobiology *s* иммунобиология
immunoblast *s* иммунобласт
lymph-borne ~ иммунокомпетентный лимфобласт
immunoblot *s* иммуноблот
immunoblotting *s* иммуноблоттинг
immunochemistry *s* иммунохимия
immunocompetence *s* иммунологическая зрелость; иммунологическая компетентность, иммунокомпетентность

misdirected ~ извращённая иммунологическая реактивность
immunocompetent *a* иимунокомпетентный
immunoconglutinin *s* иммуноконглютинин
immunoconjugate *s* иммуноконъюгат
immunocrippling *s* иммунный крипплинг
immunocrossreactivity *s* перекрёстная иммунологическая реактивность
immunocyte *s* иммуноцит
immunocytochemistry *s* иммуноцитохимия
immunocytology *s* иммуноцитология
immunocytoma *s* иммуноцитома
immunocytometry *s* иммуноцитометрия
immunodefence *s* иммунная защита; иммунитет
immunodeficiency *s* иммунодефицит, иммунодефицитное состояние, иммунологическая недостаточность
immunodetection *s* иммунологический анализ
immunodiflusion *s* иммунодиффузия
radial ~ радиальная иммунодиффузия, метод Манчини
immunodisfunction *s* нарушения иммунной функции
immunoelectroassay *s* иммуноэлектрофоретический анализ, электроиммуноанализ
immunoelectroosmophoresis *s* противоточный иммуноэлектрофорез, встречный иммуноэлектрофорез, электросинерез, электрофорез по Пезендорферу
immunoelectrophoresis *s* иммуноэлектрофорез
rocket ~ ракетный иммуноэлектрофорез по Лауреллу, однонаправленная простая электроиммунодиффузия
immunofiltration *s* иммунофильтрация
immunofixation *s* иммунофиксация
imprint ~ метод иммунофиксирующих отпечатков, иммуноимпринтинг
immunofluorescence *s* иммунофлуоресценция
double layer ~ иммунофлуоресцентный сэндвич-метод, метод двойных иммунофлуоресцирующих антител
immunogen *s* иммуноген, иммуногенное вещество; ген иммунного ответа, Ir-ген
immunogenetics *s* иммуногенетика
immunogenic *a* иммуногенный, вызывающий иммунитет
immunogenicity *s* иммуногенность
immunoglobulin *s* иммуноглобулин
immunohematology *s* иммуногематология
immunohistology *s* иммуногистология
immunoinhibition *s* иммуноторможение
immunoliposome *s* иммунолипосома
immunologic *a* иммунологический
immunologist *s* иммунолог
immunology *s* иммунология
cancer ~ иммунология опухолей, онкоиммунология
fetal ~ иммунология эмбрионального периода, иммунология плода, иммуноэмбриология
parasite ~ иммунопаразитология
immunomediator *s* иммуномедиатор
immunopathogenesis *s* иммунопатогенез
immunophenotype *s* иммунофенотип, антигенный фенотип; серотип
immunophenotyping *s* иммунофенотипирование
immunophysiology *s* иммунофизиология, физиология иммунной системы
immunopotentiation *s* потенциирование иммунного ответа, иммунопотенциирование, иммуностимулирование
immunopotentiator *s* иммунопотенциатор
immunoprecipitate *s* иммунопреципитат; образовывать иммунопреципитат
immunoprecipitation *s* иммунопреципитация
immunoquantification *s* количественный иммуноанализ
immunoreaction *s* иммунная реакция
immunoregulation *s* иммунорегуляция
faulty ~ иммунорегуляторное нарушение, иммунорегуляторный дефект
immunoresistance *s* иммунитет; иммунная резистентность

immunoresponsiveness *s* иммунологическая отвечаемость; иммунологическая реактивность
immunorestoration *s* восстановление иммунного гомеостаза
immunorheophoresis *s* иммунореофорез
immunoscreening *s* массовое иммунологическое обследование; иммунный скрининг
immunoselection *s* иммуноселекция
immunosensor *s* иммуносенсор
immunosorbent *s* иммуносорбент
immunosuppression *s* супрессия иммунного ответа, подавление иммунного ответа, иммуносупрессия
local ~ локальное ослабление иммунной реакции, местное ослабление иммунной реакции
immunosuppressor *s* иммуносупрессор
immunosurveillance *s* иммунологический надзор
immunotechnique *s* иммунологический метод
immunotopography *s* иммунотопография
immunotoxin *s* иммунотоксин
imp *s* отпрыск, побег; *bot* прививка; *v* прививать
impact *s* удар, толчок, импульс; влияние, воздействие
impairment *s* повреждение,ухудшение
impala *s* импала, чернопятая антилопа *(Aepyceros melampus)*
impalement *s bot* чашечка
impalpable *a* неосязаемый, неощутимый
impaludism *s* малярия
impar *a* непарный
imparidigitate с нечётным числом пальцев, непарнопалый
imparipinnate *a* непарноперистый
impedance *s* импеданс
impediment *s* препятствие, задержка
impedimental *a* препятствующий, задерживающий
impenetrability *s* непроницаемость
imperceptible *a* незаметный
imperception *s* нарушение восприятия, отсутствие восприятия
imperfect *a* несовершенный; неполный, незаконченный
imperfection *s* несовершенство, неполнота
~ of the geological record неполнота геологической летописи
imperforate не имеющий отверстия
impermeability *s* непроницаемость
water ~ водонепроницаемость
impermeable *a* непроницаемый
impervious *a* непроходимый; непроницаемый; водонепроницаемый; недоступный
implant *s* имплантат; *v* имплантировать, производить имплантацию
implantation *s* имплантация
implicit неявный
imponderability *s* невесомость
imponderable *a* очень лёгкий, невесомый
imporosity *s* отсутствие пористости
imporous *a* беспоровый
imported *a* занесённый,завезенный; завозной
impotence, impotency *s* импотенция, бессилие, слабость; половое бессилие
impound *v* запруживать
impounding *s* затопление
impoundment *s* запруживание; накопление воды, создание водоёмов; водоём; водохранилище
impoverished *a* обеднённый, истощённый
impoverishment *s* истощение, обеднение
impregnate *v* насыщать, наполнять, пропитывать; импрегнировать; оплодотворять
impregnation *s* пропитывание; импрегнация; оплодотворение, осеменение; зачатие
impress *s* отпечаток
impression *s* вдавление, углубление; отпечаток
imprint *s* отпечаток
imprinting *s eth* запечатление, импринтинг
improvement *s* улучшение, усовершенствование; мелиорация, окультуривание
impuberal *a* неполовозрелый; *bot* неопушённый
impulse *s* импульс
impure *a* неочищенный, загрязнённый; неоднородный
impurity *s* примесь, загрязнение
inability *s* неспособность

inactivation *s* инактивация
inactivator *s* инактиватор
inactive *a* неактивный, инертный, пассивный
inadaptability *s* неприспособленность
inadaptive *a* неприспособительный
inadequacy *s* неадекватность; недостаточность; несоответствие
inadequate *a* несоответствующий; неадекватный; неполноценный, недостаточный, нарушенный
inalbuminate *a* безбелковый
in-and-inbreeding *s* узкородственное разведение
inanimate *a* неодухотворённый, неживой, безжизненный; инертный
inanimation *s* безжизненность; инертность
inanition *s* истощение, изнурение; дефицит питания
inaperturate *a* безапертурный, беспоровый
inappetence *s* отсутствие аппетита, отсутствие влечений
inarching *s* прививка сближением
inarticulate *a zool* несочленённый; немой
inassimilable *a* неусваиваемый
inborn *a* прирождённый, врождённый
inbred *a* инбредный
inbreeding *s* инбридинг, родственное спаривание; *bot* инцухт, самоопыление
in-calf *a* стельная *(cow)*
incapacity *s* неспособность
  hybrid ~ гибридная стерильность
incapsidation *s* включение в капсид; образование капсида
incense-bearing *a* ладаноносный
incentive *s* побуждение; *a* побудительный, стимулирующий
inception *s* начало
  fruit ~ завязывание плодов
incest *s* кровосмешение
inchoate *a* зачаточный, рудиментарный
incidence *s* встречаемость
incipient *a* начинающийся, зарождающийся; начальный
incised *a* надрезанный
incisiform *a* имеющий форму резца
incision *s* разрезание, надрезание; разрез; надрез, вырез
incisive *a* режущий, острый
incisor *s* передний зуб, резец
  central ~s *pl* зацепы
  corner ~s *pl* окрайки
  intermediate ~ средниц резец
incisure *s* вырезка; насечка
incitant *s* возбудитель; стимулятор; *a* возбуждающий; стимулирующий
incitation *s* возбуждение
inciting *s eth* подстрекание
inclined *a* наклонный; пологий, отлогий
included *a* включённый; окружённый; невыступающий; заключённый
including *a* включающий; заключающий
inclusion *s* включение
incoagulability *s* несвёртываемость, некоагулируемость
incoercible *a* неукротимый
incomer *s* пришелец, особь, пришедшая со стороны, иммигрант
incompatibility *s* несовместимость
  Rh ~ резус-несовместимость
incompatible *a* несовместимый, несогласующийся
incompetence *s* недостаточность; функциональная недостаточность
incomplete *a* неполный; с просветами, редкий *(e.g., forest)*
inconceivable *a* невоспринимаемый
inconspicuous *a* незаметный, неприметный; неяркий, бледный
inconstant *a* непостоянный; нерегулярный; неустойчивый
inconvertibility *s* необратимость
incoordination *s* нарушение координации
incorporation *s* включение; внедрение; ассимиляция
incrassate, incrassated *a* утолщённый
increase *s* возрастание, рост, прирост; увеличение, прибавление; увеличиваться; размножение; *v* возрастать; расти
  ~ **in weight** весовой прирост, привес
  arithmetic ~ арифметический прирост
  body-weight ~ привес
  geometric ~ геометрический рост, геометрическое приращение
increasing *a* растущий; увеличивающийся; усиливающийся
increment *s* прирост; привес; мол. биол. инкремент
  weight ~ весовой прирост, привес
incretion *s* внутренняя секреция

incretology *s* эндокринология
incretory *a* эндокринный
incrossbred *s* инкроссбред
incrossbreeding *s* инкроссбридинг
incrossing *s* инкроссинг
incrustation *s* инкрустация, покрывание коркой; кора, корка
incrusted *a* инкустированный, покрытый коркой
incubate *v* инкубировать; культивировать, выращивать
incubation *s* инкубация; культивирование, выращивание
incubative *a* инкубационный
incubator *s* инкубатор
incubatory *a* инкубационный
incubous *a* черепицеобразно налегающий *(e.g., leaves)*
incudal *a anat* относящийся к наковальне
incudate *a* инкудатный
incumbent *a* налегающий, лежащий на
incurrent *a* вдающийся, вклинивающийся
incurvate *a* согнутый; изогнутый; сгибать; *v* изгибать; выгибать
incurvation *s* сгибание; изгиб; выгиб; *emb* инкурвация
incus *s anat* наковальня
indeciduous *a* вечнозелёный, неопадающий
indecomposable *a* неразложимый
indefinable *a* неопределимый
indefinite *a* неопределённый; нечётко ограниченный
indehiscent *a* нерастрескивающийся, нераскрывающийся
indented *a* зазубренный, насечённый
independent *a* независимый
indeterminate *a* неопределённый
index *s* индекс, указатель, показатель; коэффициент
~ **of succession** индикатор фенологических смен, фенологический индикатор, индикатор смены аспектов
cephalic ~ черепной индекс, продольно-широтный индекс, черепной укаазатель
coli ~ коли-индекс, индекс кишечной палочки
frequency ~ коэффициент встречаемости, показатель частоты
indicator индикатор, указатель
indiffusible *a* недиффундирующий
indigenous *a* аборигенный, туземный, местный, автохтонный
indigestible *a* неперевариммый
indigo *s bot* софора *(Sophora);* аморфа *(Amorpha);* баптисия *(Baptisia);* индиго
indirect *a* косвенный
indissoluble *a* нерастворимый
indistinct *a* неясный, нечёткий
individual *s* особь, индивид, индивид(уум; характерный, особенный; индивидуальный; отдельный
individualization *s* индивидуализация
individuation *s* индивидуация; формирование отдельных особей
indolent *a* вялый, неактивный; нечувствительный к боли; безболезнанный
induced *a* индуцированный, вызванный
inducer *s* индуктор
induction *s* индукция
inductor *s* индуктор
indumentum *s bot* опушение; оперение; волосяной покров
induplicate *a bot* внутрсложенный
induration *s* уплотнение; затвердение, отвердение
indusial *a ent* имеющий индузий; *bot* имеющий покрывальце
indusium *s ent* индузий, чехлик, домик; *bot* покрывальце
induviate *a* покровный
inedible *a* несъедобный
inefficient *a* неэффективный, недостаточный, неспособный
inembryonate *a* беззародышевый
inequality *s* неравенство, разница; несходство, неодинаковость; изменчивость; неравномерность
inequivalve *a* неравностворчатый
inerm, inermous *a* невооружённый; неколючий; безостый
inert *a* инертный
infallible *a* безошибочный, верный
infancy *s* ранний детский возраст, грудной возраст
infanticide *s eth* детоубийство
infantile *a* инфантильный, младенческий
infarct *s* инфаркт
infarction *s* инфаркт

infect заражать
infection *s* инфекция, заражение
infectious *a* инфекционный, заразный
infectivity *s* инфективность; инвазионная способность
infecund *a* бесплодный; неплодородный
infecundity *s* бесплодие; неплодовитость; неплодородие
inferior *a* нижний
inferoanterior *a* нижнепередний
inferolateral *a* нижнебоковой
inferomedian *a* нижнемедианный, нижнесрединный
inferoposterior *a* нижнезадний
infertile *a* бесплодный, неплодородный
infertility *s* бесплодие; неплодовитость, неплодородие
infest *v* заражать, инвазировать
infestant *s* возбудитель заболевания
infestation *s* заражение, инвазия; нападение, нашествие; засорение, засорённость
infested *a* заражённый, инвазированный; заросший сорняками
infiltrate *s* инфильтрат; просачиваться; фильтровать
infiltration *s* инфильтрация; инфильтрат; инфильтрация, просачивание; фильтрат
inflamed *a* воспалённый
inflammation *s* воспаление
inflammatory *a* воспалительный
inflate раздувать, раздуваться, вздувать, вздуваться, надувать, надуваться
inflation *s* раздутость; вздутие, вздутость
inflection *s* сгибание, изгибание
inflexion *s* изгиб
inflorescence *s* цветорасположение; соцветие; цветение
  cymose ~, definite ~, determinate ~ цимозное соцветие, определённое соцветие, верхоцветное соцветие, верхоцветник
  tassel ~ соцветие-султан
inflow *s* приток, поступление
  proprioceptive ~ проприоцептивная импульсация
  rapid ~ быстрое заполнение
influence *s* влияние, воздействие
influent *s* инфлюэнт
in-foal *a* жерёбая
information *s* информация
informofere *s* информофер
informosome *s* информосома
infra-axillary *a* подпазушный
infrabranchial *a* поджаберный
infraclavicle *s* подключичная кость
infraclavicular *a* подключичный
infraconscious *a* подсознательный
infracortical *a* подкорковый
infracostal *a* подрёберный
infraesophageal *a* подпищеводный
infragranulate *a pal* внутреннезернистый
infrahyoid *a* подподъязычный
inframarginal *a* инфрамаргинальный
inframaxillary *a* подчелюстной
infraorbital *a* подглазничный
infrapunctate *a pal* внутреннеточечный
infrared *a* инфракрасный
infrareticulate *a pal* внутреннесетчатый
infrarostral *a* инфрарострaльный
infrarugulate *a pal* внутреннеморщинистый
infrascapular *a* подлопаточный
infrasculpture *s bot* внутренняя скульптура
infraspecific *a* внутривидовой
infraspinous *a* подостный
infrasternal *a* подгрудинный
infratectal *a pal* внутреннепокровный
infratemporal *a* подвисочный
infratrochlear *a anat* подблоковый
infructescence *s bot* соплодие
infundibular *a* инфундибулярный, воронкообразный, воронковидный
infundibuliform *a* воронкообразный, воронковидный
infundibulum *s* воронка
infuriate *v* приводить в ярость, разъярять
infuse *v* вливать; настаивать, приготавливать настой
infusion *s* вливание; настой, вытяжка, экстракт; настаивание
  hay ~ сенной настой
  meat ~ мясной экстракт
  saline ~ вливание физиологического раствора
infusorian *s* инфузория *(Ciliophora)*
ingest *v* глотать, проглатывать, заглатывать
ingestion *s* глотание, проглатывание, заглатывание; заражение через рот

ingluvies *s orn* зоб; желудок *(in ruminants)*
ingrained *a* вкраплённый
ingravidate оплодотворять
ingredient *s* ингредиент, составная часть
ingrowth *s* впячивание, врастание
inguen *s* пах
inguinal *a anat* паховый
inguino-abdominal *a anat* пахово-брюшной
inhabit *v* обитать, населять
inhabitant *s* обитатель
inhabitation *s* жилище, место обитания
inhalation *s* вдыхание; всасывание, засасывание
inhale *v* вдыхать; всасывать, засасывать
inhaust *v* всасывать, засасывать
inherent *a* прирождённый, врождённый; присущий, свойственный
inherit *v* наследовать
inheritable *a* наследственный; наследуемый
inheritance *s* наследственность; наследование
inhibit *v* тормозить, задерживать; подавлять, угнетать; **to be inhibited competitively** конкурентно ингибироваться *(e.g., of enzymes)*
inhibition *s* ингибирование, торможение, задержка; подавление, угнетение; **~ to remove ~** снять ингибирование
inhibitor *s* ингибитор, задерживающий фактор
inhibitory *a* ингибирующий, тормозящий, задерживающий; подавляющий, угнетающий
inhomogeneous *a* негомогенный, неоднородный; неравномерный
inimical *a* вредный; неблагоприятный; враждебный
inion *s* инион, затылочный бугор
initial *a* начальный, первоначальный, первичный, исходный
initiate *v* начинать, положить начало; стимулировать, побуждать
initiation *s* инициирование; стимулирование; *bot* заложение; *mbiol* инициация
  ~ of culture закладка культуры
  floral ~ начало цветения
initiator *s* инициатор
inject *v* инъецировать, вводить, впрыскивать
injection *s* инъекция,введение, впрыскивание
injured *a* повреждённый
injurious *a* вредный; губительный
injuriousness *s* вредоносность
injury *s* повреждение
  immunospecific ~ иммунологическое повреждение; иммунный лизис
in-kid суягная *(of a goat)*
in-lamb суягная *(of a ewe)*
inland *a* внутриматериковый, континентальный; удалённый от моря
in-lay *a* несущаяся
inlet *s* вход, входное отверстие
in-milk *a* дойная *(cow)*
innate *a* врождённый; природный
innervate *v* иннервировать; возбуждать; раздражать
innervation *s* иннервация
innocuous *a* безвредный, неядовитый
innovation *s* обновление; *bot* побег возобновления; эволюционное новообразование
innoxious *a* безвредный, неядовитый
innutrition *s* недостаточность питания
inocomma *s* инокомма
inoculate *v* инокулировать, прививать
inoculation *s* инокуляция, засев, внесение посевного материала; прививка, заражение; прививка глазком, окулировка
  ~ of medium посев на среду
  artificial ~ искусственное заражение
  heavy ~ обильный засев
  self ~ аутовакцинация
  streak ~ посев штрихом
inoculator *s* инокулятор
inoculum *s* инокулят, посевной материал, посевная культура
inodorous *a* без запаха
inoffensive *a* безвредный
inoperculate *a* не имеющий крышечки *(of a gastropod);* не имеющий оперкулума *(e.g., of mosses)*
inophragma *s* инофрагма
inordinate *a* беспорядочный
innorganic *a* неорганический
inosculation *s bot* анастомоз
inositol *s* инозит
inoxidable *a* неокисляющийся

in-pig *a* супоросная
inpour *v* просачиваться, проникать
input *s* вход; расход, потребление
inquiline *s* инквилин, животное-сожитель; *a* поселившийся, ставший местным
inquisitive *a* пытливый, любознательный
insalivation *s* смешивание пищи со слюной
inscissus *a* разорванный
insect *s* насекомое *(Insecta)*
  apterous ~ первичнобескрылое насекомое *(Apterygota)*
  armored scale ~ щитовка *(Diaspididae)*
  blood-sucking ~s *pl* кровососущие насекомые; гнус
  cuckoo spit ~ пенница, слюнявица *(Cercopidae)*
  gall ~ орехотворка *(Cynipidae)*
  gall-producing ~ насекомое-галлообразователь
  grain ~s насекомые-вредители зерна
  homopterous ~ равнокрылое хоботное *(Homoptera)*
  Indian lac ~ лаковый червец *(Laccifer lacca)*
  injurious ~, noxious ~ вредное насекомое, насекомое-вредитель
  leaf ~ листовидка, листотел *(Phylliidae)*
  lepidopterous ~ бабочка, чешуекрылое *(Lepidoptera)*
  nerve-winged ~ сетчатокрылое *(Neuroptera)*
  orthopterous ~ прямокрылое *(Orthoptera)*
  scale ~ червец, щитовка, кокцида *(Coccidae)*
  scale-winged ~ бабочка, чешуекрылое *(Lepidoptera)*
  spittle ~ пенница, слюнявица *(Cercopidae)*
  stick ~ палочник *(Phasmoptera)*
insectary *s* инсектарий
insect-eater *s* насекомоядное *(Insectivora)*
insect-eating *a* насекомоядный
insecticide *s* инсектицид
insectifuge *s* репеллент
insection *s* надрез, насечка
insectivore *s* насекомоядное *(Insectivora)*
insectivorous *a* насекомоядный
insectology *s* энтомология
inseminate *v* осеменять, оплодотворять
insemination *n* осеменение, оплодотворение
  artificial ~ искусственное осеменение, искусственное оплодотворение
insensibility *s* нечувствительность, невосприимчивость
insensible, insensitive *a* нечувствительный, невосприимчивый
insensitivity *s* нечувствительность
inseration *s* инсерация
insert *s mbiol* вставка, инсерт
inserted *a bot* вставочный
insertion вставка, инсерция; место прикрепления; *bioch* включение
  DNA ~ ДНК-вставка
  mini ~ мини-вставка
inshore *a* прибрежный
insiccation *s* высыхание; высушивание
inside *s* внутренняя сторона; внутренний; *pl* внутренности
insight *s eth.* инсайт, №улавливание отношений¤
insinuate *v* проникать, внедряться
insipid *a* безжизненный, вялый; безвкусный
insolation *s* инсоляция
insoluble *a* нерастворимый
inspection *s* проверка
inspiration *s* вдох; вдыхание
inspissation *s* уплотнение; сгущение
instability *s* неустойчивость, нестабильность
instaminate *a bot* бестычинковый
instar *s* возрастная стадия
instep *s* плюсна
instinct *s* инстинкт
  homing ~ инстинкт возвращения к месту рождения; органотропизм
instinctive *a* инстинктивный
instratified *a* безъярусный
instrument *s* инструмент, прибор, аппарат
  sampling ~ пробоотборник, прибор для отбора проб
insufficiency *s* недостаточность
insufficient *a* недостаточный; неудовлетворительный
insufflation *s* вдувание
insula *s anat* островок
insular *a anat* инсулярный, островковый
insulation *s* изоляция

insusceptibility *s* невосприимчивость; нечувствительность
intact *a* интактный, целый, неповреждённый
intake *s* всасывание; поглощение, потребление; приём внутрь
  food ~ приём пищи
  water ~ водозабор
intectate *a pal* без толстого покрова
integral *a* существенный, неотъемлемый; полный, целый, цельный
integrase *s* интеграза
integrate *a* составной; полный, целый, цельный
integration *s* интеграция, объединение; интегрирование
integrifolious *a* цельнолистный
integument *s* интегумент, покров семяпочки; наружный покров, оболочка
  fetal ~ покровная ткань плода, плодная оболочка
integumentary *a* покровный
intellect *s* интеллект, ум
intellection *s* мышление
intelligence *s* интеллект, ум; умственные способности; смышлённость, понятливость; сведения, информация
intensificator *s* интенсификатор
intensity *s* интенсивность, напряженность
  compensation ~ компенсационная точка фотосинтеза
  population ~ плотность популяции
intention *s* намерение
interact *v* взаимодействовать
interaction *s* взаимодействие
interarticular *a* межсуставный
interatrial *a* межпредсердный
interbody *s* амбоцептор
interbrain *s* промежуточный мозг
interbranchial *a* межжаберный
interbreeding *s* интербридинг; *bot* самоопыление
intercalary *a* интеркалярный, вставочный
intercalation *s* интеркаляция, вставочный рост
intercarpal *a* межзапястный
intercarpellary *a* межплодолистиковый
intercartilaginous *a* межхрящевой
intercellular *a* межклеточный
intercept *v* перехватывать; останавливать; препятствовать; преграждать
interception *s* перехватывание, перехват; преграждение, преграда
  light ~ улавливание света
interchange *s* обмен; чередование; *v* обмениваться; чередоваться, сменяться; заменять
  segmental ~ сегментный обмен, взаимная транслокация хромосом
interchangeable *a* взаимозаменяемый
interchondral *a* межхрящевой
interclavicle *s* межключичная кость
interclavicular *a* межключичный
intercolpate *a pal* межбороздный
intercommunicate *v* сообщаться
intercommunication *s* сообщение, связь
intercompatible *a* взаимноскрещивающийся
intercondyloid *a* межмыщелковый
interconnection *s* взаимосвязь, взаимная связь; соединение
interconversion *s* взаимное превращение
intercostal *a* межрёберный
intercostobrachial *a* межрёберно-плечевой
intercourse *s* общение; половые сношения
intercross *s* интеркросс; перекрёстное опыление; перекрёстное оплодотворение
interdental *a* межзубной
interdependence *s* взаимозависимость
interdigital *a* межпальцевой
interface *s* поверхность раздела; внутренняя поверхность
interfacing *a* смежный
interfascicular *a* межпучковый
interference *s* интерференция; вмешательство; столкновение
interferon *s* интерферон
interfertile *a* способный скрещиваться
interflow *s* слияние; сливаться, соединяться; внутрипочвенный сток, подповерхностный сток
interfoliaceous *a* межлистный
interfrontal *a* межлобный
interganglionic *a* межганглионарный
intergeneric *a* межродовой
intergenic *a* межгенный

intergradation *s* интерградация; трансгрессия
intergrafting *s* внутрисортовая прививка
intergrowth *s* прорастание; срастание
interincompatible *a* взаимнонескрещиваемый, интерстерильный
interinhibitive *a* взаимотормозящий, взаимно ингибирующий
interior *s* внутренняя сторона; *a* внутренний
interjacent *a* промежуточный; переходный
interkinesis *s* интеркинез
interlacement *s* сплетение, переплетение
interlamellar *a* межпластинчатый
interlayer *s* промежуточный слой, прослойка
interleukin *s* интерлейкин
interlobar *a anat* интерлобарный, междолевой
interlobular *a anat* интерлобулярный, междольковый
interlocking *s* интерлокинг, смыкание
intermandibular *a* межчелюстной
intermaxillary *a anat* межчелюстной; резцовый
intermediary *s* посредник; промежуточный
intermediate *s* промежуточный вид, промежуточная форма; промежуточный продукт; *a* промежуточный; *v* быть посредником
intermedin *s* интермедин
intermittent *a* скачкообразный; прерывистый; перемежающийся
intermixture *s* смесь
intermolt *a* межлиночный
intermuscular *a* межмышечный
internal *a* внутренний
internalization *s* интернализация
internals *s pl* внутренние органы
interneural *a* межневральный
interneurone *s* промежуточный нейрон, вставочный нейрон
internodal *a bot* межузловой, находящийся между узлами; относящийся к междоузлию
internode *s bot* междоузлие
rachis ~ членик оси колоса
internodorsal *a* внутренний спинной
internolateral *a* внутреннебоковой
internoventral *a* внутренний брюшной
interoceptive *a* интероцептивный
interoceptor *s* интероцептор
interorbital *a* межорбитальный, межглазничный
interosseous *a* межкостный
interparietal *s* межтеменная кость; *a* межтеменной
interpenetration *s* взаимопроникновение
interpetaloid *a* межлепестковый
interpetiolar *a* межчерешковый
interphase *s* интерфаза
interplay *s* взаимодействие; *v* взаимодействовать
interpretation *s* интерпретация
interpubic *a* межлонный, межлобковый
interramal *a zool* расположенный между ветвями
interreaction *s* взаимодействие
interreduplication *s* интерредупликация
interrelation *s* взаимоотношение; взаимосвязь; соотношение
interrelationship *s* взаимное родство; взаимоотношение; взаимосвязь
interrupt *v* прерывать, нарушать; препятствовать, мешать
interrupted *a* прерванный; прерывистый; с просветами, редкий *(of forests)*
interrupter *s* прерыватель
interruption *s* прекращение; приостановка; нарушение
interscapular *a* межлопаточный
intersegmental *a* межсегментный
intersex *s* интерсекс
intersexuality *s* интерсексуальность
interspace *s* промежуток
interspecific *a* межвидовой
interspinal, **interspinous** *a* межостистый
intersterility *s* перекрёстная стерильность, стерильность при гибридизации
interstice *s* промежуток; перегородка; интерстиций
interstitial *a* интерстициальный, промежуточный
intertentacular *a* межщупальцевый
intertidal *a* литоральный, межприливный
intertrill *s* межтрелевый интервал *(bioacoustics)*
intertropic *a* межтропический; тропический

interval *s* промежуток, интервал
intervalvular *a bot* межстворчатый; *anat* межклапанный
intervarietal *a* межсортовой
intervascular *a* межсосудистый
interventricular *a* межжелудочковый
intervertebral *a* межпозвоночный
intervillous *a* межворсинковый, межворсинчатый
interzonal *a* межзональный
interzone *s* промежуточное пространство; камера *(in diatoms)*
intestinal *a* кишечный
intestine *s* кишка; *pl* кишки; кишечник; кишечный тракт; *a* внутренний, междоусобный
hind ~ задняя кишка, проктодеум
intexine *s* интэкзина
intextine *s* интэкстина
intima *s* интима
intimidate *v* пугать; запугивать
intimidation *s* запугивание, угроза, устрашение; страх; запуганность
intine *s pal* интина
intolerant *a* нетерпимый; невыносливый
intorted *a* перевитый, закрученный
intorter *s* мышца, вращающая внутрь
intoxication *s* интоксикация, отравление
intra-abdominal *a* внутрибрюшной
intra-arterial *a* внутриартериальный
intra-articular *a* внутрисуставный
intra-atrial *a* внутрипредсердный
intra-aural *a* внутриушной
intrabaculate *a pal* внутристолбчатый
intracapsular *a* внутрикапсулярный
intracardiac *a* внутрисердечный
intracartilaginous *a* внутрихрящевой
intracavitary *a* внутриполостной
intracellular *a* внутриклеточный
intraclonal *a* внутриклональный
intracranial *a* внутричерепной
intracutaneous, intradermal *a* внутрикожный
intrafoliaceous *a* внутрилистный
intragastric *a* внутрижелудочный
intrageneric *a* внутриродовой
intragranulate *a* внутризернистый
intraluminal *a* находящийся в просвете
intramolecular *a* внутримолекулярный
intramural *a* интрамуральный
intramuscular *a* внутримышечный
intranasal *a* внутриносовой
intraneous *a* внутриареальный
intranuclear *a* внутриядерный
intraocular *a* внутриглазной
intraparietal *a* внутристеночный
intraperitoneal *a* внутрибрюшинный
intrapulmonic *a* внутрилёгочный
intrarenal *a* внутрипочечный
intrareticulate *a pal* внутрисетчатый
intrarugulate *a pal* внутриморщинистый
intrasculpture *s pal* интраскульптура
intraspecific *a* внутривидовой
intraspinal *a* интраспинальный
intrasplenic *a* внутриселезёночный
intrastamineous *a* внутритычиночный
intrastelar *a* расположенный внутри стелы
intrathoracic *a* внутригрудной
intratubal *a* внутритрубный
intrauterine *a* внутриматочный
intravaginal *a* интравагинальный, внутривлагалищный
intravascular *a* внутрисосудистый
intravenous *a* внутривенный
intraventricular *a* внутрижелудочковый
intravesical *a* внутрипузырный
intravital *a* прижизненный
intravitelline *a* внутрижелточный
intrazonal *a* интразональный
intricate *a* перепутанный
intrinsic *a* внутренний; присущий; прирождённый; наследственный
introduce *v* интродуцировать, вводить, внедрять
introduction *s* интродукция, введение, внедрение
introgression *s* интрогрессия
introgressive *a* интрогрессивный
introitus *s anat* отверстие, вход в полость
intromission *s* интромиссия, введение
intromittent *a zool* приспособленный для введения
intron *s* интрон
introrse *a bot* интрорзный, обращёный внутрь, обращёный к оси
introspection *s* интроспекция
introvert *s* хобот *(of sipunculids)*
intrude *v* вторгаться, внедряться
intruder *s* вторженец
intrusion *s* вторжение, внедрение
intubation *s* интубация
inturn *v* вворачиваться

intussusception *s* интуссусцепция, врастание, проникновение, внедрение, вхождение
inundated *a* заливной, пойменный; затопляемый
inundation *s* наводнение; затопление
invade *v* заселять; вторгаться
invader *s* внедряющийся вид, вселенец; захватчик, оккупант
invaginate *v* инвагинировать, впячиваться
invagination *s* инвагинация, впячивание
invariable *a* неизменяемый, неизменный, постоянный
invasibility *s* подверженность инвазии; подверженность колонизации, подверженность вторжению
invasion *s* заселение, вторжение, инвазия
desert ~ опустынивание
forest ~ наступление леса; облесение
multiple ~ множественная инвазия
shrub ~ закустаривание
inventory *s* учёт, подсчёт
forest ~ таксация леса
invermination *s* заражение гельминтами
inverse *a* обратный, противоположный; перевёрнутый
inversion *s* инверсия; переворачивание, выворот; извращение
inversive *a* инвертируемый; инвертирующийся
invert *v* инвертировать, обращать; переворачивать; *a* инвертный
invertebrate *s* беспозвоночное животное
inverted *a* инвертированный, обращённый
inverting *s* инвертирование; *a* инвертирующий
investigation *s* исследование
investment *s* внешний покров; *eth* вклад
inviable *a* нежизнеспособный
inviscation *s* смачивание пищи слюной
invisible *a* невидимый; неразличимый
invoke *v* вызывать
involucel *s bot* вторичная обвёртка, маленькая обвёртка
involucrate *a* имеющий обвёртку
involucre *s bot* обвёртка; покров
involuntary *a* непроизвольный, невольный
involute *a* завитой, свёрнутый, закрученный, завёрнутый внутрь
involution *s* инволюция, обратное развития, дегенерация, уменьшение, сокращение; завёртывание, закручивание
iodination *s* йодирование
iodine *s* йод
chlor zinc ~ хлорцинкйод
Lugol's ~ раствор Люголя
iodophilic *a* йодофильный, окрашиваемый йодом
ionization *s* ионизация
ionophore *s* ионофор
ipecacuanha *s* ипекакуана, рвотный корень *(Psychotria ipecacuanha)*
irascible *a* раздражительный
iridescent *a* радужный; флуоресцирующий
iridocyte *s* иридоцит
iris *s* радужная оболочка; *bot* касатик *(Iris)*
irised *a* радужный; флуоресцирующий
irisroot *s* фиалковый корень *(rhizome of Iris germanica florentina)*
iron-requiring *a* железолюбивый
irradiant *a* светящийся, сияющий; излучающий
irradiate *v* освещать; испускать лучи; облучать
irradiation *s* иррадиация, облучение; лучеиспускание; распространение
irreciprocal *a* необратимый, односторонний; необоюдный
irreclaimable *a* неисправимый
irregular *a* неправильный, искривлённый; неравномерный
irreversibility *s* необратимость
irreversible *a* необратимый
irrigation *s* орошение, полив
irritability *s* раздражимость; возбудимость; раздражительность
nervous ~ нервная возбудимость; раздражимость нерва
irritable *a* раздражимый; раздражительный
irritant *s* раздражитель, раздражающее средство; раздражающий, вызывающий раздражение
irritation *s phys* ирритация, раздражение; возбуждение; *eth* раздражение, гнев
irritative *a* раздражающий, вызывающий раздражение
irruption *s* вторжение, набег, нашествие

isauxesis *s* изогония
ischiadic *a* седалищный
ischiopubic *a* седалищно-лобковый
isidiferous *a* несущий изидий
isidium *s* изидий
island *s* остров; *anat* островок
islet *s anat* островок
isoaccepting *a mbiol* изоакцепторный
isoallel *s* изоаллель
isoallely *s* изоаллелизм
isoallergen *s* изоаллерген, изологичный аллерген
isoallotype *s* изоаллотип, изотип
isoanth *s* изоанта
isoantibody *s* изоантитело
isoantigen *s* изоантиген
isoantiserum *s* изоантисыворотка, изотипическая антисыворотка
isobath *s* изобата
isocarpous *a* изокарпический
isocellular *a* равноклеточный
isochromatic *a* изохроматический, одноцветный
isochromosome *s* изохромосома
isochronic, isochronous *a* изохронный, одновременный; одинаковой продолжительности
isocoenosis *s* изоценоз
isocortex *s* изокортекс
isocytic *a* равноклеточный
isodactylous *a* равнопалый
isodem *s* изодем
isodemic *a* изодемический, равнонаселённый
isodicentric *a* изодицентрический
isodont *a* изодонтный, равнозубый
isoelectrofocusing *s* изоэлектрофокусирование
isoenzyme *s* изофермент
isogamete *s* изогамета
isogamic, isogamous *a* изогамный
isogamy *s* изогамия
isogeneity *s* изогения
isogenesis *s* изогенез
isogenetic *a* сходного происхождения
isogenic *a* изогенный
isogenomatic *a* изогеномный
isognathous *a* изогнатический
isogony *s* изогония, пропорциональный рост
isohemagglutination *s* изогемагглютинация
isoimmune *a* изоиммунный
isoimmunization *s* изоиммунизация
  Rh ~ резус-изоиммунизация
isoimmunize *v* иммунизировать изоантигеном
isolate *s* изолят; штамм; культура; изолят, изолированная группа; *v* изолировать, отделять
isolateral *a* равносторонний
isolation *s* изоляция, изолирование; разобщение, отделение; выделение
isolecithal *a* изолецитальный
isomerization изомеризация
  reductive ~ гидроизомеризация, восстановительная изомеризация
isomerous *a* изомерный
isomery *s* изомерия
isometric *a* изометрический
isometry *s* изометрия
isomorphic *a* изоморфный
isomorphism *s* изоморфизм
isonome *s* изолиния коэффициентов встречаемости
isopetalous *a* равнолепестный
isophage *s* изофаг
isophene *s* изофена
isophenic, isophenous *a* изофенный
isophyllous *a* равнолистный
isophyte *s* изофита
isoploid *s* изоплоид
isopolar *a* равнополярный
isopolyploid *s* изополиплоид
isopolyploidy *s* изополиплоидия
isopycnic *a* изопикнический
isospore *s* изоспора
isosporous *a* равноспоровый
isostasy *s* изостазия
isotachophoresis *s* изотахофорез
isotasithynic *a pal* равнонатянутый
isotherm *s* изотерма
isothermal *a* изотермический
isothiocyanate *s* изотиоцианат
  fluorescein ~ флуоресцинизотиоцианат
isotonic *a* изотонический, изотоничный
isotonicity *s* изотония
isotope *s* изотоп
isotransplantation *s* изотрансплантация
isotrisomy *s* изотрисомия
isotropy *s* изотропия, изотропность
isotype *s* изотип
isotype-matched *a* изотипически сходный

isotypy *s* изотипическое разнообразие, изотипия
isovolumic *a* изоволюмический, изоволюметрический
isozygotic *a* изозиготный
isozygoty *s* изозиготность
isozym(e) *s* изозим, изофермент
issue *s* выход, результат
  side ~ *bioch* побочный продукт
isthmian *anat* относящийся к перешейку
isthmiate *anat* соединённый перешейком
isthmus *s* перешеек, перехват, перемычка
itch *s* зуд
item *s* объект
  food ~ компонент пищи; пищевой объект
iter *s anat* проход; канал; водопровод
iteration *s* итерация, повторение
itinerary *s* маршрут, путь
itograph *s* итограф
IUCN, International Union for Conservation of Nature and Natural Resources *s* МСОП, Международный Союз охраны природы и природных ресурсов
ivory *s* дентин бивней, клык
  elephant ~ слоновая кость
ivy *s* плющ *(Hedera helix)*

J

jack *s* самец *(e.g., of ass)*
jackal *s* шакал *(Canis aureus)*
jackass *s* самец
jackdaw *s* галка *(Coloeus monedula)*
jacket *s* шкура; кожа
jackrabbit *s* заяц *(Lepus)*
jacksnipe *s* гаршнеп *(Lymnocryptes minima);* бекас *(Gallinago gallinago);* кулик-дутыш *(Calidris melanotos)*
jaculator *s* плацентарный вырост; крючковатый вырост
jaculiferous *a* меченосный
jaeger *s* поморник *(Stercorarius)*
jagged *s bot* глубокоразрезной
jagger *s* блоха песчаная *(Tunga penetrans)*
jaguar *s* ягуар *(Panthera onca)*
jaguarundi *s* ягуарунди *(Felis yagouarundi)*
janitor *s anat* привратник
janitrix *s* воротная вена
jap *s* лай; *v* лаять
jar *s* резкий звук, дребезжащий звук; дисгармония; банка, сосуд
  bell ~ стеклянный колпак
  display ~ демонстрационный сосуд
  killing ~ *энт.* морилка
  large-mouthed ~ широкогорлая склянка
  screw-cap ~ склянка с завинчивающейся пробкой
  staining ~ сосуд для окрашивания
  wide-mouthed ~ широкогорлая банка
jararaca *s herp* обыкновенная жарарака *(Bothrops jararaca)*
jasmin, jasmine *s* жасмин *(Jasminum)*
jaundice *s* желтуха
jaw *s* челюсть; челюстная конечность; пасть; *pl* мастакс *(of rotifers)*
  lower ~ нижняя челюсть; *ent* максилла
  upper ~ верхняя челюсть; *ent* жвала, мандибула
jawbone *s* челюстная кость
jawfoot *s zool* ногочелюсть
jay *s* сойка; сойка *(Garrulus);* сойка *(Garrulus glandarius)*
  Siberian ~ кукша, ронжа *(Perisoreus infaustus)*
jejunum *s* тощая кишка
jelly *s* студень, студневидная масса
  egg ~ слизистая оболочка яйца
  royal ~ *ent* маточное молочко
  Wharton's ~ *anat* вартонов студень
jellylike *a* студенистый, желеобразный
jenny *s* самка; ослица
jerboa *s* тушканчик *(Dipodi dae)*
jerk *s* рефлекс; подёргивание; вздрагивание, резкое движение
jess *s zool* опознавательный воротник
jet *s* струя
jigger *s* блоха песчаная *(Tunga penetrans)*
jimp-stemmous *a* слабостебельный
jird *s ther* песчанка; малая песчанка *(Meriones)*
John's-wood, John's-wort *s* зверобой *(Hypericum)*
joint *s* соединение; сочленение; сустав; артроз; членик; *v* соединять; сочленять
  arthrodial ~ плоский сустав, амфиартроз
  ball-and-socket ~ шаровидный сустав, артродия
  false ~ ложный сустав, псевдартроз
  gliding ~ плоский сустав, амфиартроз
  ligamentous ~ синдесмоз
  plane ~ плоский сустав, амфиартроз
  ring ~ *ent* кольцевидный членик
  spheroidal ~ шаровидный сустав, артродия
jointing *s* членистость
jordanon *s* жорданон
journey *s* миграция
jowl *s* челюсть; подгрудок *(in cattle); orn* бородка; зоб; ганаш *(in a horse);* челюстная конечность; челюстная кость
juba *s* грива; *bot* метёлка
jubate *a* гривистый, с гривой; *bot* метельчатый, имеющий метёлку
judging *s* оценка, бонитировка
jugal *a* скуловой
jug-builder *s* оса-эвмен *(Eumenes)*
jugular *a* яремный
jugulum *s* нижнее горло *(у птицы)*
juice *s* сок
  joint ~ синовиальная жидкость
juicy *a* сочный
jumbo *s* крупная особь
jump *s* прыжок, скачок; *v* прыгать, подпрыгивать

junction *s* соединение; место соединения; синапс, мионевральное соединение
exon-intron ~ экзон-интронное сочленение, экзон-интронный стык, граница сплайсинга, точка сплайсинга
gap ~ нексус, щелевидное соединение, щелевой контакт
myoneural ~ нервно-мышечный синапс
tight ~ плотный контакт, плотное соединение
uterotubal ~ маточно-трубное сочленение, соединение матки с яйцеводом
V-D ~ участок [область] стыка V- и D-сегментов
juncture *s* шов
Juneberry *s* ирга *(Amelanchier)*
jungle *s* джунгли
jute *s* джут круглоплодный *(Corchorus capsularis)*
juvenile *s* молодая особь; *a* ювенильный
juxta-articular *a* околосуставный
juxtaposition *s* сопоставление

K

kairomone *s* кайромон
kakogenesis *s* какогенез; вырождение; ненормальное развитие
kalan *s* калан, морская выдра, камчатский бобр *(Enhydra lutris)*
kale *s* капуста *(Brassica)*
kalidium *s bot* калидиум
kalymma *s* калимма
kanchil *s* канчиль, азиатский оленёк *(Tragulus)*
kangaroo *s* кенгуру; настоящий [гигантский] кенгуру *(Macropus)*
kaodzera *s* африканский трипаносомоз, сонная болезнь
kaoliang *s* гаолян, сорго японское *(Sorghum japonicum)*
kappa *s prot* каппа-частицы
karst *s* карст
karyenchyma *s* кариолимфа, кариоплазма, нуклеоплазма
karyodesma *s* кариодесма, нуклеодесма
karyogamy *s* кариогамия
karyogene *s* кариоген, ядерный ген
karyogram *s* кариограмма
karyokinesis *s* кариокинез, митоз, непрямое деление клетки
karyokinetic *a* кариокинетический
karyolemma *s* ядерная оболочка
karyological *a* кариологический
karyology *s* кариология
karyolymph *s* кариолимфа, кариоплазма, нуклеоплазма
karyolysis *s* кариолизис, растворение ядра
karyomere *s* кариомер, кариомерит
karyomitosis *s* кариокинез, митоз, непрямое деление клетки
karyon *s* клеточное ядро
karyophore *s* кариофор
karyoplasm *s* кариоплазма, кариолимфа, нуклеоплазма
karyorhexis *s* кариорексис, фрагментация ядра
karyosome *s* кариосома
karyosphere *s* кариосфера
karyotheca *s* ядерная оболочка
karyotype *s* кариотип
karyotypic *a* кариотипический
kat *s bot* кат, ката съедобная *(Catha edulis)*
katabolism *s* катаболизм, диссимиляция
katagenesis *s* катагенез
kataplexy *s* катаплексия
katathermometer *s* кататермометр
katelectrotonus *s* катэлектротон
katharobic *a* катаробный
katharometer *s* катарометр, диаферометр
kea *s orn* кеа *(Nestor notabilis)*
ked *s* кровососка *(Hippobosca);* рунец *(Melophagus)*
keel *s zool* киль; *bot* лодочка
keeled *зоол.* килеватый; имеющий киль; *bot* ладьевидный
keep *v* содержать
keeping *s* содержание
keet *s* молодая цесарка
kemp *s* ость
kenaf *s bot* кенаф, гибискус коноплёвый *(Hibiscus cannabinus)*
kendyr *s bot* кендырь *(Apocynum)*
kennel *s* конура; лисья нора; свора собак; собачий питомник
keratinization *s* кератинизация, ороговение
keratitis *s* кератит
keratogenous *a* вызывающий ороговение
keratoid *a* кератиноидный, роговой, роговидный
kernel *s* зерно; косточка; ядро *(of a nut)*
kestrel *s orn* пустельга; обыкновенная пустельга *(Falco tinnunculus)*
  red-footed ~ кобчик *(Falco vespertinus)*
key *s* ключ, определительная таблица, определитель; отмель, риф; *bot* крылатка
key-peck *s phys* рефлекс клевания ключа
key-stimulus *s* ключевой стимул
kiang *s* кианг, тибетский кулан *(Equus kiang)*
kick лягать, лягаться; брыкаться
kicking *s* удар ногой
kid *s* детёныш; козлёнок; *v* ягниться
kiddaw *s orn* чистик *(Cepphus);* кайра *(Uria)*
kidding *s* ягнение
kidney *s anat* почка
  head ~ предпочка, головная почка, пронефрос
  hind ~ метанефрос, вторичная почка

mesonephric ~, primordial ~ мезонефрос, первичная почка, туловищная почка, вольфово тело
kidney-leaved *a* почколистный
kill *s* убийство; добыча; *v* убивать; ослаблять эффект, нейтрализовать; ослаблять, успокаивать *(pain)*
killdeer *s orn* крикливый зуёк *(Charadrius vociferus)*
killer *s* хищник; *prot* «убийца»; клетка-киллер, клетка-убийца
cicada ~ оса-истребитель цикад *(Sphecius speciosus)*
false ~ малая косатка *(Pseudorca crassidens)*
insect ~ инсектицид
lymphokine-activated ~ лимфокинактивированная клетка-киллер, ЛАК-клетка
pest ~ инсектицид; пестицид
killigrew *s orn* клушица *(Pyrrocorax pyrrocorax)*
killing *s* умерщвление, убийство; убой; отстрел; цитолиз, лизис клеток; убивающий
bystander ~ неспецифический цитолиз
pup ~ *eth* детоубийство
selective ~ выборочный отстрел
kin *s* род; семья; *a* родственный, подобный
kind *s* сорт; отличительный признак; природа; качество
kindle *s* молодое животное; приплод, помёт; **in** ~ беременная *(e.g., doe rabbit)*
kindred *a* родственный; *s* сродство, схожесть
kinesiology *s* кинезиология
kinesis *s* кинез, движение
kinesthesia *s* кинестезия, мышечное чувство, мышечно-суставное чувство
kinesthesiometer *s* кинестезиометр
kinesthesis *s* кинестезия
kinesthetic *a* кинестетический
kinetics *s* кинетика
kinetid *s prot* кинетида
kinetocardiogram *s* кинетокардиограмма
kinetocardiography *s* кинетокардиография
kinetochore *s* кинетохор, центромера
kinetodesma *s* кинетодесма
kinetoplast *s* кинетопласт
kinetosome *s* кинетосома, базальное зерно, базальное тело
kinety *s* кинета
king *ent* царь; *bot* главный стебель
complemental ~, ergatoid ~ *ent* эргатоидная половая особь
sea ~ сивуч *(Eumetopias jubatus)*
substitute ~ *ent* дополнительная половая особь
kingdom *s* царство; область, сфера
kingfisher *s* зимородок; обыкновенный зимородок *(Alcedo atthis)*
kinglet *s orn* королёк *(Regulus)*
kink *s* загиб, изгиб, перегиб; извитость
kinoplasm *s* киноплазма, сократимая цитоплазма
kin-selection *s eth* отбор родичей
kinship *s* родство; близость, сходство
kion *s* нёбный язычок
kiss *s eth* поцелуй
kit *s* котёнок; детёныш; комплект, набор, кит
kite *s* коршун, сокол, осоед
kitten *s* котёнок; *v* котиться
kittiwake *s orn* моевка *(Rissa)*; обыкновенная моевка *(Rissa tridactyla)*
kiwi *s orn* киви *(Apteryx)*; *bot* киви
kleistogamous *a* клейстогамный
kleptogamy *s eth* клептогамия
kleptoparasitism *s* клептопаразитизм
klinogeotropism *s* клиногеотропизм
klinokinesis *s* клинокинез
klinostat *s* клиностат
klinotaxis *s* клинотаксис, стремление к стимулу
klon *s* клон
knag *s* сук
knaggy *a* сучковатый
knarl *s* нарост, кап
knee *s anat* колено; *orn* тарзальное соединение
kneecap, kneepan *s* коленная чашка
knellgasbacteria *s pl* водородные бактерии
knife *s* нож
knife-handle *s mal* черенок *(Ensis)*
knob *s* шишка, выпуклость; *bot* нарост, желвак; клубенёк; вздутие
quil ~ *orn* маховой бугорок
knobby *a pal* крупитчатый, мелкокрупитчатый
knoll *s* кочка, бугор

knop *s* цветковая почка, бутон
knot *s* узел; нарост, кап; сук, сучок, свиль; *orn* исландский песочник *(Calidris canutus)*
nerve ~ нервный узел, ганглий
root ~ корневой нарост
knotgrass *s bot* горец птичий, спорыш *(Polygonum aviculare);* хвостник обыкновенный *(Hippuris vulgaris)*
knotted *a* шишковатый
knotty *a* узловатый
knotweed *s bot* горец *(Polygonum)*
knuckle *s* межфаланговый сустав
kob *s* коб, болотный козёл *(Kobus kob)*
koodoo *s* лесная антилопа *(Tragelaphus)*
kookaburra *s* кукабара, гигантский зимородок *(Dacelo)*
kouprey *s* купрей, индокитайский серый бык *(Bos sauveli)*
krone *s* крона, купа
K-selection *s* К-отбор
kudu *s* лесная антилопа *(Tragelaphus)*
kwagga *s ther* квагга *(Equus quagga)*
kymograph *s* кимограф

L

label *s* этикетка, ярлык; наклейка; метка; *v* метить; этикетировать, прикреплять ярлык
label(l)ing *s* этикетирование, наклейка этикеток, прикрепление ярлыков; мечение; введение метки; метка
labelloid *a* губовидный
labial *a* губной
labiate *s* губоцветное растение; *a* губоцветный, двугубый
labiatiflorous *a* губоцветный, двугубый
labidophorous *a* клешненосный
labile *a* лабильный, неустойчивый
labiogingival *a* губодесневой
labiomental *a* губно-подбородочный
labiopalatine *a* губно-нёбный
labium *s anat* губа; *энт.* лабиум, нижняя губа; *bot* нижняя губа
labor *s* роды; *v* родить
laboratory *s* лаборатория
labral *a ent* верхнегубной
labrocyte *s* лаброцит, тучная клетка
labrum *s ent* верхняя губа
labware *s* лабораторное оборудование
labyrinth *s anat* лабиринт
labyrinthodont *a* лабиринтодонтный
lacerate разрываться
laceration *s* разрыв
lacewing *s ent* златоглазка *(Chrysopidae)*
lacinia *s ent* лациния; *bot* доля
laciniate *a* рассечённый; с неровными краями; дольчатый; надрезанный
laciniform *a* долькообразный
lacinula *s* долька
lacinulate *a* мелкодольчатый; мелконадрезной
lack *s* недостаток; отсутствие; *v* испытывать недостаток
~ **of fodder** бескормица
lackey шелкопряд кольчатый *(Malacosoma neustria)*
lacmus *s* лакмус
lacrimal *a* слёзный
lacrimation *s* слезотечение
lacrimonasal *a* слёзно-носовой, носослёзный
lactalbumin *s* лактальбумин, молочный альбумин
lactation *s* лактация, выделение молока
lacteal *s* млечный сосуд; хилёзный, млечный; *bot* млечный
lactescent *a* молочный
lactic *a* молочный
lacticifer *s bot* млечник
lactiferous *a bot* млеконосный, млечный, выделяющий молоко, проводящий молоко
lactigenic, lactigenous *a* лактогенный, продуцирующий молоко
lactobacillus *s* молочнокислая бактерия
lactoflavin *s* лактофлавин, рибофлавин, витамин B2
lactose *s* лактоза, молочный сахар
lactoserum *s* сыворотка молока
lacuna *s* лакуна, впадина, углубление, полость
lacunar *a* лакунарный
lacune *s* лакуна, впадина, углубление, полость
lacustrine *a* озёрный
ladder *s* лестница
ladder-shaped *a* ступенчатый
laden *a* налитой *(of grain)*
ladies'-slipper *s* башмачок *(Cypripedium)*
ladybird, ladybug *s* божья коровка *(Coccinellidae)*
laeotropic *a* левонаправленный; левозавёрнутый
laevorotatory *a* левовращающий
lag *s* задержка; сдвиг фаз; скрытое время, латентный период
lageniform *a* фляжковидный; бутылеобразный
lagg *s* верховое болото
lagging *s* отставание, запаздывание
lagoon *s* отстойник; лагуна
aerated ~ аэрируемый пруд
sludge ~ илонакопитель
lag-phase *s* лаг-фаза
laid *a* отложенный *(eggs);* полегший
laim *s bot* лайм настоящий *(Citrus aurantifolia)*
laika *s* лайка
lair *s* логовище, берлога; нора; *v* лежать в берлоге
lake *s* озеро
lacrimal ~ *анат.* слёзное озеро
man-made ~ водохранилище
laked *a* гемолизированный
lake-feeding *a* питающийся на озёрах

"laking" of cells *s* лизис клеток; гемолиз
lama *s* лама *(Lama)*
lamarckianism, lamarckism *s* ламаркизм
lamb *s* ягненок; *v* ягниться
  female ~ ярочка
lambing *s* окот, ягнение
lambliasis, lambliosis *s* лямблиоз
lame *a* хромой
lamella *s* пластинка; чешуйка; тонкий слой; ламелла
lamellar, lamellate *a* пластинчатый; чешуйчатый; слоистый
lamellibranch *s* пластинчатожаберный моллюск *(Lamellibranchia)*
lamellicorn *a ent* пластинчатоусый
lamelliform *a* пластинчатовидный; чешуеобразный
lamellirostral *a* пластинчатоклювый
lamellose *a* пластинчатый; чешуйчатый; слоистый
lamiation *s plsoc* ламиация
lamies *s plsoc* ламия
lamina *s* тонкая пластинка; тонкий слой; *bot* листовая пластинка
  basal ~ базальная пластинка; базальная пластинка, оболочка Бруха
  fibrous [nuclear] ~ ядерная пластинка, ламина
laminar *a* ламинарный, пластинчатый; слоистый
laminated *a* пластинчатый; слоистый; *бот.* пластинконосный
lamination *s* стратификация
lammergeier, lammergeyer *s orn* бородач, ягнятник *(Gypaetus barbatus)*
lamp *s* лампа
  spirit ~ спиртовка
lamprophyllous *a* блестящелистный
lanate *a* шерстистый; пушистый; опушённый; ворсистый
lance-leaved *a* ланцетолистный
lance-linear *a* линейно-ланцетовидный
lance-oblong *a* удлинённо-ланцетный
lanceolate *a* ланцетовидный, ланцетообразный
lance-oval *a* овально-ланцетовидный
lancet *s* ланцет; *ent* стилет
land *s* земля; угодье; местность, страна
  barren ~ пустошь
  cut-over ~ вырубка
  fall-plowed ~ зябь
  forest ~ лесное угодье
  grazing ~ пастбище, пастбищное угодье, выпас
  heathy ~ верещатник
  logged ~ вырубка
  pasture ~ пастбище, выпас
  plowing ~ пашня, пахотная земля
  virgin ~ целинная земля, целина
landing *s* посадка, приземление; улов *(e.g., of fish)*
landrail *s orn* коростель *(Crex crex)*
landscape *s* ландшафт
landscaping *s* озеленение, ландшафтная архитектура
landslide *s* оползень
langur *s* тонкотелая обезьяна, лангур *(Pygathrix);* хульман, гульман *(Presbytis entellus)*
lanner *s* средиземноморский сокол, ланнер *(Falco biarmicus)*
lanose *a* шерстистый; пушистый; опушённый; ворсистый
lantern *s* фонарь
lanuginous *a* покрытый пушком; пушистый
lanugo *s* лануго, первичный волосяной покров
lap лакать
lapidicolous *a* живущий под камнями
lappaceous *a* колючий, с колючками
lappet *s zool* лопасть; выступ; коконопряд *(Lasiocampidae)*
lapwing *s* чибис, пигалица *(Vanellus)*; чибис, пигалица *(Vanellus vanellus)*
lar *s ther* белорукий гиббон *(Hylobates lar)*
larch *s* лиственница (*Larix)*
large *a* большой, крупный; широкий; массовый, многочисленный
large-berried *a* крупноягодный
large-flowered *a* крупноцветковый
large-fruited *a* крупноплодный
large-grained *a* крупнозернистый
large-headed *a* крупноголовый, большеголовый
large-leaved *a* крупнолистный
large-seeded *a* крупносеменной
lark *s* жаворонок *(Alaudidae)*
larkspur *s* живокость, шпорник *(Delphinium)*
larva *s* личинка; гусеница
  ammocoete ~ пескоройка
  cercaria ~ церкария

coarctate ~ ложная куколка
frog ~ головастик
lepidopterous ~ гусеница
Loren's ~ трохофора
nauplius ~ науплиус
pilidium ~ пилидий
planula ~ планула
trochophore ~ трохофора
veliger ~ велигер
zoaea ~ зоэа
larva-eating *a* личинкоядный
larval *a* личиночный
larvated *a* маскированный, атипичный, скрытый
larvicide *s* ларвицид
larvina *s* безногая личинка насекомых
larviparous *a* отрождающий живых личинок
larvivorous *a* личинкоядный
laryngeal *a* гортанный
larynx *s* гортань
lash *v* хлестать; *s* хлыст
eth ~ **the tail** бить хвостом
lassitude *s* утомление; утомляемость; усталость
lata-type *s gen* лата-тип
late *a* поздний, позднеспелый
latebricolous *a* обитающий в углублении
latebrosus *a* скрытый; укрытый
late-flowering *a* поздноцветущий
latency *s* скрытое состояние, латентное состояние
lateness *s* позднеспелость
latent *a* латентный, скрытый; покоящийся, недеятельный; *bot* спящий
lateral *a* латеральный, боковой; боковой побег
lateralization *s phys* латерализация
lateriflorous *a* бокоцветковый
laterigrade *a* бокоходящий
late-ripening *a* позднеспелый
laterite *s* латерит
lateroflexion *s* латерофлексия, боковое сгибание
latex *s* латекс, млечный сок
laticiferous *a* содержащий латекс
latidentate *a* широкозубчатый
latimurate *a* широкоперегородчатый
latiporate *a* широкопоровый
latirostral *a* ширококлювый
latiseptate *a* широкоперегородчатый
lattice *s* решётка; решётчатая структура
latticed *a* решётчатый
laurel *s* лавр *(Laurus);* гардения волосистая *(Gardenia lasianthus)*
lavender *s* лаванда колосковая *(Lavandula spica)*
law *s* закон; правило; принцип
lawn *s* газон
lax *a* слабый, вялый; рыхлый; разбросанный
lay *v* откладывать *(eggs);* метать *(eggs);* ставить *(trap);* устраивать *(trap);* **to be in** ~ нестись
layer *s* слой; пласт; несушка; *bot* черенок, отводок; стелющийся побег; ярус
air ~ воздушный отводок
crown ~ полог
feeder ~ питающий подслой, фидер
germ(inal) ~ зародышевый слой, зародышевый листок
granular ~ of Ebner зернистая пластинка Эбнера
serpentine ~ змеевидный отводок
shrub ~ кустарниковый ярус; подлесок
sporiferous ~ гимений; спороносный слой
stool ~ отпрыск
layerage, layering *s* размножение отводками; ярусность
laying *s* кладка яиц; период кладки яиц; разбивка, разметка
lea *s* пастбище; залежь; залог, перелог
leach *s* рапа; *v* выщелачивать; высолаживать
leaching *s* выщелачивание; высолаживание; *a* выщелачивающийся; высолаживающийся
lead *s phys* отведение; *biom* опережение, упреждение
leader *s* главная ветвь; главный побег; вожак; доминант; *mbiol* лидерная последовательность; *orn* розовидный гребень
leadership *s eth* водительство, лидерство, руководство
leading off *s phys* отведение
leaf *s bot* лист; покрываться листвой, одеваться листьями; **in** ~ с распустившимися листьями; покрытый листвой; **to be in ~, to come**

into ~, to ~ out покрываться листьями, распускаться
acerose ~ хвоя
scale ~ чешуйчатый лист; почечная чешуя
tomato yellow ~ *phytop* желтолистность томата
leafage *s* листва; крона
leaf-bearing, leafed *a* облиственный
leaf-feeding *s* внекорневое питание
leafhopper *s ent* цикадка, кобылочка *(Cicadellidae)*
leafing *s* листосложение
leafless *a* безлистный
leaflet *s* листочек; молодой лист; *anat* листок, створка
leaf-litter *s* листовая подстилка
leaf-shaped *a* листовидный
leafstalk *s* черешок
leaftier *s* листовёртка
leafy *a bot* облиственный; листостебельный *(of liverworts);* лиственный *(of mosses)*
leakage *s* утечка, просачивание; фильтрация
leaky *a* растекающийся; неплотно соединённый, негерметичный
lean *a* тощий, худой; обеднённый *(of soil)*
leap *s* прыжок; *v* прыгать
learning *s* обучение; образование навыков; *imm* «обучение», коммитирование
relational ~ выработка навыка на отношения
reversal ~ переделка навыка
leatherback *s* кожистая черепаха *(Dermochelys); orn* ямайская савка *(Oxyura jamaicensis)*
leatherjacket *s* личинка долгоножки *(Tipulidae)*
leatherlike *a* кожистый
leathery *a* кожистый
leaved покрытый листьями, облиственный
closely ~ загущённолистный
leathery ~ кожистолистный
sparsely ~ с изреженной кроной
woolly ~ пушистолистный
leaven *s* закваска
lechwe *s* водяной козёл личи *(Kobus leche)*
lecithin *s* лецитин
lectin *s* лектин, фитогемагглютинин
lectotype *s* лектотип
LD, lethal dose *s* летальная доза
lee *s* защита, укрытие
leech *s* пиявка *(Hirudinea)*
leek *s* лук-порей *(Allium porrum)*
left-handed *a* левозарученный, левый; леворукий; левша
leg *s* нога; опора, стойка
~ of brain *ent* окологлоточный коннектив
daddy long ~ *zool* сенокосец *(Phalangida); ent* долгоножка *(Tipulidae)*
trunk ~ *zool* переопод
vestigal ~ *ent* грифелёк
leggy *a* длинноногий, высоконогий
Leghorn *s* леггорн
legume *s bot* боб; бобовое *(Leguminosae)*
leguminous *a* бобовый
leiosporous *a* имеющий гладкие споры
leishmaniasis *s* лейшманиоз
lek *s eth orn* токовище
lemma *s* нижняя цветковая чешуя
lemic *a* инфекционный; чумной
lemming *s* настоящий лемминг *(Lemmus);* копытный лемминг *(Dicrostonyx);* леммингoвая мышь *(Synaptomys)*
mole ~ слепушонка *(Ellobius)*
steppe ~ степная пеструшка *(Lagurus)*
yellow steppe ~ степная жёлтая пеструшка *(Eolagurus luteus)*
lemmocyte *s* леммоцит, шванновская клетка
lemniscus *s* лемниск, вырост гиподермы
lemon *s* лимон *(Citrus limon)*
lemur *s* лемур *(Lemuridae)*
slow ~ толстый лори *(Nycticebus)*
length *s* длина
critical day ~ критическая длина дня, критическая длительность дня
day ~ длина дня, длительность дня
lenghtened *a* удлинённый
lens *s* лупа; линза; хрусталик
crystalline ~ хрусталик глаза
lenticel *s bot* чечевичка
lenticulate *a bot* имеющий чечевички; чечевицеобразный
lentiform *a* чечевицеобразный
lentigerous *a* имеющий хрусталик

lentiginose, lentiginous *a* пятнистый, крапчатый
lentil *s* чечевица *(Lens)*
lentivirus *s* лентивирус
leonine *a* львиный
leopard *s* леопард, барс, пантера *(Panthera pardus)*
  American ~ ягуар *(Panthera onca)*
  hunting ~ гепард *(Acinonyx jubatus)*
  painted ~ оцелот *(Felis pardalis)*
lepal *s* нектарник
lepidoid *a* чешуевидный
lepidopterous *a* чешуекрылый
lepidote *a* покрытый чешуйками
lepocyte *s* лепоцит
leporin *a* заячий
leprosy *s* лепра, проказа
leptocentric *a* лептоцентрический
leptocercal *a* тонкохвостый
leptodactylous *a* тонкопалый
leptome *s* лептом
leptonema *s* лептонема
leptophloem *s* рудиментарная флоэма
leptophyllous *a* чешуйчатолистный
leptotene *s* лептотена
leptoxylem *s* рудиментарная древесина
lesion *s* повреждение, поражение
lethal *s* летальный ген, летальный фактор, леталь; *a* летальный, смертельный
lethality *s* летальность, смертность
lettuce *s bot* латук *(Lactuca)*
leucite *s* лейкопласт
leucocarpous *a bot* белоплодный
leucocytal *a* лейкоцитарный, относящийся к лейкоцитам
leucocyte *s* лейкоцит
leucocytolytic *a* разрушающий лейкоциты
leucocytopenia, leucopenia *s* лейкопения
leukocytosis *s* лейкоцитоз
leucoplast, leucoplastid *s* лейкопласт
leucopoiesis *s* лейкопоэз, образование лейкоцитов
leucotoxic *a* ядовитый для лейкоцитов; разрушающий лейкоциты
Leu-enkephalin *s* Leu-энкефалин, лейцинэнкефалин
leukoagglutination *s* лейкоагглютинация
leukoagglutinin *s* лейкоагглютинин
leukocyte *s* лейкоцит
  basophilic ~ базофильный лейкоцит, базофил
  eosinophilic ~ эозинофильный лейкоцит, эозинофил
  granular ~ зернистый лейкоцит, гранулоцит
  mononuclear ~ мононуклеарный лейкоцит, мононуклеар
  neutrophilic ~ нейтрофильный лейкоцит, нейтрофил
leukocytic *a* лейкоцитарный; относящийся к лейкоцитам
leukocytogenesis *s* образование лейкоцитов, лейкопоэз
leukocytolytic *a* разрушающий лейкоциты
leucocytopenia *s* лейкопения
leukopoiesis *s* лейкопоез, образование лейкоцитов
leucopenia *s* лейкопения
leukopoietic *a* образующий лейкоциты
leukotoxic *a* ядовитый для лейкоцитов; разрушающий лейкоциты
leukovorin *s* лейковорин, цитроворум-фактор, фолиновая кислота
levator *s* леватор, поднимающий мускул
level *s* уровень
  dietary protein ~ доля белка в пище
  prechallenge antibody ~ исходный титр антител
  stocking ~ уровень пастбищной нагрузки, пастбищная нагрузка
leveret *s* молодой заяц
lever-pressing *s* нажатие на рычаг *(в экспериментах по поведению животных)*
levigate *v* растирать в порошок; отмучивать; вымывать; выщелачивать
levotropic *a* левонаправленный; левозавёрнутый
liana, liane *s* лиана
liber *s* флоэма, луб
liberate *v* освобождать
liberatio *s* освобождение
library *s mbiol* библиотека
  epitope ~ *imm* эпитопный банк
  gene ~ клонотека генов, библиотека генов, генотека
  recombinant ~ библиотека рекомбинантных ДНК
lichen *s* лишайник *(Lichenes)*

reindeer ~ олений мох *(Cladonia rangiferina)*
lichen-covered *a* покрытый лишайниками, обросший лишайниками
lichenicole, lichenicolous *a* обитающий на лишайниках
lichenism *s* лихенизм
lichenoid *a* лишайниковидный
lichenology *s* лихенология
lick *v* лизать; облизывать; зализывать; *s* лизунец
licorice *s bot* солодка *(Glycyrrhiza)*
lid *s* веко; *bot* крышечка
lien *s* селезёнка
lienal *a* селезёночный
life *s* жизнь
average ~ средняя продолжительность жизни
shelf ~, storage ~ срок хранения
life-form жизненная форма, биологическая форма
lifeless *a* безжизненный
life-span *s* время жизни, продолжительность жизни; период полураспада
life-style *s* образ жизни
lifetime *s* время жизни; срок службы
half ~ период полужизни, время полужизни; период полураспада, время полураспада
lifting *s* поднимание; *agr* выпирание
~ **potatoes, potato** ~ сбор картошки
ligament *s anat* связка, лигамент
ligamentary *a* связочный
ligand *s* лиганд
ligand-gated *a* управляемый лигандами
ligation *s* лигатура; наложение лигатуры; *mbiol* сшивание, лигирование
ligature *s* лигатура
light *s* свет; освещение; освещать; светлый; лёгкий
infrared ~ инфракрасный свет, ИК-свет
ultraviolet ~ ультрафиолетовый свет, УФ-свет
light-absorbing *a* светопоглощающий
light-avoiding *a* светобоязливый
light-bay *a* буланый *(of a horse)*
light-diffusing *a* светорассеивающий
light-dun *a* светло-гнедой *(of a horse)*
lighted *a* изреженный, разреженный *(e.g., forest)*
light-enduring *a* световыносливый
light-footed *a* быстроногий
light-harvesting *a* светособирающий
lighting *s* освещение
lightproof *a* светоустойчивый; светонепроницаемый
light-requiring *a* светолюбивый
light-sensitive *a* светочувствительный
light-tight *a* светоустойчивый
ligneous *a* древесный, деревянистый
lignescent *a* деревенеющий, древеснеющий
lignicole, lignicolous *a* обитающий в древесине
lignification *s* лигнификация, одревеснение
lignify *v* древеснеть
lignivorous *a* питающийся древесиной
lignosa *s* древесная растительность
ligula *s* лигула, язычок
ligulate, ligulated язычковый; имеющий язычок
ligule *s* лигула, язычок
liguliflorous *a* имеющий язычковые цветки
lugulifolious *a* язычковолистный
likelihood *s biom* правдоподобие
lilac *s* сирень *(Syringa); a* сиреневый
lily *s* лилия *(Lilium)*
lily-of-the-valley *s* ландыш майский *(Convallaria majalis)*
limb *s* конечность; плечо хромосомы; *bot* отгиб; ветвь первого порядка; листовая пластинка
decaying ~ *phys* катакрота
parapodial ~ *zool* параподия
limbate *a* окаймлённый
limbic *a* краевой; лимбический
limbless *a* не имеющий конечностей
limbous *a* слегка перекрывающийся
limbus *s* край; ободок; кайма
lime *s* известь; *bot* лайм настоящий *(Citrus aurantifolia);* липа *(Tilia)*
limebody *s helm* известковое тельце
limen *s phys* предел, порог
limicolous *a* обитающий в иле
liminal *a* пороговый, относящийся к порогу
limit *s* предел, граница; *biom* лимит, предельное значение

acceptance ~s *pl biom* допустимые пределы
age ~ предельный возраст
bordering ~s *pl* пограничные области
control ~s *pl biom* контрольные пределы
pain ~ болевой порог
limitation *s* ограничение; ограниченность; лимитирование
dual ~ двойное лимитирование
space ~ пространственная ограниченность
limivorous *a* илоядный
limnetic *a* озёрный; пресноводный
limnicolous *a* озёрный
limnodium *s* болотное сообщество
limnodophilus *s* обитатель болот
limnophyte *s* лимнофит
limosphere *s bot* лимосфера
limp *s* хромота; *a* вялый, поникший; *v* хромать
limpet *s zool* блюдечко *(Patella; Acmaea)*
limp-leaved *a* склонённолистный
linden *s* липа *(Tilia)*
line *s* линия; происхождение, родословная; генеалогическая линия; черта
~ **of Owen** полоска Оуэна, параллельная дентинная полоска
~ **of Retzius** полоска Ретциуса, паралелльная эмалевая полоска
base ~ исходный материал, основной материал, исходные данные
blood ~ кровная линия, семейство
Farro-Waldeyer's ~ линия Фарро-Вальдейера, пограничный край брюшины
Hunter-Schreger ~s, incremental ~s полоски Гунтера - Шрегера, чередующиеся эмалевые полоски
Krause's ~ полоска Краузе, полоска T, телофрагма
lateral ~ боковая линия; боковой валик
lineage *s* ряд поколений; происхождение по прямой линии, родословная; линия дифференцировки, направление дифференцировки
germinal ~ последовательность по прямой линии
lineal *a* происходящий по прямой линии
lineament *s* очертание; отличительная черта
linear *a* линейный
linear-ensate *a* линейно-мечевидный
linear-fusiform *a* удлинённо-веретеновидный
linear-lanceolate *a* линейно-ланцетный
linear-oblong *a* линейно-продолговатый
linebreeding *s* линейное разведение
linecrossing *s* линейное скрещивание
lineolate *a* мелколинейчатый; исчерченный
lingua *s* язык; *ent* гипофаринкс
lingual *a* относящийся к языку, язычный
linguiform *a* языковидный
lingula *s* язычок
lingulate *a* язычковидный
lining *s* расположение в линию; выстилка
link *s* звено, связь; *v* связывать, сцеплять
linkage *s* связь, сцепление
genetic ~ сцепление генов; группа сцепленных генов
linked *a* связанный, сцепленный; соединённый
linker *s mbiol* линкер
linneon *s* линнеон, линнеевский вид
linnet *s orn* чечётка *(Acanthis);* коноплянка *(Acanthis cannabina)*
green ~ зеленушка *(Chloris)*
mountain ~ горная чечётка *(Acanthis flavirostris)*
linseed *s* льняное семя
lint *s* лён *(Linum);* лён обыкновенный *(Linum usitatissimum);* подпушек *(on seeds)*
lion *s* лев *(Panthera leo)*
Mexican [mountain] ~ пума, кугуар *(Felis concolor)*
lioness *s* львица
lip *s* губа; *a* губной
under ~ *anat* нижняя губа; *ent* лабиум
upper ~ *anat* верхняя губа; *ent* лабрум
"lip-curl" *s eth* «ощеривание»
lip-flowered *a* губоцветковый
lipid, lipide *s* липид
lipoblast *s* жировая клетка
lipocaic *s* липокаик
lipochrome *s* липохром
lipoclastic *a* жирорасщепляющий
lipocyte *s* жировая клетка

lipogenesis *s* липогенез, образование жира
lipogenetic, lipogenous *a* жирообразующий
lipogranule *s* липоидная гранула
lipoid *s* липоид; *a* липоидный
lipolysis *s* липолиз, расщепление жира
lipolytic *a* липолитический, жирорасщепляющий
lipometabolism *s* жировой обмен
lipopexia *s* отложение жира
lipophil *a* липофильный
lipophore *s* липофор
lipoprotein *s* липопротеид
liposoluble *a* жирорастворимый
liposome *s* липосома
lipotropy *s* липотропия
lipoxenous *a* липоксенный
lipoxeny *s* липоксения
liquefaction *s* разжижение
liquefy *v* разжижать
liquid *s* жидкость; *a* жидкий
liquor *s* жидкость; раствор; экстракт; настойка
lirella *s* лирелла, линейный апотеций
lisocyme *s* лизоцим
lissosphincter *s* гладкомышечный сфинктер
list *s* список, реестр, перечень; *v* вносить в список
  check ~ каталог, систематизированный перечень, список *(видов, родов)*
litharch *s* литофитная серия
lithocyst *s* литоцист
lithodesma *s* литодесма
lithodomous *a* сверлящий камни
lithophagous *a* сверлящий камни; заглатывающий камни *(of biords)*
lithophilous *a* литофильный
lithophyte *s* литофит
lithosere *s* литофитная серия
litmus *s* лакмус
litter *s* выводок, помёт, приплод; подстилка
  leaf ~ листовая подстилка, листовой опад
littling *s* детёныш в выводке
littoral *s* литораль; *a* литоральный
live *v* жить
live-bearer *s* живородящее животное
live-bearing *a* живородящий
live-born *a* живорождённый
liver *s* печень
liver-pancreas *s* гепатопанкреас
liverwort *s* печёночник *(Hepaticae)*
livery *s* одеяние, наряд, убор; покров
livestock *s* скот
lizard *s* ящерица *(Lacertilia)*
  anguine ~ веретеница *(Anguis)*
  beaded ~ ядозуб *(Heloderma)*
  flying ~ летучий дракон *(Draco volans)*
  monitor ~ варан *(Varanus)*
  ringed ~s *pl* амфисбены, двуходковые *(Amphisbaenidae)*
llama *s* лама *(Lama)*
load *s* груз, нагрузка; нагружать; взвесь; нанос, наносы
  hereditary ~ наследственная отягощённость
  mutational ~ мутационное давление
  pollen ~ *apic* обножка
loaf *s* кочан; *v* образовывать кочан
loam *s* суглинок
  sandy ~ супесь
lobar *a* долевой; лопастный
lobate *a* дольчатый
lobation *s bot* лопастность
lobe *s* доля; лопасть
  anther ~ *bot* гнездо пыльника
  calyx ~ чашелистик
  limbic ~ сводчатая извилина, лимбическая доля, доля Брока
  perianth ~ доля околоцветника, член околоцветника
  posterior ~ of the wing *ent* крылышко
  seed ~ семядоля
  side ~ боковой луч *(in hydroacoustics)*
lobed *a* дольчатый; разделённый на лопасти
  pinnately ~ перистолопастный
lobe-ears *a* вислоухий
lobopodium *s* лобоподия
lobular *a* дольчатый; дольковый
lobulate, lobulated *a* мелкодольчатый; с мелкими лопастями
lobule *s* долька; маленькая лопасть
  ~ of the ear мочка уха
  fat ~ жировая долька
local *a* местный, локальный
locale *s* местность
locality *s* местность; местообитание, местонахождение
localization *s* локализация
  germinal ~ *emb* положение зачатков

locate локализировать, определять местонахождение
location *s* местоположение
lochmium *s* сообщество зарослей
lochmodium *s* сообщество сухих зарослей
lochmodophilus *s* обитатель сухих зарослей
lociation *s* лоциация
locies *s* лоция
lock *s* пучок *(of hair);* клок *(of fur)*
genital ~ *eth* склещивание
locomotion *s* локомоция, движение, передвижение
concertina ~ гармошкообразное ползание
locomotor, locomotory *a* локомоторный, двигательный
locular, loculate *a* разделённый на полости, разделённый на камеры
locule, loculus *s zool* камера; *bot* гнездо
locus *s* очаг, фокус; локус
origin ~ точка инициации
locust *s* саранча; цикада; кобылка; *pl* настоящие саранчовые *(Acrididae)*
Italien ~ прус итальянский *(Calliptamus italicus)*
seventeen-year ~ цикада семнадцатилетняя *(Cicada septemdecim)*
lodge *s* хатка *(of beavers);* охотничий домик; полегать
lodging *s* полегание
selective ~ *cyt* избирательное заселение ниши
lodicule *s bot* лодикула, маленькая плёночка
loft *s* голубятня; *v* держать голубей
log *s* бревно; полевой журнал
loganberry *s* логанова ягода *(Rubus idaeus x Rubus ursinus)*
loggerhead *s:*
olive ~ оливковая черепаха *(Lepidochelys olivaceaa)*
red-brown ~ логгерхед, головастая морская черепаха *(Caretta caretta)*
logging *s* лесосплав
logon *s* единица содержания структурной информации
loin *s* поясничный отдел, поясница
loir *s ther* соня-полчок *(Glis glis)*
loma *s zool* перепончатая кайма
lomatocarpous *a* окаймлённоплодный
lomatophyllous *a* окаймлённолистный
loment *s bot* членистый плод
longbeaks *s ent* носатка *(Libytheidae)*
long-billed *a* длинноклювый
long-boled *a* длинноствольный, высокоствольный
long-day *a* длиннодневный
long-eared *a* длинноухий
longevity *s* долговечность, продолжительность жизни
longevous *a* долговечный; долголетний
long-flowered *a* длинноцветковый
long-haired *a* длинношёрстный
longhorn *s ent* дровосек *(Cerambycidae)*
longicorn *a* длиннорогий; *ent* длинноусый
longipennate *a* длиннокрылый
longirostral *a* длинноклювый
long-leaved *a* длиннолистный
long-legged *a* длинноногий
long-lived *a* долгоживущий
long-lobed *a* удлинённо-лопастный
long-necked *a* длинношеий
long-petaled *a* длиннолепестный
long-pistillate *a* длиннопестичный
long-sighted *a* дальнозоркий
long-sightedness *s* дальнозоркость
long-sleeved *a bot* рукавчатый
long-spiked *a* длинноколосый, крупноколосый
long-stalked *a* длинностебельчатый
long-standing *a* позднеспелый
long-stapled *a* длинноволокнистый
long-stemmed *a* длинноствольный
long-styled *a* длиннопестичный
long-tailed *a* длиннохвостый
long-term *a* долговременный; долгосрочный
long-thorned *a* длинноколючковый
long-wooled *a* длинношёрстный
look *v* смотреть
loon *s orn* гагара *(Gavia)*
loop *s* петля
looper *s ent* пяденица
loose *a* рыхлый; сыпучий; свободный, неприкреплённый
loose-hanging *a* отвислый; свободновисящий
loose-lying *a* свободнолежащий
loosestrife *s bot* вербейник *(Lysimachia);* дербенник *(Lythrum)*

lop *v* отрубать; подрезать; лесосечные отходы
lop-eared *a* вислоухий
loph *s* лоф
lophocarpous *a* гребенчатоплодный
lophodont *a* лофодонтный
lophophilus *s* обитатель холмов
lophophore *s* лофофор
lophoreticulate *a bot* гребенчатосетчатый
lophosteon *s orn* киль грудины
loquat *s bot* локва, мушмула японская *(Eriobotrya japonica)*
lorate *a* плетевидный, ремневидный
lorchel *s myc* строчок обыкновенный *(Gyromitra esculenta)*
lordosis *s* лордоз
lore *s orn* уздечка
lori *s ther* лори; тонкий лори *(Loris)*
lorica *s* панцирь
loricate *a* панцирный, имеющий панцирь
lorication *s* образование панциря
loris *s ther* лори; тонкий лори *(Loris)*
loss *s* убыток, урон, ущерб, потеря
~ **of cattle** падёж скота
~ **of flesh** похудание
blood ~ потеря крови, кровопотеря
natural ~ естественная убыль
shrinkage ~ усушка; усадка
lotus *s* лотос *(Nelumbium)*
louse *s* вошь *(Pediculus)*
book ~ книжная вошь *(Troctes divinatorius);* сеноед домовый *(Atropos pulsatoria)*
brown chicken ~ пероед куриный бурый *(Goniodes dissimilis)*
crab ~ вошь лобковая, площица *(Phthirius pubis)*
root ~ филлоксера *(Phylloxera vastatrix);* корнёвка латуковая *(Rhizobius lactucae)*
wood ~ мокрица *(Oniscidae)*
lovebird *s* неразлучник *(Agapornis)*
low *a* небольшой, недостаточный; низкий, невысокий; *s* мычание, рёв; *v* мычать, реветь
low-backed *a* с провислой спиной *(e.g., of a horse)*
low-growing *a* низкорастущий
lowing *s* мычание *(of a cow);* рёв *(of an ox or ass)*
low-set *a* коротконогий
low-stemmed *a* низкоствольный
loxocarpous *a* косоплодный
loxocolpate *a pal* кособороздный
lucern, lucerne *s* люцерна посевная *(Medicago sativa)*
lucifugous *a* светобоязливый
luciphile *a* светолюбивый
luffa *s* тыква мочальная *(Luffa cylindrica)*
lugworm *s zool* пескожил *(Arenicola)*
lumbal, lumbar *a* люмбальный, поясничный
lumbocostal *a* пояснично-рёберный
lumbodorsal *a* пояснично-спинной
lumbosacral *a* пояснично-крестцовый
lumbriciform *a* червеобразный
lumbricosis *s* аскаридоз
lumen *s* полость; просвет
luminal *a* просветный, относящийся к просвету
luminance *s* яркость
luminescence *s* люминесценция
luminescent *a* светящийся
lumpiness *s* комковатость
lumping *s* образование комьев; укрупнение
lunar, lunate *a* серповидный
lung *s* лёгкое
book ~ *zool* лёгочная «книжка»
lungless *a* не имеющий лёгких, безлёгочный
lunula *s* лунка
lunular, lunulate *a* с месяцевидными метками
lunule *s* лунка
lunulet *s* луночка
lupine *s* лупин *(Lupinus); a* волчий
lure *s* приманка; *v* приманивать
luscious *a* сочный; сладкий; приторный
lush *a* густой, буйный, пышный *(vegetation)*
luster *s* блеск
lusterless *a* матовый
lustrous *a* глянцевитый, лоснящийся; блестящий
luteinization *s* лютеинизация, образование жёлтого тела
luteosterone *s* прожестин, гормон жёлтого тела
luth *s* кожистая черепаха *(Dermochelys coriacea)*
luxuriance *s* буйный рост; луксурианс
luxuriant *a* густой, буйный, пышный *(vegetation)*

lychnidiate *a* светящийся
lying-in *s* роды
lymph *s* лимфа
intercellular ~ тканевая жидкость
lymphatic *a* лимфатический; лимфатический сосуд
lymphization *s* образование лимфы, лимфопоэз
lymphoblast *s* лимфобласт
lymphocyte *s* лимфоцит
bursa-derived ~, bursal ~ В-лимфоцит
cytotoxic T ~ цитотоксический Т-лимфоцит
delayer-type T ~ *imm* Т-лимфоцит-дилэйер
effector ~ эффекторный лимфоцит, лимфоцит-эффектор
killer ~ лимфоцит-киллер, лимфоцит-«убийца»
null ~ №нулевой¤ лимфоцит, нуль-лимфоцит
plasmacytoid ~ плазмацит
T ~ Т-лимфоцит
thymus-derived ~, thymus-processed ~ Т-лимфоцит; тимоцит
lymphocytopoiesis *s* лимфоцитопоэз
lymphocytosis *s* лимфоцитоз
lymphogenic *a* лимфогенный
lymphogenous *a* лимфообразующий; происходящий из лимфы
lymphogram *s* лимфограмма
lymphoid *a* лимфоидный
lymphokine *s* лимфокин
B-cell-derived ~ В-клеточный лимфокин, В-лимфокин
T-cell-derived ~ Т-клеточный лимфокин, Т-лимфокин
lymphology *s* лимфология
lymphotoxin *s* лимфотоксин
lynx *s* рысь *(Felis lynx)*
caracal ~, desert ~ каракал, степная рысь *(Felis caracal)*
lyocytosis *s* лиоцитоз
lyo-enzyme *s* лиофермент
lyonization *s* лайонизация
lyophilization *s* лиофилизация
lyrebird *s orn* лирохвост *(Menura)*
lyrate *a* лировидный
lyre-leaved *a* лиролистный
lysate *s* лизат
cell-free ~ бесклеточный лизат; клеточный гомогенат
phage ~ фаголизат
lyse *v* лизировать; разрушать; вызывать лизис
lysicarpous *a* лизикарпный
lysigenicous *a* лизигенный
lysis *s* лизис, распад
erythrocyte ~ гемолиз, лизис эритроцитов
lysogenesis *s* лизогенез
lysogenicity *s* лизогенность
lysogenization *s* лизогенизация
lysogeny *s* лизогения
lysosome *s* лизосома
lysotypy *s* фаготипирование
lysozyme *s* лизоцим
lytic *a* литический

M

MAC, maximum allowable concentration *s* ПДК, предельно допустимая концентрация
macaque *s* макак *(Macaca)*
rhesus ~ резус, макак-резус *(Macaca mulatta)*
mace *s* мускатный орех *(Myristica fragrans)*
macerate *v* мацерировать, размачивать; размягчать
maceration *s* мацерация, размачивание; размягчение
machine *s* машина, механизм
shaking ~ прибор для встряхивания, шюттль-аппарат
macrandrous *a bot* крупнотычинковый
macroautoradiography *s* макроавторадиография
macrobiosis *s* долговечность; долголетие, долгожительство
macrobiotic *a* долговечный, долгоживущий
macroblast *s* макробласт
macrocarpous *a* крупноплодный
macrocephalous *a* большеголовый
macrochaeta *s zool* макрохета, крупная хета
macroconidium *s* макроконидия
macroconjugant *s* макроконъюгант
macrocyst *s* макроциста
macrocyte *s* макроцит
macrocytic *a* макроцитарный
macroevolution *s* макроэволюция
macrofauna *s* макрофауна
macrofibril *s* макрофибрилла
macroflora *s* макрофлора
macrofossil *s* крупное ископаемое
macrogamete *s* макрогамета
macrogametocyte *s* макрогаметоцит
macrogamy *s* макрогамия
macroglia *s* макроглия
macroglobulin *s* макроглобулин
macrogonidium *s* макрогонидия, женский гамонт, женский гаметоцит
macrogyne *s* крупный муравей-самка
macrolide *s* макролид
macromere *s* макромер, крупный бластомер
macromolecular *a* макромолекулярный
macromolecule *s* макромолекула
macromutant *s* макромутант
macromutation *s* макромутация
macromyelon *s* продолговатый мозг
macronucleocyte *s* макронуклеоцит
macronucleus *s* макронуклеус
macronutrient *s* макронутриент
macroorganism *s* макроорганизм
macrophage *s* макрофаг
reticular ~ ретикулоцит, ретикулярная клетка
macrophanerophyte *s* макрофанерофит
macrophyllous *a* крупнолистный
macrophylogenesis *s* макроэволюция
macrophyte *s* макрофит
macropodous *a* длинночерешковый
macropolycyte *s* полисегментированный нейтрофил
macrosclere *s* макросклера
macrosclereid *s* макросклереида
macroscopic, macroscopical *a* макроскопический
macroscopy *s* макроскопия
macrosepalous *a* крупночашелистиковый
macrosome *s* макросома, мегасома
macrosomia *s* макросомия, гигантизм
macrospecies *s* линнеон, линнеевский вид
macrosporangiophore *s* макроспорангиофор
macrosporangium *s* макроспорангий
macrospore *s* макроспора
macrosporophyll *s* макроспорофилл; плодолистик
macrostomatous *a* большеротый
macrostructure *s* макроструктура
macrostylospore *s* макростилоспора
macrostylous *a* длинностолбиковый
macrotherm *s* макротерм, мегатерм, тропическое растение
macrothermophilus *s* обитатель тропиков
macrothermophyte *s* макротерм, мегатерм, тропическое растение
macrothermophytia *s* фитоценозы тропиков
macrotous *a* большеухий
macrotrichia *s* макрохеты, крупные хеты
macrovegetation *s* макрофиты
macula *s* пятно, пятнышко

maculate, maculated *a* пятнистый, покрытый пятнами
maculation *s* пятнистость
maculiferous *a* пятнистый, покрытый пятнами
magenta *s* фуксин; фуксиновый
maggot *s ent* личинка
magnetotropism *s* магнитотропизм
magnification *s* увеличение
magnificent *a* пышный
magnifier *s* лупа; увеличительное стекло
magnitude *s* величина
~ **of population** *s* численность популяции
magpie *s* сорока *(Pica)*
mahogany *s* махагониевое дерево, красное дерево, акажу *(Swietenia mahagoni)*
maiden *a* девственный
maidenhead *s* девственная плева; девственность
mail *s* панцирь; *herp* щиток
maim *s* травма; телесное повреждение; рана
mainroot *s* главный корень
maintenance *s* поддержание, сохранение, содержание
maize *s* кукуруза, маис *(Zea)*
maker *s*: **pace** ~ водитель ритма, ритмоводитель
make-up *s* конституция, сложение, натура, склад; состав
genetic ~ организация генома
malabsorption *s* нарушение всасывания
malacia *s* размягчение
malacodermous *a* мягкопокровный
malacoid *a* мягкотелый; мягкий
malacology *s* малакология
malacopterous *a* мягкопёрый
malady *s* болезнь, заболевание
malar *s* скуловая кость; *a* скуловой, щёчный
malaria *s* малярия
malassimilation *s* плохое усвоение
malaxation *s* разжёвывание, разминание; инактивация жертвы
maldevelopment *s* неправильное развитие, аномалия развития
male *s* самец; мужчина; *a* мужской
male-bee *s* трутень
malformation *s* уродство, порок развития
malfunction *s* нарушение функции, дисфункция
malignant *a* злокачественный
mallard *s* кряква *(Anas platyrhynchos)*
mallear *a anat* относящийся к молоточку
malleolar *a* лодыжечный
malleolus *s* лодыжка
malleus *s anat* молоточек
mallie, mallimauk *s orn* глупыш *(Fulmarus glacialis)*
mallow *s bot* просвирник, мальва *(Malva)*
malnutrition *s* недостаточность питания, нарушение питания
malodorous *a* зловонный, вонючий
malposed *a* неправильно расположенный
malt *s* солод
Malthusianism *s* мальтузианство
malting *s* осолаживание
maltose *s* мальтоза
mamba *s herp* мамба *(Dendroaspis)*
mamelon *s* сосочек
mamilla *s* сосок
mamma *s* молочная железа
mammal *s* млекопитающее (животное)
carnivorous ~s *pl* хищные *(Carnivora)*
cloven-hoofed ~, even-hoofed ~, even-toed ~ парнокопытное *(Artiodactyla)*
flesh-eating ~s *pl* хищные *(Carnivora)*
gnawing ~ грызун *(Rodentia)*
hoofed ~s копытные
insect-eating ~ насекомоядное *(Insectivora)*
odd-toed (hoofed) ~ непарнокопытное *(Perissodactyla)*
pouched ~ сумчатое *(Marsupialia)*
small ~ мелкое млекопитающее
web-footed ~ ластоногое *(Pinnipedia)*
mammalian *a* млекопитающий
mammalogist *s* териолог
mammalogy *s* маммология, териология
mammary *a* грудной, молочный
mammilla *s* сосок
mammillary *a* сосковый
mammillate, mammillated *a anat* усеянный бугорками; *bot* имеющий сосочки
mammoth *s* мамонт *(Elephas primigenius)*
mammotrophin *s* маммотропин, пролактин
man *s* человек *(Homo sapiens)*
management *s* управление; использование и воспроизводство;

хозяйственный режим, способ пользования; уход, содержание
cage ~ клеточное содержание
forest ~ лесоустройство, лесное хозяйство; лесопользование
nature ~ природопользование
pasture ~, range ~ организация пастбищного хозяйства, уход за пастбищем
wildlife ~ использование и воспроизводство охотничье-промысловых животных

mandarin, mandarine *s* мандарин благородный *(Citrus nobilis)*
mandible *s* мандибула; жвала; *orn* подклювье
mandibular *a* мандибулярный
mandibulohyoid *a* челюстно-подъязычный
mandrake *s bot* мандрагора *(Mandragora)*
mandrill *s ther* мандрилл *(Papio sphinx)*
manducation *s* жевание
manducatory *a* жевательный
mane *s* грива
mangabey, mangaby *s ther* мангабей *(Cercocebus)*
mange *s* чесотка
manger *s* кормушка
mangrove *s* мангровое дерево *(Rhizophora mangle)*
manicate *a bot* рукавчатый
manifestation *s* проявление; обнаружение
manifestative *a* явный, очевидный, проявляющийся
manifold *a* многократный, многочленный; *s anat* книжка
manihot *s* маниок съедобный *(Manihot esculenta)*
man-induced *a* вызванный человеком, индуцированный человеком, стимулированный человеком, антропогенный
manioc *s* маниок *(Manihot)*
manipulation *s* манипуляция
noninstrumental ~ неорудийная манипуляция, неорудийное использование предметов

man-made *a* антропогенный, созданный человеком
manna *s bot* манна, медвяная роса; манник *(Glyceria);* лишайниковая манна *(Lecanora esculenta)*
mannitol *s* маннит
man-o'-war-bird *s orn* фрегат *(Fregatidae)*
mantid *s ent* богомол *(Mantidae)*
mantle *s* мантия; загиб створки *(in diatoms)*
brain ~ мозговой плащ
chordo-mesodermal ~ хордомезодерма

manual *s* ручной приём, ручной способ; руководство, наставление, справочник
manubrium *s* манубриум, рукоятка
manul *s* манул *(Felis manul)*
manure *s* навоз, органическое удобрение; *англ.* минеральное удобрение; *v* унавоживать, удобрять
chicken ~ птичий помёт
green ~ сидеральное удобрение, сидерат

many-awned *a* многоостистый
many-bracted *a* многоприцветниковый
many-celled *a* многоклеточный
manyplies *s anat* книжка
many-seeded *a* многосемянный
many-segmented *a* многочленистый
many-stage *a* многоступенчатый, многокаскадный
manzanita *s bot* толокнянка *(Arctostaphylos)*
map *s* карта
assay ~ схема точек взятия образцов

maple *s* клён *(Acer)*
mapping *s* картирование
marabou *s* африканский марабу *(Leptoptilus crumeniferus)*
maral *s* марал *(Cervus elaphus)*
marcescent *a* увядающий
march-fly *s ent* толстоножка *(Bibionidae)*
mare *s* конематка, кобыла
margay *s* длиннохвостая кошка *(Felis wiedii)*
margin *s* край; опушка
marginal *a* маргинальный, краевой; *ent* случайный
marginate *a* окаймлённый, отороченный
marginiform *a* каёмкообразный, в виде каймы
marigold *s* календула *(Calendula);* бархатцы *(Tagetes)*

marita *s helm* марита, половозрелый сосальщик
maritime *a* морской; приморский
marjoram *s* душица *(Origanum);* майоран *(Majorana)*
mark *s* знак, метка, пятно, отметина; черта, свойство, признак; след, отпечаток
marker *s* маркёр, маркёрный ген; маркёр, метчик; маркёр, метка
markred *a* меченый; пёстрый
marking *s* маркировка, мечение; введение метки; клеймение; метка; клеймо; расцветка, окраска
scent ~ запаховое мечение, ольфакторное мечение; ольфакторная метка
marmorate *a* мраморный
marmoset *s ther* игрунка *(Callithricidae)*
plumed ~ мармозетка, уистити *(Callithrix)*
marmot *s* сурок *(Marmota)*
prairie ~ луговая собачка *(Cynomys)*
marrow *s* мозговое вещество; костный мозг; кабачок (*Cucurbita pepo*)
hair ~ сердцевина волоса
yellow bone ~ жёлтый костный мозг
marrowy *a* костномозговой
marsh *s* болото; травяное болото; *pl* марши
marsh-beet *s bot* кермек *(Limonium)*
marshbird *s* длиннохвостая камышовка *(Megalurus);* болотный кассик *(Psedoleistes)*
marshland *s* заболоченная местность
marshotter *s ther* европейская норка *(Mustela lutreola)*
marshy *a* болотистый
marsupial *s* сумчатое(животное *(Marsupialia); a* сумчатый
marten *s* куница *(Martes);* лесная куница *(Martes martes);* американская куница *(Martes americana)*
Indian ~ харза, желтогрудая куница *(Martes flavigula)*
stone ~ каменная куница, белодушка *(Martes foina)*
martin, martinet *s* ласточка *(Hirundinidae)*
common house ~ обыкновенная городская ласточка, обыкновенный воронок *(Delichon urbica)*
house ~ городская ласточка, воронок *(Delichon);* обыкновенная городская ласточка, обыкновенный воронок *(Delichon urbica)*
masculinization *s* маскулинизация
mask *v* маскировать, скрывать, прятать; *ent* маска *(of Odonata)*
masked *a* скрытый, латентный; бессимптомный; *bot* масковидный
masking *s* маскировка
mass *s* масса
egg ~ кладка, отложенные яйца
fecal ~es *pl* экскременты
pollen ~ поллиний, пыльцевой комочек
masseter *s* жевательная мышца
mass-fragmentography *s* масс-фрагментография
massula *s* поллиний, пыльцевой комочек, комочек микроспор
mast *s* плодокорм
master *s* хозяин
beach ~ секач *(fur seal)*
masticate *v* жевать, прожёвывать жвачку
mastication *s* жевание
masticatory жевательный
mastigospore *s* жгутиковая спора, флагеллиспора, зооспора
mastocyte *s* тучная клетка, лаброцит
mastoid *s* сосцевидный отросток; *a* сосцевидный
masturbation *s eth* мастурбация
mat *s* покрытие, плёнка; дернина, подстилка, войлок; ковёр; *v* переплетать, переплетаться, спутывать, спутываться
mycelial ~ мицелиальная плёнка
root ~ корневой войлок, дернина
vegetation ~ растительный мат
match *s* пара; соперник; совместимость; *v* состязаться; противостоять; спаривать, случать; подбирать пару
matching *s* проба на совместимость, перекрёстное типирование
histocompatibility ~ тканевое типирование, гистотипирование; проба на совместимость
mate *s* самец; самка; партнёр по половому процессу; партнёр; парагвайский чай *(Ilex paraguayensis); v* спаривать, спариваться; скрещивать, скрещиваться

material *s* материал; вещество
raw ~ сырьё
maternal *a* материнский
maternity *s* материнство
matgrass *s bot* белоус торчащий *(Nardus stricta)*
mating *s* спаривание; скрещивание
apolegamic ~ аполегамия
matrass *s* матрас
matrix *s* матрица; матрикс, межклеточный материал; основное вещество цитоплазмы, цитоплазматический матрикс; дичок для прививок; *анат.* матка
intercellular ~ матрикс, межклеточный материал
mitochondrial ~ матрикс митохондрии
matroclinal, **matroclinous** *a* матроклинный
matromorphic *a* сходный с материнской формой
matted *a* спутанный; задернованный
matter *s* вещество; материя
foreign ~ примесь
mattery *a* гнойный
maturation *s* созревание
early ~ скороспелость, раннее созревание
immune ~ формирование иммунокомпетентности, созревание иммунокомпетентности
mature *v* созревать; *a* зрелый, созревший; спелый
maturity *s* зрелость; спелость; **at** ~ по достижении зрелости *или* спелости
matutinal *a* утренний
maverick *s* осиротевший телёнок
mavis *s* певчий дрозд *(Turdus philomelos)*
maw живот, брюхо; *orn* зоб
maxicircle *s mbiol* максикольцо
maxilla *s ent* максилла; *orn* надклювье
second ~ вторая пара максилл; *ent* лабиум, нижняя губа
maxillary *a* максиллярный; верхнечелюстной
maxilliped, maxillipede *s* ногочелюсть
maxillofacial *a* челюстно-лицевой
maxillomandibular *a* верхне- и нижнечелюстной; *zool* челюстной
maximization *s* максимизация, доведение до максимума
mayfly *s ent* подёнка *(Ephemeroptera)*
maze *s* лабиринт
M-disk *s* полоска М, мезофрагма
meadow *s* луг
bentgrass ~ полевичник, полевицевый луг
bluegrass ~ мятличник, мятличный луг
sedge ~ осочник
short grass ~ низкотравный луг, низкозлачник
meal *s* приём пищи, еда; мука; *v* принимать пищу, есть; кормить
mealworm *s* «мучной червь», личинка хрущака мучного *(Tenebrio molitor)*
mealybug *s* червец *(Pseudococcidae)*
mean *s biom* средняя величина; *pl* средство; способ
~ **of communication** *eth* средство коммуникации
measle *s helm* цистицерк, финна
measled *a* финнозный
measles *s* корь
measure *s* мера; измерять; мероприятие
~ **of taxonomic distance** показатель таксономического расстояния
measurement *s* измерение; промер
meat *s* мясо; мякоть, сердцевина, ядро *(of a nut)*
meatal *a* относящийся к отверстию
meatfly *s* муха живородящая мясная *(Calliphora vomitoria)*
meatus *s* отверстие, проход; канал
mechanics *s* механика
developmental ~ механика развития, эпигенетика
mechanism *s* механизм
mechanoreceptor *s* механорецептор
mecism *s* гигантизм
meconium *s* меконий
mediad *a* обращённый к середине
medial *a* медиальный, срединный; средний
median *s biom* медиана; *a* средний; срединный
mediastinal *a* медиастинальный, относящийся к средостению
mediastinum *s* средостение
mediate *a* промежуточный
mediation *s* опосредование
mediator *s* медиатор, посредник; переносчик
suppressive ~ *imm* фактор супрессии
medic *s bot* люцерна *(Medicago)*

medicine *s* медицина; лекарственное средство
medine *s pal* медина, промежуточный слой
mediodorsal *a* медиодорсальный
mediolateral *a* медиолатеральный
mediopalatine *s* срединная нёбная кость
medioposterior *a* средний задний
medioventral *a* медиовентральный
medisection *s* срединный разрез
medithorax *s ent* мезоторакс, среднегрудь
medium *s* средство; способ; среда; середина; *a* средний
agar overlay ~ агаровое покрытие
contrast ~ контрастное вещество
gel ~ гелевая среда, гель
HAT ~ гипоксантин-аминоптерин-тимидиновая среда, ГАТ-среда
surrounding ~ окружающая среда, окружающие условия
medium-sized средней величины; среднего роста
medulla *s* мозговой слой, мозговое вещество; костный мозг; *a* продолговатый мозг; *bot* сердцевина
fungal ~ плектенхимная сердцевина *(of a lichen)*
spinal ~ спинной мозг
medullary *a* медуллярный, мозговой; *bot* сердцевинный
medullated *a anat* мякотный; *bot* имеющий сердцевину
medullispinal *a* спинно-мозговой
medulocell *s* миелоцит
medusa *s* медуза
medusoid *s* медузоид; *a* медузоидный
meerkat *s ther* сурикатa *(Suricata)*
megaevolution *s* макроэволюция
megafauna *s* мегафауна
megagamete *s* макрогамета
megagametocyte *s* макрогаметоцит
megagametogenesis *s* макрогаметогенез
megagametophyte *s* макрогаметофит
megakaryocyte *s* мегакариоцит
megalecithal *a* макролецитальный
megaloblast *s* мегалобласт
megalocarpous *a* крупноплодный
megalocyte *s* мегалоцит
megalogonidium *s* мегалогонидия
megalokaryocyte *s* мегакариоцит
megamere *s* макромер
megaphanerophyte *s* макрофанерофит
megaphyllous *a* крупнолистный
megapode *s* большеног *(Megapodius)*
megasclere *s* мегасклера *(of sponges)*
megasea *s* бадан толстолистный *(Bergenia crassifolia)*
megasomia *s* гигантизм
megasporangium *s* мегаспорангий
megaspore *s* мегаспора
megasporocyte *s* мегаспороцит
megasporogenesis *s* мегаспорогенез
megasporophyll *s* мегаспорофилл
megatherm *s* мегатерм, макротерм, тропическое растение
megazoospore *s* мегазооспора
meiocyte *s* мейоцит, ауксоцит
meiophylly *s* мейофилия
meiosis *s* мейоз, редукционное деление
meiosporange *s* мейоспорангий
meiospora *s* мейоспора
meiotaxy *s* мейотаксия
meiotic *a* мейотический
melancholic *a* меланхолический
melanism *s* меланизм
melanoblast *s* меланобласт
melanocarpous *a* черноплодный
melanocyte *s* меланоцит
melanogenesis *s* меланогенез, синтез меланина
melanophore *s* меланофор
melanose *s phytop* меланоз
melanosome *s* меланосома
melanospermous *a* черносемянный
melleous *a* медовый
melliferous *a* медоносный
melliphagous *a* питающий мёдом, медоядный
mellisugent *a* медососущий; питающий мёдом, медоядный
mellittophilae *s* пчелоопыляемые растения
mellivorous *a* питающий мёдом, медоядный
mellow *v* созревать, поспевать; разрыхлять *(soil); a* спелый, зрелый; плодородный, жирный *(soil);* рыхлый *(soil)*
melon *s* дыня *(Cucumis melo);* арбуз *(Citrullus)*
water ~ арбуз столовый *(Citrullus vulgaris)*
member *s* орган; член; конечность

membrana *s* мембрана, оболочка; плёнка; перепонка
membranaceous *a* мембранный; плёнчатый; перепончатый
membrane *s* мембрана, оболочка; плёнка; перепонка
abdominal ~ брюшина
adamantine ~ кутикула зуба
Bruch's ~ оболочка Бруха, базальная пластинка
germinal ~ бластодерма
gradocoll ~ градоколловый фильтр
immobilizing ~ гибридизационная подложка, блоттинг-мембрана
Krause's ~ полоска Краузе, полоска Т, телофрагма
Nasmyth's ~ оболочка Несмита, кутикула эмали, кутикула зуба
nictitant ~ третье веко, мигательная перепонка
Tenon's ~ тенонова капсула
transfer ~ гибридизационная подложка, блоттинг-мембрана
virginal ~ девственная плева
membrane-bound *a* мембраносвязанный
membrane-leaved *a* плёнчато-листный
membranella *s* мембранелла
membrane-rooted *a* плёнчато-корешковый
membraniferous *a* имеющий мембрану; образующий мембрану
membranoid *a* мембрановидный
membranous *a* мембранный; плёнчатый; перепончатый
memory *s* память
Mendelism *s* менделизм
meningoencephalitis *s* менингоэнцефалит
meninx *s* мозговая оболочка
meniscocyte *s* менискоцит, дрепаноцит
meniscus *s* мениск
~ **of touch** осязательный мениск, диск Меркеля
menopause *s* менопауза
menotaxis *s* менотаксис
mensa *s* жевательная поверхность
menses *s* менструация
menstrual *a* менструальный
menstruation *s* менструация
mental *a* умственный; психический; подбородочный
mentality *s* умственные способности; способность мышления, интеллект
mentation *s* процесс мышления
mentum *s* подбородок; *bot* базальный выступ
meospore *s* меоспора
mercury *s* ртуть
merganser *s orn* крохаль *(Mergus)*
mericarp *s* мерикарпий, дробный плод
meridian *s* меридиан
merism *s* метамерия, метамерность
meristela *s* меристела
meristem *s* меристема
meristematic *a* меристематический
meristic *a* меристический
meristogenetic *a* развивающийся из меристемы
merithallus *s* междоузлие
merle *s* чёрный дрозд *(Turdus merula)*
merlin *s orn* дербник *(Falco columbarius)*
mermithoid *a* мермитоидный
meroblast *s* меробласт
merocercus *s* мероцерк
merocrine *a* мерокринный
merocyte *s emb* мероцит; *prot* шизонт
merogamy *s* мерогамия
merogastrula *s* мерогаструла
merogenesis *s* мерогенез; сегментация
merogenote *s* мерогенота
merogony *s* мерогония
merome *s* сегмент тела, сомит, метамер
meromixis *s* меромиксис
meromorphosis *s* мероморфоз
merosome *s* сегмент тела, сомит, метамер
merospermy *s* мероспермия
merotomy *s* меротомия, разделение на части
merozoite *s* мерозоит
merozygote *s* мерозигота
merrythought *s orn* вилочка
mesarch *a* мезархный
mesaxon *s* мезаксон
mesectoderm *s* мезоэктодерма
mesembryo *s* бластула
mesencephalon *s* средний мозг; *emb* средний мозговой пузырь
mesenchymatous *a* мезенхимный
mesenchyme *s* мезенхима, мезодерма
mesendoderm *s* мезоэнтодерма
mesenterial, mesenteric *a* мезентериальный, брыжеечный

mesenteron *s* средняя кишка; гастральная полость *(of coelenterates)*
mesentery *s* мезентерий, брыжейка
mesentoblast *s* мезоэнтодерма
mesexine *s pal* мезэкзина
mesh *s* ячея
meshwork *s* сетчатая структура, ячеистая структура; сеть
mesic *a* с умеренным увлажнением
mesine *s pal* мезина
mesiophytia *s pl* внутриматериковые фитоценозы
mesoblast *s* мезобласт; мезобластема
somatic ~ соматомезобласт, мезодерма
mesoblastic *a* мезобластический; мезодермальный
mesobrochate *a* среднеячеистый
mesocardium *s* мезокардий
mesocarp *s* мезокарпий
mesocentrous *a* мезоцентральный
mesocephalic *a* мезоцефалический
mesocercaria *s helm* мезоцеркария
mesocoracoid *s* мезокоракоид
mesocotyle *s* мезокотиль
mesocycle *s* мезоцикл
mesoderm *s* мезодерма
somatic ~ соматоплевра
splanchnic ~ спланхноплевра
mesodermal *a* мезодермальный
mesodermalization, mesodermization *s* мезодермализация
mesofauna *s* мезофауна
mesogaster, **mesogastrium** *s* средняя область живота; *emb* дорсальная брыжейка желудка
mesogloea *s* мезоглея
mesognathion *s* мезогнатион
mesohydrophyte *s* мезогидрофит
mesohydrophytic *a* мезогидрофитный
mesolobus *s anat* мозолистое тело
mesomere *s* мезомер, сомит
mesomitosis *s* мезомитоз
mesomorphic, **mesomorphous** *a* мезоморфный
mesomorphy *s* мезоморфия
mesonephros *s* мезонефрос, первичная почка, вольфово тело
mesonexine *s pal* мезонэкзина
mesopetalum *s bot* губа *(in orchids)*
mesophilous *a* мезофильный
mesophloem *s* мезофлоэма
mesophragma *s ent* мезофрагма; *hist* мезофрагма, полоска М
mesophryon *s* глабелла
mesophyll *s* мезофилл
mesophyte *s* мезофит
mesophytic *a* мезофитный
mesoplastron *s* мезопластрон
mesopodial *a* плюсневой; *bot* мезоподиальный
mesopterygoid *s* мезоптеригоид
mesosaprobic *a* мезосапробный
mesoscapula *s anat* гребень лопатки, ость лопатки
mesosoma *s* мезосома
mesosperm *s* мезосперм
mesospore, mesosporium *s* мезоспорий
mesosternum *s anat* мезостернум, тело грудины; *ent* стернит среднегруди
mesotherm *s* мезотерм
mesothermophilous *a* мезотермный
mesothoracic *a ent* среднегрудной
mesothorax *s ent* мезоторакс, среднегрудь
mesotrophic *a* мезотрофный
mesotropic *a* мезотропный
mespilus *s bot* ирга *(Amelanchier)*
messenger *s* посредник, мессенджер; информационная РНК, матричная РНК, мРНК; посыльный груз
mestome *s* местом
metabasidium *s* метабазидия
metabiosis *s* метабиоз
metabolic *a* метаболический
metabolimeter *s* метаболиметр
metabolism *s* метаболизм, обмен веществ
analytic ~ катаболизм, диссимиляция
constructive ~ анаболизм, ассимиляция
crassulacean acid ~ метаболизм карбоновых кислот по типу толстянковых, САМ-метаболизм
destructive ~ катаболизм, диссимиляция
respiratory ~ газообмен
metabolite *s* метаболит
metaboly *s* метаболия
metacarpal *s* пястная кость; *a* пястный
metacercaria *s* метацеркария
metacestode *s* плероцеркоид
metachromasy *s* метахромазия
metachromatic *a* метахроматический

metachromatism *s* изменение цвета
metacontrast *s phys* метаконтраст, обратная маскировка
metadromous *a* метадромный
metaestrus *s* метаэструс
metagastrula *s* метагаструла
metagenesis *s* метагенез, чередование поколений
metagyny *s* метагиния, протандрия
metakinesis *s* метафаза
metalloenzyme *s* металлосодержащий фермент
metal-shadowed *a* напылённый металлом
metamere *s* метамер
metamerism *s* метамерия, сегментация
metamerized *a* сегментированный
metamitosis *s* метамитоз
metamorphopsia *s* метаморфопсия
metamorphosis *s* метаморфоз, превращение
metamyelocyte *s* метамиелоцит
metandry *s* метандрия
metanephros *s* метанефрос, вторичная почка
metanotum *s ent* заднеспинка
metaphase *s* метафаза
metaphloem *s* вторичная флоэма
metaphragma *s ent* метафрагма
metaphyte *s* многоклеточное растение
metaplasia *s* метаплазия
metaplasm *s* метаплазма
metapleural *a* метаплевральный; расположенный сзади и сбоку
metapleuron *s ent* метаплевра, плеврит заднегруди
metapneustic *a* метапнейстический
metapodium *s* пястье; плюсна
metapophysis *s* метапофиз
metapopulation *s* метапопуляция
metapore *s* отверстие Мажендн
metascutellum *s ent* метаскутеллум, щиток заднегруди
metascutum *s ent* метаскутум, щит заднегруди
metasoma *s* метасома; постабдомен
metasperm *s* покрытосемянное растение
metastasis *s* метастаз
metastatic *a* метастатический
metasternum *s* метастернум, мечевидный отросток; задний отдел грудины; *ent* стернит заднегруди
metastigmate *a* метастигматический
metastructure *s* ультраструктура
metasyndesis *s* метасиндезис, телосинапсис
metatarsal *a* плюсневой
metatarsophalangeal *a* плюснефаланговый
metatarsus *s* плюсна
metathalamus *s anat* метаталамус
metathorax *s ent* заднегрудь
metatroch *s* метатрох
metatrophic *a* метатрофный
metaxenia *s* метаксения
metaxylem *s bot* метаксилема
metazoan *s* многоклеточное животное *(Metazoa); a* относящийся к многоклеточным
metembryo *s* гаструла
metencephalon *s* задний мозг
Met-enkephalin *s* метионинэнкефалин, Мет-энкефалин
metepimeron *s ent* эпимер заднегруди
metepisternum *s ent* эпистернум заднегруди
metestrus *s* метэструс, послетечка
methanolysis *s* метанолиз
method *s* метод, способ; методика
  double-antibody ~ метод двойных антител, сэндвич-метод
  latex agglutination ~ реакция латекс-агглютинации
  roll immunoblot ~ роллер-иммуноблоттинг
methodization *s* приведение в систему
methosulfat *s* метосульфат
  phenazine ~ феназинметосульфат, ФМС
methoxycarbonylation *s* карбометоксилирование
methoxylation *s* метоксилирование
methylaminopterin *s* метотрексат, аметоптерин
methylation *s* метилирование
methylnitrosourea *s* метилнитрозомочевина
methysis *s* интоксикация
metopion *s* метопион
metoxenous *a par* разнохозяйный
metraterm *s helm* метратерм
mettled *a* ретивый, горячий
metula *s* метула

mew *s* чайка *(Larus);* линька птиц; сбрасывание рогов; *v* линять; сбрасывать рога; мяукать
mexine *s pal* мекзина
micella *s* мицелла
micellar *a* мицеллярный
microaerophil *s* микроаэрофил
microaerophilic *a* микроаэрофильный
microanalysis *s* микроанализ
antigen spot ~ спот-антигенный микроанализ, точечный антигенный анализ
microanatomy *s* микроскопическая анатомия; гистология
microassay *s* микроанализ
microassociation *s* микроасоциация
microautoradiograph *s* микроавторадиограф
microautoradiography *s* микроавторадиография
microbal *a* микробный
microbalance *s* микровесы
microbe *s* микроорганизм
microbeam *s* микропучок, микролуч
microbial *a* микробный
microbicidal *a* бактерицидный
microbicide *s* бактерицидное вещество
microbiologic, microbiological *a* микробиологический
microbiology *s* микробиология
microbiosis *s* микробная инфекция
microbiota *s* микрофлора и микрофауна
microbivorous *a* бактериядный
microblast *s* микробласт
microbody *s* микротельце
microburet, microburette *s* микробюретка
microburner *s* микрогорелка
microcalycular *a* мелкочашечковый
microcarpous *a* мелкоплодный
microcarrier *s* микроноситель
microcell *s lab* микроячейка; миниатюрная клетка
microcephalic *a* микроцефальный
microcercus *s* микроцерк
microchemistry *s* микрохимия
microchromosome *s* микрохромосома
microcirculation *s* микроциркуляция, капиллярное кровообращение
microclimate *s* микроклимат
micrococcus *s* микрококк
microcommunity *s* микросообщество
microconidium *s* микроконидия
microconjugant *s* микроконъюгант
microcosm *s* микрокосм
microculture *s* микрокультура
microculturing *s* микрокультивирование
microcyst *s* микроциста
microcyte *s* микроцит, мелкий эритроцит
microcytic *a* микроцитарный
microdispenser *s* микродозатор, микродиспенсер
microdissection *s* микрургия
microelectrode *s* микроэлектрод
microelement *s* микроэлемент
microenvironment *s* микроокружение, микросреда
growth-supporting ~ питающая среда
microevolution *s* микроэволюция
microfauna *s* микрофауна
microfibril *s* микрофибрилла
microfilament *s* микрофиламент
microfilaria *s* микрофилярия
microfiltration *s* микрофильтрация
microflora *s* микрофлора
microforge *s* микрокузница
microfuge *s* микрофуга, микроцентрифуга
microgamete *s* микрогамета, мужская гамета
microgametoblast *s* микрогаметобласт
microgametocyte *s* микрогаметоцит, мужской гаметоцит
microgametophyte *s* микрогаметофит
microglia *s* микроглия
microgonidium *s* микрогонидия
micrograph *s* микрофотография
microgyne *s* микрогина
microhabitat *s* микроместообитание
microhematocrit *s* микрогематокрит
microimmunofluorimetry *s* иммунофлуоресцентный микроанализ
microincineration *s* микросжигание, микросподография
micromanipulation *s* микроманипуляция
micromanipulative *a* с помощью микроманипулятора
micromanipulator *s* микроманипулятор
micromere *s* микромер
micrometer *s* микрометр
eyepiece ~, ocular ~ окуляр-микрометр
micrometry *s* микрометрия
micromutation *s* микромутация, монолокусная мутация

micromyelocyte *s* микромиелоцит
micron *s* микрометр, микрон
microneedle *s* микроигла
micronucleation *s* формирование микроядра, микронуклеация
micronucleus *s prot* микронуклеус; микроядро
micronutrient *s* микронутриент, питательный микроэлемент
microorganism *s* микроорганизм
micropaleontology *s* микропалеонтология
microparasite *s* микропаразит
micropetalous *a* микролепестный
microphotography *s* микрофотография
microphotometer *s* микрофотометр
microphthalmia *s* микрофтальмия
microphyll *s* мелкий лист
microphyllous *a* мелколистный
microphyte *s* микрофит, микроскопическое растение
microphytocoenosis *s* микрофитоценоз
micropipet, micropipette *s* микропипетка
microplania *s* пойкилоцитоз
microplate *s* микропланшет
microprobe *s* микрозонд; микропроба
microprojector *s* микропроектор
micropsia *s* микропсия
micropterism *s* микроптеризм
micropylar *a* микропилярный
micropyle *s* микропиле
microquantitation *s* микроколичественное определение
microradiography *s* микрорадиография
microreticulum *s* микросеточка
microscope *s* микроскоп
dissection ~ препаровальная лупа
microscopic, microscopical *a* микроскопический
microscopy *s* микроскопия
microsection *s* приготовление гистологических срезов
microsequencing *s* микросеквенирование
microsere *s* микросерия, серула
microslide *s* микропрепарат
microsome *s* микросома
microsomia *s* микросомия, карликовость
microspecies *s* жорданон
microspectrometry *s* микроспектрометрия
microspectrophotometer *s* микроспектрофотометр
microspectrophotometry *s* микроспектрофотометрия
microspectroscope *s* микроспектроскоп, спектральная насадка
microspermous *a* мелкосемянный
microsphere *s* микрогранула, микросфера; микроноситель
microspheric *a* микросферический
microsporangium *s* микроспорангий
microspore *s* микроспора
microsporidiosis *s* микроспоридиоз
microsporocyte *s* микроспороцит
microsporogenesis *s* микроспорогенез
microsporophore *s* микроспорофор
microsporophyll *s* микроспорофилл
microstand *s* микрогруппировка растений
microstrainer *s* микрофильтр
microstrobile *s bot* шишечка
microstroma *s* микрострома
microstructure *s* микроструктура
microstylospore *s* микростилоспора
microstylous *a* короткостолбиковый
microsubspecies *s* микроподвид, микрораса
microsurgery *s* микрургия, микрохирургия
microtherm *s* микротерм
microtiterplate *s* микротитрационный планшет
microtome *s* микротом
microtray *s* микропланшет
microtubule *s* микротрубочка
microvillus *s* микроворсинка
microzoospore *s* микрозооспора
micrurgy *s* микрургия, микрохирургия
miction *s* мочеиспускание
midbrain *s* средний мозг
midge *s ent* галлица *(Itonididae);* звонец, дергун *(Chironomidae)*
biting ~ мокрец *(Ceratopogonidae)*
chironomid ~ звонец, дергун *(Chironomus)*
fungus ~ комарик грибной *(Mycetophilidae)*
gall ~ галлица *(Itonididae)*
non-biting ~ звонец, дергун *(Chironomidae)*
midgrass *s* травы высотой 60-120 см
midgut, mid-intestine *s* средняя кишка
mid-kidney *s* мезонефрос, первичная почка, вольфово тело

midline *s* срединная линия, средняя линия
midrange *s biom* полусумма крайних значений
midrib *s* средняя жилка
midriff *s* диафрагма, грудобрюшная преграда
midsternum *s* средняя часть грудины; тело грудины
midvein *s* средняя жилка
midwood *s* лесная чаща
migraine *s* мигрень
migrant *s* мигрирующее животное; перелётная птица; *a* кочующий; перелётный
migrate *v* мигрировать, совершать переселения; **to ~ down** *ихт.* скатываться
migration *s* миграция, переселение, перемещение; перелёт
mass ~ массовая миграция; рунный ход *(of fish)*
migratory *a* мигрирующий; перелётный; проходной *(of fish)*
mild *a* мягкий; умеренный; слабый
mildew *s phytop* милдью, ложная мучнистая роса *(caused by Plasmopara)*
milfoil *s* тысячелистник *(Achillea)*
miliary *a* милиарный, просовидный
millieu *s* среда
milk *s* молоко
first ~ молозиво
milking *s* доение, дойка
milk-producing *a* образующий молоко, лактогенный
milkwort *s bot.* истод *(Polygala)*
mill *s* мельница, дробилка
millerbird *s orn* камышовка *(Acrocephalus)*
millet *s* просо *(Panicum)*
milling *s* помол, размол, дробление
milt *s* селезёнка; молоки *(of fish)*
mimesis *s* миметизм, подражание
mimetic *a* миметический, подражательный
mimetism *s* миметизм, подражание
mimic *s* животное-имитатор; подражательный; *pl gen* мимики; *v* подражать
mimicry *s* мимикрия
mince *v* измельчать, крошить; смягчать, ослаблять; *s* измельчённая ткань
mind *s* ум, разум; интеллект; психика; психическая деятельность; память
mine *s ent* мина, ход
miner *s* минёр, минирующее насекомое; *orn* печник-землекоп, минёр, пещерный поползень *(Geositta)*; *орнит.* манорина *(Manorina)*
mineralocorticoid *s* минералокортикоид
mingle смешиваться
miniantibody *s* «миниантитело»
minicell *s* миниклетка
minicircle *s mbiol* миникольцо
minigen *s* минигeн
minimax *s biom* минимакс
minimization *s* минимизация
minimus *s* мизинец
mininetwork *s* минисеть
minisatellite *s* минисателлитная ДНК
minisegment *s* минисегмент
mink *s ther* норка *(Mustela)*
minor *a* второстепенный; побочный; *bioch* минорный
"minors" *s pl* минорные антигены
mint *s* мята *(Mentha)*
miospore *s* миоспора
miracidium *s* мирацидий
mire *s* топь, трясина; болото; верховое болото
misbirth, **miscarriage** *s* выкидыш, аборт
miscible *a* смешивающийся
miscopying *s gen* «ошибка копирования»
misidentified *a* неправильно определённый
misimprinting *s eth* запечатление неадекватного объекта
mismatch *s mbiol* ошибочное спаривание
missel *s* дрозд-деряба *(Turdus viscivorus)*
misshapen *a* уродливый
missing *a* недостающий
mist *s bot* качим метельчатый, перекати-поле *(Gypsophila paniculata)*
mistletoe *s bot* омела белая *(Viscum album)*
mite *s* клещ
cheese ~ клещ сырный *(Tyroglyphus siro)*
chigger ~ клещ *Trombicula irritans*
follicle ~ железница, угрица *(Demodex folliculorum)*
itch ~ чесоточный зудень *(Sarcoptes scabiei)*

spider ~ клещик паутинный *(Tetrarhynchus)*
true ~ амбарный клещ, мучной клещ *(Acaridae)*
mite-borne *a* клещевой, клещевого происхождения
mithan *s ther* гаял *(Bos gaurus)*
miticidal *a* акарицидный
miticide *s* акарицид
mitochondrial *a* митохондральный
mitochondrion *s* митохондрия
mitogen *s* митоген
mitogenetic *a* вызывающий митоз, митогенный
mitogenicity *s* митогенные свойства, митогенность
mitoplast *s* митопласт
mitoschisis, mitosis *s* митоз, непрямое деление ядра, кариокинез
mitosome *s* митосома
mitotic *a* митотический
mitra *s bot* колпачок
mitriform *a bot* колпачковидный
mixed *a* смешанный
mixed-nerved *a bot* смешанно-жилковатый
mixer *s* мешалка,смеситель
gradient ~ градиентный смеситель, устройство для создания градиентов
vibration ~ вибрационный смеситель, вортекс
mixing *s* смешивание, смешение
mixobiosis *s* миксобиоз
mixochimaera *s* миксохимера
mixoploid *s* миксоплоид
mixoploidy *s* миксоплоидия
mixote *s* миксохимера
mixotrophic *a* миксотрофный
mixotrophy *s* миксотрофия
mixovariation *s* миксовариация
mixture *s* смесь; смешивание, смешение; *v* смешивать
mnemonic *a* мнемонический
mnemonics *s* мнемоника
mnemotaxis *s* мнемотаксис
moan *s* стон; *v* стонать
mobbing *s* совместное нападение; *orn* окрикивание
mobile *a* подвижный, мобильный
mobility *s* подвижность, мобильность
mobilization *s* мобилизация; активация
mocassin *s* щитомордник *(Agkistrodon)*
mocker, mockingbird *s* певчий пересмешник *(Mimus)*
modality *s* модальность
mode *s* образ, способ, метод; *biom* мода
~ **of existence** образ жизни; способ существования
model *s* модель
patching ~ пэтчинг-модель, модель Делизи - Перельсона
moderate *a* умеренный
modification *s* модификация, видоизменение
habit ~ переделка навыка
modificator, modifier *s* модификатор; ген-модификатор; *bioch* модулятор
immune ~, immunologic response ~ иммуномодулятор
modulation *s* модуляция
modulator *s* модулятор
moist *a* влажный
moistening *s* увлажнение
moisture *s* влажность; влага; сырость
moisture-loving *a* влаголюбивый
moistureproof *a* влагонепроницаемый
molar *s* коренной зуб, моляр
mold *s* плесень, плесневой гриб; перегной; *v* гнить, разлагаться
glume ~ альтернариоз *(caused by Alternaria)*
mucoraceous ~ мукоровый гриб *(Mucorales)*
plaster ~ плесневой налёт *(caused by Oospora fimicola and Papulospora byssina)*
slime ~ слизевик, миксомицет *(Myxomycetes)*
moldy *a* плесневой; заплесневелый
mole *s* крот *(Talpa)*
star-nosed ~ звездорыл, звездонос *(Condylura cristata)*
water ~ выхухоль *(Desmana)*
molecular *a* молекулярный
molecule *s* молекула
shuttle ~ молекула-переносчик, «челночная» молекула
mole-vole *s* слепушонка *(Ellobius)*
long-clawed ~ прометеева полёвка *(Prometheomys)*
mollusk *s* моллюск, мягкотелое *(Mollusca)*
molt *s* линька; *v* линять
moment *s* момент

monad *s* монада
monadelphous *a* однобратственный
monandrous *a* однотычинковый
monandry *s* монандрия
monanthous *a* одноцветковый
monaster *s cyt* монастер, материнская звезда
monaural *a phys* монауральный
monaxonic *a* одноосный
monecious *a bot* однодомный; обоеполый, гермафродитный
mongoose *s* мангуст *(Herpestes)*
Egyptian ~ египетский мангуст, фараонова крыса, ихневмон *(Herpestes ichneumon)*
mongrel *s* помесь; *a* нечистокровный, смешанный
monilicorn *a ent* имеющий четковидные антенны
moniliform *a* четковидный, похожий на нитку бус
monitor *s* монитор; *herp* варан *(Varanus)*
Komodo dragon ~ комодский варан *(Varanus komodoensis)*
monitoring *s* мониторинг, перманентный контроль, слежение; дозиметрический контроль; дозировка
monkey *s* обезьяна
Allen's swamp ~ чёрно-зелёная мартышка *(Allenpithecus)*
black howler ~ чёрный ревун *(Alouatta caraya)*
grass ~, green ~, grivet ~ зелёная мартышка *(Cercopithecus aethiops)*
howler ~ ревун *(Alouatta)*
leaf(-eating) ~ лангур, тонкотелая обезьяна *(Pygathrix)*
lion ~ львиная игрунка *(Leontocebus)*
long-nosed ~ носач *(Nasalis)*
New World ~ цепкохвостая обезьяна, капуцин *(Cebidae)*
rhesus ~ макак-резус *(Macaca mulatta)*
spider ~ паукообразная обезьяна, коата *(Ateles)*
squirrel ~ беличья обезьяна, саймири *(Saimiri)*
white-eyelid ~ мангабей, черномазая обезьяна *(Cercocebus)*
monoaperturate *a pal* одноапертурный
monocarp *s* монокарпное растение
monocarpellary *a* одноплодолистиковый
monocarpic, monocarpous *a* одноплодный
monocelled *a* одноклеточный
monocentric *a* моноцентрический
monocephalous *a bot* одноголовчатый
monocercus *s* моноцерк
monocerous *a* однорогий
monochasium *s* монохазий
monochlamydeous *a* однопокровный
monochro(mat)ic *a* монохроматический, одноцветный
monoclimax *s* моноклимакс
monoclinous *a* обоеполый, гермафродитный
monoclonal *a* моноклональный
monoclonality *s* моноклональность
monoclone *s* моноклональное антитело
monocoenosis *s* моноценоз
monocolpate *a pal* однобороздный
monocormic *a* одноствольный
monocotyledon *s* однодольное растение
monocotyledonous *a* односемядольный, однодольный
monocular *a* монокулярный
monoculture *s* монокультура
monocyesis *s* одноплодная беременность
monocyte *s* моноцит
monocytic *a* моноцитарный
monocytosis *s* моноцитоз
monodactylous *a* однопалый
monodactyly *s* однопалость
monoecious *a bot* однодомный; обоеполый, гермофродитный
monofactorial *a* монофакториальный
monogamous *a* моногамный, однобрачный
monogamy *s* моногамия
monogenesis *s* моногенез
monogenetic *a* моногенетический
monogenomic *a* моногеномный
monogerminal *a* однояйцевый, монозиготный
monogenous *a* моногенетический
monogeny *s* моногения; моногония, бесполое размножение
monogony *s* моногония, бесполое размножение
monogynopaedium *s* семья
monogynous *a* моногинный; однопестичный
monogyny *s* моногиния
monohaploid *s* моногаплоид

monohybrid *s* моногибрид
monolayer *s* монослой; монослойная культура
monolete *a* однощелевой, однорубцовый
monolocular *a* однокамерный
monomeric *a* мономерный, односегментный; *бот.* одночленный
monomery *s* мономерия
monomorphic *a* мономорфный
monomorphism *s* мономорфизм
monomorphous *a* мономорфный
mononuclear *a* мононуклеарный, одноядерный
mononucleotide *s* мононуклеотид
  flavine ~ флавинмононуклеотид, ФМН
  nicotinamide ~ никотинамидмононуклеотид
monoovular *a* однояйцевый, монозиготный *(of twins)*
monopetalous *a* однолепестный, спайнолепестный
monophagy *s* монофагия, одноядность
monophosphate *s*:
  adenosine ~ аденозинмонофосфат, АМФ
  cytidine ~ цитидинмонофосфат, ЦМФ
  deoxyadenosine ~ дезоксиаденозинмонофосфат, дАМФ
  deoxycytidine ~ дезоксицитидинмонофосфат, дЦМФ
  deoxyguanosine ~ дезоксигуанозинмонофосфат, дГМФ
  deoxyuridine ~ дезоксиуридинмонофосфат, дУМФ
  guanosine ~ гуанозинмонофосфат, ГМФ
  inosine ~ инозинмонофосфат
  thymidine ~ тимидинмонофосфат, ТМФ
  uridine ~ уридинмонофосфат, УМФ
monophyletic *a* монофилетический
monophyllous *a* однолистный
monophyly *s* монофилия
monoploid *s* моноплоид, истинный гаплоид; *a* моноплоидный, гаплоидный
monoplont *s* моноплонт
monopolar *a* монополюсный
monoporate *a* монопоровый
monopterous *a bot* однокрылый
monopyrenous *a bot* однокосточковый
monosaccate *a pal* одномешковый
monosaccharide *s* моносахарид
monosepalous *a* одночашелистиковый
monosome *s* моносома
monosomic *s* моносомик; *a* моносомический
monospermic, monospermous *a* моноспермический, односемянный; моноспермный
monospermy *s* моноспермия
monosporangium *s* моноспорангий
monospore *s* моноспора
monosporous *a* односпоровый
monostachyous *a* одноколосый
monostele *s* моностела, моностель
monostelic *a* имеющий одну стелу
monostichous *a* однорядный
monostigmatous *a zool* с одной стигмой; *bot* однорыльцевый
monostratal *a* однослойный
monostylous *a bot* одностолбиковый
monosulcate *a pal* однобороздный
monosymmetry *s* билатеральная симметрия
monothalamous *a zool* однокамерный; *bot* одногнёздный
monothecious *a bot* одногнёздный
monothely *s ent* монотелия, многомужество, женская полигамия
monothetic *a* монотетический
monotocous *a* employплодный
monotopic *a* монотопный
monotreme *a pal* одноапертурный
monotreme *s* однопроходное яйцекладущее *(Monotremata)*
monotrichous *a* монотрихальный, одножгутиковый
monotropic *a* монотропный
monotropy *s eth* монотропия
monotype *s* монотип
monotypic *a* монотипичный, монотипический
monoverticillate *a* одномутовчатый
monovoltine *a* моновольтинный
monozygotic *a* монозиготный
monster *s* чудовище; урод
  Gila ~ *герп.* ядозуб *(Heloderma)*
monstrosity *s* уродство
moor *s* болото
moorberry *s* клюква болотная *(Oxycoccus palustris)*
moorfowl *s* болотная дичь; шотландская куропатка, граус *(Lagopus scoticus)*

moorhen *s* камышница, болотная курица *(Gallinula);* камышница, болотная курица, водяная курочка *(Gallinula chloropus)*
moorland *s* болотистая местность
moory *a* болотистый
moose *s* лось *(Alces)*
mor *s* мор, грубый гумус, кислый гумус, низкозольный гумус
mordant *s hist* протрава
morel *s* сморчок *(Morchella);* паслён чёрный *(Solanum nigrum)*
mores *s pl* повадки
moribund *a* умирающий; отмирающий
morph *s* морфа
morphallaxis *s* морфаллаксис
morpheme *s eth.* морфема
morphism *s* морфизм
morphoclimax *s* морфоклимакс
morphogenesis *s* морфогенез, формообразование
morphogenetic *a* морфогенетический
morphogeny *s* морфогенез, формообразование
morphography *s* морфография
morphological *a* морфологический
morphology *s* морфология
morphometric *a* морфометрический
morphoplasm *s* морфоплазма, киноплазма
morphosis *s* морфоз
morphospecies *s* морфологический вид
mortal *a* смертный; смертельный
mortality *s* смертность; смертельность
  mass ~ массовая гибель
mortiferous *a* летальный, смертельный
morula *s* морула
morulation *s* образование морулы
mosaic *s gen* мозаик; мозаичный; *phytop* мозаика, мозаичность, мозаичная болезнь
  aucuba ~ аукуба-мозаика
  oat ~ мозаика овса; закукливание овса
  sandal leaf-curl ~ курчавость листьев сандала
  sex ~ мозаик по полу; интерсекс
mosaicism *s gen* мозаичность, мозаицизм
mosquito *s* комар *(Culicidae)*
moss *s* мох *(Musci);* моховое болото
  bog ~ торфяной мох, сфагнум *(Sphagnum)*
  carragheen ~ караген, ирландский мох *(Cladonia islandica)*
  common hair ~ кукушкин лён *(Polytrichum commune)*
  cup ~ ягель, олений мох *(Cladonia rangiferina)*
  hair ~ кукушкин лён *(Polytrichum commune)*
  Iceland ~ лишайник исландский мох *(Cetraria islandica)*
  Irish ~ караген, ирландский мох *(Cladonia islandica)*
  peat ~ торфяной мох, сфагнум *(Sphagnum)*
  raised ~ верховое болото
  reindeer ~ ягель, олений мох *(Cladonia rangiferina)*
  snake ~, staghorn ~ плаун булавовидный *(Lycopodium clavatum)*
  true ~ зелёный мох *(Bryales)*
moss-cap *s bot* калиптра, колпачок, чехлик
moss-capsule *sbot* спорогон, тека, урна
moss-covered *a* мшистый
mossery *s* моховое болото
mossy *a* мшистый
moth *s* моль; бабочка
  bagworm ~ мешочница *(Psychidae)*
  bella ~ медведица красивая *(Utethesia bella)*
  brown-tail ~ златогузка *(Euproctis phaeorrhaea)*
  clearwing ~ стеклянница *(Sesiidae)*
  clothes ~ моль платяная, моль мебельная *(Tineola biseliella)*
  codling ~ плодожорка яблонная *(Laspeyresia pomonella)*
  death's head hawk ~ (бражник) мёртвая [адамова] голова *(Acherontia atropos)*
  emperor ~ павлиноглазка сатурния, павлиний глаз *(Saturnia);* малый ночной павлиний глаз, сатурния малая *(Saturnia pavonia)*
  fall webworm ~ бабочка белая американская *(Hyphantria cunea)*
  geometer ~, geometrid ~ пяденица *(Geometridae)*
  hawk ~ бражник *(Sphingidae)*
  hummingbird ~ бражник *(Sphingidae)*
  lappet ~ коконопряд *(Lasiocampidae)*

large emperor ~ ночной павлиний глаз большой *(Saturnia pyri)*
leaf-roller ~ *pl* листовёртка *(Tortricidae)*
leopard ~ древоточец *(Cossidae)*
measuring-worm ~ пяденица *(Geometridae)*
naked clothes ~ моль платяная, моль мебельная *(Tineola bisseliella)*
noctuid ~ совка *(Noctuidae)*
nun ~ шелкопряд-монашенка *(Lymantria monacha)*
oak ~ коконопряд дубовый *(Lasiocampa quercus)*
oak procession ~ походный шелкопряд дубовый *(Thaumetopoea pityocampa)*
owlet ~s совки *(Noctuidae)*
pear codling ~ плодожорка грушевая *(Laspeyresia pyrivora)*
processionary ~ шелкопряд дубовый походный *(Thaumetopoea pityocampa)*
silk ~ шелкопряд *(Bombyx)*
sphinx ~ бражник *(Sphingidae)*
tiger ~ медведица *(Arctiidae)*
tortricid ~, tortrix ~ листовёртка *(Tortricidae)*
tussock ~ волнянка *(Lymantriidae)*
underwing ~ ленточница, орденская лента *(Catocala)*
wood leopard ~ древесница въедливая *(Zeuzera pyrina)*
woolen, woollen ~ моль шубная *(Tinea pellionella)*

mother *s* мать
stem ~ основательница

mother-of-pearl *s* перламутр; перламутровый слой; а перламутровый

motile *a* подвижный

motility *s* подвижность

motion *s* движение

motional *a* двигательный

motionless *a* неподвижный

motivation *s* мотивация, побуждение; побуждающий стимул

motive *s* повод, мотив; побуждение; двигательный; движущий

motivity *s* двигательная сила

motofacient обуславливающий движение

motoneuron *s* мотонейрон, двигательный нейрон

motor *s* двигательный мускул; двигательный нерв; *a* двигательный; движущий

motorial *a* двигательный; движущий

motorium *s prot* двигательный центр, моториум

mottle *s phytop* крапчатость, пятнистость; мозаика; мозаичность, мозаичная болезнь; крапинка, пятнышко
tobacco ~ крапчатость табака; стеблевая пятнистость табака

mottling *s* образование мозаик

moufflon *s* муфлон, горный баран *(Ovis ammon)*

mould *s see* **mold**

moult *s see* **molt**

mound *s* куча, холм; муравьиная куча
termite ~ термитник

mount *s* гистологический срез; препарат; заключать гистологические срезы; предметное стекло; *eth* садка; *v* монтировать препараты
slide ~ микропрепарат

mountain *s* гора; *a* горный

mounting *s* монтировка препарата; *eth* садка; ~ **in toto** изготовление тотальных препаратов

mouse *s* мышь
big-eared climbing ~ большеухий хомяк *(Ototylomys)*
birch ~ мышовка *(Sicista)*
bog ~ лемминговая мышь, мышевидный лемминг *(Synaptomys)*
bristly ~ колючий рисовый хомячок *(Neacomys)*
brown ~ бурый хомячок *(Scotinomys)*
bush ~ мышовка *(Sicista)*
common field ~ пашенная полёвка *(Microtus agrestis);* лесная мышь *(Apodemus sylvaticus)*
dawn meadow ~ южно-азиатская полёвка *(Eothenomys)*
deer ~ белоногий хомячок *(Peromyscus);* олений хомячок *(Peromyscus maniculatus)*
etho ~ акациевая крыса *(Aethomys)*
fat ~ толстый хомячок, жирная мышь *(Steatomys)*
fat-tailed ~ жирнохвостая песчанка *(Pachyuromys)*
field ~ лесная мышь *(Apodemus)*

forest ~ кустарниковая крыса *(Grammomys);* африканская мышь *(Thamnomys)*
gerbil ~ длинноухая мышь *(Malacothrix);* высокогорный хомячок *(Eligmodontia)*
golden ~ золотистый хомячок *(Ochrotomys)*
harvest ~ мышь-малютка *(Micromys minutus);* полевой хомяк *(Reithrodontomys)*
hazel ~ орешниковая соня *(Muscardinus avellanarius)*
house ~ домовая мышь *(Mus)*
lemming ~ лемминговая мышь, мышевидный лемминг *(Synaptomys)*
long-legged pouched ~ сумчатый тушканчик *(Antechinomys)*
meadow ~ серая полёвка *(Microtus)*
Old World harvest ~ мышь-малютка *(Micromys)*
opossum ~ соневидный кускус *(Cercartetus)*
pocket ~ мешотчатый прыгун *(Perognathus)*
pygmy ~ карликовый хомячок *(Baiomys)*
red-backed ~ лесная полёвка, рыжая полёвка *(Clethrionomys)*
snow ~ снежная полёвка *(Microtus nivalis)*
wood ~ лесная мышь *(Apodemus)*
woodland jumping ~ полумышь *(Napaeozapus)*

mouse-hare *s* пищуха, сеноставка *(Ochotona);* степная пищуха, малая пищуха *(Ochotona pusilla)*
mouse-shrew *s* мышевидная белозубка *(Myosorex)*
mousetrap *s* мышеловка
moustache *s* усы
mouth *s* рот, ротовое отверстие
~ **of the womb** *anat* маточный зев
cell ~ цитостом, клеточный рот
shell ~ устье раковины

mouthparts *s pl* ротовые части, ротовой аппарат
piercing-sucking ~ колюще-сосущий ротовой аппарат

movable *a* подвижный
movement *s* движение; телодвижение; переселение, кочёвка
kinetic ~ кинетизм
nastic ~ настия
vermicular ~ перистальтика

movette *s lab* мувет
mow *v* косить, жать; *s* стог; копна
mowing *s* косьба, кошение
mshara *s* мшара, багон, заболоченный сосняк
much-branched *a* сильноветвистый
muciferous *a* выделяющий слизь; содержащий слизь; заполненный слизью
mucific *a* выделяющий слизь
mucify *v* выделять слизь
mucilage *s* слизь; растительный клей
mucilaginous *a* слизистый; клейкий, вязкий, липкий
mucin *s* муцин, слизистый секрет
muck *s* навоз; перегной; болотный ил, сапропель
mucoid *s* мукоид; *a* мукоидный, слизеподобный
mucolytic *a* муколитический, растворяющий слизь
mucopeptide *s* мукопептид
mucopolysaccharide *s* мукополисахарид
mucoprotein *s* мукопротеид
mucosa *s* мукоза, слизистая оболочка
mucous *a* слизистый; покрытый слизью; секретирующий слизь; образующий слизь
mucro *s* мукро, остроконечие
mucronate, mucronated *a* заострённый, остроконечный
mucronule *s bot* небольшое остриё
muculent *a* слизеподобный
mucus *s* слизь
mud *s* ил, тина
muflon *s see* **moufflon**
mulberry *s* шелковица, тут *(Morus);* малина душистая *(Rubus odoratus)*
mulch *s* мульча, подстилка
mule *s* мул
mull *s* мулль, мягкий лесной гумус
mullein *s bot* коровяк *(Verbascum)*
multarticulate *a* многочленистый; многосуставный
multibaculate *a pal* множественно-столбчатый
multicamerate *a* многогнёздный; многокамерный

multicarinate *a* многокилевый; многожилковый
multicaulis *a* многостебельный
multicellular *a* многоклеточный
multichamber *a* многокамерный
multiciliate *a* многоресничный
multicipital *a* мутовчатый
multicolored *a* многоцветный; пёстрый
multicostate *a bot* многожилковатый
multidentate *a* многозубчатый
multidiluter *s lab* многоканальный дилютер, мультидилютер
multidish *s lab* многолуночный планшет, агглютинационный планшет
multifactorial *a* полифакториальный
multifasciculate *a bot* сильномочковатый, многопучковый
multifetation *s* множественная беременность, беременность несколькими плодами
multifid *a* многонадрезный; многораздельный
multiflagellate *a* многожгутиковый
multiflowered *a* многоцветковый
multifoliate *a* многолистный
multifoliolate *a* многолисточковый
multiformity *s* полиморфизм
multiganglionate *a* имеющий много ганглиев
multigenic *a* полифакториальный
multigyrate *a* имеющий много извилин
multiheterocolpate *a pal* множественно-разнобороздный
multijugate *a* многопарный
multilaminate *a* многопластинчатый
multilateral *a* многосторонний
multilayer *a* многослойный
multilobate *a* многодольный
multilocular *a* многогнёздный, многокамерный
multimammate *a* имеющий много сосков
multinervate *a* многожилковый
multinodal *a* многоузловой
multinuclear, multinucleate *a* многоядерный
multinucleolate *a* имеющий несколько ядрышек
multiocular *a* многоглазый
multiovulate *a* многояйцевой
multiparasitism *s* мультипаразитизм
multiparous *s* многократно рожавшая
multipartial *a imm* поливалентный
multipartite *a* многораздельный
multipipetter *s lab* многоканальная пипетка, мультисэмплер
multiple *a* множественный, сложный, со сложной структурой
multiple-aged *a* разновозрастный
multiplication *s* размножение
multiplicity *s* множественность
~ **of infection** *s* множественность заражения
multiplier *s* луковица-детка
multiply размножаться; увеличиваться в числе
multipolar *a* мультиполярный, многополюсный
multipunctate *a bot* многоточечный
multiradiate *a* многолучевой
multiramose *a* многоветвистый
multiseptate *a* многоперегородчатый
multiserial *a* многорядный
multisiliquous *a* многостручковый
multi-spike-and-wave *s phys* множественный разряд «пик – волна»
multispiral *a* многовитковый
multistaminate *a* многотычинковый
multistepper *s lab* многоканальная пипетка, мультисэмплер
multisulcate *a* многобороздчатый
multitentaculate *a* многощупальцевый
multitongued *a bot* многоязычковый
multituberculate *a* многобугорчатый
multivalency *s* поливалентность, многовалентность
multivalent *s* поливалент, мультивалент
multivalve *s* многостворчатая раковина
multivariate *a* многомерный
multivoltine *a* мультивольтинный, поливольтинный
mumia *s ent* куколка
mung *s* фасоль маш *(Phaseolus aureus)*
mungoose *s см.* **mongoose**
muntjac *s ther* мунтжак *(Muntiacus)*
murali-divided *a bot* имеющий перегородки
mureinoplast *s* муреинопласт
muricate, muricated *a* покрытый колючками
murine *a* мышиный; мышевидный
murmur *s* шум; жужжание
murre *s orn* кайра *(Uria)*
murrelet *s orn* старик *(Synthliboramphus)*
muscle *s* мышца, мускул

adductor ~ аддуктор, приводящая мышца; аддуктор, замыкающий мускул
cardiac ~ сердечная мышца, миокард
closing ~ аддуктор, замыкающий мускул
constrictor ~ сфинктер
fixation ~ мускул-фиксатор
synergistic ~ мышца-синергист
uterine ~ мышца матки, миометрий
muscoid *a* мохообразный
muscovy *s* мускусная утка *(Cairina moschata)*
muscular, musculary *a* мышечный
musculature *s* мускулатура
musculocutaneous *a* мышечно-кожный; кожно-мускульный
mushroom *s* гриб; боровик *(Boletus edulis);* **pick mushrooms, go picking mushrooms** *v* собирать грибы, ходить по грибы
aspen ~ подосиновик *(Boletus versipellis)*
common meadow ~ шампиньон обыкновенный *(Agaricus campester)*
death cap ~ бледная поганка *(Amanita phalloides)*
edible ~ съедобный гриб
field ~ шампиньон обыкновенный *(Agaricus campester)*
golden oyster ~ вешенка рожковидная *(Pleurotus cornucopiae)*
Hiratake ~ вешенка обыкновенная *(Pleurotus ostreatus)*
honey ~ опёнок настоящий [осенний] *(Armillariella mellea)*
horse ~ шампиньон полевой *(Agaricus arvensis)*
maitake ~ грифола курчавая, гриб-баран *(Grifola frondosa)*
matsutake ~ трихолома матцутаке *(Tricholoma matsutake)*
meadow ~ шампиньон обыкновенный *(Agaricus campester)*
oyster ~ вешенка обыкновенная *(Pleurotus ostreatus)*
parasol ~ гриб-зонтик пёстрый *(Lepiota procera)*
plum ~ ивишень *(Clitopilus prunulus)*
Satan's ~ сатанинский гриб *(Boletus satanus)*
musk *s* мускус
musk-deer *s* кабарга *(Moschus)*
muskeg *s* сфагновое болото, торфяное болото; верховое болото
musk-ox *s* овцебык, мускусный бык *(Ovibos moschatus)*
muskrat *s* ондатра *(Ondatra zibethica)*
Russian ~ выхухоль *(Desmana moschata)*
musquash *s см.* **muskrat**
mussel *s* двустворчатый моллюск; мидия *(Mytilus)*
must *s* плесень; покрываться плесенью; виноградное сусло
mustang *s* мустанг
mustard *s* горчица
mustelids *s pl* куньи, куницы *(Mustelidae)*
musty *a* плесневый
mutability *s* мутабильность
mutable *a* мутабильный
mutafacient *a* вызывающий мутации, мутагенный
mutagen, mutagene *s* мутаген
mutagenesis *s* мутагенез
mutagenic *a* мутагенный
mutagenicity *s* мутагенность
mutagenous *a* мутагенный
mutant *s* мутант
missense ~ миссенс-мутант
nonsense ~ нонсенс-мутант, бессмысленный мутант
"suicide" ~ мутант-«самоубийца»
mutate *v* мутировать
mutation *s* мутация
back ~ обратная мутация, реверсия
clustered point ~ точковая кластер-мутация, сгруппированные точковые мутации
missense ~ миссенс-мутация
nonsense ~ нонсенс-мутация, бессмысленная мутация
ochre ~ мутация охра, охра-мутация
opal ~ мутация опал, опал-мутация
transition ~ транзиция
transversion ~ трансверсия
mutator *s* ген-мутатор; *a* мутаторный
mutilation *s* калечение
muton *s* мутон
mutt *s* дворняжка
mutton *s* баранина
muttonbird *s* тонкоклювый буревестник *(Puffinus tenuirostris)*
mutualism *s* мутуализм

mutuality *s* обоюдность; взаимность; взаимозависимость
muzzle *s* морда; рыло; намордник
mycelioid *a* мицелиевидный
mycelium *s* мицелий, грибница
mycetocyte *s* мицетоцит
mycetogenetic, mycetogenic *a* образуемый грибами
mycetoid *a* грибовидный
mycetology *s* микология
mycetoma, mycetome *s* мицетома
mycetophage *s* мицетофаг, грибоядный организм
mycetophagous *a* грибоядный
mycobacterium *s* микробактерия
mycobiota *s* микофлора
mycocecidium *s* микоцецидий
mycocide *s* микоцид, фунгицид
mycoflora *s* микофлора
mycogenous *a* грибного происхождения
mycology *s* микология
mycophagy *s* микофагия
mycoplasm, mycoplasma *s* микоплазма
myco(r)rhiza *s* микориза
mycorrhizal *a* микоризный
mycosis *s* микоз
mycosterol *s* микостерол
mycotrophic *a* микотрофный
myelencephalon *s* продолговатый мозг
myelination *s* миелинизация
myeline *s* миелин
myeloblast *s* миелобласт
myelocyte *s* миелоцит
myelogenous *a* миелогенный
myeloid *a* миелоидный
myelomere *s* миеломер
myelon *s* спинной мозг
myeloplast *s* миелобласт
myelopoiesis *s* миелопоэз
myesthesia *s* мышечная чувствительность
myiophilous *a* опыляемый двукрылыми
myna *s orn* майна; майна *(Acridotheres)*
myoblast *s* миобласт
myocard, myocardium *s* миокард, сердечная мышца
myocinesis *s* миокинез, кинетика сердца
myoclonia, myoclonus *s* миоклонус, клонический спазм мышцы
myocoel, myocoele *s* миоцель, полость миотома
myocyte *s* миоцит, мышечная клетка
myodynamic *a* миодинамический
myoepithelial *a* миоэпителиальный, эпителиально-мышечный
myofibril, myofibrilla *s* миофибрилла
myofibrillomere *s* миофибрилломер
myofilament *s* миофиламент, мышечная нить
myogaster *s* мышечное брюшко
myogenesis *s* миогенез
myogenic *a* миогенный
myoglia *s* миоглия
myoglobin *s* миоглобин
myoglobulin *s* миоглобулин
myogram *s* миограмма
myograph *s* миограф
myography *s* миография
myokinesis *s* миокинез, кинетика мышц
myolemma *s* миолемма, сарколемма
myology *s* миология
myomere *s* миомер
myometrial *a* миометральный
myometrium *s* миометрий
myoneme *s* мионема
myoneural *a* мионевральный, нервно-мышечный
myoneurocyte *s* мионевроцит, мионевральная клетка
myophrisk *s* миофриск
myopia *s* миопия, близорукость
myoplasm *s* саркоплазма
myopolar *a* миополярный
myoseptum *s* миосепта
myosin *s* миозин
myospasm *s* миоспазм
myotactic *a* миотактический
myotatic *a* миотатический
myotome *s* миотом
myotonia *s* миотония, мышечное напряжение
myotonus *s* мышечный тонус
myrmecochorous *a* мирмекохорный, распространяющийся муравьями
myrmecology *s* мирмекология
myrmecophagous *a* поедающий муравьёв
myrmecophil *s* мирмекофил
myrmecophilous *a* мирмекофильный
myrmecophily *s* мирмекофилия
myrmecophobe *s* мирмекофоб
myrmecophobic *a* мирмекофобный
myrmecophyte *s* мирмекофит

myxobacterium *s* миксобактерия, слизистая бактерия
myxocete *s* клетка слизистой ткани
myxomycete *s* миксомицет, слизевик
myxopterygium *s* миксоптеригий, птеригоподий
myxovirus *s* миксовирус

N

nag *s* небольшая лошадь; пони
naiad *s ent* наяда
nail *s* ноготь; коготь
naked *a* голый; безволосый, бесшёрстный; беспокровный *(of flowers);* безлистный *(of trees)*
naked-flowered *a* голоцветковый
naked-fruited *a* голоплодный
naked-leaved *a* гололистный
naked-rooted *a* голокорешковый
naked-twigged *a* головетвистый
naked-winged *a* голокрылый
name *s* имя; название; давать имя; называть
nanism *s* нанизм, карликовость
nannander *s* карликовый самец
nanophanerophyte *s* нанофанерофит, низкорослый кустарник
nanopodzol *s* наноподзол, маломощная подзолистая почва
nanosomia *s* карликовость, нанизм
nanous *a* карликовый; низкорослый
nap *s* ворс; пушок
napaceous *a* репчатый
nape *s* затылок; задняя часть шеи; загривок
napiform *a* реповидный, репообразный
nappy *a* пушистый; опушённый; ворсистый
narcosis *s* наркоз
narcotic *s* наркотик; *a* наркотический
nares *s pl* ноздри
narial *a* относящийся к ноздрям
naricorn *s herp* носовой щиток
narrowed *a* суженный
narrow-flowered *a* узкоцветковый
narrowing *s* сужение, уменьшение
narrow-leaved *a* узколистный
narrow-margined *a* узко окаймлённый
narrow-mouthed *a* узкоротый
narrow-petaled *a* узколепестный
narrow-ringed *a* узкослойный
narrow-zoned *a* узкослойный
nasal *a* носовой, назальный
nascency *s* рождение, зарождение, возникновение
nascent *a* возникающий, появляющийся, рождающийся
nasofrontal *a* носолобный
nasolabial *a* носогубной
nasopalatine *a* носонёбный
nasopharynx *s* носоглотка
nasoturbinal *s* носовая раковина
nastic *a* настический
nasus *s ent* наличник, клипеус
nasute *a* носатый
natal *a* натальный, относящийся к рождению; ягодичный
natality *s* рождаемость; коэффициент рождаемости
natatorial *a* плавательный; приспособленный к плаванию
nates *s pl* ягодицы; зад, круп; крестец
native *a* врождённый; природный, естественный; аборигенный, местный; нативный
natterjack *s* камышовая жаба *(Bufo calamita)*
natural *a* природный, естественный, натуральный, настоящий; дикий, некультивированный
naturalist *s* натуралист
  young ~ юный натуралист, юннат
naturalization *s* натурализация; акклиматизация
nature *s* природа; сущность, основное свойство
nausea *s* тошнота
navel *s* пупок; пуповина; рубчик
navicular *s* ладьевидная кость; *a* ладьевидный, ладьеобразный
naviculoid *a* навикулоидный
navigation *s* навигация; ориентировка в полёте
neanic *a* юношеский; личиночный
nearctic *a* неарктический
neardesert *s* полупустыня
nearest-neighbor *s mbiol* ближайший сосед
nearshore *s* прибрежная зона
nearsighted *a* близорукий
neat *s* бык *(Bos taurus);* крупный рогатый скот; *a* чистый, неразбавленный
neb *s* клюв; нос; морда
nebulization *s* распыление
necatoriasis *s helm* некатороз
neck *s* шея; *anat* шейка
neck-fighting, necking *s eth* борьба шеями

necklace-shaped *a* чётковидный, в виде нитки бус
neck-wrestling *s eth* борьба шеями
necrobiosis *s* некробиоз
necrobiotic *a* некробиотический
necrocytosis *s* некроцитоз
necrogenous *a* живущий в трупах
necrolysis *s* неролиз
necron *s* неразложившееся растение
necrophage *s* некрофаг, трупоед
necrophagous *a* питающийся падалью, пожирающий трупы
necrophagy *s* некрофагия
necrophilous *a* некрофильный
necropsy *s* вскрытие трупа, аутопсия
necrose омертвевать; вызывать омертвение
necrosis *s* некроз, омертвение; *phytop* нектрия лиственных пород *(caused by Nectria galligena)*
necrotaxis *s* некротаксис
necrotic *a* некротический
necrotize *v* омертвевать; вызывать омертвение
necrotomy *s* вскрытие трупа, аутопсия
nectar *s* нектар
nectariferous *a* нектароносный
nectarine *s* персик голоплодный, нектарин *(Persica nucipersica)*
nectarivorous *a* питающийся нектаром
nectarostigma *s* медовая ямка
nectarotheca *s* нектаровместилище
nectarous *a* нектарный
nectary *s* нектарник
needle *s* игла, иголка; *pl* хвоя, иглы
needle-bearing *a* иглоносный
needle-leaved *a* иглолистный
needle-shaped *a* игловидный, игольчатый
negundo *s* негундо, клён ясенелистный *(Acer negundo)*
neigh *v* ржать
neighborhood *s* окрестности
ecological ~ экологическая окрестность
neighboring прилегающий, соседний
nelly *s* исполинский буревестник *(Macronectes giganteus);* альбатрос *(Phoebetria fusca)*
nelumbo *s* лотос *(Nelumbium)*
nemamere *s* немамер, сегмент хромонемы
nemathecium *s* нематеций
nematoblast *s* нематобласт
nematocyst *s* нематоциста, стрекательная клетка
nematode *s* нематода, круглый червь *(Nematoda)*
nematodosis *s helm* нематодоз
nematogene *a* нематогенный
nematogone *s bot* тонкостенная размножающаяся клетка
nematology *s* нематодология
nematophore *s* полип, несущий нематоцисты
neoantigen *s* неоантиген
neoblast *s* необласт
neocarpy *s* неокарпия
neocentromere *s* неоцентромера
neocyte *s* незрелая форма лейкоцита, неоцит
Neo-Darvinism *s* неодарвинизм
neoencephalon *s* передний мозг, большой мозг
neo-endemism *s* неоэндемизм
Neogaea *s* неогея
neogenesis *s* регенерация, восстановление ткани
neokinetic *a* неокинетический
Neo-Lamarckism *s* неоламаркизм
neomorph *s* неоморф, неоморфный ген
neomorphic *a* неоморфный
neomorphosis *s* неоморфоз
neonatal *a* новорождённый
neopallium *s* новая кора
neophilia *s eth* неофилия, стремление к новому
neophobia *s eth* неофобия, боязнь нового
neophyte *s* неофит
neoplasia *s* неоплазия
neoplasm *s* неоплазма, новообразование, опухоль
neoplastic *a* неопластический, опухолевый
neoptile, **neossoptile** *s* неоптиль, эмбриональный пух
neoteinia, neoteny *s* неотения
neotropical *a* неотропический
neotype *s* неотип; *a* неотиповой, повторно выделенный
nep *s bot* котовник кошачий *(Nepeta cataria)*
nephelometer *s* нефелометр
nephelometry *s* нефелометрия

nephric *a* почечный
nephridial *a* нефридиальный
nephridiopore *s* нефридиопор
nephridium *s* нефридий
nephrocoel *s* нефроцель, полость нефротома
nephrocyte *s* нефроцит, выделительная клетка
nephrodinic *a* имеющий мочеполовой канал
nephrogenic *a* нефрогенный
nephrogonoduct *s* мочеполовой канал
nephromere *s* нефротом
nephromixium *s* нефромиксий
nephron *s* нефрон
nephropore *s* нефридиопор
nephros *s anat* почка
nephrostome *s* нефростом
nephrotome *s* нефротом
nepionic *a* постэмбриональный; личиночный, ларвальный
nerval *a* нервный
nervate *a bot* жилковатый; *зоол.* снабжённый нервами
nervation, nervature *s bot* жилкование, расположение жилок; иннервация
nerve *s* нерв; жилка
nerve-knot *s* ганглий, нервный узел
nerveless *a* не имеющий жилок
nerveroot *s* корешок нерва
nervimotor *a* нервомоторный
nervimuscular *a* нервно-мышечный
nervisequent *a bot* следующий по ходу жилок
nervous *a* нервный, относящийся к нерву
nervule *s* маленькая жилка; нервная веточка
nervuration *s см.* **nervation**
nervure *s* нервюра, жилка
nest *s* гнездо; выводок; стая; рой; v вить гнездо; гнездиться
  termite ~ термитник
nest-building *s* гнездостроение
nester *s* гнездующая птица
nesting *s* гнездование
nestis *s* тощая кишка
nestle *v* вить гнездо; гнездиться
nestling *s* гнездарь, птенец
net *s* сеть; сетка; сетчатая структура; паутина
  mosquito ~ накомарник; противокомариная сетка, полог от комаров
  nerve ~ нервная сеть, нервное сплетение
  NHR ~ индексационная шкала гистосовместимости, шкала трансплантационных индексов, шкала Рапопорта є Дюссе
net-fruited *a* сетчатоплодный
net-leaved *a* сетчатолистный
net-seeded *a* сетчатосемянный
netted *a* сетчатый
netting *s* сетчатость
  mosquito ~ накомарник; противокомариная сетка, полог от комаров
nettle *s* крапива *(Urtica)*
nettlebead *s phytop* курчавость
nettle-leaved *a* крапиволистный
net-veined *a* сетчато-нервный
network *s* сеть, сетка
neural *a* невральный; мозговой; нервный
neuralgia *s* невралгия
  stump ~ фантомная боль
neuralization *s* нейрализация
neurapophysis *s* невральный отросток
neuration *s bot* жилкование, расположение жилок; иннервация
neuraxis *s* спинной мозг; *emb* нервная трубка
neuraxon *s* осевой цилиндр аксона
neurenteric *a* нервно-кишечный
neurilemma *s* неврилемма, оболочка нервного волокна, шванновская оболочка
neurite *s* неврит, аксон нервной клетки
neurobiology *s* нейробиология
neuroblast *s* нейробласт
neurocentrum *s* тело позвонка
neuroceptor *s* нейроцептор
neurocoel *s* невроцель, полость нервной трубки
neurocranium *s* мозговой череп
neurocrine *a* нейрокринный
neurocyte *s* невроцит, нейрон, нервная клетка
  giant pyramidal ~ гигантопирамидальная нервная клетка, клетка Бетца
  large grain ~ большой невроцит-зерно

multiform ~ многоморфный невроцит, клетка Гольджи
neurocyton *s* тело нервной клетки
neurodendron *s* древовидный отросток, невродендрон
neuroepithelium *s* нейроэпителий
neuroethology *s* нейроэтология
neurofibril, neurofibrilla *s* нейрофибрилла
neuroganglion *s* ганглий, нервный узел
neurogenesis *s* развитие нервной клетки
neurogenic, neurogenous *a* нейрогенный
neurogeny *s* развитие нервной клетки
neuroglia *s* невроглия
neurohumor *s* нейрогумор
neurohypophysis *s* нейрогипофиз
neuroimmunochemistry *s* нейроиммунохимия
neuroimmunology *s* нейроиммунология
neurolemma *s* неврилемма, оболочка нервного волокна, шванновская оболочка
neurology *s* неврология
neurolymph *s* спинномозговая жидкость
neurolysis *s* нейролиз
neuromere *s* невромер
neuromery *s* невромерия, сегментация нервной трубки
neuromion *s* невромион, нервно-мышечная единица
neuromotor *a* нейромоторный
neuromuscular *a* нервно-мышечный
neuron, neurone *s* нейрон, нервная клетка, невроцит
long ~ аксон
neuropeptide *s* нейропептид
neurophysiology *s* нейрофизиология
neuropil, neuropile, neuropilem *s* нейропиль
neuroplasm *s* нейроплазма, цитоплазма нейрона
neuroplex *s* нервное сплетение
neuropodium *s* невроподий
neuropore *s* нейропор
neuropterous *a* сетчатокрылый
neuroscience *s* нейробиология; неврология
neuroscientist *s* нейробиолог
neurosecretion *s* нейросекреция
neurosecretory *a* нейросекреторный
neurosis *s* невроз
neuroskeleton *s* внутренний скелет
neurosporaxanthin *s* каротиноидная кислота
neurosynapse *s* нейросинапс
neurotoxin *s* нейротоксин
neurotransmitter *s* нейромедиатор
neurotrophic *a* нейротрофический
neurotropism *s* нейротропизм
neurotubule *s* нервная трубка
neurovascular *a* нервно-сосудистый
neurula *s* нейрула
neurulation *s* нейруляция
neuter *a* бесполый; кастрированный
neutral *a* нейтральный, промежуточный, неопределённый; бесполый
neutralism *s* нейтрализм
neutralization *s* нейтрализация
cross ~ перекрёстная нейтрализация, реакция перекрёстной нейтрализации
neutralize *v* нейтрализовать
neutralizer *s* нейтрализатор
neutropenia *s* нейтропения
neutrophil *s* нейтрофил, нейтрофильный лейкоцит
neutrophilia *s* нейтрофилия
neutrophilic, neutrophilous *a* нейтрофильный
new-blown *a* расцветающий
newborn *a* новорождённый
newgrowth *s* неоплазма, новообразование, опухоль
new-laid *a* свежеснесенное *(egg)*
newt *s* тритон *(Triturus)*
nexin *s* нексин
nib *s* клюв; остриё, кончик
nibble *s* обкусывание; клёв
niche *s* ниша
nick, nicking *s mbiol* однонитевой разрыв, одноцепочечный разрыв, «ник»; **to seal** ~ лигировать одноцепочечный разрыв, заполнять одноцепочечный разрыв, «ник»-транслировать
nick-translation *s mbiol* «ник»-трансляция
nictitate *v* мигать, моргать
nidality *s* очаговость
nidamental *a* нидаментальный
nidation *s* нидация, имплантация
nidicolous *a* гнездовой
nidificate *v* вить гнездо, делать гнездо
nidification *s* гнездостроение
nidifugous *a* выводковый

nidus *s ent* гнездо; гнездообразная полость; полость развития спор; очаг заболевания
nighthawk *s orn* козодой *(Caprimulgus)*
night-heron *s orn* кваква *(Nycticorax)*
nightjar *s* козодой *(Caprimulgus)*
nightshade *s* паслён *(Solanum)*
nigredo *s phytop* меланодерма
nilg(h)ai *s ther* нильгау *(Boselaphus tragocamelus)*
nine-killer *s orn* сорокопут *(Laniuus)*
nip *v* щипать; прищипывать *(a plant)*
nipper *s* клешня; передний зуб, резец
nipple *s anat* сосок; *bot* сосочек
nipple-shaped *a* сосковидный
nit *s* гнида
nitex *s lab* нитекс
nitid, nitidous *a* блестящий; лоснящийся
nitration *s* нитрование
  oxidative ~ оксинитрование, окислительное нитрование
nitrification *s* нитрификация
nitrifier *s* нитрификатор
nitrify нитрифицировать
nitrobacter *s* нитробактер
nitrogen *s* азот
nitrogen-bearing *a* азотсодержащий
nitrogen-fixing *a* азотфиксирующий
nitrogen-free *a* безазотный, безазотистый
nitrogen-loved *a* нитрофильный
nitrogenous *a* азотный; азотистый
nitrolysis *s* нитролиз
nitrophil *s* нитрофил
nitrophilous *a* нитрофильный
nitrosification *s* нитрозификация
NK-lysed *a* лизированный NK-клеткой
N-mer *s* N-мер
nociceptive *a* ноцицептивный, болевой; угрожающий
nociceptor *s* ноцицептор, болевой рецептор
noctiflorous *a* цветущий ночью
noctilucent *a* фосфоресцирующий
noctule *s ther* вечерница *(Nyctalus);* рыжая вечерница *(Nyctalus noctula)*
nocturnal *a* ночной
nocturnalism *s* ночной образ жизни; ночное состояние
nodal *a* узловой; центральный
noddy *s* глупая крачка *(Anous)*)
node *s* узел; нарост, утолщение
  Hensen's ~ гензеновский узелок
nodose *a* узловатый
nodular *a* узелковый; узловатый
nodulation *s* образование клубеньков
nodule *s ent* узелок; *bot* узелок; клубенёк
noduliferous *a* несущий узелки
noise *s* шум
  "immunologic background" ~ «иммуногенный шумовой эффект», «иммуногенный фоновый шум»
nomad, nomade *s* бродячая форма
nomadic *a* бродячий, передвигающийся; кочевой
nomadism *s* номадизм, кочевой образ жизни
nomenclature *s* номенклатура
nominal *a* номинальный
nominate *a* именованный
nomium *s* сообщество пастбища
nomogenesis *s* номогенез
nomogenous *a* возникающий естественным путём
nomogram *s* номограмма
nomophilous *a* обитающий на пастбищах
nona *s* африканский трипаносомоз, сонная болезнь
nonadaptionism *s* неадаптационизм
nonadaptive *a* неприспособительный, неадаптивный
nonantigenic *a* неантигенный
nonaperturate *a* неапертурный, безапертурный; беспоровый
nonbacterial *a* безмикробный, стерильный
nonbearing *a* бесплодный, неплодоносный; неплодородный *(soil)*
nonbolting *a* нестрелкующийся *(e.g., onion)*
nonbranching *a* неветвящийся
nonbreeding *a* не выводящий молоди; негнездовой; неплеменной
noncellular *a* неклеточный
nonchlorophyllic *a* бесхлорофильный
nonconjuction *s* отсутствие конъюгации
noncooperativity *s mbiol* некооперативность
nondecidous *a* нелистопадный, вечнозелёный
nondigestible *a* неусвояемый
nondirectional *a* ненаправленный
nondisjunction *s* нерасхождение
nondividing *a* неделящийся

nonfertile *a* неплодородный
nonfilamentous *a* не имеющий нитей
nonfixiform *a* без определённой формы
nonflagellate *a* безжгутиковый
nonflowering *a* нецветковый
nongermination *s* невсхожесть
nonhairy *a* безволосый, бесшёрстный
nonhardened *a* незакалённый
nonhardy *a* невыносливый, нестойкий
nonheritable *a* ненаследуемый
nonhomogeneous *a* неоднородный
nonimmunogenic *a* неиммуногенный
noninducibility *s* неиндуцибельность
noninfected *a* незаражённый
noninjured *a* неповреждённый
noninoculated *a* незаражённый
nonliving *a* неживой
nonloricate *a* беспанцирный
nonluminous *a* несветящийся
nonmanifest *a* скрытый, латентный
nonmedullated *a* безмякотный
nonmetameric *a* неметамерный
nonmigratory *a* немигрирующий, оседлый
nonmolting *a* нелиняющий
nonmotile *a* неподвижный
nonmyelinated *a* безмякотный
nonnitrogenous *a* безазотистый
nonnodulating *a* неклубеньковый
nonoverlapping *s* неперекрывание
nonpathogenic *a* непатогенный
nonperishable *a* стойкий, непортящийся
nonpersistent *a* нестойкий
nonphotosynthetic *a* нефотосинтезирующий
nonpolar *a* неполярный, бесполюсный
nonpredatory *a* мирный, нехищный
nonprotein *a* непротеиновый, небелковый
nonreactive *a* нереагирующий
nonresistant *a* нерезистентный, невыносливый
nonresponsiveness *s* ренеагирование, отсутствие реакции; *imm* неотвечаемость; толерантность
nonreversible *a* необратимый
nonrigid *a* гибкий, податливый, нежёсткий
nonrooted *a* неукоренившийся
nonsaline *a* незасолённый *(soil)*
nonsegmented *a* несегментированный, нерасчленённый
nonselective *a* неизбирательный, неселективный
nonseptate *a* не имеющий перегородок
nonsexual *a* бесполый
nonsoluble *a* нерастворимый
nonspore-forming *a* не образующий спор
nonstaining *a* неокрашивающий, неокрашивающийся
non-survival *s* невыживание
nontapering *a* полнодревесный
nontoxic *a* нетоксичный, неядовитый
nontreated *a* необработанный; контрольный
nonvascular *a* бессосудистый
nonviable *a* нежизнеспособный
nonwater *a* безводный
nonwettability *s* несмачиваемость
nonwetting *a* несмачивающийся
nonwilting *a* неувядающий, незавядающий
norm *s* норма
normality *s* нормальное состояние
normalization *s* нормализация
normoblast *s* нормобласт
normoblastic *a* нормобластический
normochromic *a* нормохромный
normocyte *s* нормоцит, зрелый эритроцит
normotensive *a* нормотензивный
normotonic *a* нормотонический
normotropic *a* нормотропный
nose *s* нос; морда; чутьё, нюх; обоняние; *v* нюхать, обнюхивать; обонять
nosebrain *s* обонятельный мозг
nose-picking *s* ковыряние в носу
nosepiece *s lab* револьвер
nosogeny *s* патогенез
nosogeography *s* географическая медицина, нозогеография, география болезней
nosology *s* нозология
nosophyte *s* нозофит
nosopoietic *a* патогенный, болезнетворный
nostril *s* носовое отверстие, ноздря
  anterior ~ наружное носовое отверстие, ноздря, передняя ноздря
  posterior ~ заднее носовое отверстие, хоана, задняя ноздря
notal *a* дорсальный, спинной; тыльный

notation *s* обозначение; система обозначений; система изображения величин
notaula *s* парапсидальная борозда
notch *s* зарубка, надрез; *анат.* вырезка, выемка; *v* делать зарубку, производить надрез
note *s* звук *(bioacoustics)*
nothocline *s* гибридный клайн
notochord *s* спинная струна, хорда
notochordal *a* хордальный
Notogaea *s* нотогея
notopleural *a* нотоплевральный
notopodiuum *s zool* нотоподий
not-self *s imm* «не своё»
notum *s* ent нотум, спинка грудного сегмента
nourish *v* питать, кормить; растить; ухаживать
nourished *a* упитанный
nourishing *a* питательный
nourishment *s* питание, кормление; пища, корм
noxious *a* вредный, пагубный
nubilous *a* половозрелый
nucellus *s* нуцеллус
nucha *s* затылок; задняя часть шеи
nuchal *a* нухальный
nuciferous *a* орехоплодный
nuciform *a* ореховидный
nucivorous *a* питающийся орехами
nuclear *a* ядерный, относящийся к ядру; содержащий ядро
nuclease *s* нуклеаза
nucleate *v* образовывать ядро; содержащий ядро
nucleation *s* образование ядра; нуклеация
nucleic *a* нуклеиновый
nucleination *s* нуклеинизация
nucleocapsid *s* нуклеокапсид
nucleocentrosome *s* нуклеоцентросома
nucleocytoplasmic *a* ядерно-цитоплазматический
nucleodesma *s* нуклеодесма
nucleohistone *s* нуклеогистон
nucleohyaloplasm *s* нуклеоплазма, кариолимфа, ядерный сок
nucleoid *s* нуклеоид
nucleolar *a* нуклеолярный, ядрышковый
nucleolar-organizing *a* ядрышкоорганизующий
nucleolonema *s* нуклеоленома
nucleolus *s* ядрышко
nucleomixis *s* нуклеомиксис
nucleopeptide *s* нуклеопептид
nucleophilic *a* нуклеофильный
nucleoplasm *s* нуклеоплазма, кариолимфа, ядерный сок
nucleoprotein *s* нуклеопротеид
nucleoside *s* нуклеозид
nucleosoma *s* нуклеосома
nucleospindle *s* ядерное веретено
nucleotide *s* нуклеотид
  triphosphopyridine ~ трифосфопиридиннуклеотид
  "wrong" ~ «ошибочный» нуклеотид, неправильно спаренный нуклеотид
nucleus *s* ядро; косточка
nuculanium *s* мясистый плод, верхняя ягода
nudibranchiate *a* голожаберный
nudicaudate *a* голохвостый
nudicaulous *a* голостебельный
nudiflorous *a* голоцветковый
null-hypothesis *s* нулевая гипотеза
nulliparous *a* нерожавшая
nulliplex *s* нуллиплекс
nullisomic *s* нуллисомик
numb *v* вызывать онемение; вызывать окоченение; *a* онемелый, оцепенелый; окоченелый
number *s* порядковый номер; число; количество; *pl* численность
  ~ **of species** видовая насыщенность
  average sample ~ *biom* средний объём выборки
  binomial ~ *biom* биномиальный коэффициент
  color saponification ~ цветной показатель омыления
  copy ~ число копий; копийность
  field ~ полевой номер, первоначальный номер
  rejection ~ *biom* критическое количество
  saponification ~ показатель омыления, число омыления
  writing ~ *mbiol* индекс сверхспирализации
numerical *a* нумерический
nummiform, nummular *a* монетовидный; в виде монетных столбиков

nummulation *s* склеивание эритроцитов в монетные столбики
nun *s orn* лазоревка *(Parus coeruleus)*
nurse *v* кормить, выкармливать; выращивать
nursery *s* питомник, рассадник
nursling *s* сеянец; саженец; молодое растение
nurse-wood *s* материнское насаждение
nurture *s* питание, кормление; воспитание, забота о потомстве; *v* воспитывать; кормить, выкармливать
nut *s* орех
  ground ~ апиос клубневая *(Apios tuberosa);* арахис, земляной орех *(Arachis hypogaea)*
  Lambert ~ ломбардский орех, лещина крупная *(Corylus maxima)*
  palm ~ копра, ядро кокосового ореха
nutation *s* нутация
nutcracker *s* кедровка, ореховка *(Nucifraga);* кедровка, ореховка *(Nucifraga caryocatactes);* поползень *(Sitta)*
nutgall *s* орехоподобный галл; чернильный орешек
nuthatch *s* поползень *(Sitta)*
nutlet *s* орешек
nutmeg *s bot* мускатный орех *(Myristica);* торрея *(Torreya)*
nutria *s* нутрия *(Myocastor coypus)*
nutricism *s* нутрицизм
nutriculture *s* гидропоника
nutrient *s* питательное вещество, нутриент; питательный
nutriment *s* пища, корм
nutrition *s* питание, кормление; пища, корм
  extra ~ подкормка
nutritional *a* пищевой
nutritious *a* питательный
nutshell *s* скорлупа ореха
nutwood *s* лещина обыкновенная *(Corylus avellana)*
nuzzle *v* рыть рылом; водить носом *(e.g., of a dog)*
nyala *s* антилопа ньяла *(Tragelaphus angasi)*
nyctanthous *a* цветущий ночью, ночецветный
nyctinasty *s* никтинастия
nymph *s уте* нимфа
nymphal *a* нимфальный
nystagmus *s* нистагм
  postrotational ~ послевращательный нистагм, постнистагм

O

oak *s* дуб *(Quercus)*
cork ~ пробковый дуб *(Quercus suber)*
oakery, **oak-forest** *s* дубняк, дубрава
oakleaf *s phytop* дуболистость
oaklet, oakling *s* дубок
oak-wood *s* дубняк, дубрава
oasis *s* оазис
oat *s* овёс *(Avena)*
obconic *a* обратноконусовидный
obcordate *a* обратносердцевидный
obdeltoid *a* обратнодельтовидный
obduction *s* аутопсия, вскрытие
obelion *s* обелион
obex *s anat* задвижка
obimbricate *a* обратночерепитчатый
object *s* объект
visual ~ *phys* зрительный образ
objective *s* объектив
oblanceolate *a* обратноланцетовидный
oblate *a* сплющенный
obligate, obligatory *a* облигатный
oblique *a* косой, искривлённый
oblong *a* продолговатый, удлинённый, вытянутый
obovate, obovoid *a* обратнояйцевидный
obscured *a* скрытый, латентный
observation *s* наблюдение; изучение; исследование; *pl* результаты наблюдений
obsolescence *s* постепенное исчезновение; постепенная остановка
obsolete *a* атрофированный; незаметный, неясный
obstruction *s* закупорка
obtain *v* получать, приобретать; **to ~ energy requirements** удовлетворять потребность в энергии
obtected *a* покрытый
obtuse *a* тупоконечный
obtusifolious *a* туполистный
obtusion *s* притупление, снижение чувствительности
obvolute *a* окутанный; обвитый
occipital *a* затылочный
occiput *s* затылок
occlude *v* закрывать; закупоривать; заплывать, зарастать
occlusal *a* окклюзионный
occlusion *s* окклюзия, смыкание челюстей, прикус; закрытие; закупорка
occlusive *a* закрывающий, запирающий; преграждающий; окклюзионный
occlusor *s* замыкающая мышца
occurence *s* встречаемость; местонахождение, распространение; наличие
oceanarium *s* океанарий
oceanic *a* океанический
oceanology *s* океанология
ocellate, ocellated *a* имеющий простые глазки; пятнистый; глазчатый
ocellus *s* простой глазок; фасетка; глазок
ocelot *s ther* оцелот *(Felis pardalis)*
ochrea *s bot* раструб
ochthium *s* береговое сообщество
ochthophilus *s* обитатель берегов
ocrea *s bot* раструб
ocreate *a* раструбовидный
octactinal *a* восьмилучевой
octandrous *a* восьмитычиночный
octogynous *a* восьмипестичный
octopetalous *a* восьмилепестный
octoploid *s* октоплоид; октоплоидный
octopod *a* восьминогий
octosepalous *a* восьмичашелистиковый
octosporous *a* восьмиспоровый
octostichous *a* восьмирядный
ocular *a* глазной; *s* окуляр
oculate *a* имеющий глаза, имеющий глазки, глазчатый
oculofrontal *a* глазнично-лобный
oculogyration *s* движение глазного яблока
oculogyric, oculomotor *a* глазодвигательный
oculomotorius *s* глазодвигательный нерв
oculonasal *a* глазнично-носовой
oculus *s bot* глазок; *ent* фасеточный глаз
odd *a* нечётный
odd-colored *a* неопределённой масти
odd-pinnate *a* непарноперистый
odd-toed *a* непарнокопытный
odontoblast *s* одонтобласт
odontoclast *s* одонтокласт
odontogeny *s* развитие зубов
odontoid *a* зубовидный
odontology *s* одонтология
odontophore *s* одонтофор
odontoplast *s* одонтобласт

odontostomatous *a* имеющий рот; имеющий челюсти; вооружённые зубами
odor *s* запах
odoriferous *a* пахучий, ароматный; вонючий, зловонный
odorless *a* без запаха, непахнущий
odorous *a* пахучий, ароматный
odour *s* запах
oecad *s* экада
oecesis *s* эцезис
oecium *s* оеций
oecize *v* приживаться, колонизировать
oecology *s* экология
oecophene *s* экофен
oenocyte *s* эноцит
oesophageal *a* пищеводный
oesophagus пищевод
oestradiol *s* эстрадиол
oestriol *s* эстриол
oestrogen *s* эстроген
oestrone *s* эстрон
oestrous *a* эстральный
offal *s* отруби, высевки
off-center *s phys* off-центр
offense *s eth* нападение; агрессивные действия
offensive *a* устрашающий
officinal *a* лекарственный
off-response *s* реакция на выключение
offset *s* отпрыск, потомок; *bot* ответвление, отводок, боковой отросток
offshoot *s bot* ответвление, отводок, боковой отросток
offspring *s* потомство; отпрыск; помёт, приплод
offtype *a* нетипичный, не соответствующий типу
ogive *s biom* огива
oidium *s myc* оидий, артроспора
oike *s* местообитание
oikoplast *s* ойкопласт
oil *s* масло; нефть
  fish ~ рыбий жир
  joint ~ синовиальная жидкость
oilbird *s orn* гуахаро, жиряк *(Steatornis caripensis)*
oilseed *s* масличное семя
oil-soluble *a* маслорастворимый
oily *a* масличный; маслянистый, жирный
oinking *s* хрюканье
ointment *s* мазь
okapi *s ther* окапи *(Okapia);* окапи Джонстона *(Okapia johnstoni)*
okra *s bot* окра, бамия *(Abelmoschus esculentus)*
olecranon *s* локтевой отросток
oleiferous *a* масличный, маслоносный
olfact *v* обонять; нюхать
olfactic *a* обонятельный
olfaction *s* обоняние, нюх
olfactive *a* обонятельный
olfactology *s* ольфактология
olfactometer *s* ольфактометр
olfactory *a* обонятельный
oligandrous *a* малотычинковый
oliganthous *a* малоцветковый
oligobrochate *a* редкоячеистый
oligocarpous *a* малоплодный
oligodendrocyte *s* олигодендроцит, клетка олигодендроглии
oligodendroglia *s* олигодендроглия
oligodontous *a* малозубчатый
oligodynamic *a* олигодинамический
oligoforate *a* редкодырчатый
oligogene *s* олигоген
oligogenic *a* олигогенный
oligoglia *s* олигоглия
oligohaline *a* олигогалинный
oligokinesia *s* олигокинезия
oligolecithal *a* олиголецитальный
oligolectic *a* с малой избирательностью
oligomer *s* олигомер
oligomerization *s* олигомеризация
oligomerous *a* олигомерный
oligonitrophilic *a* олигонитрофильный
oligonucleotide *s* олигонуклеотид
oligophyllous *a* малолистный
oligopod *a* олигоподный; камподеовидный
oligopyrene *a* олигопиренный
oligorhizous *a* малокорневой
oligosaccharide *s* олигосахарид
oligosaprobic *a* олигосапробный
oligospermous *a* малосемянный
oligostemonous *a* малотычинковый
oligotaxy *s bot* олиготаксия
oligotroph, oligotrophic *a* олиготрофный
olive *s* маслина европейская *(Olea europeaea); anat* олива
omasum *s anat* книжка
ombrophil *s* омброфил

ombrophilous *a* приспособленный к дождям
ombrophobe *s* омброфоб
omental *a* сальниковый
omentulum *s* малый сальник
omentum *s anat* сальник
ommateum *s* сложный глаз
ommatidium *s* омматидий
ommatophore *s zool* глазной стебелёк, омматофор
omnipotent *a* омнипотентный
omnipresent *a* повсеместный
omniserum *s* полиспецифическая антисыворотка
omnivor, omnivore *s* всеядное животное
omnivorous *a* всеядный
omohyoid *s* лопаточно-подъязычный мускул; *a* лопаточно-подъязычный
omphalic *a* пупочный
omphalodium *s bot* основание рубчика семени, углубление рубчика семени
oneger *s* дикий осёл, онагр *(Equus hemionus onager)*
onchocercosis *s* онхоцеркоз
oncilla *s ther* онцилла *(Felis tigrinus)*
oncogene *s* онкоген
oncogenicity *s* онкогенность, онкогенные свойства
oncoprotein *s* онкогенный белок, опухолеспецифический белок, онкобелок
oncornavirus *s* онкорнавирус
oncosphere *s helm* онкосфера, шестикрючный зародыш
oncospore *s* спора с крючками; семя с крючками
oncotic *a* онкотический
one-celled *a* одноклеточный
one-egg *a* однояйцовый, монозиготный
one-layered *a* одноярусный; однослойный
one-row *a* однорядный
one-seeded *a* односемянный
one-segmented *a* одночленистый, одночленный
on-goings *s* поведение; образ жизни
onion *s* лук *(Allium)*
  bear's ~ лук медвежий, черемша *(Allium ursinum)*
onset *s* начало, наступление; вспышка, проявление
ontocline *s* онтоклин
ontogenesis *s* онтогенез
ontogenetic *a* онтогенетический
ontogeny *s* онтогенез
ontomutation *s* онтомутация, онтогенетическая мутация
onychogenic *a* образующий ноготь, образующий коготь
onyx *s* ноготь; коготь
ooapogamy *s* овоапогамия, ооапогамия, диплоидный партеногенез, соматический партеногенез
ooblastema *s* овобластема, oобластема, оплодотворённое яйцо
oocapt *s* овикапт
oocarp *s* ооспора, оплодотворённая яйцеклетка
oocyst *s* ооциста
oocyte *s* ооцит
ooecium *s* оэций
oogamy *s* оогамия
oogenesis *s* оогенез
oogenous *a* образующий яйцо
oogonium *s* оогоний
ooid *a* яйцевидный
ookinesis *s* оокинез
ookinete *s* оокинета
oolemma *s* оолемма
oophore *s bot* оофит, половое поколение
oophoridium *s* мегаспорангий
oophyte *s bot* оофит, половое поколение
oosome *s* оосома
oosperm *s* оплодотворённое яйцо
oosphere *s* оосфера, яйцо до оплодотворения, женская гамета
oospore *s* ооспора, оплодотворённая яйцеклетка; ооциста
oostegite *s* оостегит
ootheca *s* спорангий; оотека; яйцевой кокон; яйцевая капсула
ootid *s* оотида
ootocous *a* яйцекладущий
ootype *s* оотип
ooze *s* ил, тина; болото, топкая местность; просачивание, выделение влаги; истечение; *v* просачиваться, выделять влагу
oozy *a* илистый; болотистый, топкий
opacification *s* помутнение
opacity *s* мутность, непрозрачность
opaque *a* непрозрачный, непроницаемый

open *a* открытый; распустившийся, расцветший; разреженный, редкостойный
opening *s* отверстие, щель; просвет; распускание
  breathing ~ дыхательное отверстие; дыхальце
  gill ~ жаберная щель
  mouth ~ рот, ротовое отверстие
  mouth-like ~ *bot* устьице
  ring ~ *bioch* дециклизация, размыкание цикла
  shell ~ *mal* устье
  stomatal ~ открывание устьиц
operation *s* операция
operator *s* оператор
opercle *s* покрышка, крышечка; *pal* оперкулум
operculate, operculated *a* с крышечкой
operculum *s* покрышка, крышечка; *палин.* оперкулум
operon *s gen* оперон
ophthalmic глазной
ophthalmogyric *a* глазодвигательный
opisthaptor *s helm* задний фиксаторный аппарат
opisthe *s prot* задняя дочерняя особь
opisthion *s* опистион
opisthocoelous *a anat* опистоцельный
opisthonephros *s* опистонефрос
opisthorchiasis *s helm* описторхоз
opisthosoma *s* опистосома
opium *s* опиум
opophilous *a* питающийся соком растений
opossum *s* опоссум *(Didelphidae)*
oppilation *s* закупорка
opposite *a* супротивный; противоположный
oppositiflorous *a* супротивноцветковый
oppositifolious *a* супротивнолистный
opposition *s* оппозиция, противопоставление
oppressed *a* угнетённый
opsonization *s* опсонизация
opsonometry *s* опсонометрия
optic *a* глазной, зрительный; оптический
opticokinetic *a* глазодвигательный
optimization *s* оптимизация
optimum *s* оптимум
optional *a* произвольный; факультативный, необязательный
optogram *s* оптограмма
optometer *s* оптометр
optomotor *a* оптомоторный
orach, orache *s* лебеда *(Atriplex)*
orad *adv* по направлению ко рту
oral *a* ротовой
orally *adv* через рот, орально
orang *s* орангутан *(Pongo pygmaeus)*
orange *s* апельсин, цитрус *(Citrus);* оранжевый
  bergamot ~ цитрус бергамот *(Citrus bergamis)*
  king ~, mandarin ~ мандарин благородный *(Citrus nobilis)*
  marmelade ~ померанец, апельсин горький *(Citrus aurantium)*
  methyl ~ метиловый оранжевый
  navel ~ пупочный апельсин
  Seville ~ померанец, апельсин горький *(Citrus aurantium)*
orangutan, orangutang *s* орангутан *(Pongo pygmaeus)*
orbicular *a* кругообразный; округлый; сферический
orbit *s* орбита, глазница
orbital *s* орбиталь; орбитальный, глазничный
orbitonasal *a* глазнично-носовой
orbitosphenoid *s* глазнично-клиновидная кость
orcein *s* орсеин
orchic *a* семенниковый
orchid *s* орхидное растение *(Orchidaceae)*
orchil *s* лишайник красильный *(Roccella)*
orchis *s anat* яичко
orchitic *a* семенниковый
order *s* отряд; порядок; последовательность; расположение, размещение; ранг
  hook ~ ранг в стаде рогатых копытных
orderliness *s* упорядоченность
ordination *s* классификация, расположение
oread *s* солнцелюбивое растение, гелиофит
orgadium *s* сообщество редкого леса
orgadophilus *s* обитатель редкого леса
organ *s* орган
  ~ of Zuckerkandl поясничный аортальный параганглий, орган Цукеркандля

appendicular ~ конечность
cell ~ органоид, органелла
closed respiratory ~s закрытая дыхательная система, замкнутая дыхательная система
Cuverian ~ кювьеров орган
end ~ концевой орган; рецептор
exudate ~ *энт.* эксудаторий
flasklike sense ~ *ent* ампуллярная сенсилла
Gene's ~ орган Жене, подщитовая [головная] железа
Giraldes' ~ жиральдесов орган, парадидимис
Golgi ~ аппарат Гольджи
Golgi tendon ~ нервно-сухожильное веретено
Haller's ~ тархальный хеморецептор
Heyer-Grosser ~ орган Гейера - Гроссера, кожный клубок
motor end ~ двигательный концевой аппарат
neurotendinous ~ нервно-сухожильное веретено
open respiratory ~s открытая дыхательная система
parietal ~ париетальный орган, парапинеальный орган, теменной глаз
pineal ~ эпифиз, щитовидная железа
pituitary ~ гипофиз
Rosenmuller's ~ эпоофорон, рудимент первичной почки
Siebold's ~ *ent* слуховой гребень тимпанального органа
target ~ орган-мишень
Zuckerkandl's ~ орган Цукеркандля, пояснично-аортальный параганглий

organelle *s* органелла, органоид
organic *a* органический
organism *s* организм
causative ~ болезнетворный микроорганизм
filter-feeding ~ фильтратор, фильтрующее животное

organismal, organismic *a* организменный
organization *s* организация
organizer *s emb* организатор
organofaction, organogenesis, organogeny *s* органогенез
organography *s* органография
organoid *s* органелла, органоид
organoleptic *a* органолептический
organology *s* органология
organonymy *s* номенклатура органов
organophosphorous *a* фосфорорганический
organophyly *s* филогения органов
organotropy *s* органотропность
organule *s* клетка, элемент
orientation *s* ориентация, ориентирование, ориентировка
orifice *s* вход, отверстие, устье
cardiac ~ *anat* кардия
external ~ of uterus наружный маточный зев
vaginal ~ вход во влагалище, влагалищное отверстие

origin *s* происхождение; начало; источник; начало кооридинат, нуль шкалы
~ of replication *mbiol* репликатор, начало репликации
arbitrary ~ *biom* произвольное начало
replicon ~ точка инициации репликона

original *a* оригинальный; первоначальный
originate *v* происходить; порождать, производить
origination *s* начало; происхождение
originative *a* дающий начало, порождающий
oriole *s orn* иволга *(Oriolus);* цветной трупиал *(Icterus)*
ornamental *s* декоративное растение; *a* декоративный
ornamentation *s* орнаментация
ornis *s* авифауна
ornithic *a* птичий
ornithocoprophilous *a* предпочитающий помёт птиц
ornithologist *s* орнитолог
ornithology *s* орнитология
ornithophilous *a* орнитофильный, опыляемый птицами
oroid *a pal* апертуровидный
oroidiferous *a pal* апертуровидноносный
oroimmunity *s* пассивный иммунитет
orolochmis *s* сообщество субальпийских зарослей
orophyte *s* орофит
orrhodiagnosis *s* серодиагностика
orrhoimmunity *s* пассивный иммунитет
orrhology *s* серология
orrhoreaction *s* сывороточная реакция

orthochromasia *s* ортохромазия
orthocolpate *a* прямобороздный
orthocomplex *s* ортокомплекс, ортовезикулярный комплекс, совокупность ортопузырьков
orthodentine *s* ортодентин, остеодентин
orthodromic *a* ортодромный
orthoevolution *s* ортоэволюция
orthogenesis *s* ортогенез
orthograde *a* прямоходящий
orthokinesis *s* ортокинез, ортокинетическое движение
orthomyxovirus *s* ортомиксовирус
orthophoria *s* ортофория
orthophyte *s* ортофит
orthoploidy *s* ортоплоидия
orthopteroid *a pal* ортоптероидный
orthopterous *a* прямокрылый
orthoselection *s* ортоселекция
orthospermous *a* прямосемянный
orthostatic *a* ортостатический
orthotopic *a* ортотопический
orthotropic *a* ортотропный
orthotropism *s* ортотропизм
orthotropous *a* ортотропный
orthotubule *s* ортотрубочка
orthovesicular *a* ортовезикулярный
oryx *s* сернобык, орикс *(Oryx gazella)*
oscheal *a* мошоночный
oscillation *s* осциллярия, колебание
oscillogram *s* осциллограмма
oscillography *s* осциллография
oscillometry *s* осциллометрия
oscitation *s* зевота, зевание
oscule *s* пора; устье, оскулюм *(of sponges)*
osculiferous *a* пористый, имеющий поры
osier *s* ива (*Salix),* лоза
osier-bed *s* тальник
osmesis *s* обоняние
osmeterium *s ent* осметерия, вилкообразная мешковидная железа
osmetic *a* обонятельный
osmiophil *a* осмиофильный
osmolality *s* осмотическое давление
osmometer *s* осмометр
osmoregulation *s* осморегуляция, осмотическая регуляция
osmose, osmosis *s* осмос
osmotic *a* осмотический
osmotrophic *a* осмотрофный
osmotrophy *s* осмотрофия, осмотическое питание
osmotropism *s* осмотропизм
osprey *s orn* скопа *(Pandion haliaetus)*
ossa *s* кости, костный аппарат
osseous *a* костный
ossicle *s* косточка
  zygocardiac ~ *zool* жевательная пластинка желудка
ossific *a* остеогенный, костеобразующий
ossification *s* окостенение; образование костного вещества
ossify *v* окостеневать
osteal *a* костный
osteoarticular *a* костно-суставной
osteoblast *s* остеобласт, костеобразующая клетка
osteochondral *a* остеохондральный
osteoclast *s* остеокласт
osteocranium *s* костный череп
osteocyte *s* остеоцит, костная клетка
osteodentine *s* остеодентин
osteogen *s* остеогенная ткань
osteogenesis *s* остеогенез
osteogenetic *a* остеогенный, костеобразующий
osteogeny *s* остеогенез
osteoid *s* остеоид, костеподобное вещество
osteology *s* остеология
osteon *s* остеон, костная единица, гаверсова система
osteoplastic *a* остеопластический, костеобразующий
osteoscute *s ther* костно-роговой щиток
ostiate *a* имеющий отверстие
ostiole *s* отверстие; пора
ostium *s* отверстие; вход, устье; остия
ostrich *s* африканский страус *(Struthio camelus)*
otic *a* ушной; слуховой
otoconium *s* отоконий
otocyst *s* отоцист, слуховой пузырёк
otolith *s* отолит
oto-occipital *a* отоокципитальный, затылочно-ушной
otosalpinx *s* евстахиева труба
otostapes *s* стремечко уха
otosteal *a* относящийся к костям уха
otosteon *s* слуховая косточка
otter *s* настоящая) выдра *(Lutra);* выдры *(Lutrinae)*

ouabain *s* уабаин
ouistiti *s ther* уистити, мармозетка, обыкновенная игрунка *(Callithrix)*
ounce *s* ирбис, снежный барс *(Uncia uncia)*
ousel *s orn*: **water** ~ оляпка *(Cinclus)*
outbreak *s* взрыв, массовое появление, вспышка
outbreeding *s* аутбридинг
outcrossing *s* ауткроссинг
outdoor *a* под открытым небом
outer *a* наружный, внешний
outflow, outflux *s* отток, выделение
outgrowth *s* отросток; вырост, выпячивание
outlier *s biom* выброс, аномальное значение
outlimb *s* дистальная часть конечности
outline *s* очертание, наружный контур
outnumber *v* превосходить численностью, превосходить количеством
outpost *s* форпост
output *s* продукт, отдача, выход; выходные данные; результат вычислений
  milk ~ надой
  urine ~ диурез
outrun *s* отгонное пастбище
outwandering *s* эмиграция
outward *a* внешний, наружный
ouzel *s orn*:
  ring ~ белозобый дрозд *(Turdus torquatus)*
  water ~ оляпка *(Cinclus)*
oval *a* яйцевой; яйцевидный, овальный
ovalbumin *s* яичный белок, овальбумин
ovalocyte *s* овалоцит
ovarian *a* яичниковый
ovariole *s* овариола, яйцевая трубочка
ovariotestis *s* гермафродитная железа
ovary *s* яичник; *bot* завязь
ovate *a* яйцевидный, овальный
ovate-acuminate *a* овально-остроконечный
ovate-ellipsoidal *a* овально-эллипсоидальный
ovate-lanceolate *a* овально-ланцетовидный
ovate-oblong *a* овально-продолговатый
ovate-orbicular *a* овально-округлый
oven *s* печь
  drying ~ сушильный шкаф
  sterilizing ~ сухой стерилизатор, пастеровская печь
overadaptation *s* сверхадаптация
overactivity *s* гиперактивность, повышенная активность
overall *a* полный; общий; предельный; всеобъемлющий, всеобщий, веохватывающий
overbreathing *s phys* гипервентиляция; усиленное дыхание
overchill *v* переохлаждать
overchilled *a* переохлаждённый
overcrowding *s* перенаселение, перенаселённость
overcrusted *a* покрытый корой
overdominance *s* сверхдоминирование, сверхдоминантность
overdriven *a* загнанный *(of a horse)*
overfatigue *s* переутомление
overfeed *v* перекармливать
overfeeding *s* перекармливание
overfermentation *s* перебраживание, перебрpremium
overgrazing *s* чрезмерный выпас
overground *a* надземный
overgrow *v* зарастать; перерастать
overgrown *a* заросший; чрезмерно выросший, переросший; разросшийся
overgrowth *s* гипертрофия, чрезмерный рост; чрезмерное развитие; разрастание; заросль
overhair *s* остевой волос
overheat *s* перегрев
overhunting *s* перепромысел, чрезмерный промысел
overlapping *s* перекрывание
overlay *s*: **agar** ~ верхний слой агара
overmature *a* переспелый, перезрелый
overmaturity *s* перезрелость
overpopulate перенаселять
overpopulation *s* перенаселение, перенаселённость
overproduction *s* перепроизводство; микр. сверхсинтез
overripe *a* переспелый, перезрелый; перестойный *(forest)*
oversaturation *s* перенасыщение
overshoot *s* овершут
overshot *s* бустер-инъекция, вторичная инъекция антигена
overstaining *s* чрезмерное окрашивание

overstock *v* переуплотнять
overstocked *a* переуплотнённый
overstocking *s* переполнение; чрезмерный выпас, чрезмерная нагрузка пастбища
overstrain *s* чрезмерное напряжение, перенапряжение
overtop *v* господствовать; угнетать
overtopped *a* подчинённый, угнетённый
overweight *s* избыточный вес
overwetting *s* переувлажнение
overwinter *v* перезимовывать
ovicapt *s* овикапт
ovicell *s* яйцеклетка
oviduct *s* яйцевод, яйцепровод
ovification *s* развитие яйца
ovigerm *s* яйцеклетка
ovigerous *a* яйцеобразующий, яйцепроизводящий
ovine овечий
oviparity *s* овипария
oviparous *a* яйцекладущий
oviposit откладывать яйца
oviposition *s* откладка яиц
ovipositor *s* яйцеклад
ovisac *s* яйцевая капсула; граафов пузырёк
ovism *s* овизм
ovocenter *s* овоцентр, ооцентр, центросома яйцеклетки
ovocyst *s* ооциста
ovocyte *s* овоцит, ооцит
ovogenesis *s* овогенез, оогенез
ovoglobulin *s* овоглобулин, яичный глобулин
ovoid *a* овоидный, яйцевидный
ovotestis *s* гермафродитная железа
ovoviviparity *s* яйцеживорождение
ovoviviparous *a* яйцеживородящий
ovulation *s* овуляция
ovulatory *a* овуляционный, овуляторный
ovule *s* яйцеклетка; *bot* семяпочка, семязачаток
  cervical ~ шеечный овоид
ovuliferous *a* образующий семяпочку, образующий семязачаток
ovum *s* яйцо, женская зародышевая клетка
owl *s* сова *(Strigiformes)*
  barn ~ сипуха *(Tyto)*
  eagle ~ филин *(Bubo)*; филин *(Bubo bubo)*
  long-eared ~ ушастая сова *(Asio otus)*
  scops ~ сплюшка, обыкновенная совка *(Otus scops)*
  screech ~ *орнит.* совка *(Otus)*; североамериканская совка *(Otus asio)*
  snowy ~ белая сова, полярная сова *(Nyctea scandiaca)*
owlet *s* совёнок; домовый сыч *(Athene noctua)*; воробьиный сыч *(Glaucidium)*
own-rooted *a* корнесобственный
owre *s* арни, индийский водяной буйвол *(Bubalus bubalus)*
ox *s* вол, буйвол, бизон, бык
  Cambodian forest ~ купрей *(Bos sauveli)*
  humped ~ зебу *(Bos indicus)*
  Javanese wild ~ бантенг *(Bos javanicus)*
  musk ~ овцебык, мускусный бык *(Ovibus moschatus)*
  wild ~ тур *(Bos primigenius)*
oxidant *s* окислитель
oxidation *s* окисление
  photochemical ~ фотоокисление, фотохимическое окисление
  radiochemical ~ радиоокисление
oxidative *a* окислительный
oxidization *s* окисление
oxidize *v* окислять, оксидировать
oxidizer *s* окислитель
oxidoreduction *s* окисление – восстановление
oxidosome *s* оксидосома
oxinitration *s* оксинитрование
oxpecker *s orn* волоклюй *(Buphagus)*
oxyacanthous *a* острошиповатый
oxybiotic *a* аэробный
oxygen *s* кислород
oxygenate *v* насыщать кислородом, оксигенировать
oxygenation *s* насыщение кислородом, оксигенация
oxygenator *s* оксигенатор
oxygen-containing *a* кислородсодержащий
oxygenous *a* кислородный
oxygeophilus *s* обитатель гумуса
oxygeophytium *s* фитоценоз гумуса
oxyhemoglobin *s* оксигемоглобин
oxylophyte *s* оксилофит
oxymeter *s* оксиметр
oxyntic *a* секретирующий кислоту

oxypathia *s* оксипатия
oxyphil, oxyphile *s* оксифил
oxyphilous *a* оксифильный, кислотолюбивый
oxyphobe *s* оксифоб
oxyphobous *a* оксифобный
oxyphyte *s* оксифит
oxytocin *s* окситоцин
oxytropism *s* окситропизм
ozone *s* озон
ozonize *a* озонировать

P

pabular *a* пищевой, кормовой
pabulum *s* пища; корм
pace *s* шаг; длина шага; аллюр; иноходь
  racking ~ иноходь
pacemaker *s phys* водитель ритма, ритмоводитель
pacer *s* кардиостимулятор; иноходец
pachyderm *s* толстокожее животное
pachydermate, pachydermatous, pachydermous *a* толстокожий
pachymeninx *s anat* твёрдая мозговая оболочка
pachynema *s* пахинема
pachynosis *s* рост в толщину
pachyphyllous *a bot* толстолистный
pachypleurous *a bot* толстостенный
pachytene *s* пахитена
Pacific *a* тихоокеанский, пацифический
pacing *s phys* кардиостимуляция
pack *s* стая; свора; выводок; стадо; тампон
packaging *s mbiol* упаковка
packed *a* уплотнённый, насыщенный
packing *s* образование стаи, образование выводка, образование стада; скопление; уплотнение; *mbiol* упаковка; набивка, заполнение
pad *s* лапа; подушечка на лапе; след лапы; плавающий лист
  Seitz ~ фильтр Зейтца
  wing ~ *ent* зачаток крыла, недоразвитое крыло
paddle *s* плавник; ласт; плавательная пластинка
paddock *s* загон; выгон
paddy *s* падди, рис-сырец
paedogamy *s* педогамия
paedogenesis *s* педогенез
paedomorphic *a* педоморфный
paedomorphism *s* педоморфизм; педоморфоз
paedomorphosis *s* педоморфоз
paigle *s* первоцвет *(Primula)*
pain *s* боль
painless *a* безболезненный
paint *s* окраска; масть
painter *s* пума, кугуар *(Felis concolor)*
pair *s* пара; *v* спаривать, случать; спариваться
  royal ~ *ent* царь и царица
pairing *s* спаривание; конъюгация *(of chromosomes)*
pakchoi *s* капуста китайская *(Brassica chinensis)*
pal *s orn* тупик *(Fratercula arctica)*
pala *s ent* лопаточка, одночлениковая лапка *(in Corixidae);* лопатообразная структура
palaceous *a* краечерешковый
palaearctic *a* палеарктический
palaeform *a* расширенный в лопаточку
Palaeogaea *s* Палеогея
palama *s orn* перепонка
palatability *s* вкусовая привлекательность, вкусовые качества
palatable *a* вкусный
palatal *a* нёбный
palate *s* нёбо; *bot* выпуклая часть нижней губы
palatine *a* нёбный
palatoglossal *a* нёбно-язычный
palatonasal *a* нёбно-носовой
palatopharyngeal *a* нёбно-глоточный
palatopterygoid *a* нёбно-крыловидный
palatoquadrate *a* нёбно-квадратный, палатоквадратный
pale *a* бледный
palea *s* плёнка; чешуя; верхняя цветковая чешуя
paleaceous, paleate *a* плёнчатый; чешуйчатый
paleobiology *s* палеобиология
paleobotany *s* палеоботаника
paleocranium *s* палеокраний
paleoecology *s* палеоэкология
paleoencephalon *s* задний отдел мозга, древний мозг
paleogene *s* палеоген
paleogenesis *s* палеогенез
paleogenetics *s* палеогенетика
paleogeography *s* палеогеография
paleola *s* чешуйка
paleolate *a* имеющий чешуйки
paleolimnology *s* палеолимнология
paleontology *s* палеоонтология
paleophytology *s* палеоботаника
paleosynchorology *s* палеосинхорология
paleosynecology *s* палеосинэкология
paleozoology *s* палеозоология

palet *s* плёнка; чешуя; верхняя цветковая чешуя
palette *s ent* палетка
paliform *a* лопатообразный
palindrome *s mbiol* палиндром
palingenesis *s* палингенез
palingenetic *a* палингенетический
palisade *a* столбчатый, палисадный
palisander *s* палисандровое дерево *(Dahlbergia)*
pallah *s* пала, импала, чернопятая антилопа *(Aepyceros melampus)*
pallet *s* палетка, пластинка
palliate *a* имеющий мантию, покрытый мантией
pallid *a* бледный, бесцветный
pallium *s* паллиум, плащ головного мозга; *мал.* мантия
palm *s* пальма; ладонь
  date ~ финик пальчатый *(Phoenix dactylifera)*
palmar *a* ладонный
palmate *a* дланевидный, лапчатый, пальчатый
palmatifid *a* дланевидно-надрезный
palmatiform *a* дланевидный, пальчатовидный
palmatilobate *a* дланевидно-лопастный, пальчато-лопастный
palmatinerved *a* дланевидно-нервный, пальчато-нервный, дланевидно-жилковатый, пальчато-жилковатый
palmatipartite *a* дланевидно-раздельный, пальчато-раздельный
palmatisected *a* дланевидно-рассечённый, пальчато-рассечённый
palmella *s* пальмелла
palmelloid *a* пальмеллевидный
palmetto *s bot* сабаль *(Sabal);* пальметта
palmicolous *a* обитающий на пальмах
palmula *s* эмподиум, аролиум
palp *s* пальпа; щупик
palpate *v* пальпировать; *a* имеющий пальпы
palpation *s* пальпация, ощупывание, прощупывание
palpebra *s* веко
palpifer *s zool* пальпифер
palpiform *a* похожий на пальпу; щупикообразный
palpiger *s ent* пальпигер, основание лабиального щупика
palpitation *s* пальпитация; трепетание; дрожь
palpon *s zool* пальпон
palpulus *s* небольшой щупик
palpus *s* пальпа; щупик
palpus-bearer *s* пальпифер
palsy *s* паралич; *v* парализовать
paludal *a* болотистый, болотный
paludification *s* заболачивание
paludous *a* болотный; болотистый
palynogram *s* палинограмма
palynological *a* палинологический
palynology *s* палинология
palynomorph *s* палиноморфа
pampiniform *a* усикообразный
pamprodactylous *a* пампродактильный
pan *s* ванночка, кювета; испаритель; *v* сортировать; разделять методом пэннинга
  culture ~ вегетационный сосуд
  shallow ~ плоская кювета
panagglutination *s* панагглютинация
panaperturate *a pal* рассеянно-поровой
panarctic *a* панарктический
panclimax *s* панклимакс
pancolpate *a pal* рассеянно-бороздный
pancreas *s* поджелудочная железа
pancreatic *a* панкреатический
pancreatitis *s* панкреатит
panda *s ther* панда *(Ailurinae)*
pandemic *a* пандемичный
pandurate, pandurated, panduriform *a bot* гитарообразный
panel *s* панель, набор *(e.g., of sera)*
panformation *s* панформация
panformion *s* панформион
pangamic *a* пангамный
pangamy *s* пангамия, беспорядочное спаривание
pangen *s* панген
pangenesis *s* пангенезис
pangolin *s* панголин, ящер *(Manis)*
panhydrosis *s* пангидроз
panicle *s bot* метёлка
panicled *a* метёлконосный
panicle-shaped *a* метёлковидный
paniculate *a* метельчатый
paniculation *s* вымётывание метёлки
panmictic *a* панмиктический
panmixia, **panmixis** *s* панмиксия, свободное скрещивание
pannage *s* плодокорм

panniculus *s* поверхностный слой ткани
panning *s imm* пэннинг
pannose *a* лохматый; шерстистый; войлочный
panoistic *a ent* паноистический
panphotometric *a* панфотометрический
panporate *a pal* рассеянно-поровой
pant *s* тяжёлое дыхание; *v* тяжело дышать, задыхаться
pantanal *s* пантаналь
panther *s* пантера, леопард, барс *(Panthera pardus);* ирбис *(Felis uncia);* пума, кугуар *(Felis concolor)*
pantophagous *a* всеядный
pantropical *a* пантропический
pap *s* паста; эмульсия; мякоть *(of a fruit);* паренхима
papaya *s* папайя, дынное дерево *(Carica papaya)*
paper *s* бумага
  DBM ~ диазобензилоксиметилцеллюлозная бумага, ДБМ-бумага
  DEAE ~ диаминоэтиламиноэтилцеллюлозная бумага, ДЭАЭ-бумага
papery *a bot* бумажистый
papilionaceous *a bot* мотыльковый
papilla *s* папилла, сосочек
  nerve ~ осязательное тельце, тельце Мейснера
  optic ~ сосок зрительного нерва
papillary *a* папиллярный
papillate *a* несущий папиллы; покрытый папиллами; сосочковидный, в виде папилл
papilliferous *a* несущий папиллы
papilliform *a* сосочковидный, в виде папилл
papovaviruse *s* паповавирус
pappiferous *a bot* образующий летучки, несущий летучки, образующий хохолки, несущий хохолки
pappose *a bot* хохлатый
pappus *s bot* летучка; хохолок
pappy *a* мягкий, сочный
paprica *s* перец овощной сладкий *(Capsicum frutescens)*
papula *s* папула, прыщик; *zool* пупырышек; сосочек
papuliferous *a* покрытый пупырышками, покрытый сосочками
papyraceous *a bot* бумагообразный
papyrus *s* папирус *(Cyperus papyrus)*
para-adenovirus *s* парааденовирус
parabasal *a* парабазальный
parabiont *s* парабионт
parabiosis *s* парабиоз
parabiotic *a* парабиотический
parablast *s* парабласт
patacallus *s* паракаллюс
paracarpium *s* паракарпий
paracarpous *a* паракарпный
paracellulose *s* парацеллюлоза
paracentral *a* околоцентральный
parachordal *s* парахордаль; *a* парахордальный
parachute *s* парашют
paraclimax *s* параклимакс
paracme *s* паракме
paracoel *s* боковой желудочек мозга
paracorolla *s* привенчик
parademe *s* парадема, вторичная аподема
paraderm *s* парадерма
paradesmus *s* парадесма
paradidymis *s* парадидимис, орган Жиральде
paradoxical *a* парадоксальный
parafacial *a* парафациальный
paraffin *s* парафин
paraflocculus *s anat* наружный клочок, долька мозжечка
parafrons *s ent* парафронс
paraganglion *s* параганглий, хромаффинное тельце
  aorticolumbar ~ пояснично-аортальный параганглий, орган Цукеркандля
paragaster *s* парагастральная полость
paragastrula *s* парагаструла
parageneon *s* парагенеон
paragenesis *s* парагенез
paragglutination *s* групповая агглютинация
paragglutinin *s* коагглютинин
paraglossa *s ent* параглосса
paraheliotropism *s* парагелиотропизм
parakeet *s см.* **parrakeet**
paralimnic *a* приозёрный
parallelinervate, parallelodrome *a bot* параллельно-нервный
parallelotropism *s* параллелотропизм, ортотропизм

parallel-veined *a bot* параллельно-нервный
parallergy *s* параллергия
paralysis *s* паралич
paralyzant *a* вызывающий паралич, парализующий
paralyze *v* парализовать
paralysis *s* паралич
parameter *s* параметр, показатель
parametric *a* параметрический
paramorph *s* параморфа
paramutualism *s* парамутуализм, факультативный симбиоз
paramylon, paramylum *s* парамил
paramyotonia *s* парамиотония
paramyxovirus *s* парамиксовирус
paranasal *a* околоносовой
paranephric *a* околопочечный
paranephros *s* надпочечник
paranuclein *s* паренуклеин, пиренин
paranucleus *s* паренуклеус, побочное ядро, акцессорное ядро
parapepsin *s* парапепсин, пепсин B
parapetalous *a bot* прилепестниковый
paraphototropism *s* парафототропизм
paraphyll *s* парафилл, «прилист»
paraphysate *a* имеющий парафизы
paraphysis *s* парафиза, стерильная гифа; парафиз, вырост крыши переднего мозга; *ent* хитинизированное уплотнение
paraplasm *s* параплазма, дейтоплазма
paraplectenchyma *s* псевдопаренхима, ложная паренхима
parapodium *s* параподия
parapolar *a* расположенный около полюса
parapophysis *s* парапофиз, боковой отросток
paraproct *s* парапрокт, анальная пластинка
parapsidal *a* парапсидальный
parapteron *s* надкрыловая пластинка, параптерон
paraquadrate *s* околоквадратная кость
parareflexia *s* парарефлексия
pararosaniline *s* парарозанилин, основной фукчин
parasecretion *s* ненормальная секреция
paraselectivity *s* параселективность
parasematic *a* парасематический
parasexual *a* парасексуальный
parasite *s* паразит
  beaver ~ бобровая блоха *(Ptatypsyllidae)*
  blood ~ кровепаразит
  epizoic ~, external ~ эктопаразит, наружный паразит
  internal ~ эндопаразит, внутренний паразит
parasitemia *s* паразитемия
parasitic *a* паразитический; паразитарный
parasitism *s* паразитизм
parasitize *v* паразитировать
parasitized *a* поражённый паразитами
parasitoid *s* паразитоид
parasitologist *s* паразитолог
parasitology *s* паразитология
parasitosis *s* паразитарное заболевание, паразитоз
parasphenoid *s* парасфеноид; парасфеноидный
paraspore *s* параспора
parastemon *s* стаминодий
parasterility *s* парастерильность
parasternal *a bot* окологрудинный
parasternum *s herp* брюшное ребро
parastichy *s bot* парастиха, вторичная спираль, побочная спираль
parasympathetic *a* парасимпатический
parasynapsis, parasyndesis *s* парасинапсис, парасиндез
parasystoly *s* парасистолия
parathecium *s* паратеций
parathormone *s* паратгормон
parathyroid *s* околощитовидная железа, паращитовидная железа
paratoid *a* паратоидный, околоушный
paratomy *s* паратомия
paratonic *a* паратонический
paratope *s* антиген-связывающий активный центр, рецепторная зона, паратоп, антидетерминанта
paratrophic *a* паратрофический
paratype *s* паратип
paratypic *a* паратипический
paravariation *s* паравариация, модификация
paravertebral *a* околопозвоночный, паравертебральный
paravesical *a* околопузырный
paraxial *a* параксиальный
parcel *s bot* парцелла

pard *s* леопард,барс, пантера *(Panthera pardus)*
parencephalon *s* мозжечок
parenchyma *s* паренхима, паренхиматозная ткань
parenchymal *a* паренхимный
parenchymatous *a* паренхиматозный
parenchymula *s* паренхимула
parent *s* родитель
B-cell ~ В-клетка-предшественник, пре-В-клетка
donor ~ *gen* родитель-донор
parentage *s* происхождение; родословная; родство
parenteral *a* парэнтеральный
parhomology *s* кажущаяся гомология
paridigitate *s* парнокопытное, парнопалое *(Artiodactyla)*
parietal *a* париетальный, пристенный; теменной; *bot* стенкоположный
parietofrontal *a* лобно-теменной
parieto-occipital *a* затылочно-теменной
parietotemporal *a* височно-теменной
paripinnate *a bot* парноперистый
park *s* парк; заповедник
paroecia *s ecol* паройкия, приселение
paroophoron *s* параофорон, рудимент мезонефроса
parotic *a* околоушный
parous *a* рожавшая; беременная
parovarium *s* эпоофорон, придаток яичника
parrakeet *s* маленький попугай, попугайчик
parrot *s* попугай *(Psittacidae)*
gray ~ серый попугай, жако *(Psittacus erithacus)*
mountain ~ кеа *(Nestor notabilis)*
parsley *s* петрушка огородная *(Petroselinum crispum)*
beaked ~ купырь *(Anthriscus)*
part *s* часть, доля; компонент; деталь
mouth ~ *ent* ротовой придаток
parted *a bot* рассечённый
parthenapogamy *s* партеноапогамия, диплоидный партеногенез, соматический партеногенез
parthenita *s* партенита
parthenocarpic *a* партенокарпический
parthenocarpy *s* партенокарпия
parthenogamy *s* партеногамия
parthenogenesis *s* партеногенез, девственное размножение
parthenogenetic *a* партеногенетический
parthenomyxis *s* партеногамия
parthenospore *s* партеноспора, азигоспора
parthenote *s* партенот
partial *a bot* парциальный
particle *s* частица, корпускула
attraction ~ центриоль
kappa ~ *ген.* каппа-частица
particolored *a* пёстрый, разноцветный
particular *a* особый, частный, специфический; определённый; индивидуальный, отдельный
particulate *a* частичный; корпускулярный
partigen *s* партиген; фрагмент молекулы антигена
partite *a* раздельный, дольный; *bot* рассечённый
partition *s* разделение; расчленение
partitioning s распределение
partner *s* партнёр, член пары
fusion ~s *pl* сливающиеся клетки
myeloma ~ миеломная клетка-партнёр
partridge *s* куропатка; (серая) куропатка *(Perdix)*
monal ~ кундык *(Tetraophasis)*
painted ~ расписной турач, индийский франколин *(Francolinus pictus)*
rock ~ кеклик, каменная куропатка *(Alectoris)*
Shelly's ~ турач Шелли, франколин Шелли *(Francolinus shelleyi)*
snow ~ снежная куропатка *(Lerwa lerwa);* гималайский улар *(Tetraogallus himalayensis)*
parturiate родить; рожать; размножаться
parturient *a* рожающая; родовой; послеродовой
parturition *s* роды
parumbilical *a* околопупочный
parvicellular *a* мелкоклеточный
parviflorous *a* мелкоцветковый
parvovirus *s* парвовирус
pascual *a* пастбищный
pasqueflower *s* прострел *(Pulsatilla)*
pass *s* проход, переход
passable *a* проходимый

passage *s anat* проход, проток; перенос; перелёт; *micr* пассаж; пассирование; пересев
~ **of blood** кровотечение
~ **of strains** пересев штаммов
transplacental ~ перенос через плаценту, трансплацентарный перенос
water ~ водовод
passaging *s* пассирование; перепрививание
passante *s* пассант
passerine *a* воробьиный
passion *s*: **sexual** ~ половое влечение
passionflower *s* страстоцвет *(Passiflora)*
paste *s* клей
insect ~ клей для ловчих колец
pastern *s* путовая кость, бабка *(in a horse)*
pasteurization *s* пастеризация
pasteurize *v* пастеризовать
pasteurizer *s* пастеризатор
pasturage *s* пастбище, выгон, выпас; пастьба, выпас; подножный корм
pasture *s* пастбище, выгон, выпас; подножный корм
pastured *a* стравленный
pasturing *s* пастьба, выпас; стравливание
pat *v* погладить; ~ **a dog** погладить собаку
patabiont *s* обитатель лесной подстилки
patch *s* пятно; участок
"active" ~ антигенная детерминанта, эпитоп
bare ~ чёрная парша, ризоктониоз *(caused by Rhizoctonia solani)*
Fusarium ~ снежная плесень *(caused by Fusarium nivale)*
Payer's ~ пейерова бляшка
wheat bare ~ ризоктониоз пшеницы *(caused by Rhizoctonia solani)*
patch-clamp *s* пэтч-кламп
patchiness *s* мозаичность
patchouli *s bot* пачули *(Pogostemon cablin)*
patchy *a* пятнистый
patella *s* коленная чашка, надколенник; *bot* origпателла, блюдцевидный апотеций
patellar *a* коленный; *bot* пателлярный
patelliform *a* блюдцевидный
patellula *s bot* блюдечко
patency *s* раскрытое состояние
patent *a* открытый, доступный; явный, очевидный
paternal *a* отцовский; родственный по отцу
paternity *s* отцовство
path *s* путь; ход; тропинка, тропа; *anat* нервный путь
pathfinder *s* передовое нервное волокно
pathobolism *s* патологический обмен веществ, патологический метаболизм
pathogen, pathogene *s* болезнетворный организм; *a* патогенный, болезнетворный
fungal ~ патогенный гриб
pathogenesis *s* патогенез
pathogenetic *a* патогенетический
pathogenic *a* патогенный, болезнетворный
pathogenicity *s* патогенность, болезнетворность
pathogeny *s* патогенез
pathologic, pathological *a* патологический
pathology *s* патология
plant ~ фитопатология
pathometabolism *s* патологический обмен веществ, патологический метаболизм
pathomorphology *s* патоморфология
pathophoresis *s* патофорез, передача болезней
pathway *s* путь обмена, путь метаболизма; путь; ход; тропинка, тропа
Embden-Meyerhof-Parnes' ~ гликолиз, путь Эмбдена - Мейергофа - Парнеса
Hageman ~ *imm* каскад Хагемана
membrane attack ~ формирование мембраноатакующего комплекса
paracellular ~ трансклеточный путь, трансклеточный транспорт
patriclinous, patroclinous *a* патроклинный
patrocliny *s* патроклиния, патроклинность
patrogenesis *s* патрогенез, андрогенез, мужской партеногенез
patrogony *s* патрогония
patrolling *s eth* охрана территории
pattern *s* структура, конфигурация; особенность, характер; паттерн, образ

жизни; манера поведения; узор, рисунок; тип, способ
consummatory ~ консумматорный акт, завершающий акт
expressive ~ *eth* выразительное движение
fixed action ~ фиксированная последовательность действий; фиксированная последовательность движения
gene ~ набор генов
immune ~ иммунологический антигенный рисунок *(e.g., of a cell)*
immunoelectrophoretic ~ иммуноэлектрофоретический профиль
inherited behavior ~ наследственная форма поведения
labeling ~ паттерн мечения, распределение метки
locomotor ~ локомоторный паттерн, характер локомоции
restriction fragment ~ рестрикционная карта
two-dimensional peptide ~ двумерная пептидная карта, белковый фингерпринт
vegetation ~ мозаика растительности
patulector *s* патулектор
patulent, patulous *a* развесистый, раскидистый, разбросанный
paunch *s anat* рубец
paunchy *a zool* отвислый
pause *s* пауза
paw *s* лапа; *v* трогать лапой, скрести лапой; бить копытом
pawing *s* рытьё; роющие движения
pawpaw *s см.* **papaw**
paxilla *s* паксилла
paxillate *a* столбчатый
pea *s* горох *(Pisum);* горошина
peach *s* персик *(Persica);* персик обыкновенный *(Persica vulgaris)*
plum ~ персикослива *(Persica vulgaris x Prunus salicina)*
peachick *s* молодой павлин
peacock *s* самец павлина; павлин *(Pavo)*
peacock-pheasant *s* павлиний фазан *(Polyplectron)*
peafowl *s* павлин *(Pavo)*
peahen *s* самка павлина, пава
peak *s* пик, максимум, высшая точка
adaptive ~ адаптивный пик, максимум приспособления
peakedness *s biom* островершинность, положительный эксцесс
peanut *s* земляной орех *(Arachis)*
pear *s* грушевое дерево *(Pyrus);* груша
alligator ~ авокадо *(Persea americana)*
prickly ~ опунция *(Opuntia)*
pearl *s* жемчужина
pear-shaped *a* грушевидный
peat *s* торф
peatbog *s* торфяное болото
peat-forming *a* торфообразующий
peatmoss *s* торфяной мох, сфагнум *(Sphagnum)*
peaty *a* торфянистый; торфяной
peccant *a* болезненный; болезнетворный
peccary *s ther* пекари *(Tayassu)*
peck *s* клевок; *anat* книжка; *v* клевать; долбить клювом
pecker *s* дятел; клюв
pecking *s* клевание
peck-order, peck-right *s eth* «порядок клевания»
pecten *s* гребень, гребешок
pectic *a* пектиновый
pectinate *a* гребенчатый, напоминающий гребень
pectoral *a* грудной
pectus *s* грудная область
pedal *a* относящийся к стопе; ножной; педальный
pedate *a* имеющий ногу, имеющий ноги; *bot* лапчатый; дольчатый
pedatifid *a bot* стоповидно-надрезной
pedatiform *a bot* стоповидный
pedatilobate *a bot* стоповидно-лопастный
pedatipartite *a bot* стоповидно-раздельный
pedicel *s* ножка; стебелёк, короткая ножка; пенёк рога; *ent* стебелёк брюшка; *bot* цветоножка
pedicellaria *s* педицеллярия
pedicellate *a* имеющий ножку, имеющий стебелёк; расположенный на ножке, расположенный на стебельке
pedicle *s* ножка; стебелёк, короткая ножка; пенёк рога; *ent* стебелёк брюшка; *bot* цветоножка
pedigree *s* педигри, родословная
pedigreed *a* породистый, с родословной

pediophytium *s* горная растительная ассоциация
pedipalp, pedipalpus *s* педипальпа
pedogamy *s* педогамия
pedogenesis *s* педогенез, почвообразование
pedogenic *a* почвообразующий
pedology *s* почвоведение
pedomorphic *a see* **paedomorphic**
pedomorphosis *s see* **paedomorphosis**
peduncle *s* ножка
pedunculate, pedunculated *a* имеющий ножку, имеющий стебелёк; расположенный на ножке, расположенный на стебельке
peel *s* кожура; кожица; *v* снимать кожуру, снимать кожицу; лущить, шелушить, шелушиться
peep *s* писк; чириканье; *v* пищать; чирикать
peeper *s* квакша *(Hyla)*
peepul *s* фикус священный *(Ficus religiosa)*
peewit *s* чибис *(Vanellus vanellus);* сорочий жаворонок *(Grallina)*
pegwood *s bot* бересклет европейский *(Euonymus europaeus)*
pekan *s ther* илька, пекан *(Martes pennanti)*
pekingese *s* пекинес
pelican *s* пеликан *(Pelecanus)*
pellet *s zool* погадка; гранула
  faecal ~ фекальный пеллет, фекалия
  pollen ~ *apic* обножка
pelleted *a* гранулированный
pellicle *s* пелликула; тонкая плёнка
pelliculate *a* покрытый пелликулой
pellucid *a* прозрачный, просвечивающий
pelma *s anat* подошва
pelobiont *s* пелобионт, обитатель ила
pelogenous *a* пелогенный
pelophyte *s* пелофит
pelt *s* шкура; кожа
  sheep ~ овчина
pelta *s bot* щиток; щитовидный апотеций
peltate *a bot* щитовидный
peltatilobate *a bot* щитолопастный
peltatipartite *a bot* щитораздельный
peltatisected *a bot* щиторассечённый
peltigerous *a bot* щитоносный
pelvic *a* тазовый
pelviform *a bot* глубокоблюдчатый
pelvis *s* таз; почечная лоханка
  kidney ~ почечная лоханка
  small ~ малый таз
pen *s* загон; перо; лебедь *(самка);* средняя жилка *(листа)*
  bleeding ~ ланцет
  sheep ~ овчарня
pendent *a* свисающий, плакучий
pendulous *a* висячий, отвислый
peneclimax *s* пенеклимакс
peneplain *s* пенеплен, денудационная равнина
penetrability *s* проницаемость
penetrance *s* пенетрантность
penetrant *s zool* пенетранта; *gen* пенетрантный, проявляющийся
penetrate *v* проникать, проходить сквозь; пронизывать
penetration *s* проникновение
  stomatal ~ устьичная проницаемость
penguin *s* пингвин *(Sphenisciformes)*
  Arctic ~ бескрылая гагарка, бескрылый чистик *(Pinguinus impennis)*
penicillate, **penicilliform** *a* кистеобразный
penicillus *s* кисть, кисточка, пучок волосков; пучок щетинок
penis *s* пенис, фаллос, мужской совокупительный орган, половой член
penna *s* пенна, контурное перо
pennage *s* загон
penning *s* содержание в загонах
pennant *s ther* илька, пекан *(Martes pennanti)*
pennatularian *s* морское перо, пеннатулярия *(Pennatularia)*
penninerved *a bot* перисто-нервный
pennycress *s bot* ярутка *(Thlaspi)*
pensile *a* висячий *(a nest)*
pentacarpous *a* пятиплодный
pentachenium *s bot* пятисемянка
pentactinal *a* пятилучевой
pentadactylism *s* пятипалость
pentafluorbenzylation *s* пентафторбензилирование
pentagonal *a bot* пентагональный, пятиугольный
pentagynous *a* пятипестичный
pentamerous *a* пятичленный, разделённый на пять частей
pentandrous *a* пятитычинковый

pentapetalous *a* пятилепестный
pentaphyllous *a* пятилистный
pentaploid *s* пентаплоид; *a* пентаплоидный
pentasomic *s* пентасомик; *a* пентасомический
pentaspermous *a* пятисемянный
penultimate *a* предпоследний
peony *s* пион *(Paeonia)*
pepo *s* тыква, тыквина *(тип плода)*
pepper *s* перец *(Piper, Capsicum)*
 betel ~ бетель *(Piper betle)*
peppermint *s* мята перечная *(Mentha piperita)*
peppery *a* перечный
pepsinate *v* пепсинировать
pepsiniferous, pepsinogenous *a* продуцирующий пепсин
peptide *s* пептид
 adjuvant ~ адъювант-пептид
 autologous ~ аутологичный пептид, аутопептид
 gastrin-releasing ~ гастрин-высвобождающий пептид, ГВП
peptidoglycane *s* пептидогликан
peptidylation *s* пептидилирование
peptone *s* пептон
peptonization *s* пептонизация
peptonize *v* превращаться в пептоны, пептонизироваться
peracidity *s* избыточная кислотность
peraeopod *s zool* перейопод, грудная нога
perceivable *a* ощутимый, воспринимаемый, различимый
perceive *v* воспринимать; чувствовать; понимать
percentage *s* процент, процентное отношение
 survival ~ выживаемость
percept *s* воспринимаемый объект
perceptible *a* ощутимый, воспринимаемый, различимый
perception *s* перцепция, восприятие
perceptive *a* ощущающий, воспринимающий
perch *s* насест; жёрдочка; *v* садиться на насест
percolate *s* фильтрат; промывная вода; *v* фильтровать; процеживать; просачиваться
percolation *s* просачивание; проникновение; процеживание
percolator *s* перколятор; фильтровальная установка, фильтр
percussion *s* перкуссия, выстукивание
perdominant *s* пердоминант
peregrin, peregrine *s orn* сапсан *(Falco peregrinus)*
pereion *s* перейон, грудной отдел
pereiopod *s* перейопод, грудная нога
perennate *v* вегетировать круглый год
perennial *s* многолетник, многолетнее растение; *a* многолетний; сохраняющийся круглый год
perfoliate *a* пронзённолистный
perforate *v* перфорировать, продырявливать
perforated *a* продырявленный
perforatorium *s* перфораторий
perforin *s* каналообразующий белок, пороформирующий белок, перфорин
performance *s* деятельность; работа; действие; продуктивность, производительность, оценка продуктивности; коэффициент полезного действия
 breeding ~ плодовитость
 cardiac ~ работа сердца, деятельность сердца
 maze ~ прохождение лабиринта
 muscle ~ мышечная деятельность
 reproductive ~ репродуктивная способность, репродуктивность
perfusate *s* перфузируемая жидкость, перфузат
perfusion *s* перфузия; кровоток, кровоснабжение
perianth *s* околоцветник
periarthric *a* околосуставной
periblast *s* перибласт, наружный слой, эпибласт
periblastic *a* перибластический, поверхностный
periblastula *s* перибластула
periblem *s* периблема
pericambium *s* перикамбий
pericardiac, pericardial *a* перикардиальный
pericardium *s* перикард, перикардий, околосердечная сумка
pericarp *s* перикарпий, околоплодник
pericellular *a* окружающий клетку

pericentral, pericentric *a* перицентральный
perichaetium *s* перихеций
perichondrium *s* перихондрий, надхрящница
perichordal *a* перихордальный, окружающий нотохорд
periclinal *a* периклинальный
periclinium *s bot* общая обвёртка
pericranium *s* надкостница черепа
pericycle *s* перицикл
pericystic *a* окружающий пузырь
pericyte *s* перицит, адвентициальная клетка, клетка Руже, клетка Маршана
peridental *a* околозубной, периодонтальный
periderm *s* перидерма
peridesm *s* перидесма
perididymis *s amat* перидидимис
peridium *s* перидий
perienteric *a* околокишечный, окружающий кишку
perienteron *s* висцеральная полость
periesophageal *a* околопищеводный
periganglionic *a* периганглионарный
perigastric *a* окружающий пищеварительный тракт
perigastrium *s* полость тела, целом
perigastrula *s* перигаструла
perigonium *s* перигоний
perigynium *s* перигиний
perigynous *a* околопестичный
perihemal *a* перигемальный
peri-intestinal *a* окружающий кишечник
perikaryon *s* перикарион
peril *s* опасность
perilous *a* опасный
perilymph *s* перилимфа
perimetrium *s* периметрий, серозный покров матки
perimuscular *a* окружающий мышцу
perimysium *s* перимизий
perineal *a* промежностный, перинеальный
perineum *s* промежность
perineurium *s* периневрий
perineuronal *a* окружающий нервную ткань
period *s* период
  flowering ~ период цветения; фаза цветения
  incubation ~ период инкубации, период выращивания; инкубационный период; время инкубации; время насиживания
  lag ~ лаг-период
  reaction ~ длительность реакции
  service ~ сервис-период
periodicity *s* периодичность
  ~ **of spawning** повторность нереста
  sexual ~ половой цикл
periodont *s* периодонт, околозубная ткань
periople *s* венчик копыта
perioral *a* расположенный около рта
periost, periosteum *s* периост, надкостница
periostracum *s* периостракум
periotic *a* околоушный
peripetalous *a* окололепестный
peripharyngeal *a* окологлоточный
peripherad *adv* по направлению к периферии
peripheral *a* периферический
periphoranthium *s* общая обвёртка соцветия
periphorium *s bot* перифориум
periphysis *s* перифиза
periphytic *a* перифитонный
periphyton *s* перифитон
periplasm *s* периплазма
periplast *s* перипласт
peripneustic *a ent* перипнейстический
periproct *s* перипрокт
peripterous *a bot* нисбегающий
perisarc *s* перисарк
perish *v* погибать; умирать
perisperm *s* перисперм, питательная ткань
perispondylic *a* окружающий позвонок
perisporangium *s* периспорангий
perispore *s* периспора
perissodactyl *a* непарнопалый
perissodactyle *s* непарнокопытное *(Perissodactyla)*
perissoplidy *s* периссоплоидия
peristalsis *s* перистальтика
peristaltic *a* перистальтический
peristethium *s ent* среднегрудь
peristome *s* перистом
perithecigerous *a* несущий перитеции
perithecium *s* перитеций

peritoneal *a* брюшинный, перитонеальный
peritoneum *s* брюшина
peritreme *s* перитрема
peritrichous *a prot* кругоресничный; *micr* перитрихиальный
peritroch *s* перитроха; ресничный поясок
peritrophic *a* перитрофический
periuterine *a* околоматочный
perivascular *a* периваскулярный
perivisceral *a* первисцеральный, окружающий внутренности
perivitelline *a* перивителлиновый
periwinkle, periwinkler *s* береговая улитка, литорина *(Littorina)*; *bot* барвинок *(Vinca)*
permanent *a* постоянный
permeability *s* проницаемость
permeable *a* проницаемый
permeate *v* проникать, проходить
permissive *a* пермиссивный
permutation *s biom* перестановка
pernicious *a* пагубный, вредный, губительный, пернициозный
peroneal *a* малоберцовый
peroxidation *s* перокисление, перекисное окисление
peroxisome *s* пероксиома
perprolate *a* сверхпродолговатый
perpetual *a* цветущий круглый год
perseverance *s* нестомчивость *(of a dog)*
perseveration *s* персеверация
persimmon *s* хурма *(Diospyros)*
persist *v* оставаться, продолжать существовать, сохраняться, выживать
persistence, persistency *s* выживание, стойкость, устойчивость, выносливость; персистенция; упорство
persistent *a* стойкий, устойчивый; персистирующий; неопадающий; непрерывно возобновляющийся
persister *s* стойкий организм, устойчивый организм
person *s* самостоятельная особь, индивид; зооид
personate *a bot* масковидный
perspiration *s* потоотделение; испарина
pertusate *a bot* продырявленный
perula *s bot* чешуйка
pervious *a* проницаемый, проходимый
peryginium *s* перигиниум
pessimum *s* пессимум
pessulus *s orn* козелок
pest *s* вредитель; паразит; чума
  insect ~ насекомое-вредитель
  museum ~ кожеед музейный *(Anthrenus museorum)*
  storage ~ амбарный вредитель
pesticide *s* пестицид
pestilence *s* эпидемия
pet *s* комнатное животное; *a* комнатный, домашний, ручной
petal *s* лепесток
petaled *a* имеющий лепестки
petaliferous *a* несущий лепестки
petaliform *a* лепестковидный
petaline *a* лепестковый, лепестковидный
petalless *a* безлепестковый
petaloid, petaloideous *a* лепестковидный
petalous *a* имеющий лепестки
petal-shaped *a* лепестковидный
peter out *v* исчезать; истощаться
petiolar *a* черешковый
petiolate *a ent* стебельчатобрюхий; стебельчатый; *bot* черешчатый
petiole *s ent* стебелёк; *bot* черешок
petiolule *s* черешочек
petrel *s* буревестник *(Procellaria)*
  ashy ~ пепельная качурка *(Oceanodroma homochroa)*
petricolous *a* обитающий на скалах
petrium *s* сообщество скал
petrodium *s* сообщество валунов
petrohyoid *a* каменисто-подъязычный
petrolatun *s* вазелин
  liquid ~ вазелиновое масло
petromastoid *a* каменисто-сосцевидный
petro-occipital *a* каменисто-затылочный
petrophilous *a* обитающий на скалах
petrophyte *s* петрофит, скальное растение
petrosal *a anat* каменистый
petrosphenoidal *a anat* каменисто-клиновидный
petrosquamosal *a anat* каменисто-чешуйчатый
petrotympanic *a anat* каменисто-барабанный
petrous *a* окаменелый; каменистый
pewit *s* чибис *(Vanellus vanellus);* сорочий жаворонок *(Grallina)*

PFU, plaque-forming unit *s* бляшкообразующая единица
phacella *s* гастральная нить *(in coelenterates)*
phacoid *a* чечевицеобразный
phaeic *a* тёмноокрашенный
phaeism *s* потемнение
phaeochrous *a* тёмноокрашенный
phaeodium *s* феодий
phaeohyphomycosis *s myc* феогифомикоз
phage *s* фаг
DNA ~ ДНК-содержащий фаг
helper ~ фаг-помощник
latent латентный фаг, симбиотический фаг, профаг
wild-type ~ «дикий» тип фага, исходный тип фага
phage-typing *s* фаготипирование
phagocytal *a* фагоцитарный
phagocyte *s* фагоцит; *v* фагоцитировать
phagocytic *a* фагоцитарный
phagocytize *v* фагоцититировать
phagocytosis *s* фагоцитоз
phagolysosome *s* фаголизосома
phagoplankton *s* фагопланктон
phagosome *s* фагосома
phagotroph *s* фаготрофный организм, фаготроф
phagotrophic фаготрофный
phagotrophy *s* фаготрофия
phagotyping *s* фаготипирование
phalange *s* фаланга
phalangeal *a* относящийся к фаланге
phalanx *s* фаланга; пучок тычинок
phallus *s* пенис, фаллос, мужской совокупительный орган, половой член
phanerogam *s* явнобрачное растение
phanerogamous *a* явнобрачный
phanerophyte *s* явнобрачное растение
pharyngeal *a* глоточный
pharyngopalatine *a* нёбно-глоточный
pharynx *s* глотка
phase *s* фаза; период; стадия
coupling ~ фаза сцепления; стадия конъюгации
eclipse ~ период «затмения», период исчезнования, эклипсная фаза
lag ~ лаг-фаза, скрытое время, латентный период
"recruitment" ~ рекрут-фаза
phaser *s*: **syringe** ~ шприц-диспенсер
phasing *s mbiol* фазирование
phasmid *s* фазмида
pheasant *s* фазан *(Phasianus colchicus)*
phellem *s* феллема
phellium *s* сообщество скальных участков
phelloderm *s* феллодерма
phellogen *s* феллоген, пробковый камбий
phelloid *a* пробковидный; пробковатый
phellophilus *s* обитатель скальных участков
phene *s* фен
phengophil *a* светолюбивый
phengophobe *a* избегающий света, фотофобный
phenocline *s* феноклин
phenocontour *s* феноконтур
phenocopy *s* фенокопия
phenodeme *s* фенодим
phenodeviant *s* фенодевиант
phenogam *s* явнобрачное растение
phenogenetics *s* феногенетика
phenol *s* фенол
phenological *a* фенологический
phenology *s* фенология
phenome *s* феном
phenomenon *s* феномен
anamnestic ~ анамнестическая реакция, вторичный иммунный ответ, бустер-эффект
immunological ~ иммунологический феномен; иммунная реакция
interference ~ феномен интерференции, взаимовлияние двух различных организмов
recruitment ~ вовлечение, рекрутирование
reflex ~ рефлекторное явление
release ~ расслабление
"self-cure" ~ локальная иммунная реакция, местная иммунная реакция
veto ~ *imm* «вето»-феномен
phenon *s* фенон
phenotype *s* фенотип
antigenic ~ иммунофенотип, антигенный фенотип
HLA ~ *imm* HLA-фенотип
PLT ~ *imm* ТПЛ-фенотип
phenotypic *a* фенотипический
pheophytin *s* феофитин
pheromone *s* феромон
priming ~ мотивирующий феромон; прайминг-феромон

phialide *s* фиалида
philomel *s* соловей *(Luscinia)*
philtrum *s anat* фильтр
phlebogram *s* флебограмма; рентгенограмма вен
phlebography *s* флебография; рентгенография вен
phlebomanometer *s* флебоманометр
phlebophorous *a* жидконосный
phloem *s* флоэма, луб
phloeoderma *s bot* эндодерма
phlogistic *a* воспалительный
phobotaxis *s* фоботаксис, реакция избегания
phoebe *s orn* феб *(Sayornis)*
pholidosis *s* фолидоз
phomopsiosis *s* фомопсиоз *(caused by Phomopsis)*
phonation *s eth* издавание звуков
phoneme *s eth* фонема
phonoreceptor *s* фонорецептор
phoranthium *s bot* ложе соцветия
phoresia, phoresis, phoresy *s* форезия
phosphate *s* фосфат
nicotinamide-adenine dinucleotide ~ никотинамид-аденин-динуклеотидфосфат, НАДФ, трифосфопиридиннуклеотид
ribose ~ рибозофосфат
phosphoarginine *s* фосфоаргинин, аргининфосфат
phosphocreatine *s* креатинфосфат
phosphoenolpyruvate *s* фосфоенолпируват
phosphoenzyme *s* фосфофермент
phosphoglyceraldehyde *s* фосфоглицериновый альдегид
phospholipid *s* фосфолипид
phosphoprotein *s* фосфопротеид, фосфопротеин
phosphorescence *s* фосфоресценция
phosphorescent *a* фосфоресцирующий
phosphorolysis *s* фосфоролиз
phosphorylate *v* фосфорилировать; **to ~ on primary group** фосфорилировать по первичной группе
phosphorylation *s* фосфорилирование
photosynthetic ~ фотосинтетическое фосфорилирование, фотофосфорилирование
phosphotide *s* фосфолипид
photic *a* световой
photoaddition *s* фотоприсоединение
photoassimilation *s* фотоассимиляция
photoautotroph *s* фотоавтограф
photobiology *s* фотобиология
photocell *s* фотоэлемент
photoceptor *s* фоторецептор
photochemistry *s* фотохимия
photocinesis *s* фотокинез
photocontrol *s* фоторегуляция, влияние света
photodegradation *s* фотодеградация
photodimer *s* фотодимер
photodimerization *s* фотодимеризация
photodissociation *s* фотодиссоциация
photodynamics *s* фотодинамика
photofixation *s* фотофиксация
photogenic *a* фотогенный
photography *s* фотография
aerial ~ аэрофотосъёмка
water penetration aerial ~ аэрофотосъёмка сквозь толщу воды
photoinactivation *s* фотоинактивация
photoinduction *s* фотоиндукция
photoinhibition *s* фотоингибирование
photokinesis *s* фотокинез; фототаксис
photokinetic *a* фотокинетический
photolabeling *s* фотомечение
photolabile *a* фотолабильный
photoluminescence фотолюминесценция
photolysis *s* фотолиз, фотодиссоциация
photometabolism *s* фотометаболизм
photometer *s* фотометр
photometry *s* фотометрия
photomicrograph *s* микрофотография, микрофотоснимок
photomicrography *s* микрофотография
photonasty *s* фотонастия
photooxidation *s* фотоокисление
photoperiod *s* фотопериод, световой период
photoperiodicity *s* фотопериодичность
photoperiodism *s* фотопериодизм
photophase *s* фотофаза, световая фаза
photophilous *a* светолюбивый
photophobic *a* светобоязливый, фотофобный; избегающий света
photophobism *s* светобоязнь; избегание света
photophore *s* фотофор
photophosphorylation *s* фотофосфорилирование
photophygous *a* тенелюбивый

photopia *s* фотопия
photoplagiotropy *s* фотоплагиотропия
photoplethysmography *s* фотоплетизмография
photoproduction *s* фотообразование
photoprotection *s* фотопротекция
photopsia *s* фотопсия
photoreaction *s* фотореакция
photoreactivation *s* фотореактивация
photoreactivity *s* фотохимическая реакционная способность
photoreception *s* восприятие света, фоторецепция
photoreceptor *s* фоторецептор
photoreduction *s* фоторедукция, фотовосстановление
photorelease *s* фотовысвобождение
photorepair *s* фоторепарация
photorespiration *s* фотодыхание, световое дыхание
photoreversibility *s* фотообратимость
photosensible *a* светочувствительный
photosensitization *s* фотосенсибилизация
photostable *a* фотостабильный
photostage *s* световая стадия, фотофаза
photosynthesis *s* фотосинтез
photosystem *s* фотосистема
phototaxis *s* фототаксис
phototonus *s* чувствительность к свету
phototroph *s* фототроф, фототрофный организм
phototrophic *a* фототрофный
phototrophy *s* фототрофия
phototropism *s* фототропизм
phragma *s zool* перегородка
phragmobasidium *s* фрагмобазидия
phragmoplast *s* фрагмопласт
phragmosis *s eth* баррикадирование
phragmosome *s* фрагмосома
phragmospore *s* фрагмоспора
phrase *s* фраза *(bioacoustics)*
phratrie, phratry *s* фратрия
phreatophyte *s* фреатофит
phrenic *a* диафрагмальный, грудобрюшный
phrenology *s* френология
phycobilisome *s* фикобилисома
phycobiont *s* фикобионт, симбиотическая водоросль
phycocenology *s* фикоценология
phycology *s* альгология
phylacobiosis *s* филакобиоз
phyletic *a* филетический
phyllary *s bot* листочек обвёртки
phyllidium *s* ботридий
phylloclade *s* филлокладий, листоветвь
phyllode, phyllody *s* филлодий
phyllogenetic *a* листообразующий
phylloid *a* листообразный, листовидный
phyllomania *s* ненормальное листообразование, чрезмерное листообразование
phyllophagous *a* питающийся листьями
phyllophorous *a* листообразующий; несущий листья
phyllopod *s* листоногое ракообразное
phyllopodous *a zool* листоногий
phylloptosis *s* опадение листьев
phylloquinone *s* филлохинон
phyllosiphonic *a bot* филлосифонический
phyllotaxis, phyllotaxy *s* филлотаксис, листорасположение
phylloxanthin *s* филлоксантин, ксантофилл
phyllozooid *s* филлозооид
phylocoenogenesis *s* филоценогенез
phylogenesis *s* филогенез
phylogenetic *a* филогенетический
phylogeny *s* филогения
phylon *s* генетически родственная группа
phylum *s* тип
physical *a* физический; природный; телесный
physiogeny *s* физиогения
physiognomics *s* физиогномика
physiognomy *s* физиономия, внешний вид
physiography *s* физиография; физическая география
physiological *a* физиологический
physiology *s* физиология
bodily ~ соматофизиология
physiopathology *s* патологическая физиология, патофизиология
physocarpous *a* вздутоплодный
phytalbumin *s* растительный белок
phytium *s* растительная формация
phytivorous *a* травоядный; растительноядный
phytoactinometer *s* фитоактинометр
phytoaeron *s* фитоаэрон
phytoagglutinin *s* фитоагглютинин
phytobiology *s* биология растений

phytobiotic *a* фитобиотический, живущий внутри растений
phytocenology *s* фитоценология
phytocenosis *s* фитоценоз, растительное сообщество
phytochemistry *s* биохимия растений
phytochrome *s* фитохром
phytoclimatology *s* фитоклиматология
phytodetritus *s* растительный детрит
phytoedaphon *s* фитоэдафон, почвенная микрофлора
phytogamy *s* перекрёстное оплодотворение
phytogenesis *s* происхождение растений
phytogenetics *s* генетика растений
phytogenous *a* фитогенный, производимый растениями, растительного происхождения
phytogeny *s* происхождение растений
phytogeography *s* фитогеография, география растений
phytograph *s* фитограмма
phytography *s* фитография
phytohemagglutinin *s* фитогемагглютинин
phytohormone *s* фитогормон, регулятор роста растений
phytoid *a* напоминающий растение
phytokinin *s* фитокинин
phytolite *s* ископаемое растение
phytologic ботанический
phytology *s* ботаника
phytome *s* совокупность растений, растительность
phytomer *s* фитомер
phytometer *s* фитоиндикатор
phytonomy *s* фитономия
phytoparasite *s* растение-паразит
phytopathology *s* фитопатология
phytophage *s* фитофаг, растительноядный организм
phytophagous *a* травоядный; растительноядный
phytophagy *s* фитофагия
phytophilous *a* фитофильный
phytophysiology *s* физиология растений
phytoserology *s* фитосерология
phytosociology *s* фитосоциология, фитоценология, геоботаника
phytosterol *s* фитостерин
phytosuccivorous *a* питающийся соками растений
phytotaxonomy *s* таксономия растений
phytoteratology *s* тератология растений
phytotomy *s* анатомия растений
phytotoxic *a* фитотоксичный, токсичный для растений
phytotoxin *s* фитотоксин
phytotron *s* фитотрон
phytotrophic *a* автотрофный
pia *s* мягкая мозговая оболочка
pick *v* клевать *(grain);* снимать, собирать *(fruit);* щипать *(grass)*
picking *s* сбор *(of fruit)*
picodnavirus *s* пикоднавирус
picornavirus *s* пикорнавирус
piebald *a* пегий; *s* помесь, гибрид
piece *s* часть; компонент; фрагмент
  Fc ~ Fc-фрагмент, Fc-область *(иммуноглобулина)*
  intracellular transport [secretory tail, transport] ~ *imm* секреторный компонент, внутриклеточный транспортный пептид
  nose ~ револьвер
pied *a* пегий
piercer *s* прокалывающий орган
piesesthesia *s* барестезия, чувство давления
piesimeter *s* барестезиометр
pig *s* свинья *(Suidae);* поросёнок
  Eurasian ~ кабан, свинья *(Sus)*
  gelded ~ боров
  Guinea ~ морская свинка *(Cavia porcellus)*
  true ~s *pl* свиньи, свиные *(Suidae)*
  water ~ водосвинка *(Hydrochoerus)*
  wild ~ кабан, дикая свинья *(Sus scrofa)*
pigeon *s* голубь *(Columba)*
  Cape ~ капский буревестник *(Daption capensis)*
  kelp ~ клуша *(Larus fuscus)*
  sea ~ атлантический чистик *(Cepphus grylle)*
  wood ~ вяхирь *(Columba palumbus)*
piglet *s* поросёнок
pigment *s* пигмент
pigmental, pigmentary *a* пигментный
pigmentation *s* пигментация
pika *s* пищуха, сеноставка *(Ochotona)*
pike *s bot* шип, колючка
pileate *a bot* столбчатый; *myc* шляпочный, имеющий шляпку

pileolus *s zool* маленький зонтик, маленький колокол; *myc* маленькая шляпка
pileorhiza *s bot* корневой чехлик
pileus *s zool* зонтик, колокол; *myc* шляпка
pilidium *s zool* пилидий
piliferous *a* несущий волосы, несущий волоски; волосистый, мохнатый
piliform *a* воловидный; волосковидный
pilocystidium *s* пилоцистида
pilo-erection *s eth* пилоэрекция
pilose *a* волосистый, мохнатый
pilus *s* фимбрия
pimple *s* пупырышек
pin *s* булавка
pinbone *s* седалищная кость
pincer *s* клешня; хелицера; *pl* пинцет
pine *s* сосна *(Pinus)*
  stone ~ сосна итальянская, пиния *(Pinus pinea)*
pineal *a* шишковидный
pineapple *s* ананас *(Ananas sativus)*
pinecone *s* сосновая шишка
pinion *s* сосна съедобная *(Pinus edulis);* маховое перо; *v* подрезать крылья
pinna *s* перо; *bot* листочек
pinnate *a* перистый
  abruptly ~ прерванно-перистый
  interruptedly ~ прерывисто-перистый
  odd ~ непарноперистый
pinnatifid *a bot* перисто-надрезный
pinnatilobate *a bot* перисто-лопастный
pinnation *s* перистость
pinnatipartite *a bot* перисто-раздельный
pinnatisect, pinnatisected *a bot* перисто-рассечённый
pinniform *a* имеющий форму пера, перообразный
pinninervate *a bot* перисто-нервный
pinniped *s* ластоногое животное (*Pinnipedia); a* ластоногий
pinnule *s zool* пиннула; *bot* вторичный листочек
pinocytic *a* пиноцитозный
pinocytosis *s* пиноцитоз
pinosome *s* пиноцитозный пузырёк
pintail *s orn* шилохвость *(Anas acuta);* белобрюхий рябок *(Pterocles alchata)*
pin-toed *a* косолапый
pinworm *s helm* острица *(Enterobius vermicularis)*
piny *a* сосновый, поросший сосной
pioneer *s* пионер
pip *s* семечко; зёрнышко; шильце *(sprout);* корнеклубень; *v* пищать; чирикать; разбивать скорлупу; вылупляться
pipe *s* трубка; пение, свист; *v* петь
pipet, pipette *s* пипетка; *v* пипетировать; раскапывать; **to ~ off** отсасывать пипеткой
  applicator ~ пипетка для внесения *(образцов),* (пипетка-)аппликатор
  blood-count [blood-diluting] ~ меланжер
  gel-loading ~ пипетка для внесения образца в гель; сэмплер
  graduated ~, measuring ~ измерительная пипетка, градуированная пипетка
  piston ~ штемпель-пипетка
  syringe ~ шприц-пипетка
pipettor *s lab* устройство для пипетирования, пипеттор
pipistrelle *s* нетопырь *(Pipistrellus)*
pipit *s orn* конёк *(Anthus)*
pipping *s* проклёвывание
piscivorous *a* рыбоядный
pisiform *a* гороховидный
pistache, pistachio *s* фисташка обыкновенная *(Pistacia vera)*
pistil *s* пестик
pistillate *a bot* пестичный, женский
pistillid, pistillidium *s* архегоний
pistillode *s* пистиллодий
pit *s* ямка, углубление; *bot* пора; яма; косточка
  bitter ~ *phytop* горькая ямчатость
  rutting ~ купалка; гонная яма
pitch *s* смола; дёготь; *mbiol* шаг
pitcher-leaf *s* кувшинчик, ловчий лист
pitfall *s* яма
pith *s bot* сердцевина; *v* извлекать сердцевину
pithless *a bot* не имеющий сердцевины
pithy *a* сердцевинный
pitted *a* ямчатый, покрытый маленькими впадинками
pituicyte *s* питуицит
pituitary *s* гипофиз; *a* гипофизарный
place *s* место
  drinking ~ водопой, место водопоя
  nesting ~ гнездовье

watering ~ водопой, место водопоя
placenta *s* плацента; *bot* семяносец
placental *a* плацентарный
placentate *a* имеющий плаценту
placentation *s* плацентация; образование плаценты; тип строения плаценты
placode *s* плакода
placoid *a* плакоидный; пластинкообразный
plagioclimax *s* плагиоклимакс
plagiophyllous *a* колосистый
plagiotropic *a* плагиотропный
plagiotropism *s* плагиотропизм
plague *s* чума; массовое размножение
grasshopper ~ нашествие саранчи
plan *s* план
plan-aperturate *a pal* экваториально-апертурный
plane *s* плоскость; платан *(Platanus)*
planea *s* бластея
planeton *s* мигрирующий вид
planetous *a* мигрирующий
planigale *s* плоскоголовая сумчатая мышь *(Planigale)*
planitegillate *a* плоскопокровный
planocellular *a* плоскоклеточный
planogamete *s* планогамета
planosome *s* планосома
planospore *s* планоспора
planozygote *s* планозигота
plant *s* растение; установка; *v* расти; сажать; засевать, сеять
abased alpine ~s *pl* сниженные альпийцы
aerial ~ эпифит, надземное растение
annual ~ однолетник, однолетнее растение
autochorous ~ автохор
basiphilous ~ базифил
bee ~ медонос
calciphilous ~ кальцефил
calciphobous ~ кальцефоб
CAM ~ растение с обменом по типу толстянковых, САМ-растение
castor-oil ~ клещевина обыкновенная *(Ricinus communis)*
century ~ столетник, агава американская *(Agave americana)*
chalk ~ качим *(Hypsophila)*
chasmophilous ~ хазмофит
chiropterophilous ~ хироптерофил
compass ~ компасное растение; сильфия дольчатая *(Silphium laciniatum)*
cotton ~ хлопчатник *(Gossypium)*
cushion ~ растение-подушка, подушечник
earth ~ геофит
eurythermic ~ эвритерм
eurytopic ~ эвривалент, эвритопное растение
eutrophic ~ эвтроф
gutta-percha ~ гуттаперченос
gypsophilous ~ гипсофил
heterotrophic ~ гетеротроф
honey ~ медонос
hydrophytic гидрофит
lowered alpine ~s *pl* сниженные альпийцы
macrophytic ~s макрофлора
megatrophic ~ мегатроф
meliferous ~ медонос
money ~ лунник однолетний *(Lunaria annua)*
mycotrophic ~ микотроф
myrmecochorous ~ мирмекохор
nitrogen-loving ~, nitrophilous ~ нитрофил
oil ~ масличное растение; клещевина обыкновенная *(Ricinus communis)*
ornitophilous ~ орнитофил
paradise ~ волчеягодник обыкновенный, волчье *лыко (Daphne mezereum)*
perennial ~ многолетник, многолетнее растение
pilot ~ опытная установка
pitcher ~ саррацения *(Sarracenia)*
polster ~ растение-подушка, подушечник
purification ~ очистная станция
rubber ~ каучуконос; фикус каучуконосный *(Ficus elastica)*
salt marsh ~ галофит
sclerophyllous ~ жестколистное растение, склерофилл
seed(-bearing*)* ~ семенное растение, семенник
siliciphilous ~ кремнефил
skeet ~ борщевик обыкновенный *(Heracleum spondylium)*
snow-favored ~ хионофил
stenotrophic ~ стенотроф

succulent ~ суккулент
sugar ~ сахароносс
surface ~ хамефит
tall aerial ~ фанерофит
planta *s* подошва; *ent* планта, первый членик лапки
plantain *s* подорожник *(Plantago)*
plantar *a* подошвенный
plantation *s* плантация, насаждение
plant-feeder *s* фитофаг, травоядное животное, растительноядное животное
plant-free *a* свободный от растений
plantigrade *a* стопоходящий
planting *s* посадка; сев; насаждение; культура; ~ **out** высадка
band ~ ленточный посев, ленточная посадка
plant-inhabiting *a* живущий на растении
plantlet *s* росток, проросток, всход
plantlouce *s* тля *(Aphidiidae)*
plantling, plantule *s* росток, проросток, всход
planula *s* планула
plaque *s* пятно, бляшка; зона гемолиза; тромбоцит
plasm, plasma *s* плазма; гемолимфа; цитоплазма
blood ~ плазма; гемолимфа
ground ~ основное вещество цитоплазмы
plasmacyte *s* плазмацит, плазматическая клетка
plasmagene *s* плазмаген
plasmal *s* плазмаль
plasmalemma *s* плазмалемма
plasmalogen *s* плазмалоген
plasmasome *s* плазмосома
plasmatic *a* плазматический
plasmid *s* плазмида
Tox ~ Tox-плазмида
plasmin *s* плазмин
plasminogen *s* плазминоген, профибринолизин
plasmocyte *s* плазмоцит
plasmode *s* плазмодий
plasmodesm, plasmodesma *s* плазмодесма
plasmodiocarp *s* плазмодиокарпий
plasmodiophorous *a* образующий плазмодий
plasmodiotrophoblast *s* плазмодиотрофобласт
plasmodium *s* плазмодий
plasmogamy *s* плазмогамия
plasmokinin *s* фактор VIII, антигемофильный фактор, тромбопластический фактор плазмы
plasmolysis *s* плазмолиз
plasmolytic *a* плазмолитический
plasmon, plasmone *s* плазмон
plasmoptysis *s* плазмоптиз
plasmosome *s* плазмосома
plasmozyme *s* протромбин, тромбиноген, фактор Д
plasson *s* плассон, недифференцированная протоплазма
plastic *a* пластичный, пластический
plasticity *s* пластичность
serologic ~ нестабильность антигенных свойств
plastid *s* пластида
blood ~ клетка крови
green ~ хлоропласт
plastidome *s* пластидом
plastin *s* пластин, линин
plastocont *s* пластоконт, хондриоконт
plastodeme *s* пластодем
plastogamy *s* пластогамия
plastogene *s* пластоген
plastomere *s* пластомер
plastoquinone *s* пластохинон
plastron *s zool* пластрон, брюшной щит
platanna *s* шпорцевая лягушка *(Xenopus)*
plate *s* пластина; планшет, плашка; *zool* пластинка; чешуйка; бляшка; чашка; *v* высевать на чашки
agar ~ чашка с агаровой средой; агаровая пластинка
axial ~ *emb* осевая полоска
blood ~ кровяная пластинка, тромбоцит
clinoid ~ *anat* спинка турецкого седла
cooling ~ столик с охлаждением
equatorial ~ метафазная пластинка
genial ~ подбородочный щиток
gradient ~ градиентная чашка
hemolytic gel ~ гемолитический гель-планшет
metaphase ~ метафазная пластинка
microtiter ~ титрационный микропланшет
motor ~ моторная бляшка, двигательная точка Ранвье
slide ~ предметное стекло

streak, streaked ~ штриховая пластинка
wax ~ *apic* зеркальце восковой железы
platelet *s* пластинка; кровяная пластинка, тромбоцит
plateletpheresis *s* тромбаферез, тромбоцитоферез
platiculture *s micr* пластинчатая разводка
plating *s* посев; культивирование
dilution ~ посев методом разведения
replica ~ получение реплик, получение отпечатков
platybasic *a* платибазальный
platycarpous *a* широкоплодный
platycephalic *a* имеющий уплощённый череп
platycerous *a bot* широкорогий
platydactyl, platydacltyle *a* имеющий уплощённые пальцы
platyphyllous *a* широколистный
platypus *s* утконос *(Ornithorhynchus anatinus)*
platysma *s* платизма
platyspermous *a* широкосемянный
platystachyous *a* ширококолосый
play *s* игра
play-biting *s eth* игровой укус
play-face *s eth* «игровое лицо»
plectenchyma *s* плектенхима
plectridial *a micr* плектридиальный
pleiochasium *s bot* плейохазий
pleiocotyledony *s* многосемядольность
pleiocyclic *a* многоциклический
pleiomery *s* плейомерия
pleiopetalous *a* плейопетальный, махровый
pleiophyllous *a* плейофильный
pleiotaxis, plelotaxy *s* плейотаксия
pleiotropic *a* плейотропный
pleiotropy *s* плейотропия
pleioxeny *s par* многохозяйность
pleogamy *s* плеогамия
pleomorph *s* плеоморф
pleomorphic *a* плеоморфный
pleomorphism, pleomorphy *s* плеоморфизм
pleon *s* брюшко
pleophagous *a* питающийся разнообразной пищей; *par* имеющий несколько хозяев
pleopod *s* плеопод
pleotreme *a pal* плеотремный, более чем одноапертурный
pleotrophic *a* питающийся разнообразной пищей
plerergate *s* муравей-бочка
plerocercoid *s* плероцеркоид
plesiomorphic *a* плезиоморфный
plesiomorphy *s* плезиоморфия
plethysmogram *s* плетизмограмма
plethysmograph *s* плетизмограф
plethysmography *s* плетизмография
impedance ~ реография, импедансная плетизмография
plethysmothallus *s bot* плетизмоталлюс
pleura *s* плевра
pleural *a* плевральный
pleuranthous *a* бокоцветный
pleurapophysis *s* плеврапофиз
pleurite *s ent* бочок, плейрит, склерит боковой части
pleurocarpic, pleurocarpous *a* бокоплодный
pleurocarpy *s* бокоплодность
pleurocentrum *s anat* плевроцентр
pleurocerebral *a* плевроцеребральный
pleurocystidium *s* плевроцистида
pleurodont *a* плевродонтный
pleurogynous *a* бокопестичный
pleuromitose *s* плевромитоз
pleuron *s zool* эпимеральная пластинка
pleuropedal *a* плевропедальный
pleuroperitoneum *s* плевроперитонеум
pleurorrhizous *a* краекорешковый
pleurosporous *a* плевроспоровый
pleurovisceral *a* плевровисцеральный
plexiform сетевидный
plexus *s* сплетение
plica *s* складка
plicate *a* складчатый
plication *s* образование складок
ploidy *s* плоидность
plot *s* делянка, опытный участок, площадка; план; схема; диаграмма
calibration ~ калибровочная кривая
quadruplicated ~s четырёхкратно повторяющееся наблюдение
scatter ~ диаграмма рассеяния, диаграмма разброса данных
plotophyte *s* плотофит
plotter *s* графопостроитель, плоттер
plover *s orn* зуёк; ржанка *(Charadriidae)*

bastard ~ чибис, пигалица *(Vanellus vanellus)*
black-bellied ~ тулес *(Pluvialis squatarola)*
common ~ чибис, пигалица *(Vanellus vanellus)*
crocodile ~, Egyptian ~ египетский бегунок, крокодилов сторож *(Pluvianus aegyptius)*
golden ~ золотистая ржанка *(Pluvialis apricaria)*
gray ~ тулес *(Pluvialis squatarola)*
green ~ чибис, пигалица *(Vanellus vanellus)*
pigeon ~ тулес *(Pluvialis squatarola)*
red-legged ~ обыкновенная камнешарка *(Arenaria interpres)*
ring, ringed ~ галстучник *(Charadrius hiaticula)*
silver ~ тулес *(Pluvialis squatarola)*
sociable ~ кречётка *(Vanellus gregarius)*
plowable *a* пахотный
plowing *s* вспашка, пахота
pluck *v* собирать
plug *s* пробка
plum *s* слива *(Prunus)*
pluma *s* контурное перо
plumage *s* оперение
plumate *a* покрытый перьями, перистый
plume *s* перо; контурное перо оперение; *bot* перистый хохолок; *v* обрастать перьями
plumiped *a orn* мохноногий
plumose *a* оперённый
plump *a* выполненный *(e.g., of grain)*
plumulate *a* пушистый
plumule *s* пуховое перо; *bot* почечка зародыша
plunge нырять, погружаться
plunger *s* пружинный штифт; фиксатор; плунжер
plural-row *a* многорядный
plurannual *s* многолетник, многолетнее растение; *a* многолетний; сохраняющийся круглый год
pluriaxial *a* многоосный
pluricellular *a* многоклеточный
pluriceptor *s* плюрирецептор, плюрицептор
plurilocular *a* многокамерный, многогнёздный
plurinuclear *a* многоядерный
pluripartite *a* многолопастный, многораздельный
pluripolar *a* многополюсный
pluripotent *a* плюрипотентный
pluriserial *a* многорядный
plurivalent *a* многовалентный
plurivorous *a* многоядный
pluteus *s* плутеус
pluviisilvae *s pl* влажнотропические леса
pneumatized *a* с воздушными полостями
pneumatocyst *s* плавательный пузырь; пневматофор; *orn* воздушный мешок
pneumatophore *s* пневматофор
pneumogastric *a* лёгочно-желудочковый
pneumogram *s* пневмограмма
pneumograph *s* пневмограф
pneumography *s* пневмография
pneumonia *s* пневмония, воспаление лёгких
pneumoperitoneum *s* пневмоперитонеум
pneumostome *s zool* пневмостом, дыхательное отверстие
pneumotaxis *s* пневмотаксис
pneumothorax *s* превмоторакс
poach браконьерствовать
poacher *s* браконьер
poaching *s* браконьерство
pochard *s orn* нырок *(Netta)*
pocket *s* карман; дивертикул; очаг обитания; *bot* пазуха, сумка
air ~ *bot* воздушная полость
antigen-binding ~ антигенсвязывающий «карман»
Rathke's ~ *emb* карман Ратке
root ~ *bot* корневой кармашек
wax ~ *apic* восковая железа
pocket-shaped *bot* сумкообразный
pod *s* коробочка; боб; стручок; фасоль *(Phaseolus);* небольшое стадо *(e.g., of seals),* стайка *(of birds)*
cassia ~s кассия пучковатая, александрийские стручки *(Cassia fasciculata)*
egg ~ *ent* кубышка
podal *a* ножной
podeon *s ent* стебелёк
podetium *s* подеций
podex *s* анальная область; *ent* пигидий
podium *s* нога, ножка *(e.g., in echinoderms)*
podobranchia *s* подобранхия
pododerm *s* слой кожи копыта

podogynium *s* гиноподий
podomere *s* сегмент конечности
podotheca *s* подотека
podzol *s* подзол
poecilophyllous *a* пестролистный
poic *a* луговой
poikilocyte *s* пойкилоцит
poikiloploid *a* пойкилоплоидный, миксоплоидный
poikilotherm *s* пойкилотермное животное
poikilothermal *a* пойкилотермный
point *s* точка; отличительная черта, особенность; кончик; остриё, острый конец; ответвление оленьего рога; *pl* стати; экстерьер
apical ~ *bot* точка роста; конус нарастания
beak ~ кончик клюва
breed ~s отличительные черты породы
drip ~ удлинённый кончик листьев
equivalence ~ *imm* зона эквивалентности
fixation ~ участок связывания, сайт связывания, участок фиксации, сайт фиксации
growing ~ *bot* точка роста, конус нарастания
initiation ~ область инициации, сайт инициации
metopic ~ метопион
nasal ~ назион
root growing ~ точка роста корня, конус нарастания корня
scale ~s *pl* стати экстерьера
shoot growing ~ точка роста побега, конус нарастания побега
starting ~ исходное положение; отправная точка
watering ~ водопой
wilting ~ коэффициент завядания; влажность завядания
pointed *a* заострённый, остроконечный
pointer *s* пойнтер
pointless *a* тупоконечный
poison *s* яд
poison-bearing *a* ядовитый
poisoning *s* отравление; интоксикация
blood ~ заражение крови
poisonous *a* ядовитый, отравляющий; токсический
poium *s* луговое сообщество
polar *a* полярный
polarimetry *s* поляриметрия
polarity *s* полярность
polarization *s* поляризация
polarograph *s* полярограф
polarography *s* полярография
polder *s* польдер
pole *s* полюс
polecat *s* чёрный хорёк, обыкновенный хорёк, лесной хорёк *(Mustela putorius)*
marbled ~, mottled ~ *ther* перевязка *(Vormela peregusna)*
polioplasm *s* полиоплазма, зернистая цитоплазма
poliovirus *s* полиовирус
poll *s* затылок; комолое животное, безрогое животное; *v* срезать рога; подрезать верхушку *(of a tree)*
polled *a* комолый, безрогий; лишённый рогов; с подрезанной верхушкой *(of a tree)*
pollen *s* пыльца
pollenizer *s* опылитель
pollex *s ther* большой палец
pollinarium *s* поллинарий
pollinate опылять
pollination *s* опыление
artificial ~ искусственное опыление; *zool* искусственное осеменение
controlled ~ искусственное опыление; принудительное скрещивание
spontaneous ~ самоопыление
supplementary ~ доопыление
pollinator *s* опылитель
polliniferous *a* пыльценосный
pollinium *s* поллиний
pollinizer *s* опылитель
pollinodium *s* поллинодий
pollinoid *s bot* мужская гамета, спермацит
pollutant *s* загрязнитель, загрязняющее вещество
pollution *s* загрязнение
noise ~ шумовое загрязнение, зашумлённость
optical ~ обезображивание пейзажа
polocyte *s* полоцит, полярное тельце, направительное тельце
polster *s* подушковидное растение
polyaddition *s* полиприсоединение
polyadelphous *a* многобратственный
polyadenilation *s* полиаденилирование

polyalkylation *s* полиалкилирование
polyandrous *a* многотычинковый
polyandry *s* полиандрия
polyanthous *a* многоцветковый
polyblast *s* полибласт
polybrochate *a* многоячеистый
polycarpellary *a* многоплодолистиковый
polycarpic, polycarpous *a* поликарпический, многократно плодоносящий; многоплодный
polycephalous *a bot* многоголовчатый
polychaete *s* полихета, многощетиноковый червь *(Polychaeta)*
polychromasy *s* полихромазия
polychromatic *a* полихромный
polychromatocyte *s* полихроматоцит, полихроматофильный эритроцит
polychromatophil, polychromatophile *s* полихроматофил, полихроматофильный эритробласт
polychromatophilia *s* полихроматофилия
polychronic *a* возникающий неоднократно
polycistronic *a* полицистронный
polyclimax *s* поликлимакс
polyclonality *s* поликлональные свойства, поликлональность
polyclonal *s* поликлональный иммуноглобулин
polycotyledon *s* многосемядольное растение
polycotyledonous *a* многосемядольный
polycyesis *s* множественная беременность
polycystic *a* многогнёздный, многокамерный
polydactyl *a* многопалый
polydactylism, polydactyly *s* полидактилия, многопалость
polydelphous *a* многобратственный
polyembryony *s* полиэмбриония
polyenergid *a* полиэнергидный
polyesterification *s* полиэтерификация
polyfactorial *a* полифакториальный
polyforate *a pal* многодырчатый
polygamous *a* полигамный
polygamy *s* полигамия
polygasric *a anat* многокамерный
polygene *s* полиген
polygenesis *s* полигенез
polygenetic *a* полигенетический, полифилетический
polygenic *a* полигенный
polygenomatic *a* полигеноматический, полигеномный
polygeny *s* полигенность, полигения
polygynous *a* полигинный; многопестичный
polygyny *s* полигиния, полигамия
polyhaploid *s* полигаплоид
polyheads *s pl* полиголовки
polyhedrosis *s* полиэдроз
polyhybrid *s* полигибрид
polykaric *a* многоядерный
polykaryocyte *s* поликариоцит
polykaryon *s* полиэнергидное ядро
polykinetid *s prot* поликинетида
polykinety *s prot* поликинета
polylecithal *a* полилецитальный
polylepidous *a bot* многочешуйчатый
polylinker *s mbiol* полилинкер
polymastigote *a* многожгутиковый
polymer *s* полимер
polymerization *s* полимеризация
polymerous *a* полимерный
polymery *s* полимерия
polymitosis *s* полимитоз
polymixic *a* полимиктный
polymorph *s* полиморф; полиморфноядерный лейкоцит, сегментноядерный лейкоцит
polymorphic *a* полиморфный
polymorphism *s* полиморфизм
polymorphous *a* полиморфный
polynuclear *a* многоядерный
polynucleotide *s* полинуклеотид
polyomavirus *s* вирус полиомы
polyp *s* полип
polypeptide *s* полипептид
polypericolpate *a pal* многобороздный
polypericolporate *a pal* многобороздно-поровый
polypetalous *a* многолепестный
polyphage *s* полифаг, многоядный организм
polyphagous *a* многоядный
polyphagy *s* полифагия, многоядность
polypheny *s* полифения, плейотропия
polyphylesis *s* полифилия
polyphyletic *a* полифилетический
polyphyllous *a* многолистный
polyphyodont *a* полифиодонтный
polypide *s* полипид
polyplicate *a* многоскладчатый

polyploid *s* полиплоид; *a* полиплоидный
duplicational ~ дупликационный полиплоид, аутополиплоид
polyploidy *s* полиплоидия
polypneustic *a* полипнейстический
polypod *a* имеющий много конечностей
polypoid *a* полипоидный
polyporate *a* многопоровый
polypore *s myc* трутовик *(Laetiporus)*
polyprotodont *a* многорезцовый
polypterous *a bot* многокрылый
polypyrenous *a* многокосточковый
polyrhizal *a* многокорешковый
polyribosome *s* полирибосома, полисома
polyrugate *a bot* многоморщинистый
polysaccharide *s* полисахарид
polysaprobic *a* полисапробный
polysepalous *a* многочашелистиковый
polysheath *s* поличехол
polysomaty *s* полисоматия
polysome *s* полирибосома, полисома
polysomic *s* полисомик; *a* полисомный
polysomitic *a* многосегментный
polysomy *s* полисомия
polyspermatous, polyspermous *a* многосемянный; полиспермный
polyspermy *s* полиспермия
polyspondyly *s* полиспондилия
polysporous *a* многоспоровый
polystachyous *a* многоколосый
polystele *s* полистель, полистела
polystely *s* полистелия
polystemonous *a* многотычиночный
polystichous *a* многорядный
polystigmatic *a* многорыльцевый
polystomatous *a helm* имеющий много присосок
polystome *s helm* многоустка *(Polystomum)*
polysulcate *a pal* многобороздный
polysynaptic *a* полисинаптический, многосинаптический
polytene *a* политенный
polythalamous *a bot* многокамерный, сборный
polythetic *a* политетический
polytocous *a* плодовитый
polytopic *a* политопный
polytopy *s* политопия
polytreme *a pal* политремный, многоапертурный
polytrichous *a* густоволосистый
polytrophic *a* политрофный, многоядный
polytropic *a* политропный
polytropism *s* политропизм
polytypic *a* политипичный, политипический
polyvoltine *a* поливольтинный
pomace *s* мезга, выжимки
pome *s bot* яблоко
pomegranate *s* гранат *(Punica);* слива чёрная, слива канадская *(Punica nigra)*
pomelo *s* грейпфрут *(Citrus grandis)*
pomeridian *a* полуденный
pomiculture *s* плодоводство, садоводство
pomology *s* помология
pompe *s* насос
pond *s* пруд; водоём; бассейн; запруда; *v* запруживать; образовывать пруд
ponerid *s* муравей-бульдог *(Ponerida)*
Pontian *a* понтический
pony *s* пони
poodle *s* пудель
pool *s* группа, совокупность; пул, фонд; лужа; заводь; полынья
gene ~ генофонд
neuron ~ гомогенная группа нейронов
phage mating ~ вегетативный фонд фага
pooling of data *biom* группировка данных
poophilous *a* обитающий на лугах
poophyte *s* растение лугов
pop-eyed *a* пучеглазый
poplar *s* тополь *(Populus)*
black ~ осокорь, тополь чёрный *(Populus nigra)*
poppy *s* мак *(Papaver)*
populate населять, заселять
population *s* популяция, население; *biom* совокупность; генеральная совокупность
livestock ~ поголовье
parent ~ исходная популяция; исходная совокупность; генеральная совокупность
soil-biological ~ биологическая популяция почвы, совокупность почвенных организмов
target ~ вакцинируемое население; иммунизируемые животные; популяция клеток-мишеней
poral *a* поровый

porcine *a* свиной
porcupine *s* дикобраз *(Hystrix)*
pore *s* пора, отверстие
air ~ *bot* устьице
genital ~ половое отверстие
germ ~ микропиле, проростковая пора
passage ~ проход *(in diatoms)*
stomatal ~ устьичная щель
poriferous *a* пористый
poriform *a* порообразный
pork *s* свинина
porkworm *s helm* трихинелла *(Trichinella)*
porogamy *s bot* порогамия
poroid *s* маленькое углубление
porosity *s* порозность, пористость, скважность
porous *a* пористый
porphyrophore *s* клетка, содержащая порфирин
porta *s* ворота органа
portal *s* ворота органа; воротная вена; *a* воротный
porto-caval *a* портокавальный
position *s* положение
nursing ~ поза кормления
possum *s* поссум, кускус *(Phalangeridae)*
postabdomen *s zool* постабдомен
postadaptation *s* постадаптация
postanal *a* постанальный
postantennal *a* постантеннальный, расположенный за антеннами
postaxial *a* постаксиальный, послеосевой
postcava *s* нижняя полая вена
postcentral заднецентральный
postcentrum *s* задняя часть тела позвонка
postcephalic *a* расположенный за головой
postcerebral *a* расположенный позади мозга
postclimax *s* постклимакс
postclival *a anat* расположенный позади ската
postcranial *a* заднечерепной; затылочный
postdicrotic *a* постдикротический
postembryonic *a* постэмбриональный
posteriad *adv* назад, по направлению к задней части тела
posterior *a* задний, последующий, позднейший
posterity *s* потомство
posterodorsal *a* заднеспинной
posterolateral *a* заднелатеральный
posteromedial *a* заднемедиальный
posteroventral *a* заднебрюшной
postfixation *s* постфиксация, вторичная фиксация
postfrontal *a* заднелобный
postfurca *s* постфурка
postganglionic *a* постганглионарный, заузловой
postgena *s ent* защека
postglenoid *a* засуставной
posticous *a* задний
postjunctional *a* постсинаптический
postlarval *a* постларвальный
postmitotic *a* постмитотический
postmortal *a* посмертный
postnatal *a* послеродовой, постнатальный
postnotum *s* постскетеллум, постнотум
postorbital *a* посторбитальный, заглазничный
postotic *a* заушный
postparietal *a* заднетеменной
postpartum *a* послеродовой
postpetiole *s ent* постпетиолюс
postpharyngeal *a* заглоточный
postponement *s* откладывание; отставление подкрепления
postpubis *s* заднелобковая кость
postretinal *a* расположенный позади сетчатки
postscutellum *s* постскутеллум, постнотум
postspawning *a* посленерестовый
postsynaptic *a* постсинаптический
posttemporal *a* задневисочный
posttesticular *a* расположенный сзади семенников
posttransfusion *a* посттрансфузионный
posttrematic *a* зажаберный
posture *s eth* положение, поза
postzygapophysis *s anat* задний сочленовный отросток
pot *s* сосуд; вегетацинный сосуд; тигель
melting ~ тигель
potato *s* картофель *(Solanum tuberosum), картошка (colloquial);* **to set potatoes** сажать картошку
air ~ ямс клубненосный *(Dioscorea bulbifera)*

Jerusalem ~ земляная груша, топинамбур *(Helianthus tuberosus)*
sweet ~ батат, сладкий картофель *(Ipomoea batatas)*

potency *s* потенция; сила
biological ~ биологическая активность
immunological ~ иммуногенность
prospective ~ *emb* проспективная потенция

potent *a* сильный, сильнодействующий
potential *s* потенциал; *a* потенциальный
potentiality *s* потенциальная возможность
potentiation *s* потенцирование, потенциация
potentiometer *s* потенциометр
pothead *s ther* гринда *(Globicephala)*
potherb *s* зелень
potted *a bot* горшечный
pouch *s* мешок; карман; сумка; дивертикул; зоб; сумка; защёчный мешок
~ **of Douglas** *анат.* дугласово пространство
brood ~ выводковая камера
cheek ~ защёчный мешок
cirrus ~ сумка цирруса, мешок цирруса
egg ~ яйцевый мешок, яйцевый кокон, оотека, яйцевая капсула
hair ~ волосяной мешочек

pouch-bearing *a bot* сумконосный
pouched *a* сумчатый
poult *s* птенец
poultry *s* домашняя птица
pounce *s* коготь
pouter *s* дутыш
powder *s* порошок
powdered *a bot* припудренный
powdery *a* порошкообразный, порошковидный
power *s* сила, мощность; способность; производительность
absorbing ~ сила поглощения; абсорбционная способность, поглощающая способность
adaptation ~ приспособляемость
antigenic ~ антигенность
breeding ~ плодовитость
caloric ~ теплотворность, теплотворная способность
crop-producing ~ плодородие; продуктивность; урожайность
evaporative ~ сила испарения; испаряемость
feeding ~ питательность
germinating энергия прорастания
growing ~, growth ~ способность к росту; жизнеспособность; энергия роста
immunizing ~ иммуногенность
magnification ~ увеличение
motive ~ движущая сила, энергия
pathogenic ~ патогенность
sensitizing ~ сенсибилизирующая активность

pox *s* оспа
practicality *s* практичность, полезность
practiced *a* опытный; приобретённый на практике
praecoce *s* выводковая птица
praefloration *s* цветочное почкосложение
prairie *s* прерия; степь; луг
prairiefication *s* олуговение
prancing *s* гарцующая походка
prasoline *s* прасолин
pratal *a* луговой
pratincole *s orn* тиркушка *(Glareola)*
pratum *s* луг
preabdomen *s zool* переднебрюшие
preacetabular *a* расположенный впереди присоски; предвертлужный
preadaptation *s* преадаптация
preanal *a* пранальный
preaortic *a* расположенный перед аортой
prebloom *a* до цветения
precartilage *s* зародышевая хряшевая ткань
precasted *a* преформированный
precautionary *a* принимаемый для предосторожности; предупреждающий; предупредительный
precava *s* верхняя полая вена
prechallenged *a* предсенсибилизированный
prechordal *a* прехордальный
precinctive *a* эндемичный
precipitability *s* осаждаемость
precipitant *s* осадитель
precipitate *s* осадок; осаждать, осаждаться

precipitation *s* осаждение, преципитация; осадки; атмосферные осадки; выпадение осадков
precipitator *s* осадитель
precipitin *s* преципитин
preclimax *s* преклимакс
precocial *a* зрелорождающийся; *orn* выводковый
precocious *a* рано развившийся; ранний, скороспелый
precocity *s* преждевременное развитие, преждевременная зрелость
sexual ~ преждевременная половозрелость
precomissure *s* передняя комиссура
precuneus *s amat* предклин, предклинье
precursor *s* предшественник
vitamin ~ провитамин
precursory *a* предшествующий
predaceous *a* хищный
predation, predatism *s* хищничество
predator *s* хищник; пожиратель
predatory *a* хищный
predecessor *s* предшественник
predentary *a* предзубной
predetermination *s* предопределение
predetermine *v* заранее определять; предопределять
predetermined *a* предетерминированный
prediction *s* предсказание, предвидение, прогноз
predisposititon *s* предрасположение
predominance *s* преобладание
predominant *a* доминирующий, численно преобладающий
predominate *v* доминировать, численно преобладать
predormition *s* состояние, предшествующее сну
predorsal *a* антедорсальный
preen *v* охорашиваться; чистить перья
preening *s* прининг, уход за оперением
preexcitation *s* преждевременное возбуждение, антесистолия
preference *s* предпочтение
preferendum *s* предпочитаемые условия среды
preflight *a orn* до подъёма на крыло
prefloration *s* цветочное почкосложение
prefoliation *s* листовое почкосложение, листосложение в почках
prefontanel *s anat* большой родничок
preformation *s* преформация
prefrontal *a* предлобный
prefusion *s* фаза, предшествующая слиянию
preganglionic *a* преганглионарный
pregastrula *s* предгаструла
pregastrular *a* догаструляционный, предгаструляционный
pregeminum *s anat* передние холмы четверохолмия
pregnancy *s* беременность
pregnant *a* беременная, стельная *(cow)*, жерёбая *(mare)*, супоросная *(pig)*, суягная *(ewe)*, щенная *(dog)*
pregnenedioldional *s* альдостерон
pregnenedioldione *s* кортикостерон
pregnenedioltrione *s* кортизон
pregnenedione *s* простерон
pregnenetrioldione *s* прегнентриолдион, кортизол
pregnenoldione *s* оксипрогестерон
pregnenoltrione *s* дегидрокортикостерон
prehalter *s ent* закрыловая пластинка
preheterokinesis *s* прегетерокинез
prehyoid *a* мандибулогиоидный, предгиоидный
preimaginal *a* преимагинальный
preimmunization *s* предварительная сенсибилизация, предсенсибилизация
preincubation *s* предварительная инкубация, преинкубация
prejunctional *a* пресинаптический
prekallikrein *s* прекалликреин, фактор Флетчера
preload *s* преднагрузка
prelocalization *s* предлокализация
premature *a* недозрелый; скороспелый; недоношенный
premaxilla *s* предчелюстная кость; межчелюстная кость
premolar *s anat* премоляр; *a* премолярный, ложнокоренной
premolt *a* предлинечный
premortal *a* предсмертный
pre-mRNA *s* пре-мРНК
premunition, **premunity** *s* премуниция, инфекционный иммунитет, нестерильный иммунитет, преиммунитет
prenasal *s* предносовая кость
prenatal *a* пренатальный, внутриутробный, предродовой

preoccipital *a* предзатылочный
preocular *a* расположенный перед глазом
preoral *a* предротовой
preorbital *a* предглазничный
preparation *s* препарат; приготовление, подготовка; препарирование
preparatory *a* подготовительный, начальный
prepare *v* препарировать
preparietal *s* предтеменная кость; *a* предтеменной
prepenna *s* пуховое перо
prepharynx *s helm* предглотка
prepituitary *s* передняя доля гипофиза
prepotence, prepotency *s* препотенция
prepotent *a* препотентный
prepubis *s* предлобковая кость
prepuce *s* крайняя плоть, препуциум
prepupa *s* предкуколка
preputial *a* препуциальный
prepyloric *a* препилорический
pre-RNA *s* пре-РНК, РНК-предшественник
prescutum *s ent* предщит, прескутум, передний склерит среднеспинки
presensitization *s* предварительная сенсибилизация, предсенсибилизация
presentation *s eth* предъявление; *imm* представление, презентация, презентирование
preservation *s* сохранение; охрана; консервация
preservative *s* фиксатор, фиксирующая жидкость, консервант
preserve *s* заповедник; заказник; *v* сохранять, оберегать; фиксировать
preserved *a* фиксированный
prespermatid *s* пресперматида
presphenoid *s* пресфеноид
press *s* пресс
pressoreceptor *s* прессорецептор
pressory *a* прессорный
pressure *s* давление
  biotic ~ биотический пресс
  grazing ~ пастбищная нагрузка
  population ~ популяционный пресс, давление большой численности населения
pressure-jump *s* скачок давления
presternal *a* предгрудинный
presternum *s anat* рукоятка грудины; *zool* передняя часть грудины
presumptive *a* презумптивный
presynaptic *a* пресинаптический
pretreated *a* предварительно обработанный
pretreatment *s* предварительная обработка
prevail *v* преобладать, превалировать, господствовать
prevailing, prevalent преобладающий, широко распространённый
prevention *s* предотвращение, профилактика
preventive *a* профилактический, предупредительный
prevernal *a* ранневесенний
prevertebral *a* предпозвоночный
prevomer *s anat* предсошник
prey *s* добыча; жертва; **to fall** ~ становиться жертвой
prey-catching *s* схватывание добычи
prezygapophysis *s zool* передний сочленовный отросток
prezygotic *a* дозиготический, презиготический
prick *s* шип; колючка; укол; *v* колоть, прокалывать
prick-eared *a* имеющий остроконечные уши, со стоячими ушами
pricket *s* годовалый олень; рог молодого оленя; очиток едкий *(Sedum acre)*
prick-hedge *s* колючая изгородь
prickle *s* шип, колючка
prickly *a* шиповатый, с шипами, с колючками
prickly-seeded *a* имещий колючие семена
pride *s ther* прайд
primary *a* первоначальный, первичный; основной, главный; *s* маховое перо первого порядка
primate *s* примат *(Primates)*
  nonhuman ~s *pl* приматы кроме человека
  subhuman ~ человекообразная обезьяна *(Anthropoidea)*
prime биохим. служить затравкой
primer *s mbiol* затравка, праймер
  DNA ~ ДНК-затравка, ДНК-праймер
  RNA ~ РНК-затравка, РНК-праймер

primeval *a* первобытный, первозданный; девственный *(forest)*
priming *s imm* примирование
primiparous *a* впервые рожающая
primitive *a* первобытный, примитивный
primordial *a* примордиальный, зародышевый, зачаточный
primordium *s* зачаток; проспора, предспора, примордиальная спора; *bot* примордный
leaf ~ листовой бугорок
principle *s* правило, принцип
active ~ действующее начало
all-or-nothing ~ принцип «всё *или* ничего»
Bergmann's ~ правило Бергмана, принцип Бергмана
prink *v* чистить перья
priority *s* приоритет
prisere *s* присерия
pristine *a* первоначальный; чистый, нетронутый
privates *s pl* наружные половые органы
privet *s bot* бирючина *(Ligustrum)*
proaccelerin *s* проакцелерин
proacrosome *s* проакросома, акробласт
proamnion *s* проамнион
proangiosperm *s* предшественник покрытосеменных
proanthesis *s* раннее цветение; распускание первого цветка
probability *s* вероятность
probacteriophage *s* профаг, латентный фаг, симбиотический фаг, пробактериофаг, криптофаг
probe *s* зонд; *v* зондировать
Auro ~ *imm* аурозонд
cDNA ~ кДНК-зонд
DNA ~ ДНК-зонд
voltage ~ диагностический электрод, электрод-зонд
problem *s* задача
proboscidiform *a* хоботообразный
proboscis *s* хобот; хоботок
procambium *s* прокамбий
procarp *s bot* прокарпий
procaryotic *a* прокариотный
procedure *s* процедура, методика, приёмы проведения; операция
dilution ~ разведение
electrocution ~ умерщвление электрическим током
experimental ~ методика опыта
proceptivity *s eth* процептивность
procercoid *s* процеркоид
process *s* процесс; отросток, придаток, вырост, лопасть; *v* подвергать обработке, обрабатывать
pleural coxal ~ верхний субкоксальный мыщелок
pleural wing ~ плейральный столбик
posterior notal ~ задняя тергальная лопасть
repair ~ репарация, процесс репарации
xiphoid ~ мечевидный отросток грудины, ксифистернум
processing *s mbiol* процессинг
data ~, **information** ~ обработка информации
intrathymic ~ «обучение» в тимусе, коммитирование в тимусе
processor *s* устройство для обработки препаратов, процессор
multitip ~ многоячейковый капельный процессор
sample ~ процессор образцов
tissue ~ устройство для проводки образцов, гистопроцессор
prochorion *s* прохорион
prochromosome *s* прохромосома
procoagulant *s* прокоагулянт
procoelous *a* процельный
procoracoid *s* прокоракоид
procreate *v* порождать; производить потомство
procreation *s* воспроизведение; размножение
procrypsis *s* защитная окраска
procryptic *a* защитный *(coloration)*
proctiger *s* анальная область
proctodaeum *s emb* проктодеум, задняя кишка
procumbent *a* лежачий; стелющийся
procure *v* добывать
procurvation *s* искривление кпереди
prodromic *a* продромальный
produce *v* производить, продуцировать, вызывать
producer *s* продуцент
antibody ~ антителопродуцирующая клетка, антителопродуцент
producing *s* продуцирование; *a* продуцирующий, образующий
product *s* продукт

waste ~s *pl* отходы
production *s* продукция; продуцирование, производство; продуктивность, производительность
mass ~ крупномасштабное производство
toxin ~ токсинообразование
productional *a* продукционный
productive *a* продуктивный, производительный; урожайный; продуцирующий
productivity *s* продуктивность, производительность
proembryo *s* проэмбрион; *bot* предзародыш; предросток
proenzyme *s* профермент
proestrus *s* проэструс, предтечка
proferment *s* профермент
profile *s* профиль
pollen ~ пыльцевая диаграмма
profiling *s* анализ профиля; анализ спектра
profuse *a* обильный
progametange *s* прогаметангий
progamete *s zool* гаметоцит; *bot* прогамета
progamic *a* прогамный
progenesis *s* прогенез
progenetic *a* прогенетический
progenitor *s* прародитель, предок; предшественник; клетка-предшественник
primititve lymphoid ~ пролимфобласт
thymus-homing ~ претимоцит
progenote *s* прогенота, первичная клетка
progeny *s* потомство; потомок
progeoesthetic *a bot* отогнутый вниз
progeotropism *s* положительный геотропизм
proglottid, proglottis *s helm* проглоттида
prognathous *a* прогнатический, с выдающимися челюстями
prognosis *s* прогноз
progradation *s* проградация
program *s* программа
programming *s* составление программы; программирование
progress *s* прогресс, развитие
selection ~ прогресс при отборе; результат отбора
progression *s* продвижение, движение; последовательность; развитие
prohaptor *s helm* передний фиксаторный аппарат
prohydrotropism *s* положительный гидротропизм
projection *s* проекция; выступ; вырост
prolarva *s* предличинка
prolate *a* продолговатый
proleg *s* ложная ножка
proliferation *s* пролиферация, быстрое разрастание, быстрое размножение
clonal ~, clonogenic ~ клональная пролиферация, размножение клона
proliferous *a* пролиферирующий, быстро разрастающийся, быстро размножающийся; *bot* отпрысковый
prolific *a* плодовитый, плодородный; плодоносный; производительный
proligerous *a* пролиферирующий, быстро разрастающийся, быстро размножающийся; *bot* отпрысковый
promastigote *s prot* промастиготная форма; *a* промастиготный
promegaloblast *s* промегалобласт
promeristem *s* промеристема
prometaphase *s* прометафаза
prometatropic *a* облигатно-перекрёстно-опыляемый
prominent *a* выступающий, торчащий; выпуклый
promiscuity *s* промискуитет
promiscuous *a* беспорядочный
promontory *s anat* выступ
promorphology *s* проморфология
promote *v* стимулировать
promoter *s* стимулятор; *mbiol* промотор
promoting *s* стимуляция; активация
promunity *s* депрессионный иммунитет, промунитет
promuscis *s ent* хоботок
promycelium *s* промицелий
promyelocyte *s* промиелоцит
pronation *s* пронация
pronator *s* пронатор
prone *a* лежащий ничком, распростёртый
pronephros *s* пронефрос, предпочка
prong *s* отросток рога
pronghorn *s* вилорог, вилорогая антилопа *(Antilocapra americana)*
pronormoblast *s* пронормобласт
pronotum *s ent* пронотум, переднеспинка
pronounced *a* явный, ясно выраженный

pronucleus *s* пронуклеус
pronymph *s* пронимфа
proodophytia *s pl* начальные стадии фитоценозов
proofreading *s mbiol* исправление ошибок
propagate размножать, размножаться; распространять, распространяться; разводить
propagation *s* размножение; распространение; разведение; ~ **by cuttings** размножение черенками; ~ **by layerings** размножение отводками
~ **of errors** *biom* преобразование ошибок
artificial ~ искусственное размножение; искусственное разведение
propagative *a* размножающийся
propagator *s* культиватор *(culturing deivice)*; распространитель
shake-flask ~ *микр.* качалка, встряхиватель
yeast ~ дрожжерастительный аппарат; дрожжерастительный чан
propagule *s* побег, росток, отпрыск; соредия
proparasite *s* пропаразит
property *s* свойство
lyophilic ~ лиофильность
lyophobic ~ лиофобность
prophage *s* профаг, латентный фаг, симбиотический фаг, пробактериофаг, криптофаг
prophase *s* профаза
prophloem *s* протофлоэма
prophototropism *s* положительный фототропизм
prophyll, prophyllum *s* прицветничек; первый лист
propinquity *s* близость, родство
propleuron *s ent* проплевра, бочок переднегруди
propneustic *a ent* пропнейстический
propodium *s* проподиум
propodosoma *s* проподосома
propolis *s* прополис, пчелиный клей
propositus *s* пробанд
proprioception *s* проприорецепция, проприоцепция
proprioceptor *s* проприорецептор
propriogenic *a* проприогенный
propulsion *s* пропульсия, проталкивание вперёд; движение вперёд
propupa *s* предкуколка
propylation *s* пропилирование
proral *a ther* направленный спереди назад
proscapula *s* ключица
prosection *s* анатомирование, вскрытие
prosencephalon *s emb* передний мозговой пузырь; *anat* конечный и промежуточный мозг
prosenchyma *s* прозенхима
proserosyme *s* протромбин, тромбиноген, фактор II
prosoma *s* просома
prosome *s mbiol* просома
prosoplectenchyma *s* прозоплектенхима
prosopyle *s* прозопиль
prosorus *s* просорус
prosperous *a* процветающий
prospore *s* предспора, проспора
prostate *s* простата, предстательная железа
prostecobacteria *s pl* простекобактерии
prosternum *s ent* простернум, переднегрудь
prosthomere *a* преоральный сомит, предротовой сомит
prostomium *s* простомиум, предротовая лопасть
prostrate *a* лежащий ничком, распростёртый; обессиленный; измождённый; *bot* стелющийся
prostration *s* распростёртое положение; прострация
protandrism *s* протандрия
protandrous *a* протандрический
protandry *s* протандрия
proteantigen *s* белковый антиген
protect *v* защищать, предохранять
protection *s* защита, предохранение, охрана; иммунитет
bank ~ берегоукрепление
cross ~ перекрёстный иммунитет
environmental ~ защита окружающей среды
forest ~ лесозащита
nature ~ охрана природы
protective *a* защитный, предохранительный; покровительственный *(coloration)*
proteid, proteide *s* протеид

protein *s* белок; *a* белковый
carrier ~ белок-носитель, белок-переносчик
channel-forming ~, cytolytic pore-forming каналообразующий белок, пороформирующий белок, порфорин
defensive ~ антитело
differentiation-inducing ~ белковый индуктор дифференцировки
full-length ~ первичный продукт трансляции, непроцессированный белок
high ~ высокое содержание белка
red ~ лактоферрин, лактотрансферрин, лактосидерофилин, красный белок молока
serum ~ сывороточный белок, серопротеин
transacting ~ трансактивирующий белок, белок-трансактиватор
vegetable ~ растительный белок, глютен
whole-cell-lysate ~ клеточный белковый экстракт
proteinaceous *a* белковый
protein-binding *s* связывание с белком
protein-free *a* безбелковый
proteolysis *s* протеолиз, расщепление белка
proteolytic *a* протеолитический, расщепляющий белок
proteometabolism *s* белковый обмен
proteose *s* протеоза
proteose-peptone *s* протеозопептон
proter *s prot* передняя дочерняя особь
proterandrous *a* протандрический
proterandry *s* протандрия
proterogynous *a* протерогинный, протогинный
proterogyny *s* протогиния, протерогиния
proterosoma *s* протеросома
prothallial *a* заростковый
prothallium, **prothallus** *s* проталлиум, заросток
prothecium *s* протециум, примитивный перитеций
prothorax *s ent* проторакс, переднегрудь, первый грудной сегмент
prothrombin *s* протромбин, тромбиноген, фактор II
prothrombinase *s* протромбиназа, фактор V
prothrombinogen *s* антифибринолизин, проконвертин, фактор VII
prothrombokinase *s* антигемофильный фактор, антигемофильный глобулин, тромбопластический фактор плазмы А, фактор VIII
prothymocyte *s* протимоцит, претимусный Т-лимфоцит, про-Т-лимфоцит
protist *s* протист, одноклеточный организм
protistology *s* протистология
protobasidium *s* протобазидия
protobiont *s* протобионт
protoblast *s* протобласт
protocercal *a* протоцеркальный
protocerebrum *s* протоцеребрум
protochlorophyll *s* протохлорофилл
protocneme *s* протокнема
protoconch *s* протоконх
protocone *s* протокон
protocooperation *s* протокооперация
protocorm *s bot* протокорм
protoepiphyte *s* протоэпифит
protoerythrocyte *s* эритробласт
protofibril *s* протофибрилла
protogala *s* молозиво, колострум
protogene *s* протоген, доминантный аллель
protogenesis *s* протогенез, абиогенез
protogynous *a* протерогинный, протогинный
protogyny *s* протогиния, протерогиния
protheme *s* протогем
protokaryon *s* протокарион
protomer *s* протомер
protomerite *s* протомерит
protomitosis *s* промитоз, протомитоз
protomorphic *a* примитивный
protonation *s* протонирование
protonema *s* протонема
protonephridium *s* протонефридий
protonophore *s* протонофор
protonymph *s* протонимфа
protooncogene *s* протоонкоген
protopathic *a* протопатический
protophloem *s* протофлоэма
protophyte *s* одноклеточное растение
protoplasm *s* протоплазма
protoplasmatic, protoplasmic *a* протоплазматический
protoplast *s* протопласт

protopodite *s* протоподит
protorhabdion *s helm* ротовая капсула
protosource *s* первоисточник
protospore *s* протоспора
protostele *s* протостела
protostoma *s* первичный рот, бластопор
prototroch *s* прототрох
prototroph *s* прототроф
prototrophic *a* прототрофный
prototype *s* прототип
protoxylem *s* протоксилема
protozoan *s* простейшее животное; *a* протозойный
protozoology *s* протозоология
protozoon *s* простейшее животное
protozygote *s* протозигота
protractile *a* выдвижной
protractor *s* протрактор, вытягивающий мускул
protrude высовывать; выступать
protrudence *s* выступ
protrudent *a* выступающий; торчащий
protrusion *s* высовывание, выпячивание; выступ
protuberance *s* бугор, бугорок, возвышение
provenance *s* происхождение; место происхождения
proventriculus *s* преджелудок, *orn* железистый желудок
province *s* провинция
provinculum *s mal* первичный замок
provine *s* побег, отводок; разводить отводками
provirus *s* провирус
provisional *a* провизорный
provitamin *s* провитамин
provocative *s* раздражитель, возбудитель; *a* возбуждающий; побуждающий; провоцирующий
provoke *v* вызывать, возбуждать; побуждать; провоцировать
prowl *v* красться; бродить; рыскать
proximal *a* проксимальный
proximaloaperturate *a pal* проксимально-апертурный
proximity *s* eth близость
pruinose *a* покрытый беловатым налётом
prune *v* обрезать, подрезать; слива домашняя *(Prunus domestica)*
pruning *s* обрезка, подрезка
natural ~ самоочищение
psalterium *s anat* книжка; псалтырь, лира Давида
psammon *s* псаммон
psammophile *s* псаммофильный организм
psammophilous *a* псаммофильный, песколюбивый
psammophyte *s* псаммофит, растение песков
psammosere *s* псаммофильная серия
pseudannual *s* псевдооднолетник
pseudaposematic *a* псевдоапосематический
pseudapospory *s* псевдоапоспория
pseudaxis *s* ложная ось
pseudepisematic *a* псевдоэписематический, имеющий привлекающую окраску
pseudimago *s* субимаго
pseudoallele *s* псевдоаллель
pseudoallelism *s* псевдоаллелизм
pseudoapogamy *s* псевдоапогамия
pseudobiting *s eth* игровой укус
pseudobivalent *s* псевдобивалент
pseudobrachium *s* рукообразный плавник
pseudobranch *s* псевдобранхия, спиракулярная жабра
pseudobulb *s* ложная луковица
pseudocardinal *s zool* ложнокардинальный зуб
pseudocarp *s* ложный плод
pseudochiasma *s* псевдохиазма
pseudocilium *s bot* ложная ресничка
pseudoclimax *s* псевдоклимакс
pseudocoel *s* псевдоцелом
pseudoconch *s herp* ложная носовая раковина
pseudocone *s ent* псевдоконичечский глаз; *a* псевдоконический
pseudocopulation *s* амплексус
pseudocostate *a* имеющие ложное жилкование
pseudocyst *s prot* псевдоциста
pseudoderm *s* псевдодерма
pseudodominance *s* псевдодоминирование
pseudodont *a* имеющий роговые зубы
pseudoepiphyte *s* ложный эпифит
pseudoergate *s ent* псевдоэргат, ложный рабочий

pseudofoliaceous *a* имеющий листовидный вырост
pseudogamy *s* псевдогамия, ложногамия
pseudogene *s* псевдоген
pseudoglobulin *s* псевдоглобулин
pseudohaploid *s* псевдогаплоид
pseudoheart *s* ложное сердце, сократимый кровеносный сосуд *(in annelids);* осевой орган *(in echinoderms)*
pseudohermaphroditism *s* псевдогемафродитизм
pseudolabium *s helm* ложная губа
pseudolamina *s bot* ложная пластинка
pseudoloculus *s* псевдолокула *(in diatoms)*
pseudolysogeny *s* псевдолизогения
pseudomitotic *a* псевдомитотический
pseudomixis *s* псевдомиксис, псевдогамия
pseudomonade *s* псевдомонада
pseudomonocarpous *a* ложнооднoплодный
pseudomonocotyledonous *a* ложнооднодольный
pseudonodulus *s* ложный узелок *(in diatoms)*
pseudonotum *s ent* ложноспинка, задняя часть щитка
pseudonychium *s ent* аролиум, эмболиум, лопасть между коготками
pseudonymph *s* псевдонимфа, полукуколка
pseudoocellus *s ent* ложный глазок; ложный узелок *(in diatoms)*
pseudoparenchyma *s* псевдопаренхима
pseudoparthenogenesis *s* псевдопартеногенез
pseudoperianth *s* ложный околоцветник
pseudoperidium *s* псевдоперидий
pseudoplaque *s* «ложная» бляшка, ложноположительное пятно гемолиза
pseudoplasmodium *s* псевдоплазмодий
pseudopod, pseudopodium *s* псевдоподия, ложноножка
pseudopore *s* ложная пора
pseudopupa *s* ложнокуколка, пупарий
pseudoramulus *s* ложная ветвь
pseudoraphe *s* ложный шов
pseudorat *s* ложная мышь *(Pseudomys)*
pseudoscolex *s* ложный сколекс, псевдосколекс
pseudosematic *a* псевдосематический *(coloration)*
pseudoseptum *s bot* перфорированная перегородка, перегородка с порами; ложная септа *(in diatoms)*
pseudosessile *a ent* ложносидячебрюхий
pseudoshrub *s* ложный кустарник
pseudosleeping *s orn* ложный сон
pseudosperm *s* ложное семя; ложный плодолистик
pseudospore *s* телиоспора
pseudostele *s* ложная стель
pseudostipula *s* ложный прилистник
pseudosucker *s* ложная присоска
pseudothallus *s* ложное слоевище
pseudotrachea *s* псевдотрахея, ложная трахея
pseudovitellus *s* ложный желток
pseudovum *s* партеногенетическое яйцо
psuedoworker *s см.* **pseudoergate**
pseudoxerophilous *a* ложноксерофильный
psilad *s* растение, свойственное прерии
psilate *a pal* гладкий
psilicolous *a* обитающий в прерии
psilium *s* сообщество прерии
psilolophate *a* гладкогребенчатый
psilotegillate *a* гладкопокровный
P-site *s mbiol* Р-сайт, Р-участок
psychics *s* психика
psychologic, psychological *a* психологический
psychology *s* психология
psychophysiology *s* психофизиология
psychrometer *s* психрометр
psychrophile *s* психрофильный организм, холодолюбивый организм
psychrophilic *a* психрофильный, холодолюбивый
psychrophyte *s* психрофит
ptarmigan *s* куропатка *(Lagopus)*
ptenophyllium *s* сообщество лиственных лесов
pteridography, pteridology птеридология
pteridophyte *s* птеридофит, папоротникообразное растение
pterocarpous *a* крылоплодный
pterodium *s* крылатка, крылатый плод
pterogate *s ent* птерогат
pteroid *a* крыловидный; папоротниковидный

pteropleurite *s ent* птероплейрит, боковой склерит груди
pteropleuron *s ent* птероплевра, анэпимера среднегрудки
pteropodial *a* крылоногий
pterospermous *a* крылосемянный
pterospore *s* растение, размножающееся крылатками
pterostigma *s ent* птеростигма, крыловой глазок
pterotheca *s ent* покров крыльев *(in a pupa)*
pterygium *s* птеригиум
pterygode, pterygodum *s ent* тегула, надкрыловая пластинка
pterygoid *s* крыловидная кость
pterygophore *s* птеригофор
pterygopodium *s* птеригоподий, миксоптеригий
pteryla *s* птерилия
ptilinum *s ent* лобный пузырь
ptiloerection *s* подъём шерсти дыбом; *orn* распушение
ptyxis *s* листосложение
puberal, pubertal *a* половозрелый
puberty *s* половая зрелость
puberulent *a* короткоопушённый
pubes *s* лонная область
pubescence *s* половое созревание; опушение; опушённость
pubescent *a* половозрелый; опушённый; пушистый
pubic *a* лонный, лобковый
pubis *s* лонная кость, лобковая кость
puddle *s* лужа
pudu *s* олень пуду *(Pudu)*
puff *s* пуф, вздутие *(хромосомы)*
puffball *s* дождевик *(Lycoperdon)*; одуванчик лекарственный *(Taraxacum officinale)*; кардиоспермум *(Cardiospermum)*
puffin *s orn* тупик *(Fratercula)*
  horn-billed ~ тупик-носорог *(Cerorhinca monocerata)*
  tufted ~ топорик *(Lunda cirrhata)*
puffing *s* пуфинг, образование пуфов; *bot* выбрасывание спор
pullulate *v* прорастать; почковаться; быстро размножаться
pullulation *s* прорастание; почкование; быстрое размножение
pullet *s* молодка
pullus *s* птенец
pulmogastric *a* лёгочно-желудочный
pulmonary *a* лёгочный
pulmonate *a* имеющий лёгкие
pulp *s* пульпа; мякоть
pulpy *a* мясистый
pulsation *s* пульсация
pulsative, **pulsatory** *a* пульсирующий
pulse *s* пульс; импульс; бобовое растение; боб; *v* пульсировать
  pressure ~ пульсовые колебания давления
pulse-labeled *a* импульсно-меченый
pulverulent *a* пыльный, мучнистый
pulvillar *a ent* пульвиллярный
pulvillus *s ent* пульвилла, подушечка
pulvinar *a* подушечковидный
pulvinate *a* подушечковидный
pulvinulus *s ent* пульвилла, подушечка
pulvinus *s* листовая подушечка
puma *s* пума, ягуар *(Felis concolor)*
pump *s* насос
pumpkin *s* тыква обыкновенная *(Cucurbita pepo)*
  guinea ~ баклажан *(Solanum melongena)*
punctate *s* пунктат; пятнистый; точечный
punctated *a* пятнистый; точечный
punctitegillate *a* перфорированно-покровный
punctulate, punctulated *a* мелкоточечный; мелкоямчатый
punctum *s* точка; верхушка
puncture *s* пункция; прокол; укус
pungency *s* жгучесть *(напр. перца)*
pungent *a* острый; жгучий *(e.g., taste)*
punkies *s pl ent* мокрецы *(Heleidae)*
pup *s* щенок; молодой тюлень; *v* щениться
  gray ~, silvery ~ котик-однолеток
pupa *s* куколка
pupal *a* пупальный, куколочный
puparium *s* пупарий, ложный кокон
pupate *v* окукливаться
pupation *s* окукливание, образование куколки; *phytop* закукливание
pupigenous *a* куклородный
pupigerous *a* содержащий куколку
pupil *s* зрачок
pupillary *a* зрачковый, относящийся к зрачку

pupiparous *a* куклородный
puppy *s* щенок
 mud ~ протей *(Necturus)*; скрытожаберник *(*Cryptobranchus*)*; амбистома *(Ambystoma)*
pure *a* чистый, беспримесный
pure-blooded *a* чистокровный
purebred *s* чистопородное животное; *a* чистопородный
pure-breeding *s* чистопородное разведение
purification *s* очищение, очистка
 natural ~ самоочищение
purifier *s* очиститель
 centrifugal ~ очищающая центрифуга
 water ~ водоочистител
purine *s* пурин
purple *s* пурпур
purposiveness *s* целенаправленность
purr *s* мурлыкание; *v* мурлыкать
purret *s* лук-порей *(Allium porrum)*
PURR-PUSS *s* «кот-мурлыка»
pursue *v* преследовать, гнаться
pursuer *s* преследователь
purulent *a* гнойный
pus *s* гной
push-up *s* кормовой столик
pussycat *s* клевер пашенный *(Trifolium arvense)*
pussytoe *s bot* кошачья лапка *(Antennaria)*
pustule *s* пустула, прыщ; пузыревидное выпячивание; *bot* подушечка, споровая масса
 rust ~ телейтопустула
putative *a* предполагаемый
putrefaction *s* гниение
putrefactive *a* вызывающий гниение
putrefy *v* гнить
putrescence *s* гниение
putrescent *a* гниющий; гнилостный
putrid *a* гнилой; гнилостный
pycnidiophore *a* несущий пикниды
pycnidium, pycnium *s* пикнида, кувшинообразное вместилище
pycnoconidium *s* пикноконидия
pycnogonidium *s* пикногонидия
pycnometer *s* пикнометр
pycnosis *s* пикноз
pycnostachous *a* густоколосый
pycnotic *a* пикнотичный
pygidium *s* пигидий, анальный сегмент
pygmaean, pygmean *a* карликовый
pygmy *s* пигмей, карлик; карликовый
pygmy-owl *s* воробьиный сыч, сыч-воробей *(Glaucidium)*
pygochord *s* пигохорд
pygostyle *s* пигостиль
pyknic *a* пикнический
pylome *s* пилом
pylorus *s* пилорус
pyogenic *a* гноеродный
pyramid *s* пирамида
 Pascal ~ треугольник Паскаля
pyrene *s bot* косточка, семя
pyrenocarp *s* пиренокарпий
pyrenoid *s* пиреноид
pyrenophore *s* пиренофор
pyridoxine *s* пиридоксин
pyriform *a* грушевидный, грушеобразный
pyrimidine *s* пиримидин
pyrium *s* пирогенная сукцессия
pyrogenic *a* пирогенный
pyrogram *s* пирограмма
pyrophosphate *s* пирофосфат
 adenosine ~ аденозинпирофосфат
pyrophosphorylation *s* пирофосфорилирование
pyrophyte *s* пирофит
python *s* питон; питон *(Python)*
pyxidate *a bot* кувшиновидный
pyxidium *s bot* крыночка
pyxis *s bot* крыночка

Q

Q-disk *s* Q-диск, анизотропный диск
quack *s* кряканье; *v* крякать
quackgrass *s bot* пырей ползучий *(Agropyron repens)*
quader *s anat* предклинье
quadrate *s* квадрат
  permanent ~ постоянная пробная площадка, постоянная учётная площадка
quadrennial *a* четырёхлетний
quadrialate *a bot* четырёхкрылый
quadricorn *a bot* четырёхрогий
quadrifarious *a bot* четырёхрядный
quadrifid *a* четырёхраздельный
quadriflorous *a bot* четырёхцветковый
quadrifoliate *a bot* четырёхлистный
quadrijugate *a bot* четырёхпарный
quadrilobate *a* четырёхлопастный; четырёхдольчатый
quadrinomial *s* четверное название; *a* квадриномиальный
quadripartite *bot* четырёхраздельный
quadriphyllous *bot* четырёхлистный
quadriplex *s* квадриплекс
quadriradiare *a* четырёхлучевой
quadrivalent *s cyt* квадривалент; *a* квадривалентный
quadrivalve, quadrivalvular *a* четырёхстворчатый
quadrumanous *a zool* четверорукий
quadruped *s* четвероногое животное; *a* четвероногий
quadrupedal *a* четвероногий
quag *s* болото; трясина
quagga *s ther* квагга *(Equus quagga)*
quaggy *a* рыхлый, дряблый; отвислый
quagmire *s* сплавина; зыбун; трясина; болото
quail *s* перепел *(Coturnix)*
  button ~ трёхперстка *(Turnix)*
qualitation *s* качественное определение
qualitative *a* качественный
quality *s* качество
  ~ **of locality** бонитет
  adequate ~ доброкачественность
  market ~ товарный вид
quantification, quantitation *s* количественное определение, количественный анализ; подсчёт
quantitative *a* количественный
quantity *s* количество; величина
quantosome *s* квантисома
quarantine *s* карантин; подвергать карантину
quarter *s* место; четверть
  animal ~s виварий
  winter ~s места зимовки
quartet *s* квартет, тетрада
quasielastic квазиупругий
quasilinkage *s gen* ложное сцепление
Quaternary *s* четвертичный период
quaternate *a bot* расположенный четвёрками
quaternization *s* кватернизация
quebracho *s* квебраховое дерево, квебрахо *(Schinopsis quebracho-colorado)*
queen *s ent* матка, царица, плодущая самка
queenless *a* безматочный
queenright *a apic* имеющий матку
quench тушить, гасить; блокировать, экранировать *(незанятые сайты связывания)*
quenched *a* угасший
quenching *s* тушение, гашение
  fluorescence ~ тушение флуоресценции, гашение флуоресценции
quest *v* искать пищу
quiescence *s* покой, состояние покоя
quiescent *a* покоящийся
quiet *a* спокойный; неподвижный
quill *s* перо; стержень пера; очин пера; игла *(of porcupine)*
quinate *a bot* расположенный пятёрками
quince *s* айва *(Cydonia)*
quinine *s* хинное дерево *(Chinchona);* хинин
quinquecostate *a* пятиребристый
quinquecuspidate *a* пятивершинный
quinquedentate *a* пятизубчатый
quinquefarious *a* пятилучевой; пятирядный
quinquefid *a* пятираздельный
quinqueflorous *a* пятицветковый
quinquefoil *s bot* лапчатка *(Potentilla)*
quinquefoliate *a* пятилистный

quinquelocular *a bot* пятигнёздный
quinqueserial *a* пятирядный
quinquevalve, quinquevalvular *a* пятистворчатый
quitch *s bot* трава с ползучими корневищами; свинорой *(Cynodon);* вероника виргинская *(Veronica virginiana);* вероника трубкоцветная *(Veronica tubiflora)*
quitches *s pl* корневищные злаки
quiver *s* дрожь; *v* дрожать; подрагивать
quivering *s* дрожание; вибрация ног
quotient *s* коэффициент; индекс; показатель; часть, доля

R

rabbit *s* кролик; кролик *(Oryctolagus)*
rock ~ капский даман *(Procavia capensis);* пищуха, сеноставка *(Ochotona)*
rabbitry *s* крольчатник
rabies *s* бешенство
raccon *s* енот *(Procyon)*
race *s* раса; порода; русло; канал; мчаться; бросаться
half ~ полураса; раса, характеризующаяся гетерозиготностью
serologic ~ серологический штамм
racemate *s* рацемат
racemation, raceme *s* кисть; гроздь
racemiferous *a* кистеносный
racemization *s* рацемизация
racemose *a* кистевидный, кистеобразный
racemule *s* небольшая кисть *или* гроздь
racer *s* полоз; полоз *(Coluber)*
racerunner *s* ящурка, степная ящерица *(Eremias);* ящерица-бегун, хлыстохвостая ящерица *(Cnemidophorus)*
rachial, rachidial *a* осевой; позвоночный; спинальный, спинномозговой
rachilla *s* вторичная ось соцветия, ось колоска
rachis *s* ось; стержень; позвоночный столб, позвоночник; ось сложного колоса; главная ось цветоносного побега
rachislike *a* стержневидный
rack *s* подставка, штатив, держатель, адаптор, стеллаж; *v* сливать с осадка, декантировать; сцеживать
racoon *s* енот *(Procyon)*
racy *a* чистокровный, породистый
radial *a anat* лучевой, радиальный
radiance *s* излучение
radiate *a* радиально-симметричный; *v* излучать
radiate-veined *a* радиально-жилковатый
radiatiform *a* имеющий краевые цветки
radiating *a* лучеобразно расходящийся
radiation *s* радиация, излучение; эволюционная радиация
background ~ фоновое излучение, радиационный фон
photosynthetic active ~ фотосинтетически активное излучение, фотосинтетически активная радиация
radiation-exposed *a* облучённый
radiation-induced, radiation-initiated *a* индуцированный облучением, вызванный облучением
radical *s* радикал; радикальный; *a bot* прикорневой
radicant *a* укореняющийся от стебля
radicantium *s* сидячий организм
radication *s* укоренение
radicel *s* первичный корешок; зародышевый корень
radicicolous *a* живущий на корнях
radiciflorous *a* корнецветковый
radicivorous *a* корнеядный
radicle *s anat* корешок; *bot* первичный корешок; зародышевый корень
radicose *a* имеющий большие корни
radicular *a* корешковый
radicule *s anat* корешок; *bot* первичный корешок; зародышевый корень
radiculodium *s* корневой узелок
radioactivation *s* радиоактивация
radioactivity *s* радиоактивность
radioassay *s* радиоизотопный анализ, радиоанализ
radioautograph *s* радиоавтограф
radioautography *s* радиоавтография, авторадиография
radiobiology *s* радиобиология
radiocarbon *s* радиоактивный углерод, радиоуглерод
radiocarpal *a* лучезапястный
radiochromatography *s* радиохроматография
radioecology *s* радиоэкология
radiogenetics *s* радиационная генетика
radiograph *s* рентгенограмма
radiography *s* рентгенография, радиография
radioimmunity *s* радиорезистентность, радиационная устойчивость
radioimmunoassay *s* радиоиммуноанализ
radioimmunochromatography *s* радиоиммунохроматография
radioimmunodetection *s* радиоиммуноанализ

radioimmunoelectrophoresis *s* радиоиммуноэлектрофорез
radioimmunology *s* радиоиммунология
radioimmunoprecipitation *s* радиоиммунопреципитация
radioisotope *s* радиоизотоп, радиоактивный изотоп
radiolabeled, radiolabelled *a* меченный радиоактивным изотопом
radiolarian *prot* радиолярия, лучевик *(Radiolaria)*
radiology *s* радиология
radiolysis *s* радиолиз
radiometer *s* радиометр
radiomimetic *a* радиомиметический
radiomorphose *s* радиоморфоз
radionuclide *s* радиоизотоп, радиоактивный изотоп
radioparent *a* лучепроницаемый, рентгенопрозрачный
radiopolymerization *s* радиополимеризация
radioprotection *s* противолучевая защита
radioreaction *s* лучевая реакция
radioreceptor *s* радиорецептор, рецептор, воспринимающий излучение
radioresistance *s* радиорезистентность, радиоустойчивость
radioscopy *s* рентгеноскопия
radioselection *s* радиоселекция, радиационная селекция
radiosensibility *s* радиочувствительность
radiosensitive *a* радиочувствительный
radiosensitivity *s* радиочувствительность
radiosensitization *s* радиосенсибилизация
radiosymmetrical *a* радиально-симметричный
radiotelemetry *s* радиотелеметрия
radiotracking *s* радиопрослеживание
radioulnar *a* лучелоктевой
radish *s* редька *(Raphanus)*
  small ~ редис *(Raphanus sativus radiculata)*
radius *s* лучевая кость; *ent* радиальная жилка
radix *s* корень
radula *s zool* тёрка, радула
rafting *s* лесосплав
rage *s* ярость, гнев; *v* неистовствовать, буйствовать
ragweed *s* амброзия; амброзия высокая *(Ambrosia elatior);* крестовник-желтуха, крестовник Якова *(Senecio jacobaea)*
ragworm *s* нереида*(Nereidae)*
rail *s orn* пастушок *(Rallus)*
rain *s* дождь
rainless *a* засушливый
raise *v* выращивать; разводить; выводить
raised *a* выпуклый *(e.g., of a bacterial colony)*
raisin *s* калина, гордовина *(Viburnum)*
raising *s* выращивание; разведение; выведение
  horse ~ коневодство
  plant ~ растениеводство
  poultry ~ птицеводство
  sheep ~ овцеводство
rale *s* хрип
ram *s* баран
ramate *a* ветвистый
ramble *a bot* ползучий, вьющийся; ползти, *v* виться
Rambouillet *s* рамбулье
rame *s* ветвь; ответвление
ramellose *a* мелковетвистый
ramen *s bot* рамень
ramentaceous *a bot* покрытый шелухой *или* чешуйками
ramentum *s* шелушинка, чешуйка
ramibaculate *a* разветвлённо-столбчатый
ramicercus *s helm* рамицерк
ramicolous *a* живущий на ветвях
ramicorn *a* имеющий разветвлённые рога; *ent* имеющий разветвлённые антенны
ramification *s* разветвление; ответвление, ветвь; отросток
ramified *a* разветвлённый
  widely ~ разлапистый *(of trees)*
ramiflorous *a* с цветками на ветвях
ramify *v* ветвиться; разветвляться; давать отростки
ramiparous *a* образующий ветви; разветвляющийся
ramolissement *s* размягчение
ramose *a* разветвлённый; ветвистый
ramule *s* маленькая ветвь, веточка; веточка второго порядка
ramulose, ramulous *a* мелковетвистый; ветвящийся

ramulus *s* маленькая ветвь, веточка; веточка второго порядка
ramus *s* ветвь; ответвление
rancid *a* прогорклый
random *a* случайный; произвольный; беспорядочный; безвыборный
randomization *s* рандомизация, метод случайного отбора
range *s* ряд; сфера, зона, область; размах; диапазон, интервал; *biom* вариационный размах; ареал; область распространения; район обитания; пастбище
~ **of humidity** предел колебания влажности
adjacent ~s *pl* соприкасающиеся ареалы
alpine ~ альпийское пастбище
antibacterial ~ антибактериальный спектр действия
confidence ~ *biom* доверительный интервал
daily ~ суточная амплитуда, суточные колебания
home ~ участок обитания, индивидуальный участок
host ~ круг хозяев
mean ~ *biom* размах, полусумма крайних значений
overgrazed ~ выбитое пастбище
variation ~ диапазон изменчивости
rank *s* ранг, разряд; таксономический ранг, таксономический уровень; *v* распределять по рангу
ranking *s* ранжирование
highest ~ занимающий доминирующее положение
lower ~ занимающий подчинённое положение
rapacious *a* хищный; прожорливый
rape *s* капуста *(Brassica)*
raphe *s anat* шов; *bot* семенной шов; семенной рубчик; шов *(in diatoms)*
abdominal ~ белая линия живота
raphide *s* рафида
raphidiferous *a* содержащий рафиды
rapid *a* скорый, быстрый; скоротечный
rapid-growing *a* быстрорастущий
rapport *s* взаимопонимание
rash *s* сыпь
rasorial *a* разгребающий *(e.g., bird)*
raspberry *s* малина *(Rubus);* ежевика *(Rubus)*
Arctic ~ княженика *(Rubus arcticus)*
mountain ~ морошка приземистая *(Rubus chamaemorus)*
red ~ малина обыкновенная *(Rubus idaeus)*
raspberry-yard *s* малинник
rat *s* крыса; крыса *(Rattus)*
brown ~ серая крыса, пасюк *(Rattus norvegicus)*
common ~ серая крыса, пасюк *(Rattus norvegicus)*
infant ~ крысёнок
mole ~ слепыш *(Spalax); pl* землекоповые *(Bathyergidae);* цокор *(Myospalax)*
Norway ~ серая крыса, пасюк *(Rattus norvegicus)*
weanling ~ крысёнок-отъёмыш
rat-borne *a* распространяемый крысами
rate *s* скорость; темп; интенсивность; степень; коэффициент; пропорция; норма; расценка; частота; *v* оценивать
~ **of application** норма применения
~ **of evolution** темп эволюционного процесса, скорость эволюции
~ **of exploitation** интенсивность эксплуатации; норма эксплуатации
~ **of fertilization** скорость оплодотворения; степень оплодотворения, оплодотворяемость
~ **of mortality** показатель смертности, смертность
additive growth ~ аддитивная скорость роста
age-specific death ~ возрастная смертность
birth ~ рождаемость
conception ~ степень оплодотворения, оплодотворяемость
death ~ показатель смертности, смертность
egg-laying ~ яйценоскость
emergence ~ энергия прорастания
expiratory ~ скорость выдоха
gross reproductive ~ валовая рождаемость
growth ~ скорость роста, темп роста
infection ~ заражённость
mutation ~ частота мутаций

parasite ~ паразитарный индекс
recruitment ~ восполняемость
reproductive ~ темп размножения
specific growth ~ удельная скорость роста
stocking ~ плотность посадки
survival, survivorship ~ выживаемость

rathe *a* рано цветущий; рано созревающий; утренний, ранний

rating *s eth* занимаемый ранг
hazard ~ шкала опасности

ratio *s* отношение; соотношение; степень; коэффициент
coding ~ кодовое число
food conversion ~ оплата корма
survival ~ выживание

ration *s* рацион

ratite *a orn* бескилевой

ratoon *s* корневой отпрыск, отводок; детка; *v* давать новые отпрыски, отрастать

rattan *s* ротанговая пальма *(Calamus)*

rattle *s* дробь; стук; треск
tail ~ биение хвостом

rattler, rattlesnake *s* настоящий гремучник *(Crotalus)*

rat-trap *s* крысоловка

raucous *a* хриплый; резкий, пронзительный

raven *s* ворон *(Corvus corax); pl* врановые *(Corvidae)*

ravenous *a* хищный; прожорливый

raw *a* сырой; необработанный; чувствительный *(e.g., skin); s* ссадина; больное место; *v* сдирать кожу

ray *s* луч; *v* расходиться лучами; излучать; подвергать действию лучей, облучать
infrared ~s *pl* инфракрасные лучи, ИК-лучи
ultraviolet ~s *pl* ультрафиолетовые лучи, УФ-лучи

razorbill *s orn* гагарка *(Alca torda);* водорез *(Rhynchops)*

reabsorption *s* реабсорбция, обратное всасывание

reach *s* радиус действия; колено реки, плёс

react *v* реагировать; влиять, действовать; взаимодействовать

reactant *s* действующее вещество; реагент

reaction *s* реакция; воздействие; влияние
coupling ~ реакция связывания; *bioch* сочетание
delayed ~ задержанная реакция; запаздывающий рефлекс
exchange ~ *bioch* замещение
Gram ~ метод окраски бактерий по Граму
paraserum ~ реакция ложной агглютинации, параагглютинация
"recollection" ~ анамнестическая реакция, вторичный иммунный ответ, бустер-эффект
sexual ~ половая реакция; половая активность
soil ~ реакция почвы, pH почвы

reactivate реактивировать

reactivation *s* реактивация
UV- ~ УФ-реактивация

reactivity *s* реактивность
immune ~ иммунореактивность

reactor *s* реактор; ферментёр; тест-животное

read *s* сычуг

readout *s* считывание

reader *s* аппарат для прочтения планшетов, планшет-ридер

reading *s* считывание; отсчёт, показание; *pl* показатели, данные; *a* отсчётный
genetic code ~ считывание генетического кода, транскрипция

reafference *s* реафферентация

reafforestation *s* лесовосстановление

reagent *s* реактив; реагент

reap *v* убирать урожай, жать

rear *v* разводить, выращивать, выводить; *a* задний

rearing *s* разведение; выращивание: выведение; вертикальная стойка, подъём на задние лапы

rearrangement *s* перегруппировка; перестановка; перестройка; *mbiol* реаранжировка

rebalancing *s* восстановление равновесия

recapitulation *s* рекапитуляция

recapture, recatch *s* повторная поимка, повторная добыча
ratio ~ относительная попадаемость

recent *a* рецентный, ныне живущий

receptacle *s* ложе; цветоложе, ложе соцветия; цветоножка; рецептакула,

расширенный конец таллома; терминальный диск *(in mosses); anat* вместилище; приёмник
floral ~ цветоложе
proboscis ~ *helm* хоботковое влагалище
seminal ~ семяприёмник, сперматека
receptive *a* восприимчивый, рецептивный
receptivity *s* восприимчивость, рецептивность; поглощательная способность
sexual ~ половая восприимчивость, половая рецептивность
receptor *s* рецептор; датчик
adrenergic ~ адренорецептор, адренергический рецептор
cholinergic ~ холинорецептор, холинергический рецептор
distance ~ дистанционный рецептор, дистантный рецептор, телерецептор
gravity ~ гравирецептор
homing ~ «хоминг»-рецептор
pressure ~ прессорецептор, барорецептор
stretch ~ тензорецептор, рецептор растяжения
receptor-mediated *a* опосредуемый рецептором
receptosome *s* рецепторосома
recess *s anat* впадина, ямка, полость
cochlear ~ *anat* улитковая впадина
costodaphragmatic ~ рёберно-диафрагмальный синус, рёберно-диафрагмальное углубление
epitympanic ~ надбарабанное углубление, надбарабанное пространство
jugular ~ яремная ямка
optic ~ зрительная ямка, зрительное углубление
pineal ~ шишковидное углубление, шишковидный карман
suprapineal ~ надшишковидное угубление, надшишковидный карман
recessive *s* рецессив; рецессивный
recipient *s* реципиент; *биогеогр.* колонизуемый
reciprocal *a* реципрокный; взаимный, обоюдный; обратный; *s* обратная величина
reciprocity *s* взаимная полезность
reclamation *s* мелиорация; освоение
sylvicultural ~ агролесомелиорация
reclinate, reclinated *a* склонённый вниз; отклонённый назад
reclining *a* наклонный, неперпендикулярный
recognition *s* распознавание; узнавание
recolonization *s* повторное заселение, повторная колонизация
recombinant *s* рекомбинант; рекомбинантный
recombinase *s* рекомбиназа
recombination *s* рекомбинация
~ **by breakage and reunion** рекомбинация путём разрыва и воссоединения
DNA ~ ДНК-рекомбинация
interchromosomal ~ межхромосомная рекомбинация, межхромосомная перестройка
recon *s* рекон
reconstitution *s* восстановление, воссоздание
reconstruction *s* реконструкция
recontamination *s* повторное загрязнение *или* заражение
record *s* регистрация, запись
geological ~ геологическая летопись
recorder *s* записывающее устройство, самописец
chart ~ регистрирующее устройство, записывающее устройство, ленточный самописец
single-channel ~ одноканальный самописац
thermal ~ термограф
recording *s* регистрация; *phys* отведение
recover *v* восстанавливать; выздоравливать
recovery *s* восстановление; выздоровление; выделение *(of a microbe);* выход препарата; утилизация
waste ~ рекуперация, восстановление отходов
recreation *s* рекреация
recrudescence *s* рецидив; обострение
recruitment *s* пополнение, восстановление; *phys* феномен подкрепления; *imm* рекрутмент, рекрутинг
rectal *a* ректальный

rectangle *s* прямоугольник
rectangular *a* прямоугольный
rectification *s* ректификация, очистка
rectigradation *s* ректиградация
rectimurate *a pal* с прямыми перегородками
rectinerved *a* пряможилковатый
rectipetality *s bot* стремление к прямолинейному росту
rectirostral *a* прямоклювый, имеющий прямой носик
rectivenous *a* пряможилковатый
rectouterine *a* прямокишечно-маточный
rectrix *s* рулевое перо
rectum *s* прямая кишка
recultivation *s* рекультивация
recumbency *s* лежачее положение
recumbent *a* лежащий; лежачий
recuperation *s* вторичная индукция компетентности; восстановление; рекуперация, восстановление отходов
recurrence *s* рецидив, повторение, возврат
recurrent *a* рецидивирующий, периодически повторяющийся, возвратный
recurved *a* согнутый вниз; отогнутый назад
recycle *s* повторный цикл
recycling *s bioch* метаболический цикл, кругооборот; *imm* рециклинг
red красный
scarlet ~ судан IV
redbill *s* красноклювая шилохвость *(Anas erythrorhynchos)*
redbreast *s orn* зарянка *(Erithacus rubecula)*
reddening *s* покраснение
rede *s anat* сычуг
redeposit *v* переоткладывать
redescription *s* переописание
redia *s* редия
redifferentiation *s* повторная дифференцировка
redistribution *s* перераспределение, перегруппировка; *imm* образование «шапочки», кэппинг
redivive *s* травянистый многолетник
redolent *a* сильно пахнущий
redox *s* окислительно-восстановительный потенциал; *a* окислительно-восстановительный
redoxreaction *s* окислительно-восстановительная реакция
redpoll *s orn* чечётка *(Acanthis)*
red-ripe *a* созревший, спелый
redshank *s orn* травник *(Tringa totanus)*
dusky ~, spotted ~ щёголь *(Tringa erythropus)*
redstart *s orn* горихвостка *(Phoenicurus)*
reduce *v* восстанавливать; снижать, понижать; уменьшать; ослаблять; редуцировать; сокращать; похудеть
reducer *s* редуцент; редуцирующий фактор, редуктор
reductant *s* восстановитель
reductase *s* редуктаза
reduction *s* уменьшение; сокращение; восстановление; редукция
~ **of bacteriophage** лизогенизация фага
photochemical ~ фотовосстановление, фотохимическое восстановление
radiochemical ~ радиовосстановление
redundancy *s* избыточность; избыток, излишек
DNA ~ ДНК-избыточность
reduplication *s* редупликация
redwing *s* дрозд-белобровик *(Turdus iliacus);* красноплечий чёрный трупиал *(Agelaius phoeniceus)*
ree *s* самка турухтана
reed *s* тростник *(Phragmites)*
reedbed *s* заросль тростника
reedbuck *s* болотный козёл *(Redunca);* большой болотный козёл, большой редунка *(Redunca arundinum)*
reedhaunter *s orn* болотный печник *(Limnornis)*
reedstand *s* заросль тростника
reedy *a* заросший камышом; тростниковый
reeve *s* самка турухтана *(Philomachus pugnax)*
refamiliarize *v eth* вновь ознакомляться
refeeding *s* возобновление кормления; повторное кормление
referent *s eth* обозначаемое
reflect *v* отгибать, загибать; отражать
reflectance *s* отражение
reflected *a bot* завёрнутый, отогнутый
reflection *s* отражение; *phys.* рефлексия
reflectory *a* рефлекторный
reflex *s* рефлекс; рефлекторный

~ **of movement** рефлекторное движение
arc ~ рефлекторная дуга; спинномозговой рефлекс
axon ~ аксональный рефлекс, аксон-рефлекс
biceps ~ рефлекс двуглавой мышцы, бицепс-рефлекс

reflexed *a bot* завёрнутый
reflexion *s* отражение; *phys* рефлексия
reflexogenic *a* рефлексогенный, вызывающий рефлекс
reflexograph *s* рефлексограф
reflexometer *s* рефлексометр
reflorescence *s* повторное цветение
refluxion *s* кипячение с обратным холодильником
refoliation *s* вторичное облиствение
reforestation *s* возобновление леса
refracted *a* преломлённый; наклонённый назад под углом
refraction *s* рефракция, преломление
refractivity *s* преломляющая способность
refractometer *s* рефрактометр
refractory *a* рефрактерный, невозбудимый; невосприимчивый
refrigerate охлаждать, охлаждаться; замораживать, замораживаться
refrigeration *s* охлаждение; замораживание
refringence *s* рефракция, преломление
refuge *s* убежище, рефугий; служить убежищем; долгосрочный заказник; резерват
refugium *s* убежище, рефугий
regenerate *v* регенерировать, восстанавливать; перерождаться, обновляться
regeneration *s* регенерация, восстановление; перерождение, обновление
region *s* область, регион; зона, район; отдел тела, область тела
~ **of growth** зона роста
~ **of plant alimentation** площадь питания растений
abdominal ~ брюшной отдел
Amazonian ~ Амазония
Arctogaeic ~ Арктогея
basilar ~ основание черепа
C ~ *imm* константная область, C-область, C-домен, C-сегмент
complementarity-determining ~ *imm* гипервариабельный участок, hv-участок
constant ~ *imm* константная область, C-область, C-домен, C-сегмент
D -encoded ~ DH-область гена, DH-сегмент гена
exchanged ~ рекомбинированный участок, рекомбинировавший сегмент
hypervariable ~ *imm* гипервариабельный участок, hv-участок
immune response ~ область генов иммунного ответа, область Ir-генов, Ir-область
invariable ~ *imm* константная область, C-область, C-домен, C-сегмент
Ir ~ область генов иммунного ответа, область Ir-генов, Ir-область
source ~ материнский регион; территория, послужившая источником колонизации
switch ~ сайт переключения, область переключения, сви-сайт; s-область

regional *a* региональный, пространственный
regionalizalion *s* регионализация, районирование
regma *s bot* регма, дробная коробочка
regression *s* регрессия
regressive *a* регрессивный
regrowth *s* отрастание; отава; подрост
regular *a* правильный, нормальный; регулярный; радиально-симметричный
regulate *v* регулировать, упорядочивать, контролировать
regulation *s* регулирование, регуляция
osmotic ~ осморегуляция
temperature ~ терморегуляция

regulator *s* регулятор
growth ~ регулятор роста, ростовое вещество

regulatory *a* регуляторный
regurgitate *v* отрыгивать
rehabilitation *s* восстановление; рекультивация; оздоровление
reimmunization *s* повторная иммунизация, реиммунизация
reimplantation *s* реимплантация

reindeer *s* северный олень *(Rangifer tarandus)*
Reine Claude *s* слива ренклод *(Prunus domestica ssp. italica)*
reinfection *s* реинфекция, повторное заражение
reinfestation *s* реинвазия
reinforcement *s* подкрепление, усиление
reinoculation *s* пересев; повторная инокуляция
reintegration *s* восстановление целостности организма
reinvasion *s* реинвазия
reject *v* отвергать
rejection *s* отторжение; *pl* экскременты
rejuvenescence *s* омолаживание; обновление
relapse *s* рецидив
related *a* связанный; родственный
relation *s* отношение, связь, зависимость; родство
  consanguineous ~ кровное родство
  lock-key ~ принцип «ключа и замка»
relationship *s* отношение, взаимоотношение; родство, родственные отношения
  genetic ~ генетическое родство
  water ~**s** водный режим
  weight-length ~ соотношение между весом и длиной
relatives *s pl* близкие формы, родственные формы
relax *v* ослаблять, расслаблять
relaxation *s* релаксация, расслабление
  NMR ~, nuclear magnetic resonance ~ ЯМР-релаксация
relay *v* переключать
release *s* выброс; высвобождение; секреция; *v* высвобождать; выделять
  dormancy ~ выход из покоя
releaser *s* пусковой механизм;пусковой раздражитель; стимул, вызывающий осуществление инстинктивного действия
relegation *s* изгнание
reliable *a* достоверный, надёжный
relict *s* реликт, реликтовая форма; *a* реликтовый, остаточный
relief *s* рельеф; облегчение
relieve *v* облегчать
relocate *v* переселять, перемещать; пересаживать
relocation *s* переселение, перемещение; пересадка
rem *s* бэр, биологический эквивалент рентгена
remains *s pl* остатки
  stubble ~ пожнивные остатки, стерня
remastication *s* пережёвывание
  ~ **of cud** пережёвывание жвачки
remating *s* вторичное спаривание
remembrance *s* память
remex *s* маховое перо первого порядка
remission *s* ремиссия
remit ослабевать, стихать, уменьшаться
remittence *s* ремиссия
remittent *a* перемежающийся, ремиттирующий
  ~ **fever** перемежающаяся лихорадка
remnant *s* остаток
remontant *a* ремонтантный
remote *a* отдалённый; дистанционный
removal *s* удаление; отнятие, отщепление; передвижение, перемещение
  top ~ прищипывание; пикировка
remove *v* удалять; отнимать, отщеплять; отделять; передвигать, перемещать
ren *s* северный олень *(Rangifer tarandus)*
renal *a anat* почечный
rename *v* переименовывать
renaturation *s* ренатурация
renest *v* вновь гнездиться
renette *s helm* выделительная клетка
renew *v* обновлять; восстанавливать
renewal *s* обновление; возрождение; восстановление
reniculus *s anat* почечка; доля почки
renipericardial *a* реноперикардиальный
rennet *s* сычуг; сычужный фермент; реннин, химозин; *bot* подмаренник *(Galium)*
rennin *s* реннин, химозин
renovation *s* обновление; восстановление
rentier *s* северный олень *(Rangifer tarandus)*
reorganization *s* реорганизация, перестройка
reovirus *s* реовирус
reoxidation *s* реокисление
repair *s bioch* восстановление, репарация; заживление; *v* восстанавливать

dark ~ темновое восстановление, темновая репарация
SOS ~ SOS-репарация
repand *a* имеющий слегка волнистые края
repandodentate *a* волнисто-зубчатый
repeat *s* повторение; дупликация; *mbiol* повтор; *orn* повторное кольцевание
Alu ~ *Alu*-повтор
long terminal ~ длинный концевой повтор, LTR-последовательность
short terminal ~ короткий концевой повтор, STR-последовательность
tandem ~ тандемная дупликация
repeatability *s* повторяемость; *bioch* повторность
repel *v* отгонять, отпугивать, отталкивать
repellent *s* репеллент; *a* отталкивающий
repent *a bot* ползучий, стелющийся; *zool* пресмыкающийся, ползающий
repertoire *s* репертуар; состав; спектр
antibody ~ спектр антител
behavioral ~ *eth* поведенческий репертуар
idiotopic ~ набор идиотипических детерминант; идиотип
paratopic ~ эпитопная карта
replace *v* замещать; заменять
replacement *s* замещение; замена
competitive ~ конкурентное вытеснение
RNA base ~ замена РНК-оснований
replanting *s* пересев, пересадка, подсадка
replasmolysis *s* повторный плазмолиз
replenishment *s* повторное наполнение; пополнение
replete *a* наполненный, насыщенный; переполненный; изобилующий; *s ent* «медовая бочка»
replica *s* реплика, отпечаток
gene ~ генная реплика, копия гена
replicate воспроизводить; повторять; *bot* отогнутый назад
replication *s* повторность; *gen* репликация, ауторепродукция
theta ~ тэта-репликация *(ДНК)*
replicator *s gen* репликатор; пипетка-репликатор
multichannel ~ многоканальная пипетка
replicon *s gen* репликон
repolarization *s* реполяризация
repolymerization *s* реполимеризация
repopulate *v* вновь заселять, вновь заселяться
repopulation *s* вторичное заселение
representation *s* представительство
ipsilateral ~ представительство *(in the brain)*, одноимённое стороне раздражения
representative представитель; репрезентативный, показательный, характерный, типичный
repress *v* репрессировать, подавлять
repression *s* репрессия, подавление; **to bring about** ~ осуществлять репрессию
CFU- ~ подавление образования колоний
repressor *s* ингибитор функции; репрессор
lac ~ lac-репрессор
reproduction *s* репродукция, вопроизведение, размножение; воспроизводство; возобновление
bud ~ почкование
net ~ коэффициент размножения
reproductive *a* половой, репродуктивный
reptile *s* пресмыкающееся
repugnatorial *a* отпугивающий
repulsion *s gen* отталкивание
require требовать; нуждаться
requirement *s* потребность
quantum ~ of photosynthesis квантовый расход фотосинтеза
rescreening *s* повторный скрининг
rescue *s* спасение
research *s* исследование, анализ; исследовать
resect *v* иссекать
resection *s* иссечение, резекция
reseeding *s* пересев, повторный пересев
resemblance *s* сходство
resemble походить, иметь сходство
reservation *s* заповедник; резерват; резервация
reserve *s* запас; заповедник; запасать
food ~ запас питательных веществ, кормовая база
game ~ охотничий резерват
nature ~ заповедник

reservoir *s* резервуар, вместилище; водохранилище, водоём
reset *v* пересаживать, рассаживать; регулировка; настройка; установка на нуль
central ~ *eth* центральная перестройка
resetting *s* пересадка, рассаживание
resident *s* резидент, постоянный обитатель; оседлый, немигрирующий
residual *s* остаток, остаточная величина; остаточный; остаточная вариация, показатель статической ошибки; остаточное явление
residue *s* остаток; осадок, отстой; остаток, радикал
resilient *a* упругий, эластичный
resin *s* смола; камедь
resiniferous *a* смолоносный
resinous *a* смолистый
resistance *s* сопротивление; сопротивляемость, резистентность; устойчивость; невосприимчивость
acid ~ кислотоустойчивость
body ~ сопротивляемость организма
cold ~ холодоустойчивость
drought ~ засухоустойчивость
frost ~ морозостойкость
immune ~ иммунитет
phage ~ фагоустойчивость
respiratory ~ *phys* сопротивление дыханию
salinity ~, salt ~ солеустойчивость
thermal ~ теплоустойчивость
winter ~ зимостойкость
resistant *a* резистентный; устойчивый; невосприимчивый
disease ~ иммунный
resolution *s* растворение; рассасывание; разрушение; разрешение, разрешающая способность
resolve растворяться; рассасываться
resonance *s* резонанс
resonator *s* резонатор
resorption *s* резорбция, поглощение; всасывание
resource *s* ресурс
respiration *s* дыхание
respiratory *a* респираторный, дыхательный
respirometer *s* респирометр
respond реагировать, отвечать
responder *s* отвечающий организм; респондер
autologous ~ аутореактивный организм, аутологичный респондер; аутореактивная клетка
good ~, high ~ организм с сильным иммунитетом, высоотвечающая особь
low ~, poor ~ организм с ослабленным иммунитетом, низкоотвечающая особь
response *s* реакция, ответ
allergic ~ аллергическая реакция, аллергия
allospecific ~ иммунный ответ на аллоантиген, аллоиммунная реакция
anamnestic ~ анамнестическая реакция, вторичный иммунный ответ, бустер-эффект
antibody ~ гуморальный иммунный ответ; образование антител, антителогенез
antiself ~, autoimmune ~, autospecific ~ аутоиммунный ответ, аутоиммунная реакция
B-cell ~ гуморальная иммунная реакция, гуморальный иммунный ответ
booster ~ анамнестическая реакция, вторичный иммунный ответ, бустер-эффект
conditioned ~ условная реакция, условный ответ
conditioned emotional ~ условно-рефлекторная эмоциональная реакция
evoked ~ вызванный ответ, вызванная реакция
lysogenic ~ лизогенная реакция, лизогенизация
memory ~ анамнестическая реакция, вторичный иммунный ответ, бустер-эффект
MLC ~ реакция смешанной культуры лимфоцитов, СКЛ-реакция
off ~ *phys* off-ответ, «ответ выключения»
on ~ *phys* on-ответ, «ответ включения»
primary immune ~ первичный иммунологический ответ, первичная иммунная реакция, иммунологическая реакция
recruiting рекрутмент, рекрутинг; *phys* реакция вовлечения

respiratory ~ дыхательная реакция, реакция дыхания
secondary immune ~ вторичный иммунологический ответ, вторичная иммунная [иммунологическая] реакция
selection ~ реагирование на отбор; результат отбора
short-lived immune ~ непродолжительный иммунный ответ, транзиторная иммунная реакция
startle ~ старт-реакция, реакция вздрагивания; испуг
responsible *a* ответственный; **to be ~ for the choice** определять выбор
responsiveness *s* реактивность
rest *s* отдых, покой; перерыв; остаток; штатив; подставка
restaining *s* повторное окрашивание; восстановление окраски
resting *s* покой; период покоя; *a* покоящийся, спящий
restituion *s* возвращение к норме; восстановление
restlessness *s* беспокойство, беспокойное состояние
restocking *s* повторное заселение, возобновление *(e.g., of a forest)*
restoration *s* восстановление
restore *v* восстанавливать
restricted *a* ограниченный
absolutely ~ абсолютно верный сообществу, верный местообитанию
restriction *s* ограничение, рестрикция
genetic ~ генетическое ограничение, генетическая рестрикция
result *s* результат; *v* получать в результате, получаться в результате
resupinate *a bot* перевёрнутый
resurgence *s* возрождение
resuscitate *v* оживлять, реанимировать
resuscitation *s* оживление, реанимация
resuspend *v* ресуспендировать
resynthesis *s* ресинтез
retainer *s* садок; виварий
retardation *s* задержка, запаздывание, замедление
rete *s anat* сеть, сетка
retention *s* удерживание; задержание, задержка
water ~ влагоудержание
retentivity *s* удерживающая способность; *eth* способность к сохранению опыта
retest *s* повторная проба, повторный анализ
retial *a* относящийся к сети
reticle *s* сеточка
reticular *a* сетчатый, сеткообразный
reticulocyte *s* ретикулоцит
reticulocytosis *s* ретикулоцитоз
reticuloendothelial *a* ретикулоэндотелиальный
reticuloendothelium *s* ретикулоэндотелий
reticulosome *s* ретикулосома
reticulum *s* ретикулюм, сеть, сеточка, сетчатое образование; сетка
retiform *a* сетевидный, сетеобразный
retina *s anat* сетчатка, сетчатая оболочка
retinaculum *s anat* связка; уздечка; *энт.* зацепка, вырост; *bot* прилипальце
retinal *a* относящийся к сетчатке
retinerved *a* сетчато-жилковатый
retinoblast *s* ретинобласт
retinol *s* ретинол, витамин А
retractile *a* втягивающийся; втяжной
retraction *s* втягивание, втяжение, ретракция
retractor *s* ретрактор, втягивающий мускул
retrad *adv* в направлении кзади
retrap, retrapping *s* повторный отлов
retreat *s* убежище; логово, логовище; отступление; *v* уходить, отступать; повторно обрабатывать
retriever *s* ретривер
retroaction *s* обратное действие
retrodeviation *s* отклонение кзади
retrodisplacement *s* смещение кзади
retroflexed *a* изогнутый кзади, отогнутый назад
retrogradation *s* ретроградация; уменьшение плотности популяции
retrograde *a* ретроградный
retrogress *s* регресс, упадок; регрессировать; ухудшаться
retrogression *s* ретрогрессия, снижение уровня, дегенерация; перерождение; ухудшение; вырождение
retro-inhibition *s mbiol* негативный контроль с обратной связью
retromorphosis *s* ретроморфоз
retroposed *a* смещённый кзади

retroposon *s* ретропозон
retrorse *a* отогнутый назад, отогнутый вниз
retrostalsis *s* обратная перистальтика
retrosternal *a* загрудинный
retrotransposon *s* ретротранспозон
retroverted *a* повёрнутый *или* загнутый назад
retrovirus *s* ретровирус
return *s* возврат
  venous ~ венозный приток
reunion *s* воссоединение
reuse *s* повторное использование; повторно использовать
reutilization *s* повторное использование
revaccination *s* ревакцинация, повторная вакцинация
revegetation *s* возобновление растительного сообщества; восстановление растительного покрова
revehent *a anat* отводящий, выносящий
reverberatory *a* реверберaторный, резонансный
reversal *s* перемена, изменение
  imprinting ~ *eth* смена запечатления
  sex ~ трансформация пола
reversibility *s* реверсивность; обратимость
reversible *a* обратимый
reversion *s* реверсия, атавизм
reversionary *a* атавистический
revertant *s* ревертант
revision *s* ревизия
revival *s* возрождение; оживление
revive *v* оживать; реанимировать, оживлять
revolute *a* закрученный назад *или* вниз
reward *s* вознаграждение, подкрепление
  food ~ пищевое вознаграждение, пищевое подкрепление
Rh, rhesus factor *s* резус фактор
rhabdite *s* рабдит
rhabdolith *s* рабдолит
rhabdom *s ent* рабдом, зрительная палочка
rhabdovirus *s* рабдовирус
rhabdus *s* рабда
rhachial осевой
rhachilla *s* вторичная ось соцветия, ось колоска
rhachis *s* ось; стержень; позвоночный столб, позвоночник; ось сложного колоса; главная ось цветоносного побега
rhamnose *s* рамноза
rhamphoid *a* клювообразный
rhamphotheca *s orn* рамфотека, роговой чехол клюва
rheobase *s* реобаза, порог гальванической возбудимости
rheogram *s* реограмма
rheography *s* реография
rheometer *s* реометр
rheonome *s* электростимулятор
rheoreceptor *s* реорецептор
rheotactic *a* реотропный
rheotaxis *s* реотаксис
rheotropic *a* реотропный
rheotropism *s* реотропизм
rhesus *s* резус-фактор
rhinal *a* носовой, обонятельный
rhinencephalon *s* обонятельный мозг, обонятельные доли мозга
rhinesthesia *s* обоняние
rhinoceros *s ther* носорог *(Rhinocerotidae)*
rhinopharynx *s* носоглотка
rhinophore *s zool* ринофор
rhinosporidiosis *s phytop* риноспоридиоз *(caused by Rhinosporidium seeberi)*
rhinovirus *s* риновирус
rhipidate *a* веерообразный
rhipidium *s bot* веер, опахало
rhipidostichous *a* веерообразный
rhizanthous *a* корнецветковый
rhizine *s* ризина
rhizobia *s* микориза
rhizocarpous *a* корнеплодный
rhizocaul *s* гидрориза
rhizocollesy *s* срастание корней
rhizoctonia *s phytop* ризоктония
rhizogenic, **rhizogenous** *a* корнеобразующий
rhizoid *s* ризоид; *a* ризоидный, корневидный
rhizomatous *a* корневищный
rhizome *s* корневище
rhizomorph *s* ризоморфа
rhizomorphous *a* корнеобразный; корневищевидный
rhizomycelium *s* ризомицелий

rhizophagous *a* корнеядный, корнегрызущий
rhizophore *s* ризофор
rhizophorous *a* корненосный
rhizopodium *s* ризоподия
rhizopod *s prot* корненожка *(Rhizopoda)*
rhizosphere *s* ризосфера
rhizotaxis *s* корнерасположение
rhodamine *s* родамин
rhodopsin *s* родопсин, зрительный пурпур
rhodoxanthin *s* родоксантин
rhoium *s* сообщество ручья
rhombencephalon *s* первичный задний мозг; третий мозговой пузырь
rhopalium *s zool* ропалий
rough *a* грубый
roughage *s* грубый корм
rough-haired *a* грубошёрстный
rhubarb *s* ревень *(Rheum)*
rhyacad *s* растение стремительного потока
rhyacium *s* сообщество стремительного потока
rhynchocoel *s* ринхоцель, хоботное влагалище
rhynchodaeum *s* прецеребральная область *(in Nemertini)*
rhynchophorous *a* имеющий клюв
rhynchostome *s helm* ринхостом
rhythm *s* ритм; ритмичность, периодичность, цикличность
alpha ~ альфа-ритм
beta ~ бета-ритм
cardiac ~ ритм сердца, частота сердцебиений
delta ~ дельта-ритм
heart ~ ритм сердца, частота сердцебиений
theta ~ тета-ритм
triple ~ ритм галопа; трёхчленный ритм
rhythmicity *s* ритмичность, периодичность, цикличность
rhythmometry *s* ритмометрия
rib *s* жилка; ребро
ribbon *s* лента
lingual ~ *mal* тёрка, радула
ribby с крупными жилками
riboflavin *s* рибофлавин
ribonuclease *s* рибонуклеаза, РНКаза
ribonucleoprotein *s* рибонуклеопротеид
ribonucleoside *s* рибонуклеозид
ribonucleotide *s* рибонуклеотид
ribose *s* рибоза
riboside *s* рибозид
nicotinamide ~ никотинамидрибозид
ribosome *s* рибосома
ribulose *s* рибулоза
rice *s* рис *(Oryza)*
rich *a* изобилующий, богатый; тучный, плодородный; сильный *(odor);* яркий *(light);* сочный *(fruit)*
rickettsia *s* риккетсии
rickettsiosis *s* риккетсиоз
ricochet *s* рикошет, рикошетирующий бег
ride *s* просека; верховая езда; верховая лоошадь; отбивать животное от стада
ridge *s* гребень
supraorbital ~ надбровная дуга
urinogenital ~ урогенитальная складка
Wolffian ~ вольфова складка
right *s* право
peck ~ иерархическое положение, право вожака
rigidity *s* ригидность, окоченелость, окоченение
rigor *s* неподвижность, оцепенение; озноб
cold ~ устойчивость к холоду, холодоустойчивость
heat ~ устойчивость к высоким температурам, теплоустойчивость
rima *s* щель, трещина
rimate *a* щелевой; имеющий щели
rimoportula *s* двугубый вырост, щелевидный вырост
rimose, rimouse *a* щелевой; имеющий щели
rind *s* кора; кожица; кожура; *v* очищать кожицу; снимать кожуру; сдирать кору
ring *s* кольцо; кольцевидная структура; годичное кольцо, годичный слой; *v* кольцевать; делать кольцевой надрез;
to split the ~ open *bioch* разрывать кольцо
annual ~ годичное кольцо, годичный слой
arterial ~ артериальный круг, виллизиев круг *(of the brain)*
cabbage black ~ *phytop* чёрная кольцевая пятнистость капусты

growth ~ годичное кольцо, годичный слой

hemolytic ~ гемолитическая бляшка

ringdove *s orn* вяхирь *(Columba palumbus);* египетская горлица *(Streptopelia risoria);* кольчатая горлица *(Streptopelia decaocto)*

ringed *a* кольчатый; кольцевидный; окольцованный

ringent *a bot* раскрытый, зияющий

ringing *s* кольцевание

ring-porous *a* кольцепоровый

ringspot *s phytop* кольцевая пятнистость

ringtail *s orn* самка луня; молодой лунь; молодой беркут; *ther* кольцехвостый кускус *(Pseudocheirus);* какомицли *(Bassariscus)*

riparian *a* прибрежный, обитающий по берегам

ripe *a* спелый, зрелый, созревший

early ~ скороспелый, ранний

ripen *v* зреть, созревать; вызревать

ripeness *s* спелость, зрелость

early ~ раннеспелость

late ~ позднеспелость

ripening *s* созревание; вызревание; созревающий; приспевающий *(о лесном насаждении)*

early ~ раннеспелый, скороспелый

late ~ позднеспелый

ripple *s* пульсация

risk *s* риск

ritual *s eth* ритуал; *a* ритуальный

ritualistic *a* ритуалистический, демонстративный, условный

ritualization *s* ритуализация

ritualize *v* ритуализировать, превращать в ритуал

rival *s* конкурент, соперник; соперничать

rivalry *s* соперничество

riverine *s* обитатель рек; речной

rivose *a bot* с извилистыми каналами

RNA, ribonucleic acid *s* РНК, рибонуклеиновая кислота; ~ **minus strand** РНК минус цепь; ~ **plus strand** РНК плюс цепь

acceptor ~, adaptor ~ транспортная РНК, тРНК

chloroplast ~ хлоропластная РНК, РНК хлоропластов

heterogenous nuclear ~ гетерогенная ядерная РНК, гяРНК

informational ~ матричная РНК, информационная РНК, мРНК

messenger ~ матричная РНК, информационная РНК, мРНК

mitochondrial ~ митохондриальная РНК, РНК митохондрий

precursor ~ РНК-предшественник, пре-РНК

primer ~ РНК-затравка

ribosomal ~ рибосомная РНК, рРНК

small nuclear ~ низкомолекулярная ядерная РНК, няРНК

soluble ~, transfer ~ транспортная РНК, тРНК

translational ~ матричная РНК, информационная РНК, мРНК

roach *s* таракан

roadside *a* придорожный

roam *v* бродить, скитаться

roan *s* лошадиная антилопа *(Hippotragus equinus); a* чалый

strawberry ~ рыже-чалый

roar *s* рычание; рёв; мычание; ржание; *v* реветь

roaring *s* рычание; рёв; мычание; ржание

robber-fly *s* ктырь *(Asilidae)*

robin *s orn* зарянка *(Erithacus rubecula);* зарянковая мухоловка *(Eopsaltria);* дрозд *(Turdus)*

blue-throated ~ варакушка *(Luscinia svecica)*

rock-boring *a* сверлящий скалы

rocking *s* качание; укачивание

rock-n-roller *s* качалка с вращательным миксером, ротамиксер

rod *s* палочка; удочка; удилище

Corti's ~ кортиев столбик

Meckel's ~ меккелев хрящ

spreading ~, streaking ~ палочка для размазывания, шпатель

rodent *s* грызун *(Rodentia)*

cricetid ~ хомяк *(Cricetidae)*

double-toothed ~ зайцеобразное *(Lagomorpha)*

simple-toothed ~ грызун *(Rodentia)*

rodent-borne *a* распространяемый грызунами

rodent-mole *s* цокор *(Myospalax)*

roding *s* тяга *(of woodcocks)*

rod-like *a* палочкообразный, палочковидный

roe *s* косуля *(Capreolus);* икра; молоки

roebuck *s* самец косули
roentgenogram *s* рентгенограмма
roentgenography *s* рентгенография
rogue *s gen* уклоняющаяся форма
role *s* поведенческая роль
roll *s* трубка, рулон
leaf ~ *phytop* скручивание листьев
roll-call *s eth* перекличка
roller *s orn* сизоворонка *(Coracias); ent* листовёртка
rolling *s* перекатывание
roof *s anat* крыша; верхняя стенка
~ **of the mouth** нёбо
~ **of skull** свод черепа
rook *s* грач *(Corvus frugilegus)*
rookery *s* залёжка, лежбище *(тюленей, моржей);* колония птиц, птичий базар; гнездовье; грачовник
roost *s* насест; ночёвка, ночлег, место ночлега
rooster *s* петух; самец птицы
root *s bot* корень; пускать корни, укореняться; *anat* корешок; корень; *pl* корнеплоды; **~s in clusters** мочковатая корневая система; **to take** ~ приниматься
fungus ~ микориза
holdfast ~ корень-прицепка
pneumatophore ~ пневматофор, дыхательный корень, воздушный корень
primordial ~ первичный корешок, зародышевый корешок
storage ~ запасающий корень; корнеплод
rootage *s* укоренение; корневая система
rooting *s* укоренение; укореняющийся
rootlet *s bot* корешок
rootstock *s* корневой побег; корневище
ropy *a* липкий; вязкий, тягучий
rose *s* роза *(Rosa)*
rosefinch *s orn* чечевица *(Caprodacus)*
rosein *s* фуксин
rosellate *a bot* собранный в розетку, розеточный
rosette *s bot* розетка; *phytop* розетка, розеточная болезнь
rosette-like *a* розетковидный; розеточный
rosetting *s micr* розеткообразование
rostel *s* носик; клювик; хоботок
rostellate *a* имеющий носик, клювик *или* хоботок
rostelliform *a* клювиковидный; носиковидный; хоботковидный
rostrate *a* имеющий рострум *или* клюв
rostrulum *s* маленький рострум; клювик
rostrum *s* рострум, клювовидный выступ
rot *s* гниль; гниение; *v* гнить; портиться
blue ~ синева древесины
bottom ~ прикорневое побурение паслёновых *(caused byCorticium vagum)*
trunk ~ of chestnut «чернильная болезнь» каштана *(caused by Phytophthora cinnamomi)*
wheat crown ~ «снежная плесень» злаков *(caused by Fusarium graminearum)*
rotate *v* чередовать; сменять; *bot* колесовидный
rotate-funneled *a bot* колесовидно-ворончатый
rotation *s* вращение; чередование; севооборот, ротация, смена
crop ~ севооборот
factor ~ *biom* чередование факторов
five-course ~ пятипольный севооборот
rotator *s* мускул-вращатель
rotifer *s* коловратка *(Rotifera, Rotatoria)*
rotten *a* гнилой, испорченный, разложившийся
rotter *s* микроогранизм, вызывающий гниение, гниль
rotting *s* гниение
rotula *s* надколенник, надколенная чашечка; скобка *(in sea-urchins)*
rotundifolious *a* круглолистный
rough *a* шероховатый, неровный
rough-haired *a* грубоволосистый; грубошёрстный
roughleg *s orn* мохноногий канюк, зимняк *(Buteo lagopus)*
roughness *s* шероховатость, неровность; бугристость
rough-scaled *a* пушисточешуйчатый
rough-tongued *a* пушистоязычковый
round *s* цикл
rounding *s biom* округление; ~ **down** округление снизу; ~ **off** округление; ~ **up** округление сверху
rounded *a* округлый

round-fruited *a* круглоплодный, шаровидно-плодный
round-headed *a* шаровидно-головчатый
round-leaved *a* круглолистный
round-spiked *a* шаровидно-колосный
roundworm *s* круглый червь *(Nemathelminthes)*
large ~ аскариды *(Asscaridae)*
rousette *s* летучая собака, ночной крылан *(Rousettus)*
route *s* путь, маршрут
row *s* ряд, междурядье
rpm, revolutions per minute оборотов в минуту
RQ *s* respiratory quotient ДК, дыхательный коэффициент
rubber *s* каучук, резина
hevea ~, **Para** ~ гевея бразильская *(Hevea brasiliensis)*
rubicolous *a* живущий на малине
rubigo *s phytop* ржавчина
ruderal *a* рудеральный, растущий на пустырях
rudiment *s* зачаток; рудимент
limb ~ зачаток конечности; рудиментарная конечность
rudimentary *a* зачаточный; рудиментарный, недоразвитый
rue *s* рута *(Ruta)*
ruff *s orn* турухтан *(Philomachus pugnax);* кольцо перьев вокруг шеи
ruffed *a orn* гривастый
ruffle *v* ерошить
to ~ up взъерошивать
ruffling *s* ундуляция, колебание
rufinism *s* руфинизм, красная пигментация
ruga *s* морщина, складка
rugate *a* морщинистый, складчатый
rugged *a* шероховатый
rugoporate *a* морщинисто-поровый
rugose *a* морщинистый; складчатый
rugosity *s* морщинистость; складчатость; шероховатость
rugous *a* морщинистый, складчатый;шероховатый
rugulose *a* мелкоморщинистый, мелкоскладчатый
rule *s* правило
rumble *s* урчание; *v* урчать
rumbling *s* бурчание, урчание
rumen *s* рубец
ruminant *s* жвачное животное *(Ruminantia); a* жвачный
antlered ~ олень, плотнорогое *(Cervidae)*
hollow-horned ~ полорогое *(Bovidae)*
ruminate *v* жевать жвачку; *a* крапчатый
rumination *s* пережёвывание жвачки
rump *s orn* крестец; гузка; круп *(of a horse)*
run *s* стадо; стая *(of birds);* тропа; серия опытов; *v* бежать, бегать; преследовать, травить; течь, литься; ползти, виться
runcinate *a bot* струговидный, с загнутыми назад зубцами
runner побег; отпрыск; столон; отводок; ус; плеть; лоза; стелющееся растение; вьющиеся растение
running *s eth* побежка; текучий; проточный; ползучий, вьющийся; сыпучий
running-out *s:* **potato** ~ *phytop* «истощение» картофеля
runoff *s* сток; поверхностный сток
runt *s* малорослое животное; карликовое растение
runway *s eth* побежка
rupestrine, rupicolous *a* живущий на скалах
ruptile *a bot* неопределённо ломающийся, неопределённо растрескивающийся
rupture *s* разрыв; трещина
rusa *s* олень замбар *(Rusa)*
rush *s* стремительное движение; бросок; *bot* ситник *(Juncus); v* бросаться; мчаться
rust *s phytop* ржавчина; ржавчинный гриб; *v* быть поражённым ржавчиной, страдать от ржавчины
rusted *a phytop* поражённый ржавчиной
rustic *a* сельский
rustle *v* шелестеть, шуршать
rustproof, rust-resistant *a phytop* устойчивый против ржавчины, иммунный к ржавчине
rut *s* половая охота, течка, гон; *v* быть в половой охоте
rutabaga *s* брюква *(Brassica napus var. napobrassica)*
rutting *s* половая охота, течка, гон

ruttishness *s* половая охота, течка, гон; время спаривания
rye *s* рожь посевная *(Secale cereale)*
spurred ~ спорынья *(caused by Claviceps purpurea)*
ryegrass *s bot* плевел *(Lolium)*
rype *s* белая куропатка *(Lagopus lagopus)*

S

saber-toothed *a* саблезубый
sable *s* соболь *(Martes zibellina)*
American ~, Hudson Bay ~ американская куница *(Martes americana)*
Russian ~, Siberian ~ соболь *(Martes zibellina)*
sabulous *a* песчаный; супесчаный
sac *s* мешок; мешочек; сумка
allantoic ~ аллантоис
amniotic ~ амнион
brood ~ зародышевая камера, выводковая камера
bursal ~ *orn* фабрициева сумка
cirrus ~ сумка цирруса, мешок цирруса
conjunctival ~ конъюнктивальный мешок
egg ~ *ent* яйцевая камера
fetal ~, gestation ~ плодный мешок, плодные оболочки
heart ~ *anat* околосердечная сумка, перикард
honey ~ *apic* медовый зобик, медовый желудочек
ink ~ чернильный мешок, чернильная железа
pericardial ~ *anat* околосердечная сумка, перикард
pleural ~ плевральная полость
poison ~ резервуар ядовитой железы
pollen ~ *bot* пыльцевой мешок, гнездо пыльника
seed ~ семенной пузырёк
shell ~ скорлуповая железа
spore ~ спорангий
saccate *a* мешкообразный; *bot* мешковый; *helm* инкапсулированный
sacchariferous *a* сахароносный; содержащий сахар, сахаристый
saccharify *v* осахаривать, осахариваться; превращать в сахар, превращаться в сахар
saccharimeter *s* сахариметр
saccharimetry *s* сахариметрия
saccharobiose *s* сахароза
saccharolytic *a* сахаролитический, расщепляющий сахар
saccharometabolism *s* сахарный обмен
saccharometer *s* сахарометр
saccharose *s* сахароза
sacciform, saccular *a* мешкообразный, мешковидный
sacculation *s* образование мешочков; мешотчатое образование, группа мешочков
saccule *s* мешочек; *ent* нижняя часть вальвы; *micr* саккул; *bot* внешний покров спорангия
sacculus *s* мешочек; *bot* внешний покров спорангия; *ent* нижняя часть вальвы; саккула
sack *s* мешок; мешочек; сумка
sacrad *adv* по направлению к крестцу
sacral *a* крестцовый
sacriplex *s* крестцовое сплетение
sacrococcygeal *a* крестцово-копчиковый
sacroiliac *a* крестцово-подвздошный
sacrolumbar *a* крестцово-поясничный
sacrosciatic *a* крестцово-седалищный
sacrovertebral *a* крестцово-позвоночный
sacrum *s* крестец, крестцовый отдел позвоночника
saddle *s* седло
safety *s* безопасность
saffron *s* шафран посевной *(Crocus sativus)*
safranin *s* сафранин
sagacious *a* умный, понятливый
saginate *v* откармливать
sagitta *s* слуховой камешек, отолит
sagittal *a* сагиттальный
sagittate *a* стреловидный; стрельчатый
sagittifolious *a* стрелолистный
saiga *s* сайга, сайгак *(Saiga tatarica)*
sainfoin *s* эспарцет кормовой *(Onobrychis viciaefolia);* люцерна посевная *(Medicago lupulina)*
saki *s* саки, чёртова обезьяна *(Pithecia);* мохнатый саки, хиропот *(Chiropotes)*
salamander *s* саламандра *(Salamandra)*
blind white ~ европейский протей *(Proteus anguinus)*
giant ~ исполинская саламандра *(Andrias);* гигантская амбистома *(Dicamptodon); pl* скрытожаберники *(Cryptobranchidae)*
mole ~ амбистома *(Ambystoma)*
salient *a* выступающий вперёд, выдающийся

saliferous *a* солеобразующий; солесодержащий, соленосный
salification *s* засоление
salina *s* засолённое болото, засолённый марш; солёное озеро
saline *s* раствор солей; физиологический раствор; засолённое болото, засолённый марш; солончак, засолённая почва; *a* солевой; солёный
salinification *s* засоление
salinity *s* солёность
salinization *s* засоление
salinometry *s* солеметрия
salinous *a* солёный
salitral *s* салитраль
saliva *s* слюна
salivant *a* вызывающий слюноотделение
salivary *a* слюнной
salivate *v* выделять слюну; вызывать слюноотделение
salivation *s* выделение слюны, слюноотдеение; слюнотечение
salsify *s* козлобородник *(Trapogon)*
salsuginous *a* обитающий на засоленных почвах
salt *s* соль; *a* солёный; **to ~ out** высаливать, высаливаться
saltant *s* мутант
saltation *s* сальтация, мутация; скачкообразное движение; *phys* сальтаторное проведение возбуждения; бег прыжками
saltatory *a* прыгающий, передвигающийся прыжками; прыгательный; скачкообразный
saltfree *a* бессолевой
saltigrade *a* прыгающий, передвигающийся прыжками
salting *s* засоление
salting-out *s* высаливание
salt-loving *a* солелюбивый
salt-tolerant *a* солевыносливый, устойчивый к засолению
salvage *s* реутилизация
samara *s bot* крылатка
samaroid *a* крылатковидный
sameness *s* сходство; тождество
sample *s* проба; образец; *biom* выборочная совокупность, выборка; *v* отбирать пробу, брать пробу, брать образец
composite ~ смешанная проба; смешанный образец
puncture ~ пунктат
random ~ случайная выборка, произвольная выборка; случайная проба, произвольная проба
representative ~ репрезентативная проба; репрезентативная выборка
sampler *s* пробоотборник; прибор для взятия проб; автоматическая пипетка, сэмплер
sampling *s* отбор проб; взятие образцов; *биом.* выборка, метод выборки; ~ **on successive occasions** повторный отбор выборок
acceptance ~ *biom* выборочный контроль
multiphase ~ многофазовый выбор, метод многоступенчатой выборки
one-zero ~ метод выборки данных «да – нет»
random ~ случайная выборка, случайный отбор выборок
scan ~ метод выборки данных сканированием
systematic ~ систематический отбор проб; систематическая выборка
sanctuary *s* заповедник
sand *s* песок; песчаная почва; *pl* пески, пустыня
sandalwood *s* сандаловое дерево *(Santalum)*
sand-bathing *s eth* купание в песке
sandgrouse *s orn* рябок *(Pterocles)*
sand-loving *a* песколюбивый
sandoz *s* сандоз
sandpiper *s orn* песочник *(Calidris); pl* мелкие кулики *(Charadriiformes)*
black-bellied ~ чернозобик *(Calidris alpina)*
broad-billed ~ грязовик *(Limicola falcinellus)*
buff-breasted ~ канадский песочник *(Tryngites subruficollis)*
common ~ кулик-перевозчик *(Actitis hypoleucos)*
curlew ~ краснозобик *(Calidris ferruginea)*
gray-rumped ~ пепельный улит *(Tringa brevipes)*
green ~ черныш *(Tringa ochropus)*

least ~ песочник-крошка *(Calidris minutilla)*
marsh ~ поручейник *(Tringa stagnatilis)*
purple ~ морской песочник *(Calidris maritima)*
red-backed ~ чернозобик *(Calidris alpina)*
robin ~ исландский песочник *(Calidris canutus)*
shart-tailed ~, Sibirian pectoral ~ острохвостый песочник *(Calidris acuminata)*
solitary ~ улит-отшельник *(Tringa solitaria)*
stilt ~ ходулочниковый зуёк *(Micropalama himantopus)*
Terek ~ мородунка *(Xenus cinereus)*
white-fronted ~ белолобый зуёк *(Charadrius marginatus)*
wood ~ фифи *(Tringa glareola)*
sandwich *s*: **multiple antibody** ~ многослойный иммуносэндвич
sandworm *s zool* пескожил *(Arenicola)*
sandy *a* песчаный; песчанистый
sanguicolous *a* обитающий в крови
sanguifacient *a* кроветворный, гемопоэтический
sanguiferous *a* кровеносный
sanguification *s* кроветворение, гемопоэз
sanguimotion *s* движение крови, циркуляция крови
sanguimotor, sanguimotory *a* циркуляторный, гемодинамический
sanguine *a* кровяной; сангвинический
sanguineous *a* кровяной; содержащий кровь
sanguinivorous *a* питающийся кровью; кровососущий
sanguinopoietic *a* кроветворный, гемопоэтический
sanguivorous *a* питающийся кровью; кровососущий
sanitation *s* санация, оздоровление
santonica *s* полынь цитварная *(Artemisia cina)*
sap *s* сок
bleeding ~ пасока
exudation ~ пасока
false ~ двойная заболонь
nuclear ~ ядерный сок, кариоплазма, кариолимфа
sapajou *s* капуцин *(Cebus)*
sapful *a* сочный
saphena *s* подкожная вена ноги
sapless *a* увядший, высохший
sapling *s* побег, отпрыск, отводок; молодое деревцо
saponification *s* омыление, сапонификация
sappy *a* сочный; мясистый
saprium *s* сапрофитное сообщество
saprobe, saprobiont *s* сапробионт
saprogenic, saprogenous *a* сапрогенный, вызывающий гниение, гнилостный
sapropel *s* сапропель
sapropelic *a* сапропелевый, сапропельный
saprophage *s* сапрофаг
saprophagy *s* сапрофагия
saprophyte *s* сапрофит
saprophytic *a* сапрофитный
saprophytophagous *a* питающийся гниющими растениями
saprotroph *a* сапротрофный
saproxylobiont *s* обитатель отмершей древесины
saprozoic *a* сапрозойный
saprozoite, saprozoon *s* сапрозойный организм
sapwood *s* заболонь
sarcoblast *s* саркобласт, миобласт
sarcocarp *s* саркокарпий, мясистый мезокарпий
sarcocaul *s* мясистый стебель
sarcocyst *s* саркоциста
sarcoderm *s* саркодерма
sarcodic *a* саркодовый
sarcogenic *a* образующий мякоть; образующий мышечную ткань
sarcolemma *s* сарколемма
sarcology *s* миология
sarcomere *s* саркомер
sarcophagous *a* плотоядный, питающийся мясом
sarcoplasm *s* саркоплазма
sarcoplast *s* саркобласт, миобласт
sarcopoietic *a* образующий мышцу, миогенный
sarment *s* лоза; отпрыск; ус
sarmentaceous *a* образующий усы; в виде лозы, в виде отпрыска, в виде уса
sarmentose *a* в виде лозы, в виде отпрыска, в виде уса

sarsaparilla *s bot* аралия *(Aralia);* сассапариль сизая *(Smilax glauca)*
sassaby *s* антилопа топи, сассаби *(Damaliscus lunatus)*
sassafras *s bot* сассафрас лекарственный *(Sassafras officinale)*
satellite *s* спутник, сателлит; *a* сателлитный
satiate *v* насыщать
satiated *a* сытый, насытившийся
satiation *s* насыщение
saties *s* сатия
satiny *a* атласный, шелковисто-блестящий
sation *s* сатион
satisfaction *s* удовлетворение
saturant *s* насыщающее вещество; *a* насыщающий; поглощающий, впитывающий
saturate *v* насыщать
saturation *s* насыщение; насыщенность; поглощение, впитывание
saucer-shaped *a* чашевидный; блюдцеобразный
saurian *s* ископаемый ящер; *a* напоминающий ящерицу
savage *v eth* бить копытами; лягать
savanna, savannah *s* саванна
savory *s бот.* чабер *(Satureja)*
savoyed *a* морщинистый, сморщенный
saw *s* пила
ampule ~ пилка для ампул
sawbill *s orn* крохаль *(Mergus)*
sawfly *s ent* пилильщик
horntail ~ рогохвост *(Siricidae)*
saw-toothed *a* пильчатый
saxatile *a* обитающий на скалах
saxaul *s* саксаул *(Haloxylon)*
saxicavous *a* бурящий скалы
saxicolous *a* обитающий на скалах
scab *s* корка, струп; парша
scabby *a* мелкочешуйчатый; покрытый струпьями, шелудивый
scabrous *a* шершавый, шероховатый
scala *s* лестницеподобная структура; лестница
Loewenberg's ~ улитковый проток, канал улитки, проток улитки
scald *s* ожог; *phytop* ринхоспорозный ожог, ринхоспорозная пятнистость *(caused by Rhynchosporium secalis)*
sun ~ солнечный ожог, солнечный опал
scale *s* шкала; масштаб; чешуя; чешуйка; червец, щитовка *(Coccodea)*
micrometer eyepiece ~ окулярный микрометр, окуляр-микрометр
scale-bearing *a* чешуеносный
scaled *a* чешуйчатый, покрытый чешуёй
bristly ~ щетинисто-чешуйчатый
closely ~ густочешуйчатый
scaleless *a* бесчешуйный
scalene *s* лестничная мышца
scaling *s* выпадение осадка
scalpel *s* скальпель
scalpriform *a zool* долотообразный
scalprum *s zool* режущий край резца
scaly *a* чешуйчатый; имеющий чешуйки
scaly-barked чешуекорый
scaly-flowered *a* чешуецветковый
scaly-fruited *a* чешуеплодный
scandent *a bot* лазающий
scanner *s* сканирующее устройство, сканер
infrared ~ инфракрасный сканер, термограф, тепловизор
pH ~ система для проточной записи pH, проточный pH-метр, pH-сканер
tomographic ~ томограф
scanning *s* сканирование, разложение изображения; сканирующий, растровый
multiple view ~стереосканирование, многовекторное сканирование, многопроекционное сканирование
scintillation ~ сцинтиграфия
scansorial *a* приспособленный к лазанию
scape *s* стержень, ствол; *ent* скапус, рукоятка; стрелка, безлистный стебель
scape-bearing *a* стрелконосный
scaphoid *s* ладьевидная кость; *a* ладьевидный, ладьеобразный
scapiflorous *a* стрелкоцветный
scapiform, scapoid *a* стрелковидный
scapose *a* стебельковый
scapula *s* лопатка
scapular *a* лопаточный
scapuloclavicular *a* лопаточно-ключичный
scar *s* рубец, шрам; отпечаток
scarab *s* скарабей *(Scarabaeus)*
scarce *a* редкий, редко встречающийся
scare *v* пугать; отпугивать

scarecrow *s* пугало
scarf *s orn* баклан *(Phalacrocorax)*
scar-fruited *a* рубчикоплодный
scarfskin *s* эпидермис, кожица, кутикула
scarification *s* скарификация
scarious *a* тонкоперепончатый; *бот.* плёнчатый
scarlet *a* алый
scarred *a* рубцеватый
scarring *s* рубцевание
scat *s* экскременты, помёт
scatobiont *s* копробионт, обитатель навоза, обитатель экскрементов
scatology *s* скатология, копрология
scattergram *s* диаграмма рассеяния
scattering *s* рассеивание; разбрасывание
  light ~ светорассеяние, рассеяние света
scaup *s orn* чернеть *(Aythya);* морская чернеть *(Aythya marila)*
scavenge *v* питаться падалью
scavenger *s* падальщик; сапрофит
scavenging *s:* **superoxide** ~ утилизация супероксидных радикалов
scent *s* запах; чутьё, нюх; след; приманка; *v* чуять; охотиться по нюху; **to ~ about** обнюхивать; **to ~ out** почуять
scentless *a* непахучий, без запаха
schedule *s* режим; программа; план; расписание; таблица
scheme *s* схема
  prey ~ стереотип поиска добычи
schindylesis *s anat* зубчатый шов
schizocarp *s* схизокарпий, дробный распадающийся плод
schizocoel *s* шихоцель, схизоцель
schizogamy *s* шизогамия, схизогамия
schizogenesis *s* шизогенез, схизогенез
schizogenetic *a* шизогенетический, схизогенетический
schizognathous *a* шизогнатический, схизогнатический
shizogony *s prot.* шизогония, схизогония, синтомия, множественное деление
schizont *s prot.* шизонт, схизонт
schizophyte *s* шизофит, схизофит
schizostely *s* шизостелия, схизостелия
school *s* стая, косяк *(of fish);* стадо *(of whales); v* собираться стаями, собираться косяками, собираться стадами
schooling *s* скосячивание; *a* стайный
schou *s* тибетский олень, олень шу *(Cervus elaphus wallichi)*
sciad, sciaphyte *s* сциофит, тенелюбивое растение
sciatic седалищный
scintigraphy *s* сцинтиграфия
scintillation *s* сцинтилляция
scintillator *s* сцинтилляционный счётчик
scintiscan *s* сцинтиграмма; сканограмма
scion *s* побег, отпрыск; привой, прививочный черенок
  cleft-graft ~ привой, прививочный черенок
scion-rooted *a* корнесобственный
sciophilous *a* тенелюбивый; цветущий в тени
sciophyte *s* сциофит, тенелюбивое растение
scission *s* расщепление; рассечение; дробление; *bioch* расщепление
scissiparity *s* размножение дроблением
scissura *s anat* щель; трещина
sclera *s* склера, белочная оболочка
sclereid *s* склереида
sclerenchyma *s* склеренхима
sclerite *s* склерит
sclerobasidium *s* склеробазидия
scleroblast *s* склеробласт, скелетообразующая клетка
sclerocarp *s* склерокарпий
sclerocauly *s* жесткостебельность
sclerocorneal *a* склерокорнеальный
scleroderm *s* склеродерма
sclerodermatous *a* склеродермальный
sclerogenic, sclerogenous *a* склерогенный
scleroid *a* склероидный, жёсткий
scleromer *s* метамер скелета, сегмент скелета
sclerophyll *s* склерофилл, жестколистное растение
sclerophylly *s* склерофилия, жестколистность
sclerophyte *s* склерофит
sclerose твердеть; уплотняться; подвергаться склерозу
sclerosis *s* склероз
sclerotic *a* содержащий лигнин; отвердевший; склеральный, склеротический, плотный, твёрдый
sclerotica *s* склера, белочная оболочка
sclerotium *s* склероций

sclerotization *s* склеротизация
sclerotized *a* склеротизированный, имеющий утолщённую оболочку
sclerotome *s* склеротом
scolecospore *s* червеобразная спора, нитевидная спора
scolex *s* сколекс, головка ленточного червя
scolopale *s* штифт, окончание сколопидия
scolophore, scolopidium, scolopophore *s ent* сколопидий
scoot *s* тонкоклювая кайра *(Uria aalge);* гагарка *(Alca torda)*
scopa *s apic* щёточка
scopate, **scopiform** *a* щёткообразный
scops-owl *s* совка *(Otus);* обыкновенная совка, сплюшка*(Otus scops)*
scopula *s prot* скопула
scorch *s* ожог
scoring *s* подсчёт
scorpion *s* скорпион *(Scorpionida)*
scorpionfly *s* скорпионница *(Mecoptera)*
scoter *s orn* турпан, синьга *(Melanitta)*
black ~, common ~ синьга *(Melanitta nigra)*
scotochromogenous *a* скотохромогенный
scotophilous *a* обитающий в темноте
scotopia *s* ночное зрение, скотопическое зрение
scout *s* тонкоклювая кайра *(Uria aalge);* гагарка *(Alca torda)*
scraggy *a zool* тощий, худой; *bot* чахлый
scramble *s eth* схватка; *v* карабкаться, пробираться
scrambler *s* цепляющееся растение
scraper *s* животное-обскрёбыватель
gel ~ лезвие для геля, скребок для геля
scraping *s* соскоб
scratch царапать; чесать; чесаться, почёсываться; скрестись; рыть когтями; рыться
scratching *s* царапанье; почёсывание
scray, scraye *s* речная крачка, обыкновенная крачка *(Sterna hirundo)*
scream *s* визг; *v* визжать
screech *s* крик, визг; уханье *(of an owl);* кричать; ухать *(of an owl);* **the ~ of the owl** уханье совы
screech-owl *s orn* совка *(Otus)*
screen *s* экран; сито; массовое обследование, скриннинг; сетка от насекомых; *v* экранировать; просеивать, сортировать; отцеживать
complement ~ определение активности комплемента
screening *s* отбор, проверка, тестирование; массовое обследование, скриннинг; экранирование; защита
screw *s* винт
screwworm *s* личинка мясной мухи *(Calliphoridae)*
scrobe *s ent* бороздка
scrobiculate *a anat* ямчатый
scrobiculus *s anat* ямка, углубление
scrotal *a* мошоночный
scrotum *s* мошонка
scrub *s* кустарник
scrubby *a* покрытый кустарником; низкорослый; захудалый *(of an animal)*
scrubwood *s* низкорослое чахлое дерево
scruff *s* задняя часть шеи, выя
sculptine *s* скульптина
scum *s* пена; тина
scurf *s* налёт
black ~ *phytop* чёрная короста картофеля *(caused by Corticium solani)*
scurfy *a bot* покрытый мелкими чешуйками; мелкочешуйчатый
scut *s* хвост *(of a hare)*
scutal *a* скутальный; щитковый; щитовой; чешуйчатый
scutate *a* покрытый щитами, покрытый щитками
scute *s* щиток; чешуя, чешуйка; жучка *(of sturgeons)*
scutellar *a* щитковый; имеющий щиток
scutellate *a* щитковидный; покрытый чешуйками, покрытый щитками; имеющий щиток
scutellum *s* скутеллум, щиток
scutiferous *a* щитоносный
scutifolious *a* щитолистный
scutiform *a* щитовидный
scutigerous *a* щитоносный
scuttle *s* поспешное бегство; *v* поспешно убегать
scutum *s* скутум, щит
scypha *s bot* бокалец, сцифа
scyphiferous *a bot* имеющий бокальца, имеющий сцифы
scyphiform *a* чашеобразный, бокаловидный
scyphistoma *s* сцифистома

scyphoid *a* чашеобразный, бокаловидный
scyphulus *s bot* чашечка
sea *s* море; океан; волна; *a* морской; океанический
seabird *s* морская птица
sea-eagle *s* орлан *(Haliaeetus)*
seafoods *s pl* морепродукты
seal *s* тюлень *(Phocidae); pl* ушастые тюлени *(Otariidae)*
Alaska fur ~ северный морской котик *(Callorhinus ursinus)*
Atlantic gray ~ тевяк, серый тюлень, длинномордый тюлень *(Halichoerus grypus)*
banded ~ полосатый тюлень, крылатка *(Histriophoca fasciata)*
bay ~ обыкновенный тюлень, ларга *(Phoca vitulina)*
bearded ~ лахтак, морской заяц *(Erignathus barbatus)*
bladder-nosed ~ хохлач *(Cystophora cristata)*
Caspian ~ каспийский тюлень, каспийская нерпа *(Pusa caspia)*
common ~ обыкновенный тюлень, ларга *(Phoca vitulina)*
crab-eater ~ тюлень-крабоед *(Lobodon carcinophagus)*
crested ~ хохлач *(Cystophora cristata)*
eared ~ сивуч *(Eumetopias jubatus)*
earless ~ обыкновенный тюлень, ларга *(Phoca vitulina)*
elephant ~ морской слон *(Mirounga)*
gray ~ тевяк, серый тюлень, длинномордый тюлень *(Halichoerus grypus)*
Greenland ~ лысун, гренландский тюлень *(Histriophoca groenlandica)*
harbor ~ обыкновенный тюлень, ларга *(Phoca vitulina)*
harp ~ лысун, гренландский тюлень *(Histriophoca groenlandica)*
Hawaiian monk ~ гавайский тюлень-монах *(Monachus schauinslandi)*
hooded ~ хохлач *(Cystophora cristata)*
idle ~ безгаремный самец котик
jar ~ кольчатая нерпа *(Pusa hispida)*
larga ~ обыкновенный тюлень, ларга *(Phoca vitulina)*
leopard ~ морской леопард *(Hydrurga leptonyx)*
Mediterranean monk ~ тюлень-монах, белобрюхий тюлень *(Monachus monachus)*
monk ~ тюлень-монах *(Monachus)*
northern fur ~ северный морской котик *(Callorhinus ursinus)*
ribbon ~ полосатый тюлень, крылатка *(Histriophoca fasciata)*
sand ~ обыкновенный тюлень, ларга *(Phoca vitulina)*
true ~ обыкновенный тюлень, ларга *(Phoca vitulina)*
white-coat ~ белёк
sealer *s* охотник на тюленей; зверобойное судно; запечатывающее устройство
sealing *s* тюлений промысел; котиковый промысел
sealwort *s bot* купена изменчивая *(Polygonatum commutatum)*
seam *s* шов; рубец; морщина
search *s* исследование, изучение; поиск; *v* исследовать, изучать; искать
season *s* сезон, время года; период, время; **to be in** ~ быть в половой охоте, быть в течке; **to come in** ~ приходить в половую охоту, приходить в течку
closed ~ запретный сезон, время запрета
lean ~ сезон снижения упитанности
nonbreeding ~ анэструс
spawning ~ нерестовый период; икрометание
seasonal *a* сезонный
seat *s* местонахождение; месторасположение
seatworm *s* острица *(Enterobius vermicularis)*
sebaceous *a* сальный
sebum *s* секрет сальных желёз
secernent *s* секреторный орган
secernmet *s* секреция
secodont *a nat* секодонтный, бугорчато-режущий
secondary *a* вторичный; второстепенный; придаточный; *s* маховое перо второго порядка
secret, secreta *s* секрет
secretagogue *s* стимулятор секреции; усиливающий секрецию
secrete *v* секретировать, выделять
secretion *s* секреция, выделение; секрет

bile ~ жёлчеотделение
sweat ~ потоотделение
secreto-inhibitory *a* задерживающий секрецию
secretomotor *a* возбуждающий секрецию
secretory *a* секреторный, выделительный
sectile *a* малорассечённый; резной
section *s* рассечение; сечение; разрез; срез
sectioning *s* изготовление срезов; резка
sectorial *a* секториальный; режущий *(teeth)*
secund *a* односторонний, однобокий
secundiflorous *a* односторонне цветковый
secure *v micr* выделять
sedentary *a* сидячий; оседлый; малоподвижный; *bot* прикреплённый
sedge *s* осока *(Carex)*
sedgy *a* поросший осокой
sediment *s* осадок, отстой; нанос, отложение
sedimentation *s* седиментация, отстаивание, образование осадка; отложение наносов
see *v* видеть
seed *s* семя; *v* сеять; обсеменяться; собирать семена; наливаться; *micr* засевать, производить посев
corn ~ *bot* зерновка; семя кукурузы
dodder ~ *bot* рыжик посевной *(Camelina sativa)*
single ~ *bot* зерновка
winged ~ *bot* крылатка
seedball *s* семенная коробочка; шаровидный плод
seedbud *s* семяпочка
seedcase *s* семенная оболочка
seedeater *s* птица, питающаяся семенами; канареечный вьюрок *(Serinus)*
seeded *a* семенной
prickly ~ колючесемянный
seeding *s* обсеменение; посев
seedless *a* бессемянный
seedling *s* сеянец; проросток; саженец; *pl* рассада
seedpod *s* стручок; боб
seedstalk *s* семяножка
seedworm *s ent* вредитель семян; плодожорка *(Laspeyresia)*
seepage *s* впитывание, инфильтрация; пропитывание; просачивание
segment *s* сегмент, членик; *v* делиться на сегменты
body ~ сомит
diversity-region gene ~ D-сегмент гена
insertion ~ инсерционный сегмент; вставка, инсерция
isophene ~ псевдоаллель, изофенный сегмент
joinig-region gene ~ J-сегмент гена
muscle ~ миотом, миомер
ocular ~ *zool* первый сегмент, акрон
tapeworm ~ проглоттида
segmentation *s* сегментация, разделение на сегменты, членистость; дробление; деление
segregation *s* сегрегация, разъединение; *ген* расщепление
seismonastic *a* сейсмонастический
seismonasty *s* сейсмонастия
seismotaxis *s* сейсмотаксис
seizure *s* захват; судорога; приступ, припадок
select *v* отбирать, селекционировать
selection *s* отбор, селекция; ~ **against a dominant allele** отбор против доминантного аллеля; ~ **against a recessive allele** отбор против рецессивного аллеля
clone ~ клоновый отбор, клональная селекция
cumulative ~ накапливающий отбор, накапливающая селекция
directional ~ направленный отбор, направленная селекция
genotypic ~ отбор по генотипу, генотипическая селекция
HAT ~ селекция на среде с гипоксантин-аминоптерин-тимидином, ГАТ-селекция
intrapopulation ~ внутрипопуляционный отбор, внутрипопуляционная селекция
intrasexual ~ внутриполовой отбор, интрасексуальная селекция
line, linear ~ линейный отбор, линейная селекция
mass ~ массовый отбор, массовая селекция
messenger ~ *bioch* выбор матрицы

normalizing ~ нормализующий отбор, селекция
one-way recurrent ~ периодический односторонний отбор, селекция по линии только одного родителя
periodic ~ периодический отбор, периодическая селекция
r- ~ r-отбор
recurrent ~ повторяющийся отбор; периодический отбор, периодическая селекция
recurrent reciprocal ~ периодический реципрокный отбор, периодическая реципрокная селекция
secondary ~ вторичный отбор, вторичная селекция
sexual ~ половой отбор, половой подбор
sib ~ клоновый отбор, клональная селекция

selective *a* селективный, избирательный; отбирающий
selectivity *s* селективность, избирательность
ion ~ избирательная проницаемость для ионов
selenodont *a anat* селенодонтный, лунчатый
selenotropism *s* селенотропизм
self *s imm* «своё»; **non ~, not ~** «не своё», «чужое»
self-absorption *s* самопоглощение, самовсасывание
self-acting *a* авоматический
self-aggregation *s* аутоагрегация; аутоагрегирующий
self-aggression *s imm* аутоагрессия
self-aligning *a* саморегулирующийся
self-annealing *s mbiol* самоотжиг
self-antigen *s* аутоантиген
self-assembly *s* самосборка
self-association *s* самоассоциация
stacking ~ стекинг-самоассоциация
self-balancing *a* самоуравновешивающийся, самобалансирующийся
self-colored *a* однородной масти
self-compatibility *s* самосовместимость
self-complementary *a* самокомплементарный
self-condensation *s* автоконденсация, самоконденсация
self-conditioning *s eth* самообучение; самообусловливание
self-coupling *s* полимеризация
self-determinant *s* аутоантигенная детерминанта, аутоэпитоп, аутодетерминанта
self-development *s* саморазвитие, спонтанное развитие
self-differentiation *s* самодифференциация, аутодифференциация, самодифференцировка
self-digestion *s* аутолиз, самопереваривание
self-discrimination *s* распознавание «своего», когнатное распознавание
"self-education" *s* феномен «самообучения»
self-epitope *s* аутоантигенная детерминанта, аутоэпитоп, аутодетерминанта
self-fertile *a* автофертильный, способный к самооплодотворению
self-fertilization *s* самооплодотворение; самоопыление
self-grafting *s* самопрививка, самосрастание
self-healing *s* самозаживление
self-hemagglutination *s* неспецифическая гемагглютинация, спонтанная немагглютанация
self-idiotype *s* аутоидиотипическая детерминанта, аутоидиотип
self-immunization *s* аутоиммунизация
self-incompatibility *s* самонесовместимость; самостерильность
self-induction *s* самоиндукция
self-infection *s* аутоинфекция
selfing *s* самоопыление; сельфинг
self-inoculation *s* аутовакцинация
self-killer *s* аутологичная клетка-киллер
self-limited *a* самоограничивающийся
self-licking *s* самооблизывание
self-lysis *s* аутолиз
self-marker *s* аутоантигенный маркер, антигенный маркер аутологичной ткани, маркёр «своего»
self-mutilation *s* самокалечение
self-nourishing *a* автотрофный
self-organization *s* самоорганизация

self-peptide *s* аутологичный пептид, аутопептид
self-poisoning *s* аутоинтоксикация
self-pollination *s* самоопыление
self-preservation *s* самосохранение
self-propagating *a* саморазмножающийся
self-pruning *s* самоочищение; самоизреживание
self-purification *s* самоочищение
self-recognition *s imm* распознавание «своего», когнатное распознавание; *eth* самоузнавание
self-regulation *s* саморегулирование, саморегуляция
self-renewing *a* самовосстанавливающийся; самообновляющийся
self-repair *s* самовосстановление
self-replication *s* ауторепликация, саморепликация
self-reproduction *s* самовоспроизводство
self-senzitization *s* аутосенсибилизация
self-seeding *s* самосев
self-sown *a* выросший самосевом
self-splicing *s* автономный сплайсинг, самосплайсинг
self-sterility *s* самостерильность
self-suggestion *s* самовнушение, аутосуггестия
self-thinning *s* самоизреживание
self-tolerance *s imm* аутотолерантность
selva *s* сельва
semantization *s eth* семантизация
sematic *a* сематический, сигнализирующий об опасности
semen *s* сперма, семя
semianatropous *a* полуанатропный, с полувёрнутой семяпочкой
semiaquatic *a* полуводный
semiarid *a* семиаридный, полузасушливый
semicell *s bot* полуклетка
semicircular *a* полукружный
semicomplete *a* неполный
semiconservative *a mbiol* полуконсервативный
semidesert *s* полупустыня; *a* полупустынный
semidominance *s* неполное доминирование
semidormant *a* с пониженнной активностью
semidouble *a* полумахровый
semidwarf *a* полукарликовый
semifloret, semifloscule *s* язычковый цветок
semiflosculous *a* язычковый
semifluid *a* полужидкий
semifrutex *s* полукустарник, кустарничек
semigamy *s* семигамия
semihumid *a* полувлажный
semilethal *s* полулеталь; полулетальный
semiliquid *a* полужидкий
semilocular *a bot* полугнёздный
semilunar *a anat* полулунный; серповидный
semimaturation *s* неполное созревание
semimembranous *a anat* полуперепончатый
semimetamorphosis *s* неполный метаморфоз, неполное превращение
seminal семенной
seminary *s* рассадник; питомник
semination *s* рассеивание семян, распространение семян; осеменение, оплодотворение
seminicole, seminicolous *a* растущий на семенах
seminiferous *a bot* семеноносный; образующий семя; семявыносящий
seminific *a bot* образующий семя; образующий сперму
semiotics *s eth* семиотика
semiparasite *s* полупаразит
semipenniform *a anat* одноперистый
semipermeable *a* полупроницаемый
semiplacenta *s* семиплацента, неполная плацента
semiplume *s* пуховое перо
semipupa *s* полукуколка; псевдонимфа
semiquantitative *a* полуколичественный
semirecondite *a* полуприкрытый
semirunner *s* полустелющееся растение
semisaprophyte *s* полусапрофит
semisavanna *s* полусаванна
semishrub *s* полукустарник
  dwarf ~ полукустарничек
semispecies *s* полувид
semispinal *a anat* полуостистый
semisteril *s gen* полустериль
semistreptostylic *a anat* полустрептостильный
semisynthetic *a* полусинтетический
semithin *a* полутонкий

semituberous *a* полуклубневой
sempervirent *a* вечнозелёный
senescence *s* старение, одряхление
senile *a* сенильный, старческий
senility *s* старость
senior *s* половозрелое животное
sensation *s* ощущение, восприятие; чувство
sense *s* чувство
~ **of smell** обоняние
~ **of taste** вкус
~ **of touch** осязание
color ~ цветовое зрение, цветоощущение, цветоразличение
genesic ~ инстинкт размножения
kinesthetic ~ кинестезия, мышечное чувство
light ~ светоощущение
muscular ~ кинестезия, мышечное чувство
pressure ~ барестезия, ощущение давления
reproductive ~ инстинкт размножения
static ~ чувство равновесия, статокинетическое чувство
tactile ~ осязание
senseless *a* нечувствительный
sensibility *s* чувствительность; восприимчивость
sensibilization *s* сенсибилизация
sensibilizator, sensibilizer, sensibilizin *s* сенсибилизирующий агент, сенсибилизатор, сенсибилизин
sensiferous *a* передающий ощущение
sensillum *s* сенсилла
sensimeter *s* эстезиометр
sensing *s*: **remote** ~ дистанционное изучение, телеметрия
sensitive *a* чувствительный
sensitivity *s* чувствительность
light ~ светочувствительность
sensitization *s* сенсибилизация
sensitizer *s* сенсибилизирующий агент, сенсибилизатор, сенсибилизин
sensomotor *a* сенсомоторный, сенсорно-двигательный
sensor *s* датчик; сенсор; рецептор
sensorial *a* сенсориальный
sensorimotor *a* сенсомоторный, сенсорно-двигательный
sensorium *s* сенсориум, чувствительный центр; сознание
sensory *a* сенсорный, чувствительный
sentinel *s eth* особь-часовой
sepal *s* чашелистик
sepaled *a* имеющий чашелистики
sepaloid *a* чашелистиковидный
sepalous *a* имеющий чашелистики
separable *a* отделимый; разделимый
separate *v* отделять; разделять; разобщать; отдельный; раздельный
separation *s* отделение; разделение; разобщение; разлучение; ~ **by chromatography** хроматографическое разделение
anaphase ~ анафазное расхождение
distance ~ пространственная изоляция
foam ~ отделение путём вспенивания
separator *s* разделитель, сепаратор
centrifugal ~ разделяющая центрифуга, центробежный сепаратор
phase ~ фазовая центрифуга, фазовый разделитель, делительный сепаратор
sepsis *s* сепсис
septal *a* относящийся к перегородке
septate *a* имеющий перегородки; разделённый перегородками; септатный
septempartite *a* семираздельный
septicidal *a bot* растрескивающийся по перегородкам
septiferous *a* имеющий перегородки, разделённый перегородками
septulate *a* ложноперегородчатый
septulum *s* небольшая перегородка, вторичная перегородка
septum *s* септа, перегородка
sequenation *s* секвенирование
sequenator *s* секвенатор
sequence *s* последовательность; ряд; чередование; порядок следования
Alu ~ *Alu*-последовательность
arithmetic ~ *биом.* арифметическая последовательность, арифметическая прогрессия
coding ~ кодирующая последовательность, кодирующая область *(гена);* экзон
consensus ~ консенсус, консенсусная последовательность
generation ~s чередование поколений
geometric ~ *биом.* геометрическая последовательность, геометрическая прогрессия

intervening ~ вставочная последовательность, интрон
sequencer *s* секвенатор
sequencing *s mbiol* секвенирование
dideoxy ~ дидезокси-секвенирование
sequestration *s* секвестрация
seral *a* сукцессионный, серийный
seralbumin *s* сывороточный альбумин
serclimax *s* серклимакс
sere *s* серия
serial *a* серийный
sericate, sericeous *a bot* шелковистый
series *s* серия; ряд
ecological-edaphic ~ эдафогенная сукцессия
random ~ случайная последовательность
trial ~ опытная серия
seroconversion *s* сероконверсия
seroculture *s micr* серокультура
serodiagnosis *s* серодиагностика
serodifferentiation *s* классификация серотипов
serofibrinous *a* серозно-фибринозный
seroimmunity *s* пассивный иммунитет, сывороточный иммунитет
serology *s* серология
seromarker *s* сывороточный маркер
seromucous *a* серозно-слизистый
seronegative *a* сероотрицательный, серонегативный
seropositive *a* сероположительный, серопозитивный
seroreaction *s* серологическая реакция; серологическая проба, серотест; сывороточная реакция; сывороточная болезнь
seroresponse *s* гуморальный иммунный ответ; сывороточная реакция; сывороточная болезнь
serosa *s* серозная оболочка
serosal *a* серозный
seroscopy *s* агглютиноскопия; серологический анализ крови
serose *s* серозная оболочка, сероза
serosity *s* серозная жидкость; сыворотка; сывороточный; серозный характер
serotaxonomy *s* серологическая таксономия
serotest *s* серореакция, серологическая реакция, серотест
serotinal *a* поздний, поздноцветущий
serotine *s мам.* кожан *(Eptesicus)*
serotinous *a* поздний, поздноцветущий
serotoxin *s* серотоксин, сывороточный токсин
serotype *s* серотип, серологический тип
serotyping *s* серотипирование
serous *a* серозный
serozyme *s* фактор VII, антифибринолизин, проконвертин
serpent *s* змея
serpentine *a* змеиный; змеевидный
serra *s* пильчатая структура
serrate *a* зубчатый; пильчатый
serrate-ciliate *a* пильчато-реснитчатый
serrate-dentate *a* пильчато-зубчатый
serrate-flowered *a* пильчато-цветковый
serrate-leaved *a* пильчато-листный
serrate-spiny *a* с колючими зубчиками
serratiform *a* зубчатообразный; пиловидный
serration *s* зубчатость; зубчик
serratulate *a* мелькопильчатый, мелкозубчатый
serrature *s* зубчатость; зубчик
serriform *a* зубчатообразный; пиловидный
serrulate *a* мелкопильчатый
serule *s* серула
serum *s* сыворотка; серозная жидкость; серозный экссудат
antibody-containing ~ антисыворотка, иммунная сыворотка
hyperimmune ~ антисыворотка после реиммунизации, гипериммунная сыворотка
immune ~ антисыворотка, иммунная сыворотка
protective ~ антисыворотка, иммунная сыворотка
test ~ стандартная сыворотка, тест-сыворотка
serumal *a* сывороточный
serumless *a* бессывороточный
serval *s ther* сервал *(Felis serval)*
serve *v* случать
service *s* действие, функция; случка, половой акт, осеменение; ирга канадская *(Amelanchier canadensis)*
Service *s*:
Fish and Wildlife ~ Служба охраны рыболовства и диких животных *(США)*

Wildlife ~ Служба охраны диких животных *(США)*
serviceberry *s* ирга *(Amelanchier)*
sesame *s* кунжут *(Sesamum)*
sesamoid *s* сезамовидная кость; *a* сезамовидный
sesquidiploid *s* сесквидиплоид
sessile *a* сидячий; *zool* прикреплённый; *bot* бесчерешковый
set *s* ряд; серия; набор; *biom* множество, совокупность; кладка; саженец; севок; посадочный материал; *v* сажать, высаживать; завязь; завязываться *(of fruit);* делать стойку *(of a dog);* ставить сеть; телосложение; насиженный *(egg);* **to ~ on** натравливать *(a dog);* **to ~ out** высаживать; **to ~ up** восстанавливать силы
~ **of head** постав головы *(in a horse)*
~ **of observations** серия наблюдений
~ **of pupies** помёт щенят
~ **of samples** серия образцов; серия проб
~ **of symptomes** симптомокомплекс
chromosome ~ хромосомный набор
control ~ контрольная группа
critical ~ *biom* критическое множество
preparatory ~ установочная реакция
seta *s* щетина; щетинка; *bot* спорофор
setaceous *a* покрытый щетинками; щетинистый; щетинковидный
setback *s* регресс, задержка
crop ~ неурожай
setiferous, setigerous *a* щетинконосный
setireme *s ent* плавательная конечность
setting *s* завязывание *(e.g., of fruit);* ~ **up** закладка
settle осаждаться, отстаиваться, оседать; заселять, поселять; *apic* собираться
settled *a* осевший; оседлый; заселённый; оплодотворенная, беременная
settlement *s* поселение; колония
settler *s* отстойник
settling *s* оседание; осаждение, отстаивание; осадок, отстой
setula, setule *s* щетинка
setulose, setulous *a* покрытый щетинками; щетинконосный
setup *s* устройство, приспособление; установка, регулировка; заплеск, накат волны, нагон
sever *v* отделять; разъединять; перерезать, разрезать
sewage *s* сточные воды
sex *s* пол; *v* устанавливать пол, определять пол
sexduction *s* сексдукция
sexfid *a* шестинадрезный
sexine *s pal* сэкзина
sexing *s* установление пола, определение пола
sexless *a* бесполый
sex-limited *a* ограниченный полом
sex-linkage *s* сцепление с полом
sex-linked *a* сцепленный с полом
sexual *a* половой, сексуальный
sexuality *s* сексуальность, наличие полов; разделение полов; раздельнополость
sexupara *s* тля-полоноска
shade *s* тень; *v* затенять; защищать
shade-enduring *a* теневыносливый
shade-intolerant *a* не выносящий тени
shade-requiring *a* тенелюбивый
shade-tolerant *a* теneустойчивый
shading *s* затенение, притенение
shadow *s* тень; *v* затенять; защищать
shadow-fighting *s* «чистка» рогов о кусты
shadowing *s lab* оттенение
shaft *s* ствол; стебель; столб; стержень;стержневидная структура; ось; стержень
~ **of bone** *анат.* диафиз
~ **of femur** тело бедренной кости
~ **of fibula**
~ **of raceme** *bot* ось кисти
shag *s* баклан *(Phalacrocorax)*
shaggy *a* лохматый; мохнатый; косматый; длинношёрстный; длиннощетинистый
shake *s* встряхивание; *v* встряхивать
shaker *s* шюттель-аппарат, шейкер; качалка; *pl bot* трясунка средняя *(Briza media)*
gyratory ~, rotary ~ шюттель-аппарат с круговым вращением, ротационный шейкер
shaking *s eth* встряхивание
shallot *s* лук-шалот *(Allium ascalonicum)*
shallow *a* неглубокий, мелкий; поверхностный
sham *v* притворяться

shank *s* голень; стержень; ствол; стебель
shanking *s phytop* стеблевая гниль лука *(caused by Phytophthora)*
shape *s* форма
shapeless *a* аморфный, бесформенный
shaping *s* формирование навыка
sharebone *s* лобковая кость
sharp-fruited *a* остроплодный
sharp-haired *a* островолосистый
sharp-petaled *a* остролепестный
sharp-pointed *a* остроконечный
sharp-toothed *a* острозубчатый
shatter *v* осыпаться *(e.g., of seeds)*
shear *s* сдвиг
shearing *s mbiol* гидродинамическое фрагментирование
shearwater *s orn* буревестник *(Puffinus)*
she-ass *s* ослица
sheath *s anat* оболочка; влагалище; *ent* надкрылье; *бот.*обвёртка; *mmicr* капсула, чехол, футляр
  bundle ~ *bot* обкладка сосудистых пучков
  crural ~, femoral ~ бедренная фасция
  Henle's ~ периневрий
  medullary ~ *bot* медуллярное покрытие; кольцо протоксилемы; *бот.* сердцевинная трубка; миелиновая оболочка
  nerve ~ нервное влагалище, периневрий
  phage ~ наружный чехол фага
  pith ~ заболонь
  wing ~ *ent* элитра, надкрылье
sheathing *a* кроющий
shed линять, менять оперение; сбрасывать; терять *(teeth);* ронять *(leaves);* осыпаться *(of leaves);* **to ~ the eggs** нереститься, метать икру, откладывать икру
shedder *s* линяющее животное
she-deer *s* важенка
she-dog *s* сука, самка собаки
sheep *s* овца, баран; горный баран *(Ovis)*
  Barbary ~ гривистый баран *(Ammotragus lervia)*
  bighorn ~ снежный баран, толсторог, чубук *(Ovis canadensis)*
  blue ~ голубой баран, нахур *(Pseudois nayaur)*
  long-wooled ~ длинношёрстная овца
  Marco Polo's ~ памирский архар, памирский баран *(Ovis ammon polii)*
  mountain ~ снежный баран, толсторог, чубук *(Ovis canadensis)*
  musk ~ мускусный бык, овцебык *(Ovibos moschatus)*
  Pamir ~ памирский архар, памирский баран *(Ovis ammon polii)*
  red ~ уриал, ладакский баран *(Ovis ammon vignei)*
  wild ~ горный баран, архар *(Ovis ammon)*
sheepskin *s* овчина
sheet *s* лист; пластина; пласт
sheeting *s* расслаивание
she-goat *s* коза
shelduck *s orn* пеганка *(Tadorna)*
  common ~ пеганка, атайка *(Tadorna tadorna)*
  roody ~, ruddy ~ огарь, красная утка *(Tadorna ferruginea)*
shelgoose *s orn* пеганка *(Tadorna)*
shell *s* раковина; раковинка; панцирь, щит, домик, капсула, оболочка; скорлупа; кожура; моллюск
  butterfly ~ *мал.* донакс *(Donax)*
  cameo ~ шлемовидная улитка *(Cassididae)*
  coat-of-mail ~ панцирный моллюск *(Loricata)*
  conch ~s *pl mal* стромбиды *(Strombidae)*
  cone ~ *mal* конус *(Conidae)*
  ear ~ *mal* морское ушко *(Haliotidae)*
  horn ~ оболочка рога
shellac *s* шеллак
shellfish *s* водное животное с панцирем
shell-leaved *a* раковинолистный
shell-less *a* беспанцирный
shelly *a* имеющий раковину; похожий на раковину
shelter *s* убежище, прикрытие; укрывать, служить убежищем; защищать; домик
shelterbelt *s* защитная полоса
sheltered *a* закрытый; защищённый; богатый убежищами; ремизный
shelterwood *s* материнское насаждение
she-wolf *s* волчица
shield *s* щит; щиток; экран; защита
  embryonic ~ зародышевый щиток
  ventral ~ *herp* пластрон; брюшной щит
shieldbone *s anat* лопатка

shieldbug *s ent* щитник *(Pentatomidae)*
shielding *s* экранирование; защита
shift *s* смещение, сдвиг; смещаться; перемещение
reading frame ~ *mbiol* изменение порядка считывания, сдвиг рамки считывания
shin *s* голень
shiver *s* дрожь; *v* дрожать
shock *s* шок; удар; толчок; *v* вызвать шок
electroconvulsive ~ электрошок, электросудорожное раздражение
shoebill *s* китоглав, королевская цапля *(Balaeniceps rex)*
shoelike *a bot* башмачковидный
shoot *s* веточка; росток, побег; рост; прорастание; прирост; *pl* поросль; *v* расти; прорастать; распускаться; **to ~ leaves** покрываться листьями
~ **of apex** *bot* конус нарастания
coppice ~ придаточный побег; *pl* корневая поросль
deferred ~ росток из спящей почки
floral ~ цветоножка, цветоносный побег
resting ~ спящий побег, спящий черенок
root ~ корневой отпрыск, побег корневого происхождения
stool ~s *pl* корневая поросль; пнёвая поросль
shooting *s* отстрел; стеблевание
shore *s* берег; побережье; шор
shorn *past participle* остриженный
shortage *s* недостаток, нехватка
short-awned *a* короткоостистый
short-boled *a* короткоствольный
short-branched *a* коротковетвистый
short-day *a* короткодневный
short-eared *a* короткоколосистый; короткоухий
short-flowering *a* короткоцветковый
short-fruited *a* короткоплодный
short-haired *a* гладкошёрстный
Shorthorn *s* шортгорн
short-leaved *a* коротколистный
short-legged *a* коротконогий
short-lived *a* короткоживущий, недолговечный
short-lobed *a* коротколопастный
short-nerved *a bot* коротконервный
short-petaled *a* коротколепестный
short-stalked *a* короткостебельчатый
short-stemmed *a* короткоствольный
short-styled *a* короткопестичный
short-term *a* кратковременный; краткосрочный
short-thorned *a* короткошиповатый
short-toothed *a* короткозубчатый
short-winged *a* короткокрылый
short-wooled *a* короткошёрстный
shot *s* доза; укол; впрыскивание; инъекция
booster ~ бустер-инъекция
shoulder *s* плечо; *pl* загривок, холка
shoulder-blade лопатка
shoveler *s orn* широконоска *(Anas clypeata)*
shrew *s* землеройка *(Soricidae)*
brown-toothed water ~ водяная землеройка, кутора *(Neomys)*
Eurasian ~ водяная землеройка, кутора *(Neomys)*
shriek *s* пронзительный крик, визг; *v* пронзительно кричать, визжать
shrike *s orn* сорокопут *(Lanius)*
red-backed ~ обыкновенный жулан *(Lanius collurio)*
shrill *s* пронзительный вой; пронзительный крик; пронзительно выть; пронзительно кричать
shrill-bark *s* визгливый лай
shrink *v* съёживаться; сморщиваться; усыхать
shrinkage *s* сокращение; сжатие; сморщивание
shrivel *v* вянуть; сморщиваться; съёживаться; пожухнуть
shrub *s* кустарник
dwarf ~, low ~ кустарничек
shrubage *s* кустарниковый ярус
shrubbery *s* кустарник; *bot* клекачка перистая *(Staphylea pinnata)*
shrubby *a* кустарниковый
shrubwood *s* низкоствольный лес, низкоствольник; порослевый лес
shuck *s* шелуха; скорлупа; лущить
~ **of corn** обвёртка кукурузного початка
shudder *s* дрожь; вздрагивание; *v* дрожать; вздрагивать
shuffle *v* поворачивать яйца
shunt *s* шунт, сброс; *v* отводить, шунтировать

shutoff *s* закрытый сезон; остановка; выключение
shy *a* пугливый; настороженный
sialaden *s* слюнная железа
sialic, sialine *a* слюнный
sialogenous *a* слюнообразующий
sialosis *s* слюноотделение
sibbing *s* скрещивание сибсов
siblings *s pl* сибсы, сиблинги
sibmating *s* скрещивание сибсов
sibs *s pl* сибсы, сиблинги
sibship *s* кровное родство
siccation *s* иссушение
siccocolous *a* ксерофильный
siccostabile *a* засухоустойчивый
sick *a* больной
sickle-fruited *a* серпоплодный
sickle-leaved *a* серполистный
sickle-lobed *a* серполопастный
sicklemia *s* серповидно-клеточная анемия
sickle-shaped *a* серповидный
sicklewort *s bot* черноголовка обыкновенная *(Prunella vulgaris)*
sickling *s* образование серповидных форм
sickly *a* болезненный, хворый, хилый, нездоровый; тошнотворный
sickness *s* болезнь, заболевание
side *s* сторона; бок
lee ~ подветренная сторона
sidebone *s* подвздошная кость
side-chain *s mbiol* боковая цепь
side-flowered *a* бокоцветковый
side-leaved *a* боколистный
siderogenous *a* железообразующий
siderophile *s* сидерофил
siderophilic, siderophilous *a* сидерофильный
side-veined *a bot* боконервный
sieve *s* молекулярное сито; микрофильтр; *v* просеивать, фильтровать; подвергать гель-фильтрации
sievebone *s* решётчатая кость
sieving *s* просеивание, фильтрование; гель-фильтрация, гель-хроматография
molecular ~ гель-фильтрация, гель-хроматография
sift *v* просеивать, просеиваться; фильтровать
sight *s* зрение; видеть; **out of** ~ вне поля зрения
sighting *s* визуальное наблюдение
sightless *a* слепой
sigillate *a* печаточный
sigmoid *a* сигмовидный
sign *s* знак; символ; признак, симптом; регуляторная последовательность; *mbiol*
сигнальная последовательность; след
recognition ~ *mbiol* сайт узнавания; сигнал для узнавания
stop ~ терминирующий кодон, стоп-кодон
signal сигнал; знак; ~ **"go"** индуцирующий сигнал; сигнал активации клеток
augmentative ~ усилительный сигнал, усиливающий сигнал; *imm* хелперный фактор, хелперный сигнал, T-хелперный фактор
signaling передача сигнала; сигнализация
mitogenic ~ митогенная индукция; митогензависимая активация
significance *s* значительность, важность; значение; *biom* значимость, достоверность
significant *a* значительный, важный; *biom* значимый, достоверный
sika *s* пятнистый олень *(Cervus nippon)*
silage *s* силос; *v* силосовать
silencing *s mbiol* сайленсинг
silent *a* молчащий; бессимптомный
silica *s* окись кремния
siliceous *a* кремнезёмный
silicle *s* стручочек
silicular *a* стручковидный, стручкообразный
siliculose, siliculous имеющий стручки
silique *s* стручок
siliquiform *a* стручковидный, стручкообразный
silk *s* шёлк
silk-bearing *a* шелконосный
silkmoth *s* шелкопряд; шелкопряд *(Bombyx);* шелкопряд тутовый *(Bombyx mori)*
silkworm *s* тутовый шелкопряд *(Bombyx mori)*
silky *a* шелковистый
silt *s* ил; илистый нанос; **to** ~ **up** заиливать
bed ~ донные отложения

siltation, silting up *s* заиление, отложение ила
silva *s* лес; лесная растительность
silvan *a* лесной
silver *a* серебристый
silvereye *s orn* белоглазка *(Zosterops)*
silverfish *s ent* чешуйница *(Lepisma)*
silvicultural *a* лесоводческий
silviculture *s* лесоводство
silvics *s* лесоведение
simian *s* обезьяна; *a* обезьяний; обезьяноподобный
similarity *s* сходство, подобие
simplex *s gen* симплекс
simulation *s* симуляция; имитация; имитационное моделирование
simultaneous одновременный; синхронный
sinew *s* сухожилие; *pl* мускулатура
sinewy *a* мускульный; мускулистый
singing *s* пение
single *a* одиночный; *bot* немахровый
single-celled *a* одноклеточный
single-flowered *a* одноцветковый
single-flowering *a bot* одиночноцветущий
single-fruited *a bot* одиночноплодный
single-layered *a* однослойный
single-stranded *a mbiol* однонитевой, одноцепочечный
singlet *a* синглетный
singleton *s* одинец, одиночное животное
sinistral *a* левосторонний, левый; левозавитой; леворукий
sinistrogyration *s* поворот влево
sinistrorse *a bot* левовращающий
sink *v* погружаться, опускаться
sinks *s*: **metabolic** ~ *micr* метаболические стоки
sinoatrial *a* синусно-предсердный
sinocarotid *a* синокаротидный
sinuaperturate *a pal* выемчато-апертурный
sinuate *a bot* выемчатый
sinuation *s* извитость
sinuatrial *a* синусно-предсердный
sinuauricular *a* синоаурикулярный; *bot* выемчато-ушковатый, с выемчатыми ушками
sinuosity *s* извитость
sinuous *a* извилистый, волнистый
sinus *s anat* синус; пазуха; полость
  accessory ~es of nose придаточные пазухи носа
  basilar ~ основная пазуха; базальное сплетение
  cardinal ~ кардинальная вена, кювьеров проток
  cerebral ~ желудочек голоного мозга
  maxillary ~ гайморова полость, верхнечелюстная пазуха
  rectal ~ заднепроходная пазуха, морганиева пазуха, прямокишечная пазуха, анальный синус
  venous ~ of sclera венозный синус склеры, шлеммов канал
siotropism *s* сиотропизм
siphon *s* сифон
siphonate *a zool* имеющий один *или* несколько сифонов
siphonogamic, siphonogamous *a* сифоногамный
siphonogamy *s* сифоногамия
siphonoglyph *s* сифоноглиф
siphonostomatous *a* имеющий трубчатый рот; имеющий канал для сифона
siphonozooid *s* сифонозооид
siphorhinal *a* имеющий трубчатые ноздри
sippe *s* таксон
sire *s* производитель; *v* давать потомство; осеменять
sired *a* оплодотворённая; осеменённая
siskin *s* чиж *(Spinus)*
sisterly *a* сестринский
sit сидеть; насиживать; высиживать
site *s* место, участок; местообитание; местонахождение; *gen* участок, сайт
  ~ of action место приложения действия
  ~ of entry входные ворота
  A- ~ *см.* **aminoacyl-tRNA-binding site**
  active ~ активный участок; активный центр
  aminoacyl ribosome ~ аминоацильный центр рибосомы
  aminoacyl-tRNA-binding ~ аминоацил-тРНК-связывающий участок, А-участок, А-сайт
  assembly ~ участок сборки, сайт сборки
  binding ~ связывающий участок, сайт связывания; центр связывания
  breeding ~ гнездовье

cap ~ *mbiol* сайт кэпирования, кэп-сайт
cell-binding recongition ~ клеточносвязывающий сайт узнавания
chain-antigenic ~ секвенциальный эпитоп
complement fixation ~, complement fixing ~ комплементсвязывающий участок, сайт фиксации комплемента
consensus end ~ сигнал об окончании считывания; сайт терминации транскрипции
coupling ~ место сопряжения
ectopic ~ эктопический участок, эктопическая область
fertilization ~ место оплодотворения
hidden valence ~ скрытая антигенная детерминанта, латентная антигенная детерминанта, криптотоп
home ~ гнездовье; участок гнезда
initiation ~ участок инициации, сайт инициации
linear antigenic ~ антигенная детерминанта секвенциального типа в линейном полимере
mRNA-splicing ~ *mbiol* сигнал для сплайсинга мРНК
nesting ~ гнездовье
overlapping ~ перекрывающийся участок, перекрывающаяся последовательность
p ~, peptidyl ribosome ~, peptidyl-tRNA binding ~ пептидил-тРНК-связывающий центр рибосомы, пептидильный центр рибосомы
polyadenylation ~ сайт полиаденилирования
R ~, recognition ~ участок распознавания, R-участок, сайт узнавания
switch ~ сайт переключения, свич-сайт
unbound binding ~, unoccupied binding ~ свободный сайт связывания, незанятый сайт связывания

site-specific *a* сайт-специфичный
sitotropism *s* ситотропизм
sitter *s* наседка
situation *s* местоположение; положение
size *s* величина, размер
colony ~ размер колонии
sample ~ *biom* объём выборки
sizeable *a* значительных размеоров
skein *s* клубок
skeletal *a* скелетный
skeletogenous *a* образующий скелет, скелетогенный
skeletogeny *s* развитие скелета
skeleton *s* скелет, остов
external ~ внешний скелет, экзоскелет
facial ~ лицевой череп, кости лица
internal ~ внутренний скелет, эндоскелет
skeletonizer вредитель, скелетирующий листья
skew *a* косой; скошенный; наклонный
skewness *s biom* асимметрия; деформация
skidding *s* трелёвка
skill *s eth* умение; навык
motor ~ двигательный навык; тонкое моторное умение
skimmer *s orn* водорез *(Rhynchops)*
skin *s* кожа; шкура; шкурка; кожица, кожура; оболочка, плёнка; верхний слой; *v* снимать кожу; снимать шкуру, сдирать шкуру
inner ~ дерма, собстенно кожа
nail ~ надногтевая пластинка, эпонихий
sheep ~ овчина
true ~ дерма, собственно кожа
yolk ~ желточная оболочка
skink *s herp* сцинк *(Scincus)*
skiophilous *a* тенелюбивый; цветущий в тени
skiophyte *s* сциоцит, тенелюбивое растение
skipper *s* прыгающее насекомое; *ent* толстоголовка *(Hesperiidae)*
skirt *s* опушка, окраина
skittish *a* норовистый *(a horse)*
skototaxis *s* скототаксис
skua *s* поморник *(Stercorarius)*
skulk *v* пробираться; скрываться, прятаться
skull *s* череп
skullcap *s* свод черепа, черепной свод, крыша черепа; *bot* шлемник *(Scutellaria)*
skunk *s ther* скунс; полосатый скунс *(Mephitis)*
skylark *s* полевой жаворонок *(Alauda arvensis)*
slab *s* пластинка, полоска, блок

slant *s* скошенная плотная питательная среда, «косячок»; *a* скошенный, косой
agar ~ скошенный агар.
stock ~ исходная культура на скошенной среде
slanting, slanting of medium *s* скашивание питательной среды
slash *s* вырубка; валежник, порубочные остатки; зарубка, надрез; болотистая местность
slater *s* мокрица *(Oniscidae)*
slaty *a* слоистый; пластинчатый
slaughter *s* убой; *v* убивать, забивать
sleep *s* сон; *v* спать
frozen ~ гипотермия, искусственная зимняя спячка
winter ~ зимняя спячка, гибернация
sleeper *s ther* соня *(Gliridae)*
sleepiness *s* сонливость
sleeplessness *s* бессонница
sleepy *a* вялый; гниющий *(fruit)*
slender *a* тонкий
slender-limbed *a bot* тонкочленный
slender-stemmed *a* тонкостебельчатый
slice *s* срез; шлиф
slick *s* солевое пятно *(on soil)*
slide *s* предметное стекло; микроскопический препарат
microscope ~ предметное стекло; микроскопический препарат
slime *s* слизь
slimy *a* слизистый
sling *v* высыпать, выбрасывать
slink *s zool* выкидыш, недоносок; *v* выкинуть; красться
slip *s* скольжение; скользить; сдвиг, смещение; побег, отросток; отводок; *v* выкинуть, скинуть плод; спускать *(dogs)*
slipcover *s bot* лодочка
slippery *a* скользкий; увёртливый
slit *s anat* щель; продольный разрез; *v* разрезать в длину
slither *v* скользить *(e.g., of snakes)*
sloe *s* слива колючая, тёрн *(Prunus spinosa)*
slope *s* склон, откос; скошенная твёрдая среда, №косячок¤
agar ~ скошенный агар, «косячок»
continental ~ материковый склон, континентальный склон
curve ~ наклон кривой
serum ~ скошенная сывороточная среда
sloth *s ther* ленивец *(Bradypus)*
slough *s* выползок, сброшенная кожа *(of a snake);* влажная болотистая местность; топь, трясина; заводь; *v* осыпаться *(of fruit)*
sloughing *s* отторжение некротических масс
slow *a* медленный
slow-bolting *a bot* вяло стрелкующийся
slow-growing *a* медленно растущий, тугорослый
slow-maturing *a* позднеспелый, медленно созревающий
slowworm *s herp* ломкая веретеница, медяница *(Anguis fragilis);* чешуеног *(Pygopus)*
sludge *s* ил; тина; осадок; отстой
slug *s* слизень; *phys* компактная порция
sluggish *a* медленный, вялый; медлительный
slumber *s* сон
slurry *s* суспензия
small *a* маленький, небольшой; мелкий
~ **of the back** поясница, поясничная облать
small-celled *a* мелкоклеточный
small-combed, small-crested *a bot* мелкогребешковый
small-fruited *a* мелкоплодный
small-leaved *a* мелколистный
small-pistilled *a* мелкопестичный
smallpox *s* оспа
small-seeded *a* мелкосемянный
small-stalked *a* мелкостебельчатый
smart саднить
smear *s* мазок; *v* делать мазок
smedivirus *s* вирус Смеди
smegma *s* смегма, препуциальная смазка
smell *s* запах; обоняние; нюхать
smew *s orn* луток, малый крохаль *(Mergus albellus)*
smooth *a* гладкий; ровный
smooth-edged *a* цельнокрайний
smoothering *s* заглушение, глушение *(by weeds)*
~ **of seedlings** удушье сеянцев *(caused by Thelephora terrestris)*
smooth-flowered *a* гладкоцветковый
smooth-fruited *a* гладкоплодный
smooth-haired *a* гладкошёрстный

smoothing *s biom* сглаживание
smooth-leaved *a* гладколистный
smooth-seeded *a* гладкосемянный
smooth-shelled *a* с гладкой раковиной, с гладким панцирем
smooth-spiked *a* гладкоколосый
smooth-walled *a* с гладкой оболочкой
smut *s phytop* головня *(caused by Ustilaginales)*; головнёвый гриб; *v* заражать головнёй
smutty *a* поражённый головнёй
snag *s* сук, сучок; нарост, кап; *v* срезать сучья
snail *s* улитка *(Gastropoda)*
  heteropodous ~s *pl* киленогие *(Heteropoda)*
  sessile-eyed ~s *pl* сидячеглазые *(Basommatophora)*
snail-eating *a* питающийся улитками
snail-shaped *a* раковинообразный; улитковидный
snake *s* змея *(Serpentes)*
  Aesculapian ~ эскулапова змея *(Elaphe longissima)*
  black ~ чёрный полоз *(Coluber constrictor)*
  black and yellow ~ куроед *(Spilotes)*
  blind ~ слепозмейка *(Typhlops)*; узкоротая змея *(Leptotyphlops)*
  carpet ~ ромбический питон *(Morelia)*
  colubrid ~s ужеобразные змеи, ужовые *(Colubridae)*
  constricting ~ удав *(Boinae)*
  copperhead ~ щитомордник *(Agkistrodon)*; австралийская медноголовая змея *(Austrelaps)*
  double-ended ~, glass ~ слепозмейка *(Typhlops)*
  grass ~ обыкновенный уж *(Natrix natrix)*; песчаная змея *(Psammodynastes)*; подвязочная змея *(Thamnophis)*; песчаный уж Фитцингера *(Psammophylax)*
  gray rat ~ серый лазающий полоз, крысиная змея *(Elaphe obsoleta)*
  green ~ песчаная змея *(Psammodynastes)*; африканский зелёный уж *(Philothamnus)*
  hair ~ волосатик *(Gordiacea)*
  hog-nosed ~ свиноносая змея, крючконосый уж *(Heterodon)*
  innocuous ~ неядовитая змея
  smooth ~ медянка *(Coronella)*
snakebird *s orn* змеешейка *(Anhinga)*
snakeskin *s*: **shed** ~ выползок
snaking *s* трелёвка
snap *v* хватать *(by teeth)*, схватить; кусать, огрызаться; *s* укус; щелканье, треск; ~ **at bait** взять приманку
snap-freezing *s* быстрое замораживание, мгновенное замораживание
snare *s* петля; силок; ловушка
snarl *s* ворчание; рычание; *v* ворчать; рычать; огрызаться
sneeze *v* чихать
sniff *v* нюхать, обнюхивать, принюхиваться; чуять; сопеть; фыркать
sniffing *s* обнюхивание
snigging *s* трелёвка
snipe *s* бекас *(Gallinago)*
  great ~ дупель *(Gallinago media)*
  half ~, jack ~ гаршнеп *(Lymnocryptes minima)*
snood *s orn* серёжка
snore *s* храп; *v* храпеть
snort *s* фырканье; храпение; *v* фыркать; храпеть
snout *s* рыло, морда; *ent* хоботок; *bot* рыльце
snout-shaped *a* хоботковидный
snowballing *s* быстрый рост, интенсивное развитие
snowbreak *s* снеголом, повреждение снегом
snowcock *s* улар, горная индейка *(Tetraogallus)*
snowdrift *s* сугроб; *bot* бурачок приморский *(Alyssum maritimum)*
snuff *v* нюхать, обнюхивать; вдыхать; чуять
snuffle *v* сопеть
soak *v* впитывать, впитываться, всасывать, всасываться; пропитывать, пропитываться; просачиваться
soar *s* парение, парящий полёт; *v* парить
sobole *s bot* побег, отпрыск; *pl* поросль
soboliferous *a* дающий побеги, дающий отпрыски
sociability *s* общественность, стайность
social *a* общественный, живущий в сообществе
socialization *s* установление общественных отношений

sociation *s* социация
socies *s* соция
society *s* сообщество; социэта
layer ~, stratum ~ ярусная социэта, ламиация
sociogram *s* социограмма
sociology *s* изучение сообществ
ecological ~ синэкология
plant ~ фитоценология; геоботаника
systematic plant ~ систематика растительных сообществ
sociule *s* социула
socket *s anat* впадина; ячейка; суставная ямка
~ of the hip вертлужная впадина
antennary ~ антеннальная ямка
eye ~ глазница
tooth ~ зубная альвеола, луночка зуба
sod *s* дернина; дёрн
sodding *s* задернение
soddy *a* дерновый
soft *a* мягкий, пластичный; слабопроникающий; слабый, дряблый, вялый; легкоразлагающийся
softening *s* смягчение; размягчение
soft-finned *a* мягкопёрый
sof-fruited *a* мягкоплодный
soft-leaved *a* мягколистный
softness *s* мягкость; мучнистость
soft-rayed *a* мягкопёрый
soft-shelled *a* мягкокожий
softwood *s* хвойное дерево; *a* неодревесневший
soggy *a* болотистый, сырой
soil *s* почва, грунт, земля; почвенный слой; *a* почвенный
virgin ~ целинная почва, целина
soil-dwelling *a* почвенный, живущий в почве, обитающий в почве
soil-fixing *a* почвозакрепляющий
soil-forming *s* почвообразование; *a* почвообразующий
soil-inhabiting *a* почвенный, живущий в почве, обитающий в почве
sol *s* золь
solan *s orn* олуша *(Sula)*
solar солнечный
soldier *s ent* солдат; *bot* телорез *(Stratiotes)*
soldier's-woundwort *s см.* **sanguinary**
sole *s anat* подошва
solenocyte *s* соленоцит
soliciting *s*: **food** ~ выпрашивание пищи
solid *s* твёрдое тело; твёрдая частичка; *pl* сухой остаток, сухие вещества; *a* цельный, сплошной; чистый, беспримесный; твёрдый; плотный
solid-colored *a* со сплошной мастью
solid-hoofed *a* непарнокопытный
solidification *s* загустение, застывание, отвердение
solidify *v* застывать; твердеть
solidungulate *s* непарнокопытное животное; *a* непарнокопытный
soliped *a* непарнокопытный
solitary *s* отшельник; *a* единственный, единичный; одиночный, отдельный; растущий отдельно
solubility *s* растворимость
solubilization *s* солюбилизация
soluble *a* растворимый
solute *s* растворённое вещество
solution *s* раствор; растворение; решение
food ~ питательный раствор, питательная среда
gel ~ гель-раствор
nutrient ~ питательный раствор, питательная среда
preservative ~ фиксирующая жидкость
reference ~ контрольный раствор, референс-раствор
solvable *a* растворимый
solvation *s* сольватация
solve *v* растворять
solvent *s* растворитель; растворяющий
solvent-jump *s mbiol* скачок растворителя
solving *s* решение
soma *s* сома
somatic *a* соматический
somatoblast *s* соматобласт
somatoblastomere *s* соматобластомер
somatoderm *s* соматодерма
somatogamy *s* соматогамия, псевдоапогамия
somatogenic *a* развивающийся от соматических клеток
somatology *s* соматология
somatomammotrophin *s* соматомаммотропин
chorionic ~ хориомаммотропин, лактоген, соматомаммотропин хориона
somatome *s* сомит, сегмент тела

somatophyte *s* соматофит
somatoplasm *s* соматоплазма
somatopleure *s* соматоплевра
somatosensory *a* соматосенсорный
somatotropic *a* соматотропный
somatotropin *s* соматотропин
somatotropism *s* соматотропизм
somatotype *s* тип конституции, тип телосложения
somesthesia *s* соместезия
somite *s* сомит, сегмент тела
somniferous *a* снотворный
song *s* песня
  pair ~ парное пение
  plastic ~ пластичная песня, подпесня
songbird, songster *s* певчая птица
sonicate *v* разрушать ультразвуком
sonication *s* обработка ультразвуком
sonifier *s* ультразвуковой дезинтегратор
sonogram *s eth* сонограмма
sonometer *s* сонометр, аудиометр
soothsayer *s ent* богомол *(Mantodea)*
sappy *a* насыщенный влагой
soral *a* относящийся к сорусу
sorbitol *s* сорбит
sore *s* рана, болячка
sorediate *a* несущий соредии
soredium *s* соредия
sorghum *s* сорго *(Sorghum)*
sorocarp *s myc* сорокарп
sorophore *s* ножка соруса
sorore *s* физиологический вид
sorosis *s* сложный плод
sorption *s* сорбция
sorrel *s* щавель *(Rumex);* гнедой
sort *v* сортировать, классифицировать
sorter *s* сортировщик, сортер
sorting *s* сортировка, сортинг; классификация
sorus *s* сорус
SOS-reparation *s mbiol* SOS-репарация
sound *s* звук, тон; плавательный пузырь; *a* здоровый, крепкий; *v* издавать звук; давать сигнал; нырять ко дну
  ultrasonic ~ ультразвук
sounder *s* лот; стадо кабанов
sour *a* кислый
source *s* источник, основа, начало
souslik *s* суслик *(Citellus)*
sow *s* самка свиньи; свиноматка; медведица, имеющая медвежат; сеять, высеивать
sowing *s* посев; *a* посевной
soya *s* соя культурная *(Glycine soja)*
space *s* пространство; расстояние, интервал
  glandular ~s железистые поля
  Haversian ~ канал остеона, гаверсов канал
  intercostal ~ межрёберье, межрёберный промежуток
  interpleural ~ средостение
  lysigenous intercellular ~ *bot* лизигенный межклетник
  Nuel's ~ паратуннель, туннель Нэеля
  popliteal ~ подколенная ямка
  schizogenous intercellular ~ схизогенный межклетник
spacer *s mbiol* спейсер, спейсерная область; прокладка
spacing *s* распределение в пространстве, пространственное распределение; размещение
spadefoot *s herp* чесночница *(Pelobatidae);* лопатоног *(Scaphiopodidae)*
spadiceous *a* початковый
spadix *s* початок
spado *s* кастрированное животное, кастрат
span *s* интервал
  life ~ продолжительность жизни
spanemia *s* анемия, малокровие
spank *v* быстро бежать
spanworm *s* гусеница пяденицы
spare *s* запас; шпора; *a* запасной, резервный; *v* драться шпорами
sparrow *s* воробей *(Passer);* мелкая птица
  field ~ малая воробьиная овсянка, овсянка-крошка *(Spizella pusilla)*
  golden-crowned ~ венценосная овсянка *(Zonotrichia coronata)*
  hedge ~ лесная завирушка *(Prunella modularis)*
sparrow-hawk *s* ястреб-перепелятник *(Accipiter nisus)*
  besra ~ малый перепелятник *(Accipiter virgatus)*
sparse *a* редкий, разбросанный, рассеяный
sparsiflorous *a* рассеянно-цветковый
sparsifolious *a* рассеянно-листный
spasm *s* спазм, судорога

vasomotor ~ ангиоспазм
spasmodic *a* спастический, судорожный
spasmoytic *a* спазмолитический, антиспазматический
spathaceous, spathal *a bot* имеющий покрывало, имеющий обвёртку
spathe *s bot* покрывало, обвёртка
spathella *s bot* покрывальце, чехлик, плёнка
spatula *s* лопаточка; лопатообразный орган; шпатель
spatulate *a* лопатообразный, лопатовидный
spawn *s* икра; отпрыск, потомок; многочисленное потомство; мицелий, грибница; *v* зарождаться; порождать; плодить; метать икру; нереститься, нерестовать
spawning *s* нерест; *a* нерестующий
spay *s* овариектомирования самка; овариектомия, удаление яичников; *v* овариектомировать, удалять яичники
spear *s helm* копьё; острый шип; *bot* побег, отпрыск; стрелка; *v* выбрасывать стрелку
spear-bearing *a* копьеносный; несущий стрелку
spear-bristled *a* колючещетинистый
spear-leaved *a* копьевиднолистный
spear-lipped *a bot* копьевидногубый
specialis *s* специализированный вид
specialization *s* специализация
specialized *a* специализированный
speciation *s* видообразование
successional ~ анагенез
species *s* вид; виды
perdominant ~ пердоминант
species-specific *a* видоспецифичный
specific *a* специфический; видовой
specificity *s* специфичность; **to show a** ~ проявлять специфичность
specifier *s* спецификатор
specimen *s* образец; экземпляр; особь
speciose *a* богатый видами
speck *s phytop* пятнистость
black ~ чёрная короста картофеля *(caused by Corticium solani)*
speckled *a* крапчатый
speckled-brown *a* коричнево-пёстрая; гнедая в яблоках
spectrogram *s* спектрограмма
spectrograph *s* спектрограф
mass ~ масс-спектрограф
spectrography *s* спектрография
mass ~ масс-спектрография
spectrometer *s* спектрометр
dichroism ~ дихрограф
mass ~ масс-спектрометр
spectrometry *s* спектрометрия
mass ~ масс-спектрометрия
spectrophotometer *s* спектрофотометр
spectrophotometry *s* спектрофотометрия
spectroscope *s* спектроскоп
mass ~ масс-спектроскоп
spectroscopy *s* спектроскопия
mass ~ масс-спектроскопия
NMR ~, nuclear magnetic resonance ~ ядерная магнитно-резонансная спектроскопия, ЯМР-спектроскопия
spectrum *s* спектр
speculum *s orn* глазок
speed *s* скорость
speedwell *s bot* вероника *(Veronica)*
spelaean *a* пещерный
speleobios *s* пещерные организмы
speleology *s* спелеология
spellbone *s* малая берцовая кость
spelt *s bot* пшеница спельта, полба настоящая *(Triticum spelta)*
sperm *s* сперма, семя, сперматозоиды; сперматозоид, спермий; антерозоид
spermaceti *s* спермацет
spermaduct *s* семяпровод, семявыводящий проток, семявыносящий проток
spermagone *s* спермогоний
spermagonial *a* относящийся к спермогонию
spermagoniferous *a* имеющий спермогоний
spermagonium *s* спермогоний
spermaphyte *s* семенное растение
spermarium, spermary *s* семенник
sperm-aster *s* сперматическая звезда
spermatangium *s* сперматангий, антеридий
spermateliosis *s* сперматогенез; спермиогенез
spermatheca *s* сперматека; семеприёмник
spermatic *a* семенной; относящийся к сперме
spermatid *s* сперматида
spermatium *s* спермаций

spermatize *v* оплодотворять, осеменять
spermatoblast *s* сперматида
spermatocyst *s* семенной пузырёк
spermatocyte *s* сперматоцит
secondary ~ сперматоцит II порядка, пресперматида
spermatogenesis *s* сперматогенез
spermatogenetic, spermatogenous *a* сперматогенный, производящий сперму
spermatogonial *a* сперматогониальный
spermatogonium *s* сперматогоний
spermatoid *a* спермообразный
spermatophore *s zool* сперматофор
spermatophyte *s* семенное растение
spermatoplast *s* сперматопласт, мужская половая клетка
spermatotheca *s* сперматека, семеприёмник
spermatozoid, spermatozoon *s* сперматозоид, спермий; антерозоид
spermiducal *a* относящийся к семяпроводу
spermiduct *s* семяпровод, семявыводящий проток, семявыносящий проток
spermiogenesis *s* спермиогенез
spermism *s* спермизм
spermium *s* сперматозоид, спермий
spermoblast *s* сперматида
spermocarp *s* спермокарпий
spermoderm *s* спермодерма; семенная кожура
spermoduct *s* семяпровод, семявыводящий проток, семявыносящий проток
spermogenous *a* спермогенный
spermogonium *s* спермагоний
spermophile *s* суслик *(Citellus)*
sphacelated *a* сгнивший, разложившийся; омертвевший; усохший; увядший
sphacelia *s* сфацелия
sphaeraphis *s bot* сферическое скопление рафид
sphaerenchyma *s* сференхима, сфероидальная паренхима
sphagnicolous *a* обитающий в сфагнуме
sphagniopratum *s* сфагновое болото
sphagnous *a* сфагновый
sphenethmoid *s* клиновидно-обонятельная кость
sphenoid *s* сфеноид, клиновидная кость; *a* сфеноидальный
sphenoidal *a* сфеноидальный, относящийся к сфеноиду
sphenomandibular *a* основно-мандибулярный
sphenomaxillary *a* основно-челюстной
sphenopalatine *a* основно-нёбный
spheraster *s zool* многолучевая сферическая спикула
sphere *s* сфера
attraction ~ центросфера
immunolatex ~s латексные иммуносферы
latex ~ латексная гранула, латексный шарик, латексная сфера
spherocyte *s* сфероцит
spheromere *s* сферомер
spheroplast *s* сферопласт
spherule *s* маленькая сфера
yolk ~ желточный шар
sphincter *s* сфинктер, сжиматель
sphingophilum *s* растение, опыляемое бражниками
sphinx *s ent* бражник *(Sphingidae)*
sphragis *s ent* сфрагис
sphygmic *a* пульсовый
sphygmogram *s* сфигмограмма, пульсовая кривая
sphygmograph *s* сфигмограф
sphygmography *s* сфигмография
sphygmomanometer *s* сфигмоманометр
sphygmus *s* пульс; пульсация
spicate *a* колосовидный; имеющий колоски [колосья]; колосистый
spice *s* пряность
spiciferous *a* колосоносный
spicigerous *a* колосовидный; имеющий колосья; колосистый
spicula *s* спикула, скелетная игла *(of sponges);* маленькое остриё; шип; колючка; *bot* вторичный колосок
birotulate gemmule ~ амфидиск
spicular *a* относящийся к спикуле; игольчатый; игловидный; шиповидный, шиповатый
spiculate *a* имеющий спикулы; колючий, покрытый шипами, покрытый колючками, заострённый
spicule *s* спикула, скелетная игла *(of sponges);* маленькое остриё; шип; колючка; *bot* вторичный колосок

birotulate gemmule ~ амфидиск
spiculiferous *a* имеющий спикулы; колючий
spiculigenous *a* образующий спикулы; имеющий спикулы
spiculigerous *a* имеющий спикулы; колючий
spiculum *s* спикула, скелетная игла *(of sponges);* маленькое остриё; шип; колючка; *bot* вторичный колосок
spider *s* паук *(Araneida)*
bird ~ паук-птицеяд *(Avicularia avicularia)*
crab ~ паук-бокоход, паук-краб *(Thomisidae)*
fishing ~ паук-охотник *(Dolomedes)*
garden ~ паук-крестовик *(Araneus diadematus)*
jumping ~ паук-скакун *(Salticidae)*
raft ~ паук-охотник *(Dolomedes)*
red ~ паутинный клещик *(Tetranychidae)*
water ~ водяной паук, серебрянка *(Argyroneta aquatica);* паук-охотник *(Dolomedes);* пирата *(Pirata)*
wolf ~ тарантул *(Lycosa); pl* пауки-волки *(Lycosidae)*
spiderling *s* молодой паук
spiderweb *s* паутина
spiderwort *s* традесканция *(Tradescantia)*
spigot *s* коническая прядильная трубочка *(in spiders)*
spike *s* острый выступ; остриё; шип; колос; *phys* спайк, пик
overshoot ~ усиленный спайк, усиленный пик
spikelet *s bot* колосок; вторичный колосок
spiladophilous *a* обитающий на глине
spiladophytium *s* фитоценоз на глине
spin *v* плести
spinach, spinage *s* шпинат *(Spinacea)*
spinal *a* спинальный, позвоночный; спинной; спинномозговой
spinate *a* имеющий иглы, имеющий колючки, имеющий шипы
spination *s* наличие игл, наличие колючек, наличие шипов
spindle *s cyt* веретено
aortic ~ луковица аорты
attraction ~ веретено тяготения, «сияние», «лучистость»
spindle-shaped *a* веретенообразный, веретеновидный
spindle-stalked *a* веретеновидно-стебельчатый
spine *s* спинной хребет, позвоночник, позвоночный столб; *zool* игла; иголка *(of a hedgehog)*; ость; колючка, шип; *pal*столбик
alar ~ ость клиновидной кости
dendritic ~ дендритный шипик
dorsal ~ позвоночник, позвоночный столб
hatching ~ *ent* эмбриональный зубец
intercondyloid ~ межмыщелковое возвышение
jugular ~ яремный отросток
lumbar ~ поясничный отдел позвоночника
mobile ~ *ent* подвижный шип
nasal ~ носовая ость
obturator ~ запирательный гребень
tibial ~ бугристость большеберцовой кости
spineless *a* без колючек; неопушённый
spinelet *s* шипик
spinescent *a* колкий, колючий; имеющий иглы, колючки; остроконечной формы
spiniferous *a* имеющий иглы, имеющий колючки, имеющий шипы
spiniform *a* в форме иглы, в форме колючки, в форме шипа
spinneret *s* прядильный орган
spinnerule *s* паутинная трубочка
spinolophate *a* шиповато-гребенчатый
spinose, spinous *a* колючий, колкий; покрытый иглами, колючками *или* шипами
spinster *s* яловая самка
spinulate, spinulated *a* покрытый иголочками, покрытый шипиками
spinulation *s* защитный колючий покров; обрастание иглами, обрастание колючками, обрастание шипами
spinule *s* иголочка, мелкая колючка, шипик
spinuliferous *a* имеющий иголочки, имеющий шипики
spiny *a* колючий, покрытый иглами, покрытый шипами
spiracle *s* дыхательное отверстие; *энт.* дыхальце, стигма; брызгальце; дыхало *(of a whale)*

spiral *s* спираль; *a* спиральный
spire *s* спираль; росток; стрелка; верхушка дерева; острый конец оленьего рога; *v* прорастать *(of grain);* выходить в трубку; колоситься
spireme *s* спирема
spirilla, spirillum *s* спирилла
spirit *s* спирт; *a* спиртовой
spirituous *a* спиртовой
spirocarpous *a* спиральноплодный
spirochaete, spirochete *s* спирохета
spirogram *s* спирограмма
spirograph *s* спирограф
spirography *s* спирография
spiroid *a* имеющий форму спирали
spirolobous *a* спирально-дольный
spirometer *s* спирометр
spirometry *s* спирометрия
spirulate *a* спирально расположенный
spissitude *s* густота, плотность
spit *s* слюна; мокрота
  cuckoo ~, frog ~ «кукушкины слюнки» *(produced by spitbugs)*
spitter *s* годовалый олень
spittle *s* слюна; мокрота; «кукушкины слюнки» *(produced by spitbugs)*
spittlebug *s ent* пенница *(Cercopidae)*
splanchnic *a* внутренностный
splanchnocoel *s* спланхноцель
splanchnocranium *s* спланхнокраниум
splanchnology *s* спланхнология
splanchnopleure *s* спланхноплевра, висцеральный листок
splanchnotom *s* спланхнотом
spleen *s* селезёнка
splenetic *a* селезёночный
splenial *s* пластинчатая кость, крышечная кость
splenic *a* селезёночный
splenocyte *s* спленоцит, селезёночный макрофаг
splice *v mbiol* сплайсировать
spliceosome *s* сплайсингосома
splicing *s* сплайсинг
split *s* расщепление; раскалывание; растрескивание; *a* расщеплённый; расколотый; растрескавшийся; *v* расщеплять, расщепляться; раскалывать, раскалываться; растрескивать, растрескиваться; разлагать на компоненты, расщеплять;
  to ~ off, to ~ out отщеплять
splitting *s* деградация, расщепление; гидролиз
spondyl, spondyle, spondylus *s* позвонок
spongiocyte *s* спонгиоцит
spongioplasm *s* спонгиоплазма
spongiose, spongious *a* губчатый, рыхлый, спонгиозный
spongoid *a* губкообразный
spongy *a* губчатый; пористый; топкий, болотистый
sponk *s* гнилушка; *myc* трутовик *(Polyporus)*
spontaneous а спонтанный, самопроизвольный
spoonbill *a orn* колпица *(Platalea);* утка-широконоска *(Anas clypeata)*
spoor *s* след; *v* идти по следу
sporadic а спорадический, случайный, единичный
sporadophytium *s* открытое сообщество
sporange *s* спорангий
sporangiferous *a* спорангиеносный
sporangiocarp *s* спорангиокарпий, аскокарпий
sporangiole, sporangiolum *s* спорангиоль
sporangiophore *s* спорангионосец
sporangiospore *s* спорангиоспора
sporangium *s* спорангий
spore *s* спора
  brand ~ уредоспора, головнёвая спора
  fragmentation ~ конидия
  secondary ~ аскоспора, вторичная спора, обособленная спора
  septate ~ фрагмоспора
  swarm ~ зооспора
spore-former *s* спорообразующий микроорганизм
sporeless *a* бесспоровый
sporeling *s* молодой спорофит
sporicide *s* спороубивающее средство
sporidiferous *a* споридиеносный
sporidiole, sporidiolum *s* маленькая спора
sporidium *s* споридия
sporiferous *a* спороносный; образующий споры
sporification *s* спорообразование, формирование спор
sporiparity *s* размножение спорами
sporiparous *a* размножающийся спорами
sporoblast *s* споробласт
sporocarp *s* спорокарпий

sporocyst *s* спороциста
sporocyte *s* спороцит
sporoderm *s* спородерма
sporodochium *s* спородохий
sporoduct *s* спородукт
sporogenesis *s* спорообразование, спорогенез
sporgenous *a* спорогенный, спорообразующий
sporogonium *s* спорогоний
sporogony s спорогония
sporoid *a* спорообразный
sporont *s* спорoнт
sporophore *s* спорофор
sporophydium *s* спорофидий; споропочка, оогоний *(of Charophyta)*
sporophyll *s* спорофилл
sporophyte *s* спорофит
sporoplasm *s* спороплазма
sporosac *s zool* споросак; *bot* споровый мешок
sporozoite *s* спорозоит
sporozooid *s* зооспора
sport *s* спорт, почковая мутация
sporulation *s* споруляция
sporule *s* маленькая спора
spot *s* пятно; *phytop* пятнистость
coral ~ коралловая пятнистсть *(caused by Nectria cinnabarina)*
hot ~ природный очаг; *gen* «горячая точка»
milky ~ *anat* молочное пятно, молочный островок
nesting ~ гнездовье
pod ~ антракноз *(caused by Melanconiales)*
receptive ~ воспринимающий бугорок
spotted *a* пятнистый
spotting *s phytop* пятнистость
spawl *v* простираться, раскидываться
sprawler *s* ползающее животное
sprawling *a* раскидистый
spray *s* веточка; побег; распыляемый раствор; *v* опрыскивать, обрызгивать; разбрызгивать
spray-drying *s* сушка распылением
sprayer *s* опрыскиватель; разбрызгиватель
spraying *s* распыление, разбрызгивание; опрыскивание
spread *s* распространение; разрастание; *biom* размах
spreading *s* распространение; распластывание; *a* стелющийся; раскидистый
sprig *s* веточка; побег; давать побеги; черенок; *v* размножать черенками, сажать черенками
spring *s* весна; молодая поросль; ключ, источник, родник; прыжок; *ent* прыгательная вилка; весенний; *v* прыгать; **to ~ up** всходить *(of seeds)*
springbok, springbuck, springer *s* шпрингбок, антилопа-прыгун *(Antidorcas marsupialis)*
springhare *s ther* долгоног *(Pedetes)*
springing *a* стельная *(о корове)*
spring-planted *a* весенней посадки, весеннего посева, яровой
springtail *s ent* ногохвостка, хвостоножка *(Collembola)*
sprout *s* всход; росток; побег; почка; поросль; рассада; *v* давать ростки, удалять ростки, обламывать ростки; давать почки; распускаться
Brussel's ~s *pl* капуста брюссельская *(Brassica oleracea gemmifera)*
fruit-bearing ~ плодовое копьецо
malt ~ солодовый проросток
stump ~s *pl* пнёвая поросль
sprouting *s* проращивание, прорастание; всход, росток, побег; *a* прорастающий
spruce *s* ель (Picea)
spur *s* шип, шпора**;** *ent* шип, шпора; *bot* шпорец, брахипласт; спорынья *(Claviceps purpurea)*
fruit ~ плодоножка
spurless *a* бесшпорцевый
spurlike *a* шпорцевидный
sputterer *s* устройство для ионного напыления
sputum *s* мокрота
squab *s* неоперившийся птенец; неоперившийся голубь
squama *s anat* чешуя; *ent* закрыловая пластинка; *bot* чешуйка
squamate *a* чешуйчатый
squamation *s* чешуйчатость; расположение чешуй
squame *s anat* чешуя; *ent* закрыловая пластинка; *bot* чешуйка
squamella *s* мелкая чешуйка; плёнчатый прицветничек

squamiferous *a* имеющий чешую, чешуеносный
squamiform *a* чешуевидный
squamigerous *a* имеющий чешую, чешуеносный
squamodisk *s helm* чешуевидный диск
squamosal *s* чешуйчатая кость; *a* относящийся к чешуе; покрытый чешуйками
squamose, squamous покрытый чешуями, покрытый чешуйками; чешуйчатый; чешуевидный
squamula *s* чешуйка; *bot* лодикула, маленькая плёночка
squamulate *a* имеющий маленькие чешуйки
squamule *s* чешуйка; *bot* лодикула, маленькая плёночка
squamulose *a* имеющий маленькие чешуйки
square *s* квадрат
square-headed *a* булавовидный
squarrose *a* оттопыренный
squarrulose *a* слегка оттопыренный
squash *s* мезга, мякоть; тыква крупноплодная столовая *(Cucurbita maxima); v* раздавливать, расплющивать
squat *s* нора, берлога; *eth* припадание к земле; приседание; *a* припавший к земле; *v* лежать в норе, лежать в берлоге
squatting *s eth* припадание к земле; приседание
squawk *s* пронзительный крик; *v orn* пронзительно кричать
squeak *s* писк; *v* пищать
squeaker *s* неоперившийся птенец; неоперившийся голубь
squeeze *s* сжатие, сдавливание; *v* сжимать, сдавливать
squint *s med* косоглазие
squint-eyed *a* косоглазый
squinting *s med* косоглазие
squirrel *s* белка *(Sciurus)*
  barking ~ луговая собочка *(Cynomys)*
  ground ~ суслик *(Citellus)*
  Russian flying ~ обыкновенная летяга *(Pteromys volans)*
stab s палочка, столбик; палочкоядерный
stabile *a* стабильный, устойчивый
stability s стабильность, устойчивость
  thermal ~ теплоустойчивость, теплостойкость
stabilization *s* стабилизация
stab-inoculation s посев уколом
stable *a* стабильный, устойчивый, постоянный, неизменяемый; s стойло
stack *s* стог
stacking s *mbiol* стекинг; стогование
stag s благородный олень *(Cervus elaphus)*
stage s стадия, период, этап; столик
  ~ **of latency** латентный период, скрытый период
  booting ~ стеблевание, стадия выхода в трубку
  dead-ripe ~ мёртвая спелость; перезрелость
  dormant ~ стадия покоя, период покоя
  phenological ~ фенофаза, фенологическамя фаза
  quiescent ~ стадия покоя, период покоя
stag-headed, stag-horned *a* суховершинный
staging *s* стадийность; фазность
stagnant *a* застойный, стоячий
stagnation *s* стагнация, застой, остановка
stain *s* краситель, красящее вещество, краска; *v* красить, окрашивать
stainable *a* окрашивающийся, принимающий окраску
staining *s* окрашивание, окраска
  immunosilver ~ иммуносеребрение
  indirect ~ непрямое окрашивание
staircase *s phys* феномен лестницы
  ~ **of vestibule** *anat* лестница преддверия
stalk *s* стебель, стебелёк; ножка; чубук; *v* подкрадываться
  anther ~ тычиночная нить
  belly ~ пуповина, пупочный канатик
  cabbage ~ кочерыжка
  flower ~ цветонос, цветоножка; стрелка
  fruit ~ пордоножка
  hypopysial ~ ножка гипофиза, гипофизарная ножка
  infundibular ~, pituitary ~ воронка гипоталамуса
  root ~ корневой отпрыск
  seed ~ семяносец
  yolk ~ *emb* желточный проток

stalked *a* стебельчатый
stalk-eyed *a* имеющий стебельчатые глаза, стебельчатоглазый
stalk-shooting *s bot* выход в трубку
stallion *s* жеребец-производитель
stamen *s* тычинка
stamen-bearing *a* тычинконосный
stamenless *a* бестычинковый
staminal *a* тычиночный
staminate *a* тычинконосный; образующий тычинки
staminiferous *a* тычинконосный
staminode, staminodium *s* стаминодий
stamp *v* бить копытом, стучать лапами
stance *s* поза, положение
  hostile ~ враждебная поза
  mating ~ положение при спаривании
stand *s* насаждение; древостой; травостой; штатив
  ~ **of trees** древостой
  grass ~ травостой
  plant ~ растительный покров
  timber ~ древостой
standard *s* стандарт; образец, норма; стандартный; нормальный; образцовый; *bot* флаг, парус
stand-at-ease *s phys* свободное стояние
stand-by *a* запасной; вспомогательный
standing *s* систематический статус; стоящий; стоячий *(of water)*
standstill *s* остановка; пауза
stank *s* небольшая запруда
stapes *s anat* стремя, стремечко
staphyle *s* нёбный язычок
staple *s* волокно
star *s* звезда, звёздочка, пятно, отметина
  daughter ~ *cyt* диастер, стадия двойной звезды
starch *s* крахмал
  animal ~, **hepatic** ~, **liver** ~ животный крахмал, гликоген
starch-containing крахмалосодержащий
starch-forming крахмалообразующий
starch-free бескрахмальный
starch-splitting расщепляющий крахмал, амилолитический
starchy *a* крахмалистый
stare *s eth* настороженное выражение лица; пристальный взгляд; *v* смотреть в упор
  avert ~ отведение взгляда
staring *a* торчащий *(fur)*
starling *s* скворец *(Sturnus)*
starling-house *s* скворечник
starter *s micr* закваска
startle *v* вздрагивать
starvation *s* голодание
starve *v* голодать, умирать от истощения; **to ~ out** отмирать
starveling *s* истощённое животное
starving *a* голодающий
stasigenesis *s* стазигенез
stasis *s* стаз, закупорка; остановка роста, задержка роста
state *s* состояние
  steady ~ устойчивое состояние; *micr* стационарная фаза
stathmokinesis *s* статмокинез
static *a* статический, неподвижный; стационарный
station *s* стация, отдельный участок обитания; станция
  marking ~ пункт мечения
stationary *a* стационарный, неподвижный; стабильный, постоянный
statistics *s* статистика
  sampling ~ *biom* выборочная статистика
statoblast *s* статобласт
statocyst *s* статоцист
statocyte *s* статоцит
statolith *s* статолит
statospore *s* неподвижная спора
stature *s* телосложение
status *s* статус, статут, состояние
  ~ **of species** статут вида
  dynamic ~ динамическое состояние
  light ~ световой режим
  reproductive ~ репродуктивный статус
  social ~ ранг в стае, положение в сообществе
staurigamia *s* перекрёстное опыление
staurophyll *s* стаурофилл
steady-state *a* стационарный
steam *s* пар; *v* стерилизовать паром
steamer *s* пропариватель
steaming *s* пропаривание; стерилизация паром
steatogenesis *s* липогенез
steatolysis *s* липолиз, расщепление жира
steatolytic *a* расщепляющий жир, липолитический

steenbok *s* штейнбок, стенбок *(Raphiceros);* обыкновенный стенбок *(Raphiceros campestris)*
steep *s* замачивание; *v* замачивать
steer *s* бык
steganopodous *a* имеющий перепончатые лапы
stegocrotaphic *a* стегальный
steinbock, steinbok *s* штейнбок, стенбок *(Raphiceros);* обыкновенный стенбок *(Raphiceros campestris)*
stele *s* стель, стела
stellar, stellate *a* звездчатый; звездообразный
stellular *a* мелкозвездчатый
stelocyttarous *a ent* на ножке *(a wasp's nest)*
stem *s* стебелёк, ножка; стержень, ствол, стебель, цветонос, цветоножка, плодоножка; донце *(of a onion)*
  brain ~ стволовая часть мозга
  bulb ~ донце луковицы
  climbing ~ цепляющийся стебель, лазящий стебель, лиановидный стебель
  creeping ~ стелющийся стебель, ползучий побег, столон, плеть
  creeping underground ~ подземный стебель, корневище
  fistular ~ стебель-дудка
  floral ~ ось цветка
  flower-bearing ~ цветонос, цветоножка
  nerve ~ нервный ствол
  piping ~ стебель-дудка
  procumbent ~, prostrate ~ стелющийся стебель, ползучий побег, столон, плеть
  subterranean ~ подземный стебель, корневище
stem-clasping *a bot* стеблеобъемлющий
stemless *a* бесстебельный
stemma *s ent* стемма, простой глазок; латеральный глазок
stemness *s* стволовость
stemnode *s bot* колено
stem-root *s* стеблекорень
stenobaric *a* стенобарный
stenobotryous *a* узкокистистый
stenobulbous *a* узколуковичный
stenocarpous *a* узкоплодный
stenocephalous *a anat* узкоголовый; *bot* узкоголовчатый
stenochor *s* стенохор
stenochoric *a* стенохорный
stenocladous *a* узковетвистый
stenoecic *a* стенойкный, стенобионтный
stenogynous *a* узкопестистый
stenohaline *a* стеногалинный
stenohydric *a* стеногидрический
stenoky *s* стенойкия
stenolobate *a* узколопастный
stenonotal *ent* стенонотальный
stenopalynous *a* палинологически однотипный
stenopetalous *a* узколепестный
stenophagous *a* стенофагный
stenophotic *a* стенофотный
stenophyllous *a* узколистный
stenosepalous *a* с узкими чашелистиками
stenospermocarpy стеноспермокарпия
stenostachyous *a* узкоколосистый
stenostomatous *a* узкоротый
stenosynusic *a* органиченный в распространении
stenotherm *s* стенотермный организм
stenothermal, stenothermic *a* стенотермный
stenotopic *a* стенотопный, стенохорный
stenotropic *a* стенотропный
step *s* шаг
stephanaperturate *a pal* венечно-апертурный
stephanocarpous *a* венчикоплодный
stephanocolpate *a* венечно-бороздный
stephanoporate *a* венечно-поровый
steppe *s* степь
  wooded ~ лесостепь
steppification, steppization *s* остепнение
stereoblastula *s* стереобластула
stereocilia *s* стереоцилия
stereokinesis *s* стереокинез
stereome *s* стереом
stereomicroscope *s* стереоскопический микроскоп
stereoplasm *s* стереоплазма
stereospecificity *s* стереоспецифичность
stereospondylous *a* стереоспондильный
stereotaxis *s* стереотаксис
stereotropism *s* стереотропизм
stereotype *s* стереотип
stereotypy *s* стереотипия
sterigma *s* стеригма

sterile *a* бесплодный, стерильный
sterility *s* бесплодие, стерильность
sterilization *s* стерилизация
~ by intermittent steaming, discontinuous ~, fractional ~, intermittent ~ тиндализация, дробная стерилизация
sterilize стерилизовать; кастрировать
sterilizer *s* стерилизатор
dry-air ~, hot-air ~ сушильный шкаф, пастеровская печь, суховоздушный стерилизатор
steam ~ автоклав, паровой стерилизатор
stern *s* зад; хвост *(of a dog)*
sternal *a anat* грудинный; *ent* стернальный
sternebra *s* сегмент грудины
sternellum *s* стернеллум
sternite *s* стернит
sterno-costal *a* грудино-рёберный
sternum *s anat* грудина; *ent* стернум, стернит
sternutation *s* чихание
steroid *s* стероид; *a* стероидный
steroidogenesis *s* стероидогенез
sterol *s* стерол, стерин
sterrhad *s* болотное растение
sterrhium *s* сообщество болота
sterrhophilus *s* обитатель болот
sterrula *s zool* стеррула
sthenic *a* активный; сильный, крепкий
stichid, stichidium *s bot* стихидий
stichobasidium *s* стихобазидия
stick *s* прут; *bot* ветка; черешок; клейкость, липкость; *v* приклеиваться, прилипать, липнуть
walking ~ *ent* палочник *(Phasmoptera)*
stickiness *s* клейкость, липкость
matrix ~ слипание матрикса
sticky *a* прилипающий, липкий, клейкий
stictocarpous *a* испещрённоплодный
stictophyllous *a* испещрённолистный
stiff *a* жёсткий; тугоподвижный, ригидный; окоченевший, застывший; сильнодействующий
stiff-leaved *a* жёстколистный
stiffness *s* жёсткость; ригидность, оцепенелость, тугоподвижность
stiff-thorned *a* жёсткоколючковый
stifle *s* коленная чашка; коленный сустав
stigma *s zool* стигма; *ent* дыхальце; птеростигма, крыловой глазок; *bot* рыльце
stigmatic *a bot* рыльцевый
stigmatiferous *a bot* имеющий рыльце
stigmatoid *a bot* рыльцевидный
stile *s s* стержнеобразная спикула *(in sponges);* кристалический стебелёк *(in mollusks); ent* грифелёк; спинная щетинка; *bot* столбик
stilet *s* стилет
still *s* перегонный куб, дистиллятор; *v* перегонять, дистиллировать
stillage *s* барда
stillbirth *s* мертворождение
stillborn *a* мертворождённый
stilogonidium *s* стилогонидия, конидия
stilt *s orn* ходулочник *(Himantopus)*
stimulant *s* стимулятор; стимул; раздражитель; *a* стимулирующий
stimulate *v* стимулировать; раздражать
stimulation *s* стимулирование, стимуляция; раздражение
AC ~ электростимуляция переменным током
acoustic ~ фоностимуляция
DC ~ гальванизация
flicker ~ раздражение мелькающим светом
galvanic ~ гальванизация
photic ~ световое раздражение; фотостимуляция
punishment ~ отрицательное подкрепление
reward ~ положительное подкрепление
stimulator *s* стимулятор
acoustic ~ фоностимулятор
photic ~ фотостимулятор
stimulatory *a* стимулирующий
stimulus *s* стимул; раздражитель
conditioned ~ условный раздражитель; адекватный стимул; фактор кондиционирования среды
target ~ *eth* стимул-мишень
stimulus-response *s* стимул-реакция
sting *s* жало; ужаление; жгучий волосок; *v* жалить
stinger *s* жало; жалоносное насекомое
stinging *a* жгучий; жалящий
stinkbug *s ent* щитник *(Pentatomidae)*
stinking *a* вонючий, зловонный

stint *s* кулик-песочник *(Calidris)*
little ~ кулик-воробей *(Calidris minuta)*
long-toed ~ длиннопалый песочник *(Calidris subminuta)*
red-necked ~ песочник-красношейка *(Calidris ruficollis)*
stipe *s bot* ножка
stipel *s* прилистничек
stipellate *a* имеющий прилистнички
stipiform *a* в виде ножки; стеблевидный
stipitate *a* имеющий ножку, имеющий стебелёк; образовавшийся на ножке; стебельковый
stippling *s* зернистость, точечность
basophilic ~ базофильная зернистость, базофильная точечность
stipular *a* прилистниковый
stipulate *a* имеющий прилистники
stipule *s* прилистник
stipuliferous *a* имеющий прилистники
stipuliform *a* прилистниковидный
stir *v* взбалтывать, перемешивать, встряхивать
stirrer *s* мешалка
stirrup *s anat* стремя, стремечко
stoat *s* горностай *(Mustela erminea)*
stock *s* линия, потомство одной особи; штамм; порода; скот; поголовье скота; стадо; косяк, стая; ствол; подвой; запас; биомасса; исходный раствор
foundation ~ основной семенной материал, основной маточный материал
pedigree ~ племенное стадо
planting ~ саженцы, посадочный материал
surplus ~ излишний запас
stocked *a* заселённый; облесённый; уплотнённый *(of a forest);* беременная
stocking *s* заселение; выпас
stocky *a* приземистый, коренастый
stolon *s* столон
stolonate *a* имеющий столоны
stoloniferous *a* дающий столоны
stoma *s bot* устьице
stomach *s* желудок
first ~ рубец
fore ~ сетка
fourth ~ сычуг
honey ~ медовый зобик
honeycomb ~ сетка
miniature ~ павловский желудочек, изолированный желудочек
read ~, **rennet** ~ сычуг
third ~ книжка
stomachal, stomachic *a* желудочный, гастрический
stomatal *a* устьичный
stomate *a* имеющий устьица
stomatic *a* ротовой, оральный
stomatiferous *a* имеющий устьица
stomodaeum, stomodeum *s* стомодеум, первичный рот
stone *s* косточка *(in fruit);* камень
ear ~ отолит
island stepping ~s камни для перехода через ручьи
stonechat *s orn* чекан *(Saxicola);* черноголовый чекан *(Saxicola torquata)*
stonecrop *s* очиток *(Sedum)*
stonefly *s* веснянка *(Plecoptera)*
stoneless *a* бескосточковый
stonewort *s* хара, харовая водоросль *(Chara)*
stony *a* каменистый
stool *s* стул, испражнения; побег; отводок; маточное растение; *v* пускать побеги, куститься
stooling *s* кущение
stop *s* остановка; задержка; прекращение; *v* останавливать, останавливаться; задерживать; прекращать
stopped-flow *s mbiol* остановленная струя, остановленный поток
stopper *s* пробка
stop-response *s* стоп-ответ
storage *s* хранение; накопление; запоминающее устройство; *a* запасной; запасающий
store *s* запас; **to ~ up** накапливать, запасать, аккумулировать
storey *s* ярус растительности
storing *s* запасание
stork *s* аист *(Ciconia)*
black ~ чёрный аист *(Ciconia nigra)*
black-necked ~ азиатский ябиру *(Xenorhynchus asiaticus)*
hammerhead ~ молотоглав, молотоголовая цапля *(Scopus umbretta)*
jabiru ~ ябиру *(Jabiru mycteria)*
open-billed ~ аист-разиня *(Anastomus)*

whale-headed ~ китоглав, королевская цапля *(Balaeniceps rex)*
storksbill *s bot* аистник *(Erodium)*
storm-petrel *s orn* качурка *(Hydrobatidae)*
story *s* ярус растительности
stotting *s eth* смотровой прыжок
stove *s* печь, сушительная печь; паровая баня; теплица, оранжерея
straggle *v* одиноко расти
straight *a* прямой
straight-boled *a* прямоствольный
straight-seeded *a* прямосемянный
straight-stalked *a* прямостебельный
straight-stemmed *a* прямоствольный
straight-toothed *a* прямозубчатый
straight-veined *a* пряможилковатый
strain *s* раса, линия; штамм; растяжение; напряжение; деформация; элита; *v* растягивать, растягиваться; напрягать, напрягаться; деформировать, деформироваться; сжимать, сжиматься; процеживать
biological ~ биологическая раса
breeding ~s элита
pure ~ чистопородная линия; чистый штамм, чистая культура
reference ~ эталонный штамм, справочный штамм, референс-штамм
tester ~ линия-анализатор
strainer *s* фильтр; сито
straining *s* фильтрование; процеживание
strait *s* пролив
inferior pelvic ~ тазовый выход
superior pelvic ~ тазовый вход
strand *s* тяж; цепь, нить
Billroth's ~ тяж Бильрота, мякотный тяж, трабекула селезёнки
coding ~ кодирующая цепь
minus ~ «минус-цепь»
noncoding ~ некодирующая цепь
plus ~ «плюс-цепь»
stranded *a* выбросившийся на берег
strange *a* чуждый, заносный, интродуцированный
stranger *a* чуждый вид, заносный вид, интродуцированный вид
strangle *s* удушье; *v* душить, давить; задыхаться
strategy *s* стратегия
stratification *s* стратификация; расчленение на слои, расчленение на ярусы, расчленение на группы; создание слоёв *(in an experiment);* ярусность; *биом*. расслоение; **~ after selection** расслоение *(of a sample)*
stratify стратифицировать
stratobios *s* обитатели опада и подстилки
stratose *a* слоистый
stratum *s* пласт, слой; ярус
field ~ ярус травянистой растительности
straw *s* солома
strawberry *s* земляника *(Fragaria); a* буланая
strawworm *s* вредитель злаков
stray *s* животное, отбившееся от стада
streak *s micr* линия, штрих; штриховая культура; *phytop* стрик, полосатость; *v* делать посев штрихом
agar ~ штриховая культура на агаре
bean stripple ~ полосатая гравировка фасоли
tomato yellow ~ жёлтая полосатость томата
streaking *s micr* штриховая разводка
streak-inoculated *a micr* посеянный штрихами
streaky *a* полосатый
stream *s* течение; поток; *v* течь
blood ~ ток крови, кровоток
determination ~ *emb* детерминационный поток
streamlined *a* обтекаемый
strength *s* сила; прочность; крепость, концентрация
antigenic ~ антигенность
shear ~ сопротиление сдвигу, сопротивление срезу, прочность на сдвиг, прочность на срез
transpiration ~ интенсивность транспирации
strepsinema *s* стрепсинема
strepsitene *s* стрепситена
streptocarpous *a* скрученноплодный
streptocaulis *a* скрученностебельчатый
streptophyllous *a* скрученнолистный
streptopilose *a* скрученноволосистый
stress *s phys* стресс; напряжение; усилие; нагрузка
menntal ~ нервно-психическое напряжение; эмоциональный стресс
permissible ~ допустимая нагрузка

stressor *s phys* стрессогенный фактор, стрессор
stretch *s* участок, отрезок; *mbiol* фрагмент секвенирования
stretching *s* потягивание
stria *s* штрих, полоска; бороздка
striatal *a* стриарный
striated *a* полосатый, исчерченный, покрытый полосками, покрытый бороздками; жилковатый
striation *s* полосатость; бороздчатость; исчерченность
striato-punctate с точечными полосками, с точечными бороздками
strict *a* точный; определённый; строгий
stricture *s* стриктура, сужение канала
stridulate стрекотать, трещать
stridulation *s* стрекотание, стридуляция
strigil, strigilis *s ent* щёточка
strigillose *a* мелкощетинистый
strigose *a* щетинистый, колючий
strike *v* пускать корни, укореняться, приживаться, приниматься; сажать; нападать, бросаться, кусать, жалить *(of a snake)*; подсекать *(fish)*
string *s bot* волокно; жилка; гроздь, кисть; сухожилие
  auditory ~ слуховая струна
  eye ~ сухожилие глазной мышцы
  gene ~ генная нить, генонема
  navel ~ пуповина, пупочный канатик
string-leaved *a* струнолистный
stringless *a* безволокнистый
stringlike *a* верёвчатый, струновидный
stringy *a* волокнистый
striola *s* тонкая линия; мелкий штрих
striolate *a* мелкоштрихованный
strip *s* полоса, полоска; *v* выдавливать
  Casparian ~ поясок Каспари
stripe *s* полосатость; полоса, полоска
  leaf ~ полосатость листьев *(caused by Helminthosporium gramineum)*
  maize ~ штриховатость кукурузы
striped *a* полосатый
stripped *a bot* обнажённый
strobila *s zool* стробила; *bot* шишка
strobilaceous *a* имеющий шишки; шишковидный, шишкообразный; конусообразный
strobilation *s* стробиляция
strobile *s zool* стробила; *bot* шишка
strobiliferous *a* шишконосный; образующий шишки
strobilocercus *s* стробилоцерк
strobiloid *a* шишковидный, шишкообразный; конусообразный
strobilus *s zool* стробила; *bot* шишка
stroke *s phys* удар; *a* ударный, систолический
stroma *s* строма, основное вещество
strombuliferous *a* имеющий спиральную структуру
strombuliform *a* спиралевидный
strophiolate *a bot* имеющий семенные придатки
strophiole *s bot* семенной придаток
strophism *s* тенденция к завиванию
strophotaxis *s* завивание, закручивание; способность завиваться
structure *s* структура; строение
  fine ~ ультраструктура
  kranz ~ *bot* кранц-структура
  Porter's ~ модель Портера
  secondary sexual ~ вторичный половой признак
  stand ~ состав насаждения по группам
structuredness *s* структурированность
structureless *s* бесструктурный
struggle *s* борьба; ~ **for existence** борьба за существование; ~ **for life** борьба за жизнь; ~ **for survival** борьба за выживание
struma *s* желвак; нарост; подушкообразное вздутие; зоб
strumose *a* имеющий желваки, имеющий наросты
strychnine *s* стрихнин; *bot* чилибуха, стрихнос рвотный, рвотный орех *(Strychnos nux-vomica)*
stub *s* пень
stubble *s* стерня
stud *s* племенное стадо
studbook *s* студбук, родословная книга
study *s* исследование; изучение
  electron spin resonance ~ метод электронного спинового резонанса
  molecular orbital ~ метод молекулярных орбит
  screening ~ массовое обследование, скрининг
stubble *s* стерня; ~ **weed** пожнивной сорняк
  broken ~ лущёная стерня

stuff *s* материал; вещество
fatigue ~ вещество утомления, кенотоксин
food ~ продукт питания
stump *s* пень, пенёк; *v* срубать, корчевать
stunt *s* остановка в росте, задержка в росте; *phytop* карликовость; малорослое животное; *v* останавливать рост; останавливаться в росте, чахнуть
bushy ~ *phytop* кустистая карликовость
stunted *a* низкорослый; угнетённый; чахлый; корявый
stunting *s* низкорослость; *a* замедляющий рост
stupeous, stupose *a* паклеобразный
stupulose *a* коротковолосистый
stylar *a ent* относящийся к грифельку; *bot* относящийся к столбику
stylate *a ent* имеющий грифельки; *bot* имеющий столбики
style *s* стержнеобразная спикула *(in sponges);* кристалический стебелёк *(in mollusks); ent* грифелёк; спинная щетинка; *bot* столбик
stylet *s* стилет
stylifer *s ent* пластинка, несущая грифелёк
styliferous *a ent* имеющий грифельки; *bot* имеющий столбики
styloconic *a* стилоконический
styloid *a* шиловидный
stylomandibular *a* шилонижнечелюстной
stylomastoid *a* шилососцевидный
stylopharyngeal *a* шилоглоточный
stylopodium *s* стилоподий
stylospore *s* стилоспора
stylostegium *s* внутренний венчик
styptic *a* кровоостанавливающий
subacid *a* слабокислотный; кисловатый
subalpine *a* субальпийский
subapical *a* субапикальный, околовершинный
subacute *a* подострый
subalimentation *s* недостаточное питание
subaquatic *a* подводный
subarachnoidal *a* субарахноидальный, подпаутинный
subarborescent *a* древовидный
subarctic *a* субарктический
subassociation *s* субассоциация
subaxillary *a* подмышечный; *bot* подпазушный
subbasal *a* расположенный близ основания
subbranch *s* нижняя ветвь
subbranchial *a* поджабренный
subcartilaginous *a* неполностью охрящевевший; прохондральный
subcaudal *a* подхвостовой
subcaulescent *a* с укороченным стеблем
subcellular *a* субклеточный
subcentral *a* почти центральный, близкий к центру
subchela *s* подклешня, ложная клешня
subchromatid *s* субхроматида
subcircular *a* округлый
subclass *s* подкласс
subclimax *s* предклимакс
subclone *s* субклон; очередной пассаж
subcloning *s* пересев, субклонирование
subcordate *a* слегка сердцевидный
subcortical *a* подкорковый, подкоровый; *prot* субкортикальный
subcosta *s ent* субкоста, субкостальная жилка
subcostal *a* подрёберный
subcrenate *a* слабогородчатый
subculture *s* пересев, пассаж; пассированная культура; *v* пересевать, пассировать
subculturing *s* пересев; субкультивирование
subcutaneous подкожный
subcuticle, subcuticula *s* субкутикула
subcuticular *a* расположенный под кутикулой
subcutis *s* субкутис, подкожный слой
subdermal *a* субдермальный, подкожный
subdivision *s* подразделение
subdomain *s mbiol* субдомен
subdominant *s* субдоминант; соподчинённая особь; *a* соподчинённый
subdue смирять, подчинять, покорять; успокаивать
subdumi *s pl* полукустарники
subdural *a* субдуральный
subechinate *a* слегка шиповатый

subechinolophate *a* гребенчатый и слегка шиповатый
subelongated *a* слегка удлинённый
subepidermal *a* субэпидермальный
subepithelial *a* субэпителиальный
subequatorial *a* приэкваториальный
suber *s* пробка, пробковая ткань
suberect *a* приподнятый
subereous *a* пробковый, пробковидный
suberiferous *a* продуцирующий пробку
suberification, suberization *s* опробковение
suberose *a* пробковый, пробковидный
subesophageal *a* подпищеводный
subfamily *s* подсемейство
subformation *s* субформация
subfossil *a* субфоссильный
subgene *s* субген, мутон
subgeniculate слегка согнутый
subgenus *s* подрод
subgeocolous *a* обитающий под землёй
subglobose *a* округлый, почти шаровидный
subglossal *a* подъязычный
subgroup *s* подгруппа
subhumid *a* субгумидный
subhymenium *s* субгимениум, подгимениальный слой
subiculum *s* подушковидный мицелий
subimago *s* субимаго
subinfluent *s* субинфлюэнт
subinguinal *a* подпаховый
subinoculation *s* перевивка, пассаж
subisopolar *a* почти равнополярный
subitaneous *a* внезапный, неожиданный; субитанный, летний
subjacent *a* расположенный в основании
sublanceolate *a* почти ланцетовидный
sublayer *s* подъярус; подстилающий слой *(of soil)*
sublenticular *a* слегка чечевицеобразный, напоминающий линзу
sublethal *a* сублетальный
sublimate сублимировать, возгонять
sublimation *s* сублимация, возгонка
subliminal *a* подпороговый
sublingual *a* подъязычный
sublophate *a* почти гребенчатый
submarine *a* подводный
submaxilla *s* мандибула
submaxillary *a* подчелюстной
submental *s ent* субментум; *a* подподбородочный, относящийся к субментуму
submentum *s ent* субментум, подподбородок
submersed *a* погружённый
submetacentric *a* субметацентрический
submicroscopy *s* электронная микроскопия
submission *s eth* подчинение, покорность
submucosa *s* подслизистая оболочка
subneural *a* субневральный
subnodal *a ent* подузелковый; *bot* подузловой
subnymph *s* субнимфа
suboccipital *a* подзатылочный
subor *s* суборь
suboral *a* подротовой; околоротовой
suborbital *a* подглазничный
suborder *s zool* подотряд; *bot* подпорядок
subordination *s* субординация, подчинение
subparietal *a* подтеменной
subpharyngeal *a* подглоточный
subphrenic *a* поддиафрагмальный
subpilate *a pal* с мелкими выростами
subpleural *a* субплевральный
subpolar *a* субполярный
subpopulation *s* субпопуляция
subportion *s* субфрагмент
subprolate *a* почти продолговатый
subpsilate *a pal* почти гладкий
subramose *a* слабоветвистый, маловетвящийся
subregion *s* подобласть
subreniform *a* слегка почковидный
subsacral *a* подкрестцовый
subsample *s* часть пробы; *biom* подвыборка, выборка из выборки
subsampling *s* отбор части из пробы
subscapular *a* подлопаточный
subsere *s* субсерия; вторичная серия
subserous *a* субсерозный, подсерозный
subsessile *a* почти сидячий
subset *s* субпопуляция
subshrub *s* полукустарник
  dwarf ~ полукустарничек
subsociation *s* субсоциация
subsoil *s* подпочва
subsong *s orn* подпесня
subspeciation *s* подвидообразование

subspecies *s* подвид
substance *s* субстанция, вещество, материя
agglutinating ~ агглютинирующее вещество, агглютинин
agglutinin-stimulating ~ агглютиноген
antimitotic ~ антимитотическое вещество, антимитотик
blood group ~s *pl* группоспецифические вещества крови; групповые антигены
chromophilic ~ хромофильная субстанция, субстанция Ниссля
cortical ~ of a cell кортикальный слой клетки
dotter ~ желток
fatigue ~ вещество утомления, кенотоксин
foreign ~ примесь, постороннее вещество
gonadotropic ~ гонадотропная субстанция, гонадотропин
growth ~ ростовое вещество, ауксин
growth inhibiting ~ вещество, подавляющее рост, ингибитор роста
growth-promoting ~, growth-stimulating ~ вещество, стимулирующее рост, стимулятор роста
inhibitory ~ угнетающее вещество, ингибитор
lipid-mobilizing ~ вещество, мобилизующее липиды; жиромобилизующее вещество
oxidizing ~ окислитель
precursor ~ предшественник
respiratory ~ дыхательный материал
reticulated ~ сетчатая субстанция
trace ~ микроэлемент
substernal *a* подгрудинный
substitute *s* заменитель; *v* замещать
blood ~ заменитель крови, кровезаменитель
substitution *s* замена; замещение
substrain *s* субштамм
substrate *s* субстрат; подложка
substructure *s* субструктура
subtend *a* сопутстствующий
subterranean *a* подземный
subthreshold *a* подпороговый
subtribe *s bot* подколено
subtropical *a* субтропический
subtropics *s* субтропики
subulate *a* шиловидный
subumbrella *s* субумбрелла
subungual *a* подногтевой, подкогтевой
subunguis *s* когтевая подушка, пальцевая подушка; пальцевой мякиш
subunit *s* субъединица
subvital *a* субвитальный
subzonal *a* субзональный
subzone *s* подзона
subzygomatic *a* подскуловой
success *s* успех
succession *s* последовательность; сукцессия, смена сообществ
crop ~ плодосмен, севооборот
floral ~ последовательность развития цветков
syngenetic ~ сингенез, сингенетическая сукцессия
successional *a* относящийся к сукцессиям
succiniferous *a* янтароносный; смолоносный
succinylation *s* сукцинилирование
succise *a* обрубленный, подрезанный
succulence, succulency *s* суккулентность, сочность, мясистость
succulent *s* суккулент; *a* сочный, мясистый
succumb *v* погибать
succus *s* сок
suck сосать; всасывать, засасывать
sucker *s* сосунок; присоска; корневой побег, корневой отпрыск; волчок, боковой побег; *ent* листоблошка *(Psylla); v* выбрасывать боковые побеги
root ~ корневой побег, корневой отпрыск; корневая поросль
suckerlet *s* присасывательный клапан
suckhole *s* ботрия
suckle *v* кормить грудью; вскармливать; сосать
suckling *s* сосунок; клевер *(Trifolium)*
sucrose *s* сахароза
suction *s* сосание; всасывание, засасывание
suctorial *a* сосущий
Sudan *s* судан
sudanophil, sudanophilic *a* суданофильный, окрашивающийся суданом

sudation *s* потоотделение
sudor *s* пот
sudoriferous потовой
sufficiency *s* достаточность
  nutritional ~ удовлетворительность питания
suffocate *v* задыхаться; душить, удушать
suffocation *s* удушье; замор
suffrutescent *a* полукустарниковый, кустарничковый
suffrutex *s* полукустарник
suffruticose, suffruticous *a* полукустарниковый, кустарничковый
suffumigation *s* окуривание снизу
sugar *s* сахар
  amino ~s аминосахара
  deoxy ~ дезоксисахар
  grape ~ виноградный сахар, глюкоза
  gum ~ арабиноза
  malt ~ мальтоза
  milk ~ лактоза, молочный сахар
  starch ~ крахмальный сахар, декстроза
  wood ~ ксилоза
sugar-bearing *a* сахароносный
sugarcane *s* сахарный тростник *(Saccharum)*
sugary *a* сахаристый
sugescent *a* сосущий
suint *s* жиропот
sulcate *a anat* бороздчатый; *bot* бороздный
sulcation *s* образование борозд
sulculus *s* бороздка
sulcus *s* борозда; извилина
sulfamation *s* сульфамирование
sulfate *s* сульфат
  chondroitin ~ хондроитинсульфат
  dermatan ~ дерматансульфат
  heparitin ~ гепаритинсульфат
  mucoitin ~ мукоитинсульфат
  sodium dodecyl ~ додецилсульфат натрия
sulfation *s* сульфатирование
sulfhemoglobin *s* сульфгемоглобин
sulfination *s* сульфинирование
sulfoalkylation *s* сульфоалкилирование
sulfofication *s* сульфофикация
sulfolipid *s* сульфолипид
sulfomethylation *s* сульфометилирование
sulfonation *s* сульфирование, сульфонирование
sulfoxidation *s* сульфокисление
sulfuretum *s* сообщество серных бактерий
sulfurization *s* сульфуризация
sulfurize *v* окуривать серой
sum *s* сумма
sumac, sumach *s bot* сумах *(Rhus)*
summation *s* суммирование, суммация
summer *s* лето; *a* яровой; **in the top of** ~ в разгаре лета
summit *s* верхушка, вершина
sunblister, sunburn *s* солнечный ожог
sundew *s bot* росянка *(Drosera)*
sunflower *s* подсолнечник *(Helianthus);* очный цвет полевой *(Anagallis arvensis)*
sunleaf *s* световой лист
sun-loving *a* солнцелюбивый
sunny *a* солнечный
sunscald, sunscorch *s phytop* солнечный ожог
sunspot солнечное *s* пятно
sun-tan *s* загар
sunwatcher *s* такырная круглоголовка *(Phrynocephalus helioscopus)*
superabduction *s* чрезмерное отведение конечности
superactivity *s* повышенная активность
superalimentation *s* перекармливание
superalkalinity *s* повышенная щёлочность
superantigen *s* суперантиген
supercentrifuge *s* ультрацентрифуга
superciliary *a* надглазничный; надбровный
supercilium *s* бровь
superclass *s* надкласс
supercoil *s* суперспираль, спираль второго порядка
supercooling *s* переохлаждение
supercrescence *s* эпибиоз
supercrescent *a* эпибионтный
superdominance *s* сверхдоминирование
superexcitation *s* чрезмерное возбуждение, перевозбуждение
superfamily *s* надсемейство; суперсемейство
superfecundation *s* суперфекундация
superfemale *s* сверхсамка
superfetation *s* суперфетация
superficial *a* поверхностный
superficies *s* наружная поверхность
superflexion *s* чрезмерное сгибание

superfunction *s* гиперфункция
supergene *s* суперген
superheating *s* перегревание, перегрев
superhelicity *s mbiol* сверхспиральность
superimpregnation *s* суперфетация
superinfection *s* суперинфекция
supermale *s* сверхсамец
supermaxilla *s* верхняя челюсть
supermicroscope *s* электронный микроскоп
supermotility *s* чрезмерная подвижность
supernatant *s* надосадочная жидкость
supernumerary *a* сверхчисленный, избыточный; *s* добавочная хромосома
supernutrition *s* повышенное питание
superorder *s zool* надотряд; *bot* надпорядок
superordinate *a* соподчинённый
superorganism *s* сверхорганизм
superovulation *s* суперовуляция
superoxygenation *s* перенасыщение кислородом
superparasite *s* сверхпаразит, гиперпаразит
superparasitism *s* сверхпаразитизм, гиперпаразитизм
superposed *a* вышележащий
superregeneration *s* сверхрегенерация
supersacral *a* надкрестцовый
supersaturation *s* пересыщение
supersensitive *a* сверхчувствительный
supersex *s* суперсекс
supersound *s* ультразвук
superspecies *s* надвид
superstition *s eth* №суеверное¤ поведение
superstrain *s* элита, лучшая часть стада
superstructure *s* сверхструктура
supertemplate *s* поверхносная матрица
supertension *s* повышенное давление; повышенный тонус
superterranean, **superterraneous** *a* надземный
superthreshold *a* надпороговый
supervision *s* контроль
supination *s* супинация
supinator *s anat* супинатор
supine *a* лежащий на спине; в положении супинации
supply *s* снабжение; *v* снабжать, подавать, питать; запас; восполнять
 blood ~ кровоснабжение
 food ~ кормовая база, пищевые ресурсы
 nerve ~ иннервация
 vascular ~ кровоснабжение
support *s* опора; подложка; носитель; штатив
 activated ~ активированная подложка
 inert ~ инертная подложка; инертный носитель
 resin ~ полимерная подложка; полимерный носитель
suppress подавлять
suppression *s* супрессия, подавление
 conditioned ~ условное подавление
 nonsense ~ нонсенс-супрессия
 recessive ~ рецессивная супрессия
suppressor *s* супрессор; супрессорный ген, (ген-)супрессор
 gene ~ супрессорный ген, ген-супрессор
 missense- ~ миссенс-супрессор, супрессор миссенс-мутаций
 nonsense- ~ нонсенс-супрессор, супрессор бессмысленных мутаций
suppurate *v* гноиться, нагнаиваться
suppuration *s* нагноение
suppurative *a* гнойный; гноеродный
supra-axillary *a bot* надпазушный
suprabranchial *a* наджаберный
supracaudal *a* надхвостовой
supracellular *a* надклеточный
supraclavicular *a* надключичный
supracondylar *a* надмыщелковый
supracostal *a* надрёберный
supracranial *a* надчерепной
supraglenoid *a* надсуставной
supraliminal *a* надпороговый
supralittoral *a* супралиторальный
supramaxilla *s* верхняя челюсть
supramaxillary *a* верхнечелюстной
supramolecular *a* надмолекулярный
supraoccipital *a* верхнезатылочный
supraorbital *a* надглазничный
supraorganism *s* сверхорганизм
suprapharyngeal *a* надглоточный
suprarenal *s* надпочечник; *a* надпочечный
suprarenin *s* адреналин
suprareticulate *a* поверхностно-сетчатый
suprascapular *a* надлопаточный
supraspinal *a* супраспиральный, надостный

suprasternal *a* надгрудинный
suprastigmal *a* супрастигмальный
suprategillar *a* поверхностно-покровный
supratemporal *a* надвисочный
supraterraneous *a* надземный
supremacy *s* верховенство, доминирующее положение
~ of the herd доминирование в стаде, руководство стадом
sura *s* икра, икроножная область
sural *a* относящийся к икроножной области
suramen *s* сложная суборь, сурамень
surcingle *s anat* хвост хвостатого ядра
surculigerous, surculose *a* образующий побеги
surculus *s* присоска; корневой отпрыск; волчок, боковой побег
surdity *s* глухота
surf *s* прибой
surface *s* поверхность
surface-active *a* поверхностно-активный
surfactant *s* поверхностно-активное вещество
surfbird *s* бурунный кулик *(Aphriza virgata)*
surficial *a* поверхностный
surgery *s* хирургия
surgical *a* хирургический
surroundings *s* окружающая среда
surucucu *s herp* бушмейстер, сурукуку *(Lachesis muta)*
surveillance *s* надзор, контроль
immunological ~ иммунологический надзор
serological ~ серологический контроль
survey *s* обследование; съёмка
aerial ~ учёт с самолёта, воздушная съёмка, аэрофотосъёмка
reconnaissance ~, route ~ маршрутная съёмка
sketch ~ глазомерная съёмка
survival *s* выживание, переживание; **the ~ of the fittest** выживание наиболее приспособленных
survive *v* выживать, переживать; **to ~ to flight** сохраняться до подъёма на крыло
survivor *s* выживший организм
susceptibility *s* чувствительность, восприимчивость; подверженность
susception *s* сусцепция, восприятие
suscitate *v* повышать активность, стимулировать
suspension *s* взвесь, суспензия; приостанвка
suspensor *s* суспензор, подвесок; гиомандибулярный хрящ
suspensorium *s* подвесок; гиомандибулярный хрящ
suspensory *a zool* относящийся к подвеску, поддерживающий, подвешивающий
suspirious *a* тяжело дышащий
sustain *v* выдерживать, выносить, переносить, поддерживать
sustainable *a* устойчивый; **~ development** устойчивое развитие
sustained *a* устойчивый, постоянный, непрерывный
sustentacular *a anat* поддерживающий
sustentator *s ent* крючковидный кремастер
sutural *a* шовный
suture *s* шов
swab *s* тампон на стержне; мазок; *v* брать мазок
swallow *s* глоток; ласточка *(Hirundinidae)*; *v* глотать, проглатывать
bank ~ береговая ласточка, ласточка-береговушка *(Riparia riparia)*
barn ~, common ~, European ~ деревенская ласточка, ласточка-касатка *(Hirundo rustica)*
sand ~ береговая ласточка, ласточка-береговушка *(Riparia riparia)*
sea ~ обыкновенная крачка *(Sterna hirundo)*
swallowing *s* глотание
swallowtail *s ent* парусник, кавалер *(Papilio);* махаон *(Papilio machaon)*
swamp *s* болото; лесное болото; топь, трясина
peat ~ торфяник, торфяное болото
swamping *s* заболачивание
swampland *s* болото; лесное болото; топь, трясина
swampy *a* болотистый, заболоченный, топкий
swan *s* лебедь *(Cygnus);* лебедь-шипун *(Cygnus olor)*
mute ~ лебедь-шипун *(Cygnus olor)*

trumpeter ~ лебедь-трубач *(Cygnus buccinator)*
sward *s* травостой; травяной покров; дернина; *v* задерновывать
swarding *s* задернение
swardy *a* дернистый
swarm *s* рой; стая; косяк; скопление; *v* роиться; собираться стаями, собираться косяками; кишеть
hybrid ~ гибридная популяция
swarmer *s* зооспора; «бродяжка»
swarming *s* роение
swath *s* просека; прокос
sweat *s* пот
sweating *s* потоотделение
swede *s* брюква *(Brassica napus var. napobrassica)*
sweep *v* косить сачком
sweet *a* пресный *(water);* сладкий; незасолённый *(soil)*
sweet-fruited *a* сладкоплодный
sweet-leaved *a* сладколистный
sweet-rooted *a* сладкокорневой
swell *v* набухать; разбухать
swelling *s* набухание; разбухание; опухание; опухоль
root ~ корневой наплыв
swerve *v* отклоняться, сворачивать
swift *s orn* стриж *(Apus)*
swimmer *s* плавающее животное, пловец
swimmeret, swimmerette *s* плавательная ножка
swine *s* свинья *(Sus)*
swing *s* размах, взмах; рывок
switch *s* кисточка *(of a tail)*
swivel *s* шарнирное устройство, ветрлюг
RNA ~ РНК- шарнирное устройство
sword-bearing *a* меченосный
sword-leaved *a* мечевиднолистный
sword-shaped *a* мечевидный
swordtail *s ent* парусник, кавалер *(Papilionidae)*
syllable *s eth* слог
syllepsis *s* оплодотворение; беременность
sylva *s* лес; лесная растительность; обитатель леса
sylvan, sylvatic *a* лесной
sylvestral, **sylvestrian** *a* растущий в лесу, живущий в лесу
sylviculture *s* лесоводство
sylviduct *s anat* водопровод мозга, сильвиев водопровод
sylvula *s* лесной питомник; лесная культура
symbiont *s* симбионт
symbiontic *a* относящийся к симбионту
symbiose *s* симбиоз; сосуществовать
symbiosis *s* симбиоз
symbiote *s* симбионт
symbiotic *a* симбиотический, живущий в симбиозе
symmetry *s* симметрия
sympathectomy *s* симпатэктомия, десимпатизация
sympathetic *a* симпатический
sympathicoblast *s* симпатобласт
sympathicotonia *s* симптатикотония
sympathicotonic *a* симптатикотонический
sympathicus *s* симпатическая нервная система
sympathoblast *s* симпатобласт
sympathomimetic *a* симпатомиметический
sympatric *a* симпатрический
sympatry *s* симпатрия
sympetalous *a bot* спайнолепестный
symphily *s* симфилия
symphyantherous *a* сростнопыльниковый
symphycarpous *a* сростноплодный, синкарпный
symphyllodium *s* симфиллодий
symphyllous *a bot* спайнолистный
symphyogenesis *s* симфиогенез
symphysis *s* симфиз, гемиартроз, полусустав
symplast *s* симпласт
symplectic *s anat* симплектикум
sympodial *a* симподиальный
sympodium *s* симподий
symport *s* симпорт
symptom *s* симптом
synacme, synacmy *s bot* синакме
synandrium *s* группировка самцов
synantherous *a* сростнопыльниковый
synanthesis *s bot* синакме
synanthous *a* сростноцветковый
synanthropic *a* синантропный
synanthy *s* синантия
synapomorphy *s* синапоморфия
synaposematic *a* синапосематический

synapse *s* синапс
synapsid *a anat* синапсидный
synapsis *s* синапсис, конъюгация хромосом
synaptene *s* синаптена, зиготена, амфитена
synaptic *a* синаптический
synaptosome *s* синаптосома
synaptospermy *s bot* синаптоспермия
synarmophytous *a* гинандрический, гермафродитный
synarthrosis *s* синартроз, сочленение
synaesthesia *s* синэстезия
syncarp *s* синкарпий
syncarpous *a* синкарпный
syncarpy *s* синкарпия
syncaryon *s* синкарион
syncaryophyte *s* синкариофит, спорофит
syncephalic *a bot* сростноголовчатый
synchondrosis *s* синхондроз
synchorology *s* синхорология
synchronic *a* синхронный
synchronization *s* синхронизация
synchronizer *s* синхронизатор
time ~ синхронизатор
synchronous *a* синхронный
synclonus *s phys* общий клонус
syncolpate *a pal* меридиально-слитнобороздный
synconium *s* синконий, полая ось соцветия
syncryptic *a* синкриптический
syncyte *s* синцитий
syncytial *a* относящийся к синцитию
syncytiotrophoblast *s* синтрофобласт, синцитиотрофобласт, синцитиальный трофобласт, плазмодиотрофобласт
syncytium *s* синцитий
syndactyl *a* имеющий сросшиеся пальцы
syndactyly *s* синдактилия, сращение пальцев
syndemicolpate *a pal* меридиально-слитнополубороздный
syndesis *s* синдезис
syndesmology *s* синдесмология
syndesmosis *s* синдесмоз
syndiploidy *s* синдиплоидия
syndrome *s* синдром
synechorology *s* синхорология
synecology *s* синэкология, биоценология, экология сообществ
synema *s bot* колонка тычинок
syneresis *s* синерезис
synergetic, synergic *a* синергичный, действующий совместно
synergid *s* синергида
synergism *s* синергия
synergist *s* синергист
synergistic *a* синергичный, действующий совместно
synergy *s* синергия
synesthesia *s* синэстезия
synexine *s* синэкзина
syngameon *s* группа скрещивающихся видов
syngamic *a* сингамный
syngamy *s* сингамия
syngenesious *a* сростнопыльниковый
syngenesis *s* половое размножение; сингенез
syngenetics *s plsoc* сингенетика
syngynium *s* группировка самок
synkaryon *s* синкарион
synkaryophyte *s* диплоидное растение, спорофит
synkinesia, synkinesis *s* совместное рефлекторное движение
synmorphology *s* синморфология
synoecious *a* однодомный
synoicous *a* однодомный
synonym *s* синоним
synonymy *s* синонимика
synorrhizous *a* сростнокорешковый
synost(e)osis *s* синостоз, сращение костей
synovia *s* синовия, синовиальная жидкость
synreflexia *s* синрефлексия
synsacrum *s orn* сложный крестец
syntaxon *s* синтаксон
synthesis *s* синтез
synthesized *a* синтезированный
synthesizer *s* синтезатор
synthrophism *s* синтрофизм
syntony *s* синтония
syntrophoblast *s* синтрофобласт, синцитиотрофобласт, синцитиальный трофобласт, плазмодиотрофобласт
synusia *s* синузия
synzoospore *s* синзооспора
syringe *s* шприц
backfill ~ пипетка-шприц с обратным наполнением
hot ~ термошприц, термопипетка

loading ~ пипетка для внесения образца; инжектор

sample application ~ аппликационная пипетка, сэмплер

syrinx *s* евстахиева труба; фистула, свищ; *orn* сиринкс

syrphetobios *s* биоценоз гниющего навоза, биоценоз компоста

syrup *s* сироп

Apathy gum ~ гумми- сироп Апати

Karo corn ~ кукурузная патока Каро

system *s* система

action ~ характер поведения

computer-video microscope ~ микроскоп с телевизионным ЭВМ-корректором изображения

ecological ~ экосистема

Haversian ~ остеон, костная единица, гаверсова система

paternal-maternal ~ отцовско-материнская система, P-M-система

skeletal ~ костная система, скелет

SOS ~ SOS-система

systematic *a* систематический

systematics *s* систематика, таксономия

systematology *s* учение о системах

systemic *a* системный

systole *s phys* систола

systolic *a* систолический

sysympaedia *s pl* вторичные ассоциации, вторичные группировки

T

table *s* таблица; пластинка; стол; столик
position-specific scoring ~ *gen* карта точковых мутаций
water ~ уровень грунтовых вод
tabular *a* пластинчатый; слоистый
tabulare *s* пластинчатая кость
tabulation *s* составление таблицы, сведение в таблицы
tachyauxesis *s* быстрый рост
tachygenesis *s* тахигенез
tachytely *s* тахителия
tachytrophism *s* ускоренный обмен веществ
tackle *v* схватить; хватать
tactile *a* осязательный, тактильный; осязаемый
taction *s* чувство осязания; осязание
tactor *s* осязательный рецептор
tactual *a* осязательный, тактильный
tadpole *s* головастик
taenia *s anat* тяж; *helm* цепень *(Taenia)*
taeniate *a* лентообразный; полосатый
taenidium *s ent* тенидия
taenioid *a* лентообразный
tag *s* метка; конец хвоста *(e.g., in a fox)*; *v* метить
tagging *s* мечение
tagma *s* тагма; сегмент таллома
tagmosis *s* тагматизация
tahr *s ther* тар *(Hemitragus)*
taiga *s* тайга
tail *s* хвост, хвостовой придаток, хвостовая часть; отросток *(фага); v* купировать хвост, обрезать хвост
~ **of spleen** передний конец селезенки
main ~s *orn* шие рулевые перья
phage ~ отросток фага
tailed *a* хвостатый, имеющий хвост
tail-flicks *eth* подёргивания хвостом
tailgut *s emb* хвостовая кишка
tailhead *s* корень хвоста
tailing *s* расширение пика, образование размытых пиков; *mbiol* наращивание
homopolymer ~ *mbiol* присоединение гомополимера, гомополимерное наращивание
peak ~ расширение пика, образование размытых пиков
tailless *a* бесхвостый
tail-shaped *a* хвостовидный
taint *s* порок; порча; *v* портиться; загнивать
tainted *a* испорченный
hereditary ~ наследственно отягощённый
take *s* добыча; улов; *v* хватать; ловить; отбирать *(a sample);* снимать показания; допускать самца при спаривании; **to ~ to the air** взлететь; **to ~ a culture** *микр.* делать посев; **to ~ a dust bath** купаться в пыли; **to ~ roots** укореняться; **to ~ temperature** измерять температуру; **to ~ the earth** уходить в нору; **to ~ up** поглощать; оттягивать
take-all *s phytop* выпревание
taking-up *s* поглощение; оттягивание
takyr *s* такыр
tall *a* высокий; высокорослый
tallgrass *s* высокая трава, высокотравье
tallow-bearing *a bot* салоносный
tall-stalked *a* высокостебельный
talocrural *a* голеностопный
talon *s* коготь
talus *s* таранная кость
tame *v* приручать; дрессировать; приручаться; *a* прирученный, одомашненный, ручной; окультуренный; непуганный *(e.g., bird)*
tameless *a* дикий, неприрученный
tan *s* загар
tangle *s* сплетение, заросли
tangled *a* спутанный, перевитый
tangoreceptor *s* тангорецептор
tank *s* чан, резервуар, ванна, сборник, бассейн, тэнк; водоём; небольшое водохранилище; испаритель
aeration ~ аэротэнк; аэрируемый сосуд
agitation ~ смеситель, смесительный резервуар
clarification ~ осветлитель, отстойник
digestion ~ септик-тэнк
dosing ~ дозатор
drinking ~ поилка
earth ~ водопойный пруд
extended aeration ~ аэротэнк продолженной аэрации
fermentation ~ бродильный чан
oxidation ~ окислительный резервуар

septic ~ септик-тэнк
skimming ~ пеноотделитель
tansy *s* пижма *(Tanacetum)*
tantivy *s* быстрый галоп
tap *s* пункция; *v* делать прокол; подсачивать *(trees)*
tapered *a* оттянутый; суженный; со сбегом к кончику, конический
tapering *a* сбежистый
tapetal *a* тапетальный
tapetum *s ent* зеркальце; *bot* тапетум, выстилающий слой
tapeworm *s* ленточный червь *(Cestoda)*
dog ~ эхинококк *(Echinococcus granulosus);* тыквовидный цепень *(Dipylidium caninum)*
taphrium *s* сообщество канав
taphrophilus *s* обитатель канав
taphrophyte *s* растение канав
tapir *s* тапир *(Tapirus)*
tapping *s* подсочка *(of trees)*
taproot *s* стержневой корень, главный корень
tarantula *s* паук-птицеяд *(Theraphosidae);* апулийский тарантул *(Lycosa tarentula)*
tarbagan *s* тарбаган, монгольский сурок *(Marmota sibirica)*
tardigrade *s zool* тихоходка *(Tardigrada)*
target *s* мишень; клетка-мишень; *a* прицельный
reference ~ эталонная мишень, референс-мишень
tarn *s* крачка *(Sterna);* болотная крачка *(Chlidonias)*
tarpan *s ther* тарпан *(Equus caballus gmelini)*
tarragon *s* полынь эстрагон *(Artemisia dracunculus)*
tarrock *s orn* моевка *(Rissa tridactyla);* крачка *(Sterna)*
tarsal *a* тарзальный; предплюсневой
tarsier *s ther* долгопят *(Tarsius)*
tarsometatarsal *a* предплюсне-плюсневой
tarsometatarsus *s orn* цевка
tarsophalangeal *a* предплюсно-фаланговый
tarsus *s ent* лапка; *orn* плюсна
tartareous *a* шероховатый, с грубой поверхнстью
tasicolpate *a pal* натянутобороздный
task *s* задача
tassel *s zool* султан; *orn* борода; *bot* метёлка, султан
taste *s* вкус; вкусовое ощущение; *v* ощущать вкус
tasteless *a* безвкусный
TATA-box *s mbiol* ТАТА-блок
tautomerization *s* таутомеризация
tautonym *s* тавтоним
tautonymy *s* тавтонимия
tawny *a* рыжевато-бурый, рыжевато-коричневый
taxation *s* таксация
taxidermist *s* таксидермист
taxidermy *s* таксидермия
taxis *s* таксис
taxocene *s* таксоценоз
taxocline *s* таксономический клин
taxon *s* таксон
taxonomic *a* таксономический
taxonomist *s* систематик
taxonomy *s* таксономия
tea *s* чай *(Thea);* камелия *(Camelia)*
beef ~ мясной бульон
Paraguay ~ мате, парагвайский чай *(Ilex paraguariensis)*
teak *s* тик, джатовое дерево *(Tectona grandis)*
teal *s orn* чирок *(Anas)*
team *s* выводок; группа сходных животных
teasel *s bot* ворсянка *(Dipsacus)*
teat *s* сосок
technique *s* технический приём, способ, метод; методика
labeling ~ способ мечения
spreading ~ посев размазыванием
immunologic ~ иммунобиотехнология
non-waste ~ безотходная технология
recombinant DNA ~ генная инженерия, генетическая инженерия, метод рекомбинантных ДНК
technosphere *s* техносфера
tectology *s* тектология
tectorial *a* текториальный, покровный, кроющий
tectorium *s* текториум, покрышка
tectospondylic *a* тектоспондильный
tectrix *s* покровное перо, кроющее перо
tectum *s anat* перемычка, мостик
teem *v* быть плодовитым; изобиловать; кишеть

teeming *a* плодовитый; изобилущий; кишаший
teeth *s pl* зубы; **to bare ~, to show ~** оскаливать зубы
teethe *v* прорезываться
teething *s* прорезывание зубов
tegillate *a* тонкопокровный
tegmen *s anat* покровная структура; крыша; тегмен; *bot* покров
tegula *s ent* тегула, подкрыловая пластинка
tegular *a ent* тегулярный; черепицеобразный
tegumen *s anat* покровная структура; крыша; тегмен; *bot* покров
tegument *s* оболочка, наружный покров; *helm* тегумент
tela *s* сетевидная структура; ткань
telarian *a* прядущий паутину
teleblem *s myc* общее покрывало
teleceptor *s* телерецептор, дистантный рецептор
teledendron *s* терминальное нервное окончание
telegamic *a* телегамный
telegony *s* телегония
teleianthous *a bot* обоеполый
telemetry *s* телеметрия
telencephalon *s* конечный мозг; *emb* мозговой пузырь
teleneuron *s* нервное окончание
teleonomy *s* телеономия
teleoptile *s orn* дефинитивный наряд
telereceptor *s* телерецептор, дистантный рецептор
telescopiform *a* телескопический
telethmoid *s* предносовая кость
teleutosorus *s* телейтопустула
teleutospore *s* телейтоспора
teleutosporiferous *a* несущий телейтоспоры
teleutostage *s* телейтостадия
telianthus *a bot* обоеполый
teliosorus *s* телейтопустула
teliospore *s* телейтоспора
teliostage *s* телейтостадия
telium *s* телейтопустула
telmathium *s* фитоценоз болот
telmathology *s* болотоведение
telmathophilus *s* обитатель болот
telmicolous *a* обитающий в болотах
teloblast *s* телобласт
telocentric *a* телоцентрический
telocoel *s anat* боковой желудочек мозга
telodendrion, telodendron *s* концевое разветвление аксона
telolecithal *a* телолецитальный
telolemma *s* телолемма
telome *s bot* телом
telomere *s* теломер
telomitic *a* теломитический
telophase *s* телофаза
telophragma *s* телофрагма
teloreceptor *s* телерецептор
telosomic *s* телосомик
telosynapsis, telosyndesis *s* телосинапсис, метасиндезис
telotaxis *s* телотаксис
telotroch *s* телотрох, ресничное кольцо; телотрох, бродяжка
telson *s* тельсон
telum *s* телум
temnospondylous *a* темноспондильный
temperament *s* нрав; характер; темперамент
temperate *a* умеренный
temperature *s* температура; **to take ~** измерять температуру
accumulated ~s сумма температур
base growth ~ начальная температура роста, биологический нуль
cumulative ~s сумма температур
minimum effective ~ минимальная эффективная температура, порог развития
temperature-sensitive *a* чувствительный к температуре
tempering *s* темперирование
template *s* матрица; *a* матричный
template-directed *a* управляемый матрицей
template-RNA *s* матричная РНК, информационная РНК, мРНК
temple *s* висок
tempolabile *a* изменяющийся с течением времени
temporal *a* височный; временный, преходящий; временной
temporary *a* временный; сезонный; *s* молочный зуб
temporomalar *a* височно-скуловой
temporooccipital *a* височно-затылочный
temporoparietal *a* височно-теменной
temporozygomatic *a* височно-скуловой

tenacious *a* вязкий, липкий, клейкий; крепкий; живучий
tenaculum *s* зацепка вилочки *(in springtails); bot* прицепка
tendency *s* тенденция, стремление
tender *a* нежный, тонкий, слабый; мягкий; болезненный
tendinous *a* сухожильный
tendon *s* сухожилие
  chitinized ~ *ent* хитиновый тяж
  coronary ~ фиброзное кольцо
tendril *s bot* усик, усикообразный придаток; плодоножка *(of a watermelon)*
tendrillar *a* вьющийся
tendril-shaped *a* усиковидный
tenent *a* удерживающий, цепкий
teneral *a ent* только что полинявший
tenia *s anat* тяж; *helm* цепень *(Taenia)*
tenidium *s ent* тенидия
tenofibril *s* волокно сухожилия
tenoceptor *s* тенорецептор
tenrec *s ther* тенрек *(Tenrecidae)*
tension *s* натяжение; растяжение; напряжённое состояние, напряжение; давление
  muscle ~ напряжение мышц
  suction ~ сосущая сила
  superficial ~, **surface** ~ поверхностное натяжение
tensor *s anat* тензор, напрягающая мышца
tentacle *s* щупальце; щупик
tentacular *a* относящийся к щупальцам
tentaculiferous *a* несущий щупальца
tentaculiform *a* щупальцевидный
tentaculum *s* щупальце; щупик
tentative *a* опытный, пробный, экспериментальный; гипотетический, предполагаемый
tentorium *s:*
  ~ **of cerebellum** палатка мозжечка, намёт мозжечка, мозжечковый намёт
  ~ **of hypohysis** диафрагма турецкого седла
tenuimarginate *a* с тонкими краями
tenuitegillate *a* тонкопокровный
tepal *s* листочек околоцветника
tephrocarpous *a* пепельноплодный
teratic *a* уродливый
teratism *s* уродство, уродливое развитие
teratogen *s* тератогенный фактор
teratogenesis *s* тератогенез
teratogenic *a* тератогенный
teratology *s* тератология
tercel *s* самец сокола
terebrate *a* буравящий, имеющий буравящий орган
terete, teretial *a* почти цилиндрический, вальковатый
teretifolious *a* вальковатолистный
tergal *a* дорсальный, спинной; тыльный
tergite *s* тергит
tergosternal *a* тергостернальный
term *s* период, срок
terminal *s* конечная часть, окончание; *a* терминальный, конечный, концевой; *bot* верхушечный
  input ~ «входной участок»
  presynaptic ~ пресинаптическое окончание
terminalia *s pl ent* внешние гениталии, гипопигий
terminalization *s* терминализация
termination *s* окончание; *mbiol* терминация, окончание считывания
terminator *s mbiol* терминатор
termini *s pl*:
  blunt ~ *mbiol* «тупые» концы
  cohesive ~ *mbiol* «липкие» концы
terminiflorous *a* верхушечноцветный
terminology *s* терминология, номенклатура
termitarium *s* термитник
termite *s* термит *(Isoptera)*
termiticole, termitophilous *a* живущий в гнезде термитов
tern *s* крачка *(Sterna)*
  Caspian ~ чеграва *(Sterna caspia)*
ternate *a* тройчатый
ternatopinnate *a* триждыперистый
terrain *s* местность; территория, район
terraneous *a* наземный
terrapin *s* черепаха; *pl* пресноводные черепахи *(Emydidae)*
terrarium *s* террариум
terrestrial *a* живущий на земле; сухопутный; наземный
terrick *s orn* крачка *(Sterna);* (молодая) моевка *(Rissa);* кайра *(Uria)*
terricolous *a* почвообитающий, почвенный
territorial *a* территориальный

territorialism, territoriality *s* территориальные отношения, территориальность
territory *s* территория, участок обитания; защищаемая часть участка обитания
tertian *s* трёхдневная малярия *(caused by Plasmodium vivax)*
tertiary *a* третьего порядка; третичный
tessellated *a* клетчатый, шахматный, с прямоугольными пятнами; мозаичный
test *s* исследование; испытание; тест, проба; опыт; анализ; реакция; реактив; *biom* критерий; *v* исследовать; испытывать; проверять; производить анализ; **to put to ~** подвергать испытанию
  allium ~ проба на корешках лука, луковый тест
  anti-DNA ~ проба на антитела к ДНК
  blank ~, blind контрольный опыт, слепой опыт
  blood ~ анализ крови
  blood-sedimentation ~ проба для определения скорости осаждения эритроцитов, СОЭ-тест
  fragility ~ определение осмотической стойкости эритроцитов
  immunobead ~ иммуногранулотест
  laboratory ~ лабораторная проверка, лабораторное испытание
  latex fixation ~ *imm* реакция латексагглютинации, тест латексагглютинации, латекс-иммобилизационный тест
  linkage ~ проверка на сцепление
  mirror absorption ~ *imm* реакция взаимоабсорбции
  multiwell tray ~ метод многолуночных планшетов
  omnibus ~ омнибус-критерий
  orthostatic ~ ортостатистическая проба
  over-all ~ омнибус- критерий
  SAFA ~ гомогенный иммунофлуоресцентный анализ антигенов
  sign ~ *biom* знаковый тест, критерий знаков
  soluble antigen fluorescent antibody ~ гомогенный иммунофлуоресцентный анализ антигенов
  vitality ~ определение жизнеспособности; проверка жизнеспособности
  X-ray ~ рентгеновский анализ
testa *s bot* теста, семенная кожура; панцирь
testable *a* доступный для проверки опытом
testectomy *s* удаление семенника, кастрация
tester *s* испытательный прибор; контрольно-измерительный прибор; зонд, щуп; лаборант; испытатель
  density ~ плотномер
  lung ~ спирометр
  moisture ~ влагомер
  pH ~ pH-метр
  saline ~ солемер
testicle *s* семенник, мужская половая железа
testicular *a* тестикулярный, относящийся к семенникам
testing *s* испытание, опробование, исследование; тестирование
testis *s* семенник, мужская половая железа
test-kit *s* тест-набор
test-sensitivity *s* чувствительность метода
testudinate *a* имеющий раковину
tetania *s* тетания
tetanic *a* столбнячный, тетанический
tetanilla *s* фибрилляция мышц; тетания
tetanus *s* тетанус, тетаническое сокращение; столбняк
tetrachaenium *s bot* четырёхсемянка
tetrachoric *a biom* тетрахорический, четырёхклеточный
tetrachotomous *a* четырёхраздельный, разделённый на четыре части
tetracolporate *a pal* четырёхдистальнобороздный
tetracrepid *a* четырёхлучевой; четырёхконечный
tetract, tetractine *s* четырёхлучевая спикула
tetracyclic *a* четырёхмутовчатый; четырёхкруговой
tetrad *s* тетрада
tetradactyl *a* четырёхпалый
tetradymous *a* четырёхскладчатый

tetragonal *a* тетрагональный, четырёхугольный
tetragynous *a* с четырьмя плодолистиками
tetramer *s* тетрамер
tetramerous *a* четырёхчленный
tetramorphic *a* тетраморфный
tetrandrous *a* четырёхтычиночный
tetrapetalous *a* четырёхлепестный
tetraphyllous *a* четырёхлистный
tetraploid *s* тетраплоид; тетраплоидный
double ~ двойной тетраплоид, октоплоид
tetraplody *s* тетраплоидия
tetrapod *s* четвероногое животное; четвероногий
tetrapolar *a* тетраполярный, четырёхполюсныйюбдлщш98
tetrapterous *a* четырёхкрылый
tetraquetrous *a* четырёхгранный
tetraselenodont *a anat* тетраселенодонтный
tetrasomic *s* тетрасомик; *a* тетрасомический
tetraspermous *a* четырёхсемянный
tetrasporangium *s* тетраспорангий
tetraspore *s* тетраспора
tetrastichous *a* четырёхрядный
tetrathecal *a* четырёхгнёздный
tetravalent *s* квадривалент
tetrazolium *s* тетразолий
T-even T-чётный *(phage)*
textiform *a* сетчатый, ретикулярный
texture *s* текстура; структура, строение
thalamencephalon *s* промежуточный мозг; таламус, зрительный бугор
thalamiflorous *a* ложецветковый
thalamus *s* таламус, зрительный бугор; цветоложе
thallicolous *a* растущий на слоевище; поражающий слоевище
thalline *a* талломный, слоевищный
thallogen *s* таллофит, талломное растение, слоевищное растение
thalloid *a* талломный; талломообразный
thallome *s* таллом, слоевище
thallophyte *s* таллофит, талломное растение, слоевищное растение
thallose *a* талломный; талломообразный
thallospore *s* спора, образующаяся на талломе
thallus *s* таллом, слоевище
thamnium *s* кустистое слоевище *(of lichens)*
thanatocoenosis *s* танатоценоз
thanatoid *a* летальный, смертельный; ядовитый *(of snakes)*
thanatology *s* танатология
tanatophoric *a* летальный, смертельный
thanatosis *s* танатоз, рефлекторная неподвижность, «притворная» смерть
theca *s* тека; сумка; капсула; оболочка; аск, спорангий; пыльцевой мешок
thecaphore *s* карпофор
thecaspore *s* аскоспора
thecate, theciferous, thecigerous *a* заключённый в оболочку, покрытый текой
thecium *s* гимений
thecodont *a anat* текодонтный
thelytoky *s* телитокия
thenal *a* ладонный
thenar *s anat* тенар
theory *s* теория
theriology *s* териология, маммалиология,маммалогия
thermal, thermic *a* термический, тепловой
thermic *a* тепловой, термический
thermocleistogamy *s bot* термоклейстогамия
thermocline *s* термоклин, температурный скачок, слой температурного скачка
thermocouple *s* термопара
thermoduric *a* термостойкий, терморезистентный
thermocurrent *s* термоток
thermoesthesia *s* температурная чувствительность
thermoexchange *s* теплообмен
thermoexcitatory *a* вызывающий теплообразование
thermogenesis *s* термогенез, теплообразование
thermogenic *a* термогенный
thermogram *s* термограмма
thermograph *s* термограф
thermoinducible *a* термоиндуцируемый
thermoinhibitory *a* тормозящий образование тепла, задерживающий образование тепла
thermolabile *a* термолабильный, неустойчивый к тепловому воздействию

thermoluminescence *s* термолюминесценция
thermolysis *s* термолиз, термическое расщепление; теплоотдача
thermometer *s* термометр
 recording ~ термограф
thermometry *s* термометрия
thermonasty *s* термонастия
thermoperiodicity, thermoperiodism *s* термопериодизм
thermophil *s* термофил, теплолюбивый организм
thermophilic, thermophilous *a* термофильный, теплолюбивый
thermophyte *s* термофит, теплолюбивое растение
thermopile *s* термоэлемент
thermoreceptor *s* образование тепла температурный рецептор, терморецептор
thermoregulation *s* терморегуляция
thermoresistant *a* терморезистентный, теплоустойчивый
thermostability *s* термостабильность, термостойкость, теплоустойчивость
thermostat *s* термостат
thermotactic *a* термотаксический
thermotaxis *s* термотаксис
thermotolerant *a* терморезистый, теплоустойчивый
thermotropism *s* термотропизм
therodrymium *s* сообщество лиственного леса
therophyllous *a* листопадный
therophyte *s* терофит, однолетник, однолетнее растение
thiamine *s* тиамин, аневрин
thick *a* частый, густой; толстый; густой, вязкий
thickened *a* утолщённый
thickening *s* утолщение
thicket *s* чаща, заросль; поросль; молодой лес
thick-flat *a bot* слегка приплюснутый *(a fruit)*
thick-fleshed *a* хорошо упитанный
thick-fruited *a* толстоплодный
thickgrowing *a* густорастущий
thick-headed *a* толстоголовчатый
thick-lipped *a* толстогубый
thick-veined *a bot* толстонервный
thick-walled *a* толстостенный; с толстой оболочкой; мясистый *(of a fruit)*
thigh *s* бедро; бедренная кость; *orn* голень
 second ~ голень
tighbone *s* бедренная кость
thigmesthesia *s* осязание, тактильная чувствительность
thigmomorphosis *s* тигмоморфоз
thigmotaxis *s* тигмотаксис
thigmotropism *s* тигмотропизм
thimic-dependent *a* тимусзависимый
thimic-independent *a* тимуснезависимый
thin *a* тонкий; разбавленный; ненасыщенный; жидкий; слабый; *v* разбавлять, разжижать
thin-crowned *a* с изреженной кроной
thin-fleshed *a* плохо упитанный
thinicolous *a* обитающий на дюнах
thinium *s* сообщество дюн
thinking *s* мышление
thin-layer *a* тонкослойный
thinned *a* изреженный, разреженный
thinophilus *s* обитатель дюн
thinophyte *s* растение дюн
thin-walled *a* тонкостенный; с тонкой оболочкой
thioacylation *s* тиоацилирование
thiobacterium *s* тиобактерия, серобактерия
thiocyanation, thiocyanogenation *s* роданирование, тиоцианирование
thiogenic *a* образующий серу
thiomethylation *s* тиометилирование
thiophil *s* тиофил
thiophilic *a* тиофильный, серолюбивый
thiourea *s* тиомочевина
thirst *s* жажда; *v* испытывать жажду
thistle *s* бодяк *(Cirsium);* артишок *(Cynara);* чертополох *(Carduus)*
 sow ~ осот *(Sonchus)*
thistledown *s bot* пушок, хохолок
thoracal, thoracic *a* торакальный, грудной
thorax *s anat* торакс, грудная клетка; *ent* грудь, грудной отдел
thorn *s* шип, колючка; *bot* боярышник *(Crataegus)*
thornbush *s* колючий кустарник
thorned *a* шиповатый, колючий
 closely ~ густошиповатый
thorny *a* шиповатый, колючий

thorny-headed *a* колючеголовчатый
thoroughbred *a* чистопородный, чистокровный
thrasher *s* пересмешник *(Mimidae)*
thread *s* нить
thread-leaved *a* нитчатолистный
threadlike *a* нитевидный
thread-stalked *a* нитчатостебельчатый
threadworm *s* нематода, круглый червь *(Nematoda)*
threat *s eth* угроза
three-celled *a* трёхклеточный; трёхкамерный
three-lobed *a* трёхдольный; трёхлопастной
three-merous *a bot* трёхчленный
thremmatology *s* тремматология
threshold *s* порог, предел; мера чувствительности
above ~ надпороговый
thrive *v* буйно разрастаться
throat *s* горло; гортань; глотка; *bot* зев
throatiness *s* подгрудок
throb *s* биение, пульсация; *v* биться, пульсировать
throbbing *s* биение, пульсация; ритмическое движение
throe *s* сильная боль; агония
thromb *s* тромб; кровяной сгусток
thrombin *s* тромбин, тромбаза, фибрин-фермент
thrombinogen *s phys* фактор II, протромбин
thrombocyte *s* тромбоцит, кровяная пластинка
thrombocytopoiesis *s* тромбоцитопоэз
thrombogen *s phys* фактор II, протромбин
thromboplastin *s* тромбопластин, тромбокиназа
tissue ~ фактор III, тканевой тромбопластин, тканевая тромбокиназа
thrombosin *s* тромбин, тромбаза, фибрин-фермент
thrombotonin *s* серотонин
thrombus *s* тромб
throstle *s* певчий дрозд *(Turdus philomelos)*
throw *v* приносить потомство; менять кожу
throwback *s* регресс; возврат к прошлому; атавизм
thrush *s* дрозд *(Turdus)*
mistle ~ деряба *(Turdus viscivorus)*
thrusting *s eth* толчки тазом
thumb *s* большой палец
thumbscrew *s lab* микрометрический винт
thurl *s* тазобедренный сустав
thyme *s* тимьян, чабрец *(Thymus)*
thymocyte *s* тимоцит
thymocytopoiesis *s* лимфопоэз в тимусе, тимоцитопоэз
thymus *s* тимус, вилочковая железа, зобная железа
thyrocalcitonin *s* тиреокальцитонин, гипокальциемический фактор
thyroglossal *a* щитовидно-язычный
thyroid *s* щитовидная железа
thyroid-stimulating *a* тиреотропный, стимулирующий щитовидную железу
thyrotropin *s* тиреотропин
thyrsoid *bot* пирамидально-метельчатый, тирсоидный
tibia *s* большеберцовая кость; *ent* голень
tibial *a* большеберцовый; *ent* голенный
tibiotarsal *a ent* тибиотарзальный
tic *s med* тик
tick *s* клещ *(Ixodoidea)*
relapsing-fever ~ аргасовый клещ *Ornithodoros turicata*
sheep ~ рунец овечий *(Melophagus ovinus)*
tickbird *s* буйволовый скворец *(Buphagus)*
tick-borne *a* передаваемый клещами
tigellum *s* тигеллум, укороченный стебель
tiger *s* тигр *(Panthera tigris)*
American ~ ягуар *(Panthera onca)*
cat ~ исполинская сумчатая куница, тигровая кошка *(Dasyurus maculatus);* оцелот *(Felis pardalis);* онцилла *(Felis tigrinus)*
deer ~, **red** ~ пума, кугуар, горный лев *(Felis concolor)*
tigroid *s* тигроид, тигроидное вещество, субстанция Ниссля
till *v* возделывать; обрабатывать
tillable *a* пахотный

tiller *s* побег, росток; отросток; *v* выбрасывать побеги, давать ростки; прорастать; куститься
tillering *s* кущение, ветвление
tilt *s* наклонное положение, наклон
timbal *s ent* цимбалы
timber *s* древесина
  blue ~ синева древесины
  standing ~ лес на корню
timbered *a* лесистый
timberland *s* лес
time *s* время; период
  breeding ~ время осеменения; время спаривания, период спаривания
  circulation ~ скорость кровотока
  developmental ~ продолжительность развития
  roosting ~ ночной отдых, ночёвка *(of birds)*
time-giver *s* ритмоводитель. водитель ритма
timekeeping *s* хронометрирование, хронометраж
timid *a* робкий, пугливый
timing *s* тайминг; временной паттерн
timothy *s* тимофеевка *(Phleum)*
tinamou *s orn* тинаму *(Tinamus)*
tinction *s* окрашивание
tinctorial *a* красильный
tincture *s* тинктура, настойка; раствор; оттенок
tine *s* отросток *(of a horn)*
tingibility *s* окрашиваемость, способность окрашиваться
tint *s* оттенок; тон; *v* слегка окрашивать; подцвечивать
tintometer *s* колориметр
tiny *a* карликовый; крошечный; маленький
tip *s* конец, верхушка; конечная точка, верхушечная точка
  dripping ~, rain ~ *bot* капельное остриё
tirium *s* сообщество пустоши
tissue *s* ткань
  brown adipose ~ бурая жировая ткань, бурый жир
  cancellous ~ губчатая ткань; губчатое вещество кости
  compact ~ компактное вещество кости
  dental ~ дентин
  endothelial ~ эндотелий
  envelope ~ *bot* периблема
  epidermic ~ эпидермис
  epithelial ~ эпителий, эпителиальная ткань
  formative ~ *bot* меристема
  mesenchymal ~ мезенхима
  parenchymatous ~ паренхима, паренхиматозная ткань
  sieve ~ флоэма
  spurious ~ *myc* ложная паренхима
tissue-specific *a* тканеспецифичный
tissular *a* тканевой
tit *s* синица *(Parus);* маленькая птица
  azure ~ белая лазоревка, князёк *(Parus cyanus)*
  blue ~ лазоревка *(Parus caeruleus)*
  coal ~ московка *(Parus ater)*
  crested ~ хохлатая синица, гренадёрка *(Parus cristatus)*
  fire-capped ~ огненноголовый ремез *(Cephalopyrus flammiceps)*
  long-tailed ~ длиннохвостая синица, ополовник *(Aegithalos);* длиннохвостая синица, ополовник *(Aegithalos caudatus)*
  marsh ~ черноголовая гаичка *(Parus palustris)*
  mouse-coloured ~ мышиный ремез *(Remiz musculus)*
  penduline ~ ремез *(Remiz)*
  willow ~ буроголовая гаичка, пухляк *(Parus montanus)*
  yellow-breasted ~ желтогрудая лазоревка *(Parus flavipectus)*
titer *s* титр
  killing ~ индекс цитолиза
titlark *s orn* конёк луговой *(Anthus pratensis)*
titmouse *s* синица *(Parus)*
  black-crested ~ черношапочная гаичка *(Parus atricapillus)*
  crested ~ хохлатая синица, гренадёрка *(Parus cristatus)*
  long-tailed ~ длиннохвостая синица, ополовник *(Aegithalos caudatus)*
  marsh ~ черноголовая (болотная) гаичка *(Parus palustris)*
  penduline ~ ремез *(Remiz pendulinus)*
titrate *s* титруемый раствор; *v* титровать
titration *s* титрование
toad *s* жаба; жаба *(Bufo)*

clawed ~ шпорцевая лягушка *(Xenopus)*
dwarf ~ жабка *(Pelophryne)*
horned ~ рогатка *(Ceratophrys cornuta);* жабовидная ящерица *(Phrynosoma)*
midwife ~ жаба-повитуха *(Alytes obstetricans)*
oak ~ жаба-малютка *(Bufo quercicus)*
spade-footed ~ лопатоног *(Scaphiopus);* чесночница *(Pelobates)*
Surinam ~ суринамская пипа *(Pipa pipa)*

toadfrog *s* чесночница *(Pelobatidae)*
toadstool *s* поганка *(inedible mushroom);* ядовитый гриб
tobacco *s* табак *(Nicotiana)*
common ~, rustic ~, Syrian ~, wild ~ махорка *(Nicotiana rustica)*
tocophero1 *s* токоферол, витамин Е
tocus *s* роды, родовой акт
tod *s* лисица *(Vulpes);* серая лисица *(Urocyon)*
toe *s* палец ноги
tokostome *s* половое отверстие
tolerance *s* толерантность, устойчивость, выносливость; ~ **for winter conditions** зимоустойчивость
cold ~ холодостойкость
drought ~ засухоустойчивость
internal ~ толерантность к «своему», аутотолерантность
salt ~ солевыносливость
shade ~ теневыносливость
submersion ~ *bot* устойчивость к затоплению

tolerant *a* толерантный, терпимый, выносливый
tolerate *v* переносить, выносить
tolerization *s* приобретение толерантности, толеризация
tolerogenic *a* толерогенный, вызывающий иммунную толерантность
tolerogenicity *s* толерогенность
tom *s* самец
tomato *s* томат настоящий, помидор *(Lycopersicum esculentum)*
tomcat *s* кот
tomentose *a* опушённый, покрытый пушком, пушистый; косматый, мохнатый; войлочный
tomentulose *a* коротковойлочный
tomentum *s bot* войлок
tomography *s* томография
tomtit *s orn* синица *(Parus)*
tone *s* тонус; тон *(in bioacoustics)*
tone-modulated *a* тонально-модулированный *(bioacoustics)*
tongue *s* язык
lolling ~ свисающий язык *(e.g., in a dog)*
"tongues of fire" *phytop* «огненые языки», ржавчина боярышника *(caused by Gymnosporidium clavariiforme)*

tonic *s* тонизирующее средство; *a* тонизирующий
tonicity *s* тоничность, концентрация; тонус
tonoclonc *a* тоноклонический
tonofibril *s* тонофибрилла
tonoplast *s* тонопласт
tonotaxis *s* тонотаксис
tonsil *s* миндалина
tooth *s* зуб; *bot* зубец
auditory ~ слуховой зубчик
azzle ~ моляр, большой коренной зуб
canine ~ клык
cheek ~ моляр, большой коренной зуб
corner ~ резец *(in a horse)*
cuspid ~, cuspidate ~ клык
eye ~ глазной зуб, верхний клык
hair ~ слуховой зубчик
labial ~ резцы и клыки
laniary ~ клык
lophodont ~ лофодонтный зуб, складчатый зуб
molar ~ моляр, большой коренной зуб
multicuspid ~ моляр, большой коренной зуб
premolar ~ премоляр, малый коренной зуб
selenodont ~ селенодонтный зуб, лунчатый зуб
stomach ~ нижний клык
tricuspid ~ моляр, большой коренной зуб
wisdom ~ зуб мудрости, третий моляр

toothed *a* зубчатый
tooth-leaved *a* зубчатолистный
tooth-seeded *a* зубчатосемянный
tooth-sepaled *a* зубчаточашечный

top *s* верх, верхушка, верхняя часть; *pl* надземная масса растения; ботва; ~ **off** обрезать вершину
~ of shoulder холка
bunchy ~ *phytop* кустистость верхушки
curly ~ *phytop* курчавость, скручивание верхушки
top-dry *s* суховершинность
topical *a* местный, локальный
topinambour *s* топинамбур, земляная груша *(Helianthus tuberosus)*
topknot *s* хохолок
topline *s anat* линия верха
topochemistry *s* топохимия
topocline *s* топоклин
topodeme *s* тoподем
topogenesis *s* топогенез
topognosis *s* локализация чувствительности
topophototaxis *s* топофототаксис
topotaxis *s* топотаксис
topotropism *s* топотропизм
topotype *s* топотип
topotypical *a* топотипичный, топотипический
topshaped *a* кеглевидный
topsoil *s* верхний слой почвы
topswarm *s apic* первый рой
toque *s* макак *(Macaca)*
torfaceous *a* растущий на болотах
tormentil *s bot* лапчатка *(Potentilla)*
tornado *s* смерч, торнадо, вихрь, ураган
tornaria *s* торнария
tornote *a* с тупым краем
torose *a* имеющий большие вздутия, бугристый, шишковатый; узловатый
torpid *a* оцепеневший, онемелый; бездеятельный, апатичный
torrid *a bot* выжженный солнцем
torsion *s* торсия, спиральное закручивание
torticone *a* турбоспиральный
tortoise *s* черепаха
tortuosity *s* искривление, извитость
tortuous *a* извилистый
torularhodin *s* каротиноидная кислота
torulose, torulous *a* имеющий маленькие бугорки
torus *s ent* стебелёк брюшка; *bot* ложе, цветоложе; ложе соцветия, общее цветоложе; тор
total *s* целое; итог; *a* целый, общий
totipotency *s* тотипотентность
totipotent *a* тотипотентный
toucan *s orn* тукан *(Rhamphastos)*
touch *s* осязание; ощупывание; *v* ощупывать
touchwood *s* гниль древесины *(caused by Polyporus squamosus);* трутовик *(e.g., Fomes)*
tough *a* тягучий; вязкий
tower-shaped *a* башневидный
towlike *a* паклевидный
toxalbumin *s* токсальбумин; токсин
toxic *s* токсичное вещество, ядовитое вещество; *a* токсичный, ядовитый
toxicide *s* противоядие, антидот
toxicogenic *a* токсиногенный, образующий яд
toxicology *s* токсикология
toxicometry *s* токсикометрия
toxiferous *a* содержащий яд, токсичный, ядовитый
toxigenic *a* токсинообразующий, токсикогенный
toxin *s* токсин, яд
animal ~ зоотоксин
extracellular ~ экзотоксин,истинный токсин
fatigue ~ токсин усталости, кенотоксин
intracellular ~ эндотоксин
plant ~ фитотоксин
true ~ экзотоксин, истинный токсин
toxoid *s* токсоид, анатоксин
toxoplasmosis *s* токсоплазмоз *(caused by Toxoplasma gondii)*
Tox-plasmid *s* Tox-плазмида
trabecula *s anat* трабекула, перекладина
trabeculation *s anat* образование трабекул, образование перекладин
trace *s* след; небольшое количество
primitive ~ *emb* первичная полоска
tracer *s* меченый атом, изотопный индикатор; индикаторный; следящее устройство
radioactive ~ радиоактивный индикатор
trachea *s* трахея, дыхательное горло; *bot* древесинный сосуд
tracheal, tracheate *a* трахеальный; *bot* сосудистый
tracheation *s* трахеация

tracheid *s bot* трахеида, сосудовидная клетка
trachenchyma *s bot* сосудистая ткань
tracheole *s ent* трахеола
tracheomycosis *s phytop* трахеомикоз
trachycarpous *a* шероховато-плодный
trachyphyllous *a* шероховато-листный
trachyspermous *a* шероховато-семянный
tracing *a* стелющийся; *s* слежение, прослеживание; запись
track *s* след; путь; *v* следить, выслеживать
tracking *s* слежение; отслеживание
tract *s* тракт, путь; участок, пространство, территория
ear ~ наружный слуховой проход
feather ~s *orn* птерилии
Flechsig's ~ задний спинно-мозжечковый путь, путь Флексига, пучок Флексига
trade-off *s* трэйдоф
trail *s* след; стелющийся побег; тропа; *v* идти по следу; *bot* стелиться
trailing *a bot* стелющийся, ползучий
train *v* дрессировать, вырабатывать навыки. приучать, обучать; натаскивать *(a dog);* формировать, направлять
training *s* дрессировка, выработка навыков, обучение; натаска *(of dogs); bot* формирование, направление
wind ~ искривление под действием ветра
trait *s* признак, особенность, черта
trama *s* трама
trample *v* вытаптывать
trampling down *s* вытаптывание
transacetylation *s* переацетилирование
transacylation *s* переацилирование
transalkylation *s* переалкилирование
transamidation *s* переамидирование
transamination *s* переаминирование
transaminomethylation *s* переаминометилирование
transapical *a* трансапикальный
transcapillary *a* транскапиллярный
transcarboxylation *s* карбоксилирование
transcolpate *a pal* поперечнобороздчатый
transconfiguration *s* трансположение
transconformation *s* трансконформация, структурный переход
transcript *s mbiol* транскрипт
transcriptase *s* транскриптаза, РНК-зависимая ДНК-полимераза
reverse ~ обратная транскриптаза, ревертаза
transcription *s* транскрипция, считывание генетического кода
transdetermination *s* трансдетерминация
transducer *s phys* передатчик
transductant *s* трансдуктант
transduction *s* преобразование; трансдукция
transect *s* трансекта
transepithelial *a* трансэпителиальный
transfection *s* трансфекция
transfer *s* перенос; передача; пассаж, пееревивка; пресадка; трансплантация; *v* переносить, перемещать; передавать; перевивать; пересаживать
allogenic ~ аллотрансплантация
homologous ~ аллотрансплантация
information ~ *eth* обмен информацией
isologous ~ изотрансплантация
tissue ~ пересадка ткани
transform *v* превращать, преобразовывать
transformant *s* трансформант
transformation *s* трансформация, превращение, преобразование
blast-cell ~ бласттрансформация
neoplastic ~ злокачественное перерождение, неопластическая трансформация
transformism *s* трансформизм
transfuse *v* делать переливание
transfusion *s* переливание
transgenation *s* генная мутация; трансгеноз, перенос генов
transgene *s* трансген
transgenesis *s* трансгеноз, перенос генов
transgenic *a* трансгенный
transgenote *v* осуществлять перенос генов
transgression *s* трансгрессия
transheterozygote *s* трансгетерозигота
transhydrogenation *s* перегидрирование
transient *a* временный, преходящий, транзиторный; проходящая особь
transition *s* переход, перемещение; *mbiol* транзиция
thermal ~ термопереход, тепловой переход

transitional *a* переходный, промежуточный
translation *s mbiol* трансляция
nick ~ ник-трансляция
translocation *s* перемещение; транслокация; передвижение веществ
transmembrane *a* трансмембранный
transmethylation *s* трансметилирование, переметилирование
transmissible *a* передающийся; трансмиссивный
transmission *s* передача; перенос
aerial ~ воздушный путь передачи
insect ~ перенос насекомыми
retrograde ~ обратное проведение
transmitter *s* передатчик, переносчик; *phys* медиатор
transmural *a* трансмуральный, пронизывающий стенку
transmutation *s* преобразование, превращение
transneuronal *a* межнейронный
transovarian *a* трансовариальный
transphosphorylation *s* перефосфорилирование, трансфосфорилирование
transpiration *s* транспирация
transpirometer *s* транспирометр
transplacental *a* трансплацентарный, осуществляемый через плаценту
transplant *v* пересаживать; переселять; делать пересадку, пересаживать
transplantat *s* трансплантат
transplantation *s* трансплантация, пересадка; перевивка
transport *s* транспорт, перенос, перемещение
active ~ активное перемещение; активный транспорт
electron ~ электронный транспорт, перенос электронов
transporter *s* транспортер
transposition *s* транспозиция, перемещение; транспозиция
transposon *s gen* транспозон, Тп-элемент, перемещающийся элемент
transreplication *s gen* трансрепликация
transudate *s* транссудат; *v* просачиваться
transudation *s* транссудация, выпотевание
transvection *s* трансвекция
transverse *a* поперечный; трансверсальный
transversion *s* трансверсия
tranverso-gallop *s* диагональный галоп
trap *s* ловушка; капкан; *v* ставить ловушку, ставить капкан, ловить капканом
bacterial ~ бактериальный фильтр
break-back ~ ловушка-давилка, плашка
cannon net ~ *orn* пушечная сеть
corral ~ ловушка-корраль
enclosure ~ ловушка с захлопывающейся дверкой
live ~ живоловка
pitfall ~ ловчая яма; ловчая канавка
rat ~ крысоловка
span ~ ловушка-давилка, плашка
steel ~ капкан
trap-day *s* ловушко-сутки; ловушко-день
trap-night *s* ловушко-ночь
trauma *s* травма
traumatic *a* травматический; раневой
traumatropism *s* травматропизм
travel *s* переселение, миграция
trawl *s* трал
tray *s* лоток; планшет; пластинка
agglutination ~ многолуночный планшет, агглютинационный планшет
assay ~ аналитический столик
running ~ камера для электрофореза
tread *s* шаг; ход *(of a horse);* спаривание *(of birds); v* спариваться *(of birds)*
treat *v* обрабатывать; лечить
treatment *s* обработка; лечение
anthelminthic ~ дегельминтизация
biological ~ биологичесая очистка
photoperiodic ~ фотопериодическое воздействие
radiation ~ облучение
seed ~ протравливание семян
tree *s* дерево; *v* загнать на дерево; взбираться на дерево, влезать на дерево
apple ~ яблоня *(Malus)*
box ~ самшит вечнозелёный *(Buxus sempervirens)*
Brasil nut ~ бразильский орех *(Bertholletia excelsa)*
camphor ~ коричник камфарный, камфарный лавр *(Cinnamonium camphora)*

caoutchouc ~ гевея бразильская *(Hevea brasiliensis)*
castor-bean ~ клещевина обыкновенная *(Ricinus communis)*
cherry ~ вишня *(Cerasus)*
genealogic, genealogical ~ генеалогическое древо
gum ~ эвкалипт *(Eucalyptus)*
Judas ~ иудино дерево, багряник стручковатый *(Cercis siliquastrum)*
monkey-bread ~ баобаб, адансония *(Adansonia)*
mother ~, parent ~ семенное дерево, семенник
pomegranate ~ гранат *(Punica)*
respiratory ~s *pl zool* водные лёгкие
rowan ~ рябина обыкновенная *(Sorbus aucuparia)*
rubber ~ гевея *(Hevea)*
seed ~ семенное дерево, семенник
stunted ~ корявый подрост; угнетённый подрост
subordinate ~ подчинённое дерево
sugar ~ клён сахарный *(Acer saccharum)*
weed ~ дерево-сорняк
windfallen ~s, windthrown валежник
wolf ~s *pl* разросшиеся деревья

treecreeper *s orn* пищуха *(Certhia)*
tree-grass-steppe *s* настоящая саванна
treeless *a* безлесный
treelike *a* древовидный
tree-toad *s* квакша *(Hylidae)*
trefoil *s* клевер *(Trifolium)*
trema *s anat* отверстие, канал, проход
tremelloid *a* студенистый, киселеобразный
trend *s* направление, тенденция; *biom* тренд
feeding ~s *pl* особенности питания
trepidation *s* дрожь, дрожание, тремор; тревога, беспокойство
treppe *s phys* феномен лестницы, лестница Боудича
triactinal *a* трёхлучевой
triad *s* триада
triadelphous *a bot* трёхбратственный
trial *s* испытание, опыт, проба; испытательный, опытный
triandrous *a* трёхтычинковый
triannulate *a* трёхкольчатый
trianthous *a* трёхцветковый
triaristate *a* трёхостный
triarticulate *a* трёхчленистый
triaster *s* трёхполюсная фигура деления
tribe *s* триба; племя
tributary *s* венозная ветвь
tricarinate *a* трёхкилевой
tricephalous *a* трёхглавый
triceps *s* трицепс, трёхглавая мышца
trichidium *s bot* трихидий, стеригма
trichilium *s bot* трихилиум
trichina *s helm* трихина *(Trichinella)*
trichite *s zool* трихит
trichoblast *s* трихобласт
trichocarpous *a* волосистоплодный
trichocyst *s* трихоциста
trichogen *s* трихогенная клетка
trichogyne *s* трихогина
trichoid *a* волосообразный, волосовидный
trichome *s* трихом
trichosclereid *s* разветвлённая склереида
trichothallic *a bot* имеющий нитевидный таллом
trichotomic, trichotomous *a* трихотомический
trichotomy *s* трихотомическое ветвление
trichotrophy *s* питание волос
thichroic *a* трихроичный
trichromatic *a* трихромный
tricipital *a anat* трёхглавый
trickle *s* струйка; *v* сочиться, капать
tricolor *a* трёхцветный
tricolpate *a* трёхбороздный
tricompound *a* тройчатосложный
triconodont *a anat* триконодонтный, трёхзубчатый
tricornavirus *s* трикорнавмрус
tricrotic *a* трикротический, трёхвершинный
tricrural *a* трёхветвистый
tricuspid, tricuspidate *a* трёхстворчатый; трёхзубчатый; имеющий три острия; трёхбугорчатый
tridactyl, tridactylous *a* трёхпалый
triennial *a* трёхлетний
trifacial *s* тройничный нерв; *a* тройничный
trifarious *a bot* трёхрядный
trifid *a* трёхнадрезный
triflorous *a* трёхцветковый
trifoliate *a* трёхлистный

trifurcate *a* трёхветвистый, трёхвильчатый
trigamma *s ent* трёхзубчатое жилкование
trigamous *a bot* тригамный
trigeminal *a* тройничный
trigenic *a* тригенный, контролируемый тремя генами
trigger *s bioch* пусковой механизм; насторожка *(of a trap)*
trigonocarpous *a* треугольноплодный
trigonospermous *a* треугольносемянный
trigynous трёхпестичный
triheterozygote *s* тригетерозигота
trijugate *a bot* трёхпарный
trilabiate *a* трёхгубый
trilatiporate *a* трёхширокопоровый
trilete *a* трёхлучевой
trill *s eth* трель
trilobate *a* трёхлопастный; трёхдольчатый
trilocular *a* трёхгнёздный; трёхкамерный
trilophodont *a anat* трилофодонтный
trim обрезать, подрезать, подстригать
trimerous *a* трёхчленный; тримерный
trimming *s* обрезка, подрезка
trimorphous *a* триморфный
trinervate *a* трёхжилковый, трёхнервный
trinominal *a* триноминальный
triovulate *a* трёхсемяпочковый
tripartite *a* трёхраздельный
tripe *s anat* рубец
  plain ~ рубец
  pocket ~ *anat* сетка
  regular ~ рубец
tripetalous трёхлепестный
triphosphate *s*:
  adenosine ~ аденозинтрифосфат, АТФ
  cytidine ~ цитидинтрифосфат, ЦТФ
  deoxyadenosine ~ дезоксиаденозинтрифосфат, дАТФ
  deoxycytidine ~ дезоксицитидинтрифосфат, дЦТФ
  deoxyguanosine ~ дезоксигуанозинтрифосфат, дГТФ
  deoxyuridine ~ дезоксиуридинтрифосфат, дУТФ
  guanosine ~ гуанозинтрифосфат, ГТФ
  inosine ~ инозинтрифосфат
  thymidine ~ тимидинтрифосфат, ТТФ
  uridine ~ уридинтрифосфат, УТФ
tripinnate *a* триждыперистый
tripinnatifid *a* триждыперистонадрезный
tripinnatisect *a* триждыперисторассечённый
triple-nerved, triple-ribbed *a* трёхжилковый, трёхнервный
triplet *s* триплет; *pl* тройня
  amber ~ амбер-триплет
triplex *s gen* триплекс
tripling *s phys* тригеминия
triploblastic *a emb* трёхслойный
triploid *s* триплоид; *a* триплоидный
triploidy *s* триплоидия
triplostichous *a* трёхрядный
tripping *s* искусственное раскрывание цветка
tripterous *a* трёхкрылый
tripus *s anat* треножник
triquetrous *a bot* трёхгранный
triquetrum *s* трёхгранная кость; вставочная кость, вормиева кость
triquinate *a bot* трёх-пятираздельный
triradial, traradiate *a* трёхлучевой
trisaccharide *s* трисахарид
trischistoclastic *a* трёхрасщеплённый
trisected *a* трёхрассечённыё
triseptate *a* трёхперегородчатый; имеющий три раздела
triserial *a* трёхрядный; группами по три
trisetose *a* трёхщетинковый
trisomic *s* трисомик
trisomy *s* трисомия
trispermous *a* трёхсемянный
trisporous *a* трёхспоровый
tristichous *a* трёхрядный
tristylous *a bot* трёхстолбиковый
tristyly *s* триморфная гетеростолбчатость
triternate *a* трижды тройчатый
tritiated, tritium-labelled *a* меченный тритием
tritocerebrum *s ent* тритоцеребрум
tritonymph *s zool* тритонимфа
tritor *s anat* жевательная поверхность
tritreme *a pal* с тремя апертурами
tritubercular *a anat* тритуберкулярный; трёхбугорчатый
triturate *v* растирать, толочь
triungulin, triungulinid *s* триунгулин
trivalent *s* тривалент
trivalve *a* трёхстворчатый
trochanter *s anat* вертел; *ent* вертлуг
trochantin *s anat* малый вертел; *ent* трохантин, малый вертлуг

trochate *a* колесовидный
trochlea *s anat* блок
trochlear *a* блоковый; блоковидный
trochophore, trochosphere *s* трохофора
troglobiont *s* троглобионт, обитатель пещер
troglocolous *a* пещерный, обитающий в пещерах
trollflower *s bot* купальница *(Trollius)*
troop *s* стадо
tropeic *a* килевидный
trophallaxis *s ent* трофаллаксис, взаимное кормление
trophectoderm *s* трофобласт
trophi *s pl* ротовые части; мастакс, жевательный аппарат *(of rotifers)*
trophic *a* трофический; пищевой
trophidium *s ent* личинка первого возраста
trophobiosis *s* трофобиоз
trophoblast *s* трофобласт
trophocyte *s* трофоцит, питающая клетка
trophoderm *s* трофодерма
trophogenic, trophogenous *a* трофогенный
tropholytic *a* трофолитический
trophont *s* трофонт
trophoplasm *s* трофоплазма
trophotaxis *s* трофотаксис, пищевой таксис
trophotropism *s* трофотропизм
trophozoite, trophozooid *s* трофозоит
trophy *s* добыча
tropibasic *a* тропибазальный
tropical *a* тропический
tropicbird *s orn* фаэтон *(Phaethon)*
tropism *s* тропизм
  transverse ~ диатропизм
tropistic *a* относящийся к тропизмам
tropophyte *s* растение тропиков
tropophytic *a* тропофитный
tropotaxis *s* тропотаксис
trot *s* рысь; *v* идти рысью
trouble *s* беспокойство, волнение; тревога; забота; болезнь
trough *s* кормушка
  drinking ~ поилка
troupial *s orn* трупиал *(Icteridae)*
trowel *s bot* копалка
truffle *s myc* трюфель; трюфель *(Tuber)*
trullifolious *a bot* чашевиднодистный
trumpet *s* трубка; bot воронка
truncal *a anat* туловищный; стволовой
truncate, truncated *a* усечённый
truncation *s mbiol* усечение
trunk *s* ствол; туловище; ствол; *zool* хобот
truss *s bot* кисть; гроздь
tryma *s bot* сочная костянка
trypomastigote *s prot* трипомастиготная форма
tryptic *a* триптический
tsetse *s* муха цеце *(Glossina)*
tuatara *s herp* гаттерия, туатара *(Sphenodon punctatus)*
tuba *s anat* труба
tubal *a* трубный
tubate *a* трубкообразный
tube *s* труба, трубка; пробирка
  air ~ трахея; трахеола; дыхательная трубка
  alimentary ~ пищеварительный тракт
  auricular ~ наружный слуховой проход
  Bellini's ~ трубочка Беллини, прямой мочевой каналец
  breathing ~ трахея; трахеола; дыхательная трубка
  bronchial ~s мелкие бронхи
  collecting ~ собирательный мочевой каналец
  conjugation ~ *micr* соединительный мостик, конъюгационный мостик; конъюгационная трубка
  dermomuscular ~ кожно-мускульный мешок
  dropping ~ пипетка
  egg ~ яйцевод; *ent* овариола, яйцевая трубочка
  fermentation ~ бродильная трубка
  germ ~ *bot* проростковая ростковая трубочка, трубочка, зародышевая трубочка
  latex ~ *bot* млечник, млечный сосуд
  lung ~ *orn* лёгочная трубка, лёгочная трахея
  Malpighian ~ мальпигиев сосуд
  otopharyngeal ~ евстахиева труба
  ovarian ~ яйцевод; *ent* овариола, яйцевая трубочка
  pitch ~ *phytop* смоляная воронка
  Rainey's ~ трубочка Рейни
  test ~ пробирка

tube-builder *s ent* листовёртка *(Tortricidae)*
tube-dwelling живущий в трубках
tuber *s anat* бугор; *bot* клубень
tubercle *s anat* бугорок; *bot* мелкий клубень; клубенёк
mammillary ~ сосцевидный отросток
root ~ корневой клубенёк
tubercular, tuberculate, tuberculose *a* бугорчатый; клубеньковый
tuberidium *s* ложный клубень
tuberiferous *a* образующий клубеньки; клубненосный
tuberization *s* клубнеобразование
tuberosity *s anat* бугристость
ishial ~ седалищный бугор
tuberous *a* бугорчатый, бугристый; клубневой
tub-gardening *s* кадочная культура
tubicolous *a* живущий в трубках
tubicorn *s* полорогое животное; *a* полорогий
tubifacient *a* строящий трубки
tubilingual *a* имеющий трубчатый язык
tubing *s* разливание по пробиркам
tubular *a* трубчатый, тубулярный
tubulate, tubulated *a* имеющий трубку; трубковидный
tubulation *s* тубуляция
tubule *s anat* трубочка; сосуд; каналец
Malpighian ~ мальпигиев сосуд
Pfluger's ~ трубочка Пфлюгера, исчерченный проток
tubuliflorous *a* трубчатоцветковый
tubulose *a* трубчатый
tuft *s* пучок; хохолок; бородка; *bot* розетка
tufted *a* кустовой; пучковатый; с хохолком; с бородкой
tulip *s* тюльпан *(Tulipa)*
tumbler *s* вегетационный сосуд; *orn* турман
tumor *s* новообразование, опухоль; вздутие
tumorous *a* опухолевый
tundra *s* тундра
tung *s bot* тунг *(Aleurites)*
tunic *s zool* туника, оболочка; *бот.* туника; семенная кожура
nerve ~ оболочка нерва, нервное влагалище
nervous ~ of eyeball сетчатка, сетчатая оболочка глаза
tunicated *a* имеющий тунику; покрытый оболочкой
tunicle *s* естественный покров, интегумент
tunnel *s anat* туннель
adjoining ~ паратуннель, туннель Неэля
Corti's ~, inner ~ внутренний туннель, кортиев туннель
outer ~ паратуннель, туннель Неэля
tup *s* баран-годовик; *v* спариваться; покрывать овцу
tupping *s* спаривание, случка
tur *s* тур
turbid *a* мутный
turbidimeter *s* турбидиметр
turbidimetry *s* турбидиметрия
turbidity *s* муть, мутность; помутнение
turbinal *a* спирально закрученный; носовая раковина
turbinate *a zool* кубаревидный *(of a shell)*
turgescence *s* тургесценция, набухание; тургор
turgid *a* тургесцентный, набухший, раздутый, разбухший
turgor *s* тургор
turkey *s* индюк, индейка *(Melagris gallopavo)*
brush ~ кустарниковая курица *(Aepypodius);* сорная курица, большеног *(Talegalla);* кустарниковый большеног *(Alectura)*
tom ~ индюк
turn *s bioch* виток
turnip *s* репа *(Brassica rapa);* капуста полевая *(Brassica campestris)*
turnover *s* оборачиваемость; обмен; круговорот; обновление *(напр. белка)*
genetic ~ обновление генофонда
population ~ оборот популяции
turnstone *s orn* камнешарка *(Arenaria)*
turpentine *s* живица, скипидар; синкарпия *(Syncarpia)*
turtle *s* черепаха; наземная черепаха *(Testudo)*
hawksbill ~ бисса *(Eretmochelys)*
loggerhead ~ логгерхед, каретта *(Caretta)*
turtledove *s* горлица *(Streptopelia)*
tush, tusk *s* клык; бивень

tusker *s* слон с большими бивнями; кабан-секач
tussock *s* кочка; дерновина; полевица белая *(Agrostis alba)*
tween-brain *s* промежуточный мозг
tweezers *s* пинцет
twice-forked *a* дваждыраздвоенный
twig *s* побег; веточка; веточка
twin *s* близнец; *pl* близнецы, двойня; *a* близнецовый; двойной; спаренный, сдвоенный
twiner *s* вьющееся растение
twinflowered *a* удвоенноцветковый
twining *a* вьющийся, обвивающий
twinning *s* рождение двойни
twin-row *a* двухрядный
twist *s phytop* закрученность *(caused by Dilophospora alopecuri); ent* листовёртка
twisted-grained *a* с извилистыми волокнами
twisting *s* закручивание, скручивание
twitter *s* щебет; чириканье; песня *(of a skylark); v* щебетать, чирикать; петь *(of a skylark)*
two-celled *a* двухклеточный; двухкамерный
two-chambered *a* двухкамерный; с двумя полостями
two-combed, two-crested *a* двугребенчатый
two-dimensional *a* двухмерный; плоский
two-forked *a* двухвильчатый
two-glanded *a bot* двужелезистый
two-horned *a* двурогий
two-humped *a* двугорбый
two-layered *a* двухслойный; двухъярусный
two-nerved *a* двухжилковый
two-ranked *a* двухрядный
two-seeded *a* двусемянный
two-stage *a* двухэтапный; двухстадийный; двухступенчатый
two-storied *a* двухъярусный
two-stranded *a* двухнитевой, двухцепочечный
two-veined *a* двухжилковый
tylenchoid *a* тиленхоидный
tylosis *s* тил, каллус
tylote *s* гантелеобразная спикула
tylotic *a* относящийся к каллусу; имеющий каллус
tylus *s ent* наличник
tympanic *a* барабанный
tympanum *s* барабанная полость; *ent* тимпанальный орган; *myc* эпифрагма
tyndalization *s* тиндализация, дробная стерилизация
type *s* тип, типичный экземпляр, типичный представитель; разновидность; типовой экземпляр
  reaction ~ тип реакции; фенотип
type-species *s* типовой вид
typhlosole *s* тифлозоль
typical *a* типичный
typiform *s* типичная форма
typing *s* типирование
  antigenic ~ скрининг антигенов; антигенное картирование
  blood ~ определение группы крови
  genetic ~ генотипирование
  serologic ~ серологическое типирование, серотипирование, определение серотипа
  tissue ~ тканевое типирование, гистотипирование
typogenesis *s* типогенез
typology *s* типология
tyroid *a* казеозный
tystie, tystle *s orn* обыкновенный чистик *(Cepphus grylle)*

U

ubiquinone *s* убихинон
ubiquist *s* убиквист
udder *s* вымя
udometer *s* дождемер
ugrug, **ugruk** *s* морской заяц, лахтак *(Erignathus barbatus)*
ulcer *s* язва
ulcerate *v* изъязвлять, изъязвляться; *a pal* дистальнопоровый
uliginous болотистый; болотный; илистый; вязкий, липкий
ulna *s* локтевая кость
ulnar *a* локтевой
ulnocarpal *a* запястно-локтевой
ulotrichous *a* шерстистый
ultimate *a* последний
ultracentrifugation *s* ультрацентрифугирование
ultracentrifuge *s* ультрацентрифуга
ultracryotomy *s* ультракриотомия
ultrafilter *s* ультрафильтр
ultrafiltration *s* ультрафильтрация
ultramicrobalance *s* ультрамикровесы
ultramicrobe *s* вирус
ultramicroelectrode *s* ультрамикроэлектрод
ultramicroscope *s* ультрамикроскоп
ultramicrotom *s* ультрамикротом
  freezing ~ криоультрамикротом
ultramicrotomy *s* ультрамикротомия
ultrapasterization *s* уперизация, ультрапастеризация
ultrasonic *a* ультразвуковой
ultrasonication *s* обработка ультразвуком
ultrasound *s* ультразвук
ultrastability *s* ультраустойчивость
ultrastructure *s* ультраструктура
ultraviolet *a* ультрафиолетовый
ultravirus *s* ультравирус, фильтрующийся вирус
umbel *s bot* зонтик
umbellate *a* зонтичный; расположенный зонтиками; зонтиконосный
umbellet *s bot* вторичный зонтик
umbelliferous *a* образующий зонтик; относящийся к зонтичным
umbellula *bot* зонтичек
umbellulate *a* собранный в зонтики
umbellule *bot* зонтичек
umbilical *a* пупочный; пуповинный
umbilicate, umbilicated *a* с глубоким пупком *(of a shell)*
umbilicus *s* пупок; рубчик
umbo *s bot* выпуклость, горбина; *mal* умбо, верхушка
umbraculum *s zool* зонтиковидный отросток; зонтиковидный рот
umbrageous *a* тенистый; дающий тень
umbraticole, umbraticolous *a* теневыносливый; тенелюбивый
umbrella *s* колокол
  aerostatic ~ *bot* парашют, парус
umbrella-leaved *a* зонтиколистный
umbrous *a* тенистый; дающий тень
unaffected *a* непоражённый; нетронутый
unarmed *a zool* невооружённый
unavailable *a* недоступный
unbalance *s* отсутствие равновесия
unbiased *a* объективный*; biom* несмещённый
unblended *a* несмешанный, без примесей
unbound *a bot* отделённый
unbounded *a bioch* биохим. несвязанный
unbranched *a* неразветвлённый
unbroken *a* неповреждённый, целый
unciferous *a anat* несущий крючья
uncinate *a anat* крючковатый, крючковидный; относящийся к крючку
uncinatum *s* крючковидная кость
uncleaned *a* неочищенный
uncleaved *a* недробившийся
uncoating *s:* **virus** ~ декапсидация вируса
uncolored *a* неокрашенный
uncomplemented *a* не имеющий комплемента, бескомплементный
unconditioned *a phys* безусловный
unconjugated *a* неконъюгированный; несвязанный
unconscious *a* бессознательный
uncontaminated *a* незагрязненный; неинфицированный, незараженный
uncontrollable *a* неконтролируемый
uncoupler *s phys* разобщающий агент
uncoupling *s phys* разобщение
uncovered *a* непокрытый
uncrystallizable *a* некристаллизующийся
unculled *a* неотбракованный, небонитированный

uncultivated *a* необработанный, неосвоенный, неокультуренный *(soil);* некультивируемый
uncus *s* ункус, крючок, крючковидный отросток
uncut *a* несрезанный; неподстриженный; нескошенный
underclass *s* подкласс
undercoat *s* подшёрсток; пух
undercolor *s* подцветка
undercooling *s* переохлаждение
underdevelopment *s* недостаточное развитие
underestimation *s* недооценка
underexploitation *s* неполное использование, недостаточное использование
underfed *a* недоедающий; недокармливающий
underfeeding *s* недокармливание
underfur *s* подшёрсток
underground *a* подземный; *s* подпочва, подстилающая порода
undergrowth *s* подрост; подлесок, мелколесье
underhorn *s anat* нижний рог
underjaw *s* нижняя челюсть
underlayer *s* подслой
underline *s* нижняя часть брюха
underlip *s* нижняя губа
undernourishment *s* недокармливание; недостаточное питание; истощение
underparts *s pl* нижние участки тела
underpinning *s* нижняя часть туловища; подбрюшье
underpopulation *s* недоуплотнение популяции
underreplication *s* недорепликация
underripe *a* недозрелый
undershrub *s* кустарничек; полукустарник
underside *s* нижняя сторона
undersized *a* маломерный; низкорослый
undersoil *s* подпочва, подстилающая порода
understaining *s* недостаточная окраска
understerilization *s* недостаточная стерилизация
understocked *a* недостаточной полноты *(of a forest)*
understory *s* подрост; подлесок, мелколесье
underventilation *s* гиповентиляция
underwater *s* грунтовые воды
underwing *s ent* заднее крыло; ленточница, орденская лента *(Catocala)*
underwood *s* подлесок, мелколесье
undesirable *a* неприемлемый, нежелательный
undetectable *a* необнаруживаемый, невыявляемый
undeveloped *a* недоразвитый; необработанный; неосвоенный
undifferentiated *a* недифференцированный
undifferentiation *s* отсутствие дифференциации
undigested *a* непереваренный
undigestible *a* неперевариммый
undiluted *a* неразбавленный, неразведённый
undissociated *a* недиссоциированный
undissolving *a* нерастворяющийся
unditegillate *a pal* волнистопокровный
undocked *a* с необрезанным хвостом, с некупированным хвостом
undose, undulate *a bot* волнистый; струйчатый
undulating *a* ундулирующий, колеблющийся, волнообразный
undulation *s* ундуляция, волнообразное движение
undulipodium *s* ундулиподия
uneasiness *s* беспокойство, тревога
unencapsulated *a* бескапсульный; неинкапсулированный
uneven-aged *a* разновозрастный
unexcited *a* невозбуждённый
unfavorable *a* неблагоприятный, отрицательный
unfermentable *a* несбраживаемый, неферментируемый
unfertile *a* неплодородный
unfertilized *a* неоплодотворённый; неудобренный
unfilled *a* ненасыщенный *(biotope)*
unfit *a* негодный, неподходящий
unfledged *a* неоперённый
unfold развёртывать, развёртываться; распускаться
unfolding *s* разворачивание, развёртывание

unformalinized *a* неформалинизированный
unfruitful *a* неплодоносный; неплодородный
unfurl развёртывать, развёртываться; распускаться
ungraded *a* неотсортированный
ungual *s* ноготь; коготь; коготок; копыто; *a* ногтевой; когтевой; коготковый
unguiculate *a bot* имеющий ноготки; имеющий когти, имеющий коготки
unguiculus *a* коготкообразный придаток, коготок
unguis *s* ноготь; коготь; коготок; копыто; *bot* ноготок
ungulate *s* копытное; *a* копытный; копытообразный
  even-toed ~ парнокопытное *(Artiodactyla)*
  horned ~ полорогое *(Bovidae)*
  odd-toed ~ непарнокопытное *(Perissodactyla)*
  paridigitate ~ парнокопытное *(Artiodactyla)*
unguligrade *a* копытоходящий; копытный
unhealthy *a* вредный; болезненный, нездоровый
unialgal *a* альгологически чистый
uniarticulate *a* одночлениковый
uniaxial *a* одноосевой, одноосный
unibracteate *a* одноприцветничковый
unibranchiate *a* одножаберный
unicamerate *a* однокамерный
unicapsular *a* однокапсульный; *bot* однокоробочковый
unicell *s* одноклеточный организм
unicellate, unicellular *a* одноклеточный
unicostate *a* с одной жилкой
unicotyledonous *a* однодольный
unicuspid *a anat* однобугорчатый
unidactyl *a* однопалый
unidentate *a* однозубчатый
unifacial *a* односторонний, имеющий одну поверхность
unifactorial *a* монофакториальный
uniflagellate *a* одножгутиковый
uniflorous *a* одноцветковый
unifoliate *a* однолистный
uniforate *a* имеющий одно отверстие
uniform *a* однообразный; одинаковый; однородный
uniformity *s* единообразие, однородность
  ~ of type *s* однотипность
unijugate *a* однопарный
unilabiate *a* одногубый
unilaminate *a* однослойный
unilateral *a* односторонний
unilobar *a* однолопастный; имеющий одну долю
unilocular *a* одногнёздный; однокамерный
unimodal *a* однородный
unimucronate *a* одновершинный
uninerved *a* одножилковый
uninjurious *a* безвредный; неповреждённый
uninominal *a* униноминальный
uninuclear, uninucleate *a* одноядерный
union *s* соединение, слияние; унион
  accidental ~ случайное сосуществование
  ligamentous ~ соединение посредством связок
unioval *a* однояйцовый, монозиготный
uniparous *a* рожавшая один раз
unipetalous *a* однолепестный
unipolar *a* монополярный, униполярный
uniseptal, uniseptate *a* имеющий одну перегородку
uniserial *a* однорядный
uniserrate *a bot* однажды пильчатый
unisetose *a* несущий одну щетинку
unisexual *a* однополый
unispiral *a* односпиральный
unit *s* единица; единица измерения
  animal ~ особь; голова
  feto-placental ~ плодоплацентарная единица, комплекс плод - плацента
  functional ~ функциональный комплекс; функциональная единица
  heat ~ калория
  life ~ парцелла органического мира
  Morgan ~ морганида
  motor ~ мотонейрон, моторная клетка
  multiple ~ сложный нейрон
  nerve ~ нервная единица, нейрон
  sample ~ количество образца; *biom* элемент выборки
  single ~ одиночный нейрон
united *a* соединённый; сросшийся

unitunicate *a myc* однооболочковый
univalent *s* унивалент; *a* одновалентный; моновалентный
univalve *s* одностворчатая раковина
universe *s biom* генеральная совокупность; среда
univoltine *a* унивольтинный
unlabeled *a* немеченный, не содержащий метки
unload *s phys* разгрузка, освобождение от нагрузки, разряд
unmedullated *a* безмиелиновый
unmixed *a* несмешанный, чистый
unnucleated *a* безъядерный
unpaired *a* непарный
unplant *v* вырывать с корнем
unplanted *a* непосаженный, некультивированный, незасаженный; незаселённый
unpretentious *a* неприхотливый
unprimed *a imm* непримированный
unproductive *a* непродуктивный, непроизводительный
unprompted *a* спонтанный, непроизвольный
unpurified *a* неочищенный
unreclaimed *a* нерегенерированный
unreduced *a* невосстановленный
unrefined *a* неочищенный
unrelated *a* неродственный
unreliable *a* ненадёжный; недостоверный
unrenewable *a* невозобновимый
unresponsiveness *s* иммунологическая неотвечаемость; иммунологическая толерантность
unrest *s* беспокойство, волнение
unrestricted *a* неограниченный
unrewarded *a* неподкрепляемый
unripe *a* незрелый, недозрелый, неспелый
unrolling *a bot* развёртывающийся
unsaponifiable *a* неомыляющийся
unsaturated *a* ненасыщенный
unsegmented *a* несегментированный
unsexed *a* кастрированный; *orn* не разделенный по полу
unstable *a* нестабильный; изменчивый
unstalked *a* бесстебельный
unsterile *a* нестерильный
unsterilized *a* нестерилизованный
unstrained *a* непроцеженный, нефильтрованный
unstratified *a* безъярусный
unstriated *a* неисчерченный; гладкий *(muscle)*
unsusceptible *a* невосприимчивый; нечувствительный
untagged *a* немеченый
untaimed *a* дикий, неприрученный
unthrifty *a* хилый, болезненный, чахлый; плохо развивающийся
untreated *a* необработанный; непротравленный *(seeds);* контрольный *(plot)*
untripped *a bot* невскрытый
untrue *a* отклоняющийся от нормы, не соответствующий типу
unveined *a* безжилковый
unviable, unvital *a* нежизнеспособный
unwind *v* раскручиваться
upas *s* упас, анчар *(Antiaris toxicaria)*
updating *s* уточнение, обновление
upland *s* нагорье, возвышенность; *a* горный, нагорный; суходол, суходольный
upright *a* прямостоячий
upright-flowered *a* прямоцветковый
upright-headed *a* прямоголовчатый
upright-spiked *a* прямоколосый
upright-spined *a* прямоколючковый
uprooting *s* выкорчёвывание, корчёвка, вырывание с корнем
upset *s* нарушение, расстройство
upside *s* верхняя сторона
upsilting *s* заиление
upstanding *a* длинноногий
upstream *adv* вверх по течению; *mbiol* слева, выше
uptake *s* поглощение; потребление; ввод
urachus *s emb* мочевой проток
uraniscus *s* нёбо
urceolate *a bot* кувшинчатый
urceole *s bot* кувшинчик
urchin *s* ёж
urea *s* мочевина
urediniospore *s* урединиоспора
uredinium *s* урединиум
uredopustule *s* уредопустула
uredospore *s* уредоспора
uredostage *s* уредостадия
urema *s* урема
uresis *s* мочеиспускание

ureter *s* мочеточник
ureteric *a* мочеточниковый
urethra *s* мочеиспускательный канал, уретра
uretic *a* мочевой
urge *s* стремление; влечение; позыв
urial *s* уриал, степной баран *(Ovis ammon vignei)*
uric *a* мочевой
uridine *s* уридин
urina *s* моча
urinary *a* мочевой
urination *s* мочеиспускание
urine *s* моча
uriniferous *a* выводящий мочу
uriniparous *a* вырабатывающий мочу, мочеобразующий
urinogenital *a* мочеполовой
urinous *a* мочевой
urite *s zool* абдоминальный сегмент; анальный циррус
urkaryote *s* уркариот
urn *s bot* урна
urocarpous *a* хвостоплодный
urochord *s* урохорд
urochrome *s* урохром
urocyst *s* мочевой пузырь
urocystic *a* мочепузырный
urodaeum *s anat* уродеум
urodelous *a* хвостатый
urogenital *a* мочеполовой
urology *s* урология
uromere *s zool* брюшной сегмент
urophyllous *a* хвостолистный
uropod *s zool* уропод
uropygial *a* надхвостный, копчиковый
uropygium *s orn* уропигий, надхвостье
urosome *s zool* уросома
urosteon *s orn* медианное окостенение
urostyle *s anat* уростиль
ursine *a* медвежий
urticant *a* жгучий, стрекательный
urticator *s* стрекательная клетка
urus *s* тур, первобытный бык *(Bos primigenius)*
use *s* использование; *v* использовать, применять
  consumptive ~ суммарное потребление
  industrial water ~ промышленное водопотребление
  water ~ водопотребление
user *s* пользователь, потребитель
uterine *a* маточный
uterosacral *a* маточно-крестцовый
uterus *s anat* матка
utilization *s* утилизация, использование
utricular *a* вздутый, пузырчатый
uvea *s anat* сосудистая оболочка
uviform *a* гроздевидный
uviolizing *s* ультрафиолетовое облучение
uvula *s* язычок

V

vaccigenous *a* образующий вакцину
vaccinate *v* вакцинировать
vaccination *s* вакцинация
renewed ~ ревакцинация
vaccine *s* вакцина; вакцинный
adjuvant ~ вакцина с адъювантом, адъювант-вакцина
autogenous ~. autologous ~ аутогенная вакцина, аутовакцина
corresponding ~ аутогенная вакцина, аутовакцина
formolated ~ формол-вакцина, формалинизированная вакцина
heterologous ~, heterotypic ~ гетеровакцина
homologous ~ аутогенная вакцина, аутовакцина
mixed ~ поливакцина, ассоциированная вакцина, комбинированная вакцина, поливалентная вакцина
personal ~ аутогенная вакцина, аутовакцина
stock~ аутогенная вакцина, аутовакцина
vaccine-challenged *a* поствакцинальный, вызванный введением вакцины
vaccinia *s* коровья оспа
vaccinization *s* вакцинация
vacuolar *a* вакуолярный
vacuolate, vacuolated *a* вакуолизированный, содержащий вакуоли
vacuolation *s* образование вакуолей
vacuole *s* вакуоль
autophagic ~ аутофагосома, аутофагическая вакуоль
pinocytotic ~ пиноцитозная вакуоль, пиноцитозный пузырёк
vacuolization *s* вакуолизация
vacuome *s* вакуом
vagal *a* вагусный
vage, vagile *a* бродячий; блуждающий
vagility *s* подвижность, склонность к перемещениям
vagina *s* влагалище
vaginal *a* влагалищный
vaginate *a* одетый покровом; несущий покров; имеющий оболочку
vaginella *s bot* влагальце
vaginula *s bot* маленькое влагалище
vagotomy *s* ваготомия
vagotony *s* ваготония, парасимпатикотония
vagotropic *a* ваготропный
vagrant *a* бродячий
vague *a* неясный, нечёткий, неопределённый
vagus *s anat* блуждающий нерв
valence *s* валентность
valid *a* валидный
validity *s* валидность
valine *s* валин
vallate *a* окруженный валиком
vallecula *s* углубление, жёлоб, выемка
valleculate *a* желобчатый; желобоватый
valley *s anat* углубление; желоб; выемка
valuation *s* оценка; таксация
value *s* ценность; значение; число, величина
~ of a division цена деления
buffer ~ буферность
caloric, calorific ~ калорийность, энергетическая ценность
crossing-over ~ процент кроссинговера
desired ~ желательное значение; заданная величина
expected ~ *biom* ожидание
germination ~ степень всхожести
globular ~ цветной показатель крови
linkage ~ *gen* величина сцепления, сила сцепления
lytic ~ литический индекс, показатель литической активности
modal ~ *biom* мода
relative selective ~ относительная селекционная ценность
saturation ~ степень насыщения
selective ~ селективная ценность
swelling ~ степень набухания
theoretical ~ теоретическое значение; теоретически рассчитанная величина
valval *см.* **valvular**
valvate *a anat* клапанный; *bot* створчатый
valve *s anat* клапан; *bot* вальва, створка
Bauhin's ~ баугиниева заслонка
valveless *a anat* бесклапанный; *bot* бесстворчатый

valvelet, valvula *s anat* маленький клапан; маленькая складка; *bot* створочка
valvular *a* относящийся к вальвам, относящийся к клапанам; *anat* клапанный; содержащий клапаны; *bot* створчатый
valvule *s anat* маленький клапан; маленькая складка; *bot* створочка
vampire *s* вампир; обыкновенный вампир, десмод *(Desmodus)*
common ~ обыкновенный вампир, десмод *(Desmodus)*
hairy-legged ~ мохноногий вампир, малый кровосос *(Diphylla ecaudata)*
true ~ обыкновенный вампир, десмод, большой кровосос *(Desmodus rotundus)*
white-winged ~ белокрылый вампир *(Diaemus youngi)*
vane *s* опахало *(of a feather); bot* флаг, парус
vanish *v* исчезать
vanishing *s* исчезновение; *a* исчезающий
vapor *s* пар
vaporization *s* испарение; выпаривание
variability *s* вариабельность, изменчивость
variable *s* переменная; *a* изменчивый
variance *s* варианса; дисперсия
genotypic ~ генотипическая варианса, генотипическая дисперсия
total ~ общая дисперсия, общее среднее квадратическое
variant *s* вариант; *biom* варианта
variate *s* случайная величина, случайная переменная; изменяться; отклоняться
variation *s* вариация, изменчивость; изменение
directional ~ направленная изменчивость, ортогенез
food ~s *pl* различия в пище
idiotypic ~ разнообразие идиотипических детерминант
individual ~ индивидуальная изменчивость; *биом.* единичное отклонение
phase ~ фазовая вариация, расщепление на два варианта
stimulus parameter ~ изменение параметра раздражения
variegated *a* разноцветный; неоднородный; смешанный; разнообразный
variegate-leaved *a* разнолистный
variegation *s phytop* вирусная мозаика; пестролистность
plastid ~ пластидная пестролистность
varietal сортовой
variety *s* вариетет, разновидность; сорт
clonal ~ сорт-клон
line ~ сорт-линия
pollinating ~ сорт-опылитель
vary *v* варьировать
vasal *a anat* сосудистый
vascular *a* васкулярный, сосудистый
vascularity *s* васкулярность, наличие кровеносных сосудов
vascularization *s* васкуляризация
vasculature *s* сосудистая сеть
vasculocardiac *a* сердечно-сосудистый
vasculogenesis *s* образование и развитие сосудов
vasculomotor *a* сосудодвигательный, вазомоторный
vasculum *s* ботанизирка
vasifactive *a* сосудообразующий
vasoconstriction *s* сужение сосудов
vasoconstrictor *s* вазоконстриктор, сосудосуживающий фактор; сосудосуживающий нерв
vasodepression *s* снижение сосудистого тонуса
vasodilatation *s* расширение сосудов
vasoexcitatory *a см.* **vasohypertonic**
vasoformative *a* сосудообразующий
vasohypertonic *a* сосудосуживающий; гипертензивный
vasohypotonic *a* сосудорасширяющий; гипотензивный
vasoinhibition *s* торможение сосудосуживания
vasomotion *s* сосудодвигательная функция; сосудодвигательная реакция
vasomotor *a* вазомоторный, сосудодвигательный
vasopressor *a* сосудосуживающий
vasoreflex *s* сосудистый рефлекс
vasorelaxation *s* расширение сосудов
vault *s anat* свод
veal *s* телёнок; телятина
vection *s* перенос
vector *s* вектор; переносчик

insect ~ насекомое-переносчик
vector-borne *a* трансмиссивный, передаваемый переносчиком
vectorcardiography *s* вектор-кардиография
vegetable *s* овощ; *a* овощной; растительный
vegetable-feeder *s* растительноядное животное
vegetal *a* растительный; вегетативный
vegetalization *s* вегетализация
vegetarian *s* травоядное животное; *a* травоядный
vegetate *v* расти
vegetation *s* растительность; разрастание ткани
tall herbaceous ~ высокотравье
vegetational *a* растительный
vegetation-inhabiting *a* живущий на растениях
vegetative *a* растительный; вегетативный
vehicle *s mbiol* вектор
veil *s bot* покрывало
~ **of palate** нёбная занавеска
vein *s* вена; кровеносный сосуд; жилка *(крыла насекомого или листа)*
common cardinal ~ кювьеров проток
veined *a* имеющий вены, имеющий жилки
closely ~ *bot* тонконервный
palmately ~ *bot* пальчато-нервный
transverse-parallel ~ *бот.* поперечно-параллельно-нервный
veining *s* жилкование
veinless *a* безжилковый
veinlet *s* маленькая вена; маленькая жилка
velamen *s* веламен
veliger *s zool* велигер
vellus *s myc* ножка
velocity *s* скорость
velum *s zool* велюм, парус; *bot* покрывало
velutinate, velutinous *a* бархатистый
velvet *s* «бархат» *(on horns)*
velvety *a bot* бархатистый; пушистый
venation *s* жилкование
venenate отравлять; отравленный; ядовитый
veneniferous *a* содержащий яд
venenous *a* ядовитый
veniplex *s* венозное сплетение
venom *s* яд
venomotor *a* венодвигательный
venomous *a* ядовитый
venopressor *a* венопрессорный, повышающий венозное давление
venose *a* жилковатый
venous *a* венозный
vent *s anat* отверстие; выход; *zool* отверстие клоаки
venter *s* живот, брюшко; расширенная часть, выпуклость; *bot* брюшная сторона
ventilation *s phys* вентиляция
salt ~ обмен солей *(in soil)*
ventilatory *a* вентиляторный, дыхательный
ventrad *adv* по направлению к брюшной стороне
ventral *a* вентральный, брюшной
ventricle *s* желудочек; желудок; *orn* зоб
ventricose *a* вздутый
ventricular *a* желудочный; желудочковый
ventriculus *s* желудок
venule *s* маленькая вена; венула, маленькая жилка
Venus's-flytrap *s bot* венерина мухоловка *(Dionaea muscipula)*
verdant *a* облиственный, покрытый листьями; зелёный
verdure *s* зелень, зелёная растительность
verification *s* проверка
verifier *s* контролирующее устройство
vermicide *s* вермицид, противоглистное средство
vermicular, vermiculate *a* червеобразный
vermilion *a* алый
vermin *s* вредители, паразиты; хищники
vermination *s* заражение гельминтами
vermis *s anat* червячок мозжечка; *zool* средняя часть мозжечка
vernation *s* листосложение, листовое почкосложение
vernicose *a bot* глянцевитый
verruca *s* бородавка
verrucate, verrucose, verrrucous *a* бородавчатый, покрытый бородавками
verruculose *a* мелкобородавчатый
versatile *a zool* свободноподвижный, гибкий; *bot* подвижный
versatility *s* многосторонность

habitat ~ эвритопность
versicolored *a* разноцветный, пёстрый; способный менять окраску
version *s* вариант; загиб
vertebra *s* позвонок
odontoid ~ осевой позвонок, эпистрофей
vertebral *a* позвоночный
vertebrarium *s* позвоночник, позвоночный столб
vertebrate *s* позвоночное животное; позвоночный; *a bot* сочленённый
vertex *s* верхушка, вершина; темя; макушка
vertical *a* вертикальный; теменной
verticil *s* мутовка
verticillaster *s* полумутовка, ложная мутовка
verticillate *a* мутовчатый
verumontanum *s anat* семенной холмик, семенной бугорок
vervain *s* вербена *(Verbena)*
vesical *a* пузырный, относящийся к пузырю
vesicle *s* пузырёк
air ~ воздушный пузырёк; лёгочная альвеола
blastodermic ~ бластула
lipid ~ липосома
Polian ~ *zool* полиев пузырь
Swammerdam's ~ *mal* сперматека
Tiedemann's ~ *zool* тидеманово тельце
water-expulsion ~ сократительная вакуоль
vesicotrilete *a bot* пузырчато-трёхлучевой
vesicular *a* везикулярный, пузырчатый; ячеистый
vespertine *a orn* ночной; *bot* распускающийся вечером
vespiary *s* осиное гнездо; колония ос
vespine, vespoid *a* осиный; осообразный, осовидный
vessel *s* сосуд; сосуд, склянка; судно, корабль
bile ~ жёлчный проток
conical ~ коническая колба, колба Эрленмейера
latex ~, milk ~ *bot* млечник, млечный сосуд
minute ~ мелкий сосуд; капилляр
seed ~ околоплодник
spermatic ~ семяпровод
vestibular *a* вестибулярный
vestibulate *анат.* с преддвериями
vestibule *s anat* преддверие
vestige *s* рудимент, остаток, след
vestigial *a* рудиментарный, остаточный
vestiture *s* наружный покров тела
vetch *s* вика, горошек *(Vicia)*
vex *v* беспокоить; дразнить
vexillary, vexillate *bot* имеющий парус
vexillum *s orn* опахало пера; *bot* парус
viability *s* жизнеспособность
viable *a* жизнеспособный
vial *s* пробирка; склянка; флакон; пузырёк; ампула; сосуд
vibraculum *s* вибракулярия
vibratile *a ent* вибрирующий
vibrio *s* вибрион
vibriocarrier *s* вибриононоситель
vibrissa *s* вибрисса
vibrocardiogram *s* виброкардиограмма
vicariad, vicariant *s* викариант, замещающий вид
vicariation *s* викарирование, географическое замещение
vicarious *a* викарный, замещающий
vice *s* норов *(of a horse);* порок, дефект; недостаток
vicine *s* пришелец из соседнего сообщества
vicinism *s* вицинизм
vicissitude *s* перемена; смена; чередование
victim *s* жертва
vicugna, vicuna *s ther* викунья *(Vicugna vicugna)*
view *s* вид
viewer *s*: **agglutination** ~ агглютиноскоп
vigilance *s* настороженность; зоркость; бессонница
vigor *s* сила, мощность; энергия
heterokaryotic ~ мощность, обусловленная гетерокариозом
heterotic ~, hybrid ~ гибридная сила, гибридная мощность, гетерозис
yeast ~ подъёмная сила дрожжей
vigorous *a* сильный, мощный
villiferous *a* ворсинчатый
villiform *a* ворсинковидный
villose *a anat* покрытый ворсинками; ворсинчатый; *bot* волосистый; пушистый, покрытый пушком

villosity *s anat* ворсинчатость; *bot* волосистость
villous *a anat* покрытый ворсинками; ворсинчатый; *bot* волосистый; пушистый, покрытый пушком
villus *s* ворсинка
vimen *s* гибкий побег
vinaceous, vinous *a* винно-красный
vine *s* вьющееся растение; лиана; виноградная лоза; виноград *(Vitis)*
vinegar *s* уксус
violet *a* фиолетовый; *s* краситель фиолетового цвета; фиалка *(Viola)*
viper *s* гадюка *(Viperidae)*
viral *a* вирусный
virescence *s* позеленение
virescent *a* зеленоватый
virga *s ent* хлыст
virgate *a* стержнеобразный, прутьевидный
virgin *a* девственный; первобытный; целинный; неоплодотворённый
virion *s* вирион
viroid *s* вироид
virology *s* вирусология
virulence *s* вирулентность
virulent *a* вирулентный
viruliferous *a* содержащий вирус
virus *s* вирус
  arthropod-borne ~ арбовирус, артропонозный вирус
  bacterial ~ бактериальный вирус, бактриофаг
  enteric ~ энтеровирус, кишечный вирус
  occult ~ провирус
  oncorna ~ онкорнавирус
  Papova ~ паповавирус
  pox ~ поксвирус
vurus-carriage *s* вирусоносительство
viruses-satellite *s* вирус-сателлит, неродственный вирус
virus-resistant *a* устойчивый к вирусу
viscera *s pl* внутренности, внутренние органы
visceral *a* висцеральный, внутренностный
viscerocranium *s* висцеральный череп
visceromotor *a* висцеромоторный
viscerotropic *a* висцеротропный
viscid *a* вязкий, клейкий, липкий
viscidulous *a* клейковатый
viscosimeter *s* вискозиметр
viscosity *s* вязкость
vision *s* зрение
visitation *s* посещение; залёт
visual *a* зрительный
visualization *s* визуализация
visuognosis *s* зрительное восприятие
vital *a* жизнеспособный, жизненный
vitalism *s* витализм
vitalistic *a* виталистический
vitality *s* жизнеспособность, жизненность
vitamin *s* витамин
  primary ~ провитамин
vitamin-free *a* не содержащий витаминов
vitaminization *s* витаминизация
vitaminize витаминизировать
vitaminous *a* витаминозный
vitellicle *s* желточный мешок
vitelligenous *a* продуцирующий желток, вителлогенный
vitelline *a* желтковый; желточный
vitelloduct *s* желточный проток
vitellogenesis *s* вителлогенез, желткообразование
vitellogenous *a* продуцирующий желток, вителлогенный
vitellophag *s* вителлофаг
vitellus *s* желток яйца
viticolous *a* живущий на винограде
viticulture *s* виноградарство
vitreous *a* стекловидный, гиалиновый
vitreum *s anat* стекловидная влага
vitrification *s* витрификация; гиалинизация; застекловывание
vivarium *s* виварий
viviparity *s* вивипария, живорождение
viviparous *a* живородящий
vivipary *s* вивипария, живорождение
vivisection *s* вивисекция
vixen *s* лисица
vocal *a* голосовой
vocalization *s* голосовой сигнал; вокализация, издавание звуков
  appeasement ~ умиротворяющие звуки
voice *s* голос; **to give** ~ подать голос *(напр. залаять)*
voidings *s pl* экскременты
vola *s* вола *(of diatoms)*
volant *a* летающий

volatile *s* летучее вещество; крылатое животное; *a* изменчивый, непостоянный

vole *s* полёвка; серая полёвка *(Microtus)*

northern mole ~ слепушонка *(Ellobius talpinus)*

sagebrush ~ степная пеструшка *(Lagurus)*

volley *s* серия мышечных сокращений

voltage-clamp *s* фиксация потенциала

voltage-gated *a* потенциалозависимый

voltage-sensitive *a* потенциалочувствительный

voluble *a* способный к скручиванию; вьющийся

volume *s* объём

packed cell ~ показатель гематокрита, гематокритное число, гематокрит, объём осажденных эритроцитов

volumeter *s* объёмометр

voluntary *a* произвольный

volute *a* спиральный, закрученный спиралью; *bot* завиток

volution *s* спиральное закручивание

volva *s myc* вольва, покрывальце, влагалище

volvate *a* имеющий вольву

vomer *s anat* сошник

vomerine *анат.* сошниковый

voracious *a* прожорливый

vortex *s anat* завиток

vulnerability *s* уязвимость

vulnerable *a* ранимый, уязвимый; ~ **to scouring** подверженный размыву

vulpine лисий

vulture *s* гриф *(Aegypiidae)*

bearded ~ бородач, ягнятник *(Gypaetus barbatus)*

griffon ~ белоголовый сип *(Gyps fulvus)*

long-billed ~ индийский сип *(Gyps indicus)*

palm-nut ~ пальмовый гриф, грифовый орлан *(Gypohierax angolensis)*

turkey ~ гриф-индейка *(Cathartes aura)*

vulturous *a* хищный

vulva *s* вульва

W

wadden *s* ватты
wader *s* болотная птица; *pl* ржанкообразные *(Charadriiformes)*
wag *s* взмах; кивок; *v* махать хвостом, вилять хвостом
wagging *s* виляние хвостом
wagtail *s* трясогузка *(Motacilla)*
  yellow ~ жёлтая трясогузка, плиска *(Motacilla flava)*
wake *v* просыпаться; будить
wakeful *a* бодрствующий
walk *s* шаг; **to ~ the dog** выгуливать собаку
walking *s* ходьба
  chromosomal ~ *mnbiol* «прогулка по хромосоме»
  cosmid ~ *mnbiol* скриниг космиды, космидная «прогулка»
walking-stick *s ent* палочник *(Phasmoptera)*
wall *s* стенка, перегородка
  cell ~ оболочка клетки, клеточная оболочка; клеточная стенка
  embryo sac ~ *bot* оболочка зародышевого мешка, стенка зародышевого мешка
  nail ~ ногтевой валик
wall-bee *s* пчела-каменщица *(Chalicodoma)*
wallcreeper *s orn* краснокрылый стенолаз *(Tichodroma muraria)*
walled *a* имеющий оболочку
wallow *s eth* купалка; валяться, барахтаться *(о животных)*
wallwort *s* бузина чёрная *(Sambucus nigra)*
walnut *s* орех *(Juglans)*
walrus *s* морж *(Odobenus rosmarus)*
wand *s* тонкая ветка; побег
wanderer *s* кочующая особь, кочевник
wandering *s* миграция, странствование; *a* подвижный; нефиксированный; блуждающий
waning *s eth* сниженная реактивность; затухание реакции
wanthill *s* кротовина
wanting *a* недостаточный; недостающий
wanton *a* буйный *(growth);* резвый
wapiti *s ther* вапити *(Cervus canadensis)*
war *s* война
  lymphocyte "civil" ~ массовый аутолиз лимфоцитов; реакция «трансплантат против хозяина», РТПХ
warble *s* личинка кожного овода
warblefly *s* носоглоточный овод *(Oestridae)*
warbler *s* славка *(Sylviidae);* древесница *(Parulidae)*
  aquatic ~ вертлявая камышовка *(Acrocephalus paludicola)*
  blue-throated ~ варакушка *(Luscinia svecica)*
  booted ~ бормотушка *(Hippolais caligata)*
  brown willow ~ пеночка-теньковка *(Phylloscopus collybitus)*
  grasshopper ~ обыкновенный сверчок *(Locustella naevia)*
  olivaceous ~ бледная пересмешка, большая бормотушка *(Hippolais pallida)*
  reed ~ тростниковая камышовка *(Acrocephalus scirpaceus)*
  river ~ речной сверчок *(Locustella fluviatilis)*
  willow ~ пеночка-весничка *(Phylloscopus trochilus);* дымчатая пеночка *(Phylloscopus fulgiventer)*
  yellow-browed ~ пеночка-зарничка *(Phylloscopus inornatus)*
warden *s* смотритель
  game ~ егерь
warm-blood *s* теплокровное животное; *a* теплокровный
warm-blooded *a* теплокровный
warm-enduring *a* тепловыносливый
warmer *s* подогореватель
  slide ~ подогревательный столик
warm-requiring *a* теплолюбивый
warmwater *a* тепловодный
warning *s* предупреждение, предостережение
warren *s* охотничий заповедник
wart *s* бородавка; *bot* нарост; наплыв
  tree ~ *phytop* раковый нарост на дереве; чёрный рак *(caused by Physalospora malorum)*
warted *a* бородавчатый; бугорчатый, покрытый наростами, покрытый наплывами

warthog *s ther* бородавочник *(Phacochoerus aethiopicus)*
warty *a* бородавчатый
wary *a* осторожный; настороженный
washing *s* умывание
wasp *s* оса *(Vespa)*
bee-killer ~ пчелиный волк, филант *(Philanthus)*
fig ~ бластофага *(Blastophaga psenes)*
gall ~ орехотворка *(Cynipidae)*
sand ~ аммофила *(Ammophila*
tailed ~ стебельчатобрюхое перепончатокрылое *(Apocrita)*
tool-using ~ аммофила *(Ammophila)*
wood ~ рогохвост *(Siricidae)*
wastage *s* постепенная потеря, утечка
waste *s* пустошь; пустынный; отходы, отбросы; выделения
domestic ~ бытовые отходы; *pl* бытовые сточные воды
industrial ~ промышленные сточные воды
intestinal ~ испражнения
livestock ~ отходы животноводства; *pl* животноводческие стоки
tree ~ опад
wastewater *s* сточная вода
watch *s* наблюдение, надзор; *v* наблюдать, охранять
water *s* вода; водоём; *pl* околоплодная жидкость, плодные воды
shallow ~ мелководье
water-bearing *a* водоносный
waterbuck *s* водяной козёл *(Kobus)*
water-carrying *a* водоносный
waterer *s* поилка
waterfowl *s* водоплавающая птица, водоплавающая дичь
watering *s* водопой; полив, орошение
watering-place *s* водопой
waterless *a* безводный
waterlogged *a* заболоченный; насыщенный водой
water-loving *a* влаголюбивый. водолюбивый
watermelon *s* арбуз обыкновенный *(Citrullus vulgaris)*
water-miscible *a* водорастворимый
water-ouzel *s orn* оляпка *(Cinclus cinclus)*
waterproof *a* водонепроницаемый
water-repellent *a* гидрофобный
water-soluble *a* водорастворимый
water-stable *a* водостойкий, водопрочный
water-tight *a* водонепроницаемый
watery *a* водянистый; заболоченный
watten *s* ватты
wattle *s zool* серёжка, бородка; акация *(Acacia)*
wave *s* волна; колебание; волнистость, неровность
dicrotic ~ дикротическая волна, дикрота
population ~s *pl* популяционные волны, колебания численности популяции
recoil ~ дикротическая волна, дикрота
wave-leaved *a* волнистолистный
wave-petaled *a* волнистолепестковый
wavy *a* волнистый
wax *s* воск
ear ~ ушная сера
grafting ~ замазка для прививки
grave ~ трупный жировоск
wax-like *a* восковидный
wax-producing *a* восконосный
waxwing *s orn* свиристель *(Bombycilla)*
waxy *a* восковой
waxy-flowered *a* восковцветный
waxy-fruited *a* воскоплодный
waxy-leaved *a* восколистный
waxy-seeded *a* воскосемянный
way-grass *s bot* горец птичий, спорыш *(Polygonum aviculare)*
weak *a* слабый, хилый
wean *v* отнимать от груди; отлучать от матери
weaning *s* отнятие от груди
weanling *s* отъёмыш
weapons *s* оружие
weasand *s* трахея
weasel *s* ласка *(Mustela nivalis);* горностай *(Mustela erminea)*
gray-headed ~ тайра *(Eira barbara)*
least ~ ласка *(Mustela nivalis)*
mountain ~ солонгой, сусленник *(Mustela altaica)*
pale ~ солонгой, сусленник *(Mustela altaica)*
Siberian ~ колонок *(Mustela sibirica)*
weather *s* погода; погодный, метеорологический
weaver, weaverbird *s orn* ткач, ткачик *(Ploceidae)*

web *s* паутина; перепонка; плавательная перепонка; *anat* соединительная ткань; опахало *(of a feather)*
food ~ «паутина трофических отношений», трофические отношения
webbed *a* перепончатый
webbing *s* перепонка; плавательная перепонка
webworm *s:* **fall** ~ американская белая бабочка *(Hyphantria cunea)*
wedge-leaved *a* клинолистный
wedge-shaped *a* клиновидный
weed *s* бурьян, сорняк, сорная трава; пропалывать, уничтожать сорняки
weed-covered *a* заросший сорняками
weed-eater *s* травоядное животное
weedicide *s* гербицид
weediness *s* засорённость сорняками
weed-killing *a* гербицидный
weedy *a* заросший сорняками; поросший водорослями
weevil *s ent* долгоносик, слоник *(Curculionidae)*
alder-tree ~ скрытнохоботник ольховый *(Cryptorrhynchus lapathi)*
biscuit ~ точильщик хлебный *(Anobium paniceum)*
black vine ~ скосарь бороздчатый *(Otiorrhynchus sulcatus)*
blossom ~ цветоед *(Anthonomus)*
hazelnut ~ плодожил ореховый *(Curculio nucum)*
larger pine ~ долгоносик большой сосновый *(Hylobius abietis);* смолёвка сосновая *(Pissodes pini)*
leaf-rolling ~ трубковёрт *(Attelabidae)*
lentil ~ зерновка фасолевая *(Bruchus signaticornis)*
nut ~ плодожил, долгоносик *(Curculio)*
pea ~ зерновка гороховая *(Bruchus pisorum)*
pine ~ долгоносик большой сосновый *(Hylobius abietis);* смолёвка *(Pissodes)*
raspberry ~ скосарь одиночный *(Otiorrhynchus singularis)*
weight *s* масса, вес; **to gain** ~ прибавлять в весе; **to reduce in** ~ терять в весе
weir *s* плотина; запруда; *v* сооружать плотину; запруживать
weissmannism *s* вейсманизм
well *s* колодец
well-fed *a* упитанный
well-marked *a* хорошо выраженный
well-stocked *a* хорошо заселённый
wet мочить, смачивать; влажный; сырой; мокрый
wetland *s* заболоченное место, сырой участок
wettability *s* смачиваемость
wetting *s* увлажнение
whalebone *s* китовый ус
whalehead *s* китоглав, королевская цапля *(Balaeniceps rex)*
whaup *s orn* большой кроншнеп *(Numenius arquata)*
wheat *s* пшеница *(Triticum)*
emmer ~ полба настоящая, пшеница двузернянка *(Triticum dicoccon)*
scab ~ *phytop* корневая гниль пшеницы *(caused by Gibberella saubinetii)*
wheel *s* колесо
wheel-shaped *a* колесовидный
whelp *s* щенок; детёныш; *v* щениться; производить детёнышей
whelping *s* щенение
whimbrel *s orn* средний кроншнеп *(Numenius phaeopus)*
whimper *s* скуление; завывание; хныканье
whine *s* скуление; подвывание; писк *(of a mosquito); v* скулить
whining *s* скуление; подвывание; вой
~ **of mosquitoes** комариный писк
whiplike *a* плетевидный
whir *s* жужжание; шум; *v* жужжать; шуметь
whirlbone *s* надколенник; головка бедренной кости
whirr *v* стрекотать, трещать; жужжать
whisk *v* смахивать, отгонять
whiskers *s pl* вибриссы, усы
whistle *s* свист; пронзительный визг
whistler *s* грызун, издающий свист; *orn* свистун *(Pachycephala)*
white *a* белый; седой; прозрачный, бесцветный; *s* белок *(of an egg);* склера, белочная оболочка; *bot* заболонь; *ent* белянка *(Pieridae)*
white-cheeked *a orn* с белыми щёчками
white-eye *s orn* белоглазка *(Zosterops);* белоглазый нырок *(Aythya nyroca)*
white-faced *a* беломордый
whitefly *s ent* алейродида, белокрылка *(Aleurodidae)*

white-girdled *a* белокаёмчатый
white-nerved *a bot* белонервный
white-seeded *a* белосемянный
whitethroat *s* славка *(Sylvia);* серая славка *(Sylvia communis);* белогорлый колибри *(Leucochloris albicollis)*
lesser ~ славка-завирушка *(Sylvia curruca)*
whizzer *s* центрифуга
whizzing *s* центрифугирование
whoop *s* удод *(Upupa epops)*
whorl *s* мутовка; *mal* оборот
body ~ *mal* завиток
whorled *a* мутовчатый
wide-meshed *a* крупносетчатый; крупноячеистый
widespread *a* широко распространённый
widgeon *s orn* свиязь *(Anas penelope)*
widow *s* козодой *(Caprimulgidae)*
black ~ паук «чёрная вдова» *(Latrodectes mactans)*
wigeon *s orn* свиязь *(Anas penelope)*
wiggle *v* вилять, шевелить
wiggler, wiggle-tail *s* личинка комара, куколка комара
wild *a* дикий; дикорастущий; пугливый; **to run** ~ одичать; зарасти; **in the** ~ в природе
wilderness *s* дикая природа, дикая местность, глухомань
wildfowl *s* пернатая дичь
wilding *s* дичок, дикое растение; одичавшее растение; дикое животное
wildlife *s* живая природа; дичь, дикие животные
wildling *s* дичок, дикое растение; одичавшее растение; дикое животное
willock *s* кайра *(Uria)*
willow *s* ива *(Salix)*
brittle ~ ракита, ива ломкая *(Salix fragilis)*
Caspian ~ верба, шелюга-жёлтая, ива волчниковая *(Salix daphnoides)*
common ~ ветла, белолоз, ива серебристая, ива белая *(Salix alba)*
crack ~ ракита, ива ломкая *(Salix fragilis)*
European ~ ветла, белолоз, ива серебристая, ива белая *(Salix alba)*
glaucous ~, pussy ~ краснотал, верболаз, ива-шелюга, верба красная *(Salix acutifolia)*
red ~ ива красная *(Salix rubra);* ива длиннолистная *(Salix interior);* дёрен душистый *(Cornus amomum)*
silver ~ краснотал, верболоз, ива-шелюга, верба красная *(Salix acutifolia);* ива серебристоплодная *(Salix argyrocarpa)*
swamp ~ ива чёрная *(Salix nigra);* краснотал, ива-шелюга, верба красная *(Salix acutifolia)*
varnished ~ ракита, ива ломкая *(Salix fragilis)*
violet ~ верба, шелюга-жёлтая, ива волчниковая *(Salix daphnoides)*
willow-leaved *a* иволистный
wilt *s phytop* вилт, увядание; *v* вянуть, увядать; поникать
clover ~ рак клевера *(caused by Sclerotinia trifoliorum)*
cotton ~ вилт хлопчатника *(caused by Fusarium oxysporum vasinfectum)*
tomato spotted ~ бронзовость томата
vascular ~ фузариозное увядание, вилт *(caused byFusarium oxysporum)*
wilting *s* увядание, завядание; привядание; поникание
wind *s* ветер
dry ~ суховей
windbend *s* сгибание ветром
windborne *a* принесённый ветром
windbreak *s* бурелом, ветровал; ветрозащитная полоса
wind-disseminated *a* распространяющийся ветром
windfall *s* бурелом, ветровал
wind-firm *a* ветроустойчивый
winding *s* изгиб, извилина
window *s* окно
windpipe *s* трахея
wind-pollinated *a* ветроопыляемый
wind-pollinating *a* ветроопыление
wind-protective *a* ветрозащитный
wind-resisting *a* ветроустойчивый
wing *s* крыло; *anat* крылоподобный придаток; *bot* боковой лепесток, крыло; ветка; крылатка *(of a seed);* **to take** ~ взлетать; **on the** ~ на лету; в полёте
~ **of heart** *ent* крыловидная мышца
bastard ~ *orn* крылышко
clear ~ *pl ent* стеклянница *(Aegeriidae)*
wingbeat *s* взмах крыла

winged *a* крылатый
wing-edged *a bot* крылоугольный
wing-flowered *a* крылоцветковый
wingless *a* бескрылый
wing-span *s* размах крыльев
wing-tag *s* крыловая метка
wink *s* моргание, мигание; *v* моргать, мигать
winter *s* зима; *v* зимовать, проводить зиму; перезимовывать
winter-annual *a* озимый
wintergreen *a* зимнезелёный
winterhardy *a* зимостойкий
wintering *s* зимовка; перезимовывание
winterkill, winterkilling *s* вымерзание
wintry *a* зимний
wire *s* проволока
  planting ~ *micr* игла для посева
wireworm *s ent* проволочник *(Elateridae)*
wisent *s* зубр *(Bison bonasus)*
wisteria *s* глициния, вистерия *(Wisteria)*
witches'-broom *s phytop* ведьмина метла
wither *v* увядать, вянуть; сохнуть, засыхать
withering *s phytop* увядание, вилт
withers *s pl* холка
withstand *v* выдерживать; противостоять, сопротивляться
wobble *v* качаться, шататься
wolf *s* волк *(Canis lupus)*
  Abyssinian ~ эфиопский шакал *(Canis simensis)*
  bitch ~ волчица
  common ~ волк *(Canis lupus)*
  dog ~ самец волк
  European ~, gray ~ волк *(Canis lupus)*
  prairie ~, steppe ~ койот *(Canis latrans)*
  strand ~ полосатая гиена *(Hyaena)*
wolverene, wolverine *s* росомаха *(Gulo gulo)*
womb *s anat* матка
wombat *s ther* вомбат *(Vombatidae)*
wood *s* дерево, древесина; лес
  dead ~ сухостой
  deadfallen ~ валежник
  dwarf-pine elfin ~ сосновый стланик
  elfin ~ стланик
  fruiting ~ плодушая ветка
  overmature ~, overripe ~ перестойное насаждение
  sap ~ заболонь
  small ~ перелесок
  soft ~ мягкая древесина; хвойный
  standing ~ лес на корню
wood-boring *a* сверлящий дерево
woodchuck *s* сурок *(Marmota);* лесной сурок *(Marmota monax)*
woodcock *s* вальдшнеп*(Scolopax)*
woodcreeper *s orn* древолаз *(Dendrocolaptidae)*
wood-eating *a* питающийся древесиной
wooded *a* лесистый
woodknob *s* наплыв, кап
woodland *s* лесистая местность, лесной массив, лес
  elfin ~ криволесье; стланик
  field ~ перелесок
  open ~ редколесье
  thorn ~ колючий лес; колючее редколесье
woodless *a* безлесный
wood-meadow *s* парковый лес
woodpecker *s* дятел *(Picidae)*
  black ~ чёрный дятел, желна *(Dryocopus martius)*
woodworm *s* точильщик мебельный *(Anobium punctatum)*
woody *a* лесистый, лесной; деревянный; древесный
woody-rooted *a* древесно-корешковый
woody-spiked *a* древесно-колосый
woof *s* фырканье; лай
wool *s* шерсть
woolly *a* шерстистый, шерстный, покрытый шерстью, опушённый волосками; хлопьевидный
work *s* работа
worker *s* рабочий, рабочая особь; неоплодотворённая самка
world-wide *a* всесветный
worm *s* червь; червеобразное животное; личинка; гусеница
  bladder ~ *helm* цистицерк; финна
  brain ~ *anat* червячок мозжечка
  cotton-boll ~ *ent* гусеница совки хлопковой, коробочный червь *(Heliothis armigera, Chloridea obsoleta)*
  cut ~ подгрызающая совка
  gordian ~, horsehair ~ волосатик *(Gordiacea)*
  lug ~ пескожил *(Arenicola)*
  maw ~ аскарида *(Ascaris)*

measuring ~ гусеница пяденицы *(Geometridae)*
nematode ~ круглый червь, нематода *(Nematoda)*
palolo ~ палоло *(Eunice viridis)*
pot ~ энхитреида *(Enchytraeidae)*
proboscis ~ скребень, колючеголовый червь *(Acanthocephala)*
seat ~ острица *(Enterobius vermicularis)*
sewage ~, sludge ~ трубочник *(Tubifex)*
spiny headed ~, thorny-headed ~ скребень, колючеголовый червь *(Acanthocephala)*
thread ~ круглый червь, нематода *(Nematoda)*
timber ~ вредитель древесины
tongue ~ пятиустка, пентастомида *(Linguatulidae)*

wormhole *s* червоточина
worming *s* дегельминтизация
worm-shaped *a* червеобразный
wormwood *s* полынь *(Artemisia);* полынь обыкновенная, чернобыльник *(Artemisia vulgaris)*
wort *s* сусло; *bot* ясменник *(Asperula)*
wound *s* рана; *v* ранить
wrapping *a* завёртывающий, обвёртывающий
wreath-bearing *a* венценосный
wren *s orn* крапивник *(Troglodytidae)*
willow ~ пеночка-весничка *(Phylloscopus trochilus)*
wrestle *s eth* борьба; игровая борьба
wriggle *v* изгибаться, извиваться; пробираться ползком, ползти извиваясь
wriggler *s* личинка *или* куколка комара
wrinkle *s* морщина, складка
wrinkled *a* морщинистый
wrinkle-leaved *a* морщинистолистный
wrinkle-seeded *a* морщинистосемянный
wrist *s anat* запястье
wryneck *s orn* вертишейка *(Jynx);* обыкновенная вертишейка *(Jynx torquilla)*
wuff *s* тявканье; *v* тявкать
fund *s:* **WWF World Wildlife Fund** Всемирный фонд дикой природы

X

xanthocarpous *a* желтоплодный
xanthodermic *a* имеющий желтоватый покров
xanthophore *s* ксантофор, жёлтый хроматофор
xanthophyll *s* ксантофилл
xanthopous *a* желтостебельный
xanthospermous *a* желтосемянный
xenia *s* ксения
xeniobiosis *s* ксенобиоз
xenoaggression *s imm* ксеногенная агрессия; ксеноиммунный ответ
xenoantibody *s* ксеногенное антитело, ксеноантитело
xenoantigen *s* ксеногенный антиген, ксеноантиген
xenoantiserum *s* ксеногенная антисыворотка, ксеноантисыворотка
xenobiotic *s* ксенобиотик
xenodiagnosis *s* ксенодиагностика
xenoepitope *s* ксеноантигенная детерминанта, ксеноэпитоп
xenogamy *s* перекрёстное оплодотворение
xenogenesis *s* гетерогенез
xenogenic *a* ксеногенный
xenogenicity *s* ксеногенные свойства, ксеногенность
xenograft *s* ксенотрансплантат
xenografting *s* ксенотрансплантация
xenomorphosis *s* гетероморфоз
xenoparasit *s* ксенопаразит
xenosome *s* ксеносома
xerarch *s* ксерическая серия
xeric *a* ксерический, сухоустойчивый
xeriobole, xerobole *s bot* ксеробол
xerocarpous *a* сухоплодный
xerocephalous *bot* сухоголовчатый
xerocolous *a* сухолюбивый
xeroflorous *a* сухоцветковый
xerogeophyte *s* ксерогеофит
xerohytium *s* сообщество сухого леса
xeromorph *s* ксероморф
xeromorphic *a* ксероморфный
xeromorphosis *s* ксероморфоз
xeromorphous *a* ксероморфный
xeromorphy *s* ксероморфизм
xerophil, xerophile *s* ксерофильный организм
xerophilous *a* ксерофильный, сухолюбивый
xerophyte *s* ксерофит
xerophytic *a* ксерофитный
xerophytium *s* сообщество ксерофитов
xeropolum *s* степная растительность
xerosere *s* ксерическая серия
xerotherm *s* ксеротерм
xerothermic *a* ксеротермический
xiphiplastrom *s herp* задний парный щиток
xiphisternum *s anat* мечевидный отросток
xiphoid *a* мечеобразный
xiphophyllous *a* мечелистный
xylem *s* ксилема
xylene *s* ксилол
xylium *s* лесное сообщество
xylocarp *s* ксилокарп, жёсткий деревянистый плод
xylogen *s* ксилоген, лигнин
xylogenous *a* растущий на древесине
xylophage *s* ксилофаг
xylophagous *a* питающийся древесиной
xylophilous *a zool* предпочитающий древесину; *bot* растущий на древесине
xylophyte *s* деревянистое растение
xylosucrose *s* ксилосахароза
xylotomous *a* способный бурить древесину
X-zone *s anat* X-зона

Y

yak *s* як *(Bos mutus)*
yam *s bot* ямс, диоскорея *(Dioscorea);* торица полевая *(Spergula arvensis)*
yap *s* лай, тявканье; *v* лаять, тявкать
yarrow *s* тысячелистник *(Achillea)*
yawl *s* вой; *v* выть
yawn *s* зевок; зевота; *v* зевать
yawning *s* зевота; *bot* разрывной
yean *v* ягниться
yeanling *s* козлёнок; ягнёнок
year *s* год
year-class *s* поколение
yearling *s* годовик, первогодок, годовалое животное; однолетний саженец, однолетнее растение
year-old *a* годовалый
yeast *s* дрожжи
yeastnel *s* дрожжевой экстракт
yeasty *a* дрожжевой; бродящий
yellow *a* жёлтый; *s* краситель жёлтого цвета; *ent* желтушка *(Colias); pl phytop* желтуха
yellow-jacket *s* настоящая оса, складчатокрылая оса *(Vespidae)*
yellowlegs *s orn:*
  lesser ~ желтоногий улит *(Tringa flavipes)*
  greater ~ пёстрый улит *(Tringa melanoleuca)*
yellowtail *s ent* златогузка *(Euproctis chrysorhoea)*
yelp *s* визг, взвизгивание, скуление, лай; *v* взвизгивать, визжать; скулить
yernik *s bot* ерник
yeti *s* йети, «снежный человек»
yew *s* тис *(Taxus)*
yield *s* урожай; улов, выход продукции, полезная продуктивность
  ~ **of a generation** урожайность поклления
  ~ **to the fishery** промысловый возврат
  counting ~ результат счёта
  intermediate ~ промежуточная культура
  molar growth ~ молярный урожай биомассы
  sustainable sustained ~ устойчивый улов, постоянная продукция
yolk *s* желток
yolky *a* богатый желтком
young *s* молодняк; детёныш; молодь; *a* молодой
yusiloid *s* предлобковая кость

Z

zebra *s* зебра *(Equus zebra)*
Burchell's ~ бурчеллова зебра *(Equus burchelli)*
steppe ~ квагга *(Equus quagga)*
zebu *s* зебу *(Bos indicus)*
zenotropism *s* отрицательный геотропизм
zeolite *s* цеолит
zeren *s* монгольский дзерен, зобастая антилопа *(Procapra gutturosa)*
zero *s* нуль
zeugopodium *s* предплечье; голень
zippering *s mbiol* «застёжка-молния»
zoea *s zool* зоеа
zoetic *a* жизненный, относящийся к жизни
zoic *a* животный
zoidiogamy *s* зоидиогамия
zoidiophilous *a* зоофильный; *ent* предпочитающий животных; *bot* опыляемый животными
zoidogamous *a bot* опыляемый животными
zokor *s ther* цокор *(Myospalax)*
zonal *a* зональный
zonality *s* зональность; зональное распределение; зональная особенность
zonaperturate *a pal* зонально-апертурный
zonate *a* опоясанный, окаймлённый
zonation *s* зональное распределение, зонирование; *myc* концентрический рост
zone *s* зона; пояс; область
~ **of complete inhibition** зона полного подавления, презона
~ **of continued sampling** *biom* область продолжения выбора, область безразличия
abscission ~ *bot* отделительный слой
annual ~ *bot* годичное кольцо, годичный слой
dead ~ мёртвая зона, область нечувствительности
intertidal ~ приливно-отливная зона, литораль
life ~ жизненная зона, биом
mantle ~ область мантии
root ~ корнеобитаемый слой
spray ~ зона брызг, зона заплесков, супралитораль
tidal ~ приливно-отливная зона, литораль
zoniporate *a pal* зонально-поровый
zonisulcate *a pal* зонально-дистальнобороздный
zonociliate *a* опоясанный ресничками
zonosulcate *a pal* зонально-дистальнобороздный
zonula, zonule *s anat* малая зона, поясок
~ **of Zinn** циннова связка, ресничный поясок
zoo *s* зоопарк
zooanthroponosis *s* зооантропоноз
zoobiotic *a* зообиотический, паразитирующий на животном
zooblast *s* животная клетка
zoocarp *s bot* зооспора
zoocecidium *s* галл, вызванный животным
zoocenosis *s* зооценоз
zoochlorella *s* зоохлорелла
zoochore *s* зоохор
zoochoric *a* зоохорный
zooecium *s zool* зооэций
zooecology *s* экология животных
zoogeographical *a* зоогеографический
zoogeography *s* зоогеография
zooglea, zoogloea *s* зооглея
zoogonous *a* живородящий
zoogony *s* живорождение
zooid *s* зоооид
zoolite, zoolith *s* окаменелое животное
zoological *a* зоологический
zoology *s* зоология
zoome *s* совокупность животных; сообщество животных
zoomorphosis *s* зооморфоз
zoon *s* животное; зооид
zoonose *s* зооноз
zoonomy *s* зоономия
zoopaleontology *s* палеозоология
zooparasite *s* животный паразит, зоопаразит
zoophagous *a* питающийся животной пищей
zoophile *s* растение, опыляемое животными

zoophilous *a* зоофильный; *ent* предпочитающий животных; *bot* опыляемый животными
zoophobous *a* зоофобный
zoophysics, zoophysiology *s* физиология животных
zoophyte *s* зоофит
zoosperm *s* сперматозоид; зооспора
zoosporangiophore *s* зооспорангиофор
zoosporangium *s* зооспорангий
zoospore *s* зооспора
zoosporic *a* имеющий зооспоры
zoosporiferous *a* образующий зооспоры
zoosporogenesis *s* образование зооспор
zoosuccivorous *a* сосущий кровь животных
zootechnics, zootechny *s* зоотехника
zootomy *s* зоотомия, анатомия животных
zootoxin *s* зоотоксин
zootrophic *a* зоотрофный, питающийся животными
zootype *s* зоотип
zooxanthella *s* зооксантелла
zoozygospore *s* подвижная зигоспора
zygantrum *s herp* зигантрум
zygapophysis *s anat* сочленовный отросток
zygodactyl, zygodactyle *a anat* зигодактильный
zygodont *a* зигодонтный
zygoid *a* диплоидный
zygoma *s* скуловая дуга; скуловая кость
zygomatic *a* скуловой
zygomelous *a* имеющий парные придатки
zygomorphic, **zygomorphous** *a* зигоморфный
zygomorphy *s* зигоморфизм
zygoneury *s* зигоневрия
zygopleural *a* двусторонне-симметричный
zygosis *s* зигозис, слияние гамет
zygosperm *s* зигоспора
zygosphene *s anat* зигосфен
zygosporangium *s* зигоспорангий
zygospore *s* зигоспора
zygosporocarpic *a* зигоспорокарпный
zygosporophore *s* зигоспорофор, зигофор
zygotactism, zygotaxis *s* зиготаксис
zygote *s* зигота
zygotic *a* зиготический
zygotropism *s* зиготропизм
zygozoospore *s* зигозооспора
zyme *s* фермент
zymodeme *s* зимодем
zymogen *s* зимоген, профермент
serum ~ плазминоген
zymogenic *a* вызывающий брожение; ферментативный
zymohydrolysis *s* ферментативный гидролиз
zymogram *s* зимограмма
zymology *s* энзимология
zymolysis *s* ферментативное расщепление
zymophore *s* активный центр фермента
zymometer, zymosimeter *s* зимозиметр
zymosis *s* ферментация; ферментативное расщепление; инфекционная болезнь
zymosterol *s* зимостерин
zymotic *a* бродильный; ферментативный; инфекционный; контагиозный

Some biographical names

Bates H.W. Бейтс Г.У.
Bateson W. Бэтсон У.
Begon M. Бигон М.
Beneden E. van Бенеден Э. ван
Bernard C. Бернар К.
Beyerinck M.W. Бейеринк М.В.
Bichat M.F.X. Биша М.Ф.
Bottallo L. Боталло Л.
Boussigault J.-B. Буссенго Ж.Б.
Bowman W. Боумен У.
Buffon G.-L. L. Бюффон Ж.Л.Л.
Candolle A.P. de Декандоль О.П.
Cannon W.B. Кеннон У.Б.
Cuvier G. Кювье Ж.Л.К.Ф.Д.
Darwin C.R. Дарвин Ч.Р.
Desor E. Дезор Э.
Dohrn A. Дорн А.
Dollo L. Долло Л.
Driesch H.A.E. Дриш Х.
Eimer T.G.H. Эймер Т.Г.Г.
Elton C.S. Элтон Ч.С.
Fallopio G. Фаллопий Г.
Galton F. Гальтон Ф.
Galvani L. Гальвани Л.
Garstang W. Гарстанг У.
Gegenbaur K. Гегенбаур К.
Gesner C. Геснер К.
Goldschmidt R. Гольдшмидт Р.
Golgi C. Гольджи К.
Graaf R. de Грааф Р. де
Gram H.C.J. Грам Х.К.И.
Haeckel E. Геккель Э.
Haldane J.B.S. Холдейн Д.Б.С.
Heidenhain R.P.H. Гейденгайн Р.П.Г.
Henle F.G.J. Генле Ф.Г.Я.
Hertwig O. Гертвиг О.
Highmor N. Гаймор Н.
Hutchinson G. Хатчинсон Д.
Koch R. Кох Р.
Kupffer K.W. Купфер К.В. фон
Lamarck J.B.P.A.de Monet Ламарк Ж.Б.П.А. де М.
Langerhans P. Лангерганс П.
Leeuwenhoek A. van Левенгук А. ван
Linnaeus C. Линней К.
Lorentz K. Лоренц К.
Lotka A.J. Лотка А.Д.
Malpighi M. Мальпиги М.
Mayr E. Майр Э.У.
Mendel G.J. Мендель Г.И.
Morgan T.H. Морган Т.Х.
Nissl F. Ниссль Ф.
Odum E.P. Одум Ю.
Osborn G.F. Осборн Г.Ф.
Pacini F. Пачини Ф.
Paneth J. Панет Й.
Pauling L.K. Полинг Л.К.
Ranvier L.A. Ранвье Л.А.
Raunkiaer C. Раункиер К.
Ray J. Рей Д.
Reaumur R.-A.F. Реомюр Р.А.
Roux W. Ру В.
Saussure N.T. de Соссюр Н.Т. де
Schleiden M.J. Шлейден М.Я.
Schwann T. Шванн Т.
Shelford V.E. Шелфорд В.Э.
Simpson G.G. Симпсон Д.Г.
Spemann H. Шпеман Х.
Swammerdam J. Сваммердам Я.
Virchow R. Вирхов Р.
Waksman S. Ваксман З.
Wallace A.R. Уоллес А.Р.
Warming J.E. Варминг Й.
Weber E.H, Вебер Э.Г.
Weismann A.F. Вейсман А.

ISBN 0-444-51700-6